VITAMIN D

SECOND EDITION

NUTRITION AND HEALTH
Adrianne Bendich, PhD, FACN, Series Editor

For other titles published in this series, go to
www.springer.com/series/7659

Vitamin D

Physiology, Molecular Biology, and Clinical Applications

Second Edition

Edited by

Michael F. Holick, Ph.D., M.D.

*Boston University School of Medicine, Professor
of Medicine, Physiology and Biophysics, Director,
Vitamin D, Skin and Bone Research Laboratory,
Boston Medical Center, Program Director,
General Clinical Research Unit, Boston, MA*

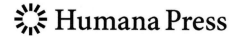 Humana Press

Editor
Michael F. Holick, Ph.D., M.D.
85 East Newton St.
Boston MA 02118
M-1013
USA
mfholick@bu.edu

Series Editor
Adrianne Bendich, PhD
GlaxoSmithKline Consumer Healthcare
Parsippany, NJ
advianne.4.bendich@gsk.com

ISBN 978-1-60327-300-8   e-ISBN 978-1-60327-303-9
DOI 10.1007/978-1-60327-303-9
Springer New York Dordrecht Heidelberg London

Library of Congress Control Number: 2010922338

Printed on acid-free paper

Humana Press is part of Springer Science+Business Media (www.springer.com)

Dedication

I dedicate this book to my best friend, colleague, and business partner for more than 37 years, my loving wife Sally who has always been supportive of my career and has been the absolute joy and love of my life.

In Memoriam

Louis V. Avioli M.D.

Last year I participated in an International Bone and Mineral Society conference for trainees and junior faculty focusing on successful career planning. My message was simple; pick the right mentor. It's a strategy I know well because any success I've enjoyed in my career reflects the wisdom and unselfishness of my mentor, Lou Avioli.

Lou was a self-described public school kid from New Jersey who earned a full scholarship to Princeton from which he graduated magna cum laude. He received his M.D. from Yale and became interested in calcium metabolism during his fellowship at University of North Carolina. In 1966, Lou came to Washington University School of Medicine as assistant professor of medicine and director of endocrinology at the Jewish Hospital of St. Louis. Lou organized the nation's first Division of Bone and Mineral Diseases and within just 6 years was appointed the Sidney and Stella Schoenberg Professor of Medicine.

At the time of Lou's arrival in St. Louis osteoporosis research was pretty much in its infancy. Because techniques of generating authentic bone cells in culture were not in hand, most studies were largely phenomenological. Lou, however, saw the potential of the field and his leadership skills and dynamism enabled him to become its father. His interests were broad but he had a particular passion for vitamin D metabolism and how it relates to the skeleton. When I first met Lou in the early 1970s John Haddad was one of his fellows and together they developed the first assay for serum 25-hydroxyvitamin D. Lou's had the capacity to consistently visualize his research in the context of patient care and publicize the significance of skeletal disease. He was prolific, publishing in excess of 300 papers and with Steve Krane, *Metabolic Bone Diseases and Related Disorders*". Before Lou Avioli, osteoporosis was a boring entity which did not lend itself to meaningful investigation.

Lou saw the big picture. He realized that progress in skeletal research demanded a first-class research society and so in the mid-1970s he convened a committee of leaders in the field to organize the American Society for Bone and Mineral Research. The ASBMR, which is now a behemoth, began with fewer than 100 members and Lou as its first president. Only one who witnessed the beginning of the society and where it stands today can appreciate the magnitude of Lou's accomplishment. Principally however, Lou was a mentor who encouraged independence. He pushed each of us to develop our own programs and yet maintain a family and sharing environment within our group. He was devoted to the concept that the welfare of the trainee always supersedes that of the mentor.

We lost Lou in 1999 after a long struggle with prostate cancer. His passing left a void in the lives of many, particularly those with whom he interacted daily. The fun has never been the same.

Steven L. Teitelbaum
Washington University School of Medicine

Series Editor Introduction

The Nutrition and Health series of books have, as an overriding mission, to provide health professionals with texts that are considered essential because each includes: (1) a synthesis of the state of the science, (2) timely, in-depth reviews by the leading researchers in their respective fields, (3) extensive, up-to-date fully annotated reference lists, (4) a detailed index, (5) relevant tables and figures, (6) identification of paradigm shifts and the consequences, (7) virtually no overlap of information between chapters, but targeted, inter-chapter referrals, (8) suggestions of areas for future research and (9) balanced, data-driven answers to patient/health professionals questions that are based upon the totality of evidence rather than the findings of any single study.

The goal of the series is to develop volumes that are adopted as the standard text in each area of nutritional sciences that the volume reviews. Evidence of the success of the series is the publication of second, third, and even fourth editions of more than half of the volumes published since the Nutrition and Health Series was initiated in 1997. The series volumes that are considered for second and subsequent editions have clearly demonstrated their value to health professionals. Second editions provide readers with updated information as well as new chapters that contain relevant up-to-date information. Each editor of new and updated volumes has the potential to examine a chosen area with a broad perspective, both in subject matter as well as in the choice of chapter authors. The international perspective, especially with regard to public health initiatives, is emphasized where appropriate. The editors, whose trainings are both research and practice oriented, have the opportunity to develop a primary objective for their book; define the scope and focus, and then invite the leading authorities from around the world to be part of their initiative. The authors are encouraged to provide an overview of the field, discuss their own research and relate the research findings to potential human health consequences. Because each book is developed de novo, the chapters are coordinated so that the resulting volume imparts greater knowledge than the sum of the information contained in the individual chapters.

"*Vitamin D: Physiology, Molecular Biology and Clinic Aspects*, Second Edition," edited by Michael F. Holick, Ph.D., M.D. fully exemplifies the Nutrition and Health Series' goals for a second edition. This volume is very timely as there have been more scientific papers published about vitamin D than any other essential nutrient over the past several years and the field of vitamin D research has expanded greatly in the decade since the first edition of this volume was published. In fact, Dr. Holick is the recognized leader in the field of vitamin D research and his innovative studies had led literally hundreds of researchers around the globe to join him in the exploration of the role of vitamin D in subcellular interactions, cellular functions, tissue and organ physiology, as well as clinical applications of naturally occurring forms of vitamin D and synthetic variants that have been developed to improve human health. Thus it is not surprising that the first volume contained 25 important chapters, and the current volume contains more than twice that number of chapters, 59 to be exact, that are either updates of the chapters included in the first edition or brand new chapters that are the result of the recent upsurge in basic as well as clinical research in vitamin D.

This important volume presents a timely review of the latest science concerning vitamin D's role in optimizing health as well as preventing many chronic diseases. The volume also examines the potential for vitamin D and its synthetic derivatives to act as valuable therapeutic agents in the treatment of serious diseases such as cancer and kidney disease. The overarching goal of the editor is to provide fully referenced information to health professionals so that they may enhance the nutritional welfare and overall health of clients and patients who may not be fully aware of the new findings of vitamin D's effects in diverse health areas from a to z: in this instance, from asthma to zygote functions. This excellent, up-to-date, and comprehensive volume will add great value to the practicing health professional as well as those professionals and students who have an interest in the latest information on the science behind vitamin D as a component of an individual's nutritional status and the consequences of low vitamin D levels on organ functions throughout the body. Chapters address the importance of the different requirements of vitamin D throughout the life span as well as the unique needs during pregnancy, lactation, and menopause for women. This timely volume reviews the ability of vitamin D to modulate the effects of the most prevalent chronic diseases and conditions that are widely seen in the majority of patient populations.

Dr. Holick is the internationally recognized leader in the field of vitamin D research with particular expertise in clinical dermatology. Michael F. Holick, Ph.D., M.D., is Professor of medicine, physiology and biophysics; Director of the General Clinical Research Unit and Director of the Bone Health Care Clinic and the Heliotherapy, Light, and Skin Research Center at Boston University Medical Center. Dr. Holick is an excellent communicator to both lay audiences as well as health professionals. He has worked tirelessly to develop the second edition of his *Vitamin D* volume; the first edition was the benchmark in the field when published in 1999. This volume continues to include extensive, in-depth chapters covering the most important aspects of the complex interactions between vitamin D and other dietary components, the ongoing debate concerning the best indicator of optimal vitamin D status and its nutrient requirements, and the impact of less than optimal status on disease risk.

The introductory 11 chapters provide readers with the basics of vitamin D biology including an introduction to the topic of photobiology, the molecular biology of each of the biological forms from pre-vitamin D through the most active form of the vitamin, as well as the molecular biology of the vitamin D receptor and related nuclear receptors. There is a comprehensive chapter on the metabolism of vitamin D, its metabolites, and clinically relevant analogs. There are individual chapters that describe the biological effects of vitamin D on bone, the kidneys, on intestinal absorption of calcium and related minerals as well as vitamin D's direct effects on the parathyroid glands. The final chapter in this section reviews the diversity of vitamin D target genes and the effects of activation.

The second section contains six chapters that examine the growing area of investigation into the non-skeletal functions of vitamin D. The field of vitamin D research was greatly enriched with the relatively recent discovery of the extra-renal sites of synthesis of the active form of vitamin D, 1,25-dihydroxyvitamin D. This historic event is chronicled in an excellent chapter written by Dr. Daniel Bilke who was one of the first investigators to elucidate the extracellular synthesis of active vitamin D. Vitamin D's role in the immune system and its role in the treatment of tuberculosis, the consequences of D's effects on immune cells that affect colon cancer risk; direct effects on cancer cells; its functions in the brain and in adipocytes are each examined in detail in separate chapters.

The third section contains the most comprehensive review of the status of vitamin D across the globe that has ever been gathered in one volume. These 14 chapters examine vitamin D status and the effects of deficiency in the United States, Canada, Northern Europe, Mediterranean countries, the Middle East, Africa, India, Asia, and New Zealand. There is a detailed chapter concerning the assays for 25-hydroxyvitamin D and another critically important chapter on vitamin D toxicity. The fourth section describes the health consequences of vitamin D deficiency and resistance on musculoskeletal health and contains eight chapters. The first chapter examines the effects of vitamin D deficiency in pregnancy and lactation. The next two chapters concentrate of the importance of vitamin D in child health and the impact of vitamin D deficiency that results in rickets. A related chapter describes the effects of inherited vitamin D-resistant rickets. Genetic defects in the metabolism of vitamin D as well as receptors are covered in three related chapters. There is also a key chapter on vitamin D's role in fall and fracture prevention especially in the elderly.

The fifth section looks at the interactions between sunlight, vitamin D, and cancer risks in seven chapters. There are several important chapters, written by the leaders in this field of research, including the Garland brothers that examine the epidemiology that links sunlight exposure and reduced as well as increased risk of specific cancers. We are reminded that aging is associated with both a decreased capacity for skin to synthesize vitamin D and at the same time, an increased risk of most cancers. The data from a recent well-controlled intervention trial that examined the effects of calcium and vitamin D supplementation on fracture risk also collected data of cancer occurrence and found a significantly lower risk of cancer in the supplemented cohort. The authors of the original study, Dr. J. Lappe and Dr. R. Heaney, have contributed this important chapter.

The sixth and seventh sections examine the clinical importance of vitamin D, its naturally occurring active forms, as well as synthetic forms of vitamin D that have been investigated as therapeutics. This section includes a review of the new data linking low vitamin D status during pregnancy and increased risk of type 1 diabetes in the offspring. Survey data have also found associations between vitamin D and certain immune-related chronic diseases including multiple sclerosis and rheumatoid arthritis. Individual, in-depth chapters on type II diabetes, cardiovascular health, blood pressure, and lung disease are included. The molecular development of novel forms and the importance of vitamin D analogues for treatment in chronic kidney disease, for example, are examined in detail. Final chapters look at the importance of vitamin D analogues in the treatment of psoriasis, therapeutic and chemoprevention of prostate cancer, and the use of vitamin D analogues in the treatment of serious autoimmune diseases. From my short descriptions of each section of this volume, it is obvious that Dr. Holick has succeeded in including the most significant areas of vitamin D research over the past decade in one single volume.

Vitamin D's role in health and disease resistance is complex, yet the editor and authors have provided chapters that balance the most technical information with practical discussions of their importance for clients and patients as well as graduate and medical students, health professionals, and academicians. Hallmarks of each of the chapters include abstracts at the beginning of each chapter as well as key words. Each chapter has complete definitions of terms with the abbreviation fully defined and there is consistent use of terms between chapters. There are over 260 relevant tables, graphs, and figures as well as more than 5,600 up-to-date references; all of the 59 chapters include a conclusion section that provides the highlights of major findings. The volume contains a highly annotated index and within chapters, readers are referred to relevant information in other chapters.

Dr. Holick has chosen 97 of the most well recognized and respected authors who are internationally distinguished researchers, clinicians, and epidemiologists. The authors provide a broad foundation for understanding the role of nutritional status, dietary intakes, route of vitamin D synthesis, life stages of patients and also disease state and the multiple effects of genetics and molecular biology on the clinical aspects of therapeutic management of chronic conditions that can be affected by vitamin D. The inventory of valuable information on the current vitamin D status of populations around the globe attests to the critical importance of adequate vitamin D especially during pregnancy, lactation, and early childhood. The volume adds further value as recommendations and practice guidelines are included at the end of relevant chapters.

In conclusion, "*Vitamin D: Physiology, Molecular Biology and Clinic Aspects*, Second Edition,", edited by Michael F. Holick, Ph.D., M.D. provides health professionals in many areas of research and practice with the most up-to-date, well-referenced volume on the importance of vitamin D and its significance in maintaining human health. This volume will again serve the reader as the benchmark in this complex area of interrelationships between vitamin D nutritional status, molecular functions, physiological functioning of organ systems, disease status, age, sex, genetic characteristics, route of administration, duration, vitamin D analogues, and the expanding knowledge of the sites and roles of the extra-renal synthesis of active vitamin D. Moreover, the interactions between genetic and environmental factors and the numerous co-morbidities seen especially in the aging population are clearly delineated so that students as well as practitioners can better understand the complexities of these interactions with regard to vitamin D. Dr. Holick is applauded for his efforts to develop the most authoritative resource in the field to date and this excellent text is a very welcome addition to the Nutrition and Health Series.

Adrianne Bendich, Ph.D., FACN

Preface

In the past 8 years since this book appeared, it is now being recognized globally that vitamin D deficiency is one of the most if not the most prevalent nutritionally related medical condition in the world. The major reason for this is the lack of appreciation that there is very little vitamin D present either naturally or in fortified foods from dietary sources and that it is exposure to sunlight that has been and continues to be the major source of vitamin D for both children and adults worldwide. Vitamin D deficiency will not only prevent children from attaining their genetically programmed maximum height but will also put them at increased risk for having a lower peak bone density as well as increased risk of fracture later in life. In utero, vitamin D deficiency not only increases the risk of the mother having preeclampsia and a caesarian section for birthing but increases the infant's risk of wheezing disorders. Vitamin D deficiency during childhood is thought to increase risk of the child developing type I diabetes, multiple sclerosis, rheumatoid arthritis, and Crohn's disease later in life. Vitamin D deficiency in adults not only will cause osteopenia and osteoporosis but increase risk of many deadly chronic diseases including common cancers, autoimmune diseases, heart disease, stroke, and infectious diseases.

As noted in the first edition of this book, vitamin D is a truly remarkable as it is both an essential nutrient and hormone that has a wide variety of biologic effects on the human body that are important for health. The reason that vitamin D plays such a crucial role in maintaining health is in part due to the fact that vitamin D receptors are present in every tissue and cell in the body. It is also known that once vitamin D is made in the skin or obtained from the diet, it undergoes sequential activation steps in the liver to 25-hydroxyvitamin D and kidney to 1,25-dihydroxyvitamin D before it can act on its vitamin D receptor in target tissues. 1,25-Dihydroxyvitamin D not only regulates calcium metabolism, but interacts with its receptor in various cells to regulate cell growth, insulin production, renin production, modulates immune function, and enhances the destruction of the infectious agents and is important for vascular tone and myocardial function.

The goal of the second edition of this book is to provide the reader with a global appreciation of how common vitamin D deficiency is in the world. It explores how vitamin D is able to maintain not only skeletal health but prevent serious chronic diseases. The book also provides a roadmap for how to evaluate patients with vitamin D deficiency and how to appropriately treat them.

As a result of a mountain of new information about the health benefits of vitamin D, the book has been expanded substantially to include many new topics including several chapters identifying vitamin D deficiency epidemics in Asia, Africa, Europe, and the United States. New chapters on the role of vitamin D in preventing type I diabetes, wheezing disorders in young children, cardiovascular disease, type II diabetes, infectious diseases, and cancers have been added to this book. As a result, the number of chapters in this book has more than doubled from the original 25 to 59 chapters.

The inspiration for the second edition of the *Vitamin D: Physiology, Molecular Biology and Clinic Aspects* came not only from discussions with medical students, house staff, health-care professionals, internists, dermatologists, basic scientists but also from my interaction with the lay press and from my global travels in response to invitations to provide the latest information concerning the growing vitamin D deficiency epidemic. The second edition brings together leading world experts in various aspects of vitamin D; the same experts who are not only doing cutting edge research in the field of vitamin D, but many are clinicians–scientists who see patients with vitamin D deficiency and all of the serious ramifications. The book is designed and organized not only to be an up-to-date review on the subject, but also to provide medical students, graduate students, health-care professionals, and even the lay public with a reference source for the most up-to-date information about the vitamin D deficiency pandemic and its clinical implications for health and disease. It is hoped that this book will not only stimulate new interest regarding the vitamin D deficiency pandemic, but ignite a grassroots effort to eliminate this insidious deficiency by encouraging worldwide food fortification programs with vitamin D and to educate the public and health-care professionals about the beneficial effect of sensible sun exposure as a major means for satisfying the body's vitamin D requirement.

ACKNOWLEDGMENTS

I am grateful to all of the hard work and efforts that were made by all of the contributing authors. My thanks to Lorrie Butler and the staff at Humana/Springer for all their help in organizing the book. A special thanks to Dr. Adrianne Bendich, Ph.D, for her helpful comments, guidance, and insightfulness in being series editor of the outstanding Nutrition and Health series.

Contents

PART I: INTRODUCTION AND BASIC BIOLOGY

PART V: SUNLIGHT, VITAMIN D AND CANCER

PART VI: VITAMIN D DEFICIENCY AND CHRONIC DISEASE

PART VII: CLINICAL USES OF VITAMIN D ANALOGUES

Contributors

JOHN S. ADAMS, MD • *Department of Orthopedic Surgery, University of California at Los Angeles, Los Angeles, CA, USA*

LUCIANO ADORINI • *Intercept Pharmaceuticals, Corciano (Perugia) 06073, Italy*

DARE AJIBADE • *Department of Biochemistry and Molecular Biology, University of Medicine and Dentistry of New Jersey, New Jersey Medical School, Newark, NJ 07103, USA*

ALBERTO ASCHERIO, MD DRPH • *Departments of Nutrition and Epidemiology, Harvard School of Public Health; Channing Laboratory, Department of Medicine, Brigham and Women's Hospital and Harvard Medical School, Boston, MA, USA*

JULIA BARSONY • *Division of Endocrinology and Metabolism, Georgetown University, Washington, DC 20007, USA*

BRYAN S. BENN • *Department of Biochemistry and Molecular Biology, University of Medicine and Dentistry of New Jersey, New Jersey Medical School, Newark, NJ 07103, USA*

ISHIR BHAN, MD, MPH • *Renal Unit, Massachusetts General Hospital, Boston, MA 02114, USA*

DOUGLASS BIBULD, MD • *Mattapan Community Health Center, Mattapan, MA 02126, USA*

DANIEL D. BIKLE, MD, PHD • *Veterans Affairs Medical Center and University of California, San Francisco, CA, USA*

N. BINKLEY • *Osteoporosis Clinical Center and Research Program, University of Wisconsin-Madison, Madison, WI 53705, USA*

HEIKE BISCHOFF-FERRARI, MD, DRPH • *Centre on Aging and Mobility, University of Zurich, Switzerland*

MARK J. BOLLAND, MBCHB, PHD • *Department of Medicine, University of Auckland, Auckland, New Zealand*

ØYVIND SVERRE BRULAND • *Department of Oncology, Rikshospitalet-Radiumhospitalet Medical Center, Montebello, 0310 Oslo, Norway*

THOMAS BURNE • *Queensland Brain Institute, University of Queensland, St Lucia, QLD 4072; Queensland Centre for Mental Health Research, The Park Centre for Mental Health, Wacol, QLD 4076, Australia*

MONA S. CALVO, PHD • *Office of Applied Research and Safety Assessment, Center for Food Safety and Applied Nutrition, US Food and Drug Administration, Laurel, MD, USA*

CARLOS A. CAMARGO JR, MD, DRPH • *Department of Emergency Medicine, Massachusetts General Hospital, Harvard Medical School, Boston, MA*

CARSTEN CARLBERG • *Life Sciences Research Unit, University of Luxembourg, L-1511 Luxembourg, Luxembourg; Department of Biochemistry, University of Kuopio, FIN-70211 Kuopio, Finland*

TAI C. CHEN • *Vitamin D, Skin and Bone Research Laboratory, Section of Endocrinology, Nutrition, and Diabetes, Department of Medicine, Boston University School of Medicine, Boston, MA, USA*

HONG CHEN, MD • *Department of Orthopedic Surgery and Department of Molecular, Cell and Developmental Biology, Orthopedic Hospital Research Center, UCLA/Orthopedic Hospital UCLA, Los Angeles, CA 90095-7358, USA*

SYLVIA CHRISTAKOS • *Department of Biochemistry and Molecular Biology, University of Medicine and Dentistry of New Jersey, New Jersey Medical School, Newark, NJ 07103, USA*

RENE F. CHUN, PHD • *Department of Orthopedic Surgery and Department of Molecular, Cell and Developmental Biology, Orthopedic Hospital Research Center, UCLA/Orthopedic Hospital UCLA, Los Angeles, CA 90095-7358, USA*

HEIDE S. CROSS • *Department of Pathophysiology, Medical University of Vienna, Austria*

FERNANDO CRUZAT • *Departamento de Bioquimica y Biologia Molecular, Facultad de Ciencias Biologicas, Universidad de Concepcion, Concepcion, Chile*

XIAOYING CUI • *Queensland Brain Institute, University of Queensland, St Lucia, QLD 4072; Queensland Centre for Mental Health Research, The Park Centre for Mental Health, Wacol, QLD 4076, Australia*

ARNE DAHLBACK • *Department of Physics, University of Oslo, 0316 Oslo, Norway*

HECTOR F. DELUCA • *Department of Biochemistry, University of Wisconsin-Madison, Madison, WI 53706, USA*

MARIE B. DEMAY • *Endocrine Unit, Massachusetts General Hospital, Boston, MA 02114, USA*

MANUEL DÍAZ-CURIEL • *Department of Internal Medicine/Bone Metabolism Disorders, Fundación Jiménez Díaz, Universidad Autónoma, RETICEF, Madrid, Spain*

AMY D. DIVASTA, MD, MMSC • *Divisions of Adolescent Medicine and Endocrinology, Children's Hospital Boston, Boston, MA 02115, USA*

DIANE R. DOWD • *Department of Pharmacology, Case Western Reserve University, Cleveland, OH 44106, USA*

ADRIANA S. DUSSO PHD • *Renal Division, Department of Internal Medicine. Washington University School of Medicine, St. Louis, MO 63110. USA*

GHADA EL-HAJJ FULEIHAN, MD, MPH • *Calcium Metabolism and Osteoporosis Program, American University of Beirut, Beirut, Lebanon*

MYRTO ELIADES, MD • *Division of Endocrinology, Diabetes and Metabolism, Tufts Medical Center, Boston, MA 02111, USA*

SAMER EL-KAISSI, MD, PHD • *Specialized Diabetes & Endocrine Center, King Fahad Medical City, Riyadh 11525, Saudi Arabia*

DARRYL EYLES • *Queensland Brain Institute, University of Queensland, St Lucia, QLD 4072; Queensland Centre for Mental Health Research, The Park Centre for Mental Health, Wacol, QLD 4076, Australia*

DAVID FELDMAN • *Division of Endocrinology, Gerontology and Metabolism, Stanford University School of Medicine, Stanford University Medical Center, Stanford, CA 94305-5103, USA*

PHILIP R. FISCHER • *Department of Pediatric and Adolescent Medicine, Mayo Clinic, Rochester, MN, USA*

FRANK C. GARLAND • *Department of Family and Preventive Medicine, University of California San Diego, La Jolla, CA 92093-0631, USA; Naval Health Research Center, 140 Sylvester Rd, San Diego, CA 92106-3521, USA*

CEDRIC F. GARLAND • *Department of Family and Preventive Medicine, University of California San Diego, La Jolla, CA 92093-0631, USA; Naval Health Research Center, 140 Sylvester Rd, San Diego, CA 92106-3521, USA*

ADIT A. GINDE, MD, MPH • *Department of Emergency Medicine, University of Colorado Denver School of Medicine Aurora, CO*

EDWARD GIOVANNUCCI, MD, SCD • *Departments of Nutrition, Epidemiology, Harvard School of Public Health; Channing Laboratory, Department of Medicine, Brigham and Women's Hospital and Harvard Medical School, Boston, MA 02115, USA*

GAIL R. GOLDBERG • *MRC Human Nutrition Research, Elsie Widdowson Laboratory, Cambridge UK; MRC Keneba, The Gambia*

DAVID GOLTZMAN, MD • *Department of Medicine, McGill University and McGill University Health Centre, Montreal, QC, Canada, H3A 1A1*

CATHERINE M. GORDON, MD, MSc • *Divisions of Adolescent Medicine and Endocrinology, Children's Hospital Boston, Boston, MA 02115, USA*

EDWARD D. GORHAM • *Department of Family and Preventive Medicine, University of California, San Diego, La Jolla, CA 92093-0631, USA; Naval Health Research Center, 140 Sylvester Rd, San Diego, CA 92106-3521, USA*

RAVINDER GOSWAMI • *Department of Endocrinology and Metabolism, All India Institute of Medical Sciences, Ansari Nagar, New Delhi, India*

WILLIAM B. GRANT, PHD • *Sunlight, Nutrition, and Health Research Center (SUNARC), San Francisco, CA 94164-1603, USA*

TIM GREEN • *Food, Nutrition and Health, The University of British Columbia, Vancouver, BC, Canada*

DAVID A. HANLEY, MD, FRCPC • *Departments of Medicine, Community Health Sciences and Oncology, Faculty of Medicine, University of Calgary, Calgary, Alberta T2N 4N1, Canada*

C.V. HARINARAYAN • *Department of Endocrinology and Metabolism, Sri Venkateswara Institute of Medical Sciences, Tirupati – 517 507, Andhra Pradesh, India*

LOUISE HARVEY • *Queensland Brain Institute, University of Queensland, QLD 4072, Australia*

ROBERT P. HEANEY, MD • *Creighton University, Omaha, NE 68131, USA*

MARTIN HEWISON, PHD • *Department of Orthopedic Surgery, University of California at Los Angeles, Los Angeles, CA, USA*

MICHAEL F. HOLICK, PHD, MD • *Vitamin D, Skin and Bone Research Laboratory, Section of Endocrinology, Nutrition, and Diabetes, Department of Medicine, Boston University School of Medicine, Boston, MA, USA*

BRUCE W. HOLLIS, PHD • *Darby Children's Research Institute, Medical University of South Carolina, Charleston, SC 29414, USA*

ELINA HYPPÖNEN, PHD, MSC, MPH • *MRC Centre of Epidemiology for Child Health, UCL Institute of Child Health, London WC1N 1EH, UK*

LANDING M.A. JARJOU • *MRC Keneba, The Gambia*

GLENVILLE JONES • *Department of Biochemistry, Queen's University, Kingston, Ontario, Canada K7L 3N6*

KERRY S. JONES • *MRC Keneba, The Gambia*

SHASHANK R. JOSHI • *Lilavati Hospital, Bhatia Hospital and Joshi Clinic; Department of Endocrinology, Seth GS Medical College and KEM Hospital, Mumbai, India*

ASTA JUZENIENE • *Department of Radiation Biology, Rikshospitalet-Radiumhospitalet Medical Center, Montebello, 0310 Oslo, Norway*

ROLFDIETER KRAUSE, MD • *Chair of Clinical Complementary Medicine[1]; Nephrological Center Moabit[2], Charité – University Medical Center[1]; KfH Kuratorium for Dialysis and Kidney Transplantation[2], Berlin, Germany*

ARUNA V. KRISHNAN • *Divisions of Endocrinology, Department of Medicine, Stanford University School of Medicine, Stanford, CA 94305, USA*

JOAN M. LAPPE, PHD, RN, FAAN • *Creighton University, Omaha NE 68131, USA*

MARÍA JESÚS LARRIBA • *Instituto de Investigaciones Biomédicas "Alberto Sols", Consejo Superior de Investigaciones Científicas-Universidad Autónoma de Madrid, Madrid E-28029, Spain.*

G. LENSMEYER • *Department of Laboratory Medicine, University of Wisconsin-Madison, Madison, WI 53705, USA*

YAN CHUN LI, PHD • *Department of Medicine, The University of Chicago, Chicago, IL 60637, USA*

JANE B. LIAN • *Department of Cell Biology, University of Massachusetts Medical School, Worcester, MA 01655, USA*

THOMAS S. LISSE, PHD • *Department of Orthopedic Surgery and Department of Molecular, Cell and Developmental Biology, Orthopedic Hospital Research Center, UCLA/Orthopedic Hospital UCLA, Los Angeles, CA 90095-7358, USA*

PHILIP LIU, PHD • *Division of Dermatology, Department of Medicine, University of California at Los Angeles, Los Angeles, CA, USA*

ZHIREN LU • *Vitamin D, Skin and Bone Research Laboratory, Section of Endocrinology, Nutrition, and Diabetes, Department of Medicine, Boston University School of Medicine, Boston, MA, USA*

PAUL N. MACDONALD • *Department of Pharmacology, Case Western Reserve University, Cleveland, OH 44106*

ALAN MACKAY-SIM • *Eskitis Institute for Cell and Molecular Therapies, Griffith University, Brisbane, QLD 4111 Australia*

PETER J. MALLOY • *Division of Endocrinology, Gerontology and Metabolism, Stanford University School of Medicine, Stanford University Medical Center, Stanford, CA 94305-5103, USA*

JONATHAN M. MANSBACH, MD • *Department of Medicine, Children's Hospital Boston, Harvard Medical School, Boston, MA, USA*

RAMAN KUMAR MARWAHA • *Department of Endocrinology and Thyroid Research, Institute of Nuclear Medicine & Allied Sciences, Timarpur, Delhi 110054, India*

JOHN MCGRATH • *Queensland Brain Institute, University of Queensland, St Lucia, QLD 4072; Queensland Centre for Mental Health Research, The Park Centre for Mental Health, Wacol, QLD 4076; Department of Psychiatry, University of Queensland, St Lucia, QLD 4072, Australia*

LINDA A. MERLINO, MS • *The University of Iowa, Iowa City, IA, USA*

JOHAN MOAN • *Department of Radiation Biology, Rikshospitalet-Radiumhospitalet Medical Center, Montebello, 0310 Oslo; Department of Physics, University of Oslo, Oslo 0316, Norway*

SHARIF B. MOHR • *Department of Family and Preventive Medicine, University of California San Diego, La Jolla, CA 92093-0631, USA; Naval Health Research Center, 140 Sylvester Rd, San Diego, CA 92106-3521, USA*

MARTIN A. MONTECINO • *Departamento de Bioquimica y Biologia Molecular, Facultad de Ciencias Biologicas, Universidad de Concepcion, Concepcion, Chile*

LEIF MOSEKILDE, MD, DMSci • *Department of Endocrinology and Metabolism C, Aarhus University Hospital, DK 8000 rhus C, Denmark*

KASSANDRA L. MUNGER, M.Sc. • *Department of Nutrition, Harvard School of Public Health, Boston, MA, USA*

ALBERTO MUÑOZ • *Instituto de Investigaciones Biomédicas "Alberto Sols", Consejo Superior de Investigaciones Científicas-Universidad Autónoma de Madrid, Madrid E-28029, Spain*

TALLY NAVEH-MANY • *Hebrew University Hadassah Medical Center, Jerusalem 91120, Israel*

MEINRAD PETERLIK • *Department of Pathophysiology, Medical University of Vienna, Austria*

JOHN M. PETTIFOR • *MRC Mineral Metabolism Research Unit, Department of Paediatrics, Chris Hani Baragwanath Hospital and the University of the Witwatersrand, Johannesburg, South Africa*

ANASTASSIOS G. PITTAS, MD, MS • *Division of Endocrinology, Diabetes and Metabolism, Tufts Medical Center, Boston, MA 02111, USA*

LORI A. PLUM • *Department of Biochemistry, University of Wisconsin-Madison, Madison, WI 53706, USA*

ALINA CARMEN POROJNICU • *Department of Radiation Biology, Rikshospitalet-Radiumhospitalet Medical Center, Montebello, 0310 Oslo, Norway*

ANN PRENTICE • *MRC Human Nutrition Research, Elsie Widdowson Laboratory, Cambridge UK; MRC Keneba, The Gambia*

JOSE MANUEL QUESADA-GOMEZ • *Unidad de Metabolismo Mineral, Servicio de Endocrinología y Nutrición, Centro CEDOS, Unidad de I+D+i Sanyres, Hospital Universitario Reina Sofía, RETICEF, Córdoba, Spain*

RAHUL RAY, PhD • *Boston University School of Medicine, Boston, MA 02118, USA*

JORG REICHRATH • *Klinik fur Dermatologie, Venerologie und Allergologie, Universitatsklinkum des Saarlandes, Homburg, Saar 66421, Germany*

IAN R. REID, MD • *Department of Medicine, University of Auckland, Auckland, New Zealand*

INEZ SCHOENMAKERS • *MRC Human Nutrition Research, Elsie Widdowson Laboratory, Cambridge UK*

GARY G. SCHWARTZ • *Departments of Cancer Biology and Epidemiology and Prevention, Comprehensive Cancer Center of Wake Forest University, Winston-Salem, NC 27157, USA*

ROBERT SCRAGG, MBBS, PhD • *School of Population Health, University of Auckland Private Bag, Auckland, New Zealand*

SUPHIA SHERBEENI, MD, FRCPE • *Specialized Diabetes & Endocrine Center, King Fahad Medical City, Riyadh 11525, Saudi Arabia*

JUSTIN SILVER • *Hebrew University Hadassah Medical Center, Jerusalem 91120, Israel*

JANET L. STEIN • *Department of Cell Biology, University of Massachusetts Medical School, Worcester, MA 01655, USA*

GARY S. STEIN • *Department of Cell Biology, University of Massachusetts Medical School, Worcester, MA 01655, USA*

XIAOCUN SUN • *Departments of Nutrition and Medicine, The University of Tennessee, Knoxville, TN 37996-1920, USA*

HECTOR TAMEZ, MD, MPH • *Renal Unit, Massachusetts General Hospital, Boston, MA 02114, USA*

SARAH N. TAYLOR, MD • *Darby Children's Research Institute, Medical University of South Carolina, Charleston, SC 29414, USA*

STEVEN L. TEITELBAUM • *Washington University School of Medicine, Saint Louis, MO, USA*

√ TOM D. THACHER • *Department of Family Medicine, Mayo Clinic, Rochester, MN, USA*

RAVI THADHANI, MD, MPH • *Renal Unit, Massachusetts General Hospital, Boston, MA 02114, USA*

MASANORI TOKUMOTO, MD, PhD • *Renal Division, Department of Internal Medicine. Washington University School of Medicine, St. Louis, MO 63110. USA*

KRISTEN K. VAN DER VEEN, BA • *Divisions of Adolescent Medicine and Endocrinology, Children's Hospital Boston, Boston, MA 02115, USA*

ANDRÉ J. VAN WIJNEN • *Department of Cell Biology, University of Massachusetts Medical School, Worcester, MA 01655, USA*

BERNARD VENN • *Department of Human Nutrition, University of Otago, Dunedin, New Zealand*

REINHOLD VIETH • *Pathology and Laboratory Medicine, Department of Laboratory Medicine and Pathobiology, Mount Sinai Hospital, University of Toronto, Toronto, ON M5G 1X5, Canada*

CAROL L. WAGNER, MD • *Darby Children's Research Institute, Medical University of South Carolina, Charleston, SC 29414, USA*

SUSAN J. WHITING, PhD • *College of Pharmacy and Nutrition, University of Saskatchewan, Saskatoon, Saskatchewan, Canada*

MICHAEL B. ZEMEL • *Departments of Nutrition and Medicine, The University of Tennessee, Knoxville, TN 37996-1920, USA*

I INTRODUCTION AND BASIC BIOLOGY

1 Vitamin D and Health: Evolution, Biologic Functions, and Recommended Dietary Intakes for Vitamin D

Michael F. Holick

Abstract   Vitamin D deficiency is now being recognized as one of the most common medical conditions worldwide. The consequences of vitamin D deficiency include poor bone development and health as well as increased risk of many chronic diseases including type I diabetes, rheumatoid arthritis, Crohn's disease, multiple sclerosis, heart disease, stroke, infectious diseases, as well as increased risk of dying of many deadly cancers including colon, prostate, and breast. The major source of vitamin D for most humans is exposure to sunlight. However, avoidance of sun exposure has resulted in an epidemic of vitamin D deficiency. Once vitamin D is made in the skin or ingested from the diet, it requires activation steps in the liver and kidney to form 25-hydroxyvitamin D [25(OH)D] and 1,25-dihydroxyvitamin D. 25(OH)D is the major circulating form of vitamin D used by clinicians to determine a patient's vitamin D status. A blood level of 25(OH)D <20 ng/ml is considered to be vitamin D deficiency, whereas a level 21–29 ng/ml is insufficient, and to maximize vitamin D's effect for health, 25(OH)D should be >30 ng/ml. Vitamin D intoxication will not occur until a blood level of 25(OH)D exceeds 150–200 ng/ml. Both the adequate intake recommendations and safe upper limits for vitamin D are woefully underestimated. For every 100 IU of vitamin D ingested, the blood level of 25(OH)D increases by 1 ng/ml. Thus, children during the first year of life need at a minimum 400 IU of vitamin D/day and 1,000 IU of vitamin D/day may be more beneficial and will not cause toxicity. The same recommendation can be made for children 1 year and older. For adults, a minimum of 1,000 IU of vitamin D/day is necessary and 2,000 IU of vitamin D/day is preferred if there is inadequate sun exposure. The safe upper limit for children can easily be increased to 5,000 IU of vitamin D/day, and for adults, up to 10,000 IU of vitamin D/day has been shown to be safe. The goal of this chapter is to give a broad perspective about vitamin D and to introduce the reader to the vitamin D deficiency pandemic and its insidious consequences on health that will be reviewed in more detail in the ensuing chapters.

Key Words: Vitamin D; sunlight; 25-hydroxyvitamin D; cancer; diabetes; adequate intake; rickets; vitamin D deficiency, 1,25-dihydroxyvitamin D, vitamin D receptor

From: *Nutrition and Health: Vitamin D*
Edited by: M.F. Holick, DOI 10.1007/978-1-60327-303-9_1,
© Springer Science+Business Media, LLC 2010

1. EVOLUTIONARY PERSPECTIVE

1.1. The Calcium Connection

Approximately 400 million years ago, as vertebrates ventured from the ocean onto land, they were confronted with a significant crisis. As they had evolved in the calcium-rich ocean environment, they utilized this abundant cation for signal transduction and a wide variety of cellular and metabolic processes. In addition, calcium became a major component of the skeleton of marine animals and provided the "cement" for structural support. However, on land, the environment was deficient in calcium; as a result, early marine vertebrates that ventured onto the land needed to develop a mechanism to utilize and process the scarce amounts of calcium in their environment in order to maintain their calcium-dependent cellular and metabolic activities and also satisfy the large requirement for calcium to mineralize their skeletons *(1)*.

For most ocean-dwelling animals that were bathed in the high calcium bath (approximately 400 nmol), they could easily extract this divalent cation from the ocean by specific calcium transport mechanisms in the gills or by simply absorbing it through their skin. However, once on land, a new strategy was developed whereby the intestine evolved to efficiently absorb the calcium present in the diet. For reasons that are unknown, an intimate relationship between sunlight and vitamin D evolved to play a critical role in regulating intestinal absorption of calcium from the diet to maintain a healthy mineralized skeleton and satisfy the body's requirement for this vital mineral.

Although vitamin D (D represents D_2 or D_3 or both) became essential for stimulating the intestine to absorb calcium from the diet, it could only do so by being activated first in the liver to 25-hydroxyvitamin D [25(OH)D] and then in the kidneys to its active form 1,25-dihydroxyvitamin D [$1,25(OH)_2D$] *(1, 2)* (Fig. 1). Once formed, $1,25(OH)_2D$ enters the circulation and travels to its principal calcium-regulating target tissues, the small intestine and bone. The small intestinal absorptive cells contain specific receptors (known as vitamin D receptors) that specifically bind $1,25(OH)_2D$ and in turn activate vitamin D-responsive genes to enhance intestinal calcium absorption *(1, 3)* (see Chapter 7 for details). However, when dietary calcium is insufficient to satisfy the body's requirement, $1,25(OH)_2D$ travels to the bone and interacts with the bone-forming cells (osteoblasts), which in turn stimulate the formation of bone-resorbing cells (osteoclasts). This process results in an increase in osteoclastic activity, which is responsible for removing calcium stores from the bone and depositing it into the blood to maintain the blood calcium in the normal range (calcium homeostasis). Thus, the

--→

Fig. 1. (continued) $1,25(OH)_2D$ enhances intestinal calcium absorption in the small intestine by stimulating the expression of the epithelial calcium channel (ECaC; also known as transient receptor potential cation channel subfamily V member 6; TRPV6) and the calbindin 9 K (calcium-binding protein; CaBP). $1,25(OH)_2D$ is recognized by its receptor in osteoblasts causing an increase in the expression of receptor activator of NFκB ligand (RANKL). Its receptor RANK on the preosteoclast binds RANKL which induces the preosteoclast to become a mature osteoclast. The mature osteoclast removes calcium and phosphorus from the bone to maintain blood calcium and phosphorus levels. Adequate calcium and phosphorus levels promote the mineralization of the skeleton and maintain neuromuscular function. Holick copyright 2007. Reproduced with permission.

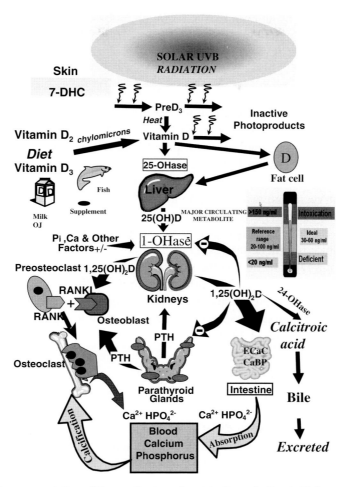

Fig. 1. Schematic representation of the synthesis and metabolism of vitamin D for regulating calcium, phosphorus, and bone metabolism. During exposure to sunlight 7-dehydrocholesterol (7-DHC) in the skin is converted to previtamin D_3 (preD$_3$). PreD$_3$ immediately converts by a heat-dependent process to vitamin D_3. Excessive exposure to sunlight degrades previtamin D_3 and vitamin D_3 into inactive photoproducts. Vitamin D_2 and vitamin D_3 from dietary sources are incorporated into chylomicrons, transported by the lymphatic system into the venous circulation. Vitamin D (D represents D_2 or D_3) made in the skin or ingested in the diet can be stored in and then released from fat cells. Vitamin D in the circulation is bound to the vitamin D-binding protein which transports it to the liver where vitamin D is converted by the vitamin D-25-hydroxylase (25-OHase) to 25-hydroxyvitamin D [25(OH)D]. This is the major circulating form of vitamin D that is used by clinicians to measure vitamin D status (although most reference laboratories report the normal range to be 20–100 ng/ml, the preferred healthful range is 30–60 ng/ml). 25(OH)D is biologically inactive and must be converted in the kidneys by the 25-hydroxyvitamin D-1α-hydroxylase (1-OHase) to its biologically active form 1,25-dihydroxyvitamin D [1,25(OH)$_2$D]. Serum phosphorus, calcium, fibroblast growth factor (FGF-23), and other factors can either increase (+) or decrease (–) the renal production of 1,25(OH)$_2$D. 1,25(OH)$_2$D feedback regulates its own synthesis and decreases the synthesis and secretion of parathyroid hormone (PTH) in the parathyroid glands. 1,25(OH)$_2$D increases the expression of the 25-hydroxyvitamin D-24-hydroxylase (24-OHase) to catabolize 1,25(OH)$_2$D and 25(OH)D to the water soluble biologically inactive calcitroic acid which is excreted in the bile.

major physiologic role of vitamin D is to maintain intra- and extracellular calcium concentrations in order to maintain essential metabolic functions, which are important for most physiologic activities *(4)* (Fig. 1).

1.2. Photosynthesis of Vitamin D in the Skin

Vitamin D is recognized as the sunshine vitamin because when the skin is exposed to sunlight, the action of the ultraviolet B (UVB = 290–315 nm) portion of sunlight causes the photolysis of 7-dehydrocholesterol to previtamin D (see Chapter 2 for details) *(5)*. The photosynthesis of vitamin D has been occurring on earth for at least 750 million years *(1)*. Evidence shows that the earliest phytoplankton that lived in the Sargasso Sea over 750 million years ago produced previtamin D_2. Although it is unknown whether this photosynthetic process played a significant role in calcium metabolism in these early life forms, there is strong evidence that most land animals from amphibians to primates retain this vital photosynthetic process *(1)*.

Once previtamin D_3 is formed in the skin, it is quickly transformed by the rearrangement of its double bonds (isomerization) into vitamin D_3 (Fig. 1). A variety of endogenous factors and environmental influences can alter the skin's production of vitamin D, including skin pigmentation, sunscreen use, clothing, latitude, season, time of day, and aging (see Chapter 2 for details) *(5)*.

1.3. Metabolism of Vitamin D

It is now well accepted that in order for vitamin D to carry out its essential biologic functions on calcium and bone metabolism it must be metabolized first in the liver mitochondria and microsomes to [25(OH)D]. This is the major circulating form of vitamin D and is used to determine a patient's vitamin D status. It, however, is not active and must be metabolized in the mitochondria of renal tubular cells to $1,25(OH)_2D$ *(1 – 4)* (Fig. 1).

From an evolutionary perspective, there is firm evidence that the metabolism of vitamin D to $1,25(OH)_2D$ occurred in bony fish, amphibians, reptiles, and most vertebrates including humans. Although it is unknown why vitamin D required two hydroxylations before it became biologically active, it is curious that both carbon 25 and carbon 1 are prone to autooxidation because of their increased reactivity, i.e., carbon 25 is a tertiary carbon and carbon 1 is allelic to the triene system. It is possible that early in evolution these carbons were subjected to autooxidation. Once this occurred, the hydroxylations imparted some unique properties to the vitamin D molecule, possibly by enhancing the membrane transport of cations such as calcium. Since terrestrial vertebrates become dependent on vitamin D for their calcified skeletons, they developed enzymes to convert vitamin D efficiently to $1,25(OH)_2D$ *(1, 6)*

2. VITAMIN D DEFICIENCY AND SOURCES OF VITAMIN D

2.1. Consequences of Vitamin D Deficiency on Skeletal Health

At the turn of the twentieth century, rickets was rampant in the industrialized cities of northern Europe and North America. Vitamin D deficiency caused severe growth retardation. The lack of calcium in the bones resulted in deformities of the skeleton, characterized by a widening at the ends of the long bones because of disorganization in the hypertrophy and maturation of chondrocytes in the epiphyseal plates. Vitamin D deficiency is also associated with a low-normal blood calcium, low or low-normal fasting blood phosphorus, and elevated parathyroid hormone (PTH) levels that cause a mineralization defect in the skeleton (1, 2). Thus, the lack of calcium in the bone leads to the classic rachitic deformities of the lower limbs: bowed legs or knock knees (7) (Fig. 2) (see Chapter 33 for details). After the skeleton has matured and the epiphyseal plates have closed, the role of vitamin D is to maintain the maximum mineral content in the skeleton. It does this by optimizing the bone remodeling process. As long as there is an adequate source of calcium, the adult skeleton can maintain its peak bone mass for several decades until menopause and/or aging alters bone remodeling so that bone resorption is greater than bone formation, leading to gradual bone loss. Vitamin D deficiency in adults causes a decrease in the efficiency of intestinal calcium absorption from 30–40% to about 10–15% and secondary hyperparathyroidism (2, 4). This results in a mineralization defect of the skeleton that is similar to rickets. However, because the epiphyseal plates are closed in adults, there is no widening of the end of the long bones. Instead, there is a subtle but significant defect in bone mineralization of the collagen matrix (osteoid) that was laid down by the osteoblasts, leading to the painful bone disease osteomalacia (unmineralized osteoid). In addition, since the body can no

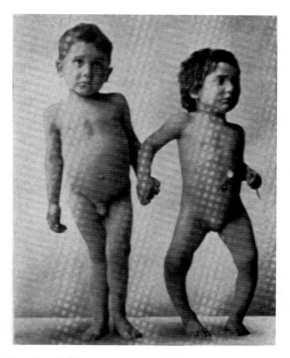

Fig. 2. Sister (*right*) and brother (*left*) ages 4 years and 6.5 years, respectively, demonstrating classic knock knees and bow legs, growth retardation, and other skeletal deformities. Holick copyright 2006. Reproduced with permission.

longer depend on the dietary calcium to satisfy its calcium requirements, it calls on the skeleton for its calcium stores. The increased PTH production mobilizes the precious calcium stores from the skeleton, thereby making the skeleton more porotic. Thus, vitamin D deficiency in adults can cause and exacerbate osteopenia and osteoporosis (see Chapter 35).

When dietary calcium intake, even in the presence of adequate vitamin D, is inadequate to satisfy the body's calcium needs, 1,25(OH)$_2$D stimulates the expression of receptor activator of NFκB (RANK) ligand (RANKL) that mobilizes osteoclast precursor stem cells and induces them to become mature osteoclasts (1, 8) (Fig. 1). These osteoclasts, in turn, enhance bone calcium resorption, thereby maintaining the blood calcium in an acceptable physiologic range (1, 2). Vitamin D does not have a direct role in the mineralization of the skeleton (9, 10). It indirectly participates in skeletal mineralization by its effects on maintaining the serum calcium and phosphorus in the normal range so they are at supersaturating levels for deposition into the collagen matrix to form calcium hydroxyapatite crystals.

2.2. Sources of Vitamin D

A major source of vitamin D for most humans comes from exposure of the skin to sunlight (4, 5). A variety of factors limit the skin's production of vitamin D$_3$. An increase in skin pigmentation (11) or the topical application of a sunscreen (12) will absorb solar UVB photons, thereby significantly reducing the production of vitamin D$_3$ in the skin by as much as 99% (12). Aging decreases the capacity of the skin to produce vitamin D$_3$ because of the decrease in the concentration of its precursor 7-dehydrocholesterol. Above the age of 65, there is fourfold decrease in the capacity of the skin to produce vitamin D$_3$ when compared with a younger adult (13). An alteration in the zenith angle of the sun caused by a change in latitude, season of the year, or time of day can dramatically influence the skin's production of vitamin D (4, 5, 14). Above and below latitudes of approximately 35° north and south, respectively, vitamin D synthesis in the skin is absent during most of the winter.

In nature, very few foods contain vitamin D (Table 1). Oily fish and oils from the liver of some fish, including cod and tuna, are naturally occurring foods containing vitamin D. Although it is generally accepted that livers from meat animals such as cows, pigs, and chickens contain vitamin D, there is no evidence for this. However, the common practice of Eskimos eating a small amount of polar bear liver did provide them with their vitamin D and vitamin A requirements because this animal concentrates both of these vitamins in its liver. Mushrooms and yeast contain ergosterol which is the provitamin D for vitamin D$_2$. When sun dried or UVB irradiated, the ergosterol in mushrooms and yeast is converted to vitamin D$_2$. When compared to vitamin D$_3$, the only differences are a double bond between carbons 22 and 23 and a methyl group on carbon 24 (Table 1).

In the United States and Canada, milk is routinely fortified with vitamin D as are some bread products, orange juices, cereals, yogurts, and cheeses (4, 15, 16). However, three surveys of the vitamin D content in milk in the United States and Canada have revealed that upward of 70% of samples tested did not contain 50% of the amount of vitamin D stated on the label (17 – 19). Approximately 10% of milk samples did not

Table 1

Various Food, Nutritional Supplement, and Pharmaceutical Forms of Vitamin D

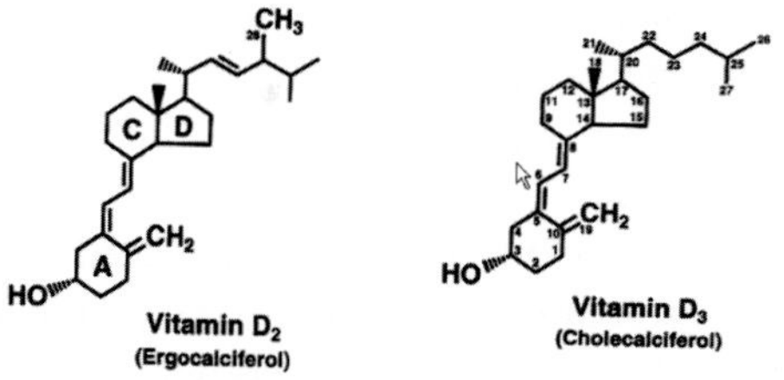

Source	*Vitamin D content IU = 25 ng*

Vitamin D$_2$ (Ergocalciferol) Vitamin D$_3$ (Cholecalciferol)

Natural sources

Cod liver oil	~400–1,000 IU/tsp vitamin D$_3$
Salmon, fresh wild caught	~600–1,000 IU/3.5 oz vitamin D$_3$
Salmon, fresh farmed	~100–250 IU/3.5 oz vitamin D$_3$, vitamin D$_2$
Salmon, canned	~300–600 IU/3.5 oz vitamin D$_3$
Sardines, canned	~300 IU/3.5 oz vitamin D$_3$
Mackerel, canned	~250 IU/3.5 oz vitamin D$_3$
Tuna, canned	236 IU/3.5 oz vitamin D$_3$
Shiitake mushrooms, fresh	~100 IU/3.5 oz vitamin D$_2$
Shiitake mushrooms, sun dried	~1,600 IU/3.5 oz vitamin D$_2$
Egg yolk	~20 IU/yolk vitamin D$_3$ or D$_2$
Sunlight/UVB radiation	~20,000 IU equivalent to exposure to one minimal erythemal dose (MED) in a bathing suit. Thus, exposure of arms and legs to 0.5 MED is equivalent to ingesting ~3,000 IU vitamin D$_3$

Fortified foods

Fortified milk	100 IU/8 oz usually vitamin D$_3$
Fortified orange juice	100 IU/8 oz vitamin D$_3$
Infant formulas	100 IU/8 oz vitamin D$_3$
Fortified yogurts	100 IU/8 oz usually vitamin D$_3$
Fortified butter	56 IU/3.5 oz usually vitamin D$_3$
Fortified margarine	429/3.5 oz usually vitamin D$_3$
Fortified cheeses	100 IU/3 oz usually vitamin D$_3$
Fortified breakfast cereals	~100 IU/serving usually vitamin D$_3$

Pharmaceutical sources in the United States

Vitamin D$_2$ (ergocalciferol)	50,000 IU/capsule
Drisdol (vitamin D$_2$) liquid	8,000 IU/cc

Supplemental sources

Multivitamin	400 IU vitamin D[a]; vitamin D$_2$ or vitamin D$_3$
Vitamin D$_3$	400, 800, 1,000, 2,000, 10,000, and 50,000 IU

[a]Vitamin D or calciferol usually means the product contains vitamin D$_2$. Ergocalciferol or vitamin D$_2$ means it only contains vitamin D$_2$.

Cholecalciferol or vitamin D$_3$ indicates the product only contains vitamin D$_3$. Reproduced with permission Holick copyright 2007.

contain any vitamin D. In Europe, most countries do not fortify milk with vitamin D because in the 1950s there was an outbreak of vitamin D intoxication in young children resulting in laws that forbade the fortification of foods with vitamin D *(4)*. However, with the recognition that vitamin D deficiency continues to be a great health concern for both children and older adults, many European countries are fortifying cereals, breads, and margarine with vitamin D. In many European countries, margarine is the major dietary source of vitamin D *(20)*. Sweden and Finland recently permitted fortifing milk with vitamin D.

Multivitamin D preparations often contain 400 IU of vitamin D_2 or vitamin D_3. Pharmaceutical preparations of vitamin D include capsules and tablets that contain 50,000 IU of vitamin D_2 and a pediatric liquid formulation (Drisdol) that contains 8,000 IU/ml. The recent recognition that vitamin D deficiency is a global health problem has led to the availability of vitamin D supplements. In the United States and Canada, 1,000 IU of vitamin D_3 is easily obtained in most pharmacies as an over-the-counter supplement (Table 1).

2.3. *Definition of Vitamin D Deficiency, Insufficiency, and Sufficiency*

There has been a lot of controversy about what the definition of vitamin D deficiency, insufficiency, and sufficiency should be. It is generally accepted that a blood level of 25(OH)D <20 ng/ml is considered to be vitamin D deficient *(4)*. This is based in part on the provocative testing of healthy adults who had blood levels of 25(OH)D between 11 and 25 ng/ml and who received 50,000 IU of vitamin D once a week for 8 weeks. The blood level of 25(OH)D increased more than 100%. On average, the blood level of PTH decreased by 55% for those who had a 25(OH)D level between 11 and 15 ng/ml and 35% decrease for those 16–20 ng/ml. Those adults who had a blood level above 20 ng/ml and who received the 50,000 IU of vitamin D/day had no change in their PTH level *(21)* (Fig. 3). When evaluating PTH levels in the general population, it was observed that the PTH levels continued to decline until the blood levels of 25(OH)D were between 30 and 40 ng/ml *(22, 23)* (Fig. 4). Furthermore, to maximize the efficiency of intestinal calcium absorption, it was reported by Heaney et al. *(24)* that in the same women who had a blood level of 20 ng/ml and was raised to on average 32 ng/ml, the efficiency of intestinal calcium absorption increased by 65%. Thus, to maximize vitamin D's effect on calcium metabolism, it is now recommended that a 25(OH)D be above 30 ng/ml. Based on these observations, it has been suggested that vitamin D deficiency be defined as a 25(OH)D <20 ng/ml, insufficiency 21–29 ng/ml, and sufficiency ≥30 ng/ml.

The recent revelations that vitamin D deficiency increases risk of chronic diseases including autoimmune diseases such as type I diabetes, multiple sclerosis, and rheumatoid arthritis, deadly cancers including prostate, colon, breast, cardiovascular disease, and stroke and may help modulate the immune system and fight infectious diseases including tuberculosis and upper respiratory tract infections, a blood level of 25(OH)D of >30 ng/ml is recommended for these health benefits and some have suggested that the blood level should be at least 40 ng/ml (Table 2). It is believed that many tissues and cells in the body have the capacity to locally produce $1,25(OH)_2D$. The local

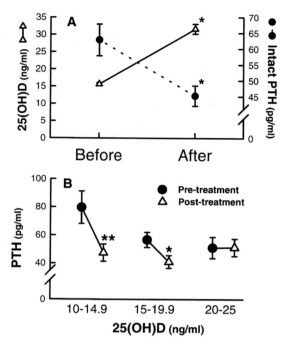

Fig. 3. (**a**) Serum levels of 25(OH)D (△) and parathyroid hormone (PTH; ●) before and after therapy with 50,000 IU of vitamin D_2 and calcium supplementation once a week for 8 weeks. (**b**) Serum levels of PTH levels in patients who had serum 25(OH)D levels between 10 and 25 ng/ml and who were stratified in increments of 5 ng/ml before and after receiving 50,000 IU of vitamin D_2 and calcium supplementation for 8 weeks, $^*P < 0.001$; $^{**}P < 0.02$. (Data from *(21)*).

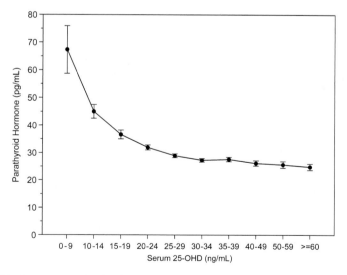

Fig. 4. Relationship between serum 25(OH)D concentrations and mean (±SE) serum concentrations of parathyroid hormone in osteoporosis patients receiving treatment. PTH = parathyroid hormone; 25(OH)D = 25-hydroxyvitamin D. Reproduced with permission *(23)*.

Table 2

Health Benefits and Disease Incidence Prevention Related to Serum 25(OH)D Level (83)

Serum 25(OH)D ng/ml	0	10	20	30	40	50	60	70
Rickets			100%					
Osteomalacia			100%					
Cancers, all combined						75%		
Breast cancer				30%		50%		67%
Ovarian cancer					20%	25%		
Colon cancer				50%		67%		
Non-Hodgkins lymphoma				25%	30%			
Kidney cancer				50%		67%		
Endometrial cancer						35%		
Type I diabetes				50%		80%		
Type 2 diabetes				50%				
Fractures, all combined				50%				
Falls, women				72%				
Multiple sclerosis					50%		66%	
Heart attack (men)				50%				
Peripheral vascular disease				80%				
Preeclampsia				50%				
Cesarean section				75%				

production of 1,25(OH)D is believed to control directly or indirectly up to 2,000 genes that are responsible for cell growth, angiogenesis, immunomodulation, vascular smooth muscle, and cardiomyocyte proliferation and inflammatory activities (Fig. 5) *(4, 5, 25 – 27)* (see Chapters 12–17, 40–51 for details).

2.4. The Vitamin D Deficiency Pandemic

It is now recognized and documented in this book that vitamin D deficiency is one of the most common medical conditions in the world *(4)*. It has been estimated that upward of 30–50% of both children and adults in the United States, Canada, Mexico, Europe, Asia, New Zealand, and Australia are vitamin D deficient. It is also extremely common in the most sunniest areas of the world including Mideast Countries including Saudi Arabia, Qatar, United Arab Emirates, and India *(28 – 36)* (see Chapters 23–28 for details).

The major reason for worldwide epidemic of vitamin D deficiency is the lack of appreciation that essentially none of our foods contain an adequate amount of vitamin D to satisfy the body's requirement which is now estimated to be 3,000–5,000 IU of vitamin D/day *(37)*. It is likely that our hunter/gather forefathers were generating several

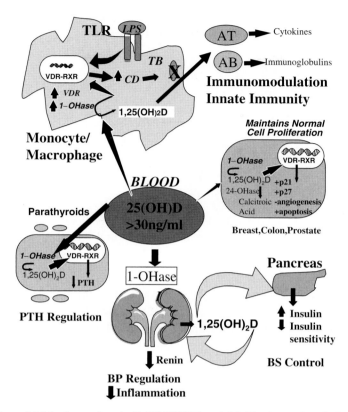

Fig. 5. Metabolism of 25-hydroxyvitamin D [25(OH)D] to 1,25 dihydroxyvitamin D 1,25(OH)$_2$D for non-skeletal functions. When a monocyte/macrophage is stimulated through its toll-like receptor 2/1 (TLR2/1) by an infective agent such as *Mycobacterium tuberculosis* (TB) or its lipopolysaccharide (LPS) the signal upregulates the expression of vitamin D receptor (VDR) and the 25-hydroxyvitamin D-1-hydroxylase (1-OHase). 25(OH)D levels >30 ng/ml provides adequate substrate for the 1-OHase to convert it to 1,25(OH)$_2$D. 1,25(OH)$_2$D returns to the nucleus where it increases the expression of cathelicidin (CD) which is a peptide capable of promoting innate immunity and inducing the destruction of infective agents such as TB. It is also likely that the 1,25(OH)$_2$D produced in the monocytes/macrophage is released to act locally on activated T (AT) and activated B (AB) lymphocytes which regulate cytokine and immunoglobulin synthesis, respectively. When 25(OH)D levels are ≈30 ng/ml, it reduces risk of many common cancers. It is believed that the local production of 1,25(OH)$_2$D in the breast, colon, prostate, and other cells regulates a variety of genes that control proliferation, including p21 and p27 as well as genes that inhibit angiogenesis and induced apoptosis. Once 1,25(OH)$_2$D completes the task of maintaining normal cellular proliferation and differentiation, it induces the 25-hydroxyvitamin D-24-hydroxylase (24-OHase). The 24-OHase enhances the metabolism of 1,25(OH)$_2$D to calcitroic acid which is biologically inert. Thus, the local production of 1,25(OH)$_2$D does not enter the circulation and has no influence on calcium metabolism. The parathyroid glands have 1-OHase activity and the local production of 1,25(OH)$_2$D inhibits the expression and synthesis of PTH. The production of 1,25(OH)$_2$D in the kidney enters the circulation and is able to downregulate renin production in the kidney and to stimulate insulin secretion in the β-islet cells of the pancreas. Holick copyright 2007. Reproduced with permission.

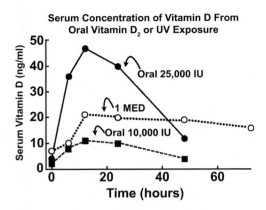

Fig. 6. Comparison of serum vitamin D_3 levels after a whole-body exposure to 1 MED (minimal erythemal dose) of simulated sunlight compared with a single oral dose of either 10,000 or 25,000 IU of vitamin D_2. Reproduced with permission *(5)*.

thousand IUs of vitamin D/day from their exposure to sunlight. An adult in a bathing suit exposed to one minimal erythemal dose of ultraviolet radiation (a slight pinkness to the skin 24 h after exposure) was found to be equivalent to ingesting between 10,000 and 25,000 IU of vitamin D (Fig. 6). This is the likely reason why humans depended on sun for their vitamin D requirement and that very few foods have ever provided humans with their vitamin D requirement with the exception of people living in the far northern and southern regions of the globe where they depended on the food chain, i.e., seal blubber and polar bear liver and other fish sources for their vitamin D needs. Based on several studies, it has been estimated that 100 IU of vitamin D_2 or vitamin D_3 ingested daily for at least 2 months will raise the blood level of 25(OH)D by 1 ng/ml *(37, 38)*. When healthy adults in Boston received 1,000 IU of vitamin D_2 or vitamin D_3 daily for 3 months, they raised the blood level of 25(OH)D by 10 ng/ml. However, since the mean blood level was ~19 ng/ml at baseline, none of the subjects raised the blood level above 30 ng/ml *(38)* (Fig. 7). Thus, in the absence of any sun exposure 2,000–3,000 IU of vitamin D/day is needed for both children and adults to sustain a blood level of above 30 ng/ml.

3. RECOMMENDED ADEQUATE DIETARY INTAKE OF VITAMIN D

During the past few years, several studies reported that the recommended adequate intake for vitamin D are woefully inadequate for all age groups. These studies are reviewed and Table 3 summarizes what the present adequate recommendations are and what is reasonable based on the most current literature.

3.1. Birth to 6 Months

In utero, the fetus is wholly dependent on the mother for vitamin D. The 25(OH)D passes from the placenta into the fetus' blood stream. Since the half-life for 25(OH)D is approximately 2–3 weeks, after birth the infant can remain vitamin D sufficient for several weeks as long as the mother was vitamin D sufficient. However, most pregnant

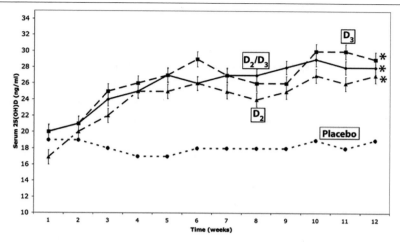

Fig. 7. Mean (±SEM) serum 25-hydroxyvitamin D levels after oral administration of vitamin D_2 and/or vitamin D_3. Healthy adults recruited at the end of the winter received either placebo [($n = 14$; (···•···)], 1,000 IU of vitamin D_3 [D_3, $n = 20$; (-■-)], 1,000 IU of vitamin D_2 [D_2, $n = 16$; (-▲-)], or 500 IU of vitamin D_2 and 500 IU of vitamin D_3 [D_2 and D_3, $n = 18$; (-♦-)] daily for 11 weeks. The total 25-hydroxyvitamin D levels are demonstrated over time. $*P = 0.027$ comparing 25(OH)D over time between vitamin D_3 and placebo. $**P = 0.041$ comparing 25(OH)D over time between 500 IU vitamin D_3 + 500 IU vitamin D_2 and placebo. $***P = 0.023$ comparing 25(OH)D over time between vitamin D_2 and placebo. Reproduced with permission *(38)*.

Table 3

Adequate Intake (AI) Tolerable Upper Limit (UL) Recommendations by IOM and Reasonable Daily Allowance and Safe Upper Levels (SUL) for Vitamin D Based on Published Literature

	IOM		*Reasonable Daily allowance (IU/day)*	*SUL [IU/day]*
	AI [IU (μg)/day]	*UL [IU (μg)/day]*		
0–6 month	200 (5)	1,000 (25)	400–1,000	2,000
6–12 months	200 (5)	1,000 (25)	400–1,000	2,000
1–18 year	200 (5)	2,000 (50)	1,000–2,000	5,000
19–50 year	200 (5)	2,000 (50)	1,500–2,000	10,000
51–70 year	400 (10)	2,000 (50)	1,500–2,000	10,000
71+ year	600 (15)	2,000 (50)	1,500–2,000	10,000
Pregnancy	200 (5)	2,000 (50)	1,500–2,000	10,000
Lactation	200 (5)	2,000 (50)	1,500–2,000 4,000–6,000 (for infant's requirement)	10,000

women are vitamin D deficient or insufficient *(39, 40)*. Lee et al. *(39)* reported at the time of birth of 40 mother–infant pairs, 76% moms and 81% of newborns had a 25(OH)D <20 ng/ml despite the fact that the pregnant women were ingesting about 600 IU of vitamin D a day from a prenatal supplement and consumption of two glasses of milk.

Infants depend on either sunlight exposure or dietary vitamin D for his/her vitamin D requirement from birth. Human breast milk and unfortified cow's milk have very little vitamin D *(41)* (see   Chapter 10). Thus, infants who are fed only human breast milk are prone to developing vitamin D deficiency especially during the winter when they and their mothers are not obtaining their vitamin D from exposure to sunlight. Conservative estimates of the length of time a human milk-fed infant in the Midwest must be exposed to sunlight to maintain serum 25(OH)D concentrations above 20 ng/ml is about 30 min/week with just a diaper on *(42, 43)*.

Human milk and colostrums contains low amounts of vitamin D on average 15.9 $\pm$ 8.6 IU/L *(44)*. There is a direct relationship between vitamin D intake and vitamin D content in human milk. However, even when women were consuming between 600 to 700 IU/day of vitamin D, the vitamin D content in their milk was only between 5 and 136 IU/l *(43, 44)*. Only when lactating women were given 4,000–6,000 IU vitamin D a day was enough vitamin D transferred in her milk to satisfy her infant's requirement *(41)* (see Chapter 32).

Vitamin D intakes between 340 and 600 IU/day have been reported to have the maximum effect on linear growth of infants *(45–47)*. When Chinese infants were given either 100, 200, or 400 IU of vitamin D/day, none of the infants demonstrated any evidence of rickets *(48)*. This observation is consistent with what Jeans *(49)* observed and was the basis for recommending that children only need 200 IU of vitamin D/day *(44)*. However, Markstad and Elzouki *(20)* reported that Norwegian infants fed infant formula containing 300 IU/day obtained blood levels of 25(OH)D of >11 ng/ml which at the time was considered the lower limit of normal. However, essentially all experts agree that the level should be at least 20 ng/ml suggesting that even 300 IU/day is not adequate for infants.

Therefore, a minimum of 100 IU of vitamin D a day is satisfactory for preventing rickets. However, in the absence of any exposure to sunlight, it is likely that infants require a much larger amount of vitamin D to optimize bone health. Infants require at least 400 IU of vitamin D to maximize bone health in the absence of sunlight exposure. The American Academy of Pediatrics *(50)* and the Canadian Pediatric Association both recommended 400 IU/day which is twice what the Institute of Medicine (IOM) of the US National Academy of Sciences recommended in 1997 *(44)*. Whether 400 IU/day is enough to provide all the health benefits associated with vitamin D is not known at this time. However, infants who received 2,000 IU of vitamin D/day for the first year of life in Finland reduced their risk of developing type 1 diabetes in the ensuing 31 years by 78% *(51)*. Thus, the safe upper limit should be at least 2,000 IU/day.

3.2. Ages 6–12 Months

As for infants, it was recommended by the IOM *(21)* that the adequate dietary intake for ages 6–12 months be 200 IU (5 μg)/day. This was based on the observation that in the absence of sun-mediated vitamin D synthesis, approximately 200 IU/day of vitamin D maintained circulating concentrations of 25(OH)D above 11 ng/ml of older infants but below circulating concentrations attained in similar aged infants in the summer *(20)*. It was also suggested that twice this amount, i.e., 400 IU/day, would not be excessive *(45)* and is in line with what the American Academy of Pediatrics and the Canadian Pediatric Society has recommended *(50)*. However, because 2,000 IU was safe and effective in Finnish children, *(51)* the safe upper limit should be at least 2,000 IU/day.

3.3. Ages 1–8 Years

Children aged 1–8 years of all races used to obtain most of their vitamin D from exposure to sunlight, and, therefore, did not normally need to ingest vitamin D *(52)*. However, children are spending more time indoors, and when they go outside, they often wear sun protection limiting their ability to make vitamin D in their skin. There is overwhelming evidence that children of all ages are at high risk for vitamin D deficiency *(53 – 56)* and its insidious health consequences. There was no data on how much vitamin D is required to prevent vitamin D deficiency in children of ages 1–8 years. Extrapolating from available data in slightly older children of ages 10–17 years (who were presumably exposed to an adequate amount of sun-mediated vitamin D since they lived in Lebanon) who ingested either 1,400 IU vitamin D_3 once a week or 14,000 IU once a week for 1 year provides an insight into the vitamin D requirement for children in this age group. The children who received 1,400 IU/week increased their blood level of 25(OH)D from 14 ± 9 to 17 ± 6 ng/ml while the children who received 14,000 IU/week for 1 year increased their blood levels from 14 ± 8 to 38 ± 31 ng/ml. No intoxication was noted (see Chapter 24). Thus, this age group needs at least 1,000 IU/day. Therefore, the IOM recommended 200 IU/day for this age group should be increased to between 1,000 and 2,000 IU/day *(44)*. The safe upper limit should be increased to at least 5,000 IU/day.

3.4. Ages 9–18 Years

Children aged 9–18 have a rapid growth spurt that requires a marked increase in their requirement of calcium and phosphorus to maximize skeletal mineralization. During puberty, the metabolism of 25(OH)D to $1,25(OH)_2D$ increases *(57)*. In turn, the increased blood levels of $1,25(OH)_2D$ enhances the efficiency of the intestine to absorb dietary calcium and phosphorus to satisfy the growing skeleton's requirement for these minerals during its rapid growth phase. However, despite the increased production of $1,25(OH)_2D$, there is no scientific evidence demonstrating an increased requirement for vitamin D in this age group probably because circulating concentrations of $1,25(OH)_2D$ are approximately 500–1,000 times lower than those of 25(OH)D (i.e., 15–60 pg/ml vs. 20–100 ng/ml, respectively).

A few studies conducted in children during the pubertal years show that children maintained a serum 25(OH)D of at least >11 ng/ml with dietary vitamin D intakes of 2.5–10 µg/day *(57)*. When intakes were <2.5 µg/day, Turkish children aged 12–17 years had 25(OH)D levels consistent with vitamin D deficiency, i.e., <11 ng/ml *(58)*. Based on Maalouf et al. *(53)*, this age group needs 2,000 IU vitamin D/day to maintain a blood level above 30 ng/ml. This is an order of magnitude higher compared to what the IOM *(44)* recommended as the AI for vitamin D of 200 IU (5 μg)/day. The safe upper limit should be increased to at least 5,000 IU/day.

3.5. Ages 19–50 Years

Most young and middle-aged adults used to obtain their vitamin D from casual everyday exposure to sunlight. This age group is now at equal risk of vitamin D deficiency because of decreased outdoor activities and aggressive sun protection. Very few studies have evaluated the vitamin D requirement in this age group. An evaluation of 67 white and 70 black premenopausal women ingesting 138 ± 84 and 145 ± 73 IU/day, respectively, revealed that serum 25(OH)D levels were in the insufficient or deficient range (circulating concentrations of 21.4 ± 4 and 18.3 ± 5 ng/ml, respectively) *(59)*.

A study conducted in submariners provided an insight as to how much vitamin D was required for this age group in the absence of exposure to sunlight. Young groups of male submariners during a 3-month voyage received either a placebo or 600 IU of vitamin D a day. Circulating concentrations of 25(OH)D decreased by 50% over the 3 months in the submariners who did not receive vitamin D supplementation, whereas 25(OH)D levels were maintained at about 20 ng/ml in the group that received 600 IU/day. After the voyage when the submariners were exposed to sunlight, there was a dramatic more than 50% increase in 25(OH)D levels that were >30 ng/ml *(5)*.

In a cross-sectional study, 67 white (age 36.5 ± 6.4 year) and 70 black (age 37 ± 6.7 year) premenopausal women living in the New York City area who were active and exposed to sunlight had average daily vitamin D intakes of 138 ± 84 IU/day and 145 ± 73 IU/day, respectively. Both groups of women had insufficient circulating concentrations of 25(OH)D concentrations *(59)*. During the winter months (November through May) in Omaha, Nebraska, 6% of a group of young women aged 25–35 years ($n = 52$) maintained serum concentrations of 25(OH)D >20 ng/ml but <30 ng/ml when the daily vitamin D intake was estimated to be between 131 and 135 IU/day *(60)*. Healthy adults aged 18–84 years who received 1,000 IU vitamin D_3/day for 3 months during the winter increased their 25(OH)D from 19.6 ± 11.1 to 28.9 ± 7.7 ng/ml *(38)*.

Therefore, based on all of the available literature, both sunlight and diet play an essential role in providing vitamin D to this age group. The IOM *(44)* recommended AI for this age group of 200 IU (5 μg)/day is totally inadequate. At least 1,000 IU vitamin D/day and probably 1,500–2,000 IU/day is needed to maintain the 25(OH)D >30 ng/ml. The safe upper limit should be increased to at least 10,000 IU/day *(73)*.

3.6. Ages 51–70 Years

Men and women aged 51–70 are very dependent on sunlight for most of their vitamin D requirement. However, this age group is much more health conscious and is especially concerned about their skin health as it relates to skin cancer and wrinkles. Therefore, there is increased use of clothing and sunscreen over sun-exposed areas, which decreases the amount of ultraviolet B radiation penetrating the skin and thereby decreasing the cutaneous production of vitamin D_3 by 95–100% *(4, 5, 11)*. In addition, age decreases the capacity of the skin to produce vitamin D_3 *(12)*. Although it has been suggested that aging may decrease the ability of the intestine to absorb dietary vitamin D, two studies have revealed that aging does not alter the absorption of physiologic or pharmacologic doses of vitamin D *(38, 61, 62)*.

It was thought that men and women in this age group require twice the amount of vitamin D compared with younger adults. This is in part based on the data of Krall and Dawson-Hughes *(63)* who observed in 66 healthy postmenopausal women (mean age 60 ± 5 year) a seasonal variation in calcium retention that positively correlated with serum 25(OH)D levels in women on low calcium diets. Furthermore, an evaluation of bone loss in 247 postmenopausal women (mean age 64 ± 5 year) who consumed an average of 100 IU/day found that women who received an additional supplement of 700 IU of vitamin D/day lost less bone than women who were on 200 IU/day *(64)*. A similar observation was made in women who were placed on placebo or 400 IU/day *(65)*. It was also observed that when the vitamin D intake was up to 220 IU/day, women had higher PTH values between March and May than women studied between August and October *(66)*. This suggested that in the absence of sunlight-mediated vitamin D synthesis, 220 IU/day may be inadequate to maximize bone health. Vitamin D deficiency in this age group should not be minimized. A study of 169 men and women ages 49–83 years suggested that 37% were vitamin D deficient *(67)*. However, now that it is appreciated that to maximize intestinal calcium transport that the serum 25(OH)D needs to be >30 ng/ml, *(24)* this age group needs to be on at least 3–5 times more vitamin D (i.e., 1,500–2,000 IU/day) than that recommended by the IOM *(44)* for the AI of 400 IU ($10\,\mu g$)/day. The safe upper limit should be increased to at least 10,000 IU/day *(73)*.

3.7. *Age 71 Years and Older*

Those who are 71 years and older are at high risk of developing vitamin D deficiency *(4, 5, 63–67)* owing to a decrease in outdoor activity, which reduces cutaneous production of vitamin D. Many studies have looked at dietary supplementation of vitamin D in older men and women and its influence on serum 25(OH)D, PTH and bone health as measured by bone mineral density and fracture risks. Several randomized, double-blind clinical trials of elderly men and women who had an intake of 400 IU/day showed insufficient 25(OH)D levels *(64, 68)*. When men and women were supplemented with 400–1,000 IU/day, they had a significant reduction in bone resorption. When a group of elderly French women were supplemented with calcium and 800 IU of vitamin D, there was a significant decrease in vertebral and nonvertebral fractures *(69)*. A similar observation was made in free-living men and women 65 years and older who received 500 mg of calcium and 700 IU of vitamin D *(70)*.

A comparison of the vitamin D status of elderly people in Europe and North America revealed that the elderly are prone to vitamin D deficiency and associated abnormalities in blood chemistries and bone density *(4)*. When elders up to 84 years took 1,000 IU vitamin D_3/day for 3 months their blood levels did not raise above 30 ng/ml *(38)*. Thus, even though the IOM *(44)* recommended an AI for this age group of 600 IU ($15\,\mu g$)/day it is still 2–3 times lower than what is needed to maintain blood levels above 30 ng/ml. (Table 3). Thus, the AI should be increased to 1,500–2,000 IU/day. The safe upper limit should be increased to at least 10,000 IU/day.

3.8. Pregnancy

It has always been assumed that during pregnancy the vitamin D requirement is increased because of fetal utilization of vitamin D and the requirement of vitamin D to increase the maternal intestinal calcium absorption. During the first and second trimesters, the fetus is developing most of its organ systems and laying down the collagen matrix for its skeleton. During the last trimester, the fetus begins to calcify the skeleton thereby increasing its maternal demand for calcium. This demand is met by increased production of $1,25(OH)_2D$ by the mother's kidneys and placental production of $1,25(OH)_2D$. Circulating concentrations of $1,25(OH)_2D$ gradually increase during the first and second trimesters owing to an increase in vitamin D-binding protein concentrations in the maternal circulation. However, the free levels of $1,25(OH)_2D$ which are responsible for enhancing intestinal calcium absorption are only increased during the third trimester. Although there is ample evidence that the placenta converts $25(OH)D$ to $1,25(OH)_2D$ and transfers $1,25(OH)_2D$ to the fetus, these quantities are relatively small and appear not to affect the overall vitamin D status of pregnant women. It is now recognized that pregnant women are at high risk for vitamin D. Furthermore, vitamin D deficiency increases risk of having preeclampsia *(71)* and a caesarian section *(72)*. Thus, since 600 IU/day does not prevent vitamin D deficiency in pregnant women, the IOM *(44)* recommendation for an AI for pregnant women of 200 IU (5 µg)/day is inadequate. They should be taking at least a prenatal vitamin containing 400 IU vitamin D with a supplement that contains 1,000 IU vitamin D (Table 3). Thus, they need between 1,500 and 2,000 IU/day. The safe upper limit should be raised to at least 10,000 IU/day.

3.9. Lactation

During lactation, the mother needs to increase the efficiency of dietary absorption of calcium to ensure an adequate calcium content in her milk. The metabolism of $25(OH)D$ to $1,25(OH)_2D$ is enhanced in response to this new demand. However, because circulating concentrations of $1,25(OH)_2D$ are 500–100 times less than $25(OH)D$, the increased metabolism probably does not significantly alter the daily requirement for vitamin D. Therefore, the IOM recommended *(44)* that the AI for lactating women is the same as that for nonlactating women, i.e., 200 IU (5 µg)/day. However, to satisfy their requirement to maintain a $25(OH)D >30$ ng/ml, they should take at least a multivitamin containing 400 IU vitamin D along with a 1,000 IU vitamin D supplement. To satisfy the infants requirement if they receive their sole nutrition from breast milk Hollis et al *(41)* reported 4,000–6,000 IU/day is what is required to transfer enough vitamin D into human milk to satisfy the infants needs. Thus, at a minimum, lactating women should be on 1,500–2,000 IU/day and to satisfy the requirement of their infants, 4,000–6,000 IU/day may be needed. The safe upper limit should be increased to at least 10,000 IU/day.

3.10. Tolerable Upper Intake Levels

In the 1950s, there was an outbreak of presumed vitamin D intoxication in infants and young children. It was assumed that this was due to over fortification of milk with vitamin D. However, this was never proven and the clinical cases that reported infants with hypercalcemia had physical characteristics consistent with Williams syndrome. This syndrome is associated with increased sensitivity to vitamin D. Vitamin D intoxication has also been associated with prolonged intakes (months to years) of pharmacologic amounts usually 50,000–1 million units of vitamin D a day of vitamin D (4). It is associated with hypercalcemia, hyperphosphatemia, and markedly elevated levels of 25(OH)D usually >200 ng/ml. This can lead to nephrocalcinosis with renal failure, calcification of soft tissues including large and small blood vessels and kidney stones. It was assumed that infants aged 0–12 months may be more sensitive to pharmacologic amounts of vitamin D, and, therefore, the IOM (44) recommended an upper safe level of 1,000 IU (40 µg)/day. For children and adults, the recommended upper safe level was 2,000 IU (80 µg)/day (Table 3). However, the Finnish study of infants receiving 2,000 IU vitamin D/day during the first year of life did not demonstrate any toxicity. For adults ingesting up to 10,000 IU of vitamin D/day for 5 months did not cause any toxicity. Thus, the tolerable safe upper intake levels for children 0–12 months should be at least 2,000 IU/day and for children over 1–18 years should be at least 5,000 IU/day, and for all adults the level should be at least 10,000 IU/day.

4. CAUSES OF AND TREATMENT STRATEGIES FOR VITAMIN D DEFICIENCY

4.1. Causes

The major cause of vitamin D deficiency is both inadequate exposure to sunlight along with an inadequate intake of vitamin D from dietary sources. Patients with fat malabsorption problems are at high risk because of their inability to absorb dietary and supplemental vitamin D. A silent cause of vitamin D deficiency is celiac disease. Often the first indication that the patient has a intestinal malabsorption problem is based on the observation that patients who are vitamin D deficient receiving pharmacologic doses of vitamin D do not raise their blood levels of 25(OH)D. There are also many other causes for vitamin D deficiency as well as syndromes that appear to be like vitamin D deficiency but are caused by acquired or hereditary disorders in vitamin D metabolism or in vitamin D recognition. These are summarized in Table 4 and found in a recent review (4) (Fig. 8).

4.2. Strategies for Preventing and Treating Vitamin D Deficiency

In the United States, there is only one pharmaceutical form of vitamin D, vitamin D$_2$. It comes either as a liquid for infants and young children and contains 8,000 IU of vitamin D$_2$ in 1 ml. For older children and adults, there is a vitamin D capsule that contains an oil that has 50,000 IU of vitamin D$_2$. To treat vitamin D deficiency, I typically

Table 4
Strategies to Treat and Prevent Vitamin D Deficiency

	Cause/mechanism	Prevention and safety	Treatment (Tx) and maintenance (M)
Children 0–1 year	Breast fed without supplementation	400 IU vitamin D/day Sensible sun exposure 2,000 IU vitamin D/day is safe	Tx-200,000 IU vitamin D_3 q 3 months or M-1,000–2,000 IU vitamin D_2 or vitamin D_3/day Achieve 25(OH)D ~30–100 ng/ml
Children 1–18	Inadequate sun exposure/always use sun protection Children of color Inadequate supplementation	400–2,000 IU vitamin D/day Sensible sun exposure 2,000–5,000 IU vitamin D_3/day is safe	Tx-50,000 IU vitamin D_2 q week × 8 M-1,000–2,000 vitamin D/day Achieve 25(OH)D ~30–100 ng/ml
Adults 19–71+	Inadequate sun exposure Adults of color Inadequate supplementation Decreased 7-DHC in skin (over age 50 years)	1,500–2,000 IU vitamin D_3/day 50,000 IU vitamin D_2 q 2 weeks or 1 month Sensible sun exposure Up to 10,000 IU vitamin D/day is safe	Tx-50,000 IU vitamin D_2 q week × 8 Repeat x1 if 25(OH)D <30 ng/ml M-50,000 IU q 2 weeks or month Achieve 25(OH)D ~30–100 ng/ml
Pregnancy	Fetal utilization Inadequate sun exposure inadequate supplementation	1,500–2,000 IU vitamin D/day or 50,000 IU vitamin D_2 q 2 weeks for providing enough vitamin D for infant's need 4,000–6,000 IU/day Up to 10,000 IU vitamin D_3/day is safe	Tx-50,000 IU vitamin D_2 q week × 8 Repeat x1 if 25(OH)D <30 ng/ml M-50,000 IU vitamin D_2 q2 or 4 weeks achieve 25(OH)D ~30–100 ng/ml

Table 4
(continued)

	Cause/mechanism	Prevention and safety	Treatment (Tx) and maintenance (M)
Lactation	Infant utilization Inadequate sun exposure inadequate supplementation Secreted into milk	1,500–2,000 IU vitamin D/day or 50,000 IU vitamin D_2 q 2 weeks Up to 10,000 IU vitamin D_3/day is safe	Tx-50,000 IU vitamin D_2 q week × 8 Repeat x1 if 25(OH)D <30 ng/ml M-50,000 IU vitamin D_2 q2 or 4 weeks achieve 25(OH)D ~30–100 ng/ml
Malabsorption syndromes	Malabsorption of vitamin D Inadequate sun exposure Inadequate supplementation	Adequate exposure to sun or UVB radiation 50,000 IU vitamin D_2 qd, qod, or q week 2,000–6,000 IU vitamin D/day Achieve 25(OH)D ~30–60 ng/ml	Tx-UVB irradiation (tanning bed) 50,000 IU vitamin D_2 q d M-50,000 IU vitamin D_2 q week Achieve serum 25(OH)D ~30–100 ng/ml
Drugs that activate SXR Transplant patients	Enhanced destruction of 1,25(OH)$_2$D and 25(OH)D	50,000 IU vitamin D_2 qod or q week 2,000–4,000 IU vitamin D/day Achieve 25(OH)D ~30–60 ng/ml	Tx-50,000 IU vitamin D_2 q 2/week × 8–12 weeks Repeat x1 if 25(OH)D <30 ng/ml M-50,000 IU vitamin D_2 q 2 week or 4 weeks Achieve 25(OH)D ~30–100 ng/ml

Table 4
(continued)

	Cause/mechanism	Prevention and safety	Treatment (Tx) and maintenance (M)
Obesity	Fat sequestration	2,000–4,000 IU vitamin D/day 50,000 IU vitamin D_2 1 or 2/week Achieve 25(OH)D ~30–60 ng/ml	Tx-50,000 IU vitamin D_2 2/week × 8–12 weeks Repeat x1 if 25(OH)D <30 ng/ml M-50,000 IU vitamin D_2 q 2 week or 4 weeks Achieve 25(OH)D ~30–100 ng/ml
Nephrotic syndrome	Loss 25(OH)D into urine attached to DBP	2,000–4,000 IU vitamin D/day 50,000 IU vitamin D_2 1 or 2/×/week Achieve 25(OH)D ~30–60 ng/ml	Tx-50,000 IU vitamin D_2 2/week × 8–12 weeks Repeat x1 if 25(OH)D <30 ng/ml M-50,000 IU vitamin D_2 q 2 week or 4 weeks Achieve 25(OH)D ~30–100 ng/ml
Chronic kidney disease GFR >30 cc/ml/1.73 M^2	1,25$(OH)_2$D production reduced due to elevated serum Pi and fibroblast growth factor 23	Control serum Pi and 1,000–2,000 IU vitamin D_3/day 50,000 IU vitamin D_2 q 2 weeks Maintain 25(OH)D ~30–60 ng/ml	Tx-50,000 IU vitamin D_2 q week × 8 Repeat x1 if 25(OH)D <30 ng/ml M-50,000 IU vitamin D_2 q 2 or 4 weeks Achieve 25(OH)D ~30–100 ng/ml

Table 4
(continued)

	Cause/mechanism	Prevention and safety	Treatment (Tx) and maintenance (M)
Chronic kidney disease GFR < ~30% of normal	Loss of 1-OHase resulting in inadequate production 1,25(OH)$_2$D	1,000–2,000 IU vitamin D/day 50,000 IU vitamin D$_2$ q 2 weeks and maintain 25(OH)D ~30–60 ng/ml	Tx-1,25(OH)$_2$D$_3$ (calcitriol) 0.25= 1.0 µg po BID 1–2 µg IV q 3 days Paricalcitol (Zemplar) 0.04–0.1 µg/kg IV qod initial may increase to 0.24 µg/kg 2–4 µg po 3 ×/week Doxecalciferol (Hectorol) 10–20 µg po 3 ×/week
Primary or tertiary hyperparathyroidism	Increased metabolism of 25(OH)D to 1,25(OH)$_2$D by PTH and increased destruction of 25(OH)D	1,000–2,000 IU vitamin D$_3$/day maintain 25(OH)D ~30–60 ng/ml	50,000 IU vitamin D$_2$ q week × 8 weeks Achieve 25(OH)D ~30–60 ng/ml
Granulomatous disorders	Increased metabolism 25(OH)D to 1,25(OH)$_2$D by macrophages	800–1,000 IU vitamin D/day Maintain 25(OH)D ~20–35 ng/ml and <40 ng/ml to prevent hypercalciuria and hypercalcemia in some patients	50,000 IU of vitamin D$_2$ once a month achieve 25(OH)D ~20–35 ng/ml

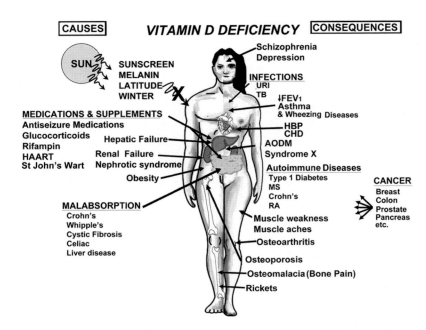

Fig. 8. A schematic representation of the major causes for vitamin D deficiency and potential health consequences. Holick copyright 2007. Reproduced with permission.

give my patients 50,000 IU of vitamin D_2 once a week for 8 weeks to fill the empty vitamin D tank followed by 50,000 IU of vitamin D_2 every 2 weeks. I have followed patients up to 5 years on this regimen and most patients have blood levels between 30 and 50 ng/ml within 2–3 months after initiating treatment. For patients who have a blood level of 25(OH)D of >30 ng/ml in order to maintain this level, I give my patients 50,000 IU of vitamin D_2 once every 2 weeks. In more than 100 patients on these treatment regimens, I have not observed any untoward toxicity including either kidney stones or hypercalcemia. For neonates who are vitamin D deficient, they can receive an average of 1,000 IU of vitamin D/day until their blood level is >30 ng/ml. There are other various strategies that have been used using very high pharmacologic doses of vitamin D that have been both safe and effective and are outlined in Table 4.

5. MYTHS

5.1. Vitamin D Treatment Will Worsen Hypercalcemia in Hyperparathyroid Patients

Patients who suffer from hyperparathyroidism have an elevated level of PTH and calcium in their serum. These patients are often vitamin D deficient, and physicians are reluctant to treat the vitamin D deficiency because there is concern that the patients will become more hypercalcemic. From my own experience and published literature, there is no evidence that correcting vitamin D deficiency in patients who suffer from primary hyperparathyroidism will increase their blood calcium *(74)*. I have observed in my clinic that these patients who suffer from hyperparathyroidism and I correct their vitamin D deficiency often have improvement in muscle weakness and aches and pains in their bones and muscles. Furthermore, if the patient is sent for parathyroid surgery, they are less likely to develop the hungry bone syndrome and become hypocalcemic after the surgery. Patients with hyperparathyroidism are more likely to develop vitamin D deficiency because PTH stimulates the kidneys to produce more $1,25(OH)_2D$, thus using up the $25(OH)D$ substrate. However, the major reason is that the higher serum level of $1,25(OH)_2D$ stimulates the 24-OHase, which increases the destruction of both $1,25(OH)_2D$ and $25(OH)D$.

5.2. Treatment with Vitamin D Will Increase Risk of Kidney Stones

Although there are some published studies that have suggested that increasing calcium and vitamin D will increase the risk of kidney stones, these studies were not well controlled for either calcium or vitamin D intake and other causes of kidney stone development were not evaluated *(75)*. The major cause for kidney stones is the increased absorption of dietary oxalate that is often present in dark green leafy vegetables including spinach. This is the reason why it has been observed that increasing calcium intake reduces risk of developing kidney stones. However, patients who have a history of kidney stones do need to be cautious about their calcium intake. For patients with kidney stones who are not getting an adequate amount of calcium in their diet and need to take a calcium supplement, I recommend either calcium citrate or calcium citrate malate. The reason is that the citrate and malate will chelate the calcium in the urine decreasing risk of kidney stone development. Treating vitamin D deficiency and increasing vitamin D intake to raise blood levels of $25(OH)D$ of >30 and <100 ng/ml will not increase the risk of developing kidney stones unless there is some other underlying calcium or bone metabolic disorder *(4)*.

5.3. Patients with Chronic Kidney Disease Do Not Need to Be Treated with Vitamin D Since They Cannot Activate It

It is true that the kidneys are the major source for the production of $1,25(OH)_2D$ that acts as an endocrine hormone to regulate calcium and bone metabolism. As a result, patients with chronic kidney disease (CKD) often require either $1,25(OH)_2D$ or one of its active analogues to help prevent an increase in their PTH levels and associated metabolic bone disease *(4, 76)*. However, the National Kidney Foundation recommendation known as the KDOQI guidelines recommends that all patients with CKD maintain a blood level of $25(OH)D$ of >30 ng/ml. The reason for this is that it is now recognized that the parathyroid glands have the ability to produce locally $1,25(OH)_2D$ which may help suppress the overproduction of PTH *(78)*. Furthermore, the $25(OH)D$ may be also converted to $1,25(OH)_2D$ locally within noncalcemic tissues such as colon, prostate, breast which may help regulate cell growth and prevent malignancy *(4, 5)*.

5.4. Treatment with 50,000 IU of Vitamin D Long-Term Will Cause Vitamin D Intoxication

I have found it to be very effective in treating vitamin D deficiency and insufficiency with the only pharmaceutical form of vitamin D available in the United States, i.e., 50,000 IU of vitamin D_2 once a week for 8 weeks. For patients who are severely vitamin D deficient, i.e., a 25(OH)D <20 ng/ml and for obese patients, I usually extend the treatment for an additional 4–8 weeks. (Table 4) On average, I observe an increase in their blood level by ∼100% after 8 weeks. There was no untoward toxicity with this treatment regimen. However, treating vitamin D deficiency does not prevent the reoccurrence of vitamin D deficiency. Thus, patients who have their vitamin D deficiency corrected either need to take 1,000–2,000 IU of vitamin D/day or 50,000 IU of vitamin D_2 once every 2 weeks (79). A review of my patients medical charts for up to 5 years revealed that most patients who had an average blood level of 19 ng/ml were able to raise and maintain their blood level between 40 and 50 ng/ml. In more than 100 patients that I have had on this regimen, none have developed any untoward toxicity including hypercalcemia or kidney stones (79).

6. CONCLUSION

Vitamin D is taken for granted and is not usually considered an important nutrient. Because most humans obtain their vitamin D passively from exposure to sunlight, it would normally not be necessary to recommend a dietary allowance for vitamin D. Everyday casual exposure to sunlight will provide active children with their vitamin D requirement. Children who live in far northern and southern latitudes and who make vitamin D during the summer months store the excess in their body fat and utilize these stores during the winter when they are unable to make vitamin D in the skin. Young, middle aged, and older adults who are active outdoors obtain their vitamin D requirement from sunlight (80, 81). Outdoor workers in Brazil in the summer had achieved serum 25(OH)D levels of ∼40 ng/ml that decreased to ∼25 ng/ml in the winter (80, 82). In sunrich environments, including Africa where there is no restriction to sun exposure from clothing or sunscreen use, serum 25(OH)D levels range from 54 to 90 ng/ml (82). However, both children and adults who are not outdoors or who wear sun protection before going outdoors and in the winter are at high risk of vitamin D deficiency (48, 53, 54, 80, 81).

Vitamin D deficiency is the most common medical condition in the world. Looker et al (78) reported from the National Health and Nutrition examination surveys, a comparison of blood levels of 25(OH)D from 1988, 1994 when compared to data collected from 2000 to 2004 revealed aged adjusted mean serum 25(OH)D levels were 2–8 ng/ml lower in NHANES 2000–2004 than NHANES III. Both children and adults in Australia who have either avoided sun exposure or always wore sun protection have now reported to be at risk of vitamin D deficiency. Children and adults living in the Middle East, and even in Florida, are at risk of vitamin D deficiency because of sun protection practices.

There is a great need to have the adequate recommendations by the Institute of Medicine revise substantially upward so that all children should receive at least 400 and preferably 1,000–2,000 IU of vitamin D/day. All teenagers and adults should ingest at least 1,000 and up to 2,000 IU of vitamin D/day. The urgency for these increased recommendations is based on the fact that many countries use the adequate intake recommendations from the United States for their own recommendations. Most importantly, food manufacturers are limited as to how much vitamin D can be added to foods and the number of foods that can be fortified with vitamin D since these adequate intake recommendations and the safe upper limits are easily attained with the amount of vitamin D present in fortified foods in the United States. However, if the adequate intake recommendations were increased by 5–10 times of what is now recommended, the manufacturers will have the ability to not only increase the amount of vitamin D per serving, but also increase the number of foods fortified with vitamin D and not be at risk of reaching what is considered to be safe upper limits, which should be increased to 2,000 IU/day of vitamin D for neonates, 5,000 IU/day for children, and 10,000 IU/day for all adults.

ACKNOWLEDGMENTS

This work was supported in part by NIH grant UL1RRO 25771 and the UV Foundation.

REFERENCES

1. Holick MF (1989) Phylogenetic and evolutionary aspects of vitamin D from phytoplankton to humans. In: Pang PKT and Schreibman MP (eds) Verebrate endocrinology: fundamentals and biomedical implications, Vol. 3. Academic Press, Inc. (Harcourt Brace Jovanovich), Orlando, FL
2. Holick MF, Garabedian M (2006) Vitamin D: photobiology, metabolism, mechanism of action, and clinical applications. In: Favus MJ (ed) Primer on the metabolic bone diseases and disorders of mineral metabolism, 6th edn. American Society for Bone and Mineral Research, Washington, DC.
3. Strugnell SA, DeLuca HF (1997) The vitamin D receptor – structure and transcriptional activation. Proc Soc Exp Biol 215:223–228
4. Holick MF (2007) Vitamin D deficiency. N Engl J Med 357:266–281
5. Holick MF (2004) Vitamin D: importance in the prevention of cancers, type 1 diabetes, heart disease, and osteoporosis. Robert H. Herman Memorial Award in Clinical Nutrition Lecture, 2003. Am J Clin Nutr 79:362–371
6. Holick MF (1997) The evolution of vitamin D from phytoplankton to man. Norman AW, Bouillon R, Thomasset M (eds) Vitamin D: chemistry, biology and clinical applications of the steroid hormone (Proceedings of the Tenth Workshop on Vitamin D). University of California Press, Riverside, pp 771–776
7. Rajakumar K, Greenspan SL, Thomas SB, Holick MF (2007) Solar ultraviolet radiation and vitamin D: a historical perspective. Am J Public Health 97(10):1746–1754
8. Khosla S (2001) The OPG/RANKL/RANK system. Endocrinology 142(12):5050–5055
9. Holtrop ME, Cox KA, Cares DL, Holick MF (1986) Effects of serum calcium and phosphorus on skeletal mineralization in vitamin D-deficient rats. Am J Physiol 251:E234–E240
10. Underwood JL, DeLuca HF (1984) Vitamin D is not directly necessary for bone growth an mineralization. Am J Physiol 246:E493–E498
11. Clemens TL, Henderson SL, Adams JS, Holick MF (1982) Increased skin pigment reduces the capacity of skin to synthesis vitamin D$_3$. Lancet 1(8263):74–76
12. Matsuoka LY, Ide L, Wortsman J, MacLaughlin J, Holick MF (1987) Sunscreens suppress cutaneous vitamin D$_3$ synthesis. J Clin Endocrinol Metab 64:1165–1168
13. Holick MF, Matsuoka LY, Wortsman J (1989) Age, vitamin D, and solar ultraviolet. Lancet 2: 1104–1105

14. Webb AR, Kline L, Holick MF (1988) Influence of season and latitude on the cutaneous synthesis of vitamin D_3: exposure to winter sunlight in Boston and Edmonton will not promote vitamin D_3 synthesis in human skin. J Clin Endocrinol Metab 67:373–378

15. Holick MF, Chen TC (2008) Vitamin D deficiency: a worldwide problem with health consequences. Am J Clin Nutr 87(4):1080S–1086S

16. Tangpricha V, Koutkia P, Rieke SM, Chen TC, Perez AA, Holick MF (2003) Fortification of orange juice with vitamin D: a novel approach to enhance vitamin D nutritional health. Am J Clin Nutr 77:1478–1483

17. Tanner JT, Smith J, Defibaugh P, Angyal G, Villalobos M, Bueno M, McGarrahan E (1988) Survey of vitamin content of fortified milk. J Assoc Off Analyt Chem 71:607–610

18. Holick MF, Shao Q, Liu WW, Chen TC (1992) The vitamin D content of fortified milk and infant formula. N Engl J Med 326:1178–1181

19. Chen TC, Heath H, Holick MF (1993) An update on the vitamin D content of fortified milk from the United States and Canada. N Engl J Med 329:1507

20. Markestad T, Elzouki Ay (1991) Vitamin D deficiency rickets in northern Europe and Libya. In: Glorieux FH (ed). Rickets nestle nutrition workshop series. Raven, New York

21. Malabanan A, Veronikis IE, Holick MF (1998) Redefining vitamin D insufficiency. Lancet 351:805–806

22. Thomas KK, Lloyd-Jones DH, Thadhani RI, Shaw AC, Deraska DJ, Kitch BT, Vamvakas EC, Dick IM, Prince RL, Finkelstein JSL (1988) Hypovitaminosis D in medical inpatients. N Engl J Med 338:777–783

23. Holick MF, Siris ES, Binkley N, Beard MK, Khan A, Katzer JT, Petruschke RA, Chen E, de Papp AE (2005) Prevalence of vitamin D inadequacy among postmenopausal North American women receiving osteoporosis therapy. J Clin Endocrinol Metab 90:3215–3224

24. Heaney RP, Dowell MS, Hale CA, Bendich A (2003) Calcium absorption varies within the reference range for serum 25-hydroxyvitamin D. J Am Coll Nutr 22(2):142–146

25. Moan J, Porojnicu AC, Dahlback A, Setlow RB (2008) Addressing the health benefits and risks, involving vitamin D or skin cancer, of increased sun exposure. Proc Natl Acad Sci USA 105(2):668–673

26. Bischoff-Ferrari HA, Giovannucci E, Willett WC, Dietrich T, Dawson-Hughes B (2006) Estimation of optimal serum concentrations of 25-hydroxyvitamin D for multiple health outcomes. Am J Clin Nutr 84:18–28

27. Grant WB, Holick MF (2005) Benefits and requirements of vitamin D for optimal health: a review. Alter Med Rev 10:94–111

28. Holick MF (2006) High prevalence of vitamin D inadequacy and implications for health. Mayo Clin Proc 81(3):353–373

29. Sedrani SH (1984) Low 25-hydroxyvitamin D and normal serum calcium concentrations in Saudi Arabia: Riyadh region. Ann Nutr Metab 28:181–185

30. Chapuy MC, Preziosi P, Maaner M, Arnaud S, Galan P, Hercberg S, Meunier PJ (1997) Prevalence of vitamin D insufficiency in an adult normal population. Osteopor Int 7:439–443

31. Boonen S, Bischoff-Ferrari A, Cooper C, Lips P, Ljunggren O, Meunier PJ, Reginster JY (2006) Addressing the musculoskeletal components of fracture risk with calcium and vitamin D: a review of the evidence. Calcif Tissue Int 78(5):257–270

32. Lips P (2001) Vitamin D deficiency and secondary hyperparathyroidism in the elderly: consequences for bone loss and fractures and therapeutic implications. Endocr Rev 22:477–501

33. Bakhtiyarova S, Lesnyak O, Kyznesova N, Blankenstein MA, Lips P (2006) Vitamin D status among patients with hip fracture and elderly control subjects in Yekaterinburg, Russia. Osteoporos Int 17(3):441–446

34. McKenna MJ (1992) Differences in vitamin D status between countries in young adults and the elderly. Am J Med 93:69–77

35. Larsen ER, Mosekilde L, Foldspang A (2004) Vitamin D and calcium supplementation prevents osteoporotic fractures in elderly community dwelling residents: a pragmatic population-based 3-year intervention study. J Bone Miner Res 19:370–378

36. Lips P, Hosking D, Lippuner K et al (2006) The prevalence of vitamin D inadequacy amongst women with osteoporosis: an international epidemiological investigation. J Intern Med 260:245–254
37. Heaney RP, Davies KM, Chen TC, Holick MF, Barger-Lux MJ (2003) Human serum 25-hydroxycholecalciferol response to extended oral dosing with cholecalciferol. Am J Clin Nutr 77:204–210
38. Holick MF, Biancuzzo RM, Chen TC, Klein EK, Young A, Bibuld D, Reitz R, Salameh W, Ameri A, Tannenbaum AD (2008) Vitamin D_2 is as effective as vitamin D_3 in maintaining circulating concentrations of 25-hydroxyvitamin D. J Clin Endocrinol Metab 93(3):677–681
39. Lee JM, Smith JR, Philipp BL, Chen TC, Mathieu J, Holick MF (2007) Vitamin D deficiency in a healthy group of mothers and newborn infants. Clin Pediatr 46:42–44
40. Bodnar LM, Simhan HN, Powers RW, Frank MP, Cooperstein E, Roberts JM (2007) High prevalence of vitamin D insufficiency in black and white pregnant women residing in the northern United States and their neonates. J Nutr 137:447–452
41. Hollis BW, Wagner CL (2004) Assessment of dietary vitamin D requirements during pregnancy and lactation. Am J Clin Nut 79:717–726
42. Specker BL, Valanis B, Hertzberg V, Edwards N, Tsang RC (1985) Cyclical serum 25-hydroxyvitamin D concentrations paralleling sunshine exposure in exclusively breast-fed infants. J Pediatr 110:744–747
43. Specker BL, Valanis B, Hertzberg V, Edwards N, Tsang RC (1985) Sunshine exposure and serum 25-hydroxyvitamin D. J Pediatr 107:372–376
44. Food and Nutrition Board Institurte of Medicine (1997) Dietary reference intakes for calcium, phosphorus, magnesium, vitamin D, and Fluoride. IOM National Academy Press, Washington, DC, pp 7–30
45. Nakao H (1998) Nutritional significance of human milk vitamin D in neonatal period. Kobe J Med Sci 34:21–128
46. Feliciano ES, Ho ML, Specker BL, Falciglia G et al (1994) Seasonal and geographical variations in the growth rate of infants in China receiving increasing dosages of vitamin D supplements. J Trop Pediatr 40:162–165
47. Foman SJ, Younoszai K, Thomas L (1966) Influence of vitamin D on linear growth of normal full-term infants. J Nutr 88:345–350
48. Specker B, Ho M, Oestreich A, Yin T et al (1992) Prospective study of vitamin D supplementation and rickets I China. J Pediatr 120:733–739
49. Jeans PC (1950) Vitamin D. JAMA 1243:177–181
50. Wagner CL, Greer FR and the Section on Breast Feeding and Committee on Nutrition (2008) Prevention of rickets and vitamin D deficiency in infants, children, and adolescents. Pediatrics 122:1142–1152
51. Hypponen E, Laara E, Jarvelin M-R (2001) Virtanen SM. Intake of vitamin D and risk of type 1 diabetes: a birth-cohort study. Lancet 358:1500–1503
52. Pettifor JM, Ross FP, Moodley G, Wang J et al (1978) Serum calcium, magnesium, phosphorus, alkaline, phosphatase and 25-hydroxyvitamin D concentrations in children. S Afr Med J 53:751–754
53. Maalouf J, Nabulsi M, Vieth R, Kimball S, El-Rassi R, Mahfoud Z, Fuleihan GE (2008) Short term and long term safety of weekly high dose of vitamin D_3 supplementation in school children. J Clin Endocrin Metab First published ahead of print April 29, 2008 as doi: 10.1210/jc.2007-2530
54. Gordon CM, Feldman HA, Sinclair L, Williams AL, Kleinman PK, Perez-Rossello J, Cox JE (2008) Prevalence of vitamin D deficiency among healthy infants and toddlers. Arch Pediatr Adolesc Med 162(6):505–512
55. Gordon CM, Williams AL, Feldman HA, May J, Sinclair L, Vasquez A, Cox JE (2008) Treatment of hypovitaminosis D in infants and toddlers. J Clin Endocrinol Metab 93(7):2716–2721
56. Rajakumar K, Fernstrom JD, Holick MF, Janosky JE, Greenspan SL (2008) Vitamin D status and response to vitamin D_3 in obese vs. non-obese African American children. Obesity 16:90–95
57. Aksnes L, Aarskog D (1982) Plasma concentrations of vitamin D metabolites in puberty: effect of sexual maturation and implications for growth. J Clin Endocrinol Metab 55:94–101

58. Gultekin A, Ozalp I, Hasanoglu A, Unal A (1987) Serum-25-hydroxycholecalciferol levels in children and adolescents. Turk J Pediatr 29:155–162

59. Meier DE, Luckey MM, Wallenstein S, Clements TL et al (1991) Calcium, vitamin D, and parathyroid hormone status in young white and black women: association with racial differences in bone mass. J Clin Endocrinol Metab 72:703–710

60. Kinyamu HK, Gallagher JC, Galhorn KE, Petranick KM, Rafferty KA (1997) Serum vitamin D metabolites and calcium absorption in normal young and elderly free-living women and in women living in nursing homes. Am J Clin Nutr 65:790–797

61. Holick MF (1986) Vitamin D requirements for the elderly. Clin Nutr 5:121–129

62. Clemens TL, Zhou X, Myles M, Endres D, Lindsay R (1986) Serum vitamin D_2 and vitamin D_3 metabolite concentrations and absorption of vitamin D_2 in elderly subjects. J Clin Endocrinol Metab 63:656–660

63. Krall EA, Dawson-Hughes B (1991) Relation of fractional ^{47}Ca retention to season and rates of bone loss in healthy postmenopausal women. J Bone Miner Res 6:1323–1329

64. Dawson-Hughes B, Harris SS, Krall EA, Dallal GE, Falconer G, Green CL (1995) Rates of bone loss in postmenopausal women randomly assigned to one of two dosages fo vitamin D. Am J Clin Nutr 61:1140–1145

65. Dawson-Hughes B, Dallal GE, Krall EA, Harris S, Sokoll LJ, Falconer G (1991) Effect of vitamin D supplementation on wintertime and overall bone loss in healthy postmenopausal women. Ann Intern Med 115:505–512

66. Krall WE, Sahyoun N, Tannenbaum S, Dallal G, Dawson-Hughes B (1989) Effect of vitamin D intake on seasonal variations in parathyroid hormone secretion in postmenopausal women. N Engl J Med 321:1777–1783

67. Tangpricha V, Pearce EN, Chen TC, Holick MF (2002) Vitamin D insufficiency among free-living healthy young adults. Am J Med 112(8):659–662

68. Lips P, Wiersinga A, van Ginkel FC et al (1988) The effect of vitamin D supplementation on vitamin D status and parathyroid function in elderly subjects. J Clin Endocrinol Metab 67:644–650

69. Chapuy MC, Arlot ME, Duboeuf F, Brun J, Crouzet B, Arnaud S, Delmas PD, Meunier PJ (1992) Vitamin D_3 and calcium to prevent hip fractures in elderly women. N Engl J Med 327(23):1637–1642

70. Dawson-Hughes B, Harris SS, Krall EA, Dallal GE (1997) Effect of calcium and vitamin D supplementation on bone density in men and women 65 years of age or older. N Engl J Med 337:670–676

71. Bodnar LM, Catov JM, Simhan HN, Holick MF, Powers RW, Roberts JM (2007) Maternal vitamin D deficiency increases the risk of preeclampsia. J Clin Endocrinol Metab 92(9):3517–3522

72. Merewood A, Mehta SD, Chen TC, Holick MF, Bauchner H (2009) Association between severe vitamin D deficiency and primary caesarean section. J Clin Endo Metab 94(3):940–945

73. Vieth R, Garland C, Heaney R et al (2007) The urgent need to reconsider recommendations for vitamin D nutrition intake. Am J Clin Nutr 85:649–650

74. Grey A, Lucas J, Horne A, Gamble G, Davidson JS, Reid IR (2005) Vitamin D repletion in patients with primary hyperparathyroidism and coexistent vitamin D insufficiency. J Clin Endocrinol Metab 90:2122–2126

75. Jackson RD, LaCroix AZ, Gass M et al (2006) Calcium plus vitamin D supplementation and the risk of fractures. N Engl J Med 354(7):669–683

76. K/DOQI (2003) Clinical practice guidelines for bone metabolism and disease in chronic kidney disease. Am J Kidney Dis 42(Suppl 3):S1-S201

77. Dusso AS, Sato T, Arcidiacono MV et al (2006) Pathogenic mechanisms for parathyroid hyperplasia. Kidney Int 70:S8–S11

78. Looker AC, Pfeiffer CM, Lacher DA, Schleicher RL, Picciano MF, Yetley EA (2008) Serum 25-hydroxyvitamin D status of the US population: 1988–1994 compared with 2000–2004. Am J Clin Nutr 88(6):1519–1527

79. Pietras SM, Obayan BK, Cai MH, Holick MF (2009) Vitamin D_2 treatment for vitamin D deficiency and insufficiency for up to 6 years. Arch Int Med 169: 1806–1807

80. Brot C, Vestergaard P, Kolthoff N, Gram J, Hermann AP, Sorensen OH (2001) Vitamin D status and its adequacy in healthy Danish perimenopausal women: relationships to dietary intake, sun exposure and serum parathyroid hormone. Br J Nutr 86(1):S97–S103

81. Maeda SS, Kunii IS, Hayashi L, Lazaretti-Castro M (2007) The effect of sun exposure on 25-hydroxyvitamin D concentrations in young healthy subjects living in the city of Sao Paulo, Brazil. Braz J Med Biol Res 40(12):1653–1659

82. Hollis BW (2005) Circulating 25-hydroxyvitaminD levels indicative of vitamin D sufficiency: implications for establishing a new effective dietary intake recommendation for vitamin D. J Nutr 135: 317–322

83. http://www.grassrootshealth.org (adapted from)

2 Photobiology of Vitamin D

Tai C. Chen, Zhiren Lu,
and Michael F. Holick

Abstract   The major function of vitamin D (either vitamin D_2 or D_3) is to maintain healthy bone. Most humans obtain their vitamin D requirement through casual exposure of the skin to solar ultraviolet B and from dietary intake. The cutaneous synthesis of vitamin D is a function of 7-dehydrocholesterol concentration in epidermis, melanin pigmentation, and the solar zenith angle which depends on latitude, season, and time of day. Our recent study also indicates that altitude may influence the production of previtamin D_3. One area which has shown more progress during the past decade is the use of simulated sunlamp to improve vitamin D production in patients with intestinal malabsorption and elderly who were infirmed or living in northern latitude. Vitamin D deficiency is common in infants, children, and adults worldwide. The major cause of vitamin D deficiency globally is an underappreciation of the crucial role of sunlight in providing humans with their vitamin D requirement. The association between vitamin D deficiency and the increased risk of cancers, autoimmune diseases, infectious diseases, and cardiovascular disease indicates the importance of sunlight, vitamin D, and overall health and well-being of the general population.

Key Words: Sunlight; vitamin D; previtamin D; latitude; season; sunscreen; melanin; black; tanning; ultraviolet B radiation

1. INTRODUCTION

Vitamin D (represents either D_2 or D_3) is essential for maintaining healthy bone *(1, 2)*. The major source of vitamin D for most humans comes from the exposure of the skin to sunlight *(1, 2)*. During exposure to sunlight, 7-dehydrocholesterol (provitamin D_3) in the skin absorbs high-energy solar ultraviolet B photons (UVB; 290–315 nm) and is photolyzed to previtamin D_3 and then thermoisomerized to vitamin D_3 in the bilayer of the plasma membrane. After vitamin D is made in the skin or ingested in the diet, it is transported to the liver where it is hydroxylated on C-25 to form 25-hydroxyvitamin D [25(OH)D], the major circulating form of vitamin D that has been used as an index to determine vitamin D nutritional status *(2)*. To become active, 25(OH)D is further hydroxylated in the kidney to $1\alpha,25$-dihydroxyvitamin D [$1,25(OH)_2D$]. One of the major functions of vitamin D is the stimulation of the small intestine to increase its absorption of dietary calcium and phosphorus to maintain normal levels of calcium and

From: *Nutrition and Health: Vitamin D*
Edited by: M.F. Holick, DOI 10.1007/978-1-60327-303-9_2,
© Springer Science+Business Media, LLC 2010

phosphorus in the circulation *(2)*. When dietary calcium is inadequate to satisfy the body's calcium requirement, $1,25(OH)_2D$, in concert with parathyroid hormone (PTH), mobilizes monocytic stem cells in the bone marrow to become mature osteoclasts, which in turn enhances the removal of calcium from the bone into the circulation to maintain normal serum calcium levels *(2)*. In addition, $1,25(OH)_2D$ has other biologic actions, unrelating to calcium metabolism, in many tissues or cells that possess the $1,25(OH)_2D$ receptor. These include enhancement of cellular differentiation and/or inhibition of cellular proliferation, maintenance of normal neuromuscular, immunomodulatory, and cardiovascular function *(3)*.

2. HISTORICAL PERSPECTIVE

Rickets, a bone-deforming disease, is synonymous with vitamin D deficiency. It was first described in the mid-seventeenth century in Northern Europe during the industrial revolution as a major health problem for the young children when people began to migrate to city centers and live in an environment that was devoid of direct exposure to sunlight. The first observation into the potential cause of this bone-deforming disease and the implication of exposure to the sun in the prevention and cure of this disease was reported by Sniadecki in 1822 *(4)*. He reported that children who lived in Warsaw, Poland, had a high incidence of rickets, while children who lived on the farms surrounding Warsaw were free of the disease. He concluded that it was the lack of sunlight that was the likely cause of rickets. Later in 1890, Palm investigated factors that might associate with rickets in an epidemiological survey study and concluded that the common denominator for rickets in children was the lack of sunlight exposure *(2)*. Therefore, he suggested a systematic sunbathing as a means for preventing and curing rickets. However, these insightful observations went unnoticed just as was Sniadecki's earlier finding.

Using another approach, Bretonneau in the mid-1800s administered cod liver oil to treat a 15-month-old child with acute rickets and noted an incredible speedy recovery of the patient *(5)*. Later, Trousseau, a student of Bretonneau, used liver oils from a variety of fish and marine animals for the treatment of rickets and osteomalacia. In his monograph on therapeutics he advocated the use of fish liver oil preferably accompanied by exposure to sunlight to rapidly cure both rickets and osteomalacia *(5)*. These clinical observations led many to believe that rickets was caused by some type of nutritional deficiency. Unfortunately, these important observations were also disregarded.

By the turn of the twentieth century rickets had became epidemic in Northern Europe as well as in the industrialized northeastern region of the United States. However, it was not until 1918 when Mellanby fed rachitic beagle puppies cod liver oil and found that their rickets was cured that the scientific community began to consider rickets as a nutritional deficiency disease *(6)*. He concluded that cod liver oil possessed a fat soluble nutritional factor that he called the antirachitic factor. Originally it was thought that the antirachitic factor was the newly discovered vitamin A. However, McCollum et al. *(7)* found that by heating and aerating cod liver oil with oxygen they were able to destroy vitamin A activity without affecting its antirachitic activity. The finding therefore established a new fat soluble principle in cod liver oil which was subsequently

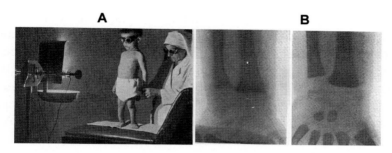

Fig. 1. (**a**) A rachitic child was irradiated to a mercury vapor arc lamp. (**b**) A wrist bone X-ray picture before (*left panel*) and after (*right panel*) 4 months of mercury vapor arc lamp therapy. (Reproduced from *(89)*).

named vitamin D by McCollum. About the same time as Mellanby's observations *(6)*, Huldschinsky exposed four rachitic children to radiation from a mercury vapor arc lamp and demonstrated by X-ray analysis that the rickets was cured after 4 months of therapy *(8)* (Fig. 1). He further demonstrated that the effect was not localized at the site of irradiation because exposing one arm to the radiation resulted in healing of both arms. In 1921, Hess and Unger exposed seven rachitic children in New York City to sunlight and, based on X-ray examination, reported marked improvement in rickets of each child *(9)*. The finding prompted Goldblatt and Soames *(10, 11)*, Hess and Weinstock *(12)*, and Steenbock and Black *(13)* to expose a variety of foods, such as wheat, lettuce, olive, linseed oils, and rat chow and other substances, such as human and rat plasma, to ultraviolet radiation. Both groups found that the ultraviolet irradiation imparted antirachitic activity to the substances. This finding led Steenbock to recommend that the irradiation of milk might be an excellent method to provide vitamin D to children and to prevent rickets. This suggestion was followed first by the addition of ergosterol (provitamin D_2) and its subsequent irradiation to impart antirachitic activity *(14)*, and, ultimately, the addition of synthetic vitamin D directly to milk. This simple concept led to the eradication of rickets as a significant health problem in the United States and other countries that used this practice. Thus, nearly one century after Sniadecki had first suggested the importance of sunlight exposure for prevention of rickets it was finally demonstrated that exposure to sunlight alone or ingesting ultraviolet irradiated foods or substances could prevent and cure this crippling bone disease.

3. VITAMIN D FORTIFICATION IN MILK

In the 1950s, there was an outbreak of neonatal hypercalcemia. In Great Britain that was thought to be due to over-fortification of milk of vitamin D. Although there was little evidence that this was the cause, most European countries reacted by banning the fortification of dairy products and other foods with vitamin D. As a result, vitamin D fortification was severely restricted in Europe. Consequently, the incidence of rickets became more frequent and this childhood disease continues to be a health problem in Europe *(15, 16)*. Realizing this problem, many European countries have begun adding vitamin D to various foods, including cereals and margarine. Only recently Sweden and Finland have permitted milk to be fortified with vitamin D. In the United States, the American Medical Association's Council on Foods and Nutrition recommended in 1957 that milk should contain 400 IU (10 μg) per quart and that the vitamin D content be measured at least twice yearly by an independent laboratory. Accordingly, the FDA modified its guidelines for the safe levels of vitamin D in milk and stipulated that one quart should contain 400 IU and no more than 600 IU of vitamin D. Fortification above 800 IU/qt might cause a public health threat and should be prohibited. A survey of vitamin D content in fortified milk revealed that 80 and 73% of the milk samples from the United States and Canada, respectively, did not contain 80–120% of label claim *(17–19)*. Some samples even had undetectable amount of vitamin D and some had 2- to 3-fold excess vitamin D.

The fact that antirachitic activity could be produced by exposing skin or foods to ultraviolet (UV) radiation led scientists to identify the precursor of vitamin D. The first one identified was ergosterol from yeast. Ergosterol is a major sterol in the fungal and plant kingdoms. The product from the irradiated ergosterol was first mistakenly believed to be a single product and was named vitamin D_1. The term vitamin D_1 was dropped after the preparation was found to be a mixture of several compounds. The vitamin D

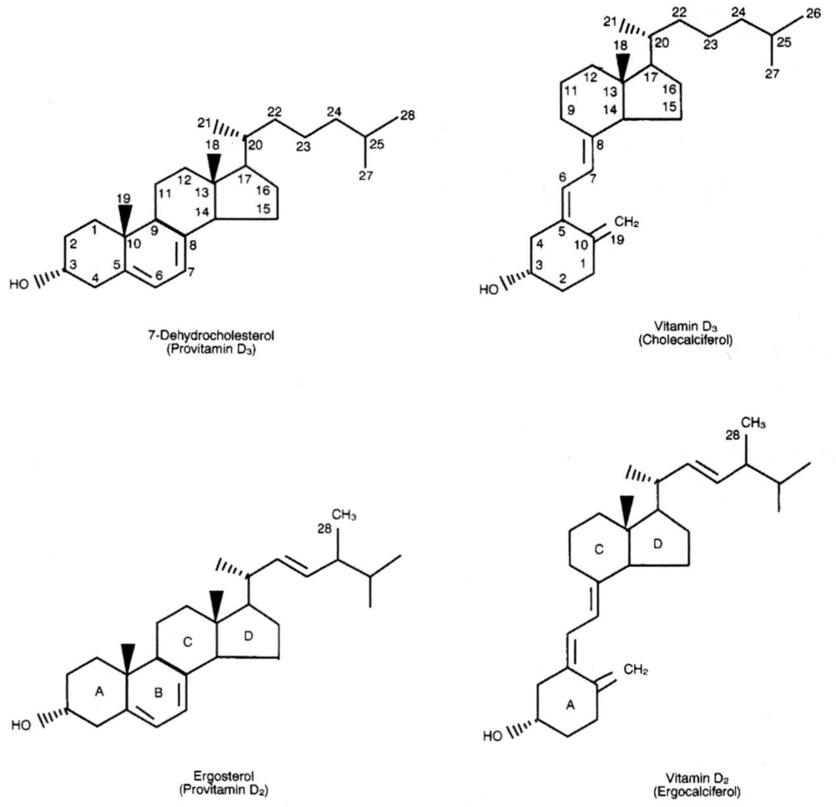

Fig. 2. Chemical structures of vitamin D_2 and vitamin D_3 and their respective precursors, ergosterol, and 7-dehydrocholesterol.

which was isolated in pure form from the irradiated ergosterol was named ergocalciferol or vitamin D_2. Its structure was identified after it was synthesized chemically by Windaus and colleagues *(20)* (Fig. 2). Originally it was thought that vitamin D_2 was the same product produced in the skin after sunlight exposure. However, it was found that vitamin D_2 was not as effective as the vitamin D from pig skin in curing rickets in chickens. The observations suggested that vitamin D_2 and the vitamin D isolated from the UV irradiated pig skin might not be the same. This suspicion was proven to be correct when Windaus's laboratory isolated a new compound with a side chain structure of cholesterol and showed that this cholesterol-like compound had the same antirachitic potency as those isolated from pig skin (Fig. 2). This finding led Windaus and Bock in 1937 to isolate 7-dehydrocholesterol (7-DHC) from pig skin. They demonstrated that the irradiation of 7-DHC produced a vitamin D which was named vitamin D_3 or cholecalciferol *(21)*. Subsequently, the presence of 7-DHC was demonstrated in human cadaver skin *(22)*, and the skin of fish, amphibians, reptiles, birds, and mammals including rats, hamsters, sheep, and polar bears *(23–31)*.

4. PHOTOSYNTHESIS OF PREVITAMIN D_3 IN THE SKIN

After the discovery of vitamin D_3, it was initially thought that 7-DHC was directly converted to vitamin D_3 after sunlight exposure. However, Velluz et al. were unable to detect any vitamin D_3 from a 7-DHC solution when they exposed the solution to ultraviolet radiation (UVR) at 0°C. Instead, they found an unknown compound and named this compound as previtamin D_3 *(32)*. They also reported that previtamin D_3 was not stable at room temperature and it slowly isomerized to vitamin D_3 *(33)*.

4.1. Photoconversion of 7-DHC to Previtamin D_3

When 7-DHC absorbs UV (290–315 nm) radiation, it causes a bond cleavage between carbon 9 and carbon 10 and an isomerization of 5,7-diene to form the s-*cis*, s-*cis*-previtamin D_3. Because there is a steric interaction between the C ring and the carbon 19 methyl group of s-*cis*, s-*cis*-previtamin D_3, this conformer of previtamin D_3 is energetically unfavorable and therefore is less stable. As a result, a rotation around the carbon 5 and carbon 6 single bond occurs and an energetically more stable s-*trans*, s-*cis*-previtamin D_3 is formed (Fig. 3). Alternately, the unstable s-*cis*, s-*cis*-previtamin D_3 conformer can undergo an intramolecular hydrogen rearrangement to form vitamin D_3.

7-DHC is present in all layers of human skin. Approximately 65% of 7-DHC per unit area is found in the epidermis and the remaining 35% is in the dermis. The highest concentration per unit area is found in the stratum basale and stratum spinosum *(34)*. These epidermal layers, therefore, have the greatest potential for previtamin D_3 synthesis. However, the amount of previtamin D_3 produced also depends on the number and energy of the photons reaching each layer of skin. Bunker and Harris reported in 1937 that the most effective wavelength for curing rickets in rats was 297 nm *(35)*. They and their colleagues fed irradiated 7-DHC to rachitic rats and concluded that 7-DHC

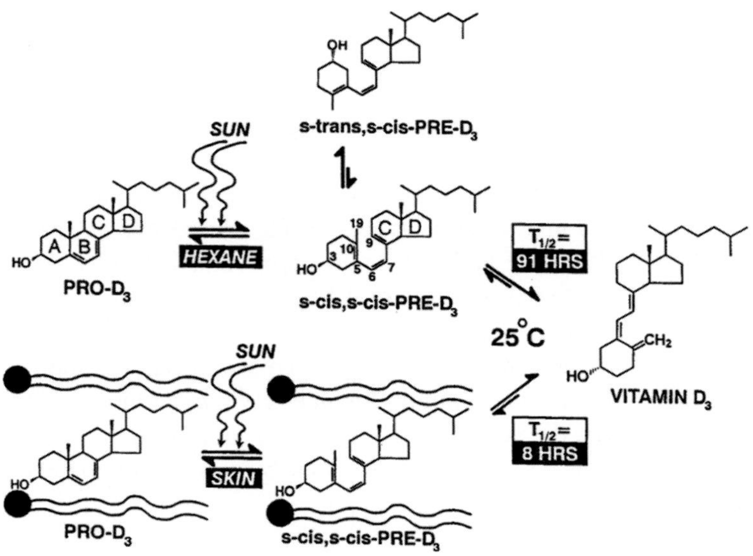

Fig. 3. Photolysis of 7-dehydrocholesterol (pro-D$_3$ or 7-DHC) to previtamin D$_3$, its thermal isomerization to vitamin D$_3$ in n-hexane and in the skin, and the stereo-structure of previtamin D$_3$ in the fatty acid bilayers of skin after exposure to sunlight (Reproduced with permission from *(45)*).

irradiated at 297 nm had the greatest antirachitic activity, while solutions irradiated at wavelengths longer than 313 nm showed no activity. Later, Kobayashi and Yasumara *(36)* showed that irradiation of an ergosterol solution with 295 nm radiation gave the maximum yield of previtamin D$_2$. MacLaughlin et al. examined the photosynthesis of previtamin D$_3$ from 7-DHC in human skin after exposing the tissue to narrow-band radiation or simulated solar radiation *(37)*. They reported that the optimum wavelengths for the production of previtamin D$_3$ were between 295 and 300 nm *(37)*. In Caucasians with skin type II, 20–30% of the radiation of 295 nm is transmitted through the epidermis; the majority of the UVB photons (290–320 nm) are absorbed by the stratum spinosum of the epidermis. In Blacks with skin type V only about 2–5% of the UVB photons penetrate through the epidermis *(38, 39)*. As the UVR penetrates the epidermis, it is absorbed by a variety of molecules including DNA, RNA, proteins, as well as 7-DHC. The 5,7-diene of 7-DHC absorbs solar radiation between 290 and 315 nm causing it to isomerize resulting in a bond cleavage between carbon 9 and 10 to form a 9,10-*seco*-sterol previtamin D$_3$ *(24, 34, 37)*. Because most of the radiation responsible for producing previtamin D$_3$ is absorbed in the epidermis, greater than 95% of the previtamin D$_3$ that is produced is in the epidermis *(34)*. Once previtamin D$_3$ is synthesized in the skin, it can undergo either a photoconversion to lumisterol, tachysterol, and 7-DHC or a heat-induced isomerization to vitamin D$_3$ (Fig. 4).

Determination of the subcellular localization of 7-DHC and previtamin D$_3$ in human epidermal tissue revealed that most 7-DHC and previtamin D$_3$ were in the membrane fraction, while only 20% was in the cytosol. Based on this finding, it has been postulated that most 7-DHC is entrapped in membrane. It is likely that the 3β-hydroxyl group of the

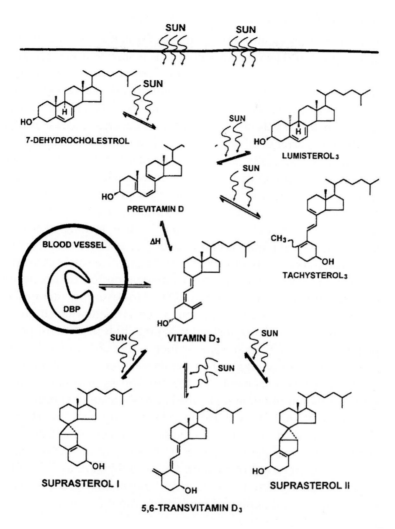

Fig. 4. Photosynthesis of vitamin D_3 from 7-dehydrocholesterol (pro-D_3 or 7-DHC), its photoisomerization and photodegradation, and the photoisomers of previtamin D_3 (Holick copyright 2006 with permission).

7-DHC molecule is near the polar head group of the membrane fatty acids and interacts with it through hydrophilic forces, while the nonpolar rings and side chain are associated with the nonpolar tail of the fatty acid by hydrophobic van der Waals interactions. Thus, when 7-DHC in the skin's plasma membrane is exposed to UVB radiation the thermodynamically less favorable conformation of the s-*cis*, s-*cis*-previtamin D_3 is preserved through hydrophobic and hydrophilic interactions with the bilayer lipid membrane fatty acids. Thus, a rotation around carbon 5 and carbon 6 to form the thermodynamically more stable s-*trans*, c-*cis* conformer is presented in the membrane and is only permissible in an organic solvent, such as hexane.

4.2. Conversion of Previtamin D₃ to Vitamin D₃

The isomerization of previtamin D_3 to vitamin D_3 is the last step in the synthesis of vitamin D_3 in human skin. The reaction rate of this isomerization is temperature dependent and is enhanced by raising the temperature. Earlier studies found that this process was not affected by acids, bases, catalysts, or inhibitors of radical chain processes (40). Furthermore, no intermediate was detected during the isomerization. This led to the conclusion that the reaction was an intramolecular concerted process involving a [1,7]-sigmatropic hydrogen rearrangement (40), which is an antarafacially (opposite side of a plane) allowed and a suprafacially (same side of a plane) forbidden process (41). Much of the information about the previtamin D_3 isomerization has been obtained from experiments using organic solvents for the conversion (42, 43) and assumed to be the same in human skin. There is no evidence for the existence of an enzymatic process in the skin that can convert previtamin D_3 to vitamin D_3. It is postulated that in an organic solvent, such as hexane, previtamin D_3 preferentially exists in a s-*trans*, s-*cis* conformation, which is thermodynamically more stable and cannot be easily converted to vitamin D_3. Therefore, it takes several days for the isomerization reaction between previtamin D_3 and vitamin D_3 to reach equilibrium at 37°C. However, previtamin D_3 in the skin is maintained in the s-*cis*, s-*cis* conformation, a conformation which greatly facilitates its conversion to vitamin D_3. Thus, instead of taking 30 h for 50% of previtamin D_3 to convert to vitamin D_3 at 37°C in hexane, it took only 2.5 h in the human skin at the same temperature. The results suggest that the interaction of previtamin D_3 with membrane fatty acids in skin is responsible for the efficient formation of vitamin D_3 in the skin. During the formation of vitamin D_3, the hydrophilic and hydrophobic interactions of the s-*cis*, s-*cis*-previtamin D_3 with the membrane fatty acids are disrupted, thereby facilitating the ejection of vitamin D_3 from the skin cell membrane into the extracellular space. By diffusion it enters the circulation by binding to the serum vitamin D binding protein (DBP). The removal of vitamin D_3 from the skin as it is being produced, thereby, changes the isomerization reaction from a reversible process to an irreversible process (34, 44). This would explain the relatively rapid rise in the serum vitamin D_3 concentration after UVB exposure.

A comparative study of the kinetic and thermodynamic properties of the isomerization reaction in human skin and in an organic solvent revealed that not only the equilibrium of the reaction was shifted in favor of vitamin D_3 synthesis in human skin (equilibrium constant K at 37°C = 11.44) compared to hexane ($K = 6.15$) but also the rate of the reaction was increased by more than 10-fold in human skin ($T_{1/2}$ at 37°C = 2.5 h) when compared to hexane ($T_{1/2} = 30$ h) (44). This accelerated rate of isomerization was also observed in chicken, frog, and lizard skin (44, 45). The enthalpy ($H°$) change for the reaction was −21.58 and −15.60 kJ mol^{-1} in human skin and in hexane, respectively. The activation energy for both the forward and the reverse reactions was lower in human skin than in hexane. Thus, human skin profoundly changed the rate constant and equilibrium constant in favor of vitamin D_3 formation.

The importance of membrane microenvironments on previtamin $D_3 \rightleftarrows$ vitamin D_3 isomerization received further support from a kinetic study in an aqueous solution of β-cyclodextrin (46). Cyclodextrins, a group of naturally occurring, truncated cone-shaped oligosaccharides, have an unique ability to complex a variety of foreign molecules including steroids into their hydrophobic cavities in aqueous solution *(47)* and catalyze reactions of a wide variety of guest molecules *(48)*. Among the various cyclodextrins, β-cyclodextrin has been shown to form 2:1 (host/guest) inclusion complexes with vitamin D_3 *(49)*. Using this model, Tian and Holick demonstrated that, at 5°C, the forward (κ_1) and reversed (κ_2) rate constants for previtamin $D_3 \rightleftarrows$ vitamin D_3 isomerization were increased by more than 40 and 600 times, respectively, compared with those in *n*-hexane *(46)*. The equilibrium constant of the reaction was significantly reduced by more than 12-fold when compared to that in *n*-hexane at 5°C, and the percentage of vitamin D_3 at equilibrium was increased as the temperature was increased. When complexed with β-cyclodextrin, the previtamin $D_3 \rightleftarrows$ vitamin D_3 isomerization became endothermic ($H° = 13.05$ kJ mol^{-1}) in contrast to being exothermic in *n*-hexane, suggesting that the thermodynamically unfavorable s-*cis*, s-*cis*-previtamin D_3 conformers are stabilized by β-cyclodextrin; and therefore, the rate of the isomerization is increased *(46)*.

4.3. Translocation of Vitamin D₃ from the Skin into the Circulation

After vitamin D_3 is formed from the thermally induced isomerization of previtamin D_3 in the epidermis, it is transported into the dermal capillary bed beneath the dermoepidermal junction. Little is known about the mechanism of this translocation process. In an attempt to understand this event, Tian et al. studied the kinetics of vitamin D_3 formation and the time course of appearance of vitamin D_3 in the circulation after exposure of chicken to UVB radiation. Their data indicate a much faster rate of formation of vitamin D_3 from previtamin D_3 than the reverse reaction (return back to previtamin D_3 from vitamin D_3) and a relatively fast rate of translocation from skin to circulation *(42)*. By examining the time course of appearance of vitamin D_3 in circulation, they found a rapid phase of vitamin D_3 appearance from 8 to about 30 h post-irradiation and a relatively slower phase of its disappearance after the circulating concentration of vitamin D_3 reached its peak. No previtamin D_3 could be detected 1 h after UVB irradiation. Thus, only vitamin D_3 is preferentially removed from skin into circulation, leaving behind previtamin D_3 in the epidermis for the continued thermoisomerization to vitamin D_3.

The role of DBP in the translocation of vitamin D_3 into circulation is implicated from the work of Haddad et al. *(50)*. They investigated the transport of cutaneously synthesized vitamin D_3 into circulation in seven healthy volunteers who received whole-body irradiation with 27 mJ/cm^2 dosage of UVB light (290–320 nm) by comparing the time course distribution of plasma protein-bound vitamin D_3 in high- (>1.3 g/ml) and low-density (<1.3 g/ml) fractions after UVB irradiation. They found that plasma vitamin D concentration began to increase 10 h after irradiation, peaked at 24 h, and lasted for a week in the high-density layer where all the hDBP was present. When actin affinity chromatography was used to specifically bind DBP, it also removed vitamin D_3 from the plasma of irradiated subjects. These observations indicate that the endogenously photosynthesized vitamin D_3 circulates in serum almost exclusively on DBP, which differs from the orally administered vitamin D_2 *(50)*, which is evenly distributed between the high- and low-density layers at 4, 8, and 24 h after the ingestion. Thus, DBP is important in the translocation of vitamin D from skin into the circulation.

4.4. Photodegradation of Vitamin D$_3$

Once vitamin D is formed in the skin from previtamin D$_3$, it is transported into the circulation by a diffusion process from the epidermis into the dermal capillary bed. It is believed that this process is prompted by its attraction to the DBP *(34)*. If vitamin D$_3$ in the skin is exposed to sunlight prior to its transfer into the circulation, the triene system of vitamin D$_3$ structure will absorb solar UVR and photolyze to three major photoproducts, 5,6-*trans*-vitamin D$_3$, suprasterol 1, and suprasterol 2 *(51)*. Exposure to as little as 10 min of sunlight in Boston in the summer resulted in the photodegradation of 30% of [^{3}H]-vitamin D$_3$ in a test-tube model system *(51)*. Likewise, exposure to 0.5, 1, and 3 h of the summer sunlight caused 50, 75, and 95% degradation of the original [^{3}H]-vitamin D$_3$, respectively. Although winter sunlight in Boston does not promote vitamin D$_3$ synthesis, the longer wavelength UVR such as UVA present in the winter sunlight could potentially photodegrade vitamin D stores in the skin and in the circulation.

4.5. Photoisomers of Vitamin D$_3$

Previtamin D$_3$ photosynthesized in the skin can isomerize either to vitamin D$_3$ or to a variety of products, including tachysterol$_3$ and lumisterol$_3$. For example, during the first 10 min of simulated equatorial solar radiation, about 10–15% of the epidermal 7-DHC in Caucasian skin was converted to previtamin D$_3$ without any detectable amounts of tachysterol$_3$ or lumisterol$_3$ *(52)*. After 1 h of exposure, 5 and 30% of the original 7-DHC was converted to tachysterol$_3$ and lumisterol$_3$, respectively, whereas the amount of previtamin D$_3$ remained at about 15%. The concentrations of lumisterol increased with increasing exposure times, reaching 60% by 8 h. When black skin was irradiated under the same conditions, longer exposure times were required to reach the maximal previtamin D$_3$ formation *(52)*. Thus, it is the photochemical degradation of previtamin D$_3$ rather than melanin pigmentation that is most responsible for limiting the production of previtamin D$_3$ in human skin.

5. REGULATION OF THE CUTANEOUS SYNTHESIS OF PREVITAMIN D$_3$

5.1. Role of Melanin Pigmentation

In 1967, Loomis *(53)* suggested that melanin pigmentation evolved for protection from vitamin intoxication because of excessive exposure to sunlight. He further promoted the concept that as people migrated away from the equator, they gradually lost their skin pigmentation in order to synthesize adequate amounts of vitamin D$_3$ in their skin to maintain a healthy skeleton. However, it is unlikely that this is the major mechanism that prevents excessive vitamin D$_3$ formation in the skin. It is now recognized that sunlight itself is the most important factor for regulating the total production of vitamin D$_3$ in the skin. Sunlight can destroy any excess previtamin D$_3$ and vitamin D$_3$ in the skin *(51)*.

It is well known that melanin effectively absorbs solar radiation from 290 to 700 nm, including the portion that is responsible for previtamin D$_3$ synthesis (290–315 nm).

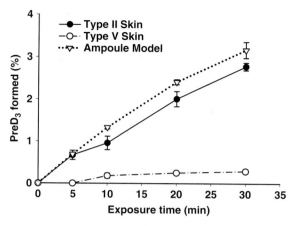

Fig. 5. The conversion of 7-dehydrocholesterol (pro-D_3 or 7-DHC) to previtamin D_3 in an ampoule model, type II and type V skin after exposing to noon sunlight in June at Boston (42°N), Massachusetts. The data represent the mean ± SEM of duplicate determinations (Reproduced with permission from *(56)*).

Thus, melanin pigmentation can diminish the production of previtamin D_3 in the skin. The effects of skin pigmentation on previtamin D_3 synthesis in humans were studied in two in vitro models by exposing type II and type V skin along with ampoules containing 7-DHC solution to noon sunlight in June on a cloudless day in Boston, Massachusetts. It was observed that in June, $0.67 \pm 0.11\%$ of 7-DHC in epidermis was converted to previtamin D_3 in type II skin, but no detectable amount was found in type V skin samples after 5 min of sunlight exposure (Fig. 5). A small amount ($0.18 \pm 0.06\%$) of epidermal 7-DHC was converted to previtamin D_3 in type V skin after 10 min of sun exposure, whereas in the same time period, 0.95% of 7-DHC was converted to previtamin D_3 in type II skin. The synthesis of previtamin D_3 in epidermis continued to increase to 2.01 ± 0.18 and $2.78 \pm 0.09\%$ after 20 and 30 min sunlight exposure, respectively, in type II skin. Unlike type II skin, the conversion of epidermal 7-DHC to previtamin D_3 in type V skin samples only increased modestly to 0.25 ± 0.04 and $0.29 \pm 0.05\%$ after 20 and 30 min of exposure, respectively. The production of previtamin D_3 in the ampoule model *(14)* was about 0.4% more than that found in type II skin between 10 and 30 min of exposure.

Studies were performed to determine the effect of increased skin pigmentation on the cutaneous production of vitamin D_3 by measuring the circulating concentrations of vitamin D_3 in Caucasian, Indian, Pakistani, and Black volunteers *(54, 55)*. Exposure of Caucasian subjects to 1.5 MED (0.054 J/cm^2) of UVR greatly increased serum vitamin D_3 concentration by more than 50-fold 24–28 h after the exposure, whereas the same dose had no effect on Black volunteers (Fig. 6) *(54)*. However, when a 6-fold greater dose of UV radiation (0.32 J/cm^2) was applied to one of the Black volunteers, there was an increase in serum vitamin D_3. In this study, the time course of appearance and disappearance of vitamin D_3 was similar to that found in Caucasian volunteers, but that the peak concentration was still lower than those seen in Caucasian volunteers exposed to 0.054 J/cm^2 UV radiation by 40%. These results indicate that increased

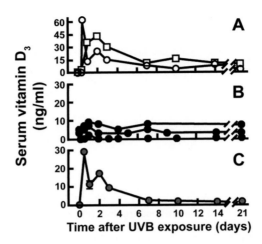

Fig. 6. Increase in serum vitamin D_3 after total-body exposure to 0.054 J/cm^2 of UVB of (**a**) two lightly pigmented Caucasians and (**b**) three heavily pigmented Blacks. (**c**) Serum level of vitamin D_3 in one black subject in **panel b** after exposure to a 0.32 J/cm^2 dose of UV radiation (Reproduced with permission from *(54)*).

skin pigmentation can greatly reduce the UVR-mediated synthesis of vitamin D_3. When people with intermediate skin color, such as Indians and Pakistanis, were exposed to 1.5 MED (0.073–0.2 J/cm^2) of whole-body UVR, there was a similar time course of appearance and disappearance of serum vitamin D_3 as that seen in Caucasian controls exposed to 1.5 MED (0.046–0.072 J/cm^2) of UVR *(55)*. Thus, people with darker skin need longer exposure to UVR than fair-skinned people to photosynthesize the same amount of vitamin D. However, the capacity to produce vitamin D is no different among the groups.

The effect of skin pigmentation on the serum levels of 25(OH)D after UVB irradiation was investigated in volunteers with different skin types *(56)*. Healthy volunteers with different skin types (II, III, IV, and V) were recruited in the beginning of winter. Based on the manufacturer recommendation each volunteer received a total of 0.75 MED (minimal erythema dose) in each session. To achieve his goal, the skin types II, III, IV, and V received an average of 6, 8, 11, and 12 min of UV irradiation during the first session, respectively. The sessions were held three times a week for 12 weeks for a total of 36 sessions. The blood was drawn for 25(OH)D measurement at the baseline visit before UVB exposure, during and at the conclusion of the study. Figure 7 demonstrates that UVB irradiation greatly increased the serum 25(OH)D in all skin types. The percentage increase in the 25(OH)D level at the end of the study for skin type II, III, IV, and V were 210 ± 53, 187 ± 64, 125 ± 55, and 40%, respectively.

5.2. Influence of Altitude, Latitude, Time of Day, and Weather Conditions on Previtamin D_3 Production

It was first observed in 1897 by Kassowitz that incidence of rickets was markedly increased during the winter months and decreased during summer and fall *(57)*. Later, Schmorl *(58)* demonstrated from his autopsy studies that the highest incidence of rickets

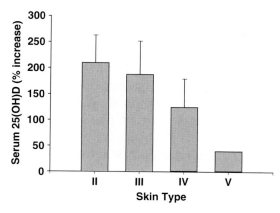

Fig. 7. The serum 25(OH)D levels in volunteers with different skin types after weekly exposure to simulated sunlight for 12 weeks (0.75 MED per session). The data represent the mean ± SEM of 4–5 volunteers for skin type II, III, and IV. A single determination was obtained for skin type IV (Reproduced with permission from *(56)*).

occurred between November and May. Hansemann *(59)* also noted that almost all children who were born in the autumn and died in the spring showed marked manifestations of rickets while those born in the spring and during the summer were essentially free of the disease. During the winter months in the Northern hemisphere, people wear more clothing and go outdoors less often that prevents their skin from direct exposure to sunlight *(60)*. Furthermore, the zenith angle of the sun is increased in the winter, and the amount of UVB radiation reaching the earth's surface is significantly diminished because most of it is absorbed by the stratospheric ozone layer. Therefore, the amount of solar UVB radiation reaching the biosphere is a function of the solar zenith angle and depends on latitude, season, and time of day *(61, 62)*. Several studies were conducted to determine the effect of latitude, season, and time of day on the cutaneous production of previtamin D_3 *(61–63)*. A more comprehensive study was conducted utilizing a model consisting of 7-DHC solution (50 μg/ml ethanol) sealed in borosilicated glass ampoules. On a cloudless day, the ampoules were exposed to sunlight on an hourly basis throughout the day once a month in seven cities located in both the Northern and Southern hemispheres. Because of the absorption properties of human skin and the effect of skin pigmentation on the absorption of solar UVB radiation, the results obtained from the ampoules represented the maximal conversion of 7-DHC to previtamin D_3. In order to correlate the synthesis of previtamin D_3 in ampoules with human skin, 1 cm^2 of Caucasian skin samples with skin type III *(64)*, which sometimes burns and always tans, were exposed to simulated sunlight along with the ampoules containing 50 μg 7-DHC in 1 ml of absolute alcohol. It was found that 0.8% of 7-DHC was converted to previtamin D_3 in the ampoules before any previtamin D_3 appeared in the skin samples. Based on this correlation, we have estimated the capacity of type III skin to synthesize vitamin D_3 from natural sunlight each month at seven locations with different latitudes in both the Northern and Southern hemispheres. It was found that following exposure to noon sunlight for 1 h, previtamin D_3 formation was negligible during the months of December through February in Boston, USA (42°N); from November through March in Edmonton,

Canada (52°N); and from October through March in Bergen, Norway (61°N). Similarly, people with the same skin type cannot synthesize previtamin D$_3$ from April through September in Ushuaia, Argentina (55°S). However, individuals with this skin type can synthesize previtamin D$_3$ year round in Cape Town, South Africa (35°S); Johannesburg, South Africa (26°S); and Buenos Aires, Argentina (34°S) (Fig. 8). Thus, previtamin D$_3$ synthesis increased from the spring to summer months and decreased thereafter. These changes are reflected in the seasonal variation of serum 25(OH)D$_3$ concentrations in children and adults *(65–68)*. Correlation studies between the synthesis of previtamin D$_3$ in ampoules and in skin type V samples indicated that 1.8% of 7-DHC was converted to previtamin D$_3$ in ampoules before any previtamin D$_3$ could be detected in the skin samples. Therefore, individuals with skin type V would require a considerably longer period of sun exposure to make the same amount of vitamin D in their skin compared to those with skin type III.

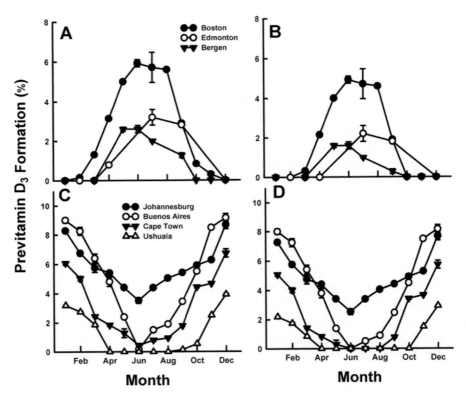

Fig. 8. Influence of season and latitude on the synthesis of previtamin D$_3$ in Blacks (skin type V) and Caucasians (skin type II). **Panels a** and **c** represent the calculated formation of previtamin D$_3$ in the Northern and Southern hemispheres in the skin of Caucasians and **panels b** and **d** represent the calculated formation of previtamin D$_3$ in the Northern and Southern hemispheres in the skin of Blacks. The cities in the Northern hemisphere include Boston (42°N), Edmonton (52°N), and Bergen (61°N), and the cities in the Southern hemisphere are Johannesburg (26°S), Buenos Aires (34°S), Cape Town (35°S), and Ushuala (55°S). The data represent the means ± SEM of duplicate determinations. It should be noted that the data for May and June in Edmonton are missing (Copyright Chen and Holick, 2009 with permission).

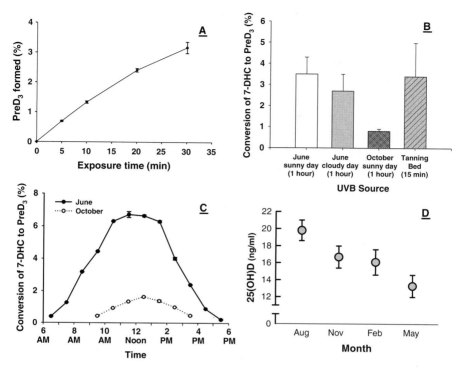

Fig. 9. (**a**) Time-dependent formation of previtamin D$_3$ in ampoules containing 7-DHC in ethanol after exposure to sunlight at noon in June in Boston. (**b**) A comparison of previtamin D$_3$ production after exposure to sunlight in June on a sunny day and cloudy day for 1 h, on a sunny day in October for 1 h, and tanning bed for 15 min. (**c**) Conversion of 7-DHC to previtamin D$_3$ at various times throughout the day in June and in October on a sunny day in Boston. (**d**) Circulating levels of 25(OH)D in healthy free-living nursing home residents at various seasons of the year (Reproduced with permission from *(69)*).

The time-dependent conversion of 7-DHC to previtamin D$_3$ was also studied at noon time in June in Boston (Fig. 9). The figure indicates that after 5 min, ~0.8% of 7-DHC was converted to previtamin D$_3$, and by 35 min, ~3.3% of 7-DHC was photolyzed to previtamin D$_3$ and its photoproducts (Fig. 9a). This showed that previtamin D$_3$ production occurred when 7-DHC was exposed to sunlight and that the efficiency of conversion was almost linear as a function of time over a period of 30 min. Because the zenith angle during a day is dependent on the time of a day and is much more oblique in the early morning and late afternoon that will result in a longer path length for the solar UVB photons to pass through, it is important to evaluate the effect of time of day on previtamin D$_3$ synthesis. As shown in Fig. 9c, greater than 6% of 7-DHC was converted to previtamin D3 in June in Boston, whereas no previtamin D$_3$ production was observed before 7:00 a.m. or after 6:00 p.m. on the same day. The previtamin D$_3$ production between the hours of 8:00 and 10:00 a.m. and 4:00 and 6:00 p.m. was <20% of that produced at noon time. The figure also shows that much less previtamin D$_3$ was produced in October even at noon time; there was an 80% reduction in the efficiency of conversion of 7-DHC to previtamin D_3 at noon time compared to noon time in June. Likewise, the length of no-conversion periods was found to be longer in the morning and in the afternoon in October comparing to June. The length of these periods will also increase in people with higher degrees of skin pigmentation *(53)*.

When the efficiency of 7-DHC conversion to previtamin D_3 on a cloudy day was compared to a cloudless day, the efficiency was reduced by about 20%. Furthermore, the conversion efficiency on a cloudless day in June for 1 h is equivalent to that exposed to tanning bed radiation for 15 min ($\sim$30 mJ/cm^2) (Fig. 9b).

Recently, we investigated the effects of altitude on the production of previtamin D_3 in Nepal, India, and Mount Everest using an ampoule model *(69)*. The ampoules containing 7-DHC in ethanol and sealed under argon were exposed to sunlight on cloudless days at 27°N in Nepal and India during the last week of October and during the first 2 weeks of November, 2006. The lowest altitude was in Agra at 169 m and the highest altitude was at Mount Everest base camp at 5,350 m. After exposure for 1 h between 11:30 a.m. and 12:30 p.m., the samples were stored in the dark before they were evaluated by high-performance liquid chromatography for the conversion of 7-DHC to previtamin D_3 and its other photoproducts, including lumisterol and tachysterol. A dramatic influence of altitude on the synthesis of previtamin D_3 and its photoproducts at the same latitude of 27°N during the period of October (last 2 weeks) and November (first 2 weeks) was observed (Fig. 10). In Agra (169 m) and Katmandu (1,400 m), about 0.5% of 7-DHC was converted to previtamin D_3 and its photoproducts. There was an almost linear increase in the production of previtamin D_3 and its photoproducts with increasing altitude that was about 400% higher at the base camp of Everest at 5,350 m compared to Agra at 169 m.

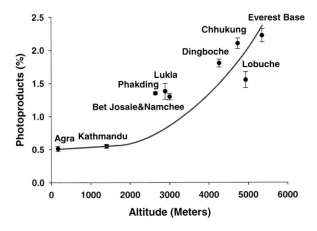

Fig. 10. Effect of altitudes on the production of previtamin D_3. Ampoules containing 7-DHC in ethanol were exposed for 1 h between 11:30 a.m. and 12:30 p.m. at 27°N in India and Nepal at various attitudes (Reproduced with permission from *(69)*).

5.3. *Effect of Aging on the Cutaneous Production of Previtamin D₃*

Abundant studies have suggested that osteomalacia caused by vitamin D deficiency is becoming an epidemic in Asia, Europe, and the United States *(70–76)*. It has been estimated that about 30–40% of elderly people with hip fracture in the United States and Great Britain are vitamin D deficient. Lamberg-Allard reported that circulating concentrations of 25(OH)D were low in long-stay geriatric patients and in residents at the old peoples home due to lack of exposure to sunlight and low vitamin D intake *(76)*. Similarly, a study conducted at a nursing home in the Boston area demonstrated that approximately 60% of the nursing home residents were vitamin D deficient during the winter months *(77)*. Even near the end of the summer we found that in the Boston area 30, 43, and 80% of the free-living elderly Caucasian, Hispanic, and Black subjects, respectively, had 25(OH)D levels of less than 20 ng/ml.

Furthermore, Lester et al. *(68)* reported that elderly subjects had lower circulating levels of 25(OH)D than healthy young adults. These data suggested that aging decreased the capacity of human skin to produce vitamin D_3. It has been known that skin thickness decreases linearly with age after the age of 20 years *(78)*. This is correlated with an age-dependent decrease in the epidermal concentrations of 7-DHC *(79)*. Exposure of human skin samples from various age groups to simulated sunlight showed that the skin from 8- and 18-year-old subjects produced 2- to 3-fold greater amounts of previtamin D_3 than the skin from 77- and 82-year-old subjects. Whole-body exposure of healthy young and older adults to the same amount of UV radiation confirmed the in vitro finding. When the circulating concentrations of vitamin D were determined before and at various times after the exposure, the young volunteers (age range 22–30 years) raised their circulating concentrations of vitamin D_3 to a maximum of 30 ng/ml within 24 h while the elderly subjects (62–80 years) were able to reach a maximum concentration of only 8 ng/ml *(80)*. In one study, we exposed 7-DHC in ethanol in an ampoule to tanning bed irradiation (MedSun by Wolff System Technology Corporation, Atlanta, GA.) for 10 min. We observed a linear increase of 7-DHC conversion to previtamin D_3 from about 1% at 1 min to about 10% at 10 min (Fig. 11 a). We also determined the circulating concentrations of 25(OH)D in 15 healthy young adults (skin types II and III, 20–53 year) after most of their body were exposed to the same tanning bed three times a week (0.75 MED per week) for 7 weeks (total radiation was approximately 4 MED for the period). After 1 week, there was a 50% increase in 25(OH)D that continued to increase for 5 weeks before reaching a plateau of about 150% above baseline values (Fig. 11b). Using the same tanning bed, we also exposed a 76-year-old male volunteer to 0.75 MED three times a week for 7 weeks. His serum 25(OH)D level increased from 29 to 47 ng/ml after 7 weeks of exposure. However, unlike the young adults (Fig. 11b), his serum 25(OH)D reached plateau with about 60% increase after 2 weeks of exposure (Fig. 11c). Because there was no significant increase in this subject's skin pigment throughout the study, it is likely that the photochemical synthesis and degradation of vitamin D_3 might have reached equilibrium after 2 weeks.

5.4. Effect of Sunscreen Use and Clothing on Previtamin D_3 Formation

There is a great concern about the damaging effects of chronic exposure to sunlight on the skin. Long-term exposure to sunlight can cause dry wrinkled skin as well as increase risk of non-melanoma skin cancer such as squamous cell and basal cell carcinoma. As a result of this concern, there has been a major effort by dermatologists to encourage

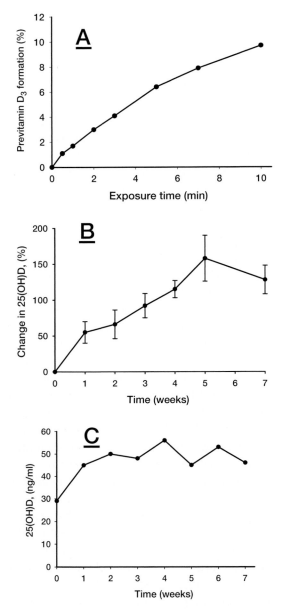

Fig. 11. Production of previtamin D$_3$ and serum level of 25(OH)D after the exposure of 7-DHC solution in ampoules and human volunteers to a tanning bed lamp. (**a**) Ampoules containing 7-DHC were placed and exposed to a tanning bed lamp. At various times, an ampoule was removed and the conversion of 7-DHC to previtamin D$_3$ was measured by HPLC. (**b**) Healthy adults were exposed to 0.75 MED in a tanning bed three times a week for 7 weeks. Circulating concentrations of 25(OH)D were determined at baseline and once a week thereafter. (**c**) A healthy 76-year-old man was exposed to tanning bed radiation equivalent to 0.75 MED three times a week for 7 weeks. His circulating concentrations of 25(OH)D were obtained at weekly intervals (Reproduced with permission from *(69)*).

people to apply sunscreens on their body before going outdoors *(81)*. There is no question of the benefit of sunscreen use for the protection of the skin from the damaging effects of sunlight. However, the radiation that is responsible for causing sunburn, skin wrinkling, and skin cancer is the same radiation that is responsible for producing previtamin D_3. Thus, the sunscreen use can also prevent the beneficial effect of sunlight, the photoconversion of 7-DHC to previtamin D_3 *(82, 83)*. Matsuoka et al. studied the effects of *para*-aminobenzoic acid (PABA) on previtamin D_3 formation in skin samples in vitro and on the cutaneous vitamin D synthesis in vivo. They found that 5% PABA solution totally blocked the photoisomerization of 7-DHC to previtamin D_3 in human skin specimens. When the volunteers who were applied with 5% PABA, which had a sun protection factor of 8, did not raise their circulating concentrations of vitamin D_3 after exposure to 1 MED UVR. The same dose of UVR increased the serum level of vitamin D_3 10-fold in control subjects (without PABA) 24 h after exposure. Therefore, any UVB blocking agent that prevents the damaging effects of sunlight will also prevent the cutaneous production of previtamin D_3. The impact of chronic sunscreen use on the vitamin D status of the elderly has been evaluated *(83)*. It was found that the 25(OH)D concentrations in serum of long-term PABA users were significantly lower than the non-users. Among the 19 long-term PABA users, 4 had borderline to overt vitamin D deficiency based on low circulating concentrations of 25(OH)D.

Clothing also blocks transmission of nonvisible UV radiation *(60, 84)*. Robson and Diffey *(84)* tested 60 different fabrics and observed that the nature and type of the textures and the structure of the fabric affect the sun protection factor. For example, a polyester blouse would have a SPF-2, while cotton twill jeans would have a SPF of 1,000 or complete protection against extreme sunlight exposures. Matsuoka and her colleagues examined the effect of the commonly used fabrics including cotton, wool, and polyester in black and white colors, on UVB transmission properties; the photoproduction of previtamin D_3 from 7-DHC; and the elevation of serum vitamin D_3 after irradiation with 1 MED of UVB in volunteers wearing jogging garments made of these fabrics *(60)*. They found that direct transmission of UVB was attenuated the most by black wool and the least by white cotton. None of the fabrics allowed the conversion of 7-DHC to previtamin D_3 after irradiation with up to 40 min of simulated sunlight or the increase in serum vitamin D_3 in volunteers. Increasing the whole-body irradiation dose to 6 MEDs still failed to elevate their serum vitamin D_3 levels in garment-clad subjects. It was concluded that clothing significantly impairs the formation of vitamin D_3 after 6 MEDs of UVB photostimulation. They also studied the effect of regular (seasonal) street clothing on serum vitamin D_3 response to whole-body irradiation with suberythematous dose of UVB (27 mJ/cm^2) radiation in seven healthy subjects. They found that summer clothing partially and autumn clothing totally prevented an elevation of the vitamin D_3 in response to UVB radiation 24 h later *(60)*.

5.5. *Influence of Season on 25(OH)D Levels in Nursing Home and Home Care Elderly*

The influence of season on vitamin D production in ampoule model was confirmed with a study conducted in a nursing home in Boston *(69, 85)*. Forty-five residents in this nursing home, who were taking a multivitamin that contained 400 IU of vitamin D_2, showed a dramatic decline in their 25(OH)D levels from the end of the summer to the beginning of the following summer (Fig. 9d). Based on the new definition of vitamin D deficiency [25(OH)D <20 ng/ml], 49, 67, 74, and 78% of the nursing home residents were vitamin D deficient in August, November, February, and May, respectively, as the mean serum 25(OH)D levels declined. Results obtained from free-living White elders cared for by Boston Medical Center also showed similar seasonal effects. Among the 69 White patients, 62 and 40% had serum 25(OH)D concentrations less than 20 ng/ml during February/March and September/October, respectively. In the same study, no significant difference in vitamin D deficiency was observed between February/March and September/October in 46 Black patients.

5.6. Tanning Bed Irradiation Enhances Vitamin D Status and Bone Mineral Density

In patients with intestinal disease, resection, or bypass, there is impaired absorption of both vitamin D and calcium. Therefore, patients with short-bowel syndrome and malabsorption are prone to vitamin D deficiency and nutritional osteomalacia. One alternative to improve their vitamin D nutrition is through cutaneous synthesis of vitamin D. We reported that a 57-year-old woman with a long history of Crohn's disease and short-bowel syndrome who had only 2 ft of small intestine remaining after three bowel resections remained to be vitamin D deficient (serum 25(OH)D, <20 ng/ml) after 36 months of treatment with a daily 400 IU of oral vitamin D supplement and 200 IU of vitamin D through TPN infusion (86). She was then exposed to UVB radiation in a tanning bed wearing a one-piece bathing suit for 10 min, three times a week for 6 months. After 4 weeks of exposure, her serum 25(OH)D level increased by 357% from 7 to 32 ng/ml, parathyroid hormone level decreased by 52% from 92 to 44 pg/ml, and the serum calcium level increased from 7.8 to 8.5 mg/dl (Fig. 12). After 6 months of UVB, her serum 25(OH)D level was maintained above 30 ng/ml and was free of muscle weakness and bone and muscle pain, those are common symptoms due to vitamin D deficiency. In another study, two short-bowel syndrome patients were irradiated twice a week for 8 weeks (6 min each session) with a commercial portable UV indoor tanning lamp that has a spectral output that mimics natural sunlight (87). These two patients were able to maintain their serum 25(OH)D levels during the winter month. In a follow-up study, an increased frequency of UV lamp exposure (five times per week for 8 weeks, 5–10 min each session depending on the skin type) was applied to five cystic fibrosis patients at their lower backs in a seated position (87). Their serum 25(OH)D levels increased from 21 ± 3 ng/ml at the baseline to 27 ± 4 ng/ml at the end of 8 weeks of exposure ($p = 0.05$).

Tangpricha et al. (88) investigated the serum levels of 25(OH)D and bone mineral density of subjects who regularly used a tanning bed and compared those biomarkers to subjects who did not use a tanning bed. They found that the subjects who used a tanning bed had serum 25(OH)D concentrations 90% higher than those of non-user group (46.2 ± 3.2 and 24.12 ± 1.2 ng/ml, respectively; $p < 0.001$). In addition, tanning bed users had lower PTH levels compared to the tanning bed non-users (21.4 ± 1.0 and

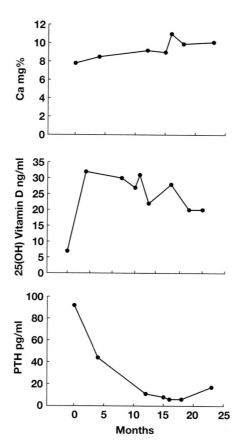

Fig. 12. Serum calcium, 25(OH)D, and PTH levels in a patient with Crohn's disease who had whole-body UVB exposure for 10 min, three times a week for 6 months (Reproduced with permission from *(86)*).

25.3 ± 0.8 pg/ml, respectively; $p = 0.01$). Consequently, tanners had significantly higher bone mineral density and z scores at the total hip than did nontanners.

6. SUMMARY

Vitamin D deficiency is synonymous with rickets. The initial description of the bone-deforming disease, rickets, in the mid-seventeenth century by two English physicians, Francis Glisson and Daniel Whistler, led to the observation by Sniadecki in 1822 that lack of sunlight exposure was likely a cause of rickets and the discovery of vitamin D a century later by Elmer McCollum in 1922. It is well established that most vertebrates, including amphibians, birds, reptiles, and primates, depend on sun exposure for their vitamin D requirement, because very few foods naturally contain vitamin D.

The first step leading to the cutaneous synthesis of vitamin D_3 is the absorption of solar UVB photons with energies 290–315 nm by 7-DHC in the skin that leads to the opening of its C-9 and C-10 bond and the formation of previtamin D_3. Previtamin D_3

then undergoes a rapid thermoisomerization within the plasma membrane to form vitamin D_3. However, excessive exposure to sunlight will not result in vitamin D intoxication, because both previtamin D_3 and vitamin D_3 can be photolyzed to noncalcemic or inactive photoproducts. During the winter, there is a minimal previtamin D_3 production in the skin at latitudes above 35°N. Altitude also significantly influences vitamin D_3 production. For example, in November at 27°N, very little previtamin D_3 synthesis was detected in Agra (169 m) and Katmandu (1,400 m). Production of previtamin D_3 increased 2- and 4-fold at 3,400 m and at Everest base camp (5,300 m), respectively. In addition, increased skin pigmentation, application of a sunscreen, aging, and clothing have a dramatic effect on previtamin D_3 production in the skin. It is estimated that exposure in one-piece bathing suit to 1 MED is equivalent to ingesting between 10,000 and 25,000 IU of vitamin D_2. The importance of sunlight exposure for providing most humans with their vitamin D requirement is well documented by the seasonal variation in circulating levels of 25(OH)D.

Vitamin D deficiency [serum 25(OH)D < 20 ng/ml] is common in infant, children, and adults worldwide. Exposure to simulated sunlamps that produce UVB radiation is an excellent source for producing vitamin D_3 in the skin and is especially efficacious in patients with fat malabsorption syndromes. The major cause of vitamin D deficiency globally is an underappreciation of the crucial role of sunlight in providing humans with their vitamin D requirement. The association regarding increased risk of common deadly cancers, autoimmune diseases, infectious diseases, and cardiovascular disease with living at higher latitudes and being prone to vitamin D deficiency should alert all health-care professionals about the importance of vitamin D for overall health and well-being.

ACKNOWLEDGMENTS

The authors would like to thank Andrew Tannenbaum for the graphics. This work was supported in part by a grant (MO1RR 00533) from the National Institutes of Health and UV Foundation.

REFERENCES

1. DeLuca HF (2008) Evolution of our understanding of vitamin D. Nutr Rev 66(10 Suppl 2):S73–S87
2. Holick MF (2008) Vitamin D: a D-Lightful health perspective. Nutr Rev 66(10 Suppl 2): S182–S194
3. Holick MF (2008) The vitamin D deficiency pandemic and consequences for nonskeletal health: mechanisms of action. Mol Aspects Med 29(6):361–368
4. Sniadecki J (1939) Cited by W Mozolowski: "Jedrzej Sniadecki (1768–1883) on the cure of rickets.". Nature 143:121
5. Holick MF (1990) Vitamin D and the skin: photobiology, physiology and therapeutic efficacy for psoriasis. In: Heersche JNM and Kanis JA (eds) Bone and mineral research, vol 7. Elsevier, Amsterdam, pp 313–366
6. Mellanby T (1918) The part played by an 'accessory factor' in the production of experimental rickets. J Physiol 52:11–14
7. McCollum EF, Simmonds N, Becker JE, Shipley PG (1922) Studies on experimental rickets and experimental demonstration of the existence of a vitamin which promotes calcium deposition. J Biol Chem 53:293–312

8. Huldschinsky K (1919) Heilung von Rachitis durch kunstliche Honensonne. Dtsch Med Wochenschr 45:712–713

9. Hess AF, Unger LJ (1921) Cure of infantile rickets by sunlight. JAMA 77:39

10. Goldblatt H, Soames KN (1924) A study of rats on a normal diet irradiated daily by the mercury vapor quartz lamp or kept in darkness. Biochem J 17:294–297

11. Goldblatt H (1924) A study of the relation of the quantity of fat-soluble organic factor in the diet to the degree of calcification of the bones and the development of experimental rickets in rats. Biochem J 17:298–326

12. Hess AF, Weinstock M (1924) Antirachitic properties imparted to inert fluids and green vegetables by ultraviolet irradiation. J Biol Chem 62:301–313

13. Steenbock H, Black A (1924) The induction of growth-promoting and calcifying properties in a ration by exposure to ultraviolet light. J Biol Chem 61:408–422

14. Steenbock H (1924) The induction of growth-promoting and calcifying properties in a ration by exposure to light. Science 60:224–225

15. Stamp TC, Walker PG, Perry W, Jenkins MV (1980) Nutritional osteomalacia and late rickets in greater London, 1974–1979: clinical and metabolic studies in 45 patients. Clin Endocrinol Metab 9: 81–105

16. Marksted T, Halvorsen S, Halvorsen KS et al (1984) Plasma concentrations of vitamin D metabolites before and during treatment of vitamin D deficiency rickets in children. Acta Paediatr Scand 73: 225–231

17. Tanner JT, Smith J, Defibaugh P et al (1988) Survey of vitamin content of fortified milk. J Assoc Off Anal Chem 71:607–610

18. Holick MF, Shao Q, Liu WW, Chen TC (1992) The vitamin D content of fortified milk and infant formula. N Engl J Med 326:1178–1181

19. Chen TC, Shao Q, Heath H, Holick MF (1993) An update on the vitamin D content of fortified milk from the United states and Canada. N Eng J Med 329:1507

20. Fieser LD, Fieser M (1959) Vitamin D. In: Steroids. Reinhold, New York, pp 90–168

21. Windaus A, Bock F (1937) Uber das provitamin aus dem sterin der schweineschwarte. Hoppe-Seylers Z Physiol Chem 245:168

22. Rauschkolb EW, Winston D, Fenimore DC, Black HS, Fabre LF (1969) Identification of vitamin D_3 in human skin. J Invest Dermatol 53:289–293

23. Okano T, Yasumura M, Mizuno K, Kobayashi T (1977) Photochemical conversion of 7-dehydrocholesterol into vitamin D_3 in rat skins. J Nutr Sci Vitaminol (Tokyo) 23:165–168

24. Esvelt RR, Schnoes HK, DeLuca HF (1978) Vitamin D_3 from rat skins irradiated in vitro with ultraviolet light. Arch Biochem Biophys 188:282–286

25. Pask-Hughes PA, Calam DH (1982) Determination of vitamin D_3 in cod-liver oil by high performance liquid chromatography. J Chromatogr 246:95–104

26. Muller-Mulot VW, Rohrer G, Schwarzbauer K (1979) Zur auffindung naturilich vorkommender vitamin D_3-ester im lebertran chemische bestimmung des freien, verester-ten und gesamt-vitamin D_3. Fette Seifen Anstrichm 81:38–40

27. St Lezin MA. Phylogenetic occurrence of vitamin D and provitamin D sterols. M.S. Dissertation, MIT Press, Cambridge, MA, 1983.

28. Koch EM, Koch FC (1941) The provitamin D of the covering tissues of chickens. J Poult Sci 20:33

29. Wheatley RH, Sher DW (1969) Studies of the lipids of dog skin. J Invest Dermatol 36:169–170

30. Kenny DE, Irlbeck NA, Chen TC, Lu Z, Holick MF (1998) Determination of vitamins D, A and E in sera and vitamin D in milk from captive and free-ranging polar bears (*Ursus maritimus*) and 7-dehydrocholesterol levels in skin from captive polar bears. Zoo Biol 17:285–293.

31. Morris JG (1997) Ineffective vitamin D synthesis of vitamin D in kittens exposed to sun and ultraviolet light is reversed by an inhibitor of 7-dehydrocholesterol-Δ^7-reductase. In: Norman AW, Bouillon R, Thomasset M (eds) Proceedings of the tenth workshop on vitamin D, Strasbourg, France, 24–29 May, 1997, 721–722.

32. Velluz L, Petit A, Amiard G (1948) Sur un stage non photochimique dans la formation des calciferols: essais d'interpretation. Bull Soc Chim Fr 15:1115–1120

33. Velluz L, Amiard G, Petit A (1949) Le precalciferol-ses relations d'equilibre avec le calciferol. Bull Soc Chim Fr 16:501–508
34. Holick MF, MacLaughlin JA, Clark MB, Holick SA, Potts JT Jr, Anderson RR, Blank IH, Parrish JA, Elias P (1980) Photosynthesis of vitamin D_3 in human skin and its physiologic consequences. Science 210:203–205
35. Bunker JWM, Harris RS, Mosher ML (1940) Relative efficiency of active wavelengths of ultraviolet light in activation of 7-dehydrocholesterol. J Am Chem Soc 62:508–511
36. Kobayashi T, Yasumara M (1973) Studies on the ultraviolet irradiation of previtamin D and its related compounds effect of wavelength on the formation of potential vitamin D_2 in the irradiation of ergosterol by monochromatic ultraviolet rays. J Nutr Sci Vitamino 119:123–128
37. MacLaughlin JA, Anderson RR, Holick MF (1982) Spectral character of sunlight modulates photosynthesis of previtamin D_3 and its photoisomers in human skin. Science 216:1001–1003
38. Anderson RR, Parrish JA (1982) Optical properties of human skin. In: Regan JD, Parrish JA (eds) The science of photomedicine. Plenum Press, New York, pp 147–194
39. MacLaighlin JA, Holick MF (1983) Photobiology of vitamin D in the skin. In: Goldsmith LA (ed) Biochemistry and physiology of the skin. Oxford University Press, New York, pp 734–754.
40. Havinga E (1973) Vitamin D, example and challenge. Experientia 29:1181–1193
41. Woodward RB, Hoffmann R (1965) Selection rules for sigmatropic reactions. J Am Chem Soc 87:2511–2513
42. Buisman JA, Hanewald KH, Mulder FJ, Roborgh JR, Keuning KJ (1968) Evaluation of the effect of isomerization on the chemical and biological assay of vitamin D. Analysis of fat-soluble vitamins X. J Pharm Sci 57:1326–1329.
43. Tian XQ, Chen TC, Matsuoka LY, Wortsman J, Holick MF (1993) Kinetic and thermodynamic studies of the conversion of previtamin D_3 to vitamin D_3 in human skin. J Biol Chem 268:14888–14892
44 Tian XQ, Chen TC, Lu Z, Shao Q, Holick MF (1994) Characterization of the translocation process of vitamin D_3 from the skin into the circulation. Endocrinology 135:655–661
45. Holick MF, Tian XQ, Allen M (1995) Evolutionary importance for the membrane enhancement of the production of vitamin D_3 in the skin of poikilothermic animals. Proc Natl Acad Sci USA 92:3124–3126
46. Tian X, Holick MF (1995) Catalyzed thermal isomerization between previtamin D_3 and vitamin D_3 via β-cyclodextrin complexation. J Biol Chem 270:8706–8711
47. Albers E, Muller BW (1992) Complexation of steroid hormones with cyclodextrin derivatives: substituent effects of the guest molecule on solubility and stability in aqueous solution. Pharm Sci 81:756–761
48. Chen ET, Pardue HL (1993) Analytical applications of catalytic properties of modified cyclodextrins. Anal Chem 65:2563–2567
49. Bogoslovsky NA, Kurganov BI, Samochvalova NG, Isaeva TA, Sugrobova NP, Gurevich VM, Valashek IE, Samochvalov GI (1988) Vitamin D: molecular, cellular and clinical endocrinology. Walter de Gruyter & Co, Berlin, pp 1021–1023
50. Haddad JG, Matsuoko LY, Hollis BW, Hu YZ, Wortsman J (1993) Human plasma transport of vitamin D after its endogenous synthesis. J Clin Invest 91:2552–2555
51. Webb AR, de Costa B, Holick MF (1989) Sunlight regulates the cutaneous production of vitamin D_3 by causing its photodegradation. J Clin Endocrinol Metab 68:882–887
52. Holick MF, MacLaughlin JA, Dopplet SH (1981) Regulation of cutaneous previtamin D_3 photosynthesis in man: skin pigment is not an essential regulator. Science 211:590–593
53. Loomis F (1967) Skin-pigment regulation of vitamin D biosynthesis in man. Science 157:501–506
54. Clemens TL, Henderson SL, Adams JS, Holick MF (1982) Increased skin pigment reduces the capacity of skin to synthesize vitamin D_3. Lancet 1:74–76
55. Lo C, Paris PW, Holick MF (1986) Indian and Pakistani immigrants have the same capacity as Caucasians to produce vitamin D in response to ultraviolet irradiation. Am J Clin Nutr 44:683–685
56. Chen TC, Chimeh F, Lu Z, Mathieu J, Person KS, Zhang A, Kohn N, Martinello S, Berkowitz R, Holick MF (2007) Factors that influence the cutaneous synthesis and dietary sources of vitamin D. Arch Biochem Biophys 460:213–217

57. Kassowitz M (1897) Tetanie and autointoxication in kindersalter. Wien Med Presse XXXViii 97:139

58. Schmorl G (1909) Die pathologigische anatomie de rachitischen knochenerkrankung mit besonderer ber ucksichtigung imer histologie und pathogenese. Ergeb d Inn Med Kinderheilkd IV 403

59. Hansemann D (1906) Veber den einfluss der domestikation auf die entstehung der krankheiten. Berl Klin Wochenschr Xliii 629:670

60. Matsuoko LY, Wortsman J, Dannenberg MJ, Hollis B, Lu Z, Holick MF (1992) Clothing prevents ultraviolet-B radiation-dependent photosynthesis of vitamin D_3. J Clin Endocrinol Metab 75: 1099–1103

61. Webb AR, Kline L, Holick MF (1988) Influence of season and latitude on the cutaneous synthesis of vitamin D_3: exposure to winter sunlight in Boston and Edmonton will not promote vitamin D_3 synthesis in human skin. J Clin Endocrinol Metab 67:373–378

62. Lu Z, Chen TC, Holick MF (1992) Influence of season and time of day on the synthesis of vitamin D_3. In: Holick MF and Kligman AM (eds) Biological effects of light. Walter de Gruyter, Berlin/New York, pp 57–61.

63. Ladizesky M, Lu Z, Oliver B, Roman NS, Diaz S, Holick MF, Mautalen C (1995) Solar ultraviolet B radiation and photoproduction of vitamin D_3 in Central and Southern areas of Argentina. J Bone Miner Res 10:545–549

64. Fitzpatrick TB (1988) The validity and practicality of sun-reactive skin types I through VI. Arch Dermatol 124:869–871

65. Ala-Houhala M, Parviainen MT, Pyykko K, Visakorpi JK (1984) Serum 25-hydroxyvitamin D levels in Finnish children aged 2–17 years. Acta Paediatr Scand 73:232–236

66. Oliveri MB, Ladizesky M, Mautalen CA, Alonso A, Martinez L (1993) Seasonal variations of 25-hydroxyvitamin D and parathyroid hormone in Ushuaia (Argentina), the southernmost city of the world. Bone Miner 20:99–108

67. Sherman SS, Hollis BW, Tobin JD (1990) Vitamin D status and related parameters in a healthy population: the effects of age, sex and season. J Clin Endocrinol Metab 71:405–413

68. Lester E, Skinner RK, Wills MR (1977) Seasonal variation in serum 25-hydroxyvitamin D in the elderly in Britain. Lancet 1:979–980

69. Holick MF, Chen TC, Lu Z, Sauter E (2007) Vitamin D and skin physiology: a D-lightful story. J Bone Miner Res Supply 2:V28–V33

70. Chalmers J, Conacher DH, Gardner DL, Scott PJ (1967) Osteomalacia- a common disease in elderly women. J Bone Joint Surg (Br) 49B:403–423

71. Jenkins DH, Roberts JG, Webster D, Williams EO (1973) Osteomalacia in elderly patients with fracture of the femoral neck. J Bone Joint Surg (Br) 55B:575–580

72. Doppelt SH, Neer RM, Daly M, Bourret L, Schiller A, Holick MF (1983) Vitamin D deficiency and osteomalacia in patients with hip fractures. Orthop Trans 7:512–513

73. Sokoloff L (1978) Occult osteomalacia in American patients with fracture of the hip. Am J Surg Pathol 2:21–30

74. Whitelaw GP, Abramowitz AJ, Kavookjian H, Holick MF (1991) Fractures and vitamin D deficiency in the elderly. Complications Ortho 6:70–80

75. Omdahl JL, Garry PJ, Hunsaker LA, Junt WC, Goodwin JS (1982) Nutritional status in a healthy elderly population: vitamin. Am J Clin Nutr 36:1225–1233

76. Lamberg-Allard T (1984) Vitamin D intake, sunlight exposure, and 25-hydroxyvitamin D levels in elderly during one year. Ann Nutr Metab 28:144–150

77. Webb AR, Pilbeam C, Hanafin N, Holick MF (1990) An evaluation of the relative contributions of exposure to sunlight and diet on the circulating concentrations of 25-hydroxyvitamin D in an elderly nursing home population in Boston. Am J Clin Nutr 51:1075–1081

78. Tan CY, Strathum B, Marks R (1982) Skin thickness measurement by pulsed ultrasound: its reproducibility, validation and variability. Br J Dermatol 106:657–667

79. MacLaughlin JA, Holick MF (1985) Aging decreases the capacity of human skin to produce vitamin D_3. J Clin Invest 76:1536–1538

80. Holick MF, Matsuoka LY, Wortsman J (1989) Age, vitamin D, and solar ultraviolet radiation. Lancet ii:1104–1105

81. Gilchrest BA (1993) Sunscreens – a public health opportunity. N Engl J Med 329(16):1193–1194

82. Matsuoko L, Ide L, Wortsman J, MacLaughlin JA, Holick MF (1987) Sunscreens suppress cutaneous vitamin D$_3$ synthesis. J Clin Endocrinol Metab 64:1165–1168

83. Matsuoko LY, Wortsman J, Hanifan N, Holick MF (1988) Chronic sunscreen use decreases circulating concentrations of 25-hydroxyvitamin D: a preliminary study. Arch Dermatol 124:1802–1804

84. Robson J, Diffey BL (1990) Textiles and sun protection. Photodermatol Photoimmunol Photomed 7:32–34

85. Tangpricha V, Pearce EN, Lu Z, Mathieu JS, Flanagan JN, Chen TC, Holick MF (2002) Vitamin D insufficiency among free-living young healthy adults. Am J Med 112:659–662

86. Koutkia P, Lu Z, Chen TC, Holick MF (2001) Treatment of vitamin D deficiency due to Crohn's disease with tanning bed ultraviolet B radiation. Gastroenterology 121:1485–1488

87. Chandra P, Wolfenden LL, Ziegler TR, Tian J, Luo M, Stecenko AA, Chen TC, Holick MF, Tangpricha V (2007) Treatment of vitamin D deficiency with UV light in patients with malabsorption syndromes: a case series. Photodermato, Photoimmunol Photomed 23:179–185

88. Tangpricha V, Turner A, Spina C, Decastro S, Chen TC, Holick MF (2004) Tanning is associated with optimal vitamin D status (serum 25-hydroxyvitamin D concentration) and higher bone mineral density. Am J Clin Nutr 80:1645–1649

89. Gamgee KML (1927) The artificial light treatment of children in rickets, anaemia & malnutrition. P.B. Hoeber Inc., New York, 172pp.

3 The Functional Metabolism and Molecular Biology of Vitamin D Action

Lori A. Plum and Hector F. DeLuca

Abstract   The evolution of our understanding of the biological impact of vitamin D is briefly reviewed, with a focus on the physiology and endocrinology of the vitamin D system. This chapter attempts to bring the molecular discoveries in vitamin D metabolism and mechanisms of action in to focus on known physiology and endocrinology. The latest developments on metabolism of vitamin D, the enzymes involved, and the genes responsible are presented. The impact of the molecular discoveries on current views of the importance of vitamin D in public health is also presented.

Key Words: Vitamin D; vitamin D metabolism; calcium; intestine; bone; 25-hydroxyvitamin D; 1,25-dihydroxyvitamin D; vitamin D receptor; calcium transport; bone mobilization

1. INTRODUCTION

The dietary requirement for vitamin D and/or the necessity for exposure to the sun's ultraviolet rays were first recognized in the prevention or cure of the disease rickets *(1–4)*. This disease is the result of a failure to mineralize the organic matrix of bone. The resultant decrease in bone strength causes skeletal malformations and may result in death *(5)*. In adults, the disease osteomalacia occurs in which undermineralized osteoid seams appear and bones are easily fractured *(6)*.

Rickets in children and osteomalacia in adults became prevalent at the time of the industrial revolution because smoke and urbanization deprived the industrialized population of sunlight *(5)*. Steenbock and Black found that ultraviolet light could be used to induce vitamin D activity in the lipid portions of food *(7)*. This process was then used to fortify food, which eliminated rickets as a major medical problem. Nutritional rickets then became rare in the developed countries. However, forms of rickets and osteomalacia not cured by nutritional levels of vitamin D remained. Of these, two result from genetic defects in the vitamin D system. Vitamin D-dependent rickets type I occurs when the kidneys are unable to metabolize 25-hydroxyvitamin D_3 [$25(OH)D_3$] to the more polar

From: *Nutrition and Health: Vitamin D*
Edited by: M.F. Holick, DOI 10.1007/978-1-60327-303-9_3,
© Springer Science+Business Media, LLC 2010

and hormonally active 1,25-dihydroxyvitamin D_3 [1,25$(OH)_2D_3$] (8, 9). Thirty different mutations in the 25(OH)D-1-hydroxylase gene that result in this disease have been characterized (10–18). The inability to produce 1,25$(OH)_2D_3$ greatly impairs skeletal mineralization since 1,25$(OH)_2D_3$ is the functional or hormonal form of vitamin D (19). The more recently discovered form of rickets, vitamin D-dependent rickets type II, occurs when the vitamin D receptor (VDR) protein is nonfunctional; thus, a tissue resistance to 1,25$(OH)_2D_3$ occurs (20–22). Many genetic mutations in the VDR gene have been identified which render the protein nonfunctional or dramatically less functional (23–26). The absence of functional VDR precludes the transcriptional regulation of vitamin D-regulated genes involved in calcium and phosphorus homeostasis. Therefore, calcium and phosphorus are not maintained at sufficient levels in the serum to support bone mineralization. These two types of rickets illustrate both the functional metabolism and molecular biology of vitamin D action.

2. OVERALL ROLE OF THE VITAMIN D HORMONE IN CALCIUM AND PHOSPHORUS HOMEOSTASIS

The classical function of the vitamin D hormone is to increase serum calcium and serum phosphorus levels required for skeletal mineralization and, therefore, the prevention of rickets and osteomalacia (27–29). The hormone accomplishes this through its actions on the intestine, bone, and kidney. In the intestine, 1,25$(OH)_2D_3$ stimulates the active transport of calcium and independently phosphate from the lumen of the intestine to the blood (30–39). In the presence of parathyroid hormone, 1,25$(OH)_2D_3$ acts on the distal nephron to improve reabsorption of calcium (40) and on bone to increase resorption (41, 42), providing an additional source of mineral for new bone formation (41–43). When calcium and phosphate are very low in the diet, the effect of vitamin D on kidney and bone results in retaining calcium and phosphate to support new mineralization (44, 45). When adequate calcium and phosphate are present in the diet, the role of vitamin D on intestine dominates because parathyroid hormone secretion is suppressed. This suppression is due to normal serum calcium acting on the calcium receptor in the parathyroid gland and the direct action of 1,25$(OH)_2D_3$ in the parathyroid gland (46–48). The absence of parathyroid hormone turns off bone resorption (42) and bone calcium is conserved. In this way, environmental calcium is utilized for both neuromuscular (discussed below) and mineralization functions (49, 50).

Control of phosphate is under different but related mechanisms. In general, dietary phosphate is rarely limiting. Furthermore, 1,25$(OH)_2D_3$ stimulates intestinal phosphate absorption (36, 37). However, parathyroid hormone blocks phosphate reabsorption in the kidney causing a phosphate diuresis. Additionally, FGF23 that is elevated by 1,25$(OH)_2D_3$ (51) also results in a phosphate diuresis. Keeping phosphate in the normal range while calcium is elevated, FGF23 also suppresses 1,25$(OH)_2D_3$ synthesis (52). However, a full physiological understanding of the interaction of 1,25$(OH)_2D_3$ and FGF23 is not available. For example, under phosphate limitation, 1,25$(OH)_2D_3$ elevates serum phosphate and little or no phosphate is found in the urine (53). Either 1,25$(OH)_2D_3$ does not elevate FGF23 under hypophosphatemic conditions or the FGF23/Klotho system is unable to cause phosphaturia under this circumstance.

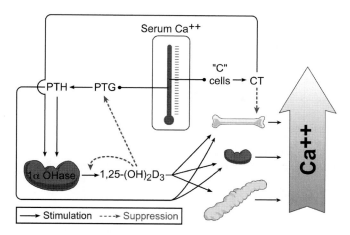

Fig. 1. Pathways of 1,25(OH)$_2$D$_3$, PTH, and calcitonin (CT) actions for altering blood calcium levels.

The purpose of 1,25(OH)$_2$D$_3$ and PTH under dietary calcium restriction and hypocalcemia is to provide calcium in the serum for neuromuscular function (49, 50). Extreme hypocalcemia places one in grave danger of death resulting from hypocalcemic tetany. To avoid this situation, 1,25(OH)$_2$D$_3$ and PTH increase serum calcium levels, without increasing serum phosphorus levels, through actions on kidney and bone. Figure 1 depicts the actions of 1,25(OH)$_2$D$_3$ and PTH on their classical target tissues in order to maintain serum calcium.

3. FUNCTIONAL METABOLISM

Chemical synthesis of [^{3}H]-vitamin D$_3$ of high specific activity was a major breakthrough in elucidating the functional metabolism of vitamin D (54). This permitted experiments with truly physiological doses of [^{3}H]-vitamin D$_3$. Upon administration of [^{3}H]-vitamin D$_3$, polar metabolites of vitamin D were quickly observed in the serum, intestine, kidney, and bone (55, 56). The chemical identification of these metabolites, the chemical synthesis of these metabolites, and the testing of these metabolites for biological activity were vital to our present knowledge of the vitamin D endocrine system and have been reviewed (57).

Vitamin D may be obtained by dietary means from only a few sources, i.e., fish liver oils or fortified foods. Otherwise, the principal source of vitamin D is skin (58). The epidermis contains a pool of 7-dehydrocholesterol (59) (Fig. 2). This compound contains a chromophore because of the conjugated double bond system that absorbs 280–310 nm ultraviolet light. Upon absorption of the UV light, 7-dehydrocholesterol undergoes a photochemical isomerization to form previtamin D$_3$ (60, 61). Previtamin D$_3$ exists in equilibrium with its isomer, vitamin D$_3$ (62). Vitamin D$_3$ is transported in the blood bound to the vitamin D-binding protein or albumin to the liver (63). In the liver, the vitamin D$_3$ molecule is enzymatically hydroxylated on carbon 25 to form 25(OH)D$_3$ (64, 65). The enzyme responsible for this step has not yet been identified. There is no doubt that it is a CYP-450 enzyme (66). A mitochondrial CYP27A1 has been cloned and is capable of 25-hydroxylating vitamin D when the substrate is present

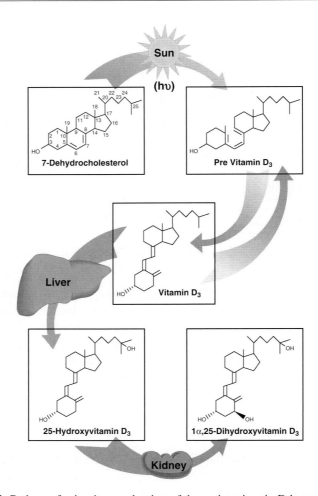

Fig. 2. Pathway for in vivo production of the native vitamin D hormone.

in high concentrations *(67)*. However, its primary function is in bile acid metabolism and its null mutation does not result in a disturbance of vitamin D metabolism *(68–70)*. Multiple cytochrome P450 enzymes have been cloned that are likely candidates for playing a role in 25-hydroxylation but final confirmation of any one of them remains *(66, 71)*. The true enzyme responsible for production of 25(OH)D$_3$ therefore has not yet been conclusively demonstrated. 25-Hydroxyvitamin D$_3$ is transported to the kidney where it is again enzymatically hydroxylated, but this time on carbon 1, to form 1,25(OH)$_2$D$_3$ *(72, 73)*. 1,25(OH)$_2$D$_3$ is the biologically active form of vitamin D$_3$ *(74–76)*. The enzyme responsible has been clearly determined and is CYP27B1 *(66)*. Null mutant mice develop rickets, hypocalcemia, and hypophosphatemia in exact parallel to the human disease, vitamin D-dependency rickets type I *(77, 78)*. This disease in humans has been clearly shown to be the result of mutations in the CYP27B1 gene *(10–18)*. 1,25(OH)$_2$D$_3$ is classified as a hormone because its production is at a site distant from the targets of action, and the production of 1,25(OH)$_2$D$_3$ is feedback regulated by rising serum calcium and phosphorus levels *(57)*. Other factors that control levels of this hormone and its precursors are shown in Table 1.

Table 1
Factors Affecting the Levels of Vitamin D$_3$, 25(OH)D$_3$, and 1,25(OH)$_2$D$_3$

Compound in serum	*Factor*	*Direction of change*	*References*
Vitamin D$_3$	Increasing age	Decrease	*(205, 206)*
	Increased skin pigment	Decrease	*(207–209)*
	Obesity	Decrease	*(210, 211)*
	Para-aminobenzoic acid	Decrease	*(212)*
25(OH)D$_3$	Administration of native hormone	Decrease	*(213–218)*
	Anticonvulsant drugs	Decrease	*(219–225)*
	Glucocorticoids	Decrease	*(226)*
	Increasing age	Decrease	*(206, 227–230)*
	Increased skin pigment	Decrease	*(207, 231, 232)*
	Kidney failure	Decrease	*(233–235)*
	Low dietary calcium	Decrease	*(217, 236–238)*
	Low dietary phosphorus	Decrease	*(239)*
	Obesity	Decrease	*(210, 211, 230, 240–242)*
	Pregnancy	Decrease	*(243–246)*
	Strontium	Increase	*(247)*
	Winter season	Decrease	*(230, 248–257)*
1,25(OH)$_2$D$_3$	4-Aminoquinoline drugs	Decrease	*(258–260)*
	Calcitonin	Increase	*(261)*
	Estrogens	Increase	*(262, 263)*
	Glucocorticoids	Decrease	*(226)*
	Increasing age	Decrease	*(264–276)*
	Ketoconazole	Decrease	*(277–279)*
	Low dietary calcium	Increase	*(237, 280–287)*
	Low dietary phosphorus	Increase	*(217, 270, 281–285, 288–297)*
	Obesity	Increase	*(210, 240)*
	Pregnancy	Increase	*(243, 244, 265, 272, 298–301)*
	PTH	Increase	*(107, 108, 282, 302–304)*
	Renal failure	Decrease	*(267, 272, 282, 291, 305–316)*
	Strontium	Decrease	*(247)*
	Verapamil	Decrease	*(236, 317)*

More recent developments include the idea that the CYP27B1 is expressed in a paracrine or autocrine manner in extrarenal tissue *(79–81)*. These are based either on CYP27B1 mRNA measurements or on 1,25(OH)$_2$D$_3$ produced by cells in culture.

While there is no doubt about extrarenal production of $1,25(OH)_2D_3$ in disease states such as sarcoidosis *(82)*, and certain malignancies *(83)*, the idea of normal production of $1,25(OH)_2D_3$ by extrarenal tissue (except placenta) is not supported by several in vivo experiments *(84–86)*. Whether extrarenal production of $1,25(OH)_2D_3$ occurs in biologically significant amounts under normal conditions remains uncertain.

The catabolism of $1,25(OH)_2D_3$ also proceeds by a cytochrome P-450 pathway. The major enzyme in this pathway is the 25-OH-D_3-24-hydroxylase or CYP24. The 24-hydroxylase enzyme is present in highest concentrations in kidney mitochondria *(87)*. Its substrates are $25(OH)D_3$ and $1,25(OH)_2D_3$, with most reports indicating that the enzyme has a higher affinity for $1,25(OH)_2D_3$ *(88)*. The gene responsible is the CYP24 and the encoded enzyme is able to execute all enzymatic reactions in the catabolism of $1,25(OH)_2D_3$ by the C23/C24 pathway *(89, 90)*. Purified 24-hydroxylase enzyme has been shown to possess 23-hydroxylase activity in vitro *(89, 90)*. The final product from $1,25(OH)_2D_3$ metabolism is calcitroic acid *(91, 92)*. Calcitroic acid has been found in the bile; therefore, it is thought to be the eventual excretory product of $1,25(OH)_2D_3$ *(93, 94)*. Although $24,25$-$(OH)_2D_3$ has been postulated as another active metabolite of vitamin D *(95–99)*. This has been clearly ruled out by experiments with 24,24-difluoro-$1,25(OH)_2D_3$. This lab has maintained vitamin D-deficient rats on $25(OH)D_3$ difluorinated at the 24-position for two consecutive generations *(19, 100)*. The fluoro groups on the 24-position preclude metabolism of $1,25(OH)_2D_3$ to 24-hydroxylated metabolites *(101)*. When vitamin D-deficient rats were maintained on 24,24-difluoro-$25(OH)D_3$ as their sole source of vitamin D_3, growth, reproduction, and skeletal mineralization remained normal when compared to rats maintained on $25(OH)D_3$ or $1,25(OH)_2D_3$ *(19, 100, 102)*. Thus, it is clear that 24-hydroxylation of vitamin D plays no significant role in the functions of vitamin D. The CYP24 null mouse has been produced with a phenotype of high plasma levels of $1,25(OH)_2D_3$ and defective mineralization *(103)*. The defect in mineralization was later shown to be a toxic consequence of high levels of $1,25(OH)_2D_3$ *(104)*. Thus, the null mutant CYP24 mice confirm the conclusion that the 24-hydroxylation of vitamin D compounds has no functional role except as an intermediate in the catabolism and elimination of vitamin D compounds, especially $1,25(OH)_2D_3$ *(103, 104)*.

4. REGULATION OF THE PRODUCTION AND CATABOLISM OF $1,25(OH)_2D_3$

Calcium and phosphorus levels in the plasma are tightly regulated by the vitamin D/PTH/calcitonin endocrine system. As a result, in vivo there are three situations that result in differential regulation of $1,25(OH)_2D_3$ production (Fig. 3). First, serum calcium levels may be low, while serum phosphorus levels are normal. Second, serum phosphorus levels may be low, while serum calcium levels are normal. And third, serum calcium and serum phosphorus levels may both be low. In each situation, $1,25(OH)_2D_3$ either alone or in conjunction with PTH normalizes serum calcium and serum phosphorus levels.

Under conditions of low serum calcium, the calcium receptor on parathyroid glands activates the secretion of parathyroid hormone *(105)*. The parathyroid hormone acts on the kidney to induce 25(OH)D-1-hydroxylase *(75, 76, 106–108)*. The increased activity of the 25(OH)D-1-hydroxylase increases the circulating levels of 1,25(OH)$_2$D$_3$ *(107)*.

Hypocalcemia, Normaphosphatemia

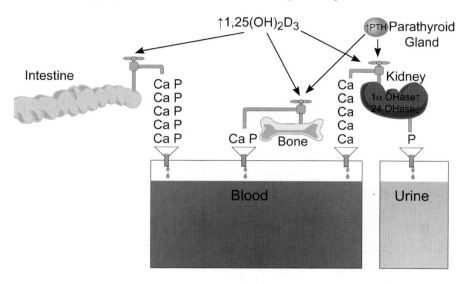

Normacalcemia, Hypophosphatemia

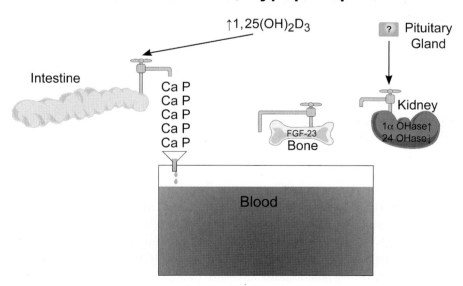

Fig. 3. Diagrams showing three different blood mineral scenarios and the mechanisms in place to normalize the plasma calcium and phosphorus levels.

Hypocalcemia, Hypophosphatemia

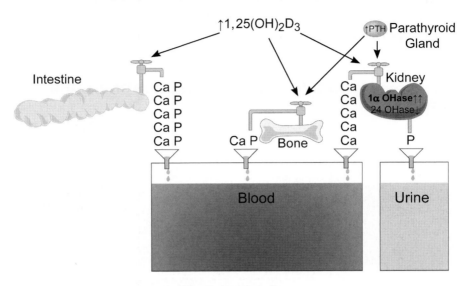

Fig. 3. (continued)

1,25-Dihydroxyvitamin D_3 acts upon the intestine to stimulate the active transport of calcium and phosphorus from the intestinal lumen to the serum *(30–38)*. In the bone, parathyroid hormone and $1,25(OH)_2D_3$ activate bone calcium mobilization *(39, 41, 42, 109)*. In addition, $1,25(OH)_2D_3$ and parathyroid hormone act upon the kidney to stimulate the active reabsorption of calcium from the urine to blood *(42)*. Parathyroid hormone also acts upon the kidney to reduce phosphorus reabsorption *(110)*. Therefore, in this situation, serum calcium levels will increase to normal, while serum phosphorus remains unchanged. The function of $1,25(OH)_2D_3$ and PTH in this situation is to increase serum calcium levels to avoid hypocalcemic tetany. Once serum calcium levels have returned to normal, parathyroid hormone secretion is no longer stimulated and, thus the production of $1,25(OH)_2D_3$ is no longer activated.

Under conditions of low serum phosphorus and normal serum calcium, the secretion of parathyroid hormone is not stimulated since the serum calcium levels are normal or elevated. But low serum phosphorus levels can trigger an increase in the activity of the 25(OH)D-1-hydroxylase enzyme by an unknown mechanism *(111–113)*. Growth hormone (GH) and insulin-like growth factor-1 (IGF-1) have been proposed as mediators of 1-hydroxylase activity in response to hypophosphatemia *(114–116)*. Although IGF-1 and GH do appear to stimulate 1-hydroxylase activity at low serum phosphorus levels in hypophysectomized rats, they are not likely to mediate the 1-hydroxylase response to serum phosphorus because IGF-1 and GH do not increase 1-hydroxylase activity when administered to hypophysectomized rats with normal serum phosphorus levels *(115)*. If IGF-1 and GH were mediators of serum phosphorus effects on the 1-hydroxylase then administration of these compounds to rats should be able to increase 1-hydroxylase activity independent of serum phosphorus. Regardless, the 1-hydroxylase response to serum phosphorus seems to be mediated by some pituitary gland hormone since hypophysectomy eliminates both the 1-hydroxylase response to serum phosphorus and decreases serum $1,25(OH)_2D_3$ levels *(117–119)*. In hypophosphatemia, increased activity of the 1-hydroxylase enzyme increases circulating levels of $1,25(OH)_2D_3$. Without PTH present, $1,25(OH)_2D_3$ will act solely upon the intestine to stimulate the active absorption of calcium and phosphorus from the lumen of the intestine to blood *(109)*. In addition, $1,25(OH)_2D_3$ stimulates the secretion of FGF23 from osteocytes *(120)*. However, FGF23 causes a phosphate diuresis in the kidney. It is not clear what role FGF23 plays in the physiologic role of $1,25(OH)_2D_3$ in phosphate metabolism. Interestingly, FGF23 also suppresses production of $1,25(OH)_2D_3$ *(52)*. Although vitamin D lowers serum phosphate slightly under conditions of dietary excess of phosphate *(44)*, it does exactly the opposite under hypophosphatemic conditions *(45)*. Further, $1,25(OH)_2D_3$ healing of rickets in rats requires an elevation of serum phosphate *(121)*. A number of in vivo experiments will be required before the involvement of FGF23 in the physiologic actions of vitamin D is understood. The action of hypophosphatemia in increasing the activity of the 1α-hydroxylase ultimately will increase both serum calcium and serum phosphorus levels. Once serum phosphorus is increased, the activity of 25(OH)D-1-hydroxylase is no longer stimulated.

When both serum calcium and serum phosphorus levels are low, the 25(OH)D-1-hydroxylase is super-stimulated. Low serum phosphorus increases 1-hydroxylase activity *(111–113)*. In addition, low serum calcium stimulates PTH secretion which also activates 1-hydroxylase activity *(75, 76, 106–108)*. The result is extremely high circulating levels of $1,25(OH)_2D_3$. 1,25-Dihydroxyvitamin D_3 will act alone in the intestine to increase calcium and phosphorus absorption in the intestine. The high PTH levels in combination with $1,25(OH)_2D_3$ will increase bone calcium and phosphorus mobilization and calcium reabsorption from the urine, while PTH itself will increase renal phosphate excretion. The result of $1,25(OH)_2D_3$ action will be a net increase in serum calcium and phosphorus *(109)*.

The catabolism of $1,25(OH)_2D_3$ is also strictly regulated to prevent hypercalcemia and hyperphosphatemia. Thus, the parathyroid hormone, $1,25(OH)_2D_3$, and serum phosphorus regulate the principal catabolic enzyme in the vitamin D endocrine system, the 25(OH)D-24-hydroxylase (CYP24) *(122, 123)*. Parathyroid hormone is secreted under conditions of hypocalcemia and not when calcium levels are adequate. Therefore, PTH decreases 25(OH)D-24-hydroxylase activity and mRNA levels during hypocalcemia *(122, 124)*. As long as serum calcium remains low, PTH will continue to stimulate the production of $1,25(OH)_2D_3$, and repress the production of $1,24,25(OH)_3D_3$ and $24,25(OH)_2D_3$. Thus, while low serum calcium persists, $1,25(OH)_2D_3$ catabolism to biologically inactive metabolites will be low. In addition, $1,25(OH)_2D_3$ activates its own breakdown by stimulating transcription of the 24-hydroxylase gene, so that once serum calcium is normalized, $1,25(OH)_2D_3$ will return to basal levels *(123, 125, 126)*. Low serum phosphorus decreases the mRNA levels of the 24-hydroxylase to prevent catabolism of $1,25(OH)_2D_3$ under hypophosphatemic conditions *(127)*. The PTH hormone, since its synthesis is directly responsive to serum calcium, has the major control of vitamin D catabolism and 24-hydroxylase activity, providing for increased synthesis and decreased destruction of $1,25(OH)_2D_3$. Unfortunately, the mechanism(s) whereby

PTH controls 1α- and 24-hydroxylase activity is not entirely clear at the present time, but it involves transcriptional regulation *(124, 128)*. The PTH-sensitive site has been located in the promotor of the 1α-hydroxylase and overlaps with the CAAT box of that gene *(129)*. Another important factor is VDR expression in kidney, parathyroid gland, and perhaps other tissues but not intestine *(130, 131)*. Hypocalcemia and vitamin D deficiency eliminate VDR expression rendering these organs unresponsive to $1,25(OH)_2D_3$ *(132–134)*. Thus, a calcium need prevents induction of the CYP24 by $1,25(OH)_2D_3$ and prevents PTH suppression in the parathyroid gland by $1,25(OH)_2D_3$ *(131, 135)*.

5. MOLECULAR MECHANISM OF VITAMIN D ACTION

By using radiolabeled $1,25(OH)_2D_3$, Stumpf and colleagues demonstrated that this compound localized in the nucleus of target cells *(136)*. Multiple investigators demonstrated a nuclear VDR *(137, 138)*. Subsequent isolation of the VDR cDNA revealed that it was homologous to other steroid hormone receptors *(139, 140)*. The VDR is a member of the steroid hormone receptor superfamily and, more specifically, the thyroid/vitamin D/retinoic acid receptor subfamily *(141)*. Therefore, the VDR possesses both a DNA binding domain (C) and a ligand binding domain (E) in addition to other regions of the protein denoted A/B, D, and F *(142)*. As stated above, the autosomal recessive disease, vitamin D-dependency rickets type II, results from mutations in the VDR gene. The different VDR mutations that result in either a truncated VDR protein or the inability of the VDR protein to bind DNA demonstrate that the VDR is a necessary requirement in the vitamin D signal transduction cascade *(14–16)*.

Because of the ability of the VDR to influence gene transcription directly in the presence of its ligand, the correct term for the VDR is ligand-induced transcription factor. The sequences of DNA in the target genes to which the VDR binds are called VDREs or vitamin D response elements *(141)*. The sequences usually consist of two hexameric repeats of the consensus sequence AGGTCA separated by three nucleotides. The shorthand notation for the VDRE is direct repeat 3 or DR3 *(143)*. Specificity for responsiveness to $1,25(OH)_2D_3$ is determined by the three nucleotide spacer *(143)*. When the spacer is modified to four or five nucleotides, the response element becomes responsive to thyroid hormone and retinoic acid, respectively *(143)*. VDREs are found in the promoter regions of genes which regulate calcium and phosphorus homeostasis as well as those genes involved in vitamin D metabolism and other functions.

As a result of ChIP/chip assays, the binding of VDR to sites either one or more kilobases upstream from the start site or in introns has been detected *(144)*. Reporter gene assays have been used to confirm that at least some are functional VDREs. Exactly how they function as enhancers has not been clearly established. At least four or five such sites have been found in the VDR gene itself, only one of which appears functional *(144)*. Similar findings have been reported for the RANKL gene *(145)*.

It is very clear that VDR binds to the DRE only if a retinoid X receptor (RXR) protein is available *(141)*. The RXR binds to the $5'$ arm, while the VDR binds to the $3'$ arm *(141)*. The binding of the VDR–RXR complex is markedly enhanced if not dependent on $1,25(OH)_2D_3$ *(141)*. It is now believed that several additional proteins bind to the complex formed on the VDRE. Certainly TFIIB, RNA Pol II, and several

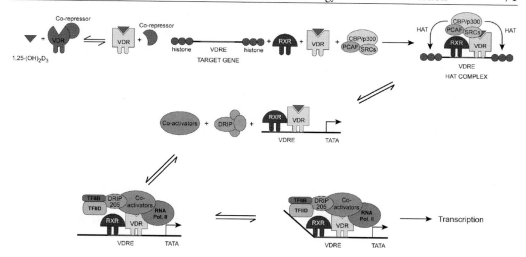

Fig. 4. A proposed model for vitamin D-influenced target gene expression.

other proteins are found in the complex *(141)*. The first wave of proteins that binds to DNA-bound liganded receptor exerts histone acetyltransferase activity to "loosen" the chromatin structure *(146)*. The first group of proteins leave, and a second group of proteins, including in some cases DRIP 205 along with coactivator proteins and Pol II *(141)* bind to the complex. This complex initiates transcription. A diagrammatic sketch of this postulated mechanism of $1,25(OH)_2D_3$ in transcription is provided only as a guide of currently available information (Fig. 4). Many questions are left unanswered by the above hypothesis. For example, how are certain genes suppressed by the action of $1,25(OH)_2D_3$? Microarrays of intestinal genes have shown some 50 are upregulated and an equally impressive number are downregulated to provide the full phenotype of vitamin D action *(147, 148)*. Although some genes controlled by $1,25(OH)_2D_3$ such as CYP24 may have more than one VDRE in the proximal promoter *(149, 150)*, usually a single VDRE is found. There are other genes where VDRE cannot be found in the proximal promotor such as RANKL and VDR, yet they are clearly regulated by $1,25(OH)_2D_3$ *(131, 151–155)*. ChIP/chip assay reveal VDREs either in the distal promoter (1–5 kb upstream) or in the introns *(144, 145)*. How these "enhancer" sites are used in regulation of the target gene expression is largely speculative. Thus, much more information can be expected in the coming years to provide insight into the mechanisms of regulation.

Many cellular experiments have been described that reveal a "non" genomic mechanism of action of $1,25(OH)_2D_3$ *(156, 157)*. These usually but not always involve high concentrations of $1,25(OH)_2D_3$ (10^{-8} M or higher) and are both nonphysiological and lack structural specificity. In vivo no difference between vitamin D-deficient VDR –/– and vitamin D supplemented VDR –/– mice could be found *(158)* and, further, VDR –/– animals show no toxicity or other response to high levels of $1,25(OH)_2D_3$ *(158)*. The reported transcaltachia or rapid intestinal calcium absorption 14 minutes after a high dose of $1,25(OH)_2D_3$ *(159)* could not be repeated in rats or chicks in the authors' laboratory. The evidence for in vivo non-genomic actions of vitamin D is not convincing.

There have been many genes identified in cell culture or in vivo that are regulated by 1,25(OH)$_2$D$_3$ *(150, 160)*. Exactly how they are regulated and what they provide to describe the in vivo function of 1,25(OH)$_2$D$_3$ will be left for discussion when the various functions of vitamin D are described in the several chapters in this volume.

6. ANALOGS: MECHANISMS OF ACTION

Since the discovery of 1,25(OH)$_2$D$_3$, hundreds of analogs have been synthesized with the distinct hope of preparing analogs that will selectively express specific functions of vitamin D. The first major effort was made to prepare "non-calcemic" analogs that would carry out functions of vitamin D in differentiation, in immune function, or on cell proliferation without raising serum calcium levels. This has been largely disappointing but some understanding has resulted. Unfortunately, analogs were tested in vitro for receptor binding, transcription activity, cell differentiation and cell proliferation, and in vivo for calcemic action. This generated several "non" calcemic compounds *(160–162)*. Multiple compounds seemed to be very active in vitro as an anti-proliferative agent but failed in vivo because of rapid elimination *(162)*. This is also true of MC903 *(163)* which is effective against psoriasis when applied topically but ineffective orally because it is rapidly eliminated *(164)*.

However, selective vitamin D compounds have been synthesized that favor bone formation *(165–167)* or suppression of secondary hyperparathyroidism *(168–170)*. However, these compounds are NOT devoid of calcemic activity. All vitamin D analogs, if given in high enough dose, will cause hypercalcemia. However, the dose at which they are effective in vivo versus the dose that causes hypercalcemia may differ. Thus, 19-nor-1α,25-dihydroxyvitamin D$_2$ is as effective at suppression of secondary hyperparathyroidism as 1,25(OH)$_2$D$_3$, but at a dose that has much less effect on serum calcium levels *(171)*. Similarly, 2-methylene-19-nor-(20S)-1α,25-dihydroxyvitamin D$_3$ (2MD) is 30–50 times more effective on bone while being equally active as 1,25(OH)$_2$D$_3$ on intestine *(155)*. The mechanism of selectivity is not understood in these cases.

In attempting to assess the molecular basis for analog selectivity, Moras and colleagues *(172–174)* and Vanhooke et al. *(175, 176)* crystallized the ligand binding domain (LBD) of VDR (minus a 20 amino acid fragment in the flexible portion) with various analogs in the binding pocket. The ligands did not induce a structural change in the LBD, but the ligands were more or less distorted in the binding pocket. It is possible that the free energy of crystallization drives this phenomenon, while in solution the ligand may induce a three-dimensional change that may cause differential binding of certain coactivators. It is of some interest that some analogs may alter susceptibility of VDR to proteolytic digestion *(177–181)*. Much remains unknown in this area that could be of therapeutic importance.

7. FUNCTION OF VITAMIN D NOT RELATED TO CALCIUM AND PHOSPHATE HOMEOSTASIS

Although the primary driving force of the vitamin D system is to regulate serum calcium and phosphorus for mineralization of the skeleton, there is no doubt that functions of the vitamin D hormone in systems not related to calcium and phosphorus take place. The VDR and its ligand are clearly found in cells and tissues not playing a role in calcium and phosphate homeostasis (Table 2). In particular, the VDR is found in the islet cells of the pancreas, in activated cells of the immune system, in the macrophage, in epithelial cells of the intima of blood vessels, in cells of the stomach, in keratinocytes of skin, in epithelial cells of the colon, and in cells of the placenta (141, 182–185). These localizations have in part driven scientists to determine whether there might not be a function of vitamin D in those cells. There is now little doubt that the vitamin D hormone has an impact on the keratinocytes. Both oral and topical administrations of 1,25(OH)$_2$D$_3$ will suppress psoriasis (186, 187). Yang et al. demonstrated that the immune system is clearly affected by the vitamin D hormone (188). Both in

Table 2
Tissue Distribution of Vitamin D$_3$, 25(OH)D$_3$, and 1,25(OH)$_2$D$_3$

Compound	Species	Tissues	References
[^{14}C]-vitamin D$_3$	Human	Bile	(318–320)
		Blood	
		Feces	
[^{3}H]-vitamin D$_3$	Rat	Adipose	(321–326)
		Blood	
		Bone	
		Fetus	
		Intestine	
		Kidney	
		Liver	
		Mammary gland	
		Placenta	
		Skeletal muscle	
[^{3}H, ^{14}C]-vitamin D$_3$	Rat	Fetal intestine	(327)
		Fetal bone	
[^{3}H]-25(OH)D$_3$	Human	Bile	(319, 320)
		Blood	
		Feces	
		Urine	
	Rat	Bile	(322, 323, 325, 328)
		Blood	
		Liver	
		Feces	
		Fetus	
		Kidney	
		Liver	
		Placenta	
		Urine	

Table 2
(continued)

Compound	Species	Tissues	References
[^{3}H]-1,25(OH)$_2$D$_3$	Rat	Bone	
		– Chondroblasts	
		– Chondrocytes	
		– Osteoblasts	
		Bone marrow	
		– Reticular cells	
		Brain	
		– Amygdala	
		– Stria terminalis	
		– Thalamus	
		– Medulla	
		oblongata	
		Colon	
		Duodenum	
		Esophagus	
		Heart muscle	
		Ileum	
		Jejunum	
		Kidney	
		– Distal tubule	
		– Glomeruli	
		podocytes	
		– Collecting	
		duct cells	
		– Proximal	
		tubule	
		Liver	
		– Lipocytes	
		Mammary gland	
		Pancreas	
		– Islet B cells	
		Parathyroid	
		– Parenchymal	
		cells	
		Pituitary	
		– Thyrotropes	
		Placenta	
		Pyloric muscle	

Table 2
(continued)

Compound	Species	Tissues	References
		Salivary gland	*(329–348)*
		– Epithelial cells of striated ducts	
		– Granular convoluted tubules	
		– Myoepithelial cells	
		Skin	
		– Epidermis	
		– Hair shafts	
		– Sebaceous glands	
		Spinal cord	
		Stomach	
		– Endocrine cells	
		Teeth	
		Testis	
		Thymus	
		– Reticular cells	
		Thyroid	
		Trachea	
		Yolk Sac	

vitamin D deficiency and in $1,25(OH)_2D_3$ excess, a marked change in delayed hypersensitivity is found *(188, 189)*. Furthermore, there is clear evidence in autoimmune diseases such as experimental autoimmune encephalomyelitis, a model of multiple sclerosis, that the vitamin D hormone can clearly prevent the development of this disease or can prevent progression of the disease *(190, 191)*. Furthermore, the same can be said for type I diabetes *(191, 192)*, arthritis *(160, 191)*, and lupus *(160, 191, 193)*. It is not clear how $1,25(OH)_2D_3$ suppresses these model diseases but it certainly can do so. Unfortunately the price is hypercalcemia because $1,25(OH)_2D_3$ retains its strong effectiveness on the elevation of plasma calcium and phosphorus levels. These studies not only suggest that the vitamin D hormone plays an important role in bodily functions not related to calcium but also illustrate promise that vitamin D derivatives might become useful in the prevention and/or treatment of these diseases.

8. VITAMIN D IN PUBLIC HEALTH

For many years following the elimination of nutritional rickets as a major medical problem with vitamin D, an attitude persisted that once the mineralization problem is corrected, vitamin D is no longer needed for public health. This has recently been dramatically changed by the discovery of the VDR and the distribution of the native hormone (Table 2) in a variety of non-calcium-related cell types. The treatment of psoriasis by the topical application of a vitamin D derivative is now well established *(164)*. Certain clinical experiences that utilize vitamin D as co-treatments for malignancy have also been reported *(194)*. However, a major awareness of vitamin D has come through epidemiological studies that point to a possible relationship to a variety of health problems. A reduction in risk for colorectal cancers and breast cancers is associated with higher blood levels of 25(OH)D$_3$ *(195, 196)*. A relationship between increased sunlight exposure and a reduction of colorectal and breast cancers has also been reported *(195, 196)*. In continuing the epidemiological discourse, reports have now appeared that the incidence of multiple sclerosis is inversely related to sunlight exposure *(190)* and 25(OH)D$_3$ levels in blood *(197)*. A similar relationship has been reported with type I diabetes *(192)*. More recently, an epidemiological study has appeared reporting that higher blood levels of 25(OH)D$_3$ are associated with a reduced risk of heart attacks *(198)*. Thus, some scientists in the vitamin D field have been lobbying to increase the recommended daily allowance level of vitamin D to levels that will raise blood 25(OH)D$_3$ in the range that reduces the risk of cancers and autoimmune diseases *(199, 200)*. An additional debate has taken place between scientists interested in skin production of vitamin D and suppression of both cancers and autoimmune diseases. Dermatologists believe that sunlight exposure is damaging and contributes to melanoma, squamous cell carcinoma, and basal cell carcinoma, thus warning against increased sunlight exposure for production of vitamin D for prevention of the diseases *(201)*. However, there is a clear alternative, namely the intake of supplemental vitamin D to raise blood levels. Thus, vitamin D has become an important public health consideration and it is likely that recommended vitamin D intakes in the future will be increased to 2,000–4,000 international units per day. However, it must be remembered that vitamin D, like its derivatives, is potentially toxic; raising the intakes of vitamin D much above this level carries with it the risk of accumulation of ordinary vitamin D in the adipose tissues *(202)* and eventual saturation, leading to high conversions to 25(OH)D$_3$ that then may cause intoxication (Table 3). It is believed that 25(OH)D$_3$ is the form of vitamin D responsible for intoxication caused by vitamin D$_3$ *(203, 204)*. A true safety study assessing at what levels the population is completely safe must be carried out before recommendations are made for levels of vitamin D intake above 4,000 units per day.

A partial compilation of estimates of the biological lifetime of vitamin D$_3$ and its principle metabolites in rat and man is provided in Table 3. Certainly the body lifetime of vitamin D$_3$ and its first metabolite, 25(OH)D$_3$, are measured in weeks and days. However, as might be expected for the active hormonal form, the lifetime of 1,25(OH)$_2$D$_3$ is measured in hours. The prolonged residence of vitamin D$_3$ in adipose tissue explains the difficulty in producing vitamin D-deficient experimental animals and signals a concern of accumulation of higher doses in the adipose tissue that might later cause a problem.

Table 3
Pharmacokinetics of Vitamin D_3, 25(OH)D_3, and 1,25(OH)$_2D_3$

Drug administered	Route	Amount	Species	Blood C_{max}	Blood T_{max}	Blood $t_{1/2}$ or $T_{50\% of max}$	Adipose $t_{1/2}$	References
[^{3}H]-vitamin D_3	IV, single	110 mcg	Human			20–30 h		(318)
[1,2-^{3}H or 4-^{14}C]-vitamin D_3	IV, single	110–216 mcg	Human			1–2 days (Phase I) 18–32 days (Phase II)		(349)
[1,2-^{3}H]-vitamin D_3	IV, single	6 mcg or 80 mcg	Human			12 h		(350)
[1,2-^{3}H]-vitamin D_3	IV, single	0.25 mcg	Rat (D-deficient)		< 1 h	~1–2 days (Phase I) ~14 days (Phase II)		(351)
Vitamin D_3	Oral, chronic	0.2 mcg/kg	Rat			1.4 days (Phase I?)	81 days (s.c.) 97 days (perirenal)	(352)
[4-^{14}C]-vitamin D_3	Oral, chronic	0.5 mcg–125 mcg	Rat (D-deficient)			21 days (Phase II)	81 days	(202)
[26,27-^{3}H]-25 (OH)D_3	IV, single	10 μCi	Human			13–19 days (normal) 41–46 days (anephric)		(353)
[26,27-^{3}H or 23,24-^{3}H]-25 (OH)D_3	IV, single	2–5 μCi	Human			28 days		(354)

Table 3 (continued)

Drug administered	Route	Amount	Species	Blood C_{max}	Blood T_{max}	Blood $t_{1/2}$ or $T_{50\% of max}$	Adipose $t_{1/2}$	References
[26,27-^{3}H]-25(OH)D$_3$	IV, single	Unknown	Human			23 days (normal) 42 days (anephric)		(355)
[26,27-^{3}H]-25(OH)D$_3$	IV, single	5 μCi	Human			13 days		(355)
[26,27-^{3}H]-25(OH)D$_3$	IV, single	5 μCi	Human			26 days		(356)
[6,19,19-^{2}H]-25(OH)D$_3$	IV, single	Enough to label 5% of total body pool	Human (dialysis patients)			10 days		(357)
25(OH)D$_3$	Oral, single	1.5–15 mcg/kg	Human	30–150 ng/ml	4–8 h	22 days		(358)
1,25(OH)$_2$D$_3$	IP, single	60 ng/kg	Human (dialysis patients)	~100 pg/ml	3 h	~12 h		(359)
[^{3}H]-1,25(OH)$_2$D$_3$	IP, single	0.38 μCi	Rat			12 h		(360)
1,25(OH)$_2$D$_3$	IV, single	1 mcg + 1 μCi radiolabel	Human			14 h		(361)
1,25(OH)$_2$D$_3$	IV, single	0.07 mcg/kg	Human	365 pg/ml		11 h		(362)

Compound	Route	Dose	Species	Cmax	Tmax	Half-life	Ref
$1,25(OH)_2D_3$	IV, single	60 ng/kg	Human (dialysis patients)	~475 pg/ml	2 min	17 h	(359)
$1,25(OH)_2D_3$	IV, single	2 mcg	Human (dialysis patients)	~175 pg/ml	<1 h	19–26 h	(363)
$1,25(OH)_2D_3$	IV, single	0.03 mcg/kg	Human (dialysis patients)	NA	5 min	15 min	(364)
$1,25(OH)_2D_3$	IV, single	1.5 mcg/m^2	Human	297 pg/ml	5 min	~4 h	(365)
$1,25(OH)_2D_3$	IV, single	0.7 mcg/kg	Human	413 pg/ml		26 h	(366)
$1,25(OH)_2D_3$	IV, single	0.17–1.6 mcg/kg	Human	460–6,680 pg/ml		12–21 h	(367)
[26,27-^{3}H]-$1,25(OH)_2D_3$	IV, single	59.6 pmol	Rat (D-deficient)			7–12 h	(368)
$1,25(OH)_2D_3$	IV, single	10–50 mcg/kg	Rat			2–4 h	(163)
$1,25(OH)_2D_3$	Oral, single	1 mcg + 1 μCi radiolabel	Human		4 h	13 h	(361)
$1,25(OH)_2D_3$	Oral, single	0.07 mcg/kg	Human	420 pmol/l	4 h	8–12 h	(369)
$1,25(OH)_2D_3$	Oral, single	0.03 mcg/kg	Human	100	4		(370)

Table 3 (continued)

Drug administered	Route	Amount	Species	Blood C_{max}	Blood T_{max}	Blood $t_{1/2}$ or $T_{50\% \, of \, max}$	Adipose $t_{1/2}$	References
1,25(OH)$_2$D$_3$	Oral, single	2 mcg	Human	235–351 pmol/L (normal) 108–212 pmol/L	1–7 h (normal) 3–7 h (uremic)	5–8 h (normal) 18–44 h (uremic)		(309)
1,25(OH)$_2$D$_3$	Oral, single	0.5 mcg	Human	~60 pg/ml	4 h	3.5 h		(371)
1,25(OH)$_2$D$_3$	Oral, single	4 mcg	Human	133 pg/ml	4 h	12–24 h		(372)
1,25(OH)$_2$D$_3$	Oral, single	60 ng/kg	Human (dialysis patients)	~80 pg/ml	6 h	~12 h		(359)
1,25(OH)$_2$D$_3$	Oral, chronic	0.25 mcg	Human (dialysis patients)			18 h		(373)
1,25(OH)$_2$D$_3$	Oral	0.25–0.5 mcg	Human			3–6 h		(374)
1,25(OH)$_2$D$_3$	Oral, single	2 mcg	Human (dialysis patients)	~50–100 pg/ml	2–4 h	25–38 h		(363)
1,25(OH)$_2$D$_3$	Oral, single	1.5 mcg/m^2	Human	101 pg/ml	4 h	~24 h		(365)
1,25(OH)$_2$D$_3$	Oral, single	1.6 mcg/kg	Human	~1,800 pg/ml	4–8 h	6 h		(375)

Drug	Route, schedule	Dose	Species	Concentration	T_{max}	Half-life	Reference
1,25(OH)$_2$D$_3$	Oral, days 1, 2, and 3 each week for 6 weeks	0.07–0.63 mcg/kg[a]	Human	210–1,100 pg/ml	1.5–10	18–34 h	(376)
1,25-OH)$_2$D$_3$ [^{3}H]-1,25(OH)$_2$D$_3$	Oral, single Oral, single	0.07 mcg/kg 0.38 μCi	Human Rat	223 pg/ml	2 h	28 h 13 h	(366) (360)
1,25(OH)$_2$D$_3$	Subcutaneous, single	25–40 ng/kg	Human			12–14 h	(377)
1,25(OH)$_2$D$_3$	Subcutaneous, single	0.03 mcg/kg	Human (dialysis patients)		5 min	~4 h	(364)
1,25(OH)$_2$D$_3$	Subcutaneous, single	2–10 mcg	Human (cancer patients)	112–301 pg/ml	1–3 h	11–19 min	(378)

[a] 60 kg

BIBLIOGRAPHY

1. Mellanby E (1919) An experimental investigation on rickets. Lancet 1:407–412
2. McCollum EV, Simmonds N, Becker JE et al (1922) Studies on experimental rickets. XXI. An experimental demonstration of the existence of a vitamin which promotes calcium deposition. J Biol Chem 53:293–312
3. Huldshinsky K (1919) Heilung von rachitis durch kunstlickhe hohensonne. Deut Med Wochschr 45:712–713
4. Chick H, Palzell EJ, Hume EM (1923) Studies of rickets in Vienna 1919–1922. Medical Research Council.
5. Hess A ed. (1929) In: Rickets, including osteomalacia and tetany. Lee & Febiger, Philadelphia, pp. 22–37
6. Sebrell WH, Harris RS (1954) Vitamin D group. The vitamins. Academic Press, New York, pp 1131–1266
7. Steenbock H, Black A (1924) Fat-soluble vitamins. XVII. The induction of growth-promoting and calcifying properties in a ration by exposure to ultraviolet light. J Biol Chem 61:405–422
8. Scriver CR, Reade TM, DeLuca HF et al (1978) Serum 1,25-$(OH)_2D_3$ levels in normal subjects and in patients with hereditary rickets or bone disease. N Engl J Med 299:976–979
9. Fraser D, Kooh SW, Kind HP et al (1973) Pathogenesis of hereditary vitamin D-dependent rickets: An inborn error of vitamin D metabolism involving defective conversion of 25-hydroxyvitamin D to 1,25-dihydroxyvitamin D. N Engl J Med 289:817–822
10. Kim CJ, Kaplan LE, Perwad F et al (2007) Vitamin D 1α-hydroxylase deficiency. J Clin Endocrinol Metab 92(8):3177–3182
11. Porcu L, Meloni A, Casula L et al (2002) A novel splicing defect (IVS6+1G—>T) in a patient with pseudovitamin D deficiency rickets. J Endocrinol Invest 25:557–560
12. Wang X, Zhang MYH, Miller WL et al (2002) Novel gene mutations in patients with 1α-hydroxylase deficiency that confer partial enzyme activity in vitro. J Clin Endocrinol Metab 87(6):2424–2430
13. Wang JT, Lin C-J, Burridge SM et al (1998) Genetics of vitamin D 1α-hydroxylase deficiency in 17 families. Am J Hum Genet 63:1694–1702
14. Fu GK, Lin D, Zhang MYH et al (1997) Cloning of human 25-hyroxyvitamin D-1α-hydroxylase and mutations causing vitamin D-dependent rickets type 1. Mol Endo 11:1961–1970
15. Kitanaka S, Murayama A, Sakaki T et al (1999) No enzyme activity of 25-hydroxyvitamin D_3 1α-hydroxylase gene product in pseudovitamin D deficiency rickets, including that with mild clinical manifestation. J Clin Endocrinol Metab 84:4111–4117
16. Kitanaka S, Takeyama K-I, Murayama A et al (1998) Inactivating mutations in the 25-hydroxyvitamin D_3 1α-hydroxylase gene in patients with pseudovitamin D-deficiency rickets. N Engl J Med 338: 653–661
17. Yoshida T, Monkawa T, Tenenhouse HS et al (1998) Two novel 1α-hydroxylase mutations in French-Canadians with vitamin D dependency rickets type I. Kidney Int 54:1437–1443
18. Smith SJ, Rucka AK, Berry JL et al (1999) Novel mutations in the 1α-hydroxylase (P450c1) gene in three families with pseudovitamin D-deficiency rickets resulting in loss of functional enzyme activity in blood-derived macrophages. J Bone Miner Res 14:730–739
19. Brommage R, Jarnagin K, DeLuca HF et al (1983) 1- hydroxylation But not 24- hydroxylation of vitamin D is required for skeletal mineralization in rats. Am J Physiol 244:E298–E304
20. Eil C, Lieberman UA, Rosen JF et al (1981) A cellular defect in hereditary vitamin D-dependent rickets Type II: defective nuclear uptake of 1,25-dihydroxyvitamin D in cultured skin fibroblasts. N Engl J Med 304:1588–1591
21. Bell NH, Hamstra AJ, DeLuca HF (1978) Vitamin D-dependent rickets Type II: Resistance of target organs to 1,25-dihydroxyvitamin D. N Engl J Med 298:996–999
22. Rosen JF, Fleischman AR, Finberg L et al (1979) Rickets with alopecia: An inborn error of vitamin D metabolism. J Pediatrics 94:729–735
23. Marx SJ, Liberman UA, Eil C et al (1984) Hereditary resistance to 1,25-dihydroxyvitamin D. Rec Prog Horm Res 40:589–620

24. Wiese RJ, Goto H, Prahl JM et al (1993) Vitamin D-dependency rickets type II: truncated vitamin D receptor in three kindreds. Mol Cell Endocrinol 90:197–201

25. Liberman UA, Marx SJ (1990) Vitamin D dependent rickets. In: Favus MJ (ed) Primer on the metabolic bone diseases and disorders of mineral metabolism, 1st edn. William Byrd Press, Richmond, 178–182

26. Liberman UA (2007) Vitamin D-resistant diseases. J Bone Miner Res 22(S2):V105–V107

27. Underwood JL, DeLuca HF (1983) Vitamin D is not directly necessary for bone growth and mineralization. Am J Physiol 246:E493–E498

28. DeLuca HF (1967) Mechanism of action and metabolic fate of vitamin D. Vitam Horm 1967(25): 315–367

29. DeLuca HF, Schnoes HK (1983) Vitamin D: recent advances. Ann Rev Biochem 52:411–439

30. Schachter D, Rosen SM (1959) Active transport of Ca45 by the small intestine and its dependence on vitamin D. Am J Physiol 196:357–362

31. Higaki M, Takahashi M, Suzuki T et al (1965) Metabolic activities of vitamin D in animals. III. Biogenesis of vitamin D sulfate in animal tissues. J Vitaminol 11:261–265

32. Martin DL, DeLuca HF (1969) Calcium transport and the role of vitamin D. Arch Biochem Biophys 134:139–148

33. Walling MW, Rothman SS (1969) Phosphate-independent, carrier-mediated active transport of calcium by rat intestine. Am J Physiol 217:1144–1148

34. Wasserman RH, Kallfelz FA, Comar CL (1961) Active transport of calcium by rat duodenum in vivo. Science 133:883–884

35. Schachter D (1963) Vitamin D and the active transport of calcium by the small intestine. In: Wasserman RH (ed) The transfer of calcium and strontium across biological membranes. Academic Press, New York, 197–210

36. Chen TC, Castillo L, Korycka-Dahl M et al (1974) Role of vitamin D metabolites in phosphate transport of rat intestine. J Nutr 104:1056–1060

37. Walling MW (1977) Effects of 1,25-dihydroxyvitamin D3 on active intestinal inorganic phosphate absorption. In: Norman AW, Schaefer K, Coburn JW (eds) Vitamin D: biochemical, chemical, and clinical aspects related to calcium metabolism. Walter de Gruyter, Berlin, 321–330

38. Harrison HE, Harrison HC (1962) Intestinal transport of phosphate: Action of vitamin D, calcium, and potassium. Am J Physiol 201:1007–1012

39. Nicolaysen R, Eeg-Larsen N (1956) The mode of action of vitamin D. In: Wolstenholme GWE, O'Connor CM (eds) Ciba foundation symposium on bone structure and metabolism. Little, Brown, and Co., Boston, 175–186

40. Yamamoto M, Kawanobe Y, Takahashi H et al (1984) Vitamin D deficiency and renal calcium transport in the rat. J Clin Invest 74:507–513

41. Carlsson A (1952) Tracer experiments on the effect of vitamin D on the skeletal metabolism of calcium and phosphorus. Acta Physiol Scand 26:212–220

42. Rasmussen H, DeLuca H, Arnaud C et al (1963) The relationship between vitamin D and parathyroid hormone. J Clin Invest 42:1940–1946

43. Morii H, Lund J, Neville PF et al (1967) Biological activity of a vitamin D metabolite. Arch Biochem Biophys 120(3):508–512

44. Steenbock H, Herting DC (1955) Vitamin D and growth. J Nutr 57:449–468

45. Cramer JW, Steenbock H (1956) Calcium metabolism and growth in the rat on a low phosphorus diet as affected by vitamin D and increases in calcium intake. Arch Biochem Biophys 63:9–13

46. Darwish HM, DeLuca HF (1996) Analysis of binding of the 1,25-dihydroxyvitamin D3 receptor to positive and negative vitamin D response elements. Arch Biochem Biophys 334: 223–234

47. Demay MB, Kiernan MS, DeLuca HF et al (1992) Sequences in the human parathyroid hormone gene that bind the 1,25-dihydroxyvitamin D3 receptor and mediate transcriptional repression in response to 1,25-dihydroxyvitamin D3. Proc Natl Acad Sci USA 89:8097–8101

48. Silver J, Naveh-Many T, Mayer H et al (1986) Regulation of vitamin D metabolites of parathyroid hormone gene transcription in vivo in the rat. J Clin Invest 78:1296–1301

49. DeLuca HF (1981) The transformation of a vitamin into a hormone – the vitamin D story. Harvey Lect 75:333–379
50. DeLuca HF (1985) The vitamin D-calcium axis – 1983. In: Rubin RP, Weiss GB, Putney JW Jr (eds) Calcium in biological systems, vol 53. Plenum Publishing Corporation, New York, 491–511
51. Liu S, Tang W, Zhou J et al (2006) Fibroblast growth factor 23 is a counter-regulatory phosphaturic hormone for vitamin D. J Am Soc Nephrol 17(5):1305–1315 (Epub 2006, April 5)
52. Shimada T, Kakitani M, Yamazaki Y et al (2004) Targeted ablation of Fgf23 demonstrates an essential physiological role of FGF23 in phosphate and vitamin D metabolism. J Clin Invest 113(4):562–658
53. Bellin SA, Herting DC, Cramer JW et al (1954) The effect of vitamin D on urinary citrate in relation to calcium phosphorus and urinary phosphorus. Arch Biochem Biophys 80:18–23
54. Neville PF, DeLuca HF (1966) The synthesis of [1,2-^{3}H]-vitamin D3 and the tissue localization of a 0.25 µg (10 IU) dose per rat. Biochemistry 5:2201–2207
55. Norman AW, Lund J, DeLuca HF (1964) Biologically active forms of vitamin D3 in kidney and intestine. Arch Biochem Biophys 108:12–21
56. Lund J, DeLuca HF (1966) Biologically active metabolites of vitamin D_3 from bone, liver, and blood serum. J Lipid Res 7:739–744
57. DeLuca HF (1974) Vitamin D: the vitamin and the hormone. Fed Proc 33:2211–2219
58. Vieth R (1999) Vitamin D supplementation, 25-hydroxyvitamin D concentrations, and safety. Am J Clin Nutr 69:842–856
59. Windus A, Bock F (1937) Uber das provitamin aus der sterin der schweineschwarte. Z Physiol Chem 245:168–170
60. Esvelt RP, Schnoes HK, DeLuca HF (1978) Vitamin D3 from rat skins irradiated in vitro with ultraviolet light. Arch Biochem Biophys 188:282–286
61. Windus A, Schenck F, Weder Fv (1936) Uber das antirachitisch wirksame bestrahlungs-produkt aus 7-dehydro-cholesterin. Hoppe-Seylers Z Physiol Chem 241:100–103
62. Velluz L, Amiard G (1949) Chimie organique-le precalciferol. Compt Rend 228:692–694
63. Holick MF, Clark MB (1978) The photobiogenesis and metabolism of vitamin D. Fed Proc 37: 2567–2574
64. Ponchon G, DeLuca HF, Suda T (1970) Metabolism of (1,2)3H-vitamin D3 and (26,27)3H-25-hydroxyvitamin D3 in rachitic chicks. Arch Biochem Biophys 141:397–408
65. Horsting M, DeLuca HF (1969) In vitro production of 25-hydroxycholecalciferol. Biochem Biophys Commun 36:251–256
66. Prosser DE, Jones G (2004) Enzymes involved in the activation and inactivation of vitamin D. Trends Biochem Sci 29(13):664–673
67. Guo Y-D, Strugnell S, Back DW et al (1993) Transfected human liver cytochrome P-450 hydroxylates vitamin D analogs at different side-chain positions. Proc Natl Acad Sci USA 90:8668–8672
68. Pikuleva IA, Bjorkhem I, Waterman MR (1997) Expression, purification, and enzymatic properties of recombinant human cytochrome P450c27 (CYP27). Arch Biochem Biophys 343(1):123–130
69. Rosen H, Reshef A, Maeda N et al (1998) Markedly reduced bile acid synthesis but maintained levels of cholesterol and vitamin D metabolites in mice with disrupted sterol 27-hydroxylase gene. J Biol Chem 273(24):14805–14812
70. Repa JJ, Mangelsdorf DJ (2000) The role of orphan nuclear receptors in the regulation of cholesterol homeostasis. Ann Rev Cell Dev Biol 16:459–481
71. Ohyama Y, Yamasaki T (2004) Eight cytochrome P450s catalyze vitamin D metabolism. Front Biosci 9:3007–3018
72. Fraser DR, Kodicek E (1970) Unique biosynthesis by kidney of a biologically active vitamin D metabolite. Nature 228:764–766
73. Gray R, Boyle I, DeLuca HF (1971) Vitamin D metabolism: the role of kidney tissue. Science 172:1232–1234
74. Boyle IT, Miravet L, Gray RW et al (1972) The response of intestinal calcium transport to 25-hydroxy and 1,25-dihydroxyvitamin D in nephrectomized rats. Endocrinology 90:605–608
75. Holick MF, Garabedian M, DeLuca HF (1972) 1,25-Dihydroxycholecalciferol: metabolite of vitamin D3 active on bone in anephric rats. Science 176:1146–1147

76. Wong RG, Norman AW, Reddy CR et al (1972) Biologic effects of 1,25-dihydroxycholecalciferol (a highly active vitamin D metabolite) in acutely uremic rats. J Clin Invest 51:1287–1291

77. Dardenne O, Prud'Homme J, Arabian A (2001) Targeted inactivation of the 25-hyroxyvitamin D$_3$-1α-hydroxylase gene (CYP27B1) creates an animal model of pseudovitamin D-deficiency rickets. Endocrinology 142:3135–3141

78. Panda DK, Miao D, Tremblay ML et al (2001) Targeted ablation of the 25-hydroxyvitamin D 1α-hydroxylase enzyme: Evidence for skeletal, reproductive, and immune dysfunction. Proc Natl Acad Sci USA 98(13):7498–7503

79. Norman AW (2008) From vitamin D to hormone D: fundamentals of the vitamin D endocrine system essential for good health. Am J Clin Nutr 88:491S–499S

80. Bikle DD, Chang S, Crumrine D et al (2004) 25 Hydroxyvitamin D 1α-hydroxylase is required for optimal epidermal differentiation and permeability barrier homeostasis. J Invest Dermatol 122: 984–992

81. Hewison M, Adams JS (2005) Extra-renal 1α-hydroxylase activity and human disease. In: Feldman D, Pike JW, Glorieux FH (eds) Vitamin D, vol 79, 2nd edn. Elsevier Academic Press, San Diego, CA, 1379–1400

82. Barbour GL, Coburn JW, Slatopolsky E et al (1981) Hypercalcemia in an anephric patient with sarcoidosis: Evidence for extrarenal generation of 1,25-dihydroxyvitamin D. N Engl J Med 305(8): 440–443

83. Jones G, Ramshaw H, Zhang A et al (1999) Expression and activity of vitamin D-metabolizing cytochrome P450s (CYP1α and CYP24) in human nonsmall cell lung carcinomas. Endocrinology 140(7):3303–3310

84. Reeve L, Tanaka Y, DeLuca HF (1983) Studies on the site of 1,25-dihydroxyvitamin D$_3$ synthesis in vivo. J Biol Chem 258(6):3615–3617

85. Shultz TD, Fox J, Heath H 3rd et al (1983) Do tissues other than the kidney produce 1,25-dihydroxyvitamin D$_3$ in vivo? A reexamination. Proc Natl Acad Sci USA 80(6):1746–1750

86. Vanhooke JL, Prahl JM, Kimmel-Jehan C et al (2006) CYP27B1 null mice with *LacZ* reporter gene display no 25-hydroxyvitamin D$_3$-1α-hydroxylase promoter activity in the skin. Proc Natl Acad Sci USA 103(1):75–80

87. Pedersen JI, Shobaki HH, Holmberg I et al (1983) 25-Hydroxyvitamin D$_3$-24-hydroxylase in rat kidney mitochondria. J Biol Chem 258:742–746

88. Omdahl JL, Morris HA, May BK (2002) Hydroxylase enzymes of the vitamin D pathway: Expression, function, and regulation. Annu Rev Nutr 22:139–166

89. Akiyoshi-Shibata M, Sakaki T, Ohyama Y (1994) Further oxidation of hydroxycalcidiol by calcidiol 24-hydroxylase. Eur J Biochem 224:335–343

90. Beckman MJ, Tadikonda P, Werner E et al (1996) Human 25-hydroxyvitamin D$_3$-24-hydroxylase, a multicatalytic enzyme. Biochemistry 35:8465–8472

91. Makin G, Lohnes D, Byford V et al (1989) Target cell metabolism of 1,25-dihydroxyvitamin D$_3$ to calcitroic acid. Evidence for a pathway in kidney and bone involving 24-oxidation. Biochem J 262(1):173–180

92. Reddy GS, Tserng KY (1989) Calcitroic acid end product of renal metabolism of 1,25-dihydroxyvitamin D$_3$ through C-24 oxidation pathway. Biochemistry 28(4):1763–1769

93. Esvelt RP, Rivizzani MA, Paaren HE (1981) Synthesis of calcitroic acid, a metabolite of 1,25-dihydroxycholecalciferol. J Org Chem 46:456–458

94. Onisko BL, Esvelt RP, Schnoes HK et al (1980) Metabolites of 1,25-dihydroxyvitamin D3 in rat bile. Biochemistry 19:4124–4130

95. Rasmussen H, Bordier P (1978) Vitamin D and bone. Metab Bone Dis Rel Res 1:7–13

96. Ornoy A, Goodwin D, Noff D, Edelstein S (1978) 24,25-dihydroxyvitamin D is a metabolite of vitamin D essential for bone formation. Nature 276:517–519

97. Henry HL, Taylor AN, Norman AW (1977) Response of chick parathyroid glands to the vitamin D metabolites 1,25-dihydroxyvitamin D3 and 24,25-dihydroxyvitamin D3. J Nutr 107:1918–1926

98. Garabedian M, Lieberherr M, Nguyen TM et al (1978) In vitro production and activity of 24,25-dihydroxycholecalciferol in cartilage and calvarium. Clin Orthop Relat Res 135:241–248

 99. Henry HL, Norman AW (1978) Vitamin D: two dihydroxylated metabolites are required for normal chicken egg hatchability. Science 201:835–837
100. Jarnagin K, Brommage R, DeLuca HF (1983) 1-But not 24-hydroxylation of vitamin D is required for growth and reproduction in rats. Am J Physiol 244:E290–E297
101. Halloran BP, DeLuca HF, Barthell E (1981) An examination of the importance of 24-hydroxylation to the function of vitamin D during early development. Endocrinology 108:2067–2071
102. Miller SC, Halloran BP, DeLuca HF (1981) Studies on the role of 24-hydroxylation of vitamin D in the mineralization of cartilage and bone of vitamin D-deficient rats. Calcif Tissue Int 33: 489–497
103. St-Arnaud R, Arabian A, Glorieux FH (1996) Abnormal bone development in mice deficient for the vitamin D 24-hydroxylase gene. ASBMR 18th Annual Meeting. Seattle, WA, p S126
104. St-Arnaud R, Arabian A, Travers R (2000) Deficient mineralization of intramembranous bone in vitamin D-24-hydroxylase-ablated mice is due to elevated 1,25-dihydroxyvitamin D and not to the absence of 24,25-dihydroxyvitamin D. Endocrinology 141(7):2658–2666
105. Brown EM, Gamba G, Riccardi D (1993) Cloning and characterization of an extracellular Ca+2-sensing receptor from bovine parathyroid. Nature 366:575–580
106. Omdahl JL, Gray RW, Boyle IT et al (1972) Regulation of metabolism of 25-hydroxycholecalciferol metabolism by kidney tissue in vitro by dietary calcium. Nat N Biol 237:63–64
107. Garabedian M, Holick MF, DeLuca HF et al (1972) Control of 25-hydroxycholecalciferol metabolism by the parathyroid glands. Proc Natl Acad Sci USA 69:1673–1676
108. Fraser DR, Kodicek E (1973) Regulation of 25-hydroxycholecalciferol-1-hydroxylase activity in kidney by parathyroid hormone. Nat N Biol 241:163–166
109. Garabedian M, Tanaka Y, Holick MF et al (1974) Response of intestinal calcium transport and bone calcium mobilization to 1,25-dihydroxyvitamin D_3 in thyroparathyroidectomized rats. Endocrinology 94:1022–1027
110. Forte LR, Nickols GA, Anast CS (1976) Renal adenylate cyclase and the interrelationship between parathyroid hormone and vitamin D in the regulation of urinary phosphate and adenosine cyclin $3',5'$ monophosphate excretion. J Clin Invest 57:559–568
111. Tanaka Y, DeLuca HF (1973) The control of 25-hydroxyvitamin D metabolism by inorganic phosphorus. Arch Biochem Biophys 154:566–574
112. Baxter LA, DeLuca HF (1976) Stimulation of 25-hydroxyvitamin D_3-1-hydroxylase by phosphate depletion. J Biol Chem 251:3158–3161
113. Hughes MR, Brumbaugh PF, Haussler MR (1975) Regulation of serum 1,25-dihydroxyvitamin D3 by calcium and phosphate in the rat. Science 190:578–580
114. Gray RW (1987) Evidence that somatomedins mediate the effect of hypophosphatemia to increase serum 1,25-dihydroxyvitamin D3 levels in rats. Endocrinology 121:504–512
115. Halloran BP, Spencer EM (1988) Dietary phosphorus and 1,25-dihydroxyvitamin D metabolism: Influence of insulin-like growth factor-1. Endocrinology 123:1225–1229
116. Spencer EM, Tobiassen O (1981) The mechanism of the action of growth hormone on vitamin D metabolism in the rat. Endocrinology 108:1064–1070
117. Gray RW (1981) Control of plasma 1,25-$(OH)_2$-vitamin D concentrations by calcium and phosphorus in the rat: effects of hypophysectomy. Calcif Tissue Int 33:485–488
118. Pahuja DN, DeLuca HF (1981) Role of the hypophysis in the regulation of vitamin D metabolism. Mol Cell Endocrinol 23:345–350
119. Brown DJ, Spanos E, MacIntyre I (1980) Role of pituitary hormones in regulating renal vitamin D metabolism in man. Br Med J 280:277
120. Liu S, Zhou J, Tang W (2006) Pathogenic role of Fgf23 in Hyp mice. Am J Physiol Endocrinol Metab 291(1):E38–E49
121. Tanaka Y, Frank H, DeLuca HF (1973) Biological activity of 1,25-dihydroxyvitamin D_3 in the rat. Endocrinology 92:417–422
122. Tanaka Y, Lorenc RS, DeLuca HF (1975) The role of 1,25-dihydroxyvitamin D3 and parathyroid hormone in the regulation of chick renal 25-hydroxyvitamin D_3-24-hydroxylase. Arch Biochem Biophys 171:521–526

123. Tanaka Y, DeLuca HF (1974) Stimulation of 24,25-dihydroxyvitamin D3 production by 1,25-dihydroxyvitamin D3. Science 183:1198–1200

124. Shinki T, Jin CH, Nishimura A (1992) Parathyroid hormone inhibits 25-hydroxyvitamin D_3-24-hydroxylase mRNA expression stimulated by 1,25-dihydroxyvitamin D3 in rat kidney but not in intestine. J Biol Chem 267:13757–13762

125. Zierold C, Darwish HM, DeLuca HF (1994) Identification of a vitamin D-response element in the rat calcidiol (25-hydroxyvitamin D3) 24-hydroxylase gene. Proc Natl Acad Sci USA 91:900–902

126. Ohyama Y, Ozono K, Uchida M (1994) Identification of a vitamin D-responsive element in the 5′-flanking region of the rat 25-hydroxyvitamin D3 24-hydroxylase gene. J Biol Chem 269: 10545–10550

127. Wu SX, Finch J, Zhong M (1996) Expression of the renal 25-hydroxyvitamin D-24-hydroxylase gene-regulation by dietary phosphate. Am J Phys 40:F203–F208

128. Brenza HL, Kimmel-Jehan C, Jehan F (1998) Parathyroid hormone activation of the 25-hydroxyvitamin D_3-1α-hydroxylase gene promoter. Proc Natl Acad Sci USA 95:1387–1391

129. Brenza HL (2002) Regulation of 25-hydroxyvitamin D_3-1α-hydroxylase gene expression. Ph.D. Thesis. University of Wisconsin-Madison.

130. Strom M, Sandgren ME, Brown TA et al (1989) 1,25-Dihydroxyvitamin D3 up-regulates the 1,25-dihydroxyvitamin D3 receptor in vivo. Proc Natl Acad Sci USA 86:9770–9773

131. Healy KD, Zella JB, Prahl JM et al (2003) Regulation of the murine renal vitamin D receptor by 1,25-dihydroxyvitamin D_3 and calcium. Proc Natl Acad Sci USA 100(17):9733–9737

132. Sandgren ME, DeLuca HF (1990) Serum calcium and vitamin D regulate 1,25-dihydroxyvitamin D_3 receptor concentration in rat kidney in vivo. Proc Natl Acad Sci USA 87(11):4312–4314

133. Goff JP, Reinhardt TA, Beckman MJ et al (1990) Contrasting effects of exogenous 1,25-dihydroxyvitamin D [1,25-$(OH)_2$D] versus endogenous 1,25-$(OH)_2$D, induced by dietary calcium restriction, on vitamin D receptors. Endocrinology 126(2):1031–1035

134. Beckman MJ, DeLuca HF (2002) Regulation of renal vitamin D receptor is an important determinant of 1α,25-dihydroxyvitamin D_3 levels in vivo. Arch Biochem Biophys 401(1):44–52

135. Naveh-Many T, Silver J (1990) Regulation of parathyroid hormone gene expression by hypocalcemia, hypercalcemia, and vitamin D in the rat. J Clin Invest 86:1313–1319

136. Stumpf WE, Sar M, DeLuca HF (1981) Sites of action of 1,25(OH)$_2$ vitamin D3 identified by thaw-mount autoradiography. In: Cohn DV, Talmage RV, Matthews JL (eds) Hormonal control of calcium metabolism. Excerpta Medica, Amsterdam-Oxford-Princeton, pp 222–229

137. Brumbaugh PF, Haussler MR (1975) Nuclear and cytoplasmic binding components for vitamin D metabolites. Life Sci 16:353

138. Kream BE, Reynolds RD, Knutson JC (1976) Intestinal cytosol binders of 1,25-dihydroxyvitamin D3 and 25-hydroxyvitamin D3. Arch Biochem Biophys 176:779–787

139. Baker AR, McDonnell DP, Hughes M (1988) Cloning and expression of full-length cDNA encoding human vitamin D receptor. Proc Natl Acad Sci USA 85:3294–3298

140. Burmester JK, Wiese RJ, Maeda N et al (1988) Structure and regulation of the rat 1,25-dihydroxyvitamin D3 receptor. Proc Natl Acad Sci USA 85:9499–9502

141. Pike JW, Shevde NK (2005) The vitamin D receptor. In: Feldman D, Pike JW, Glorieux FH (eds) Vitamin D, vol 11, 2nd edn. Elsevier Academic Press, San Diego, CA, pp 167–191

142. McDonnell DP, Scott RA, Kerner SA et al (1989) Functional domains of the human vitamin D3 receptor regulate osteocalcin gene expression. Mol Endocrinol 3:635–644

143. Umesono K, Murakami KK, Thompson CC (1991) Direct repeats as selective response elements for the thyroid hormone, retinoic acid, and vitamin D3 receptors. Cell 65:1255–1266

144. Zella LA, Kim S, Shevde NK et al (2006) Enhancers located within two introns of the vitamin D receptor gene mediate transcriptional autoregulation by 1,25-dihydroxyvitamin D_3. Mol Endocrinol 20(6):1231–1247

145. Kim S, Yamazaki M, Zella LA (2006) Activation of receptor activator of NF-kappaB ligand gene expression by 1,25-dihydroxyvitamin D_3 is mediated through multiple long-range enhancers. Mol Cell Biol 26(17):6469–6486

146. Carlberg C, Seuter S (2007) The vitamin D receptor. Dermatol Clin 25:515–523

147. Kutuzova GD, DeLuca HF (2004) Gene expression profiles in rat intestine identify pathways for 1,25-dihydroxyvitamin D_3 stimulated calcium absorption and clarify its immunomodulatory properties. Arch Biochem Biophys 432(2):152–166

148. Kutuzova GD, DeLuca HF (2007) 1,25-Dihydroxyvitamin D_3 regulates genes responsible for detoxification in intestine. Toxicol Appl Pharmacol 218(1):37–44

149. Chen KS, DeLuca HF (1995) Cloning of the human 1α,25-dihydroxyvitamin D_3 24-hydroxylase gene promoter and identification of two vitamin D-responsive elements. Biochim Biophys Acta 1263(1): 1–9

150. Carlberg C, Dunlop TW, Frank C (2005) Molecular basis of the diversity of vitamin D target genes. In: Feldman D, Pike JW, Glorieux FH (eds) Vitamin D, vol 18, 2nd edn. Elsevier Academic Press, San Diego, CA, pp 313–325

151. Nagai M, Sato N (1999) Reciprocal gene expression of osteoclastogenesis inhibitory factor and osteoclast differentiation factor regulates osteoclast formation. Biochem Biophys Res Commun 257: 719–723

152. Strom M, Sandgren ME, Brown TA et al (1989) 1,25-Dihydroxyvitamin D_3 up-regulates the 1,25-dihydroxyvitamin D_3 receptor in vivo. Proc Natl Acad Sci USA 86(24):9770–9773

153. Naveh-Many T, Marx R, Keshet E (1990) Regulation of 1,25-dihydroxyvitamin D_3 receptor gene expression by 1,25-dihydroxyvitamin D_3 in the parathyroid in vivo. J Clin Invest 86(6):1968–1975

154. Huang L, Xu J, Wood DJ et al (2000) Gene expression of osteoprotegerin ligand, osteoprotegerin, and receptor activator of NF-kappaB in giant cell tumor of bone. Possible involvement in tumor cell-induced osteoclast-like cell formation. Am J Pathol 156(3):761–767

155. Shevde NK, Plum LA, Clagett-Dame M et al (2002) A potent analog of 1α,25-dihydroxyvitamin D_3 selectively induced bone formation. Proc Natl Acad Sci USA 99(21):13487–13491

156. Fleet JC (2004) Rapid, membrane-initiated actions of 1,25 dihydroxyvitamin D: What are they and what do they mean? J Nutr 134:3215–3218

157. Norman AW, Mizwicki MT, Norman DPG (2004) Steroid-hormone rapid actions, membrane receptors and a conformational ensemble model. Nat Rev Drug Discov 3(1):27–41

158. Demay MB (2005) Mouse models of vitamin D receptor ablation. In: Feldman D, Pike JW, Glorieux FH (eds) Vitamin D, vol 20, 2nd edn. Elsevier Academic Press, San Diego, CA, pp 341–349

159. Nemere I, Yoshimoto Y, Norman AW (1984) Calcium transport in perfused duodena from normal chicks: enhancement within fourteen minutes of exposure to 1,25-dihydroxyvitamin D_3. Endocrinology 115(4):1476–1483

160. Nagpal S, Na S, Rathnachalam R (2005) Noncalcemic actions of vitamin D receptor ligands. Endocrine Rev 26(5):662–687

161. Dusso AS, Negrea L, Gunawardhana S et al (1991) On the mechanisms for the selective action of vitamin D analogs. Endocrinology 128(4):1687–1692

162. Binderup L, Binderup E, Godtfredsen WO (2005) Development of new vitamin D analogs. In: Feldman D, Pike JW, Glorieux FH (eds) Vitamin D, vol 84, 2nd edn. Elsevier Academic Press, San Diego, CA, pp 1489–1510

163. Kissmeyer A-M, Binderup L (1991) Calcipotriol (MC 903): pharmacokinetics in rats and biological activities of metabolites. A comparative study with $1,25(OH)_2D_3$. Biochem Pharmacol 41(11): 1601–1606

164. Segaert S, Duvold LB (2006) Calcipotriol cream: a review of its use in the management of psoriasis. J Dermatolog Treat 17(6):327–337

165. Sicinski RR, Prahl JM, Smith CM (1998) New 1α,25-dihydroxy-19-norvitamin D_3 compounds of high biological activity: synthesis and biological evaluation of 2-hydroxymethyl, 2-methyl, and 2-methylene analogues. J Med Chem 41(23):4662–4674

166. Ke HZ, Qi H, Crawford DT et al (2005) A new vitamin D analog, 2MD, restores trabecular and cortical bone mass and strength in ovariectomized rats with established osteopenia. J Bone Miner Res 20:1742–1755

167. Plum LA, Fitzpatrick LA, Ma X et al (2006) 2MD, a new anabolic agent for osteoporosis treatment. Osteoporosis Int 17(5):704–715

168. Slatopolsky E, Finch JL, Brown AJ (2007) Effect of 2-methylene-19-nor(20S)-1α-hydroxy-bishomopregnacalciferol (2MbisP), an analog of vitamin D, on secondary hyperparathyroidism. J Bone Miner Res 22:686–694

169. DeLuca HF, Plum LA, Clagett-Dame M (2007) Selective analogs of 1α,25-dihydroxyvitamin D_3 for the study of specific functions of vitamin D. J Steroid Biochem Mol Biol 103(3–5):263–268

170. Brown AJ, Slatopolsky E (2007) Drug insight: vitamin D analogs in the treatment of secondary hyperparathyroidism in patients with chronic kidney disease. Nat Clin Pract Endocrinol Metab 3(2): 134–144

171. Slatopolsky E, Finch J, Ritter C et al (1995) A new analog of calcitriol, 19-nor-1,25$(OH)_2D_2$, suppresses parathyroid hormone secretion in uremic rats in the absence of hypercalcemia. Am J Kidney Dis 26(5):852–860

172. Tocchini-Valentini G, Rochel N, Wurtz JM et al (2001) Crystal structures of the vitamin D receptor complexed to superagonist 20-epi ligands. Proc Natl Acad Sci USA 98(10):5491–5496

173. Tocchini-Valentini G, Rochel N, Wurtz J-M et al (2004) Crystal structures of the vitamin D nuclear receptor liganded with the vitamin D side chain analogues calcipotriol and seocalcitol, receptor agonists of clinical importance. Insights into a structural basis for the switching of calcipotriol to a receptor antagonist by further side chain modification. J Med Chem 47:1956–1961

174. Rochel N, Wurtz JM, Mitschler A et al (2000) The crystal structure of the nuclear receptor for vitamin D bound to its natural ligand. Molec Cell 5:173–179

175. Vanhooke JL, Benning MM, Bauer CB et al (2004) Molecular structure of the rat vitamin D receptor ligand binding domain complexed with 2-carbon-substituted vitamin D_3 hormone analogues and a LXXLL-containing coactivator peptide. Biochemistry 43(14):4101–4110

176. Vanhooke JL, Tadi BP, Benning MM et al (2007) New analogs of 2-methylene-19-nor-(20S)-1,25-dihydroxyvitamin D_3 with conformationally restricted side chains: Evaluation of biological activity and structural determination of VDR-bound conformations. Arch Biochem Biophys 460:161–165

177. Van den Bemd GC, Pols HA, Birkenhäger JC et al (1996) Conformational change and enhanced stabilization of the vitamin D receptor by the 1,25-dihydroxyvitamin D_3 analog KH1060. Proc Natl Acad Sci USA 93(20):10685–10690

178. Väisänen S, Juntunen K, Itkonen A et al (1997) Conformational studies of human vitamin-D receptor by antipeptide antibodies, partial proteolytic digestion and ligand binding. Eur J Biochem 248(1):156–162

179. Castillo AI, Sánchez-Martinez R, Jiménez-Lara AM et al (2006) Characterization of vitamin D receptor ligands with cell-specific and dissociated activity. Mol Endocrinol 20(12):3093–3104

180. Yamamoto H, Shevde NK, Warrier A et al (2003) 2-Methylene-19-nor-(20S)-1,25-dihydroxyvitamin D_3 potently stimulates gene-specific DNA binding of the vitamin D receptor in osteoblasts. J Biol Chem 278(34):31756–31765

181. Peleg S, Sastry M, Collins ED (1995) Distinct conformational changes induced by 20-epi analogues of 1α,25-dihydroxyvitamin D_3 are associated with enhanced activation of the vitamin D receptor. J Biol Chem 270(18):10551–10558

182. Christakos S, Norman AW (1979) Studies on the mode of action of calciferol. XVIII. Evidence for a specific high affinity binding protein for 1,25 dihydroxyvitamin D_3 in chick kidney and pancreas. Biochem Biophys Res Commun 89(1):56–63

183. Veldman CM, Cantorna MT, DeLuca HF (2000) Expression of 1,25-dihydroxyvitamin D_3 receptor in the immune system. Arch Biochem Biophys 374(2):334–338

184. Evans KN, Bulmer JN, Kilby MD et al (2004) Vitamin D and placental-decidual function. J Soc Gynecol Investig 11(5):263–271

185. Merke J, Milde P, Lewicka S et al (1989) Identification and regulation of 1,25-dihydroxyvitamin D_3 receptor activity and biosynthesis of 1,25-dihydroxyvitamin D_3. Studies in cultured bovine aortic endothelial cells and human dermal capillaries. J Clin Invest 83(6):1903–1915

186. Perez A, Raab R, Chen TC (1996) Safety and efficacy of oral calcitriol (1,25-dihydroxyvitamin D_3) for the treatment of psoriasis. Br J Dermatol 134(6):1070–1078

187. Pèrez A, Chen TC, Turner A (1996) Efficacy and safety of topical calcitriol (1,25-dihydroxyvitamin D_3) for the treatment of psoriasis. Br J Dermatol 134(2):238–246

188. Yang S, Smith C, Prahl JM et al (1993) Vitamin D deficiency suppresses cell-mediated immunity in vivo. Arch Biochem Biophys 303(1):98–106

189. Yang S, Smith C, DeLuca HF (1993) 1α,25-Dihydroxyvitamin D_3 and 19-nor-1α,25-dihydroxyvitamin D_2 suppress immunoglobulin production and thymic lymphocyte proliferation in vivo. Biochim Biophys Acta 1158(3):279–286

190. Niino M, Fukazawa T, Kikuchi S (2008) Therapeutic potential of vitamin D for multiple sclerosis. Curr Med Chem 15:499–505

191. DeLuca HF, Cantorna MT (2001) Vitamin D: its role and uses in immunology. FASEB J 15(14): 2569–2585

192. Tai K, Need AG, Horowitz M, Chapman IM (2008) Vitamin D, glucose, insulin, and insulin sensitivity. Nutrition 24:269–285

193. Abe J, Nakamura K, Takita Y (1990) Prevention of immunological disorders in MRL/l mice by a new synthetic analogue of vitamin D_3: 22-oxa-1α,25-dihydroxyvitamin D_3. J Nutr Sci Vitaminol (Tokyo) 6(1):21–31

194. Deeb K, Trump DL, Johnson CS (2007) Vitamin D signaling pathways in cancer: potential for anticancer therapeutics. Nat Rev Cancer 7(9):684–700

195. Garland CF, Gorham ED, Mohr SB (2007) Vitamin D and prevention of breast cancer: pooled analysis. J Steroid Biochem Mol Biol 103(3–5):708–711

196. Gorham ED, Garland CF, Garland FC (2007) Optimal vitamin D status for colorectal cancer prevention: a quantitative meta analysis. Am J Prev Med 32(3):210–216

197. Munger KL, Levin LI, Hollis BW (2007) Elevated serum 25-hydroxyvitamin D predicts a decreased risk of MS. Mult Scler 13:280–307

198. Giovannucci E, Liu Y, Hollis BW et al (2008) 25-Hydroxyvitamin D and risk of myocardial infarction in men: a prospective study. Arch Intern Med 168(11):1174–1180

199. Sayre RM, Dowdy JC, Shepherd JG (2007) Reintroduction of a classic vitamin D ultraviolet source. J Steroid Biochem Mol Biol 103(3–5):686–688

200. Rajakumar K, Greenspan SL, Thomas SB et al (2007) SOLAR ultraviolet radiation and vitamin D a historical perspective. Am J Public Health 97(10):1746–1754

201. Lim HW, Carucci JA, Spencer JM et al (2007) Commentary: A responsible approach to maintaining adequate serum vitamin D levels. J Am Acad Dermatol 57:594–595

202. Rosenstreich S, Rich C, Volwiler W (1971) Deposition in and release of vitamin D_3 from body fat: evidence for a storage site in the rat. J Clin Invest 50:679–687

203. Vieth R (1990) The mechanisms of vitamin D toxicity. Bone Miner 11(3):267–272

204. Shepard RM, DeLuca HF (1980) Determination of vitamin D and its metabolites in plasma. Methods Enzymol 67:393–413

205. MacLaughlin J, Holick MF (1985) Aging decreases the capacity of human skin to produce vitamin D_3. J Clin Invest 76(4):1536–1538

206. Aksnes L, Rodland O, Aarskog D (1988) Serum levels of vitamin D_3 and 25-hydroxyvitamin D_3 in elderly and young adults. Bone Min 3:351–357

207. Clemens TL, Adams JS, Henderson SL et al (1981) Increased skin pigment reduces the capacity of skin to synthesise vitamin D_3. Lancet 1(8263):74–76

208. Matsuoka LY, Wortsman J, Haddad JG et al (1990) Skin types and epidermal photosynthesis of vitamin D_3. J Am Acad Dermatol 23:525–526

209. Matsuoka LY, Wortsman J, Haddad JG et al (1991) Racial pigmentation and the cutaneous synthesis of vitamin D. Arch Dermatol 127:536–538

210. Liel Y, Ulmer E, Shary J et al (1988) Low circulating vitamin D in obesity. Calcif Tissue Int 43:199–201

211. Wortsman J, Matsuoka LY, Chen TC et al (2000) Decreased bioavailability of vitamin D in obesity. Am J Clin Nutr 72:690–693 (Erratum: Am J Clin Nutr 2003;77:1342.)

212. Matsuoka LY, Ide L, Wortsman J et al (1987) Sunscreens suppress cutaneous vitamin D_3 synthesis. J Clin Endocrinol Metab 64:1165–1168

213. Loré F, Di Cairano G, Periti P et al (1982) Effect of the administration of 1,25-dihydroxyvitamin D_3 on serum levels of 25-hydroxyvitamin D in postmenopausal osteoporosis. Calcif Tissue Int 34: 539–541

214. Baran DT, Milne ML (1983) 1,25 Dihydroxyvitamin D-induced inhibition of [^{3}H]-25 hydroxyvitamin D production by the rachitic rat liver in vitro. Calcif Tissue Int 35(4–5):461–464

215. Bell NH, Shaw S, Turner RT (1984) Evidence that 1,25-dihydroxyvitamin D$_3$ inhibits the hepatic production of 25-hydroxyvitamin D in man. J Clin Invest 74:1540–1544

216. Halloran BP, Bikle DD, Levens MJ et al (1986) Chronic 1,25-dihydroxyvitamin D$_3$ administration in the rat reduces the serum concentration of 25-hydroxyvitamin D by increasing metabolic clearance rate. J Clin Invest 78:622–628

217. Berlin T, Björkhem I (1987) On the regulatory importance of 1,25-dihydroxyvitamin D$_3$ and dietary calcium on serum levels of 25-hydroxyvitamin D$_3$ in rats. Biochem Biophys Res Commun 144(2):1055–1058

218. Halloran BP, Castro ME (1989) Vitamin D kinetics in vivo: effect of 1,25-dihydroxyvitamin D administration. Am J Physiol 256:E686–E691

219. Hahn TJ, Birge SJ, Scharp CR et al (1972) Phenobarbital-induced alterations in vitamin D metabolism. J Clin Invest 51(4):742–748

220. Hahn TJ, Hendin BA, Scharp CR et al (1972) Effect of chronic anticonvulsant therapy on serum 25-hydroxycalciferol levels in adults. N Engl J Med 287(18):900–904

221. Hahn TJ, Hendin BA, Scharp CR (1975) Serum 25-hydroxycalciferol levels and bone mass in children on chronic anticonvulsant therapy. N Engl J Med 292:550–554

222. Stamp TCB, Round JM, Rowe DJF et al (1972) Plasma levels and therapeutic effect of 25-hydroxycholecalciferol in epileptic patients taking anticonvulsant drugs. Br Med J 4:9–12

223. Bouillon R, Reynaert J, Claes JH (1975) The effect of anticonvulsant therapy on serum levels of 25-hydroxy-vitamin D$_3$ calcium, and parathyroid hormone. J Clin Endocrinol Metab 41:1130–1135

224. Jubitz W, Haussler MR, McCain TA (1977) Plasma 1,25-dihydroxyvitamin D levels in patients receiving anticonvulsant drugs. J Clin Endocrinol Metab 44(4):617–621

225. Gascon-Barré M, Delvin EE, Glorieux FH et al (1981) Influence of vitamin D$_3$ status, phenobarbital, and diphenylhydantoin treatment on the plasma 25-hydroxyvitamin D$_3$ concentrations in the rat. Can J Physiol Pharmacol 59(10):1073–1081

226. Sambrook P (2005) Glucocorticoids and vitamin D. In: Feldman D, Pike JW, Glorieux FH, (eds) Vitamin D, vol 73, 2nd edn. Elsevier Academic Press, San Diego, CA, pp 1239–1251

227. Preece MA, Tomlinson S, Ribot CA et al (1975) Studies of vitamin D deficiency in man. Quart J Med, New Series XLIV(176):575–589

228. Baker MR, Peacock M, Nordin BEC (1980) The decline in vitamin D status with age. Age Ageing 9:249–252

229. Omdahl JL, Garry PJ, Hunsaker LA (1982) Nutritional status in a healthy elderly population: vitamin D. Am J Clin Nutr 36:1225–1233

230. Need AG, Morris HA, Horowitz M et al (1993) Effects of skin thickness, age, body fat, and sunlight on serum 25-hydroxyvitamin D. Am J Clin Nutr 58:882–885

231. Preece MA, Ford JA, McIntosh WB (1973) Vitamin D deficiency among Asian immigrants to Britain. Lancet i(7809):907–910

232. Bell NH, Greene A, Epstein S et al (1985) Evidence of alteration of the vitamin D-endocrine system in blacks. J Clin Invest 76:470–473

233. Pietrek J, Kokot F, Kuska J (1978) Kinetics of serum 25-hydroxyvitamin D in patients with acute renal failure. Am J Clin Nutr 31:1919–1926

234. Hidiroglou M, Williams CJ, Ivan M (1979) Pharmacokinetics and amounts of 25-hydroxycholecalciferol in sheep affected by osteodystrophy. J Dairy Sci 62:567–571

235. Khamiseh G, Vaziri ND, Oveisi F (1991) Vitamin D absorption, plasma concentration and urinary excretion of 25-hydroxyvitamin D in nephritic syndrome. Proc Soc Exp Biol Med 196:210–213

236. Fox J, Della-Santina CP (1989) Oral verapamil and calcium and vitamin D metabolism in rats: effect of dietary calcium. Am J Physiol 257:E632–E638

237. Clements MR, Johnson L, Fraser DR (1987) A new mechanism for induced vitamin D deficiency in calcium deprivation. Nature 325:62–65

238. Vieth R, Fraser D, Kooh SW (1987) Low dietary calcium reduces 25-hydroxycholecalciferol in plasma of rats. J Nutr 117:914–918

239. Dominguez JH, Gray RW, Lemann J Jr. (1976) Dietary phosphate deprivation in women and men: Effects on mineral and acid balances, parathyroid hormone and the metabolism of 25-OH-vitamin D. J Clin Endocrinol Metab 45(5):1056–1068

240. Bell NH, Epstein S, Greene A (1985) Evidence for alteration of the vitamin D-endocrine system in obese subjects. J Clin Invest 76:370–373

241. Compston JE, Vedi S, Ledger JE (1981) Vitamin D status and bone histomorphometry in gross obesity. Am J Clin Nutr 34:2359–32363

242. Hey H, Stokholm KH, Lund BJ (1982) Vitamin D deficiency in obese patients and changes in circulating vitamin D metabolites following jejunoileal bypass. Int J Obes 6:473–479

243. Kubota M, Ohno J, Shiina Y et al (1982) Vitamin D metabolism in pregnant rabbits: Differences between the maternal and fetal response to administration of large amounts of vitamin D_3. Endocrinology 110(6):1950–1956

244. Delvin EE, Gilbert M, Pere MC et al (1988) In vivo metabolism of calcitriol in the pregnant rabbit doe. J Dev Physiol 10:451–459

245. Paulson SK, DeLuca HF, Battaglia F (1987) Plasma levels of vitamin D metabolites in fetal and pregnant ewes. Proc Soc Exp Biol Med 185(3):267–271

246. Paulson SK, Ford KK, Langman CB (1990) Pregnancy does not alter the metabolic clearance of 1,25-dihydroxyvitamin D in rats. Am J Physiol 258:E158–E162

247. Omdahl JL, Jelinek G, Eaton RP (1977) Kinetic analysis of 25-hydroxyvitamin D_3 metabolism in strontium-induced rickets in the chick. J Clin Invest 60:1202–1210

248. Gupta MM, Round JM, Stamp TCB (1974) Spontaneous cure of vitamin-D deficiency in Asians during summer in Britain. Lancet 1(7858):586–588

249. Haddad JG, Stamp TCB (1974) Circulating 25-hydroxyvitamin D in man. Am J Med 57:57–62

250. Stamp TCB, Round JM (1974) Seasonal changes in human plasma levels of 25-hydroxyvitamin D. Nature 247:563–565

251. McLaughlin M, Raggatt PR, Brown DJ et al (1974) Seasonal variations in serum 25-hydroxycholecalciferol in healthy people. Lancet 1(7857):536–538

252. Pettifor JM, Ross FP, Solomon L (1978) Seasonal variation in serum 25-hydroxycholecalciferol concentrations in elderly South African patients with fractures of femoral neck. Br Med J 1(6116): 826–827

253. Hidiroglou M, Proulx JG, Roubos D (1979) 25-Hydroxyvitamin D in plasma of cattle. J Dairy Sci 62:1076–1080

254. Juttmann JR, Visser TJ, Buurman C et al (1981) Seasonal fluctuations in serum concentrations of vitamin D metabolites in normal subjects. Br Med J 282:1349–1352

255. Chesney RW, Rosen JF, Hamstra AJ (1981) Absence of seasonal variation in serum concentrations of 1,25-dihydroxyvitamin D despite a rise in 25-hydroxyvitamin-D in summer. J Clin Endocrinol Metab 53(1):139–142

256. Smith BS, Wright H (1984) Relative contributions of diet and sunshine to the overall vitamin D status of the grazing ewe. Vet Rec 115:537–538

257. Van der Klis FRM, Jonxis JHP, van Doormaal JJ et al (1996) Changes in vitamin-D metabolites and parathyroid hormone in plasma following cholecalciferol administration to pre- and postmenopausal women in the Netherlands in early spring and to postmenopausal women in Curaçao. Br J Nutr 75:637–646

258. O'Leary TJ, Jones G, Yip A et al (1986) The effects of chloroquine on serum 1,25-dihydroxyvitamin D and calcium metabolism in sarcoidosis. N Engl J Med 315(12):727–730

259. Barré PE, Gascon-Barré M, Meakins JL et al (1987) Hydroxychloroquine treatment of hypercalcemia in a patient with sarcoidosis undergoing hemodialysis. Am J Med 82(6): 1259–1262

260. Adams JS, Diz MM, Sharma OP (1989) Effective reduction in the serum 1,25-dihydroxyvitamin D and calcium concentration in sarcoidosis-associated hypercalcemia with short-course chloroquine therapy. Ann Intern Med 111(5):437–438

261. Henry HL (2005) The 25-hydroxyvitamin D 1α-hydroxylase. In: Feldman D, Pike JW, Glorieux FH (eds) Vitamin D, vol 5, 2nd edn. Elsevier Academic Press, San Diego, CA, pp 69–83

262. Baksi SN, Kenny AD (1981) Vitamin D metabolism in Japanese quail: gonadal hormones and dietary calcium effects. Am J Physiol 241(4):E275–E280

263. Tanaka Y, Castillo L, DeLuca HF (1976) Control of renal vitamin D hydroxylases in birds by sex hormones. Proc Natl Acad Sci USA 73(8):2701–2705

264. Haussler MR, Hughes MR, McCain TA et al (1977) 1,25-Dihydroxyvitamin D_3: Mode of action in intestine and parathyroid glands, Assay in humans and isolation of its glycoside from *Solanum Malacoxylon*. Calcif Tissue Res 22(Suppl):1–18

265. Pike JW, Toverud S, Boass A et al (1977) Circulating $1\alpha,25$-$(OH)_2D$ during physiological states of calcium stress. In: Norman A, Schaefer K, Coburn J, DeLuca H, Fraser D, Grigoleit HG, Herrath DV (eds) Vitamin D: biochemical, chemical, and clinical aspects related to calcium metabolism (Proceedings of the third workshop on vitamin D). De Gruyter, New York, pp 187–189

266. Gallagher JC, Riggs BL, Eisman J et al (1979) Intestinal calcium absorption and serum vitamin D metabolites in normal subjects and osteoporotic patients – Effect of age and dietary calcium. J Clin Invest 64(3):729–736

267. Chesney RW, Rosen JF, Hamstra AJ et al (1980) Serum 1,25-dihydroxyvitamin D levels in normal children and in vitamin D disorders. Am J Dis Child 134(2):135–139

268. Lund B, Clausen N, Lund B et al (1980) Age-dependent variations in serum 1,25-dihydroxyvitamin D in childhood. Acta Endocrinol 94:426–429

269. Seino Y, Shimotsuji T, Yamaoka K et al (1980) Plasma 1,25-dihydroxyvitamin D concentrations in cords, newborns, infants, and children. Calcif Tissue Int 30:1–3

270. Gray RW (1981) Effects of age and sex on the regulation of plasma 1,25-$(OH)_2D$ by phosphorus in the rat. Calcif Tissue Int 33(5):477–484

271. Gray RW, Gambert SR (1982) Effect of age on plasma 1,25-$(OH)_2$ vitamin D in the rat. Age 5(2): 54–56

272. Manolagas SC, Culler FL, Howard JE et al (1983) The cytoreceptor assay for 1,25-dihydroxyvitamin D and its application to clinical studies. J Clin Endcrinol Metab 56:751–760

273. Armbrecht HJ, Forte LR, Halloran BP (1984) Effect of age and dietary calcium on renal 25(OH)D metabolism, serum 1,25$(OH)_2D$, and PTH. Am J Physiol 246:E266–E270

274. Epstein S, Bryce G, Hinman JW et al (1986) The influence of age on bone mineral regulating hormones. Bone 7:421–425

275. Buchanan JR, Myers CA, Greer RBIII (1988) Effect of declining renal function on bone density in aging women. Calcif Tissue Int 43:1–6

276. Fox J (1990) Production and metabolic clearance rates of 1,25-dihydroxyvitamin D_3 during maturation in rats: Studies using a rapid, primed-infusion technique. Horm Metab Res 22:278–282

277. Glass AR, Eil C (1986) Ketoconazole-induced reduction in serum 1,25-dihydroxyvitamin D. J Clin Endocrinol Metab 63(3):766–769

278. Glass AR, Eil C (1988) Ketoconazole-induced reduction in serum 1,25-dihydroxyvitamin D and total serum calcium in hypercalcemic patients. J Clin Endocrinol Metab 66(5):934–938

279. Saggese G, Bertelloni S, Baroncelli GI et al (1993) Ketoconazole decreases the serum ionized calcium and 1,25-dihydroxyvitamin D_3 levels in tuberculosis-associated hypercalcemia. Am J Dis Child 147(3):270–273

280. Boyle IT, Gray RW, DeLuca HF (1971) Regulation by calcium of in vivo synthesis of 1,25-dihydroxycholecalciferol and 21,25-dihydroxycholecalciferol. Proc Natl Acad Sci USA 68(9): 2131–2134

281. Morrissey RL, Wasserman RH (1971) Calcium absorption and calcium-binding protein in chicks on differing calcium and phosphorus intakes. Am J Physiol 220(5):1509–1515

282. Haussler MR, Baylink DJ, Hughes MR (1976) The assay of $1\alpha,25$-dihydroxyvitamin D_3: Physiologic and pathologic modulation of circulating hormone levels. Clin Endocrinol 5: 151s–165s

283. Hughes MR, Baylink DJ, Jones PG et al (1976) Radioligand receptor assay for 25-hydroxyvitamin D_2/D_3 and $1\alpha,25$-dihydroxyvitamin D_2/D_3. J Clin Invest 58:61–70

284. Taylor CM, Caverzasio J, Jung A (1983) Unilateral nephrectomy and 1,25-dihydroxyvitamin D_3. Kidney Int 24:37–42

285. Fox J, Ross R (1985) Effects of low phosphorus and low calcium diets on the production and metabolic clearance rates of 1,25-dihydroxycholecalciferol in pigs. J Endocr 105:169–173

286. Paulson SK, Kenny AD (1985) Effect of dietary mineral and vitamin D content and parathyroidectomy on the plasma disappearance rate of 1,25-dihydroxyvitamin D_3 in rats. Biopharm Drug Dispos 6:359–372

287. Jongen MJ, Bishop JE, Cade C et al (1987) Effect of dietary calcium, phosphate and vitamin D deprivation on the pharmacokinetics of 1,25-dihydroxyvitamin D_3 in the rat. Horm Metab Res 19: 481–485

288. Baxter LA, DeLuca HF (1976) Stimulation of 25-hydroxyvitamin D_3-1α-hydroxylase by phosphate depletion. J Biol Chem 251(10):3158–3161

289. Gray RW, Wilz DR, Caldas AE et al (1977) The importance of phosphate in regulating plasma 1,25-$(OH)_2$-vitamin D levels in humans: Studies in healthy subjects, in calcium-stone formers and in patients with primary hyperparathyroidism. J Clin Endocrinol Metab 45:299–306

290. Gray RW, Garthwaite TL, Phillips LS (1983) Growth hormone and triiodothyronine permit an increase in plasma 1,25$(OH)_2$D concentrations in response to dietary phosphate deprivation in hypophysectomized rats. Calcif Tissue Int 35:100–106

291. Llach F, Massry SG (1985) On the mechanism of secondary hyperparathyroidism in moderate renal insufficiency. J Clin Endocrinol Metab 61:601–606

292. Rader JI, Baylink DJ, Hughes MR et al (1979) Calcium and phosphorus deficiency in rats: effects on PTH and 1,25-dihydroxyvitamin D_3. Am J Physiol 236(2):E118–E122

293. Insogna KL, Broadus AE, Gertner JM (1983) Impaired phosphorus conservation and 1,25 dihydroxyvitamin D generation during phosphorus deprivation in familial hypophosphatemic rickets. J Clin Invest 71:1561–1569

294. Lufkin EG, Kumar R, Heath HIII (1983) Hyperphosphatemic tumoral calcinosis: Effects of phosphate depletion on vitamin D metabolism, and of acute hypocalcemia on parathyroid hormone secretion and action. J Clin Endocrinol Metab 56(6):1319–1322

295. Maierhofer WJ, Gray RW, Lemann J Jr. (1984) Phosphate deprivation increases serum 1,25-$(OH)_2$-vitamin D concentrations in healthy men. Kidney Int 25:571–575

296. Portale AA, Booth BE, Halloran BP et al (1984) Effect of dietary phosphorus on circulating concentrations of 1,25-dihydroxyvitamin D and immunoreactive parathyroid hormone in children with moderate renal insufficiency. J Clin Invest 73:1580–1589

297. Portale AA, Halloran BP, Murphy MM et al (1986) Oral intake of phosphorus can determine the serum concentration of 1,25-dihydroxyvitamin D by determining its production rate in humans. J Clin Invest 77:7–12

298. Halloran BP, Barthell EN, DeLuca HF (1979) Vitamin D metabolism during pregnancy and lactation in the rat. Proc Natl Acad Sci USA 76(11):5549–5553

299. Kumar R, Cohen WR, Silva P et al (1979) Elevated 1,25-dihydroxyvitamin D plasma levels in normal human pregnancy and lactation. J Clin Invest 63:342–344

300. Steichen JJ, Tsang RC, Gratton TL et al (1980) Vitamin D homeostasis in the perinatal period: 1,25-dihydroxyvitamin D in maternal, cord, and neonatal blood. N Engl J Med 302(6):315–319

301. Wieland P, Fischer JA, Trechsel U et al (1980) Perinatal parathyroid hormone, vitamin D metabolites, and calcitonin in man. Am J Physiol 239(5):E385–E390

302. Mawer EB, Backhouse J, Hill LF et al (1975) Vitamin D metabolism and parathyroid function in man. Clin Sci Mol Med 48:349–365

303. Kaplan RA, Haussler MR, Deftos LJ et al (1977) The role of 1α,25-dihydroxyvitamin D in the mediation of intestinal hyperabsorption of calcium in primary hyperparathyroidism and absorptive hypercalciuria. J Clin Invest 59:756–760

304. Lambert PW, Hollis BW, Bell NH et al (1980) Demonstration of a lack of change in serum 1α,25-dihydroxyvitamin D in response to parathyroid extract in pseudohypoparathyroidism. J Clin Invest 66:782–791

305. Piel CF, Doorf BS, Avioli LV (1973) Metabolism of tritiated 25-hydroxycholecalciferol in chronically uremic children before and after successful renal homotransplantation. J Clin Endocrinol Metab 37:944–948

306. Eisman JA, Hamstra AJ, Kream BE et al (1976) A sensitive, precise, and convenient method for determination of 1,25-dihydroxyvitamin D in human plasma. Arch Biochem Biophys 176(1): 235–243

307. Christiansen C, Christensen MS, Melsen F et al (1981) Mineral metabolism in chronic renal failure with specific reference to serum concentration of $1,25(OH)_2D$ and $24,25(OH)_2D$. Clin Nephrol 15(1):18–22

308. Juttmann JR, Buurman CJ, De Kam E et al (1981) Serum concentrations of metabolites of vitamin D in patients with chronic renal failure (CRF). Consequences for the treatment with 1α-hydroxy derivatives. Clin Endocrinol (Oxf) 14(3):225–236

309. Papapoulos SE, Clemens TL, Sandler LM et al (1982) The effect of renal function on changes in circulating concentrations of 1,25-dihydroxycholecalciferol after an oral dose. Clin Sci 62:427–429

310. Pitts TO, Piraino BH, Mitro R (1988) Hyperparathyroidism and 1,25-dihydroxyvitamin D deficiency in mild, moderate, and severe renal failure. J Clin Endocrinol Metab 67:876–881

311. Dusso A, Lopez-Hilker S, Lewis-Finch J et al (1989) Metabolic clearance rate and production rate of calcitriol in uremia. Kidney Int 35:860–864

312. Patel S, Simpson RU, Hsu CH (1989) Effect of vitamin D metabolites on calcitriol metabolism in experimental renal failure. Kidney Int 36:234–239

313. Portale AA, Booth BE, Tsai HC et al (1982) Reduced plasma concentration of 1,25-dihydroxyvitamin D in children with moderate renal insufficiency. Kidney Int 21:627–643

314. Wilson L, Felsenfeld A, Drezner MK et al (1985) Altered divalent ion metabolism n early renal failure: Role of $1,25(OH)_2D$. Kidney Int 27:565–573

315. St. John A, Thomas MB, Davies CP et al (1992) Determinants of intact parathyroid hormone and free 1,25-dihydroxyvitamin D levels in mild and moderate renal failure. Nephron 61:422–427

316. Salusky IB, Goodman WG, Horst R et al (1990) Pharmacokinetics of calcitriol in continuous ambulatory and cycling peritoneal dialysis patients. Am J Kidney Dis XVI(2):126–132

317. Fox J (1988) Verapamil induces PTH resistance but increases duodenal calcium absorption in rats. Am J Physiol 255:E702–E707

318. Avioli LV, Lee SW, McDonald JE et al (1967) Metabolism of vitamin D_3 ^3H in human subjects – Distribution in blood, bile, feces, and urine. J Clin Invest 46(6):983–992

319. Gray RW, Weber HP, Dominguez JH et al (1974) The metabolism of vitamin D_3 and 25-hydroxyvitamin D_3 in normal and anephric humans. J Clin Endocrinol Metab 39:1045–1056

320. Arnaud SB, Goldsmith RS, Lambert PW et al (1975) 25-Hydroxyvitamin D_3: Evidence of an entero-hepatic circulation in man. Proc Soc Exp Biol Med 149:570–572

321. Norman AW, DeLuca HF (1963) The preparation of $[^3H]$-vitamin D_2 and D_3 – Their localization in the rat. Biochemistry 2:1160–1168

322. Haddad JG Jr, Boisseau V, Avioli LV (1971) Placental transfer of vitamin D_3 and 25-hydroxycholecalciferol in the rat. J Lab Clin Med 77(6):908–915

323. Rojanasathit S, Haddad JG (1976) Hepatic accumulation of vitamin D_3 and 25-hydroxyvitamin D_3. Biochim Biophys Acta 421:12–21

324. Weisman Y, Vargas A, Duckett G et al (1978) Synthesis of 1,25-dihydroxyvitamin D in the nephrectomized pregnant rat. Endocrinology 103(6):1992–1996

325. Weisman Y, Sapir R, Harell A et al (1976) Maternal-perinatal interrelationships of vitamin D metabolism in rats. Biochim Biophys Acta 428:388–395

326. Dueland S, Pedersen JI, Helgerud P et al (1983) Absorption, distribution, and transport of vitamin D_3 and 25-hydroxyvitamin D_3 in the rat. Am J Physiol 245:E463–E467

327. Noff D, Edelstein S (1978) Vitamin D and its hydroxylated metabolites in the rat. Placental and lacteal transport, subsequent metabolic pathways and tissue distribution. Horm Res 9:292–300

328. Larsson S-E, Lorentzon R (1977) Excretion of active metabolites of vitamin D in urine and bile of the adult rat. Clin Sci Mol Med 53:373–377

329. Stumpf WE, O'Brien LP (1987) Autoradiographic studies with $[^3H]$-1,25 dihydroxyvitamin D_3 in thyroid and associated tissues of the neck region. Histochemistry 87(1):53–58

330. Stumpf WE, Hayakawa N (2007) Salivary glands epithelial and myoepithelial cells are major vitamin D targets. Eur J Drug Metab Pharmacokinet 32(3):123–129

331. Stumpf WE, Sar M, O'Brien LP (1987) Vitamin D sites of action in the pituitary studied by combined autoradiography-immunohistochemistry. Histochemistry 88(1):11–16

332. Frolik CA, DeLuca HF (1973) Stimulation of 1,25-dihydroxycholecalciferol metabolism in vitamin D-deficient rats by 1,25-dihydroxycholecalciferol treatment. J Cin Invest 52(3):543–548

333. Stumpf WE, Sar M, Reid FA et al (1979) Target cells for 1,25-dihydroxyvitamin D_3 in intestinal tract, stomach kidney, skin, pituitary, and parathyroid. Science 206:1188–1190

334. Stumpf WE, Sar M, Narbaitz R et al (1980) Cellular and subcellular localization of 1,25-$(OH)_2$ Vitamin D_3 in rat kidney – Comparison with localization of parathyroid-hormone and estradiol. Proc Natl Acad Sci USA 77(2):1149–1153

335. Stumpf WE, Sar M, Reid FA et al (1981) Autoradiographic studies with [^{3}H]-1,25-$(OH)_2$ vitamin D_3 and [^{3}H]-25-OH-vitamin D_3 in rat parathyroid glands. Cell Tissue Res 221(2):333–338

336. Stumpf WE, Sar M, Clark SA et al (1982) Brain target sites for 1,25-dihydroxyvitamin D_3. Science 215(4538):1403–1405

337. Stumpf WE, Narbaitz R, Huang S et al (1983) Autoradiographic localization of 1,25-dihydroxyvitamin D_3 in rat placenta and yolk sac. Horm Res 18:215–220

338. Sar M, Stumpf WE, DeLuca HF (1980) Thyrotropes in the pituitary are target cells for 1,25 dihydroxy vitamin D_3. Cell Tissue Res 209:161–166

339. Simpson RU, DeLuca HF (1980) Characterization of a receptor-like protein for 1,25-dihydroxyvitamin D_3 in rat skin. Proc Natl Acad Sci USA 77(10):5822–5826

340. Clark SA, Stumpf WE, Sar M (1980) Target cells for 1,25-dihydroxyvitamin D_3 in the pancreas. Cell Tissue Res 209(3):515–520

341. Clark SA, Dame MC, Kim YS et al (1985) 1,25-Dihydroxyvitamin D_3 in teeth of rats and humans: receptors and nuclear localization. Anat Rec 212(3):250–254

342. Narbaitz R, Stumpf W, Sar M (1981) The role of autoradiographic and immunocytochemical techniques in the clarification of sites of metabolism and action of vitamin D. J Histochem Cytochem 29(1):91–100

343. Rhoten WB, Christakos S (1981) Immunocytochemical localization of vitamin D-dependent calcium binding protein in mammalian nephron. Endocrinology 109(3):981–983

344. Gascon-Barré M, Huet PM (1982) Role of the liver in the homeostasis of calciferol metabolism in the dog. Endocrinology 110(2):563–570

345. Merke J, Kreusser W, Bier B (1983) Demonstration and characterization of a testicular receptor for 1,25-dihydroxycholecalciferol in the rat. Eur J Biochem 130(2):303–308

346. Levy FO, Eikvar L, Jutte NHPM (1985) Appearance of the rat testicular receptor for calcitriol (1,25-dihydroxyvitamin D_3) during development. J Steroid Biochem 23(1):51–56

347. Stumpf WE, O'Brien LP (1987) 1,25 $(OH)_2$ vitamin D_3 sites of action in the brain. An autoradiographic study. Histochemistry 87(5):393–406

348. Narbaitz R, Stumpf WE, Sar M et al (1983) Autoradiographic localization of target cells for 1,25-dihydroxyvitamin D_3 in bones from fetal rats. Calcif Tissue Int 35(2):177–182

349. Mawer EB, Lumb GA, Stanbury SW (1969) Long biological half-life of vitamin D_3 and its polar metabolites in human serum. Nature 222:482–483

350. Smith JE, Goodman D (1971) The turnover and transport of vitamin D and of a polar metabolite with the properties of 25-hydroxycholecalciferol in human plasma. J Clin Invest 50:2159–2167

351. Ponchon G, DeLuca HF (1969) Ethanol-induced artifacts in the metabolism of [^{3}H]-vitamin D_3. Proc Soc Exp Biol Med 131:727–731

352. Brouwer DA, van Beek J, Ferwerda H et al (1998) Rat adipose tissue rapidly accumulates and slowly releases an orally-administered high vitamin D dose. Br J Nutr 79(6):527–532

353. Bec P, Bayard F, Louvet JP (1972) 25-Hydroxycholecalciferol dynamics in human plasma. Rev Europ Etudes Clin Et Biol XVII:793–796

354. Batchelor AJ, Compston JE (1983) Reduced plasma half-life of radio-labeled 25-hydroxyvitamin D_3 in subjects receiving a high-fibre diet. Br J Nutr 49:213–216

355. Davie MW, Lawson DEM, Emberson C (1982) Vitamin D from skin: contribution to vitamin D status compared with oral vitamin D in normal and anticonvulsant-treated subjects. Clin Sci 63:461–472

356. Clements MR, Davies M, Hayes ME (1991) The role of 1,25-dihydroxyvitamin D in the mechanism of acquired vitamin D deficiency. Clin Endocrinol 37(1):17–27

357. Vicchio D, Yergey A, O'Brien K (1993) Quantification and kinetics of 25-hydroxyvitamin D₃ by isotope dilution liquid chromatography/thermospray mass spectrometry. Biol Mass Spectrometry 22: 53–58

358. Haddad JG Jr, Rojanasathit S (1976) Acute administration of 25-hydroxycholecalciferol in man. J Clin Endocrinol Metab 42:284–290

359. Salusky IB, Goodman WG, Horst R (1990) Pharmacokinetics of calcitriol in continuous ambulatory and cycling peritoneal dialysis patients. Am J Kidney Dis XVI(2):126–132

360. Vieth R, Kooh SW, Balfe JW (1990) Tracer kinetics and actions of oral and intraperitoneal 1,25-dihydroxyvitamin D₃ administration in rats. Kidney Int 38:857–861

361. Mawer EB, Backhouse J, Davies M et al (1971) Metabolic fate of administered 1,25-dihydroxycholecalciferol in controls and in patients with hypoparathyroidism. Lancet i:1203–1206

362. Salusky I, Goodman WG, Horst R et al (1988) Plasma kinetics of intravenous calcitriol in normal and dialysed subjects and acute effect on serum PTH levels. In: Norman A, Schaefer K, Grigoleti HG, Herrath DV (eds) Vitamin D: molecular, cellular, and clinical endocrinology (Proceedings of the seventh workshop on vitamin D). De Gruyter, New York, pp 781–782

363. Levine BS, Song M (1996) Pharmacokinetics and efficacy of pulse oral versus intravenous calcitriol in hemodialysis patients. J Am Soc Nephrol 7:488–496

364. Torregrosa JV, Campistol JM, Más M et al (1996) Usefulness and pharmacokinetics of subcutaneous calcitriol in the treatment of secondary hyperparathyroidism. Nehrol Dial Transplant 11(3):54–57

365. Bianchi ML, Ardissino GL, Schmitt CP et al (1999) No difference in intestinal strontium absorption after an oral or an intravenous 1,25(OH)₂D₃ bolus in normal subjects. J Bone Miner Res 14: 1789–1795

366. Brandi L, Egfjord M, Olgaard K (2002) Pharmacokinetics of 1,25(OH)₂D₃ and 1α(OH)D₃ in normal and uraemic men. Nephrol Dial Transplant 17(5):829–842

367. Fakih MG, Trump D, Muindi JR (2007) A phase I pharmacokinetic and pharmacodynamic study of intravenous calcitriol in combination with oral Gefitinib in patients with advanced solid tumors. Clin Cancer Res 13(4):1216–1223

368. Frolik CA, DeLuca HF (1972) Metabolism of 1,25-dihydroxycholecalciferol in the rat. J Clin Invest 51(11):2900–2906

369. Mason RS, Lissner D, Posen S (1980) Blood concentrations of dihydroxylated vitamin D metabolites after an oral dose. Br Med J 280:449–450

370. Ohno J, Kubota M, Hirasawa Y et al (1982) Clinical evaluation of 1α-hydroxycholecalciferol and 1α,25-dihydroxycholecalciferol in the treatment of renal osteodystrophy. In: Norman A, Schaefer K, Herrath DV, Grigoleit HG (eds) Vitamin D, chemical, biochemical and clinical endocrinology of calcium metabolism. W. De Gruyter, New York, pp 847–852

371. Levine BS, Singer FR, Bryce GF et al (1985) Pharmacokinetics and biologic effects of calcitriol in normal humans. J Lab Clin Med 105:349–357

372. Seino Y, Tanaka H, Yamaoka K et al (1987) Circulating 1α,25-dihydroxyvitamin D levels after a single dose of 1α,25-dihydroxyvitamin D₃ or 1α-hydroxyvitamin D₃ in normal men. Bone Miner 2:469–485

373. Kimura Y, Nakayama M, Kuriyama S et al (1991) Pharmacokinetics of active vitamin D3, 1α-hydroxyvitamin D₃ and 1α,25-dihydroxyvitamin D₃ in patients on chronic hemodialysis. Clin Nephrol 35(2):72–77

374. Dechant KL, Goa KL (1994) Calcitriol. A review of its use in the treatment of postmenopausal osteoporosis and its potential in corticosteroid-induced osteoporosis. Drugs Aging 5(4):300–312

375. Beer TM, Munar M, Henner WD (2001) A phase I trial of pulse calcitriol in patients with refractory malignancies. Pulse dosing permits substantial dose escalation. Cancer 91(12):2431–2439

376. Muindi JR, Peng Y, Potter DM et al (2002) Pharmacokinetics of high-dose oral calcitriol: Results from a phase 1 trial of calcitriol and paclitaxel. Clin Pharmacol Ther 72:648–659

377. Selgas R, Martinez M-E, Miranda B et al (1993) The pharmacokinetics of a single dose of calcitriol administered subcutaneously in continuous ambulatory peritoneal dialysis patients. Perit Dial Int 13:122–125

378. Smith DC, Johnson CS, Freeman CC et al (1999) A Phase I trial of calcitriol (1,25-dihydroxycholecalciferol) in patients with advanced malignancy. Clin Cancer Res 5:1339–1345

4 Metabolism and Catabolism of Vitamin D, Its Metabolites and Clinically Relevant Analogs

Glenville Jones

Abstract This chapter discusses the current state of knowledge of vitamin D metabolism and the specific enzymes involved. Vitamin D_3 undergoes a two-step metabolic activation involving sequential hydroxylations at 25- and 1α-carbons by cytochrome P450-based hydroxylases (CYP2R1 and CYP27B1) to give first the main circulating form 25-hydroxyvitamin D_3 and then a hormonally active form 1α,25-dihydroxyvitamin D_3. The plant-derived vitamin D_2 undergoes the same activation steps. This review highlights the recent finding of extra-renal sites of CYP27B1 expression and the physiological implications of this discovery. 1α,25-Dihydroxyvitamin D_3 is inactivated by another cytochrome P450 enzyme (CYP24A1) which produces a series of metabolic products culminating in either a side chain-truncated biliary excretory form, calcitroic acid, or a 26,23-lactone derivative. This chapter also discusses the current knowledge of the metabolism of the clinically relevant analogs of vitamin D, ranging from prodrug forms (e.g. 1(OH)D) that require a step or more of activation to produce a biologically active form to the calcitriol analogs, which are active as administered. Differences between metabolism-sensitive and metabolism-resistant vitamin D analogs are discussed in the context of evaluating the relative importance of analog metabolism in their mechanism of action. The review ends by attempting to predict future directions in the field, focussing on determination of CYP structure, the knowledge gained from mouse CYP knockouts and future vitamin D drug design.

Key Words: Vitamin D metabolism; vitamin D analogs; 25-hydroxyvitamin D; 1α,25-dihydroxyvitamin D; 25-hydroxylase; 1α-hydroxylase; 24-hydroxylase; cytochrome P450

1. METABOLISM OF VITAMIN D_3 AND 25(OH)D_3

The elucidation of the metabolism of vitamin D_3 is arguably one of the most important developments in nutritional sciences over the latter half of the twentieth century. An appreciation that vitamin D_3 represents a precursor to the functionally active form and that two steps of activation are necessary to produce the hormone 1α,25-dihydroxyvitamin D_3 (1,25(OH)$_2D_3$) constitute historical landmarks in modern vitamin research *(1)*. These developments not only spawned detailed studies of the biological properties of vitamin D metabolites produced and the regulation of cytochrome

From: *Nutrition and Health: Vitamin D*
Edited by: M.F. Holick, DOI 10.1007/978-1-60327-303-9_4,
© Springer Science+Business Media, LLC 2010

P450-containing enzymes involved in their production but also provided the stimulus for the chemical synthesis of a plethora of vitamin D analogs (~thousands at last count). Furthermore, it appears that susceptibility to vitamin D catabolic pathways together with other important parameters such as binding to the vitamin D receptor (VDR) functional complex and binding to the vitamin D-binding protein (DBP) are probably key elements in dictating the differences in the actions of so-called calcaemic and non-calcaemic vitamin D analogs. Therefore, from the perspective of its historical significance and relevance, it seems entirely logical to consider the metabolism of vitamin D and its analogs together at this stage of a general text on vitamin D. It should be stated that since the publication of the first edition of this book, there has been significant progress made on the nature of the cytochromes P450 involved in vitamin D metabolism including the release of the first X-crystal structure, namely of CYP2R1.

1.1. 25- and 1α-Hydroxylation

Vitamin D_3 can be synthesized in the skin (see Chapter 1) or be derived from dietary sources. As such, this precursor does not circulate for long in the bloodstream, but, instead, is immediately taken up by adipose tissue or liver for storage. In humans, tissue storage of vitamin D can last for months or even years. Ultimately, the vitamin D_3 undergoes its first step of activation, namely 25-hydroxylation in the liver (Fig. 1). Over

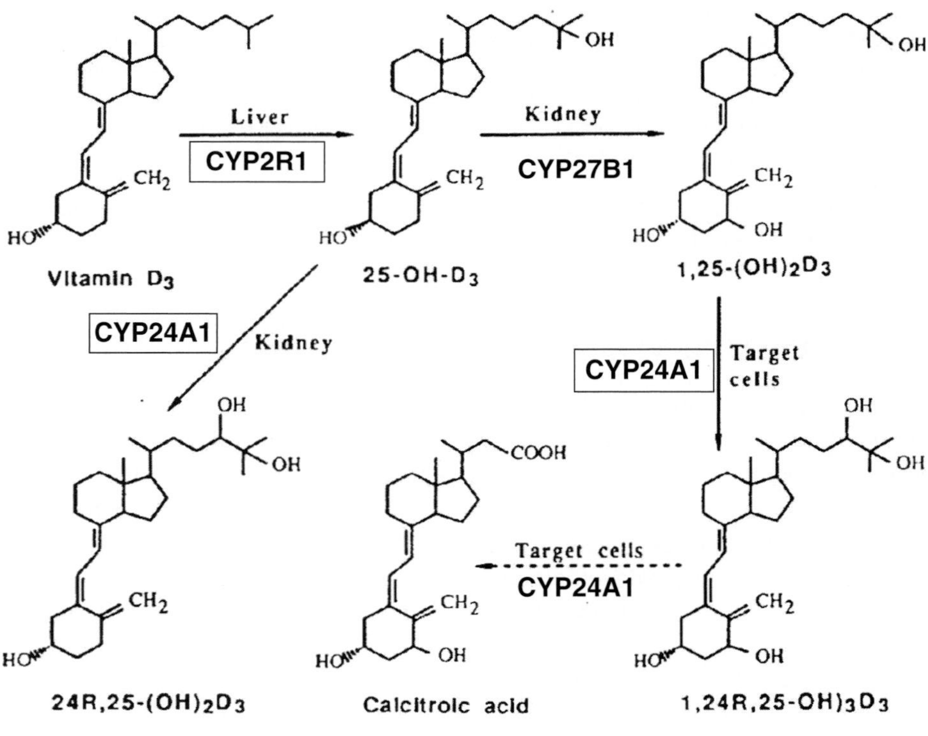

Fig. 1. Metabolism of vitamin D_3.

the years, there has been much controversy over whether 25-hydroxylation is carried out by one enzyme or two and whether this cytochrome P450-based enzyme is found in the mitochondrial or microsomal fractions of liver *(2)*. In the past few years, this has been clarified substantially by work which suggests that the mitochondrial enzyme is CYP27A1 and the microsomal enzyme is CYP2R1. The mitochondrial form CYP27A1 has been purified to homogeneity, subsequently cloned from several species *(3–5)*, and appears to be a bifunctional cytochrome P450 which in addition to synthesis of 25-hydroxylating vitamin D_3 also carries out the side chain hydroxylation of intermediates involved in bile acid biosynthesis. Even though 25-hydroxylation of vitamin D_3 has been clearly demonstrated in cells transfected with CYP27A1, there was continuing scepticism in the vitamin D field that a single cytochrome P450 could explain all the metabolic findings observed over the past two decades of research. These unexplained observations include

(A) Using the perfused rat liver, Fukushima et al. *(6)* demonstrated *two* 25-hydroxylase enzyme activities: a high-affinity, low-capacity form (presumably microsomal) and a low-affinity, high-capacity form (presumably mitochondrial; CYP27A1).

(B) Regulation, albeit weak, of the liver 25-hydroxylase in animals given normal dietary intakes of vitamin D after a period of vitamin D deficiency *(7)* is not explained by a transcriptional mechanism since the gene promoter of CYP27A1 lacks a VDRE.

(C) No obvious $25(OH)D_3$ or $1,25(OH)_2D_3$ deficiency in patients suffering from the genetically inherited disease, cerebrotendinous xanthomatosis, where CYP27A1 is mutated. Although a subset of these patients can suffer from osteoporosis, this is more likely due to biliary defects leading to altered enterohepatic circulation of $25(OH)D$ *(8)*.

(D) CYP27A1 does not appear to 25-hydroxylate vitamin D_2.

Despite the persistent doubts, one thing that the existence of CYP27A1 does explain is the occasional reports of extra-hepatic 25-hydroxylation of vitamin D_3 *(9)*. CYP27A1 mRNA has been detected in a number of extra-hepatic tissues including kidney and bone (osteoblast) *(10, 11)*.

Over the past 5 years, the "missing" liver *microsomal* 25-hydroxylase has been identified as CYP2R1 and cloned from several species *(12)*. Work has shown that it fits the activity profile predicted above in the points A–D. First, CYP2R1 has a high substrate affinity for vitamin D_2, vitamin D_3 and its 1α-hydroxylated counterparts, $1(OH)D_2$ and $1(OH)D_3$ *(12–15)*, making it a strong candidate to be called the physiologically relevant 25-hydroxylase. Second, Chen et al. *(13)* demonstrated that there is a human mutation of CYP2R1 L99P which results in null enzyme activity and in vitamin D-dependent rickets (VDDR type 1a). Lastly, CYP2R1 shows a mainly liver tissue distribution *(12)*. A slightly modified, fully active version of human CYP2R1 was recently crystallized with vitamin D_3 docked in its active site and its X-ray crystal structure determined to 2.7 Å *(15)*. The product of the 25-hydroxylation step, 25-hydroxyvitamin D_3 [$25(OH)D_3$], is the major circulating form of vitamin D_3 and in humans is present in plasma at concentrations in the range 10–40 ng/ml (25–125 nM). The main reason for the stability of this metabolite is its strong affinity for the vitamin D-binding (globulin) protein of blood (DBP) which is around 5×10^{-8} M.

The second step of activation, 1α-hydroxylation, occurs in the kidney *(12)* and the synthesis of the *circulating* hormone in the normal mammal appears to be the *exclusive* domain of that organ. The main evidence for this comes from clinical medicine, where patients with chronic kidney disease gradually deplete their circulating levels of $1,25(OH)_2D_3$ over the five-stage history of their disease and if left untreated go on to develop frank renal osteodystrophy caused by a deficiency of $1,25(OH)_2D_3$ caused by lack of 1α-hydroxylase and secondary hyperparathyroidism. This condition is prevented by $1,25(OH)_2D_3$ administration. In 1997, as the result of a tremendous amount of attention over 25 years of research, the cytochrome P450, CYP27B1, representing the 1α-hydroxylase enzyme was finally cloned from a rat renal cDNA library by St Arnaud's group in Montreal *(16)*. This was rapidly followed by cloning of cDNAs representing mouse and human CYP27B1 *(17–19)* as well as the human and mouse genes *(16, 18–20)*. It had been known for some time that the mitochondrially located 1α-hydroxylase enzyme comprises three proteins: a cytochrome P450, ferredoxin and ferredoxin reductase for activity and is strongly down-regulated by $1,25(OH)_2D_3$ & FGF23 and upregulated by PTH *(21–23)*. It is now evident that the promoter for the CYP27B1 gene contains the transcriptional regulatory elements necessary to explain the observed physiological up-regulation by PTH *(24)* and down-regulation by $1,25(OH)_2D_3$ *(25)*. The human CYP27B1 gene co-localizes to the chromosome 12q14 where vitamin D dependency rickets type 1 (VDDR-Ib), a human disease state, was first proposed to be caused by a mutation of the 1α-hydroxylase 35 years ago *(17, 21, 26, 113)*.

Over the past 20 years, suggestions of extra-renal 1α-hydroxylase expression have gone from incidental findings to a new concept that vitamin D works through an autocrine/paracrine mechanism in addition to the long-standing endocrine mechanism involving circulating calcitriol *(27, 28)*. Extra-renal expression of the 1α-hydroxylase enzyme activity in various tissues (e.g. placenta, bone and skin *(29)*) has been placed on a firmer footing by the development of highly specific antibodies to CYP27B1 protein and PCR-based detection techniques for CYP27B1 mRNA after the cloning of the CYP27B1 gene. Also, Adams and Gacad *(30)* had demonstrated the pathological significance of extra-renal 1α-hydroxylase expression by showing

(a) the existence of a $25\text{-}OH\text{-}D_3\text{-}1\alpha$-hydroxylase in sarcoid tissue in sarcoidosis patients;
(b) that this poorly regulated enzyme results in elevated plasma $1,25(OH)_2D_3$ levels; and
(c) that this causes hypercalciuria and hypercalcaemia in such sarcoidosis patients.

The availability of new specific CYP27B1 tools has confirmed the molecular basis of the disease as an over-expression of CYP27B1 *(31)*. The regulation of this 1α-hydroxylase in macrophages by cytokines and other growth factors has been demonstrated to involve interferon-gamma *(32)*. But the real value of the probes for CYP27B1 has been emergence of the widespread distribution of extra-renal 1α-hydroxylase in normal tissues including osteoblast, keratinocyte, monocyte and colonic, prostatic and breast epithelial cells *(27, 28, 31, 33–35)*. The existence of CYP27B1 in various extra-renal sites around the body implies that the enzyme boosts local $1,25(OH)_2D_3$ production allowing for higher intracellular concentrations of $1,25(OH)_2D_3$ in order to trigger transcriptional regulation of genes unaffected by the plasma concentrations of the hormone *(27, 28, 31)*. However, it should be noted that this hypothesis is currently unproven at the level of measurement of intracellular $1,25(OH)_2D_3$ concentrations.

1.2. 24-Hydroxylation

24-Hydroxylation of both $25(OH)D_3$ and $1,25(OH)_2D_3$ has been shown to occur in vivo *(36, 37)*. The importance of this step has been immersed in controversy since it has been claimed that 24-hydroxylated metabolites might play a role in (a) bone mineralization *(38, 39)* and (b) egg hatchability *(40)*. Experimental evidence favours a different function for 24-hydroxylation, namely *inactivation* of the vitamin D molecule.

This concept comes from four main lines of evidence:

(i)   The levels of $24,25(OH)_2D_3$ do not appear to be regulated, reaching >100 ng/ml in hypervitaminotic animals *(41)*.
(ii) There is no apparent $24,25(OH)_2D_3$ receptor similar to VDR within the orphan class of the steroid receptor superfamily.
(iii) Synthesis of vitamin D analogs blocked with fluorine atoms at the various carbons of the side chain (e.g. $24F_2$-$1,25(OH)_2D_3$) results in molecules with the *full* biological activity of vitamin D in vivo *(42)*.
(iv) 24-Hydroxylation appears to be the first step in a degradatory pathway demonstrable in vitro *(43, 44)* (Fig. 2), which culminates in a biliary excretory form, calcitroic acid, observed in vivo *(45)*.

The 25-OH-D_3-24-hydroxylase was originally characterized as a P450-based enzyme over 35 years ago *(46)* and more recently the cytochrome P450 species purified and

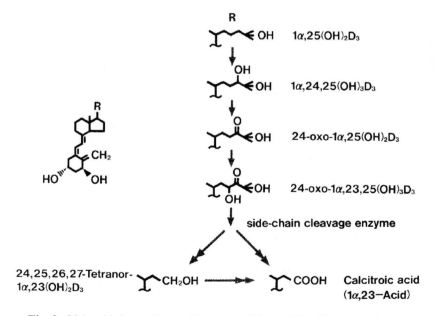

Fig. 2. C24-oxidation pathway. (Reproduced from *(43)* with permission).

cloned by Okuda's group *(47)*. The enzyme appears to 24-hydroxylate both $25(OH)D_3$ and $1,25(OH)_2D_3$, the latter with a 10-fold higher efficiency *(48, 49)*. However, since the circulating level of $25(OH)D_3$ is ~1,000 times higher than $1,25(OH)_2D$, the role of the enzyme in vivo is not clear. The enzyme, particularly the renal form which appears to be expressed at high constitutive levels in the normal animal, may be involved in the inactivation and clearance of excess $25(OH)D_3$ in the circulation *(43)*. On the other hand, the 24-hydroxylase may be involved in target cell destruction of $1,25(OH)_2D_3$. This topic will be discussed further under Section 2.1.

1.3. 26-Hydroxylation and 26,23-Lactone Formation

$25,26-(OH)_2D_3$ was the first dihydroxylated metabolite to be identified back in the late 1960s *(50)* and yet it is still the most poorly understood. The metabolite is readily detectable in the plasma of animals given large doses of vitamin D_3 and it retains strong affinity for DBP *(51)*. However, its biological activity is inferior to other endogenous vitamin D compounds and it is presumed to be a minor catabolite. The knowledge that CYP27A1 is involved in vitamin D_3 activation and that $26(27)(OH)D$ and $1,27(OH)_2D_3$ are formed from vitamin D_3 and $1(OH)D$, respectively, in CYP27A1 transfection systems *(5)* when taken together suggests that 26(27)-hydroxylation may be a consequence of errant side chain hydroxylation.

The most abundant 26-hydroxylated analog appearing in vivo is the 26,23-lactone derivative of $25(OH)D_3$. $25-OH-D_3-26,23$-Lactone accumulates in hypervitaminotic animals in vivo because of its extremely strong affinity for DBP *(51)*. The route of synthesis of this metabolite is depicted in Fig. 3 and research indicates that 26-hydroxylation follows 23-hydroxylation in this process *(52)* and a wealth of evidence *(53, 54)* now suggests that CYP24A1 is responsible for 23- as well as 24-hydroxylation and is also involved in 26,23-lactone formation. Indeed, mutational analysis of human CYP24A1 *(54)* has shown that a A326G modification changes the enzyme from a 24-hydroxylating enzyme synthesizing predominantly calcitroic acid to a 23-hydroxylating enzyme making 26,23-lactone (Fig. 4). The role of $25-OH-D_3-$ 26,23-lactone or its $1,25(OH)_2D_3$ counterpart, which has also been reported *(54, 55)*, is currently unknown but members of the synthetic 26,23-lactone family are strong VDR antagonists *(56)* with potential therapeutic value in Paget's disease *(57)*. Since some species (e.g. opossum, guinea pig) make predominantly 26,23-lactone over calcitroic acid, it is theorized that 26,23-lactone formation represents a backup degradatory pathway for $25(OH)D_3$ and/or $1,25(OH)_2D_3$ or helps to reinforce the inactivation of the vitamin D endocrine system by antagonizing the further action of remaining $1,25(OH)_2D_3$ at the VDR *(54)*.

2. CATABOLISM OF $1,25(OH)_2D_3$

2.1. C-24 Oxidation Pathway to Calcitroic Acid

As described in Section 1.2 above, $1,25(OH)_2D_3$ is a very good substrate for the 24-hydroxylase. Using a variety of cell lines representing specific vitamin D target organs (intestine: CaCo2 cells; osteosarcoma: UMR-106 cells; kidney: LLC-PK1

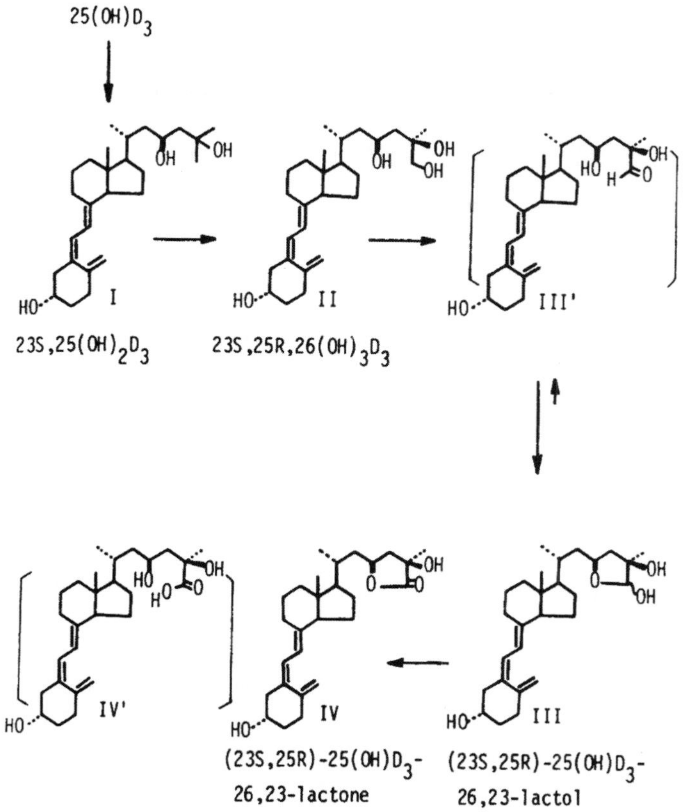

Fig. 3. 26,23-Lactone pathway. The pathway shown is for 25-OH-D$_3$. An analogous pathway exists for 1α,25-(OH)$_2$D$_3$. (Reproduced from *(52)* with permission).

cells; keratinocyte: HPK1A and HPK1A-ras) a number of researchers have shown that 24-hydroxylation is the first step in the C-24 oxidation pathway, a 5-step, vitamin D-inducible, ketoconazole-sensitive pathway which changes the vitamin D molecule to water-soluble truncated products (Figs. 2 and 4) *(43, 44)*. In most biological assays, the intermediates and truncated products of this pathway possess lower or negligible activity. Furthermore, many of these compounds have little or no affinity for DBP making their survival in plasma tenuous at best. The cloning of the cytochrome P450 component (CYP24) of the 24-hydroxylase enzyme has led to detection of CYP24mRNA in a wide range of tissues *(58)*, corroborating the earlier studies reporting widespread 24-hydroxylase enzyme activity in most, if not all, vitamin D target cells *(59)*. Additional studies have shown that mRNA transcripts for CYP24 are virtually undetectable in naive target cells not exposed to 1,25(OH)$_2$D$_3$ but increase dramatically by a VDR-mediated mechanism within hours of exposure to 1,25(OH)$_2$D$_3$ *(58)*. It is therefore attractive to propose that 24-hydroxylation is not only an important step in inactivation of excess 25(OH)D$_3$ in the circulation but also involved in the inactivation of 1,25(OH)$_2$D$_3$ inside target cells. As such one can hypothesize that C-24 oxidation is a target cell attenuation or desensitization process which constitutes a molecular switch

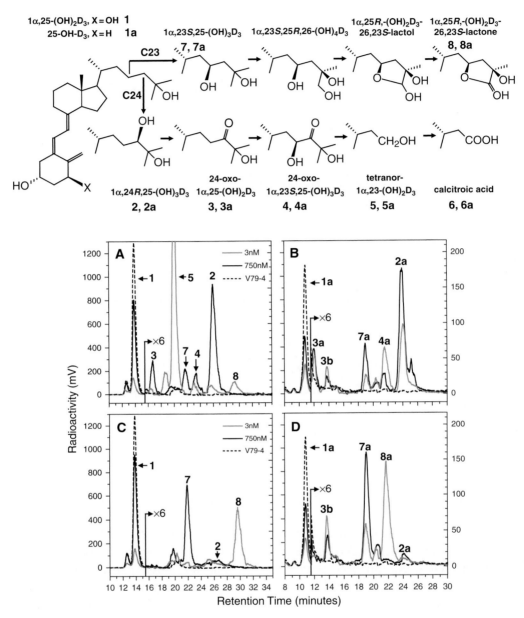

Fig. 4. Effect of A326G mutation on the hydroxylation properties of human CYP24A1. Chromatograms of the lipid extracts of V79 cells expressing either human CYP24A1 (**a** and **b**) or A326G human CYP24A1 (**c** and **d**) incubated with either 25-OH-D$_3$ (**b** and **d**) or 1α,25-(OH)$_2$D$_3$ (**a** and **c**). Note that human CYP24A1 makes mainly 24-hydroxylated metabolites and calcitroic acid whereas A326G human CYP24A1 makes mainly 23-hydroxylated metabolites and 26,23-lactones. The numbered metabolites are shown on the pathways above the chromatograms. (Reproduced from *(54)* with permission).

to turn off vitamin D responses inside target cells *(59)*. The recent development of a CYP24A1 knockout mouse *(60)* resulting in hypercalcaemia, hypercalciuria, nephrocalcinosis and premature death in 50% of null animals seems to support this hypothesis. On the other hand, surviving animals have changes in bone morphology which could suggest an alternative role for 24-hydroxylase in bone mineralization although these cyp24–/– animals are rescued by crossing them with vdr–/– animals, suggesting that abnormal bone is more likely due to $1,25(OH)_2D_3$ toxicity *(60, 61)*. Surviving CYP24 null animals exhibit altered pharmacokinetics of exogenously administered $1,25(OH)_2D_3$ and appear to have much reduced ability to get rid of the hormone, suggesting that the mammal has little in the way of backup systems to catabolize $1,25(OH)_2D_3$ when CYP24A1 is absent *(53)*. Nonetheless, recent work by Thummel's group has shown that $1,25(OH)_2D_3$ is subject to glucuronidation by UGT1A4 *(62)*.

Calcitroic acid, the final product of $1,25(OH)_2D_3$ catabolism, is probably not synthesized in the liver because C-24 oxidation does not occur in hepatoma cells and therefore must presumably be transferred from target cells to liver via some plasma carrier. Though calcitroic acid has been found in various tissues in vivo *(63)*, the nature of any transfer mechanism has not been elucidated.

3. METABOLISM AND CATABOLISM OF THE ANALOGS OF VITAMIN D

3.1. *Activation of Prodrugs*

Prodrugs are synthetic analogs of vitamin D_3 requiring *one or more step(s)* of activation by endogenous enzyme systems before they are biologically active (e.g. *one step*: $1(OH)D_2$, $1(OH)D_3$ or $25(OH)D_3$; *multiple steps*: vitamin D_2 and dihydrotachysterol).

3.2. *Vitamin D_2*

Though vitamin D_2 (for structure see Table 1) can be synthesized naturally by irradiation of ergosterol, little finds its way into the human diet unless it is provided as a dietary supplement. Indeed, vitamin D_2 has been used as a dietary supplement in lieu of vitamin D_3 since the 1930s. Since vitamin D_2 is an artificial form of the vitamin and it is dependent on the same activation steps as vitamin D_3 to become biologically active, one could make a strong case for considering it as a prodrug. In many ways the metabolism of vitamin D_2 is analogous to that of vitamin D_3. For instance, it has been established that vitamin D_2 gives rise to a similar series of metabolites in the form of $25(OH)D_2$ *(64)*, $1,25(OH)_2D_2$ *(65)* and $24,25(OH)_2D_2$ *(66)*. The formation of these metabolites suggests that the enzymes involved in side chain metabolism namely 25-, 1α- and 24-hydroxylases do not discriminate against compounds bearing the vitamin D_2 side chain. However, metabolic studies have also revealed the formation of several additional metabolites including $24(OH)D_2$ *(67)*, $1,24S(OH)_2D_2$ *(68)*, $24,26(OH)_2D_2$ *(69)* and $1,25,28(OH)_3D_2$ *(70)*. Pharmaceutical companies have exploited these subtle differences in the metabolism of vitamin D_2 by synthesizing molecules incorporating the features of the vitamin D_2 side chain, namely the C22=C23 double bond or the C-24 methyl group (see Table 1), into the structure of other analogs (e.g. calcipotriol).

Table 1
Vitamin D Prodrugs

Vitamin D prodrug [ring structure][a]	Side chain structure	Company	Status	Possible target diseases	Mode of delivery	References
1α-OH-D_3 [3]	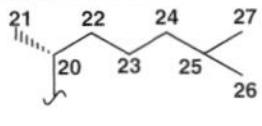	Leo	In use Europe	Osteoporosis	Systemic	Barton *et al.* (85)
1α-OH-D_2 [3]	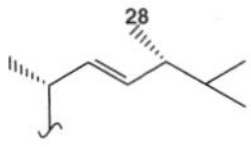	Genzyme	In use USA	Secondary hyperparathyroidism	Systemic	Paaren *et al.* (89)
Dihydrotachysterol [2]	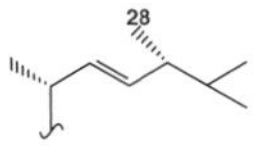	Duphar	Withdrawn	Renal failure	Systemic	Jones *et al.* (33)
Vitamin D_2 [1]	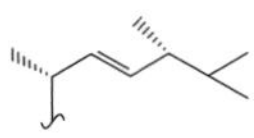	Various	In use USA	Rickets Osteomalacia	Systemic Systemic	Fraser *et al.* (151)
1α-OH-D_5 [3]	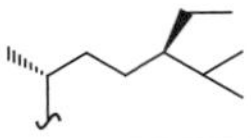	NCI	Clinical trials	Cancer	Systemic Systemic	Mehta *et al.* (160)

[a] Structure of the vitamin D nucleus (secosterol ring structure).

Vitamin D nucleus

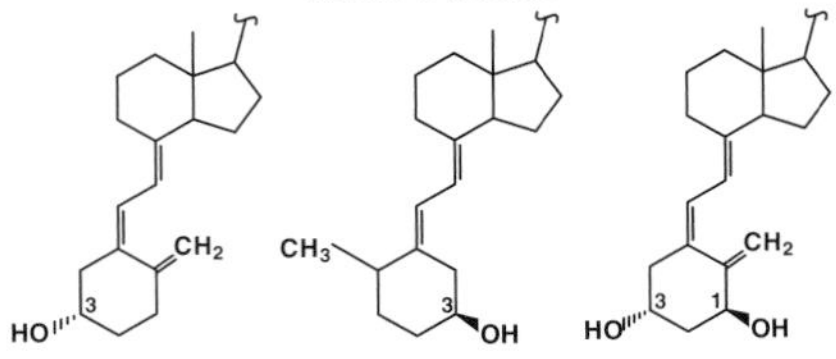

[1]   [2]   [3]

Biologically active vitamin D_2 compounds, such as $1,25(OH)_2D_2$ and $1,24S(OH)_2D_2$, are also subject to further metabolism although it differs from that of $1,25(OH)_2D_3$, essentially because the modifications in the vitamin D_2 side chain prevent the C23C24 cleavage observed during calcitroic acid production. Instead, the principal products are more polar tri- and tetra-hydroxylated metabolites such as $1,24,25(OH)_3D_2$, $1,25,28(OH)_3D_2$ and $1,25,26(OH)_3D_2$ from $1,25(OH)_2D_2$ (70–72) and $1,24,26(OH)_3D_2$ from $1,24S(OH)_2D_2$ (73), all likely produced by the action of CYP24A1. In this latter case, the rate of $1,24S(OH)_2D_2$ metabolism appears slower than that of $1,25(OH)_2D_3$ (73). Some catabolites retain considerable biological activity and at least one, $1,25,28(OH)_3D_2$, is patented for use as a drug.

3.2.1. DIHYDROTACHYSTEROL

This example of a vitamin D prodrug represents the oldest vitamin D analog and was developed in the 1930s as a method of stabilizing the triene structure of one of the photoisomers of vitamin D. The structure of dihydrotachysterol$_2$ shown in Table 1 contains an A-ring rotated through 180°, a reduced C10=C19 double bond and the side chain structure of ergosterol/vitamin D_2. The side chain depicted is that of vitamin D_2 because the clinically approved drug version of dihydrotachysterol is dihydrotachysterol$_2$.

However, it should be noted that dihydrotachysterol$_3$ (DHT$_3$) can also be chemically synthesized with the side chain of vitamin D$_3$. The metabolism of both dihydrotachysterol$_2$ (DHT$_2$) and dihydrotachysterol$_3$ (DHT$_3$) has been extensively studied over the past four decades *(74–77)*. Initial studies performed in the early 1970s showed that DHT is efficiently converted to its 25-hydroxylated metabolite *(78)*.

The effectiveness of DHT to relieve hypocalcaemia of chronic renal failure in the absence of functional renal 1α-hydroxylase led to the hypothesis *(79)* that

> 25-OH-DHT might represent the biologically-active form of DHT, by virtue of its 3ß-hydroxy group being rotated 180° into a "pseudo 1α-hydroxyl position".

It was thus believed that 1α-hydroxylation of 25(OH)DHT was unnecessary. This viewpoint prevailed for at least a decade but debate was renewed when Bosch et al. *(75)* were able to provide evidence for the existence of a mixture of 1α- and 1β-hydroxylated products of 25(OH)DHT$_2$ in the blood of rats dosed with DHT$_2$. Studies involving the perfusion of kidneys from vitamin D-deficient rats with an incubation medium containing 25(OH)DHT$_3$ and using diode array spectrophotometry to analyse the extracts showed this molecule to be subject to extensive metabolism by renal enzymes but failed to give the expected 1-hydroxylated metabolites (Fig. 5), opening up the possibility that the 1α- and 1β-hydroxylated metabolites observed by Bosch et al. *(75)* might be formed by a putative extra-renal 1-hydroxylase activity *(80)*. Following the synthesis of appropriate authentic standards, subsequent research *(77)* has confirmed the in vivo formation and identity of 1α,25(OH)$_2$DHT and 1β,25(OH)$_2$DHT in both rat and human. The ability of these 1α- and 1β-hydroxylated forms of both DHT$_2$ and DHT$_3$ to stimulate a VDRE-inducible growth hormone reporter system exceeded that of 25(OH)DHT and in the process established 1α,25(OH)$_2$DHT and 1β,25(OH)$_2$DHT as the most potent derivatives of DHT identified to date. The importance of the "pseudo 1α-hydroxyl group" hypothesis now stands in question though current findings do not rule out that the biological activity of DHT might be due to the collective action of a group of metabolites including 25(OH)DHT, 1α,25(OH)$_2$DHT and 1β,25(OH)$_2$DHT. The enzymatic origin of these dihydroxylated metabolites is still undetermined despite the availability of recombinant CYP27A1, CYP27B1 and CYP24A1 although work suggests that extra-renal hydroxylases of bone marrow origin might be involved *(80, 81)*.

Though the enzymes involved in the activation of DHT, especially the 1-hydroxylation step, have an altered specificity towards this molecule, the enzymes involved in the catabolism of DHT$_3$ appear to treat the molecule as they would 25(OH)D$_3$ or 1,25(OH)$_2$D$_3$. Side chain hydroxylated derivatives of both 25(OH)DHT$_3$ and 1,25(OH)$_2$DHT$_3$ have been identified and appear to be analogous to intermediates of the C-24 oxidation and 26,23-lactone pathways of vitamin D$_3$ metabolism *(82, 83)*.

3.2.2. 1(OH)D$_2$ AND 1(OH)D$_3$

The prodrug, 1(OH)D$_3$, was developed in the early 1970s *(84, 85)* following the discovery of the hormone, 1α,25-(OH)$_2$D$_3$, and the realization that the kidney was the main site of its synthesis *(21, 22, 86)*. The rationale behind its use was to circumvent the 1α-hydroxylation step involved in vitamin D activation thereby providing a molecule

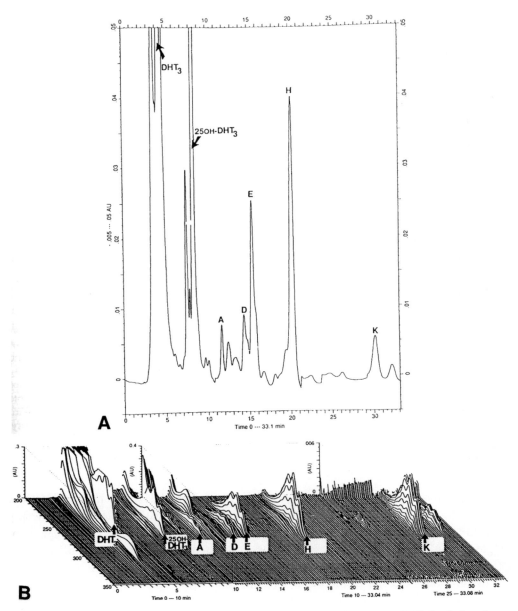

Fig. 5. In vivo metabolism of dihydrotachysterol3 in the rat. Diode array HPLC of the plasma extract of a rat administered 1 mg DHT3 18 h prior to sacrifice. Metabolites are labelled 25-OH-DHT3 and peaks A–L. All possess the distinctive tricuspid UV spectrum (λ_{max} 242.5, 251 and 260.5 nm). Metabolites A–L were subsequently identified as side chain-modified compounds analogous to vitamin D metabolites of the C-24 oxidation and 26,23-lactone pathways depicted in Figs. 1 and 2. (Reproduced from *(76)* with permission).

which could still be activated even in the absence of a functional kidney. It soon became an alternative drug therapy to synthetic 1,25(OH)2D3 in renal osteodystrophy and other hypocalcaemic conditions. Aside from the advantage of a reduced cost of synthesis, 1(OH)D3 offers the potential biological edge of requiring a step of activation in the form of 25-hydroxylation to produce an active molecule. It was believed that the requirement for an activation step might alter the pharmacokinetics of the drug compared to $1,25(OH)_2D_3$, delaying slightly its initial effects and extending its duration of action thereby making the drug less likely to cause acute hypercalcaemia. The 25-hydroxylation of $1(OH)D_3$ was first investigated using the isolated perfused rat liver *(6)* and confirmed that the liver represents the main site of activation. More recent work has confirmed that both CYP27A1 and CYP2R1 are able to 25-hydroxylate $1(OH)D_3$ efficiently *(6, 15)*, and there are no data to support promoter-mediated regulation of these enzymes by $1,25(OH)_2D_3$. It is widely assumed that either 25-hydroxylase enzyme is only loosely regulated and therefore constitutes an insignificant barrier to drug activation. The theoretical advantages of $1(OH)D_3$ over $1,25(OH)_2D_3$ have not materialized in clinical practice *(87)*.

A prodrug based on vitamin D_2 has also been synthesized in the form of $1(OH)D_2$ *(88, 89)*. Although developed as a potential anti-osteoporosis drug and currently approved for the treatment of secondary hyperparathyroidism and renal osteodystrophy as a result of chronic kidney disease *(90, 91)*, this molecule has proved to be a valuable tool in studying hydroxylation reactions in the liver. At low substrate concentrations, $1(OH)D_2$, like $1(OH)D_3$, is 25-hydroxylated by liver hepatomas, Hep3B and HepG2, producing the well-established, biologically active compound $1,25(OH)_2D_2$ *(92)*. However, when the substrate concentration is increased to micromolar values the principal site of hydroxylation of $1(OH)D_2$ becomes the C-24 position, the product being $1,24S(OH)_2D_2$ (Fig. 6), another compound with significant biological activity in several calcaemia and cell proliferation assay systems *(68, 73, 92)*. This metabolite has been previously reported in cows receiving massive doses of vitamin D_2 *(68)*. Transfection studies using the liver cytochrome P450, CYP27A1, expressed in COS-1 cells suggest that $1,24S(OH)_2D_2$ is a product of this cytochrome *(6)*. Whether the formation of this unique metabolic product of $1(OH)D_2$ is the reason for the relative lower toxicity of $1(OH)D_2$ as compared to $1(OH)D_3$ *(93)* has not been established definitively. Recent data *(14, 15)* suggest that CYP2R1 is responsible for the 25-hydroxylation of $1(OH)D_2$ at physiologically relevant concentrations of substrate (Fig. 6).

3.3. Metabolism-Sensitive Analogs

These synthetic analogs of $1,25(OH)_2D_3$ require *no* activation in vivo but are susceptible to attack by catabolic enzyme systems, in most cases rendering them biologically inactive (e.g. calcipotriol, OCT, KH1060).

3.3.1. CYCLOPROPANE RING CONTAINING ANALOGS OF VITAMIN D

These analogs are modified in their side chains such that C-26 is joined to C-27 to give a cyclopropane ring consisting of C-25, C-26 and C-27. The best-known member of this group of compounds is MC 903 or calcipotriol *(94)*, the structure of which is shown in Table 2. In addition to the cyclopropane ring, calcipotriol features a C22=C23 double bond and a 24*S*-hydroxyl group which has been proposed to act as a surrogate C-25 hydroxyl in interactions of the molecule with the VDR. Calcipotriol was the first

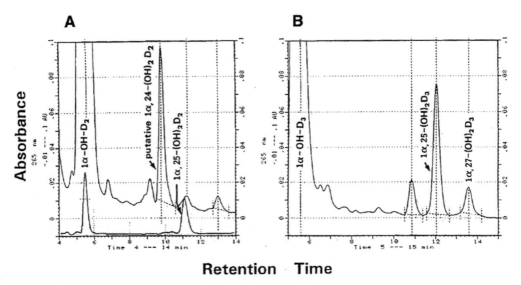

Fig. 6. In vitro metabolism of 1α-OH-D$_2$ and 1α-OH-D$_3$ by the hepatoma Hep3B. (**a**) HPLC trace using diode array detector at 265 nm of an extract of Hep3B cells incubated with 10 μM 1α-OH-D$_2$. The peak at 9.79 min was later conclusively identified by GC-MS and co-migration with authentic standard as 1α,24S-(OH)$_2$D$_2$. Note that inset is a trace of standards: 1α-OH-D$_2$, 5.6 min; 1α,25-(OH)$_2$D$_2$, 11.1 min. (**b**) HPLC trace using diode array detector at 265 nm of an extract of Hep3B cells incubated with 10 μM 1α-OH-D$_3$. Note that the peak at 12.04 min comigrated with authentic 1α,25-(OH)$_2$D$_3$. Subsequent work with standard 1α,26(27)-(OH)$_2$D$_3$ (synthesized by Martin Calverley, Leo Pharmaceuticals) has confirmed its identity. (From *(92)* with permission).

vitamin D analog to be approved for topical use in psoriasis and is currently used world-wide for the successful control of this skin lesion *(95, 96)*.

Pharmacokinetic data acquired for calcipotriol showed that it had a very short $t_{1/2}$, in the order of minutes, results that are consistent with the lack of a hypercalci-uric/hypercalcaemic effect when administered in vivo *(97)*. The first metabolic studies *(98)* revealed that calcipotriol was rapidly metabolized by a variety of different liver preparations from rat, minipig and human to two novel products. These workers *(99)* were able to isolate and identify the two principal products as a C22=C23 unsaturated, 24-ketone (MC1046) and a C22—C23 reduced, 24-ketone (MC1080). These results were confirmed and extended by others *(99)* who showed that calcipotriol metabolism was not confined to liver tissue, but could be carried out by a variety of cells including those cells exposed to topically administered calcipotriol in vivo, namely keratinocytes. Fur-thermore, these workers *(99)* proposed further metabolism of the 24-ketone in these vitamin D target cells to side chain cleaved molecules including calcitroic acid (Fig. 7). The main implications of this work are that calcipotriol is subject to rapid metabolism initially by non-vitamin D-related enzymes, then by vitamin D-related pathways prob-ably including CYP24A1 *(14)* to a side chain cleaved molecule *(99)*. Catabolites are produced in a variety of tissues and appear to have lower biological activity than the parent molecule. Since calcipotriol is administered topically the work suggests that it acts and is broken down locally and may never reach detectable levels in the

Table 2
Analogs of 1α,25-(OH)$_2$D$_3$

Vitamin D analog [ring structure][a]	Side chain structure	Company	Status	Possible target diseases	Mode of delivery	Reference
Calcitriol, 1α,25-(OH)$_2$D$_3$ [3]	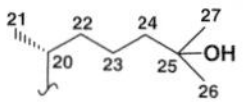	Roche, Duphar	In use worldwide	Hypocalcemia Psoriasis	Systemic Topical	Baggiolini et al. (152)
26,27-F$_6$-1α,25-(OH)$_2$D$_3$ [3]	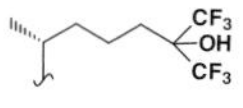	Sumitomo-Taisho	In use Japan	Osteoporosis Hypoparathyroidism	Systemic Systemic	Kobayashi et al. (112)
19-Nor-1α,25-(OH)$_2$D$_2$ [5]	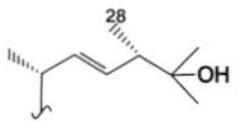	Abbott	In use USA	Secondary hyperparathyroidism	Systemic	Perlman et al. (153)
22-Oxacalcitriol (OCT) [3]	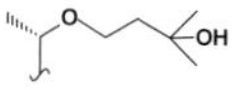	Chugai	In use Japan	Secondary hyperparathyroidism Psoriasis	Systemic Topical	Murayama et al. (100)
Calcipotriol (MC903) [3]	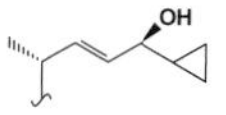	Leo	In use worldwide	Psoriasis Cancer	Topical Topical	Calverley (94)
1α,25-(OH)$_2$-16-ene-23-yne-D$_3$ (Ro 23–7553) [6]		Roche	Pre-clinical	Leukemia	Systemic	Baggiolini et al. (154)
EB1089 [3]	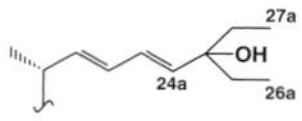	Leo	Clinical trials	Cancer	Systemic	Binderup et al. (121)
20-epi-1α,25-(OH)$_2$D$_3$ [3]	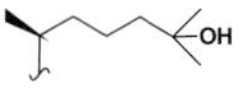	Leo	Pre-clinical	Immune diseases	Systemic	Calverley et al. (155)
2-methylene-19-nor-20-epi-1α,25-(OH)$_2$D$_3$ (2MD) [7]	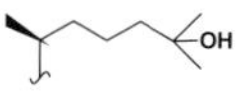	Deltanoids	Pre-clinical	Osteoporosis	Systemic	Shevde et al. (161)
BXL-628 (formerly Ro-269228) [8]	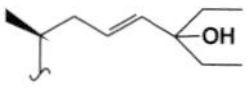	Bioxell	Clinical trials	Prostate Cancer	Systemic	Marchiani et al. (159)
ED71 [4]	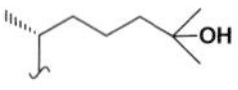	Chugai	Clinical trials	Osteoporosis	Systemic	Nishii et al. (156)
1α,24(S)-(OH)$_2$D$_2$ [3]	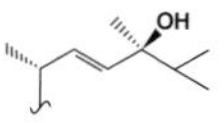	Genzyme	Pre-clinical	Psoriasis	Topical	Strugnell et al. (92)
1α,24(R)-(OH)$_2$D$_3$ (TV-02) [3]	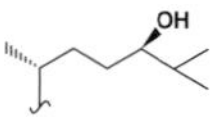	Teijin	In use Japan	Psoriasis	Topical	Morisaki et al. (157)

[a]Structure of the vitamin D nucleus (secosterol ring structure).

Vitamin D nucleus

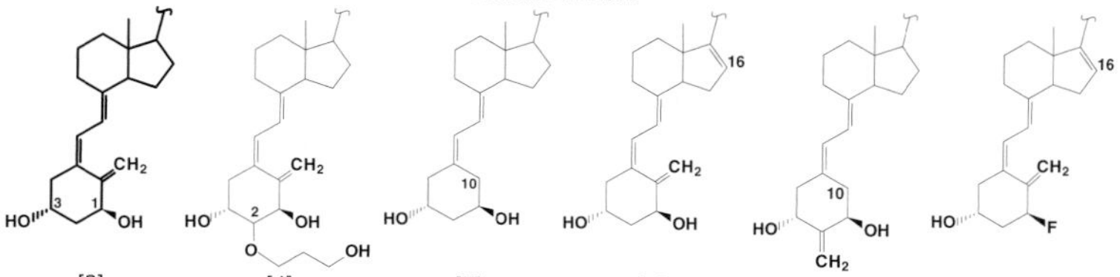

[3] [4] [5] [6] [7] [8]

bloodstream. Should calcipotriol enter the circulation, the ability of liver and target cells to breakdown calcipotriol provides a backup system to prevent hypercalcaemia.

The reduction of the C22=C23 double bond during the earliest phase of calcipotriol catabolism was an unexpected event given that the C22=C23 double bond in vitamin D$_2$

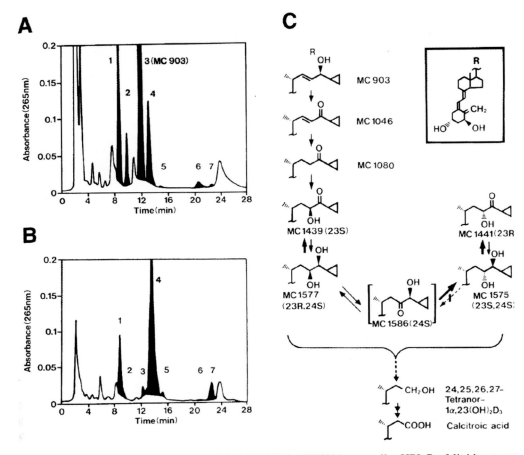

Fig. 7. In vitro metabolism of calcipotriol (MC903) by HPK1A-ras cells. HPLC of lipid extracts following incubation of MC903 with (**a**) HPK1A human keratinocytes and (**b**) HPK1A-ras human keratinocytes. Peak 1 = MC1080; Peak 2 = MC1046; Peak 3 = MC903 (calcipotriol); Peak 4 = mixture of MC1439 and MC1441; Peak 5 = Tetranor-1α,23(OH)$_2$D$_3$; Peak 6 = MC1577; Peak 7 = MC1575. 7C: Proposed Pathway of calcipotriol metabolism in cultured keratinocytes. (Reproduced from *(99)* with permission).

compounds is extraordinarily stable to metabolism. It thus appears that metabolism of calcipotriol provides evidence that the C-24 methyl group in the vitamin D$_2$ side chain must play a stabilizing role, preventing the formation of the 24-ketone which facilitates the reduction of the C22=C23 double bond. However, it is still unknown which enzyme is responsible for this reduction in the side chain of calcipotriol, although it is widely suspected that the extremely versatile CYP24A1 may once again be involved.

3.3.2. OXA-GROUP-CONTAINING ANALOGS

These compounds involve the replacement of a carbon atom (usually in the side chain) with an oxygen atom. The best known of these are the 22-oxa-analogs including 22-oxa-calcitriol (OCT) and KH1060 *(100, 101)*. Both of these molecules are metabolically fascinating to study because *the oxa-atom makes the molecule inherently unstable should it be hydroxylated at the adjacent carbon atom.* The hydroxylation at an adjacent carbon generates an unstable hemi-acetal which spontaneously breaks down to eliminate the carbons distal to the oxa-group. In the case of the 22-oxa-compounds the expected product(s) would be C-20 alcohol/ketone.

The metabolism of OCT has been extensively studied in a number of different biological systems including primary parathyroid *(102)*, primary keratinocyte cells *(103)* as well as cultured osteosarcoma, hepatoma and keratinocyte cell lines *(104)*. In all these systems, OCT is rapidly broken down. Judicious use of two different radioactive labels in the form of [26-^{3}H]OCT and [2β-^{3}H]OCT enabled Brown et al. *(102)* to suggest that the side chain was truncated, though definitive proof of the identity of the products was not immediately forthcoming. It was not until later work *(104)* that the principal metabolites were unequivocably identified by GC-MS as 24(OH)OCT, 26(OH)OCT and hexanor-1α,20-dihydroxyvitamin D$_3$ (Fig. 8). In the case of the keratinocyte-derived cell line, HPK1A-ras, an additional product, hexanor-20-oxo-1α-hydroxyvitamin D$_3$, is also formed. These latter two truncated products are suggestive of hydroxylation of OCT at

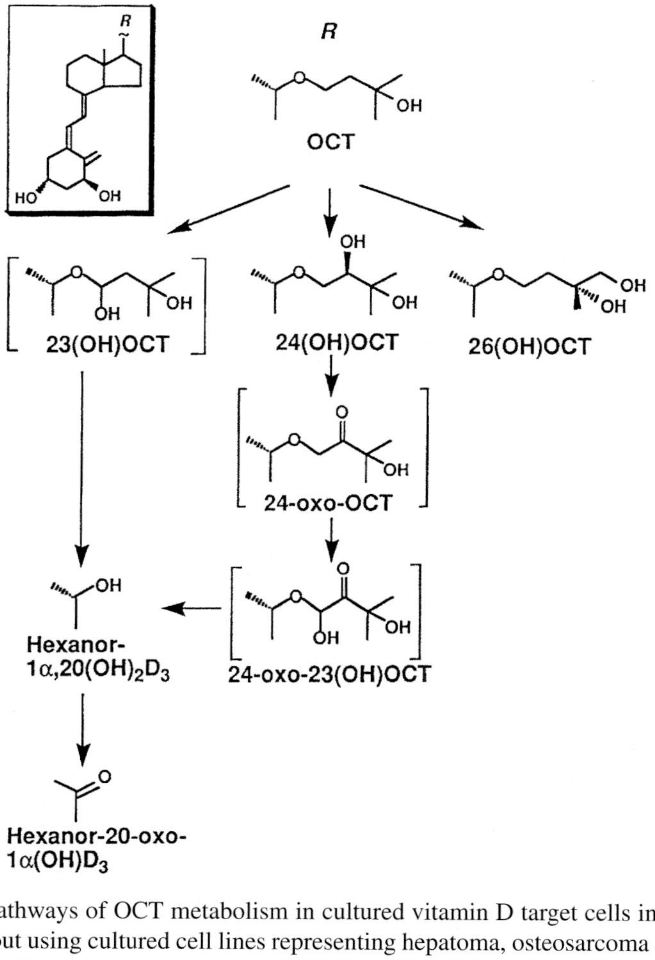

Fig. 8. Proposed pathways of OCT metabolism in cultured vitamin D target cells in vitro. Metabolic pathways worked out using cultured cell lines representing hepatoma, osteosarcoma and keratinocyte. (Taken from *(104)* with permission).

the C-23 position to give the theoretical unstable intermediate. Though all of these products were isolated from in vitro systems, there is evidence that the processes also occur in vivo because Kobayashi et al. *(105)* have generated data which suggest that the biliary excretory form of OCT in the rat is a glucuronide ester of the truncated 20-alcohol.

The above example of a simple oxa-analog provides useful knowledge which can help in predicting the metabolic fate of a complex oxa-analog such as KH1060. This highly potent compound which possesses in vitro cell-differentiating activity exceeding that of any other analog synthesized to date has four different modifications to the side chain of $1,25(OH)_2D_3$, namely (1) 22-oxa-group, (2) the 20-epi side chain stereochemistry, (3) 24a-homologation and (4) 26- and 27-dimethyl homologation (see Table 2 for structure).

Since all of these changes are known to separately affect biological activity in vitro and in vivo as well as side chain metabolism *(106, 107)*, it comes as no surprise that the metabolism of KH1060 is extremely complex. KH1060 has a very short $t_{1/2}$ in pharmacokinetic studies in vivo *(108)* giving a metabolic profile with at least 16 unknown metabolites *(109)*. Dilworth et al. *(109)* reported the first in vitro study using micromolar concentrations of KH1060 incubated with the keratinocyte-derived cell line HPK1A-ras in which these workers were able to discern 22 different metabolites after multiple HPLC steps and assigned structures to 12 of these metabolites (see Fig. 9). As would be expected from consideration of the studies of other oxa-compounds, two of these were truncated products and were identical to the molecules formed from another 22-oxa compound, OCT. As would be expected from consideration of the studies of other homologated compounds, other products are hydroxylated at specific carbons of the side chain including C-26 and C-26a. As with the metabolism-resistant analog, EB1089 (see following section 3.3.2) and $26,27$-dimethyl-$1,25(OH)_2D_3$ (Leo code: MC1548), the presence of dimethyl groups in the terminus of the side chain appears to attract hydroxylation to these sites in KH1060. One novel metabolite found only for KH1060 is 24a-OH-KH1060, observed both in broken cell and intact cell models *(109, 110)*.

An important facet of this complex metabolic profile is that rather than simplifying our understanding of the mechanism of action of KH1060, this data complicates it. This is because biological assays performed on each of the metabolic products of KH1060 have shown that several of the principal and long-lived metabolites retain significant vitamin D-dependent gene inducing activity in reporter gene expression systems *(111)*. While current published assays have demonstrated a high biological activity for KH1060, these assays are performed in whole cell assay systems (cell culture, organ culture, transfected cell systems) over extended time periods (usually 24–72 h) where metabolism is known to occur. Yet analysts do not employ inhibitors of metabolism and often assume that the biological effects observed are due to the parent compound, not to its metabolic products. In the case of KH1060 where metabolism is rapid, it would seem to be prudent to assess the rate of metabolism in the bioassay model or else attempt to block metabolism by the use of appropriate inhibitors (e.g. ketoconazole).

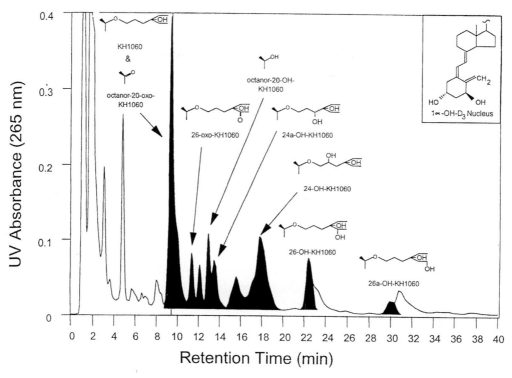

Fig. 9. In vitro metabolism of KH1060 by HPK1A-ras cells. HPLC of lipid extracts following incubation of KH1060 (10 μM) with the human keratinocyte, HPK1A-ras for 72 h. Nine peaks (*darkened*) possessing the characteristic UV chromophore of vitamin D (λ_{max} 265 nm, λ_{min} 228 nm) are visible in the HPLC profile reproduced here. Rechromatography of these peaks on a second HPLC system resulted in the further resolution of these 9 peaks into 22 separate metabolites. Many of these metabolites were identified *(81)* by comparison to synthetic standards on HPLC and GC-MS. Some examples of the types of structures corresponding to each peak are provided. These include Peak at 13.39 min = 24a-OH-KH1060; Peak at 22.14 min= 26-OH-KH1060. (Reproduced with permission from *(109)*).

3.4. Metabolism-Resistant Analogs

These synthetic analogs of 1,25(OH)$_2$D$_3$ require *no* activation in vivo and are resistant to attack by catabolic enzyme systems because of blocking groups in metabolically sensitive regions (e.g. 26,27-F$_6$-1,25(OH)$_2$D$_3$; 1,25(OH)$_2$D$_3$-16-ene-23-yne; EB1089).

3.4.1. F$_6$-1,25-(OH)$_2$D

This analog was first synthesized in the early 1980s *(112)*, along with a number of other side chain fluorinated analogs, to test the importance of certain key hydroxylation sites (e.g. C-23, C-24, C-25, C-26(27), C-1) to biological activity. It was noted immediately that 26,27-F$_6$-1,25(OH)$_2$D$_3$ was extremely potent (10-fold higher than 1,25(OH)$_2$D$_3$) in assays to measure calcaemic activity both in vitro and in vivo *(114–116)*. Lohnes and Jones *(60)* presented evidence using a bone cell line, UMR106, that 26,27-F$_6$-1,25(OH)$_2$D$_3$ had a longer $t_{1/2}$ inside target cells due to the apparent lack of 24-hydroxylation of 26,27-F$_6$-1,25(OH)$_2$D$_3$. At around the same time, Morii's group noted the appearance of a metabolite of $26,27\text{-}F_6\text{-}1,25(OH)_2D_3$ which they have identified as $26,27\text{-}F_6\text{-}1,23,25(OH)_3D_3$ *(117)*. This compound possesses excellent calcaemic activity in its own right *(117)* but whether this derivative is in part responsible for the biological activity of $26,27\text{-}F_6\text{-}1,25(OH)_2D_3$ is not conclusively proven. Nonetheless, $26,27\text{-}F_6\text{-}1,25(OH)_2D_3$ has undergone clinical trials for hypocalcaemia associated with hypoparathyroidism and uraemia *(118, 119)*.

3.4.2. UNSATURATED ANALOGS

The idea of introducing double bond(s) into the side chain of vitamin D analogs arose from experience with vitamin D_2. Vitamin D_2 metabolites have similar biological activity to those of vitamin D_3, so that the introduction of the double bond is not deleterious. As mentioned earlier the metabolism of the side chain is significantly altered by this relatively minor change.

The modification has not been confined to the introduction of a C22=C23 double bond. Roche has developed molecules with two novel modifications:

(a) Introduction of a C16=C17 double bond and (b) introduction of a C23≡C24 triple bond that when combined produce the well-studied 16-ene, 23-yne analog of $1,25(OH)_2D_3$ *(120)* (see Table 2 for structure). Leo Pharmaceuticals has introduced the unsaturated analog EB1089 which contains a conjugated double bond system at C22=C23 and C24=C24a, in addition to both main side chain and terminal dimethyl types of homologation *(121)* (see Table 2 for structure). These two series of Roche and Leo compounds have shown strong anti-proliferative activity both in vitro and in vivo *(120, 122, 123)*.

The metabolism of the 16-ene compound by the perfused rat kidney has been studied *(124)*. These workers *(124)* found that the introduction of the C16=C17 double bond reduces 23-hydroxylation of the molecule and the implication is that the D-ring modification must alter the conformation of the side chain sufficiently to subtly change the site of hydroxylation by CYP24, the cytochrome P450 thought to be responsible for 23- and 24-hydroxylation. Both Dilworth et al. *(106)* studying the metabolism of $20\text{-epi-}1,25(OH)_2D_3$ and Shankar et al. studying the metabolism of $20\text{-methyl-}1,25(OH)_2D_3$ have noted the absence or much reduced 23-hydroxylation reinforcing the view that modifications around the C17-C20 bond profoundly influence the rate of 23-hydroxylation.

The metabolism of the 16-ene,23-yne analog of $1,25(OH)_2D_3$ by WEHI-3 myeloid leukaemic cells has recently been reported *(126)*. Because this molecule is blocked in the C-23 and C-24 positions one might predict that this molecule might be stable to C-24 oxidation pathway enzyme(s); however, it was found experimentally that the 16-ene,23-yne analog has the same $t_{1/2}$ as $1,25(OH)_2D_3$ when incubated with this cell line ($t_{1/2}=6.8$ h). The main product of $[25\text{-}^{14}C]1,25(OH)_2\text{-}16\text{-ene},23\text{-yne-}D_3$ was not identified by these workers but appeared to be more polar than the starting material. Similar work *(127)* used the perfused rat kidney and GC-MS to identify the main metabolite as $1,25,26(OH)_3\text{-}16\text{-ene},23\text{-yne-}D_3$. As with $24\text{-}F_2\text{-}1,25(OH)_2D_3$, it appears that the C-26 becomes vulnerable to attack when the C-23 and C-24 positions are metabolically blocked. As in the case of calcipotriol, not all vitamin D analogs containing unsaturation are resistant to metabolism suggesting that the type of unsaturation, exact position and context (neighbouring groups) must also be considered.

Another unsaturated analog which one might predict would be relatively metabolically stable is EB1089 with its conjugated double bond system. However, as pointed out earlier EB1089 contains three structural modifications: the conjugated double bond system is accompanied by two types of side chain homologation. Nevertheless, as expected the conjugated double bond system dominates the metabolic fate of EB1089, there being no C-24 oxidation activity due to the blocking action of the conjugated diene system. When metabolism is studied with either in vitro liver cell systems or the cultured keratinocyte cell line, HPK1A-ras, disappearance of EB1089 is much slower than that of $1,25(OH)_2D_3$ *(128, 129)*. Such data are consistent with the fairly long $t_{1/2}$ observed in pharmacokinetic studies in vivo *(108)*. Since the conjugated system of EB1089 blocks C-24 oxidation reactions, it is not surprising that a different site in the molecule becomes

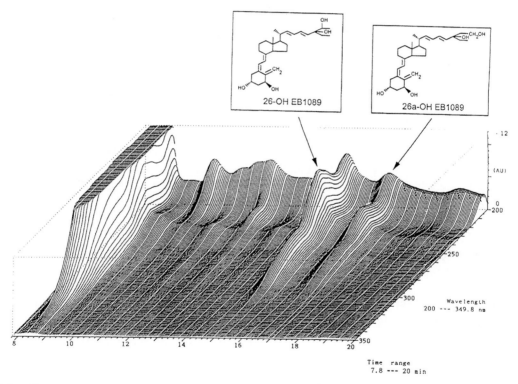

Fig. 10. In vitro metabolism of EB1089 by HPK1A-ras cells. Diode array HPLC of a lipid extract following incubation of EB1089 (10 μM) with the human keratinocyte, HPK1A-ras, for 72 h. In addition to the substrate at 8.5 min, two metabolites showing the distinctive UV chromophore of EB1089 (λ_{max} 235 nm, shoulder 265 nm) are visible in the part of the HPLC profile reproduced here (8–20 min). Metabolite peaks at 15.03 min and 16.55 min were isolated by extensive HPLC and identified *(128, 129)* by comparison to synthetic standards on HPLC, GC-MS and NMR. The identifications are Peak A at 15.03 min= 26-OH-EB1089; Peak B at 16.55 min= 26a-OH-EB1089. The structures of each of the metabolite peaks are depicted in the insets. (Reproduced from *(128)* with permission).

the target for hydroxylation, albeit at a much reduced rate. Diode array spectrophotometry has allowed for the identification of the principal metabolic products of EB1089 as 26- and 26a-hydroxylated metabolites *(128, 129)* (Fig. 10). These metabolites of EB1089 have been chemically synthesized and, as in the case of KH1060, have been shown to retain significant biological activity in cell differentiation and anti-proliferative assays *(129)*.

Again, it is interesting to note that with EB1089 and other molecules blocked in the C-23 and C-24 positions such as $1,24S(OH)_2D_2$ *(73, 92)*, the terminal carbons C-26 and C-26a become the sites of further hydroxylation. However, it should also be considered that even in molecules not blocked in the C-23 and C-24 positions but containing the terminal 26- and 27-dimethyl homologation such as $26,27$-dimethyl-$1,25(OH)_2D_3$ (MC1548) *(107)*, there seems to be significant terminal 26a-hydroxylation occurring. Thus, the hydroxylation of EB1089 at C-26 and C-26a may be in part a consequence of the introduction of the conjugated double bond system and only in part a consequence of the introduction of the terminal homologation.

When the C22=C23 double bond is present in the side chain in the absence of a C-24 methyl group, as in calcipotriol, the double bond appears vulnerable to reduction. As pointed out earlier, the *principal metabolites of calcipotriol are reduced in the C22–C23 bond* except for one, the C22=C23 unsaturated, 24-ketone (MC1046) *(98, 99)*. This suggests a C-24 ketone must be present to allow for reduction of the double bond to occur. Work using the Roche compound, Δ^{22}-$1,25(OH)_2D_3$, an analog which contains the C22=C23 double bond but lacks a C-24 substituent, tends to indirectly support this theory *(130)*. When incubated with the chronic myelogenous leukaemic cell line, RWLeu-4, this molecule, like $1,25(OH)_2D_3$, is converted, presumably via metabolites analogous to intermediates in the C-24 oxidation pathway, to the side chain-truncated product $24,25,26,27$-tetranor-$1,23(OH)_2D_3$, a molecule which lacks the C22=C23 double bond.

4. IMPORTANT IMPLICATIONS DERIVED FROM METABOLISM STUDIES

4.1. Relative Importance of Metabolism in the Mechanism of Action of Vitamin D Analogs

There is currently tremendous interest in explaining the mechanism of action of vitamin D analogs at the molecular level, in particular clarification of the difference between "calcaemic" and "non-calcaemic" analogs.

The susceptibility of a specific vitamin D analog to metabolism and excretion undoubtedly plays a significant role in determining the biological activity of the analog in vivo. In practice, the rate of metabolism of an analog can be studied in cultured cells in vitro, and several cell lines from liver or target cell sources are available to act as valuable tools to reflect this process occurring in vivo *(131)*. In the 10 years since the publication of the first edition of this book, all the individual vitamin D-related CYPs (CYP2R1, CYP27A1, CYP27B1, CYP24A1) have been introduced into appropriate expression vectors (for *E. coli*, yeast, insect and mammalian cell expression) to enable the study of the metabolism of any vitamin D analog reproducibly and outside of the influence of hormonal or ionic regulatory factors *(132)*. In addition, the emergence of CYP24 inhibitors as well as general CYP inhibitors such as ketoconazole *(133, 134)* has opened up the possibility of studying the action of any analog without metabolic complications. Perhaps the ultimate degree of sophistication in current metabolic studies is the use of mice genetically engineered to exhibit over-expression or ablation of any of the vitamin D-related CYPs *(53, 135, 136)*.

Use of the above metabolic approaches allows for the assessment of the relative importance of metabolism within the framework of the full list of biological parameters:

(1) Susceptibility to metabolism.
(2) *Affinity of the analog for plasma DBP* which dictates transport, cell entry and plasma clearance *(137)*.
(3) *Affinity of the analog for target cell VDR–RXR heterodimeric complex* and the resultant affinity of this complex for the VDRE found in the promoter of target genes *(138, 139)*.
(4) Possible *tissue-specific recruitment of downstream coactivators (140)* to the VDR–transactivation complex which might allow for action in cell proliferative or differentiation roles without actions at classical calcium and phosphate homeostatic tissues.

Data for each one of these parameters can be collected in various different ways. Binding assays for DBP and VDR *(137)* and VDRE-mediated transactivation assays have been available for some time now *(138, 139)*. Patterns of coactivator (DRIP) recruitment are now being added in recent analog VDR transcriptional analyses *(140–143, 158)*.

The more detailed studies of the post-VDR steps of the transactivation process are now starting to reveal differences in the affinity of different coactivators for the liganded VDR–RXR heterodimeric complex *(140–145)*. Furthermore, the data are suggestive that analogs which form stable complexes are more active than 1,25(OH)$_2$D$_3$ in vitro. However, taking the reports together no consistent pattern of coactivator recruitment has emerged that might explain analog selectivity on a general scale *(140–145)*.

Rather this author concludes that the application of these assays has shown that none of the parameters, VDR binding, DBP binding *(146, 147)*, recruitment of specific coactivators, transactivation activity *or* rate of metabolism, *when considered separately,* is able to fully explain why some analogs have superior biological activity as compared with 1,25(OH)$_2$D$_3$. However, when these metabolic VDR and DBP parameters *are considered together* for a given analog such as 20-epi-1,25(OH)$_2$D$_3$ *(106)* as in Fig. 11 then one begins to understand the complexity and the fact that many components might contribute to explain the apparent overall superiority of 20-epi-1,25(OH)$_2$D$_3$ over 1,25(OH)$_2$D$_3$ observed in gene transactivation models in vitro (Fig. 10d). One might anticipate this complexity to be even greater in vivo when additional pharmacokinetic parameters (e.g. target cell or hepatic clearance) are added to the in vitro picture.

In the whole animal in vivo, pharmacokinetic data can be acquired for each analog which probably reflects more than one of these parameters to different degrees. Pharmacokinetic data reflect the following important parameters including

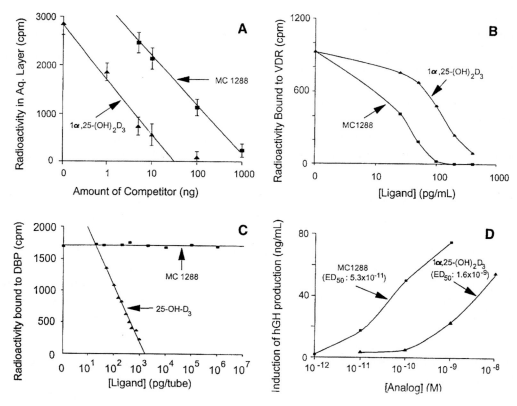

Fig. 11. Biological parameters for 20-epi-1,25(OH)$_2$D$_3$ (MC1288) **a**: Ability of 20-epi-1, 25(OH)$_2$D$_3$ to compete for C24-oxidation pathway enzymes. **b**: VDR-binding affinity of 20-epi-1, 25(OH)$_2$D$_3$ compared to 1,25(OH)$_2$D$_3$. **c**: DBP-binding affinity of 20-epi-1,25(OH)$_2$D$_3$ compared to 1,25(OH)$_2$D$_3$. **d**: Gene transactivation (VDRE placed upstream of GH reporter gene) by 20-epi-1,25(OH)$_2$D$_3$ as compared to 1,25(OH)$_2$D$_3$ in the COS-1 cell line. (Reproduced from *(106)* with permission).

(a) The affinity of the vitamin D analog for DBP in the bloodstream.
(b) The rate of target cell uptake and metabolism by target cell enzymes.
(c) The rate of liver cell uptake, hepatic metabolism and biliary clearance.
(d) The rate of storage depot uptake and release.

In the case of some of the analogs shown in Tables 1 and 2, pharmacokinetic data *(81, 109, 110, 128, 129)* are available and can be compared to the data provided by in vitro metabolic studies. It is apparent from perusal of pharmacokinetic and metabolic data that the analogs which we have defined as *metabolically resistant are usually "calcaemic"* analogs and those termed *metabolically sensitive are "non-calcaemic"*. In fact, this classification can be refined along the lines suggested by Kissmeyer et al. *(81)*, all of their compounds segregate into at least two groups (perhaps more) on the basis of their pharmacokinetic parameters:

Calcaemic Analogs (Strong or Weak): Those analogs with a long $t_{1/2}$ which is either a function of strong DBP binding *or* a reduced rate of metabolism (or both). There appear to be a group of analogs in which a long $t_{1/2}$ is correlated with a slower rate of metabolism (e.g. 26,27-F_6-1,25(OH)$_2$D$_3$, EB1089, 2-MD and ED-71). With the exception of ED-71, which has a strong affinity for DBP, most of these active analogs bind DBP poorly.

Non-calcaemic Analogs: Those analogs with a short $t_{1/2}$ which is either a function of poor DBP binding *or* a rapid rate of metabolism (or both) (e.g. calcipotriol, KH1060 and OCT).

It should be noted that certain classifications used in the vitamin D literature are somewhat artificial since analogs that are *purely* "non-calcaemic" have not been created. All "non-calcaemic" analogs will stimulate in vitro "calcaemic gene" expression and will eventually cause hypercalcaemia in vivo if their concentration is raised sufficiently. The crucial question to the development of anti-cancer or immune-suppressive analogs is whether systemically administered, "weakly calcaemic" or "non-calcaemic" analogs can (e.g. 1(OH)D$_5$, 19-nor-1,25(OH)$_2$D$_2$ and BXL-628) produce their anti-cell proliferation/pro-cell differentiation effects in vivo at concentrations lower than that required to produce calcaemia. Various in vivo clinical trials currently underway will be the acid test for this question.

When considering molecular mechanisms of action at the target cell level, metabolism is often disregarded or given too little emphasis. Furthermore, certain invalid metabolic assumptions made during biological activity testing include (i) the analog is biologically active as administered and (ii) the analog is stable in the in vitro target cell model used, whether organ culture, cultured cell or transfected cell. The validity of this approach is made more tenuous *when data acquired with different in vitro models where metabolic considerations may or may not apply are compared to data acquired in vivo where metabolic considerations definitely apply.* Invalid comparisons of in vivo and in vitro data have led to frequent false expectations when taking promising analogs from in vitro assays to pre-clinical animal testing.

In summary, from studies performed thus far, metabolism appears to be one of a number of key parameters which dictates the survival and hence biological activity of the analog when administered topically or systemically in vivo. At this point in time, a pattern is emerging that suggests "non-calcaemic" or "calcaemic" analogs are metabolically sensitive and metabolically resistant, respectively. However, it seems unlikely that permutations of these biochemical parameters will translate into pure "non-calcaemic" and "calcaemic" analogs, but rather therapeutic agents with improved cell-differentiating or calcaemic activities suitable for different applications.

4.2. Future Directions

4.2.1. STRUCTURAL ASPECTS OF VITAMIN D-RELATED CYTOCHROME P450S

One of the highlights of the past 10 years has been the emergence of X-ray crystal structures for the ligand-binding domain of VDR, DBP and certain CYPs. The recent over-expression, crystallization and X-ray crystallographic determination of the structure of CYP2R1 will spur the study of the mitochondrial vitamin D-related CYPs 24A1, 27A1 and 27B1. As alluded to earlier, mutagenesis studies of human CYP24A1 have already revealed the important information that explains why some species synthesize calcitroic acid via a 24-hydroxylation pathway and others 26,23-lactones via a 23-hydroxylation pathway *(54)*. This study has also revealed that computer homology modelling of the mitochondrial CYPs has also reached a new level of sophistication that has allowed the prediction of active site lining residues *(54, 148)*. One anticipates that emerging polymorphisms of any of the human vitamin D-related CYPs will be tested through such in vitro expression systems such as the Chinese hamster lung fibroblast cell line, V-79 and rationalized using these new CYP structures and computer models.

The complex regulation of the 1α-hydroxylase (CYP27B1) in kidney and extra-renal sites by PTH, FGF23, $1,25(OH)_2D_3$, IFN-γ and potentially other factors is of central interest to physiologists trying to understand the role of CYP27B1 around the body *(27, 28)*. Over the coming years, the relative importance of the endocrine $1,25(OH)_2D_3$ produced by the kidney CYP27B1 and the paracrine/autocrine $1,25(OH)_2D_3$ produced by the extra-renal CYP27B1 should emerge. Perhaps equally important will be to clarify the roll of the cell membrane receptors megalin/cubilin in the uptake of 25-OH-D_3 bound to its DBP carrier into vitamin D target cells. The fact that the megalin knockout mouse *(149)* suffers from vitamin D deficiency supports the view that a special mechanism exists in the kidney and there is evidence that extra-renal tissues also need to express this receptor protein in order to concentrate and metabolize $25(OH)D_3$ *(150)*.

4.2.2. HYDROXYLASE GENE KNOCKOUTS

The advent of the genetically engineered mouse has opened up new possibilities of studying the influence of metabolism on the biological activity of an analog within the in vivo context. In particular, the CYP24A1 knockout mouse *(53, 60, 61)* offers the advantage of comparing biological activity or alternate catabolism of a particular analog in the presence and absence of CYP24A1. Despite the availability of this CYP24A1-null model, few analogs except $1,25(OH)_2D_3$ have been studied to date in this way, though we might expect more in the years to come.

Since the first edition of this book, the emergence of CYP2R1 as the best candidate for physiologically relevant 25-hydroxylase *(14)* also opens up the possibility of engineering a CYP2R1 null mouse and testing the veracity of this hypothesis in a whole animal context in the near future. There has been no shortage of other 25-hydroxylases, i.e. CYP27A1, CYP3A4, CYP2J1/2, so that the deletion of CYP2R1 will allow us to observe the degree of redundancy among these liver-related CYPs. One might predict from various kinetic data for these alternative liver CYPs that none is able to operate effectively at the nanomolar concentrations of substrate found in vivo. The study of the molecular structure, properties and regulation of vitamin D-related cytochrome P450s will continue over the coming decade.

4.2.3. FUTURE VITAMIN D ANALOG DESIGN AND DEVELOPMENT

The success of vitamin D analogs in a variety of clinical applications, particularly chronic kidney disease, hypocalcaemic conditions and psoriasis, will continue to fuel interest in the development of more effective vitamin D analogs for use in these conditions as well as in osteoporosis, cancer and immunosuppression. Over the immediate future we can anticipate (1) a continuing search for novel synthetic modifications to the vitamin D molecule (e.g. 2-MD); (2) combination of "useful" modifications in order to fine-tune the best analogs (e.g. BXL-628); (3) synthesis of coactivator-specific molecules based on emerging transactivation and structure–activity information gained from earlier generations of molecules; (4) clinical testing of a new class of vitamin D analogs, the CYP24-inhibitors which have already been used with promising results in Phase 11B psoriasis and are also going into Phase 11B for testing in the treatment of secondary hyperparathyroidism associated with chronic kidney disease.

In particular, one can envision that VDR ligand-binding pocket studies and cytochrome P450 substrate-binding pocket studies just beginning to emerge will provide valuable information for the design of further generations of non-steroidal vitamin D analogs. With the application of molecular modelling techniques to the study of vitamin D-related proteins for the first time, we will be in the position to view the fit of vitamin D analog to active site topology and fine-tune this fit. The future of vitamin D metabolism remains bright while the pursuit of novel vitamin D analogs continues to be attractive to the pharmaceutical industry.

ACKNOWLEDGEMENTS

The author wishes to acknowledge the help of F. Jeffrey Dilworth and David Prosser in assembly of the tables and figures contained within this document. Some of the work cited here is supported through grants to the author from the Medical Research Council of Canada/Canadian Institutes of Health Research. Over the past 20 years, the research described here has involved some excellent research trainees: David Lohnes, Fuad Qaw, Stephen Strugnell, Sonoko Masuda, Jeffrey Dilworth, VN Shankar, David Prosser and Martin Kaufmann as well as the interdisciplinary collaborations from talented scientists from around the world from Hugh LJ Makin, Martin Calverley, Joyce Knutson, Charles Bishop, Noboru Kubodera, Anne-Marie Kissmeyer, Richard Kremer, Rene St-Arnaud, Mark R. Haussler and Hector F.DeLuca. I sincerely thank them all.

REFERENCES

1. Holick MF (1995) Vitamin D: Photobiology, Metabolism and Clinical Applications. In: Degroot L (ed) Endocrinology, vol 2, 3rd edn. Saunders, Philadelphia, 990–1014
2. Bhattacharyya MH, DeLuca HF (1973) The regulation of the rat liver calciferol-25-hydroxylase. J Biol Chem 248:2969–2973
3. Andersson S, Davis DL, Dahlback H, Jornvall H, Russell DW (1989) Cloning, structure and expression of the mitochondrial cytochrome P450 sterol 26-hydroxylase, a bile acid biosynthetic enzyme. J Biol Chem 246:8222–8229
4. Okuda KI, Usui E, Ohyama Y (1995) Recent progress in enzymology and molecular biology of enzymes involved in vitamin D metabolism. J Lipid Res 36:1641–1652
5. Guo Y-D, Strugnell S, Back DW, Jones G (1993) Transfected human liver cytochrome P-450 hydroxylates vitamin D analogs at different side-chain positions. Proc Natl Acad Sci USA 90: 8668–8672
6. Fukushima M, Suzuki Y, Tohira Y, Nishii Y, Suzuki M, Sasaki S, Suda T (1976) 25-Hydroxylation of 1α-hydroxyvitamin D_3 in vivo and in the perfused rat liver. FEBS Lett 65:211–214

7. Baran DT, Milne ML (1986) 1,25-Dihydroxyvitamin D increases hepatocyte cytosolic calcium levels: a potential regulator of vitamin D-25-hydroxylase. J Clin Invest 77:1622–1626

8. Berginer VM, Shany S, Alkalay D, Berginer J, Dekel S, Salen G, Tint GS, Gazit D (1993) Osteoporosis and increased bone fractures in cerebrotendinous xanthomatosis. Metabolism 42:69–74

9. Tucker G, Gagnon RE, Haussler MR (1973) Vitamin D_3-25-hydroxylase: tissue occurrence and lack of regulation. Arch Biochem Biophys 155:47–57

10. Axen E, Postlind H, Wikvall K (1995) Effects of CYP27 mRNA expression in rat kidney and liver by 1α,25-dihydroxyvitamin D_3, a suppressor of renal 25-hydroxyvitamin D_3-1α-hydroxylase activity. Biochem Biophys Res Commun 215:136–141

11. Ichikawa F, Sato K, Nanjo M, Nishii Y, Shinki T, Takahashi N, Suda T (1995) Mouse primary osteoblasts express vitamin D_3 25-hydroxylase mRNA and convert 1α-hydroxyvitamin D_3 into 1α,25-dihydroxyvitamin D_3. Bone 16:129–135

12. Cheng JB, Motola DL, Mangelsdorf DJ, Russell DW (2003) Deorphanization of cytochrome P450 2R1: a microsomal vitamin D 25-hydroxylase. J Biol Chem 278:38084–38093

13. Cheng JB, Levine MA, Bell NH, Mangelsdorf DJ, Russell DW (2004) Genetic evidence that the human CYP2R1 enzyme is a key vitamin D 25-hydroxylase. Proc Natl Acad Sci U S A 101: 7711–7715

14. Jones G, Byford V, West S, Masuda S, Ibrahim G, Kaufmann M, Knutson J, Strugnell S, Mehta R (2006) Hepatic activation & inactivation of clinically-relevant vitamin D analogs and prodrugs. Anticancer Res 26:2589–2596

15. Strushkevich N, Usanov SA, Plotnikov AN, Jones G, Park H-W (2008) Structural analysis of CYP2R1 in complex with vitamin D_3. J Mol Biol 380:95–106

16. St-Arnaud R, Messerlian S, Moir JM, Omdahl JL, Glorieux FH (1997) The 25-hydroxyvitamin D 1α-hydroxylase gene maps to the pseudovitamin D-deficiency rickets (PDDR) disease locus. J Bone Miner Res 12:1552–1559

17. Takeyama K-I, Kitanaka S, Sato T, Kobori M, Yanagisawa J, Kato. S (1997) 25-Hydroxyvitamin D_3 1α-hydroxylase & vitamin D synthesis. Science 277:1827–1830

18. Monkawa T, Yoshida T, Wakino S, Shinki T, Anazawa H, DeLuca HF, Suda T, Hayashi M, Saruta T (1997) Molecular cloning of cDNA and genomic DNA for human 25-hydroxyvitamin D_3 1α-hydroxylase. Biochem Biophys Res Commun 239:527–533

19. Fu GK, Lin D, Zhang MY, Bikle DD, Shackleton CH, Miller WL, Portale AA (1997) Cloning of human 25-hydroxyvitamin D-1α-hydroxylase and mutations causing vitamin D dependent rickets type 1. Mol Endocrinol 11:1961–1970

20. Fu GK, Portale AA, Miller WL (1997) Complete structure of the human gene for the vitamin D 1α-hydroxylase, P450c1. DNA Cell Biol 16:1499–1507

21. Gray RW, Omdahl JL, Ghazarian JG, DeLuca HF (1972) 25-Hydroxycholecalciferol-1-Hydroxylase: subcellular location and properties. J Biol Chem 247:7528–7532

22. Henry HL (1979) Regulation of the hydroxylation of 25-hydroxyvitamin D_3 in vivo and in primary cultures of chick kidney cells. J Biol Chem 254:2722–2729

23. Stubbs J, Liu S, Quarles LD (2007) Role of fibroblast growth factor 23 in phosphate homeostasis and pathogenesis of disordered mineral metabolism in chronic kidney disease. Semin Dial 20:302–308

24. Brenza HL, Kimmel-Jehan C, Jehan F, Shinki T, Wakino S, Anazawa H, Suda T, DeLuca HF (1998) Parathyroid hormone activation of the 25-hydroxyvitamin D_3-1α-hydroxylase gene promoter. Proc Natl Acad Sci USA 95:1387–1391

25. Murayama A, Takeyama K, Kitanaka S, Kodera Y, Hosoya T, Kato S (1998) The promoter of the human 25-hydroxyvitamin D_3 1α-hydroxylase gene confers positive and negative responsiveness to PTH, calcitonin, and 1α,25$(OH)_2D_3$. Biochem Biophys Res Commun 249:11–16

26. Kitanaka S, Takeyama K, Murayama A, Sato T, Okumura K, Nogami M, Hasegawa Y, Niimi H, Yanigisawa J, Tanaka T, Sato K (1998) Inactivating mutations in the 25-hydroxyvitamin D_3-1α-hydroxylase gene in patients with pseudovitamin D deficiency rickets. N Engl J Med 338:653–661

27. Jones G (2007) Expanding role for vitamin D in chronic kidney disease: Importance of blood 25-OH-D levels & extra-renal 1α-hydroxylase in the classical and non-classical actions of 1α,25-dihydroxyvitamin D_3. Semin Dial 20:316–324

28. Holick MF (2007) Vitamin D deficiency. N Engl J Med 357:266–281

29. Lester GE, Gray TK, Williams ME (1981) In vitro 1α-hydroxylation of ^{3}H-25-hydroxyvitamin D_3 by isolated cells from rat kidneys and placentae. In: Cohn DV, Talmage RV, Matthews JL (eds) Hormonal control of calcium metabolism. Excerpta Medica, Amsterdam, 376

30. Adams JS, Gacad MA (1985) Characterization of 1α-hydroxylation of vitamin D_3 sterols by cultured alveolar macrophages from patients with sarcoidosis. J Exp Med 161:755–765

31. Hewison M, Burke F, Evans KN, Lammas DA, Sansom DM, Liu P, Modlin RL, Adams JS (2007) Extra-renal 25-hydroxyvitamin D_3-1α-hydroxylase in human health and disease. J Steroid Biochem Mol Biol 103:316–321

32. Stoffels K, Overbergh L, Giulietti A, Verlinden L, Bouillon R, Mathieu C (2006) Immune regulation of 25-hydroxyvitamin-D_3-1α-hydroxylase in human monocytes. J Bone Miner Res 21: 37–47

33. Jones G, Ramshaw H, Zhang A, Cook R, Byford V, White J, Petkovich M (1998) Expression and activity of vitamin D-metabolizing cytochrome P450 s (CYP1α & CYP24) in human non-small cell lung carcinomas. Endocrinology 140:3303–3310

34. Bises G, Kállay E, Weiland T, Wrba F, Wenzl E, Bonner E, Kriwanek S, Obrist P, Cross HS (2004) 25-hydroxyvitamin D_3-1α-hydroxylase expression in normal and malignant human colon. J Histochem Cytochem 52:985–989

35. Chen TC, Wang L, Whitlatch LW, Flanagan JN, Holick MF (2003) Prostatic 25-hydroxyvitamin D-1α-hydroxylase and its implication in prostate cancer. J Cell Biochem 88:315–322

36. Holick MF, Schnoes HK, DeLuca HF, Gray RW, Boyle IT, Suda T (1972) Isolation and identification of 24,25-dihydroxycholecalciferol: A metabolite of vitamin D_3 made in the kidney. Biochemistry 11:4251–4255

37. Holick MF, Kleiner-Bossaller A, Schnoes HK, Kasten PM, Boyle IT, DeLuca HF (1973) 1,24,25-Trihydroxyvitamin D_3. A metabolite of vitamin D_3 effective on intestine. J Biol Chem 248: 6691–6696

38. Ornoy A, Goodwin D, Noff D, Edelstein S (1978) 24,25-Dihydroxyvitamin D is a metabolite of vitamin D essential for bone formation. Nature 276:517–519

39. Rasmussen H, Bordier P (1978) Vitamin D & bone. Metab Bone Dis Rel Res 1:7–13

40. Henry HL, Norman AW (1978) Vitamin D: two dihydroxylated metabolites are required for normal chicken egg hatchability. Science 201:835–837

41. Jones G, Vriezen D, Lohnes D, Palda V, Edwards NS (1987) Side chain hydroxylation of vitamin D_3 and its physiological implications. Steroids 49:29–55

42. Brommage R, Jarnagin K, DeLuca HF, Yamada S, Takayama H (1983) 1- but not 24-hydroxylation of vitamin D is required for skeletal mineralization in rats. Am J Physiol 244:E298–E304

43. Makin G, Lohnes D, Byford V, Ray R, Jones G (1989) Target cell metabolism of 1,25-dihydroxyvitamin D_3 to calcitroic acid. Evidence for a pathway in kidney and bone involving 24-oxidation. Biochem J 262:173–180

44. Reddy GS, Tserng K-Y (1989) Calcitroic acid, end product of renal metabolism of 1,25-dihydroxyvitamin D_3 through C-24 oxidation pathway. Biochemistry 28:1763–1769

45. Esvelt RP, Schnoes HK, DeLuca HF (1979) Isolation and characterization of 1α-hydroxy-23-carboxytetranorvitamin D: a major metabolite of 1,25-dihydroxyvitamin D_3. Biochemistry 18: 3977–3983

46. Knutson JC, DeLuca HF (1974) 25-Hydroxyvitamin D_3-24-hydroxylase. Subcellular location and properties. Biochemistry 13:1543–1548

47. Ohyama Y, Noshiro M, Okuda K (1991) Cloning and expression of cDNA encoding 25-hydroxyvitamin D_3 24-hydroxylase. FEBS Lett 278:195–198

48. Ohyama Y, Okuda K (1991) Isolation and characterization of a cytochrome P450 from rat kidney mitochondria that catalyzes the 24-hydroxylation of 25-hydroxyvitamin D3. J Biol Chem 266: 8690–8695

49. Tomon M, Tenenhouse HS, Jones G (1990) Expression of 25-hydroxyvitamin D_3-24-hydroxylase activity in CaCo-2 cells. An In Vitro model of intestinal vitamin D catabolism. Endocrinology 126:2868–2875

50. Suda T, DeLuca HF, Schnoes HK, Tanaka Y, Holick MF (1970) 25,26-dihydroxyvitamin D_3, a metabolite of vitamin D_3 with intestinal transport activity. Biochemistry 9:4776–4780

51. Horst RL (1979) 25-OH-D_3-26,23-Lactone: a metabolite of vitamin D_3 that is 5 times more potent than 25-OH-D_3 in the rat plasma competitive protein binding radioassay. Biochem Biophys Res Commun 89:286–293

52. Yamada S, Nakayama K, Takayama H, Shinki T, Takasaki Y, Suda T (1984) Isolation, identification and metabolism of (23S,25R)-25-hydroxyvitamin D_3-26,23-lactol: a biosynthetic precursor of (23S,25R)-25-hydroxyvitamin D_3-26,23-lactone. J Biol Chem 259:884–889

53. Masuda S, Byford V, Arabian A, Sakai Y, Demay MB, St-Arnaud R, Jones G (2005) Altered Pharmacokinetics of 1α,25-dihydroxyvitamin D_3 and 25-hydroxyvitamin D_3 in the blood and tissues of the 25-hydroxyvitamin D-24-hydroxylase (CYP24A1) null mouse. Endocrinology 146: 825–834

54. Prosser D, Kaufmann M, O'Leary B, Byford V, Jones G (2007) Single A326G mutation converts hCYP24A1 from a 25-OH-D_3-24-hydroxylase into -23-hydroxylase generating 1α,25-$(OH)_2D_3$-26,23-lactone. Proc Natl Acad Sci USA 104:12673–12678

55. Ishizuka S, Ishimoto S, Norman AW (1984) Isolation and identification of 1α,25-dihydroxy-24-oxo-vitamin D_3, 1α,25-dihydroxyvitamin D_3-26,23-lactone, 1α,24(S),25-trihydroxy-vitamin D_3: in vivo metabolites of 1α,25-dihydroxyvitamin D_3. Biochemistry 23:1473–1478

56. Toell A, Gonzalez MM, Ruf D, Steinmeyer A, Ishizuka S, Carlberg C (2001) Different molecular mechanisms of vitamin D_3 receptor antagonists. Mol Pharmacol 59:1478–1485

57. Ishizuka S, Kurihara N, Reddy SV, Cornish J, Cundy T, Roodman GD (2005) (23S)-25-Dehydro-1α-hydroxyvitamin D_3-26,23-lactone, a vitamin D receptor antagonist that inhibits osteoclast formation and bone resorption in bone marrow cultures from patients with Paget's disease. Endocrinology 146:2023–2030

58. Shinki T, Jin CH, Nishimura A, Nagai Y, Ohyama Y, Noshiro M, Okuda K, Suda T (1992) Parathyroid hormone inhibits 25-hydroxyvitamin D_3-24-hydroxylase mRNA expression stimulated by 1α,25-dihydroxyvitamin D_3 in rat kidney but not in intestine. J Biol Chem 267:13757–13762

59. Lohnes D, Jones G (1992) Further metabolism of 1α,25-dihydroxyvitamin D_3 in target cells. J Nutr Sci Vitaminol Special Issue:75–78

60. St-Arnaud R, Arabian A, Travers R, Glorieux FH (1997) Abnormal intramembranous ossification in mice deficient for the vitamin D 24-hydroxylase gene. In: Norman AW, Bouillon R, Thomasset M (eds) Vitamin D. Chemistry, biology and clinical applications of the steroid hormone. Vitamin D Workshop Inc., Riverside, CA, pp 635–639

61. St-Arnaud R, Arabian A, Travers R, Glorieux FH (1997) Partial rescue of abnormal bone formation in 24-hydroxylase knock-out mice supports a role for 24,25-$(OH)_2D_3$ in intramembranous ossification. J Bone Miner Res 12:33 (abstract S111)

62. Hashizume T, Xu Y, Mohutsky MA, Alberts J, Hadden C, Kalhorn TF, Isoherranen N, Shuhart MC, Thummel KE (2008) Identification of human UDP-glucuronosyltransferases catalyzing hepatic 1α,25-dihydroxyvitamin D_3 conjugation. Biochem Pharmacol 75:1240–1250

63. Esvelt RP, DeLuca HF (1980) Calcitroic acid: biological activity and tissue distribution studies. Arch Biochem Biophys 206:404–413

64. Suda T, DeLuca HF, Schnoes HK, Blunt JW (1969) Isolation and identification of 25-hydroxyergocalciferol. Biochemistry 8:3515–3520

65. Jones G, Schnoes HK, DeLuca HF (1975) Isolation and identification of 1,25-dihydroxyvitamin D_2. Biochemistry 14:1250–1256

66. Jones G, Rosenthal A, Segev D, Mazur Y, Frolow F, Halfon Y, Rabinovich D, Shakked Z (1979) Isolation and identification of 24,25-dihydroxyvitamin D_2 using the perfused rat kidney. Biochemistry 18:1094–1101

67. Jones G, Schnoes HK, Levan L, DeLuca HF (1980) Isolation and identification of 24-hydroxyvitamin D_2 and 24,25-dihydroxyvitamin D_2. Arch Biochem Biophys 202:450–457

68. Horst RL, Koszewski NJ, Reinhardt TA (1990) 1α-Hydroxylation of 24-hydroxyvitamin D_2 represents a minor physiological pathway for the activation of vitamin D_2 in mammals. Biochemistry 29:578–582

69. Koszewski NJ, Reinhardt TA, Napoli JL, Beitz DC, Horst RL (1988) 24,26-Dihydroxyvitamin D_2: a unique physiological metabolite of vitamin D_2. Biochemistry 27:5785–5790

70. Reddy GS, Tserng K-Y (1986) Isolation and identification of 1,24,25-trihydroxyvitamin D_2, 1,24,25,28-tetrahydroxyvitamin D_2, 1,24,25,26-tetrahydroxyvitamin D_2: new metabolites of 1,25-dihydroxyvitamin D_2 produced in the rat kidney. Biochemistry 25:5328–5336

71. Clark JW, Reddy GS, Santos-Moore A, Wankadiya KF, Reddy GP, Lasky S, Tserng K-Y, Uskokovic MR (1993) Metabolism and biological activity of 1,25-dihydroxyvitamin D_2 and its metabolites in a chronic myelogenous leukemia cell line, RWLEU-4. Bioorg Med Chem Lett 3:1873–1878

72. Masuda S, Strugnell S, Knutson JC, St-Arnaud R, Jones G (2006) Evidence for the activation of 1α-hydroxyvitamin D_2 by 25-hydroxyvitamin D-24-hydroxylase: delineation of pathways involving 1α,24-dihydroxyvitamin D_2 & 1α,25-dihydroxyvitamin D_2. Biochim Biophys Acta (Mol Cell Biol Lipids) 1761:221–234

73. Jones G, Byford V, Kremer R, Makin HLJ, Rice RH, deGraffenreid LA, Knutson JC, Bishop CA (1996) Anti-proliferative activity and target cell catabolism of the vitamin D analog, 1α,24(S)-dihydroxyvitamin D_2 in normal and immortalized human epidermal cells. Biochem Pharmacol 52:133–140

74. Suda T, Hallick RB, DeLuca HF, Schnoes HK (1970) 25-hydroxydihydrotachysterol$_3$ Synthesis and biological activity. Biochemistry 9:1651–1657

75. Bosch R, Versluis C, Terlouw JK, Thijssen JHH, Duursma SA (1985) Isolation and identification of 25-hydroxydihydrotachysterol$_2$, 1α,25-dihydroxydihydrotachysterol$_2$ and 1β,25-dihydroxy-dihydrotachysterol$_2$. J Steroid Biochem 23:223–229

76. Jones G, Edwards N, Vriezen D, Porteous C, Trafford DJH, Cunningham J, Makin HLJ (1988) Isolation and identification of seven metabolites of 25-hydroxy-dihydrotachysterol$_3$ formed in the isolated perfused rat kidney: A model for the study of side-chain metabolism of vitamin D. Biochemistry 27:7070–7079

77. Qaw F, Calverley MJ, Schroeder NJ, Trafford DJH, Makin HLJ, Jones G (1993) *In vivo* metabolism of the vitamin D analog, dihydrotachysterol. Evidence for formation of 1α,25- and 1β,25-dihydroxydihydrotachysterol metabolites and studies of their biological activity. J Biol Chem 268:282–292

78. Bhattacharyya MH, DeLuca HF (1973) Comparative studies on the 25-hydroxylation of vitamin D_3 and dihydrotachysterol$_3$. J Biol Chem 248:2974–2977

79. Wing RM, Okamura WH, Pirio MP, Sine SM, Norman AW (1974) Vitamin D in solution: conformations of vitamin D_3, 1,25-dihydroxyvitamin D_3 and dihydrotachysterol$_3$. Science 186:939–941

80. Shany S, Ren S-Y, Arbelle JE, Clemens TL, Adams JS (1993) Subcellular localization and partial purification of the 25-hydroxyvitamin D-1-hydroxylation reaction in the avian myelomonocytic cell line HD-11. J Bone Miner 8:269–276

81. Qaw F, Schroeder NJ, Calverley MJ, Maestro M, Mourino A, Trafford DJH, Makin HLJ, Jones G (1992). In vitro synthesis of 1,25-dihydroxydihydrotachysterol in the myelomonocytic cell line, HD-11. J Bone Miner Res 7:S161 (Abstract 274)

82. Qaw FS, Makin HLJ, Jones G (1992) Metabolism of 25-hydroxy-dihydrotachysterol$_3$ in bone cells *in vitro*. Steroids 57:236–243

83. Schroeder NJ, Qaw F, Calverley MJ, Trafford DJH, Jones G, Makin HLJ (1992) Polar metabolites of dihydrotachysterol$_3$ in the rat: Comparison with *in vitro* metabolites of 1α,25-dihydroxy dihydrotachysterol$_3$. Biochem Pharm 43:1893–1905

84. Holick MF, Semmler E, Schnoes HK, DeLuca HF (1973) 1α-Hydroxy derivative of vitamin D_3: a highly potent analog of 1α,25-dihydroxyvitamin D_3. Science 180:190–191

85. Barton DH, Hesse RH, Pechet MM, Rizzardo E (1973) A convenient synthesis of 1α-hydroxy-vitamin D_3. J Am Chem Soc 95:2748–2749

86. Fraser DR, Kodicek E (1970) Unique biosynthesis by kidney of a biologically active vitamin metabolite. Nature 228:764–766

87. Gallagher JC, Goldgar D (1990) Treatment of postmenopausal osteoporosis with high doses of synthetic calcitriol. A randomized control study. Ann Intern Med 113:649–655

88. Lam HY, Schnoes HK, DeLuca HF (1974) 1α-Hydroxyvitamin D_2: a potent synthetic analog of vitamin D_2. Science 186:1038–1040

89. Paaren HE, Hamer DE, Schnoes HK, DeLuca HF (1978) Direct C-1 hydroxylation of vitamin D compounds: convenient preparation of 1α-hydroxyvitamin D_3, 1α,25-dihydroxyvitamin D_3 and 1α-hydroxyvitamin D_2. Proc Natl Acad Sci USA 75:2080–2081

90. Gallagher JC, Bishop CW, Knutson JC, Mazess RB, DeLuca HF (1994) Effects of increasing doses of 1α-hydroxyvitamin D_2 on calcium homeostasis in post-menopausal osteopenic women. J Bone Miner Res 9:607–614

91. Tan AU Jr, Levine BS, Mazess RB, Kyllo DM, Bishop CW, Knutson JC, Kleinman KS, Coburn JW (1997) Effective suppression of parathyroid hormone by 1α-hydroxy-vitamin D_2 in hemodialysis patients with moderate to severe secondary hyperparathyroidism. Kidney Int 51:317–323

92. Strugnell S, Byford V, Makin HLJ, Moriarty RM, Gilardi R, LeVan LW, Knutson JC, Bishop CW, Jones G (1995) 1α,24(S)-dihydroxyvitamin D_2: A biologically active product of 1α-hydroxyvitamin D_2 made in the human hepatoma, Hep3B. Biochem J 310:233–241

93. Sjoden G, Smith C, Lindgren V, DeLuca HF (1985) 1α-hydroxyvitamin D_2 is less toxic than 1α-hydroxyvitamin D_3 in the rat. Proc Soc Exp Biol Med 178:432–436

94. Calverley MJ (1987) Synthesis of MC-903, a biologically active vitamin D metabolite analog. Tetrahedron 43:4609–4619

95. Kragballe K, Gjertsen BT, De Hoop D, Karlsmark T, van de Kerkhof PC, Larko O, Nieboer C, Roed-Petersen J, Strand A, Tikjob G (1991) Double-blind, right/left comparison of calcipotriol and betamethasone valerate in treatment of psoriasis vulgaris. Lancet 337:193–196

96. Jones G, Calverley MJ (1993) A dialogue on analogues: newer vitamin-D drugs for use in bone disease, psoriasis, and cancer. Trends Endocrinol Metab 4:297–303

97. Binderup L (1988) MC903 - A novel vitamin D analogue with potent effects on cell proliferation and cell differentiation. In: Norman AW, Schaefer K, Grigoleit H-G, von Herrath D (eds) Vitamin D. Molecular, cellular and clinical endocrinology. De Gruyter, Berlin, pp 300–309

98. Sorensen H, Binderup L, Calverley MJ, Hoffmeyer L, Rastrup Anderson N (1990) In vitro metabolism of calcipotriol (MC 903), a vitamin D analogue. Biochem Pharmacol 39:391–393

99. Masuda S, Strugnell S, Calverley MJ, Makin HLJ, Kremer R, Jones G (1994) In vitro metabolism of the anti-psoriatic vitamin D analog, calcipotriol, in two cultured human keratinocyte models. J Biol Chem 269:4794–4803

100. Murayama E, Miyamoto K, Kubodera N, Mori T, Matsunaga I (1986) Synthetic studies of vitamin D analogues. VIII. Synthesis of 22-oxavitamin D_3 analogues. Chem Pharm Bull (Tokyo) 34:4410–4413

101. Hansen K, Calverley MJ, Binderup L (1991) Synthesis and biological activity of 22-oxa vitamin D analogues. In: Norman AW, Bouillon R, Thomasset M (eds) Vitamin D: gene regulation, structure-function analysis and clinical application. De Gruyter, Berlin, pp 161–162

102. Brown AJ, Berkoben M, Ritter C, Kubodera N, Nishii Y, Slatopolsky E (1992) Metabolism of 22-oxacalcitriol by a vitamin D-inducible pathway in cultured parathyroid cells. Biochem Biophys Res Commun 189:759–764

103. Bikle DD, Abe-Hashimoto J, Su MJ, Felt S, Gibson DFC, Pillai S (1995) 22-Oxa-calcitriol is a less potent regulator of keratinocyte proliferation and differentiation due to decreased cellular uptake and enhanced catabolism. J Invest Dermatol 105:693–698

104. Masuda S, Byford V, Kremer R, Makin HLJ, Kubodera N, Nishii Y, Okazaki A, Okano T, Kobayashi T, Jones G (1996) In vitro metabolism of the vitamin D analog, 22-oxacalcitriol, using cultured osteosarcoma, hepatoma and keratinocyte cell lines. J Biol Chem 271:8700–8708

105. Kobayashi T, Tsugawa N, Okano T, Masuda S, Takeuchi A, Kubodera N, Nishii Y (1994) The binding properties with blood proteins and tissue distribution of 22-oxa-1α,25-dihydroxyvitamin D_3, a noncalcemic analogue of 1α,25-dihydroxyvitamin D_3 in rats. J Biochem 115:373–380

106. Dilworth FJ, Calverley MJ, Makin HLJ, Jones G (1994) Increased biological activity of 20-epi-1,25-dihydroxyvitamin D_3 is due to reduced catabolism and altered protein binding. Biochem Pharmacol 47:987–993

107. Dilworth FJ, Scott I, Green A, Strugnell S, Guo Y-D, Roberts EA, Kremer R, Calverley MJ, Makin HLJ, Jones G (1995) Different mechanisms of hydroxylation site selection by liver and kidney

cytochrome P450 species (CYP27 and CYP24) involved in vitamin D metabolism. J Biol Chem 270:16766–16774

108. Kissmeyer A-M, Mathiasen IS, Latini S, Binderup L (1995) Pharmacokinetic studies of vitamin D analogues: relationship to vitamin D binding protein (DBP). Endocrine 3:263–266

109. Dilworth FJ, Williams GR, Kissmeyer A-M, Løgsted-Nielsen J, Binderup E, Calverley MJ, Makin HLJ, Jones G (1997) The vitamin D analog, KH1060 is rapidly degraded both in vivo and in vitro via several pathways: principal metabolites generated retain significant biological activity. Endocrinology 138:5485–5496

110. Rastrup-Anderson N, Buchwald FA, Grue-Sorensen G (1992) Identification and synthesis of a metabolite of KH1060, a new potent $1\alpha,25$-dihydroxyvitamin D_3 analogue. Bioorg Med Chem Lett 2:1713–1716

111. Van Den Bemd GJ-CM, Dilworth FJ, Makin HLJ, Prahl JM, DeLuca HF, Jones G, Pols HAP, Van Leeuwen JPTM (2000) Contribution of several metabolites of the vitamin D analog 20-epi-22-oxa-24a,26a,27a-trihomo-$1,25$-$(OH)_2$vitamin D_3 (KH1060) to the overall biological activity of KH1060 by a shared mechanism of action. Biochem Pharmacol 59:621–627

112. Kobayashi Y, Taguchi T, Mitsuhashi S, Eguchi T, Ohshima E, Ikekawa N (1982) Studies on organic fluorine compounds. XXXIX. Studies on steroids. LXXIX. Synthesis of $1\alpha,25$-dihydroxy-26,26,26,27,27,27-hexafluorovitamin D_3. Chem Pharm Bull (Tokyo) 30:4297–4303

113. Koeffler HP, Armatruda T, Ikekawa N, Kobayashi Y, DeLuca HF (1984) Induction of macrophage differentiation of human normal and leukemic myeloid stem cells by $1\alpha,25$-dihydroxyvitamin D_3 and its fluorinated analogs. Cancer Res 44:6524–6528

114. Inaba M, Okuno S, Nishizawa Y, Yukioka K, Otani S, Matsui-Yuasa I, Morisawa S, DeLuca HF, Morii H (1987) Biological activity of fluorinated vitamin D analogs at C-26 and C-27 on human promyelocytic leukemia cells, HL-60. Arch Biochem Biophys 258:421–425

115. Kistler A, Galli B, Horst R, Truitt GA, Uskokovic MR (1989) Effects of vitamin D derivatives on soft tissue calcification in neonatal and calcium mobilization in adult rats. Arch Toxicol 63:394–400

116. Inaba M, Okuno S, Nishizawa Y, Imanishi Y, Katsumata T, Sugata I, Morii H (1993) Effect of substituting fluorine for hydrogen at C-26 and C-27 on the side chain of $1\alpha,25$-dihydroxyvitamin D_3. Biochem Pharmacol 45:2331–2336

117. Sasaki H, Harada H, Hanada Y, Morino H, Suzawa M, Shimpo E, Katsumata T, Masuhiro Y, Matsuda K, Ebihara K, Ono T, Matsushige S, Kato S (1995) Transcriptional activity of a fluorinated vitamin D analog on VDR-RXR-Mediated gene suppression. Biochemistry 34:370–377

118. Nakatsuka K, Imanishi Y, Morishima Y, Sekiya K, Sasao K, Miki T, Nishizawa Y, Katsumata T, Nagata A, Murakawa S (1992) Biological potency of a fluorinated vitamin D analogue in hypoparathyroidism. Bone Miner 16:73–81

119. Nishizawa Y, Morii H, Ogura Y, DeLuca HF (1991) Clinical trial of 26,26,26,27,27,27-hexafluoro-$1\alpha,25$-dihydroxyvitamin D_3 in uremic patients on hemodialysis: preliminary report. Contrib Nephrol 90:196–203

120. Zhou J-Y, Norman AW, Chen D-L, Sun G, Uskokovic M, Koeffler HP (1990) 1,25-Dihydroxy-16-ene-23-yne-vitamin D_3 prolongs survival time of leukemic mice. Proc Natl Acad Sci USA 87:3929–3932

121. Binderup E, Calverley MJ, Binderup L (1991) Synthesis and biological activity of 1α-hydroxylated vitamin D analogues with poly-unsaturated side chains. In: Norman AW, Bouillon R, Thomasset M (eds) Vitamin D: gene regulation, structure-function analysis and clinical application. de Gruyter, Berlin, pp 192–193

122. Colston KW, Mackay AG, James SY, Binderup L, Chandler S, Coombes RC (1992) EB1089: A new vitamin D analogue that inhibits the growth of breast cancer cells in vivo and in vitro. Biochem Pharmacol 44:2273–2280

123. James SY, Mackay AG, Binderup L, Colston KW (1994) Effects of a new synthetic analogue, EB1089, on the oestrogen-responsive growth of human breast cancer cells. J Endocr 141:555–563

124. Reddy GS, Clark JW, Tserng K-Y, Uskokovic MR, McLane JA (1993) Metabolism of 1,25-$(OH)_2$-16-ene D_3 in kidney: influence of structural modification of D-ring on side chain metabolism. Bioorg Med Chem Lett 3:1879–1884

125. Shankar VN, Byford V, Prosser DE, Schroeder NJ, Makin HLJ, Wiesinger H, Neef G, Steinmeyer A, Jones G (2001) Metabolism of a 20-methyl substituted series of vitamin D analogs by cultured human cells: Apparent reduction of 23-hydroxylation of the side chain by 20-methyl group. Biochem Pharmacol 61:893–902

126. Satchell DP, Norman AW (1996) Metabolism of the cell differentiating agent 1,25-(OH)$_2$-16-ene-23-yne vitamin D$_3$ by leukemic cells. J Steroid Biochem Molec Biol 57:117–124

127. Dantuluri PK, Haning C, Uskokovic MR, Tserng K-Y, Reddy GS (1994) Isolation and identification of 1,25,26-(OH)$_3$-16-ene-23-yne D$_3$, a metabolite of 1,25-(OH)$_2$-16-ene-23-yne D$_3$ produced in the kidney. In: Ninth workshop on vitamin D abstract book, Orlando, May 28–June 2 1994, abstract #43, p 32.

128. Shankar VN, Dilworth FJ, Makin HLJ, Schroeder NJ, Trafford DAJ, Kissmeyer A-M, Calverley MJ, Binderup E, Jones G (1997) Metabolism of the vitamin D analog EB1089 by cultured human cells: redirection of hydroxylation site to distal carbons of the side chain. Biochem Pharmacol 53:783–793

129. Kissmeyer A-M, Binderup E, Binderup L, Hansen CM, Andersen NR, Schroeder NJ, Makin HLJ, Shankar VN, Jones G (1997) The metabolism of the vitamin D analog EB 1089: Identification of in vivo and in vitro metabolites and their biological activities. Biochem Pharmacol 53:1087–1097

130. Wandkadiya KF, Uskokovic MR, Clark J, Tserng K-Y, Reddy GS (1992) Novel evidence for the reduction of the double bond in Δ^{22}-1,25-dihydroxyvitamin D$_3$. J Bone Miner Res 7:S171 (Abstract 315)

131. Jones G, Lohnes D, Strugnell S, Guo Y-D, Masuda S, Byford V, Makin HLJ, Calverley MJ (1994) Target cell metabolism of vitamin D and its analogs. In: Norman AW, Bouillon R, Thomasset M (eds) Vitamin D. A pluripotent steroid hormone: structural studies, molecular endocrinology and clinical applications. deGruyter, Berlin, pp 161–169

132. Masuda S, Kaufmann M, Byford V, Gao M, St-Arnaud R, Arabian A, Makin HLJ, Knutson JC, Strugnell S, Jones G (2004) Insights into vitamin D metabolism using CYP24 over-expression and knockout systems in conjunction with liquid chromatography/mass spectrometry (LC/MS). J Steroid Biochem & Mol Biol 89–90:149–153

133. Schuster I, Egger H, Astecker N, Herzig G, Schüssler M, Vorisek G (2001) Selective inhibitors of CYP24: mechanistic tools to explore vitamin D metabolism in human keratinocytes. Steroids 66: 451–462

134. Posner GH, Crawford KR, Yang HW, Kahraman M, Jeon HB, Li H, Lee JK, Suh BC, Hatcher MA, Labonte T, Usera A, Dolan PM, Kensler TW, Peleg S, Jones G, Zhang A, Korczak B, Saha U, Chuang SS (2004) Potent low-calcemic selective inhibitors of CYP24 hydroxylase: 24-sulphone analogs of the hormone 1α,25-dihydroxyvitamin D$_3$. J Steroid Biochem & Mol Biol 89–90: 5–12

135. Dardenne O, Prud'homme J, Arabian A, Glorieux FH, St-Arnaud R (2001) Targeted inactivation of the 25-hydroxyvitamin D$_3$-1α-hydroxylase gene (CYP27B1) creates an animal model of pseudovitamin D-deficiency rickets. Endocrinology 142:3135–3141

136. Panda DK, Miao D, Tremblay ML, Sirois J, Farookhi R, Hendy GN, Goltzman D (2001) Targeted ablation of the 25-hydroxyvitamin D 1α-hydroxylase enzyme: evidence for skeletal, reproductive, and immune dysfunction. Proc Natl Acad Sci USA 98:7498–7503

137. Bouillon R, Okamura WH, Norman AW (1995) Structure-function relationships in the vitamin D endocrine system. Endocr Rev 16:200–257

138. Terpening CM, Haussler CA, Jurutka PW, Galligan MA, Komm BS, Haussler MR (1991) The vitamin D-responsive element in the rat bone Gla protein gene is an imperfect direct repeat that cooperates with other cis-elements in 1,25-dihydroxyvitamin D$_3$- mediated transcriptional activation. Mol Endocrinol 5:373–385

139. Cheskis B, Lemon BD, Uskokovic MR, Lomedico PT, Freedman LP (1995) Vitamin D$_3$-retinoid X receptor dimerization, DNA binding, and transactivation are differentially affected by analogs of 1,25-dihydroxyvitamin D$_3$. Mol Endocrinol 9:1814–1824

140. Freedman LP, Reszka AA (2005) Vitamin D receptor cofactors: Function, regulation and selectivity. In: Feldman D, Pike JW, Glorieux FH (eds) Vitamin D, 2nd edn. Elsevier-Academic Press, New York, pp 263–279

141. Issa LL, Leong GM, Sutherland RL, Eisman JA (2002) Vitamin D analogue-specific recruitment of vitamin D receptor coactivators. J Bone Miner Res 17:879–890

142. Takeyama K, Masuhiro Y, Fuse H, Endoh H, Murayama A, Kitanaka S, Suzawa M, Yanagisawa J, Kato S (1999) Selective interaction of vitamin D receptor with transcriptional coactivators by a vitamin D analog. Mol Cell Biol 19:1049–1055

143. Peleg S, Ismail A, Uskokovic MR, Avnur Z (2003) Evidence for tissue- and cell-type selective activation of the vitamin D receptor by Ro-26-9228, a noncalcemic analog of vitamin D_3 . J Cell Biochem 88:267–273

144. Peleg S, Sastry M, Collins ED, Bishop JE, Norman AW (1995) Distinct conformational changes induced by 20-epi analogues of 1α,25-dihydroxyvitamin D_3 are associated with enhanced activation of the vitamin D receptor. J Biol Chem 270:10551–10558

145. Nayeri S, Danielsson C, Kahlen J, Schräder M, Mathiasen IS, Binderup L, Carlberg C (1995) The anti-proliferative effect of vitamin D_3 analogues is not mediated by inhibition of the AP-1 pathway, but may be related to promoter selectivity. Oncogene 11:1853–1858

146. Bouillon R, Allewaert K, Xiang DZ, Tan BK, Van Baelen H (1991) Vitamin D analogs with low affinity for the vitamin D binding protein: Enhanced *in vitro* and decreased *in vivo* activity. J Bone Miner Res 6:1051–1057

147. Dusso AS, Negrea L, Gunawardhana S, Lopez-Hilker S, Finch J, Mori T, Nishii Y, Slatopolsky E, Brown AJ (1991) On the mechanisms for the selective action of vitamin D analogs. Endocrinology 128:1687–1692

148. Prosser DE, Guo Y-D, Geh KR, Jia Z, Jones G (2006) Molecular modelling of CYP27A1 and site-directed mutational analyses affecting vitamin D hydroxylation. Biophys J 90:1–21

149. Willnow TE, Nykjaer A (2005) Chapter 10: Endocytic Pathways for 25-hydroxyvitamin D_3. In: Feldman D, Pike JW, Glorieux FH (eds) Vitamin D, 2nd edn. Elsevier Academic Press, New York, pp 153–163

150. Rowling MJ, Kemmis CM, Taffany DA, Welsh J (2006) Megalin-mediated endocytosis of vitamin D binding protein correlates with 25-hydroxycholecalciferol actions in human mammary cells. J Nutr 136:2754–2759

151. Fraser D, Kooh SW, Kind P, Holick MF, Tanaka Y, DeLuca HF (1973) Pathogenesis of hereditary vitamin D dependency rickets. N Engl J Med 289:817–822

152. Baggiolini EG, Wovkulich PM, Iacobelli JA, Hennessy BM, Uskokovic MR (1982) Preparation of 1-alpha hydroxylated vitamin D metabolites by total synthesis. In: Norman AW, Schaefer K, von Herrath D, Grigoleit H-G (eds) Vitamin D: chemical, biochemical and clinical endocrinology of calcium metabolism. DeGruyter, Berlin, pp 1089–1100

1523. Perlman KL, Sicinski RR, Schnoes HK, DeLuca HF (1990) 1α,25-Dihydroxy-19-nor-vitamin D_3, a novel vitamin D-related compound with potential therapeutic activity. Tetrahedron Lett 31: 1823–1824

154. Baggiolini EG, Partridge JJ, Shiuey S-J, Truitt GA, Uskokovic MR (1989) Cholecalciferol 23-yne derivatives, their pharmaceutical compositions, their use in the treatment of calcium-related diseases, and their antitumor activity, US 4,804,502 [Abstract]. Chem Abstr 111:58160d

155. Calverley MJ, Binderup E, Binderup L (1991) The 20-epi modification in the vitamin D series: Selective enhancement of "non-classical" receptor-mediated effects. In: Norman AW, Bouillon R, Thomasset M (eds) Vitamin D: gene regulation, structure-function analysis and clinical application. de Gruyter, Berlin, pp 163–164

156. Nishii Y, Sato K, Kobayashi T (1993) The development of vitamin D analogues for the treatment of osteoporosis. Osteoporosis Int 1:S190–S193 (Suppl)

157. Morisaki M, Koizumi N, Ikekawa N, Takeshita T, Ishimoto S (1975) Synthesis of active forms of vitamin D. Part IX. Synthesis of 1α,24-dihydroxycholecalciferol. J Chem Soc Perkin Trans 1(1): 1421–1424

158. Whitfield GK, Jurutka PW, Haussler C et al. (2005) Chapter 13: Nuclear Receptor: Structure-Function, Molecular Control of gene Transcription and Novel Bioactions. In: Feldman D, Pike JW, Glorieux FH (eds) Vitamin D, 2nd edn. Elsevier Academic Press, New York, pp. 219–262

159. Marchiani S, Bonaccorsi L, Ferruzzi P, Crescioli C, Muratori M, Adorini L, Forti G, Maggi M, Baldi E (2006) The vitamin D analogue BXL-628 inhibits growth factor-stimulated proliferation and invasion of DU145 prostate cancer cells. J Cancer Res Clin Oncol 132:408–416

160. Mehta R, Hawthorne M, Uselding L, Albinescu D, Moriarty R, Christov K, Mehta R (2000) Prevention of N-methyl-N-nitrosourea-induced mammary carcinogenesis in rats by 1α-hydroxyvitamin D_5. J Natl Cancer Inst 92:1836–1840

161. Shevde NK, Plum LA, Clagett-Dame M, Yamamoto H, Pike JW, DeLuca HF (2002) A potent analog of 1α,25-dihydroxyvitamin D3 selectively induces bone formation. Proc Natl Acad Sci USA 99:13487–13491

5 The Molecular Biology of the Vitamin D Receptor

Diane R. Dowd and Paul N. MacDonald

Abstract   The biological effects of 1,25-dihydroxyvitamin D_3 [1,25(OH)$_2$D] are mediated through a soluble receptor protein termed the vitamin D receptor (VDR). The VDR binds 1,25(OH)$_2$D with high affinity and high selectivity. In the target cell, the interaction of the 1,25(OH)$_2$D hormone with VDR initiates a complex cascade of molecular events culminating in alterations in the rate of transcription of specific genes or gene networks. This chapter discusses the molecular biology of the VDR and focuses on various aspects of VDR function with an emphasis on the macromolecular interactions that are required for the transcriptional regulatory activity of the VDR. These macromolecular interactions include the association of VDR with the 1,25(OH)$_2$D ligand, the mechanisms required for specific, high-affinity interaction of VDR with DNA, the heterodimeric interaction of VDR with retinoid X receptor (RXR), and protein–protein contacts that comprise the communication links between the VDR and the transcriptional machinery. This chapter also touches on some recent data that suggest that the VDR has transcriptional activity independent of the 1,25(OH)$_2$D ligand in the hair follicle and in the skin. This last aspect demonstrates a novel role for the VDR and its implications in the transcriptional mechanism of the VDR are profound.

   Key Words: Vitamin D receptor (VDR); 1,25-dihydroxyvitamin D_3; vitamin D; cholicalciferol; transcription; steroid hormone receptor; nuclear receptor

1. INTRODUCTION

Vitamin D was discovered as a micronutrient that is essential for normal skeletal development and for maintaining bone integrity. However, vitamin D is more appropriately classified as a hormone and it is the vitamin D endocrine system that regulates skeletal homeostasis. Its predominant role is to preserve skeletal calcium by ensuring that adequate absorption of dietary calcium and phosphorous takes place. In addition to this calciotropic role, vitamin D functions in a plethora of cellular actions, perhaps the most fundamental of which is cellular differentiation *(1)*. In skeletal tissue, the hormonal form of vitamin D, 1,25-dihydroxyvitamin D_3 [1,25(OH)$_2$D], increases osteoclast number *(2)* possibly by inducing the differentiation of preosteoclasts into mature bone-resorbing cells *(3)*. Vitamin D also acts directly on the osteoblast wherein one well-established effect is stimulating the synthesis of several bone matrix proteins including

From: *Nutrition and Health: Vitamin D*
Edited by: M.F. Holick, DOI 10.1007/978-1-60327-303-9_5,
© Springer Science+Business Media, LLC 2010

osteocalcin and osteopontin. Thus, it is via an integrated series of diverse effects that vitamin D is thought to preserve and maintain the integrity of the bony tissues.

The biological effects of $1,25(OH)_2D$ are mediated through a soluble receptor protein termed the vitamin D receptor (VDR). The VDR binds $1,25(OH)_2D$ with high affinity and high selectivity. In the target cell, the interaction of the $1,25(OH)_2D$ hormone with VDR initiates a complex cascade of molecular events culminating in alterations in the rate of transcription of specific genes or gene networks. Central to this mechanism is the requisite interaction of VDR with retinoid X receptor (RXR) to form a heterodimeric complex. This complex binds to specific DNA sequence elements (VDREs) in vitamin D-responsive genes and ultimately influences the rate of RNA polymerase II-mediated transcription. Thus, the VDR–RXR heterodimer serves as the functional transcriptional enhancer in vitamin D-activated transcription. Following VDR–RXR interaction with the VDRE, protein–protein interactions between the VDR–RXR heterodimer and the transcription machinery are essential for the mechanism of vitamin D-mediated gene expression.

This chapter discusses the molecular biology of the VDR and focuses on various aspects of VDR function with an emphasis on the macromolecular interactions that are required for the transcriptional regulatory activity of the VDR. These macromolecular interactions include the association of VDR with the $1,25(OH)_2D$ ligand, the mechanisms required for specific, high-affinity interaction of VDR with DNA, the heterodimeric interaction of VDR with retinoid X receptor (RXR), and protein–protein contacts that comprise the communications links between VDR and the transcriptional machinery. This chapter also touches on some recent data that suggest that the VDR has transcriptional activity independent of the $1,25(OH)_2D$ ligand in the hair follicle and in the skin. This last aspect demonstrates a novel role for the VDR and its implications in the transcriptional mechanism of the VDR are profound.

2. THE VITAMIN D RECEPTOR GENE

The location of the human VDR (hVDR) gene on chromosome 12 was originally determined using Southern blot analysis of DNA from human-Chinese hamster cell hybrids (4). This was further refined to the 12q13-14 region using somatic cell hybrid mapping (5) and in situ hybridization and linkage analysis (6). Human chromosome 12q13.3 is also the location of the 1α-hydroxylase gene which is involved in pseudovitamin D-deficient rickets (PDDR) (6, 7). This autosomal recessive disorder, caused by impaired activity of the renal 1α-hydroxylase, results in insufficient levels of serum $1,25(OH)_2D$. It is intriguing that the two most crucial components of the vitamin D endocrine system, namely VDR and the 1α-hydroxylase, map close to each other on the same region of human chromosome 12.

The gene encoding the hVDR was originally isolated from a human liver genomic DNA library (8) and subsequent reports characterized the hVDR gene and its promoter (8)–(12) (Fig. 1). The hVDR gene spans over 70 kb of genomic DNA and is a complex structure consisting of 14 exon sequences interrupted by intronic sequences ranging in size from 0.2 to 13 kb. A GC-rich, TATA-less promoter directs the transcription of at least three VDR mRNA transcripts (11). The noncoding 5'-end of the gene includes

Fig. 1. Human VDR chromosomal gene. Exons are represented by *boxes*. *Filled boxes* represent noncoding sequences; *open boxes* are coding sequences.

exons 1a–f, while eight additional exons (exons 2–9) encode the VDR protein *(11, 12)*. Exon 2 encodes the two known translation initiation codons. The first initiation codon encodes a protein of 427 amino acids. A T to C transition in the initiation codon creates a polymorphic *FokI* site and results in initiation of translation at an alternative ATG start codon beginning at the tenth nucleotide downstream, encoding a receptor protein of 424 amino acids. Exon 2 also encodes sequences for the first zinc finger of the DNA-binding domain (see below). Exon 3 contains the sequence for the second zinc finger. The observation that the two motifs of the zinc-finger DNA-binding domain are encoded by separate exons is a characteristic trait of the steroid receptor superfamily. The majority of the C-terminal ligand-binding domain is encoded by exons 8 and 9.

The human VDR cDNA or mRNA transcript consists of 4,605 bp containing a 115 bp noncoding leader sequence, a 1,281-bp open reading frame, and 3,209-bp of 3′-noncoding sequence *(13)*. The functional significance of this rather long 3′-untranslated region is unknown, but it is a characteristic feature of the steroid receptor superfamily. The size of this cDNA agrees well with the predominant mRNA species observed in human extracts (approximately 4.6 kb). While cDNAs were identified which varied in their 5′-untranslated region, most use either the Met1 or polymorphic Met4 translation initiation codon, thereby producing a 48-kDa VDR protein. Recent studies have reported that a small proportion of transcripts produce a 54 kDa protein with a slightly longer N-terminus, suggesting that expression of the VDR is under the complex control of multiple promoters *(12, 14)*. This larger protein, VDRB1, was co-expressed at a 1:3 ratio with the 48 kDa VDR protein (VDRA), had 60% of the transcriptional activity of VDRA, and was localized differently in the cell.

Sequence comparison of the VDR cDNAs showed striking similarity with the members of the superfamily of nuclear receptors for steroid and thyroid hormones. An area of high sequence relatedness is found in the DNA-binding domain (DBD) of these receptors. This 70 amino acid domain is rich in cysteine, lysine, and arginine residues and it is the region of the receptor that is responsible for high-affinity interaction with specific DNA sequence elements. An area of more limited homology is in the large C-terminal domain which is responsible for high-affinity interaction with the various hormones, termed the ligand-binding domain (LBD). Interestingly, distinct regions of high similarity are present in the LBDs of the various receptors despite the obvious structural non-relatedness of their individual ligands. This may be due to the fact that this C-terminal domain functions in other related areas of receptor function (e.g., in protein–protein interactions; see below).

The deduced amino acid sequences of the VDR from several species are illustrated in Fig. 2. A comparison of the coding regions of the human, rat, mouse, bovine, and avian VDRs shows a high degree of sequence identity particularly within the N-terminal portion (aa 16–116 of hVDR) that constitutes the DBD (97% identity) and the C-terminal domain (aa 226–427 of hVDR) which comprises most of the 1,25(OH)$_2$D-binding domain (85% identity). The sequences tend to diverge in the hinge region of the

Fig. 2. Deduced amino acid sequence comparison of vitamin D receptors (VDRs) from several species. The *asterisks* denote conserved residues.

receptors which is the region located between the DBD and the LBD. The high degree
of sequence conservation of VDR between species supports the fundamental roles that
these domains serve in VDR function.

3. MOLECULAR ANALYSIS OF THE FUNCTIONAL DOMAINS OF THE VDR

The nuclear receptor superfamily is characterized by a modular structure consisting
of regions required for specific functions. The amino terminus is of a variable length
and contains a transactivation domain termed AF-1. In the VDR, this region is very
short. The central DBD has two zinc-finger motifs that are responsible for protein–DNA
interactions. The carboxy-terminal domain contains the ligand-binding domain (LBD)
and the AF-2 domain. These regions are discussed in more detail below.

3.1. The DNA-Binding Domain (DBD)

The location of the DBD of VDR was originally mapped to the N-terminal 113 amino
acids of the receptor (15, 16). Sone et al. showed that the N-terminal 21 amino acids
immediately preceding the first cysteine residue could be removed without affecting
DNA binding or the transcriptional activation potential of the VDR (17). Thus, the minimal domain of the VDR that mediates VDR–DNA interactions was determined to reside
between amino acid residues 22 and 113 in the human sequence. There are nine cysteine
residues within the DBD that are conserved throughout the members of the superfamily
of receptor proteins. The first eight of these cysteines (counting from the N-terminus)
tetrahedrally coordinate two zinc atoms to form two zinc-finger DNA-binding motifs
(Fig. 3a). In mutagenesis studies, mutation of the first eight of the nine cysteine residues
to serines eliminated VDR binding to both nonspecific and specific DNA sequences and
eliminated VDR-mediated transactivation (17). A serine mutation at the ninth cysteine
residue (C84S) had little effect on VDR function suggesting that this residue is not functionally analogous to the first eight cysteines. These data showed the essential nature of
the first eight cysteines in the overall organization and structural integrity of the DBD
zinc fingers. Later crystallization of the VDR DBD confirmed the structural predictions
(18).

While the VDR–RXR heterodimer is the active transcription factor, efforts to co-
crystallize their DBDs with DNA have been unsuccessful. However, Shaffer and
Gewirth successfully crystallized the VDR DBD homodimer bound to a DR3 response
element (18). Thus, much of what is known of the nature of VDR–RXR–DNA interactions is modeled after functional and structural data of the homodimers bound to DNA as
well as of other related nuclear receptors bound to their recognition sites. The common
structural feature of the DBDs for all these receptors is the folding of two α-helices in
the carboxyl terminal portion of the each zinc finger into a single DNA-binding domain.
The first α-helix in the amino terminal finger (denoted helix 1 in Fig. 3b) lies across the
major groove of DNA making specific contacts with the DNA-binding site and it is this
region that contains the crucial amino acids that determine response element specificity
for some receptors. The second α-helix (denoted helix 2 in Fig. 3b) folds across the first

a.

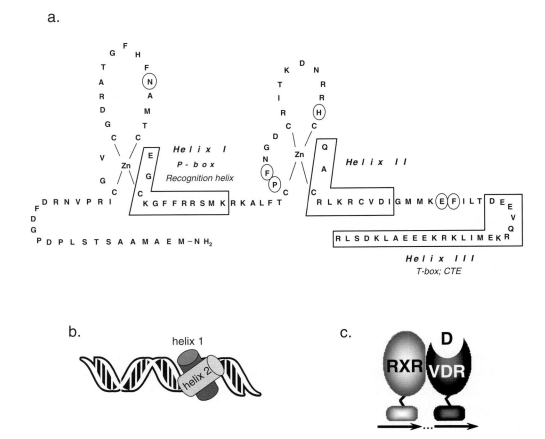

Fig. 3. Binding of VDR to its response element. (**a**) The N-terminal, zinc-finger DNA-binding motifs of human vitamin D receptor. Residues *circled* are involved in the homodimers interface when bound to DNA. (**b**) Orientation of helix 1 and helix 2 when bound to DNA. (**c**) Orientation of the VDR–RXR heterodimers when bound to a VDRE.

in a perpendicular arrangement. The DBD is rich in the positively charged amino acids, lysine and arginine, several of which form favorable electrostatic interactions with the negatively charged phosphate backbone of the DNA helix.

The VDR response element, VDRE, consists of two hexameric half-sites that are arranged as a direct repeat separated by a spacer of three nucleotides (DR3). Unlike some of the other nuclear receptors, VDR does not show any preference for particular sequences outside of the half-site to which it is bound *(19)*. When bound to a DR3, the VDR and RXR each binds to half-site in a head-to-tail dimer with RXR occupying the 5′ or upstream half-site in each element (Fig. 3c) *(20, 21)*. The VDR–VDR homodimer also binds to DR3 response elements in a head-to-tail arrangement *(18)*. VDR dimerization contacts involve the side chains of Pro61, Phe62, and His75 of the upstream subunit and residues Asn37, Glu92, and Phe93 of the downstream subunit *(18)*. These sequences are conserved among the nine known VDRs. Mutation of the core of this hydrophobic dimer interface abolished cooperative assembly on DR3 elements *(22)* and reporter gene activation *(23)*.

Helix I of the VDR DBD is also referred to as the recognition helix, the proximal-box, or the P-box. The recognition helix confers target sequence selectivity for binding of the VDR homodimer to its element (see Fig. 3a). The key interactions occur with four conserved residues, Glu42, Lys45, Arg49, and Arg50, and each residue makes sequence-specific base contacts in the major groove of the half-site *(18)*. Helix III, referred to as the T-box or C-terminal extension (CTE), resides just C-terminal to the second zinc finger and mediates homodimer interaction and interaction of nuclear receptors with DNA *(24, 25)*. Mutations in the CTE of the VDR resulted in a dramatic reduction in VDR binding to DNA and in transactivation indicating an important role for this α-helical domain in VDR function. However, the crystal structure of VDR demonstrated that the CTE of VDR only contacted the DNA at two positions *(18)*. This is in stark contrast to the thyroid hormone receptor CTE which makes 15 contacts with DNA. Shaffer and Gewirth's studies further suggest that the primary role of the VDR CTE is to provide not additional protein–DNA contacts but the mechanism by which the protein discriminates the spacing between the half elements. Thus, the CTE in VDR most likely prevents dimerization on incorrectly spaced response elements.

3.2. The Multifunctional C-terminal Domain

Although the large C-terminal domain of VDR (aa 116 – 427 of hVDR) clearly functions as the binding motif for the $1,25(OH)_2D$ ligand, it also fulfills several other critical roles in VDR function (reviewed in *(26)*). A prominent role for the LBD is that of a protein–protein interaction surface through which VDR contacts its heterodimeric partner RXR for high-order binding to DNA. It is through this C-terminal domain that VDR also interacts with other proteins such as TFIIB and coactivators that are important for the mechanism of VDR-mediated transcription. Finally, key serine residues in this domain serve as sites of phosphorylation that may be important in regulating the transcriptional activity of the VDR.

The LBD is responsible for high-affinity binding of $1,25(OH)_2D$, exhibiting equilibrium-binding constants on the order of 10^{-10}–10^{-11} M *(27)*–*(30)*. Both the 1-hydroxyl and 25-hydroxyl moieties are crucial for efficient recognition by VDR. For example, the monohydroxylated vitamin D_3 compounds, $1(OH)D$ and $25(OH)D$, bind with approximately 100-fold less affinity than $1,25(OH)_2D$. The nonhydroxylated parent vitamin D_3 compound does not bind significantly to the VDR. Modifications in both the side chain structure and in the A ring of the secosteroid also dramatically reduce the affinity of the LBD for the analog.

The specific amino acid residues that are directly involved in $1,25(OH)_2D$ interactions were identified when the LBD was crystallized bound to its natural ligand *(31)*. While earlier attempts to crystallize the VDR failed, Rochel et al. achieved success upon removing a region that decreased protein solubility. The removal of this highly variable region did not affect hormone binding or transactivation and is unlikely to disrupt structural integrity. The structure of the VDR LBD is organized similar to other nuclear receptors into 13 α-helices and 3 β-sheets that together form a hydrophobic pocket that binds ligand. Helix 12, containing the ligand-dependent AF-2 region, contacts the ligand directly while creating a coactivator-binding surface. The VDR ligand-binding cavity is large compared to other receptors and the $1,25(OH)_2D$ ligand occupies only about one half of the volume. Rochel et al. *(31)* conducted ligand-docking modeling of synthetic analogs to begin to understand the differences in activities of the various ligands. They observed variations in conformation of the ligand and receptor as well as different contacts in the binding pockets. These differences may result in changes in half-lives and transcriptional activities. The information obtained from the crystallization of the LBD and these docking studies will aid in the development of synthetic analogs that mimic the advantageous effects of $1,25(OH)_2D$ without the hypercalcemic side effects.

In addition to hormone binding, the LBD is required for several other aspects of receptor function, including an importance in mediating protein–protein interactions. One important protein–protein contact is the heterodimerization of VDR with the retinoid X receptor (RXR) family of receptors. By drawing analogies to other related receptors, the crystal structure of the RXR–RXR homodimeric complex provides some insight into what may be the crucial heterodimerization surface of the VDR–RXR complex. The RXRα crystal structure revealed an α-helical-rich LBD (65%) consisting of 11 α-helices organized into what has been termed "a three-layer antiparallel sandwich" *(32)*. The RXR dimer is symmetrically arranged with the interaction surface being formed mainly by helix 10 and, to a lesser extent, by helix 9. Helix 10 of RXR corresponds to the C-terminal region of VDR identified by Nakajima et al. *(33)* as being crucial for heterodimer formation. Like RXR, the VDR LBD also forms a sandwich in three layers of β-sheets *(31)*. Thus, based on the structural relatedness between VDR and RXR, the C-terminal region of hVDR that includes residues 382–403 (helix 10) may directly contact helix 10 of RXR to comprise the major interaction surface which would generate a structurally symmetrical VDR–RXR heterodimeric complex.

The C-terminal domain also functions as a transactivation domain. Transcriptional activation domains are generally defined through mutations that selectively affect the transcriptional activity of the receptor without disrupting other receptor functions such as ligand binding, response element interaction, or nuclear localization. These domains are often transferable; that is, fusing the activation domain itself to a heterologous DNA-binding domain imparts high-order transcription to the heterologous fusion protein. The extreme C-terminus of the nuclear receptors has a highly conserved, constitutive, hormone-dependent activation domain known as the AF2 domain. Nakajima et al. showed that removing the 25 amino acid AF2 domain from the C-terminus of hVDR (Δ403-427) resulted in a complete loss of $1,25(OH)_2D$-/VDR-activated transcription *(33)*. This loss of function was not due to altered binding of RXR, VDRE, or hormone, and the mutant receptor was appropriately targeted to the cell nucleus. Later studies demonstrated that the AF2 domain resides within helix 12 and the main structural feature is that of an amphipathic α-helix *(31)*. Upon binding ligand, the AF-2 domain, located in helix 12, folds over the top of the LBD and creates a platform for the docking of nuclear receptor coactivators required for nuclear receptor-dependent transcription. Communication between the VDR and the transcriptional machinery is discussed below.

4. MOLECULAR MECHANISM OF TRANSCRIPTIONAL CONTROL BY VDR

4.1. VDR Interaction with Vitamin D-Responsive Elements

4.1.1. VITAMIN D-RESPONSIVE ELEMENTS

Nuclear receptors modulate transcription by binding to specific DNA elements in the promoter regions of hormone-responsive genes. The specific VDR-interactive promoter sequences are termed vitamin D-responsive elements (VDREs). VDREs from a variety of vitamin D-responsive genes have been identified (Fig. 4) on the basis of several functional criteria: (1) deletion or mutation of the element resulted in the loss of promoter responsiveness to $1,25(OH)_2D$; (2) the sequence alone conferred vitamin D responsiveness to an otherwise unresponsive, heterologous promoter; and (3) the element served as a high-affinity binding site for VDR in vitro.

By comparing the limited number of natural response elements identified to date, the VDRE is generally described as an imperfect direct repeat of a core hexanucleotide sequence, *G/A G G T G/C A*, with a spacer region of three nucleotides separating each half element (also termed DR-3 for direct repeat with a three nucleotide spacer). This direct repeat motif is analogous to DNA elements that mediate retinoic acid and thyroid hormone responsiveness (RAREs and TREs) and it contrasts with responsive elements that mediate glucocorticoid- or estrogen-responsive genes (GREs and EREs) which are generally palindromic or inverted repeat sequences (Fig. 4, lower panel). It is apparent that some degree of plasticity exists in the sequence of each half element of the VDRE,

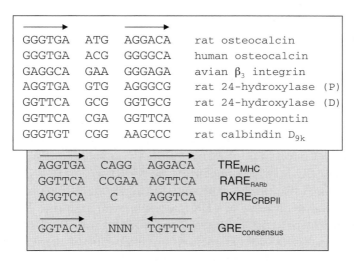

Fig. 4. Comparison of vitamin D response elements (VDREs) from several vitamin D-responsive genes and other hormone receptor response elements.

suggesting that there is flexibility in the precise sequence that will mediate vitamin D responsiveness. However, there is little question that the nucleotide sequence of the element is critical for receptor-mediated transcription. Modest changes in the nucleotides of either half element in the rat osteocalcin VDRE disrupted VDR–VDRE interactions and compromised VDR-dependent transactivation *(34)*.

An additional determinant of response specificity for this class of nuclear receptors is the length of the spacer region *(35)*. In general, VDR, thyroid hormone receptor (TR), and retinoic acid receptor (RAR) all recognize similar direct repeat sequences, but with differing spacer regions of 3, 4, and 5 nucleotides, respectively. This phenomenon, termed the "3-4-5 rule" and later expanded to the "1-5 rule" *(36, 37)*, illustrates that half-site spacing plays a major role in determining selective hormonal response. With regard to the VDRE, it is clear that the actual nucleotide sequence of the spacer region is also important since mutations in the spacer of the rat osteocalcin VDRE disrupt VDR binding and transactivation *(34)*.

The natural VDREs identified thus far provide only a snapshot of the DNA sequences that mediate the transcriptional effects of the VDR and the elements in Fig. 4 are certainly a limited sample. Variations on the DR-3 motif for VDREs have been identified in the elements that mediate vitamin D responsiveness in the calbindin D9k and calbindin D28k genes *(38, 39)*. Moreover, several synthetic elements with large spacer regions and inverted arrangements can mediate vitamin D responsiveness under certain conditions *(40)*. It is likely that the affinity of VDR for these atypical elements may vary from that of the classic DR-3 motif adding yet another level of regulatory complexity to the process of VDR-mediated gene expression. Interestingly, VDREs and other NR regulatory elements may function over very long distances. For example, Nerenz et al. recently identified a $1,25(OH)_2D$-responsive enhancer in the human RANKL promoter that resides -20 kb upstream of the transcriptional start site *(41)*. This region contains a VDRE to which the VDR/RXR heterodimer can bind and which is responsible for induction by $1,25(OH)_2D$.

4.1.2. A ROLE FOR THE $1,25(OH)_2D$ LIGAND

Insight into a role for the $1,25(OH)_2D$ ligand in the transactivation process has emerged from studies examining VDR, RXR, and VDRE interactions. Nanomolar amounts of the $1,25(OH)_2D$ ligand dramatically enhance VDR–RXR heterodimerization, both the direct interaction of VDR with RXR in solution *(17, 42)* and the interaction of the VDR–RXR heterodimer with the VDRE *(17, 43, 44)*. Surface plasmon resonance quantitated the binding constants for these interactions and showed a clear $1,25(OH)_2D$-dependent decrease in VDR interaction with itself (i.e., VDR homodimers) and a concomitant increase in VDR heterodimerization with RXR *(45)*. These ligand-induced changes in VDR–RXR interactions are likely due to altered conformations of the VDR in the absence and presence of $1,25(OH)_2D$ *(46)*. Thus, one putative role for the $1,25(OH)_2D$ in VDR-mediated transcription may be to induce a distinct conformational change in VDR that disrupts weak homodimers of unliganded VDR and promotes liganded VDR heterodimerization with RXR. The interaction of VDR and RXR generates a heterodimeric complex that is highly competent to bind DR-3 VDREs and subsequently affect the transcriptional process.

The possibility exists that other natural compounds may serve as activators of VDR-mediated responses. Makishima et al. *(47)* screened classical nuclear receptors to identify those that were activated by the bile acid lithocholic acid (LCA). They found that micromolar concentrations of LCA and its metabolites directly bind to VDR and activate transcription. Moreover, LCA- or $1,25(OH)_2D$-liganded VDR also stimulates the expression of endogenous CYP3A, the P450 enzyme responsible for degradation of LCA in the liver and the intestine. While LCA is implicated as a toxin that promotes colorectal carcinogenesis *(48, 49)*, $1,25(OH)_2D$ protects against colon cancer *(50)*. Thus, induction of CYP3A by $1,25(OH)_2D$ and LCA may represent a detoxification pathway for LCA and explain the potential preventative effects of $1,25(OH)_2D$ in colon cancer.

4.1.3. GENE REGULATION THROUGH NEGATIVE RESPONSE ELEMENTS

The role of VDR as a transcriptional activator is well known, but numerous studies demonstrate that VDR also directly down-regulates the transcription of some genes. The genes encoding the parathyroid hormone (PTH), the parathyroid hormone-related peptide (PTHrP), and CYP27B1 are all down-regulated by VDR *(51)–(53)*. Both the rat PTH promoter and the human PTHrP promoter contain negative VDREs (nVDREs) which closely resemble the consensus VDRE sequence and which are bound by either VDR homodimers or VDR/RXR heterodimers. However, negative regulation may not necessarily require both VDRE half-sites.

A second type of nVDRE, composed of E-box-type motifs (5′-CATCTG-3′), was identified in the promoters of the human CTP27B1 gene, the human PTH gene, and the human PTHrP gene *(54, 55)*. These nVDREs are transcriptionally active in the absence of $1,25(OH)_2D$ and are bound by a helix-loop-helix transcription factor known as VDR-interacting repressor (VDIR) *(56)*. In the absence of $1,25(OH)_2D$, VDIR recruits CBP/p300 HAT activity to promote active transcription. However, in the presence of $1,25(OH)_2D$, ligand-induced coregulator switching occurs which is responsible for transcriptional repression *(57)*. Basically, liganded VDR/RXR heterodimers associate with the VDRE-bound VDIR, not with the DNA itself. This results in the dissociation of the CBP/p300 HAT coactivator and the recruitment of histone deacetylases and NCoR/SMRT corepressors. Thus, the chromatin is subsequently remodeled and transcription inhibited.

4.1.4. A ROLE FOR THE UNLIGANDED VDR IN TRANSCRIPTIONAL REGULATION

Studies of the VDR knockout (VDRKO) mice and the $25(OH)D$-1α-hydroxylase knockout mice suggest a role for the unliganded receptor in controlling transcription *(58, 59)*. The VDRKO mice exhibit an impairment of calcium absorption and progressive hypocalcemia, hypophosphatemia, and compensatory hyperparathyroidism. These metabolic imbalances result in growth retardation, severe skeletal defects including decreased bone mineral density, thinned bone cortex, and widened undermineralized growth plates. These bone defects are secondary to the malabsorption of calcium, since VDRKO mice fed a rescue diet rich in calcium and phosphorus develop normally without bone abnormalities. The VDRKO mice also develop alopecia which, in contrast to the skeletal phonotype, is not corrected by the rescue diet. This indicates a direct role for VDR in hair follicle cycling. Interestingly, mice that lack the 25(OH)D-1α(OH)ase enzyme and that are totally devoid of the 1,25(OH)$_2$D ligand do not exhibit alopecia *(59)*. Additional studies using VDR transgenes with mutations in the ligand-binding domain demonstrated that these VDR mutants were able to restore hair growth *(60)*. VDR ablation also sensitizes mice to chemically and UV-induced tumorigenesis *(61, 62)*, a response that was not displayed in the 25(OH)D-1α(OH)ase knockout. Thus, the VDR, but not its ligand, is required for protection against carcinogenesis *(61)*. Taken together with the alopecia data, these studies suggest that in the epidermis, the VDR has functions that are independent of the 1,25(OH)$_2$D ligand but require interactions with nuclear factors. There is precedent for this, as in vitro studies show that VDR can associate with various transcription factors and induce select genes in the absence in ligand *(63)–(66)*. Indeed, VDR has been shown to activate target promoters in the absence of 1,25(OH)$_2$D selectively in primary keratinocyte cell culture models *(67)*. Alternatively, unliganded VDR may repress a subset of target genes in a manner analogous to other nuclear receptors through corepressor interactions.

4.2. Communication Between VDR and the Transcriptional Machinery

The past decade has witnessed significant progress in understanding the sequence of events that follow VDR–RXR heterodimer binding to the VDRE and lead to RNA polymerase II-directed transcription. Central to the process of activated transcription are the general transcription factors and the ordered assembly of the preinitiation complex (PIC), beginning with TATA-binding protein (TBP, a subunit of TFIID), binding to the TATA element of class II promoters in a process that is facilitated by TFIIA [reviewed in *(68)*]. Then, TFIIB enters the complex by direct interaction with TBP. RNA Pol II, in association with TFIIF, binds to this early complex by contacting TFIIB. The further association with TFIIE and other general factors results in a complex capable of accurately initiating RNA synthesis. Transcription initiated by these minimal components represents basal-level transcription which can be stimulated (or repressed) by sequence-specific, trans-acting factors such as the VDR.

The nuclear receptor superfamily contains two regions responsible for transcriptional activation: AF1, located in the extreme N-terminus, and AF2, located in the C-terminus. The N-terminus of the VDR is truncated compared to other nuclear receptors and thus it was thought to be unlikely that an analogous constitutive AF-1 domain exists N-terminal to the DBD in VDR. Indeed, as previously mentioned, removing the N-terminal 22 amino acids preceding the DBD had no affect on VDR-activated transcription. However, mutation of Arg-18 and Arg-22 compromised hVDR transcriptional activity presumably by compromising interaction with transcription factor TFIIB *(69)*. Moreover, a polymorphic variant, lacking the first three amino acids at the N-terminus has increased transactivation potency due to better interaction with TFIIB *(69)*. Thus, the short N-terminus of VDR may act as a docking site for TFIIB. VDR also interacts with TFIIB via its C-terminus *(42, 70)*, an interaction that leads to a functional increase in vitamin D-mediated transcription *(70)*. These studies clearly indicate that the VDR–TFIIB interaction is functionally significant and they suggest that the VDR–RXR heterodimer may communicate with the PIC, in part, through specific protein–protein contacts between VDR and TFIIB.

In addition to contacting components of the PIC, VDR and other nuclear receptors are known to interact with a variety of coregulators, i.e., coactivator (CoA) and corepressor (CoR) proteins that also function in the mechanism of steroid-regulated gene transcription *(26)*. Coregulators can be classified into two main groups. The first group contains factors that are recruited by the nuclear receptor to the target promoter to modulate the chromatin architecture. This is done by covalently modifying histones by such processes as acetylation/deacetylation and methylation/demethylation. The second group of coregulators includes ATP-dependent chromatin remodeling factors that modulate promoter accessibility to transcription factors and to the basal transcriptional machinery.

Basically, in the absence of ligand, the nuclear receptor may interact with CoR proteins such as nuclear corepressor (NCoR), silencing mediator of retinoic acid and thyroid hormone receptor (SMART), hairless, and Alien, which in turn associate with histone deacetylases leading to a locally increased chromatin packaging and decrease in gene expression. The binding of ligand induces the dissociation of the CoR and the association of a CoA of the p160/SRC family. Ligand binding results in conformational changes that create a hydrophobic cleft composed of helices 3, 4, 5, and 12. The hydrophobic cleft serves as a docking surface for the p160/SRC family of proteins through their LXXLL domains, and this provides a scaffold for further recruitment of coactivators including histone acetytransferases (HATs) such as p300/CBP and histone methyltransferases (HMTs). By covalently modifying histones, these enzymes cause the relaxation in the chromatin architecture and provide a signal for recruitment of additional coregulatory proteins. Ligand-activated nuclear receptors then change rapidly from interacting with the p160/SRC family to interacting with mediator/VDR-interacting protein (DRIP) complexes, such as Med1 *(71)*. The mediator/DRIP complexes consist of approximately 15–20 proteins and build a bridge to the basal transcription machinery *(71, 72)*. In fact, liganded VDR binds to DRIP/mediator and this ternary complex can then bind directly to the Pol II holoenzyme and recruit the transcriptional machinery to the promoter *(73)*. In this way, ligand-activated nuclear receptors execute two tasks, modification of chromatin and regulation of transcription.

In addition, several other proteins that potentiate VDR-mediated transcription have been described. One example is NCoA-62/ski-interacting protein (SKIP), a coactivator which binds VDR simultaneously with SRC-1 to form a ternary complex that synergistically enhances VDR-stimulated transcription *(63, 74)*. Subsequent studies identified NCoA62/SKIP in subcomplexes of the spliceosome *(75, 76)*. This, combined with NCoA62/SKIP's ability to contact varied transcription factors such as the VDR, suggests a potentially important role in coupling nuclear receptor-mediated transcription with mRNA splicing *(77, 78)*.

A current model for VDR-mediated transcription is illustrated in Fig. 5. This model incorporates numerous properties of VDR that were discussed in this chapter. The initial event in this model is high-affinity binding of the $1,25(OH)_2D$ ligand to the VDR.

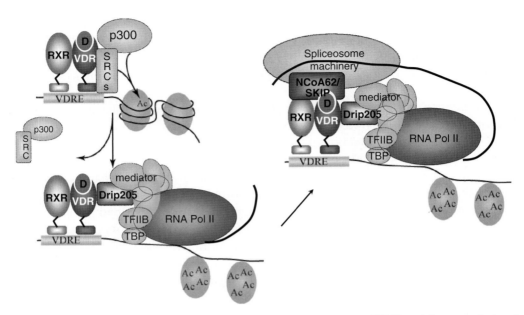

Fig. 5. Putative molecular communications between vitamin D receptor (VDR) and the transcriptional complex.

Ligand binding induces VDR/RXR heterodimerization and the heterodimer specifically binds VDREs in the promoter regions of vitamin D-responsive genes. The VDR–RXR heterodimer serves as a platform for the sequential binding of a wide variety of proteins needed for efficient transcriptional regulation. These include basal transcriptional components such as TFIIB and mediator/DRIP complexes. Coactivator proteins form additional contacts between the VDR and the PIC and it is the interaction with and the communication between VDR, RXR, TFIIB, and other ligand-dependent coactivator proteins such as SRC-1 that may determine the overall transcriptional activity of a vitamin D-responsive gene. It also incorporates proteins such as NCoA62/SKIP, which may enter later to act at more distal steps such as the processing of the nascent RNA transcript. Understanding the complex interplay that occurs between these various factors is crucial to unraveling the complexities of activated or repressed transcription mediated by vitamin D and the VDR.

REFERENCES

1. Abe E, Miyaura C, Sakagami H et al (1981) Differentiation of mouse myeloid leukemia cells induced by 1α,25-dihydroxyvitamin D_3. Proc Natl Acad Sci USA 78:4990–4994
2. Holtrop ME, Cox KA, Clark MB et al (1981) 1,25-dihydroxycholecalciferol stimulates osteoclasts in rat bones in the absence of parathyroid hormone. Endocrinology 108:2293–2301
3. Bar-Shavit Z, Teitelbaum SL, Reitsma P et al (1983) Induction of moncytic differentiation and bone resorption by 1,25(OH)$_2$D$_3$. Proc Natl Acad Sci USA 80:5908–5911
4. Faraco JH, Morrison NA, Baker A et al (1989) ApaI dimorphism at the human vitamin D receptor gene locus. Nucleic Acids Res 17:2150
5. Szpirer J, Szpirer C, Riviere M et al (1991) The Sp1 transcription factor gene (*SP1*) and the 1,25-dihydroxyvitamin D$_3$ receptor gene (*VDR*) are colocalized on human chromosome arm 12q and rat chromosome 7. Genomics 11:168–173

6. Labuda M, Fujiwara TM, Ross MV et al (1992) Two hereditary defects related to vitamin D metabolism map to the same region of human chromosome 12q13-14. J Bone Miner Res 7:1447–1453

7. Kitanaka S, Takeyama K, Murayama A et al (1998) Inactivating mutations in the 25-hydroxyvitamin D3 1alpha-hydroxylase gene in patients with pseudovitamin D-deficiency rickets. N Engl J Med 338:653–661

8. Pike JW, Kesterson RA, Scott RA et al (1988) Vitamin D₃ receptors: molecular structure of the protein and its chromosomal gene. In: Norman AW, Schaefer K et al (ed) Vitamin D: molecular, cellular and clinical endocrinology. Walter de Gruyter & Co., Berlin, New York

9. Sone T, Marx SJ, Liberman UA et al (1990) A unique point mutation in the human vitamin D receptor chromosomal gene confers hereditary resistance to 1,25-dihydroxyvitamin D₃. Mol Endocrinol 4: 623–631

10. Hughes MR, Malloy PJ, Kieback DG et al (1988) Point mutations in the human vitamin D receptor gene associated with hypocalcemic rickets. Science 242:1702–1705

11. Miyamoto K, Kesterson RA, Yamamoto H et al (1997) Structural organization of the human vitamin D receptor chromosomal gene and its promoter. Mol Endocrinol 11:1165–1179

12. Crofts L, Hancock M, Morrison N et al (1998) Multiple promoters direct the tissue-specific expression of novel N-terminal variant human vitamin D receptor gene transcripts. Proc Natl Acad Sci USA 95:10529–10534

13. Baker AR, McDonnell DP, Hughes M et al (1988) Cloning and expression of full-length cDNA encoding human vitamin D receptor. Proc Natl Acad Sci USA 85:3294–3298

14. Sunn K, Cock T, Crofts L et al (2001) Novel N-terminal variant of human VDR. Mol Endocrinol 15:1599–1609

15. Allegretto EA, Pike JW, Haussler MR (1987) Immunochemical detection of unique proteolytic fragments of the chick 1,25-dihydroxyvitamin D₃ receptor. Distinct 20 kDa DNA-binding and 45 kDa hormone-binding species. J Biol Chem 262:1312–1319

16. McDonnell DP, Scott RA, Kerner SA et al (1989) Functional domains of the human vitamin D₃ receptor regulate osteocalcin gene expression. Mol Endocrinol 3:635–644

17. Sone T, Kerner S, Pike JW (1991) Vitamin D receptor interaction with specific DNA: association as a 1,25-dihydroxyvitamin D₃-modulated heterodimer. J Biol Chem 266:23296–23305

18. Shaffer PL, Gewirth DT (2002) Structural basis of VDR–DNA interactions on direct repeat response elements. EMBO J 21:2242–2252

19. Freedman LP, Arce V, Perez Fernandez R (1994) DNA sequences that act as high affinity targets for the vitamin D3 receptor in the absence of the retinoid X receptor. Mol Endocrinol 8:265–273

20. Jin CH, Pike JW (1996) Human vitamin D receptor-dependent transactivation in Saccharomyces cerevisiae requires retinoid X receptor. Mol Endocrinol 10:196–205

21. Schrader M, Nayeri S, Kahlen JP et al (1995) Natural vitamin D₃ response elements formed by inverted palindromes: polarity-directed ligand sensitivity of vitamin D₃ receptor-retinoid X receptor heterodimer-mediated transactivation. Mol Cell Biol 15:1154–1161

22. Towers TL, Luisi BF, Asianov A et al (1993) DNA target selectivity by the vitamin D3 receptor: mechanism of dimer binding to an asymmetric repeat element. Proc Natl Acad Sci USA 90: 6310–6314

23. Quack M, Szafranski K, Rouvinen J et al (1998) The role of the T-box for the function of the vitamin D receptor on different types of response elements. Nucleic Acids Res 26:5372–5378

24. Lee MS, Kliewer SA, Provencal J et al (1993) Structure of the retinoid X receptor α DNA binding domain: a helix required for homodimeric DNA binding. Science 260:1117–1121

25. Wilson TE, Paulsen RE, Padgett KA et al (1992) Participation of non-zinc finger residues in DNA binding by two nuclear orphan receptors. Science 256:107–110

26. Sutton AL, MacDonald PN (2003) Vitamin D: more than a "bone-a-fide" hormone. Mol Endocrinol 17:777–791

27. Brumbaugh PF, Haussler MR (1975) Specific binding of 1α,25-dihydroxycholecalciferol to nuclear components of chick intestine. J Biol Chem 250:1588–1594

28. Brumbaugh PF, Haussler MR (1974) 1,25-dihydroxycholecalciferol receptors in the chick intestine. II. Temperature dependent transfer of the hormone to chromatin via a specific cytosol receptor. J Biol Chem 249:1258–1262

29. Wecksler WR, Norman AW (1980) A kinetic and equilibrium binding study of 1α,25-dihydroxyvitamin D_3 with its cytosol receptor from chick intestinal mucosa. J Biol Chem 255:3571–3574

30. Mellon W, DeLuca HF (1979) An equilibrium and kinetic study of 1,25-dihydroxyvitamin D_3 binding to chicken intestinal cytosol employing high specific activity 1,25-dihydroxy[^{3}H-26,27]vitamin D_3. Arch Biochem Biophys 197:90–95

31. Rochel N, Wurtz JM, Mitschler A et al (2000) The crystal structure of the nuclear receptor for vitamin D bound to its natural ligand. Mol Cell 5:173–179

32. Bourguet W, Ruff M, Chambon P et al (1995) Crystal structure of the ligand-binding domain of the human nuclear receptor RXR-α. Nature 375:377–382

33. Nakajima S, Hsieh J-C, MacDonald PN et al (1994) The C-terminal region of the vitamin D receptor is essential to form a complex with a receptor auxiliary factor required for high affinity binding to the vitamin D-responsive element. Mol Endocrinol 8:159–172

34. Demay MB, Kiernan MS, DeLuca HF et al (1992) Characterization of 1,25-dihydroxyvitamin D_3 receptor interactions with target sequences in the rat osteocalcin gene. Mol Endocrinol 6:557–562

35. Umesono K, Evans RM (1989) Determinants of target gene specificity for steroid/thyroid hormone receptors. Cell 57:1139–1146

36. Leid M, Kastner P, Chambon P (1992) Multiplicity generates diversity in the retinoic acid signalling pathways. Trends Biochem Sci 17:427–433

37. Mangelsdorf DJ, Umesono K, Kliewer SA et al (1991) A direct repeat in the cellular retinol-binding protein type II gene confers differential regulation by RXR and RAR. Cell 66:555–561

38. Darwish HM, DeLuca HF (1992) Identification of a 1,25-dihydroxyvitamin D_3-response element in the 5′-flanking region of the rat calbindin D-9 k gene. Proc Natl Acad Sci USA 89:603–607

39. Gill RK, Christakos S (1993) Identification of sequence elements in mouse calbindin-$D_{28 k}$ gene that confer 1,25-dihydroxyvitamin D_3- and butyrate-inducible responses. Proc Natl Acad Sci USA 90:2984–2988

40. Carlberg C, Bendik I, Wyss A et al (1993) Two nuclear signalling pathways for vitamin D. Nature 361:657–660

41. Nerenz RD, Martowicz ML, Pike JW (2008) An enhancer 20 kilobases upstream of the human receptor activator of nuclear factor-κB ligand gene mediates dominant activation by 1,25-dihydroxyvitamin D_3. Mol Endocrinol 22:1044–1056

42. MacDonald PN, Sherman DR, Dowd DR et al (1995) The vitamin D receptor interacts with general transcription factor IIB. J Biol Chem 270:4748–4752

43. Liao J, Ozono K, Sone T et al (1990) Vitamin D receptor interaction with specific DNA requires a nuclear protein and 1,25-dihydroxyvitamin D_3. Proc Natl Acad Sci USA 87:9751–9755

44. MacDonald PN, Dowd DR, Nakajima S et al (1993) Retinoid X receptors stimulate and 9-cis retinoic acid inhibits 1,25-dihydroxyvitamin D_3-activated expression of the rat osteocalcin gene. Mol Cell Biol 13:5907–5917

45. Cheskis B, Freedman LP (1996) Modulation of nuclear receptor interactions by ligands: kinetic analysis using surface plasmon resonance. Biochemistry 35:3309–3318

46. Peleg S, Sastry M, Collins ED et al (1995) Distinct conformational changes induced by the 20-epi analogues of 1α,25-dihydroxyvitamin D_3 are associated with enhanced activation of the vitamin D receptor. J Biol Chem 270:10551–10558

47. Makishima M, Lu TT, Xie W et al (2002) Vitamin D receptor as an intestinal bile acid sensor. Science 296:1313–1316

48. Debruyne PR, Bruyneel EA, Karaguni IM et al (2002) Bile acids stimulate invasion and haptotaxis in human colorectal cancer cells through activation of multiple oncogenic signaling pathways. Oncogene 21:6740–6750

49. Kozoni V, Tsioulias G, Shiff S et al (2000) The effect of lithocholic acid on proliferation and apoptosis during the early stages of colon carcinogenesis: differential effect on apoptosis in the presence of a colon carcinogen. Carcinogenesis 21:999–1005

50. Lamprecht SA, Lipkin M (2001) Cellular mechanisms of calcium and vitamin D in the inhibition of colorectal carcinogenesis. Ann N Y Acad Sci 952:73–87

51. Demay MB, Kiernan MS, DeLuca HF et al (1992) Sequences in the human parathyroid hormone gene that bind the 1,25-dihydroxyvitamin D_3 receptor and mediate transcriptional repression in response to 1,25-dihydroxyvitamin D_3. Proc Natl Acad Sci USA 89:8097–8101

52. Falzon M (1996) DNA sequences in the rat parathyroid hormone-related peptide gene responsible for 1,25-dihydroxyvitamin D3-mediated transcriptional repression. Mol Endocrinol 10:672–681

53. Russell J, Ashok S, Koszewski NJ (1999) Vitamin D receptor interactions with the rat parathyroid hormone gene: synergistic effects between two negative vitamin D response elements. J Bone Miner Res 14:1828–1837

54. Kim MS, Fujiki R, Murayama A et al (2007) 1Alpha,25(OH)2D3-induced transrepression by vitamin D receptor through E-box-type elements in the human parathyroid hormone gene promoter. Mol Endocrinol 21:334–342

55. Murayama A, Takeyama K, Kitanaka S et al (1998) The promoter of the human 25-hydroxyvitamin D3 1 alpha-hydroxylase gene confers positive and negative responsiveness to PTH, calcitonin, and 1 alpha,25(OH)2D3. Biochem Biophys Res Commun 249:11–16

56. Murayama A, Kim MS, Yanagisawa J et al (2004) Transrepression by a liganded nuclear receptor via a bHLH activator through co-regulator switching. EMBO J 23:1598–1608

57. Fujiki R, Kim MS, Sasaki Y et al (2005) Ligand-induced transrepression by VDR through association of WSTF with acetylated histones. EMBO J 24:3881–3894

58. Sakai Y, Kishimoto J, Demay MB (2001) Metabolic and cellular analysis of alopecia in vitamin D receptor knockout mice. J Clin Invest 107:961–966

59. Panda DK, Miao D, Tremblay ML et al (2001) Targeted ablation of the 25-hydroxyvitamin D 1alpha -hydroxylase enzyme: evidence for skeletal, reproductive, and immune dysfunction. Proc Natl Acad Sci USA 98:7498–7503

60. Skorija K, Cox M, Sisk JM et al (2005) Ligand-independent actions of the vitamin D receptor maintain hair follicle homeostasis. Mol Endocrinol 19:855–862

61. Ellison TI, Smith MK, Gilliam AC et al (2008) Inactivation of the vitamin D receptor enhances susceptibility of murine skin to UV-induced tumorigenesis. J Invest Dermatol

62. Zinser GM, Sundberg JP, Welsh J (2002) Vitamin D(3) receptor ablation sensitizes skin to chemically induced tumorigenesis. Carcinogenesis 23:2103–2109

63. Baudino TA, Kraichely DM, Jefcoat SC Jr. (1998) Isolation and characterization of a novel coactivator protein, NCoA-62, involved in vitamin D-mediated transcription. J Biol Chem 273: 16434–16441

64. Masuyama H, Jefcoat SC, MacDonald PN (1997) The N-terminal domain of transcription factor IIB is required for direct interaction with the vitamin D receptor and participates in vitamin D-mediated transcription. Mol Endocrinol 11:218–228

65. Lavigne AC, Mengus G, Gangloff YG et al (1999) Human TAF(II)55 interacts with the vitamin D(3) and thyroid hormone receptors and with derivatives of the retinoid X receptor that have altered transactivation properties [In Process Citation]. Mol Cell Biol 19:5486–5494

66. Tolon RM, Castillo AI, Jimenez-Lara AM et al (2000) Association with Ets-1 causes ligand- and AF2-independent activation of nuclear receptors. Mol Cell Biol 20:8793–8802

67. Ellison TI, Eckert RL, MacDonald PN (2007) Evidence for 1,25-dihydroxyvitamin D3-independent transactivation by the vitamin D receptor: uncoupling the receptor and ligand in keratinocytes. J Biol Chem 282:10953–10962

68. Zawel L, Reinberg D (1992) Advances in RNA polymerase II transcription. Curr Opin Cell Biol 4: 488–495

69. Jurutka PW, Remus LS, Whitfield GK et al (2000) The polymorphic N terminus in human vitamin D receptor isoforms influences transcriptional activity by modulating interaction with transcription factor IIB. Mol Endocrinol 14:401–420

70. Blanco JCG, Wang I-M, Tsai SY et al (1995) Transcription factor TFIIB and the vitamin D receptor cooperatively activate ligand-dependent transcription. Proc Natl Acad Sci USA 92:1535–1539

71. Rachez C, Suldan Z, Ward J et al (1998) A novel protein complex that interacts with the vitamin D3 receptor in a ligand-dependent manner and enhances VDR transactivation in a cell- free system. Genes Dev 12:1787–1800

72. Rachez C, Lemon BD, Suldan Z et al (1999) Ligand-dependent transcription activation by nuclear receptors requires the DRIP complex. Nature 398:824–828

73. Chiba N, Suldan Z, Freedman LP et al (2000) Binding of liganded vitamin D receptor to the vitamin D receptor interacting protein coactivator complex induces interaction with RNA polymerase II holoenzyme. J Biol Chem 275:10719–10722

74. Zhang C, Baudino TA, Dowd DR et al (2001) Ternary complexes and cooperative interplay between NCoA-62/Ski-interacting protein and steroid receptor coactivators in vitamin D receptor-mediated transcription. J Biol Chem 276:40614–40620

75. Makarov EM, Makarova OV, Urlaub H et al (2002) Small nuclear ribonucleoprotein remodeling during catalytic activation of the spliceosome. Science 298:2205–2208

76. Zhou Z, Licklider LJ, Gygi SP et al (2002) Comprehensive proteomic analysis of the human spliceosome. Nature 419:182–185

77. Auboeuf D, Honig A, Berget SM et al (2002) Coordinate regulation of transcription and splicing by steroid receptor coregulators. Science 298:416–419

78. Zhang C, Dowd DR, Staal A et al (2003) Nuclear coactivator-62 kDa/Ski-interacting protein is a nuclear matrix-associated coactivator that may couple vitamin D receptor-mediated transcription and RNA splicing. J Biol Chem 278:35325–35336

6 VDR and RXR Subcellular Trafficking

Julia Barsony

Abstract   During the past decade, a new dynamic perspective of vitamin D receptor (VDR) and retinoid X receptor (RXR) functions has emerged. The ability to monitor receptor movement in living cells by fluorescent techniques in real time has led to the realization that VDR, RXR, and most other nuclear receptors (NRs) and transcription factors constantly shuttle between the cytoplasm and the nucleus as well as between subnuclear compartments, and revealed the transient nature of receptor–DNA interactions. In this review, the significance of receptor trafficking is first highlighted, along with diseases associated with abnormal receptor localization. The significance of spatial and temporal control of transcription for the regulation of cell growth and differentiation is emphasized. Next, our current knowledge of the nuclear import and export machinery is summarized. Regulation of NR transport is discussed at the level of the receptor (nuclear localization sequence and nuclear export sequence modifications), at the level of import and export receptor expression, and at the level of signal-dependent changes in nuclear pore complex conformation. An understanding of how nuclear architecture and intranuclear NR mobility contribute to gene regulation concludes the review of the general aspects of NR trafficking. Then, information specific for VDR and RXR import, export, and intranuclear trafficking is presented in detail. Conclusions emphasize that understanding the spatial and temporal aspects of VDR functions is important, and express hope that rapid initial progress in this area will not be halted by economic and ideological pressure.

Key Words: Vitamin D; vitamin D receptor; vitamin D response elements; transcription; 1,25-dihydroxyvitamin D; hair; skin; RXR; VDR

1. SIGNIFICANCE OF RECEPTOR LOCALIZATION

The vitamin D receptor (VDR) is a member of the nuclear receptor (NR) superfamily of DNA-binding transcription factors. VDR is a ligand-induced regulator of gene transcription, part of a dynamic regulatory network, controlling directly or indirectly the expression of more than a thousand genes involved in cell division, cell growth and differentiation, cellular defense, and cellular adaptation to changes in the metabolism of the organism. The intricate combinatory network of regulatory interactions is just as complex and dynamic as the world economy, hence the use of large-scale computerized mathematical and simulation methods to describe both (1–3), but until recently this description remained static and did not take into account the spatial and temporal organization of transcription control.

From: *Nutrition and Health: Vitamin D*
Edited by: M.F. Holick, DOI 10.1007/978-1-60327-303-9_6,
© Springer Science+Business Media, LLC 2010

Over the past decade, considerable progress has been made to improve our understanding of the intracellular transport of NRs and the diseases caused by the mislocalization of NRs and other transcriptional regulators. The physiological importance of VDR localization became first apparent in the late 1980s, when we traced back the reason for hereditary vitamin D-resistant rickets in two families to the inability of the receptor to translocate from the cytoplasm to the nucleus *(4)*. Similarly, mislocalization of other nuclear receptors, coregulators, and transcription factors has been identified in association with several human diseases. Such diseases include Rubinstein–Taybi syndrome (mental and growth retardation and increased risk of tumors) due to defective transport of the coactivator CBP *(5)*, androgen insensitivity syndrome due to defective transport of the androgen receptor (AR) *(6)*, disorders of sex determination due to defective transport of the SRY and SOX9 transcription factors *(7)*, the hereditary neurological disease triple A syndrome due to defective transport of DNA repair proteins *(8)*, and heterotaxy, a disease characterized by abnormal left–right arrangement of organs due to the defective transport of ZIC3 transcription factor *(9)*. These and other diseases attest to the functional significance of intracellular transport for transcriptional regulation.

Spatial and temporal control of transcription is particularly important for the appropriate responses to mitogenic signals. Much insight into this control derived from studies demonstrating that the progression of the cell cycle in *Caulobacter* requires a coordinated succession of events that control chromosome replication, differentiation, and cell division *(10)*. These studies showed that differential control of the amount and location of three master transcriptional regulators controls the temporal and spatial expression of cell cycle-regulated genes *(10)*. Moreover, changes in transcription factor localization are often required for responses to mitogenic stimulation in mammalian cells. For example, mitogenic stimulation suppresses PPARgamma's genomic activity by inducing nuclear export of PPARgamma, through inducing its binding to ERK cascade component MAPK/ERK-kinases *(11)*. Other well-characterized examples for signal-specific control of transcription factor localization are NF-κB *(12)*, MAP kinases *(13)*, STAT *(14)*, Smad1 *(15)*, and p53 *(16, 17)*. Not surprisingly, defects in transcription factor targeting have been detected in many different types of cancers. For example, cellular stress blocks the export of the tumor suppressor p53 *(18, 19)* and defects in p53 export are often associated with neoplasms *(20, 21)*. Furthermore, viral oncoproteins often elicit their tumorogenic action by altering trafficking of transcriptional regulators. One well-known example is the v-ErbA, an oncogenic derivative of the thyroid hormone receptor α 1 (TRα), which is carried by the avian erythroblastosis virus (AEV). v-ErbA interferes with the action of TRα in a dominant negative fashion, dimerizes with TRα and the retinoid X receptor (RXR), and sequesters a significant fraction of the two NRs in the cytoplasm *(22)*. These examples demonstrate the importance of understanding the mechanisms that regulate subcellular trafficking of transcription factors for understanding the mechanisms of cell cycle control and tumor development.

In summary, intact subcellular targeting is essential for the proper functioning of NRs and other transcriptional regulators. Elucidating the roles of receptor nucleocytoplasmic trafficking gives insight into the spatial and temporal organization and mechanisms of transcriptional control, leads to a better understanding of the regulation of hormone actions, provides mechanistic insights into the pathophysiology of several diseases, and may reveal new molecular targets for treatment.

2. SPATIAL AND TEMPORAL CONTROL OF NUCLEAR RECEPTOR FUNCTIONS

Regulation of NR functions is part of the intricate combinatory network of regulatory interactions that influence cellular division, growth and differentiation, morphogenesis, and adaptive responses to environmental changes by regulating gene transcription in the nucleus and by rapid, nongenomic actions outside the nucleus. Regulatory checkpoints include NR expression and degradation, NR interactions with other proteins, and post-transcriptional modifications of NRs (such as phosphorylation and acetylation). Gaining insight into the spatial and temporal organization of NR functions elucidates the interplay between protein and gene networks in regulatory circuits, which allows novel level of integration to transcriptional control.

During the last decade, several kinetic studies with fluorescent chimeras of NRs demonstrated that most NRs shuttle constantly between cytoplasm and nucleus *(23)*. This shuttling through the nuclear pore complex (NPC) is mediated by import and export receptors and is influenced by other components of the transport pathway (Fig. 1).

Adapter proteins, subtypes of *importin α*, recognize nuclear localization signals (NLSs) on NRs in the cytoplasm. One or two bipartite NLSs have been found in the DNA-binding domain (DBD) of most nuclear receptors. In addition to these NLSs in the DBD, a constitutive NLS was identified in the hinge region of the androgen receptor (AR) *(24)*. A ligand-regulated NLS was found in the ligand-binding domain (LBD) of the glucocorticoid receptor (GR) *(25)* and the AR *(26)*. The localization of NRs to the nucleus depends on the interplay between multiple NLS domains, and the roles of these domains in ligand-dependent and ligand-independent import of NRs is the subject of ongoing studies *(27)*. The importin α/NR complex is then stabilized by the binding of the cargo-loaded importin α to *importin β*. This triple import complex docks to the cytoplasmic side of the NPC via importin β and translocates to the nuclear side of the NPC. Ran-GTP gradient confers directionality of nuclear import and export receptor movement across the NPC. Direct binding of the arginine-rich NLS of certain transcription factors to importin β has also been described and may represent an alternate mechanism for nuclear import *(28)*. In addition, NRs can use an indirect, piggyback mechanism for nuclear import via binding to other NLS-containing proteins *(29)*.

Importin binding of Ran-GTP terminates translocation, leading to the displacement of importin α-cargo protein from the import complex and the formation of an importin β-Ran-GTP complex. The cargo-free importin α binds to the export receptor, cellular apoptosis susceptibility factor (CAS), cooperatively with Ran-GTP and is then recycled back to the cytoplasm. The Ran-GTP-bound importin β translocates back to the cytoplasm either alone or with cargos that can directly bind to importin β. In the cytoplasm, Ran-binding protein 1 (RanBP1) and Ran-specific GTPase-activating protein 1 (Ran-GAP) activities cause GTP hydrolysis. This leads to the disassembly of export complexes. CAS recycles back to the nucleus by passing through the NPC. Another import factor, NTF2, facilitates Ran transport into the nucleus, ending the import process.

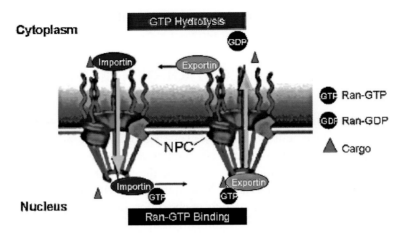

Fig. 1. Schematic representation of nucleo-cytoplasmic transport of proteins through the nuclear pore complex (NPC). Ran-GTP gradient confers directionality of nuclear import and export receptor movement across the NPC. Adapter proteins, like importin α, recognize nuclear localization signals (NLSs) on their cargo in the cytoplasm. This complex is stabilized by the binding of the import receptor importin β, which interacts with the importin β-binding (IBB) domain of the cargo-loaded importin α. This triple import complex then docks to the cytoplasmic side of the NPC via importin β and moves to the nuclear side of the NPC. There, binding of Ran-GTP terminates translocation, leading to the displacement of importin α-cargo protein from the import complex and the formation of an importin β/Ran-GTP complex. The C-terminal IBB-domain of the cargo-free importin α binds to the export receptor CAS cooperatively with Ran-GTP and is recycled back to the cytoplasm. The Ran-GTP-bound importin β translocates back to the cytoplasm either alone or with cargos that can directly bind to importin β. In the cytoplasm, the activities of Ran-binding protein 1 (RanBP1) and the Ran-specific GTPase-activating protein 1 (Ran-GAP) result in the hydrolysis of GTP on Ran, which leads to the disassembly of export complexes. CAS recycles back to the nucleus by passing through the NPC. Another import receptor, NTF2, together with its dimerization partner, takes Ran into the nucleus. Export receptors bind with their cargo and with Ran-GTP in the nucleus. After this complex crosses the NPC, hydrolysis of GTP leads to the disassembly of the export complex, and exportin recycles back to the nucleus.

The mechanisms of NR *export* from the nucleus into the cytoplasm are less well understood. Three export receptors are likely to play significant roles in NR export: the export receptor called chromosomal region maintenance 1 protein (Crm-1) or exportin-1 *(30–33)*, calreticulin *(30, 34)*, and exportin-7 *(35, 36)*. The best-characterized export pathway uses Crm-1. Cargos for Crm-1 are either proteins that carry a leucine-rich nuclear export signal (NES) *(37)* or RNAs *(38)*. Leptomycin B (LMB) is a specific inhibitor of Crm-1-mediated export; it prevents binding of cargo proteins to Crm-1 *(39)*. Crm-1 binds with the nuclear export sequence of the cargo and with Ran-GTP in the nucleus. After this complex crosses the NPC, hydrolysis of GTP leads to the disassembly of the export complex, and exportin recycles back to the nucleus. Both calreticulin-mediated export and exportin-7-mediated export of glucocorticoid receptor (GR) have been shown to affect transcriptional activities. Calreticulin directly binds to and exports GR, whereas exportin-7 influences the export of GR by exporting the 14-3-3 interacting protein. The NES-containing RIP140 corepressor also interacts with the GR and exports it from the nucleus *(40)*. Because RIP140 binds to both VDR *(41)* and RXR *(42)*, this mechanism may be important for the regulation of calcitriol localization. Current research is likely to elucidate novel export pathways of nuclear receptors and transcriptional control by the interplay of multiple export pathways.

The complexities of nuclear import and export mechanisms for NRs provide versatile, hierarchical *regulatory* mechanisms for the spatial control of hormone responses *(43)*. Regulation of transport is most specific at the level of the cargo NRs, whereas regulation of transport receptor expression and recycling influences multiple cargoes. The least specific, large-scale regulatory mechanisms influence the transport capacity of the NPC.

The primary sequence and the secondary or tertiary structure of an NLS or NES domain within specific NRs influence their affinities for transport receptors. Affinities for transport receptors govern transport efficiency *(44)* and thereby significantly alter NR abundance in the cytoplasm and in the nucleus. Signal-induced changes in the secondary or tertiary structure of NLS and NES domains therefore allow cyclical and timely responses to cellular states. For instance, the secondary and tertiary structures of NRs are influenced by post-transcriptional modifications (phosphorylation *(45, 46)*, methylation *(47)*, ubiquitination *(48)*) in a way that involve the transport domains. Furthermore, protein interactions can mask the NLS or NES *(49)* or serve to dock NRs in the cytoplasm or in the nucleus *(26, 50)*.

Regulation at the level of transport receptors is more general than regulations at individual NRs, affecting multiple cargoes. Alterations of the levels of importins and exportins play important roles in development, differentiation, and transformation *(51, 52)*. Relative specificity may be achieved by selectively influencing expression of one or another importin α isoform *(53)*. Such regulatory mechanisms have recently been described using simulation and experimental validation *(54)*. These studies have indicated that nuclear import rate is limited principally by importin α and Ran availability, but is also influenced by NTF2 *(54)*. The model has also revealed that increased concentrations of importin β and the exchange factor RCC1 actually inhibit rather than stimulate import.

The signal-dependent global regulation of NPC conformation influences the transport of multiple protein complexes and therefore has broad consequences on cellular functions. An interesting example for this type of regulation is the triggering of changes in NPC permeability and structure by calcium release *(55, 56)*. It is not yet clear how changes in NPC permeability regulate NR transcriptional activities.

Overall, these three hierarchical mechanisms control cellular localization of NRs. The spatial control mechanisms act in concert with temporal control mechanisms to regulate NR functions.

Temporal control of NR functions includes coordinated sequential changes in NR interactions with coregulators and with DNA (reviewed in *(57)*). Ligand binding induces conformational changes of NRs, exposing new surfaces for protein interactions; hence, the unliganded and liganded NRs form different complexes with coregulators. Moreover, agonists and antagonists of NRs induce formation of distinct coregulator complexes, and these differences lead to selective NR functions *(58, 59)*. Complex temporal kinetics govern the recruitment of multiple coactivators *(60)*, but a consensus on how the rules of patterning regulate transcription rate of specific genes has not been developed.

Recent studies revealed that even the liganded NRs/coactivator complexes are subject to dynamic rearrangements *(61)*.

Studies using advanced photobleaching imaging methods and chromatin immuno-precipitation (ChIP)-based methods with microarrays and with DNA sequencing have collectively indicated that the interaction of NRs with target DNA sequences is also dynamic (recently reviewed in *(62)*).

For all practical purposes, the spatial and temporal controls of NR functions are insep-arable. While nucleo-cytoplasmic partitioning of NRs and coregulators play roles in the temporal control of transcription, NR intranuclear mobility and targeting into distinct nuclear subcompartments are likely to be even more important. Over the past decade, nuclear organization and intranuclear mobility of NRs and their coregulators have been extensively studied using green fluorescent protein chimeras. These pioneering stud-ies revealed structural elements of nuclear architecture, particularly lamin proteins and short actin filaments *(63, 64)*, and linked nuclear functions to nuclear subcompartments and chromosomal territories, which are maintained by scaffolding proteins (recently reviewed in *(65, 66)*). There is increasing evidence that the rapid movements of genomic regions are important for the regulation of gene expression *(67)*. Transcriptional activa-tion is frequently associated with movements of genes from the nuclear periphery to the nuclear interior. Actin and myosin might have roles in the repositioning of genes and move nucleosomes and chromatin, using RNA-polymerase holoenzymes as power-ful ATP-dependent molecular motors *(64, 68)*. Initial insights into the mechanisms that govern how NRs find their target sequences came from photobleaching experiments, demonstrating that NRs move around *(1)* by rapid (10–100 μm^2/s) diffusion, limited by steric constraints from chromatin and nuclear bodies *(2)*, by facilitated diffusion *(69)*, and *(3)* by active ATP-dependent motion *(70)*. It is also well established that ligand binding reduces intranuclear mobility of NRs *(71–74)*. Overall, an understanding of how nuclear architecture and NR mobility contribute to gene regulation is now emerg-ing. In the forefront of this rapid progress are studies related to VDR mobility, described in more detail below.

Advances in our understanding of nucleo-cytoplasmic and intranuclear transport of NRs provide exciting opportunities for *pharmaceutical applications*. Among the most promising applications are monitoring ligand-induced NR translocation for high-throughput drug screening *(73, 75)* and selective modulation of aberrant NR localization for the treatment of cancer and inflammation *(76)*. Multiple approaches emerged for the modulation of NR localization. Hawiger and his coworkers designed cell-permeable peptides containing the NLS from NF-κB and demonstrated that this peptide inhibits inflammatory response by inhibiting the import of NF-κB in vitro and in vivo *(77, 78)*. Furthermore, recombinant protein constructs have been designed with multiple NLSs and NESs to be targeted into the cytoplasm or the nucleus in a ligand-dependent fash-ion, aiming to treat diseases caused by mislocalization of NRs in certain types of cancer *(79)*. Others developed small molecular inhibitors that specifically and selectively target abnormal subnuclear organization in cancer *(65)*. These efforts have not yet been aimed to control aberrant VDR and RXR localization, but the roles of RXR mislocalization in acute promyelocytic leukemia *(80)* and renal-cell carcinoma *(81)* have already been described and may be targeted in the near future.

3. NUCLEAR IMPORT MECHANISMS FOR VDR AND RXR

Vitamin D receptors (VDRs) belong to the type two nuclear receptor family of proteins, which all act as heterodimers with retinoid X receptors (RXRs) to regulate transcription of their target genes. Similar to other NRs, VDR and RXR functions are under spatial and temporal control. The best understood aspect of this control is the nuclear import for VDR and RXR.

The *physiological importance* of VDR import became evident when our studies on patients with hereditary vitamin D-resistant rickets found that the VDR defect in some of the patients is characterized by the inability of the receptor to translocate from the cytoplasm into the nucleus *(82)*. Using wheat germ agglutinin in permeabilized cells, we also demonstrated that VDR interacts with the NPC during the early phase of activation by calcitriol treatment, strongly supporting the roles of nuclear import in the receptor activation process *(82)*. We made the next major step in understanding VDR import when we succeeded in generating transcriptionally active fluorescent protein chimeras of VDR and RXR *(83, 84)*. Using these fluorescent protein chimeras of VDR and RXR and live cell imaging, we visualized both constitutive and hormone-dependent nuclear import of these receptors *(84, 83)*. Later, other laboratories also carried out localization studies with fluorescent protein-tagged VDR *(85–87)* and RXR *(88, 89)*.

We detected the unliganded GFP–VDR in both the cytoplasm and the nucleus (Fig. 2a). This *distribution* was in agreement with our previous immunocytochemical VDR localization finding that utilized microwave fixation *(82)*, and with findings of cell fractionation experiments *(90)*, but was contrary to immunocytochemical VDR localization that utilized chemical fixatives *(91)*. The cytoplasmic proportion of VDR in our studies and in studies by other laboratories was variable, depending on the cellular environment (cell type, cell cycle, metabolic activity, cellular stress, etc.) *(87–94)*. All the studies agreed that calcitriol treatment induces rapid redistribution of VDR, reaching a predominantly nuclear steady-state distribution within 5–30 min *(84–85)* (Fig. 2b). The RXR steady-state distribution was predominantly nuclear in both immunocytochemical studies on fixed cells and in fluorescent protein-tagged live cell experiments *(95, 83)*.

Accumulating evidence indicated that both VDR and RXR interact with nuclear *import receptors* via multiple binding sites. Mutational analyses defined the bipartite NLS segments within the VDR and RXR that interact with the nuclear import receptors (Fig. 3a). First, Haussler's laboratory identified NLS1 of the VDR within the DNA-binding domain *(96)*. This bipartite NLS between the two zinc fingers of the DBD is homologous to the NLS found in most NRs (NLS1). We later confirmed that this is a functional NLS in living cells by generating mutations in NLS1 of the GFP–VDR (K53Q, R54G, K55E) *(83)*. Mutations of this region resulted in cytoplasmic retention of both unliganded and liganded VDR *(96)*. Subsequent studies identified several putative nuclear localization sequences within the VDR (Fig. 3a) and explored their functional significance. A cluster of basic amino acids (NLS2) in the second zinc finger of VDR is homologous to the NLS found in the orphan nuclear receptor TR2 *(97)*. Mutations of these basic amino acids to lysine did not cause cytoplasmic retention of GFP–VDR (Barsony J. and Prufer K., unpublished), thus this segment is unlikely to function as an obligatory NLS. Two more clusters of basic amino acids within the VDR

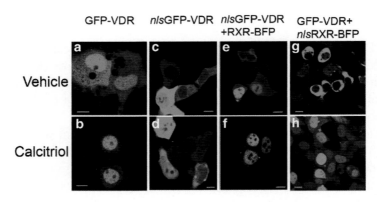

Fig. 2. Effects of calcitriol and dimerization on VDR and RXR import. HEK293 cells expressed wild-type GFP–VDR, the NLS1 mutant GFP–VDR (*nls*GFP–VDR), co-expressed the *nls*GFP–VDR with RXR–BFP, or co-expressed the GFP–VDR with the NLS1 mutant of YFP–RXR (*nls*RXR–BFP). Images were taken of living cells using a Carl Zeiss Axiovert 100 fluorescent microscope with a LSM 410 laser-scanning unit. The 488-nm line of a krypton-argon laser with a band-pass 510–525-nm emission filter was used for GFP detection. The blue (BFP) signal is not depicted. Unliganded GFP–VDR partitioned between the cytoplasm and the nucleus (*a*). Alone, both *nls*GFP–VDR (*c*) and *nls*RXR–BFP (not shown) sequestered in the cytoplasm, and the presence of *nls*RXR–BFP caused cytoplasmic retention of GFP–VDR (*g*). The presence of RXR–BFP allowed nuclear transport of the *nls*GFP–VDR (*e*). Calcitriol addition induced nuclear accumulation of GFP–VDR (*b*), even in the presence of *nls*RXR–BFP (*h*). Calcitriol failed to induce nuclear accumulation of *nls*GFP–VDR (*d*), but the presence of RXR–BFP allowed nuclear accumulation of *nls*GFP–VDR (*f*). These findings suggest that RXR and VDR form dimmers in the cytoplasm. Furthermore, these data indicate that RXR dominates the import of the unliganded VDR, and liganded VDR dominates the import of RXR. Bars, 10 μm.

appeared initially to function as NLSs. A peptide, representing amino acids 102–111 of VDR, has been shown to enable nuclear accumulation of the fluorescein-labeled IgG *(98)*. We deleted these amino acids 102–110 (NLS3) of the GFP–VDR without significant impairment of the nuclear import of the unliganded VDR, but this deletion of NLS3 region abolished the hormone-dependent nuclear accumulation of GFP–VDR. Michigami and his colleagues showed that another short segment of 20 amino acids in the hinge region of VDR (NLS4; 154–173) enabled GFP-tagged alkaline phosphatase to appear in the nucleus *(85)*, but using mutational analysis we were unable to confirm that this segment functions as an NLS in the context of the intact receptor: mutations at amino acids 154–158 (R154G, P155A, P156A, R158G) and at amino acids169–173 (R169G, P170A, R173G) did not prevent nuclear import of GFP–VDR (Barsony J. and Prufer K., unpublished).

RXR also interacts with import receptors. Figure 3b shows the NLS1 of RXR in the first zinc finger of RXR, which is a homologue of the NLS1 of VDR. We introduced point mutations at amino acids 160, 161, 164, and 165 (K160Q, R161G, R164G, K165Q) of this NLS1 of RXR to generate *nls*1YFP–RXR *(99)*. Microscopy showed that the *nls*1YFP–RXR was retained in the cytoplasm (Fig. 3b). An additional NLS is located between amino acids 181 and 186 within the second zinc finger of RXR (NLS2).

a VDR NLS Segments

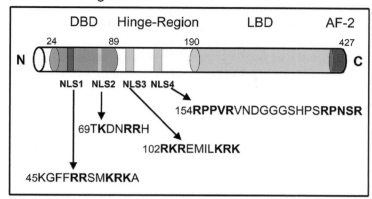

b RXR NLS Segments

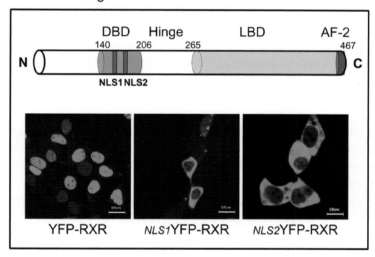

Fig. 3. Schematic representation of VDR and RXR nuclear import domains. *Colored bars*, flanked by the bordering amino acid numbers on *top*, depict VDR and RXR functional domains: DNA-binding domain (DBD), hinge region, and ligand-binding domain (LBD). (**a**) Amino acids representing the putative nuclear localization sequences (NLSs) of VDR are shown in *bold letters*. In living cells only the NLS1 mutations caused cytoplasmic retention of receptors (Fig. 2). (**b**) Two NLSs within the DNA-binding domain of RXR are indicated with the *red bars*. Images are shown of live HEK293 cells stably expressing wild-type YFP–RXR or its mutants. The NLS1 mutations were K160Q, R161G, and R164G, and the NLS2 mutations were R182E and R184E. Mutations caused cytoplasmic retention. Images were taken using a Zeiss LSM410 confocal microscope. Bars, 10 μm.

Mutations in this segment (R182E, R184E) also caused cytoplasmic retention of the NLS2YFP–RXR (Fig. 3b).

In addition to the segments that interact with import receptors, VDR segments involved in *other protein–protein interactions* proved to be essential for the calcitriol-induced nuclear accumulation. We have shown that ligand-dependent acceleration of

GFP–VDR import is disturbed by deletion or mutations in the activation function 2 (AF2) domain *(84, 99)*. This finding indicated that coactivators could play a role in VDR import. Additionally, residues 105–109 within the putative NLS3 of VDR were also essential for calcitriol-induced nuclear accumulation of VDR. The same amino acids bind to hsp70 *(100)*. The loss of hsp70 binding to VDR has no effect on hormone-binding and coactivator interactions *(100)*. Therefore, this finding could indicate a role for hsp70 in the regulation of nuclear import of the liganded VDR.

Significantly, *dimerization* with RXR also influences nuclear import of VDR (Fig. 2). Dimerization with RXR is essential for both constitutive- and calcitriol-dependent functions of VDR. Not surprisingly, dimerization also affects the mobility of these receptors. The ability to generate multicolor fluorescent protein chimeras of nuclear receptors provided the means to study protein–protein interactions and receptor dynamics in living cells by microscopy in real time. Fluorescence resonance energy transfer (FRET) signal is generated when a higher wavelength acceptor protein comes into close proximity (less than 50 Å) with a complementary lower wavelength fluorescent protein. Increase in the ratio of intensities thus marks the interaction of donor and acceptor proteins in real time. Our FRET experiments using green VDR (acceptor) and blue RXR (donor) demonstrated the presence of VDR/RXR heterodimers with and without calcitriol treatment (Fig. 4). FRET detected VDR/RXR binding both in the cytoplasm and in the nucleus without calcitriol and mostly in the nucleus after calcitriol treatment *(83)* (Fig. 4).

We used co-expression experiments with wild-type and NLS mutants of VDR and RXR to study the impact of dimerization on receptor nuclear import (Fig. 2). Co-expression of wild-type RXR–BFP restored nuclear localization of *nls1*GFP–VDR, whereas co-expression of *nls1*RXR–BFP retained the wild-type GFP–VDR in the cytoplasm. Moreover, calcitriol treatment induced nuclear accumulation of *nls1*RXR–BFP when it was co-expressed with GFP–VDR. These data indicated that RXR dominates the nuclear import of unliganded VDR, whereas liganded VDR dominates the nuclear import of RXR. In addition, these data strongly supported the notion that VDR dimerizes with RXR in the cytoplasm.

Immunoprecipitation experiments elucidated the interactions of VDR and RXR with import receptors in more detail *(89)*. Findings in this study indicated that RXR binds to importin β, whereas VDR binds to importin α. Calcitriol treatment stabilized the interaction of VDR/RXR heterodimers with importin α, but inhibited RXR binding to importin β *(89)*. In another study, yeast two-hybrid screening used VDR-(4-232) as the bait to identify interacting proteins. Findings indicated that importin (α) 4 is the import receptor for VDR *(101)*. This finding was confirmed in digitonin permeabilized cells, as the addition of recombinant importin 4 facilitated nuclear import of both liganded and unliganded VDR, but importin beta failed to support the nuclear import of either liganded or unliganded VDR *(101)*.

To date, very little is known about the *regulation* of VDR or RXR import. This area is likely to attract considerable interest in the near future. Studies are needed to elucidate the roles of multiple receptor import mechanisms in the adaptation to diverse environmental stimuli.

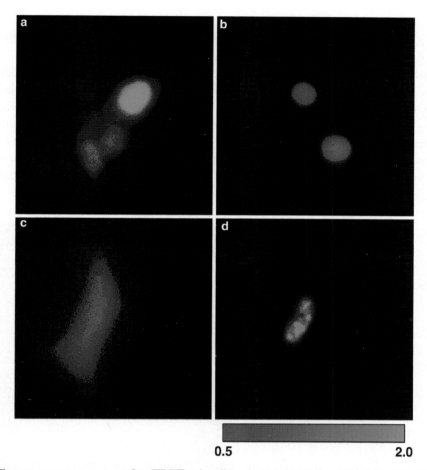

0.5   2.0

Fig. 4. Fluorescence energy transfer (FRET) visualizes interactions of unliganded GFP–VDR and RXR–BFP. CV-1 cells were cotransfected with GFP–VDR and RXR–BFP (*a* and *d*), GFP–VDR and *hd*RXR–BFP (*b*), or with BFP alone (*c*). Representative ratio images are shown from FRET experiments. Ratio images were generated from the images taken with a 370–390-nm excitation filter and with either a 515–555 or a 435–485-nm emission filter. Positive FRET signal is generated when the green GFP–VDR and blue RXR–BFP are closer than 50 Å and are labeled with green pseudocolor, whereas the low ratios are labeled with blue pseudocolor (see scale on *bottom*). Positive FRET signal indicate heterodimers; most of them are in the nuclei, and a few of them are in the cytoplasm (*a*). The ratio image from cells expressing the GFP–VDR and the *hd*RXR–BFP shows ratio values in the blue range because the BFP emission is stronger than the GFP emission (*b*). The ratio image from cells expressing only the BFP shows similar ratio values (*c*). After treating the GFP–VDR/RXR–BFP-expressing cells with 10 nM calcitriol, ratio images indicate the strongest green signal for heterodimers within intranuclear foci (*d*). *Bar*, 10 μm.

4. NUCLEAR EXPORT MECHANISMS FOR VDR AND RXR

Perhaps the most surprising finding about VDR and RXR localization was their short stay in the nucleus. A model that pictures these receptors as residing in the nucleus, once imported, most easily fits into a model that assumes a predominantly nuclear steady-state localization and the relatively long-lasting transcriptional regulatory effects of VDR and RXR. Unfortunately, this simplistic model was inconsistent with experimental data.

Photobleaching experiments demonstrated that both VDR and RXR rapidly shuttle between the cytoplasm and the nucleus (Fig. 5). Results indicated that both VDR and RXR frequently interact with import and export receptors, as the recovery half-times were 15–30 min for the unliganded VDR and 20–30 min for the unliganded RXR *(99)*. This rapid exchange is not unique to VDR and RXR; several subsequent kinetic studies with fluorescent chimeras of nuclear receptors demonstrated that most of them shuttle constantly between cytoplasm and nucleus with similar exchange rates *(23)*.

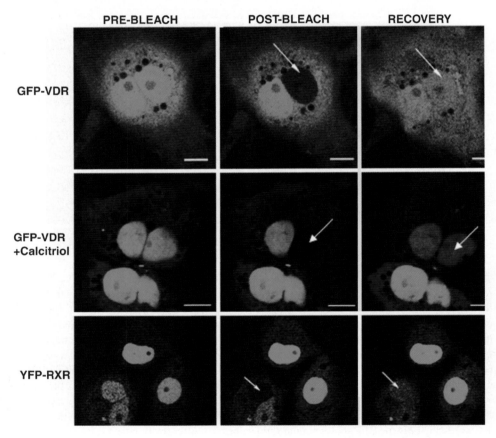

Fig. 5. Fluorescence recovery after photobleaching demonstrates that VDR and RXR shuttle between the cytoplasm and the nucleus. COS-7 cells expressing GFP–VDR (*upper and middle panels*) were treated with vehicle (0.1% ethanol; *upper*) or calcitriol (5 nM for 1 h; *middle*) before microscopy. COS7 cells expressing YFP–RXR (*lower panels*) were treated with vehicle. Cells with two nuclei were selected (*left*), and a 2 μm² spot (4 × 4 pixel) within one of the nuclei (*arrow*) was exposed to a focused laser beam at full power for 1 s/pixel. This intense illumination caused irreversible photobleaching (*darkening*) of all the receptors within the exposed nucleus, whereas the receptors in the other nucleus remained *bright*. After photobleaching, serial images were taken for measurement of fluorescence recovery. The images taken immediately after photobleaching are shown in the center panels and images taken at complete recovery are shown on the right panels. Bars, 10 μm.

Time-resolved confocal microscopy and photobleaching experiments demonstrated that the kinetics of VDR shuttling are hormone dose dependent *(87, 84)*. Our photobleaching experiments showed that an index for shuttling rate, the recovery half-time in the nucleus, was 5–15 min after treatment with 10-nM calcitriol *(99)*. Another photobleaching study monitored VDR import rates and found that calcitriol treatment increased the rate even in CaCo cells, in which hormone-induced nuclear accumulation of GFP–VDR was not apparent at steady state *(87)*.

The interactions of VDR with *export receptors* have been demonstrated using an inhibitor, mutational analysis, and microscopy on intact cells and experiments on permeabilized cells.

Leptomycin B (LMB) is a specific inhibitor of Crm-1 mediated export; it prevents binding of cargo proteins to Crm-1 *(39)*. The nuclear export of unliganded GFP–VDR was inhibited by LMB treatment, resulting in an exclusively nuclear receptor accumulation in intact cells (Fig. 6) and diminished loss of GFP–VDR from the nucleus in permeabilized cells *(99)*. The export of RXR was not sensitive to LMB *(99)*. After calcitriol treatment, the rate of GFP–VDR export tripled and this faster export was no longer sensitive to LMB, indicating that the liganded VDR uses an export mechanism different from the unliganded VDR, in that it no longer uses Crm-1 for export.

We searched for the NES, the segment of VDR that binds Crm-1, using mutational analysis. Results revealed that mutations of the leucines 320–325 to alanines in the ligand-binding domain of VDR abrogated LMB sensitivity *(99)*. We then explored the involvement of calreticulin in VDR and RXR export using mutations of the conserved calreticulin-binding sites in the context of fluorescent protein chimeras of the receptors, as well as permeabilized cell export assays. These studies indicated that mutations of

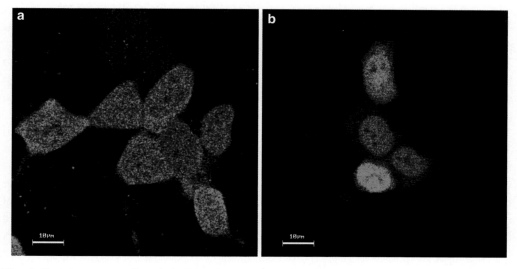

Fig. 6. Crm-1 exports unliganded VDR. GL48 cells (a HEK293-derived cell line stably expressing GFP–VDR) were treated with vehicle (DMSO) or LMB (2 nM for 3 h) before microscopy. Treatment with LMB caused nuclear accumulation of GFP–VDR by preventing its Crm-1-mediated export. Images of living cells were taken by confocal microscope (Zeiss LSM410). Bars, 10 μm.

this region in both GFP–VDR (F47A, F48A) and YFP–RXR (F158A, F159A) inhibited nuclear export *(99)*.

VDR and RXR export may also involve *indirect* mechanisms, through "piggybacking" on binding protein partners. The contribution of this mechanism was indicated by Katagiri and his colleagues, who showed that treatment with NGF induced phosphorylation of orphan nuclear receptor NGFI-B (also called nur77) at amino acid 105 and that this phosphorylation induced NGFI-B export together with the export of the dimerization partner, RXR *(46)*. Similar "piggybacking" could occur using a corepressors partner that interacts with Crm-1 *(102)*.

The *physiological importance* of VDR and RXR export is not yet understood. We found that inhibition of VDR export decreased calcitriol-induced transcriptional activity. ROS-AI cells, which stably express a 24-hydroxylase-luciferase reporter, were first treated with either vehicle (DMSO) or with LMB (2 nM for 3 h), and then treated with 10-nM calcitriol for 24 h. Calcitriol induced 29-fold increase in luciferase reporter expression in vehicle pretreated cells, whereas calcitriol induced only a 13-fold induction in LMB pretreated cells (Barsony, Prufer, unpublished). This indicated a role for receptor export in transcriptional activity. Moreover, VDR is likely to have cytoplasmic functions; perhaps cytoplasmic VDR participates in the "nongenomic" or ligand-independent actions. It is also possible that cytoplasmic localization is necessary for efficient receptor degradation and rapid receptor turnover.

5. INTRANUCLEAR TRAFFICKING OF VDR AND RXR

One of the long ignored aspects of VDR function was the intranuclear dynamics, addressing issues ranging from the mechanisms by which the receptor finds target sites within the nucleus to the kinetics of receptor binding and release from DNA target sequences, as well as interactions with the nuclear matrix and with the intranuclear proteasomal degradation machinery. Progress in this area required a change of thinking. The generally accepted model of VDR functions stipulated that VDR, together with RXR, remains bound to the promoter regions of target genes for the entire duration of hormone-induced transcriptional control. This concept obscured the complexity and dynamic nature of intranuclear receptor functions.

Results from our photobleaching experiments demonstrated that GFP–VDR and YFP–RXR are in rapid motion within the cell nucleus. For FRAP, a 0.2 μm^2 area (less than 5% of total nuclear area) corresponding to a discrete bright spot within the nucleus was subjected to intense laser illumination for 0.5 s, which caused the receptors in the illuminated area to completely lose fluorescence. Following this photobleaching, serial images were taken and fluorescence recovery was calculated in the area that had been bleached. Data were obtained from at least 20 nuclei of COS-7 cells. The residence times for the unliganded GFP–VDR and the YFP–RXR in randomly selected locations were about 2 s, consistent with a facilitated diffusion mechanism. Since then, several studies described the intranuclear motion of various steroid receptors based on photobleaching experiments and found similar residence time *(103–107)*. We found that calcitriol binding decreased the intranuclear mobility of GFP–VDR; the residence time for the liganded receptor was 5–7 s (Barsony et al. unpublished). This finding is also consistent with studies on other NRs.

Hormone treatment induces focal accumulation of most nuclear receptors, including the VDR and RXR and their dimer (Figs. 4 and 7) *(83)*. Mutational analysis indicated that the focal accumulation of GFP–VDR signifies binding to regulatory DNA sequences *(83)* (Fig. 7). A mutation that prevented DNA binding of VDR prevented the formation of discrete bright foci in the nuclei of GFP–VDR-expressing cells. In addition, colocalization experiments showed that after calcitriol treatment, GFP–VDR and RXR–BFP bind to the same DNA-binding sites (foci) *(108)*. We used fluorescence loss in photobleaching (FLIP) measurements to study the kinetics of GFP–VDR exchange between foci and areas of diffuse signal. For FLIP, a 0.4 μm^2 area corresponding to diffuse nuclear signal distant from the selected bright spot was subjected to intense laser illumination for increasing duration (0.5 s to 1 min) in 1 s increments. Loss of fluorescence was measured over the selected bright spot to detect the time-dependent decrease in fluorescence intensities, as more and more receptors turned black in the nucleus due to photobleaching. As expected, shorter photoexposure caused loss of fluorescence outside the foci (6.5 s), and longer photoexposure caused complete loss of fluorescence even in the foci (42 s). From this difference, the exchange rate between receptor and DNA-binding sites and residence time at foci was calculated. Interestingly, the residence time is not uniform among foci, indicating variability in the affinity of DNA-binding sites for the receptor. Similar FLIP experiments in BFP–VDR-expressing COS-7 cells characterized the residence time of YFP–RXR at the foci. Results indicated that the residence time of VDR and RXR is not identical at every focus (Barsony et al. unpublished).

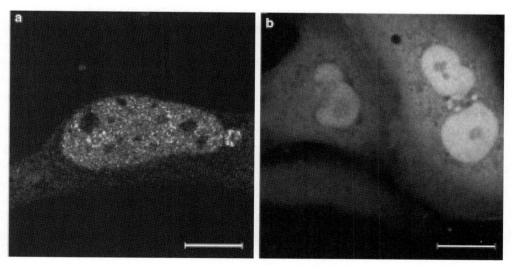

Fig. 7. Calcitriol-induced nuclear GFP–VDR foci formation requires intact DNA binding. A mutation (R77Q) was introduced in the DNA-binding region of GFP–VDR to create *dna*GFP–VDR *(83)*. GFP–VDR and *dna*GFP–VDR were stably expressed in CV-1 cells. (**a**) Treatment with calcitriol (10 nM calcitriol for 1 h) caused formation of bright foci. (**b**) Calcitriol did not cause focal accumulation of *dna*GFP–VDR. Images of living cells were taken by confocal microscope (Zeiss LSM410). Bars, 10 μm.

Published results from ChIP assays on osteoblastic cells after calcitriol treatment also indicated that there is a rapid and cyclical association of VDR with the osteopontin promoter *(109)*. The kinetic analysis, though, was limited to binding assessed in 15-min intervals over a period of 165 min after treatment with calcitriol. Despite this limitation, the transient nature of VDR binding to DNA target sites was evident.

Larger aggregates of unliganded GFP–VDR formed in cells that overexpressed the recombinant receptor. There was no rapid exchange of GFP–VDR at these sites. Very similar aggregates of GFP–VDR formed in the nucleus after treatment with inhibitors of proteasomal degradation. These aggregates likely signify nuclear domains for the interaction of VDR with active proteolytic centers *(110–112)*.

Last, but not least, several studies indicated that VDR interacts with the nuclear matrix *(113, 114)*. ChIP assays indicated that VDR interaction with matrix-bound coregulators is cyclical *(113)*. This interaction may contribute to the mechanisms that direct the receptors and other transcription factors to the sites of active transcription or to domains of RNA splicing. The kinetics of VDR interaction with the matrix-bound proteins await to be explored in living cells.

6. SUMMARY AND CONCLUSION

This review of our current understanding of the rules and regulators of VDR and RXR subcellular trafficking was presented in the context of the general mechanisms of protein transport across the nuclear membrane and the mechanisms of subcellular trafficking of other members of the NR protein superfamily. Along this line of investigation, data regarding the subcellular trafficking of coregulators are also accumulating. Interactions between receptors and protein partners can now be described by taking into account space and time, using multiple-color fluorescent protein chimeras and FRET, but these studies have yet to be extended to VDR and to be integrated into the model of transcriptional control by VDR. Results from ongoing biochemical analysis of transcriptional regulation are still easier to fit into a static model, but we must understand that the multitude of possible protein interactions seldom takes place within the compartmentalized environment of cells. Therefore, live cell studies are essential for the detection of protein interactions.

We described receptor import and export mechanisms and the mutational analysis that defined the regions of VDR and RXR that interact with known import and export receptors. The mechanisms that regulate VDR and RXR import and export in response to cellular events have yet to be elucidated. Advanced microscopy techniques generated an abundance of experimental results that argued against the belief that steroid receptor complexes are statically bound to regulatory sites in hormone-responsive promoters and suggested that receptors and their cofactors must interact with the response element repeatedly during transcriptional activation. A new model depicting a transient association of NRs with target sites is gaining broad acceptance, and available data suggest that VDR and RXR are not exceptions in this regard. As we learn more about the nuclear architecture, chromatin remodeling and global transcriptional regulation, our insight into the mechanisms specific for VDR and RXR mobility will be even more essential. This field of study developed rapidly until recently. As resources are dwindling for basic science, concerns about the applicability of new findings on receptor trafficking and feasibility of experimental plans emerged. It appears sometimes that instead of live cell experiments, we should favor doing studies on the "train" while it is resting in the station (test tubes). However, change is in the air to bring back enthusiasm to this field.

REFERENCES

1. Lee SI, Pe'er D, Dudley AM, Church GM, Koller D (2006) Identifying regulatory mechanisms using individual variation reveals key role for chromatin modification. *Proc Natl Acad Sci USA* 103:14062–14067
2. Brockmann R, Beyer A, Heinisch JJ, Wilhelm T (2007) Posttranscriptional expression regulation: what determines translation rates? *PLoS Comput Biol* 3:e57
3. Liu ET (2005) Systems biology, integrative biology, predictive biology. *Cell* 121:505–506
4. Barsony J, McKoy W, DeGrange DA, Liberman UA, Marx SJ (1989) Selective expression of a normal action of the 1,25-dihydroxyvitamin D3 receptor in human skin fibroblasts with hereditary severe defects in multiple actions of that receptor. *J Clin Invest* 83:2093–2101
5. Bentivegna A *et al* (2006) Rubinstein-Taybi Syndrome: spectrum of CREBBP mutations in Italian patients. *BMC. Med. Genet.* 7:77
6. Cutress ML, Whitaker HC, Mills IG, Stewart M, Neal DE (2008) Structural basis for the nuclear import of the human androgen receptor. *J Cell Sci* 121:957–968
7. Nikolova G, Vilain E (2006) Mechanisms of disease: Transcription factors in sex determination–relevance to human disorders of sex development. *Nat Clin Pract Endocrinol Metab* 2:231–238
8. Kiriyama T *et al* (2008) Restoration of nuclear-import failure caused by triple A syndrome and oxidative stress. *Biochem Biophys Res Commun* 249:745–753
9. Bedard JE, Purnell JD, Ware SM (2007) Nuclear import and export signals are essential for proper cellular trafficking and function of ZIC3. *Hum Mol Genet* 16:187–198
10. Collier J, Shapiro L (2007) Spatial complexity and control of a bacterial cell cycle. *Curr Opin Biotechnol* 18:333–340
11. Burgermeister E, Seger R (2007) MAPK kinases as nucleo-cytoplasmic shuttles for PPARgamma. *Cell Cycle* 6:1539–1548
12. Birbach A *et al* (2002) Signaling molecules of the NF-kappa B pathway shuttle constitutively between cytoplasm and nucleus. *J Biol. Chem* 277:10842–10851
13. Aplin AE, Hogan BP, Tomeu J, Juliano RL (2002) Cell adhesion differentially regulates the nucleo-cytoplasmic distribution of active MAP kinases. *J Cell Sci* 115:2781–2790
14. Meyer T, Marg A, Lemke P, Wiesner B, Vinkemeier U (2003) DNA binding controls inactivation and nuclear accumulation of the transcription factor Stat1. *Genes Dev* 17:1992–2005
15. Xiao Z, Brownawell AM, Macara IG, Lodish HF (2003) A novel nuclear export signal in Smad1 is essential for its signaling activity. *J Biol Chem* 278:34245–34252
16. O'Keefe K, Li H, Zhang Y (2003) Nucleocytoplasmic shuttling of p53 is essential for MDM2-mediated cytoplasmic degradation but not ubiquitination. *Mol Cell Biol* 23:6396–6405
17. Wadhwa R et al (2002) Hsp70 family member, mot-2/mthsp70/GRP75, binds to the cytoplasmic sequestration domain of the p53 protein. *Exp Cell Res* 274:246–253
18. Stommel JM *et al* (1999) A leucine-rich nuclear export signal in the p53 tetramerization domain: regulation of subcellular localization and p53 activity by NES masking. *EMBO J* 18:1660–1672
19. Kanai M *et al* (2007) Inhibition of Crm1-p53 interaction and nuclear export of p53 by poly(ADP-ribosyl)ation. *Nat. Cell Biol.* 9:1175–1183
20. Foo RS *et al* (2007) Regulation of p53 tetramerization and nuclear export by ARC. *Proc Natl Acad Sci USA* 104:20826–20831
21. Nie L, Sasaki M, Maki CG (2007) Regulation of p53 nuclear export through sequential changes in conformation and ubiquitination. *J Biol Chem* 282:14616–14625
22. Bonamy GM, Allison LA (2006) Oncogenic conversion of the thyroid hormone receptor by altered nuclear transport. *Nucl Recept Signal* 4:e008

23. Bunn CF *et al* (2001) Nucleocytoplasmic shuttling of the thyroid hormone receptor alpha. *Mol Endocrinol* 15:512–533

24. Zhou ZX, Sar M, Simental JA, Lane MV, Wilson EM (1994) A ligand-dependent bipartite nuclear targeting signal in the human androgen receptor. Requirement for the DNA-binding domain and modulation by NH2-terminal and carboxyl-terminal sequences. *J Biol Chem* 269:13115–13123

25. Picard D, Yamamoto KR (1987) Two signals mediate hormone-dependent nuclear localization of the glucocorticoid receptor. *EMBO J* 6:3333–3340

26. Poukka H *et al* (2000) The RING finger protein SNURF modulates nuclear trafficking of the androgen receptor. *J Cell Sci* 113(Pt 17):2991–3001

27. Kaku N, Matsuda K, Tsujimura A, Kawata M (2008) Characterization of nuclear import of the domain-specific androgen receptor in association with the importin alpha/beta and Ran-guanosine 5'-triphosphate systems. *Endocrinology* 149:3960–3969

28. Palmeri D, Malim MH (1999) Importin beta can mediate the nuclear import of an arginine-rich nuclear localization signal in the absence of importin alpha. *Mol Cell Biol* 19:1218–1225

29. Guo D *et al* (2006) Peroxisome proliferator-activated receptor (PPAR)-binding protein (PBP) but not PPAR-interacting protein (PRIP) is required for nuclear translocation of constitutive androstane receptor in mouse liver. *Biochem Biophys Res Commun* 347:485–495

30. Grespin ME *et al* (2008) Thyroid hormone receptor alpha 1 follows a cooperative CRM1/calreticulin-mediated nuclear export pathway. *J Biol Chem* 283(16):10568–10580

31. Nonclercq D *et al* (2007) Effect of nuclear export inhibition on estrogen receptor regulation in breast cancer cells. *J Mol Endocrinol* 39:105–118

32. Zimmerman TL, Thevananther S, Ghose R, Burns AR, Karpen SJ (2006) Nuclear export of retinoid X receptor alpha in response to interleukin-1beta-mediated cell signaling: roles for JNK and SER260. *J Biol Chem* 281:15434–15440

33. Amazit L *et al* (2003) Subcellular localization and mechanisms of nucleocytoplasmic trafficking of steroid receptor coactivator-1. *J Biol Chem* 278:32195–32203

34. Holaska JM *et al* (2001) Calreticulin Is a receptor for nuclear export. *J Cell Biol* 152:127–140

35. Mingot JM, Bohnsack MT, Jakle U, Gorlich D (2004) Exportin 7 defines a novel general nuclear export pathway. *EMBO J* 23:3227–3236

36. Kino T et al (2003) Protein 14-3-3sigma interacts with and favors cytoplasmic subcellular localization of the glucocorticoid receptor, acting as a negative regulator of the glucocorticoid signaling pathway. *J Biol Chem* 278:25651–25656

37. Stade K, Ford CS, Guthrie C, Weis K (1997) Exportin 1 (Crm1p) is an essential nuclear export factor. *Cell* 90:1041–1050

38. Fornerod M, Ohno M (2002) Exportin-mediated nuclear export of proteins and ribonucleoproteins. *Results Probl Cell Differ* 35:67–91

39. Ossareh-Nazari B, Bachelerie F, Dargemont C (1997) Evidence for a role of CRM1 in signal-mediated nuclear protein export. *Science* 278:141–144

40. Zilliacus J *et al* (2001) Regulation of glucocorticoid receptor activity by 14-3-3-dependent intracellular relocalization of the corepressor rip140. *Mol Endocrinol* 15:501–511

41. Masuyama H, Jefcoat SCJ, MacDonald PN (1997) The N-terminal domain of transcription factor IIB is required for direct interaction with the vitamin D receptor and participates in vitamin D-mediated transcription. *Mol Endocrinol* 11:218–228

42. Wiebel FF, Steffensen KR, Treuter E, Feltkamp D, Gustafsson JA (1999) Ligand-independent coregulator recruitment by the triply activatable OR1/retinoid X receptor-alpha nuclear receptor heterodimer [In Process Citation]. *Mol Endocrinol* 13:1105–1118

43. Terry LJ, Shows EB, Wente SR (2007) Crossing the nuclear envelope: hierarchical regulation of nucleocytoplasmic transport. *Science* 318:1412–1416

44. Heger P, Lohmaier J, Schneider G, Schweimer K, Stauber RH (2001) Qualitative highly divergent nuclear export signals can regulate export by the competition for transport cofactors in vivo. *Traffic* 2:544–555

45. Amazit L *et al* (2007) Regulation of SRC-3 intercompartmental dynamics by estrogen receptor and phosphorylation. *Mol Cell Biol* 27:6913–6932

46. Katagiri Y *et al* (2000) Modulation of retinoid signalling through NGF-induced nuclear export of NGFI-B. *Nat Cell Biol* 2:435–440

47. Mostaqul H *et al* (2006) Suppression of receptor interacting protein 140 repressive activity by protein arginine methylation. *EMBO J* 25:5094–5104

48. Li M *et al* (1998) Nuclear inclusions of the androgen receptor protein in spinal and bulbar muscular atrophy. *Ann Neurol* 44:249–254

49. Ruden DM, Xiao L, Garfinkel MD, Lu X (2005) Hsp90 and environmental impacts on epigenetic states: a model for the trans-generational effects of diethylstibesterol on uterine development and cancer. *Hum Mol Genet* 14(Spec No 1):R149–R155

50. Carrigan A *et al* (2007) An active nuclear retention signal in the glucocorticoid receptor functions as a strong inducer of transcriptional activation. *J Biol Chem* 282:10963–10971

51. Yasuhara N *et al* (2007) Triggering neural differentiation of ES cells by subtype switching of importin-alpha. *Nat Cell Biol* 9:72–79

52. Gluz O *et al* (2008) Nuclear karyopherin alpha2 expression predicts poor survival in patients with advanced breast cancer irrespective of treatment intensity. *Int J Cancer* 123:1433–1438

53. Fagerlund R, Kinnunen L, Kohler M, Julkunen I, Melen K (2005) NF-{kappa}B is transported into the nucleus by importin {alpha}3 and importin {alpha}4. *J Biol Chem* 280:15942–15951

54. Riddick G, Macara IG (2005) A systems analysis of importin-{alpha}-{beta} mediated nuclear protein import. *J Cell Biol* 168:1027–1038

55. Erickson ES, Mooren OL, Moore D, Krogmeier JR, Dunn RC (2006) The role of nuclear envelope calcium in modifying nuclear pore complex structure. *Can J Physiol Pharmacol* 84:309–318

56. Sweitzer TD, Hanover JA (1996) Calmodulin activates nuclear protein import: a link between signal transduction and nuclear transport. *Proc Natl Acad Sci USA* 93:14574–14579

57. Metivier R, Reid G, Gannon F (2006) Transcription in four dimensions: nuclear receptor-directed initiation of gene expression. *EMBO Rep* 7:161–167

58. Han SJ, Tsai SY, Tsai MJ, Omalley BW (2007) Distinct temporal and spatial activities of RU486 on progesterone receptor function in reproductive organs of ovariectomized mice. *Endocrinology* 148:2471–2486

59. Hall JM, McDonnell DP (2005) Coregulators in nuclear estrogen receptor action: from concept to therapeutic targeting. *Mol Interv* 5:343–357

60. Schaufele F *et al* (2000) Temporally distinct and ligand-specific recruitment of nuclear receptor-interacting peptides and cofactors to subnuclear domains containing the estrogen receptor. *Mol Endocrinol* 14:2024–2039

61. Lonard DM, O'malley BW (2007) Nuclear receptor coregulators: judges, juries, and executioners of cellular regulation. *Mol Cell* 27:691–700

62. Aoyagi S, Archer TK (2008) Dynamics of coactivator recruitment and chromatin modifications during nuclear receptor mediated transcription. *Mol Cell Endocrinol* 280:1–5

63. Gruenbaum Y, Margalit A, Goldman RD, Shumaker DK, Wilson KL (2005) The nuclear lamina comes of age. *Nat Rev Mol Cell Biol* 6:21–31

64. Lanctot C, Cheutin T, Cremer M, Cavalli G, Cremer T (2007) Dynamic genome architecture in the nuclear space: regulation of gene expression in three dimensions. *Nat Rev Genet* 8:104–115

65. Zaidi SK *et al* (2007) Nuclear microenvironments in biological control and cancer. *Nat Rev Cancer* 7:454–463

66. Misteli T (2008) Physiological importance of RNA and protein mobility in the cell nucleus. *Histochem Cell Biol* 129:5–11

67. Gasser SM (2002) Visualizing chromatin dynamics in interphase nuclei. *Science* 296:1412–1416

68. Dundr M *et al* (2007) Actin-dependent intranuclear repositioning of an active gene locus in vivo. *J Cell Biol* 179:1095–1103

69. Pratt WB, Galigniana MD, Harrell JM, DeFranco DB (2004) Role of hsp90 and the hsp90-binding immunophilins in signalling protein movement. *Cell Signal* 16:857–872

70. Matsuda K, Nishi M, Takaya H, Kaku N, Kawata M (2008) Intranuclear mobility of estrogen receptor alpha and progesterone receptors in association with nuclear matrix dynamics. *J Cell Biochem* 103:136–148

71. Maruvada P, Baumann CT, Hager GL, Yen PM (2003) Dynamic shuttling and intranuclear mobility of nuclear hormone receptors. *J Biol Chem* 278:12425–12432

72. Prufer K, Boudreaux J (2007) Nuclear localization of liver X receptor alpha and beta is differentially regulated. *J Cell Biochem* 100:69–85

73. Marcelli M et al (2006) Quantifying effects of ligands on androgen receptor nuclear translocation, intranuclear dynamics, and solubility. *J Cell Biochem* 98:770–788

74. Meijsing SH, Elbi C, Luecke HF, Hager GL, Yamamoto KR (2007) The ligand binding domain controls glucocorticoid receptor dynamics independent of ligand release. *Mol Cell Biol* 27:2442–2451

75. Zhu PJ *et al* (2008) A miniaturized glucocorticoid receptor translocation assay using enzymatic fragment complementation evaluated with qHTS. *Comb Chem High Throughput Screen* 11:545–559

76. Davis JR, Kakar M, Lim CS (2007) Controlling protein compartmentalization to overcome disease. *Pharm Res* 24:17–27

77. Torgerson TR, Colosia AD, Donahue JP, Lin YZ, Hawiger J (1998) Regulation of NF-kappa B, AP-1, NFAT, and STAT1 nuclear import in T lymphocytes by noninvasive delivery of peptide carrying the nuclear localization sequence of NF-kappa B p50. *J Immunol* 161:6084–6092

78. Liu D *et al* (2004) Nuclear import of proinflammatory transcription factors is required for massive liver apoptosis induced by bacterial lipopolysaccharide. *J Biol Chem* 279:48434–48442

79. Kakar M, Davis JR, Kern SE, Lim CS (2007) Optimizing the protein switch: altering nuclear import and export signals, and ligand binding domain. *J Control Release* 120:220–232

80. Dong S, Stenoien DL, Qiu J, Mancini MA, Tweardy DJ (2004) Reduced intranuclear mobility of APL fusion proteins accompanies their mislocalization and results in sequestration and decreased mobility of retinoid X receptor alpha. *Mol Cell Biol* 24:4465–4475

81. Buentig N *et al* (2004) Predictive impact of retinoid X receptor-alpha-expression in renal-cell carcinoma. *Cancer Biother Radiopharm* 19:331–342

82. Barsony J, Pike JW, DeLuca HF, Marx SJ (1990) Immunocytology with microwave-fixed fibroblasts shows 1 alpha,25-dihydroxyvitamin D3-dependent rapid and estrogen-dependent slow reorganization of vitamin D receptors. *J Cell Biol* 111:2385–2395

83. Prufer K, Racz A, Lin GC, Barsony J (2000) Dimerization with retinoid X receptors promotes nuclear localization and subnuclear targeting of vitamin D receptors. *J Biol Chem* 275:41114–41123

84. Racz A, Barsony J (1999) Hormone-dependent translocation of vitamin D receptors is linked to transactivation. *J Biol Chem* 274:19352–19360

85. Michigami T *et al* (1999) Identification of amino acid sequence in the hinge region of human vitamin D receptor that transfers a cytosolic protein to the nucleus. *J Biol Chem* 274:33531–33538

86. Sunn KL, Cock TA, Crofts LA, Eisman JA, Gardiner EM (2001) Novel N-terminal variant of human VDR. *Mol Endocrinol* 15:1599–1609

87. Klopot A, Hance KW, Peleg S, Barsony J, Fleet JC (2006) Nucleo-cytoplasmic cycling of the vitamin D receptor in the enterocyte-like cell line, Caco-2. *J Cell Biochem*

88. Lin XF *et al* (2004) RXRalpha acts as a carrier for TR3 nuclear export in a 9-cis retinoic acid-dependent manner in gastric cancer cells. *J Cell Sci* 117:5609–5621

89. Yasmin R, Williams RM, Xu M, Noy N (2005) Nuclear import of the retinoid X receptor, the vitamin D receptor, and their mutual heterodimer. *J Biol Chem* 280:40152–40160

90. Putkey JA, Wecksler WR, Norman AW (1978) The interaction of 1,25-dihydroxyvitamin D3 with its intestinal mucosa receptor: kinetic parameters and structural requirements. *Lipids* 13:723–729

91. Clemens TL et al (1988) Immunocytochemical localization of the 1,25-dihydroxyvitamin D3 receptor in target cells. *Endocrinology* 122:1224–1230

92. Dornas RA *et al* (2007) Distribution of vitamin D3 receptor in the epididymal region of roosters (Gallus domesticus) is cell and segment specific. *Gen. Comp Endocrinol* 150:414–418

93. Prufer K, Schroder CRABJ (2000) Cell cycle dependence of vitamin D receptor expression. In: Norman AW, Boillon R, Thomasset M (eds) Vitamin D Endocrine System: structural, biological, genetic and clinical aspects. Vitamin D Workshop, Inc., Riverside, CA, 379–382

94. Wu-Wong JR, Nakane M, Ma J, Ruan X, Kroeger PE (2007) VDR-mediated gene expression patterns in resting human coronary artery smooth muscle cells. *J Cell Biochem* 100:1395–1405

95. Reichrath J, Mittmann M, Kamradt J, Muller SM (1997) Expression of retinoid-X receptors (-alpha,-beta,-gamma) and retinoic acid receptors (-alpha,-beta,-gamma) in normal human skin: an immuno-histological evaluation. *Histochem J* 29:127–133

96. Hsieh JC *et al* (1998) Novel nuclear localization signal between the two DNA-binding zinc fingers in the human vitamin D receptor. *J Cell Biochem* 70:94–109

97. Yu Z, Lee CH, Chinpaisal C, Wei LN (1998) A constitutive nuclear localization signal from the second zinc-finger of orphan nuclear receptor TR2. *J Endocrinol* 159:53–60

98. Luo Z, Rouvinen J, Maenpaa PH (1994) A peptide C-terminal to the second Zn finger of human vitamin D receptor is able to specify nuclear localization. *Eur J Biochem* 223:381–387

99. Prufer K, Barsony J (2002) Retinoid X receptor dominates the nuclear import and export of the unliganded vitamin D receptor. *Mol Endocrinol* 16:1738–1751

100. Swamy N, Mohr SC, Xu W, Ray R (1999) Vitamin D receptor interacts with DnaK/heat shock protein 70: identification of DnaK interaction site on vitamin D receptor. *Arch Biochem Biophys* 363:219–226

101. Miyauchi Y *et al* (2005) Importin 4 is responsible for ligand-independent nuclear translocation of vitamin D receptor. *J Biol Chem* 280:40901–40908

102. Hong SH, Privalsky ML (2000) The SMRT corepressor is regulated by a MEK-1 kinase pathway: inhibition of corepressor function is associated with SMRT phosphorylation and nuclear export. *Mol Cell Biol* 20:6612–6625

103. Stenoien DL *et al* (2000) Subnuclear trafficking of estrogen receptor-alpha and steroid receptor coactivator-1. *Mol Endocrinol* 14:518–534

104. Stenoien DL *et al* (2001) FRAP reveals that mobility of oestrogen receptor-alpha is ligand- and proteasome-dependent. *Nat Cell Biol* 3:15–23

105. Akiyama TE, Baumann CT, Sakai S, Hager GL, Gonzalez FJ (2002) Selective intranuclear redistribution of PPAR isoforms by RXR alpha. *Mol Endocrinol* 16:707–721

106. Becker M *et al* (2002) Dynamic behavior of transcription factors on a natural promoter in living cells. *EMBO Rep* 3:1188–1194

107. McNally JG, Muller WG, Walker D, Wolford R, Hager GL (2000) The glucocorticoid receptor: rapid exchange with regulatory sites in living cells. *Science* 287:1262–1265

108. Barsony J, Prufer K (2002) Vitamin D receptor and retinoid X receptor interactions in motion. *Vitam Horm* 65:345–376

109. Kim S, Shevde NK, Pike JW (2005) 1,25-Dihydroxyvitamin D3 stimulates cyclic vitamin D receptor/retinoid X receptor DNA-binding, co-activator recruitment, and histone acetylation in intact osteoblasts. *J Bone Miner Res* 20:305–317

110. Masuyama H, MacDonald PN (1998) Proteasome-mediated degradation of the vitamin D receptor (VDR) and a putative role for SUG1 interaction with the AF-2 domain of VDR. *J Cell Biochem* 71:429–440

111. Prufer K, Schroder C, Hegyi K, Barsony J (2002) Degradation of RXRs influences sensitivity of rat osteosarcoma cells to the antiproliferative effects of calcitriol. *Mol Endocrinol* 16:961–976

112. Chen M, Singer L, Scharf A, von Mikecz A (2008) Nuclear polyglutamine-containing protein aggregates as active proteolytic centers. *J Cell Biol* 180:697–704

113. Zhang C *et al* (2003) Nuclear coactivator-62 kDa/Ski-interacting protein is a nuclear matrix-associated coactivator that may couple vitamin D receptor-mediated transcription and RNA splicing. *J Biol Chem* 278:35325–35336

114. Lian JB *et al* (2001) Contributions of nuclear architecture and chromatin to vitamin D-dependent transcriptional control of the rat osteocalcin gene. *Steroids* 66:159–170

7

Mechanism of Action of 1,25-Dihydroxyvitamin D$_3$ on Intestinal Calcium Absorption and Renal Calcium Transport

Dare Ajibade, Bryan S. Benn, and Sylvia Christakos

Abstract   In the intestine 1,25-dihydroxyvitamin D$_3$ [1,25(OH)$_2$D] induces various aspects of the transcellular active calcium transport system including calbindin, the basolateral plasma membrane pump, and the epithelial calcium channel, TRPV6. Calcium can enter the enterocyte through the epithelial calcium channel and then bind to calbindin and move through the cytosol. At the basolateral membrane calcium is transported actively by the plasma membrane calcium pump into the extracellular space. There is increasing evidence that, in the intestine, 1,25(OH)$_2$D can also enhance paracellular calcium diffusion.

In the distal nephron of the kidney 1,25(OH)$_2$D induces TRPV5 and the calbindins and affects calcium transport, at least in part, by enhancing the action of PTH. A role for the Na$^+$/Ca^{2+} exchanger in the distal tubule in vitamin D-dependent calcium reabsorption has also been suggested. In the kidney, besides enhancement of calcium transport in the distal nephron, 1,25(OH)$_2$D also modulates the 25(OH)D hydroxylases. Effects on renal phosphate reabsorption have also been suggested.

Key Words: 1,25-dihydroxyvitamin D; calcium transport; intestine; kidney; renal calcium transport; TRPV6; calmodulin; calbindin; VDR; claudin; phosphate

1. INTRODUCTION

In mammals, the plasma calcium concentration under normal conditions is maintained at 2.5mM or 10 mg/dL. Plasma calcium may be further divided into ionized calcium (45%), calcium bound to proteins (45%), and calcium complexed with small anions (10%). Vitamin D is the principle factor that maintains normal plasma calcium concentrations, and its three target sites of action, the intestine, kidney, and bone, are primarily responsible for this maintenance of calcium homeostasis (1). This chapter focuses on how vitamin D, specifically its active form, 1,25-dihydroxyvitamin D$_3$ [1,25(OH)$_2$D], acts at times of increased calcium demand to increase the efficiency of

From: *Nutrition and Health: Vitamin D*
Edited by: M.F. Holick, DOI 10.1007/978-1-60327-303-9_7,
© Springer Science+Business Media, LLC 2010

calcium absorption from the intestine and to enhance the tubular reabsorption of calcium from the kidney.

2. INTESTINAL CALCIUM ABSORPTION

2.1. Overview

As the fifth most abundant element in the human body, calcium is an essential player in many physiological processes (2). In order to serve these multiple functions, calcium from the diet must be ingested and absorbed by the body. Previous studies have shown that $1,25(OH)_2D$ is the major controlling hormone of intestinal calcium absorption (1, 3). As the body's demand for calcium increases, the synthesis of $1,25(OH)_2D$ is increased, thus stimulating the rate of calcium absorption (1). Furthermore, it has been suggested that the antirachitic action of $1,25(OH)_2D$ is due, at least in part, to its ability to increase intestinal calcium absorption, resulting in the increased availability of calcium for incorporation into bone (1).

The generally accepted view of intestinal calcium absorption is that it is a phenomenon comprised of two different modes of calcium transport: a saturable process and a nonsaturable process. Saturable calcium transport involves the rapid, active transport of calcium, predominates at calcium concentrations between 1 and 10 mM, and is mainly transcellular. In contrast, nonsaturable calcium transport involves the slower, diffusional transport of calcium, predominates at much higher intraluminal calcium concentrations (10–50 mM), and is mainly paracellular. Transcellular calcium transport is believed to be comprised of three vitamin D-dependent component steps that result in net calcium absorption: (1) the entry of calcium from the intestinal lumen across the brush border membrane, (2) the transceullar movement of calcium through the cytosol of the enterocyte, and (3) the energy-requiring extrusion of calcium against a concentration gradient at the basolateral membrane of the enterocyte into the lamina propria and eventually into the plasma (1, 3–5). Paracellular calcium transport is believed to be due to passive movement of calcium across tight junctions and intracellular spaces and is directly related to the concentration of calcium in the lumen of the small intestine (6). Furthermore, it is thought that while active transcellular calcium absorption is observed predominately in the proximal portion of the intestine (but can occur at a slower rate in other parts of the intestine; (3)), paracellular calcium absorption occurs at a similar rate throughout the length of the small intestine (duodenum, jejunum, and ileum). $1,25(OH)_2D$ has been reported to affect both the transcellular, saturable process (3–5) and the paracellular path (3, 6) (Fig. 1).

2.2. Effect of 1,25-Dihydroxyvitamin D₃ on Saturable, Transcellular Intestinal Calcium Absorption

To clarify further the process of intestinal calcium absorption, each event in the saturable process (calcium entry, transcellular movement of calcium, and calcium extrusion) is discussed in relation to the role of $1,25(OH)_2D$.

Transcellular   **Paracellular**

Vitamin D adequate- Duodenum

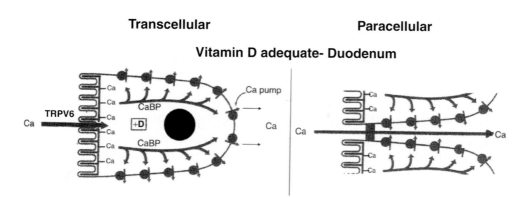

Fig. 1. Models of vitamin D-mediated processes of intestinal calcium absorption. *Left panel*: Transcellular intestinal calcium absorption. In the duodenum and jejunum calcium can enter the enterocyte through the epithelial calcium channel, TRPV6 (TRPV6 is not present in the ileum). Calcium can bind to calbindin that can modulate TRPV6 channel activity and/or act as an intracellular calcium buffer to prevent toxic levels of calcium from accumulating in the intestinal cell. It has also been suggested that calbindin can act as a facilitator of calcium diffusion through the interior of the cell to the basolateral membrane. In the vicinity of the calcium pump, whose synthesis, similar to calbindin and TRPV6, is induced by 1,25(OH)$_2$D, calcium is transported actively into the extracellular space. *Right panel*: Paracellular pathway. There is increasing evidence that, in the intestine, 1,25(OH)$_2$D can enhance paracellular calcium diffusion by regulating tight junction proteins.(Reproduced with permission from *(3)*).

2.2.1. ROLE OF 1,25(OH)$_2$D ON THE ENTRY OF CALCIUM ACROSS THE BRUSH BORDER MEMBRANE

The first phase of the calcium absorptive process involves entry of calcium into the enterocyte, which has been shown to be increased by 1,25(OH)$_2$D *(3, 4)*. Recently, the apical epithelial calcium channel TRPV6, which may potentially be important in the control of intestinal calcium absorption, was cloned from rat duodenum, suggesting a mechanism for calcium entry *(7)*. In the intestine TRPV6 is localized in the duodenum (in highest concentration) and in jejunum but is not present in the ileum *(3)*. TRPV6 activity has been shown to be regulated by interaction with calmodulin *(8)*. Association of TRPV6 with the S100A10-annexin 2 complex and Rab11a has been reported to be required for targeting and retention of TRPV6 to the plasma membrane and recycling of TRPV6 to the plasma membrane, respectively *(9, 10)*. In addition, it has been shown that the expression of TRPV6 is regulated by 1,25(OH)$_2$D and low dietary calcium *(11, 12)*, supporting its proposed role as a vitamin D-inducible epithelial channel involved in cellular calcium entry. Furthermore, in vitamin D receptor (VDR) knockout (KO) mice, where the major defect is in intestinal calcium absorption, TRPV6 was found to be markedly decreased in the intestine, suggesting that the expression of TRPV6 may be a rate-limiting step in the process of vitamin D-dependent intestinal calcium absorption *(13)*. TRPV6 null mutant mice were recently generated, allowing, for the first time, in vivo studies of the effect of complete ablation of TRPV6 on 1,25(OH)$_2$D-mediated active intestinal calcium absorption *(14)*. Although these mice have normal serum calcium levels, they were reported to have increased PTH and 1,25(OH)$_2$D, suggesting a disturbance in calcium homeostasis *(14, 15)*. Further studies are needed to determine the exact role of TRPV6 in $1,25(OH)_2D$-mediated entry of calcium and whether other calcium channels, yet to be identified, are also involved in vitamin D-mediated calcium entry.

2.2.2. ROLE OF $1,25(OH)_2D$ ON TRANSCELLULAR MOVEMENT OF CALCIUM

The second phase of the calcium absorptive process occurs more slowly and involves the movement of calcium through the interior of the enterocyte. One of the most pronounced known effects of $1,25(OH)_2D$ is increased synthesis of the calcium-binding protein calbindin, the first identified target of $1,25(OH)_2D$ action *(16)*. Two major subclasses of calbindin exist: calbindin-D_{9k}, an approximately 9,000 molecular weight protein present in mammalian intestine and in bovine, mouse, and neonatal rat kidney, and calbindin-D_{28k}, an approximately 28,000 molecular weight protein present in avian intestine, mammalian and avian kidney, mammalian and avian pancreas, and mammalian to molluskan brain *(17–20)*. While calbindin-D_{9k} contains two calcium-binding domains, calbindin-D_{28k} has four functional high-affinity calcium-binding sites *(17, 19)*. Furthermore, no amino acid sequence homology exists between calbindin-D_{28k} and calbindin-D_{9k} *(17, 19)*. In intestine, only one calbindin is present: calbindin-D_{9k} in mammalian intestine and calbindin-D_{28k} in avian intestine. Early studies in chicks established a strong correlation between the level of calbindin and an increase in intestinal calcium transport *(20–22)*. In addition, intestinal calbindin is induced by $1,25(OH)_2D$ and low dietary calcium *(17, 19)* and is markedly reduced in VDR KO mice which have a defect in intestinal calcium absorption *(13, 23)*, supporting the proposed role of calbindin as a facilitator of calcium diffusion *(3)*. However, studies using analogs of $1,25(OH)_2D$ have shown that the induction of calbindin-D_{9k} does not always correlate with an increase in intestinal calcium absorption *(24, 25)*. These studies, as well as recent findings that indicate normal serum calcium levels and active intestinal calcium absorption in calbindin-D_{9k} KO mice *(15, 26, 27)*, indicate that calbindin may be compensated for by another factor. Calbindin may also have another role in the intestine, for example, as a modulator of TRPV6 calcium channel activity and/or as an intracellular calcium buffer to prevent toxic levels of calcium from accumulating in the intestinal cell during $1,25(OH)_2D$-dependent transcellular calcium transport.

2.2.3. ROLE OF $1,25(OH)_2D$ ON CALCIUM EXTRUSION FROM THE INTESTINAL CELL

The third phase of $1,25(OH)_2D$-dependent intestinal calcium absorption is calcium extrusion from the intestinal cell that involves calcium transport against a concentration gradient by the intestinal plasma membrane calcium pump PMCA1b. Previous studies have shown that PMCA1b activity and expression are stimulated by $1,25(OH)_2D$, suggesting that the intestinal calcium absorptive process may involve a direct effect of $1,25(OH)_2D$ on calcium pump expression *(28–31)*. While it has been suggested that the sodium/calcium exchanger, which is also present at the basolateral membrane of the enterocyte, may play a role in calcium extrusion, it should be noted that this cotransporter in the intestine is not $1,25(OH)_2D$ inducible *(32)*.

2.3. Other Models of Intestinal Calcium Transport – Nonsaturable, Paracellular Intestinal Calcium Absorption

Intestinal calcium absorption, being the only source of new calcium in the body, determines the availability of calcium for different functions in the body (3–5, 33–35). The intestinal epithelium is a continuous layer of individual cells with very narrow intercellular space that allow for selective permeability diffusion of small molecules and ions (3). This paracellular pathway, that functions throughout the entire length of the intestine but predominates in the more distal regions when dietary calcium is adequate or high, is driven by the luminal electrochemical gradient and the integrity of intercellular tight junctions (36). Tight junctions are specialized membrane domains located in the apical region of the enterocyte, which constitute a barrier to the movement of molecules and ions by maintaining charge and size selectivity (36, 37). The molecular mechanisms driving the paracellular calcium diffusion and its vitamin D dependency, which have been a matter of debate, remain much less defined compared to the vitamin D-mediated transcellular transport of calcium. It has been suggested that 1,25(OH)$_2$D can promote paracellular calcium diffusion by increasing junction ion permeability. Recent studies have shown that the tight junction proteins claudin 2 and claudin 12, that form paracellular channels in the intestinal epithelia, are regulated by 1,25(OH)$_2$D (38). In addition, studies from Dr. DeLuca's lab recently showed that 1,25(OH)$_2$D downregulates cadherin 17 (important for cell to cell contact) and aquaporin 8 (a tight junction channel) in the intestine, further suggesting that transjunctional movement of calcium can be regulated by 1,25(OH)$_2$D (39). Future studies examining different regions of the intestine as well as novel 1,25(OH)$_2$D-regulated proteins involved in both transcellular and paracellular calcium absorption are needed.

2.4. 1,25(OH)$_2$D and Intestinal Phosphorus Absorption

It should be noted that although the major biologic function of vitamin D is to maintain calcium homeostasis, 1,25(OH)$_2$D can also enhance the intestinal absorption of dietary phosphorus, principally from the proximal intestine (40–42). It has been suggested that 1,25(OH)$_2$D acts by affecting sodium-dependent phosphorus influx into the brush border membrane by regulating the transcription of the intestinal type IIb sodium phosphate co-transporter (43). However this has been a matter of debate (41). Under conditions of high dietary phosphorus intake 1,25(OH)$_2$D does not stimulate jejunal phosphate absorption (40). It has been suggested that the concentration of dietary phosphorus is the major determinant of net phosphorus absorption.

3. RENAL CALCIUM TRANSPORT

3.1. Overview

In addition to the intestine, the kidney is also a major target tissue involved in the regulation by 1,25(OH)$_2$D of calcium homeostasis. Almost 98% of the calcium filtered by the glomerulus is reabsorbed along the nephron. Only plasma calcium that is not bound to proteins is filtered in the glomerulus. The different regions of the nephron involved in the reabsorption process and the hormonal sites of action are schematically

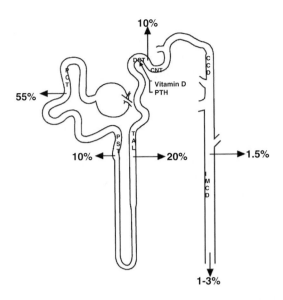

Fig. 2. Calcium absorption at different nephron sites and hormonal site of action. The percentages refer to the percentage of filtered calcium absorbed at different nephron segments. In the proximal convoluted tubule (PCT) and proximal straight tubule (PST) approximately 65% of the filtered calcium is absorbed. In these parts of the nephron, calcium absorption is passive and proceeds via a paracellular path. In the thick ascending loop of Henle (TAL) both active and passive transport pathways contribute to calcium reabsorption. Of the filtered calcium 10% is absorbed by the distal convoluted tubule (DCT) and connecting tubules (CNT). In this part of the nephron calcium absorption involves active transport, is hormonally regulated, and proceeds via a transcellular pathway. CCD cortical collecting duct; MCD inner medullary collecting duct. [Reproduced with permission *(76)*.]

represented in Fig. 2. The proximal tubule absorbs about 65% of the filtered calcium and the loop of Henle absorbs another 20% *(35, 44, 45)*. Calcium absorption in the proximal tubule is passive and proceeds via a paracellular pathway. Both passive and active transport pathways have been reported to contribute to calcium absorption from Henle's loop (the ascending medullary and cortical thick limbs are the parts of Henle's loop where calcium absorption occurs) *(35, 45)*. The remaining 10% of the filtered calcium is absorbed by the distal convoluted tubule and the connecting tubules *(35, 45)*. In the distal nephron calcium absorption involves active transport and proceeds via a transcellular pathway *(35, 45)*. It is the distal nephron that is believed to be the key site for calcium regulation. It is here that the calcium is dually regulated by $1,25(OH)_2D$ and parathyroid hormone (PTH) *(35, 45)*. In this part of the chapter, the role of vitamin D in the kidney is discussed, including the factors involved in this process: the epithelial calcium channel TRPV5, the vitamin D-dependent calcium-binding proteins calbindin-D_{28K} and calbindin-D_{9K} and the plasma membrane calcium pump, PMCA1b (Fig. 3). Besides the role of $1,25(OH)_2D$ in the tubular reabsorption of calcium, other effects in the kidney also discussed are the effect of $1,25(OH)_2D$ on the production of renal vitamin D hydroxylases and the effect of $1,25(OH)_2D$ on phosphate reabsorption.

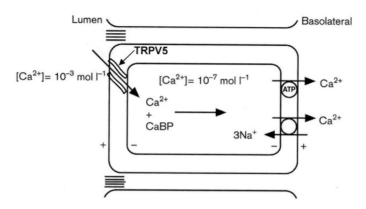

Fig. 3. Model of transcellular calcium transport in the distal nephron. Transcellular calcium transport occurs in the distal nephron and consists of influx through an apical calcium channel TRPV5, binding to calbindin, diffusion through the cytosol and active extrusion at the basolateral membrane. [Modified and reproduced with permission *(77)*.]

3.2. Effect of 1,25(OH)$_2$D on Renal Calcium Transport

3.2.1. CURRENT UNDERSTANDING OF THE ROLE OF 1,25(OH)$_2$D

Data obtained using discrete nephron segments as well as studies using renal tubule cells have indicated that vitamin D metabolites enhances calcium transport in the distal nephron *(46, 47)*. In addition, vitamin D deficiency has been reported to decrease the distal tubular reabsorption of calcium and to decrease the stimulatory effect of PTH on calcium reabsorption *(48, 49)*. Using mouse distal convoluted tubule cells, 1,25(OH)$_2$D has been shown to accelerate PTH-dependent calcium uptake significantly *(47)*. In the presence of 1,25(OH)$_2$D, the time required for PTH to induce membrane hyperpolarization (required for the stimulation of calcium entry into the distal tubule cells) as well to increase intracellular calcium and ^{45}Ca uptake is significantly reduced *(47)*. In addition, 1,25(OH)$_2$D increases PTH receptor mRNA and binding activity, suggesting that this effect may be involved in the acceleration of PTH-dependent calcium entry *(50)* by 1,25(OH)$_2$D. In the kidney, active calcium reabsorption comprises a sequence of processes restricted to the distal convoluted tubule (DCT) and the connecting tubule (CNT). Similar to studies in the intestine, apical entry of calcium is facilitated through the renal epithelial cell via the highly calcium selective channel TRPV5 as a result of the inward electrochemical gradient across the apical membrane *(35)*. Calbindin-D$_{28K}$ has been reported to serve as a calcium buffer thereby maintaining intracellular calcium concentration at a low nontoxic concentration. Calbindin has also been reported to regulate calcium by direct association with TRPV5 and to act as a shuttle for calcium from the apical side to the basolateral side where the Na$^+$/Ca^{2+}-exchanger (NCX1) and the plasma membrane ATPase (PMCA1b) extrude Ca^{2+} into the blood compartment *(35, 45)*. Since transcellular calcium transport in the distal tubule cell is a multiple-step process, similar to transcellular calcium transport in the enterocyte, 1,25(OH)$_2$D enhances calcium transport by affecting various steps in this process as outlined below.

3.2.2. ROLE OF VITAMIN D-DEPENDENT CALCIUM-BINDING PROTEINS AND TRPV5 IN RENAL CALCIUM TRANSPORT

Consistent with the calcium transport studies localizing the site of action of $1,25(OH)_2D$ to the distal nephron *(46, 47)*, autoradiographic data have demonstrated that the nuclear uptake of $[^3H]$ $1,25(OH)_2D$ is localized predominately in the distal nephron *(51)*. In addition, the exclusive localization of the vitamin D-dependent calcium-binding proteins calbindin-D_{28K} and calbindin-D_{9K} is in the distal nephron [the distal convoluted tubule, the connecting tubule, and the cortical collecting duct *(52–54)*]. It has been suggested that the two proteins affect renal calcium absorption by different mechanisms. Calbindin-D_{28k} has been reported to stimulate calcium transport from the apical membrane and calbindin-D_{9k} was reported to enhance ATP-dependent calcium transport of the basolateral membrane *(55, 56)*. These two calcium-binding proteins are induced by $1,25(OH)_2D$ in the kidney *(57)*. The expression of calbindin-D_{9K} but not calbindin-D_{28K} in the kidney is inhibited under high dietary calcium conditions, further suggesting different functions for these proteins in the kidney *(57)*. Studies using VDR/calbindin-D_{28k} double KO mice showed that, compared to the VDR KO mice, the double KO mice have higher urinary calcium excretion and more severe secondary hyperparathyroidism and rachitic skeletal phenotype, providing in vivo evidence for a role for calbindin-D_{28k} in renal calcium handling *(58)*. However a decrease in calbindin-D_{28k} does not always correlate with an increase in urinary calcium excretion, suggesting that calbindin-D_{28k} is not the limiting factor in all cases of hypercalciuria *(59, 60)*.

Similar to studies in the intestine, a $1,25(OH)_2D$-regulated apical calcium channel TRPV5 (that shares 73.4% sequence homology with TRPV6) has also been identified in the distal convoluted tubule and connecting tubule and suggested to be a facilitator of calcium entry *(11, 35)*. The S100A10-annexin 2 complex was found to facilitate the translocation of TRPV5 (as well as TRPV6 in the intestine) to the plasma membrane *(9)*. The anti-aging protein klotho activates TRPV5 by hydrolyzing sugar residues from the extracellular domain of TRPV5, resulting in prolonged expression of TRPV5 at the plasma membrane *(61)*. In addition, direct association of calbindin-D_{28k} and TRPV5 has been reported *(62)*. It has been suggested that renal calbindin-D_{28k} acts as a dynamic calcium buffer, regulating the calcium concentration surrounding the TRPV5 pore by a direct association with the channel and facilitating renal calcium transport by preventing calcium channel inactivation *(62)*. TRPV5 KO mice have impaired renal calcium reabsorption, hypercalciuria, compensatory calcium hyperabsorption in the intestine, and disturbances in bone structure, supporting the role of TRPV5 as a facilitator of calcium entry during active calcium reabsorption in the distal tubule *(63)*. Thus, $1,25(OH)_2D$ affects calcium transport in the distal tubule by enhancing the action of PTH and by inducing TRPV5 and the calbindins.

3.2.3. ROLE OF THE PLASMA MEMBRANE CALCIUM PUMP AND THE Na^+/Ca^{2+} EXCHANGER

In addition to the calbindins and TRPV5, the plasma membrane calcium pump (PMCA1b) and the Na^+/Ca^{2+} exchanger have also been localized to the distal nephron *(64, 65)*. The interrelationship between the basolateral extrusion system in the distal nephron and 1,25(OH)$_2$D is not clear at this time. Consistent with the finding that calbindin-D$_{9K}$ enhances ATP-dependent calcium transport at the renal basolateral membrane, a calbindin-D$_{9K}$-binding domain has been identified in PMCA1b *(66)*. Thus, 1,25(OH)$_2$D may not directly regulate the activity of the renal calcium pump but may do so indirectly by increasing calbindin-D$_{9K}$ *(67)*.

A model of distal tubule renal calcium transport, similar but not identical to the model of intestinal calcium transport, is proposed (Fig. 3). Presumably TRPV5 facilitates calcium entry at the distal nephron, calbindin-D$_{28K}$ increases the influx of calcium by preventing calcium channel inactivation and acts as a diffusional carrier of calcium to the basolateral membrane where calbindin-D$_{9K}$ binds calcium and stimulates the basolateral extrusion of calcium via PMCA1b. It should be noted that in the 25(OH)D 1αhydroxylase KO mouse, that is defective in the synthesis of 1,25(OH)$_2$D, besides calbindin and TRPV6, the renal Na$^+$/Ca^{2+} calcium exchanger is also markedly decreased, suggesting a role for the Na$^+$/Ca^{2+} exchanger in vitamin D-dependent renal calcium reabsorption *(68)*. Further studies of the exact role of the Na$^+$/Ca^{2+} exchanger in the 1,25(OH)$_2$D regulation of renal calcium transport are needed.

3.3. Other Effects of 1,25(OH)$_2$D in the Kidney

Besides the role of 1,25(OH)$_2$D in enhancing the tubular reabsorption of calcium, another important effect of 1,25(OH)$_2$D in the kidney is its ability to regulate its own production. 1,25(OH)$_2$D decreases its own production by inhibiting the renal 25(OH)D 1α hydroxylase enzyme, which hydroxylates 25(OH)D at the α position of carbon 1 of the A-ring, resulting in the formation of 1,25(OH)$_2$D. 1,25-Dihydroxyvitamin D also stimulates the renal 24-hydroxylase enzyme, which hydroxylates the 24 position of both 25(OH)D and 1,25(OH)$_2$D *(69)*. The 24-hydroxylation of 1,25(OH)$_2$D is thought to be the first step in the catabolism of 1,25(OH)$_2$D *(70)*. The predominant localization of these two enzymes is in the proximal convoluted tubule *(71)*. The mechanism by which 1,25(OH)$_2$D reciprocally regulates these two enzymes is transcriptional *(72, 73)*. Besides enhancement of calcium transport in the distal nephron and modulation of 25(OH)D hydroxylases, effects of 1,25(OH)$_2$D on renal phosphate reabsorption have also been suggested (an increase or a decrease has been reported depending on the PTH status and experimental conditions). 1,25(OH)$_2$D has been reported to downregulate the PHEX gene (mutations in the PHEX gene are responsible for X-linked hypophosphatemia), further suggesting a role for 1,25(OH)$_2$D in renal phosphate transport *(75)*. The mechanisms involved in effects of 1,25(OH)$_2$D on phosphate transport are topics of current investigation.

4. SUMMARY

In the intestine 1,25(OH)$_2$D induces various aspects of the transcellular active calcium transport system including calbindin, the basolateral plasma membrane pump and the epithelial calcium channel, TRPV6. Calcium can enter the enterocyte through the epithelial calcium channel and then bind to calbindin and move through the cytosol. At the basolateral membrane calcium is transported actively by the plasma membrane calcium pump into the extracellular space. There is increasing evidence that, in the intestine, 1,25(OH)$_2$D can also enhance paracellular calcium diffusion.

In the distal nephron of the kidney 1,25(OH)$_2$D induces TRPV5 and the calbindins and affects calcium transport, at least in part, by enhancing the action of PTH. A role for the Na$^+$/Ca^{2+} exchanger in the distal tubule in vitamin D-dependent calcium reabsorption has also been suggested. In the kidney, besides enhancement of calcium transport in the distal nephron, 1,25(OH)$_2$D also modulates the 25(OH)D hydroxylases. Effects on renal phosphate reabsorption have also been suggested.

Further research will provide new insight concerning mechanisms of 1,25(OH)$_2$D action and additional effects and targets of 1,25(OH)$_2$D action in kidney and intestine.

REFERENCES

1. DeLuca HF (2004) Overview of general physiologic features and functions of vitamin D. Am J Clin Nutr 80:1689S–1696S
2. Carafoli E, Santella L,Branca D, Brini M (2001) Generation, control and processing of cellular calcium signals. Crit Rev Biochem Mol Biol 36:107–260
3. Wasserman RH (2004) Vitamin D and the Intestinal Absorption of Calcium: A View and Overview. In Feldman D, Pike JW, Glorieux FH (eds) Vitamin D, 2nd edn. Academic Press, San Diego, pp 411–428
4. Pérez AV, Picotto G, Carpentieria AR, Rivoira MA, López ME, De Talamoni NG (2008) Regulation of intestinal calcium absorption: emphasis on molecular mechanisms of transcellular pathway. Digestion 77:22–34
5. Bronner F (2003) Mechanisms and functional aspects of intestinal calcium absorption. J Exp Zoolog A Comp Exp Biol 300(1):47–52
6. Karbach U (1992) Paracellular calcium transport across the small intestine. J Nutr 122:672–677
7. Peng JB, Chen XZ, Berger UV, Vassilev PM, Tsukaguchi H, Brown EM, Hediger MA (1999) Molecular cloning and characterization of a channel-like transporter mediating intestinal calcium absorption. J Biol Chem 274(32):22739–22746
8. Lambers TT, Weidema AF, Nilius B, Hoenderop JG, Bindels RJ (2004) Regulation of the mouse epithelial Ca2(+) channel TRPV6 by the Ca(2+)-sensor calmodulin. J Biol Chem 279(28): 28855–28861
9. van de Graaf SF, Hoenderop JG, Gkika D, Lamers D, Prenen J, Rescher U, Gerke V, Staub O, Nilius B, Bindels RJ (2003) Functional expression of the epithelial Ca(2+) channels (TRPV5 and TRPV6) requires association of the S100A10-annexin 2 complex. EMBO J 22(7):1478–1487
10. van de Graaf SF, Chang Q, Mensenkamp AR, Hoenderop JG, Bindels RJ (2006) Direct interaction with Rab11a targets the epithelial Ca^{2+} channels TRPV5 and TRPV6 to the plasma membrane. Mol Cell Biol 26(1):303–312
11. Song Y, Peng X, Porta A, Takanaga H, Peng JB, Hediger MA, Fleet JC, Christakos S (2003) Calcium transporter 1 and epithelial calcium channel messenger ribonucleic acid are differentially regulated by 1,25 dihydroxyvitamin D3 in the intestine and kidney of mice. Endocrinology 144:3885–3894
12. van Abel M, Hoenderop JG, van der Kemp AW, van Leeuwen JP, Bindels RJ (2003) Regulation of the epithelial Ca^{2+} channels in small intestine as studied by quantitative mRNA detection. Am J Physiol Gastrointest Liver Physiol 285:G78–G85
13. Van Cromphaut SJ, Dewerchin M, Hoenderop JG, Stockmans I, Van Herck E, Kato S, Bindels RJ, Collen D, Carmeliet P, Bouillon R, Carmeliet G (2001) Duodenal calcium absorption in vitamin D receptor-knockout mice: functional and molecular aspects. Proc Natl Acad Sci USA 98: 13324–13329
14. Bianco SD, Peng JB, Takanaga H, Suzuki Y, Crescenzi A, Kos CH, Zhuang L, Freeman MR, Gouveia CH, Wu J, Luo H, Mauro T, Brown EM, Hediger MA (2007) Marked disturbance of calcium homeostasis in mice with targeted disruption of the Trpv6 calcium channel gene. J Bone Miner Res 22(2):274–285

15. Benn BS, Ajibade D, Porta A, Dhawan P, Hediger M, Peng JB, Jiang Y, Oh GT, Jeung EB, Lieben L, Bouillon R, Carmeliet G, Christakos S (2008) Active intestinal calcium transport in the absence of transient receptor potential vanilloid type 6 and calbindin-D9k. Endocrinology 149(6):3196–3205

16. Wasserman RH, Taylor AN (1966) Vitamin D3-induced calcium-binding protein in chick intestinal mucosa. Science 152(3723):791–793

17. Christakos S, Gabrielides C, Rhoten WB (1989) Vitamin D-dependent calcium binding proteins: chemistry, distribution, functional considerations, and molecular biology. Endocr Rev 10:3–26

18. Christakos S (1995) Vitamin D-dependent calcium binding proteins: chemistry, distribution, functional considerations, and molecular biology. Endocr Rev Monograph 4:208–210

19. Christakos S, Liu Y, Dhawan P, Peng X (2005) The calbindins: calbindin D9k and calbindin D28K. In: Feldman D, Pike JW and Glorieux F, (eds) Vitamin D, 2nd edn. Academic Press, San Diego, pp 721–735

20. Taylor AN, Wasserman RH (1969) Correlations between the vitamin D-induced calcium binding protein and intestinal absorption of calcium. Fed Proc 28(6):1834–1838

21. Feher JJ, Wasserman RH (1979) Calcium absorption and intestinal calcium-binding protein: quantitative relationship. Am J Physiol 236:E556–E561

22. Corradino RA, Fullmer CS, Wasserman RH (1976) Embryonic chick intestine in organ culture: stimulation of calcium transport by exogenous vitamin D-induced calcium-binding protein. Arch Biochem Biophys 174:738–743

23. Li YC, Amling M, Pirro AE, Priemel M, Meuse J, Baron R, Delling G, Demay MB (1998) Normalization of mineral ion homeostasis by dietary means prevents hyperparathyroidism, rickets, and osteomalacia, but not alopecia in vitamin D receptor-ablated mice. Endocrinology 139:4391–4396

24. Wang YZ, Li H, Bruns ME, Uskokovic M, Truitt GA, Horst R, Reinhardt T, Christakos S (1993) Effect of 1,25,28-trihydroxyvitamin D2 and 1,24,25-trihydroxyvitamin D3 on intestinal calbindin-D9K mRNA and protein: is there a correlation with intestinal calcium transport?. J Bone Miner Res 8:1483–1490

25. Krisinger J, Strom M, Darwish HD, Perlman K, Smith C, DeLuca HF (1991) Induction of calbindin-D 9k mRNA but not calcium transport in rat intestine by 1,25-dihydroxyvitamin D3 24-homologs. J Biol Chem 266(3):1910–1913

26. Kutuzova GD, Akhter S, Christakos S, Vanhooke J, Kimmel-Jehan C, DeLuca HF (2006) Calbindin D(9k) knockout mice are indistinguishable from wild-type mice in phenotype and serum calcium level. Proc Natl Acad Sci U S A 103(33):12377–12381

27. Akhter S, Kutuzova GD, Christakos S, DeLuca HF (2007) Calbindin D9k is not required for 1,25-dihydroxyvitamin D3-mediated Ca^{2+} absorption in small intestine. Arch Biochem Biophys 460(2):227–232

28. Zelinski JM, Sykes DE, Weiser MM (1991) The effect of vitamin D on rat intestinal plasma membrane Ca-pump mRNA. Biochem Biophys Res Commun 179(2):749–755

29. Wasserman RH, Smith CA, Brindak ME, De Talamoni N, Fullmer CS, Penniston JT, Kumar R (1992) Vitamin D and mineral deficiencies increase the plasma membrane calcium pump of chicken intestine. Gastroenterology 102(3):886–894

30. Cai Q, Chandler JS, Wasserman RH, Kumar R, Penniston JT (1993) Vitamin D and adaptation to dietary calcium and phosphate deficiencies increase intestinal plasma membrane calcium pump gene expression. Proc Natl Acad Sci U S A 90(4):1345–1349

31. Pannabecker TL, Chandler JS, Wasserman RH (1995) Vitamin-D-dependent transcriptional regulation of the intestinal plasma membrane calcium pump. Biochem Biophys Res Commun 213(2):499–505

32. Ghijsen WE, De Jong MD, Van Os CH (1983) Kinetic properties of Na^+/Ca^{2+} exchange in basolateral plasma membranes of rat small intestine. Biochim Biophys Acta 730(1):85–94

33. Wasserman RH (2004) Vitamin D and the dual processes of intestinal calcium absorption. J Nutr 134:3137–3139

34. Bouillon R, Van Cromphaut S, Carmeliet G (2003) Intestinal calcium absorption: molecular vitamin D mediated mechanisms. J Cell Biochem 88:332–339

35. Hoenderop JG, Nilius B, Bindels RJ (2005) Calcium absorption across epithelia. Physiol Rev 85: 373–422

36. Tsukita S, Furuse M, Itoh M (2001) Multifunctional strands in tight junctions. Nat Rev Mol Cell Biol 2:285–293

37. Tang VW, Goodenough DA (2003) Paracellular ion channel at the tight junction. Biophys J 84: 1660–1673

38. Fujita H, Sugimoto K, Inatomi S, Maeda T, Osanai M, Uchiyama Y, Yamamoto Y, Wada T, Kojima T, Yokozaki H, Yamashita T, Kato S, Sawada N, Chiba H (2008) Absorption between enterocytes tight junction proteins claudin-2 and -12 are critical for vitamin D-dependent Ca^{2+} absorption between enterocytes. Mol Biol Cell 19(5):1912–1921

39. Kutuzova GD, DeLuca HF (2004) Gene expression profiles in rat intestine identify pathways for 1,25-dihydroxyvitamin D3 stimulated calcium absorption and clarify its immunomodulatory properties. Arch Biochem Biophys 432:152–166

40. Favus MJ, Bushinsky DA, Lemann J (2006) Regulation of calcium, magnesium and phosphate metabolism. In: Favus MJ (ed) Primer on Metabolic Bone Diseases and Disorders of Mineral Metabolism, 6th ed. ASBMR, Washington DC, 76–83

41. Williams KB, DeLuca HF (2007) Characterization of intestinal phosphate absorption using a novel in vivo method. Am J Physiol Endocrinol Metab 292:E1917–E1921

42. Marks J, Srai SK, Biber J, Murer H, Unwin RJ, Debnam ES (2006) Intestinal phosphate absorption and the effect of vitamin D: a comparison of rats with mice. Exp Physiol 91:531–537

43. Xu H, Bai L, Collins JF, Ghishan FK (2002) Age dependent regulation of rat intestinal type IIb sodium-phosphate cotransporter by 1,25(OH)2vitamin D3. Am J Physiol Cell Physiol 282:C487–C493

44. Lassiter WE, Gottschalk CW, Mylle M (1963) Micropuncture study of tubular reabsorption of calcium in normal rodents. Am J Physiol 204:771–775

45. Friedman PA (2000) Mechanisms of renal calcium transport. Exp Nephrol 8:343–350

46. Winaver J, Sylk DB, Robertson JS, Chen TC, Puschett JB (1980) Micropuncture study of the acute renal tubular transport effect of 25-hydroxyvitamin D3 in the dog. Miner Electrol Metab 4:178–188

47. Friedman PA, Gesek FA (1993) Vitamin D3 accelerates PTH-dependent calcium transport in distal convoluted tubule cells. Am J Physiol Renal Fluid Electrolyte Physiol 265:F300–F308

48. Bouhtiauy I, Lajeunesse D, Brunette MG (1993) Effect of vitamin D depletion on calcium transport by luminal and basolateral membranes of the proximal and distal nephrons. Endocrinology 132(115-):120

49. Bindels RJM, Hartog A, Timmermans J, Van Os CH (1991) Active Ca++ transport in primary cultures of rabbit kidney CCD: stimulation by 1,25-dihydroxyvitamin D3 and PTH. Am J Physiol 261: F799–F807

50. Sneddon WB, Barry EL, Coutermarsh BA, Gesek FA, Liu F, Friedman PA (1998) Regulation of renal parathyroid hormone receptor expression by 1,25-dihydroxyvitamin D3 and retinoic acid. Cell Physiol Biochem 8:261–277

51. Stumpf WE, Sarr M, Narbaitz R, Reid FA, Deluca HF, Tanaka Y (1982) Cellular and subcellular localization of 1,25-dihydroxyvitamin D3 in rat kidney: comparison with Roth J, Brown D, Norman AW, Orci L. Localization of the vitamin D- dependent calcium protein in mammalian kidney. Am J Physiol 243:F243–F252

52. Rhoten WB, Christakos S (1981) Immunocytochemical localization of vitamin D-dependent calcium binding protein in mammalian nephron. Endocrinology 109:981–983

53. Roth J, Brown D, Norman AW, Orci L (1982) Localization of the vitamin D- dependent calcium protein in mammalian kidney. Am J Physiol 243:F243–F252

54. Rhoten WB, Bruns ME, Christakos S (1985) Presence and localization of two vitamin D-dependent calcium binding proteins in Kidneys of higher vertebrates. Endocrinology 117:674–683

55. Bouhtiauy L, Lajeunesse D, Christakos S, Brunnette MG (1994) Two vitamin D- dependent calcium binding proteins increase calcium reabsorption by different mechanisms. II. Effect of CaBP28K. Kidney Int 45:461–468

56. Bouhtiauy L, Lajeunesse D, Christakos S, Brunnette MG (1994) Two vitamin D- dependent calcium binding proteins increase calcium reabsorption by different mechanisms. II. Effect of CaBP9K. Kidney Int 45:469–474

57. Sooy J, Kohut J, Christakos S (2000) The role of calbindin and 1,25-dihydroxyvitamin D3 in the kidney. Curr Opin Nephrol Hyperten 9:341–347

58. Zheng W, Xie Y, Li G, Kong J, Feng JQ, Li YC (2004) Critical role of calbindin-D28k in calcium homeostasis revealed by mice lacking both vitamin D receptor and calbindin-D28k. J Biol Chem 279:52401–52413
59. Caride AJ, Chini EN, Penniston JT, Dousa TP (1999) Selective decrease of mRNAs encoding plasma membrane calcium pump isoforms 2 and 3 in rat kidney. Kidney Int 56:1818–1825
60. Gkika D, Hsu Y, van der Kemp AW, Christakos S, Bindels RJ, Hoenderop JG (2006) Critical role of the epithelial Ca^{2+} Channel TRPV5 in active Ca^{2+} reabsorption as revealed by TRPV5/calbindin-D28K knockout mice. J Am Soc Nephrol 17:3020–3027
61. Chang Q, Hoefs S, van der Kemp AW, Topala CN, Bindels RJ, Hoenderop JG (2005) The beta-glucuronidase klotho hydrolyzes and activates the TRPV5 channel. Science 310:490–493
62. Lambers TT, Mahieu F, Oancea E, Hoofd L, de Lange F, Mensenkamp AR, Voets T, Nilius B, Clapham DE, Hoenderop JG, Bindels RJ (2006) Calbindin D 28K dynamically controls TRPV5-mediated Ca^{2+} transport. EMBO J. 25:2978–2988
63. Hoenderop JG, van Leeuwen JP, wan der Eerden BC, Kersten FF, van der Kemp AW, Merillat AM, Waarsing JH, Rossier BC, Vallon U, Hummler E, Bindles RJ (2003) Renal Ca^{2+} wasting, hyperabsorption and reduced bone thickness in mice lackingTPRV5. J Clin Invest 112:1906–1914
64. Borke JL, Caride A, Verma AK, Penniston JT, Kumar R (1989) Plasma membrane calcium pump and 28kDa calcium binding protein in cells of rat kidney distal tubules. Am J Physiol 257:F842–F849
65. Bourdeau JE, Taylor AN, Iacopino AM (1993) Immunocytochemical localization of sodium-calcium exchanger in canine nephron. J Am Soc Nephrol 4(1):105–110
66. James P, Vorherr T, Thulin E, Forsen S, Carafoli E (1991) Identification and primary structure of a calbindin 9K binding domain in the plasma membrane Ca^{2+} pump. FEBS Lett 278(2):155–159
67. Walters JR, Howard A, Charpin MV, Gniecko KC, Brodin P, Thulin E, Forsén S (1990) Stimulation of intestinal basolateral membrane calcium-pump activity by recombinant synthetic calbindin-D9k and specific mutants. Biochem Biophys Res Commun 170(2):603–608
68. Hoenderop JGJ, Bardenne O, van Abel M, van der Kemp AW, van Os CH, St-Arnaud R, Bindels RJM (2002) Modulation of renal calcium transport protein genes by dietary Ca^{2+} and 1,25-dihydroxyvitamin D3 in 25-hydroxyvitamin D3 1-α-hydroxylase knockout mice. FASEB J 16:1398–1406
69. Omdahl JL, Bobrovnikova EV, Annalora A, Chen P, Serda R (2003) Expression, structure-function and molecular modeling of vitamin D P450s. J Cell Biochem 88:356–362
70. Shinki T, Jin CH, Nishimura A, Nagai Y, Ohyama Y, Noshiro M, Okuda K, Suda T (1992) Parathyroid hormone inhibits 25-hydroxyvitamin D3-24-hydroxylase mRNA expression stimulated by 1 alpha,25-dihydroxyvitamin D3 in rat kidney but not in intestine. J Biol Chem 267(19):13757–13762
71. Kawashima H, Torikai S, Kurokawa K (1981) Localization of 25-hydroxyvitamin D3 1 alpha-hydroxylase and 24-hydroxylase along the rat nephron. Proc Natl Acad Sci USA 78(2):1199–1203
72. Chen KS, DeLuca HF (1995) Cloning of the human 1 alpha,25-dihydroxyvitamin D-3 24-hydroxylase gene promoter and identification of two vitamin D-responsive elements. Biochim Biophys Acta 1263(1):1–9
73. Meyer MB, Zella LA, Nerenz RD, Pike JW (2007) Characterizing early events associated with activation of target genes by 1,25-dihydroxyvitamin D3 in mouse kidney and intestine in vivo. J Biol Chem 282:22344–22352
74. Murayama A, Kim MS, Yanagisawa J, Takeyama K, Kato S (2004) Transrepression by a liganded nuclear receptor via bHLH activator through coregulator switching. EMBO J 23:1598–1608
75. Hines ER, Kolek OI, Jones MD, Serey SH, Sirjani NB, Kiela PR, Jurutka PW, Haussler MR, Collins JF, Ghishan FK (2004) 1,25-dihydroxyvitamin D3 downregulation of PHEX gene expression is mediated by apparent repression of a 110kDa transfactor that binds to a polyadenine element in the promoter. J Biol Chem 279:46406–46414
76. Friedman PA, Gesek FA (1993) Calcium transport in renal epithelial cells. Am J Physiol 264: F181–F198
77. Bindels RJM (1993) Calcium handling by the mammalian kidney. J Exp Biol 184:89–104

8 Biological and Molecular Effects of Vitamin D on Bone

Martin A. Montecino, Jane B. Lian, Janet L. Stein, Gary S. Stein, André J. van Wijnen, and Fernando Cruzat

Abstract   The physiological activities of 1,25-dihydroxyvitamin D_3 [1,25$(OH)_2$D], the active hormone of vitamin D_3, in the skeleton are far-reaching and include development and turnover of bone, differentiation and survival of distinct bone cell populations, and maintaining calcium and bone homeostasis through positive and negative control of gene expression. Here we describe these functional activities within the context of the molecular mechanisms established for the bone tissue-specific osteocalcin gene, involving interactions of the vitamin D receptor transcriptional complexes that contribute to various 1,25$(OH)_2$D activities in the skeleton.

   Key Words: Bone; VDR; osteocalcin; 1,25-dihydroxyvitamin D; skeleton; osteoblast; osteocyte; alkaline phosphatase; 24-hydroxylase; mineralization

1. INTRODUCTION

Bone is the connective tissue characterized by an extensive extracellular matrix that is mineralized. Calcium and phosphate in the form of hydroxyapatite constitute by weight up to 90% of adult bone. Crystals of hydroxyapatite $[Ca_{10}(PO4)_6(OH)_2]$ deposit within the bone, a process which is in part dependent on vitamin D_3 (VD3) for adequate intake, absorption, and retention of dietary calcium and phosphate for producing a mineralized skeleton that functions as a connective tissue. The skeleton provides rigid mechanical support to the body, protects vital organs, and serves as a reservoir of ions, especially for calcium and phosphate required for serum homeostasis. This bone mineral reservoir supports the cellular functions of all cells in the body and releases ions into the circulation for maintaining normal calcium homeostasis. In response to reduced serum calcium levels, calcium transport is stimulated across the gut and from the renal tubular lumen into the bloodstream. At the same time, calcium is mobilized from bone through the coordinated activities of osteoblast and osteoclast lineage cells to resorb

From: *Nutrition and Health: Vitamin D*
Edited by: M.F. Holick, DOI 10.1007/978-1-60327-303-9_8,
© Springer Science+Business Media, LLC 2010

the bone matrix through regulated expression of genes by 1,25-dihydroxyvitamin D_3 [1,25(OH)$_2$D]. A broad spectrum of biological activities of VD3 on distinct bone cell populations facilitates this metabolic function of bone tissue, as well as maintaining a balance between bone resorption and bone formation (bone homeostasis). The active hormone ligand, 1,25(OH)$_2$D, binds the vitamin D receptor (VDR) in osteoblasts to activate a group of genes for promoting osteoclast differentiation and bone resorption. Replacement of this resorbed bone with new bone tissue synthesized by osteoblasts is mediated by the ability of 1,25(OH)$_2$D to target genes that represent bone matrix proteins. Thus VD3 has a physiological role in the mobilization of mineral ions to support calcium homeostasis and the temporal events of bone remodeling for maintaining bone homeostasis, all through its activities in kidney, intestine, and the skeleton.

This chapter focuses on an expanding knowledge of the multiple cellular activities that are regulated by vitamin D in bone and the molecular mechanisms by which 1,25(OH)$_2$D can regulate bone tissue functions through direct activation or repression of the target genes in bone cells. The many possibilities for interactions of the vitamin D receptor transcriptional complex with other factors for regulating gene expression is provided by the example of 1,25(OH)$_2$D control of the osteoblast-specific osteocalcin gene. Among the factors interacting with the VDR include chromatin remodeling factors, tissue-specific transcription factors, and a group of coregulator proteins that facilitate either positive or negative regulation of the gene in response to physiologic signals and in different cellular contexts.

It has long been appreciated that insufficient calcium levels produce rickets in growing children when normal growth plate maturation of chondrocytes to hypertrophic cells for long bone growth becomes severely impaired (see Chapter 21). In the adult, osteomalacia occurs from impaired mineralization of new bone tissue during bone turnover. In early studies of vitamin D-deficient rodents, supplementation with calcium normalized the rachitic syndrome and direct roles for VD3 on osteoblast activities were not appreciated until discovery of the first vitamin D response element (VDRE) in the bone restricted and developmentally expressed osteocalcin gene. Through characterization of genetic mouse models, cellular pathways, and molecular components of the vitamin D pathway, a new level of understanding of the many facets of 1,25(OH)$_2$D control of bone homeostasis has been reached that impacts on the importance of daily requirements of vitamin D for general health.

2. A SPECTRUM OF CELLULAR ACTIVITIES OF 1,25(OH)$_2$D CONTRIBUTING TO BONE FORMATION

2.1. Bone Tissue Organization, Vitamin D, and Osteoblasts

The cellular composition of bone is quite heterogeneous with two distinct cell lineages giving rise to the bone forming osteoblast from mesenchymal stem cells and the bone resorbing osteoclast which arises from the hematopoietic lineage. Each lineage is represented by subpopulations of cells at different stages of maturation and organized in relation to bone architecture. Progenitor cells or immature osteoblasts are found near the outer bone surface, either in the periosteum or along the endosteum (Fig. 1). Active bone forming surfaces are lined with cuboidal osteoblasts. When bone formation is

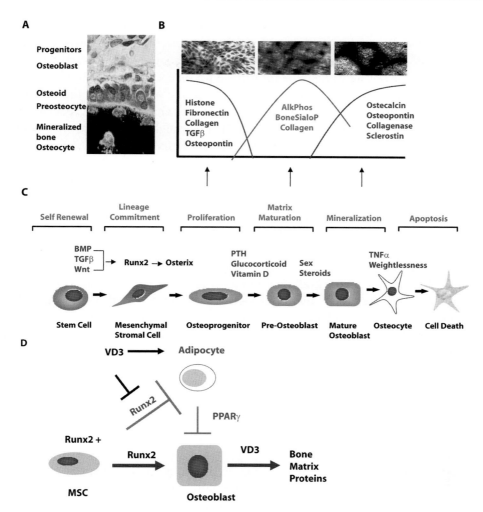

Fig. 1. Stages and markers of osteoblast growth and differentiation. (**a**) Bone surface in vivo shows organization of osteoblast lineage cells as they mature in relation to the mineralizing bone. (**b**) Primary calvarial osteoblasts cultured for 3 weeks showing cell layers at three principle stages of maturation in vitro: growth (Toluidine blue); matrix maturation (alkaline phosphatase histochemical staining, marker of differentiated osteoblast); mineralization stage (von Kossa stain for phosphate in hydroxyapatite mineral). *Lower panel* – Expression of genes most characteristic of the stages. Note osteopontin is related to proliferation and increases again during mineralization. Histone is a marker of DNA synthesis. (**c**) Schematic of stages of osteoblast differentiation with signaling factors that promote commitment to the chondro- and osteoblast phenotype include BMP2, TGFβ, and Wnt/β-catenin pathways. The transcription factors Runx2 and Osterix are essential for bone formation in vivo (established in knockout mice). Indicated hormones have effects on both cell growth and expression of bone matrix proteins in vitro and can be anabolic at low doses or catabolic at high doses in vivo. (**d**) Illustrates the differentiating promoting properties of VD3 and Runx2 on mesenchymal cells toward the bone and fat cell lineages.

triggered some of the pre-osteoblasts can divide, migrate into the interior of the bone, and differentiate. Bone lining cells, which are inactive osteoblasts, are responsive to elevated PTH and VD3 levels and must retract from the bone surface to facilitate osteoid degradation and exposure of the mineral surface for osteoclast adhesion in order to initiate bone resorption. Finally, when osteoblasts become embedded in the mineralized matrix, their functional and morphologic characteristics change to the osteocyte. Osteocytes have multiple cellular processes that reach out through canaliculi in bone tissue for connectivity and maintaining bone viability. Osteocytes have a mechanosensor function in bone *(1)*.

The bone matrix is composed largely of collagen type I (with associated minor collagens) accounting for as much as 90% of the total protein in adult bone. In addition, numerous non-collagenous proteins reside in the bone matrix that function in either promoting or inhibiting mineralization *(2)*. Many of these are specialized calcium and phosphate binding proteins that are vitamin D regulated and include the bone sialoprotein, osteopontin, which is a protein containing *O*-phosphoserine and osteocalcin, a vitamin K-dependent calcium binding protein characterized by γ-carboxyglutamic acid residues. Also important to the mineralization process are elevated levels of the plasma membrane-bound enzyme alkaline phosphatase. These genes can be negatively or positively regulated by $1,25(OH)_2D$ dependent on multiple factors as will be appreciated throughout this chapter. Considerations for gene regulation must include the physiologic levels of the hormone, the cellular context (e.g., differentiation stage of the cell), and the coregulatory protein interactions with the VDR complex on target genes, as will be described. While physiologic levels of the hormone contribute to normal bone formation, physiologic doses of VD3 that lead to bone resorption can downregulate bone matrix protein synthesis and mineralization-related enzymes until signals for bone formation are activated.

All osteoblast lineage cells express the VDR to respond to paracrine effects of the hormone. However, osteoblasts have the ability to synthesize $1,25(OH)_2D$ and express CYP27B1, the 25D 1α-hydroxylase, and CYP24, the 25(OH) vitamin D-24-hydroxylase, which is a catabolic regulator of the hormones $1,25(OH)_2D$ *(3, 4)*. VD3 modulates expression of many dependent genes in osteoblasts that mediate both bone formation through regulation of bone matrix genes *(5, 6)* and bone resorption genes that facilitate a spectrum of activities *(7, 8)*. Gene profiling studies have led to characterization of factors identified with roles in bone metabolic activities, among which are semaphorin 3B that affects osteoblast and osteoclast activity *(9)*, and FGF2 *(10)*, a key regulator of the Pi axis of mineral homeostasis that are transcriptionally regulated by $1,25(OH)_2D$.

2.2. Skeletal Development

In the last decade, mouse mutations in the VDR or the 1α-hydroxylase enzyme that is essential for the $1,25(OH)_2D$ synthesis have revealed that the vitamin D pathway is not critical for embryonic skeletal development. Mice deficient in VD3 signaling did form a normal skeleton in utero, but after birth developed rickets postnatally which could be corrected by supplementation with calcium. However, other studies clearly demonstrate direct effects of 1,25(OH)$_2$D on the regulation of genes that mediate key developmental signaling pathways for bone formation. One of these is the well-established canonical Wnt signaling pathway that regulates numerous cell differentiation lineages through multiple Wnt ligands that interact with frizzled receptor complexes *(11)*. Canonical Wnt signaling is transduced through intracellular β-catenin/TCF transcriptional complexes that regulate target genes to promote mesenchymal cell commitment to either chondrogenesis or osteogenesis, dependent on cellular levels of β-catenin. Nuclear hormone receptors commonly interact with Wnt ligands, β-catenin, and TCF complexes. VD3 induces expression of the Wnt frizzled receptor coregulator LRP5 *(12)* and in stromal cells VDR inhibits DKK1 and SFRP2 which are Wnt antagonists. Together these events enhance canonical Wnt signaling which has an anabolic effect on bone formation *(11)*.

The negative regulation of Wnt antagonists by 1,25(OH)$_2$D has also been linked to suppression of adipogenic differentiation in favor of osteoblast differentiation. An interesting observation from the vitamin D receptor null mice was a higher mRNA level of PPARγ, the transcription factor essential for adipocyte differentiation, and these cells expressed higher levels of the Wnt antagonist DKK1, which inhibits the bone promoting properties of the WNT signaling pathway. Thus, in early skeletal development, VD3 might be contributing to the bone forming properties by repressing adipogenesis of mesenchymal stem cells. The Wnt and VD3 pathways are also related to each other for hair follicle development *(13)*. While VD3 promotes differentiation of cells in the absence of VDR, β-catenin can induce tumors resembling basal cell carcinomas *(14, 15)*. Thus, identifying molecular mechanisms by which Wnt signaling converges with 1,25(OH)$_2$D regulation of target genes has important implications for skeletal development and treating tumors arising from deregulated Wnt signaling.

The bone morphogenetic protein 2 (BMP2) is highly osteogenic and a recent global gene profiling study of embryonic stem cells identified a complex temporal interplay on gene effects by VD3, BMP2, and β-catenin activities. BMP2 induction of Wnt signaling is opposed by VD3 blocking β-catenin activity, the intracellular mediator of canonical Wnt signaling that translocates to the nucleus to participate in gene regulation. Depending on the cellular context, there are differences in effects of 1,25(OH)$_2$D in influencing lineage determination to multiple phenotypes. Not until more definitive studies for these changes can be established and in vivo lineage mapping studies carried out in response to VD3 can we obtain a better understanding of the role of VD3 in contributing as a determinant of embryonic skeletal development.

2.3. Vitamin D3 Promotes Cell Differentiation at Multiple Levels

Vitamin D has a key role in influencing lineage direction of a pluripotent stem cell to multiple phenotypes *(13, 16–20)*. Highly relevant to skeletal homeostasis is the ability of 1,25(OH)$_2$D to promote adipocyte differentiation as well as activate genes that are the bone matrix components of osteoblast differentiation. As we age, our marrow becomes more "fatty" and in a way, a contest prevails in directing marrow stromal cells to either the osteoblast or adipocyte lineage. Several studies have shown that subpopulations of multipotential mesenchymal cells or committed fetal rat calvarie-derived cells can be bipotential forming either osteoblast or adipocyte colonies under appropriate culture conditions. This phenomenon occurs in cultures chronically treated with vitamin D immediately after plating of cells. In early studies, it had been clearly established that short-term (up to 24 h) treatment of osteoblast cells in vitro with $1,25(OH)_2D$, induced expression of vitamin D-regulated genes, including osteopontin and osteocalcin that represent the mineralization stage of osteoblast differentiation. However, other bone matrix-related genes were downregulated as alkaline phosphatase and the bone sialoprotein that represents the pre-mineralization stage. These findings suggest that VD3 may regulate progression through stages of osteoblast differentiation. Not until the precise functions of these bone matrix, structure-related proteins are better defined in relation to bone metabolic activities will we be able to fully interpret these activities of $1,25(OH)_2D$ in osteoblast lineage cells. As will be discussed in Section 4 of this chapter, there are mechanisms to assure that the VDR complex promotes osteogenic lineage gene expression through direct interactions with Runx2, the essential transcription factor for bone formation. Thus, tissue-specific activities of VD3 for driving differentiation appear to reside in the ligand-induced VDR interactions with other cell type phenotypic transcriptional regulators. An understanding of these molecular events mediating these changes is providing new insights for bone renewal.

The relationship between proliferation and differentiation of all skeletal cells must be maintained and is stringently regulated through the activity of cytokines, growth factors, and hormones. It is the potent antiproliferative effects of $1,25(OH)_2D$ that contribute to cell differentiation and are the basis for treatment of certain cancers (see Chapters 12, 40, 47, 50). The mechanisms for VD3 inhibition of proliferation involve $1,25(OH)_2D$-mediated upregulation of two classes of cell cycle inhibitors, p21 and p27 cyclin-dependent kinase inhibitors and the tumor suppressor pRB, thereby impeding cell cycle progression *(21, 22)*. The human p21 (WAF1/CIP1 gene) is regulated by p53 and the VDR gene has multiple p53 and VDR positive regulatory regions *(23)*. VD3 also affects progression through the cell cycle by modulating the cell cycle checkpoint by downregulating expression of the Chk1 and Claspin proteins through interactions of the VDR with E2F recognition motifs in these genes. Thus, VD3 directly interfaces with multiple regulatory mechanisms that inhibit proliferation and stimulate differentiation.

The antiproliferative effects of VD3 in normal osteoblasts are observed in pre-confluent cells and can be so potent that they have the effect of inhibiting osteoblast maturation in vitro, as primary osteoblasts in culture require multi-layering to produce bone matrix for mineralization *(24–26)*. However, if VD3 is added to post confluent committed osteoblasts, the induction of osteoblast-related genes is observed. These genes, e.g., osteocalcin and osteopontin, contribute to forming the bone extracellular matrix and become significantly upregulated enhancing further maturation of the osteoblasts and osteoblast mineralization. One must always keep in mind that vitamin D functions as an "enhancer" or "repressor" of genes related to differentiation and does not have the ability to induce a phenotype per se, as the morphogen BMP2. Its cell differentiation properties are coupled to antiproliferative properties and the ability of the vitamin D receptor complex to interact with other transcriptional regulators.

In contrast to inhibitory effects of vitamin D on osteoblasts in vitro at pharmacologic doses due to antiproliferative effects, in vivo studies have identified bone anabolic effects of vitamin D. A VDR transgene was expressed in mice in mature osteoblast under control of the osteocalcin gene promoter and this resulted in an increase in both cortical and trabecular bone. This study highlights the importance of delineating the effect of the vitamin D ligand on specific subpopulations of osteoblast lineage cells for their selective effects. Our current understanding is that VD3 is an important determinant for regulating distinct activities within cells of the osteoblast lineage, for bone homeostasis which maintains the critical balance between bone resorption, and for bone formation activities *(5, 6, 27, 28)*.

Apoptosis has been proposed to play a key role in controlling osteoblast homeostasis, particularly in the formation of new bone where only a limited number of osteoblasts on the bone surface synthesizing the osteoid will survive as osteocytic cells in the mineralized tissue. Vitamin D has well-known pro-apoptotic properties in cancer cells (see Chapters 41 and 47), but not in normal osteoblasts *(29)*. In human osteoblasts, apoptosis was shown to occur through activation of FAS ligand. Treating the cells with $1,25(OH)_2D$ exerted an anti-apoptotic effect on this pathway through downregulation of components of the mitochondrial- and FAS-related pathways. Expression of the pro-apoptotic BAX protein was decreased with a complimentary increase in the anti-apoptotic BCL protein *(30)*. Thus, vitamin D may also be contributing to osteoblast maturation at the mature osteoblast–pre-osteocyte stage by protecting cells for the transition to osteocytes in a mineralized matrix.

3. THE ROLE OF VITAMIN D IN COUPLING OSTEOBLAST ACTIVITY TO OSTEOCLAST DIFFERENTIATION FOR BONE RESORPTION

An essential calcitrophic hormone function of VD3 is the regulation of bone resorbing osteoclasts indirectly through hormone effects on the osteoblast that secrete factors essential for osteoclastogenesis, as well as directly regulating activity of the differentiated osteoclast (Fig. 2). Vitamin D contributes to the coupling of osteoclast and osteoblast activities at two stages of the remodeling process, thereby mediating completion of the bone remodeling sequence. Initially, VD3 targets osteoblasts and lining cells to retract from the bone surface and secrete VD3-responsive factors which induce osteoclast activity. At the same time, VD3 directly promotes osteoclast formation from mononuclear cells; for example, $1,25(OH)_2D$ regulates carbonic anhydrase II *(31, 32)*. Following the resorption phase, VD3 can stimulate synthesis of cytokines for pre-osteoblast recruitment and growth and expression of osteoblast proteins which form the bone matrix. In this manner, VD3 contributes to the completion of the bone remodeling sequence.

Osteoclasts are of hematopoietic origin and the progenitors can be recruited from marrow, spleen, and blood *(33, 34)*. Immature hematopoietic cells, circulating monocytes, and some tissue macrophages are capable of differentiating into osteoclasts. Characteristic features of the actively resorbing osteoclast include many unique morphological features essential for their bone resorbing functions, the clear sealing zone and ruffled border *(35, 36)*. The clear zone serves to attach osteoclasts to the bone surface and separates the bone resorption area (the Howship lacunae) from the unresorbed bone to create an acidic compartment. A multi-component complex is involved

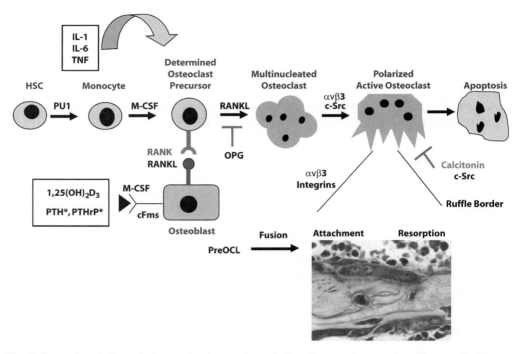

Fig. 2. Osteoclast differentiation, activation, and regulation. Stages of maturation illustrated with regulatory factors driving the differentiation stages and final adhesions and activity of the multinucleated osteoclasts onto the bone surface. Cytokines and hormones contributing to RANKL–RANK interactions between osteoblast lineage cells and mononuclear osteoclast precursors that is the critical step for differentiation. A histologic section of a bone spicule is shown with the osteoclast stained for tartrate-resistant acid phosphatase enzyme activity. On the opposing side of the trabecular bone are cuboidal osteoblasts.

in the attachment of the osteoclast to bone that includes osteopontin and the integrin $\alpha v\beta 3$ and synthesis of these proteins are increased in response to vitamin D *(37–39)*. The ruffled border, a structure of deeply infolded plasma membrane, has a large surface area for secretion of protons and enzymes to dissolve the bone and returns ions into the cell for entering the circulation. During bone resorption, besides release of the mineral, chemotactic and mitogenic factors which are stored in the bone matrix become active (for example, TGFβ) *(40)* and are important for the coupling of bone resorption to later bone formation after resorption has normalized calcium levels in the circulation. These released factors can recruit osteogenic precursors and inhibit osteoclast differentiation, thereby providing a negative feedback mechanism for bone resorption *(24–26)*. For osteoclasts to adhere to bone, osteoid on the bone surface must be removed to expose a mineralized surface. Vitamin D increases expression of MMP13, an enzyme secreted from osteoblast to carry out this function *(41)*.

To accomplish this program, $1,25(\text{OH})_2\text{D}$ directly regulates genes essential for each stage of osteoclastogenesis. RANKL in osteoblasts and other cells is required for fusion of mononuclear hematopoietic precursors for osteoclast differentiation. RANKL, anchored on osteoblast cell surfaces, interacts with RANK on the surface of osteoclast precursors, a coupling reaction that leads to the differentiated multinucleated osteo-clasts. The RANK receptor expressed in monocytes and osteoclasts is increased by 1,25(OH)$_2$D *(42, 43)*. A well-characterized VDRE and cooperating CCAAT box in the RANKL promoter respond to 1,25(OH)$_2$D by recruitment of the VDR, chromatin remodeling factors, and RNA polymerase II *(44)*. Runx2 also participates 1,25(OH)$_2$D-dependent activation of RANKL by remodeling chromatin *(45)*.

Also present in osteoblasts is an inhibitor of osteoclast differentiation, osteo-protegerin (OPG), which is a soluble form of RANKL that binds to RANK to block RANKL interactions *(46)* and thereby prevents fusion of mononuclear pre-cursors. This OPG "decoy receptor" inhibitor of osteoclast differentiation is down-regulated by 1,25(OH)$_2$D in mature osteoblasts, thus favoring bone resorption *(47)*. Notably, in MC3T3 pre-osteoblasts, OPG levels are undetectable, further suggesting that vitamin D preferentially regulates the mature osteoblast to resorb bone *(48, 49)*. Inter-estingly, the OPG gene does not have a consensus negative VDRE, yet is clearly regulated by vitamin D. The suppression is via an AP-1 site which is influenced by 1,25(OH)$_2$D *(46)*.

4. VITAMIN D REGULATION OF GENE EXPRESSION DURING BONE FORMATION

Vitamin D regulation is principally mediated through modulation of transcription. Recent reports indicate that vitamin D controls the expression of at least 913 genes and it may possibly affect, directly or indirectly, the expression of as many as 27,000 genes *(50)*. Vitamin D binds to VDR, which then interacts with specific elements located within the regulatory regions of target genes. Combinatorial- and context-dependent protein–protein interactions with other transcription factors or cofactors bound at spe-cific promoter elements may further modify transcription. Here, physiological control requires that coregulatory proteins determine specificity of biological responsiveness to regulatory cues. It is becoming increasingly evident that organization and assembly of VDR-regulatory complexes are dynamic rather than static *(51, 52)*. Modifications in the composition of these regulatory complexes provide a mechanism for integrating regula-tory signals to support positive and negative control through synergism and antagonism, respectively.

4.1. Components of Vitamin D-Dependent Regulatory Complexes

The role that vitamin D plays in bone metabolism provides a paradigm for under-standing molecular mechanisms that operate in vitamin D action. Vitamin D directly regulates the expression of genes that support bone formation during development and bone remodeling throughout life. Therefore, osteoblast differentiation is a model for understanding developmental responsiveness to vitamin D *(53, 28)*.

Vitamin D exerts its genomic effects through the VDR which is a member of the superfamily of nuclear receptors *(51, 54)*. As in other nuclear receptors, binding of the ligand induces conformational changes in the C-terminal ligand binding domain (LBD) of the VDR. The changes establish competency for VDR interaction with coactivators of the p160/SRC family, including SRC-1/NCoA-1, SRC-2/NCoA-2/GRIP/TIF2, and SRC-3/ACTR. These complexes are critical for transcriptional activation (51, 53, 54). p160/SRC coactivators form high molecular weight complexes by interacting with other coactivator proteins including p300, its related homologue CBP, and P/CAF (55). Moreover, p160/SRC coactivators have been shown to recruit CBP/p300 and P/CAF to ligand-bound nuclear receptors. Multiprotein complexes containing different activities are functionally linked to ligand-dependent transcriptional regulation (51). Coactivators such as SRC-3/ACTR, SRC-1/NCoA-1, CBP/p300, and P/CAF contain intrinsic histone acetyl transferase (HAT) activity. Therefore, protein complexes including independent HAT activities can be recruited to gene promoters by nuclear receptors in a ligand-dependent manner (51). Once bound to these promoters, the HAT activities contribute to chromatin remodeling events that increase access of additional regulatory factors to their cognate elements (56). The coactivator NCoA62/SKIP can also interact with VDR in a ligand-dependent manner. However, this protein–protein interaction occurs through a domain of VDR that is different from that recognized by p160/SRC or DRIP205 coactivators (57). Moreover, NCoA62/SKIP can form a ternary complex with VDR and SRC-1 to cooperatively stimulate VDR-mediated transcription activation.

The multisubunit DRIP (VDR-interacting protein) complex also binds to VDR in response to the ligand vitamin D (58, 59). This interaction occurs through the LBD of VDR in the same manner as the p160/SRC coactivators, resulting in transcriptional enhancement (60). In contrast to p160/SRC coactivators, DRIP is devoid of HAT and other chromatin remodeling activities and interacts with nuclear receptors through a single subunit designated DRIP205, which anchors other subunits to the receptor LBD. Several of these subunits are also present in the Mediator complex, which interacts with the C-terminal domain (CTD) of RNA polymerase II, forming the holoenzyme complex (61). Therefore, the DRIP complex appears to function as a transcriptional coactivator by forming a molecular bridge between the VDR and the basal transcription machinery, reflecting the importance of three-dimensional promoter organization to regulatory activity.

An ATP-dependent chromatin remodeling complex that binds to VDR and that is named WINAC (for WSTF [Williams Syndrome Transcription Factor] including nucleosome assembly complex) was recently reported (62). WINAC shares components with two other chromatin remodeling complexes, SWI/SNF and ISWI, and has been proposed to mediate recruitment of unliganded VDR to target genes. Nevertheless, subsequent interaction of the targeted VDR with transcriptional corepressors requires the presence of vitamin D. This complex has been reported to be involved in transcriptional repression (63, 64) and in controlling DNA replication (62).

In the last few years various investigators have shown that coactivator complexes including p160/SRC and DRIP are recruited to steroid hormone-regulated genes by nuclear receptors in a sequential and mutually exclusive manner (65–68). The ordered association of transcriptional regulators exhibits binding kinetics with periods of 40–60 min. These results provided the basis for a model in which cyclical association of different coactivator complexes reflects the dynamics of the transcription activation process of nuclear receptor-regulated genes *(69)*. Alternatively, recent reports also indicate that occupancy at the target gene regulatory regions by nuclear receptor-associated coactivator complexes may also occur gradually and at a significantly lower rate *(70, 71)*. Thus, the molecular mechanisms by which VDR-associated coactivator complexes are recruited to target genes may be context dependent, that is, directly related to the nature of the regulatory factors bound to each particular promoter (see below).

4.2. Vitamin D-Mediated Gene Expression Within the Three-Dimensional Context of Nuclear Structure in Bone Cells

Evidence is accumulating that the architectural organization of nucleic acids and regulatory proteins within the nucleus supports functional interrelationships between nuclear structure and gene expression. There is increasing acceptance that components of nuclear structure are functionally linked to the organization and sorting of regulatory information in a manner that permits utilization (reviewed in Zaidi et al. *(72)*). The primary level of organization, the representation and ordering of genes and promoter elements, provides alternatives for physiological control. The molecular organization of regulatory elements, the overlap of regulatory sequences within promoter domains, and the multipartite composition of regulatory complexes increase options for responsiveness. Chromatin structure and nucleosome organization reduce distances between regulatory sequences, facilitate cross talk between promoter elements, and render elements competent for interactions with positive and negative regulatory factors. The components of higher order nuclear architecture, including nuclear pores *(73)*, the nuclear matrix, and subnuclear domains, contribute to the subnuclear distribution and activities of genes and regulatory factors *(72, 74)*. Compartmentalization of regulatory complexes is illustrated by focal organization of PML bodies *(75)*, Runx bodies *(76, 77)*, the nucleolus *(78)*, and chromosomes *(79)*, as well as by the punctuate intranuclear distribution of sites for replication *(80)*, DNA repair *(81)*, transcription *(82)*, and the processing of gene transcripts *(83–85)*.

There is emerging recognition that nuclear structure and function are causally related. Interestingly, it has been reported that in several mammalian cells, including osteoblastic cells, VDR exhibits a punctate nuclear distribution that is enhanced upon vitamin D stimulation *(86–88)*, raising the possibility that VDR is associated with components of the nuclear architecture. Furthermore, recent results from our group indicate that VDR is bound to the nuclear matrix fraction of several osteoblastic cells in a ligand-dependent manner (Arriagada et al., Unpublished results). Similarly, Zhang et al. *(87)* have reported that the NCoA62/Skip protein, which binds to VDR and functions as a transcriptional coactivator, is also bound to the nuclear matrix in osteoblastic cells. Therefore, association of VDR with components of the nuclear architecture is dynamic rather than static. The bone-specific OC gene and skeletal-restricted Runx2 transcription factor serve as examples of obligatory relationships between nuclear structure and vitamin D-mediated physiological control of skeletal gene expression *(52, 89)*. It appears that there are similar relationships between nuclear organization and other bone-related vitamin D-responsive genes (e.g., osteopontin and 24-hydroxylase).

4.3. *Vitamin D Receptor Coregulatory Factors Provide Gene-Specific Regulation*

The rat OC gene encodes a 10 kDa bone-specific protein that is induced in osteoblasts with the onset of mineralization at late stages of differentiation *(90)*. Modulation of OC gene expression during bone formation and remodeling requires physiologically responsive accessibility of proximal and upstream promoter sequences to regulatory and coregulatory proteins, as well as protein–protein interactions that integrate independent promoter domains *(91)*. The chromatin organization of the OC gene illustrates dynamic remodeling of a promoter to accommodate requirements for phenotype-related developmental and vitamin D-responsive activity *(92)*.

Transcription of the OC gene is controlled by modularly organized basal and hormone-responsive promoter elements (see Fig. 2a) located within two DNase I-hypersensitive sites (Distal site, positions −600 to −400; proximal site, positions −170 to −70) that are only nuclease accessible in bone-derived cells expressing this gene *(91)*. A key regulatory element that controls OC gene expression is recognized by the VDR complex upon ligand stimulation. This vitamin D-responsive element (VDRE) is located in the distal region (Fig. 3a) of the OC promoter (positions −465 to −437) and functions as an enhancer to increase OC gene transcription *(92)*. Another key regulator of OC gene expression is the nuclear matrix-associated transcription factor Runx2, a member of the Runt homology family of proteins which has been shown to contribute to the control of skeletal gene expression *(89)*. Runx2 proteins serve as a scaffold for the assembly and organization of coregulatory proteins that mediate biochemical and architectural control of promoter activity. The rat OC gene promoter contains three recognition sites for Runx2 interactions, site A (−605 to −595), site B (−438 to −430), and site C (−138 to −130). Mutation of all three Runx2 sites results in significantly reduced OC expression in bone-derived cells *(93)*. The retention of a nucleosome between the proximal and the upstream enhancer domains reduces the distance between the basal regulatory elements and the VDRE and supports a promoter configuration that is conducive to protein–protein interactions between VDR-associated proteins and components of the RNA polymerase II-bound complex (Fig. 2b). Interaction of the VDR at the distal promoter region of the OC gene requires nucleosomal remodeling *(94, 95)*.

We have recently shown that within the OC gene promoter context there is a tight functional relationship between Runx2 and the vitamin D-dependent pathway *(88)*. Runx2 and VDR are components of the same nuclear complexes, colocalize at punctate foci within the nucleus of osteoblastic cells, and interact directly in protein–protein binding assays in vitro *(88)*. Additionally, mutation of the distal Runx2 sites A and B (which flank the VDRE, see Fig. 2a) abolishes vitamin D-enhanced OC promoter activity *(88)*. In contrast to most nuclear receptors, the VDR does not contain an N-terminal AF-1 transactivation domain and thus is unable to interact with coactivators through this region *(51)*. Therefore, Runx2 plays a key role in the vitamin D-dependent stimulation of the OC gene promoter in osteoblastic cells by directly stabilizing binding of the VDR to the VDRE. Runx2 also allows recruitment of the coactivator p300 to the OC promoter (Fig. 3a), which results in upregulation of both basal and vitamin D-enhanced OC gene transcription *(96)*. Based on these results, we have postulated that

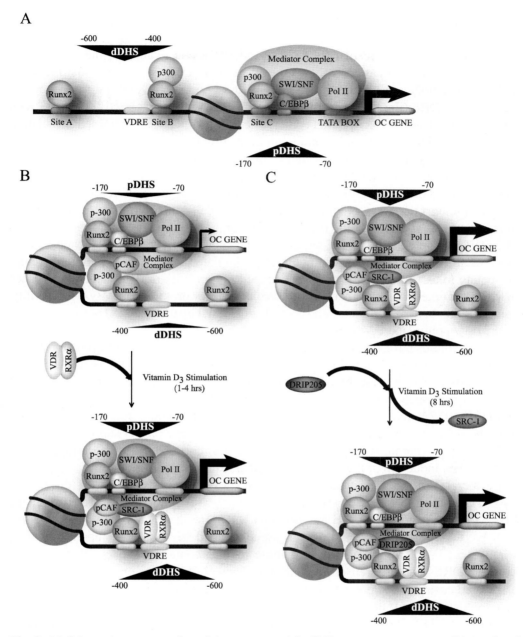

Fig. 3. (a) Schematic representation of the rat osteocalcin (OC) gene promoter transcribing at basal levels. The *circle* in the *middle* represents a positioned nucleosome flanked by a distal and proximal DNase I-hypersensitive sites (dDHS and pDHS, respectively). The different regulatory elements and the transcription factors that bind them in osteoblastic cells expressing the OC gene are also indicated. **(b)** Proposed three-dimensional organization of the OC promoter transcribing at basal levels or enhanced by vitamin D. The stimulatory effect of the VDR-associated coactivators on the general transcription machinery is represented by the size of the *arrows* at the transcription start site. The positioned nucleosome facilitates DNA bending and the functional interactions between distal and proximal promoter regulatory elements that are bound by the cognate factors. **(c)** After several hours of vitamin D treatment, DRIP205 is recruited to the OC promoter and SRC-1 is released. VDRE, vitamin D-responsive element; Pol II, RNA polymerase II.

Runx2-mediated recruitment of p300 may facilitate the subsequent interaction of p300 with the VDR upon ligand stimulation (88).

The rate of recruitment of p160/SCR-1 and DRIP coactivators to the OC gene in response to vitamin D has recently been studied (71). It has been found that the VDR and SRC-1 rapidly and stably interact with the distal region of the OC promoter encompassing the VDRE (Fig. 2b). The interaction of SRC-1 and VDR directly correlates with vitamin D-mediated transcriptional enhancement of the OC gene, increased association of the RNA polymerase complex and vitamin D-stimulated histone H4 acetylation (71, 97). Interestingly, DRIP205 was found to bind to the OC promoter only after several hours of continuous treatment with vitamin D, concomitant with release of SRC-1 (see Fig. 3c). Based on these results it has been postulated that this preferential recruitment of SRC-1 to the OC gene promoter is based on the specific distribution of regulatory elements at the distal region of the promoter. This organization may lead to the formation of a stable complex at the distal region that includes Runx2, p300, VDR, and SRC-1. Once established, this complex may directly stimulate the basal transcription machinery bound to an OC promoter actively engaged in transcription. The general relevance of vitamin D-mediated chromatin-based mechanisms of promoter activity, accessibility, and cross talk between regulatory domains is illustrated by the vitamin D responsiveness of the osteopontin and 24-hydroxylase genes.

Recent studies have established that vitamin D induces a rapid and cyclical association of the VDR/RXR heterodimer with the proximal mouse 24-hydroxylase gene promoter in osteoblastic cells (68). Vitamin D treatment also induces a rapid recruitment of coactivators such as p160/SRC and p300/CBP, which leads to acetylation of histone H4. DRIP205/Mediator is also recruited to the proximal promoter region concomitantly with the interaction of RNA polymerase II. Together, these results support a model in which highly dynamic association of the VDR with chromatin occurs during vitamin D-dependent induction of the 24(OH)ase gene in osteoblasts (68).

Carlberg et al. have also monitored the spatio-temporal regulation of the human 24(OH)ase gene (98). They have evaluated 25 contiguous genomic regions spanning the first 7.7 kb of the human 24(OH)ase promoter and found that in addition to the proximal VDREs, three further upstream regions are associated with the VDR upon vitamin D stimulation. Interestingly, only two of these regions contain sequences that resemble known VDREs that are transcriptionally responsive to this hormone. The other VDR-associated upstream promoter region does not contain any recognizable classical VDRE that could account for the presence of the VDR protein. However, simultaneous association of the VDR, RXR, p160/SRC, and DRIP/Mediator coactivators, as well as RNA polymerase, was detected in all four vitamin D-responsive sequences after the addition of the ligand (98). Remarkably, despite participating in the same process, all four chromatin regions displayed individual vitamin D-dependent patterns of interacting proteins. Based on these results, the authors propose that these upstream vitamin D-responsive regions may have a role in the implementation of gene activation, as they raise their vitamin D-dependent histone H4 acetylation status earlier than that of the proximal promoter VDREs (98). It has also been suggested that the simultaneous communication of the individual promoter regions with the RNA polymerase II complex occurs through a particular three-dimensional organization of the chromatin at the 24(OH)ase promoter.

This arrangement could be facilitating close contact between distal and proximal regulatory regions.

On the other hand, Pike et al. have described that upon vitamin D treatment of osteoblastic cells, there is a rapid and cyclical association of the VDR with the OP promoter *(68)*. This increased binding of the VDR parallels vitamin D-mediated transcriptional enhancement of the OP gene and additionally involves cyclical, sequential, and mutually exclusive recruitment of the coactivators p160/SRC, p300/CBP, and DRIP/Mediator. Interestingly and in contrast to the OC and 24(OH)ase genes, p160/SRC-p300/CBP binding does not result in increased histone H4 acetylation. These results further confirm that in osteoblastic cells different promoters are regulated by distinct mechanisms in response to vitamin D.

Recent reports have also described the molecular components of VDR-containing nuclear complexes that repress transcription. Thus, Kato and colleagues have identified a bHLH-type factor, the VDR-interacting repressor (VDIR), which directly binds to an inhibitory VDRE located in the human 1α-hydroxylase gene promoter. Binding of VDIR in the absence of ligand results in transcriptional activation of the 1α-hydroxylase gene *(99)*. Interestingly, vitamin D-induced association between VDR and VDIR leads to ligand-dependent transrepression of the 1α-hydroxylase gene. The mechanism involves an exchange in the VDR-associated coregulatory complex from the coactivator p300 to an HDAC-containing corepressor complex. This exchange is accompanied by a significant decrease in histone acetylation at the 1α-hydroxylase gene promoter and by a release of the RNA polymerase II basal transcriptional machinery *(64)*. The same group has indicated that the complex WINAC functions in this VDR-mediated transrepression of the 1α-hydroxylase gene through its ability to efficiently bind the acetylated nucleosomes present in the 1α-hydroxylase promoter region in the absence of vitamin D *(64)*.

5. CONCLUSIONS AND FUTURE DIRECTIONS

Vitamin D serves as a principal modulator of skeletal gene transcription, thus necessitating further understanding of interfaces between activity of this steroid hormone with regulatory cascades that are functionally linked to regulation of skeletal genes *(53, 52)*. There is growing appreciation for the repertoire of factors that influence gene expression for commitment to the osteoblast lineage. It is well documented that sequentially expressed genes support progression of osteoblast differentiation through developmental transition points where responsiveness to phosphorylation-mediated regulatory cascades determines competency for establishing and maintaining the structural and functional properties of bone cells *(100, 101)*. The catalog of promoter elements and cognate regulatory proteins that govern skeletal gene expression offers essential but insufficient insight into mechanisms that are operative in intact cells. Gene promoters serve as regulatory infrastructure by functioning as blueprints for responsiveness to the flow of cellular regulatory signals. However, access to the specific genetic information requires transcriptional control of skeletal genes within the context of the subnuclear organization of nucleic acids and regulatory proteins. Explanations are required for (i) convergence of multiple regulatory signals at promoter sequences; (ii) the integration of regulatory information at independent promoter domains; (iii) selective utilization of redundant regulatory pathways; (iv) thresholds for initiation or downregulation of transcription with limited intranuclear representation of promoter elements and regulatory factors; (v) mechanisms that render the promoters of cell growth and phenotypic genes competent for protein–DNA and protein–protein interactions in a physiologically responsive manner; (vi) the composition, organization, and assembly of sites within the nucleus that support transcription, and (vii) the intranuclear trafficking of regulatory proteins to transcriptionally active foci. From a regulatory perspective compartmentalization of components of VD3 control supports the integration of regulatory activities, perhaps by establishing thresholds for protein activity in time frames that are consistent with the execution of regulatory signaling.

It is necessary to understand how the regulatory machinery for vitamin D responsiveness is retained during mitosis to support gene expression in post-mitotic progeny cells. Key components of nuclear architecture are dismantled and must be re-organized to support tissue-specific transcription. In this context, we have recently reported that within the condensed mitotic chromosomes Runx2 is retained in large discrete foci at nucleolar organizing regions where ribosomal RNA genes reside *(102)*. These Runx2 chromosomal foci are associated with open chromatin, colocalize with RNA polymerase I transcription factor UBF1, and transition into nucleoli at sites of rRNA synthesis during interphase. Runx2 forms complexes with UBF1 and SL1 (another RNA pol I factor), cooccupies the rRNA gene promoter with these factors in vivo, and affects local chromatin histone modifications at rDNA regulatory regions as well as competency for transcription. Hence, these findings indicate that Runx2 not only controls lineage commitment and cell proliferation by regulating genes transcribed by RNA pol II, but modulates also RNA pol I-mediated rRNA synthesis *(102)*. A similar focal retention of Runx transcription factors with mitotic chromosomal loci of RNA polymerase II target genes that are functionally consequential has recently been observed *(103)*. Taken together, it appears that in addition to DNA methylation and post-translational modifications of histones, the mitotic occupancy of genes that support proliferation, cell growth, and phenotype by transcription factors and coregulatory proteins is an additional dimension to epigenetic control.

ACKNOWLEDGMENTS

This work was supported by grants from CONICYT-PBCT ACT044 (to M.M), FONDECYT 1095075 (TOMM) NIH PO1 AR48818 (to G.S.S), and DE12528 (to J. B. L.). The contents are solely the responsibility of the authors and do not necessarily represent the official views of the National Institutes of Health.

REFERENCES

1. Bonewald LF (2006) Mechanosensation and transduction in osteocytes. IBMS BoneKEy-Osteovision 3:7–15
2. Robey PG, Boskey AL (2006) Extracellular matrix and biomineralization of bone. In: Favus MJ (ed) Primer on the metabolic bone diseases and disorders of mineral metabolism, 6th edn. American Society for Bone and Mineral Research, Washington, DC, pp 12–19

3. Atkins GJ, Anderson PH, Findlay DM et al (2007) Metabolism of vitamin D3 in human osteoblasts: evidence for autocrine and paracrine activities of 1 alpha,25-dihydroxyvitamin D3. Bone 40: 1517–1528

4. Van DM, Koedam M, Buurman CJ et al (2006) Evidence that both 1alpha,25-dihydroxyvitamin D3 and 24-hydroxylated D3 enhance human osteoblast differentiation and mineralization. J Cell Biochem 99:922–935

5. St-Arnaud R (2008) The direct role of vitamin D on bone homeostasis. Arch Biochem Biophys 473:225–230

6. Van Leeuwen JP, van Driel M, van den Bemd GJ et al (2001) Vitamin D control of osteoblast function and bone extracellular matrix mineralization. Crit Rev Eukaryot Gene Expr 11:199–226

7. Goltzman D, Miao D, Panda DK et al (2004) Effects of calcium and of the vitamin D system on skeletal and calcium homeostasis: lessons from genetic models. J Steroid Biochem Mol Biol 89–90:485–489

8. Ross FP (2006) Osteoclast biology and bone resorption. In: Favus MJ (ed) Primer on the metabolic bone diseases and disorders of mineral metabolism, 6th edn. American Society for Bone and Mineral Research, Washington, DC, pp 30–35

9. Sutton AL, Zhang X, Dowd DR et al (2008) Semaphorin 3B is a 1,25-Dihydroxyvitamin D3-induced gene in osteoblasts that promotes osteoclastogenesis and induces osteopenia in mice. Mol Endocrinol 22:1370–1381

10. Barthel TK, Mathern DR, Whitfield GK et al (2007) 1,25-dihydroxyvitamin D3/VDR-mediated induction of FGF23 as well as transcriptional control of other bone anabolic and catabolic genes that orchestrate the regulation of phosphate and calcium mineral metabolism. J Steroid Biochem Mol Biol 103:381–388

11. Bodine PV, Komm BS (2006) Wnt signaling and osteoblastogenesis. Rev Endocr Metab Disord 7: 33–39

12. Fretz JA, Zella LA, Kim S et al (2007) 1,25-dihydroxyvitamin D3 induces expression of the Wnt signaling co-regulator LRP5 via regulatory elements located significantly downstream of the gene's transcriptional start site. J Steroid Biochem Mol Biol 103:440–445

13. Palmer HG, Njos-Afonso F, Carmeliet G et al (2008) The vitamin D receptor is a Wnt effector that controls hair follicle differentiation and specifies tumor type in adult epidermis. PLoS ONE 3: e1483

14. Chrysogelos S, Riley DE, Stein G et al (1985) A human histone H4 gene exhibits cell cycle-dependent changes in chromatin structure that correlate with its expression. Proc Natl Acad Sci USA 82: 7535–7539

15. Jande SS, Bélanger LF (1973) The life cycle of the osteocyte. Clin Orthop 94:281–305

16. Zhang S, Chan M, Aubin JE (2006) Pleiotropic effects of the steroid hormone 1,25-dihydroxyvitamin D3 on the recruitment of mesenchymal lineage progenitors in fetal rat calvaria cell populations. J Mol Endocrinol 36:425–433

17. Bellows CG, Wang Y-H, Heersche JNM et al (1994) 1,25-dihydroxyvitamin D3 stimulates adipocyte differentiation in cultures of fetal rat calvaria cells: comparison with the effects of dexamethasone. Endocrinology 134:2221–2229

18. Cianferotti L, Demay MB (2007) VDR-mediated inhibition of DKK1 and SFRP2 suppresses adipogenic differentiation of murine bone marrow stromal cells. J Cell Biochem 101:80–88

19. Pendas-Franco N, Garcia JM, Pena C et al (2008) DICKKOPF-4 is induced by TCF/beta-catenin and upregulated in human colon cancer, promotes tumour cell invasion and angiogenesis and is repressed by 1alpha,25-dihydroxyvitamin D3. Oncogene 27:4467–4477

20. Cianferotti L, Cox M, Skorija K et al (2007) Vitamin D receptor is essential for normal keratinocyte stem cell function. Proc Natl Acad Sci USA 104:9428–9433

21. Ryhanen S, Jaaskelainen T, Mahonen A et al (2003) Inhibition of MG-63 cell cycle progression by synthetic vitamin D3 analogs mediated by p27, Cdk2, cyclin E, and the retinoblastoma protein. Biochem Pharmacol 66:495–504

22. Matsumoto T, Sowa Y, Ohtani-Fujita N et al (1998) p53-independent induction of WAF1/Cip1 is correlated with osteoblastic differentiation by vitamin D3. Cancer Lett 129:61–68

23. Saramaki A, Banwell CM, Campbell MJ et al (2006) Regulation of the human p21(WAF1/Cip1) gene promoter via multiple binding sites for p53 and the vitamin D3 receptor. Nucleic Acids Res 34:543–554

24. Mundy GR, Boyce B, Hughes D et al (1995) The effects of cytokines and growth factors on osteoblastic cells. Bone 17:71S–75S

25. Gerstenfeld LC, Zurakowski D, Schaffer JL et al (1996) Variable hormone responsiveness of osteoblast populations isolated at different stages of embryogenesis and its relationship to the osteogenic lineage. Endocrinology 137:3957–3968

26. Owen TA, Aronow MS, Barone LM et al (1991) Pleiotropic effects of vitamin D on osteoblast gene expression are related to the proliferative and differentiated state of the bone cell phenotype: dependency upon basal levels of gene expression, duration of exposure, and bone matrix competency in normal rat osteoblast cultures. Endocrinology 128:1496–1504

27. Gardiner EM, Baldock PA, Thomas GP et al (2000) Increased formation and decreased resorption of bone in mice with elevated vitamin D receptor in mature cells of the osteoblastic lineage. FASEB J 14:1908–1916

28. van Driel M, Pols HA, Van Leeuwen JP (2004) Osteoblast differentiation and control by vitamin D and vitamin D metabolites. Curr Pharm Des 10:2535–2555

29. Trump DL, Hershberger PA, Bernardi RJ et al (2004) Anti-tumor activity of calcitriol: pre-clinical and clinical studies. J Steroid Biochem Mol Biol 89–90:519–526

30. Duque G, El AK, Henderson JE et al (2004) Vitamin D inhibits Fas ligand-induced apoptosis in human osteoblasts by regulating components of both the mitochondrial and Fas-related pathways. Bone 35:57–64

31. David JP, Rincon M, Neff L et al (2001) Carbonic anhydrase II is an AP-1 target gene in osteoclasts. J Cell Physiol 188:89–97

32. Quelo I, Jurdic P (2000) Differential regulation of the carbonic anhydrase II gene expression by hormonal nuclear receptors in monocytic cells: identification of the retinoic acid response element. Biochem Biophys Res Commun 271:481–491

33. Suda T, Takahashi N, Etsuko A (1992) Role of vitamin D in bone resorption. J Cell Biochem 49: 53–58

34. Roodman GD (1996) Advances in bone biology: the osteoclast. Endocr Rev 17:308–332

35. Roodman GD, Ibbotson KJ, MacDonald BR et al (1985) 1,25-dihydroxyvitamin D3 causes formation of multinucleated cells with several osteoclast characteristics in cultures of primate marrow. Proc Natl Acad Sci USA 82:8213–8217

36. Baron R, Neff L, Louvard D et al (1985) Cell-mediated extracellular acidification and bone resorption: evidence for a low pH in resorbing lacunae and localization of a 100-kD lysosomal membrane protein at the osteoclast ruffled border. J Cell Biol 101:2210–2222

37. Bruzzaniti A, Baron R (2006) Molecular regulation of osteoclast activity. Rev Endocr Metab Disord 7:123–139

38. Noda M, Vogel RL, Craig AM et al (1990) Identification of a DNA sequence responsible for binding of the 1,25-dihydroxyvitamin D_3 receptor and 1,25-dihydroxyvitamin D_3 enhancement of mouse secreted phosphoprotein 1 (Spp-1 or osteopontin) gene expression. Proc Natl Acad Sci USA 87: 9995–9999

39. Cao X, Ross FP, Zhang L et al (1993) Cloning of the promoter for the avian integrin beta 3 subunit gene and its regulation by 1,25-dihydroxyvitamin D3. J Biol Chem 268:27371–27380

40. Oursler MJ (1994) Osteoclast synthesis and secretion and activation of latent transforming growth factor beta. J Bone Miner Res 9:443–452

41. Uchida M, Shima M, Chikazu D et al (2001) Transcriptional induction of matrix metalloproteinase-13 (collagenase-3) by 1alpha,25-dihydroxyvitamin D3 in mouse osteoblastic MC3T3-E1 cells. J Bone Miner Res 16:221–230

42. Kido S, Inoue D, Hiura K et al (2003) Expression of RANK is dependent upon differentiation into the macrophage/osteoclast lineage: induction by 1alpha,25-dihydroxyvitamin D3 and TPA in a human myelomonocytic cell line, HL60. Bone 32:621–629

43. Kitazawa R, Kitazawa S (2002) Vitamin D(3) augments osteoclastogenesis via vitamin D-responsive element of mouse RANKL gene promoter. Biochem Biophys Res Commun 290:650–655

44. Kabe Y, Yamada J, Uga H et al (2005) NF-Y is essential for the recruitment of RNA polymerase II and inducible transcription of several CCAAT box-containing genes. Mol Cell Biol 25:512–522

45. Kitazawa R, Mori K, Yamaguchi A et al (2008) Modulation of mouse RANKL gene expression by Runx2 and vitamin D(3). J Cell Biochem [Epub ahead of print]

46. Kondo T, Kitazawa R, Maeda S et al (2004) 1 alpha,25 dihydroxyvitamin D3 rapidly regulates the mouse osteoprotegerin gene through dual pathways. J Bone Miner Res 19:1411–1419

47. Baldock PA, Thomas GP, Hodge JM et al (2006) Vitamin D action and regulation of bone remodeling: suppression of osteoclastogenesis by the mature osteoblast. J Bone Miner Res 21:1618–1626

48. Atkins GJ, Kostakis P, Pan B et al (2003) RANKL expression is related to the differentiation state of human osteoblasts. J Bone Miner Res 18:1088–1098

49. Varga F, Spitzer S, Klaushofer K (2004) Triiodothyronine (T3) and 1,25-dihydroxyvitamin D3 (1,25D3) inversely regulate OPG gene expression in dependence of the osteoblastic phenotype. Calcif Tissue Int 74:382–387

50. Wang TT, Tavera-Mendoza LE, Laperriere D et al (2005) large scale in silico and microarray-based identification of direct 1,25-dihydroxy vitamin D3 target genes. Mol Endocrinol 19:2685–2695.

51. Rachez C, Freedman LP (2000) Mechanisms of gene regulation by vitamin D(3) receptor: a network of coactivator interactions. Gene 246:9–21

52. Montecino M, Stein GS, Cruzat F et al (2007) An architectural perspective of vitamin D responsiveness. Arch Biochem Biophys 460:293–299

53. Christakos S, Dhawan P, Liu Y et al (2003) New insights into the mechanisms of vitamin D action. J Cell Biochem 88:695–705

54. Xu J, O'Malley BW (2002) Molecular mechanisms and cellular biology of the steroid receptor coactivator (SRC) family in steroid receptor function. Rev Endocr Metab Disord 3:185–192

55. Goodman RH, Smolik S (2000) CBP/p300 in cell growth, transformation, and development. Genes Dev 14:1553–1577

56. Narlikar GJ, Fan HY, Kingston RE (2002) Cooperation between complexes that regulate chromatin structure and transcription. Cell 108:475–487

57. Barry JB, Leong GM, Church WB et al (2003) Interactions of SKIP/NCoA-62, TFIIB, and retinoid X receptor with vitamin D receptor helix H10 residues. J Biol Chem 278:8224–8228

58. Rachez C, Suldan Z, Ward J et al (1998) A novel protein complex that interacts with the vitamin D3 receptor in a ligand-dependent manner and enhances VDR transactivation in a cell-free system. Genes Dev 12:1787–1800

59. Rachez C, Lemon BD, Suldan Z et al (1999) Ligand-dependent transcription activation by nuclear receptors requires the DRIP complex. Nature 398:824–828

60. Rachez C, Gamble M, Chang CP et al (2000) The DRIP complex and SRC-1/p160 coactivators share similar nuclear receptor binding determinants but constitute functionally distinct complexes. Mol Cell Biol 20:2718–2726

61. Kornberg RD (2005) Mediator and the mechanism of transcriptional activation. Trends Biochem Sci 30:235–239

62. Kitagawa H, Fujiki R, Yoshimura K et al (2003) The chromatin-remodeling complex WINAC targets a nuclear receptor to promoters and is impaired in Williams syndrome. Cell 113:905–917

63. Fujiki R, Kim MS, Sasaki Y et al (2005) Ligand-induced transrepression by VDR through association of WSTF with acetylated histones. EMBO J 24:3881–3894

64. Kato S, Fujiki R, Kim MS et al (2007) Ligand-induced transrepressive function of VDR requires a chromatin remodeling complex, WINAC. J Steroid Biochem Mol Biol 103:372–380

65. Burakov D, Crofts LA, Chang CP et al (2002) Reciprocal recruitment of DRIP/mediator and p160 coactivator complexes in vivo by estrogen receptor. J Biol Chem 277:14359–14362

66. Sharma D, Fondell JD (2002) Ordered recruitment of histone acetyltransferases and the TRAP/Mediator complex to thyroid hormone-responsive promoters in vivo. Proc Natl Acad Sci USA 99:7934–7939

67. Metivier R, Penot G, Hubner MR et al (2003) Estrogen receptor-alpha directs ordered, cyclical, and combinatorial recruitment of cofactors on a natural target promoter. Cell 115:751–763

68. Kim S, Shevde NK, Pike JW (2005) 1,25-Dihydroxyvitamin D3 stimulates cyclic vitamin D receptor/retinoid X receptor DNA-binding, co-activator recruitment, and histone acetylation in intact osteoblasts. J Bone Miner Res 20:305–317

69. Metivier R, Reid G, Gannon F (2006) Transcription in four dimensions: nuclear receptor-directed initiation of gene expression. EMBO Rep 7:161–167

70. Oda Y, Sihlbom C, Chalkley RJ et al (2003) Two distinct coactivators, DRIP/mediator and SRC/p160, are differentially involved in vitamin D receptor transactivation during keratinocyte differentiation. Mol Endocrinol 17:2329–2339

71. Carvallo L, Henriquez B, Paredes R et al (2008) 1alpha,25-dihydroxy vitamin D3-enhanced expression of the osteocalcin gene involves increased promoter occupancy of basal transcription regulators and gradual recruitment of the 1alpha,25-dihydroxy vitamin D3 receptor-SRC-1 coactivator complex. J Cell Physiol 214:740–749

72. Zaidi SK, Young DW, Choi JY et al (2005) The dynamic organization of gene-regulatory machinery in nuclear microenvironments. EMBO Rep 6:128–133

73. Iborra FJ, Jackson DA, Cook PR (2000) The path of RNA through nuclear pores: apparent entry from the sides into specialized pores [In Process Citation]. J Cell Sci 113(Pt 2):291–302

74. Misteli T (2000) Cell biology of transcription and pre-mRNA splicing: nuclear architecture meets nuclear function. J Cell Sci 113:1841–1849

75. Dyck JA, Maul GG, Miller WH et al (1994) A novel macromolecular structure is a target of the promyelocyte-retinoic acid receptor oncoprotein. Cell 76:333–343

76. Zeng C, McNeil S, Pockwinse S et al (1998) Intranuclear targeting of AML/CBFα regulatory factors to nuclear matrix-associated transcriptional domains. Proc Natl Acad Sci USA 95:1585–1589

77. Zaidi SK, Javed A, Choi J-Y et al (2001) A specific targeting signal directs Runx2/Cbfa1 to subnuclear domains and contributes to transactivation of the osteocalcin gene. J Cell Sci 114:3093–3102

78. Olson MO, Hingorani K, Szebeni A (2002) Conventional and nonconventional roles of the nucleolus. Int Rev Cytol 219:199–266

79. Ma H, Siegel AJ, Berezney R (1999) Association of chromosome territories with the nuclear matrix. Disruption of human chromosome territories correlates with the release of a subset of nuclear matrix proteins. J Cell Biol 146:531–542

80. Cook PR (1999) The organization of replication and transcription. Science 284:1790–1795

81. Scully R, Livingston DM (2000) In search of the tumour-suppressor functions of BRCA1 and BRCA2. Nature 408:429–432

82. Verschure PJ, van Der Kraan I, Manders EM et al (1999) Spatial relationship between transcription sites and chromosome territories. J Cell Biol 147:13–24

83. Misteli T, Spector DL (1999) RNA polymerase II targets pre-mRNA splicing factors to transcription sites in vivo. Mol Cell 3:697–705

84. Smith KP, Moen PT, Wydner KL et al (1999) Processing of endogenous pre-mRNAs in association with SC-35 domains is gene specific. J Cell Biol 144:617–629

85. Wagner S, Chiosea S, Nickerson JA (2003) The spatial targeting and nuclear matrix binding domains of SRm160. Proc Natl Acad Sci USA 100:3269–3274

86. Barsony J, Renyi I, McKoy W (1997) Subcellular distribution of normal and mutant vitamin D receptors in living cells. Studies with a novel fluorescent ligand. J Biol Chem 272:5774–5782

87. Zhang C, Dowd DR, Staal A et al (2003) Nuclear coactivator-62 kDa/Ski-interacting protein is a nuclear matrix-associated coactivator that may couple vitamin D receptor-mediated transcription and RNA splicing. J Biol Chem 278:35325–35336

88. Paredes R, Arriagada G, Cruzat F et al (2004) The bone-specific transcription factor RUNX2 interacts with the 1α,25-dihydroxyvitamin D3 receptor to up-regulate rat osteocalcin gene expression in osteoblastic cells. Mol Cell Biol 24:8847–8861

89. Lian JB, Javed A, Zaidi SK et al (2004) Regulatory controls for osteoblast growth and differentiation: role of Runx/Cbfa/AML factors. Crit Rev Eukaryot Gene Expr 14:1–41

90. Owen TA, Aronow M, Shalhoub V et al (1990) Progressive development of the rat osteoblast pheno-type in vitro: reciprocal relationships in expression of genes associated with osteoblast proliferation and differentiation during formation of the bone extracellular matrix. J Cell Physiol 143:420–430

91. Montecino M, Lian J, Stein G et al (1996) Changes in chromatin structure support constitutive and developmentally regulated transcription of the bone-specific osteocalcin gene in osteoblastic cells. Biochemistry 35:5093–5102

92. Montecino M, Stein JL, Stein GS et al (2007) Nucleosome organization and targeting of SWI/SNF chromatin-remodeling complexes: contributions of the DNA sequence. Biochem Cell Biol 85: 419–425

93. Javed A, Gutierrez S, Montecino M et al (1999) Multiple Cbfa/AML sites in the rat osteocalcin promoter are required for basal and vitamin D responsive transcription and contribute to chromatin organization. Mol Cell Biol 19:7491–7500

94. Montecino M, Frenkel B, van Wijnen AJ et al (1999) Chromatin hyperacetylation abrogates vitamin D-mediated transcriptional upregulation of the tissue-specific osteocalcin gene *in vivo*. Biochemistry 38:1338–1345

95. Paredes R, Gutierrez J, Gutierrez S et al (2002) Interaction of the 1alpha,25-dihydroxyvitamin D3 receptor at the distal promoter region of the bone-specific osteocalcin gene requires nucleosomal remodelling. Biochem J 363:667–676

96. Sierra J, Villagra A, Paredes R et al (2003) Regulation of the bone-specific osteocalcin gene by p300 requires Runx2/Cbfa1 and the vitamin D3 receptor but not p300 intrinsic histone acetyltransferase activity. Mol Cell Biol 23:3339–3351

97. Shen J, Montecino MA, Lian JB et al (2002) Histone acetylation in vivo at the osteocalcin locus is functionally linked to vitamin D dependent, bone tissue-specific transcription. J Biol Chem 277:20284–20292

98. Vaisanen S, Dunlop TW, Sinkkonen L et al (2005) Spatio-temporal activation of chromatin on the human CYP24 gene promoter in the presence of 1alpha,25-Dihydroxyvitamin D3. J Mol Biol 350:65–77

99. Murayama A, Kim MS, Yanagisawa J et al (2004) Transrepression by a liganded nuclear receptor via a bHLH activator through co-regulator switching. EMBO J 23:1598–1608

100. Stein GS, Lian JB, Montecino M et al (2002) Involvement of nuclear architecture in regulating gene expression in bone cells. In: Bilezikian JP, Raisz LG, Rodan GA (eds) Principles of bone biology, 2nd edn. Academic Press, San Diego, pp 169–188

101. Schinke T, Karsenty G (2002) Transcriptional control of osteoblast differentiation and function. In: Bilezikian JP, Raisz LG, Rodan GA (eds) Principles of bone biology. Academic Press, San Diego, pp 83–91

102. Young DW, Hassan MQ, Pratap J et al (2007) Mitotic occupancy and lineage-specific transcriptional control of rRNA genes by Runx2. Nature 445:442–446

103. Young DW, Hassan MQ, Yang X-Q et al (2007) Mitotic retention of gene expression patterns by the cell fate determining transcription factor Runx2. Proc Natl Acad Sci USA 104:3189–3194

9 Biological and Molecular Effects of Vitamin D on the Kidney

Adriana S. Dusso and Masanori Tokumoto

Abstract   The kidney is essential for the integrity of the vitamin D endocrine system: Normal kidney function ensures adequate serum levels of 1,25-dihydroxyvitamin D ([1,25(OH)$_2$D], the hormonal form of vitamin D) and of its precursor, 25-hydroxyvitamin D [25(OH)D]. In turn, normal serum 1,25(OH)$_2$D and 25(OH)D levels are critical in maintaining normal kidney function and in ameliorating the progression of kidney disease. Research in the last three decades has firmly established that renal conversion of 25(OH)D to 1,25(OH)$_2$D plays a key role in the maintenance of calcium and phosphate homeostasis and skeletal health. 1,25(OH)$_2$D endocrine actions include the coordinated regulation of the synthesis of parathyroid hormone (PTH), the bone phosphatonin FGF23, and its own serum concentrations, and consequently of the hormonal loops and ion fluxes between the parathyroid gland, the intestine, the kidney, and the bone. 1,25(OH)$_2$D tight control of this multi-organ endocrine system is central not only for bone integrity but also in preventing an excess of serum calcium and phosphate predisposing to over-mineralization of bone or ectopic calcification. Importantly, recent epidemiological evidence also suggests that the kidney is essential for the maintenance of normal serum 25(OH)D levels, a requirement for 1,25(OH)$_2$D autocrine actions. Renal uptake of 25(OH)D from the glomerular ultrafiltrate is an active process necessary for both renal and extrarenal 1,25(OH)$_2$D production. Normal renal uptake of 25(OH)D ensures the appropriate delivery of 25(OH)D for its conversion to 1,25(OH)$_2$D by an increasing number of nonrenal cells. Autocrine 1,25(OH)$_2$D actions in 1,25(OH)$_2$D producing cells, which include parathyroid cells, osteoblasts, and cells of the immune and cardiovascular system, appear to mediate the multiple health benefits and the survival advantage conferred to the general population by a normal vitamin D status. This chapter presents the current understanding of the mechanisms mediating renal control of vitamin D metabolism and 1,25(OH)$_2$D endocrine and autocrine actions important in disease prevention, with a special focus on the renoprotective actions of the vitamin D endocrine system that prevent/ameliorate the onset and progression of kidney disease.

Key Words: Kidney; 1,25-dihydroxyvitamin D; 25-hydroxyvitamin D; calcium; phosphate; GFR; 25-hydroxyvitamin D-1α-hydroxylase; parathyroid hormone; FGF23; klotho

1. THE KIDNEY, THE VITAMIN D ENDOCRINE SYSTEM, AND DISEASE PREVENTION

The essential role of vitamin D in mineral homeostasis and skeletal health has been recognized since the 1930s, when vitamin D fortification of milk eradicated rickets. In the last two decades, an impressive body of in vitro, animal, clinical, and

From: *Nutrition and Health: Vitamin D*
Edited by: M.F. Holick, DOI 10.1007/978-1-60327-303-9_9,
© Springer Science+Business Media, LLC 2010

epidemiological evidence has extended vitamin D actions beyond bone health to support "nontraditional" roles of vitamin D which are critical in disease prevention *(1)*. Indeed, in normal individuals, vitamin D deficiency has been associated with increased risk of osteoporosis, hypertension, glucose intolerance, certain infectious diseases, multiple sclerosis, cardiovascular disease, cancer *(1)*, and, importantly for this chapter, with proteinuria, a hallmark of an abnormal kidney function *(2)*.

The discovery of the hormonal form of vitamin D [1,25-dihydroxyvitamin D, $1,25(OH)_2D$ or calcitriol] in 1970 *(3)* has firmly established the essential role of the kidney in vitamin D biological actions. On its own, vitamin D has little intrinsic activity and must be metabolized to calcitriol to exert its biological actions [reviewed in *(4, 1)* and in Chapter 3]. The first step in vitamin D bioactivation is its hydroxylation of carbon 25 to 25-hydroxyvitamin D [25(OH)D], which occurs primarily in the liver. Serum levels of 25(OH)D correlate directly with vitamin D concentrations, and because they are easier to measure than serum vitamin D levels, they are used to monitor vitamin D status. The proximal convoluted tubule of the kidney is the principal site for the final and most critical step in vitamin D bioactivation: 1α-hydroxylation of 25-hydroxyvitamin D to calcitriol, a potent calcitropic steroid hormone. This reaction is catalyzed by mitochondrial renal 1α-hydroxylase (1-hydroxylase). Thus, renal calcitriol production ensures adequate systemic levels of the vitamin D hormone to exert its biological actions.

Renal calcitriol production is critical for the actions of vitamin D as it is the main regulator of the expression of its high-affinity receptor, the vitamin D receptor (VDR), in vitamin D target cells *(4)*. Most calcitriol actions are mediated by the VDR which acts as a ligand-activated transcription factor [reviewed in *(4)* and discussed in Chapter 2 and Chapter 5]. Calcitriol binding to the VDR activates the receptor to translocate from the cytosol to the nucleus where it heterodimerizes with its partner the retinoid X receptor, RXR. The VDR/RXR complex then binds specific sequences in the promoter region of vitamin D target genes, called vitamin D response elements (VDREs), and recruits basal transcription factors and co-regulator molecules to either increase or suppress the rate of transcription of vitamin D target genes by RNA-polymerase II.

From the plethora of vitamin D biological actions, it is clear that the calcitriol/VDR complex regulates the expression of numerous genes other than those implicated in the tight control of mineral homeostasis and bone health, in an increasing number of nontraditional target cells. In this chapter, special focus will be directed to vitamin D responsive genes that regulate specific cell functions in resident and infiltrating cell types along nephron segments, which are responsible for the maintenance of normal kidney function. Thus, through a simultaneous regulation of systemic calcitriol levels and VDR content in vitamin D target cells including its own, the kidney determines the capacity of traditional and nontraditional targets to respond to vitamin D. In fact, in the course of chronic kidney disease (CKD), a disorder that affects 11% of the adult US population, the degree of calcitriol deficiency/insufficiency and the concomitant VDR reduction directly associate not only with the severity of secondary hyperparathyroidism and skeletal abnormalities but also with lesions to the renal parenchyma *(5)*. The latter generates a vicious cycle for further reductions in renal calcitriol production, VDR levels, and calcitriol/VDR renoprotection, thereby accelerating the progression of renal damage to end-stage kidney disease.

Interestingly, however, the enhanced susceptibility to health disorders including albuminuria in vitamin D-deficient individuals is not caused by a reduction in serum calcitriol levels but by deficient/insufficient serum levels of the calcitriol precursor, 25-hydroxyvitamin D *(1)*. These findings suggest a key role in human health of the "local" activation of 25(OH)D to calcitriol by numerous nonrenal cells that express 1-hydroxylase activity [reviewed in *(6)* and in Chapter 7] and have uncovered another critical contribution of the kidney to vitamin actions: the uptake of 25(OH)D. Indeed, renal proximal tubular cells are also responsible for the uptake of 25(OH)D bound to its carrier in the circulation, the vitamin D-binding protein (DBP), from the glomerular filtrate, a process mediated via the endocytic receptor megalin *(7)*. Megalin-mediated endocytosis of 25(OH)D, by preventing the urinary loss of filtered 25(OH)D, ensures the adequate delivery of the calcitriol precursor to renal and nonrenal cells bearing 1-hydroxylase activity. Importantly, renal megalin expression is induced by the calcitriol/VDR complex *(8)*. Thus, impaired renal calcitriol production in CKD contributes to reduce renal megalin levels *(9)* and, consequently, to a vicious cycle of defective 25(OH)D uptake for renal and local calcitriol synthesis.

In summary, reciprocal interactions occur between the kidney and the vitamin D endocrine system that are vital for human health: Normal kidney function is required for the maintenance of normal serum levels of 25(OH)D and calcitriol, VDR content and function. In turn, the integrity of the vitamin D endocrine system is essential in maintaining normal kidney function, and in ameliorating the progression of renal damage to end-stage kidney disease. This chapter presents the current understanding of the mechanisms regulating renal control of vitamin D biological actions, namely, renal calcitriol production, the uptake of 25(OH)D and VDR expression and function, as well as those mediating vitamin D renoprotection, and their contribution to ameliorate disease progression and to the survival advantage conferred by vitamin D therapy to CKD patients.

2. ESSENTIAL ROLE OF THE KIDNEY IN VITAMIN D BIOLOGICAL ACTIONS

2.1. Renal Regulation of Systemic Calcitriol

2.1.1. CALCITRIOL SYNTHESIS

Renal 1-hydroxylase, also known as CYP27B1, is a cytochrome P450 monooxygenase which localizes in the inner mitochondrial membrane of the proximal convoluted tubule, and the main contributor to systemic calcitriol levels *(10)*. The 1-hydroxylase gene maps to chromosomal locus 12q13.1–q13.3 *(11)*. Mutations in the coding region of this gene result in the expected inability of calcitriol synthesis and the phenotype of vitamin D-dependent rickets type I *(12–14)*.

The potent calcitropic- and phosphate-mobilizing actions of circulating calcitriol require a tight control of renal 1-hydroxylase expression and activity. Parathyroid hormone (PTH), hypocalcemia, and hypophosphatemia are the major inducers, whereas hyperphosphatemia, hypercalcemia, and calcitriol repress its activity. In spite of the pathophysiological significance of these hormonal loops in mineral and skeletal homeostasis, the molecular mechanisms regulating renal 1-hydroxylase expression remain poorly understood. Specifically, PTH, the most potent inducer of 1-hydroxylase activity *(15)*, directly increases 1-hydroxylase mRNA *(16, 17)* levels in renal proximal tubular cells via increases in cAMP *(18)* and stimulation of 1-hydroxylase gene transcription *(19, 20)*. The 1-hydroxylase gene promoter contains three cAMP response elements (CRE). However, the PTH-cAMP pathway appears to act at a CRE-less sequence located near the transcription start site for this gene *(19)*.

Dietary calcium (Ca) can regulate renal 1-hydroxylase directly through changes in serum Ca and indirectly by altering PTH levels *(21)*. Indeed, parathyroidectomy severely blunts but does not eliminate the stimulation of the 1-hydroxylase by hypocalcemia *(22)*. A direct suppression of the 1-hydroxylase activity and mRNA by calcium has been demonstrated in a human proximal tubule cell line *(23)*. However, it is unclear whether this effect is mediated by the Ca-sensing receptor.

It is also unclear how dietary phosphate (P) regulates renal 1-hydroxylase activity *(24)* and mRNA *(16)* independently of low P-driven changes in serum PTH *(25)* or Ca *(26)*. The lack of a direct effect of P on renal 1-hydroxylase activity in cell culture suggests that dietary phosphate regulation of 1-hydroxylase may be mediated by a systemic hormone. Whereas the induction of 1-hydroxylase by P restriction appears to involve a growth hormone-mediated mechanism *(27)*, likely candidates for the suppressive effects of high P are the recently discovered phosphaturic factors or phosphatonins, fibroblast growth factor 23 (FGF23), frizzled-related protein 4 (FRP4) and matrix extracellular phosphoglycoprotein (MEPE) [reviewed in *(28)*]. Transgenic mice constitutively expressing FGF23 have reduced calcitriol levels in spite of low plasma phosphate *(29)*. Similarly, FRP4 administration and MEPE overexpression in vivo produce hypophosphatemia and reduce serum calcitriol levels *(30, 31)*. High P directly induces FGF23 expression, and FGF23 was shown to suppress 1-hydroxylase mRNA levels dose dependently *(32)*.

The klotho gene product, which encodes a β-glucuronidase and is implicated in aging and in facilitating FGF receptor activation by FGF23, is also a potent negative regulator of 1-hydroxylase *(33)*. Similar to the FGF23-null mice, the klotho-null mice have elevated calcitriol levels, high plasma calcium and phosphate, and die prematurely due to ectopic calcifications *(34)*. In these mice, basal 1-hydroxylase mRNA levels are increased in spite of hypercalcemia, hyperphosphatemia, and low PTH *(33)*.

Calcitriol is a key determinant of its own circulating levels. Calcitriol feedback inhibition of 1-hydroxylase minimizes the potential for vitamin D intoxication. Calcitriol directly suppresses 1-hydroxylase activity in kidney cell culture *(35, 36)* and reduces 1-hydroxylase mRNA *(37, 16, 17, 38)*. The latter does not involve direct suppression of the 1α-hydroxylase gene promoter by the calcitriol/VDR complex but calcitriol-mediated inhibition of the PTH/cAMP induction of promoter activity *(19)*. In addition, calcitriol inhibits 1-hydroxylase in vivo through several indirect mechanisms including calcitriol-mediated increases in serum Ca and P levels, decreases in serum PTH, and the recently identified induction of bone FGF23 synthesis *(39, 40)* and renal klotho expression *(33)*.

2.1.2. CALCITRIOL CATABOLISM

Calcitriol induction of its own degradation is also a key determinant of the tight regulation of serum calcitriol levels. In fact, in normal individuals, the metabolic clearance rate of calcitriol is accelerated when its production rates increases *(41)*. This is accomplished within virtually all target cells by the calcitriol-inducible vitamin D 24-hydroyxlase, which catalyzes a series of oxidation reactions at carbons 24 and 23 leading to side chain cleavage and inactivation. Mice lacking a functional 24-hydroxylase gene have high serum calcitriol levels due to the decreased capacity to degrade it *(42)*. At least two distinct vitamin D response elements in the 24-hydroxylase promoter mediate calcitriol/VDR induction of the transcription of the 24-hydroxylase gene *(43, 44)*. The 24-hydroxylase is also regulated by P and PTH in a reciprocal manner to their control of 1-hydroxylase, i.e., its activity and expression are increased by P *(24, 45)* and reduced by PTH *(15)*.

2.2. Abnormal Calcitriol Production in CKD

In the course of kidney disease, the progressive reduction in systemic calcitriol levels *(46–49)* caused by the reduction in functional renal mass is further aggravated by (a) a blunted induction of 1-hydroxylase activity in response to the high circulating PTH, which has been attributed to either uncoupling of the PTH receptor/PKA axis *(50)* or the inhibition of 1-hydroxylase activity by N-terminally truncated PTH fragments or C-terminal PTH fragments *(51)*; (b) the suppression of renal 1-hydroxylase expression by P-mediated increases in serum FGF23 levels *(32)*; and (c) direct inhibition of the activity of remnant renal 1-hydroxylase by acidosis and/or the accumulation of uremic toxins *(52)* .

In relation to calcitriol catabolism, reduced catabolic rates were expected in CKD from the low serum calcitriol and the high PTH levels. In fact, reduced metabolic clearance rates were reported in CKD patients *(53)* and in a rat model of CKD *(54)* and were considered as compensatory mechanisms to maintain serum calcitriol levels at early stages of kidney disease. However, no changes in catabolic rates were found in dogs with mild to severe renal damage, in whom serum calcitriol levels reflected the decrease in production rates by the failing kidney *(55)*. In spite of these controversial reports, it is obvious from the low serum calcitriol levels that reduced production rate is the main contributor to the calcitriol deficiency/insufficiency in CKD patients.

3. RENAL UPTAKE OF 25(OH)D: RELEVANCE IN THE MAINTENANCE OF NORMAL VITAMIN D STATUS

3.1. Essential Contribution of Megalin-Mediated Endocytosis to Calcitriol Production

Normal renal 1-hydroxylase activity requires the uptake of 25(OH)D by renal proximal tubular cells and its delivery to the inner mitochondrial membrane for its hydroxylation to calcitriol. Studies by Nykjaer and collaborators in the megalin-null mice *(7)* challenged the prior concept that 25(OH)D diffuses into renal proximal tubular cells at the basolateral membrane upon dissociation from DBP. Instead, the 25(OH)D/DBP complex in the circulation is filtered through the glomerulus and endocytosed into the proximal tubular cell via the apical-membrane receptor, megalin *(7)*, the largest member of the LDL receptor superfamily [summarized in Fig. 1]. Megalin-mediated endocytosis of 25(OH)D/DBP also requires the receptor-associated protein (RAP) *(56)* and cubilin *(57)*, a protein required for sequestering DBP on the cell surface prior to its internalization by megalin. Ablation of these proteins results in urinary excretion of DBP. Once inside the cells, DBP is degraded, apparently by legumain *(58)*, releasing 25(OH)D back to the circulation or delivering it for metabolism to either 1-hydroxylase for its activation to calcitriol or 24-hydroxylase for its metabolic inactivation. The 25(OH)D translocation to the mitochondria may also be facilitated rather than passive and appears to involve the interaction of the intracellular carboxyterminus of megalin with the intracellular vitamin D-binding proteins IDBP-1 and IDBP-3 *(59)*.

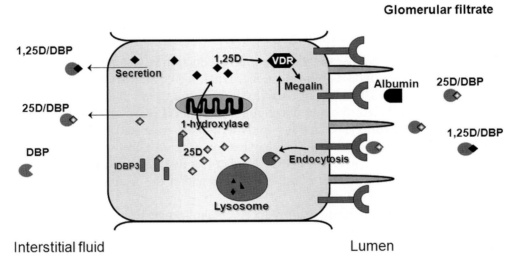

Fig. 1. Renal uptake of 25(OH)D. Circulating 25(OH)D (25D) bound to its carrier vitamin D-binding protein (DBP) is filtered by the kidney and internalized into proximal tubular cells via megalin-mediated endocytosis. Upon its release from DBP, 25(OH)D is either delivered to 1-hydroxylase for its bioactivation to calcitriol (1,25(OH)$_2$D) by interacting vitamin D-binding protein 3 (IDBP3) or it re-enters the circulation. Calcitriol induces renal megalin expression thereby generating a cycle that ensures normal systemic 25(OH)D and calcitriol levels as well as the reabsorption of low molecular weight proteins from the glomerular filtrate, including albumin.

3.2. Abnormal Uptake of 25(OH)D in CKD

In the course of CKD, the combination of progressive decreases in glomerular filtration rates (GFR) and renal megalin content *(9)* markedly impairs the uptake of 25(OH)D by proximal tubular cells. Decreased GFRs reduce the amount of filtered 25(OH)D/DBP complex, which in turn limits the amount of 25(OH)D available for uptake, thus reducing its intracellular content for conversion to calcitriol by the remnant renal 1-hydroxylase. Appropriate vitamin D supplementation with ergocalciferol or cholecalciferol, by elevating serum 25(OH)D, increases the proportion of 25(OH)D/DBP complex in serum and, consequently, the amount filtered and available for bioactivation to calcitriol. Conclusive evidence of the importance of correcting the impaired 25(OH)D availability in CKD was reported in 1984 *(60)*. Patients with a GFR below 25 ml/min and, therefore, very limiting renal 1-hydroxylase activity and low serum calcitriol levels normalized serum calcitriol when serum 25(OH)D concentrations were increased to supraphysiological levels through oral supplementation *(60)* (see Fig. 2, left panel). This suggested impaired substrate availability for the renal 1-hydroxylase. In fact, in advanced human and experimental CKD, serum levels of substrate and product of renal 1-hydroxylase strongly correlate. This serum 25(OH)D/calcitriol association does not occur in individuals with normal renal function. On the contrary, calcitriol levels are kept within the narrow limits required for normal Ca homeostasis even upon elevations in serum 25-hydroxyvitamin D up to 70-fold above normal in cases of vitamin D intoxication *(61)*.

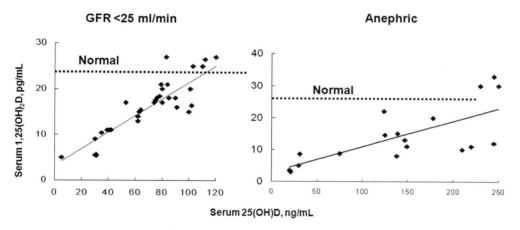

Fig. 2. Impaired availability to renal and nonrenal 1-hydroxylases in CKD. The strong correlation between serum levels of 25(OH)D and calcitriol [1,25(OH)$_2$D] in severely uremic patients (GFR < 25 ml/min) [*Left panel*] as well as in bilaterally nephrectomized patients (anephrics) [*Right panel*] demonstrates impaired 25(OH)D availability for its bioactivation to calcitriol by the remnant renal and nonrenal 1-hydroxylases in CKD. Only supraphysiological concentrations of 25(OH)D normalize serum calcitriol levels. Adapted from references *(60)* and *(65)*.

In CKD, reduced renal content of megalin *(9)* further aggravates the low 25(OH)D availability for the renal 1-hydroxylase caused by the decreases in GFR. Indeed, in rats, renal megalin-mRNA levels are reduced by 2 weeks after the induction of renal failure *(9)*. Defective induction of renal megalin by the calcitriol/VDR complex *(8)* could contribute to the reduced megalin levels in CKD and could partially account for the failure of exclusive vitamin D supplementation in correcting serum 25(OH)D levels in CKD patients [reviewed in *(62)*]. The latter raises important considerations when designing a strategy to correct vitamin D deficiency in these patients. According to the current recommendations, in CKD stages 2 or 3, high serum PTH is an indicator to evaluate vitamin D status, a bolus of 50,000 IU of either ergocalciferol (vitamin D2) or cholecalciferol (vitamin D3) should be given to increase serum 25(OH)D levels above 30 ng/ml *(62)*. However, upon a bolus administration of 50,000–100,000 IU, the initial increases in serum 25(OH)D levels are identical only over the first 3 days *(63)*. Whereas serum 25(OH)D levels continued to rise in cholecalciferol-treated subjects and peak at 14 days, serum 25(OH)D levels at day 14 were not different from baseline in ergocalciferol-treated individuals. Instead, daily oral administration of 2,000 IU of either metabolite results in identical increases in serum 25(OH)D from baseline *(64)*. The markedly lower potency and shorter duration of ergocalciferol action relative to cholecalciferol in correcting 25(OH)D deficiency *(63)* and consequently renal and local calcitriol production could be simply resolved through using daily oral administration of either metabolite, and this would also prove to be at a lower cost.

3.3. Abnormal Renal 25(OH)D Uptake in CKD Contributes to Impaired Extrarenal Calcitriol Production

Kidney disease also impairs the delivery of 25-hydroxyvitamin D to extrarenal sources of calcitriol *(65)*. As mentioned, 1-hydroxylase is expressed in numerous non-renal cells *(6)*, and their capacity to produce calcitriol was conclusively demonstrated in bilaterally nephrectomized patients undergoing hemodialysis *(66)*. In these patients, appropriate 25(OH)D supplementation could normalize serum calcitriol levels. Serum calcitriol concentrations strongly correlate with the levels of its precursor 25(OH)D (see Fig. 2, right panel), suggesting an impaired substrate availability to extrarenal sources similar to that demonstrated for the renal 1-hydroxylase in severely uremic patients *(60)*.

The physiological relevance of impaired 25(OH)D availability to nonrenal 1-hydroxylase in CKD was first suggested from studies in primary cultures of parathyroid cells, which express 1-hydroxylase *(67, 68)*. These cells appear to suppress PTH in an autocrine/paracrine manner through local activation of 25(OH)D to calcitriol *(67)*. Accordingly, in early stages of CKD, ergocalciferol administration can correct serum PTH exclusively in patients achieving serum 25-hydroxyvitamin D levels above 30 ng/ml *(69–71)*. These findings support a role for impaired 25(OH)D availability for the parathyroid 1-hydroxylase in limiting the suppression of PTH by endogenously produced calcitriol.

As mentioned earlier, in the case of the renal calcitriol production of CKD, reduced induction of megalin expression by the calcitriol/VDR complex and the resulting impairment of 25(OH)D uptake further aggravate the reduced substrate availability due to low serum 25(OH)D levels. The impact of kidney disease on 25(OH)D uptake by "nonrenal" 1-hydroxylase-bearing cells was examined in peripheral blood monocytes from hemodialysis patients as the monocyte–macrophage 1-hydroxylase is identical to the renal enzyme though more readily accessible.

Figure 3 summarizes the findings. Not only the uptake of 25(OH)D by monocytes from hemodialysis patients was markedly lower than that of monocytes from normal individuals, but more significantly, the defective uptake could be corrected through the normalization of the low serum calcitriol levels in these hemodialysis patients by intravenous calcitriol supplementation *(65)*. Megalin expression in macrophages is not firmly established, and therefore, neither the mechanisms mediating impaired 25-hydroxyvitamin D uptake nor those involved in its correction by calcitriol treatment are known.

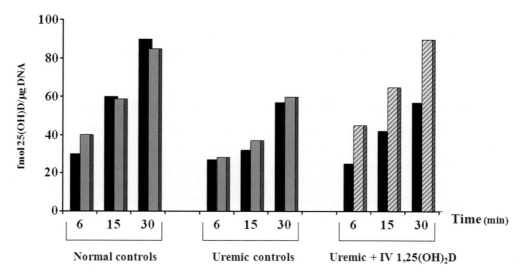

Fig. 3. Calcitriol administration in CKD corrects the defective 25(OH)D uptake by peripheral blood monocyte–macrophages. Time course for 25(OH)D uptake at day 0 (*left bar*) and at day 15 (*right bar*) by peripheral blood monocytes from normal individuals (Normal) and hemodialysis patients (Uremic) receiving either vehicle (Controls) or intravenous calcitriol (i.v. 1,25(OH)₂D) three times a week for 15 days. Modified from reference *(65)*.

The potent immunomodulatory properties of the calcitriol/VDR complex *(72)* suggest that, in CKD, reduced local calcitriol production by monocytes could partially account for the impaired calcitriol anti-inflammatory properties, as well as the higher macrophage infiltration to the renal parenchyma that aggravates renal lesions.

These findings underscore the importance of an appropriate correction of vitamin D and calcitriol deficiency in CKD to ensure vitamin D actions resulting from local calcitriol production.

4. RENAL REGULATION OF VDR EXPRESSION AND FUNCTION

4.1. Calcitriol Regulation of VDR Expression

Most if not all of calcitriol biological actions are mediated by the VDR. Since calcitriol is the best-known regulator of VDR expression in target tissues, an adequate renal calcitriol production is vital for the maintenance of normal responses to the calcitriol/VDR complex. Calcitriol upregulation of VDR expression involves dual mechanisms: It increases VDR mRNA levels and/or VDR protein stability *(73, 74)*. The latter is the most common process involved in calcitriol upregulation of VDR content. Calcitriol binding to the VDR prevents receptor degradation by the proteasome complex, thus enhancing the half-life of the ligand-bound VDR compared to that of the unliganded VDR *(74)*.

Calcitriol is known to upregulate VDR mRNA levels in the parathyroid glands and the kidney through mechanisms that are unclear since there are no VDRE in the VDR promoter. The recent demonstrations that calcitriol induces the expression of the transcription factor C/EBPβ in numerous cell types *(75–77)*, and that C/EBPβ in turn induces VDR gene expression via a C/EBP-binding site in the VDR promoter *(75)*, suggest that calcitriol-mediated increases in C/EBPβ could partially account for calcitriol upregulation of VDR mRNA levels.

4.2. Abnormal Regulation of VDR Expression and Function in CKD

In CKD, the progressive reduction in serum calcitriol results in a parallel reduction in VDR expression responsible for the onset of resistance to calcitriol replacement therapy. In fact, calcitriol-binding studies in the parathyroid glands of humans *(78, 79)*, rats *(80)*, and dogs *(81)* with hyperparathyroidism secondary to CKD support a reduction in VDR number without changes in the affinity of the receptor for calcitriol. Furthermore, not only a strong correlation exists between serum levels of calcitriol and parathyroid VDR content but more significantly, calcitriol/analog therapy effectively corrects the reduced parathyroid VDR content *(82)*.

The recent characterization of the pathogenesis of VDR reduction with the severity of parathyroid gland hyperplasia in experimental and human CKD supports a role for calcitriol induction of C/EBPβ in calcitriol regulation of VDR gene expression *(83)*. The lowest parathyroid VDR mRNA and protein expression occurs in areas of nodular growth *(83)*, the most aggressive form of parathyroid hyperplasia *(78)*. Enhanced TGF-α/EGFR expression induces the synthesis of LIP, a potent mitogen and the dominant negative isoform of C/EBPβ, which antagonizes C/EBPβ induction of VDR gene expression *(83)*. Similar mechanisms could mediate the downregulation of renal VDR with the progression of renal damage. In fact, enhanced TGF-α/EGFR activation also occurs in the kidney upon nephron reduction, prolonged renal ischemia, or prolonged exposure to angiotensin II and is a main determinant of the severity of lesions to the renal parenchyma in these conditions *(84, 85)*. A similar TGF-α/EGFR-driven induction of LIP could mediate a reduction in renal VDR gene expression, thereby aggravating the impaired VDR content resulting from low serum calcitriol levels.

4.3. Impaired VDR Function in CKD

CKD also impairs calcitriol/VDR signaling. Upon nephron reduction, a reduction in the renal expression of the VDR partner, the RXR, reduces RXR/VDR heterodimerization *(86)*. The accumulation of uremic toxins impairs the binding of the VDR/RXR heterodimer to DNA *(87)*. In addition, hypocalcemia-driven induction of calreticulin, a molecule that binds the DNA-binding domain of the VDR, competes with the VDR for DNA binding, further decreasing calcitriol/VDR regulation of gene expression *(88)*.

Kidney disease also affects the complex interactions of DNA-bound VDR/RXR with co-regulators of VDR transactivation or transrepression of vitamin D responsive genes. Megalin is one of such regulators. Megalin modulates VDR-mediated transactivation through sequestration of Skp *(89)*, a component of the VDR transcriptional complex. Thus, the reduced renal megalin expression associated with kidney disease could partially account for abnormalities in calcitriol/VDR transcriptional activity. The reduced serum calcitriol of CKD could affect either the expression of essential coactivator or co-repressor molecules *(90)* or the proper activation by phosphorylation of transcriptional co-regulators and, consequently, their recruitment to the VDR transcription pre-initiation complex. Indeed, selective recruitment of co-repressor molecules by calcitriol analogs contributes to their diverse potency in suppressing PTH gene expression "ex-vivo" *(91)*.

The marked impairment of VDR function in CKD patients prompted an assessment of the potential contribution of differences in the frequency of VDR gene polymorphisms, which are normal genetic variants of the human VDR gene *(92, 93)*. Indeed, a higher frequency of certain VDR polymorphisms is associated with changes in bone density *(94)*, the propensity to hyperparathyroidism *(95, 96)*, or resistance to vitamin D therapy *(97)* in Western and Japanese CKD patients *(98–101)*. However, the findings were controversial, which may reflect the inadequacy in the choice of the VDR polymorphism under analysis as none of the studied VDR gene variants affects either the levels of VDR gene or protein expression or the function of the translated protein *(102)*.

5. VITAMIN D REGULATION OF KIDNEY FUNCTION

This section presents an update on actions of the calcitriol/VDR complex in the kidney that are critical for the maintenance of mineral homeostasis, novel actions that contribute to the prevention of cardiovascular disorders, and with a special emphasis on those implicated in the protection of kidney function that ameliorates the progression of renal lesions to end-stage renal disease.

5.1. Mineral Homeostasis

Figure 4 summarizes the central role of renal calcitriol production and calcitriol-coordinated regulation of PTH, FGF23, and its own levels in the powerful, complex, and tightly regulated hormonal loops between the parathyroid glands, intestine, and bone responsible for maintaining Ca and P homeostasis (reviewed in *(4)*), while preventing the excesses of these minerals that could lead to over-mineralization of bone or ectopic calcification *(39)*. Briefly, the ability of calcitriol to promote intestinal absorption and renal reabsorption of calcium and phosphate could potentially lead to an excess of these minerals causing over-mineralization of bone and ectopic calcification *(39)*. This is normally avoided by calcitriol inhibition of PTH synthesis, which reduces bone resorption, and calcitriol induction of FGF23 synthesis *(40)*, which prevents hyperphosphatemia by inhibiting renal P reabsorption *(28)*, and feedbacks repress renal calcitriol production *(39)* It is clear from this current model that the most important endocrine action of the calcitriol/VDR complex in the kidney that affects mineral metabolism is the tight control of its own homeostasis.

5.1.1. CALCITRIOL/VDR REGULATION OF CALCITRIOL HOMEOSTASIS

Calcitriol/VDR induction of megalin expression in cells of the proximal convoluted tubule *(8)* plays an important role in renal calcitriol synthesis. However, calcitriol/VDR suppression of 1α-hydroxylase and stimulation of 24-hydroxylase are the key determinants of net renal calcitriol production.

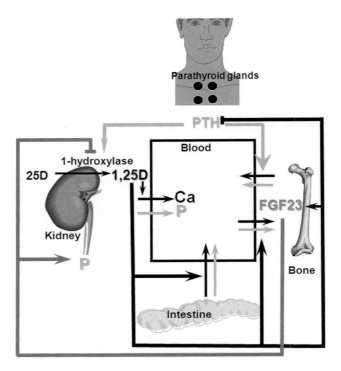

Fig. 4. Central role of renal calcitriol production in Ca, P, and skeletal homeostasis. Renal calcitriol (1,25(OH)$_2$D) production is a central integrator of hormonal feedback loops between PTH and FGF23, as well as Ca (*black arrows*) and P (*gray arrows*) fluxes between the intestine, bone, and kidney that ensure normal mineral metabolism and skeletal integrity, while preventing the excess of both ions predisposing to ectopic calcifications. Hypocalcemia stimulates PTH secretion by the parathyroid glands, which induces Ca and P resorption from bone and stimulates renal 1-hydroxylase to induce calcitriol production thus promoting intestinal and renal Ca and P absorption/reabsorption. Upon Ca normalization, both Ca and calcitriol close the loop through simultaneous suppression of PTH synthesis and renal 1-hydroxylase. Calcitriol induction of the synthesis of FGF23 in bone helps prevent hyperphosphatemia through FGF23-mediated increases in renal P excretion and suppression of 1-hydroxylase.

5.1.2. CALCITRIOL/VDR CONTROL OF RENAL HANDLING OF CALCIUM AND PHOSPHATE

Calcitriol involvement in the renal handling of calcium and phosphate continues to be controversial due to the systemic effects of the calcitriol/VDR complex through suppression of serum PTH (Chapter 14), as well as intestinal calcium and phosphate absorption (Chapter 9), the latter affecting the filter load of both ions.

Direct effects of calcitriol in enhancing renal calcium reabsorption involve the induction of trans-epithelial calcium transport through direct transactivation of calbindin expression and the epithelial calcium channel TRPV5. Several putative VDR-binding sites have been located in the human promoter of the renal epithelial calcium channel, and decreases in circulating levels of calcitriol concentrations resulted in a marked decline in the expression of the channel at the protein and mRNA levels *(103)*. Also, calcitriol indirectly increases renal calcium channel activity through inhibition of the expression of SBPRY (B-box and SPRY domain-containing protein), a TRPV5 and TRPV6 interacting protein, which inhibits TRVP5 calcium channel activity at the cell surface *(104)*. Calcitriol regulation of Ca channel activity confers two important properties to renal Ca influx, namely a high calcium selectivity and a negative feedback regulation to prevent calcium overload during trans-epithelial Ca transport *(105)*. Calcitriol also accelerates PTH-dependent calcium transport in the distal tubule *(106)*, the main determinant of the final excretion of calcium into the urine and the site with the highest VDR content.

Calcitriol also elicits direct effects on the renal handling of P through the induction of NPT2c, a renal sodium-phosphate co-transporter that facilitates renal P reabsorption. Although this transporter plays a minor role in renal phosphate reabsorption in rodents, alterations in the gene coding for NPT2c are the cause of the human phosphate wasting disorder of hereditary hyperphosphatemic rickets with hypercalciuria *(107, 108)*. This gene appears to be under direct transcriptional control by the calcitriol/VDR complex through binding to a consensus VDRE sequence in the human promoter, which is partially conserved in mammals *(39)*. Calcitriol induction of FGF23 levels, which in turn suppress renal 1-hydroxylase, raises questions on the net effect of calcitriol on renal P reabsorption.

Calcitriol induction of renal megalin may indirectly control the renal handling of P and the phosphaturic response to PTH. Megalin is essential for the maintenance of adequate steady-state expression of the sodium/P co-transporter NaPiIIa and the capacity of proximal tubular cells to react to PTH-driven inactivation of NaPiIIa by endocytosis and intracellular translocation *(109)*. Indeed, kidney-specific inactivation of the megalin gene results in markedly reduced phosphaturia due to enhanced steady-state levels of NaPiIIa in the brush border membrane, and a defective retrieval and impaired degradation of NaPiIIa in response to PTH *(109)*. Similarly, calcitriol induction of Klotho expression in the kidney could enhance the phosphaturic effects of FGF23, as well as FGF23 control of 1-hydroxylase.

5.2. Renal Calcitriol/VDR Actions Affecting the Cardiovascular System

Two critical nonclassical renal actions of the calcitriol/VDR complex could mediate the reported beneficial effects of vitamin D therapy in the cardiovascular system, the suppression of renin and the induction of the type A natriuretic peptide receptor (NPR-A) in cells of the inner medullary collecting duct *(110)*, key in determining urinary sodium concentration *(111)*.

The importance of calcitriol transcriptional regulation of the renin-angiotensin system will be discussed below in relation to its contribution to the renoprotective effects of active vitamin D therapy.

Calcitriol induction of NPR-A expression involves transactivation of the NPR-A gene by the calcitriol/VDR complex at a typical vitamin D responsive element located upstream from the gene transcription start site. Calcitriol also induces the expression of the NPR-A ligand, the atrial natriuretic peptide *(110)*. Activation of this system mediates vasodilation, urinary excretion of sodium and water, and suppression of myocardial hypertrophy and fibrosis *(112)*.

In addition, defective calcitriol/VDR induction of renal megalin could also contribute to cardiovascular disorders. Renal megalin mediates not only 25-hydroxyvitamin D uptake but also the reabsorption of low molecular weight proteins by renal proximal tubular cells *(113)*. Indeed, reduced megalin expression in renal proximal tubular cells and/or abnormalities in the recycling of cytosolic megalin to the brush border membrane mediate the low molecular weight proteinuria of Dent's disease *(114)* and Fanconi syndrome *(115)*. The contribution of reduced megalin to the strong epidemiological association between vitamin D deficiency and albuminuria in healthy individuals *(2)* is unknown at present. However, in spite of normal glomerular filtration rates, the low 25(OH)D levels of vitamin D-deficient individuals will result in a marked reduction in the filtered 25(OH)D available for uptake by renal proximal tubular cells, thereby decreasing calcitriol synthesis, which in turn could compromise calcitriol induction of megalin for protein reabsorption. The induction of renal megalin by calcitriol or its analogs could partially account for the efficacy of active vitamin D therapy in correcting low molecular weight proteinuria at early stages of kidney disease *(116)*, which could contribute to the survival advantage conferred by active vitamin D therapy to CKD patients as urinary protein is a recognized determinant of renal disease progression and cardiovascular complications *(117)*.

5.3. Calcitriol/VDR Renoprotective Actions

Irrespective of the nature of the initial renal injury, tubular damage and interstitial fibrosis are common final pathways leading to end-stage renal disease (ESRD). Once established, neither pathway can be reversed or ameliorated by currently available treatment. Interstitial fibrosis is a complex process triggered by a plethora of pathogenic mechanisms, including increased proteinuria, activation of the renin-angiotensin system (RAS), inflammation, and myofibroblast activation, the latter resulting from epithelial to mesenchymal transition (EMT). Therefore, the identification and effective targeting of these key inducers of glomerulosclerosis and tubulo-interstitial fibrosis are mandatory to attenuate the progression of renal lesions to ESRD.

Clinical trials and experimental data from several animal models support a renoprotective role for active vitamin D therapy in chronic kidney disease (CKD) *(118–121)*. In fact, active vitamin D was proven effective in attenuating glomerular injury, the activation of the renin-angiotensin (RAS) system, renal inflammation, and interstitial fibrosis through a direct control of specific mesangial cell, podocyte, tubular cell, fibroblast, or macrophage function.

In the rat model of remnant kidney after subtotal nephrectomy (SNX), a classic CKD model characterized by primary glomerular lesions, administration of active vitamin D effectively decreases albuminuria, glomerulosclerosis, and glomerular cell proliferation *(122, 118, 119)*. Active vitamin D therapy also reduces serum creatinine, which suggests a role for calcitriol (analog) in preserving renal function *(122)*. Importantly, the use of parathyroidectomized rats in models of kidney disease has conclusively demonstrated that the renal benefits of active vitamin D therapy are independent of calcitriol (analog) reduction of serum PTH *(119)*.

In primary cultures of human mesangial cells, which elicit a high-affinity binding of active vitamin D metabolites to the VDR, calcitriol treatment inhibits mesangial cell proliferation *(123)*. More significantly, in the rat anti-thy-1 mesangial proliferative glomerulonephritis model, administration of calcitriol or its less calcemic vitamin D analog, 22-oxa-calcitriol, not only inhibits mesangial cell proliferation but also prevents albuminuria, inflammatory cell infiltration, and the extracellular matrix (ECM) accumulation that accompanies myofibroblast activation *(124, 125)*. Calcitriol (analog) inhibition of the expression of TGFβ1, a well-known mediator of the onset and progression of various forms of renal lesions can partially account for calcitriol renoprotection *(126)*.

Podocytes, critical cells in the evolution of proteinuria in CKD, are also targets of calcitriol actions. Podocytes express the VDR and respond to calcitriol by reducing podocyte injury in immune and non-immune rat models of mesangial proliferative glomerulonephritis *(118, 126)*. Also, in the SNX rat model, calcitriol administration effectively reduces podocyte loss and hypertrophy, improves podocyte ultrastructure, and suppresses the expression of desmin, a marker of podocyte injury, with less activation of the cyclin-dependent proliferative cascades. Consequently, the anti-proteinuric effects of active vitamin D could be partly accounted for its ability to preserve podocyte integrity.

Also important for renoprotection is the role of calcitriol as a direct endocrine regulator of RAS, through VDR-mediated repression of the expression of the renin gene *(127, 128)*. This finding has offered some mechanistic insights into the role of vitamin D in renal protection and homeostasis as well as the molecular basis to explore the efficacy of vitamin D analogs in CKD as renin inhibitors, in modulating RAS, and in preventing glomerular hemodynamic adaptation or fibrogenic responses.

It is well established that glomerular hemodynamic changes, podocyte abnormalities, and mesangial activation are associated with proteinuria. Active vitamin D protection of glomerular structures and hyperfiltration contributes to attenuate proteinuria and glomerulosclerosis, the hallmarks of primary glomerular diseases. In fact, a diabetic VDR-null mouse model develops a more severe form of albuminuria and glomerulosclerosis, with reduced nephrin expression in podocytes, and enhanced fibronectin content in mesangial cells *(121)*. Thus, by reducing proteinuria, active vitamin D therapy may attenuate protein-dependent interstitial inflammation in nephropathies.

Another key pathogenic mechanism in the development and progression of CKD is chronic inflammation, a disorder characterized by infiltration of inflammatory cells into the glomeruli and tubulointerstitium. Indeed, the decline in renal function in patients with CKD correlates closely with the extent of inflammation. Inflammatory cells elicit their effects through the production of reactive oxygen species or the release of pro-inflammatory cytokines that modulate the response of renal resident cells to injurious stimuli. In addition, infiltrating macrophages produce pro-fibrotic cytokines such as transforming growth factor-β (TGFβ), which in turn induces matrix-producing myofibroblast activation, and tubular cell epithelial to mesenchymal transition (EMT), thereby enhancing the propensity for the fibrogenic process. The calcitriol/VDR complex has long been recognized for its immunomodulatory properties *(72)*. The VDR is present in most cells: the immune system, in particular, antigen-presenting cells including macrophages, dendritic cells, and both CD4+ and CD8+ T cells *(129, 130)*. A detailed discussion of calcitriol/VDR immunomodulatory properties is the topic of a different chapter.

A recent study reported that the vitamin D analog paricalcitol inhibited macrophage infiltration via suppression of macrophage chemoattractant protein 1 (MCP-1) in the SNX rat model *(131)*. Similarly, treatment with calcitriol almost completely abrogated the glomerular infiltration of neutrophils in the anti-thy-1 model *(125)*. By inhibiting inflammatory infiltration, active vitamin D essentially diminishes many of the detrimental effects of the glomerular and/or interstitial inflammation caused by injuries of various origins.

Active vitamin D is also effective in preventing renal interstitial lesions. However, unlike glomerulosclerosis, much less was known until recently about the mechanisms underlying vitamin D inhibition of tubulo-interstitial fibrosis. Evidence from cultures of interstitial fibroblasts, which express the VDR, has demonstrated vitamin D direct inhibition of interstitial fibrosis, a final common pathway for diverse types of causes of CKD. Also important, besides its systemic impact on RAS and inflammation, calcitriol was shown to suppress myofibroblast activation, a critical event in the diseased kidney, which involves the generation of α-SMA-positive, matrix-producing effector cells from interstitial fibroblasts. Indeed, in cultures of rat renal interstitial fibroblasts, calcitriol administration suppresses TGFβ1-induced α-SMA expression and the resulting increases in type I collagen and thrombospondin-1, in a dose-dependent manner, through upregulation of mRNA levels and protein secretion of hepatocyte growth factor (HGF) *(132)*. More significantly, in the obstructed kidney, paricalcitol administration reduces interstitial volume and the deposition of interstitial matrix components in a dose-dependent manner. Paricalcitol antifibrotic properties involve inhibition of renal mRNA expression of fibronectin, type I and III collagen, and fibrogenic TGFβ1 *(120)*.

In the fibrotic kidney that follows sustained injury, tubular epithelial cells undergo a phenotypic transition known as EMT to become myofibroblasts. In vivo models of obstructive nephropathy have shown that paricalcitol preserves the integrity of the tubular epithelium by restoring E-cadherin and VDR expression, and attenuating interstitial fibrosis *(120)*. Moreover, paricalcitol directly blocks the tubular EMT induced by TGFβ1 in cultured tubular epithelial cells, as demonstrated by restoring E-cadherin content, and suppressing α-SMA and fibronectin expression. Regardless of the underlying mechanism, the observations that paricalcitol blocks TGFβ1-mediated tubular EMT, together with its ability to inhibit myofibroblast activation, suggest that paricalcitol may effectively suppress interstitial fibrosis in pathological conditions.

The recent characterization of renal EGFR transactivation as the main determinant of the accelerated progression of renal lesions upon nephron reduction, prolonged renal ischemia *(85)*, and prolonged exposure to angiotensin II *(84)* also suggests a role for calcitriol inhibition of signaling from activated EGFR *(133)* in calcitriol renoprotective properties. As summarized in Fig. 5, Ang II binding to and activation of Ang II-receptor 1 (AT-R1) enhances the expression and activity of tumor necrosis factor converting enzyme, TACE [also called disintegrin and ADAM17 (a metalloproteinase domain 17)].

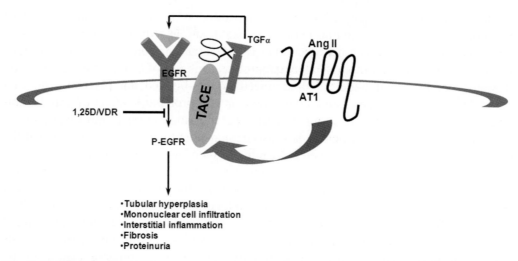

Fig. 5. TACE-dependent EGF receptor activation mediates angiotensin II-driven renal lesions in CKD. The binding of angiotensin II (Ang II) to its heptahelical AT1 receptor (AT1) activates the metallo-proteinase TACE to release TGF-α which binds to and activates the EGF receptor (EGFR) causing renal lesions. Calcitriol (1,25(OH)$_2$D)/VDR potent inhibition of EGFR signaling could contribute to vitamin D renoprotection.

Increases in TACE expression and activity, in turn, enhance the release of mature TGF-α from its transmembrane precursor, which then binds to and activates the EGFR *(84)*. More significantly, Ang II-induced renal lesions, including proteinuria, glomeruloscle-rosis, tubular cell hyperplasia, fibrosis, mononuclear cell infiltration, and interstitial inflammation, are substantially reduced in spite of high blood pressure in transgenic mice that either lack TGF-α or express an inactivated EGFR in renal tubular cells or in wild-type mice receiving specific TACE inhibitors *(84)*. Importantly, in addition to Ang II, other molecules that circulate at high levels in CKD, including PTH and endothe-lin 1, transactivate the EGFR upon binding to their cognate G protein-coupled receptors (GPCR) through a TACE-mediated mechanism *(134–136)*. Although the contribution of endothelin I and PTH transactivation of the renal EGFR to EGFR-driven renal lesions is unknown, combined therapy with active vitamin D metabolites and angiotensin II antag-onists could be more effective in preventing EGFR-driven renal lesions than exclusive suppression of the AngII-AT/R1 pathway. In fact, in the rat model of CKD, combination therapy with the ACE inhibitor enalapril and the calcitriol analog paricalcitol resulted in a better control of macrophage infiltration and the levels of MCP-1, thus synergiz-ing with renin–angiotensin blockage to ameliorate renal damage *(131)*. It is unclear at present whether paricalcitol inhibition of EGFR activation or a more effective suppres-sion of RAS by combined therapy contributed to the reported synergy.

In the past, however, several reports raised concerns on the potential nephrotoxicity of vitamin D therapy *(137, 138)*. Undoubtedly, high doses of active vitamin D, causing hypercalcemia and hypercalciuria, could certainly contribute to decrease GFR. The fear of accelerating the decline in renal function has limited for many years the use of active vitamin D or its analogs in early stages of KD. Recent studies, including prospective randomized trials in patients with mild-to-moderate CKD, have shown the benefits of moderate doses of active vitamin D on bone health and in reducing serum PTH levels with no adverse impact on renal function *(139–141)*. Furthermore, several clinical trials in CKD patients have corroborated the renoprotective effects of active vitamin D therapy demonstrated in animal models. A randomized, double-blinded, and placebo-controlled clinical trial in CKD stages 3 and 4 has shown an anti-proteinuric effect of oral paricalcitol administration *(116)*. In addition, a retrospective study in renal transplant patients with chronic allograft nephropathy has also demonstrated that calcitriol prolonged the duration of graft survival and lowered the rate of loss of renal function *(142)*.

In summary, clinical trials and data from in vivo and in vitro experiments suggest that active vitamin D and/or its analogs are renoprotective. Active vitamin D treatment results in attenuated glomerulosclerosis and interstitial fibrosis and improved kidney function. Depending on the nature and etiology of the renal injury, numerous and distinct mechanisms appear to mediate the renoprotective properties of active vitamin D or its analogs in human and experimental kidney disease, including downregulation of the RAS system and inflammation, podocyte protection, HGF induction, and preservation of tubular epithelium integrity via blockage of EMT, and possibly inhibition of renal EGFR activation. Early supplementation with either vitamin D or active vitamin D emerges as a proper strategy to halt the vicious cycle between vitamin D/calcitriol deficiency and the decline in kidney function in CKD.

ACKNOWLEDGMENTS

This work was supported in part by a grant from the National Institutes of Health (DK062713) and a grant from the Center for D-receptor Activation Research, Massachusetts General Hospital, Boston, MA, USA. The authors thank Dr. Biju Marath for his careful proofreading of the manuscript and valuable comments and suggestions and Dr. Maria Vittoria Arcidiacono for her assistance in the design of figures.

REFERENCES

1. Holick MF (2007) Vitamin D deficiency. N Engl J Med 357:266–281
2. de Boer IH, Ioannou GN, Kestenbaum B et al (2007) 25-Hydroxyvitamin D levels and albuminuria in the Third National Health and Nutrition Examination Survey (NHANES III). Am J Kidney Dis 50:69–77
3. Fraser DR, Kodicek E (1970) Unique biosynthesis by kidney of a biological active vitamin D metabolite. Nature 228:764–766
4. Dusso AS, Slatopolsky E (2005) Vitamin D. Am J Physiol Renal Physiol 289:F8-F28
5. Andress DL (2006) Vitamin D in chronic kidney disease: a systemic role for selective vitamin D receptor activation. Kidney Int 69:33–43
6. Townsend K, Evans KN, Campbell MJ et al (2005) Biological actions of extra-renal 25-hydroxyvitamin D-1alpha-hydroxylase and implications for chemoprevention and treatment. J Steroid Biochem Mol Biol 97:103–109
7. Nykjaer A, Dragun D, Walther D et al (1999) An endocytic pathway essential for renal uptake and activation of the steroid 25-(OH) vitamin D3. Cell 96:507–515
8. Liu W, Yu WR, Carling T et al (1998) Regulation of gp330/megalin expression by vitamins A and D. Eur J Clin Invest 28:100–107

9. Takemoto F, Shinki T, Yokoyama K et al (2003) Gene expression of vitamin D hydroxylase and megalin in the remnant kidney of nephrectomized rats. Kidney Int 64:414–420

10. Ghazarian JG, Jefcoate CR, Knutson JC et al (1974) Mitochondrial cytochrome p450. A component of chick kidney 25-hydrocholecalciferol-1alpha-hydroxylase. J Biol Chem 249:3026–3033

11. Labuda M, Glorieux FH (1990) Mapping autosomal recessive vitamin D dependency type I to chromosome 12q14 by linkage analysis. Am J Hum Genet 47:28–36

12. Fu GK, Lin D, Zhang MYH et al (1997) Cloning of human 25-hydroxyvitamin D-1alpha-hydroxylase and mutations causing vitamin D-dependent rickets type I. Mol Endocrinol 11: 1961–1970

13. Kitanaka S, Takeyama K, Murayama A et al (1998) Inactivating mutations in the 25-hydroxyvitamin D_3 1a-hydroxylase gene in patients with pseudovitamin D-deficiency ricket. N Engl J Med 338: 653–661

14. Yoshida T, Monkawa T, Tenenhouse HS et al (1998) Two Novel 1-Alpha-Hydroxylase Mutations in French-Canadians With Vitamin D Dependency Rickets Type I. Kidney Int 54: 1437–1443

15. Henry HL, Norman AW (1984) Vitamin D: metabolism and biological actions. Annu Rev Nutr 4: 493–520

16. Shinki T, Shimada H, Wakino S et al (1997) Cloning and expression of rat 25-hydroxyvitamin D3-1alpha-hydroxylase cDNA. Proc Natl Acad Sci USA 94:12920–12925

17. St-Arnaud R, Messerlian S, Moir JM et al. (1997) The 25-hydroxyvitamin D 1-alpha-hydroxylase gene maps to the pseudovitamin D-deficiency rickets (PDDR) disease locus. J Bone Miner Res 12:1552–1559

18. Rost CR, Kaplan RA (1981) In vitro stimulation of 25-hydroxycholecalciferol 1 alpha-hydroxylation by parathyroid hormone in chick kidney slices: evidence for a role for adenosine $3',5'$-monophosphate. Endocrinology 108:1002–1006

19. Brenza HL, Kimmeljehan C, Jehan F et al (1998) Parathyroid hormone activation of the 25-hydroxyvitamin D_3 1a-hydroxylase gene promoter. Proc Natl Acad Sci USA 95:1387–1391

20. Murayama A, Takeyama K, Kitanaka S et al (1998) The promoter of the human 25-hydroxyvitamin D-3 1-alpha-hydroxylase gene confers positive and negative responsiveness to Pth, calcitonin, and 1-alpha,25(Oh)(2)D-3. Biochem Biophys Res Commun 249:11–16

21. Omdahl JL, Gray RW, Boyle IT et al (1972) Regulation of metabolism of 25-hydroxycholecalciferol by kidney tissue in vitro by dietary calcium. Nat New Biol 237:63–64

22. Garabedian M, Holick MF, Deluca HF et al (1972) Control of 25-hydroxycholecalciferol metabolism by parathyroid glands. Proc Natl Acad Sci USA 69:1673–1676

23. Bland R, Hughes SV, Stewart PM et al (1998) Direct regulation of 25-hydroxyvitamin D_3 1a-hydroxylase by calcium in a human proximal tubule cell line. Bone 23:S260

24. Tanaka Y, Deluca HF (1973) The control of 25-hydroxyvitamin D metabolism by inorganic phosphorus. Archives Biochem Biophys 154:566–574

25. Hughes MR, Brumbaugh PF, Hussler MR et al (1975) Regulation of serum 1alpha,25-dihydroxyvitamin D3 by calcium and phosphate in the rat. Science 190:578–580

26. Bushinsky DA, Favus MJ (1989) Elevated Ca^{2+} does not inhibit the 1,25(OH)2D3 response to phosphorus restriction. Am J Physiol 256:F285-F289

27. Gray RW (1981) Control of plasma 1,25-(OH)2-vitamin D concentrations by calcium and phosphorus in the rat: effects of hypophysectomy. Calcif Tissue Int 33:485–488

28. Schiavi SC, Kumar R (2004) The phosphatonin pathway: new insights in phosphate homeostasis. Kidney Int 65:1–14

29. Shimada T, Urakawa I, Yamazaki Y et al (2004) FGF-23 transgenic mice demonstrate hypophosphatemic rickets with reduced expression of sodium phosphate cotransporter type IIa. Biochem Biophys Res Commun 314:409–414

30. Berndt T, Craig TA, Bowe AE et al (2003) Secreted frizzled-related protein 4 is potent tumor-derived phosphaturic agent. J Clin Invest 112:785–794

31. Rowe PSN, Kumagai Y, Garrett R et al (2002) CHO-cells expressing MEPE, PHEX and co-expressing MEPE/PHEX cause major changes in BMD, Pi and serum alkaline phosphatase in nude mice. J Bone Miner Res 17:S211

32. Perwad F, Azam N, Zhang MY et al (2005) Dietary and serum phosphorus regulate fibroblast growth factor 23 expression and 1,25-dihydroxyvitamin D metabolism in mice. Endocrinology 146: 5358–5364

33. Tsujikawa H, Kurotaki Y, Fujimori T et al (2003) Klotho, a gene related to a syndrome resembling human premature aging, functions in a negative regulatory circuit of vitamin D endocrine system. Mol Endocrinol 17:2393–2403

34. Yoshida T, Nabeshima Y (2002) Mediation of unusually high concentrations of 1,25-dihydroxyvitamin D in homozygous klotho mutant mice by increased expression of renal 1alpha-hydroxylase gene. Endocrinology 143:683–689

35. Henry HL (1979) Regulation of the hydroxylation of 25-hydroxyvitamin D3 in vivo and in primary cultures of chick kidney cells. J Biol Chem 254:2722–2729

36. Trechsel U, Fleisch H (1979) Regulation of the metabolism of 25-hydroxyvitamin D3 in primary cultures of chick kidney cells. J Clin Invest 64:206–217

37. Monkawa T, Yoshida T, Wakino S et al (1997) Molecular cloning of cDNA and genomic DNA for human 25-hydroxyvitamin D3 1 alpha-hydroxylase. Biochem Biophys Res Commun 239:527–533

38. Takeyama K, Kitanaka S, Sato T et al (1997) 25-Hydroxyvitamin D3 1alpha-hydroxylase and vitamin D synthesis. Science 277:1827–1830

39. Barthel TK, Mathern DR, Whitfield GK et al (2007) 1,25-Dihydroxyvitamin D3/VDR-mediated induction of FGF23 as well as transcriptional control of other bone anabolic and catabolic genes that orchestrate the regulation of phosphate and calcium mineral metabolism. J Steroid Biochem Mol Biol 103:381–388

40. Kolek OI, Hines ER, Jones MD et al (2005) 1alpha,25-Dihydroxyvitamin D3 upregulates FGF23 gene expression in bone: the final link in a renal-gastrointestinal-skeletal axis that controls phosphate transport. Am J Physiol Gastrointest Liver Physiol 289:G1036–G1042

41. Maierhofer WJ, Gray RW, Adams ND et al (1981) Synthesis and metabolic clearance of 1,25-dihydroxyvitamin D as determinants of serum concentrations: a comparison of two methods. J Clin Endocrinol Metab 53:472–475

42. St-Arnaud R, Arabian A, Glorieux FH (1996) Abnormal bone development in mice deficient for the vitamin D-24-hydroxylase gene. J Bone Miner Res 11:S126

43. Chen KS, DeLuca HF (1995) Cloning of the human 1 alpha,25-dihydroxyvitamin D-3 24-hydroxylase gene promoter and identification of two vitamin D-responsive elements. Biochimica et Biophysica Acta 1263:1–9

44. Ohyama Y, Ozono K, Uchida M et al (1994) Identification of a vitamin D-responsive element in the 5′-flanking region of the rat 25-hydroxyvitamin D3 24-hydroxylase gene. J Biol Chem 269: 10545–10550

45. Wu S, Finch J, Zhong M et al (1996) Expression of the renal 25-hydroxyvitamin D-24-hydroxylase gene: regulation by dietary phosphate. Am J Physiol 271:F203–F208

46. Christiansen C, Christensen MS, Melsen F et al (1981) Mineral metabolism in chronic renal failure with special reference to serum concentrations of 1.25(OH)2D and 24.25(OH)2D. Clin Nephrol 15:18–22

47. Juttmann JR, Buurman CJ, De Kam E et al (1981) Serum concentrations of metabolites of vitamin D in patients with chronic renal failure (CRF). Consequences for the treatment with 1-alpha-hydroxy-derivatives. Clin Endocrinol (Oxf) 14:225–236

48. Mason RS, Lissner D, Wilkinson M et al (1980) Vitamin D metabolites and their relationship to azotaemic osteodystrophy. Clin Endocrinol (Oxf) 13:375–385

49. Tessitore N, Venturi A, Adami S et al (1987) Relationship between serum vitamin D metabolites and dietary intake of phosphate in patients with early renal failure. Miner Electrolyte Metab 13: 38–44

50. Bellorin-Font E, Humpierres J, Weisinger JR et al (1985) Effect of metabolic acidosis on the PTH receptor-adenylate cyclase system of canine kidney. Am J Physiol 249:F566–F572

51. Usatii M, Rousseau L, Demers C et al (2007) Parathyroid hormone fragments inhibit active hormone and hypocalcemia-induced 1,25(OH)2D synthesis. Kidney Int 72:1330–1335

52. Wilson L, Felsenfeld A, Drezner MK et al (1985) Altered divalent ion metabolism in early renal failure: role of 1,25(OH)2D. Kidney Int 27:565–573

53. Hsu CH, Buchsbaum BL (1991) Calcitriol metabolism in patients with chronic renal failure. Am J Kidney Dis 17:185–190

54. Hsu CH, Patel S, Young EW et al (1987) Production and degradation of calcitriol in renal failure rats. Am J Physiol 253:F1015–F1019

55. Dusso A, Lopez-Hilker S, Lewis-Finch J et al (1989) Metabolic clearance rate and production rate of calcitriol in uremia. Kidney Int 35:860–864

56. Birn H, Vorum H, Verroust PJ et al (2000) Receptor-associated protein is important for normal processing of megalin in kidney proximal tubules. J Am Soc Nephrol 11:191–202

57. Nykjaer A, Fyfe JC, Kozyraki R et al (2001) Cubilin dysfunction causes abnormal metabolism of the steroid hormone 25(OH) vitamin D(3). Proc Natl Acad Sci USA 98:13895–13900

58. Yamane T, Takeuchi K, Yamamoto Y et al (2002) Legumain from bovine kidney: its purification, molecular cloning, immunohistochemical localization and degradation of annexin II and vitamin D-binding protein. Biochimica et Biophysica Acta 1:108–120

59. Adams JS, Chen H, Chun RF et al (2003) Novel regulators of vitamin D action and metabolism: Lessons learned at the Los Angeles zoo. J Cell Biochem 88:308–314

60. Halloran BP, Schaefer P, Lifschitz M et al (1984) Plasma vitamin D metabolite concentrations in chronic renal failure: effect of oral administration of 25-hydroxyvitamin D3. J Clin Endocrinol Metab 59:1063–1069

61. Hughes MR, Baylink DJ, Jones PG et al (1976) Radioligand receptor assay for 25-hydroxyvitamin D2/D3 and 1 alpha, 25- dihydroxyvitamin D2/D3. J Clin Invest 58:61–70

62. Al-Badr W, Martin KJ (2008) Vitamin D and kidney disease. Clin J Am Soc Nephrol 3(5): 1555–1560

63. Armas LA, Heaney RP (2004) Vitamin D2 is much less effective than vitamin D3 in humans. J Clin Endocrinol Metab 89:5387–5391

64. Holick MF, Biancuzzo RM, Chen TC et al (2008) Vitamin D2 is as effective as vitamin D3 in maintaining circulating concentrations of 25-hydroxyvitamin D. J Clin Endocrinol Metab 93:677–681

65. Gallieni M, Kamimura S, Ahmed A et al (1995) Kinetics of monocyte 1 alpha-hydroxylase in renal failure. Am J Physiol 268:F746–F753

66. Dusso A, Lopez-Hilker S, Rapp N et al (1988) Extra-renal production of calcitriol in chronic renal failure. Kidney Int 34:368–375

67. Ritter CS, Armbrecht HJ, Slatopolsky E et al (2006) 25-Hydroxyvitamin D(3) suppresses PTH synthesis and secretion by bovine parathyroid cells. Kidney Int 70:654–659

68. Segersten U, Correa P, Hewison M et al (2002) 25-hydroxyvitamin D(3)-1alpha-hydroxylase expression in normal and pathological parathyroid glands. J Clin Endocrinol Metab 87:2967–2972

69. Al-Aly Z, Qazi RA, Gonzalez EA et al (2007) Changes in serum 25-hydroxyvitamin D and plasma intact PTH levels following treatment with ergocalciferol in patients with CKD. Am J Kidney Dis 50:59–68

70. Chandra P, Binongo JN, Ziegler TR et al (2008) Cholecalciferol (vitamin D3) therapy and vitamin D insufficiency in patients with chronic kidney disease: a randomized controlled pilot study. Endocr Pract 14:10–17

71. Zisman AL, Hristova M, Ho LT et al (2007) Impact of ergocalciferol treatment of vitamin D deficiency on serum parathyroid hormone concentrations in chronic kidney disease. Am J Nephrol 27: 36–43

72. Mathieu C, Adorini L (2002) The coming of age of 1,25-dihydroxyvitamin D(3) analogs as immunomodulatory agents. Trends Mol Med 8:174–179

73. Costa EM, Feldman D (1986) Homologous up-regulation of the 1,25 (OH)2 vitamin D3 receptor in rats. Biochem Biophys Res Commun 137:742–747

74. Wiese RJ, Uhland-Smith A, Ross TK et al (1992) Up-regulation of the vitamin D receptor in response to 1,25-dihydroxyvitamin D3 results from ligand-induced stabilization. J Biol Chem 267: 20082–20086

75. Dhawan P, Peng X, Sutton AL et al (2005) Functional cooperation between CCAAT/enhancer-binding proteins and the vitamin D receptor in regulation of 25-hydroxyvitamin D3 24-hydroxylase. Mol Cell Biol 25:472–487

76. Gutierrez S, Javed A, Tennant DK et al (2002) CCAAT/enhancer-binding proteins (C/EBP) beta and delta activate osteocalcin gene transcription and synergize with Runx2 at the C/EBP element to regulate bone-specific expression. J Biol Chem 277:1316–1323

77. Ji Y, Studzinski GP (2004) Retinoblastoma protein and CCAAT/enhancer-binding protein beta are required for 1,25-dihydroxyvitamin D3-induced monocytic differentiation of HL60 cells. Cancer Res 64:370–377

78. Fukuda N, Tanaka H, Tominaga Y et al (1993) Decreased 1,25-dihydroxyvitamin D3 receptor density is associated with a more severe form of parathyroid hyperplasia in chronic uremic patients. J Clin Invest 92:1436–1443

79. Korkor AB (1987) Reduced binding of [3H]1,25-dihydroxyvitamin D3 in the parathyroid glands of patients with renal failure. N Engl J Med 316:1573–1577

80. Merke J, Hugel U, Zlotkowski A et al (1987) Diminished parathyroid 1,25(OH)2D3 receptors in experimental uremia. Kidney Int 32:350–353

81. Brown AJ, Dusso A, Lopez-Hilker S et al (1989) 1,25-(OH)2D receptors are decreased in parathyroid glands from chronically uremic dogs. Kidney Int 35:19–23

82. Denda M, Finch J, Brown AJ et al (1996) 1,25-dihydroxyvitamin D3 and 22-oxacalcitriol prevent the decrease in vitamin D receptor content in the parathyroid glands of uremic rats. Kidney Int 50: 34–39

83. Arcidiacono MV, Sato T, Alvarez-Hernandez D et al (2008) EGFR activation increases parathyroid hyperplasia and calcitriol resistance in kidney disease. J Am Soc Nephrol 19:310–320

84. Lautrette A, Li S, Alili R et al (2005) Angiotensin II and EGF receptor cross-talk in chronic kidney diseases: a new therapeutic approach. Nat Med 11:867–874

85. Terzi F, Burtin M, Hekmati M et al (2000) Targeted expression of a dominant-negative EGF-R in the kidney reduces tubulo-interstitial lesions after renal injury. J Clin Invest 106:225–234

86. Sawaya BP, Koszewski NJ, Qi Q et al (1997) Secondary hyperparathyroidism and vitamin D receptor binding to vitamin D response elements in rats with incipient renal failure. J Am Soc Nephrol 8: 271–278

87. Patel SR, Ke HQ, Vanholder R et al (1995) Inhibition of calcitriol receptor binding to vitamin D response elements by uremic toxins. J Clin Invest 96:50–59

88. Sela-Brown A, Russell J, Koszewski NJ et al (1998) Calreticulin inhibits vitamin D's action on the PTH gene in vitro and may prevent vitamin D's effect in vivo in hypocalcemic rats. Mol Endocrinol 12:1193–1200

89. May P, Bock HH, Herz J (2003) Integration of endocytosis and signal transduction by lipoprotein receptors. Sci STKE 2003:PE12

90. Dunlop TW, Vaisanen S, Frank C et al (2004) The genes of the coactivator TIF2 and the corepressor SMRT are primary 1alpha,25(OH)2D3 targets. J Steroid Biochem Mol Biol 89–90:257–260

91. Takeyama K, Masuhiro Y, Fuse H et al (1999) Selective interaction of vitamin D receptor with transcriptional coactivators by a vitamin D analog. Mol Cell Biol 19:1049–1055

92. Koshiyama H, Nakao K (1995) Vitamin-D-receptor-gene polymorphism and bone loss. Lancet 345:990–991

93. Morrison NA, Qi JC, Tokita A et al (1994) Prediction of bone density from vitamin D receptor alleles. Nature 367:284–287

94. Eisman JA (1999) Genetics of osteoporosis. Endocr Rev 20:788–804

95. Carling T, Kindmark A, Hellman P et al (1995) Vitamin D receptor genotypes in primary hyperparathyroidism. Nat Med 1:1309–1311

96. Gomez Alonso C, Naves Diaz ML, Diaz-Corte C et al (1998) Vitamin D receptor gene (VDR) polymorphisms: effect on bone mass, bone loss and parathyroid hormone regulation. Nephrol Dial Transplant 13:73–77

97. Kontula K, Valimaki S, Kainulainen K et al (1997) Vitamin D receptor polymorphism and treatment of psoriasis with calcipotriol. Br J Dermatol 136:977–978

98. Fernandez E, Fibla J, Betriu A et al (1997) Association between vitamin D receptor gene polymorphism and relative hypoparathyroidism in patients with chronic renal failure. J Am Soc Nephrol 8:1546–1552

99. Akiba T, Ando R, Kurihara S et al (1997) Is the bone mass of hemodialysis patients genetically determined?. Kidney Int Suppl 62:S69–S71

100. Borras M, Torregrossa V, Oliveras A et al (2003) BB genotype of the vitamin D receptor gene polymorphism postpones parathyroidectomy in hemodialysis patients. J Nephrol 16:116–120

101. Nagaba Y, Heishi M, Tazawa H et al (1998) Vitamin D receptor gene polymorphisms affect secondary hyperparathyroidism in hemodialyzed patients. Am J Kidney Dis 32:464–469

102. Whitfield GK, Remus LS, Jurutka PW et al (2001) Functionally relevant polymorphisms in the human nuclear vitamin D receptor gene. Mol Cell Endocrinol 177:145–159

103. Hoenderop JG, Muller D, Van Der Kemp AW et al (2001) Calcitriol controls the epithelial calcium channel in kidney. J Am Soc Nephrol 12:1342–1349

104. van de Graaf SF, van der Kemp AW, van den Berg D et al (2006) Identification of BSPRY as a novel auxiliary protein inhibiting TRPV5 activity. J Am Soc Nephrol 17:26–30

105. Hoenderop JG, Chon H, Gkika D et al (2004) Regulation of gene expression by dietary Ca2+ in kidneys of 25-hydroxyvitamin D3-1 alpha-hydroxylase knockout mice. Kidney Int 65: 531–539

106. Friedman PA, Gesek FA (1993) Vitamin D3 accelerates PTH-dependent calcium transport in distal convoluted tubule cells. Am J Physiol 265:F300–F308

107. Bergwitz C, Roslin NM, Tieder M et al (2006) SLC34A3 mutations in patients with hereditary hypophosphatemic rickets with hypercalciuria predict a key role for the sodium-phosphate cotransporter NaPi-IIc in maintaining phosphate homeostasis. Am J Hum Genet 78:179–192

108. Lorenz-Depiereux B, Benet-Pages A, Eckstein G et al (2006) Hereditary hypophosphatemic rickets with hypercalciuria is caused by mutations in the sodium-phosphate cotransporter gene SLC34A3. Am J Hum Genet 78:193–201

109. Bachmann S, Schlichting U, Geist B et al (2004) Kidney-specific inactivation of the megalin gene impairs trafficking of renal inorganic sodium phosphate cotransporter (NaPi-IIa). J Am Soc Nephrol 15:892–900

110. Chen S, Olsen K, Grigsby C et al (2007) Vitamin D activates type A natriuretic peptide receptor gene transcription in inner medullary collecting duct cells. Kidney Int 72:300–306

111. Zeidel ML (1993) Hormonal regulation of inner medullary collecting duct sodium transport. Am J Physiol 265:F159–F173

112. Knowles JW, Esposito G, Mao L et al (2001) Pressure-independent enhancement of cardiac hypertrophy in natriuretic peptide receptor A-deficient mice. J Clin Invest 107:975–984

113. Gekle M (2005) Renal tubule albumin transport. Annu Rev Physiol 67:573–594

114. Guggino SE (2007) Mechanisms of disease: what can mouse models tell us about the molecular processes underlying Dent disease?. Nat Clin Pract Nephrol 3:449–455

115. Norden AG, Lapsley M, Igarashi T et al (2002) Urinary megalin deficiency implicates abnormal tubular endocytic function in Fanconi syndrome. J Am Soc Nephrol 13:125–133

116. Agarwal R, Acharya M, Tian J et al (2005) Antiproteinuric effect of oral paricalcitol in chronic kidney disease. Kidney Int 68:2823–2828

117. Jerums G, Panagiotopoulos S, Tsalamandris C et al (1997) Why is proteinuria such an important risk factor for progression in clinical trials?. Kidney Int Suppl 63:S87–S92

118. Kuhlmann A, Haas CS, Gross ML et al (2004) 1,25-Dihydroxyvitamin D3 decreases podocyte loss and podocyte hypertrophy in the subtotally nephrectomized rat. Am J Physiol Renal Physiol 286:F526–F533

119. Schwarz U, Amann K, Orth SR et al (1998) Effect of 1,25 (OH)2 vitamin D3 on glomerulosclerosis in subtotally nephrectomized rats. Kidney Int 53:1696–1705

120. Tan X, Liu Y (2006) Paricalcitol attenuates renal interstitial fibrosis in obstructive nephropathy. J Am Soc Nephrol 17:3382–3393

121. Zhang Z, Sun L, Wang Y et al (2008) Renoprotective role of the vitamin D receptor in diabetic nephropathy. Kidney Int 73:163–171

122. Hirata M, Makibayashi K, Katsumata K et al (2002) 22-Oxacalcitriol prevents progressive glomerulosclerosis without adversely affecting calcium and phosphorus metabolism in subtotally nephrectomized rats. Nephrol Dial Transplant 17:2132–2137

123. Weinreich T, Merke J, Schonermark M et al (1991) Actions of 1,25-dihydroxyvitamin D3 on human mesangial cells. Am J Kidney Dis 18:359–366
124. Makibayashi K, Tatematsu M, Hirata M et al (2001) A vitamin D analog ameliorates glomerular injury on rat glomerulonephritis. Am J Pathol 158:1733–1741
125. Panichi V, Migliori M, Taccola D et al (2001) Effects of 1,25(OH)2D3 in experimental mesangial proliferative nephritis in rats. Kidney Int 60:87–95
126. Migliori M, Giovannini L, Panichi V et al (2005) Treatment with 1,25-dihydroxyvitamin D3 preserves glomerular slit diaphragm-associated protein expression in experimental glomerulonephritis. Int J Immunopathol Pharmacol 18:779–790
127. Li YC, Kong J, Wei M et al (2002) 1,25-Dihydroxyvitamin D(3) is a negative endocrine regulator of the renin-angiotensin system. J Clin Invest 110:229–238
128. Li YC, Qiao G, Uskokovic M et al (2004) Vitamin D: a negative endocrine regulator of the renin-angiotensin system and blood pressure. J Steroid Biochem Mol Biol 89–90:387–392
129. Provvedini DM, Tsoukas CD, Deftos LJ et al (1983) 1,25-dihydroxyvitamin D3 receptors in human leukocytes. Science 221:1181–1183
130. Veldman CM, Cantorna MT, DeLuca HF (2000) Expression of 1,25-dihydroxyvitamin D(3) receptor in the immune system. Arch Biochem Biophys 374:334–338
131. Mizobuchi M, Morrissey J, Finch JL et al (2007) Combination therapy with an Angiotensin-converting enzyme inhibitor and a vitamin d analog suppresses the progression of renal insufficiency in uremic rats. J Am Soc Nephrol 18:1796–1806
132. Li Y, Spataro BC, Yang J et al (2005) 1,25-dihydroxyvitamin D inhibits renal interstitial myofibroblast activation by inducing hepatocyte growth factor expression. Kidney Int 68:1500–1510
133. Cordero JB, Cozzolino M, Lu Y et al (2002) 1,25-Dihydroxyvitamin D down-regulates cell membrane growth- and nuclear growth-promoting signals by the epidermal growth factor receptor. J Biol Chem 277:38965–38971
134. Ahmed I, Gesty-Palmer D, Drezner MK et al (2003) Transactivation of the epidermal growth factor receptor mediates parathyroid hormone and prostaglandin F2 alpha-stimulated mitogen-activated protein kinase activation in cultured transgenic murine osteoblasts. Mol Endocrinol 17:1607–1621
135. Iwasaki H, Eguchi S, Marumo F et al (1998) Endothelin-1 stimulates DNA synthesis of vascular smooth-muscle cells through transactivation of epidermal growth factor receptor. J Cardiovasc Pharmacol 31(Suppl 1):S182–S184
136. Qin L, Tamasi J, Raggatt L et al (2005) Amphiregulin is a novel growth factor involved in normal bone development and in the cellular response to parathyroid hormone stimulation. J Biol Chem 280:3974–3981
137. Christiansen C, Rodbro P, Christensen MS et al (1978) Deterioration of renal function during treatment of chronic renal failure with 1,25-dihydroxycholecalciferol. Lancet 2:700–703
138. Tougaard L, Sorensen E, Brochner-Mortensen J et al (1976) Controlled trial of 1apha-hydroxycholecalciferol in chronic renal failure. Lancet 1:1044–1047
139. Coburn JW, Maung HM, Elangovan L et al (2004) Doxercalciferol safely suppresses PTH levels in patients with secondary hyperparathyroidism associated with chronic kidney disease stages 3 and 4. Am J Kidney Dis 43:877–890
140. Ritz E, Kuster S, Schmidt-Gayk H et al (1995) Low-dose calcitriol prevents the rise in 1,84-iPTH without affecting serum calcium and phosphate in patients with moderate renal failure (prospective placebo-controlled multicentre trial). Nephrol Dial Transplant 10:2228–2234
141. Rix M, Olgaard K (2004) Effect of 18 months of treatment with alfacalcidol on bone in patients with mild to moderate chronic renal failure. Nephrol Dial Transplant 19:870–876
142. Sezer S, Uyar M, Arat Z et al (2005) Potential effects of 1,25-dihydroxyvitamin D3 in renal transplant recipients. Transplant Proc 37:3109–3111

10 Vitamin D and the Parathyroids

Justin Silver and Tally Naveh-Many

Abstract   The administration of 1,25-dihydroxyvitamin D_3 [1,25$(OH)_2$D] results in a dramatic decrease in PTH mRNA levels in rats with no change in serum calcium levels. The administration of 1,25$(OH)_2$D or its analogs has become an important part of the management of secondary hyperparathyroidism due to chronic renal failure. 1,25$(OH)_2$D acts to decrease PTH gene transcription. Two mechanisms may play a role in the transrepression. First, a VDRE has been described in the PTH promoter, which mediates PTH gene transrepression in transfected cells. In addition, an alternative E-box-like motif has been identified as another class of nVDRE in the human 1α-hydroxylase promoter. Negative regulation of the hPTH gene by liganded 1,25$(OH)_2$D receptor (VDR) is mediated by VDR-interacting repressor, the VDIR, directly binding to the E-box-type nVDRE at the promoter. Post-transcriptional mechanisms of PTH mRNA regulation in experimental kidney failure and after changes in dietary calcium and phosphorus are pertinent to the understanding of PTH expression. These involve the binding of *trans*-acting factors to a defined *cis*-acting instability element in the PTH mRNA 3′-untranslated region. The bone-derived hormone, fibroblast growth factor (FGF) 23 is increased by phosphate and 1,25$(OH)_2$D and acts on the kidney to cause a phosphaturia and decrease 1,25$(OH)_2$D synthesis. FGF23 acts on the parathyroid to decrease PTH expression.

   Key Words: Parathyroid gland; parathyroid hormone; VDR; 1,25-dihydroxyvitamin D; VDRE; FGF23; PTH gene; calcium; phosphate; vitamin D analogs

1. INTRODUCTION

The action of 1,25-dihydroxyvitamin D_3 [1,25$(OH)_2$D] or its analogs to decrease PTH secretion is now a well-established axiom in clinical medicine for the suppression of the secondary hyperparathyroidism of patients with chronic kidney disease. So much so, that it is worthwhile to reflect upon its scientific basis. That is the purpose of the present review. There is ongoing academic and commercial activity in the development of drugs that may have more selective actions on the parathyroid while leading to less hypercalcemia. These attempts are of great clinical and pharmaceutical interest but still remain to be proven by rigorous scientific testing and therefore despite the extensive literature on the subject, the final word on the analogs awaits prospective outcome studies *(1, 2)* and are not discussed in detail in this chapter.

From: *Nutrition and Health: Vitamin D*
Edited by: M.F. Holick, DOI 10.1007/978-1-60327-303-9_10,
© Springer Science+Business Media, LLC 2010

2. THE PARATHYROID HORMONE GENE

2.1. The PTH Gene

The human parathyroid hormone (PTH) gene is localized on the short arm of chromosome 11 at 11p15 *(3, 4)*. The human and bovine genes have two functional TATA transcription start sites, and the rat only one. The two homologous TATA sequences flanking the human PTH gene direct the synthesis of two human PTH gene transcripts both in normal parathyroid glands and in parathyroid adenomas *(5)*. The PTH genes in all species that have been cloned have two introns or intervening sequences and three exons *(6)*. Strikingly, even though fish do not have discrete parathyroid glands, they do synthesize PTH using two distinct genes that share the same exon–intron pattern found in tetrapod PTH genes *(7, 8)*. The locations of the introns are identical in each case *(9)*. Intron A splits the 5′ untranslated sequence of the mRNA five nucleotides before the initiator methionine codon. Intron B splits the fourth codon of the region that codes for the prosequence of preProPTH. The three exons that result, thus, are roughly divided into three functional domains. Exon 1 contains the 5′ untranslated region. Exon 2 codes for the presequence or signal peptide and exon 3 codes for PTH as well as the 3′ untranslated region. It is interesting that the human gene is considerably longer in both intron A and the 3′ untranslated region of the cDNA compared to the bovine, rat, and mouse. The genes for PTH and PTHrP (PTH-related protein) are located in similar positions on sibling chromosomes 11 and 12. It is therefore likely that they arose from a common precursor by chromosomal duplication.

2.2. The PTH mRNA

Complementary DNA encoding for human *(10, 11)*, bovine *(12, 13)*, rat *(14)*, mouse *(15)*, pig *(14)*, chicken *(16, 17)*, dog *(18)*, cat *(19)*, horse *(20)*, macaca *(21)*, fugu fish *(7)*, and zebrafish *(8)* PTH have all been cloned *(9)*. The PTH gene is a typical eukaryotic gene with consensus sequences for initiation of RNA synthesis, RNA splicing, and polyadenylation. The primary RNA transcript consists of RNA transcribed from both introns and exons, and then RNA sequences derived from the introns are spliced out. The product of this RNA processing, which represents the exons, is the mature PTH mRNA, which will then be translated into preproPTH. There is considerable identity among mammalian PTH genes, which is reflected in an 85% identity between human and bovine proteins and 75% identity between human and rat proteins. There is less identity in the 3′-noncoding region. A more extensive review of the structure and sequences of the PTH gene has been published elsewhere *(9)* in the book *Molecular Biology of the Parathyroid (22)*.

3. DEVELOPMENT OF THE PARATHYROID AND TISSUE-SPECIFIC EXPRESSION OF THE PTH GENE

The thymus, thyroid, and parathyroid glands in vertebrates develop from the pharyngeal region, with contributions both from pharyngeal endoderm and from neural crest cells in the pharyngeal arches. Studies of gene knockout mice have shown that the hoxa3, pax 1, pax 9, and Eya1 transcription factors are needed to form parathyroid glands as well as many other pharyngeal pouch derivatives, such as the thymus. Glial cells missing2 (Gcm2), a mouse homologue of *Drosophila* Gcm, is a transcription factor whose expression is restricted to the parathyroid glands *(23)*. A human patient with a defective Gcm B gene, the human equivalent of Gcm-2, exhibited hypoparathyroidism and complete absence of PTH from the bloodstream *(24)*. The parathyroid gland of tetrapods and the gills of fish both express Gcm-2 and require this gene for their formation *(25)*. They also showed that the gill region expresses mRNA encoding the two PTH genes found in fish, as well as mRNA encoding the calcium-sensing receptor.

4. PROMOTER SEQUENCES

Regions upstream of the transcribed structural gene often determine tissue specificity and contain many of the regulatory sequences for the gene. For PTH, analysis of this region has been hampered by the lack of a parathyroid cell line. It has been shown that the 5 kb of DNA upstream of the start site of the human PTH gene was able to direct parathyroid gland-specific expression in transgenic mice *(26)*. Analysis of the human PTH promoter region identified a number of consensus sequences by computer analysis *(27)*. These included a sequence resembling the canonical cAMP-responsive element 5′-TGACGTCA-3′ at position −81 with a single residue deviation. This element was fused to a reporter gene (CAT) and then transfected into different cell lines. Pharmacological agents that increase cAMP led to an increased expression of the CAT gene, suggesting a functional role for the cAMP-responsive element (CRE). Specificity protein (Sp) and the nuclear factor-Y (NF-Y) complex are thought to be ubiquitously expressed transcription factors associated with basal expression of a host of gene products. Sp family members and NF-Y can cooperatively enhance transcription of a target gene. There is a highly conserved Sp1 DNA element present in mammalian PTH promoters *(28)*. Coexpression of Sp proteins and NF-Y complex leads to synergistic transactivation of the hPTH promoter, with alignment of the Sp1 DNA element essential for full activation *(28)*. The presence of a proximal NF-Y-binding site in the hPTH promoter highlights the potential for synergism between distal and proximal NF-Y DNA elements to strongly enhance transcription *(29)*.

Several groups have identified DNA sequences that might mediate the negative regulation of PTH gene transcription by 1,25-dihydroxyvitamin D [1,25(OH)$_2$D]. Demay et al. *(30)* identified DNA sequences in the human PTH gene that bind the 1,25(OH)$_2$D receptor. Nuclear extracts containing the 1,25(OH)$_2$D receptor were examined for binding to sequences in the 5′-flanking region of the hPTH gene. A 25-bp oligonucleotide containing sequences from −125 to −101 from the start of exon 1 bound nuclear proteins that were recognized by monoclonal antibodies against the 1,25(OH)$_2$D receptor. The sequences in this region contained a single copy of a motif (AGGTTCA) that is homologous to the motifs repeated in the upregulatory 1,25(OH)$_2$D response element of the osteocalcin gene. When placed upstream to a heterologous viral promoter, the sequences contained in this 25-bp oligonucleotide mediated transcriptional repression in response to 1,25(OH)$_2$D in GH4C1 cells but not in ROS 17/2.8 cells. Therefore, this downregulatory element differs from upregulatory elements both in sequence composition and in the requirement for particular cellular factors other than the 1,25(OH)$_2$D receptor (VDR) for repressing PTH transcription *(30)*. Russell et al. *(31)* have shown that there are two negative VDREs in the rat PTH gene. One is situated at –793 to –779 and bound a VDR/RXR heterodimer with high affinity and the other at –60 to –746 bound the heterodimer with a lower affinity. Transfection studies with VDRE-CAT constructs showed that they had an additive effect. Liu et al. *(32)* have identified such sequences in the chicken PTH gene and demonstrated their functionality after transfection into the opossum kidney (OK) cell line. They converted the negative activity imparted by the PTH VDRE to a positive transcriptional response through selective mutations introduced into the element. They showed that there was a p160 protein that specifically interacted with a heterodimer complex bound to the wild-type VDRE, but was absent from complexes bound to response elements associated with positive transcriptional activity. Thus, the sequence of the individual VDRE appears to play an active role in dictating transcriptional responses that may be mediated by altering the ability of a VDR/RXR heterodimer to interact with accessory factor proteins. Further work is needed to demonstrate that any of these differing negative VDREs function in this fashion in parathyroid cells.

The transrepression by $1,25(OH)_2D$ has also been shown to be dependent upon another promoter element. Kato's laboratory have identified an E-box (CANNTG)-like motif as another class of nVDRE in the human 1α-hydroxylase promoter *(33, 34)*. In sharp contrast to the previously reported DR3-like motif in the hPTH gene promoter, a basic helix-loop-helix factor, designated VDR interacting repressor (VDIR), transactivates through direct binding to this E-box-type element (1nVDRE). However, the VDIR transactivation function is transrepressed through ligand-induced protein–protein interaction of VDIR with VDR/RXR. In the absence of $1,25(OH)_2D$, VDIR appears to bind to 1nVDRE for transactivation through the histone acetylase (HAT) coactivator, p300/CBP. Binding of $1,25(OH)_2D$ to VDR induces interaction with VDIR and dissociation of the HAT coactivator, resulting in recruitment of histone deacetylase (HDAC) corepressor for ligand-induced transrepression *(34)*. They have also characterized the functions of VDIR and E-box motifs in the human (h) PTH and hPTHrP gene promoters *(35)*. They identified E-box-type elements acting as nVDREs in both the hPTH promoter (hPTHnVDRE –87 to –60 bp) and in the hPTHrP promoter (hPTHrPnVDRE –850 to –600 bp, –463 to –104 bp) in a mouse renal tubule cell line. The hPTHnVDRE alone was enough to direct ligand-induced transrepression mediated through VDR/retinoid X receptor and VDIR. Direct DNA binding of hPTHnVDRE to VDIR, but not VDR/retinoid X receptor, was observed and ligand-induced transrepression was coupled with recruitment of VDR and histone deacetylase 2 (HDAC2) to the hPTH promoter. They concluded that negative regulation of the hPTH gene by liganded VDR is mediated by VDIR directly binding to the E-box-type nVDRE at the promoter, together with recruitment of an HDAC corepressor for ligand-induced transrepression *(35)*. These studies were specific to a mouse proximal tubule cell line and await the development of a parathyroid cell line to confirm them in a homologous cell system.

5. REGULATION OF PTH GENE EXPRESSION

5.1. 1,25-Dihydroxyvitamin D

PTH regulates serum concentrations of calcium and phosphate, which, in turn, regulate the synthesis and secretion of PTH. $1,25(OH)_2D$ has independent effects on calcium and phosphate levels and also participates in a well-defined feedback loop between 1,25(OH)$_2$D and PTH *(36)*.

1,25(OH)$_2$D potently decreases transcription of the PTH gene (Fig. 2). This action was first demonstrated in vitro in bovine parathyroid cells in primary culture, where 1,25(OH)$_2$D led to a marked decrease in PTH mRNA levels *(37, 38)* and a consequent decrease in PTH secretion *(39–42)*. The physiological relevance of these findings was established by in vivo studies in rats *(43)*. The localization of 1,25(OH)$_2$D receptor mRNA (VDR mRNA) to parathyroids was demonstrated by in situ hybridization studies of the thyroparathyroid and duodenum (Fig. 1). VDR mRNA was localized to the parathyroids in the same concentration as in the duodenum, the classic target organ of 1,25(OH)$_2$D *(44)*. Rats injected with amounts of 1,25(OH)$_2$D that did not increase serum calcium had marked decreases in PTH mRNA levels, reaching <4% of control at 48 h. This effect was shown to be transcriptional both in in vivo studies in rats *(43)* and in in vitro studies with primary cultures of bovine parathyroid cells *(45)*. When 684 bp of the 5′-flanking region of the human PTH gene was linked to a reporter gene and transfected into a rat pituitary cell line (GH4C1), gene expression was lowered by

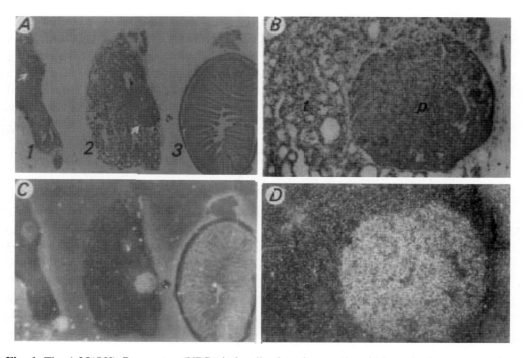

Fig. 1. The 1,25(OH)$_2$D receptor (VDR) is localized to the parathyroid in a similar concentration to that found in the duodenum, indicating that the parathyroid is a physiological target organ for 1,25(OH)$_2$D. In situ hybridization with the VDR probe in rat parathyroid–thyroid sections. (**a**) and (**c**) Parathyroid–thyroid tissue from a control rat (*left section*) and from a 1,25(OH)$_2$D-treated rat (100 pmol at 24 h) (*middle preparation*). Duodenum from a 1,25(OH)$_2$D-treated rat. *White arrows* point to parathyroid glands. (**b**) and (**d**) Higher power view of the parathyroids in (**a**) and (**c**). *Top figures* were photographed under bright-field illumination, whereas *bottom figures* show dark-field illumination of the same sections. Reproduced with permission from Naveh-Many et al. *(44)*.

1,25(OH)$_2$D *(46)*. These studies suggest that 1,25(OH)$_2$D decreases PTH transcription by acting on the 5′-flanking region of the PTH gene, probably at least partly through interactions with the vitamin D receptor-binding sequences and/or the E box that binds VDIR noted earlier. The effect of 1,25(OH)$_2$D may involve heterodimerization with the retinoid acid receptor. This is because 9 *cis*-retinoic acid, which binds to the retinoic acid receptor, when added to bovine parathyroid cells in primary culture, led to a decrease in PTH mRNA levels *(47)*. Moreover, combined treatment with 1 × 10^{-6} M retinoic acid and 1 × 10^{-8} M 1,25(OH)$_2$D decreased PTH secretion and preproPTH mRNA more effectively than either compound alone *(47)*. Alternatively, retinoic acid receptors might synergize with VDRs through actions on distinct sequences.

A further level at which 1,25(OH)$_2$D might regulate the PTH gene would be at the level of the 1,25(OH)$_2$D receptor. 1,25(OH)$_2$D acts on its target tissues by binding to the 1,25(OH)$_2$D receptor, which regulates the transcription of genes with the appropriate recognition sequences. Concentration of the VDR in 1,25(OH)$_2$D target sites could allow a modulation of the 1,25(OH)$_2$D effect, with an increase in receptor concentration leading to an amplification of its effect and a decrease in receptor concentration dampening the 1,25(OH)$_2$D effect. Naveh-Many et al. *(44)* injected 1,25(OH)$_2$D into rats and measured the levels of VDR mRNA and PTH mRNA in the parathyroid tissue. They showed that 1,25(OH)$_2$D in physiologically relevant doses led to an increase in VDR mRNA levels in the parathyroid glands in contrast to the decrease in PTH mRNA levels. Weanling rats fed a diet deficient in calcium were markedly hypocalcemic at 3 weeks and had very high serum 1,25(OH)$_2$D levels. Despite the chronically high serum 1,25(OH)$_2$D levels, there was no increase in VDR mRNA levels. Furthermore, PTH mRNA levels did not fall and were increased markedly. The low calcium in the bloodstream may have prevented the increase in parathyroid VDR levels, which may partially explain PTH mRNA suppression.

5.1.1. CALRETICULIN AND THE ACTION OF 1,25(OH)$_2$D ON THE PTH GENE

Calreticulin is a calcium-binding protein present in the endoplasmic reticulum of the cell and may also have a nuclear function. It regulates gene transcription via its ability to bind a protein motif in the DNA-binding domain of nuclear hormone receptors of sterol hormones. It has been shown to prevent vitamin D's binding and action on the osteocalcin gene in vitro *(48)*. Sela-Brown et al. *(49)* showed that calreticulin might inhibit the action of vitamin D on the PTH gene. Both rat and chicken VDRE sequences of the PTH gene were incubated with recombinant VDR and retinoic acid receptor (RXR) proteins in a gel retardation assay and showed a clear retarded band. Purified calreticulin inhibited binding of the VDR–RXR complex to the VDREs in gel retardation assays. This inhibition was due to direct protein–protein interactions between VDR and calreticulin. OK cells were transiently cotransfected with calreticulin expression vectors and either rat or chicken PTH gene promoter-CAT constructs. The cells were then assayed for 1,25(OH)$_2$D-induced CAT gene expression. 1,25(OH)$_2$D decreased PTH promoter-CAT transcription. Cotransfection with sense calreticulin, which increases calreticulin protein levels, completely inhibited the effect of 1,25(OH)$_2$D on the PTH promoters of both rat and chicken. Cotransfection with the antisense calreticulin construct did not interfere with the effect of vitamin D on PTH gene transcription. Calreticulin expression had no effect on basal CAT mRNA levels. In order to determine a physiological role for calreticulin in regulation of the PTH gene, levels of calreticulin protein were determined in the nuclear fraction of rat parathyroids. The rats were fed either a control diet or a low calcium diet, which leads to increased PTH mRNA levels, despite high serum $1,25(OH)_2D$ levels that would be expected to inhibit PTH gene transcription *(49)*. It was postulated that high calreticulin levels in the nuclear fraction would prevent the effect of $1,25(OH)_2D$ on the PTH gene. In fact, hypocalcemic rats had increased levels of calreticulin protein, as measured by Western blots, in their parathyroid nuclear faction. This may help explain why hypocalcemia leads to increased PTH gene expression, despite high serum $1,25(OH)_2D$ levels, and may also be relevant to the refractoriness of the secondary hyperparathyroidism of many chronic renal failure patients to $1,25(OH)_2D$ treatment. These studies, therefore, indicate a role for calreticulin in regulating the effect of vitamin D on the PTH gene and suggest a physiological relevance to these studies *(49)*.

Russell et al. *(50)* studied the parathyroids of chicks with vitamin D deficiency and confirmed that $1,25(OH)_2D$ regulates PTH and VDR gene expression in the avian parathyroid gland. Brown et al. *(51)* studied vitamin D-deficient rats and confirmed that $1,25(OH)_2D$ upregulated parathyroid VDR mRNA. Rodriguez et al. *(52)* showed that administration of the calcimimetic R-568 resulted in increased VDR expression in parathyroid tissue. In vitro studies of the effect of R-568 on VDR mRNA and protein were conducted in cultures of whole rat parathyroid glands. Incubation of rat parathyroid glands in vitro with R-568 resulted in a dose-dependent decrease in PTH secretion and an increase in VDR expression. Together with previous work on the effect of extracellular calcium to increase parathyroid VDR mRNA in vitro *(53)*, they concluded that activation of the CaR upregulates the parathyroid VDR mRNA.

All these studies show that $1,25(OH)_2D$, and calcium in certain circumstances, increases the expression of the VDR gene in the parathyroid gland, which would result in increased VDR protein synthesis and increased binding of $1,25(OH)_2D$. This ligand-dependent receptor upregulation would lead to an amplified effect of $1,25(OH)_2D$ on the PTH gene and might help explain the dramatic effect of $1,25(OH)_2D$ on the PTH gene.

Vitamin D may also amplify its effect on the parathyroid by increasing the activity of the calcium receptor (CaR). Canaff et al. *(54)* showed that in fact there are VDREs in the human CaR's promoter. The calcium-sensing receptor (CaR), expressed in parathyroid chief cells, thyroid C-cells, and cells of the kidney tubule, is essential for maintenance of calcium homeostasis. They showed that parathyroid, thyroid, and kidney CaR mRNA levels increased twofold at 15 h after intraperitoneal injection of $1,25(OH)_2D$ in rats. Functional VDREs have been identified in the CaR gene and probably provide the mechanism whereby $1,25(OH)_2D$ upregulates parathyroid, thyroid C-cell, and kidney CaSR expression.

The use of $1,25(OH)_2D$ is limited by its hypercalcemic effect, and therefore a number of $1,25(OH)_2D$ analogs have been synthesized that are biologically active but are less hypercalcemic than 1,25(OH)$_2$D (55). These analogs usually involve modifications of the 1,25(OH)$_2$D side chain, such as 22-oxa-1,25(OH)$_2$D, which is the chemical modification in oxacacitriol (56), or a cyclopropyl group at the end of the side chain in calcipotriol (57, 58). Brown et al. (59) showed that oxacalcitriol in vitro decreased PTH secretion from primary cultures of bovine parathyroid cells with a similar dose response to that of 1,25(OH)$_2$D. In vivo the injection of both vitamin D compounds led to a decrease in rat parathyroid PTH mRNA levels (59). However, detailed in vivo dose–response studies showed that in vivo 1,25(OH)$_2$D is the most effective analog for decreasing PTH mRNA levels, even at doses that do not cause hypercalcemia (60). Oxacalcitriol and calcipotriol are less effective for decreasing PTH RNA levels but have a wider dose range at which they do not cause hypercalcemia. This property might be useful clinically. Paricalcitol was shown to be effective at reducing PTH concentrations without causing significant hypercalcemia or hyperphosphatemia as compared to placebo. Paricalcitol treatment was shown to reduce PTH concentrations more rapidly with fewer sustained episodes of hypercalcemia and increased Ca × P product than 1,25(OH)$_2$D therapy (61). The marked activity of 1,25(OH)$_2$D analogs in vitro as compared to their modest hypercalcemic actions in vivo probably reflects their rapid clearance from the circulation (62). Despite the great interest in the development and marketing of new 1,25(OH)$_2$D analogs to decrease PTH gene expression and serum PTH levels without causing hypercalcemia, there have been few rigorous comparisons of their biological effects compared to those of 1,25(OH)$_2$D itself (63, 64). There has been a lot of debate about the relative advantages of the so-called "nonhypercalcemic" vitamin D analogs over 1,25(OH)$_2$ vitamin D$_3$. Drueke (1) has analyzed the clinical trials that have been performed and concluded that all clinical studies were retrospective in nature and suffered from the limitations of retrospective data analysis. The question is still open.

The ability of 1,25(OH)$_2$D to decrease PTH gene transcription is used therapeutically in the management of patients with chronic renal failure. They are treated with 1,25(OH)$_2$D or its prodrug 1(OH)D in order to prevent the secondary hyperparathyroidism of chronic renal failure. The poor response in some patients who do not respond may well result from poor control of serum phosphate, decreased VDR concentration in the patients' parathyroids (65), an inhibitory effect of a uremic toxin(s) on VDR-VDRE binding (66), or tertiary hyperparathyroidism with monoclonal parathyroid tumors (67). The development of calcimimetic drugs, which act to directly activate the CaR has provided a significant advance in the treatment that we can offer patients with secondary hyperparathyroidism and in clinical use these drugs are frequently used in combination with 1,25(OH)$_2$D or its analogs.

5.1.2. STUDIES ON MICE WITH VDR GENE DELETION

The VDR total knockout phenotype is characterized by high PTH levels, hypocalcemia, hypophosphatemia, bone malformations, rickets, and alopecia (68, 69). Most, but not all of the phenotypes, were reversed by correcting the serum calcium concentration with a high calcium–lactose diet (70). In order to investigate PTH regulation by the VDR in as close as possible to the physiological conditions, we have ongoing work where we have generated parathyroid-specific VDR knockout mice (PT-VDR$^{-/-}$), by crossing PTH promoter-Cre mice with total body floxed-VDR mice. VDR expression was decreased specifically in the parathyroid glands of the *PT-VDR*$^{-/-}$ mice. These mice had a normal phenotype but their serum PTH levels were significantly increased with no change in serum calcium or phosphorus. The sensitivity of the parathyroid glands of the *PT-VDR*$^{-/-}$ mice to calcium was intact as measured by serum PTH levels after changes in serum calcium. Serum type I collagen C-telopeptides (CTX), a marker of bone resorption, was increased ($\times 2.5$) in the *PT-VDR*$^{-/-}$ mice with no change in the bone formation marker, serum osteocalcin, consistent with a resorptive effect of the increased serum PTH levels in the *PT-VDR*$^{-/-}$ mice. Therefore, deletion of the VDR specifically in the parathyroid decreases parathyroid CaR expression and only moderately increases basal PTH levels, suggesting that the VDR has a limited role in parathyrodid physiology *(71)*.

5.2. Calcium

A remarkable characteristic of the parathyroid is its sensitivity to small changes in serum calcium, which leads to large changes in PTH secretion. This remarkable sensitivity of the parathyroid to increase hormone secretion after small decreases in serum calcium levels is unique to the parathyroid. All other endocrine glands increase hormone secretion after exposure to a high extracellular calcium. This calcium sensing is also expressed at the levels of PTH gene expression and parathyroid cell proliferation (Fig. 2).

Calcium and phosphate both have marked effects on the levels of PTH mRNA in vivo *(72)*. The major effect is for low calcium to increase PTH mRNA levels and low phosphate to decrease PTH mRNA levels. Naveh-Many et al. *(73)* studied rats in vivo. They showed that a small decrease in serum calcium from 2.6 to 2.1 mmol/l led to large increases in PTH mRNA levels, reaching threefold that of controls at 1 and 6 h. A high serum calcium had no effect on PTH mRNA levels even at concentrations as high as 6.0 mmol/l. Yamamoto et al. *(74)* also studied the in vivo effect of calcium on PTH mRNA levels in rats. They showed that hypocalcemia induced by a calcitonin infusion for 48 h led to a sevenfold increase in PTH mRNA levels. Rats made hypercalcemic (2.9–3.4 mM) for 48 h had the same PTH mRNA levels as controls that had received no infusion (2.5 mM). Therefore, hypercalcemia in vivo has a limited effect to decrease PTH mRNA levels. These results emphasize that the gland is geared to respond to hypocalcemia and not hypercalcemia.

5.2.1. MECHANISMS OF REGULATION OF PTH mRNA BY CACLIUM

The mechanism whereby calcium regulates PTH gene expression is particularly interesting. Changes in extracellular calcium are sensed by a calcium sensor that then regulates PTH secretion *(75, 76)*. Signal transduction from the CaSR involves activation of phospholipase C, D, and A$_2$ enzymes *(77)*. The response to changes in serum calcium involves the protein phosphatase type 2B, calcineurin *(78)*. In vivo and in vitro studies demonstrated that inhibition of calcineurin by genetic manipulation or pharmacologic agents affected the response of PTH mRNA levels to changes in extracellular calcium *(78)*.

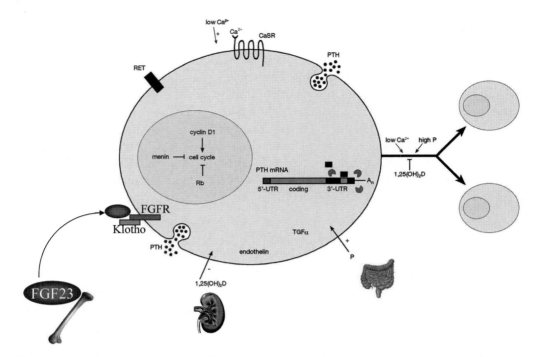

Fig. 2. Regulation of parathyroid proliferation, gene expression, and secretion. Cyclin D1 driven by the PTH promoter and activating mutations of the menin gene are known to cause parathyroid adenomas. Germ-line mutations of the latter cause MEN 1. The very rare parathyroid carcinomas show lack of expression of the retinoblastoma protein (pRb). Activating mutations of the RET proto-oncogene result in MEN 2a. A low serum calcium leads to a decreased activation of the CaR and results in increased PTH secretion, PTH mRNA stability, and parathyroid cell proliferation. A high serum phosphate leads to similar changes in all these parameters. $1,25(OH)_2D$ decreases PTH gene transcription markedly and decreases parathyroid cell proliferation. A high serum phosphate and chronic kidney disease lead to an increase in FGF23 secretion by osteocytes, which then acts on the kidney to decrease phosphate reabsorption and inhibit the synthesis of $1,25(OH)_2D$ and on the parathyroid to decrease PTH expression. Endothelin and TGFα are increased in the parathyroids of proliferating parathyroid glands. PTH mRNA stability is regulated by parathyroid cytosolic proteins (*trans* factors) binding to a short defined *cis* sequence in the PTH mRNA 3′-UTR and preventing degradation by ribonucleases in a process involving KSRP recruitment of the exosome. The defined protective proteins are AUF1 and Unr. In hypocalcemia there is more binding of the stabilizing *trans* factors to the *cis* sequence leading to a more stable transcript. A low serum phosphate leads to much less binding and a rapidly degraded PTH transcript. (Modified from Silver with permission *(99)*).

Moallem et al. *(79)* have performed in vivo studies on the effect of hypocalcemia on PTH gene expression. The effect is post-transcriptional in vivo and involves protein–RNA interactions at the 3′-untranslated region of the PTH mRNA *(79)*. A similar mechanism is involved in the effect of phosphate on PTH gene expression, so the mechanisms involved will be discussed after the independent effect of phosphate on the PT is considered.

5.3. Phosphate

5.3.1. PHOSPHATE REGULATES THE PARATHYROID INDEPENDENTLY OF CALCIUM AND 1,25(OH)$_2$D

The demonstration of a direct effect of high phosphate on the parathyroid in vivo has been difficult. One of the reasons is that the various maneuvers used to increase or decrease serum phosphate invariably leads to a change in the ionized calcium concentration. A number of careful clinical and experimental studies suggested that the effect of phosphate on serum PTH levels was independent of changes in both serum calcium and 1,25(OH)$_2$D levels. In dogs with experimental chronic renal failure, Lopez-Hilker et al. *(80)* have shown that phosphate restriction corrected their secondary hyperparathyroidism independent of changes in serum calcium and 1,25(OH)$_2$D levels. They did this by placing the uremic dogs on diets deficient in both calcium and phosphate. This led to lower levels of serum phosphate and calcium, with no increase in the low levels of serum 1,25(OH)$_2$D. Despite this, there was a 70% decrease in PTH levels. This study suggested that, at least in chronic renal failure, phosphate affected the parathyroid cell by a mechanism independent of its effect on serum 1,25(OH)$_2$D and calcium levels *(80)*. Therefore, phosphate plays a central role in the pathogenesis of secondary hyperparathyroidism, both by its effect on serum 1,25(OH)$_2$D and calcium levels and possibly independently. A raised serum phosphate also stimulates the secretion of FGF23 which in turn decreases PTH gene expression and serum PTH levels *(81, 82)*. This effect would act as a counterbalance to the stimulatory effect of phosphate on the parathyroid and is discussed separately in this chapter.

Kilav et al. *(83)* bred second-generation vitamin D-deficient rats and then placed the weanling vitamin D-deficient rats on a diet with no vitamin D, low calcium, and low phosphate. After one night of this diet, serum phosphate had decreased markedly with no changes in serum calcium or 1,25(OH)$_2$D. These rats with isolated hypophosphatemia had marked decreases in PTH mRNA levels and serum PTH. However, the very low serum phosphates in these in vivo studies may have no direct relevance to possible direct effects of high phosphate in renal failure. It is necessary to separate nonspecific effects of very low phosphate from true physiologic regulation. To establish that the effect of serum phosphate on the parathyroid was indeed a direct effect, in vitro confirmation was needed, which was provided by three groups. Rodriguez was the first to show that increased phosphate levels increased PTH secretion from isolated parathyroid glands in vitro. The effect required maintainance of tissue architecture *(84–86)*. The effect was found in whole glands or tissue slices but not in isolated cells. The requirement for intact tissue suggests either that the sensing mechanism for phosphate is damaged during the preparation of isolated cells or that the intact gland structure is important to the phosphate response.

Parathyroid responds to changes in serum phosphate at the level of secretion, gene expression, and cell proliferation, although the mechanism of these effects is unknown (Fig. 2). The effect of high phosphate to increase PTH secretion may be mediated by phospholipase A$_2$-activated signal transduction. Bourdeau et al. *(87, 88)* showed that arachidonic acid and its metabolites inhibit PTH secretion. Almaden *(89)* showed in vitro that a high phosphate medium increased PTH secretion, which was prevented by the addition of arachidonic acid.

5.4. Protein–PTH mRNA Interactions Determine the Regulation of PTH Gene Expression by Serum Calcium and Phosphate

The clearest rat in vivo models for effects of calcium and phosphate on PTH gene expression are diet-induced hypocalcemia with a large increase in PTH mRNA levels and diet-induced hypophosphatemia with a large decrease in PTH mRNA levels. In both instances the effect was post-transcriptional, as shown by nuclear transcript run-on experiments (72). Parathyroid cytosolic proteins were found to bind in vitro transcribed PTH mRNA (79). Interestingly, this binding was increased with parathyroid proteins from hypocalcemic rats (with increased PTH mRNA levels) and decreased with parathyroid proteins from hypophosphatemic rats (with decreased PTH mRNA levels). Proteins from other tissues bound to PTH mRNA, but this binding is regulated by calcium and phosphate only with parathyroid proteins. Intriguingly, binding requires the presence of the terminal 60 nucleotides of the PTH transcript.

Naveh-Many and colleagues utilized an in vitro degradation assay to study the effects of hypocalcemic and hypophosphatemic parathyroid proteins on PTH mRNA stability (79). This assay reproduced the differences in PTH mRNA levels observed in vivo. Moreover, the difference in RNA stability by the parathyroid extracts was totally dependent on an intact 3′-untranslated region (UTR) and, in particular, on the terminal 60 nucleotides. Proteins from other tissues in these rats were not regulated by calcium or phosphate. Therefore, calcium and phosphate change the properties of parathyroid cytosolic proteins, which bind specifically to the PTH mRNA 3′-UTR and determine its stability (Fig. 2). What are these proteins?

5.4.1. IDENTIFICATION OF THE PTH mRNA 3′-UTR-BINDING PROTEINS THAT DETERMINE PTH mRNA STABILITY

5.4.1.1. AU-rich Binding Factor (AUF1).
Sela-Brown et al. and Dinur et al. have utilized affinity chromatography using the PTH RNA 3′-UTR to isolate these RNA-binding proteins (90, 91). The proteins, which bind the PTH mRNA, are also present in other tissues, such as brain, but only in the parathyroid their binding is regulated by calcium and phosphate. A major band from the eluate of a PTH 3′-region RNA affinity chromatography was identical to AU-rich binding factor (AUF1) (91). Recombinant AUF1 bound the full-length PTH mRNA and the 3′-UTR. Added recombinant AUF1 also stabilized the PTH transcript in the in vitro degradation assay. These results showed that AUF1 is a protein that binds to the PTH mRNA 3′-UTR and stabilizes the PTH transcript.

The regulation of protein–PTH mRNA binding involves post-translational modification of AUF1. AUF1 levels are not regulated in PT extracts from rats fed calcium and phosphorus depleted diets. However, two-dimensional gels showed post-translational modification of AUF1 that included phosphorylation (78). There is no parathyroid cell line, but a PTH mRNA *cis* acting 63-nt element *(92)* is recognized in HEK 293 cells as an instability element. RNA interference for AUF1 decreased human PTH secretion in cotransfection experiments *(78)*.

Most patients with chronic kidney disease develop secondary hyperparathyroidism with disabling systemic complications. Calcimimetic agents are effective tools in the management of secondary hyperparathyroidism, acting through allosteric modification of the calcium-sensing receptor (CaR) on the parathyroid gland to decrease PTH secretion and parathyroid cell proliferation. R-568 decreased both PTH mRNA and serum PTH levels in adenine high phosphorus-induced chronic kidney disease *(93)*. The effect of the calcimimetic on PTH gene expression was post-transcriptional and correlated with differences in protein–RNA binding and post-translational modifications of the *trans*-acting factor AUF1 in the parathyroid. The AUF1 modifications as a result of uremia were reversed to those of normal rats by treatment with R-568. Therefore, uremia and activation of the CaR mediated by calcimimetics modify AUF1 post-translationally. These modifications in AUF1 correlate with changes in protein–PTH mRNA binding and PTH mRNA levels *(93)*.

5.4.1.2. Unr (Upstream of N-*Ras*). A second parathyroid cytosolic protein which is part of the stabilizing PTH mRNA 3'-UTR-binding complex was shown to be Unr by affinity chromatography *(90)*. Depletion of Unr by small interfering RNA decreased PTH mRNA levels in HEK293 cells transiently cotransfected with the human PTH gene. Overexpression of Unr increased the rat full-length PTH mRNA levels but not a PTH mRNA lacking the terminal 60-nucleotide *cis*-acting protein-binding region. Therefore, Unr binds to the PTH *cis* element and increases PTH mRNA levels, as does AUF1. Unr, together with the other proteins in the RNA-binding complex, determines PTH mRNA stability *(90)*. Recent findings have identified an additional decay promoting protein, KSRP, that differentially interacts with PTH mRNA to recruit the degradatory machinery *(94)*. The balance between the stabilizing and destabilizing proteins determines PTH mRNA levels in response to physiological stimuli *(95)*.

5.4.1.3. KSRP (K-homology Splicing Regulator Protein). We have shown that mRNA decay promoting protein KSRP binds to PTH mRNA in intact parathyroid glands and in transfected cells *(94)*. This binding of KSRP is decreased in glands from calcium depleted or experimental chronic kidney failure rats where PTH mRNA is more stable, compared to parathyroid glands from control and phosphorus depleted rats where PTH mRNA is less stable. The differences in KSRP-PTH mRNA binding counter those of AUF1. PTH mRNA decay depends on the KSRP-recruited exosome in parathyroid extracts. In transfected cells, KSRP over-expression and knockdown experiments show that KSRP decreases PTH mRNA stability and steady-state levels through the PTH mRNA ARE. Over-expression of isoform p45 of the PTH mRNA stabilizing protein AUF1 blocks KSRP-PTH mRNA binding and partially prevents the KSRP mediated decrease in PTH mRNA levels. Therefore, calcium or phosphorus depletion, as well as chronic kidney failure, regulate the interaction of KSRP and AUF1 with PTH mRNA and its half-life. The balance between the stabilizing and destabilizing proteins determines PTH mRNA levels in response to physiological stimuli.

5.4.2. A Conserved Sequence in the PTH mRNA 3'-UTR Binds Parathyroid Cytosolic Proteins and Determines mRNA Stability in Response to Changes in Calcium and Phosphate

We have identified the minimal sequence for protein binding in the PTH mRNA 3'-UTR and determined its functionality *(92)*. A minimum sequence of 26 nucleotides was sufficient for PTH RNA–protein binding and competition (Fig. 2). Significantly, this sequence was preserved among species *(9)*. To study the functionality of the sequence in the context of another RNA, a 63-bp cDNA PTH sequence consisting of the 26 nucleotide and flanking regions was fused to growth hormone (GH) cDNA. The conserved PTH RNA protein-binding region was necessary and sufficient for responsiveness to calcium and phosphate and determines PTH mRNA stability and levels *(92)*.

The PTH mRNA 3'-UTR-binding element is AU rich and is a type III AU-rich element (ARE). Sequence analysis of the PTH mRNA 3'-UTR of different species revealed a preservation of the 26-nt protein-binding element in rat, murine, human, macaque, feline, and canine 3'-UTRs *(9)*. In contrast to protein-coding sequences that are highly conserved, UTRs are less conserved. The conservation of the protein-binding element in the PTH mRNA 3'-UTR suggests that this element represents a functional unit that has been evolutionarily conserved. The *cis*-acting element is at the 3'-distal end in all species where it is expressed.

6. FIBROBLAST GROWTH FACTOR 23 AND THE PARATHYROID

Phosphate homeostasis is maintained by a counterbalance between efflux from the kidney and influx from intestine and bone. Fibroblast growth factor-23 (FGF23) is a bone-derived phosphaturic hormone that acts on the kidney to increase phosphate excretion and suppress biosynthesis of $1,25(OH)_2$vitamin D. FGF23 signals through fibroblast growth factor receptors (FGFR) bound by the transmembrane protein Klotho *(96)*. Since most tissues express FGFRs, expression of Klotho virtually determines FGF23 target organs. Takeshita et al. *(97)* were the first to show that Klotho protein is expressed not only in the kidney but also in the parathyroid, pituitary, and sino-atrial node. In addition, Urakawa et al. *(98)* injected rats with FGF23 and demonstrated increased Egr-1 (early growth response gene-1) mRNA levels in the parathyroid, suggesting that the parathyroid may be a further FGF23 target organ. Phosphate homeostasis is maintained by a counterbalance between efflux from the kidney and influx from intestine and bone. FGF23 is a bone-derived phosphaturic hormone that acts on the kidney to increase phosphate excretion and suppress biosynthesis of vitamin D. FGF23 signals with highest efficacy through several FGF receptors (FGFRs) bound by the transmembrane protein Klotho as a coreceptor. Since most tissues express FGFR, expression of Klotho determines FGF23 target organs. We have identified the parathyroid as a target organ for FGF23 in rats *(81)*. We showed that the parathyroid gland expressed Klotho and two FGFRs. The administration of recombinant FGF23 led to an increase in parathyroid Klotho levels. In addition, FGF23 activated the MAPK pathway in the parathyroid through ERK1/2 phosphorylation and increased early growth response 1 mRNA levels. Using both rats and in vitro rat parathyroid cultures, we showed that FGF23 suppressed both parathyroid hormone (PTH) secretion and PTH gene expression.

The FGF23-induced decrease in PTH secretion was prevented by a MAPK inhibitor. These data indicate that FGF23 acts directly on the parathyroid through the MAPK pathway to decrease serum PTH (Fig. 2). This bone–parathyroid endocrine axis adds a new dimension to the understanding of mineral homeostasis. Krajisnik et al. *(82)* showed similar results using bovine parathyroid cells in primary culture. Interestingly, they also showed that FGF23 led to a dose-dependent increase in the expression of the 1α-(OH)ase enzyme in the parathyroid. The increased $1,25(OH)_2D$ might then act autocrinely to decrease PTH gene transcription.

7. SUMMARY

The PTH gene is regulated by a number of factors (Fig. 2). $1,25(OH)_2D$ acts on the PTH gene to decrease its transcription, and this action is used in the management of patients with chronic kidney disease. The major effect of calcium on PTH gene expression in vivo is for hypocalcemia to increase PTH mRNA levels, and this is mainly post-transcriptional. Phosphate also regulates PTH gene expression in vivo, and this effect appears to be independent of the effect of phosphate on serum calcium and $1,25(OH)_2D$. The effect of phosphate is also post-transcriptional. *Trans*-acting parathyroid cytosolic proteins bind to a defined *cis* element in the PTH mRNA 3′-UTR. This binding determines the degradation of PTH mRNA by degrading enzymes and thereby PTH mRNA half-life. The post-transcriptional effects of calcium and phosphate are the result of changes in the balance of these stabilizing and degrading factors on PTH mRNA. These interactions also regulate PTH mRNA levels in experimental uremia. In diseases such as chronic renal failure, secondary hyperparathyroidism involves abnormalities in PTH secretion and synthesis. FGF23 also acts on its receptors, the Klotho-FGFR1, 2c, and 3c to decrease PTH mRNA levels and secretion. An understanding of how the parathyroid is regulated at each level will help devise rational therapy for the management of such conditions, as well as treatment for diseases, such as osteoporosis, in which alterations in PTH may have a role.

ACKNOWLEDGMENTS

This work was supported in part by grants from the Israel Science Foundation.

REFERENCES

1. Drueke TB (2005) Which vitamin D derivative to prescribe for renal patients. Curr Opin Nephrol Hypertens 14:343–349
2. Steddon SJ, Schroeder NJ, Cunningham J (2001) Vitamin D analogues: how do they differ and what is their clinical role? Nephrol Dial Transplant 16:1965–1967
3. Antonarakis SE, Phillips JA, Mallonee RL, Kazazian HHJ, Fearon ER, Waber PG, Kronenberg HM, Ullrich A, Meyers DA (1983) Beta-globin locus is linked to the parathyroid hormone (PTH) locus and lies between the insulin and PTH loci in man. Proc Natl Acad Sci USA 80:6615–6619
4. Zabel BU, Kronenberg HM, Bell GI, Shows TB (1985) Chromosome mapping of genes on the short arm of human chromosome 11: parathyroid hormone gene is at 11p15 together with the genes for insulin, c-Harvey-ras 1, and beta-hemoglobin. Cytogenet Cell Genet 39:200–205
5. Igarashi T, Okazaki T, Potter H, Gaz R, Kronenberg HM (1986) Cell-specific expression of the human parathyroid hormone gene in rat pituitary cells. Mol Cell Biol 6:1830–1833

6. Kronenberg HM, Igarashi T, Freeman MW, Okazaki T, Brand SJ, Wiren KM, Potts JT Jr (1986) Structure and expression of the human parathyroid hormone gene. Recent Prog Horm Res 42:641–663

7. Danks JA, Ho PM, Notini AJ, Katsis F, Hoffmann P, Kemp BE, Martin TJ, Zajac JD (2003) Identification of a parathyroid hormone in the fish Fugu rubripes. J Bone Miner Res 18:1326–1331

8. Gensure RC, Ponugoti B, Gunes Y, Papasani MR, Lanske B, Bastepe M, Rubin DA, Juppner H (2004) Identification and characterization of two parathyroid hormone-like molecules in zebrafish. Endocrinology 145:1634–1639

9. Bell O, Silver J, Naveh-Many T (2005) Parathyroid hormone, from gene to protein. In: Naveh-Many T (ed) Molecular biology of the parathyroid. Landes Bioscience and Kluwer Academic/Plenum Publishers, New York, pp 8–28

10. Hendy GN, Kronenberg HM, Potts JTJ, Rich A (1981) Nucleotide sequence of cloned cDNAs encoding human preproparathyroid hormone. Proc Natl Acad Sci USA 78:7365–7369

11. Vasicek TJ, McDevitt BE, Freeman MW, Fennick BJ, Hendy GN, Potts JTJ, Rich A, Kronenberg HM (1983) Nucleotide sequence of the human parathyroid hormone gene. Proc Natl Acad Sci USA 80:2127–2131

12. Kronenberg HM, McDevitt BE, Majzoub JA, Nathans J, Sharp PA, Potts JT Jr., Rich A (1979) Cloning and nucleotide sequence of DNA coding for bovine preproparathyroid hormone. Proc Natl Acad Sci USA 76:4981–4985

13. Weaver CA, Gordon DF, Kemper B (1982) Nucleotide sequence of bovine parathyroid hormone messenger RNA. Mol Cell Endocrinol 28:411–424

14. Schmelzer HJ, Gross G, Widera G, Mayer H (1987) Nucleotide sequence of a full-length cDNA clone encoding preproparathyroid hormone from pig and rat. Nucleic Acids Res 15:6740–6741

15. He B, Tong TK, Hiou-Tim FF, Al Akad B, Kronenberg HM, Karaplis AC (2002) The murine gene encoding parathyroid hormone: genomic organization, nucleotide sequence and transcriptional regulation. J Mol Endocrinol 29:193–203

16. Khosla S, Demay M, Pines M, Hurwitz S, Potts JTJ, Kronenberg HM (1988) Nucleotide sequence of cloned cDNAs encoding chicken preproparathyroid hormone. J Bone Miner Res 3:689–698

17. Russell J, Sherwood LM (1989) Nucleotide sequence of the DNA complementary to avian (chicken) preproparathyroid hormone mRNA and the deduced sequence of the hormone precursor. Mol Endocrinol 3:325–331

18. Rosol TJ, Steinmeyer CL, McCauley LK, Grone A, DeWille JW, Capen CC (1995) Sequences of the cDNAs encoding cannine parathyroid hormone-related protein and parathyroid hormone. Gene 160:241–243

19. Toribio RE, Kohn CW, Chew DJ, Capen CC, Rosol TJ (2002) Cloning and sequence analysis of the complementary DNA for feline preproparathyroid hormone. Am J Vet Res 63:194–197

20. Caetano AR, Shiue YL, Lyons LA, O'Brien SJ, Laughlin TF, Bowling AT, Murray JD (1999) A comparative gene map of the horse (Equus caballus). Genome Res 9:1239–1249

21. Malaivijitnond S, Takenaka O, Anukulthanakorn K, Cherdshewasart W (2002) The nucleotide sequences of the parathyroid gene in primates (suborder Anthropoidea). Gen Comp Endocrinol 125:67–78

22. Naveh-Many T, (2005) Molecular biology of the parathyroid. Landes Bioscience/Eurekah.com and Kluwer Academic Press, Georgetown, Texas and New York

23. Gunther T, Chen ZF, Kim J, Priemel M, Rueger JM, Amling M, Moseley JM, Martin TJ, Anderson DJ, Karsenty G (2000) Genetic ablation of parathyroid glands reveals another source of parathyroid hormone. Nature 406:199–203

24. Ding C, Buckingham B, Levine MA (2001) Familial isolated hypoparathyroidism caused by a mutation in the gene for the transcription factor GCMB. J Clin Invest 108:1215–1220

25. Okabe M, Graham A (2004) The origin of the parathyroid gland. Proc Natl Acad Sci USA 101: 17716–17719

26. Imanishi Y, Hosokawa Y, Yoshimoto K, Schipani E, Mallya S, Papanikolaou A, Kifor O, Tokura T, Sablosky M, Ledgard F, Gronowicz G, Wang TC, Schmidt EV, Hall C, Brown EM, Bronson R, Arnold A (2001) Dual abnormalities in cell proliferation and hormone regulation caused by cyclin D1 in a murine model of hyperparathyroidism. J Clin Invest 107:1093–1102

27. Kel A, Scheer M, Mayer H (2005) *In silico* analysis of regualtory sequences in the human parathyroid hormone gene. In: Naveh-Many T (ed) Molecular biology of the parathyroid. Landes Bioscience and Kluwer Academic/Plenum Publishers, New York, pp 68–83

28. Alimov AP, Park-Sarge OK, Sarge KD, Malluche HH, Koszewski NJ (2005) Transactivation of the parathyroid hormone promoter by specificity proteins and the nuclear factor Y complex. Endocrinol 146:3409–3416

29. Koszewski NJ, Alimov AP, Park-Sarge OK, Malluche HH (2004) Suppression of the human parathyroid hormone promoter by vitamin D involves displacement of NF-Y binding to the vitamin D response element. J Biol Chem 279:42431–42437

30. Demay MB, Kiernan MS, DeLuca HF, Kronenberg HM (1992) Sequences in the human parathyroid hormone gene that bind the l,25-dihydroxyvitamin D-3 receptor and mediate transcriptional repression in response to l,25-dihydroxyvitamin D-3. Proc Natl Acad Sci USA 89:8097–8101

31. Russell J, Ashok S, Koszewski NJ (1999) Vitamin D receptor interactions with the rat parathyroid hormone gene: synergistic effects between two negative vitamin D response elements. J Bone Miner Res 14:1828–1837

32. Liu SM, Koszewski N, Lupez M, Malluche HH, Olivera A, Russell J (1996) Characterization of a response element in the 5'-flanking region of the avian (chicken) parathyroid hormone gene that mediates negative regulation of gene transcription by 1,25-dihydroxyvitamin D_3 and binds the vitamin D_3 receptor. Mol Endocrinol 10:206–215

33. Fujiki R, Kim MS, Sasaki Y, Yoshimura K, Kitagawa H, Kato S (2005) Ligand-induced transrepression by VDR through association of WSTF with acetylated histones. EMBO J 24:3881–3894

34. Murayama A, Kim MS, Yanagisawa J, Takeyama K, Kato S (2004) Transrepression by a liganded nuclear receptor via a bHLH activator through co-regulator switching. EMBO J 23:1598–1608

35. Kim MS, Fujiki R, Murayama A, Kitagawa H, Yamaoka K, Yamamoto Y, Mihara M, Takeyama K, Kato S (2007) 1Alpha,25(OH)2D3-induced transrepression by vitamin D receptor through E-box-type elements in the human parathyroid hormone gene promoter. Mol Endocrinol 21:334–342

36. Naveh-Many T, Silver J (2006) Regulation of parathyroid hormone gene expression by 1,25-dihydroxyvitamin D. In: Naveh-Many T (ed) Molecular biology of the parthyroid. Landes Bioscience and Kluwer Academic/Plenum Publishers, New York, pp 84–94

37. Russell J, Silver J, Sherwood LM (1984) The effects of calcium and vitamin D metabolites on cytoplasmic mRNA coding for pre-proparathyroid hormone in isolated parathyroid cells. Trans Assoc Am Phys 97:296–303

38. Silver J, Russell J, Sherwood LM (1985) Regulation by vitamin D metabolites of messenger ribonucleic acid for preproparathyroid hormone in isolated bovine parathyroid cells. Proc Natl Acad Sci USA 82:4270–4273

39. Cantley LK, Ontjes DA, Cooper CW, Thomas CG, Leight GS, Wells SAJ (1985) Parathyroid hormone secretion from dispersed human hyperparathyroid cells: increased secretion in cells from hyperplastic glands versus adenomas. J Clin Endocrinol Metab 60:1032–1037

40. Cantley LK, Russell J, Lettieri D, Sherwood LM (1985) 1,25-dihydroxyvitamin D3 suppresses parathyroid hormone secretion from bovine parathyroid cells in tissue culture. Endocrinology 117:2114–2119

41. Chan YL, McKay C, Dye E, Slatopolsky E (1986) The effect of 1,25 dihydroxycholecalciferol on parathyroid hormone secretion by monolayer cultures of bovine parathyroid cells. Calcif Tissue Int 38:27–32

42. Karmali R, Farrow S, Hewison M, Barker S, O'Riordan JL (1989) Effects of 1,25-dihydroxyvitamin D3 and cortisol on bovine and human parathyroid cells. J Endocrinol 123:137–142

43. Silver J, Naveh-Many T, Mayer H, Schmelzer HJ, Popovtzer MM (1986) Regulation by vitamin D metabolites of parathyroid hormone gene transcription in vivo in the rat. J Clin Invest 78:1296–1301

44. Naveh-Many T, Marx R, Keshet E, Pike JW, Silver J (1990) Regulation of 1,25-dihydroxyvitamin D3 receptor gene expression by 1,25-dihydroxyvitamin D3 in the parathyroid in vivo. J Clin Invest 86:1968–1975

45. Russell J, Lettieri D, Sherwood LM (1986) Suppression by 1,25(OH)2D3 of transcription of the pre-proparathyroid hormone gene. Endocrinology 119:2864–2866

46. Okazaki T, Igarashi T, Kronenberg HM (1988) 5'-flanking region of the parathyroid hormone gene mediates negative regulation by 1,25-(OH)2 vitamin D3. J Biol Chem 263:2203–2208
47. MacDonald PN, Ritter C, Brown AJ, Slatopolsky E (1994) Retinoic acid suppresses parathyroid hormone (PTH) secretion and PreproPTH mRNA levels in bovine parathyroid cell culture. J Clin Invest 93:725–730
48. Wheeler DG, Horsford J, Michalak M, White JH, Hendy GN (1995) Calreticulin inhibits vitamin D3 signal transduction. Nucleic Acids Res 23:3268–3274
49. Sela-Brown A, Russell J, Koszewski NJ, Michalak M, Naveh-Many T, Silver J (1998) Calreticulin inhibits vitamin D's action on the PTH gene in vitro and may prevent vitamin D's effect in vivo in hypocalcemic rats. Mol Endocrinol 12:1193–1200
50. Russell J, Bar A, Sherwood LM, Hurwitz S (1993) Interaction between calcium and 1,25-dihydroxyvitamin D3 in the regulation of preproparathyroid hormone and vitamin D receptor messenger ribonucleic acid in avian parathyroids. Endocrinology 132:2639–2644
51. Brown AJ, Zhong M, Finch J, Ritter C, Slatopolsky E (1995) The roles of calcium and 1,25-dihydroxyvitamin D3 in the regulation of vitamin D receptor expression by rat parathyroid glands. Endocrinology 136:1419–1425
52. Rodriguez ME, Almaden Y, Canadillas S, Canalejo A, Siendones E, Lopez I, Aguilera-Tejero E, Martin D, Rodriguez M (2007) The calcimimetic R-568 increases vitamin D receptor expression in rat parathyroid glands. Am J Physiol Renal Physiol 292:F1390–F1395
53. Garfia B, Canadillas S, Canalejo A, Luque F, Siendones E, Quesada M, Almaden Y, Aguilera-Tejero E, Rodriguez M (2002) Regulation of parathyroid vitamin D receptor expression by extracellular calcium. J Am Soc Nephrol 13:2945–2952
54. Canaff L, Hendy GN (2002) Human calcium-sensing receptor gene. Vitamin D response elements in promoters P1 and P2 confer transcriptional responsiveness to 1,25-dihydroxyvitamin D. J Biol Chem 277:30337–30350
55. Brown AJ (2005) Vitamin D analogs for the treatment of secondary hyperparathyroidsim in chronic renal failure. In: Naveh-Many T (ed) Molecular biology of the parathyoid. Landes Bioscience and Kluwer Academic/Plenum Publishers, New York, pp 95–112
56. Nishii Y, Abe J, Mori T, Brown AJ, Dusso AS, Finch J, Lopez-Hilker S, Morrissey J, Slatopolsky E (1991) The noncalcemic analogue of vitamin D, 22-oxacalcitriol, suppresses parathyroid hormone synthesis and secretion. Contrib Nephrol 91:123–128
57. Evans DB, Thavarajah M, Binderup L, Kanis JA (1991) Actions of calcipotriol (MC 903), a novel vitamin D3 analog, on human bone-derived cells: comparison with 1,25-dihydroxyvitamin D3. J Bone Miner Res 6:1307–1315
58. Kissmeyer AM, Binderup L (1991) Calcipotriol (MC 903): pharmacokinetics in rats and biological activities of metabolites. A comparative study with 1,25(OH)2D3. Biochem Pharmacol 41: 1601–1606
59. Brown AJ, Ritter CR, Finch JL, Morrissey J, Martin KJ, Murayama E, Nishii Y, Slatopolsky E (1989) The noncalcemic analogue of vitamin D, 22-oxacalcitriol, suppresses parathyroid hormone synthesis and secretion. J Clin Invest 84:728–732
60. Naveh-Many T, Silver J (1993) Effects of calcitriol, 22-oxacalcitriol and calcipotriol on serum calcium and parathyroid hormone gene expression. Endocrinology 133:2724–2728
61. Sprague SM, Llach F, Amdahl M, Taccetta C, Batlle D (2003) Paricalcitol versus calcitriol in the treatment of secondary hyperparathyroidism. Kidney Int 63:1483–1490
62. Bouillon R, Allewaert K, Xiang DZ, Tan BK, Van Baelen H (1991) Vitamin D analogs with low affinity for the vitamin D binding protein in vitro and decreased in vivo activity. J Bone Miner Res 6:1051–1057
63. Brown AJ (1998) Vitamin D analogues. Am J Kidney Dis 32:S25–S39
64. Verstuyf A, Segaert S, Verlinden L, Casteels K, Bouillon R, Mathieu C (1998) Recent developments in the use of vitamin D analogues. Curr Opin Nephrol Hypertens 7:397–403
65. Fukuda N, Tanaka H, Tominaga Y, Fukagawa M, Kurokawa K, Seino Y (1993) Decreased 1,25-dihydroxyvitamin D3 receptor density is associated with a more severe form of parathyroid hyperplasia in chronic uremic patients. J Clin Invest 92:1436–1443

66. Patel S, Rosenthal JT (1985) Hypercalcemia in carcinoma of prostate. Its cure by orchiectomy. Urology. 25:627–629
67. Arnold A, Brown MF, Urena P, Gaz RD, Sarfati E, Drueke TB (1995) Monoclonality of parathyroid tumors in chronic renal failure and in primary parathyroid hyperplasia. J Clin Invest 95:2047–2053
68. Li YC, Pirro AE, Amling M, Delling G, Baron R, Bronson R, Demay MB (1997) Targeted ablation of the vitamin D receptor: an animal model of vitamin D-dependent rickets type II with alopecia. Proc Natl Acad Sci USA 94:9831–9835
69. Van Cromphaut SJ, Dewerchin M, Hoenderop JG, Stockmans I, Van Herck E, Kato S, Bindels RJ, Collen D, Carmeliet P, Bouillon R, Carmeliet G (2001) Duodenal calcium absorption in vitamin D receptor-knockout mice: functional and molecular aspects. Proc Natl Acad Sci USA 98: 13324–13329
70. Li YC, Amling M, Pirro AE, Priemel M, Meuse J, Baron R, Delling G, Demay MB (1998) Normalization of mineral ion homeostasis by dietary means prevents hyperparathyroidism, rickets, and osteomalacia, but not alopecia in vitamin D receptor-ablated mice. Endocrinology 139:4391–4396
71. Meir T, Levi R, Lieben L, Libutti S, Carmeliet G, Bouillon R, Silver J, Naveh-Many T (2009) Deletion of the vitamin D receptor specifically in the parathyroid demonstrates a limited role for the VDR in parathyroid physiology. Am J Physiol 297:F1192–F1198.
72. Kilav R, Silver J, Naveh-Many T (2005) Regulation of parathyroid hormone mRNA stability by calcium and phosphate. In: Naveh-Many T (ed) Molecular biology of the parathyroid. Landes Bioscience and Kluwer Academic/Plenum Publishers, New York, pp 57–67
73. Naveh-Many T, Friedlaender MM, Mayer H, Silver J (1989) Calcium regulates parathyroid hormone messenger ribonucleic acid (mRNA), but not calcitonin mRNA in vivo in the rat. Dominant role of l,25-dihydroxyvitamin D. Endocrinology 125:275–280
74. Yamamoto M, Igarashi T, Muramatsu M, Fukagawa M, Motokura T, Ogata E (1989) Hypocalcemia increases and hypercalcemia decreases the steady-state level of parathyroid hormone messenger RNA in the rat. J Clin Invest 83:1053–1056
75. Brown EM, Gamba G, Riccardi D, Lombardi M, Butters R, Kifor O, Sun A, Hediger MA, Lytton J, Hebert J (1993) Cloning and characterization of an extracellular Ca^{2+}-sensing receptor from bovine parathyroid. Nature 366:575–580
76. Yano S, Brown EM (2005) The calcium sensing receptor. In: Naveh-Many T (ed) Molecular biology of the parathyroid. Landes Bioscience and Kluwer Academic/Plenum Publishers, New York, pp 44–56
77. Kifor O, Diaz R, Butters R, Brown EM (1997) The Ca^{2+}-sensing receptor (CaR) activates phospholipases C, A2, and D in bovine parathyroid and CaR-transfected, human embryonic kidney (HEK293) cells. J Bone Miner Res 12:715–725
78. Bell O, Gaberman E, Kilav R, Levi R, Cox KB, Molkentin JD, Silver J, Naveh-Many T (2005) The protein phosphatase calcineurin determines basal parathyroid hormone gene expression. Mol Endocrinol 19:516–526
79. Moallem E, Silver J, Kilav R, Naveh-Many T (1998) RNA protein binding and post-transcriptional regulation of PTH gene expression by calcium and phosphate. J Biol Chem 273:5253–5259
80. Lopez-Hilker S, Dusso AS, Rapp NS, Martin KJ, Slatopolsky E (1990) Phosphorus restriction reverses hyperparathyroidism in uremia independent of changes in calcium and calcitriol. Am J Physiol 259:F432–F437
81. Ben Dov IZ, Galitzer H, Lavi-Moshayoff V, Goetz R, Kuro-o M, Mohammadi M, Sirkis R, Naveh-Many T, Silver J (2007) The parathyroid is a target organ for FGF23 in rats. J Clin Invest 117: 4003–4008
82. Krajisnik T, Bjorklund P, Marsell R, Ljunggren O, Akerstrom G, Jonsson KB, Westin G, Larsson TE (2007) Fibroblast growth factor-23 regulates parathyroid hormone and 1alpha-hydroxylase expression in cultured bovine parathyroid cells. J Endocrinol 195:125–131
83. Kilav R, Silver J, Naveh-Many T (1995) Parathyroid hormone gene expression in hypophosphatemic rats. J Clin Invest 96:327–333
84. Almaden Y, Canalejo A, Hernandez A, Ballesteros E, Garcia-Navarro S, Torres A, Rodriguez M (1996) Direct effect of phosphorus on parathyroid hormone secretion from whole rat parathyroid glands in vitro. J Bone Miner Res 11:970–976

85. Nielsen PK, Feldt-Rasmusen U, Olgaard K (1996) A direct effect of phosphate on PTH release from bovine parathyroid tissue slices but not from dispersed parathyroid cells. Nephrol Dial Transplant 11:1762–1768

86. Slatopolsky E, Finch J, Denda M, Ritter C, Zhong A, Dusso A, MacDonald P, Brown AJ (1996) Phosphate restriction prevents parathyroid cell growth in uremic rats. High phosphate directly stimulates PTH secretion in vitro. J Clin Invest 97:2534–2540

87. Bourdeau A, Moutahir M, Souberbielle JC, Bonnet P, Herviaux P, Sachs C, Lieberherr M (1994) Effects of lipoxygenase products of arachidonate metabolism on parathyroid hormone secretion. Endocrinology 135:1109–1112

88. Bourdeau A, Souberbielle J-C, Bonnet P, Herviaux P, Sachs C, Lieberherr M (1992) Phospholipase-A_2 action and arachidonic acid in calcium-mediated parathyroid hormone secretion. Endocrinology 130:1339–1344

89. Almaden Y, Canalejo A, Ballesteros E, Anon G, Rodriguez M (2000) Effect of high extracellular phosphate concentration on arachidonic acid production by parathyroid tissue In vitro. J Am Soc Nephrol 11:1712–1718

90. Dinur M, Kilav R, Sela-Brown A, Jacquemin-Sablon H, Naveh-Many T (2006) In vitro evidence that upstream of N-ras participates in the regulation of parathyroid hormone messenger ribonucleic acid stability. Mol Endocrinol 20:1652–1660

91. Sela-Brown A, Silver J, Brewer G, Naveh-Many T (2000) Identification of AUF1 as a parathyroid hormone mRNA 3'-untranslated region binding protein that determines parathyroid hormone mRNA stability. J Biol Chem 275:7424–7429

92. Kilav R, Silver J, Naveh-Many T (2001) A conserved cis-acting element in the parathyroid hormone 3'-untranslated region is sufficient for regulation of RNA stability by calcium and phosphate. J Biol Chem 276:8727–8733

93. Levi R, Ben Dov IZ, Lavi-Moshayoff V, Dinur M, Martin D, Naveh-Many T, Silver J (2006) Increased parathyroid hormone gene expression in secondary hyperparathyroidism of experimental uremia is reversed by calcimimetics: correlation with posttranslational modification of the trans acting factor AUF1. J Am Soc Nephrol 17:107–112

94. Nechama M, Ben Dov IZ, Briata P, Gherzi R, Naveh-Many T (2008) The mRNA decay promoting factor K-homology splicing regulator protein post-transcriptionally determines parathyroid hormone mRNA levels. FASEB J 22:3458–3468

95. Naveh-Many T, Nechama M (2007) Regulation of parathyroid hormone mRNA stability by calcium, phosphate and uremia. Curr Opin Nephrol Hypertens 16:305–310

96. Kurosu H, Ogawa Y, Miyoshi M, Yamamoto M, Nandi A, Rosenblatt KP, Baum MG, Schiavi S, Hu MC, Moe OW, Kuro-o M (2006) Regulation of fibroblast growth factor-23 signaling by klotho. J Biol Chem 281:6120–6123

97. Takeshita K, Fujimori T, Kurotaki Y, Honjo H, Tsujikawa H, Yasui K, Lee JK, Kamiya K, Kitaichi K, Yamamoto K, Ito M, Kondo T, Iino S, Inden Y, Hirai M, Murohara T, Kodama I, Nabeshima Y (2004) Sinoatrial node dysfunction and early unexpected death of mice with a defect of klotho gene expression. Circulation 109:1776–1782

98. Urakawa I, Yamazaki Y, Shimada T, Iijima K, Hasegawa H, Okawa K, Fujita T, Fukumoto S, Yamashita T (2006) Klotho converts canonical FGF receptor into a specific receptor for FGF23. Nature 444:770–774

99. Silver J (2001) Cycling with the parathyroid. J Clin Invest 107:1079–1080

11 Diversity of Vitamin D Target Genes

Carsten Carlberg

Abstract   The vitamin D receptor (VDR) is a ligand-inducible transcription factor, whose target genes play key roles in cellular metabolism, bone formation, cellular growth, differentiation, and in controlling inflammation. Many of these VDR target genes are also involved in dysregulated pathways leading to common human diseases, such as cancer, osteoporosis, or the metabolic syndrome. The activation of VDR by natural and synthetic ligands may improve such pathological conditions. On a genomic level these pathways converge on regulatory modules, some of which contain VDR-binding sites, so-called vitamin D response elements (VDREs). Transcriptome analysis, chromatin immunoprecipitation scans and in silico screening approaches already identified many genomic targets of the VDR. Important regulatory modules with VDREs should have a major impact on understanding the role and potential therapeutic value of VDR and its ligands.

   Key Words: Vitamin D; vitamin D receptor; vitamin D response elements; in silico screening; gene regulation; nuclear receptor

Abbreviations

$1,25(OH)_2D$	$1\alpha,25$-dihydroxyvitamin D_3
3C	chromosome-conformation-capture
ALOX5	5-lipoxygenase
CDK	cyclin-dependent kinase
ChIP	chromatin immunoprecipitation
CCNC	cyclin C
CDKN1	p21
CYP24A1	24-hydroxylase
CYP27B1	25-dihydroxyvitamin D_3 1α-hydroxylase
DR	direct repeat
ER	everted repeat
HAT	histone acetyltransferase
IGFBP	insulin-like growth factor binding protein
MAR	matrix attachment region
PPAR	peroxisome proliferator-activated receptor
RXR	retinoid X receptor
TSS	transcription start site
VDR	vitamin D receptor
VDRE	vitamin D response element

From: *Nutrition and Health: Vitamin D*
Edited by: M.F. Holick, DOI 10.1007/978-1-60327-303-9_11,
© Springer Science+Business Media, LLC 2010

1. INTRODUCTION

The biologically active form of vitamin D, 1α,25-dihydroxyvitamin D_3 [1,25$(OH)_2D_3$], acts as a ligand to the transcription factor vitamin D receptor (VDR). In order to directly activate a gene by 1,25$(OH)_2$D at least one VDR molecule has to bind in sufficient vicinity to the gene's transcription start site (TSS) *(1)*. However, "vicinity" could in some cases be a distance of up to 100 kB, irrespective if upstream or downstream of the TSS. Moreover, there are a number of evidences that most primary VDR target genes use multiple VDR-binding sites, the so-called vitamin D response elements (VDREs), for their full functionality *(2)*. The complete sequence of the human genome and also that of other mammalian species, such as chimp, dog, mouse, and rat, is now available, so that we are able to screen for all putative VDREs. However, the constant packaging of genomic DNA into chromatin provides a repressive environment, which in most cases denies the access to putative VDREs *(3)*. Fortunately, new experimental techniques for genome-wide analyses of chromatin modifications and transcription factor binding, such as chromatin immunoprecipitation (ChIP)-chip and massively parallel sequencing, are now available *(4)*. This will revolutionize our understanding of the genome-wide effects of the VDR and of the diversity of 1,25$(OH)_2$D target genes as outlined in this chapter.

2. VDR IS A NUCLEAR RECEPTOR

2.1. *The Nuclear Receptor Superfamily*

Nuclear receptors are the best-characterized representatives of approximately 3,000 different mammalian proteins that are involved in transcriptional regulation in human tissues *(5)*. They form a superfamily with 48 human members, of which the most have the special property to be ligand-inducible *(6, 7)*. Nuclear receptors modulate genes that affect processes as diverse as reproduction, development, inflammation, and general metabolism. They can be classified based on ligand sensitivity *(6)*, evolution of the nuclear receptor genes *(8)*, and their physiological role as interpreted from tissue-specific expression patterns *(9)*.

The ligand sensitivity approach suggests three classes of nuclear receptors. Class I contains the endocrine receptors with high-affinity hormonal lipids, such as the receptors for the steroid hormones estradiol (estrogen receptors α and β), progesterone (progesterone receptor), testosterone (androgen receptor), cortisol (glucocorticoid receptor), and aldosterol (mineralocorticoid receptor), for thyroid hormones (thyroid hormone receptors α and β), and for the biologically active forms of the fat-soluble vitamins A and D, all-*trans* retinoic acid (retinoid acid receptors α, β, and γ) and 1,25$(OH)_2$D (VDR). In class II are adopted orphan receptors that bind to dietary lipids and xenobiotics with low affinity, such as peroxisome proliferator-activated receptors (PPARs) α, δ, and γ, constitutive androstane receptor, and pregnane X receptor. Finally, in class III orphan receptors are placed, such as estrogen-related receptors, for which a physiological ligand has not yet been identified.

When the sequences of nuclear receptors are compared on DNA and protein level, the grouping significantly differs from the ligand-centered view. For example, VDR is in the same group with PPARs and the highly ligand-sensitive estrogen receptors and the orphan estrogen-related receptors are together in another group.

On the basis of mRNA expression of all nuclear receptor genes in 39 different tissues in two different mouse strains, nuclear receptors are divided into clades with distinct physiological roles. In this classification, for example VDR is grouped to bile acid and xenobiotic metabolism based on its high expression in gastroentric tissues and PPARs are linked to lipid metabolism and energy homeostasis.

2.2. Modular Structure of Nuclear Receptors

Nuclear receptors have a modular structure, onto which certain functions can be ascribed. The amino-terminus is of variable length and sequence in the different family members. It contains a transactivation domain, termed AF-1, which is recognized by co-activator proteins and/or other transcription factors, often in a ligand-independent fashion. The central DNA-binding domain has two zinc-finger motifs that are common to the entire family. The carboxy-terminal ligand-binding domain, whose overall architecture is well conserved between the various family members, nonetheless diverges sufficiently to guarantee selective ligand recognition as well as accommodate the broad spectrum of nuclear receptor ligand structures. The ligand-binding domain consists of 250–300 amino acids in 11–13 α-helices (10). Ligand binding causes a conformational change within the ligand-binding domain, whereby, at least in the case of endocrine nuclear receptors, helix 12, the most carboxy-terminal α-helix (also called AF-2 domain), closes the ligand-binding pocket via a "mouse-trap like" intramolecular folding event (11). The ligand-binding domain is also involved in a variety of interactions with nuclear proteins, such as other members of the nuclear receptor superfamily and co-regulator proteins.

2.3. The VDR

VDR is an endocrine member of the nuclear receptor superfamily (7), because it is the only nuclear protein that binds 1,25(OH)$_2$D with high affinity ($K_d = 0.1$ nM) (12, 13). VDR has been shown to form homodimers (14, 15) and heterodimers with thyroid hormone receptors (16, 17) and retinoid acid receptors (18), but by far the strongest binding partner of VDR is one of the three retinoid X receptors (RXRs) α, β, and γ (19). In mammals, the highest VDR expression is found in metabolic tissues, such as intestine and kidney, as well as in skin and the thyroid gland, but moderate expression is found in nearly all tissues (9). Moreover, the receptor is also expressed in many malignant tissues (20). Mice lacking a functional VDR gene develop alopecia (likewise found in many patients with mutations in the VDR) (21); these mice also exhibit a defect in epidermal differentiation. Moreover, VDR-null mice also show an increased susceptibility to tumor formation (22). More details on the receptor have been provided in the previous chapter.

3. VDR-BINDING SITES

3.1. DR3-Type VDREs

The binding of VDR–RXR heterodimers is achieved through the specific binding of the DNA-binding domain of the VDR to the major grove of a hexameric DNA sequence,

RXR VDR

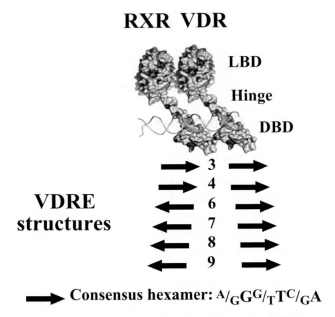

LBD

Hinge

DBD

VDRE structures

3
4
6
7
8
9

➡ **Consensus hexamer:** $^A/_GG^G/_TTC^C/_GA$

Fig. 1. VDR–RXR heterodimer on VDREs: VDR–RXR heterodimer binding to different types of VDREs is schematically depicted.

referred to as core-binding motif, with the consensus sequence RGKTSA (R = A or G, K = G or T, S = C or G) *(13)* (Fig. 1). Numerous studies (for example *(14, 23)*) have confirmed Umesomo's suggestion *(24)* that VDR–RXR heterodimers bind well to response elements (REs) that are formed by a direct repeat of the hexameric sequences with three spacing nucleotides. These DR3-type REs are therefore widely accepted as the classical structure of a VDRE.

Every $1,25(OH)_2D$ target gene has to contain at least one VDRE in its promoter region and the first VDREs have been identified rather close to the TSS of the genes. The strongest DR3-type VDREs has been identified within the *rat atrial natriuretic factor* promoter *(25)*, the mouse and pig *osteopontin* promoter *(26, 27)*, and the chicken *carbonic anhydrase II* promoter *(28)*. A number of other DR3-type VDREs have been published and were based on their in vitro binding affinity categorized into different classes *(29)*. The REs with the lowest affinity show a significant deviation from the RGKTSA consensus and may not be functional. However, it is possible that these VDREs may gain responsiveness to $1,25(OH)_2D$ in their natural promoter context through the help of flanking partner proteins (see also Section 3.4.). Moreover, the functionality of a $1,25(OH)_2D$ responding gene will also depend on a potential cooperative action of two or more VDREs, such as in the case of the *24-hydroxylase (CYP24A1)* gene *(30)*. Nevertheless, VDR–RXR heterodimers form identical complexes on all DR3-type VDREs and display no significant differences in their interaction with a given co-activator or co-repressor protein *(29)*.

3.2. Other Types of VDREs

There are also other VDRE structures, such as direct repeats with four intervening nucleotides (DR4). Effective VDR binding has also been observed on everted repeat (ER)-type REs with 6–9 spacing nucleotides (ER6, ER7, ER8, ER9) *(17, 31)* (Fig. 1). A VDRE classification according to the affinity for VDR–RXR heterodimers suggests that the degree of deviation from the core-binding motif consensus sequence RGKTSA *(32)* is proportional to the loss of in vitro functionality *(29)*. Interestingly, the DR4-type RE of the rat *pit1* gene *(33)*, which contains two perfect core-binding motifs, was found to be stronger than any known natural DR3-type VDRE *(29)*. However, one has to consider that a DR4-type REs is also recognized by the heterodimeric complexes of thyroid hormone receptor, constitutive active receptor and pregnane X receptor and other orphan nuclear receptors with RXR *(19, 34)*, whereas the same complexes bind to DR3-type REs less tightly than VDR–RXR heterodimers. The competitive situation on DR4-type REs may therefore be the reason why in vivo VDR–RXR heterodimers still prefer DR3-type REs. Moreover, VDR–RXR heterodimers bind to DR4-type REs in the same conformation as to DR3-type REs *(34)*, i.e., there seem to be no differential action of VDR on these elements due to a differential complex formation with RXR.

The VDRE of the human *osteocalcin* promoter was the first identified natural binding site for the VDR *(35, 36)*. It was initially described as a DR6-type structure but later a third cryptic hexamer was identified at a distance of 3 nucleotides, so that the whole VDRE is more likely a complex DR6/DR3-type RE (see Section 3.4.). The DR6 part of the VDRE has been shown to bind VDR homodimers *(14)* and VDR–RAR heterodimers *(18)*, whereas the DR3 part weakly binds VDR–RXR heterodimers. Other examples of a DR6-type VDREs that bind VDR homodimers and VDR–RAR heterodimers have been identified in the promoters of mouse *fibronectin (37)*, rat *CYP24A1 (38)*, and human *phospholipase C (39)*. Their functionality remain to be determined.

3.3. Chromatin and Co-factors

The major protein constituents of chromatin are the four different histones that form a nucleosome, around which DNA is wound. Covalent modifications of the lysines at the amino-terminal tails of these histone proteins neutralize their positive charge and thus their attraction for the negatively charged DNA backbone is diminished *(40)*. As a consequence, the association between the histone and the DNA becomes less stable. This influences the degree of chromatin packaging and regulates the access of transcription factors to their potential binding sites. When nuclear receptors are bound to REs in the regulatory regions of their target genes, they recruit positive and negative co-regulatory proteins, referred to as co-activators *(41)* and co-repressors *(42)*, respectively.

In a simplified view of nuclear receptor signaling, in the absence of ligand, the nuclear receptor interacts with co-repressor proteins, such as NCoR1, SMRT, hairless, and Alien, which in turn associate with histone deacetylases leading to a locally increased chromatin packaging *(43, 44)*. The binding of ligand induces the dissociation of the co-repressor and the association of a co-activator of the p160 family, such as SRC-1, TIF2, or RAC3 *(45)*. Some co-activators have histone acetytransferase (HAT) activity

or are complexed with proteins harboring such activity and this results in the net effect
of local chromatin relaxation *(46)*. In a subsequent step, ligand-activated nuclear receptors change rapidly from interacting with the co-activators of the p160 family to those
of the mediator complex, such as Med1 *(47)*. The mediator complex, which consists
of approximately 15–20 proteins, builds a bridge to the basal transcriptional machinery
(48). In this way ligand-activated nuclear receptors execute two tasks, the modification
of chromatin and the regulation of transcription.

Cell- and time-specific patterns of relative protein expression levels of some coregulators can distinctly modulate nuclear receptor transcriptional activity. This aspect
may have some diagnostic and therapeutic value in different types of cancer *(49)*. Concerning skin cancer it was postulated that the stoichiometric ratio between co-activators
of the p160 family and Med1 might regulate a $1,25(OH)_2D$-dependent balance between
proliferation and differentiation of keratinocytes *(50)*. However, the switch between
gene repression and activation is more complex than a simple alternative recruitment
of two different regulatory complexes *(51)*. Most co-regulators are co-expressed in the
same cell type at relatively similar levels, which raises the possibility of their concomitant recruitment to a specific promoter. This has been resolved by the mutually exclusive
binding of co-activators and co-repressors to ligand-bound and ligand-unbound nuclear
receptors, respectively. Therefore, repression and activation are more likely achieved
by a series of sequential multiple enzymatic reactions that are promoter and cell-type
specific. Transcriptional regulation is a highly dynamic event of rapid association and
dissociation of proteins and their modifications, including proteolytic degradation and
de novo synthesis. A pattern of recruitment and release of cohorts of co-regulatory complexes was demonstrated on a single region of the *trefoil factor-1* promoter in breast
cancer cells *(52)*. This study revealed detailed and coordinated patterns of co-regulator
recruitment and preferential selectivity for factors that have similar enzymatic activities.
Interestingly, similar cyclic behavior was also observed for the VDR *(53)*.

3.4. VDREs in the Chromatin Context

It is assumed that matrix attachment regions (MARs) subdivide genomic DNA into
units of an average length of 100 kB containing the coding region of at least one gene
(54) (Fig. 2). DNA looping should be able to bring any DNA site within the same
chromatin unit close to the basal transcriptional machinery that is assembled on the
TSS. This model suggests that also very distant sequences can serve as VDREs and
that even sequences downstream of the TSS could serve as functional VDR-binding
sites.

Due to its optimized 5'-flanking dinucleotide and core-binding motif sequences the
DR4-type RE of the rat *pit-1* gene is the most efficient known VDRE in vitro *(29, 34)*.
However, the chromatin in the region of the *pit-1* gene promoter containing this RE
seems to be closed in the adult rat, so that the responsiveness of the gene to $1,25(OH)_2D$
is lower than expected *(46)*. This indicates that a high in vitro binding affinity of VDR–
RXR heterodimers for a VDRE is not sufficient for responsiveness to $1,25(OH)_2D$.
When the promoter region that contains the VDRE is covered by condensed chromatin,
VDR–RXR heterodimers are unable to bind there. This makes sufficiently decondensed

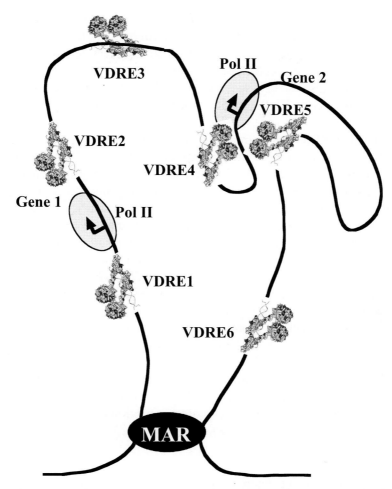

Fig. 2. 1,25(OH)$_2$D responsive chromatin units: A chromatin unit carrying two genes and six VDREs is schematically depicted. The loop may be formed by some 100 kB of genomic sequence, which is kept together by a matrix attachment region (MAR) complex. For clarity the some 500 nucleosome covering this genomic region are not shown. In this scenario VDREs 4 and 5 do sub-looping to the basal transcriptional machinery (Pol II) on the TSS (arrow) of gene 2 and induce transcription. Similarly all six VDREs have the potential to activate genes 1 and 2, i.e., both genes are supposed to be primary VDR target genes.

chromatin an essential prerequisite for a functional VDRE. Chromatin decondensation is achieved by the activity of HATs, which are recruited to their local chromatin target by co-activator proteins. In turn, these co-activators are transiently attracted to a promoter region by ligand-activated nuclear receptors and other active transcription factors. Therefore, the more transcription factor binding sites a given genomic region has and the more of these transcription factors are expressed in the respective cell, the higher is the chance that this area of the promoter gets locally decondensed. One example is the VDRE of the rat *osteocalcin* gene, which is flanked on both sides with a binding site for the transcription factor Runx2/Cbfa1 *(55)*. By contacting co-activator proteins and HATs Runx2/Cbfa1 seems to mediate the opening of chromatin locally, which allows efficient binding of VDR–RXR heterodimers to this decondensed region to occur. This mechanism suggests that VDREs are better targets for VDR–RXR heterodimers, if other transcription factors are bound to the same chromatin region. In this respect, promoter context and cell-specific expression of other transcription factors may be of greater importance to VDRE functionality and specificity than its in vitro binding profile.

The DR6/DR3 and ER9/DR3 structures of the VDREs of the human and rat *osteocalcin* genes, respectively, are the first examples of complex VDREs. These two complex VDREs show only limited homology to each other, although they are derived from orthologous genes. This suggests that for an important primary 1,25(OH)$_2$D responding gene, such as *osteocalcin*, there may be limited evolutionary pressure for a specific VDRE structure. It seems to be more important to guarantee an efficient binding of the VDR to a promoter in competition with the tight packaging of nucleosomes. It is interesting to note that the complex VDRE of the human *osteocalcin* gene is overlaid by a binding site of the transcription factor AP-1 *(56)*, which provides the RE with an increased activity. These types of REs are also observed for other nuclear receptors and often referred to as composite REs *(57)*. Another interesting example of a complex/composite VDRE has been reported in the mouse c-*fos* promoter *(58)*. Within this VDRE three hexameric core-binding motifs are forming a DR7/DR7 structure, which contains an internal binding site for the transcription factor NF-1. Additional examples are the VDRE of the human and mouse *fibronectin* gene, which contains an internal binding site for the transcription factor CREB *(37)*, or the VDRE of the rat *bone sialo protein*, which also seems to bind the general transcription factor TBP *(59)*.

3.5. Negative VDREs

Expression profiling using microarray technology indicates that comparable numbers of genes are down-regulated by 1,25(OH)$_2$D as are up-regulated by the hormone *(60)*. In general, the mechanisms of the down-regulation of genes by 1,25(OH)$_2$D are much less understood, but they also seem to require the binding of an agonist to the VDR. It is obvious that only genes, which show basal activity, can be down-regulated, i.e., these genes exhibit basal activity due to other transcription factors binding to their promoter. There are several different models that attempt to explain, how 1,25(OH)$_2$D and the VDR can mediate down-regulation of genes, but the common theme is that VDR counteracts the activity of specific transcription factors. For the physiologically important down-regulation of the *25-dihydroxyvitamin D$_3$ 1α-hydroxylase* (*CYP27B1*) gene by 1,25(OH)$_2$D a negative VDRE located at position –0.5 kB has been proposed, where VDR–RXR heterodimers do not bind directly, but via the transcription factor VDR-interacting repressor (VDIR, also called TCF3) *(61)*. In addition, two positive VDREs are located –2.6 and –3.2 kB upstream from the TSS and modulate the cell-specific activity of the negative VDRE *(62)*. Association of VDR–RXR heterodimers to TCF3-binding sites may also occur through ligand-dependent chromatin looping from more distal regions that directly bind the VDR *(62)*. In situations where these activating transcription factors are other nuclear receptors or transcription factors that bind to composite nuclear receptor REs, VDR could simply compete for DNA-binding sites *(63, 64)*. In a similar way, VDR could also compete for binding to partner proteins, such as RXR, or for common co-activators, such as SRC-1 or CBP *(65)*. In all these situations the down-regulating effects of the VDR should be of general impact, i.e., the mechanism could apply to other genes in the same way. So far, however, no general down-regulating effects of $1,25(OH)_2D$ have been reported.

The concept that multiple VDREs (see also Fig. 2) together with other transcription factor binding sites regulate primary $1,25(OH)_2D$ responding genes (see Section 5.2.) suggests that a promoter may contain both negative and positive VDREs. The activities of the different VDREs are determined by the promoter context and may not be simultaneously active. One might imagine that prior to stimulation with $1,25(OH)_2D$ only the negative VDREs bind the VDR and recruit co-repressors. This would actively condense the chromatin on a particular promoter region. The addition of ligand induces the release of co-repressor proteins and reduces chromatin density. The VDR may then be transiently released from the negative VDRE and bind to a positive VDRE, which may be uncovered through $1,25(OH)_2D$-dependent local nucleosome acetylation. The VDR then interacts with the mediator protein complex on this positive VDRE leading to transient transcriptional activation. After a certain period of time, newly synthesized, unliganded VDR again binds to the negative VDRE, which initiates chromatin closing and inactivation of the positive VDRE *(62)*. In this or even more complex scenarios, the balance between negative and positive VDREs could explain the time course of the activation of primary $1,25(OH)_2D$ responding genes.

4. VDR TARGET GENES

4.1. Classical VDR Targets

The most striking effect of severe vitamin D deficiency is rickets. Rickets can also by inflicted by mutations in the *CYP27B1* or the *VDR* gene. $1,25(OH)_2D$ is essential for adequate Ca^{2+} and P_i absorption from the intestine and hence for bone formation *(66)*. Liganded VDR has been shown to induce expression of the gene encoding for the major Ca^{2+} channel in intestinal epithelial cell, *TRPV6*, by direct binding to a VDRE at -1.2 kB from the TSS *(67)*. The phosphate co-transporter *NaP(i)-IIb* gene was also found to be induced by $1,25(OH)_2D$ but no VDREs have yet been identified for this gene *(68)*. $1,25(OH)_2D$ also down-regulates the expression of the *PTH* gene that opposes $1,25(OH)_2D$ in regulation of serum Ca^{2+} and P_i levels, but up-regulates the *FGF23* gene, which, like PTH, lowers serum P_i levels *(69)*. The induction of the *RANKL* gene by liganded VDR via multiple distant VDREs (up to 70 kB from the TSS) leads to stimulation of osteoclast precursors to fuse and form new osteoclasts, resulting in enhanced resorption of bone *(70)*.

4.2. VDR Targets in Cell Cycle Regulation

The main anti-proliferative effect of $1,25(OH)_2D$ on cells is a cell cycle block at the G_1 phase. This can be explained by changed expression of multiple cell cycle regulator genes. Among the first targets described, expression of cyclin-dependent kinase inhibitors (CDKIs) *p21* (*CDKN1A*) and *p27* were found to be up-regulated by ligand treatment *(71, 72)*. For the *CDKN1A* gene a VDRE in the proximal promoter was characterized, thus establishing *CDKN1A* as a direct 1,25(OH)$_2$D target gene *(73)*. Later on it has been questioned, whether *CDKN1A* truly is a primary target or a secondary target via up-regulation of TGFβ or of the insulin-like growth factor binding protein (IGFBP) 3, and whether the described VDRE is truly functional *(74, 75)*. However, by screening 7 kB of the *CDKN1A* promoter with overlapping ChIP regions three novel regions with 1,25(OH)$_2$D-induced VDR enrichment, two of which harvested also p53 were identified *(76)*. The direct role of the characterized 1,25(OH)$_2$D-responsive regions on regulation of *CDKN1A* expression is illustrated by the their association with VDR-RNA polymerase II complexes as well as by their ligand-dependent looping to the *CDKN1A* TSS *(51)*. Also other CDKIs, such as *p15, p16, p18,* and *p19,* show transcriptional response to 1,25(OH)$_2$D, but for *p16* the response is secondary as it can be blocked by inhibition of protein synthesis *(73, 77)*. In addition, *cyclin E, cyclin D1,* and *CDK2* were found to be down-regulated by 1,25(OH)$_2$D *(74)*. It remains to be elucidated, whether these effects are primary and occur via functional VDREs on regulatory regions of these genes.

Another interesting 1,25(OH)$_2$D-target gene is *cyclin C* (*CCNC*). The cyclin C-CDK8 complex was found to be associated with the RNA polymerase II basal transcriptional machinery *(78)* and is considered as a functional part of those mediator protein complexes that are involved in gene repression *(79)*. The fact that the *CCNC* gene, being located in chromosome 6q21, is deleted in a subset of acute lymphoblastic leukemias, suggests its involvement in tumorigenesis *(80)*. In addition, *GADD45A* and members of the *IGFBP* gene family respond to 1,25(OH)$_2$D *(81, 82)*. *GADD45A* plays an essential role in DNA repair and GADD45 proteins displace cyclin B1 from Cdc2 and thus inhibit the formation of M phase-promoting factor that is essential for G$_2$/M transition *(83)*. *GADD45A* has been shown to be a direct transcriptional target of 1,25(OH)$_2$D with a functional VDRE within the fourth exon of the gene *(84)*. IGFBPs modulate the activity of the circulating insulin-like growth factors (IGF) I and II. The *IGFBP-3* gene was first discovered to be up-regulated by 1,25(OH)$_2$D and contains a functional VDRE *(85)*. Later also *IGFBP-1* and *IGFBP-5* have been characterized as primary 1,25(OH)$_2$D target genes *(82)*. Another interesting primary 1,25(OH)$_2$D target is the *PPARD* gene, which carries a potent DR3-type VDRE in close proximity to its TSS *(86)*. PPARδ and VDR proteins are widely expressed and an apparent overlap in the physiological action of the two nuclear receptors is their involvement in the regulation of cellular growth, particularly in neoplasms. High *PPARD* expression in tumor seems to be positive for the prognosis of the respective cancer *(87)*.

Overall, 1,25(OH)$_2$D restricts cell cycle progression in several phases via multiple and partially redundant targets on parallel pathways, that when combined, provide robustness for its anti-proliferative effect.

4.3. Relative Expression of VDR Target Genes

The steady-state mRNA expression levels of some VDR target genes, such as that of the *CYP24A1* gene, are very low in the absence of ligand, but are up to 1,000-fold induced by stimulation with 1,25(OH)$_2$D *(88)*. Most other known primary 1,25(OH)$_2$D target genes, such as *CCNC* and *CDKN1A*, often show an inducibility of twofold or less after short-term treatment with 1,25(OH)$_2$D *(89, 90)*. However, both genes have 10,000- to 100,000-fold higher basal expression levels compared to that of the *CYP24A1* gene. Therefore, when the relative levels are taken into account, 2- to 20-fold more *CCNC* and *CDKN1A* than *CYP24A1* mRNA molecules are produced after induction with 1,25(OH)$_2$D.

5. VDR TARGET GENE ANALYSIS

5.1. Transcriptome Analysis

There are a number of modern methods for the identification and characterization of VDR target genes. The effect of 1,25(OH)$_2$D on the mRNA expression, i.e., 1,25(OH)$_2$D-induced changes of the transcriptome, has been assayed by multiple microarray experiment in cellular models (either an established cell line or primary cells) or in in vivo models (mostly rodents). In case the focus is on the identification of primary VDR target genes, the stimulation times are short (2–6 h), but when the overall physiological effects are the center of the study, longer treatment times are used (24–72 h). For a limited number of putative VDR target genes quantitative PCR can be applied, but for a whole genome perspective on VDR signaling, microarrays have to be used.

A few years ago mostly cDNA arrays with an incomplete number of genes were used and rather short lists of VDR target genes from colon *(90)*, prostate *(49, 91–93)*, breast *(89)*, and osteoblasts *(94, 95)*. In squamous cell carcinoma more than 900 genes respond to 1,25(OH)$_2$D after 12 h treatment in the presence of the protein synthesis inhibitor cycloheximide *(96)*. However, the number of overlapping VDR target genes in these lists was low. Since the setups of these microarray analyses were different in treatment times and probe sets, this suggests that most VDR target genes reply to 1,25(OH)$_2$D in a very tissue-specific fashion and may have only a rather transient response. However, on the basis of these results the total number of convincing primary 1,25(OH)$_2$D target genes is in the order of 250. In addition, secondary 1,25(OH)$_2$D-responding genes contribute to the physiological effects of 1,25(OH)$_2$D, but their induction is delayed by a few hours or even days and are probably mediated by primary 1,25(OH)$_2$D-responding gene products, such as transcription factors or co-regulator proteins *(90)*. For a more detailed meta-analysis of VDR target genes, standardized microarray procedures performed on whole genome chips from Affymetrix, Illumina or other commercial suppliers are essential. Results from such approaches will be published soon.

5.2. ChIP Analysis

For a detailed analysis of the regulatory regions of primary VDR target genes since a couple of years the method of ChIP became the golden standard. For the genes *CYP24A1 (88)*, *CYP27B1 (62)*, *CCNC (97)*, and *CDKN1A (76)* 7.1–8.4 kB of their promoter regions were investigated by using in each case a set of 20–25 overlapping genomic region. The spatio-temporal, 1,25(OH)$_2$D-dependent chromatin changes in the four gene promoters were studied by ChIP assays with antibodies against acetylated histone 4, VDR, RXR, and RNA polymerase II. Promising promoter regions were then screened *in silico* for putative VDREs, whose functionality was analyzed sequentially with gel shift, reporter gene, and re-ChIP assays. This approach identified four VDREs for both the *CYP24A1* and *CCNC* genes, three in the *CDKN1A* promoter and two in the *CYP27B1* gene. However, most of them are simultaneously under the control of other transcription factors, such as p53 in case of the *CDKN1A* gene *(76)*, and therefore possess significant basal levels of transcription.

An alternative approach to the identification of primary 1,25(OH)$_2$D target genes was performed with the six members of the *IGFBP* gene family. Here first an *in silico* screen was performed, which was then followed by the analysis of candidate 1,25(OH)$_2$D-responsive sequences by gel shift, reporter gene, and re-ChIP assays *(82)*. Induction of gene expression was confirmed independently using quantitative PCR. By using this approach, the genes *IGFBP1, IGFBP3,* and *IGFBP5* were demonstrated to be primary 1,25(OH)$_2$D target genes. The *in silico* screening of the 174 kB of genomic sequence surrounding all six *IGFBP* genes identified 15 candidate VDREs, 10 of which were shown to be functional in ChIP assays. Importantly, the *in silico* screening approach was not restricted to regulatory regions that comprise only maximal 2 kB of sequence up- and downstream of the TSS, as in a recent whole genome screen for regulatory elements *(98)*, but involved up to 10 kB of flanking sequences as well as intronic and intergenic sequences. In a similar approach the *5-lipoxygenase (ALOX5)* gene has been analyzed and confirmed to be a primary 1,25(OH)$_2$D target gene. From the 22 putative VDREs identified in the whole *ALOX5* gene sequence (–10 kB to +74 kB) by *in silico* screening, at least two have been validated to be functional in vitro and in the living cells. One of these VDREs is located far downstream of the TSS (+42 kB) and is one of the strongest known VDREs of the human genome *(99)*. No functional VDRE had been reported for the *ALOX5* gene before, since previous studies had been restricted to the proximal promoter region *(100, 101)*. Therefore, this approach revealed candidate VDREs that are located more than 30 kB distant from their target gene's TSS. Based on the present understanding of enhancers, DNA looping and chromatin units being flanked by insulators or MARs, these distances are not limiting *(102)*.

5.3. ChIP-Chip Assays

The combination of ChIP assays with hybridization of the resulting chromatin fragments on microarrays, the so-called ChIP-chip assays, provides an additional step for a larger scale analysis of VDR target genes. The ChIP-chip technology has been applied for the analysis of the *VDR* gene itself *(103)*, the intestinal calcium ion channel gene *TRPV6 (67)*, the Wnt signaling co-regulator *LRP5 (104),* and the TNF-like factor *RANKL* that promotes the formation of calcium-resorbing osteoclasts *(105)*. For all those genes a number of VDR-associated chromatin regions were identified, some of which were far upstream of the gene's TSS. These studies confirmed that many, if not all, VDR target genes have multiple VDR-associated regions. However, not all of these VDREs may be functional, i.e., they may not contact the gene's TSS via DNA looping. Therefore, it is necessary to apply an additional method, the so-called chromosome-conformation-capture (3C) assay. So far, 3C assays confirmed the functionality of the VDREs in the *CYP27B1 (62)* and the *CDKN1A (51)* genes.

The next step in genome-wide association studies will be massively parallel sequencing of genomic fragments obtained after ChIP assays, also referred to as ChIP-Seq, with antibodies against VDR and its partner proteins. Results will be published soon.

5.4. In Silico Screening of VDREs

The specificity of VDR for its DNA-binding sites allows constructing a model to describe the VDRE properties that can be used to predict potential binding sites in genomic sequences. For this the VDR-binding preference, often expressed as position weight matrix, has to be described on the basis of experimental data, such as series of gel shift assays with a large number of natural binding sites *(106–109)*. However, VDR–RXR heterodimers do not only recognize a pair of the nuclear receptor consensus binding motifs AGGTCA, but also a number of variations to it. Dependent on the individual position matrix description this leads to a prediction of VDREs every 1,000–10,000 bp of genomic sequence. This probably contains many false-positive predictions, which is mainly due to scoring methodology and the limitations that are imposed by the available experimental data. Wang et al. combined microarray analysis and *in silico* genome-wide screens for DR3- and ER6-type VDREs *(96)*. This approach identified several novel VDREs and VDR target genes, but most of the VDREs await a confirmation by ChIP and 3C assays.

In a position frequency matrix the quantitative characteristics of a transcription factor, i.e., its relative binding strength to a number of different binding sites, are neglected, since simply the total number of observations of each nucleotide is recorded for each position. Moreover, in the past there was a positional bias of transcription factor binding sites upstream in close vicinity to the TSS. This would be apparent from the collection of identified VDREs *(13)*, but is in contrast with a multi-genome comparison of nuclear receptor binding site distribution *(98)* and other reports on wide-range associations of distal regulatory sites *(110)*.

Internet-based software tools, such as TRANSFAC *(111)*, screen DNA sequences with databases of matrix models. The accuracy of such methods can be improved by taking the evolutionary conservation of the binding site and that of the flanking genomic region into account. Moreover, cooperative interactions between transcription factors, i.e., regulatory modules, can be taken into account by screening for binding site clusters. The combination of phylogenetic footprinting and position weight matrix searches applied to orthologous human and mouse gene sequences reduces the rate of false predictions by an order of magnitude, but leads to some reduction in sensitivity *(112)*. Recent studies suggest that a surprisingly large fraction of regulatory sites may not be conserved but yet are functional, which suggests that sequence conservation revealed by alignments may not capture some relevant regulatory regions *(113)*.

The recently published classifier method for the *in silico* screening of transcription factor binding sites *(114)* showed at the example of PPAR–RXR heterodimers, how a set of in vitro binding preferences of the three PPAR subtypes can be used as an experimental data set. Single nucleotide variants were sorted into three classes, where in class I the PPAR subtypes are able to bind the sequence with a strength of $75 \pm 15\%$ of that of the consensus PPAR-binding site, in class II with $45 \pm 15\%$, and in class III with $15 \pm 15\%$. Additional 130 PPREs were sorted on the basis of counting increasing number of variations from the consensus and taking into account the single nucleotide variant binding strength. Those variants that alone decrease the binding only modestly (class I) could be combined with even three deviations from consensus still resulting in more than 20% binding relative to consensus. Other combinations resulted in faster loss of binding detailed in 11 categories, where such combinations still resulted in more than 1% relative binding. The main advantage, when comparing the classifier to position weight matrix methods, is a clear separation between weak PPREs and those of medium and strong strength *(114)*. With this method the gene-dense human chromosome 19 (63.8 MB, 1,445 known genes) and its syntenic mouse regions (956 genes have known orthologs) were screened. Twenty percent of genes of chromosome 19 were found to contain a strong RE and additional 4% have more than two medium REs or one proximal medium RE. These numbers suggest a total of 4,000–5,000 targets for PPARs in the human genome. Presently, the same approach is used for a genome-wide screening of VDREs and the results will be published soon. First results already indicated that the number of putative VDR target genes is in the same order as that of possible PPAR targets. Certainly not all sites will be accessible and the human genome also contains weak binding sites that could gain function via interaction with other transcription factors.

In effect, these approaches and tools are still insufficient and there has to be a focus on the creation of bioinformatics resources that include more directly the biochemical restraints to regulate gene transcription. One important aspect is that most putative transcription factor binding sites are covered by nucleosomes, so that they are not accessible to the transcription factor. This repressive environment is found in particular for those sequences that are either contained within interspersed sequences, are located isolated from transcription factor modules, or lie outside of insulator sequences marking the border of chromatin loops *(115)*. This perspective strongly discourages the idea that isolated, simple VDREs may be functional in vivo. In turn, this idea implies that the more transcription factor binding sites a given promoter region contains and the more of these transcription factors are expressed, the higher is the chance that this area of the promoter becomes locally decondensed.

6. CONCLUSIONS

The sequencing of the complete human genome and the genome of other species, i.e., the availability of all regulatory sequences, enable a more mature understanding of the diversity of $1,25(OH)_2D$ target genes. Perhaps the idea of simple isolated VDREs should shift to the concept of complex VDREs, of which a simple DR3-, DR4- or ER-type VDRE represents the core. Depending on the temporal presence of cell-specific transcription factors, these complex REs may act positively or negatively in respect to $1,25(OH)_2D$. The coordinated action of these different types of VDREs could explain the individual response of target genes to $1,25(OH)_2D$.

Methods incorporating both experimental- and informatics-derived evidence to arrive at a more reliable prediction of VDR targets and binding modules can bring all available data together with the aim to predict the outcome in a specific context. We envision that in future the emphasis will shift from target genes to target regulatory modules to alter a physiological response and from individual genes to whole genome response. Therefore, a much larger challenge lies ahead when we would be confronted with the higher order of regulated networks of genes, where the sum effect of ligand treatment may reveal itself. In an effort to study this, we have started applying systems biology to the field of nuclear receptor biology, through an EU-funded Marie Curie Research Training Network, NUCSYS (www.uku.fi/nucsys).

ACKNOWLEDGMENTS

The University of Luxembourg, the Academy of Finland, the Juselius Foundation, and the EU (Marie Curie RTN NucSys) supported the work.

REFERENCES

1. Carlberg C, Dunlop TW (2006) The impact of chromatin organization of vitamin D target genes. Anticancer Res 26:2637–2645
2. Carlberg C, Dunlop TW, Saramäki A, Sinkkonen L, Matilainen M, Väisänen S (2007) Controlling the chromatin organization of vitamin D target genes by multiple vitamin D receptor binding sites. J Steroid Biochem Mol Biol 103:338–343
3. Demeret C, Vassetzky Y, Mechali M (2001) Chromatin remodelling and DNA replication: from nucleosomes to loop domains. Oncogene 20:3086–3093
4. Valouev A, Johnson DS, Sundquist A et al (2008) Genome-wide analysis of transcription factor binding sites based on ChIP-Seq data. Nat Methods
5. Maglich JM, Sluder A, Guan X et al (2001) Comparison of complete nuclear receptor sets from the human, *Caenorhabditis elegans* and *Drosophila* genomes. Genome Biol 2:0029
6. Chawla A, Repa JJ, Evans RM, Mangelsdorf DJ (2001) Nuclear receptors and lipid physiology: opening the X-files. Science 294:1866–1870
7. Nuclear-Receptor-Committee (1999) A unified nomenclature system for the nuclear receptor superfamily. Cell 97:161–163
8. Bertrand S, Brunet FG, Escriva H, Parmentier G, Laudet V, Robinson-Rechavi M (2004) Evolutionary genomics of nuclear receptors: from twenty-five ancestral genes to derived endocrine systems. Mol Biol Evol 21:1923–1937
9. Bookout AL, Jeong Y, Downes M, Yu RT, Evans RM, Mangelsdorf DJ (2006) Anatomical profiling of nuclear receptor expression reveals a hierarchical transcriptional network. Cell 126:789–799
10. Rochel N, Wurtz JM, Mitschler A, Klaholz B, Moras D (2000) Crystal structure of the nuclear receptor for vitamin D bound to its natural ligand. Mol Cell 5:173–179
11. Moras D, Gronemeyer H (1998) The nuclear receptor ligand-binding domain: structure and function. Curr Opin Cell Biol 10:384–391
12. Bikle DD, Gee E, Pillai S (1993) Regulation of keratinocyte growth, differentiation, and vitamin D metabolism by analogs of 1,25-dihydroxyvitamin D. J Invest Dermatol 101:713–718
13. Carlberg C, Polly P (1998) Gene regulation by vitamin D_3. Crit Rev Eukaryot Gene Expr 8:19–42
14. Carlberg C, Bendik I, Wyss A et al (1993) Two nuclear signalling pathways for vitamin D. Nature 361:657–660
15. Cheskis B, Freedman LP (1994) Ligand modulates the conversion of DNA-bound vitamin D_3 receptor (VDR) homodimers into VDR-retinoid X receptor heterodimers. Mol Cell Biol 14:3329–3338
16. Schräder M, Carlberg C (1994) Thyroid hormone and retinoic acid receptors form heterodimers with retinoid X receptors on direct repeats, palindromes, and inverted palindromes. DNA Cell Biol 13: 333–341

17. Schräder M, Müller KM, Nayeri S, Kahlen JP, Carlberg C (1994) VDR-T$_3$R receptor heterodimer polarity directs ligand sensitivity of transactivation. Nature 370:382–386
18. Schräder M, Bendik I, Becker-Andre M, Carlberg C (1993) Interaction between retinoic acid and vitamin D signaling pathways. J Biol Chem 268:17830–17836
19. Mangelsdorf DJ, Evans RM (1995) The RXR heterodimers and orphan receptors. Cell 83:841–850
20. Norman AW (1998) Receptors for 1,25(OH)$_2$D: past, present, and future. J Bone Miner Res 13:1360–1369
21. Kato S, Takeyama K-I, Kitanaka S, Murayama A, Sekine K, Yoshizawa T (1999) In vivo function of VDR in gene expression-VDR knock-out mice. J Steroid Biochem Mol Biol 69:247–251
22. Welsh J (2004) Vitamin D and breast cancer: insights from animal models. Am J Clin Nutr 80: 1721S–1724S
23. MacDonald PN, Dowd DR, Nakajima S et al (1993) Retinoid X receptors stimulate and 9-*cis* retinoic acid inhibits 1,25-dihydroxyvitamin D$_3$-activated expression of the rat osteocalcin gene. Mol Cell Biol 13:5907–5917
24. Umesono K, Murakami KK, Thompson CC, Evans RM (1991) Direct repeats as selective response elements for the thyroid hormone, retinoic acid, and vitamin D$_3$ receptors. Cell 65:1255–1266
25. Kahlen JP, Carlberg C (1996) Functional characterization of a 1,25-dihydroxyvitamin D$_3$ receptor binding site found in the rat atrial natriuretic factor promoter. Biochem Biophys Res Commun 218:882–886
26. Zhang Q, Wrana JL, Sodek J (1992) Characterization of the promoter region of the porcine opn (osteopontin, secreted phosphoprotein 1) gene. Eur J Biochem 207:649–659
27. Noda M, Vogel RL, Craig AM, Prahl J, DeLuca HF, Denhardt DT (1990) Identification of a DNA sequence responsible for binding of the 1,25-dihydroxyvitamin D$_3$ receptor and 1,25-dihydroxyvitamin D$_3$ enhancement of mouse secreted phosphoprotein 1 (*Spp-1* or osteopontin) gene expression. Proc Nat Acad Sci USA 87:9995–9999
28. Quelo I, Machuca I, Jurdic P (1998) Identification of a vitamin D response element in the proximal promoter of the chicken carbonic anhydrase II gene. J Biol Chem 273:10638–10646
29. Toell A, Polly P, Carlberg C (2000) All natural DR3-type vitamin D response elements show a similar functionality in vitro. Biochem J 352:301–309
30. Kerry DM, Dwivedi PP, Hahn CN, Morris HA, Omdahl JL, May BK (1996) Transcriptional synergism between vitamin D-responsive elements in the rat 25-hydroxyvitamin D$_3$ 24-hydroxylase (CYP) promoter. J Biol Chem 271:29715–29721
31. Schräder M, Nayeri S, Kahlen JP, Müller KM, Carlberg C (1995) Natural vitamin D$_3$ response elements formed by inverted palindromes: polarity-directed ligand sensitivity of vitamin D$_3$ receptor-retinoid X receptor heterodimer-mediated transactivation. Mol Cell Biol 15:1154–1161
32. Carlberg C (1995) Mechanisms of nuclear signalling by vitamin D$_3$. Interplay with retinoid and thyroid hormone signalling. Eur J Biochem 231:517–527
33. Rhodes SJ, Chen R, DiMattia GE et al (1993) A tissue-specific enhancer confers Pit-1-dependent morphogen inducibility and autoregulation on the *pit-1* gene. Genes Dev 7:913–932
34. Quack M, Carlberg C (2000) Ligand-triggered stabilization of vitamin D receptor/retinoid X receptor heterodimer conformations on DR4-type response elements. J Mol Biol 296:743–756
35. Morrison NA, Shine J, Fragonas J-C, Verkest V, McMenemey ML, Eisman JA (1989) 1,25-dihydroxyvitamin D-responsive element and glucocorticoid repression in the osteocalcin gene. Science 246:1158–1161
36. Ozono K, Liao J, Kerner SA, Scott RA, Pike JW (1990) The vitamin D-responsive element in the human osteocalcin gene. Association with a nuclear proto-oncogene enhancer. J Biol Chem 265:21881–21888
37. Polly P, Carlberg C, Eisman JA, Morrison NA (1996) Identification of a vitamin D$_3$ response element in the fibronectin gene that is bound by a vitamin D$_3$ receptor homodimer. J Cell Biochem 60: 322–333
38. Kahlen JP, Carlberg C (1994) Identification of a vitamin D receptor homodimer-type response element in the rat calcitriol 24-hydroxylase gene promoter. Biochem Biophys Res Commun 202: 1366–1372

39. Xie Z, Bikle DD (1997) Cloning of the human phospholipase C-γ1 promoter and identification of a DR6-type vitamin D-responsive element. J Biol Chem 272:6573–6577
40. Jenuwein T, Allis CD (2001) Translating the histone code. Science 293:1074–1080
41. Aranda A, Pascual A (2001) Nuclear hormone receptors and gene expression. Physiol Rev 81: 1269–1304
42. Burke LJ, Baniahmad A (2000) Co-repressors 2000. FASEB J 14:1876–1888
43. Polly P, Herdick M, Moehren U, Baniahmad A, Heinzel T, Carlberg C (2000) VDR-Alien: a novel, DNA-selective vitamin D₃ receptor-corepressor partnership. FASEB J 14:1455–1463
44. Privalsky ML (2004) The role of corepressors in transcriptional regulation by nuclear hormone receptors. Annu Rev Physiol 66:315–360
45. Leo C, Chen JD (2000) The SRC family of nuclear receptor coactivators. Gene 245:1–11
46. Castillo AI, Jimenez-Lara AM, Tolon RM, Aranda A (1999) Synergistic activation of the prolactin promoter by vitamin D receptor and GHF-1: role of coactivators, CREB-binding protein and steroid hormone receptor coactivator-1 (SRC-1). Mol Endocrinol 13:1141–1154
47. Rachez C, Suldan Z, Ward J et al (1998) A novel protein complex that interacts with the vitamin D₃ receptor in a ligand-dependent manner and enhances transactivation in a cell-free system. Genes Dev 12:1787–1800
48. Rachez C, Lemon BD, Suldan Z et al (1999) Ligand-dependent transcription activation by nuclear receptors requires the DRIP complex. Nature 398:824–828
49. Khanim FL, Gommersall LM, Wood VH et al (2004) Altered SMRT levels disrupt vitamin D₃ receptor signalling in prostate cancer cells. Oncogene 23:6712–6725
50. Oda Y, Sihlbom C, Chalkley RJ et al (2003) Two distinct coactivators, DRIP/mediator and SRC/p160, are differentially involved in vitamin D receptor transactivation during keratinocyte differentiation. Mol Endocrinol 17:2329–2339
51. Malinen M, Saramäki A, Ropponen A, Degenhardt T, Väisänen S, Carlberg C (2008) Distinct HDACs regulate the transcriptional response of human cyclin-dependent kinase inhibitor genes to Trichostatin A and 1α,25-dihydroxyvitamin D₃. Nucl Acids Res 36:121–132
52. Metivier R, Penot G, Hubner MR et al (2003) Estrogen receptor a directs ordered, cyclical, and combinatorial recruitment of cofactors on a natural target promoter. Cell 115:751–763
53. Kim S, Shevde NK, Pike JW (2005) 1,25-Dihydroxyvitamin D₃ stimulates cyclic vitamin D receptor/retinoid X receptor DNA-binding, co-activator recruitment, and histone acetylation in intact osteoblasts. J Bone Miner Res 20:305–317
54. Hancock R (2000) A new look at the nuclear matrix. Chromosoma 109:219–225
55. Sierra J, Villagra A, Paredes R et al (2003) Regulation of the bone-specific osteocalcin gene by p300 requires Runx2/Cbfa1 and the vitamin D₃ receptor but not p300 intrinsic histone acetyltransferase activity. Mol Cell Biol 23:3339–3351
56. Schüle R, Umesono K, Mangelsdorf DJ, Bolado J, Pike JW, Evans RM (1990) Jun-Fos and receptors for vitamins A and D recognize a common response element in the human osteocalcin gene. Cell 61:497–504
57. Miner JN, Yamamoto KR (1992) The basic region of AP-1 specifies glucocorticoid receptor activity at a composite response element. Genes Dev 6:2491–2501
58. Candeliere GA, Jurutka PW, Haussler MR, St-Arnaud R (1996) A composite element binding the vitamin D receptor, retinoid X receptor a, and a member of the CTF/NF-1 family of transcription factors mediates the vitamin responsiveness of the c-*fos* promoter. Mol Cell Biol 16: 584–592
59. Kim RH, Li JJ, Ogata Y, Yamauchi M, Freedman LP, Sodek J (1996) Identification of a vitamin D₃-response element that overlaps a unique inverted TATA box in the rat bone sialoprotein gene. Biochem J 318:219–226
60. White JH (2004) Profiling 1,25-dihydroxyvitamin D₃-regulated gene expression by microarray analysis. J Steroid Biochem Mol Biol 89–90:239–244
61. Murayama A, Kim MS, Yanagisawa J, Takeyama K, Kato S (2004) Transrepression by a liganded nuclear receptor via a bHLH activator through co-regulator switching. EMBO J 23: 1598–1608

62. Turunen MM, Dunlop TW, Carlberg C, Väisänen S (2007) Selective use of multiple vitamin D response elements underlies the 1α,25-dihydroxyvitamin D₃-mediated negative regulation of the human CYP27B1 gene. Nucleic Acids Res 35:2734–2747

63. Towers TL, Freedman LP (1998) Granulocyte-macrophage colony-stimulating factor gene transcription is directly repressed by the vitamin D₃ receptor: implications for allosteric influences on nuclear receptor structure and function by a DNA element. J Biol Chem 273:10338–10348

64. Towers TL, Staeva TP, Freedman LP (1999) A two-hit mechanism for vitamin D₃-dediates transcriptional repression of the granulocyte-macrophage colony-stimulating factor gene: vitamin D receptor completes for DNA binding with NFAT1 and stabilizes c-Jun. Mol Cell Biol 19:4191–4199

65. Polly P, Carlberg C, Eisman JA, Morrison NA (1997) 1α,25-dihydroxyvitamin D₃ receptor as a mediator of transrepression of retinoid signaling. J Cell Biochem 67:287–296

66. Renkema KY, Alexander RT, Bindels RJ, Hoenderop JG (2008) Calcium and phosphate homeostasis: concerted interplay of new regulators. Ann Med 40:82–91

67. Meyer MB, Watanuki M, Kim S, Shevde NK, Pike JW (2006) The human transient receptor potential vanilloid type 6 distal promoter contains multiple vitamin D receptor binding sites that mediate activation by 1,25-dihydroxyvitamin D₃ in intestinal cells. Mol Endocrinol 20:1447–1461

68. Xu H, Bai L, Collins JF, Ghishan FK (2002) Age-dependent regulation of rat intestinal type IIb sodium-phosphate cotransporter by 1,25-(OH)₂ vitamin D₃. Am J Physiol Cell Physiol 282: C487–C493

69. Saito H, Maeda A, Ohtomo S et al (2005) Circulating FGF-23 is regulated by 1α,25-dihydroxyvitamin D₃ and phosphorus in vivo. J Biol Chem 280:2543–2549

70. Kim S, Yamazaki M, Zella LA et al (2007) Multiple enhancer regions located at significant distances upstream of the transcriptional start site mediate RANKL gene expression in response to 1,25-dihydroxyvitamin D₃. J Steroid Biochem Mol Biol 103:430–434

71. Jiang H, Lin J, Su Z-z, Collart FR, Huberman E, Fisher PB (1994) Induction of differentiation in human promyelotic HL-60 leukemia cells activates p21, WAF1/CIP1, expression in the absence of p53. Oncogene 9:3397–3406

72. Wang QM, Jones JB, Studzinski GP (1996) Cyclin-dependent kinase inhibitor p27 as a mediator of the G1-S phase block induced by 1,25-dihydroxyvitamin D₃ in HL60 cells. Cancer Res 56:264–267

73. Liu M, Lee M-H, Cohen M, Bommakanti M, Freedman LP (1996) Transcriptional activation of the Cdk inhibitor p21 by vitamin D₃ leads to the induced differentiation of the myelomonocytic cell line U937. Genes Dev 10:142–153

74. Verlinden L, Verstuyf A, Convents R, Marcelis S, Van Camp M, Bouillon R (1998) Action of 1,25(OH)₂D₃ on the cell cycle genes, cyclin D1, p21 and p27 in MCF-7 cells. Mol Cell Endocrinol 142:57–65

75. Danielsson C, Mathiasen IS, James SY et al (1997) Sensitive induction of apoptosis in breast cancer cells by a novel 1,25-dihydroxyvitamin D₃ analogue shows relation to promoter selectivity. J Cell Biochem 66:552–562

76. Saramäki A, Banwell CM, Campbell MJ, Carlberg C (2006) Regulation of the human p21$^{(waf1/cip1)}$ gene promoter via multiple binding sites for p53 and the vitamin D₃ receptor. Nucleic Acids Res 34:543–554

77. Tavera-Mendoza L, Wang TT, Lallemant B et al (2006) Convergence of vitamin D and retinoic acid signalling at a common hormone response element. EMBO Rep 7:180–185

78. Rickert P, Seghezzi W, Shanahan F, Cho H, Lees E (1996) Cyclin C/CDK8 is a novel CTD kinase associated with RNA polymerase II. Oncogene 12:2631–2640

79. Bourbon HM, Aguilera A, Ansari AZ et al (2004) A unified nomenclature for protein subunits of mediator complexes linking transcriptional regulators to RNA polymerase II. Mol Cell 14:553–557

80. Li H, Lahti JM, Valentine M et al (1996) Molecular cloning and chromosomal localization of the human cyclin C (CCNC) and cyclin E (CCNE) genes: deletion of the CCNC gene in human tumors. Genomics 32:253–259

81. Akutsu N, Lin R, Bastien Y et al (2001) Regulation of gene Expression by 1α,25-dihydroxyvitamin D₃ and Its analog EB1089 under growth-inhibitory conditions in squamous carcinoma Cells. Mol Endocrinol 15:1127–1139

82. Matilainen M, Malinen M, Saavalainen K, Carlberg C (2005) Regulation of multiple insulin-like growth factor binding protein genes by 1α,25-dihydroxyvitamin D₃. Nucleic Acids Res 33: 5521–5532

83. Zhan Q, Antinore MJ, Wang XW et al (1999) Association with Cdc2 and inhibition of Cdc2/Cyclin B1 kinase activity by the p53-regulated protein Gadd45. Oncogene 18:2892–2900

84. Jiang F, Li P, Fornace AJ Jr., Nicosia SV, Bai W (2003) G2/M arrest by 1,25-dihydroxyvitamin D3 in ovarian cancer cells mediated through the induction of GADD45 via an exonic enhancer. J Biol Chem 278:48030–48040

85. Peng L, Malloy PJ, Feldman D (2004) Identification of a functional vitamin D response element in the human insulin-like growth factor binding protein-3 promoter. Mol Endocrinol 18:1109–1119

86. Dunlop TW, Väisänen S, Frank C, Molnar F, Sinkkonen L, Carlberg C (2005) The human peroxisome proliferator-activated receptor d gene is a primary target of 1α,25-dihydroxyvitamin D₃ and its nuclear receptor. J Mol Biol 349:248–260

87. Reed KR, Sansom OJ, Hayes AJ et al (2004) PPARδ status and Apc-mediated tumourigenesis in the mouse intestine. Oncogene 23:8992–8996

88. Väisänen S, Dunlop TW, Sinkkonen L, Frank C, Carlberg C (2005) Spatio-temporal activation of chromatin on the human CYP24 gene promoter in the presence of 1α,25-dihydroxyvitamin D₃. J Mol Biol 350:65–77

89. Swami S, Raghavachari N, Muller UR, Bao YP, Feldman D (2003) Vitamin D growth inhibition of breast cancer cells: gene expression patterns assessed by cDNA microarray. Breast Cancer Res Treat 80:49–62

90. Palmer HG, Sanchez-Carbayo M, Ordonez-Moran P, Larriba MJ, Cordon-Cardo C, Munoz A (2003) Genetic signatures of differentiation induced by 1α,25-dihydroxyvitamin D₃ in human colon cancer cells. Cancer Res 63:7799–7806

91. Peehl DM, Shinghal R, Nonn L et al (2004) Molecular activity of 1,25-dihydroxyvitamin D₃ in primary cultures of human prostatic epithelial cells revealed by cDNA microarray analysis. J Steroid Biochem Mol Biol 92:131–141

92. Krishnan AV, Shinghal R, Raghavachari N, Brooks JD, Peehl DM, Feldman D (2004) Analysis of vitamin D-regulated gene expression in LNCaP human prostate cancer cells using cDNA microarrays. Prostate 59:243–251

93. Ikezoe T, Gery S, Yin D et al (2005) CCAAT/enhancer-binding protein delta: a molecular target of 1,25-dihydroxyvitamin D3 in androgen-responsive prostate cancer LNCaP cells. Cancer Res 65: 4762–4768

94. Eelen G, Verlinden L, van Camp M et al (2004) The effects of 1α,25-dihydroxyvitamin D₃ on the expression of DNA replication genes. J Bone Miner Res 19:133–146

95. Eelen G, Verlinden L, Van Camp M et al (2004) Microarray analysis of 1α,25-dihydroxyvitamin D₃-treated MC3T3-E1 cells. J Steroid Biochem Mol Biol 89–90:405–407

96. Wang TT, Tavera-Mendoza LE, Laperriere D et al (2005) Large-scale in silico and microarray-based identification of direct 1,25-dihydroxyvitamin D₃ target genes. Mol Endocrinol 19:2685–2695

97. Sinkkonen L, Malinen M, Saavalainen K, Väisänen S, Carlberg C (2005) Regulation of the human cyclin C gene via multiple vitamin D₃-responsive regions in its promoter. Nucleic Acids Res 33: 2440–2451

98. Xie X, Lu J, Kulbokas EJ et al (2005) Systematic discovery of regulatory motifs in human promoters and 3′ UTRs by comparison of several mammals. Nature 434:338–345

99. Seuter S, Väisänen S, Radmark O, Carlberg C, Steinhilber D (2007) Functional characterization of vitamin D responding regions in the human 5-lipoxygenase gene. Biochim Biophys Acta 1771: 864–872

100. Klan N, Seuter S, Schnur N, Jung M, Steinhilber D (2003) Trichostatin A and structurally related histone deacetylase inhibitors induce 5-lipoxygenase promoter activity. Biol Chem 384:777–785

101. Sorg BL, Klan N, Seuter S et al (2006) Analysis of the 5-lipoxygenase promoter and characterization of a vitamin D receptor binding site. Biochim Biophys Acta 1761:686–697

102. Ogata K, Sato K, Tahirov TH (2003) Eukaryotic transcriptional regulatory complexes: cooperativity from near and afar. Curr Opin Struct Biol 13:40–48

103. Zella LA, Kim S, Shevde NK, Pike JW (2006) Enhancers located within two introns of the vitamin D receptor gene mediate transcriptional autoregulation by 1,25-dihydroxyvitamin D_3. Mol Endocrinol 20:1231–1247

104. Fretz JA, Zella LA, Kim S, Shevde NK, Pike JW (2007) 1,25-Dihydroxyvitamin D3 induces expression of the Wnt signaling co-regulator LRP5 via regulatory elements located significantly downstream of the gene's transcriptional start site. J Steroid Biochem Mol Biol 103:440–445

105. Kim S, Yamazaki M, Zella LA, Shevde NK, Pike JW (2006) Activation of receptor activator of NF-kappaB ligand gene expression by 1,25-dihydroxyvitamin D_3 is mediated through multiple long-range enhancers. Mol Cell Biol 26:6469–6486

106. Mader S, Leroy P, Chen J-Y, Chambon P (1993) Multiple parameters control the selectivity of nuclear receptors for their response elements. J Biol Chem 268:591–600

107. Schräder M, Müller KM, Becker-André M, Carlberg C (1994) Response element selectivity for heterodimerization of vitamin D receptors with retinoic acid and retinoid X receptors. J Mol Endocrinol 12:327–339

108. Schräder M, Müller KM, Carlberg C (1994) Specificity and flexibility of vitamin D signalling. Modulation of the activation of natural vitamin D response elements by thyroid hormone. J Biol Chem 269:5501–5504

109. Schräder M, Becker-Andre M, Carlberg C (1994) Thyroid hormone receptor functions as monomeric ligand-induced transcription factor on octameric half-sites. Consequences also for dimerization. J Biol Chem 269:6444–6449

110. Barski A, Cuddapah S, Cui K et al (2007) High-resolution profiling of histone methylations in the human genome. Cell 129:823–837

111. Matys V, Fricke E, Geffers R et al (2003) TRANSFAC: transcriptional regulation, from patterns to profiles. Nucleic Acids Res 31:374–378

112. Wasserman WW, Sandelin A (2004) Applied bioinformatics for the identification of regulatory elements. Nat Rev Genet 5:276–287

113. Odom DT, Dowell RD, Jacobsen ES et al (2007) Tissue-specific transcriptional regulation has diverged significantly between human and mouse. Nat Genet 39:730–732

114. Heinäniemi M, Uski JO, Degenhardt T, Carlberg C (2007) Meta-analysis of primary target genes of peroxisome proliferator-activated receptors. Genome Biol 8:R147

115. Burns JL, Jackson DA, Hassan AB (2001) A view through the clouds of imprinting. FASEB J 15:1694–1703

II   NON-SKELETAL/FUNCTIONS OF VITAMIN D

12 Extrarenal Synthesis of 1,25-Dihydroxyvitamin D and Its Health Implications

Daniel D. Bikle

Abstract   Although the kidney was initially thought to be the sole organ responsible for the production of 1,25-dihydroxyvitamin D_3 [1,25(OH)$_2$D] via its enzyme CYP27B1, it is now appreciated that the expression of CYP27B1 in tissues other than the kidney is widespread. However, the kidney is the major source for circulating 1,25(OH)$_2$D. Therefore the existence of the capacity for extrarenal 1,25(OH)$_2$D production begs the question why, and in particular whether the extrarenal production of 1,25(OH)$_2$D has physiologic importance. In this chapter this question will be discussed. First a compilation of the extrarenal sites for CYP27B1 expression is provided. This is followed by a discussion of the regulation of CYP27B1 expression and activity in extrarenal tissues, pointing out that such regulation is tissue specific and different from that of CYP27B1 in the kidney. Finally the physiologic significance of extrarenal 1,25(OH)$_2$D production is examined, with special focus on the role of CYP27B1 in regulation of cellular proliferation and differentiation, hormone secretion, and immune function. At this point the data do not clearly demonstrate an essential role for CYP27B1 expression in any tissue outside the kidney, but several examples pointing in this direction are provided. With the availability of the mouse enabling tissue-specific deletion of CYP27B1, the role of extrarenal CYP27B1 expression can now be addressed definitively.

   Key Words: CYP27B1; proliferation; differentiation; parathyroid hormone; insulin; FGF23; immune function; cancer; keratinocytes; macrophages

1. INTRODUCTION

Fraser and Kodicek *(1)* identified the kidney as the source of what was subsequently shown to be the active metabolite of vitamin D, namely 1,25-dihydroxyvitamin D [1,25(OH)$_2$D]. Subsequent studies demonstrated that the placenta could also convert 25-hydroxyvitamin D to 1,25(OH)$_2$D *(2, 3)*. Initially it was thought that in the non-pregnant individual the kidney was the only source of this hormone based on lower, if not undetectable levels of 1,25(OH)$_2$D in anephric patients in early assays *(4)* and failure to detect radiolabeled 1,25(OH)$_2$D following administration of radiolabeled 25-hydroxyvitamin D [25(OH)D] to acutely nephrectomized rats *(5, 6)*. However, other studies with anephric humans *(7, 8)* and pigs *(9)* identified low circulating

From: *Nutrition and Health: Vitamin D*
Edited by: M.F. Holick, DOI 10.1007/978-1-60327-303-9_12,
© Springer Science+Business Media, LLC 2010

levels of 1,25(OH)$_2$D that could be increased with 25(OH)D administration, and the report by Barbour et al. *(10)* of an anephric patient with sarcoidosis with clearly detectable 1,25(OH)$_2$D levels demonstrated a disease state in which extrarenal 1,25(OH)$_2$D production occurred. At about the same time, a number of investigators were finding 1,25(OH)$_2$D production by bone cells *(11)*, melanocytes *(12)*, activated macrophages *(13)*, and epidermal keratinocytes *(14)*. With the cloning of the 25(OH)D 1α hydroxylase (CYP27B1) in 1997 by four groups *(15–18)* and the subsequent development of antibodies to CYP27B1 *(19)*, CYP27B1 expression was readily demonstrated in a wide variety of tissues. In that the vitamin D receptor (VDR) is also found in a wide variety of tissues, the question of the physiologic significance of these observations must be addressed. The development of a mouse model in which CYP27B1 has been deleted *(20)*, and more specifically a mouse model in which CYP27B1 can be deleted in a tissue-specific fashion *(21)* puts us in the position to answer this question for the many tissues in which CYP27B1 expression has been identified. This chapter will address the tissue distribution of CYP27B1, the regulation of CYP27B1 in these cells, and the potential significance of these extrarenal locations in health and disease.

2. TISSUE DISTRIBUTION OF CYP27B1

Prior to the cloning of CYP27B1, identification of extrarenal sources of 1,25(OH)$_2$D production relied on demonstration that the product formed from the substrate 25(OH)D co-migrated with authentic 1,25(OH)$_2$D over several different HPLC columns, was equivalent to authentic 1,25(OH)$_2$D in the receptor-binding assay, and had the identical mass spectrum as authentic 1,25(OH)$_2$D. This standard was achieved for only a few tissues including placenta *(22)*, bone *(23)*, macrophages *(24)*, and keratinocytes *(14)*. Moreover, it was not clear that the enzyme producing 1,25(OH)$_2$D in these tissues was the same for each tissue in general and for the renal 1α-hydroxylase in particular. This situation changed in 1997 with the cloning of the 1α-hydroxylase. First, our cloning of the human 1α-hydroxylase was from keratinocytes *(15)*, and we demonstrated in a patient with pseudovitamin D deficiency rickets (PDDR) not only the mutations in the 1α-hydroxylase that caused this disease but also that the keratinocytes from that patient failed to produce 1,25(OH)$_2$D. Similar results were obtained from the placenta of a patient with PDDR *(25)*. Thus at least for the 1α-hydroxylase in keratinocytes and placenta and no doubt for most if not all other extrarenal sites, the enzyme is the same as that in the kidney. Second, access to the sequence of 1α-hydroxylase (which was renamed CYP27B1 after its cloning) has enabled investigators to use the very sensitive method of PCR to look for CYP27B1 expression in a wide variety of tissues, and in many cases the findings of PCR have been supported by immunohistochemistry (IHC) using antibodies prepared against the deduced sequence of the cloned CYP27B1. Table 1 lists tissues, both normal and malignant, in which CYP27B1 expression has been identified by PCR, IHC, and/or activity [i.e., 1,25(OH)$_2$D production]. The list is growing, and is likely not complete. The biologic function of CYP27B1 in many of these tissues is not clear.

Epithelia form the largest group of cells that have been shown to produce 1,25(OH)$_2$D and/or express CYP27B1. These cells provide the barrier between the

Table 1
Tissue Distribution of CYP27B1

Cell type	mRNA	Protein	Activity	References
Epithelia				
Epidermis	+	+	+	(14, 19)
Prostate	+	+	+	(30, 31)
Colon	+	+	+	(19, 32–34, 44, 143)
Breast	+	+		(35)
Cervix	+	+		(36)
Endometrium	+	+		(37)
Macrophages				
Pulmonary alveolar	+	+	+	(24)
Peritoneal			+	(49)
Synovial			+	(50)
Blood monocytes			+	(53, 54)
Lymph nodes	+	+		(19)
Dendritic cells	+	+	+	(55)
Testes	+			(15)
Ovary	+		+	(66)
Bone cartilage				
Osteoblasts	+	+	+	(23, 136)
Chondrocytes	+			(56)
Endocrine glands				
Parathyroid gland	+	+		(59)
Pancreatic islets	+	+	+	(19, 58)
Thyroid	+	+		(64)
Adrenal medulla		+		(19)
Placenta				
Decidua	+	+	+	(19, 22, 39)
Trophoblasts	+	+	+	(19, 38, 39)
Liver			+	(68)
Brain	+	+		(15, 19)
Tumors/cancer				
Lymphoma			+	(137–139)
Lung	+		+	(140, 141)
Glioblastomas	+			(70, 142)
Parathyroid	+			(60)
Basal cell	+	+		(42)
Squamous cell	+	+		(43)
Pancreas	+	+		(62)
Papillary thyroid	+	+		(64)
Breast	+	+		(35)
Colon	+	+		(32, 33, 44)
Prostate	+	+	+	(30, 31)
Endometrium	+	+		(45)
Cervix	+	+		(36)
Dysgerminomas	+		+	(66)

inside environment of the body and the outside, a feature which will be discussed subsequently provides one potential function for CYP27B1 in these cells. The archetype cell is the epidermal keratinocyte, which we *(14)* showed over 20 years ago to be a very high expresser of CYP27B1, higher in fact than the kidney. Based on calculations from $1,25(OH)_2D$ production by perfused pig skin we *(26)* determined that epidermal production of $1,25(OH)_2D$ could account for all the circulating $1,25(OH)_2D$ found in anephric patients (or pigs) in the basal state or after supplementation with vitamin D or 25(OH)D. However, it was also apparent from these early studies that most if not all $1,25(OH)_2D$ produced by the epidermis was used for autocrine or paracrine purposes, and that the kidney under normal conditions provided most of the circulating $1,25(OH)_2D$. This is likely because of the rapid and extensive induction of CYP24 in these cells, which limits the amount of secreted $1,25(OH)_2D$. This may also be the explanation why the increased expression of CYP27B1 in psoriasis *(19)* does not result in increased circulating levels of $1,25(OH)_2D$, unlike that seen in sarcoidosis (see below). In vitro, CYP27B1 activity as well as VDR levels decline as the keratinocytes differentiate *(27)*. IHC localization of CYP27B1 and VDR confirm these observations in that the highest expression of both CYP27B1 *(19)* and VDR *(28, 29)* is found in the stratum basale, although both are found in the more differentiated layers as well. Such observations are consistent with a role for CYP27B1 in keratinocyte proliferation and differentiation. Subsequent to these initial observations, CYP27B1 expression and activity have been found in most epithelia where it has been sought. These tissues include the prostate *(30, 31)*, colonic mucosa *(19, 32–34)*, mammary epithelium *(35)*, cervical epithelium *(36)*, endometrium *(37)*, and as a special case the decidua and trophoblasts of the placenta *(19, 22, 38, 39)*. Endometriosis is associated with increased expression of CYP27B1 in the endometrium *(37)*; preeclampsia and early pregnancy are associated with increased CYP27B1 in the placenta *(39–41)*. These examples suggest a role for CYP27B1 in hyperproliferative and/or distressed tissue. Moreover, CYP27B1 expression is also found and in many cases is increased when these epithelia become malignant. This observation further suggests a role for CYP27B1 in regulating proliferation and differentiation of these tissues. Examples include both basal cell *(42)* and squamous cell *(43)* carcinoma, breast cancer *(35)*, colon cancer *(32, 33, 44)*, prostate cancer (in which malignancy appears to reduce CYP27B1 expression and activity) *(30, 31)*, cervical cancer *(36)*, and endometrial cancer *(45)*.

Macrophages comprise a second large group of cells in which CYP27B1 expression and activity are well established. Unlike keratinocytes, CYP27B1 expression is found only when the macrophage is activated. Furthermore, as will be discussed subsequently, CYP24 activity in activated macrophages is less effective in regulating the $1,25(OH)_2D$ produced such that diseases associated with macrophage activation can and do lead to hypercalcemia and hypercalciuria as a result of increased circulating levels of $1,25(OH)_2D$. Following the initial case report of an anephric patient with sarcoidosis and readily detectable circulating $1,25(OH)_2D$ levels *(10)*, Adams and his colleagues *(13)* demonstrated $1,25(OH)_2D$ production by pulmonary alveolar macrophages from patients with sarcoidosis. This result has been confirmed by others *(46)* who have shown a correlation between CYP27B1 expression in pulmonary alveolar macrophages from patients with sarcoidosis and circulating $1,25(OH)_2D$ levels.

Furthermore, Adams and colleagues *(47)* noted that other granulomatous diseases such as tuberculosis were associated with elevated 1,25(OH)$_2$D levels due to increased production of 1,25(OH)$_2$D in the involved lung. Subsequently, CYP27B1 expression has been found in pulmonary alveolar macrophages from patients with lung cancer and expression appears to increase in patients with advanced disease *(48)*. Activated macrophages from other sources such as the peritoneum in patients with peritonitis *(49)*, from the synovium of patients with inflammatory arthritis *(50)*, granulomata in the colon of patients with Crohn's disease *(51)*, granulomata in the skin of a patient with slack skin disease, a T-cell lymphoproliferative disorder *(52)*, granulomata within lymph nodes *(19)*, circulating monocytes from patients with chronic renal failure *(53)*, and normal monocytes activated in vitro *(54)*. Dendritic cells, although not macrophages, are from the same lineage and like macrophages require activation to express CYP27B1 *(55)*.

Osteoblasts were among the first non-renal cells to be shown to have CYP27B1 activity *(11)* and at least low levels of expression have been found in chondrocytes *(56)*. Osteosarcoma cell lines likewise express CYP27B1 *(57)*, although no cases of osteosarcoma have been reported with hypercalcemia due to elevated 1,25(OH)$_2$D levels to my knowledge.

Several endocrine glands, the parathyroid gland and pancreatic islets in particular, whose products are regulated by 1,25(OH)$_2$D also express CYP27B1 *(19, 58, 59)*. Expression of CYP27B1 appears to be greater in parathyroid adenomas, whether primary or secondary to renal failure, compared to normal glands *(59, 60)*. When sequenced, none of these adenomas expressed inactivating mutations in CYP27B1 to suggest that CYP27B1 was functioning as a tumor suppressor gene *(61)*. Malignant pancreatic tissue and cell lines derived from such tissue express CYP27B1 at levels comparable to normal tissue *(62)*. Polymorphisms in CYP27B1 have been associated with increased risk of type 1 diabetes mellitus *(63)*, although it is unclear whether this link is due to regulation of insulin secretion or due to the immune response leading to this form of diabetes. The thyroid has also been shown to express CYP27B1 and this expression is increased in papillary carcinoma of this gland *(64)*. However, unlike the parathyroid gland and pancreatic islet, it is unclear whether 1,25(OH)$_2$D influences hormone secretion from this gland. CYP27B1 protein has been detected in the adrenal medulla by immunohistochemistry, although no reports of 1,25(OH)$_2$D regulated catecholamine production or altered CYP27B1 expression in patients with pheochromocytoma have been published to my knowledge.

Following the cloning of CYP27B1, we performed a tissue screen and were surprised to find a strong signal in the testes *(15)*. Using the promoter of CYP27B1 to drive a luciferase promoter, Anderson et al. *(65)* were able to localize CYP27B1 expression in both Leydig and Sertoli cells of the testes. CYP27B1 expression has also been found in the ovary *(66)*. Dysgerminomas of the ovary have a much greater expression of CYP27B1 than normal ovarian tissue *(66)*. This has significance in that hypercalcemia is known to occur with dysgerminomas, and a recent case report attributed the hypercalcemia to increased 1,25(OH)$_2$D levels in a patient with an ovarian dysgerminoma *(67)*.

The liver, well known to be the major site for 25(OH)D production has also been shown to express CYP27B1 *(68)*, which at least in the rat appears to disappear postnatally *(69)*. This observation has received little further investigation following these initial and early publications.

Expression of CYP27B1 in the brain was also a surprise result of our initial tissue screen for CYP27B1 *(15)*. Subsequent localization studies have demonstrated expression in the cerebellum and cerebral cortex *(19)*, and in particular in Purkinje cells *(65)*. Glioblastomas show increased CYP27B1 expression along with spliced variants of CYP27B1, the significance of which is uncertain *(70)*.

3. REGULATION OF EXTRARENAL CYP27b1 EXPRESSION

One of the initial observations of the regulation of extrarenal CYP27B1 is that it differed from that in the kidney, or in the proximal convoluted tubule (PCT) to be precise. The CYP27B1 in the PCT is controlled principally by three hormones, parathyroid hormone (PTH), FGF23, and $1,25(OH)_2D$ itself responding at least in part to changes in ambient calcium and phosphate levels. PTH stimulates CYP27B1 expression principally by acting on its membrane receptor to promote cyclic AMP formation, PKA activation, and enhanced binding of CCAAT box-binding protein to its proximal response element in the CYP27B1 promoter *(71)*. FGF23 inhibits CYP27B1 expression. FGF23 binds to the FGF receptor 1(IIIc) in association with Klotho, but the signaling events leading to inhibition of CYP27B1 expression are not well defined *(72)*. Low serum calcium stimulates PTH secretion, whereas low serum phosphate inhibits FGF23, and so renal production of $1,25(OH)_2D$ is stimulated by the reduction of these ions. The $1,25(OH)_2D$ produced in turn inhibits PTH secretion and stimulates FGF23 secretion thus providing a negative feedback loop. $1,25(OH)_2D$ also directly inhibits CYP27B1 expression in the kidney through a complex mechanism involving VDR that brings both histone deacetylases (HDAC) and DNA methyl transferases to the promoter of CYP27B1 inhibiting its transcription *(73)*. These feedback loops provide very tight regulation of $1,25(OH)_2D$ production by the proximal convoluted tubule of the kidney, control that differs from that of CYP27B1 in other cell types including that of distal renal tubule cells where PTH has little effect *(74)*. Most attention regarding regulation of extrarenal CYP27B1 has focused on keratinocytes and macrophages, and regulation of CYP27B1 in those two cells will now be considered (Fig. 1).

Keratinocytes respond to PTH with increased $1,25(OH)_2D$ production, but these cells do not have the classic PTH receptor and do not respond to cyclic AMP *(75)*. The mechanism by which PTH stimulates $1,25(OH)_2D$ production in these cells remains unclear. However, using a CYP27B1 promoter/luciferase reporter assay, Flanagan et al. *(76)* demonstrated PTH-stimulated expression in a kidney cell line but not in keratinocytes, suggesting that the effect of PTH may be post-transcriptional. The effect of FGF23 on keratinocyte CYP27B1 expression or function has not been reported and the expression of either FGFR 1(IIIc) or Klotho has not been demonstrated. Unlike the kidney, $1,25(OH)_2D$ does not directly affect CYP27B1 expression in keratinocytes. Rather, $1,25(OH)_2D$ regulates its own levels in the keratinocyte by inducing CYP24, the catabolic enzyme for $1,25(OH)_2D$ *(77)*. Tumor necrosis factor-α (TNF) *(78)* and

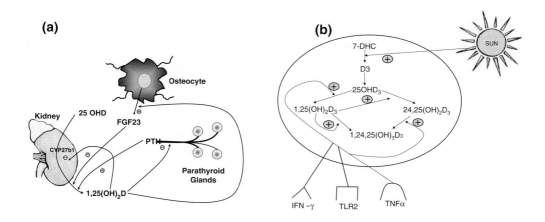

Fig. 1. Comparison of the regulation of CYP27B1 in the kidney to that in the keratinocyte. A. CYP27B1 in the kidney is regulated principally by three hormones: PTH, FGF23, and its product 1,25(OH)$_2$D. PTH stimulates, while FGF23 and 1,25(OH)$_2$D inhibit CYP27b. 1,25(OH)$_2$D in turn inhibits PTH production while stimulating that of FGF23. Calcium and phosphate likewise regulate PTH and FGF23 production providing feedback loops that tightly control CYP27B1 activity and maintain normal calcium and phosphate homeostasis. B. In the keratinocyte and other extrarenal sites of CYP27B1 expression 1,25(OH)$_2$D production is controlled primarily by cytokines such as IFN-γ and TNFα and activation of toll-like receptors (TLR). Unlike the kidney, 1,25(OH)$_2$D regulates its own levels within the cell primarily by induction of CYP24, which catabolizes both the substrate [25(OH)D] and product [1,25(OH)$_2$D] of CYP27B1. In the macrophage, this latter mechanism is lax, and conditions of increased macrophage activation can lead to excess 1,25(OH)$_2$D production and hypercalcemia. Figure 1a is adapted from Fig. 2 in Bikle *(144)*.

interferon-γ (IFN) *(79)*, on the other hand, are potent inducers of CYP27B1 activity in the keratinocyte.

Splice variants of CYP27B1 have been identified in normal and transformed keratinocytes *(42)*. As described below such variants may alter the synthesis and activity of wild-type CYP27B1, but this has not been demonstrated at least in the keratinocyte. However, it is conceivable that the apparent increase in CYP27B1 expression during the early stages of malignant transformation of various epithelia could reflect increased expression of variant forms that impair rather than increase CYP27B1 activity.

Pulmonary alveolar macrophage production of 1,25(OH)$_2$D requires activation by IFN or TNF and is inhibited by dexamethasone *(13, 80)*. The production of 1,25(OH)$_2$D by circulating monocytes can be stimulated by IFN and other cytokines including TNF, IL-1, and IL-2 *(54)*. In other studies of circulating monocytes, the combination of IFN and either phorbol ester (PMA) or lipopolysaccharide (LPS) induced CYP27B1 *(81)*. LPS stimulates through specific toll-like receptors (TLR) in association with the coreceptor CD14 *(81)*, a point that will become important when the functional role of extrarenal production of 1,25(OH)$_2$D is discussed. These investigators *(81)* observed a requirement for signaling through the JAK/STAT, p38 MAPK, and NFkB pathways, and their results implicated CEBPβ as a key transcription factor. Similar results were obtained with the macrophage cell line THP-1 *(82)*. In a v-myc-transformed myelomonocytic cell line (HD-11), Adams et al. *(83)* demonstrated stimulation of 1,25(OH)$_2$D production by LPS and IFN, and failure of PTH or calcium and phosphate to regulate CYP27B1 activity. The ability of FGF23 to regulate CYP27B1 in these cells has not been tested to my knowledge, but unless these cells express both FGFR 1 (IIIc) and Klotho, FGF23 will be ineffective.

As noted previously, CYP27B1 expression by activated macrophages can and does lead to increased circulating levels of 1,25(OH)$_2$D with the associated hypercalcemia and hypercalciuria. The mechanisms for this lack of feedback control are several. First, the major drivers for CYP27B1 expression and activity in these cells are cytokines, not PTH, and cytokines are not regulated by calcium and phosphate. In this regard the situation is not different from that in keratinocytes. Second, CYP24 induction and/or function in macrophages in response to 1,25(OH)$_2$D is blunted and does not provide the safety valve found in keratinocytes *(13)*. The mechanism appears to involve the expression of a truncated form of CYP24, which includes the substrate-binding domain but not the mitochondrial targeting sequence. This truncated form is postulated to act as a dominant negative form of CYP24, binding 1,25(OH)$_2$D within the cytoplasm and preventing its catabolism *(84)*. Both intact CYP24 and the truncated variant are induced in macrophage cell lines (HD-11 and THP-1), but CYP24 activity is not increased. A similar story may exist for CYP27B1 itself. Several non-coding splice variants of CYP27B1 have been identified in THP-1 cells as well as in the human proximal tubule kidney cell line HKC-8. At least in HCK-8 cells knockdown of the expression of several of these variant CYP27B1 forms did not alter wild-type CYP27B1 mRNA levels but did increase CYP27B1 protein and activity levels *(85)* suggesting inhibition of translation of wild-type CYP27B1 by these splice variants. It remains to be seen whether this mechanism functions in normal macrophages.

4. FUNCTION OF EXTRARENAL CYP27B1

With the demonstration that CYP27B1 like VDR is found in many if not most cells, the question of why needs to be addressed. In particular, why does a cell need to make its own 1,25(OH)$_2$D, when the kidney can provide all it needs. Until the results from tissue-specific deletions of CYP27B1 become generally available, this question will remain mostly unanswered. However, existing data support the role of extrarenal 1,25(OH)$_2$D production in a number of the nonclassic actions of 1,25(OH)$_2$D. These actions fall into three broad categories: the regulation of cellular proliferation and differentiation, the regulation of hormone secretion, and the regulation of immune function both innate and adaptive. Within a given cell these actions can overlap. The following discussion will be organized into these three categories.

4.1. Regulation of Proliferation and Differentiation

The epidermis, and the keratinocytes that form the epidermis, provides an excellent model to evaluate the ability of 1,25(OH)$_2$D to regulate proliferation and differentiation in a normal cell (Fig. 2). The epidermis is unique, in that under physiologic condi-

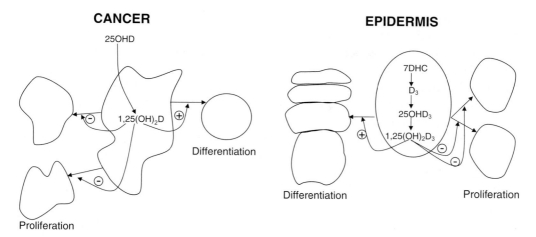

Fig. 2. The antiproliferative, prodifferentiating functions of 1,25(OH)$_2$D. One potential role for extrarenal CYP27B1 is to produce sufficient 1,25(OH)$_2$D within the cell to promote its differentiation and limit its proliferation. Two examples in this figure are premalignant cells and keratinocytes. In each case endogenous production of 1,25(OH)$_2$D is shown as limiting proliferation and promoting differentiation.

tions it is capable not only of making vitamin D but of converting it to 1,25(OH)$_2$D in the same cell and that cell is fully capable of responding to the 1,25(OH)$_2$D produced. 1,25(OH)$_2$D promotes the differentiation of keratinocytes while inhibiting their proliferation *(86)*. In the epidermis proliferation occurs in the basal layer, and as the keratinocytes move out of the basal layer differentiation is initiated. As the keratinocytes move from one layer of epidermis to the next, differentiation proceeds in a sequential fashion ultimately resulting in the enucleated corneocyte enmeshed in a lipid-rich matrix that provides the barrier function. 1,25(OH)$_2$D is involved in all steps of this process in that it limits proliferation in the basal layer while inducing in a sequential pattern the expression of genes whose products ultimately produce the permeability barrier. The ability of 1,25(OH)$_2$D to act sequentially on gene expression as the differentiation process unfolds is due to the differential distribution of coactivators (DRIP205 and SRC3) within the epidermis as a function of differentiation *(87)* and the differential utilization of these coactivators by genes involved in the early and late stages of differentiation *(88)*. CYP27B1 like VDR is found throughout the epidermis, although expression appears to be higher in the basal layer of the epidermis *(19, 28, 29)*. In our evaluation of the CYP27B1 null mouse *(89)*, we demonstrated increased keratinocyte proliferation, decreased differentiation, failure to form the calcium gradient within the epidermis, and reduced ability to reform the permeability barrier following disruption indicating a role of CYP27B1 in these processes.

Psoriasis is a chronic, generalized, and scaly erythematous dermatosis thought to be due to a Th1 or Th17-mediated immune reaction to as yet unidentified antigens in the skin that may cause or at least is accompanied by increased proliferation and decreased differentiation of the keratinocytes in the epidermis. As noted previously psoriatic skin overexpresses CYP27B1. Analogs of 1,25(OH)$_2$D, as well as calcitriol itself have proved effective therapy for moderate forms of this disease *(90)*, but topical 25(OH)D may provide an alternative approach with potentially less risk of systemic effects. This form of therapy likely works by inhibiting the inflammatory component via a direct action on the T cells *(91)* as well as by reducing keratinocyte proliferation and enhancing their differentiation *(92)*. The potential role of CYP27B1 in the immune functions of skin will be further discussed subsequently.

1,25(OH)$_2$D has been evaluated for its potential anticancer activity in animal and cell studies for over 25 years. The list of malignant cells that express VDR and CYP27B1 is now quite extensive as shown in Table 1, and many animal studies have demonstrated a reduction in growth in such cancers following administration of 1,25(OH)$_2$D or its analogs. Epidemiologic evidence is consistent with a role for vitamin D in preventing many of these cancers in humans *(93–97)*. Mechanisms are likely to include the ability of 1,25(OH)$_2$D to stimulate the expression of cell cycle inhibitors p21 and p27 *(98)* and the expression of the cell adhesion molecule E-cadherin *(99)* while inhibiting the transcriptional activity of β-catenin *(99–101)*. With respect to the skin, both BCC and SCC overexpress CYP27B1 and VDR. Mice lacking the VDR are predisposed to developing skin tumors induced either chemically *(102)* or by UV radiation *(103)*. The analog of 1,25(OH)$_2$D, EB1089, protects against skin tumors formed by overexpression of β-catenin *(104)*. Topical 1,25(OH)$_2$D protects and/or accelerates the repair of cyclobutane dimers, the signature lesion of UVB DNA damage, following UV radiation *(105)*. Part of this protection may be due to the increased expression of several critical DNA repair genes by 1,25(OH)$_2$D *(106)*. Breast, prostate, and colon cancers also express CYP27B1 and VDR. The mechanisms listed for the skin are likely to apply in these epithelia as well. A prospective 4-year trial with vitamin D and calcium showed a reduction in both breast and colon cancers *(107)*. A prospective study of patients with prostate cancer treated with high doses of 1,25(OH)$_2$D in combination with docetaxel demonstrated a survival advantage *(108)*. Although these studies support the potential role of 1,25(OH)$_2$D in the prevention and treatment of cancer, none of the studies to date can distinguish between the benefit of local production of 1,25(OH)$_2$D within the tissue and the benefit of elevated circulating 1,25(OH)$_2$D. Such distinctions will only come from the evaluation of animals with tissue-specific deletions of CYP27B1.

4.2. Regulation of Hormone Secretion

Parathyroid hormone (PTH). Circulating PTH levels are better correlated with 25(OH)D levels than with 1,25(OH)$_2$D *(109)* even though it is 1,25(OH)$_2$D that inhibits the synthesis and secretion of PTH *(110)* and prevents the proliferation of the parathyroid gland *(110, 111)*. The expression of CYP27B1 within the parathyroid gland could account for this observation. The parathyroid gene contains a negative VDRE through which 1,25(OH)$_2$D exerts its suppression *(110)*. 1,25(OH)$_2$D also induces the calcium sensing receptor (CaR) in the parathyroid gland *(112)*, which sensitizes the parathyroid gland to calcium inhibition. Although the focus of the pharmaceutical industry is on

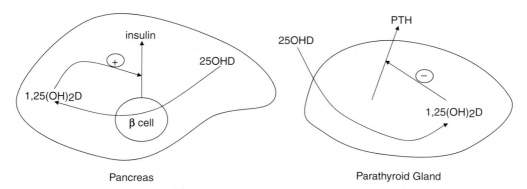

Fig. 3. The regulation of hormone secretion by endogenous production of 1,25(OH)$_2$D. Two examples are shown. CYP27B1 expression in pancreatic islets may serve to produce sufficient 1,25(OH)$_2$D to facilitate insulin secretion. On the other hand production of 1,25(OH)$_2$D in the parathyroid gland will inhibit PTH production.

developing analogs of 1,25(OH)$_2$D that selectively inhibit PTH synthesis/secretion, the potential for analogs that serve as a substrate for the CYP27B1 in the parathyroid gland needs to be explored (Fig. 3).

Insulin. A number of mostly case control and observational studies have suggested that vitamin D deficiency contributes to increased risk for type 2 (as well as type 1) diabetes mellitus *(113)*. In animal studies 1,25(OH)$_2$D has been shown to stimulate insulin secretion, although the mechanism is not well defined *(114, 115)*. Nor is it clear what role the CYP27B1 in the pancreatic islet plays in this process, although as for PTH secretion, insulin secretion may be amenable to regulation by the substrate rather than the product of CYP27B1 acting within the islet.

Fibroblast Growth Factor 23 (FGF23). FGF23 is produced primarily by bone, in particular by osteoblasts and osteocytes. 1,25(OH)$_2$D stimulates this process, but the mechanism is not clear *(116)*. Osteoblasts express CYP27B1 as noted previously, and conceivably locally produced 1,25(OH)$_2$D could impact this process. Whether FGF23 can feedback on the expression of CYP27B1 in bone remains to be demonstrated. Inasmuch as FGF23 inhibits 1,25(OH)$_2$D production by the kidney, and circulating 1,25(OH)$_2$D suppresses FGF23 production in bone, the role of the osteoblast CYP27B1 is unclear, and will remain so until osteoblast-specific deletion of CYP27B1 is accomplished and evaluated for this potential function.

4.3. Regulation of Immune Function

Adaptive immunity. The adaptive immune response involves the ability of T and B lymphocytes to produce cytokines and immunoglobulins, respectively, to specifically combat the source of the antigen presented to them by cells such as macrophages and dendritic cells. As noted previously, CYP27B1 expression is increased when these cells are activated. 1,25(OH)$_2$D exerts an inhibitory action on the adaptive immune system by suppressing the proliferation and differentiation of B-cell precursors into plasma cells *(117)*, inhibiting T-cell proliferation and function *(118)*, in particular Th1

(119) and Th17 *(120)* cells, and shifting the balance to favor Th2 cell *(121)* and regulatory T-cell (Treg) *(122)* function. At least in part these actions on T-cell proliferation and differentiation stem from actions of 1,25(OH)$_2$D on dendritic cells to reduce their antigen-presenting capability. The ability of 1,25(OH)$_2$D to suppress the adaptive immune system appears to be beneficial for a number of conditions in which the immune system is directed at self – i.e., autoimmunity. In a number of experimental models *(123, 124)* including inflammatory arthritis, autoimmune diabetes, experimental allergic encephalitis (a model for multiple sclerosis), and inflammatory bowel disease 1,25(OH)$_2$D administration has prevented and/or treated the disease process. Epidemiologic evidence supports such possibilities in humans. It has been observed that vitamin D deficiency and/or living at higher latitudes (with less sunlight) is associated with a number of autoimmune diseases including type 1 diabetes mellitus, multiple sclerosis, and Crohn's disease *(125)*. In a large Finnish study, providing infants with 2,000 IU vitamin D for their first year of life reduced the incidence of type 1 diabetes mellitus by 80% *(126)*. Other studies have linked vitamin D deficiency to increased risk of multiple sclerosis *(127)*, asthma *(128)*, and other immunologic diseases. Local production of 1,25(OH)$_2$D in synovial macrophages, pancreatic islets, neurons in the brain, lung, and intestinal mucosa, all tissues which express CYP27B1, may provide a mechanism to control the destructive immune process that is the etiology for these disease processes (Fig. 4).

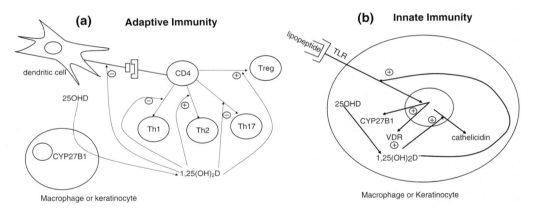

Fig. 4. 1,25(OH)$_2$D regulates both adaptive (**a**) and innate (**b**) immunity. CYP27B1 in either the macrophage or keratinocyte is increased by cytokines. The 1,25(OH)$_2$D produced then serves to inhibit the adaptive response by suppressing Th1 and Th17 proliferation and function while promoting Th2 and Treg functions. On the other hand 1,25(OH)$_2$D promotes innate immunity by increasing TLR and cathelicidin expression. This figure is adapted from Fig. 3 in Bikle *(144)*.

Innate immunity. Innate immune responses involve the activation of toll-like receptors (TLRs) in polymorphonuclear cells (PMNs), monocytes, and macrophages as well as in a number of epithelial cells including those of the epidermis, gingiva, intestine, vagina, bladder, and lungs. TLRs are transmembrane pathogen-recognition receptors that interact with specific membrane patterns (PAMP) shed by infectious agents that trigger the innate immune response in the host *(129)*. Activation of TLRs leads to the induction of antimicrobial peptides and reactive oxygen species, which kill the organism. Among those antimicrobial peptides is cathelicidin. The expression of this antimicrobial peptide is induced by 1,25(OH)$_2$D in both myeloid and epithelial cells *(130,131)*, and as noted previously these cells express both VDR and CYP27B1. The innate immune response of the macrophage and keratinocyte provide the best examples of the role of local CYP27B1 activity to meet a local challenge. Vitamin D deficiency has long been associated with increased risk of tuberculosis *(132)*, and 1,25(OH)$_2$D has long been recognized to potentiate the killing of mycobacteria by monocytes *(133)*. The mechanism underlying these observations has recently been determined by the observation that the monocyte, when activated by mycobacterial lipopeptides, expresses CYP27B1, producing 1,25(OH)$_2$D from circulating 25(OH)D, and in turn inducing cathelicidin that enhances killing of the mycobacterium. Inadequate circulating 25(OH)D levels fail to support this process despite presumably normal 1,25(OH)$_2$D levels *(134)*. In keratinocytes the expression of TLR2 and CD14, critical receptors for a number of pathogenic organisms activating the innate immune response, is increased by 1,25(OH)$_2$D as is the expression of cathelicidin *(135)*. When the skin is wounded expression of CYP27B1 increases along with that of TLR2, CD14, and cathelicidin *(135)*. These changes do not occur in mice lacking CYP27B1 in their epidermis *(135)*. Other epithelia of tissues facing the outside environment must provide a protective barrier, and it is anticipated that the innate immune system in these tissues will likewise be regulated by local 1,25(OH)$_2$D. However this needs to be demonstrated with tissue-specific deletion of CYP27B1.

5. SUMMARY

CYP27B1 is expressed widely and generally in conjunction with VDR. It seems unlikely that this is fortuitous, although at this point there are limited data to support a specific role for local production of 1,25(OH)$_2$D in most tissues where the CYP27B1 is expressed. Definitive data on this point will only come from studies with tissue-specific deletion of CYP27B1. Nevertheless, if local expression of CYP27B1 can be shown to be important, the potential to manipulate 1,25(OH)$_2$D production in the tissue of choice without incurring the hypercalcemia/hypercalciuria that limits administration of 1,25(OH)$_2$D and its analogs systemically will provide an important advance in the treatment of a variety of diseases. In particular, such therapy could be directed at regulating the uncontrolled proliferation and disordered differentiation in cancer, inhibiting excess PTH secretion in primary and secondary hyperparathyroidism, promoting insulin secretion in diabetes, limiting tissue destruction in autoimmune diseases, and enhancing the resistance to infectious organisms such as tuberculosis.

REFERENCES

1. Fraser DR, Kodicek E (1970) Unique biosynthesis by kidney of a biological active vitamin D metabolite. Nature 228(5273):764–766
2. Gray TK et al (1979) Evidence for extra-renal 1 alpha-hydroxylation of 25-hydroxyvitamin D3 in pregnancy. Science 204(4399):1311–1313
3. Weisman Y et al (1979) 1 alpha, 25-Dihydroxyvitamin D3 and 24,25-dihydroxyvitamin D3 in vitro synthesis by human decidua and placenta. Nature 281(5729):317–319
4. Brumbaugh PF et al (1974) Filter assay for 1alpha, 25-dihydroxyvitamin D3. Utilization of the hormone's target tissue chromatin receptor. Biochemistry 13(20):4091–4097

5. Reeve L et al (1983) Studies on the site of 1,25-dihydroxyvitamin D3 synthesis in vivo. J Biol Chem 258(6):3615–3617

6. Shultz TD et al (1983) Do tissues other than the kidney produce 1,25-dihydroxyvitamin D3 in vivo? A reexamination. Proc Natl Acad Sci USA 80(6):1746–1750

7. Lambert PW et al (1982) Evidence for extrarenal production of 1 alpha,25-dihydroxyvitamin D in man. J Clin Invest 69(3):722–725

8. Dusso A et al (1988) Extra-renal production of calcitriol in chronic renal failure. Kidney Int 34(3):368–375

9. Littledike ET, Horst RL (1982) Metabolism of vitamin D3 in nephrectomized pigs given pharmacological amounts of vitamin D3. Endocrinology 111(6):2008–2013

10. Barbour GL et al (1981) Hypercalcemia in an anephric patient with sarcoidosis: evidence for extrarenal generation of 1,25-dihydroxyvitamin D. N Engl J Med 305(8):440–443

11. Turner RT et al (1980) In vitro synthesis of 1 alpha,25-dihydroxycholecalciferol and 24,25-dihydroxycholecalciferol by isolated calvarial cells. Proc Natl Acad Sci USA 77(10):5720–5724

12. Frankel TL et al (1983) The synthesis of vitamin D metabolites by human melanoma cells. J Clin Endocrinol Metab 57(3):627–631

13. Adams JS, Gacad MA (1985) Characterization of 1 alpha-hydroxylation of vitamin D3 sterols by cultured alveolar macrophages from patients with sarcoidosis. J Exp Med 161(4):755–765

14. Bikle DD et al (1986) Neonatal human foreskin keratinocytes produce 1,25-dihydroxyvitamin D3. Biochemistry 25(7):1545–1548

15. Fu GK et al (1997) Cloning of human 25-hydroxyvitamin D-1 alpha-hydroxylase and mutations causing vitamin D-dependent rickets type 1. Mol Endocrinol 11(13):1961–1970

16. Takeyama K et al (1997) 25-Hydroxyvitamin D3 1alpha-hydroxylase and vitamin D synthesis. Science 277(5333):1827–1830

17. St-Arnaud R et al (1997) The 25-hydroxyvitamin D 1-alpha-hydroxylase gene maps to the pseudovitamin D-deficiency rickets (PDDR) disease locus. J Bone Miner Res 12(10):1552–1559

18. Shinki T et al (1997) Cloning and expression of rat 25-hydroxyvitamin D3-1alpha-hydroxylase cDNA. Proc Natl Acad Sci USA 94(24):12920–12925

19. Zehnder D et al (2001) Extrarenal expression of 25-hydroxyvitamin d(3)-1 alpha-hydroxylase. J Clin Endocrinol Metab 86(2):888–894

20. Panda DK et al (2001) Targeted ablation of the 25-hydroxyvitamin D 1alpha -hydroxylase enzyme: evidence for skeletal, reproductive, and immune dysfunction. Proc Natl Acad Sci USA 98(13):7498–7503

21. Dardenne O et al (2001) Targeted inactivation of the 25-hydroxyvitamin D(3)-1(alpha)- hydroxylase gene (CYP27B1) creates an animal model of pseudovitamin D- deficiency rickets. Endocrinology 142(7):3135–3141

22. Delvin EE et al (1985) In vitro metabolism of 25-hydroxycholecalciferol by isolated cells from human decidua. J Clin Endocrinol Metab 60(5):880–885

23. Turner RT et al (1983) Calvarial cells synthesize 1 alpha,25-dihydroxyvitamin D3 from 25-hydroxyvitamin D3. Biochemistry 22(5):1073–1076

24. Adams JS et al (1985) Isolation and structural identification of 1,25-dihydroxyvitamin D3 produced by cultured alveolar macrophages in sarcoidosis. J Clin Endocrinol Metab 60(5):960–966

25. Glorieux FH et al (1995) Pseudo-vitamin D deficiency: absence of 25-hydroxyvitamin D 1 alpha-hydroxylase activity in human placenta decidual cells. J Clin Endocrinol Metab 80(7): 2255–2258

26. Bikle DD et al (1994) Production of 1,25 dihydroxyvitamin D3 by perfused pig skin. J Invest Dermatol 102(5):796–798

27. Pillai S et al (1988) 1,25-Dihydroxyvitamin D production and receptor binding in human keratinocytes varies with differentiation. J Biol Chem 263(11):5390–5395

28. Milde P et al (1991) Expression of 1,25-dihydroxyvitamin D3 receptors in normal and psoriatic skin. J Invest Dermatol 97(2):230–239

29. Stumpf WE et al (1984) Topographical and developmental studies on target sites of 1,25 (OH)2 vitamin D3 in skin. Cell Tissue Res 238(3):489–496

30. Hsu JY et al (2001) Reduced 1alpha-hydroxylase activity in human prostate cancer cells corre-lates with decreased susceptibility to 25-hydroxyvitamin D3-induced growth inhibition. Cancer Res 61(7):2852–2856

31. Chen TC (2008) 25-Hydroxyvitamin D-1 alpha-hydroxylase (CYP27B1) is a new class of tumor suppressor in the prostate. Anticancer Res 28(4A):2015–2017

32. Bareis P et al (2001) 25-hydroxy-vitamin d metabolism in human colon cancer cells during tumor progression. Biochem Biophys Res Commun 285(4):1012–1017

33. Tangpricha V et al (2001) 25-hydroxyvitamin D-1alpha-hydroxylase in normal and malignant colon tissue. Lancet 357(9269):1673–1674

34. Bises G et al (2004) 25-hydroxyvitamin D3-1alpha-hydroxylase expression in normal and malignant human colon. J Histochem Cytochem 52(7):985–989

35. Segersten U et al (2005) 25-Hydroxyvitamin D3 1alpha-hydroxylase expression in breast cancer and use of non-1alpha-hydroxylated vitamin D analogue. Breast Cancer Res 7(6):R980–R986

36. Friedrich M et al (2002) Analysis of 25-hydroxyvitamin D3-1alpha-hydroxylase in cervical tissue. Anticancer Res 22(1A):183–186

37. Agic A et al (2007) Relative expression of 1,25-dihydroxyvitamin D3 receptor, vitamin D 1 alpha-hydroxylase, vitamin D 24-hydroxylase, and vitamin D 25-hydroxylase in endometriosis and gyne-cologic cancers. Reprod Sci 14(5):486–497

38. Diaz L et al (2000) Identification of a 25-hydroxyvitamin D3 1alpha-hydroxylase gene transcription product in cultures of human syncytiotrophoblast cells. J Clin Endocrinol Metab 85(7):2543–2549

39. Zehnder D et al (2002) The ontogeny of 25-hydroxyvitamin D(3) 1alpha-hydroxylase expression in human placenta and decidua. Am J Pathol 161(1):105–114

40. Fischer D et al (2007) Metabolism of vitamin D3 in the placental tissue of normal and preeclampsia complicated pregnancies and premature births. Clin Exp Obstet Gynecol 34(2):80–84

41. Vigano P et al (2006) Cycling and early pregnant endometrium as a site of regulated expression of the vitamin D system. J Mol Endocrinol 36(3):415–424

42. Mitschele T et al (2004) Analysis of the vitamin D system in basal cell carcinomas (BCCs). Lab Invest 84(6):693–702

43. Reichrath J et al (2004) Analysis of the vitamin D system in cutaneous squamous cell carcinomas. J Cutan Pathol 31(3):224–231

44. Ogunkolade BW et al (2002) Expression of 25-hydroxyvitamin D-1-alpha-hydroxylase mRNA in individuals with colorectal cancer. Lancet 359(9320):1831–1832

45. Becker S et al (2007) Expression of 25 hydroxyvitamin D3-1alpha-hydroxylase in human endometrial tissue. J Steroid Biochem Mol Biol 103(3-5):771–775

46. Inui N et al (2001) Correlation between 25-hydroxyvitamin D3 1 alpha-hydroxylase gene expression in alveolar macrophages and the activity of sarcoidosis. Am J Med 110(9):687–693

47. Barnes PF et al (1989) Transpleural gradient of 1,25-dihydroxyvitamin D in tuberculous pleuritis. J Clin Invest 83(5):1527–1532

48. Yokomura K et al (2003) Increased expression of the 25-hydroxyvitamin D(3)-1alpha-hydroxylase gene in alveolar macrophages of patients with lung cancer. J Clin Endocrinol Metab 88(12):5704–5709

49. Hayes ME et al (1987) Peritonitis induces the synthesis of 1 alpha,25-dihydroxyvitamin D3 in macrophages from CAPD patients. FEBS Lett 220(2):307–310

50. Hayes ME et al (1992) Differential metabolism of 25-hydroxyvitamin D3 by cultured synovial fluid macrophages and fibroblast-like cells from patients with arthritis. Ann Rheum Dis 51(2):220–226

51. Abreu MT et al (2004) Measurement of vitamin D levels in inflammatory bowel disease patients reveals a subset of Crohn's disease patients with elevated 1,25-dihydroxyvitamin D and low bone mineral density. Gut 53(8):1129–1136

52. Karakelides H et al (2006) Vitamin D-mediated hypercalcemia in slack skin disease: evidence for involvement of extrarenal 25-hydroxyvitamin D 1alpha-hydroxylase. J Bone Miner Res 21(9):1496–1499

53. Dusso AS et al (1991) Extrarenal production of calcitriol in normal and uremic humans. J Clin Endocrinol Metab 72(1):157–164

54. Gyetko MR et al (1993) Monocyte 1 alpha-hydroxylase regulation: induction by inflammatory cytokines and suppression by dexamethasone and uremia toxin. J Leukoc Biol 54(1):17–22
55. Fritsche J et al (2003) Regulation of 25-hydroxyvitamin D3-1 alpha-hydroxylase and production of 1 alpha,25-dihydroxyvitamin D3 by human dendritic cells. Blood 102(9):3314–3316
56. Weber L et al (2003) Cultured rat growth plate chondrocytes express low levels of 1alpha-hydroxylase. Recent Results Cancer Res 164:147–149
57. Atkins GJ et al (2007) Metabolism of vitamin D3 in human osteoblasts: evidence for autocrine and paracrine activities of 1 alpha,25-dihydroxyvitamin D3. Bone 40(6):1517–1528
58. Bland R et al (2004) Expression of 25-hydroxyvitamin D3-1alpha-hydroxylase in pancreatic islets. J Steroid Biochem Mol Biol 89-90(1-5):121–125
59. Correa P et al (2002) Increased 25-hydroxyvitamin D3 1alpha-hydroxylase and reduced 25-hydroxyvitamin D3 24-hydroxylase expression in parathyroid tumors–new prospects for treatment of hyperparathyroidism with vitamin d. J Clin Endocrinol Metab 87(12):5826–5829
60. Segersten U et al (2002) 25-hydroxyvitamin D(3)-1alpha-hydroxylase expression in normal and pathological parathyroid glands. J Clin Endocrinol Metab 87(6):2967–2972
61. Lauter K, Arnold A (2008) Analysis of CYP27B1, encoding 25-hydroxyvitamin D-1alpha-hydroxylase, as a candidate tumor suppressor gene in primary and severe secondary/tertiary hyperparathyroidism. J Bone Miner Res
62. Schwartz GG et al (2004) Pancreatic cancer cells express 25-hydroxyvitamin D-1 alpha-hydroxylase and their proliferation is inhibited by the prohormone 25-hydroxyvitamin D3. Carcinogenesis 25(6):1015–1026
63. Bailey R et al (2007) Association of the vitamin D metabolism gene CYP27B1 with type 1 diabetes. Diabetes 56(10):2616–2621
64. Khadzkou K et al (2006) 25-hydroxyvitamin D3 1alpha-hydroxylase and vitamin D receptor expression in papillary thyroid carcinoma. J Histochem Cytochem 54(3):355–361
65. Anderson PH et al (2008) Co-expression of CYP27B1 enzyme with the 1.5 kb CYP27B1 promoter-luciferase transgene in the mouse. Mol Cell Endocrinol 285(1-2):1–9
66. Evans KN et al (2004) Increased expression of 25-hydroxyvitamin D-1alpha-hydroxylase in dysgerminomas: a novel form of humoral hypercalcemia of malignancy. Am J Pathol 165(3):807–813
67. Hibi M et al (2008) 1,25-dihydroxyvitamin D-mediated hypercalcemia in ovarian dysgerminoma. Pediatr Hematol Oncol 25(1):73–78
68. Hollis BW (1990) 25-Hydroxyvitamin D3-1 alpha-hydroxylase in porcine hepatic tissue: subcellular localization to both mitochondria and microsomes. Proc Natl Acad Sci USA 87(16): 6009–6013
69. Takeuchi A et al (1994) The enzymatic formation of 1 alpha,25-dihydroxyvitamin D3 from 25-hydroxyvitamin D3 in the liver of fetal rats. Comp Biochem Physiol C Pharmacol Toxicol Endocrinol 109(1):1–7
70. Diesel B et al (2003) Gene amplification and splice variants of 25-hydroxyvitamin D3 1,alpha-hydroxylase (CYP27B1) in glioblastoma multiforme–a possible role in tumor progression? Recent Results Cancer Res 164:151–155
71. Gao XH et al (2002) Basal and parathyroid hormone induced expression of the human 25-hydroxyvitamin D 1alpha-hydroxylase gene promoter in kidney AOK-B50 cells: role of Sp1, Ets and CCAAT box protein binding sites. Int J Biochem Cell Biol 34(8):921–930
72. Strom TM, Juppner H (2008) PHEX, FGF23, DMP1 and beyond. Curr Opin Nephrol Hypertens 17(4):357–362
73. Kim MS et al (2007) 1alpha,25(OH)2D3-induced DNA methylation suppresses the human CYP27B1 gene. Mol Cell Endocrinol 265-266:168–173
74. Bajwa A et al (2008) Specific regulation of CYP27B1 and VDR in proximal versus distal renal cells. Arch Biochem Biophys 477(1):33–42
75. Bikle DD et al (1986) 1,25-Dihydroxyvitamin D3 production by human keratinocytes. Kinetics and regulation. J Clin Invest 78(2):557–566
76. Flanagan JN et al (2003) Regulation of the 25-hydroxyvitamin D-1alpha-hydroxylase gene and its splice variant. Recent Results Cancer Res 164:157–167

77. Xie Z et al (2002) The mechanism of 1,25-dihydroxyvitamin D(3) autoregulation in keratinocytes. J Biol Chem 277(40):36987–36990

78. Bikle DD et al (1991) Tumor necrosis factor-alpha regulation of 1,25-dihydroxyvitamin D production by human keratinocytes. Endocrinology 129(1):33–38

79. Bikle DD et al (1989) Regulation of 1,25-dihydroxyvitamin D production in human keratinocytes by interferon-gamma. Endocrinology 124(2):655–660

80. Pryke AM et al (1990) Tumor necrosis factor-alpha induces vitamin D-1-hydroxylase activity in normal human alveolar macrophages. J Cell Physiol 142(3):652–656

81. Stoffels K et al (2006) Immune regulation of 25-hydroxyvitamin-D3-1alpha-hydroxylase in human monocytes. J Bone Miner Res 21(1):37–47

82. Overbergh L et al (2006) Immune regulation of 25-hydroxyvitamin D-1alpha-hydroxylase in human monocytic THP1 cells: mechanisms of interferon-gamma-mediated induction. J Clin Endocrinol Metab 91(9):3566–3574

83. Adams JS et al (1994) Regulated production and intracrine action of 1,25-dihydroxyvitamin D3 in the chick myelomonocytic cell line HD-11. Endocrinology 134(6):2567–2573

84. Ren S et al (2005) Alternative splicing of vitamin D-24-hydroxylase: a novel mechanism for the regulation of extrarenal 1,25-dihydroxyvitamin D synthesis. J Biol Chem 280(21):20604–20611

85. Wu S et al (2007) Splice variants of the CYP27B1 gene and the regulation of 1,25-dihydroxyvitamin D3 production. Endocrinology 148(7):3410–3418

86. Bikle DD, Pillai S (1993) Vitamin D, calcium, and epidermal differentiation. Endocrine Rev 14:3–19

87. Oda Y et al (2003) Two distinct coactivators, DRIP/Mediator and SRC/p160, are differentially involved in VDR transactivation during keratinocyte differentiation. Mol Endocrinol 17:2329–2339

88. Hawker NP et al (2007) Regulation of Human Epidermal Keratinocyte Differentiation by the Vitamin D Receptor and its Coactivators DRIP205, SRC2, and SRC3. J Invest Dermatol 127:874

89. Bikle DD et al (2004) 25 Hydroxyvitamin D 1 alpha-hydroxylase is required for optimal epidermal differentiation and permeability barrier homeostasis. J Invest Dermatol 122(4):984–992

90. Bruce S et al (1994) Comparative study of calcipotriene (MC 903) ointment and fluocinonide ointment in the treatment of psoriasis. J Am Acad Dermatol 31(5 Pt 1):755–759

91. Bagot M et al (1994) Immunosuppressive effects of 1,25-dihydroxyvitamin D3 and its analogue calcipotriol on epidermal cells. Br J Dermatol 130(4):424–431

92. Kragballe K, Wildfang IL (1990) Calcipotriol (MC 903), a novel vitamin D3 analogue stimulates terminal differentiation and inhibits proliferation of cultured human keratinocytes. Arch Dermatol Res 282(3):164–167

93. Garland C et al (1985) Dietary vitamin D and calcium and risk of colorectal cancer: a 19-year prospective study in men. Lancet 1(8424):307–309

94. Bostick RM et al (1993) Relation of calcium, vitamin D, and dairy food intake to incidence of colon cancer among older women. The Iowa Women's Health Study. Am J Epidemiol 137(12):1302–1317

95. Kearney J et al (1996) Calcium, vitamin D, and dairy foods and the occurrence of colon cancer in men. Am J Epidemiol 143(9):907–917

96. Garland FC et al (1990) Geographic variation in breast cancer mortality in the United States: a hypothesis involving exposure to solar radiation. Prev Med 19(6):614–622

97. Hanchette CL, Schwartz GG (1992) Geographic patterns of prostate cancer mortality. Evidence for a protective effect of ultraviolet radiation. Cancer 70(12):2861–2869

98. Ingraham BA et al (2008) Molecular basis of the potential of vitamin D to prevent cancer. Curr Med Res Opin 24(1):139–149

99. Palmer HG et al (2001) Vitamin D(3) promotes the differentiation of colon carcinoma cells by the induction of E-cadherin and the inhibition of beta-catenin signaling. J Cell Biol 154(2):369–387

100. Shah S et al (2003) Trans-repression of beta-catenin activity by nuclear receptors. J Biol Chem 278(48):48137–48145

101. Shah S et al (2006) The molecular basis of vitamin D receptor and beta-catenin crossregulation. Mol Cell 21(6):799–809

102. Zinser GM et al (2002) Vitamin D(3) receptor ablation sensitizes skin to chemically induced tumorigenesis. Carcinogenesis 23(12):2103–2109

103. Ellison TI et al (2008) Inactivation of the Vitamin D Receptor Enhances Susceptibility of Murine Skin to UV-Induced Tumorigenesis. J Invest Dermatol 128:2508–2517

104. Palmer HG et al (2008) The Vitamin D Receptor Is a Wnt Effector that Controls Hair Follicle Differentiation and Specifies Tumor Type in Adult Epidermis. PLoS ONE 3(1): e1483

105. Dixon KM et al (2005) Skin cancer prevention: a possible role of 1,25dihydroxyvitamin D3 and its analogs. J Steroid Biochem Mol Biol 97(1-2):137–143

106. Moll PR et al (2007) Expression profiling of vitamin D treated primary human keratinocytes. J Cell Biochem 100(3):574–592

107. Lappe JM et al (2007) Vitamin D and calcium supplementation reduces cancer risk: results of a randomized trial. Am J Clin Nutr 85(6):1586–1591

108. Beer TM et al (2007) Double-blinded randomized study of high-dose calcitriol plus docetaxel compared with placebo plus docetaxel in androgen-independent prostate cancer: a report from the ASCENT Investigators. J Clin Oncol 25(6):669–674

109. Vieth R et al (2003) Age-related changes in the 25-hydroxyvitamin D versus parathyroid hormone relationship suggest a different reason why older adults require more vitamin D. J Clin Endocrinol Metab 88(1):185–191

110. Demay MB et al (1992) Sequences in the human parathyroid hormone gene that bind the 1,25- dihydroxyvitamin D3 receptor and mediate transcriptional repression in response to 1,25-dihydroxyvitamin D3. Proc Natl Acad Sci USA 89(17):8097–8101

111. Martin KJ, Gonzalez EA (2004) Vitamin D analogs: actions and role in the treatment of secondary hyperparathyroidism. Semin Nephrol 24(5):456–459

112. Canaff L, Hendy GN (2002) Human calcium-sensing receptor gene. Vitamin D response elements in promoters P1 and P2 confer transcriptional responsiveness to 1,25-dihydroxyvitamin D. J Biol Chem 277(33):30337–30350

113. Pittas AG et al (2007) The role of vitamin D and calcium in type 2 diabetes. A systematic review and meta-analysis. J Clin Endocrinol Metab 92(6):2017–2029

114. Kadowaki S, Norman AW (1985) Demonstration that the vitamin D metabolite 1,25(OH)2-vitamin D3 and not 24R,25(OH)2-vitamin D3 is essential for normal insulin secretion in the perfused rat pancreas. Diabetes 34(4):315–320

115. Lee S et al (1994) 1,25-Dihydroxyvitamin D3 and pancreatic beta-cell function: vitamin D receptors, gene expression, and insulin secretion. Endocrinology 134(4):1602–1610

116. Kolek OI et al (2005) 1alpha,25-Dihydroxyvitamin D3 upregulates FGF23 gene expression in bone: the final link in a renal-gastrointestinal-skeletal axis that controls phosphate transport. Am J Physiol Gastrointest Liver Physiol 289(6):G1036–G1042

117. Chen S et al (2007) Modulatory effects of 1,25-dihydroxyvitamin D3 on human B cell differentiation. J Immunol 179(3):1634–1647

118. Rigby WF et al (1984) Inhibition of T lymphocyte mitogenesis by 1,25-dihydroxyvitamin D3 (calcitriol). J Clin Invest 74(4):1451–1455

119. Lemire JM et al (1995) Immunosuppressive actions of 1,25-dihydroxyvitamin D3: preferential inhibition of Th1 functions. J Nutr 125(6 Suppl):1704S–1708S

120. Daniel C et al (2008) Immune modulatory treatment of trinitrobenzene sulfonic acid colitis with calcitriol is associated with a change of a T helper (Th) 1/Th17 to a Th2 and regulatory T cell profile. J Pharmacol Exp Ther 324(1):23–33

121. Boonstra A et al (2001) 1alpha,25-Dihydroxyvitamin d3 has a direct effect on naive CD4(+) T cells to enhance the development of Th2 cells. J Immunol 167(9):4974–4980

122. Penna G, Adorini L (2000) 1 Alpha,25-dihydroxyvitamin D3 inhibits differentiation, maturation, activation, and survival of dendritic cells leading to impaired alloreactive T cell activation. J Immunol 164(5):2405–2411

123. Adorini L (2005) Intervention in autoimmunity: the potential of vitamin D receptor agonists. Cell Immunol 233(2):115–124

124. Deluca HF, Cantorna MT (2001) Vitamin D: its role and uses in immunology. FASEB J 15(14): 2579–2585

125. Ponsonby AL et al (2002) Ultraviolet radiation and autoimmune disease: insights from epidemiological research. Toxicology 181–182:71–78
126. Hypponen E et al (2001) Intake of vitamin D and risk of type 1 diabetes: a birth-cohort study. Lancet 358(9292):1500–1503
127. Munger KL et al (2006) Serum 25-hydroxyvitamin D levels and risk of multiple sclerosis. JAMA 296(23):2832–2838
128. Litonjua AA, Weiss ST (2007) Is vitamin D deficiency to blame for the asthma epidemic? J Allergy Clin Immunol 120(5):1031-1035
129. Medzhitov R (2007) Recognition of microorganisms and activation of the immune response. Nature 449(7164):819–826
130. Gombart AF et al (2005) Human cathelicidin antimicrobial peptide (CAMP) gene is a direct target of the vitamin D receptor and is strongly up-regulated in myeloid cells by 1,25-dihydroxyvitamin D3. FASEB J 19(9):1067–1077
131. Wang TT et al (2004) Cutting edge: 1,25-dihydroxyvitamin D3 is a direct inducer of antimicrobial peptide gene expression. J Immunol 173(5):2909–2912
132. Ustianowski A et al (2005) Prevalence and associations of vitamin D deficiency in foreign-born persons with tuberculosis in London. J Infect 50(5):432–437
133. Rook GA et al (1986) Vitamin D3, gamma interferon, and control of proliferation of Mycobacterium tuberculosis by human monocytes. Immunology 57(1):159–163
134. Liu PT et al (2006) Toll-like receptor triggering of a vitamin D-mediated human antimicrobial response. Science 311(5768):1770–1773
135. Schauber J et al (2007) Injury enhances TLR2 function and antimicrobial peptide expression through a vitamin D-dependent mechanism. J Clin Invest 117(3):803–811
136. van Driel M et al (2006) Evidence for auto/paracrine actions of vitamin D in bone: 1alpha-hydroxylase expression and activity in human bone cells. FASEB J 20(13):2417–2419
137. Davies M et al (1985) Abnormal vitamin D metabolism in Hodgkin's lymphoma. Lancet 1(8439):1186–1188
138. Adams JS et al (1979) Hypercalcemia, hypercalciuria, and elevated serum 1,25 dihydroxyvitamin D concentrations in patients with AIDS and non-AIDS associated lymphoma. Blood 73:235–239
139. Seymour JF, Gagel RF (1993) Calcitriol: the major humoral mediator of hypercalcemia in Hodgkin's disease and non-Hodgkin's lymphomas. Blood 82(5):1383–1394
140. Jones G et al (1999) Expression and activity of vitamin D-metabolizing cytochrome P450s (CYP1alpha and CYP24) in human nonsmall cell lung carcinomas. Endocrinology 140(7):3303–3310
141. Mawer EB et al (1994) Constitutive synthesis of 1,25-dihydroxyvitamin D3 by a human small cell lung cancer cell line. J Clin Endocrinol Metab 79(2):554–560
142. Maas RM et al (2001) Amplification and expression of splice variants of the gene encoding the P450 cytochrome 25-hydroxyvitamin D(3) 1,alpha-hydroxylase (CYP 27B1) in human malignant glioma. Clin Cancer Res 7(4):868–875
143. Lechner D et al (2007) 1alpha,25-dihydroxyvitamin D3 downregulates CYP27B1 and induces CYP24A1 in colon cells. Mol Cell Endocrinol 263(1–2):55–64
144. Bikle DD (2009) Nonclassical actions of vitamin D. J Endocrinol Metab 94:26–34

13 Vitamin D and the Innate Immunity

Philip T. Liu, Martin Hewison, and John S. Adams

Abstract   This chapter will examine the role of vitamin D in the innate immune system as a mediator of human host defense mechanisms against microbial disease, focusing on tuberculosis. The first section will examine tuberculosis and the innate immune response to the intracellular pathogen, *Mycobacterium tuberculosis* (*M. tuberculosis*), the causative agent of tuberculosis. This is followed by a discussion of the known associations, genetic and mechanistic, between the vitamin D pathway and tuberculosis susceptibility. Finally, the chapter will conclude with a discussion on the potential for adjuvant treatment of tuberculosis with vitamin D.

Key Words: Extrarenal; 1,25-dihydroxyvitamin D; CYP27B1; proliferation; differentiation; parathyroid hormone; FGF23; immune function; cancer; keratinocytes

1. TUBERCULOSIS

1.1. Tuberculosis Overview

Tuberculosis has plagued humans throughout history with fossil evidence indicating tuberculosis infection of early hominids, such as the *Homo erectus,* and recordings of the disease by man as far back as ancient Egyptian and Chinese manuscripts *(1)*. The bacterium that causes tuberculosis, *M. tuberculosis*, was first described by Robert Koch in 1882. The bacterium primarily infects lung macrophages leading to pathogenesis of the disease. More than a century later, tuberculosis remains as a leading cause of morbidity and mortality worldwide, with one-third of the world's population infected and eight million new cases of tuberculosis each year *(2)*. Tuberculosis is one of the leading causes of death worldwide in women of reproductive age and in individuals infected with HIV *(3, 4)*. Even developed countries are not spared by this pandemic; estimates are that 10–15 million people residing in the United States are infected with *M. tuberculosis (5, 6)*. And, like the situation worldwide, mycobacterial infection is a leading cause of death among patients with AIDS in the United States *(5)*. The recent emergence of extensively drug-resistant (XDR) TB in HIV-infected individuals in KwaZulu Natal and its high mortality are an additional and urgent concern *(7)*.

From: *Nutrition and Health: Vitamin D*
Edited by: M.F. Holick, DOI 10.1007/978-1-60327-303-9_13,
© Springer Science+Business Media, LLC 2010

In addition to its importance with respect to global health, tuberculosis provides an important model for investigation of the human immune response to an intracellular pathogen and studies on tuberculosis have led to many basic immunological findings; these include [1] that the innate immune system recognizes microbial lipoproteins via Toll-like receptor 2 (TLR2) *(8)*; [2] that activation, via TLR2, of monocytes leads to instruction of the adaptive immune response via release of IL-12 *(8)*, dendritic cell differentiation *(9)*, and maturation *(10)*; and [3] that the activation of monocytes via TLR2 triggers (i) macrophage differentiation *(9)*, (ii) a nitric oxide-dependent antimicrobial pathway in mice *(11)*, and (iii) a vitamin D-dependent antimicrobial pathway in humans *(12)*. TLR2 has been shown to be important for resistance to *M. tuberculosis* in mouse models *(13–15)* and polymorphisms in both the vitamin D receptor (VDR) and TLR2 are associated with susceptibility to TB in humans *(16–23)*. The role of the innate immune response and vitamin D will be discussed in detail below.

1.1.1. INNATE IMMUNITY

Metchnikoff originally described the key direct functions of cells of the innate immune system: (1) rapid detection of microbes; (2) phagocytosis of those microbes; and (3) antimicrobial activity. Charles Janeway advanced our thinking about how the mammalian innate immune recognizes microbial pathogens, proposing that it must involve evolutionarily primitive receptors that bind conserved microbial constituents, termed pattern recognition receptors *(24)*. *M. tuberculosis* is known to activate at least two different families of pattern recognition receptors: TLRs and the nucleotide oligomerization domain (NOD)-like receptors on macrophages. The TLR2 and TLR1 heterodimer recognizes a triacylated lipoprotein derived from *M. tuberculosis*, which results in activation of NF-κB leading to the production of inflammatory cytokines and direct antimicrobial activity *(8, 11, 12)*. NOD2 recognizes muramyl dipeptide (MDP), which is a peptidoglycan present on *M. tuberculosis (25, 26)*. Triggering NOD2 similarly leads to a NF-κB-mediated inflammatory response; however, in contrast to TLRs, NOD2 also results in activation of the inflammasome *(27)*. The inflammasome is a protein complex whose function is to cleave and activate the pro-IL-1β protein into the active IL-1β cytokine through the enzymatic actions of caspase-1.

1.1.2. TOLL-LIKE RECEPTORS

Toll was first studied as a part of the dorsal ventral patterning system in *Drosophila melanogaster* embryogenesis. In 1996, Lemaitre et al. *(28)* reported that Toll-deficient adult *Drosophila* were more susceptible to fungal infection. Activation of Toll in flies results in production of antimicrobial peptides *(29)*, thus implicating Toll as a player in a primitive immune system. One year later, Medzhitov et al. *(30)* demonstrated that a constitutively active human Toll homologue, or a Toll-like receptor (TLR), modulates the adaptive immune response by inducing cytokine secretion as well as expression of co-stimulatory molecules. Together, these reports first established the importance of Toll and TLRs in host defense.

TLRs have been shown to have specificity in recognition of microbial ligands and mediate immune functions of the innate immune system. To date, 11 mammalian TLRs have been identified in both the human and murine genomes. Although, TLR1-9 are conserved between humans and mice, the murine *Tlr10* gene is nonfunctional, and the human *TLR11* gene harbors a premature stop codon which prevents its expression [31]. All the mammalian TLRs share a highly similar cytosolic Toll/IL-1 receptor (TIR) domain, which triggers several signaling pathways including the transcription factor: NF-κB [32]. The extracellular TLR domains include multiple leucine-rich repeat motifs and is responsible for recognition of conserved pathogen-associated molecular patterns (PAMPs). TLR2 heterodimerizes with TLR1 or TLR6, and the dimers mediate recognition of triacylated and diacylated bacterial lipoproteins, respectively [33]. The remainder of the known TLR ligands are as follows: viral dsRNA (TLR3), lipopolysaccharide (LPS) (TLR4), bacterial flagellin (TLR5), viral single-stranded RNA (ssRNA) (TLR7 and TLR8), bacterial unmethylated CpG DNA (TLR9), and protozoan profilin-like molecule (TLR11) [31]. The ligand for TLR10 is still unclear. Thus, TLRs provide a rapid first line of defense against a variety of microbial pathogens through the recognition of a milieu of pathogen-associated molecules.

TLR activation induces a variety of effects, including enhancement of macrophage phagocytosis [34], endosomal/lysosomal fusion [35], production of antimicrobial peptides [36, 37], as well as induction of direct antibacterial [11, 36] and antiviral activity [38–40]. *M. tuberculosis*-infected macrophages can induce a direct antimicrobial activity upon TLR2/1 activation. In a murine macrophage cell line, this activity was dependent on the generation of nitric oxide (NO) through inducible nitric oxide synthase (iNOS) activity. Addition of the iNOS inhibitors L-NIL and L-NAME ablated the murine TLR2/1-mediated antimicrobial activity; however, neither had an effect on human monocytes, suggesting that human TLR2/1-induced antimicrobial activity is fundamentally different from murine cells [11]. This correlated with the finding that, upon TLR2/1 activation, human monocytes do not generate detectable levels of NO [41]. Accordingly, the mechanism by which human macrophages kill intracellular *M. tuberculosis* intrigued immunologists for many years; the surprising role of the vitamin D synthetic/metabolic pathway in this mechanism is detailed below.

1.1.3. IMMUNOACTIVITY OF 1,25-DIHYDROXYVITAMIN D

There have been many studies on the role of active vitamin D, 1,25-dihydroxyvitamin D [1,25(OH)$_2$D], on innate and adaptive immune responses [42–44]. Insight into vitamin D-induced antimicrobial activity by human monocytes and macrophages against *M. tuberculosis* was first suggested by experiments in the labs of Rook in 1986 [45] and Crowle in 1987 [46]. These experiments were performed adding 1,25(OH)$_2$D to the extracellular medium of *M. tuberculosis*-infected human monocytes and macrophages in vitro with a resultant reduction of the intracellular bacterial load. Yet Crowle writes "concentrations of 1,25(OH)$_2$D near 4 µg/ml were needed for good protection, these levels seemed unphysiologically high compared with 26–70 pg/ml being in the normal circulating range." Nevertheless, these studies opened new questions regarding [1] the role of vitamin D in the physiological response to *M. tuberculosis* and [2] the identity of the vitamin D-dependent antimicrobial effectors.

It would be nearly a decade later before the molecular mechanism of the vitamin D-induced antimicrobial activity in macrophage began to be elucidated. One study by Sly et al. *(47)* reported 1,25(OH)$_2$D-induced antimicrobial activity to be regulated by phosphatidylinositol 3-kinase and mediated through the generation of oxygen intermediates via NADPH-dependent phagocyte oxidase. Interestingly, the observed 1,25(OH)$_2$D-induced oxidative burst occurred at a different time point than the antimicrobial activity, thus leading the authors to postulate that there had to be another key factor *(47)*. Another mechanism was proposed by Anand et al.; their study *(48), (49)* demonstrated that the 1,25(OH)$_2$D-induced antimicrobial activity was associated with downregulated transcription of the host protein, tryptophan-aspartate-containing coat protein (TACO); this protein plays an important role in *M. tuberculosis* entry and survival in human macrophages. In 2005, using a genome-wide scan for vitamin D response elements (VDREs), Wang et al. *(50)* reported that the genes encoding antimicrobial peptides, cathelicidin and hBD2(DEFB4), were regulated by the VDR. Prior to this study, human macrophages were not thought to utilize antimicrobial peptides as a defense mechanism; however, it was demonstrated in the same year that human monocytes expressed cathelicidin at both the mRNA and protein levels when stimulated with 1,25(OH)$_2$D *(12, 50, 51)*. Although it was apparent that monocytes could express cathelicidin, whether or not it played a role in host defense against intracellular *M. tuberculosis* infection was not clear. Two years later, a critical role for cathelicidin in the 1,25(OH)$_2$D-induced antimicrobial activity against intracellular *M. tuberculosis* was demonstrated in human monocytic cells using siRNA knockdowns *(52)*.

In contrast to its effects on macrophages, many studies have reported that 1,25(OH)$_2$D induces immunosuppressive effects, including but not limited to [1] inhibition of IL-12 secretion, [2] inhibition of lymphocyte proliferation and immunoglobulin synthesis, and [3] impairment of dendritic cell maturation, leading to the generation of tolerogenic dendritic cells and T-cell anergy *(53–56)*. In particular, it was suggested that 1,25(OH)$_2$D produced by the macrophage in granuloma-forming diseases, like tuberculosis and sarcoidosis, exerted a paracrine immunoinhibitory effect on neighboring, activated lymphocytes that express the VDR, and that this acts to slow an otherwise "overzealous" immune response that may be detrimental to the host *(57)*. The physiological significance of this has been highlighted by the recent development of 1,25(OH)$_2$D-deficient mouse models in which the gene for the vitamin D-activating enzyme CYP27B1 has been knocked out *(58, 59)*. A notable feature of these animals is that they present with enhanced adaptive immunity signified by multiple enlarged lymph nodes.

1.1.4. ANTIMICROBIAL PEPTIDES

Antimicrobial peptides consist of a highly diverse family of small peptides that can function as chemoattractants *(60, 61)*, dendritic cell activators *(62)*, and importantly, direct antimicrobial effectors *(63, 64)*. They exert microbicidal activity by disrupting the pathogen membrane through electrostatic interactions with the polar head groups of membrane lipids *(65)*, or the creation of membrane pores *(63)*; as such, they exhibit a wide range of microbial targets including bacteria *(66)*, fungi *(67, 68)*, parasites *(69, 70)*, and enveloped virii *(71)*. Although epithelial cells at the interface between the outside and inside environment of the host express antimicrobial peptides *(72)*, it is the population of innate immune cells that buttress that external–internal barrier in the host, such as neutrophils *(73)*, mast cells *(74)*, and monocytes/macrophages *(51, 75)*, that are recognized to be the major producers of antimicrobial peptides. Several antimicrobial peptides produced by macrophages have been demonstrated to have direct antimicrobial activity against *M. tuberculosis*, including but most likely not limited to LL-37 (cathelicidin) *(12, 76)*, hBD2 (DEFB4) *(77)*, and hepcidin *(78)*. In humans, cathelicidin and DEFB4 were found to contain activating VDREs in their promoter regions; whether or not hepcidin is vitamin D-regulated at the level of transcription is unknown *(50)*. Activation of the VDR in monocytes/macrophages results in the expression of cathelicidin at both the mRNA and protein levels *(12, 50, 76)*. siRNA knockdown of $1,25(OH)_2D$-induced cathelicidin in human monocytic cells resulted in complete loss of antimicrobial activity *(52)*, suggesting that antimicrobial peptides represent a major human macrophage host defense mechanism. Furthermore, macrophages can obtain and utilize neutrophil granules which can concentrate a variety of antimicrobial peptides against *M. tuberculosis (79, 80)*.

1.2. Vitamin D Pathway and Tuberculosis

Many studies have identified genes that may confer some degree of susceptibility to tuberculosis, including: HLA-DR alleles *(81–83)*, NRAMP-1 *(84)*, interferon-γ signaling *(85)*, SP110 *(86)*, complement receptor-1 *(87)*, and notably, the VDR *(19–23)*. However, these studies did not identify a clear cut host defense mechanism. Several studies have linked serum levels of the major circulating form of vitamin D, 25-hydroxyvitamin D [25(OH)D], to both tuberculosis disease progression and susceptibility *(23, 88)*. In 1985, a study reported that of 40 Indonesian patients with active tuberculosis and treated with anti-tuberculosis chemotherapy, 10 patients with the highest 25(OH)D levels at the outset of therapy had "less active pulmonary disease" *(88)*.

Another aspect of the vitamin D pathway that has been extensively studied is the VDR itself. There are two major VDR polymorphisms that have been studied in terms of tuberculosis susceptibility with conflicting results: TaqI *(20–22)* and FokI *(22, 89)*, located in exons nine and two of the VDR coding sequence, respectively *(90)*. Bellamy et al. conclude that the tt allele of the TaqI polymorphism protects against TB, however, studies by two other groups report no such association *(21, 89)*. Liu et al. *(22)* report that the FokI ff allele is associated with active TB among the Chinese Han population, but there are no other reports concluding an association for FokI ff and TB in any other population. These associations became clearer in a study examining the relationship between vitamin D deficiency and VDR polymorphisms with tuberculosis in the Gujarati Asians living in West London in the year 2000 *(23)*. The study reported that both the TaqI (Tt/TT) and FokI (ff) alleles were associated with tuberculosis only when the individual exhibited serum 25(OH)D deficiency *(23)*. Collectively, these studies have demonstrated that vitamin D plays an important role to host defense against *M. tuberculosis* in vivo. The problem in drawing conclusions between the studies in vitro with human inflammatory cells and these observations in vivo in humans with tuberculosis resides in the fact that previous in vitro studies used the active, $1,25(OH)_2D$ metabolite to affect antimicrobial activity, while the association to tuberculosis was with serum levels of the $1,25(OH)_2D$ substrate, $25(OH)D$.

Relatively little is known about the direct effects of $25(OH)D$ on innate immunity. Hewison et al. *(91)* found that $25(OH)D$ at physiologic levels (100 nM) suppressed CD40L-induced IL-12 production in day-7 GM-CSF/IL-4-derived DCs in vitro. Other studies in vitro have shown that intracrine metabolism of $25-1,25(OH)_2D$ via endogenous expression of CYP27B1 is a more efficient mechanism for modulating the phenotype of either DCs or monocytes compared to the exogenous addition of active $1,25(OH)_2D$ itself *(92)*. In contrast to these in vitro analyses, there are little data on the effects of altering the $25(OH)D$ status in vivo on the immune status of the host. Yang et al. *(93)* showed that profound reduction in the serum $25(OH)D$ in mice resulted in significant blunting of the cell-mediated immune response to cutaneous dinitrofluorobenzene (DNFB) challenge. Administration of $25(OH)D$ to humans with head and neck squamous cell carcinoma increases plasma IL-12 and IFN-γ levels and improves T-cell blastogenesis *(94)*. In more recent studies, we have shown that the ability of monocytes from human subjects to mount a cathelicidin response following TLR challenge is directly proportional to circulating levels of $25(OH)D$ but not $1,25(OH)_2D$ *(95)*. Importantly, this study also showed that TLR-induction of cathelicidin was enhanced in subjects supplemented with vitamin D (500,000 IU vitamin D_2 over 5 weeks), indicating that the immunomodulatory effects of $25(OH)D$ also occur in vivo.

1.2.1. ROLE OF 25-HYDROXYVITAMIN D ON THE INNATE IMMUNE RESPONSE

In 2006, a potential mechanism by which the $25(OH)D$ status of an individual may alter their ability to mount an innate immune response against *M. tuberculosis* was reported. In humans, activation of TLR2/1 results in the induction of key genes in the vitamin D pathway (Fig. 1), including the VDR and CYP27B1. Under conditions where the extracellular concentration of $25(OH)D$ is present at sufficient levels, TLR2/1 activation of monocytes results in a CYP27B1- and VDR-dependent expression of the antimicrobial peptide, cathelicidin, and direct microbicidal activity against intracellular *M. tuberculosis*. The induction of CYP27B1 and VDR in monocytes was subsequently demonstrated to be mediated through the actions of TLR2/1-induced IL-15 expression *(96)*. Interestingly, the human but the not murine cathelicidin promoter contains an activating VDRE *(51)*, perhaps suggesting a point of divergent evolution between mice and humans in the antimicrobial effectors used by the TLR-mediated innate immune response. Inhibition of the VDR resulted in ablation of the TLR2/1-induced antimicrobial activity, implicating that VDR activation is a critical step in the innate immune response against *M. tuberculosis* and potentially explaining the association of $25(OH)D$ serum levels with susceptibility to tuberculosis; e.g., where low $25(OH)D$ levels in the circulation cannot provide sufficient substrate $25(OH)D$ for CYP27B1-mediated production of $1,25(OH)_2D$ to activate the VDR-dependent antimicrobial response.

This requirement of adequate $25(OH)D$ in the extracellular environment of the human macrophage for the induction of host defense mechanisms via TLR2/1 provided a link between two well-documented clinical observations: compared to lightly pigmented

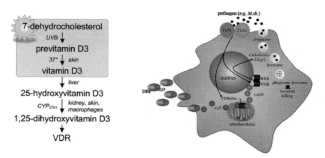

Fig. 1. Synthesis and innate immunoaction of vitamin D. The *left panel* shows synthetic/metabolic pathway of vitamin D3 beginning in the skin. The initial step is the UVB-mediated, non-enzymatic conversion of 7-dehydrocholesterol to previtamin D3. This is followed by thermal isomerization of previtamin D3 to vitamin D_3. Vitamin D_3 gains access to the circulation for oxidative metabolism first to 25-hydroxyvitamin D_3 in the liver followed by conversion via the CYP27B1 hydroxylase in the kidney, skin, and macrophage to hormone 1,25-dihydroxyvitamin D3. The *right panel* schematic recapitulates the events post interaction of the human TLR2/1 dimer with pathogen-associated membrane patterns (PAMPs) from *Mycobacterium tuberculosis* (*M.tb.*). TLR activation results in (1) induction of expression of the CYP27B1 hydroxylase and VDR genes; (2) intracrine generation of 1,25-dihydroxyvitamin D [1,25(OH)D], if and only if sufficient substrate 25-hydroxyvitamin D (25(OH)D) has been delivered to the macrophage bound to the serum vitamin D-binding protein (DBP); (3) transactivation of the cathelicidin gene via interaction of the 1,25(OH)$_2$D-VDR-retinoid X receptor (RXR) with an enhancer element in its promoter; (4) induction of expression of the cathelicidin gene product LL37; and (5) killing of ingested mycobacteria.

human populations, darkly pigmented black individuals are [1] more susceptible to virulent infections with tuberculosis and [2] have lower circulating, serum 25(OH)D levels owing to their relatively diminished capacity to synthesize vitamin D in their skin during sunlight exposure. The biosynthetic pathway of 25(OH)D in humans (Fig. 1) involves the absorption of ultraviolet B (UVB) photons from sunlight by 7-dehydrocholesterol (7HDC) in the basal layer of the epidermis and its non-enzymatic conversion to a pre-vitamin D_3 precursor in the skin; in fact, the melanin in pigmented skin will competitively absorb these UVB rays preventing this photoreaction *(97)*. In human monocytes cultured in sera from pigmented African American subjects and stimulated with a TLR2/1 ligand, there was no upregulation of cathelicidin mRNA, whereas the same human monocytes conditioned in sera from lightly pigmented subjects did *(12)*. Moreover, supplementation of the African American sera with exogenous 25(OH)D restored the induction of cathelicidin mRNA. This implies that an individual's serum 25(OH)D level may affect their ability to combat infection and that returning circulating levels of 25(OH)D to normal could potentially restore their host defense mechanisms *(12)*.

Through reostatic regulation of CYP27B1 activity and conversion of substrate 25(OH)D to product 1,25(OH)$_2$D, the macrophage directly controls its intracellular level of 1,25(OH)$_2$D *(98)*. It is also now recognized that TLR-induced antimicrobial activity can be inhibited by blocking CYP27B1 activity *(12)*. These data suggest that it is serum 25(OH)D and not the 1,25(OH)$_2$D concentration that [1] controls the intracellular 1,25(OH)$_2$D level and [2] is essential for the TLR-induced antimicrobial activity. This explains why in previous experiments in vitro, a supraphysiologic concentration of

1,25(OH)$_2$D in the conditioning extracellular media was required to generate sufficient intracellular levels of the metabolite to affect the VDR and to achieve an antimicrobial effect in human macrophages.

1.2.2. HISTORY OF VITAMIN D, SUNSHINE, AND TUBERCULOSIS

Establishment of vitamin D's role in host defense against tuberculosis provides new insights into the historical understanding of tuberculosis treatment prior to the advent of antibiotics. In the late nineteenth century, two young physicians, who themselves had contracted tuberculosis, were instructed by their physicians to travel to mountainous regions of Europe during the summertime as part of their attempt to recover. Their trek into this high UVB environment led to the "remission" of their disease. As a consequence of this success, Hermann Brehmer built the world's first high-altitude tuberculosis sanitorium in Germany, designed to allow patients to be exposed to "fresh air and sunlight." At about the same time in the United States, Edward Livingston Trudeau of New York published his original scientific finding that rabbits infected with tuberculosis had a more severe course of disease if caged indoors in the dark as opposed to being kept outdoors on a remote island. These experimental observations led him to build the first sanitorium at Saranac Lake, NY. In fact, it was the success of treatment facilities like these that paved the way to the 1903 Nobel Prize in Medicine awarded to the Danish physician, Niels Ryberg Finsen, for demonstrating that UV light was beneficial to patients with lupus vulgaris, a form of cutaneous *M. tuberculosis* infection. Despite widespread skepticism about the value of sanitoria at the time and since then, it is quite possible that the prolonged exposure to sunlight increased cutaneous vitamin D production, increased substrate 25(OH)D levels, and enhanced innate immunity to combat tuberculosis. Clearly, the harmful effects of sunlight have been well documented and emphasized, particularly its strong association with melanomas and squamous cell carcinomas *(99, 100)*. However, there is also convincing epidemiologic evidence that vitamin D has a positive association with lower incidences of colorectal and prostate cancers *(101)*. While heliotherapy is not likely to re-emerge as a useful intervention for human disease, other than for perhaps seasonal depression, vitamin D supplementation could represent an inexpensive adjuvant therapeutic approach to correcting the worldwide prevalence of vitamin D insufficiency and enhancing innate immunity to microbial infections, especially in individuals of pigmented African and South Asian descent in whom tuberculosis is rampant.

1.2.3. TREATMENT OF TUBERCULOSIS WITH VITAMIN D

There is a long history of using vitamin D to treat mycobacterial infections with apparent success. In 1946, Dowling et al. *(102)* reported the treatment of patients with lupus vulgaris with oral vitamin D. Eighteen of 32 patients appeared to be cured, nine improved. Morcos et al. *(103)* treated 24 newly diagnosed cases of tuberculosis in children with standard chemotherapy with and without vitamin D; they noted more profound clinical and radiological improvement in the group treated with vitamin D. Nursyam et al. *(104)* administered vitamin D or placebo to 67 tuberculosis patients following the sixth week of standard treatment. Out of 60 total patients, the group with vitamin D had higher sputum conversion rate and radiological improvement (100%) than placebo group (76.7%). This difference was statistically significant ($p = 0.002$). Despite the clear benefits of vitamin D treatment for tuberculosis, the mechanism of action had not been elucidated. The fact that TLR-activated macrophages can convert vitamin D to produce antimicrobial peptides could be a possible mechanism by which supplementation of patients with inactive vitamin D leads to a positive therapeutic outcome.

Progress in curtailing the human death rate from tuberculosis has been hampered by access to, cost and effectiveness of current antibiotic regimens (105). Some of these problems could potentially be overcome by adding vitamin D to the treatment regimen of tuberculosis. Although the currently published studies on the effects of vitamin D supplementation are generally inadequate to evaluate the efficacy of such treatment (106), a single oral dose of 50,000 IU of vitamin D has been shown to enhance killing of mycobacteria by whole blood of healthy volunteers (107). As such, knowledge of the role of human vitamin D metabolism and action in the basic innate immune defense mechanisms against mycobacterial infection provides hope in the development of safe, simple, and cost-effective strategies in the near future to prevent and treat tuberculosis.

REFERENCES

1. Kappelman J, Alcicek MC, Kazanci N, Schultz M, Ozkul M, Sen S (2008) First Homo erectus from Turkey and implications for migrations into temperate Eurasia. Am J Phys Anthropol 135:110
2. Raviglione MC (2003) The TB epidemic from 1992 to 2002. Tuberculosis (Edinb) 83:4
3. Fawzi WW, Msamanga GI, Spiegelman D, Wei R, Kapiga S, Villamor E, Mwakagile D, Mugusi F, Hertzmark E, Essex M, Hunter DJ (2004) A randomized trial of multivitamin supplements and HIV disease progression and mortality. N Engl J Med 351:23
4. Quinn TC, Overbaugh J (2005) HIV/AIDS in women: an expanding epidemic. Science 308:1582
5. Corbett EL, Watt CJ, Walker N, Maher D, Williams BG, Raviglione MC, Dye C (2003) The growing burden of tuberculosis: global trends and interactions with the HIV epidemic. Arch Intern Med 163:1009
6. Dye C, Scheele S, Dolin P, Pathania V, Raviglione MC (1999) Consensus statement. Global burden of tuberculosis: estimated incidence, prevalence, and mortality by country. WHO Global Surveillance and Monitoring Project. JAMA 282:677
7. Gandhi NR, Moll A, Sturm AW, Pawinski R, Govender T, Lalloo U, Zeller K, Andrews J, Friedland G (2006) Extensively drug-resistant tuberculosis as a cause of death in patients co-infected with tuberculosis and HIV in a rural area of South Africa. Lancet 368:1575
8. Brightbill HD, Libraty DH, Krutzik SR, Yang RB, Belisle JT, Bleharski JR, Maitland M, Norgard MV, Plevy SE, Smale ST, Brennan PJ, Bloom BR, Godowski PJ, Modlin RL (1999) Host defense mechanisms triggered by microbial lipoproteins through toll-like receptors. Science 285:732
9. Krutzik SR, Tan B, Li H, Ochoa MT, Liu PT, Sharfstein SE, Graeber TG, Sieling PA, Liu YJ, Rea TH, Bloom BR, Modlin RL (2005) TLR activation triggers the rapid differentiation of monocytes into macrophages and dendritic cells. Nat Med 11:653
10. Hertz CJ, Kiertscher SM, Godowski PJ, Bouis DA, Norgard MV, Roth MD, Modlin RL (2001) Microbial lipopeptides stimulate dendritic cell maturation via Toll-like receptor 2. J Immunol 166:2444
11. Thoma-Uszynski S, Stenger S, Takeuchi O, Ochoa MT, Engele M, Sieling PA, Barnes PF, Rollinghoff M, Bolcskei PL, Wagner M, Akira S, Norgard MV, Belisle JT, Godowski PJ, Bloom BR, Modlin RL (2001) Induction of direct antimicrobial activity through mammalian toll-like receptors. Science 291:1544
12. Liu PT, Stenger S, Li H, Wenzel L, Tan BH, Krutzik SR, Ochoa MT, Schauber J, Wu K, Meinken C, Kamen DL, Wagner M, Bals R, Steinmeyer A, Zugel U, Gallo RL, Eisenberg D, Hewison M, Hollis

BW, Adams JS, Bloom BR, Modlin RL (2006) Toll-like receptor triggering of a vitamin D-mediated human antimicrobial response. Science 311:1770

13. Reiling N, Holscher C, Fehrenbach A, Kroger S, Kirschning CJ, Goyert S, Ehlers S (2002) Cutting edge: Toll-like receptor (TLR)2- and TLR4-mediated pathogen recognition in resistance to airborne infection with Mycobacterium tuberculosis. J Immunol 169:3480

14. Drennan MB, Nicolle D, Quesniaux VJ, Jacobs M, Allie N, Mpagi J, Fremond C, Wagner H, Kirschning C, Ryffel B (2004) Toll-like receptor 2-deficient mice succumb to Mycobacterium tuberculosis infection. Am J Pathol 164:49

15. Bafica A, Scanga CA, Feng CG, Leifer C, Cheever A, Sher A (2005) TLR9 regulates Th1 responses and cooperates with TLR2 in mediating optimal resistance to Mycobacterium tuberculosis. J Exp Med 202:1715

16. Ogus AC, Yoldas B, Ozdemir T, Uguz A, Olcen S, Keser I, Coskun M, Cilli A, Yegin O (2004) The Arg753GLn polymorphism of the human toll-like receptor 2 gene in tuberculosis disease. Eur Respir J 23:219

17. Ben-Ali M, Barbouche MR, Bousnina S, Chabbou A, Dellagi K (2004) Toll-like receptor 2 Arg677Trp polymorphism is associated with susceptibility to tuberculosis in Tunisian patients. Clin Diagn Lab Immunol 11:625

18. Yim JJ, Lee HW, Lee HS, Kim YW, Han SK, Shim YS, Holland SM (2006) The association between microsatellite polymorphisms in intron II of the human Toll-like receptor 2 gene and tuberculosis among Koreans. Genes Immun 7:150

19. Bornman L, Campbell SJ, Fielding K, Bah B, Sillah J, Gustafson P, Manneh K, Lisse I, Allen A, Sirugo G, Sylla A, Aaby P, McAdam KP, Bah-Sow O, Bennett S, Lienhardt C, Hill AV (2004) Vitamin D receptor polymorphisms and susceptibility to tuberculosis in West Africa: a case-control and family study. J Infect Dis 190:1631

20. Bellamy R, Ruwende C, Corrah T, McAdam KP, Thursz M, Whittle HC, Hill AV (1999) Tuberculosis and chronic hepatitis B virus infection in Africans and variation in the vitamin D receptor gene. J Infect Dis 179:721

21. Selvaraj P, Narayanan PR, Reetha AM (2000) Association of vitamin D receptor genotypes with the susceptibility to pulmonary tuberculosis in female patients & resistance in female contacts. Indian J Med Res 111:172

22. Liu W, Cao WC, Zhang CY, Tian L, Wu XM, Habbema JD, Zhao QM, Zhang PH, Xin ZT, Li CZ, Yang H (2004) VDR and NRAMP1 gene polymorphisms in susceptibility to pulmonary tuberculosis among the Chinese Han population: a case-control study. Int J Tuberc Lung Dis 8:428

23. Wilkinson RJ, Llewelyn M, Toossi Z, Patel P, Pasvol G, Lalvani A, Wright D, Latif M, Davidson RN (2000) Influence of vitamin D deficiency and vitamin D receptor polymorphisms on tuberculosis among Gujarati Asians in west London: a case-control study. Lancet 355:618

24. Janeway CA Jr (1989) Approaching the asymptote? Evolution and revolution in immunology. Cold Spring Harb Symp Quant Biol 54(Pt 1):1

25. Yang Y, Yin C, Pandey A, Abbott D, Sassetti C, Kelliher MA (2007) NOD2 pathway activation by MDP or Mycobacterium tuberculosis infection involves the stable polyubiquitination of Rip2. J Biol Chem 282:36223

26. Girardin SE, Boneca IG, Viala J, Chamaillard M, Labigne A, Thomas G, Philpott DJ, Sansonetti PJ (2003) Nod2 is a general sensor of peptidoglycan through muramyl dipeptide (MDP) detection. J Biol Chem 278:8869

27. Delbridge LM, O'Riordan MX (2007) Innate recognition of intracellular bacteria. Curr Opin Immunol 19:10

28. Lemaitre B, Nicolas E, Michaut L, Reichhart JM, Hoffmann JA (1996) The dorsoventral regulatory gene cassette spatzle/Toll/cactus controls the potent antifungal response in Drosophila adults. Cell 86:973

29. Meister M, Lemaitre B, Hoffmann JA (1997) Antimicrobial peptide defense in Drosophila. Bioessays 19:1019

30. Medzhitov R, Preston-Hurlburt P, Janeway CA Jr (1997) A human homologue of the Drosophila Toll protein signals activation of adaptive immunity. Nature 388:394

31. Zhang D, Zhang G, Hayden MS, Greenblatt MB, Bussey C, Flavell RA, Ghosh S (2004) A toll-like receptor that prevents infection by uropathogenic bacteria. Science 303:1522

32. Dunne A, O'Neill LA (2005) Adaptor usage and Toll-like receptor signaling specificity. FEBS Lett 579:3330

33. Takeuchi O, Sato S, Horiuchi T, Hoshino K, Takeda K, Dong Z, Modlin RL, Akira S (2002) Cutting edge: role of Toll-like receptor 1 in mediating immune response to microbial lipoproteins. J Immunol 169:10

34. Doyle SE, O'Connell RM, Miranda GA, Vaidya SA, Chow EK, Liu PT, Suzuki S, Suzuki N, Modlin RL, Yeh WC, Lane TF, Cheng G (2004) Toll-like receptors induce a phagocytic gene program through p38. J Exp Med 199:81

35. Blander JM, Medzhitov R (2004) Regulation of phagosome maturation by signals from toll-like receptors. Science 304:1014

36. Hertz CJ, Wu Q, Porter EM, Zhang YJ, Weismuller KH, Godowski PJ, Ganz T, Randell SH, Modlin RL (2003) Activation of Toll-like receptor 2 on human tracheobronchial epithelial cells induces the antimicrobial peptide human beta defensin-2. J. Immunol 171:6820

37. Birchler T, Seibl R, Buchner K, Loeliger S, Seger R, Hossle JP, Aguzzi A, Lauener RP (2001) Human Toll-like receptor 2 mediates induction of the antimicrobial peptide human beta-defensin 2 in response to bacterial lipoprotein. Eur J Immunol 31:3131

38. Doyle S, Vaidya S, O'Connell R, Dadgostar H, Dempsey P, Wu T, Rao G, Sun R, Haberland M, Modlin R, Cheng G (2002) IRF3 mediates a TLR3/TLR4-specific antiviral gene program. Immunity 17:251

39. Kawai T, Takeuchi O, Fujita T, Inoue J, Muhlradt PF, Sato S, Hoshino K, Akira S (2001) Lipopolysaccharide stimulates the MyD88-independent pathway and results in activation of IFN-regulatory factor 3 and the expression of a subset of lipopolysaccharide-inducible genes. J Immunol 167: 5887

40. Kawai T, Sato S, Ishii KJ, Coban C, Hemmi H, Yamamoto M, Terai K, Matsuda M, Inoue J, Uematsu S, Takeuchi O, Akira S (2004) Interferon-alpha induction through Toll-like receptors involves a direct interaction of IRF7 with MyD88 and TRAF6. Nat Immunol 5:1061

41. Schmidt HH, Hofmann H, Schindler U, Shutenko ZS, Cunningham DD, Feelisch M (1996) No NO from NO synthase. Proc Natl Acad Sci USA 93:14492

42. Cantorna MT (2006) Vitamin D and its role in immunology: multiple sclerosis, and inflammatory bowel disease. Prog Biophys Mol Biol 92:60

43. Deluca HF, Cantorna MT (2001) Vitamin D: its role and uses in immunology. FASEB J 15:2579

44. Holick MF (2006) Resurrection of vitamin D deficiency and rickets. J Clin Invest 116: 2062

45. Rook GA, Steele J, Fraher L, Barker S, Karmali R, O'Riordan J, Stanford J (1986) Vitamin D3, gamma interferon, and control of proliferation of Mycobacterium tuberculosis by human monocytes. Immunology 57:159

46. Crowle AJ, Ross EJ, May MH (1987) Inhibition by 1,25(OH)2-vitamin D3 of the multiplication of virulent tubercle bacilli in cultured human macrophages. Infect Immun 55:2945

47. Sly LM, Lopez M, Nauseef WM, Reiner NE (2001) 1alpha,25-Dihydroxyvitamin D3-induced monocyte antimycobacterial activity is regulated by phosphatidylinositol 3-kinase and mediated by the NADPH-dependent phagocyte oxidase. J Biol Chem 276:35482

48. Anand PK, Kaul D (2005) Downregulation of TACO gene transcription restricts mycobacterial entry/survival within human macrophages. FEMS Microbiol Lett 250:137

49. Anand PK, Kaul D (2003) Vitamin D3-dependent pathway regulates TACO gene transcription. Biochem Biophys Res Commun 310:876

50. Wang TT, Nestel FP, Bourdeau V, Nagai Y, Wang Q, Liao J, Tavera-Mendoza L, Lin R, Hanrahan JW, Mader S, White JH (2004) Cutting edge: 1,25-dihydroxyvitamin D3 is a direct inducer of antimicrobial peptide gene expression. J Immunol 173:2909

51. Gombart AF, Borregaard N, Koeffler HP (2005) Human cathelicidin antimicrobial peptide (CAMP) gene is a direct target of the vitamin D receptor and is strongly up-regulated in myeloid cells by 1,25-dihydroxyvitamin D3. FASEB J 19:1067

52. Liu PT, Stenger S, Tang DH, Modlin RL (2007) Cutting edge: vitamin D-mediated human antimicrobial activity against Mycobacterium tuberculosis is dependent on the induction of cathelicidin. J Immunol 179:2060

53. Adorini L, Penna G, Giarratana N, Uskokovic M (2003) Tolerogenic dendritic cells induced by vitamin D receptor ligands enhance regulatory T cells inhibiting allograft rejection and autoimmune diseases. J Cell Biochem 88:227

54. D'Ambrosio D, Cippitelli M, Cocciolo MG, Mazzeo D, Di LP, Lang R, Sinigaglia F, Panina-Bordignon P (1998) Inhibition of IL-12 production by 1,25-dihydroxyvitamin D3. Involvement of NF-kappaB downregulation in transcriptional repression of the p40 gene. J Clin Invest 101:252

55. Griffin MD, Lutz W, Phan VA, Bachman LA, McKean DJ, Kumar R (2001) Dendritic cell modulation by 1alpha,25 dihydroxyvitamin D3 and its analogs: a vitamin D receptor-dependent pathway that promotes a persistent state of immaturity in vitro and in vivo. Proc Natl Acad Sci USA 98:6800

56. Hewison M, Gacad MA, Lemire J, Adams JS (2001) Vitamin D as a cytokine and hematopoetic factor. Rev Endocr Metab Disord 2:217

57. Lemire JM (1995) Immunomodulatory actions of 1,25-dihydroxyvitamin D3. J. Steroid Biochem Mol Biol 53:599

58. Panda DK, Miao D, Tremblay ML, Sirois J, Farookhi R, Hendy GN, Goltzman D (2001) Targeted ablation of the 25-hydroxyvitamin D 1alpha -hydroxylase enzyme: evidence for skeletal, reproductive, and immune dysfunction. Proc Natl Acad Sci USA 98:7498

59. Dardenne O, Prud'homme J, Arabian A, Glorieux FH, St-Arnaud R (2001) Targeted inactivation of the 25-hydroxyvitamin D(3)-1(alpha)-hydroxylase gene (CYP27B1) creates an animal model of pseudovitamin D-deficiency rickets. Endocrinology 142:3135

60. Hoover DM, Boulegue C, Yang D, Oppenheim JJ, Tucker K, Lu W, Lubkowski J (2002) The structure of human macrophage inflammatory protein-3alpha /CCL20. Linking antimicrobial and CC chemokine receptor-6-binding activities with human beta-defensins. J Biol Chem 277: 37647

61. Niyonsaba F, Ogawa H, Nagaoka I (2004) Human beta-defensin-2 functions as a chemotactic agent for tumour necrosis factor-alpha-treated human neutrophils. Immunology 111:273

62. Biragyn A, Ruffini PA, Leifer CA, Klyushnenkova E, Shakhov A, Chertov O, Shirakawa AK, Farber JM, Segal DM, Oppenheim JJ, Kwak LW (2002) Toll-like receptor 4-dependent activation of dendritic cells by beta-defensin 2. Science 298:1025

63. Ganz T (2003) Defensins: antimicrobial peptides of innate immunity. Nat Rev Immunol 3:710

64. Oppenheim JJ, Biragyn A, Kwak LW, Yang D (2003) Roles of antimicrobial peptides such as defensins in innate and adaptive immunity. Ann Rheum Dis 62(Suppl 2):ii17–ii21

65. Hoover DM, Rajashankar KR, Blumenthal R, Puri A, Oppenheim JJ, Chertov O, Lubkowski J (2000) The structure of human beta-defensin-2 shows evidence of higher order oligomerization. J Biol Chem 275:32911

66. Zanetti M (2004) Cathelicidins, multifunctional peptides of the innate immunity. J Leukoc Biol 75:39

67. Ahmad I, Perkins WR, Lupan DM, Selsted ME, Janoff AS (1995) Liposomal entrapment of the neutrophil-derived peptide indolicidin endows it with in vivo antifungal activity. Biochim Biophys Acta 1237:109

68. Shin SY, Kang SW, Lee DG, Eom SH, Song WK, Kim JI (2000) CRAMP analogues having potent antibiotic activity against bacterial, fungal, and tumor cells without hemolytic activity. Biochem Biophys Res Commun 275:904

69. Giacometti A, Cirioni O, Barchiesi F, Caselli F, Scalise G (1999) In-vitro activity of polycationic peptides against Cryptosporidium parvum, Pneumocystis carinii and yeast clinical isolates. J Antimicrob Chemother 44:403

70. Cirioni O, Giacometti A, Barchiesi F, Scalise G (1998) In-vitro activity of lytic peptides alone and in combination with macrolides and inhibitors of dihydrofolate reductase against Pneumocystis carinii. J Antimicrob Chemother 42:445

71. Tamamura H, Murakami T, Horiuchi S, Sugihara K, Otaka A, Takada W, Ibuka T, Waki M, Yamamoto N, Fujii N (1995) Synthesis of protegrin-related peptides and their antibacterial and anti-human immunodeficiency virus activity. Chem Pharm Bull (Tokyo) 43:853

72. Meyer T, Stockfleth E, Christophers E (2007) Immune response profiles in human skin. Br J Dermatol 157(Suppl 2):1
73. Sorensen O, Arnljots K, Cowland JB, Bainton DF, Borregaard N (1997) The human antibacterial cathelicidin, hCAP-18, is synthesized in myelocytes and metamyelocytes and localized to specific granules in neutrophils. Blood 90:2796
74. Di NA, Vitiello A, Gallo RL (2003) Cutting edge: mast cell antimicrobial activity is mediated by expression of cathelicidin antimicrobial peptide. J Immunol 170:2274
75. Agerberth B, Charo J, Werr J, Olsson B, Idali F, Lindbom L, Kiessling R, Jornvall H, Wigzell H, Gudmundsson GH (2000) The human antimicrobial and chemotactic peptides LL-37 and alpha-defensins are expressed by specific lymphocyte and monocyte populations. Blood 96:3086
76. Martineau AR, Wilkinson KA, Newton SM, Floto RA, Norman AW, Skolimowska K, Davidson RN, Sorensen OE, Kampmann B, Griffiths CJ, Wilkinson RJ (2007) IFN-gamma- and TNF-independent vitamin D-inducible human suppression of mycobacteria: the role of cathelicidin LL-37. J Immunol 178:7190
77. Rivas-Santiago B, Schwander SK, Sarabia C, Diamond G, Klein-Patel ME, Hernandez-Pando R, Ellner JJ, Sada E (2005) Human {beta}-defensin 2 is expressed and associated with Mycobacterium tuberculosis during infection of human alveolar epithelial cells. Infect Immun 73:4505
78. Sow FB, Florence WC, Satoskar AR, Schlesinger LS, Zwilling BS, Lafuse WP (2007) Expression and localization of hepcidin in macrophages: a role in host defense against tuberculosis. J Leukoc Biol 82:934
79. Tan BH, Meinken C, Bastian M, Bruns H, Legaspi A, Ochoa MT, Krutzik SR, Bloom BR, Ganz T, Modlin RL, Stenger S (2006) Macrophages acquire neutrophil granules for antimicrobial activity against intracellular pathogens. J Immunol 177:1864
80. Silva MT, Silva MN, Appelberg R (1989) Neutrophil-macrophage cooperation in the host defence against mycobacterial infections. Microb Pathog 6:369
81. Ravikumar M, Dheenadhayalan V, Rajaram K, Lakshmi SS, Kumaran PP, Paramasivan CN, Balakrishnan K, Pitchappan RM (1999) Associations of HLA-DRB1, DQB1 and DPB1 alleles with pulmonary tuberculosis in south India. Tuber Lung Dis 79:309
82. Mehra NK, Rajalingam R, Mitra DK, Taneja V, Giphart MJ (1995) Variants of HLA-DR2/DR51 group haplotypes and susceptibility to tuberculoid leprosy and pulmonary tuberculosis in Asian Indians. Int J Lepr Other Mycobact Dis. 63:241
83. Amirzargar AA, Yalda A, Hajabolbaghi M, Khosravi F, Jabbari H, Rezaei N, Niknam MH, Ansari B, Moradi B, Nikbin B (2004) The association of HLA-DRB, DQA1, DQB1 alleles and haplotype frequency in Iranian patients with pulmonary tuberculosis. Int J Tuberc Lung Dis 8:1017
84. Liu J, Fujiwara TM, Buu NT, Sanchez FO, Cellier M, Paradis AJ, Frappier D, Skamene E, Gros P, Morgan K (1995) Identification of polymorphisms and sequence variants in the human homologue of the mouse natural resistance-associated macrophage protein gene. Am J Hum Genet 56:845
85. Jouanguy E, Lamhamedi-Cherradi S, Lammas D, Dorman SE, Fondaneche MC, Dupuis S, Doffinger R, Altare F, Girdlestone J, Emile JF, Ducoulombier H, Edgar D, Clarke J, Oxelius VA, Brai M, Novelli V, Heyne K, Fischer A, Holland SM, Kumararatne DS, Schreiber RD, Casanova JL (1999) A human IFNGR1 small deletion hotspot associated with dominant susceptibility to mycobacterial infection. Nat Genet 21:370
86. Pan H, Yan BS, Rojas M, Shebzukhov YV, Zhou H, Kobzik L, Higgins DE, Daly MJ, Bloom BR, Kramnik I (2005) Ipr1 gene mediates innate immunity to tuberculosis. Nature 434:767
87. Fitness J, Floyd S, Warndorff DK, Sichali L, Mwaungulu L, Crampin AC, Fine PE, Hill AV (2004) Large-scale candidate gene study of leprosy susceptibility in the Karonga district of northern Malawi. Am J Trop Med Hyg 71:330
88. Grange JM, Davies PD, Brown RC, Woodhead JS, Kardjito T (1985) A study of vitamin D levels in Indonesian patients with untreated pulmonary tuberculosis. Tubercle 66:187
89. Delgado JC, Baena A, Thim S, Goldfeld AE (2002) Ethnic-specific genetic associations with pulmonary tuberculosis. J Infect Dis 186:1463
90. Uitterlinden AG, Fang Y, Van Meurs JB, Pols HA, Van Leeuwen JP (2004) Genetics and biology of vitamin D receptor polymorphisms. Gene 338:143

 91. Hewison M, Freeman L, Hughes SV, Evans KN, Bland R, Eliopoulos AG, Kilby MD, Moss PA, Chakraverty R (2003) Differential regulation of vitamin D receptor and its ligand in human monocyte-derived dendritic cells. J Immunol 170:5382

 92. Hewison M, Burke F, Evans KN, Lammas DA, Sansom DM, Liu P, Modlin RL, Adams JS (2007) Extra-renal 25-hydroxyvitamin D3-1alpha-hydroxylase in human health and disease. J Steroid Biochem Mol Biol 103:316

 93. Yang S, Smith C, Prahl JM, Luo X, Deluca HF (1993) Vitamin D deficiency suppresses cell-mediated immunity in vivo. Arch Biochem Biophys 303:98

 94. Lathers DM, Clark JI, Achille NJ, Young MR (2004) Phase 1B study to improve immune responses in head and neck cancer patients using escalating doses of 25-hydroxyvitamin D3. Cancer Immunol Immunother 53:422

 95. Adams JS, Ren S, Liu PT, Chun RF, Lagishetty V, Gombart AF, Borregaard N, Modlin RL, Hewison M (2009) Vitamin D-directed rheostatic regulation of monocyte antibacterial responses. J Immunol 182:4289.

 96. Krutzik SR, Hewison M, Liu PT, Robles JA, Stenger S, Adams JS, Modlin RL (2008) IL-15 links TLR2/1-induced macrophage differentiation to the vitamin D-dependent antimicrobial pathway. J Immunol 181:7115

 97. Hagenau T, Vest R, Gissel TN, Poulsen CS, Erlandsen M, Mosekilde L, Vestergaard P (2009) Global vitamin D levels in relation to age, gender, skin pigmentation and latitude: an ecologic meta-regression analysis. Osteoporos Int 20:133

 98. Barnes PF, Modlin RL, Bikle DD, Adams JS (1989) Transpleural gradient of 1,25-dihydroxyvitamin D in tuberculous pleuritis. J Clin Invest 83:1527

 99. Berwick M, Armstrong BK, Ben-Porat L, Fine J, Kricker A, Eberle C, Barnhill R (2005) Sun exposure and mortality from melanoma. J Natl Cancer Inst 97:195

100. Halliday GM (2005) Inflammation, gene mutation and photoimmunosuppression in response to UVR-induced oxidative damage contributes to photocarcinogenesis. Mutat Res 571:107

101. Giovannucci E (2005) The epidemiology of vitamin D and cancer incidence and mortality: a review (United States). Cancer Causes Control 16:83

102. Dowling GB, Thomas EW (1946) Treatment of lupus vulgaris with calciferol. Lancet 1:919–922

103. Morcos MM, Gabr AA, Samuel S, Kamel M, el Baz M, Michail RR (1998) Vitamin D administration to tuberculous children and its value. Boll Chim Farm 137:157

104. Nursyam EW, Amin Z, Rumende CM (2006) The effect of vitamin D as supplementary treatment in patients with moderately advanced pulmonary tuberculous lesion. Acta Med Indones 38:3

105. Zahrt TC, Deretic V (2001) Mycobacterium tuberculosis signal transduction system required for persistent infections. Proc Natl Acad Sci USA 98:12706

106. Martineau AR, Wilkinson RJ, Wilkinson KA, Newton SM, Kampmann B, Hall BM, Packe GE, Davidson RN, Eldridge SM, Maunsell ZJ, Rainbow SJ, Berry JL, Griffiths CJ (2007) A single dose of vitamin D enhances immunity to mycobacteria. Am J Respir Crit Care Med 176:208

107. Martineau AR, Honecker FU, Wilkinson RJ, Griffiths CJ (2007) Vitamin D in the treatment of pulmonary tuberculosis. J Steroid Biochem Mol Biol 103:793

14 Vitamin D and Colon Cancer

Heide S. Cross and Meinrad Peterlik

Abstract   Protection from colon cancer by vitamin D can be traced to the role of 1,25-dihydroxyvitamin D_3 [1,25-$(OH)_2D$] in controlling proliferation, differentiation, and apoptosis of colonic epithelial cells. Human colonocytes express the *CYP27B1*-encoded 25(OH)D-1α-hydroxylase and therefore are able to convert 25-hydroxyvitamin D[25(OH)D] to 1,25-$(OH)_2D$. In vitamin D insufficiency, availability of 25(OH)D is low, so that CYP27B1 activity in colonocytes may be not high enough to achieve tissue concentrations of 1,25-$(OH)_2D$ necessary to prevent tumor cell growth. Progression of colon tumors to well and moderately differentiated cancers is associated with a considerable increase in *CYP27B1* mRNA expression. Thus, colon carcinoma cells may still produce enough 1,25-$(OH)_2D$ to halt or retard further tumor growth. However, during progression to higher grades of malignancy, the efficiency of growth control by 1,25-$(OH)_2D$ is diminished due to low *CYP27B1* activity and by the predominance of the catabolic *CYP24A1*-encoded 25(OH)D-24-hydroxylase. There is evidence that the differentiation-dependent change in *CYP27B1* and *CYP24A1* expression is the result of differential epigenetic regulation of gene activity.

Results from our recent studies suggest that the chemopreventive potential of vitamin D can be augmented by factors that induce appropriate changes in *CYP27B1* and/or *CYP24A1* expression. In mice on a typical "Western style" diet, an increase in dietary calcium from 0.04 to 0.90% was accompanied by an up to 90% reduction of *CYP24A1* mRNA expression. 17β-Estradiol induced *CYP27B1* in the rectal mucosa of postmenopausal women with adenomatous polyps. The phytoestrogen genistein not only reduced *CYP24A1* but also up-regulated *CYP27B1* mRNA in mice as well as in a human colon cancer cell line. Finally, we identified folate as a potent suppressor of *CYP24A1* activity in the mouse colon. These findings provide a rationale for advocating adequate intake levels of calcium, folate, and phytoestrogens as an effective measure for prevention of colorectal cancer in humans.

   Key Words: Colorectal cancer; 1,25-dihydroxyvitamin D; CYP27B1; CYP24A1; chemoprevention; calcium; estrogen; phytoestrogens; folate

1. INTRODUCTION

In 1980 Garland and Garland proposed that sunlight and vitamin D can protect against colon cancer *(1)*. This hypothesis gained strong support when Garland et al. *(2)* in 1985 published the results of a 19-year prospective trial, showing that low dietary intakes of vitamin D and of calcium are associated with a significant risk of colorectal cancer. Since then many other observational studies reported a strong association between incidence or mortality rates of colon cancer and a low vitamin D status *(3)* or, respectively, low calcium intake *(4)*.

From: *Nutrition and Health: Vitamin D*
Edited by: M.F. Holick, DOI 10.1007/978-1-60327-303-9_14,
© Springer Science+Business Media, LLC 2010

Although an adequate supply of vitamin D reduces the incidence of colorectal cancer, the preventive effect of vitamin D consumption on colorectal cancer is a modest one *(5)*. This is plausible because dietary intake of vitamin D typically supplies only 10–20% of the daily requirement, whereas up to 80–90% come from UVB-mediated synthesis of vitamin D_3 in the epidermis *(6)*. Therefore, the impact of low sunshine exposure on vitamin D insufficiency and consequently on colorectal cancer incidence becomes that much greater. For example, an ecological study by Grant *(7)*, using DNA-weighted solar UVB exposure as an index of vitamin D_3 photoproduction in the skin, found a highly significant ($p < 0.001$) inverse association between low exposure to UVB and premature deaths from colon cancer. This association persisted even after additional ecologic risk factors (smoking, dietary factors, urban residence, poverty, etc.) were included in the analysis *(8)*. Premature mortality due to insufficient UVB exposure among white Americans was estimated at 12% of the total number of deaths from colon cancer *(7)*. To maintain an adequate vitamin D status for protection from colon cancer, vitamin D supply from UVB-mediated synthesis has to be complemented by vitamin D from food sources or even supplements. From a meta-analysis of 16 observational studies, Gorham et al. *(9)* concluded that ingestion of $\geq 1,000$ IU vitamin D_3 per day could reduce the risk of colon cancer incidence by 50%.

It must be taken into consideration that the efficiency of vitamin D_3 in reducing the risk of colon cancer depends very much on the calcium status of an individual, because calcium and vitamin D deficits act together in the pathogenesis of colorectal cancer: In a study on the effect of vitamin D and calcium supplementation on recurrence of colorectal adenomas, Grau et al. *(10)* found that 25(OH)D levels were associated with a reduced risk of adenoma recurrence only among subjects with high calcium intake. This notion is supported by findings from an interventional trial conducted by Holt et al. *(11)*, in which adenomatous polyp patients received high doses of supplemental calcium in combination with vitamin D as a chemopreventive regimen. After 6 months of treatment, a significant reduction in the rate of polyp formation was seen that was accompanied by an increase in expression of apoptotic markers. Cho et al. *(4)* concluded from an analysis of pooled primary data from 10 cohort studies, in which more than half a million individuals were followed up for 6–16 years, that optimal risk reduction for colorectal cancer necessitates high intake levels of both vitamin D and calcium.

1.1. Colorectal Cancer Prevention by Vitamin D and Calcium: Molecular and Cellular Mechanisms

1,25-Dihydroxyvitamin D_3 [1,25(OH)$_2$D] controls growth of normal and neoplastic cells by modulating the transcriptional activity of key genes involved in control of cellular proliferation, differentiation, and apoptosis (for review, see *(12)*). Studies from our laboratory *(13–16)* had identified c-Myc proto-oncogene and cyclin D1 expression as key targets of growth-inhibitory signaling from 1,25(OH)$_2$D/VDR. Since a number of intracellular proliferative pathways, viz., the Raf-1/MEK1/ERK and STAT-3, converge at c-myc and engage cyclin D1, a key element in cell cycle control, as a common downstream effector, 1,25(OH)$_2$D, must be considered a potent inhibitor of mitogenesis whatever the nature of cellular growth-promoting factors is. Another anti-proliferative mechanism of $1,25(OH)_2D$ in human colon carcinoma cells involves direct interaction with growth factor receptor-activated pathways: $1,25(OH)_2D$ not only reduces the number of ligand-occupied epidermal growth factor receptors (EGF-R) *(17)* but also blocks signal transduction downstream of EGF-R activation at c-myc *(18)*. Palmer et al. *(19)* analyzed in great detail the complicated network of intracellular pathways on which signaling from $1,25(OH)_2D$/VDR is transduced into cellular differentiation: Induction of E-cadherin and inactivation of the β-catenin/T cell transcription factor-4 (TCF-4) complex are the key actions of $1,25(OH)_2D$ in promoting differentiation of human colon carcinoma cells.

The observation that the efficiency of the anti-mitogenic action of $1,25(OH)_2D$ depends on adequate calcium nutrition can be explained on a molecular basis, at least in part, as follows: Human neoplastic colonocytes express the parathyroid-type extracellular calcium-sensing receptor (CaR) *(20)* at the mRNA and protein level as long as they retain a certain degree of differentiation *(21, 22)*. The CaR is a G protein-coupled membrane receptor which transduces minute changes in extracellular fluid Ca^{2+} concentration ($[Ca^{2+}]_o$) into various intracellular signaling pathways. Activation of the CaR in human colon adenocarcinoma-derived cells causes inhibition of phospholipase A_2 activity *(23)*, which, in turn, would reduce the amount of arachidonic acid available for the synthesis of proliferation-stimulating prostaglandins. Further downstream, CaR-activated anti-proliferative signaling targets the canonical Wnt/β-catenin pathway and thereby induces down-regulation of T cell transcription factor (TCF)-4 and induction of E-cadherin expression *(24, 25)*. Activation of the CaR will thus amplify anti-proliferative and pro-differentiating signaling from $1,25(OH)_2D$/VDR transduced along the Wnt/β-catenin pathway *(19, 24)*.

2. RELEVANCE OF ENDOGENOUS $1,25(OH)_2D$ SYNTHESIS FOR GROWTH CONTROL OF NORMAL AND NEOPLASTIC COLON EPITHELIAL CELLS

Conversion of $25(OH)D$ to $1,25(OH)_2D$ is catalyzed by the *CYP27B1*-encoded enzyme $25(OH)D$-1α-hydroxylase and occurs primarily in the kidney. However, many extra-renal cells also biosynthesize $1,25(OH)_2D$. Our laboratory was the first to show that normal and neoplastic human colon epithelial cells are endowed with a functional 25-hydroxyvitamin D-1α-hydroxylase and can thus convert $25(OH)D$ to $1,25(OH)_2D$ *(26–28)*.

Renal CYP27B1 activity is tightly regulated by serum Ca^{2+} and parathyroid hormone (PTH), as well as by feedback inhibition from $1,25(OH)_2D$. In contrast, *CYP27B1* expression in colonocytes is relatively insensitive to modulation via the PTH/$[Ca^{2+}]_o$ axis *(29)*. However, *CYP27B1* mRNA and protein levels can be significantly raised, for example, by epidermal growth factor *(30)* or 17β-estradiol *(31)*.

Thus, intracellular synthesis of $1,25(OH)_2D$ at extra-renal sites depends largely on ambient $25(OH)D$ levels and is not influenced by plasma levels of $1,25(OH)_2D$ *(32)*. This may explain why the incidence of vitamin D-dependent cancers, e.g., of the colorectum, *(33)* breast *(34)* and prostate gland *(35)*, is correlated with low serum 25(OH)D rather than with 1,25(OH)D. Strong support for the importance of intracellularly produced over circulating $1,25(OH)_2D$ for regulation of cell functions comes from the elegant study of Rowling et al. *(36)* who have shown that in mammary gland cells VDR-mediated actions depended more on megalin-mediated endocytosis of 25(OH)D than on ambient $1,25(OH)_2D$. Also Lechner et al. *(37)* could induce the characteristic anti-mitogenic effect of $1,25(OH)_2D$ when human colon carcinoma cells were treated with 25(OH)D though only when they were CYP27B1-positive. Altogether, at low serum levels of 25(OH)D, CYP27B1 activity in colonocytes may be not high enough to achieve steady-state tissue concentrations of $1,25(OH)_2D$ necessary to maintain normal cellular growth and differentiation.

3. EXPRESSION OF *CYP27B1* DURING INFLAMMATION AND TUMOR PROGRESSION

The relevance of the biologically active vitamin D compound for the maintenance of normal epithelial cell turnover in the large intestine is highlighted by results from studies with mice, which were genetically altered to block $1,25(OH)_2D$/VDR signaling: The colon mucosa of VDR null ($VDR^{-/-}$) mice show a pattern of increased DNA damage and cell division, the former probably due to formation of reactive oxygen species, which resembled the typical focal inflammatory and hyperproliferative lesions in incipient ulcerative colitis *(38)*, a condition which frequently develops into colon cancer. Interestingly, the large intestine reacts to inflammatory and hyperproliferative conditions with up-regulation of the VDR and of its ligand-synthesizing enzyme, CYP27B1: Liu et al. *(39)* reported that in a mouse model of ulcerative colitis, expression of CYP27B1 was increased fourfold compared with controls. From the observation that Cyp27b1 null mice show aggravated symptoms of the disease, it can be inferred that increased synthesis of $1,25(OH)_2D$ in colonocytes could have beneficial effects on severity and on progression of ulcerative colitis also in humans.

With respect to human colon cancer, we had shown that expression of *CYP27B1* rises about fourfold in the course of progression from adenomas to well and moderately differentiated (G1 and G2) cancers, and then substantially declines during further tumor progression *(40)*. Expression of the *VDR* showed the same dependence on tumor cell differentiation *(40)*. The majority of cancer cells in poorly differentiated (G3) lesions, however, are devoid of immunoreactivity for VDR and CYP27B1, while, at the same time, EGF receptor mRNA can be detected by in situ hybridization in almost any cancer cell *(41)*. We suggested therefore that the $1,25(OH)_2D$/VDR system can be activated in colon epithelial cells in response to mitogenic stimulation, e.g., by EGF *(17, 41)*. A strong autocrine/paracrine anti-mitogenic action of $1,25(OH)_2D$ would retard further tumor growth as long as cancer cells retain a certain degree of differentiation and high levels of CYP27B1 activity and of VDR expression. However, during progression to high-grade malignancy, signaling from the $1,25(OH)_2D$/VDR system would be too weak to effectively counteract proliferative effects from, for example, EGF-R activation *(41)*.

4. REGULATION OF VITAMIN D HYDROXYLASES, *CYP27B1* AND *CYP24A1*

It must be taken into account that the effective tissue concentration of $1,25(OH)_2D$ is determined not only by substrate availability but also by additional factors: (i) in colonocytes, $1,25(OH)_2D$ down-regulates *CYP27B1* and the *VDR*; (ii) $1,25(OH)_2D$ at the same time induces *CYP24A1*-encoded 25(OH)D-24-hydroxylase, the enzyme that initiates stepwise degradation of the hormone *(37)*; and (iii) expression of *CYP24A1* increases dramatically when colon cancers progress from a moderately to poorly differentiated state (G3) *(42)*. All this will further attenuate the anti-proliferative efficiency of locally produced 1,25(OH)D.

Prevention of colon cancer by vitamin D must therefore take a two-pronged approach: (i) consumption of vitamin D in the general population ought to be increased to achieve average 25(OH)D levels at and above at least 40 nM and (ii) changes in extra-renal vitamin metabolism in the direction of higher steady-state tissue concentrations of 1,25(OH)D could be achieved by nutrient factors like calcium, phytoestrogens, and folate.

4.1. Regulation of Vitamin D Metabolism in the Gut Mucosa by Calcium

It is intriguing that vitamin D in combination with high intake of calcium from dietary sources or supplements apparently is much more effective in reducing the risk of colorectal cancer than when given alone *(4, 10, 11)*. To find a viable explanation for this phenomenon, we studied mechanisms of vitamin D/calcium interaction on regulation of colonocyte growth and differentiation in a mouse model. When both male and female mice were fed an AIN76 minimal diet containing 0.04% calcium, they expressed colonocyte *CYP24A1* mRNA at a sixfold to eightfold higher level than their counterparts on a 0.9% calcium diet *(29)*.

At present it is not clear whether signals from low luminal calcium into elevated *CYP24A1* gene activity are transduced by the CaR. Alternatively, a lack of calcium is known to increase concentrations of free bile acids in the gut lumen. Of these, lithocholic acid, by binding to the VDR, can induce expression of *CYP24A1* *(43)*.

Our results suggest that also in humans calcium supplementation could lower the risk of colorectal cancer, because high dietary calcium suppresses vitamin D catabolism and thus favors accumulation of $1,25(OH)_2D$ in the colon mucosa. Furthermore, $1,25(OH)_2D$ would increase expression of the CaR by binding to a vitamin D responsive element in its promoter region *(44)*.

Interestingly, when *CYP27B1* expression in mouse colon was evaluated by real-time PCR in dependence of gender, it became apparent that it was significantly up-regulated in animals on 0.04% compared to 0.9% calcium though in female mice only *(45)*. Importantly, measurement of $1,25(OH)_2D$ concentrations in mucosal homogenates by a newly developed assay *(45)* indicates that up-regulation of *CYP27B1* by low calcium is translated into increased CYP27B1 protein activity causing accumulation of $1,25(OH)_2D$ in colonic mucosal cells. In parallel, in these cells apoptotic pathways, i.e., expression of the downstream effector proteases, caspase-3 and of caspase 7, are stimulated. This strongly suggests that enhanced synthesis of $1,25(OH)_2D$

in the ascending colon of female mice apparently overrides the gender-independent stimulatory effect of low calcium on CYP24A1-mediated vitamin D catabolism, thereby providing protection against incipient hyperproliferation induced by inadequate calcium nutrition.

4.2. Regulation of the Vitamin D System by Estrogenic Compounds

Although men and women suffer from similar rates of colorectal cancer deaths in their lifetime, the age-adjusted risk for colorectal cancer is less for women than for men *(46)*. This strongly indicates a protective role of female sex hormones, particularly of estrogens, against colorectal cancer (e.g., *(47, 48)*). Observational studies have further suggested that postmenopausal hormone therapy is associated with a lower risk for colorectal cancer and a lower death rate in women *(49)*. A meta-analysis of studies showed a 34% reduction in the incidence of this tumor in postmenopausal women receiving hormone replacement therapy *(50)*.

The mechanism of action of estrogens in lowering colon cancer risk is not understood. Since estrogen receptors are present in both normal intestinal epithelium and in colorectal cancers, the hormone is probably protective through these receptors and resultant post-receptor cellular activities. While the colon cannot be considered an estrogen-dependent tissue, it is part of a long list of estrogen-responsive organs. Expression of estrogen receptor (ER) subtypes α and β have been detected in cancer cell lines. The ER-α/ER-β ratio has been identified as a possible determinant of the susceptibility of a tissue to estrogen-induced carcinogenesis. Thus, binding of estrogen to ER-α induces cancer-promoting effects, whereas binding to ER-β exerts a protective action *(51)*. Normal human colon mucosa expresses primarily the ER-β type and very little ER-α, regardless of gender *(52)*.

Estrogens may exert an indirect anti-tumorigenic effect by changing VDR expression or vitamin D metabolism in colonic epithelial cells. Liel et al. *(53)* have reported that estrogen increased VDR activity in epithelial cells of the gastrointestinal tract. In the colon adenocarcinoma-derived cell line Caco-2, which is ER-β positive, but negative for ER-α, we demonstrated an increase of *CYP27B1* mRNA expression and also of enzymatic activity after treatment with 17β-estradiol *(31)*. Based on these findings a clinical pilot trial was designed, in which postmenopausal women with a past history of rectal adenomas were given 17β-estradiol daily for 1 month to reach premenopausal serum levels. Rectal biopsies were obtained at the beginning and end of trial. A predominant result was the elevation of *VDR* mRNA *(54)*. We also observed significant induction of *CYP27B1* mRNA in parallel to a decrease in *COX-2* mRNA expression in those patients who had particularly high levels of the inflammatory marker at the beginning of the trial (unpublished). From this it can be inferred that in human colonocytes estrogenic compounds have positive effects on endogenous synthesis of $1,25(OH)_2D$ and consequently on VDR-mediated anti-inflammatory and anti-mitogenic actions.

In this context it is of interest that in East Asian populations the risk of cancers of sex hormone-responsive organs, breast and prostate gland, as well as of the colorectum is clearly lower than elsewhere. This has been traced to the typical diet in this part of the world, which is rich in soy products and therefore contains high amounts of phytoestrogens. Of these, genistein induced *CYP27B1* and reduced *CYP24A1* expression and activity in a mouse model and in human colon adenocarcinoma-derived cell lines *(55)*, while daidzein, another phytoestrogen prominent in soy, and, importantly, its metabolite equol, which is strongly active in other biological systems, did not affect any of the colonic vitamin D hydroxylases *(31)*.

Genistein could also have anti-inflammatory properties in the colon: When mice were fed 0.04% dietary calcium, *COX-2* mRNA and protein were increased in the colon mucosa by about twofold in females and to a lesser extent in males. Supplementation of genistein to the diet lowered *COX-2* expression to control levels (0.5% dietary calcium) in both genders *(56)*. This suggests that genistein could have a beneficial effect on colonic inflammation similar to that seen with 17β-estradiol in the human pilot study described before.

Since genistein preferentially activates ER-β *(57, 58)*, which is equally expressed in the colon of women and men, low rates of colorectal cancer incidence in both genders in soy-consuming populations could be due to appropriate modulation of the anti-inflammatory and anti-cancer potential of vitamin D by the phytoestrogen.

4.3. Effect of Folate on CYP24A1 Expression

Folate, a water-soluble vitamin of the B family, is essential for synthesis, repair, and methylation of DNA. Evidence is increasing that a low folate status predisposes to development of several common malignancies including colorectal cancer. Giovannucci et al. *(59)* among others demonstrated that prolonged intake of folate above currently recommended levels significantly reduced the risk of colorectal cancer.

To investigate the relevance of folate for regulation of the vitamin D system *(60)*, we used C57/BL6 mice on the semisynthetic AIN76A diet, which contained, among others, 5% fat, 0.025 µg/g vitamin D_3, 5 mg/g calcium, and 2 µg/g folic acid. When this basal diet was modified to contain high fat, low calcium, low vitamin D_3, and low folic acid, mice exhibited signs of hyperplasia and hyperproliferation in the colon mucosa *(61)*, which were accompanied by a more than 2.5-fold elevated *CYP24A1* mRNA expression. When calcium and vitamin D_3 intake was optimized, while fat was still high and folic acid low, *CYP24A1* mRNA expression fell by 50%, but was still higher than in the colon mucosa of mice fed the basal (control) diet. Finally, when the diet contained high fat, low calcium, and low vitamin D, but folic acid was optimized, only then any increment in colonic *CYP24A1* due to dietary manipulations was completely abolished *(60)* (for details on diets see Fig. 1).

4.4. Epigenetic Regulation of CYP24A1 and CYP27B1 Activity

As a methyl donor, folate plays an important role in epigenetic regulation of gene expression. DNA methylation of cytosine residues of CpG islands in the promoter region of genes is associated with transcriptional silencing of gene expression in mammalian cells. This could be underlying the observed inhibitory effect of folate on CYP24A1 activity in the mouse colon *(60)*. Conversely, decreased methylation of CpG islands enhances gene activity, for example, of the VDR in a mouse model of chemically induced colon cancer *(62)*. We therefore wanted to know whether epigenetic

Content of	AIN76A	NWD1	NWD2	NWD3
Fat (corn oil), %	5	20	20	20
Vitamin D$_3$, IU /g	1	0.11	2.30	0.11
Calcium, mg/g	5	0.50	7.00	0.50
Folic acid, µg/g	2	0.23	0.23	2.00

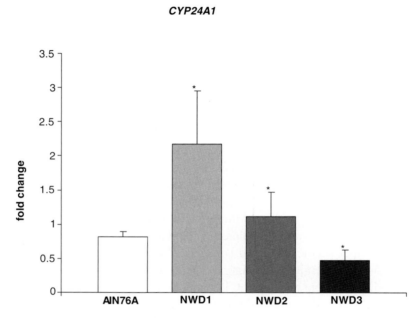

Fig. 1. Influence of diet on *CYP24A1* mRNA expression in mouse proximal colon. *CYP24A1* was evaluated by quantitative RT-PCR and is expressed as fold change over internal control. Values are mean ± SE, $n \geq 6$ mice (triplicate analysis was performed). *Asterisk* indicates statistically significant difference from AIN76 control, $p \leq 0.05$ by Student's *t* test.

regulation of gene activity through methylation/demethylation processes could account also for the differences in expression of the vitamin D hydroxylases in course of tumor progression, as observed previously in colon cancer patients *(40, 42)*. In fact, further analysis of tumor biopsies from those patients showed that in poorly differentiated cancerous lesions, regions of the *CYP24A1* promoter were demethylated and those of CYP27B1 were methylated, whereas the reverse was observed in rather differentiated cancers (unpublished). Hence, as a result of epigenetic regulation, expression of *CYP27B1* and of *CYP24A1* should change in opposite directions during progression to a highly malignant state. Figure 2 shows that this is actually the case: Transition from low- to high-grade cancers is associated with a further rise in

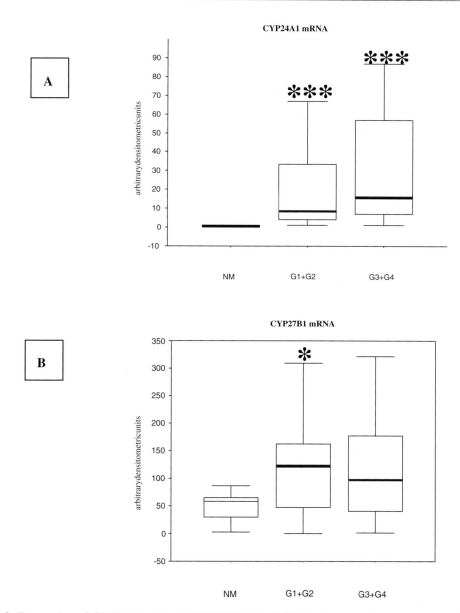

Fig. 2. Expression of CYP24A1 (**a**) and CYP27B1 (**b**) mRNA in human colon tumor tissue. Semi-quantitative RT-PCR was evaluated by densitometry. NM, normal colonic mucosa obtained from diverticulitis patients ($n = 10$) after stoma reoperation. G1, G2: biopsies from well and moderately differentiated cancerous lesions ($n = 59$). G3, G4: biopsies from poorly and undifferentiated cancers ($n = 46$). Values are mean ± SE from a total of 105 tumor patient samples. *Asterisks* indicate statistically significant difference from NM; $*p \leq 0.05$, $***p \leq 0.001$.

CYP24A1 mRNA expression (Fig. 2a) and a simultaneous decline of *CYP27B1* mRNA (Fig. 2b).

4.5. Regulation of CYP24A1 and CYP27B1 and Prevention of Colon Cancer

As mentioned before, CpG islands in the *CYP24A1* promoter are methylated in well to moderately differentiated colon cancers, and gene activity should therefore be suppressed. However, a small rise of *CYP24A1* activity is observed even at this stage of tumor progression (Fig. 2a) *(42)*. To explain this, one has to take into account that in low-grade cancerous lesions, *CYP27B1* expression, probably due to demethylation of relevant promoter regions, is exceedingly high compared to normal mucosa in non-cancer patients (Fig. 2b). Accumulation of 1,25(OH)$_2$D in the colon mucosa would thus be responsible for up-regulation of transcriptional activity of *CYP24A1 (37)* but, more importantly, also for autocrine/paracrine inhibition of tumor cell growth. In contrast, in highly malignant cancers, an efficient anti-mitogenic effect of 1,25(OH)$_2$D is unlikely, because expression of the catabolic vitamin D hydroxylase, CYP24A1, by far exceeds that of CYP27B1 (cf. Fig. 2a). We therefore suggest that in well and moderately differentiated cancers, vitamin D hydroxylase expression levels could be appropriately manipulated to halt further progression to highly malignant stages. Our findings that this could be achieved by, for example, calcium, phytoestrogens, or folate provide a rationale for advocating adequate intake levels of these nutrients as an effective measure for colon cancer prevention.

REFERENCES

1. Garland CF, Garland FC (1980) Do sunlight and vitamin D reduce the likelihood of colon cancer? Int J Epidemiol 9:227–231
2. Garland C, Shekelle RB, Barrett-Connor E, Criqui MH, Rossof AH, Paul O (1985) Dietary vitamin D and calcium and risk of colorectal cancer: a 19-year prospective study in men. Lancet 1: 307–309
3. Garland CF, Garland FC, Gorham ED et al (2006) The role of vitamin D in cancer prevention. Am J Public Health 96:252–261
4. Cho E, Smith-Warner SA, Spiegelman D et al (2004) Dairy foods, calcium, and colorectal cancer: a pooled analysis of 10 cohort studies. J Natl Cancer Inst 96(13):1015–1022
5. Grant WB, Garland CF (2004) A critical review of studies on vitamin D in relation to colorectal cancer. Nutr Cancer 48:115–123
6. Holick MF (2003) Vitamin D: a millenium perspective. J Cell Biochem 88(2):296–307
7. Grant WB (2002) An estimate of premature cancer mortality in the US due to inadequate doses of solar ultraviolet-B radiation. Cancer 94:1867–1875
8. Grant WB (2003) Ecologic studies of solar UV-B radiation and cancer mortality rates. Recent Results Cancer Res 164:371–377
9. Gorham ED, Garland CF, Garland FC et al (2005) Vitamin D and prevention of colorectal cancer. J Steroid Biochem Mol Biol 97:179–194
10. Grau MV, Baron JA, Sandler RS et al (2003) Vitamin D, calcium supplementation, and colorectal adenomas: results of a randomized trial. J Natl Cancer Inst 95:1765–1771
11. Holt PR, Bresalier RS, Ma CK et al (2006) Calcium plus vitamin D alters preneoplastic features of colorectal adenomas and rectal mucosa. Cancer 106:287–296
12. Deeb KK, Trump DL, Johnson CS (2007) Vitamin D signalling pathways in cancer: potential for anticancer therapeutics. Nat Rev Cancer 7:684–700

13. Cross HS, Pavelka M, Slavik J, Peterlik M (1992) Growth control of human colon cancer cells by vitamin D and calcium in vitro. J Natl Cancer Inst 84:1355–1357

14. Cross HS, Hulla W, Tong WM, Peterlik M (1995) Growth regulation of human colon adenocarcinoma-derived cells by calcium, vitamin D and epidermal growth factor. J Nutr 125(7 Suppl):2004S–2008S

15. Hulla W, Kallay E, Krugluger W, Peterlik M, Cross HS (1995) Growth control of human colon-adenocarcinoma-derived Caco-2 cells by vitamin-D compounds and extracellular calcium in vitro: relation to c-myc-oncogene and vitamin-D-receptor expression. Int J Cancer 62:711–716

16. Tong WM, Hofer H, Ellinger A, Peterlik M, Cross HS (1999) Mechanism of antimitogenic action of vitamin D in human colon carcinoma cells: relevance for suppression of epidermal growth factor-stimulated cell growth. Oncol Res 11:77–84

17. Tong WM, Kallay E, Hofer H et al (1998) Growth regulation of human colon cancer cells by epidermal growth factor and 1,25-dihydroxyvitamin D3 is mediated by mutual modulation of receptor expression. Eur J Cancer 34:2119–2125

18. Hulla W, Kallay E, Krugluger W, Peterlik M, Cross HS (1995) Growth control of human colon-adenocarcinoma-derived Caco-2 cells by vitamin-D compounds and extracellular calcium in vitro: relation to c-myc-oncogene and vitamin-D-receptor expression. Int J Cancer 62:711–716

19. Palmer HG, Gonzalez-Sancho JM, Espada J et al (2001) Vitamin D(3) promotes the differentiation of colon carcinoma cells by the induction of E-cadherin and the inhibition of beta-catenin signaling. J Cell Biol 154:369–387

20. Brown EM, Gamba G, Riccardi D et al (1993) Cloning and characterization of an extracellular Ca(2+)-sensing receptor from bovine parathyroid. Nature 366:575–580

21. Kállay E, Bajna E, Wrba F, Kriwanek S, Peterlik M, Cross HS (2000) Dietary calcium and growth modulation of human colon cancer cells: role of the extracellular calcium-sensing receptor. Cancer Detect Prev 24:127–136

22. Sheinin Y, Kallay E, Wrba F, Kriwanek S, Peterlik M, Cross HS (2000) Immunocytochemical localization of the extracellular calcium-sensing receptor in normal and malignant human large intestinal mucosa. J Histochem Cytochem 48:595–602

23. Kallay E, Bonner E, Wrba F, Thakker RV, Peterlik M, Cross HS (2003) Molecular and functional characterization of the extracellular calcium-sensing receptor in human colon cancer cells. Oncol Res 13:551–559

24. Chakrabarty S, Wang H, Canaff L, Hendy GN, Appelman H, Varani J (2005) Calcium sensing receptor in human colon carcinoma: interaction with Ca(2+) and 1,25-dihydroxyvitamin D(3). Cancer Res 65:493–498

25. MacLeod RJ, Hayes M, Pacheco I (2007) Wnt5a secretion stimulated by the extracellular calcium-sensing receptor inhibits defective Wnt signaling in colon cancer cells. Am J Physiol Gastrointest Liver Physiol 293:G403–G411

26. Cross HS, Peterlik M, Reddy GS, Schuster I (1997) Vitamin D metabolism in human colon adenocarcinoma-derived Caco-2 cells: expression of 25-hydroxyvitamin D3-1alpha-hydroxylase activity and regulation of side-chain metabolism. J Steroid Biochem Mol Biol 62:21–28

27. Bareis P, Bises G, Bischof MG, Cross HS, Peterlik M (2001) 25-hydroxy-vitamin d metabolism in human colon cancer cells during tumor progression. Biochem Biophys Res Commun 285:1012–1017

28. Bises G, Kallay E, Weiland T et al. (2004) 25-hydroxyvitamin D3-1alpha-hydroxylase expression in normal and malignant human colon. J Histochem Cytochem 52:985–989

29. Kallay E, Bises G, Bajna E et al (2005) Colon-specific regulation of vitamin D hydroxylases – a possible approach for tumor prevention. Carcinogenesis 26:1581–1589

30. Bareis P, Kallay E, Bischof MG et al (2002) Clonal differences in expression of 25-hydroxyvitamin D(3)-1alpha-hydroxylase, of 25-hydroxyvitamin D(3)-24-hydroxylase, and of the vitamin D receptor in human colon carcinoma cells: effects of epidermal growth factor and 1alpha, 25-dihydroxyvitamin D(3). Exp Cell Res 276:320–327

31. Lechner D, Bajna E, Adlercreutz H, Cross HS (2006) Genistein and 17beta-estradiol, but not equol, regulate vitamin D synthesis in human colon and breast cancer cells. Anticancer Res 26:2597–2603

32. Anderson PH, O'Loughlin PD, May BK, Morris HA (2005) Modulation of CYP27B1 and CYP24 mRNA expression in bone is independent of circulating 1,25(OH)(2)D(3) levels. Bone 36:654–662

33. Feskanich D, Ma J, Fuchs CS et al (2004) Plasma vitamin D metabolites and risk of colorectal cancer in women. Cancer Epidemiol Biomarkers Prev 13:1502–1508

34. Bertone-Johnson ER, Chen WY, Holick MF et al (2005) Plasma 25-hydroxyvitamin D and 1,25-dihydroxyvitamin D and risk of breast cancer. Cancer Epidemiol Biomarkers Prev 14:1991–1997

35. Tuohimaa P, Tenkanen L, Ahonen M et al (2004) Both high and low levels of blood vitamin D are associated with a higher prostate cancer risk: a longitudinal, nested case-control study in the Nordic countries. Int J Cancer 108:104–108

36. Rowling MJ, Kemmis CM, Taffany DA, Welsh J (2006) Megalin-mediated endocytosis of vitamin D binding protein correlates with 25-hydroxycholecalciferol actions in human mammary cells. J Nutr 136:2754–2759

37. Lechner D, Kallay E, Cross HS (2007) 1alpha,25-dihydroxyvitamin D3 downregulates CYP27B1 and induces CYP24A1 in colon cells. Mol Cell Endocrinol 263:55–64

38. Kallay E, Pietschmann P, Toyokuni S et al (2001) Characterization of a vitamin D receptor knock-out mouse as a model of colorectal hyperproliferation and DNA damage. Carcinogenesis 22: 1429–1435

39. Liu N, Nguyen L, Chun RF et al (2008) Altered endocrine and autocrine metabolism of vitamin D in a mouse model of gastrointestinal inflammation. Endocrinology 149:4799–4808

40. Cross HS, Bareis P, Hofer H et al (2001) 25-Hydroxyvitamin D(3)-1alpha-hydroxylase and vitamin D receptor gene expression in human colonic mucosa is elevated during early cancerogenesis. Steroids 66:287–292

41. Sheinin Y, Kaserer K, Wrba F et al (2000) In situ mRNA hybridization analysis and immunolocalization of the vitamin D receptor in normal and carcinomatous human colonic mucosa: relation to epidermal growth factor receptor expression. Virchows Arch 437:501–507

42. Cross H, Bises G, Lechner D, Manhardt T, Kallay E (2005) The Vitamin D endocrine system of the gut – its possible role in colorectal cancer prevention. J Steroid Biochem Mol Biol 97: 121–128

43. Nehring JA, Zierold C, DeLuca HF (2007) Lithocholic acid can carry out in vivo functions of vitamin D. Proc Natl Acad Sci U S A 104:10006–10009

44. Canaff L, Hendy GN (2002) Human calcium-sensing receptor gene. Vitamin D response elements in promoters P1 and P2 confer transcriptional responsiveness to 1,25-dihydroxyvitamin D. J Biol Chem 277:30337–30350

45. Nittke T, Kallay E, Manhardt T, Cross HS (2009) Parallel elevation of colonic 1,25-dihydroxyvitamin D3 levels and apoptosis in female mice on a calcium-deficient diet. Anti cancer Res 29:3727–3732.

46. Calle EE, Miracle-McMahill HL, Thun MJ, Heath CW Jr (1995) Estrogen replacement therapy and risk of fatal colon cancer in a prospective cohort of postmenopausal women. J Natl Cancer Inst 87: 517–523

47. DeCosse JJ, Ngoi SS, Jacobson JS, Cennerazzo WJ (1993) Gender and colorectal cancer. Eur J Cancer Prev 2:105–115

48. Jemal A, Murray T, Samuels A, Ghafoor A, Ward E, Thun MJ (2003) Cancer statistics, 2003. CA Cancer J Clin 53:5–26

49. Grodstein F, Newcomb PA, Stampfer MJ (1999) Postmenopausal hormone therapy and the risk of colorectal cancer: a review and meta-analysis. Am J Med 106:574–582

50. Chlebowski RT, Wactawski-Wende J, Ritenbaugh C et al (2004) Estrogen plus progestin and colorectal cancer in postmenopausal women. N Engl J Med 350:991–1004

51. Campbell-Thompson M, Lynch IJ, Bhardwaj B (2001) Expression of estrogen receptor (ER) subtypes and ERbeta isoforms in colon cancer. Cancer Res 61:632–640

52. Foley EF, Jazaeri AA, Shupnik MA, Jazaeri O, Rice LW (2000) Selective loss of estrogen receptor beta in malignant human colon. Cancer Res 60:245–248

53. Liel Y, Shany S, Smirnoff P, Schwartz B (1999) Estrogen increases 1,25-dihydroxyvitamin D receptors expression and bioresponse in the rat duodenal mucosa. Endocrinology 140:280–285

54. Protiva P, Cross H, Hopkins ME, Kallay E, Bises G, Dreyhaupt E, Augenlicht L, Lipkin M, Lesser M, Livote E, Holt PR (2008) Chemoprevention of colorectal neoplasia by estrogen: potential role of vitamin D. Cancer Prev 29(9):1788–1793

55. Cross HS, Kallay E, Lechner D, Gerdenitsch W, Adlercreutz H, Armbrecht HJ (2004) Phytoestrogens and vitamin D metabolism: a new concept for the prevention and therapy of colorectal, prostate, and mammary carcinomas. J Nutr 134:1207S–1212S

56. Bises G, Bajna E, Manhardt T, Gerdenitsch W, Kallay E, Cross HS (2007) Gender-specific modulation of markers for premalignancy by nutritional soy and calcium in the mouse colon. J Nutr 137: 211S–215S

57. An J, Tzagarakis-Foster C, Scharschmidt TC, Lomri N, Leitman DC (2001) Estrogen receptor beta-selective transcriptional activity and recruitment of coregulators by phytoestrogens. J Biol Chem 276:17808–17814

58. Kuiper GG, Lemmen JG, Carlsson B et al (1998) Interaction of estrogenic chemicals and phytoestrogens with estrogen receptor beta. Endocrinology 139:4252–4263

59. Giovannucci E, Stampfer MJ, Colditz GA et al (1998) Multivitamin use, folate, and colon cancer in women in the Nurses' Health Study. Ann Intern Med 129:517–524

60. Cross HS, Lipkin M, Kallay E (2006) Nutrients regulate the colonic vitamin D system in mice: relevance for human colon malignancy. J Nutr 136:561–564

61. Newmark HL, Yang K, Lipkin M et al (2001) A western-style diet induces benign and malignant neoplasms in the colon of normal C57Bl/6 mice. Carcinogenesis 22:1871–1875

62. Smirnoff P, Liel Y, Gnainsky J, Shany S, Schwartz B (1999) The protective effect of estrogen against chemically induced murine colon carcinogenesis is associated with decreased CpG island methylation and increased mRNA and protein expression of the colonic vitamin D receptor. Oncol Res 11:255–264

15 Mechanisms of Resistance to Vitamin D Action in Human Cancer Cells

María Jesús Larriba and Alberto Muñoz

Abstract   Initial clinical trials in cancer patients with vitamin D compounds have shown acceptable toxicity but low activity. A number of mechanisms responsible for resistance to their action in cancer cells have been recently reported. They include reduced intracellular availability of $1\alpha,25$-dihydroxyvitamin D_3 $[1,25(OH)_2D]$, loss of vitamin D receptor (VDR) expression and deregulation of transcription corepressors that modulate VDR action. Here, we summarize the data in the literature on the altered activity of the enzymes (CYP27B1, CYP24A1) that controls $1,25(OH)_2D$ levels, the repression of VDR by the transcription factor Snail1 and the overexpression of several VDR corepressors (NCoR, SMRT) in cancer cells. A better understanding of these processes must contribute to improved protocols for the clinical use of vitamin D compounds.

Key Words: Vitamin D resistance; CYP27B1; CYP24A1; VDR repression; VDR corepressors

The classical consideration of vitamin D as a regulator of calcium and phosphate metabolism and bone biology began to change in 1981, when David Feldman's and Tatsuo Suda's groups showed that the most active vitamin D metabolite $1\alpha,25$-dihydroxyvitamin D_3 $(1,25(OH)_2D$, calcitriol) inhibited the proliferation of melanoma cells and induced the differentiation of leukemic cells [1, 2]. These seminal, pioneer findings opened a new era in the study of $1,25(OH)_2D$, which is now seen as a hormone with pleiotropic effects in the organism.

A large number of epidemiological and experimental studies performed in cultured cells and animal models over the last three decades support a cancer preventive and, perhaps, a therapeutic role for $1,25(OH)_2D$ and a series of synthetic vitamin D analogs [3, 4, reviews]. The initial phase I and phase II cancer clinical studies showed acceptable toxicity but low activity of these compounds [5, review]. As is usual in the development of new antitumoral drugs, patients enrolled in these trials were not responding to any other therapy, and they were unselected in terms of putative responsiveness to vitamin D compounds. Recent data on the physiology of the vitamin D system improve our understanding of the action of $1,25(OH)_2D$ at the molecular and cellular level, which

From: *Nutrition and Health: Vitamin D*
Edited by: M.F. Holick, DOI 10.1007/978-1-60327-303-9_15,
© Springer Science+Business Media, LLC 2010

may in turn help us to design future clinical trials more rationally. In this chapter, we summarize current knowledge on the mechanisms that dictate the response or resistance to 1,25(OH)$_2$D in human cancer cells. The mechanisms have been grouped into three classes: (a) the bioavailability of 1,25(OH)$_2$D in the cell, (b) the integrity and level of expression of the vitamin D receptor (VDR); and (c) the pattern of expression of the transcription coregulators (coactivators and corepressors) that modulate VDR action.

1,25(OH)$_2$D action in human cancer cells. The anticancer action of 1,25(OH)$_2$D relies on several mechanisms at the cell level: inhibition of proliferation, invasion and angiogenesis, sensitization to apoptotic stimuli, induction of differentiation, and modulation of the immune system *(3–5)*. In each type of cancer, the combined effect of these mechanisms on tumor and stromal cells may determine 1,25(OH)$_2$D action.

1,25(OH)$_2$D has two types of effect on target cells: the regulation of transcription rate of a large number of genes (genomic effects), and the rapid transcription-independent modulation of the activity of membrane ion channels and cytosolic kinases, phosphatases, and phospholipases (non-genomic effects). Most studies indicate that both types of effect are mediated by 1,25(OH)$_2$D binding to, and activation of VDR. This is a member of the superfamily of nuclear receptors that is expressed in many cell types, and it acts as a ligand-modulated transcription factor regulating gene expression. A few studies have proposed the existence of 1,25(OH)$_2$D receptors other than VDR, but their confirmation as physiological mediators of 1,25(OH)$_2$D action is pending *(3, 5)*.

The current model for gene activation by 1,25(OH)$_2$D–VDR predicts that unliganded VDR bound (forming a heterodimer with RXR, the retinoid X receptor) to regulatory sequences (vitamin D response elements or VDRE) in target genes represses their transcription by recruiting corepressors (NCoR, SMRT, etc.) and histone deacetylases. 1,25(OH)$_2$D induces a conformational change in VDR that results in the replacement of corepressors by coactivators (SRC1, NCoA2, etc.) and increased histone acetylase activity. This results in the opening of chromatin structure thus allowing entry of the basal RNA polymerase II transcription machinery (Fig. 1) *(6, review)*. The mechanism of gene repression by 1,25(OH)$_2$D is less known, although this process is probably equally important for its action, as around one-third of the target genes are inhibited. In addition to the direct blockade of RNA polymerase II activity by binding to VDREs located close to the transcription initiation site, gene repression may be due to competition for DNA binding with other transcription factors or due to interference of their transcription regulatory function (by protein–protein interaction or modulation of their phosphorylation) *(5)*.

These considerations indicate that 1,25(OH)$_2$D acts by modulating VDR action and that such action is fine-tuned by its interaction with coregulators (coactivators and corepressors). 1,25(OH)$_2$D is the most active vitamin D metabolite because it has the highest affinity for VDR binding. 1,25(OH)$_2$D is synthesized from 25-hydroxyvitamin D$_3$ [25(OH)D] by the action of the 25-hydroxyvitamin D$_3$ 1α-hydroxylase (CYP27B1). As discussed below, this has long been thought to take place only in the kidney, but it is now known to occur in several other cell types including colon, breast, and prostate normal and transformed epithelial cells. Intracellularly, 1,25(OH)$_2$D is degraded by further hydroxylation at position 24 by the ubiquitous enzyme 24-hydroxylase (CYP24A1) *(3, 5)*.

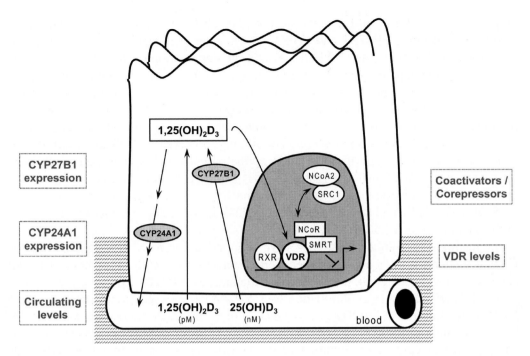

Fig. 1. Schematic representation of 1,25(OH)$_2$D metabolism and action in human cancer cells and of the mechanisms of resistance to 1,25(OH)$_2$D. Cancer epithelial cells receive 25(OH)D and 1,25(OH)$_2$D (in the nM and pM range, respectively) from the bloodstream and synthesize 1,25(OH)$_2$D from 25(OH)D by CYP27B1 action. 1,25(OH)$_2$D binding to VDR induces a conformational change that leads to the replacement of corepressors (SMRT, NCoR) by coactivators (SRC1, NCoA2) and finally induces the transcriptional activation of many genes involved in different cellular functions related to 1,25(OH)$_2$D antitumoral action. Alteration of the expression and activity of the enzymes regulating 1,25(OH)$_2$D synthesis (CYP27B1) and catabolism (CYP24A1) have been found in cancer, and they cause the reduction of intracellular 1,25(OH)$_2$D levels. Similarly, VDR repression and aberrant expression of transcriptional coactivators or corepressors have been found in cancer and they lead to 1,25(OH)$_2$D resistance.

In summary (Fig. 1), cell responsiveness to 1,25(OH)$_2$D primarily relies on the expression of VDR. Second, VDR activity depends on cellular 1,25(OH)$_2$D levels that result from the balance between the circulating concentrations of 25(OH)D and 1,25(OH)$_2$D, which cross the plasma membrane by either passive or facilitated (carrier-mediated) diffusion, and the cellular activity of CYP27B1 and CYP24A1 enzymes that control 25(OH)D conversion into 1,25(OH)$_2$D and subsequent degradation of this molecule. Third, the pattern of expression of coactivators and corepressors modulates VDR activity, thus establishing the precise mode and strength of the effects of 1,25(OH)$_2$D on the transcription of target genes. An additional overimposed level of regulation comes from other signals that may trigger molecular pathways affecting the post-translational (phosphorylation) modification, intracellular location, or half-life (polyubiquitylation) of VDR or any of its coregulators.

Resistance due to reduced bioavailability of 1,25(OH)$_2$D within cancer cells. Ligand occupancy and activation of VDR depends on the levels of circulating 25(OH)D (nM range) and $1,25(OH)_2D$ (pM range) that enter the cell from the bloodstream and on the cellular activity of CYP27B1 and CYP24A1 enzymes that regulate, respectively, the conversion of $25(OH)D$ into $1,25(OH)_2D$ and the degradation of this molecule. Obviously, vitamin D deficiency due to insufficient vitamin D_3 synthesis or diet contribution will determine low circulating $25(OH)_2D$ levels. Decreased renal CYP27B1 expression or activity has the same consequence. Kidney CYP27B1 was long time considered the only source of $1,25(OH)_2D$, which would then reach the bloodstream and act on target tissues in an endocrine fashion. The finding, however, of CYP27B1 expression and activity in normal and tumoral colon epithelial cells and later in other cell types (prostate, breast) demonstrated that $1,25(OH)_2D$ has paracrine and autocrine activities that may be involved in a defense mechanism against cancer progression (7, 8). $1,25(OH)_2D$ rapidly and strongly induces CYP24A1 expression, leading to an increase in the catabolism of $1,25(OH)_2D$ (8).

Interestingly, a number of studies have reported differences in CYP27B1 and/or CYP24A1 expression/activity between normal and tumoral cells. This has led to the currently accepted idea that, in cancer cells, attenuated CYP27B1 expression or activity or accelerated $1,25(OH)_2D$ elimination by CYP24A1 overexpression leads to reduced VDR activation and $1,25(OH)_2D$ resistance. Colon, prostate, and breast cancer cells in culture show reduced CYP27B1 expression, which in some cases appears to be a consequence of epigenetic alteration of the genetic locus, as treatment of cells with methylation or deacetylation inhibitors increased CYP27B1 expression (9). One study has proposed that the mechanism of decreased CYP27B1 activity in prostate cancer cell lines is via decreased gene expression while in primary cultures and tissues it is post-translational (10). Moreover, the finding that CYP27B1 is also present in tumor infiltrating macrophages suggests an immunomodulatory component of $1,25(OH)_2D$ production in some types of cancer (8).

CYP24A1 overexpression is commonly held responsible for the partial or total resistance of colon, breast, and prostate cancer cells to $1,25(OH)_2D$ effects. In line with this, combination of $1,25(OH)_2D$ with inhibitors of vitamin D_3 catabolizing enzymes or antisense inhibition of CYP24A1 caused a greater inhibition of proliferation of human prostate and breast cancer cells than $1,25(OH)_2D$ alone (11, 12). In colon cancer cells, $1,25(OH)_2D$ downregulates CYP27B1 and induces CYP24A1 in such a way that CYP27B1 is active only if CYP24A1 is not maximally functional, and that the malignancy of tumor cells determines the extent of $1,25(OH)_2D$ catabolism (13). Additionally, CYP24A1 polymorphisms or splicing variants (modulated by $1,25(OH)_2D$) may render different levels of constitutive and inducible CYP24A1 activity at least in prostate cancer cell lines (14). The relevance of CYP24A1 regulation has recently been extended by the finding that epigenetic silencing of CYP24A1 contributes to the selective growth inhibition that $1,25(OH)_2D$ induces in tumor-derived endothelial cells as compared to endothelial cells of non-tumor origin (15).

Taken together, these data indicate that enzymes involved in vitamin D metabolism may be important targets for cancer prevention and treatment. This is supported by in vitro studies that use epigenetic inhibitors or vitamin D analogs that prevent the increase of CYP24A1 activity (16). Furthermore, antineoplastic agents of clinical use such as daunorubicin hydrochloride, etoposide, and vincristine sulfate increase the intracellular level of 1,25(OH)$_2$D by decreasing the stability of *CYP24A1* mRNA *(17)*. This action would prolong the bioavailability of 1,25(OH)$_2$D and may thus form the basis for putative additive or synergistic anticancer treatments. Importantly, during aging the activity of CYP27B1 decreases while that of CYP24A1 increases, which contributes to partial resistance to vitamin D in the elderly.

Resistance to 1,25(OH)$_2$D due to VDR repression. Cellular response to 1,25(OH)$_2$D mainly depends on VDR expression levels. VDR is expressed in almost all cell types and tissues. Epithelial cells from gastrointestinal tract, breast, kidney, prostate, bladder, and liver contain VDR. Moreover, VDR is expressed in bone, muscle, and skin cells as well as some activated cells of the immune system *(3)*. Tumor cells derived from these tissues usually express VDR and maintain the capacity to respond to 1,25(OH)$_2$D. Remarkably, elevated VDR expression is associated with high differentiation, absence of node involvement, and favorable prognosis in colorectal cancer *(18–20)*, and with late development of lymph node metastases and longer disease-free survival in breast cancer *(21, 22)*. However, downregulation of VDR expression has been found in several tumors (colon, breast, and lung) compared with their corresponding normal tissue *(7, 23)*. In colon cancer, VDR expression is induced in the early stages of tumorigenesis (polyps and adenomas) and decreases during colon cancer progression *(7)*. Clinical response to 1,25(OH)$_2$D analogs requires the expression of VDR in tumoral cells. Therefore, the downregulation of VDR found in tumors will lead to 1,25(OH)$_2$D unresponsiveness and resistance not only to the therapy with 1,25(OH)$_2$D analogs but also to the antitumoral effects of endogenous 1,25(OH)$_2$D.

Deletions, rearrangements, or point mutations affecting the coding region of the *VDR* gene have not been found in cancer. Several polymorphisms have been described in the *VDR* gene, some of which have been associated with increased risk of breast, prostate, and colon cancer. However, their consequences for VDR expression or functionality, and therefore their implication in the development of 1,25(OH)$_2$D resistance remain to be established *(24)*.

A mechanism responsible for VDR downregulation, at least in colon cancer, has emerged in the last years. Our group has revealed that the transcription factor Snail1 binds to the human *VDR* gene promoter and represses its expression in human colon cancer cells. In addition, Snail1 also reduces *VDR* mRNA half-life *(25)*. Accordingly, overexpression of Snail1 in human colon cancer cells blocks the prodifferentiation action of 1,25(OH)$_2$D and its inhibitory effect on cell proliferation and migration. Snail1 also represses VDR expression and abrogates the antitumoral effect of EB1089 (a 1,25(OH)$_2$D analog) in xenografted mice (Fig. 2) *(25, 26)*. Upregulation of Snail1 has been found in approximately 60% of colon tumors and it has been significantly associated with diminished VDR expression *(25, 27)*. These data indicate that Snail1 induction is probably responsible for VDR downregulation during colon cancer progression and indicate that tumors with high Snail1 expression would be resistant to treatment with 1,25(OH)$_2$D or its analogs. Since Snail1 is upregulated in advanced tumors (associated with acquisition of migratory and invasive properties), 1,25(OH)$_2$D and its analogs should preferentially be used as chemopreventive drugs, particularly for high risk patients, and as chemotherapeutic agents to be administered during the early stages of carcinogenesis *(28)*. In addition, as Snail1 upregulation has been reported

SW480-ADH

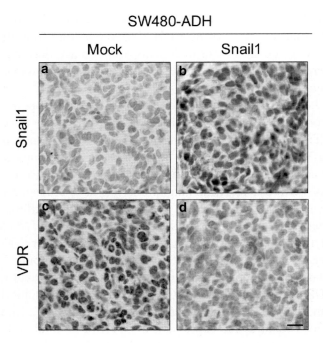

Fig. 2. Immunohistochemistry analysis of the expression of VDR and Snail1 in xenografted tumors generated in immunodeficient mice by human SW480-ADH colon cancer cells that were infected with retroviruses coding for the hemmaglutinin (HA)-tagged mouse *Snail1* cDNA or an empty virus (Mock). As shown by the use of anti-VDR and anti-HA antibodies, Snail1 overexpression downregulates VDR levels.

in cancers other than colon (synovial sarcoma and breast, gastric, and hepatocellular carcinomas) *(29, review)* this mechanism could be responsible for VDR downregulation and vitamin D resistance in other tumors. It has also been shown that *H-ras* transformation of HC11 mammary cells or NIH 3T3 fibroblasts provokes a reduction of *VDR* mRNA levels leading to $1,25(OH)_2D$ resistance. In these cases, transcriptional repression was ruled out and VDR downregulation seemed to be due to a decrease in *VDR* mRNA stability *(30, 31)*. As the Ras-Raf-MAPK pathway is one of the inducers of Snail1 *(29)*, VDR downregulation by oncogenic *H-ras* is probably mediated by Snail1. Remarkably, ectopic VDR re-expression diminished *H-ras*-induced transformation in NIH 3T3 cells *(31)*. Recent results indicate that the Snail1 homologous Snail2 (also named Slug) represses VDR expression and consequently inhibits $1,25(OH)_2D$ action in human colon cancer cells *(32)*. Snail2 cooperates with Snail1 for VDR inhibition in cultured cells, and analysis of human biopsies shows that Snail2 is upregulated in colon tumors and suggests that efficient VDR downregulation requires expression of both Snail1 and Snail2 *(32)*.

Several studies have identified compounds that increase VDR expression and/or $1,25(OH)_2D$ level in cancer and, therefore, could overcome the $1,25(OH)_2D$ resistance of tumors. $1,25(OH)_2D$ synthesis and VDR expression can be regulated by estrogens in colon cancer. 17β-Estradiol and certain phytoestrogens induce VDR and CYP27B1 expressions while they reduce that of CYP24A1. Therefore, the treatment of tumors with estrogens would increase VDR levels and overcome the resistance to $1,25(OH)_2D$ analogs *(33)*. In addition, the transcription factor p53 and its family members p63γ, p73α, and p73β activate human *VDR* gene transcription. These transcription factors are activated in turn by genotoxic stresses such as certain chemotherapeutic drugs. It is therefore expected that combined treatment with chemotherapy may enhance cell sensitivity to the antitumoral effects of $1,25(OH)_2D$ because the increase in the levels of p53 family members by chemotherapy could induce VDR *(33)*.

Resistance to $1,25(OH)_2D$ due to altered expression of VDR coregulators. Interestingly, tumoral resistance to $1,25(OH)_2D$ appears to be unrelated to VDR levels in several situations, because there is no clear relation between VDR expression and $1,25(OH)_2D$ action *(24)*. Therefore, various groups began to examine whether epigenetic mechanisms were disrupting VDR signalling, and they found that these mechanisms caused decreased responsiveness to $1,25(OH)_2D$ in prostate and breast cancer *(34)*. This epigenetic corruption results from an abnormal pattern of expression of corepressors or coactivators. In addition, altered post-translational modifications of VDR or RXR may also diminish VDR transcriptional activity.

Elevated expression of the corepressors SMRT and NCoR has been reported in prostate and breast cancer cell lines, respectively *(35, 36)*. Moreover, an increased ratio of corepressor/*VDR* RNA is also observed in matched primary tumor and normal breast cells, particularly associated with estrogen receptor negativity *(36)*. This alteration causes resistance to the antiproliferative effect of $1,25(OH)_2D$, which can be restored with inhibitors of histone deacetylases (trichostatin) and DNA methyltransferases (5-aza-2′-deoxycytidine) *(35–37)*. This also explains the synergistic inhibitory effect of combined treatment with $1,25(OH)_2D$ and sodium butyrate or trichostatin on the proliferation of prostate cancer cells *(38)*. The abnormal expression of coregulators may explain why VDR content does not predict the antiproliferative effects of $1,25(OH)_2D$ in some cancer cells *(24)*. Interestingly, in line with the increasing tendency to consider stroma as a critical regulator of tumor progression, altered VDR-mediated transcriptional activity due to abnormal VDR DNA binding and SMRT recruitment has been described in prostate cancer stroma *(39)*.

In contrast to what happens with corepressors, there is no report of alterations in the expression of VDR coactivators. The lack of such alterations in cancer cells may be a consequence of a dominant role of corepressor overexpression, which upsets the equilibrium between the two types of coregulators in normal cells. Bikle and collaborators hypothesized that the reason why squamous cell carcinoma cells fail to respond to the prodifferentiating action of $1,25(OH)_2D$, in spite of having normal levels of VDR and normal binding of VDR to VDREs, is a failure of the sequential binding of coactivator complexes to VDR *(40)*.

Finally, although data in the literature are scarce, deregulation of the translational modifications of VDR and RXR are believed to affect the gene regulatory activity of $1,25(OH)_2D$. In the transformed human keratinocyte cell line HPK1Aras, RXRα is phosphorylated at serine 260, which attenuates the transcriptional activity of VDR–RXR heterodimers and results in resistance to the growth inhibitory action of $1,25(OH)_2D$ *(41)*. HPK1Aras cells overexpress the *H-ras* oncogene and consequently exhibit an activated Ras-Raf-MAPK pathway. As this pathway is usually activated in many human cancer cells due to mutation or overexpression of membrane tyrosine kinase receptors (EGFR, MET) or downstream components such Ras or Raf, the importance of this type of alteration may be greater than suspected. VDR is known to be phosphorylated at several residues, although the significance of this remains to be assessed.

Concluding remarks. $1,25(OH)_2D$ is a pleiotropic hormone with a complex variety of actions in the organism which are mediated by its high-affinity receptor VDR present in most cell types. VDR is expressed by many tumor cells. However, partial-to-total resistance to the antitumoral effects of $1,25(OH)_2D$ arises during tumor progression. This is due to the reduced levels of $1,25(OH)_2D$ in cancer cells, loss of VDR expression, or deregulation of VDR transcriptional activity. The molecular mechanisms responsible for these alterations are, respectively, the deregulation of the enzymes synthesizing and degrading $1,25(OH)_2D$ intracellularly (CYP27B1 and CYP24A1), the upregulation of Snail1 and perhaps other transcriptional repressors of VDR, and alterations in the synthesis or post-translational modifications of VDR corepressors or partners. Better understanding of these mechanisms of resistance may allow us to improve the anticancer therapy with $1,25(OH)_2D$ and its analogs.

ACKNOWLEDGMENTS

We apologize to colleagues whose work were either inadvertently overlooked or not cited due to space limitations. We thank Robin Rycroft for his help with the English manuscript. The work in authors' laboratory is supported by the Ministerio de Ciencia e Innovación of Spain, Comunidad de Madrid and European Union.

REFERENCES

1. Abe E, Miyaura C, Sakagami H et al (1981) Differentiation of mouse myeloid leukemia cells induced by 1alpha,25-dihydroxyvitamin D_3. Proc Natl Acad Sci USA 78:4990–4994
2. Colston K, Colston MJ, Feldman D (1981) 1,25-dihydroxyvitamin D_3 and malignant melanoma: the presence of receptors and inhibition of cell growth in culture. Endocrinology 108:1083–1086
3. Hansen CM, Binderup L, Hamberg KJ et al (2001) Vitamin D and cancer: effects of $1,25(OH)_2D$ and its analogs on growth control and tumorigenesis. Front Biosci 6:D820–D848
4. Ordóñez-Morán P, Larriba MJ, Pendás-Franco N et al (2005) Vitamin D and cancer: an update of in vitro and in vivo data. Front Biosci 10:2723–2749
5. Deeb KK, Trump DL, Johnson CS (2007) Vitamin D signalling pathways in cancer: potential for anticancer therapeutics. Nat Rev Cancer 7:684–700
6. Sutton ALM, MacDonald PN (2003) Vitamin D: more than a "bone-a-fide" hormone Mol Endocrinol. 17:777–791
7. Cross HS, Kállay E, Khorchide M et al (2003) Regulation of extrarenal synthesis of 1,25-dihydroxyvitamin D_3-relevance for colonic cancer prevention and therapy. Mol Aspects Med 24:459–465
8. Townsend K, Evans KN, Campbell MJ et al 2005 Biological actions of extra-renal 25-hydroxyvitamin D-1alpha-hydroxylase and implications for chemoprevention and treatment. J Steroid Biochem Mol Biol 97(1–2):103–109
9. Khorchide M, Lechner D, Cross HS (2005) Epigenetic regulation of vitamin D hydroxylase expression and activity in normal and malignant human prostate cells. J Steroid Biochem Mol Biol 93:167–172
10. Ma JF, Nonn L, Campbell MJ et al (2004) Mechanisms of decreased Vitamin D 1alpha-hydroxylase activity in prostate cancer cells. Mol Cell Endocrinol 221:67–74

11. Townsend K, Banwell CM, Guy M et al (2005) Autocrine metabolism of vitamin D in normal and malignant breast tissue. Clin Cancer Res 11:3579–3586
12. Yee SW, Campbell MJ, Simons C (2006) Inhibition of Vitamin D_3 metabolism enhances VDR signalling in androgen-independent prostate cancer cells. J Steroid Biochem Mol Biol 98:228–235
13. Lechner D, Kallay E, Cross HS (2007) 1alpha,25-dihydroxyvitamin D_3 downregulates CYP27B1 and induces CYP24A1 in colon cells. Mol Cell Endocrinol 263:55–64
14. Muindi JR, Nganga A, Engler KL et al (2007) CYP24 splicing variants are associated with different patterns of constitutive and calcitriol-inducible CYP24 activity in human prostate cancer cell lines. J Steroid Biochem Mol Biol 103:334–337
15. Chung I, Karpf AR, Muindi JR et al (2007) Epigenetic silencing of CYP24 in tumor-derived endothelial cells contributes to selective growth inhibition by calcitriol. J Biol Chem 282:8704–8714
16. Lechner D, Manhardt T, Bajna E et al (2007) A 24-phenylsulfone analog of vitamin D inhibits 1alpha,25-dihydroxyvitamin D_3 degradation in vitamin D metabolism-competent cells. J Pharmacol Exp Ther 320:1119–1126
17. Tan J, Dwivedi PP, Anderson P et al (2007) Antineoplastic agents target the 25-hydroxyvitamin D_3 24-hydroxylase messenger RNA for degradation: implications in anticancer activity. Mol Cancer Ther 6:3131–3138
18. Vandewalle B, Adenis A, Hornez L et al (1994) 1,25-dihydroxyvitamin D_3 receptors in normal and malignant human colorectal tissues. Cancer Lett 86:67–73
19. Cross HS, Bajna E, Bises G et al (1996) Vitamin D receptor and cytokeratin expression may be progression indicators in human colon cancer. Anticancer Res 16:2333–2337
20. Evans SRT, Nolla J, Hanfelt J et al (1998) Vitamin D receptor expression as a predictive marker of biological behavior in human colorectal cancer. Clin Cancer Res 4:1591–1595
21. Eisman JA, Suva LJ, Martin TJ (1986) Significance of 1,25-dihydroxyvitamin D_3 receptor in primary breast cancers. Cancer Res 46:5406–5408
22. Berger U, McClelland RA, Wilson P et al (1991) Immunocytochemical determination of estrogen receptor, progesterone receptor, and 1,25-dihydroxyvitamin D_3 receptor in breast cancer and relationship to prognosis. Cancer Res 51:239–244
23. Anderson MG, Nakane M, Ruan X et al (2006) Expression of VDR and CYP24A1 mRNA in human tumors. Cancer Chemother Pharmacol 57:234–240
24. Campbell MJ, Adorini L (2006) The vitamin D receptor as a therapeutic target. Expert Opin Ther Targets 10:735–748
25. Pálmer HG, Larriba MJ, García JM et al (2004) The transcription factor SNAIL represses vitamin D receptor expression and responsiveness in human colon cancer. Nat Med 10:917–919
26. Larriba MJ, Valle N, Pálmer HG et al (2007) The inhibition of Wnt/beta-catenin signalling by 1alpha,25-dihydroxyvitamin D_3 is abrogated by Snail1 in human colon cancer cells. Endocr Relat Cancer 14:141–151
27. Peña C, García JM, Silva J et al (2005) E-cadherin and vitamin D receptor regulation by SNAIL and ZEB1 in colon cancer: clinicopathological correlations. Hum Mol Genet 14:3361–3370
28. Larriba MJ, Muñoz A (2005) SNAIL vs vitamin D receptor expression in colon cancer: therapeutics implications. Br J Cancer 92:985–989
29. Peinado H, Olmeda D, Cano A (2007) Snail, Zeb and bHLH factors in tumour progression: an alliance against the epithelial phenotype? Nat Rev Cancer 7:415–428
30. Rozenchan PB, Folgueira MAAK, Katayama MLH et al (2004) Ras activation is associated with vitamin D receptor mRNA instability in HC11 mammary cells. J Steroid Biochem Mol Biol 92:89–95
31. Agudo-Ibáñez L, Núñez F, Calvo F et al (2007) Transcriptomal profiling of site-specific Ras signals. Cell Signal 19:2264–2276
32. Larriba MJ, Martín-Villar E, García JM et al (2009) Snail2 cooperates with Snail1 in the repression of vitamin D receptor in colon cancer. Carcinogenesis 30:1459–1468.
33. Larriba MJ, Valle N, Álvarez S et al (2008) Vitamin D_3 and colorectal cancer. Adv Exp Med Biol 617:271–280
34. Abedin SA, Banwell CM, Colston KW et al (2006) Epigenetic corruption of VDR signalling in malignancy. Anticancer Res 26:2557–2566

35. Khanim FL, Gommersall LM, Wood VHJ et al (2004) Altered SMRT levels disrupt vitamin D_3 receptor signalling in prostate cancer cells. Oncogene 23:6712–6725
36. Banwell CM, MacCartney DP, Guy M et al (2006) Altered nuclear receptor corepressor expression attenuates vitamin D receptor signaling in breast cancer cells. Clin Cancer Res 12:2004–2013
37. Banwell CM, O'Neill LP, Uskokovic MR et al (2004) Targeting 1alpha,25-dihydroxyvitamin D_3 antiproliferative insensitivity in breast cancer cells by co-treatment with histone deacetylation inhibitors. J Steroid Biochem Mol Biol 89–90:245–249
38. Rashid SF, Moore JS, Walker E et al (2001) Synergistic growth inhibition of prostate cancer cells by 1alpha,25 Dihydroxyvitamin D_3 and its 19-nor-hexafluoride analogs in combination with either sodium butyrate or trichostatin A. Oncogene 20:1860–1872
39. Hidalgo AA, Paredes R, García VM et al (2007) Altered VDR-mediated transcriptional activity in prostate cancer stroma. J Steroid Biochem Mol Biol 103:731–736
40. Bikle DD, Xie Z, Ng D et al (2003) Squamous cell carcinomas fail to respond to the prodifferentiating actions of 1,25(OH)$_2$D: why? Recent Results Cancer Res 164:111–122
41. Solomon C, White JH, Kremer R (1999) Mitogen-activated protein kinase inhibits 1,25-dihydroxyvitamin D_3-dependent signal transduction by phosphorylating human retinoid X receptor alpha. J Clin Invest 103:1729–1735

16 Vitamin D and the Brain: A Neuropsychiatric Perspective

Louise Harvey, Thomas Burne, Xiaoying Cui, Alan Mackay-Sim, Darryl Eyles, and John McGrath

Abstract   Based on clues from epidemiology, it has been proposed that low prenatal vitamin D may be a risk factor for schizophrenia. In order to explore this hypothesis, our group has undertaken an integrated research program linking analytic epidemiology and rodent experiments. There is consistent evidence from rodents that offspring exposed to low developmental vitamin D deficiency have altered brain structure and function as adults. This chapter provides a concise summary of the evidence linking vitamin D to brain development and function. In addition, the epidemiological evidence linking hypovitaminosis D and various neuropsychiatric disorders is outlined.

Key Words: Vitamin D; animal experiments; neuroscience; schizophrenia; developmental neuroscience; epidemiology

1. INTRODUCTION

From an historical perspective, the links between vitamin D and neuropsychiatric disorders have only received attention in recent decades. The first indirect clue that vitamin D may play a role in brain development and function was when vitamin D and its metabolites were discovered in the cerebrospinal fluid of healthy adults *(1)*. However, the work of Walter Stumpf catalysed the modern interest in the impact of vitamin D on brain function. Stumpf mapped 1,25-dihydroxyvitamin D [$1,25(OH)_2D$] binding in rodent brains using radio-labelled $1,25(OH)_2D$ and autoradiography *(2, 3)*. The presence of this receptor in the central nervous system (CNS) provided the first real clues that vitamin D may have a role in brain function. Later studies using immunohistochemistry for the vitamin D receptor (VDR), in the neonatal and adult rat central nervous system, showed evidence for the VDR in multiple brain regions (e.g. temporal, orbital and cingulate cortices, thalamus, accumbens, amygdala, olfactory system and pyramidal neurons of the hippocampus), thus adding further weight to the hypothesis that vitamin D

From: *Nutrition and Health: Vitamin D*
Edited by: M.F. Holick, DOI 10.1007/978-1-60327-303-9_16,
© Springer Science+Business Media, LLC 2010

signalling is involved in brain function *(4–6)*. The later discovery of 1-hydroxylase in the human brain suggests that the CNS can synthesise 1,25(OH)$_2$D (the active form of vitamin D) from 25-hydroxyvitamin D [25(OH)D], which is the precursor inactive form (this is the "storage" form of vitamin D used to monitor serum levels). Thus, serum 25(OH)D levels may influence the autocrine and paracrine production of 1,25(OH)$_2$D in the CNS *(7–10)*, challenging the assumption that the brain is wholly reliant on circulating 25(OH)D crossing the blood–brain barrier *(11, 12)*.

The chapter has two aims. We will provide a concise summary of the evidence linking vitamin D to brain development and brain function. Mostly, this evidence is based on animal experiments. In particular, we will focus on an animal model developed by the authors to explore the impact of transient low prenatal vitamin D on brain development. Finally, we will summarize the epidemiological clues that link vitamin D to adult neuropsychiatric disorders such as schizophrenia and depression.

2. VITAMIN D AND BRAIN DEVELOPMENT

Expression of the VDR is temporally regulated in the developing rat CNS *(4)*. The earliest expression of VDR in the developing brain is in neural epithelium from day 12 of gestation. The VDR continues to be expressed in differentiating areas of the brain throughout gestation *(4)*. Specifically, the VDR emerges at days 19–21 in the pyramidal cells of the hippocampus, the same time at which the cells begin to cease division and commence differentiation in this brain region *(4)*. There is also a strong correlation between the degree of VDR staining and the degree of mitotic activity seen in the developing neuroepithelium of the developing rat CNS *(6)*. The temporal expression of VDR protein and messenger ribonucleic acid (mRNA) in the developing rat brain actually coincides with a decrease in mitotic activity and an increase in apoptosis, and this has led to the hypothesis that vitamin D could be regulating neuronal cell cycle events *(6)*.

Given nerve growth factor (NGF) is essential for the growth and survival of cholinergic basal forebrain neurons which project to the hippocampus *(13)* and 1,25(OH)$_2$D is a potent regulator of NGF *(14, 15)*, it is possible that vitamin D can effect hippocampal development by modulating NGF production *(4)*. The addition of 1,25(OH)$_2$D has been shown to increase neurite outgrowth in embryonic hippocampal explant cultures, an effect which is most likely due to the induction of NGF *(16)*. The addition of 1,25(OH)$_2$D to cultured hippocampal neurons also reduces the number of mitotic cells present and increases the amount of free NGF protein produced *(16)*. NGF is capable of binding to the neurotrophin receptor p75^{NTR} *(17)*. The promoter region of this receptor contains a vitamin D response element; hence, vitamin D can regulate the expression of the p75^{NTR} receptor in glioma cells *(18)*. NGF and p75^{NTR} are essential for programmed cell death *(19)*. As vitamin D regulates both proteins, it is possible that vitamin D could modulate neuronal survival and differentiation during development.

Vitamin D may also indirectly influence neuronal development by altering neurotrophic factor production by non-neuronal cells. The addition of 1,25(OH)$_2$D to rat primary glial cell cultures has been shown to increase the synthesis of NGF mRNA and protein *(14)*, neurotrophin-3 (NT-3) mRNA *(20)* and down-regulate neurotrophin-4 (NT-4) mRNA *(20)*. The addition of 1,25(OH)$_2$D can also increase the synthesis of glial cell line-derived neurotrophin factor (GDNF) mRNA in C6 glioma cells *(21)* but does not regulate GDNF production in primary glial cell cultures *(22)*. While GDNF is produced by non-neuronal cells, it is integral to the development of the dopaminergic *(23)* and noradrenergic systems *(24)*. This evidence therefore suggests that vitamin D could directly alter cellular development in the brain (both neuronal and non-neuronal).

Further evidence that vitamin D can effect brain development comes from studies on non-neuronal cell lines. $1,25(OH)_2D$ is able to induce cell death pathways in rat and human gliomas *(25)* and this effect extends to multiple glioma cell lines and primary cultures from surgical specimens *(26)*. Microglial astrocytes also respond to $1,25(OH)_2D$ in models of lipopolysaccharide (LPS)-induced brain inflammation and experimental autoimmune encephalitis (EAE) *(27, 28)*. Finally, it has been shown in vitro that activated microglia can metabolise $25(OH)D$ and produce biologically active $1,25(OH)_2D$ *(29)*. This evidence, taken in light of the evidence that the enzyme 1 α-hydroxylase is present in glial *(9)*, suggests that non-neuronal cells could synthesise vitamin D locally within the brain.

3. BRAIN DEVELOPMENT IN THE DEVELOPMENTAL VITAMIN D (DVD)-DEFICIENT RODENT

In order to explore the impact of transient hypovitaminosis D on brain development and subsequent adult brain structure and function, the authors have developed the developmental vitamin D (DVD) deficiency model. To obtain vitamin D depletion, female Sprague-Dawley rats are kept on a vitamin D-deficient diet. Animals are housed on a 12-h light/dark cycle (lights on at 06:00 h) using incandescent lighting to avoid ultraviolet radiation within the vitamin D action spectrum. These conditions are maintained for 6 weeks prior to mating and throughout gestation. Control animals are kept under similar conditions except they receive a vitamin D-replete diet and are housed under standard lighting conditions. After the dams have littered, all dams (and corresponding litters) are placed on vitamin D-replete diet. The vitamin D-deficient dams and DVD-deficient offspring remain normocalcemic (i.e. neither the dams nor their offspring have the rickets-like phenotype that would result from more chronic vitamin D depletion). It is important to stress that this is only a *developmental* exposure, because from birth all maternal animals and offspring receive a diet containing normal levels of vitamin D. Offspring are vitamin D replete by 2 weeks of age *(30)*, and vitamin D levels, calcium, and parathyroid hormone (PTH) levels are all normal when the animals are tested as adults (i.e. 10 weeks).

As neonates, the DVD-deficient offspring have a number of anatomical changes in the developing brain, including an increase in brain and lateral ventricle size and a decrease in thickness of the neocortex *(31)*. There are also other changes in markers of neuronal development, as DVD-deficient rats showed increased numbers of mitotic cells and decreased levels of p^{75} protein *(31)*. Further studies confirmed that cellular differentiation was profoundly altered in the DVD-deficient developing brain *(32)*. The level of apoptosis in the developing brains of DVD-deficient rats follows a different trajectory compared to controls; at embryonic day 19 (E19) there was no difference; however, at E21 and birth DVD-deficient rats had fewer apoptotic cells than controls

(32). This difference was normalised by post-natal day 7 (P7) *(32)*. It should be noted, however, that these rats were maintained on a vitamin D-deficient diet until being tested at P7. In line with a previous study *(31)*, DVD-deficient rats also showed increased rates of mitosis at significant pre- and post-natal time points (E19, E21, P0 and P7) *(32)*. More recently the VDR was shown to be present in neurospheres cultured from the subventricular zone (SVZ) of neonatal rats *(33)*. When neurosphere cultures are made from the brains of DVD-deficient neonates, neurosphere number was increased, suggesting greater cellular division *(33)*. This same study also revealed that the addition of 1,25(OH)$_2$D decreased neurosphere number. Thus, both the presence and the absence of vitamin D are capable of manipulating cellular proliferation in developing brain cells.

The timing of the reintroduction of vitamin D appears to be important in the persistence of these developmental changes into adulthood *(34)*. The enlarged lateral ventricles seen in the DVD-deficient neonates *(31)* only persist into adulthood if the introduction of the vitamin D-replete diet is delayed until weaning. Thus, re-addition of vitamin D to the diet from birth appears to partially ameliorate the lateral ventricle changes in DVD-deficient rats *(34)*.

The adult DVD-deficient rats also have altered brain expression of genes involved in cytoskeleton maintenance (MAP2, NF-L) and neurotransmission (GABA-Aα4) *(34)*. Gene array and proteomics analysis have been used to explore gene expression in the whole brain and protein expression in the prefrontal cortex and hippocampus of adult DVD-deficient rats *(35, 36)*. These studies found that DVD deficiency resulted in significantly altered expression of 36 proteins and 74 genes involved with cytoskeleton maintenance, calcium homeostasis, synaptic plasticity and neurotransmission, oxidative phosphorylation, redox balance, protein transport, chaperoning, cell cycle control and post-translational modifications *(35, 36)*. A recent study has examined protein expression in the nucleus accumbens in the DVD rat *(37)*. While this study found small fold-changes, it identified significant alterations in several proteins involved in calcium binding proteins (calbindin1, calbindin2 and hippocalcin). Other proteins associated with DVD deficiency related to mitochondrial function and the dynamin-like proteins.

4. BEHAVIOUR IN THE DVD-DEFICIENT RAT AND MOUSE

By adulthood, male DVD-deficient rats show a complex behavioural phenotype. DVD-deficient rats have disrupted latent inhibition *(38)*; however, they have normal pre-pulse inhibition *(39)* and working memory *(38)*. In the open field, DVD-deficient rats display a novelty-induced hyperlocomotion *(40)* that can be abated by injection or restraint *(39, 41)*, even though they display a normal stress response in the hypothalamic pituitary axis *(42)*. Adult, male DVD-deficient rats also exhibit a sensitivity to the *N*-methyl-D-aspartic acid receptor (NMDA-R) antagonist MK-801 *(39)*. Treatment with MK-801 induces hyperlocomotion in the open field in control rats, and DVD-deficient rats responded with an even greater enhancement in locomotor activity *(39)*. The later period of gestation appeared to be most relevant for this later effect *(30)*, because rats experiencing a DVD deficiency during late gestation also showed this effect, whereas DVD-deficient rats exposed during early gestation did not *(30)*.

A mouse model of DVD deficiency has also recently been produced in two strains of mice (129/SvJ and C57BL/6 J) *(43)*. As with the rat model, one strain exhibited spontaneous hyperlocomotion in the open field arena *(43)*. However, both strains demonstrated increased frequency of head dips in a hold board arena, indicative of increased exploratory behaviour *(43)*. This is in contrast to findings from DVD-deficient rats on the hole board test *(38)*. DVD-deficient mice were also assessed on a comprehensive screen of behavioural tests, including the elevated plus maze, forced swim test, prepulse inhibition and social interaction test *(43)*. There was no effect of maternal diet or strain on performance in any of these four tests *(43)*.

5. VITAMIN D AND NEUROPSYCHIATRIC DISORDERS: CLUES FROM EPIDEMIOLOGY

Based on clues from epidemiology, it has been proposed that low prenatal vitamin D may be associated with an increased risk of several adult-onset disorders. For example, McGrath *(12)* drew attention to diseases such as multiple sclerosis and schizophrenia that had both season of birth effects (e.g. those born in winter and spring have an increased risk of developing schizophrenia) and latitude effects (e.g. the prevalence of schizophrenia and multiple sclerosis is higher at higher latitudes).

Many studies have shown that those born in winter and spring have a significantly increased risk of developing schizophrenia *(44)* and that those born at higher latitudes are also at increased risk *(45)*, with both the incidence and the prevalence of schizophrenia being significantly greater in sites from higher latitudes *(46)*. Furthermore, based on data from cold climates, the incidence of schizophrenia is significantly higher in dark-skinned migrants compared to the native born *(47)*. Given that hypovitaminosis D is more common (a) during winter and spring, (b) at high latitudes and (c) in dark-skinned individuals *(48)*, low prenatal vitamin D "fits" these key environmental features. Preliminary evidence from analytical epidemiology studies also links low prenatal vitamin D with schizophrenia risk. For example, vitamin D supplements in the first year of life significantly reduced the risk of schizophrenia in males in a large Finnish Birth Cohort *(49)*. In addition, 25(OH)D serum levels in 26 mothers whose children developed schizophrenia were numerically (but not significantly) lower than that of 51 control mothers whose children did not develop the disease *(50)*. There was a trend-level association between low maternal vitamin D levels and schizophrenia in a subgroup of children of African-American mothers *(50)*, who would be at greatest risk of vitamin D deficiency because of their increased skin pigmentation.

There is some epidemiological data linking developmental vitamin D status and MS. The concordance rate of MS in monozygotic twins is only 30%, indicating a strong environmental influence on the development of the disease. Two northern hemisphere studies have described a season of birth effect for MS. A small Danish study described an excess of MS births in spring *(51)*, and a pooled analysis of patients from Canada, Britain, Denmark and Sweden described an increase in the percentage of people with MS born in May (spring) and a decrease in the percentage born in November (autumn) *(52)*. Other epidemiological links between vitamin D and multiple sclerosis are covered elsewhere in this book.

There is a growing body of evidence linking hypovitaminosis D and depression. First, the higher prevalence of seasonal affective disorder in high latitudes was noted by Stumpf *(53)*; however, results of randomized controlled trials of vitamin D supplements in this group have been mixed *(54–57)*. There is some evidence linking 25(OH)D levels to scores on mood scales *(55, 56)*, and hypotheses have been outlined linking vitamin D to neurotransmitters implicated in depression *(58)*. Observational studies suggest an association between low 25(OH)D levels and depression. Two studies designed primarily to examining 25(OH)D levels and performance on cognitive function have also reported an association between low 25(OH)D levels and depressed mood *(59, 60)*. However, neither of these studies included adjustments for physical activity; thus, any apparent association between 25(OH)D and mood may simply reflect that depressed individuals are less likely to go outside and thus access ultraviolet radiation.

These issues have been examined in a larger observation study *(61)*. Based on 1,283 community-based elderly residents (65–95 years), it was reported that 25(OH)D levels were 14% lower in 169 persons with minor depression and 14% lower in 26 persons with major depressive disorder compared with levels in 1,087 control individuals. Depression severity was significantly associated with decreased serum 25(OH)D levels and increased serum PTH levels. These associations persisted when controlled for activity levels. Adequately powered randomized controlled trials will be required to progress this interesting research field.

With respect to vitamin D and cognition, the results are also mixed. Several clinical studies have examined 25(OH)D levels and performance on neurocognitive measures using case–control samples. Based on patients with secondary hyperparathyroidism ($n = 21$) and matched controls ($n = 63$), Jorde and colleagues reported no significant association between 25(OH)D levels and various cognitive measures *(59)*. Another study, based on a mixed sample of 80 elderly individuals (60 years or older; half with mild Alzheimer's disease), reported no association between 25(OH)D levels and performance on a factor score derived from a large battery of neurocognitive tests *(60)*. Finally, a study based on a sample of 80 elderly individuals referred to a memory clinic reported a significant positive correlation between 25(OH)D levels and performance on the Mini Mental State Examination *(62)*. However, these studies lacked sufficient power to confidently detect small effect sizes, were not community-based and were not able to address the important potential confound of reverse causality (i.e. those with impaired cognitive ability may be less likely to go outside, and thus may develop hypovitaminosis D as a consequence of impaired cognition).

Some of these issues were addressed in a recently published study based on the large population-based NHANES III survey. The study examined the association between 25(OH)D levels and several different neurocognitive measures (including measures of attention and memory) in an adolescent group ($n = 1676$, age range 12–17 years), adult group ($n = 4747$, 20–60 years) and elderly group ($n = 4809$, 60–90 years). The study controlled for physical activity, as a proxy measure related to outdoor activity and possible ultraviolet light exposure. However, in the adolescent and adult groups, none of the psychometric measures were associated with 25(OH)D levels. In the elderly group there was a significant difference between 25(OH)D quintiles performance on a learning and memory task; however, paradoxically those with the highest quintile of 25(OH)D were most impaired on the task, contrary to the hypotheses. Lower 25(OH)D levels were not associated with impaired performance on various psychometric measures. While this cross-sectional study indicates that 25(OH)D levels are not associated with neurocognitive performance, it remains to be seen if chronic exposure to low 25(OH)D levels alters brain function in the long term. Curiously, in vitro experiments have clearly shown that vitamin D has neuroprotective qualities *(63–67)*. For example, pretreatment with 1,25(OH)$_2$D reduces the impact of glucocorticoid-induced neuronal changes *(68)*. Thus, it may be feasible that low vitamin D reduces the ability of the brain to recover after various adverse events (e.g. hypoxia, infection/inflammation, high glucocorticoid levels related to stress). Chronic hypovitaminosis D may exacerbate the resultant neuropsychiatric impairment. A longitudinal, prospective study would be better suited to exploring this hypothesis. While the animal experimental data supports an association between low developmental vitamin D and altered brain development, the impact of hypovitaminosis D during adulthood on brain function remains to be clarified. Recently, a detailed narrative review of the field reached similar conclusions *(69)*. More animal experimental work should help clarify the role of vitamin D on adult brain function; however, more focussed analytic epidemiological experiments are also required.

6. CONCLUSIONS

The list of neuropsychiatric disorders with possible links to vitamin D continues to grow. Recently, developmental vitamin D deficiency has also been linked to autism spectrum disorders *(70)*. While the results from the DVD-deficient animal experiments indicate that brain structure and function is altered in rodents, it remains to be seen if this deficiency is associated with schizophrenia in humans. How vitamin D links to cognition and depression also warrants further scrutiny. Much work remains to be done in order to understand how developmental vitamin D can influence brain development.

REFERENCES

1. Balabanova S, Richter HP, Antoniadis G, Homoki J, Kremmer N et al (1984) 25-Hydroxyvitamin D, 24, 25-dihydroxyvitamin D and 1,25-dihydroxyvitamin D in human cerebrospinal fluid. Klin Wochenschr 62:1086–1090
2. Stumpf WE, Bidmon HJ, Li L, Pilgrim C, Bartke A et al (1992) Nuclear receptor sites for vitamin D-soltriol in midbrain and hindbrain of Siberian hamster (Phodopus sungorus) assessed by autoradiography. Histochemistry 98:155–164
3. Stumpf WE, O'Brien LP (1987) 1,25 (OH)2 vitamin D3 sites of action in the brain. An Autoradiographic Study. Histochemistry 87:393–406
4. Veenstra TD, Prufer K, Koenigsberger C, Brimijoin SW, Grande JP et al (1998) 1,25-Dihydroxyvitamin D3 receptors in the central nervous system of the rat embryo. Brain Res 804: 193–205
5. Prufer K, Veenstra TD, Jirikowski GF, Kumar R (1999) Distribution of 1,25-dihydroxyvitamin D3 receptor immunoreactivity in the rat brain and spinal cord. J Chem Neuroanat 16:135–145
6. Burkert R, McGrath J, Eyles D (2003) Vitamin D receptor expression in the embryonic rat brain. Neurosci Res Commun 33:63–71
7. Hosseinpour F, Wikvall K (2000) Porcine microsomal vitamin D(3) 25-hydroxylase (CYP2D25). Catalytic properties, tissue distribution, and comparison with human CYP2D6. J Biol Chem 275: 34650–34655

8. Zehnder D, Bland R, Williams MC, McNinch RW, Howie AJ et al (2001) Extrarenal expression of
    25-hydroxyvitamin d(3)-1 alpha-hydroxylase. J Clin Endocrinol Metab 86:888–894
9. Eyles DW, Smith S, Kinobe R, Hewison M, McGrath JJ (2005) Distribution of the vitamin D receptor
    and 1 alpha-hydroxylase in human brain. J Chem Neuroanat 29:21–30
10. Sutherland MK, Somerville MJ, Yoong LK, Bergeron C, Haussler MR et al (1992) Reduction of
    vitamin D hormone receptor mRNA levels in Alzheimer as compared to Huntington hippocampus:
    correlation with calbindin-28 k mRNA levels. Brain Res Mol Brain Res 13:239–250
11. Gascon-Barre M, Huet PM (1983) Apparent [3H]1,25-dihydroxyvitamin D3 uptake by canine and
    rodent brain. Am J Physiol 244:E266–E271
12. McGrath J (2001) Does 'imprinting' with low prenatal vitamin D contribute to the risk of various adult
    disorders? Med Hypotheses 56:367–371
13. Korsching S, Auburger G, Heumann R, Scott J, Thoenen H (1985) Levels of nerve growth factor
    and its mRNA in the central nervous system of the rat correlate with cholinergic innervation. Embo J
    4:1389–1393
14. Neveu I, Naveilhan P, Jehan F, Baudet C, Wion D et al (1994) 1,25-dihydroxyvitamin D3 regulates the
    synthesis of nerve growth factor in primary cultures of glial cells. Brain Res Mol Brain Res 24:70–76
15. Wion D, MacGrogan D, Neveu I, Jehan F, Houlgatte R et al (1991) 1,25-Dihydroxyvitamin D3 is a
    potent inducer of nerve growth factor synthesis. J Neurosci Res 28:110–114
16. Brown J, Bianco JI, McGrath JJ, Eyles DW (2003) 1,25-dihydroxyvitamin D3 induces nerve growth
    factor, promotes neurite outgrowth and inhibits mitosis in embryonic rat hippocampal neurons. Neu-
    rosci Lett 343:139–143
17. Chao MV, Hempstead BL (1995) p75 and Trk: a two-receptor system. Trends Neurosci 18:321–326
18. Naveilhan P, Neveu I, Baudet C, Funakoshi H, Wion D et al (1996) 1,25-Dihydroxyvitamin D3 regu-
    lates the expression of the low-affinity neurotrophin receptor. Brain Res Mol Brain Res 41:259–268
19. Chao MV (1994) The p75 neurotrophin receptor. J Neurobiol 25:1373–1385
20. Neveu I, Naveilhan P, Baudet C, Brachet P, Metsis M (1994) 1,25-dihydroxyvitamin D3 regulates
    NT-3, NT-4 but not BDNF mRNA in astrocytes. Neuroreport 6:124–126
21. Naveilhan P, Neveu I, Wion D, Brachet P (1996) 1,25-Dihydroxyvitamin D3, an inducer of glial cell
    line-derived neurotrophic factor. Neuroreport 7:2171–2175
22. Remy S, Naveilhan P, Brachet P, Neveu I (2001) Differential regulation of GDNF, neurturin, and their
    receptors in primary cultures of rat glial cells. J Neurosci Res 64:242–251
23. Granholm AC, Reyland M, Albeck D, Sanders L, Gerhardt G et al (2000) Glial cell line-derived
    neurotrophic factor is essential for postnatal survival of midbrain dopamine neurons. J Neurosci 20:
    3182–3190
24. Quintero EM, Willis LM, Zaman V, Lee J, Boger HA et al (2004) Glial cell line-derived neurotrophic
    factor is essential for neuronal survival in the locus coeruleus-hippocampal noradrenergic pathway.
    Neuroscience 124:137–146
25. Naveilhan P, Berger F, Haddad K, Barbot N, Benabid AL et al (1994) Induction of glioma cell
    death by 1,25(OH)2 vitamin D3: towards an endocrine therapy of brain tumors? J Neurosci Res 37:
    271–277
26. Zou J, Landy H, Feun L, Xu R, Lampidis T et al (2000) Correlation of a unique 220-kDa protein with
    vitamin D sensitivity in glioma cells. Biochem Pharmacol 60:1361–1365
27. Garcion E, Sindji L, Montero-Menei C, Andre C, Brachet P et al (1998) Expression of inducible
    nitric oxide synthase during rat brain inflammation: regulation by 1,25-dihydroxyvitamin D3. Glia 22:
    282–294
28. Nataf S, Garcion E, Darcy F, Chabannes D, Muller JY et al (1996) 1,25 Dihydroxyvitamin D3 exerts
    regional effects in the central nervous system during experimental allergic encephalomyelitis. J Neu-
    ropathol Exp Neurol 55:904–914
29. Neveu I, Naveilhan P, Menaa C, Wion D, Brachet P et al (1994) Synthesis of 1,25-dihydroxyvitamin
    D3 by rat brain macrophages in vitro. J Neurosci Res 38:214–220
30. O'Loan J, Eyles DW, Kesby J, Ko P, McGrath JJ et al (2007) Vitamin D deficiency during various
    stages of pregnancy in the rat; its impact on development and behaviour in adult offspring. Psychoneu-
    roendocrinology 32:227–234

31. Eyles D, Brown J, Mackay-Sim A, McGrath J, Feron F (2003) Vitamin D3 and brain development. Neuroscience 118:641–653
32. Ko P, Burkert R, McGrath J, Eyles D (2004) Maternal vitamin D3 deprivation and the regulation of apoptosis and cell cycle during rat brain development. Brain Res Dev Brain Res 153:61–68
33. Cui X, McGrath JJ, Burne TH, Mackay-Sim A, Eyles DW (2007) Maternal vitamin D depletion alters neurogenesis in the developing rat brain. Int J Dev Neurosci 25:227–232
34. Feron F, Burne TH, Brown J, Smith E, McGrath JJ et al (2005) Developmental Vitamin D3 deficiency alters the adult rat brain. Brain Res Bull 65:141–148
35. Almeras L, Eyles D, Benech P, Laffite D, Villard C et al (2007) Developmental vitamin D deficiency alters brain protein expression in the adult rat: implications for neuropsychiatric disorders. Proteomics 7:769–780
36. Eyles D, Almeras L, Benech P, Patatian A, Mackay-Sim A et al (2007) Developmental vitamin D deficiency alters the expression of genes encoding mitochondrial, cytoskeletal and synaptic proteins in the adult rat brain. J Steroid Biochem Mol Biol 103:538–545
37. McGrath J, Iwazaki T, Eyles D, Burne T, Cui X et al (2008) Protein expression in the nucleus accumbens of rats exposed to developmental vitamin D deficiency. PLoS One 3(6):e2383
38. Becker A, Eyles DW, McGrath JJ, Grecksch G (2005) Transient prenatal vitamin D deficiency is associated with subtle alterations in learning and memory functions in adult rats. Behav Brain Res 161:306–312
39. Kesby JP, Burne TH, McGrath JJ, Eyles DW (2006) Developmental vitamin D deficiency alters MK 801-induced hyperlocomotion in the adult rat: an animal model of schizophrenia. Biol Psychiatry 60:591–596
40. Burne TH, Becker A, Brown J, Eyles DW, Mackay-Sim A et al (2004) Transient prenatal Vitamin D deficiency is associated with hyperlocomotion in adult rats. Behav Brain Res 154:549–555
41. Burne TH, O'Loan J, McGrath JJ, Eyles DW (2006) Hyperlocomotion associated with transient prenatal vitamin D deficiency is ameliorated by acute restraint. Behav Brain Res 174:119–124
42. Eyles DW, Rogers F, Buller K, McGrath JJ, Ko P et al (2006) Developmental vitamin D (DVD) deficiency in the rat alters adult behaviour independently of HPA function. Psychoneuroendocrinology 31:958–964
43. Harms LR, Eyles DW, McGrath JJ, Mackay-Sim A, Burne TH (2007) Developmental vitamin D deficiency alters adult behaviour in 129/SvJ and C57BL/6 J mice. Behav Brain Res 187(2):343–350
44. Torrey EF, Miller J, Rawlings R, Yolken RH (1997) Seasonality of births in schizophrenia and bipolar disorder: a review of the literature. Schizophr Res 28:1–38
45. Saha S, Chant DC, Welham JL, McGrath JJ (2006) The incidence and prevalence of schizophrenia varies with latitude. Acta Psychiatr Scand 114:36–39
46. Davies G, Welham J, Chant D, Torrey EF, McGrath J (2003) A systematic review and meta-analysis of Northern Hemisphere season of birth studies in schizophrenia. Schizophr Bull 29:587–593
47. Cantor-Graae E, Selten JP (2005) Schizophrenia and migration: a meta-analysis and review. Am J Psychiatry 162:12–24
48. Holick MF (1995) Environmental factors that influence the cutaneous production of vitamin D. Am J Clin Nutr 61:638S–645S
49. McGrath J, Saari K, Hakko H, Jokelainen J, Jones P et al (2004) Vitamin D supplementation during the first year of life and risk of schizophrenia: a Finnish birth cohort study. Schizophr Res 67:237–245
50. McGrath J, Eyles D, Mowry B, Yolken R, Buka S (2003) Low maternal vitamin D as a risk factor for schizophrenia: a pilot study using banked sera. Schizophr Res 63:73–78
51. Templer DI, Trent NH, Spencer DA, Trent A, Corgiat MD et al (1992) Season of birth in multiple sclerosis. Acta Neurol Scand 85:107–109
52. Willer CJ, Dyment DA, Sadovnick AD, Rothwell PM, Murray TJ et al (2005) Timing of birth and risk of multiple sclerosis: population based study. BMJ 330:120
53. Stumpf WE, Privette TH (1989) Light, vitamin D and psychiatry. Role of 1,25 dihydroxyvitamin D3 (soltriol) in etiology and therapy of seasonal affective disorder and other mental processes. Psychopharmacology (Berl) 97:285–294

54. Dumville JC, Miles JN, Porthouse J, Cockayne S, Saxon L et al (2006) Can vitamin D supplementation prevent winter-time blues? A randomised trial among older women. J Nutr Health Aging 10:151–153
55. Gloth FM 3rd, Alam W, Hollis B (1999) Vitamin D vs broad spectrum phototherapy in the treatment of seasonal affective disorder. J Nutr Health Aging 3:5–7
56. Lansdowne AT, Provost SC (1998) Vitamin D3 enhances mood in healthy subjects during winter. Psychopharmacology (Berl) 135:319–323
57. Partonen T, Vakkuri O, Lamberg-Allardt C, Lonnqvist J (1996) Effects of bright light on sleepiness, melatonin, and 25-hydroxyvitamin D(3) in winter seasonal affective disorder. Biol Psychiatry 39: 865–872
58. Berk M, Sanders KM, Pasco JA, Jacka FN, Williams LJ et al (2007) Vitamin D deficiency may play a role in depression. Med Hypotheses 69:1316–1319
59. Jorde R, Waterloo K, Saleh F, Haug E, Svartberg J (2006) Neuropsychological function in relation to serum parathyroid hormone and serum 25-hydroxyvitamin D levels : The Tromso study. J Neurol 253:464–470
60. Wilkins CH, Sheline YI, Roe CM, Birge SJ, Morris JC (2006) Vitamin D deficiency is associated with low mood and worse cognitive performance in older adults. Am J Geriatr Psychiatry 14:1032–1040
61. Hoogendijk WJ, Lips P, Dik MG, Deeg DJ, Beekman AT et al (2008) Depression is associated with decreased 25-hydroxyvitamin D and increased parathyroid hormone levels in older adults. Arch Gen Psychiatry 65:508–512
62. Przybelski RJ, Binkley NC (2007) Is vitamin D important for preserving cognition? A positive correlation of serum 25-hydroxyvitamin D concentration with cognitive function. Arch Biochem Biophys 460(2):202–205
63. Oermann E, Bidmon HJ, Witte OW, Zilles K (2004) Effects of 1alpha,25 dihydroxyvitamin D3 on the expression of HO-1 and GFAP in glial cells of the photothrombotically lesioned cerebral cortex. J Chem Neuroanat 28:225–238
64. Wang JY, Wu JN, Cherng TL, Hoffer BJ, Chen HH et al (2001) Vitamin D(3) attenuates 6-hydroxydopamine-induced neurotoxicity in rats. Brain Res 904:67–75
65. Brewer LD, Thibault V, Chen KC, Langub MC, Landfield PW et al (2001) Vitamin D hormone confers neuroprotection in parallel with downregulation of L-type calcium channel expression in hippocampal neurons. J Neurosci 21:98–108
66. Moore ME, Piazza A, McCartney Y, Lynch MA (2005) Evidence that vitamin D3 reverses age-related inflammatory changes in the rat hippocampus. Biochem Soc Trans 33:573–577
67. Losem-Heinrichs E, Gorg B, Redecker C, Schleicher A, Witte OW et al (2005) 1alpha,25-dihydroxy-vitamin D3 in combination with 17beta-estradiol lowers the cortical expression of heat shock protein-27 following experimentally induced focal cortical ischemia in rats. Arch Biochem Biophys 439:70–79
68. Obradovic D, Gronemeyer H, Lutz B, Rein T (2006) Cross-talk of vitamin D and glucocorticoids in hippocampal cells. J Neurochem 96:500–509
69. McCann JC, Ames BN (2008) Is there convincing biological or behavioral evidence linking vitamin D deficiency to brain dysfunction? FASEB J 22:982–1001
70. Cannell JJ (2008) Autism and vitamin D. Med Hypotheses 70:750–759

17 Vitamin D Modulation of Adipocyte Function

Michael B. Zemel and Xiaocun Sun

Abstract   Calcitriol has recently been demonstrated to play an important role in modulating adipocyte function by regulating adipocyte lipid metabolism and energy homeostasis via both genomic and non-genomic actions. Physiological concentrations of calcitriol dose-dependently inhibit adipocyte apopto-sis, although supra-physiological concentrations stimulate adipocyte apoptosis; the former is mediated by inhibition of mitochondrial uncoupling and the latter by mitochondrial calcium overload. Calcitriol also regulates adipose tissue fat depot location and expansion by promoting glucocorticoid production and release. Finally, calcitriol also modulates the cross talk between adipose tissue and both skeletal muscle and macrophages. Calcitriol modulation of adipocyte–macrophage cross talk results in a synergistic increase in expression and release of reactive oxygen species and inflammatory cytokines from both cell types, while calcitriol regulation of adipocyte–skeletal muscle cross talk results in inhibition of skeletal muscle fatty acid oxidation and preferential energy storage in adipocytes. Accordingly, conditions which chronically increase calcitriol levels, such as low-calcium diets, increase obesity risk, decrease metabolic flexibility, and increase oxidative and inflammatory stress.

Key Words: Adipocyte; adipokine; calcitriol; calcium; energy metabolism; obesity

1. INTRODUCTION

Although vitamin D and its metabolites are not generally considered to have a major role in the control of energy metabolism, evidence accumulated over the past 8 years indicates that calcitriol does play a key regulatory role in adipocyte lipid metabolism and thereby modulates energy homeostasis and obesity risk *(1)*. This concept emerged from the search for a mechanism to explain an apparent "anti-obesity" effect of dietary calcium first noted in epidemiological studies and subsequently confirmed in some clin-ical trials. Data from the 1987–1988 US Department of Agriculture's Nationwide Food Consumption Survey showed an inverse relationship between dietary calcium intake and body weight *(2, 3)*; notably, when stratified by ethnicity, non-Hispanic blacks had the lowest mean daily calcium intake and the highest obesity prevalence. Similarly, in the data from the first National Health and Nutrition Examination Survey (NHANES I) McCarron demonstrated a significant inverse association between calcium intake and

From: *Nutrition and Health: Vitamin D*
Edited by: M.F. Holick, DOI 10.1007/978-1-60327-303-9_17,
© Springer Science+Business Media, LLC 2010

body weight *(3)*, and data from NHANES III demonstrated a strong inverse association between calcium intake and relative risk of obesity *(4)*. Subsequent support emerged from retrospective and prospective epidemiological and observational studies, secondary analysis of past clinical trials originally conducted with other primary endpoints (e.g., skeletal, cardiovascular), and prospective clinical trials; these effects are supported by a clear mechanistic framework based upon emerging data demonstrating a key role for calcitriol in modulating energy metabolism *(5)*; this role is mediated through both genomic and non-genomic actions of calcitriol, as discussed in the remainder of this chapter.

2. Ca^{2+} SIGNALING

Calcitriol is well recognized to modulate Ca^{2+} signaling in numerous cell types *(6–10)*, including adipocytes, and earlier work on the *agouti* gene demonstrated that Ca^{2+} signaling plays a pivotal role in adipocyte lipid metabolism *(11, 12)*. *Agouti* is normally expressed in mouse melanocytes, where it is involved in modulation of hair pigmentation via melanocortin-1 receptor antagonism *(13)*; normal expression of *agouti* produces the characteristic wild-type pigmentation pattern of mouse hair, with a predominately black hair shaft with a subapical yellow segment. However, dominant mutations in the mouse *agouti* gene confer a pleiotropic syndrome characterized by obesity and insulin resistance, and expressing the wild-type *agouti* cDNA under control of either a ubiquitous promoter or an adipose tissue-specific promoter (aP2) recapitulates this syndrome *(14, 15)*. Notably, the human homolog of this gene is primarily expressed in white adipose tissue. Agouti protein exerts significant paracrine/autocrine effects by targeting ion channels, thereby causing an increase in adipocyte intracellular free calcium ($[Ca^{2+}]_i$) *(16, 17)*. Strong correlation between the degree of agouti expression and both $[Ca^{2+}]_i$ levels and body weight has been demonstrated in mice, indicating that agouti may modulate adiposity via a $[Ca^{2+}]_i$-dependent mechanism *(18)*. Indeed, an agouti/Ca^{2+} response sequence has been mapped to the fatty acid synthase (FAS) promoter region *(19)*, and increasing $[Ca^{2+}]_i$ in adipocytes via either receptor- or voltage-mediated Ca^{2+} channel activation stimulated FAS gene expression and consequently resulted in stimulation of FAS activity *(20)*; notably, there is a significant association between adipose tissue agouti expression and both FAS expression and body mass index in normal humans *(21)*. The role of Ca^{2+} signaling in obesity was confirmed by the observation that calcium channel inhibition resulted in significant decreases in adipose tissue mass and adipocyte lipogenesis in obese *agouti*-transgenic mice *(22)*. Agouti protein also plays a role in regulation of adipocyte lipolysis via a Ca^{2+}-dependent mechanism *(23, 24)*. Recombinant agouti protein inhibits both basal and agonist-stimulated lipolysis in human adipocytes. Increasing Ca^{2+} influx through either voltage- or receptor-operated Ca^{2+} channels also inhibits lipolysis, and this effect is blocked by Ca^{2+} channel antagonists, indicating that this anti-lipolytic effect is mediated by calcium signaling. The mechanism underlying the anti-lipolytic effect of Ca^{2+} has been demonstrated to be mediated by increased activation of phosphodiesterase 3B, resulting in reduced cAMP levels and consequently inhibition of hormone sensitive lipase activity. Although calcitriol and

Ca^{2+} signaling inhibit the early stages of adipogenesis, they serve to accelerate late-stage differentiation and lipid filling of existing adipocytes by stimulating expression of peroxisome proliferator-activated receptor γ (PPAR-γ) and key downstream genes, such as aP2, stearoyl-CoA desaturase (SCD-1), phosphoenolpyruvate carboxykinase (PEPCK), and FAS *(25)*. Thus, increased Ca^{2+} signaling favors a smaller number of hypertrophic adipocytes.

This regulation of lipid metabolism by $[Ca^{2+}]_i$ provides the framework for calcitriol modulation of adiposity and, consequently, the link between dietary calcium and obesity. We found that calcitriol induces rapid Ca^{2+} influx, while a specific membrane vitamin D receptor antagonist (1β,25-dihydroxyvitamin D_3) blocked this effect *(26)*. This indicated a non-genomic action of calcitriol via a putative membrane vitamin receptor, which later was identified as the membrane-associated rapid response to steroid (1,25(OH)$_2$D-MARRS) *(27)*, in modulating $[Ca^{2+}]_i$. Dietary calcium supplementation has been demonstrated to decrease the $[Ca^{2+}]_i$ concentration in various cell types including adipocytes *(4, 28–30)*, and this effect is largely mediated by suppression of calcitriol. We have demonstrated that the increased calcitriol produced in response to low-calcium diets stimulates adipocyte Ca^{2+} influx and, consequently, promotes adiposity by both inhibiting lipolysis and stimulating lipogenesis *(31, 32)*. Accordingly, suppressing calcitriol levels by increasing dietary calcium is an attractive target for obesity interventions. In support of this concept, transgenic mice expressing the agouti gene specifically in adipocytes (a human-like pattern) respond to low-calcium diets with accelerated weight gain and fat accretion, whereas high-calcium diets markedly inhibit lipogenesis, accelerate lipolysis, increase thermogenesis, and suppress fat accretion and weight gain in animals maintained at identical caloric intakes *(31)*. Further, low-calcium diets impede body fat loss whereas high-calcium diets markedly accelerate fat loss in transgenic mice subjected to caloric restriction *(32)*. These concepts are confirmed by both epidemiological and clinical data *(33, 34)*, which demonstrate that increasing dietary calcium results in significant reductions in adipose tissue mass in obese humans in the absence of caloric restriction and markedly accelerates the weight and body fat loss secondary to caloric restriction, whereas dairy products exert significantly greater effects.

3. ROLE OF THE NUCLEAR VITAMIN D RECEPTOR

In addition to regulating adipocyte metabolism via $[Ca^{2+}]_i$ through the non-genomic 1,25(OH)$_2$D-MARRS, calcitriol also exerts a genomic act via the adipocyte nuclear vitamin D receptor (nVDR) to inhibit the expression of uncoupling protein 2 (UCP2) *(35)*. Polymorphisms in the nVDR are associated with the susceptibility to obesity in humans *(36, 37)*, and several lines of evidence demonstrate alterations in the vitamin D endocrine system in obese humans *(38, 39)*. Interestingly, the nVDR is expressed at very low level in preadipocytes, but is transiently stimulated during adipogenesis and then returns to low levels. However, we have recently shown nVDR expression in mature adipocytes to be subjected to regulation by multiple compounds, including calcitriol, resulting in a positive feedback pathway in response to both agonist (calcitriol) and glucocorticoid *(40)*, as discussed in a subsequent section of this chapter.

Calcitriol acts via the nVDR to inhibit UCP2 expression and to suppress UCP2 responses to both isoproterenol and fatty acids *(35)*. The role of the nVDR in this antagonism is demonstrated by its reversal by antisense oligonucleotide-mediated nVDR knockout in adipocytes and by the failure of 1,25(OH)$_2$D-MARRS agonist and antagonist to either mimic or prevent the calcitriol inhibition of UCP2 expression. Uncoupling proteins (UCPs) are mitochondrial transporters present in the inner membrane of mitochondria and have been shown to stimulate mitochondrial proton leak and therefore exhibit a potential role in thermogenesis and energy metabolism *(41)*. Unlike the other isoforms of UCPs, UCP2 is ubiquitously expressed, with the highest level in white adipose tissue while UCP3 is dominantly expressed in skeletal muscles of human and rodents. Suppression of calcitriol by feeding high-calcium diets to energy-restricted mice results in increased adipose tissue UCP2 and skeletal muscle UCP3 expression and attenuates the decline in core temperature which otherwise occurs with energy restriction *(42)*. The high level of homology of UCP2 and UCP3 with UCP1 suggests a possible uncoupling activity which has been shown to stimulate mitochondria proton leak and therefore exhibit a potential role in thermogenesis and energy metabolism *(43–45)*; both are associated with fatty acid transport across the inner mitochondrial membrane and subsequent β-oxidation, and polymorphisms of anonymous markers encompassing the UCP2–UCP3 locus in humans are strongly linked to resting metabolic rate. Accordingly, calcitriol suppression of UCP2 expression may be anticipated to increase energetic efficiency and thereby increase obesity risk.

4. CALCITRIOL REGULATION OF ADIPOCYTE APOPTOSIS

Calcitriol regulation of both UCP2 and [Ca^{2+}]$_i$ appears to exert an additional role in energy metabolism by affecting adipocyte apoptosis *(46)*. Although calcitriol has previously been shown to exert a pro-apoptotic effect in several tissues *(47–49)*, these effects are generally observed with supra-physiological levels of the hormone (≥ 100 nM), and our data in human adipocytes also demonstrate a pro-apoptotic role of such high concentrations. However, we have also shown that lower doses of calcitriol (0.1–10 nm) dose-dependently inhibit apoptotic gene expression such as caspase-1 and caspase-3 expression but stimulate anti-apoptotic gene expression such as BCL-2 and increase the BCL-2/Bax ratio in wild-type adipocytes. Furthermore, calcitriol dose-dependently induced an increase in mitochondrial potential ($\Delta \psi$) and ATP production, while overexpressing UCP2 in adipocytes exerted the opposite effect, indicating that suppression of UCP2 expression and consequent increases in mitochondrial potential and ATP production may contribute to the anti-apoptotic effect of calcitriol. Notably, high doses of calcitriol also induced a markedly increase in mitochondrial calcium ([Ca^{2+}]$_m$) load while lower, more physiological doses of calcitriol exerted the opposite effect, indicating that the increased [Ca^{2+}]$_m$ is associated with the induction of apoptosis by calcitriol. Mitochondria are often located close to endoplasmic reticulum (ER) and are thereby exposed to the Ca^{2+} released by the inositol-1,4,5-triphosphate receptor (IP3R) and ryanodine receptor (RyR). The high Ca^{2+} levels achieved at these contact sites favor Ca^{2+} uptake into mitochondria. Because of their tight coupling to ER Ca^{2+} stores, mitochondria are highly susceptible to abnormalities in Ca^{2+} signaling *(50)*. Recent evidence suggests that the amount of Ca^{2+} going through mitochondria is crucial in triggering a Ca^{2+}-dependent apoptotic response, probably by the opening of a sensitized state of permeability transition pore (PTP) *(51)*. Thus, the anti-apoptotic effect of physiological concentrations of calcitriol appears to be mediated primarily by suppression of UCP2, while the pro-apoptotic effects observed with pharmacological concentrations are mediated by mitochondrial Ca^{2+} overload (Fig. 1). The effects of calcitriol on adipocyte apoptosis were further supported by our recent microarray study of human adipocytes *(52)*, as physiological concentrations of calcitriol suppressed the pro-apoptotic gene stanniocalcin 2 (STC2) but stimulated anti-apoptotic gene STC1. Further in vivo data provide additional supporting evidence for a role of calcitriol and of dietary calcium in adipocyte apoptosis *(46)*, with suppression of calcitriol using high-calcium diets resulted in significant, substantial increases in white adipose tissue apoptosis in diet-induced obesity.

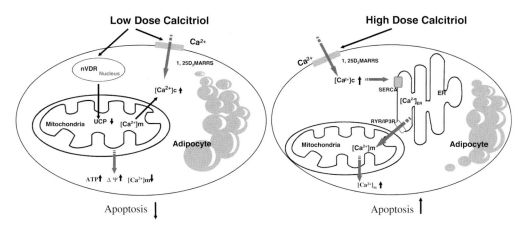

Fig. 1. Proposed mechanism for calcitriol modulation of adipocyte apoptosis.

5. CALCITRIOL MODULATION OF ADIPOCYTE GLUCOCORTICOID PRODUCTION

Calcitriol may also participate in the energy metabolism by regulating adipose tissue fat depot location and expansion. Excessive central fat deposition in obesity may result from the greater capacity for regeneration of active glucocorticoids in the visceral fat depot. Local adipose tissue glucocorticoid levels and intracellular glucocorticoid availability are controlled by the activity of 11β-hydroxysteroid dehydrogenase type 1 (11β-HSD 1) to generate active cortisol from inactive cortisone *(53)*. Overexpression of 11β-HSD 1 specifically in white adipose tissue of transgenic mice recapitulates features of the metabolic syndrome, including central obesity, hypertension, dyslipidemia, and insulin resistance *(54, 55)*, suggesting that tissue-specific dysregulation of glucocorticoid metabolism may promote expansion of adipose tissue stores. Previous studies from this laboratory demonstrate that the anti-obesity effect of dietary calcium is associated with preferential loss of central adipose tissue *(56)*, and we recently demonstrated that calcitriol directly up-regulates adipocyte 11β-HSD 1 expression and cortisol release and, consequently, correspondingly affects local cortisol levels, indicating a potential role for calcitriol in visceral adiposity *(57, 58)*. This effect is attributable to the rapid non-genomic action of calcitriol mediated through the $1,25(OH)_2D$-MARRS because this response is prevented by $1,25(OH)_2D$, which antagonizes the rapid, membrane-associated signaling events resulting from exposure to calcitriol. These findings are further supported by our recent microarray study *(52)*, as well as by data demonstrating that dietary calcium-induced suppression of calcitriol attenuated adipose tissue 11β-HSD 1 expression in diet-induced obese mice *(59)*.

Recent data from this laboratory also demonstrated an interesting positive feedback regulation of glucocorticoids on calcitriol in adipocytes *(40)*; the cortisol precursor cortisone or the synthetic glucocorticoid dexamethasone each increased nVDR expression, while 11β-HSD 1 knockdown attenuated this effect. We thus propose that the increased cortisol, which results from liganded adipocyte nVDR stimulation of 11β-HSD 1, upregulates nVDR expression, resulting in further potential binding of calcitriol by the nVDR and consequent further stimulation of active glucocorticoid generation (Fig. 2). Notably, calcitriol also exerts additional effect on stimulation of glucocorticoid production by increasing cortisol release. This effect, however, appears to be independent of the nVDR and is instead mediated by intracellular calcium signaling because knockdown 11β-HSD 1 did not affect short-term corticosterone release while modulating calcium influx by KCl, BAYK8644, and $1,25(OH)_2D$-MARRS agonist lumisterol markedly increased corticosterone release.

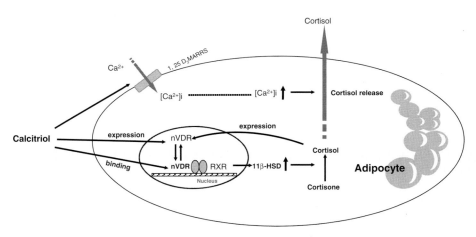

Fig. 2. Calcitriol regulation of adipocyte nVDR expression: interaction with glucocorticoids.

6. CALCITRIOL REGULATION OF CROSS TALK BETWEEN ADIPOCYTE AND SKELETAL MUSCLE IN ENERGY METABOLISM

Calcitriol may also modulate energy metabolism via regulation of the expression and production of multiple adipokines, as discussed later in this chapter. In addition to its role as a fuel reservoir, adipose tissue serves as an active endocrine organ which synthesizes and secretes a variety of biological molecules *(60–63)*. The more recent recognition that skeletal muscle may also assume a similar role in response to various metabolic stimuli suggests a potential interaction between skeletal muscle and adipose tissue *(64, 65)*. Energy partitioning between adipose tissue and skeletal muscle has been previously demonstrated. Animals lacking myostatin exhibit markedly increased skeletal muscle mass *(66, 67)* and reduced body fat accumulation, and stimulation of beta-adrenergic receptor produces a dramatic increase in skeletal muscle mass and a corresponding reduction body fat content *(68, 69)*. Consistent with this concept, data from our laboratory demonstrated an independent role of leucine and calcium antagonism in regulating energy partitioning between adipocytes and muscle cells *(70)*, with both favoring fatty acid oxidation and UCP3 expression in C2C12 muscle cells while calcitriol suppressed fatty acid oxidation and attenuated the effects of leucine and nifedipine, indicating that this increase in muscle fat oxidation is coupled with decreased fat storage and increased fat catabolism in adipocytes. Indeed, we found that leucine inhibited FAS and PPAR gamma expression in differentiated 3T3-L1 adipocytes while calcitriol stimulated the expression of both genes and attenuated the effect of leucine on FAS expression. In addition, we have demonstrated that calcitriol decreases mitochondrial biogenesis and associated regulatory gene expression in both myocytes and adipocytes while leucine exerts the opposite effect *(71)*, indicating that calcitriol modulates both adipocyte and muscle metabolism by antagonizing mitochondrial biogenesis. Moreover, muscle cells treated with conditioned medium derived from adipocytes or co-cultured with adipocytes exhibited suppressed fatty acid oxidation *(70)* and decreased mitochondria biogenesis *(71)*, indicating that one or more factors derived from adipocytes regulate skeletal muscle energy metabolism. Indeed, leucine, nifedipine, and calcitriol also modulate adiponectin production, with leucine and nifedipine increasing adiponectin production while calcitriol exerts the opposite effect *(70)*. Consequently, we further evaluated the role of adiponectin in mediating the response to leucine, nifedipine, and calcitriol *(70)*. Our data to date suggest calcitriol modulation of adipocyte–skeletal muscle cross talk is mediated, in part, by alterations in adipocyte adiponectin and skeletal muscle interleukin-15 (IL-15) and IL-6 *(70)*; however, additional studies are required to clarify the role of specific adipokines in mediating this cross talk between adipose tissue and skeletal muscle (Fig. 3).

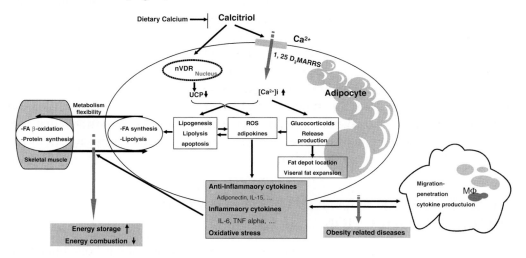

Fig. 3. Calcitriol regulation of adipocyte function and cross talk with bother organs.

7. CALCITRIOL REGULATION OF ADIPOCYTE OXIDATIVE STRESS

Adipose tissue excess is an important pathogenic mechanism underlying the obesity-related disorders, and this effect is attributable, in part, to the increased production of reactive oxygen species (ROS) and inflammatory cytokines by adipose tissue (72–76). It has been postulated that hyperglycemia and hyperlipidemia, key clinical manifestation of obesity and diabetes, may promote ROS production through several pathways (77). Indeed, the generation of ROS as byproducts of the mitochondrial electron transport chain has long been attributed to the high rates of glucose and lipid metabolism. Additional mechanism may be operated to produce ROS under high-glucose and high-lipid conditions, including the formation of advanced glycation end products (78), altered polyol pathway activity (79), activation of oxidases (80, 81), and/or reduction of antioxidant enzymes (82, 83). The interaction between ROS and calcium/uncoupling has been intensively investigated. There is a bidirectional interaction wherein ROS regulates cellular calcium signaling while manipulation of calcium signaling may also regulate cellular ROS production. Calcium signaling is essential for production of ROS (84), and elevated $[Ca^{2+}]_i$ activates ROS-generating enzymes, such as NADPH oxidase and myeloperoxidase, as well as the formation of free radicals by the mitochondrial respiratory chain. Increased ROS production also stimulates $[Ca^{2+}]_i$ by activating calcium channels on both the plasma membrane and the endoplasmic reticulum (85). Mitochondrial respiration is associated with production of ROS, and mitochondria produce a large fraction of the total ROS made in cells (86). Mild uncoupling of respiration thus diminishes mitochondrial ROS formation by dissipating mitochondrial proton gradient and potential (87).

We have recently shown that ROS production is modulated by mitochondrial uncoupling status and cytosolic calcium signaling (88) and that calcitriol regulates ROS production in cultured murine and human adipocytes via its modulation of both Ca^{2+} signaling and mitochondrial uncoupling. Consistent with this, our recent in vivo data showed that dietary calcium-induced suppression of calcitriol reduced adipose tissue ROS production (89). Interestingly, animals on a low-calcium diet showed markedly higher visceral fat gain than subcutaneous fat versus mice on a high-calcium diet and exhibited strikingly enhanced ROS production and NADPH oxidase expression in visceral fat versus subcutaneous fat, indicating that higher visceral fat predisposes to enhanced ROS production. We also demonstrated that 11β-HSD 1 expression in visceral fat was markedly higher than subcutaneous fat in mice on basal low-calcium group whereas no difference was observed between the fat depots in mice on the high-calcium diet. We also found the high-calcium diet suppressed 11β-HSD 1 expression in visceral adipose tissue compared to mice on the low-calcium diet. These findings demonstrated that suppression of ROS production by dietary calcium may be mediated, at least in part, by the regulation of glucocorticoid-associated fat distribution.

8. CALCITRIOL REGULATION OF ADIPOCYTE INFLAMMATORY CYTOKINE PRODUCTION

Adipocytes produce a variety of biological molecules, including both inflammatory cytokines such as tumor necrosis factor alpha (TNFα), IL-6, and IL-8 and anti-inflammatory factors such as adiponectin and IL-15 *(60–63)*. Dysregulated production of these adipocytokines contributes to the pathogenesis of obesity-associated metabolic syndrome. Given that obesity and related disorders are associated with low-grade systemic inflammation and oxidative stress, and that both Ca^{2+} signaling and ROS production modulate cytokine expression and release, we considered the possibility that calcitriol, and by extension dietary calcium, may also play a role in modulating adipose tissue cytokine production and systemic inflammatory stress. We found that calcitriol stimulated TNFα, IL-6, and IL-8 production in cultured human and murine adipocytes and that this effect was completely blocked by a calcium channel antagonist *(90)*, suggesting that dietary calcium suppresses inflammatory factor production in adipocyte and that calcitriol-induced Ca^{2+} influx may be a key mediator of this effect. Moreover, dietary calcium decreased production of pro-inflammatory factors such as TNFα and IL-6 and increased anti-inflammatory molecules such as IL-15 and adiponectin in visceral fat, as well as systemic biomarkers of both oxidative and inflammatory stress *(90, 91)*. Our recent clinical evidence support these observations *(91)*, as high-calcium diets suppressed circulating C-reactive protein and increased adiponectin levels during both weight loss and weight maintenance in obese subjects. In a recent follow-up randomized crossover study in overweight and obese subjects, we found that suppressing calcitriol levels with high- versus low-calcium diets suppressed both oxidative and inflammatory stress within 7 days of initiation of supplementation and that these effects increased in magnitude with increased duration of supplementation *(92)*. Collectively, these findings indicate an important role of both calcitriol in inducing oxidative and inflammatory stress and, consequently, of dietary calcium in attenuating these risk factors.

9. CALCITRIOL REGULATION OF ADIPOCYTE–MACROPHAGE CROSS TALK IN INFLAMMATION

Adipose tissue also includes a stromal vascular fraction that contains blood cells, endothelial cells, and macrophages *(93–95)*. Although adipocytes directly generate inflammatory mediators, adipose tissue-derived cytokines also originate substantially from these non-fat cells *(96)*, among which the infiltrated macrophages play a prominent role. Infiltration and differentiation of adipose tissue-resident macrophages are under the local control of chemokines, many of which are produced by adipocytes. Accordingly, cross talk between adipocytes and macrophages may be a key factor in mediating inflammatory and oxidative changes in obesity. Calcitriol stimulates production of adipokines associated with macrophage function and increases inflammatory cytokine expression in both macrophages and adipocytes *(97)*; these include CD14, migration inhibitory factor (MIF), macrophage colony-stimulating factor (M-CSF), macrophage inflammatory protein (MIP), TNFα, IL-6, and monocyte chemotactic protein-1 (MCP-1) in adipocytes, and TNFα and IL-6 in macrophages. Moreover, a cytokine protein array identified multiple additional inflammatory cytokines that were up-regulated by calcitriol in both macrophages and adipocytes *(97)*. Further, calcitriol also regulated cross talk between macrophages and adipocytes, as shown by augmentation of expression and production of inflammatory cytokines from adipocytes and macrophages in co-culture versus individual culture. These effects were attenuated by either calcium channel antagonism or increasing mitochondrial uncoupling, indicating that the pro-inflammatory effect of calcitriol is mediated by calcitriol-induced stimulation of Ca^{2+} signaling and attenuation of mitochondrial uncoupling.

10. CONCLUSION

Accumulating evidence suggests a key role for calcitriol in regulating adipocyte function. Calcitriol favors lipogenesis and inhibits lipolysis via non-genomic modulation of Ca^{2+} influx. In addition, calcitriol suppresses UCP2 expression via the nVDR and thereby increases energy efficiency and directly stimulates nVDR expression in mature adipocyte to induce a positive feedback loop between calcitriol and its receptor. Calcitriol also exerts a dose-dependent impact on adipocyte apoptosis and regulates adipose tissue fat depot location and expansion by promoting glucocorticoid production and release. Recent data also demonstrate a pivotal role of calcitriol in modulation of adipokine production resulting in significant roles in cross talk between adipocytes and macrophages in oxidative and inflammatory stress and between adipocytes and skeletal muscle in metabolic flexibility (Fig. 3).

REFERENCES

1. Zemel MB (2005) Calcium and dairy modulation of obesity risk. Obes Res 13:192–193
2. Fleming KH, Heimbach JT (1994) Consumption of calcium in the U.S.: food sources and intake levels. J Nutr 124(Suppl):1426S–1430S
3. Anonymous (1994) Optimal calcium intake. NIH Consensus Statement. National Institutes of Health, Bethesda, MD
4. Zemel MB, Shi H, Greer B, Dirienzo D, Zemel PC (2000) Regulation of adiposity by dietary calcium. FASEB J 14:1132–1138
5. Zemel MB (2007) Dairy Foods, calcium and weight management. In: B Bagchi and H Preuss (eds) Obesity epidemiology, pathophysiology and prevention. CRC Press, Boca Raton, FL
6. Xiaoyu Z, Payal B, Melissa O, Zanello LP (2007) 1alpha,25(OH)2-vitamin D3 membrane-initiated calcium signaling modulates exocytosis and cell survival. J Steroid Biochem Mol Biol 103:457–461
7. Picotto G (2001) Rapid effects of calciotropic hormones on female rat enterocytes: combined actions of 1,25(OH)2-vitamin D3, PTH and 17beta-estradiol on intracellular Ca^{2+} regulation. Horm Metab Res 33:733–738
8. Capiati DA, Vazquez G, Tellez Iñón MT, Boland RL (2000) Role of protein kinase C in 1,25(OH)(2)-vitamin D(3) modulation of intracellular calcium during development of skeletal muscle cells in culture. J Cell Biochem 77:200–212
9. Van Baal J, Yu A, Hartog A, Fransen JA, Willems PH, Lytton J, Bindels RJ (1993) Localization and regulation by vitamin D of calcium transport proteins in rabbit cortical collecting system. Am J Physiol 271:F985–F993
10. Gascon-Barre M, Haddad P, Provencher SJ, Bilodeau S, Pecker F, Lotersztajn S, Vallieres S (1994) Chronic hypocalcemia of vitamin D deficiency leads to lower intracellular calcium concentrations in rat hepatocytes. J Clin Invest 93:2159–2167
11. Shi H, Norman AW, Okamura WH, Sen A, Zemel MB (2001) 1 alpha, 25-Dihydroxyvitamin D3 modulates human adipocyte metabolism via nongenomic action. FASEB J 15:2751–2753
12. Xue B, Moustaid N, Wilkison WO, Zemel MB (1998) The agouti gene product inhibits lipolysis in human adipocytes via a Ca^{2+}-dependent mechanism. FASEB J 12:1391–1396
13. Silvers WK (1958) An experimental approach to action of genes at the agouti locus in the mouse. III. Transplants of newborn Aw-, A-and at-skin to Ay-, Aw-, A-and aa hosts. J Exp Zool 137:189–196

14. Michaud EJ, Mynatt RL, Miltenberger RJ, Klebig ML, Wilkinson JE, Zemel MB, Wilkison WO, Woychik RP (1997) Role of the agouti gene in obesity. J Endocrinol 155:207–209

15. Zemel MB, Kim JH, Woychik RP, Michaud EJ, Kadwell SH, Patel IR, Wilkison WO (1995) Agouti regulation of intracellular calcium: role in the insulin resistance of viable yellow mice. Proc Natl Acad Sci U S A 92:4733–4737

16. Kim JH, Kiefer LL, Woychik RP, Wilkison WO, Truesdale A, Ittoop O, Willard D, Nichols J, Zemel MB (1997) Agouti regulation of intracellular calcium: role of melanocortin receptors. Am J Physiol 272:E379–E384

17. Xue BZ, Wilkison WO, Mynatt RL, Moustaid N, Goldman M, Zemel MB (1999) The agouti gene product stimulates pancreatic [beta]-cell Ca^{2+} signaling and insulin release. Physiol Genomics 1: 11–19

18. Xue B, Zemel MB (2000) Relationship between human adipose tissue agouti and fatty acid synthase (FAS). J Nutr 130:2478–2481

19. Moustaid N, Sakamoto K, Clarke S, Beyer RS, Sul HS (1993) Regulation of fatty acid synthase gene transcription. Sequences that confer a positive insulin effect and differentiation-dependent expression in 3T3-L1 preadipocytes are present in the 332 bp promoter. Biochem J 292:767–772

20. Zemel MB (1998) Nutritional and endocrine modulation of intracellular calcium: implications in obesity, insulin resistance and hypertension. Mol Cell Biochem 188:129–136

21. Smith SR, Gawronska-Kozak B, Janderova L, Nguyen T, Murrell A, Stephens JM, Mynatt RL (2003) Agouti expression in human adipose tissue: functional consequences and increased expression in type 2 diabetes. Diabetes 52:2914–2922

22. Kim JH, Mynatt RL, Moore JW, Woychik RP, Moustaid N, Zemel MB (1996) The effects of calcium channel blockade on agouti-induced obesity. FASEB J 10:1646–1652

23. Xue B, Greenberg AG, Kraemer FB, Zemel MB (2001) Mechanism of intracellular calcium ($[Ca^{2+}]i$) inhibition of lipolysis in human adipocytes. FASEB J 15:2527–2529

24. Pohl SL (1981) Cyclic nucleotides and lipolysis. Int J Obes 5:627–633

25. Shi H, Halvorsen YD, Ellis PN, Wilkison WO, Zemel MB (2000) Role of intracellular calcium in human adipocyte differentiation. Physiol Genomics 3:75–82

26. Koster HP, Hartog A, Van Os CH, Bindels RJ (1995) Calbindin-D28K facilitates cytosolic calcium diffusion without interfering with calcium signaling. Cell Calcium 18:187–196

27. Rohe B, Safford SE, Nemere I, Farach-Carson MC (2005) Identification and characterization of 1,25D3-membrane-associated rapid response, steroid (1,25D3-MARRS)-binding protein in rat IEC-6 cells. Steroids 70:458–463

28. Petrov V, Lijnen P (1999) Modification of intracellular calcium and plasma renin by dietary calcium in men. Am J Hypertens 12:1217–1224

29. Rao RM, Yan Y, Wu Y (1994) Dietary calcium reduces blood pressure, parathyroid hormone, and platelet cytosolic calcium responses in spontaneously hypertensive rats. Am J Hypertens 7: 1052–1057

30. Otsuka K, Watanabe M, Yue Q, McCarron DA, Hatton D (1997) Dietary calcium attenuates platelet aggregation and intracellular Ca^{2+} mobilization in spontaneously hypertensive rats. Am J Hypertens 10:1165–1170

31. Zemel MB (2001) Calcium modulation of hypertension and obesity: mechanisms and implications. J Am Coll Nutr 20:428S–435S

32. Shi H, Dirienzo D, Zemel MB (2001) Effects of dietary calcium on adipocyte lipid metabolism and body weight regulation in energy-restricted aP2-agouti transgenic mice. FASEB J 15: 291–293

33. Zemel MB (2004) Role of calcium and dairy products in energy partitioning and weight management. Am J Clin Nutr 79:907S–912S

34. Zemel MB, Thompson W, Milstead A, Morris K, Campbell P (2004) Calcium and dairy acceleration of weight and fat loss during energy restriction in obese adults. Obes Res 12: 582–590

35. Shi H, Norman AW, Okamura WH, Sen A, Zemel MB (2002) 1alpha,25-dihydroxyvitamin D3 inhibits uncoupling protein 2 expression in human adipocytes. FASEB J 16:1808–1810

36. Ye WZ, Reis AF, Dubois-Laforgue D, Bellanne-Chantelot C, Timsit J, Velho G (2001) Vitamin D receptor gene polymorphisms are associated with obesity in type 2 diabetic subjects with early age of onset. Eur J Endocrinol 145:181–186

37. Barger-Lux MJ, Heaney RP, Hayes HF, DeLuca HF, Johnson ML, Gong G (1995) Vitamin D receptor gene polymorphism, bone mass, body size, and vitamin D receptor density. Calcif Tissue Int 57: 161–162

38. Andersen T, McNair P, Hyldstrup L, Fogh AN, Nielsen TT, Astrup A, Transbol I (1988) Secondary hyperparathyroidism of morbid obesity regresses during weight reduction. Metabolism 37:425–428

39. Bell NH, Epstein S, Greene A, Shary J, Oexmann MJ, Shaw S (1985) Evidence for alteration of the vitamin D-endocrine system in obese subjects. J Clin Invest 76:370–373

40. Sun X, Zemel MB (2008) 1-alpha, 25-Dihydroxyvitamin D and corticosteroid regulate adipocyte nuclear vitamin D receptor. Int J Obes (Lond) 32:1305–1311

41. Ricquier D (2005) Respiration uncoupling and metabolism in the control of energy expenditure. Proc Nutr Soc 64:47–52

42. Sun XC, Zemel MB (2004) Calcium and dairy products inhibit weight and fat regain during ad libitum consumption following energy restriction in Ap2-agouti transgenic mice. J Nutr 134:3054–3060

43. Boss O, Hagen T, Lowell BB (2000) Uncoupling proteins 2 and 3: potential regulators of mitochondrial energy metabolism. Diabetes 49:143–156

44. Dulloo AG, Samec S (2001) Uncoupling proteins: their roles in adaptive thermogenesis and substrate metabolism reconsidered. Br J Nutr 86:123–139

45. Saleh MC, Wheeler MB, Chan CB (2002) Uncoupling protein-2: evidence for its function as a metabolic regulator. Diabetologia 45:174–187

46. Sun X, Zemel MB (2004) Role of uncoupling protein 2 (UCP2) expression and 1alpha, 25-dihydroxyvitamin D3 in modulating adipocyte apoptosis. FASEB J 18:1430–1432

47. Elias J, Marian B, Edling C, Lachmann B, Noe CR, Rolf SH, Schuster I (2003) Induction of apoptosis by vitamin D metabolites and analogs in a glioma cell line. Recent Results Cancer Res 164:319–332

48. Wagner N, Wagner KD, Schley G, Badiali L, Theres H, Scholz H (2003) 1,25-dihydroxyvitamin D3-induced apoptosis of retinoblastoma cells is associated with reciprocal changes of Bcl-2 and bax. Exp Eye Res 77:1–9

49. Narvaez CJ, Byrne BM, Romu S, Valrance M, Welsh J (2003) Induction of apoptosis by 1,25-dihydroxyvitamin D3 in MCF-7 Vitamin D3-resistant variant can be sensitized by TPA. J Steroid Biochem Mol Biol 84:199–209

50. Rizzuto R, Pinton P, Carrington W, Fay FS, Fogarty KE, Lifshitz LM, Tuft RA, Pozzan T (1998) Close contacts with the endoplasmic reticulum as determinants of mitochondrial Ca2+ responses. Science 280:1763–1766

51. Smaili SS, Hsu YT, Youle RJ, Russell JT (2000) Mitochondria in Ca^{2+} signaling and apoptosis. J Bioenerg Biomembr 32:35–46

52. Sun X, Morris K, Zemel MB (2008) Microarray analysis of the effects of calcitriol and cortisone on human adipocyte gene expression. J Nutrigenet Nutrigenomics 1:30–48

53. Jamieson PM, Chapman KE, Edwards CR, Seckl JR (1995) 11ß-hydroxysteroid dehydrogenase is an exclusive 11ß-reductase in primary cultures of rat hepatocytes: effect of physiochemical and hormonal manipulations. Endocrinology 136:4754–4761

54. Masuzaki H, Paterson J, Shinyama H, Morton NM, Mullins JJ, Seckl JR et al (2001) A transgenic model of visceral obesity and the metabolic syndrome. Science 294:2166–2170

55. Masuzaki H, Yamamoto H, Kenyon CJ, Elmquist JK, Morton NM, Paterson JM et al (2003) Transgenic amplification of glucocorticoid action in adipose tissue causes high blood pressure in mice. J Clin Invest 112:83–90

56. Sun X, Zemel MB (2006) Dietary calcium regulates ROS production in aP2-agouti transgenic mice on high-fat/high-sucrose diets. Int J Obes (Lond) 30:1341–1346

57. Zemel MB, Richards J, Mathis S, Milstead A, Gebhardt L, Silva E (2005) Dairy augmentation of total and central fat loss in obese subjects. Int J Obes (Lond) 29:391–397

58. Morris KL, Zemel MB (2005) 1,25-dihydroxyvitamin D3 modulation of adipocyte glucocorticoid function. Obes Res 13:670–677

59. Zemel MB, Herweyer A (2007) Role of branch chain amino acids and ACE inhibition in the anti-obesity effect of milk. FASEB J 21:A328

60. Coppack SW (2001) Pro-inflammatory cytokines and adipose tissue. Proc Nutr Soc 60:349–356

61. Trayhurn P, Beattie JH (2001) Physiological role of adipose tissue: white adipose tissue as an endocrine and secretory organ. Proc Nutr Soc 60:329–339

62. Ajuwon KM, Jacobi SK, Kuske JL, Spurlock ME (2004) Interleukin-6 and interleukin-15 are selectively regulated by lipopolysaccharide and interferon-gamma in primary pig adipocytes. Am J Physiol (Reg Int Comp Physiol) 286:R547–R553

63. Okamoto Y, Kihara S, Funahashi T, Matsuzawa Y, Libby P (2006) Adiponectin: a key adipocytokine in metabolic syndrome. Clin Sci (Lond) 110:267–278

64. Chan MH, Carey AL, Watt MJ, Febbraio MA (2004) Cytokine gene expression in human skeletal muscle during concentric contraction: evidence that IL-8, like IL-6, is influenced by glycogen availability. Am J Physiol (Reg Int Comp Physiol) 287:R322–R327

65. Steensberg A, Keller C, Starkie RL, Osada T, Febbraio MA, Pedersen BK (2002) IL-6 and TNF-alpha expression in, and release from, contracting human skeletal muscle. Am J Physiol (Endocrinol Metab) 283:E1272–E1278

66. Tobin JF, Celeste AJ (2005) Myostatin, a negative regulator of muscle mass: implications for muscle degenerative diseases. Curr Opin Pharmacol 5:328–332

67. McPherron AC, Lee SJ (2002) Suppression of body fat accumulation in myostatin-deficient mice. J Clin Invest 109:595–601

68. Yang YT, McElligott MA (1989) Multiple actions of beta-adrenergic agonists on skeletal muscle and adipose tissue. Biochem J 261:1–10

69. Spurlock ME, Cusumano JC, Ji SQ, Anderson DB, Smith CK 2nd, Hancock DL, Mills SE (1994) The effect of ractopamine on beta-adrenoceptor density and affinity in porcine adipose and skeletal muscle tissue. J Anim Sci 72:75–80

70. Sun X, Zemel MB (2007) Leucine and calcium regulate fat metabolism and energy partitioning in murine adipocytes and muscle cells. Lipids 42:297–305

71. Zemel MB, Sun X (2008) Leucine and calcitriol modulation of mitochondrial biogenesis in muscle cells and adipocytes. FASEB J 22(305):2

72. Fantuzzi G (2005) Adipose tissue, adipokines, and inflammation. J Allergy Clin Immunol 115: 911–919

73. Furukawa S, Fujita T, Shimabukuro M, Iwaki M, Yamada Y, Nakajima Y, Nakayama O, Makishima M, Matsuda M, Shimomura I (2004) Increased oxidative stress in obesity and its impact on metabolic syndrome. J Clin Invest 114:1752–17561

74. Weiss R, Caprio S (2005) The metabolic consequences of childhood obesity. Best Pract Res Clin Endocrinol Metab 19:405–419

75. Pickup JC (2004) Inflammation and activated innate immunity in the pathogenesis of type 2 diabetes. Diabetes Care 27:813–823

76. Cottam DR, Mattar SG, Barinas-Mitchell E, Eid G, Kuller L, Kelley DE, Schauer PR (2004) The chronic inflammatory hypothesis for the morbidity associated with morbid obesity: implications and effects of weight loss. Obes Surg 14:589–600

77. Inoguchi T, Li P, Umeda F, Yu HY, Kakimoto M, Imamura M, Aoki T, Etoh T, Hashimoto T, Naruse M, Sano H, Utsumi H, Nawata H (2000) High glucose level and free fatty acid stimulate reactive oxygen species production through protein kinase C – dependent activation of NAD(P)H oxidase in cultured vascular cells. Diabetes 49:1939–1945

78. Shangari N, O'Brien PJ (2004) The cytotoxic mechanism of glyoxal involves oxidative stress. Biochem Pharmacol 68:1433–1442

79. Chung SS, Ho EC, Lam KS, Chung SK (2003) Contribution of polyol pathway to diabetes-induced oxidative stress. J Am Soc Nephrol 14:S233–S236

80. Inoguchi T, Sonta T, Tsubouchi H, Etoh T, Kakimoto M, Sonoda N, Sato N, Sekiguchi N, Kobayashi K, Sumimoto H, Utsumi H, Nawata H (2003) Protein kinase C-dependent increase in reactive oxygen species (ROS) production in vascular tissues of diabetes: role of vascular NAD(P)H oxidase. J Am Soc Nephrol 14:S227–S232

81. Talior I, Tennenbaum T, Kuroki T, Eldar-Finkelman H (2005) Protein kinase C-{delta} dependent acti-
    vation of oxidative stress in adipocytes of obese and insulin resistant mice: role for NADPH oxidase.
    Am J Physiol (Endocrinol Metab) 288:E405–E411

82. Fiordaliso F, Bianchi R, Staszewsky L, Cuccovillo I, Doni M, Laragione T, Salio M, Savino C, Melucci
    S, Santangelo F, Scanziani E, Masson S, Ghezzi P, Latini R (2004) Antioxidant treatment attenuates
    hyperglycemia-induced cardiomyocyte death in rats. J Mol Cell Cardiol 37:959–968

83. Manea A, Constantinescu E, Popov D, Raicu M (2004) Changes in oxidative balance in rat pericytes
    exposed to diabetic conditions. J Cell Mol Med 8:117–126

84. Gordeeva AV, Zvyagilskaya RA, Labas YA (2003) Cross-talk between reactive oxygen species and
    calcium in living cells. Biochemistry (Mosc) 68:1077–1080

85. Volk T, Hensel M, Kox WJ (1997) Transient Ca^{2+} changes in endothelial cells induced by low doses
    of reactive oxygen species: role of hydrogen peroxide. Mol Cell Biochem 171:11–21

86. Casteilla L, Rigoulet M, Penicaud L (2001) Mitochondrial ROS metabolism: modulation by uncou-
    pling proteins. IUBMB Life 52:181–188

87. Miwa S, Brand MD (2003) Mitochondrial matrix reactive oxygen species production is very sensitive
    to mild uncoupling. Biochem Soc Trans 31:1300–1301

88. Sun X, Zemel MB (2007) 1α, 25-dihydroxyvitamin D3 modulation of reactive oxygen species produc-
    tion. Obesity (Silver Spring) 15:1944–1953

89. Sun X, Zemel MB (2006) Dietary calcium regulates ROS production in aP2-agouti transgenic mice on
    high-fat/high-sucrose diets. Int J Obes (Lond) 30:1341–1346

90. Sun X, Zemel MB (2007) Calcium and 1,25-dihydroxyvitamin D3 regulation of adipokine expression.
    Obesity (Silver Spring) 15:340–348

91. Sun X, Zemel MB (2008) Dietary calcium and dairy modulation of oxidative and inflammatory stress
    in mice and humans. J Nutr 138:1047–1052

92. Zemel MB, Sun X, Sobhani T, Wilson B (2010) Effects of dairy compared with soy on oxidative
    and inflammatory stress in overweight and obese subjects. Am J Clin Nutr e-pub ahead of print:
    doi:10.3945/ajcn2009.28468.

93. Weisberg SP, McCann D, Desai M, Rosenbaum M, Leibel RL, Ferrante AW Jr (2003) Obesity is
    associated with macrophage accumulation in adipose tissue. J Clin Invest 112:1796–1808

94. Xu H, Barnes GT, Yang Q, Tan G, Yang D, Chou CJ, Sole J, Nichols A, Ross JS, Tartaglia LA, Chen
    H (2003) Chronic inflammation in fat plays a crucial role in the development of obesity-related insulin
    resistance. J Clin Invest 112:1821–1830

95. Curat CA, Miranville A, Sengenes C, Diehl M, Tonus C, Busse R, Bouloumie A (2004) From
    blood monocytes to adipose tissue-resident macrophages: induction of diapedesis by human mature
    adipocytes. Diabetes 53:1285–1292

96. Bouloumie A, Curat CA, Sengenes C, Lolmede K, Miranville A, Busse R (2005) Role of macrophage
    tissue infiltration in metabolic diseases. Curr Opin Clin Nutr Metab Care 8:347–354

97. Sun X, Zemel MB (2008) Calcitriol and calcium regulate cytokine production and adipocyte-
    macrophage cross-talk. J Nutr Biochem 19:392–399

III VITAMIN D STATUS – GLOBAL ANALYSIS

18 Determinants of Vitamin D Intake

Mona S. Calvo and Susan J. Whiting

Abstract   The objective of our chapter is to provide convincing evidence of how changes in food consumption patterns, judicious fortification of food staples, and targeted supplementation of at-risk groups could be effective public health strategies to help increase vitamin D intake, maintain bone health, and potentially prevent chronic disease. We demonstrate the limitations of the Canadian and American food supply to provide sufficient vitamin D to meet increased dietary needs when cutaneous synthesis of vitamin D is compromised. Vitamin D deficiency as measured by low circulating 25-hydroxyvitamin D [25(OH)D] and its link to increased risk of chronic disease is a significant global reality and threat to general public health, yet dietary intakes of vitamin D remain lower than the recommended dietary guidelines for the majority of individuals experiencing the lowest levels of 25(OH)D.

Key Words: Cholecalciferol (vitamin D_3); ergocalciferol (vitamin D_2); 25-hydroxyvitamin D (25(OH)D); dietary guidelines, dietary supplements, food fortification

1. INTRODUCTION

The circulating level of 25-hydroxyvitamin D [25(OH)D] is the most commonly used measure of vitamin D nutritional status and is the combined product of cutaneous synthesis made during solar exposure and dietary intake (1–3). Among people whose exposure to sunlight may be limited for whatever reason, dietary vitamin D becomes the major determinant of serum 25(OH)D. Maintaining adequate circulating 25(OH)D concentration is now considered critical to overall health maintenance and the function of the immune, integumentary, reproductive, and musculoskeletal system of men and women of all ages and races (4). Worldwide, even in geographic areas with abundant sunshine, studies show vitamin D status is sub-optimal relative to the level of 25(OH)D associated with lower risk of chronic and infectious diseases (4), mortality from all causes (5–8), and factors known to contribute to longevity (9). Recent compelling longitudinal studies now link higher serum 25(OH)D levels to reduced incidence of hip

Disclaimer: The findings and conclusions presented in this chapter are those of the authors and do not necessarily represent the views or opinions of the US Food and Drug Administration. Mention of trade names, product labels, or food manufacturers does not constitute endorsement or recommendation for use by the US Food and Drug Administration.

From: *Nutrition and Health: Vitamin D*
Edited by: M.F. Holick, DOI 10.1007/978-1-60327-303-9_18,
© Springer Science+Business Media, LLC 2010

fracture *(10, 11)*. A recently documented north–south European gradient of higher mean 25(OH)D levels in the northernmost countries compared to southern countries with available sunlight year round demonstrates that efforts to increase dietary intake of vitamin D through supplementation, fortification, and changes in food consumption patterns are effective approaches to maintaining adequate circulating 25(OH)D under any geographic condition *(12)*.

The long held assumption that the vitamin D content of the food supply is sufficient to maintain healthy circulating levels of 25(OH)D when adequate sun exposure is compromised is a fallacy held worldwide *(13)*. In this chapter, we examine the factors that influence vitamin D intake of the general population of North America. Specifically, we address the major determinants of vitamin D intake in the United States and Canada: dietary guidelines, food consumption patterns, food fortification practices and policies, and trends in supplementation practices. Appropriate modification of these four determinants to local population needs has global relevance to the development of effective and safe public health strategies to optimize vitamin D intake and status.

2. DIETARY GUIDELINES: PAST, PRESENT, AND FUTURE CONSIDERATIONS

Considerable controversy exists over the threshold value of 25(OH)D designating optimal levels for health maintenance and reduced risk of disease *(13, 14, 15)*. An optimal level suggested to be as much as 100 nmol/l has been proposed largely based on findings from cross-sectional studies showing strong associations between functional biomarkers and low disease risk or incidence *(16, 17)*. The threshold value used in establishing dietary guidelines by the United States and Canada of 27.5 nmol/l (11 ng/ml) and the United Kingdom of 25 nmol/l (10 ng/ml) relates only to bone health and is just above the range indicative of frank vitamin D deficiency observed in rickets or osteomalacia *(18, 19)*. Despite the conservative nature of this cutoff value for estimating vitamin D deficiency, a high prevalence of frank vitamin D deficiency has been demonstrated globally *(20)*. Before we can begin to implement changes in national dietary guidelines for vitamin D to improve public health, we first must agree on an optimal range of 25(OH)D.

Three lines of evidence are needed in addition to agreement on a cutoff for adequate 25(OH)D: (1) evidence that increasing vitamin D intake will safely increase circulating 25(OH)D, the main status indicator, (2) quantification of the amount of dietary vitamin D needed to raise 25(OH)D to the specified optimal range, and (3) evidence that the changes in the status indicator result in quantifiable changes in disease incidence or adverse events such as reduction in bone fractures. Providing this evidence is no easy task, given the number of potential confounding factors including source of vitamin D (supplement or food), completeness of the nutrient composition database, form of the supplement or fortificant (ergocalciferol, cholecalciferol, or metabolite 25(OH)D), dose or level of intake, frequency of intake, initial vitamin D status, age, gender, race/ethnicity, physiologic state (obese, pregnant, or lactating), and functional endpoint measured.

Studies published on or before June 2006, which could potentially provide needed evidence for dietary guideline reassessment, were recently reviewed in a systematic evidence-based review that specifically focused on the relation of vitamin D to bone and muscle health *(21)*. Cranney et al. examined the effect of vitamin D supplementation and food fortification on 25(OH)D concentrations, reporting a significant association between dose (vitamin D_2 or vitamin D_3) and 25(OH)D concentration. They determined a 1–2 nmol/l increase in 25(OH)D for each additional 100 IU of vitamin D; nonetheless, they were not able to estimate an adequate intake of vitamin D due to the lack of standardization and calibration of 25(OH)D measures used in the 16 vitamin D trials that comprised their meta-analysis. Cranney and associates found good evidence of a positive effect of vitamin D-fortified food which varied in magnitude from 15 to 40 nmol/l; however, they could not determine if the positive effect varied by age, body mass index, or race/ethnicity *(21)*. They also reported fair evidence that adults tolerated vitamin D at doses above current dietary reference intake levels, but had no data on the association between long-term adverse events and higher doses. With respect to bone health, Cranney et al. reported inconsistent evidence of an association between serum 25(OH)D concentrations and bone mineral content in infants and fair evidence in older children and older adults, but variable or inconsistent evidence for a beneficial association between 25(OH)D concentration and clinical outcomes or performance measures (fractures and falls) *(21)*. This recent systematic review provides important evidence of the efficacy and safety of using diet to increase 25(OH)D, but strong evidence relating 25(OH)D levels to specific functional outcomes is needed to justify changes in the dietary guidelines or threshold ranges for vitamin D nutritional status assessment. Further research is needed to fully understand the relation between 25(OH)D threshold values and functional outcomes of other diseases such as cancer, cardiovascular disease, and others *(22)*. This information is forthcoming, given the number of studies published after June 2006 that were not included in the systematic evidence-based review and the approximate 257 ongoing or completed clinical trials involving cholecalciferol (vitamin D_3) or ergocalciferol (vitamin D_2) dietary intervention in a variety of conditions including vitamin D deficiency reported by the National Institutes of Health (http://clinicaltrials.gov/ct2/results?intr=Cholecalciferol+ORErgocalciferol&pg=1, last accessed 9/24/2008).

3. HISTORY AND USE OF DIETARY REFERENCE INTAKES FOR VITAMIN D

Since the early 1940s, the United States and Canada have set dietary intake recommendations for nutrients. The first recommendation for vitamin D for Americans in 1941 gave the value of 400 IU (i.e., the lower value of a range for infants at the time), for adults in a footnote only, that stated "When not available from sunshine, [vitamin D] should be provided up to the minimal amounts recommended for infants" *(23)*. This value of 400 IU (10 µg) was derived from an observation that this amount of vitamin D activity, found in a teaspoon of cod liver oil, was sufficient to prevent rickets *(24)*. In the mid-1990s, the Dietary Reference Intakes (DRI) process saw Canada and the United States set nutrient recommendations together *(18)*. These recommendations are usually attributed to the Institute of Medicine (IOM) but may also be referred to as coming from the Food and Nutrition Board (FNB); in Canada, these values are attributed to Health Canada. The current (1997) recommendations for vitamin D, shown in Table 1, are

Table 1
Dietary Recommendations and Expert Guidelines for Health Professional for Vitamin D Intake

Organization and date	Age group	Recommendation	Notes
Institute of Medicine (IOM), 1997 http://www.iom.edu	1–50 years 51–70 years 71+ years	5.0 μg (200 IU) 10 μg (400 IU) 15 μg (600 IU)	Many experts indicate a need for revision as these intakes will not maintain optimal 25OHD levels. These values will be revisited and revised by 2010
Osteoporosis Society of Canada 2002 http://www.osteoporosis.ca	19–50 years 51+ years	10 μg (400 IU) 20 μg (800 IU)	For osteoporosis prevention (with calcium); and for osteoporosis treatment (adjunct, with calcium)
American College of Obstetricians & Gynecologists, 2003 http://www.acog.org	Women 51+ years	10–20 μg (400–800 IU)	For prevention of osteoporosis in at-risk women (poor intake or sun exposure)
Dietary Guidelines for Americans, 2005 http://www.health. gov/dietaryguidelines	Men and women	25 μg (1,000 IU)	"Older adults, people with dark skin, and people exposed to insufficient ultraviolet band radiation (i.e., sunlight) should consume extra vitamin D from vitamin D-fortified foods and/or supplements"
Health Canada Canada Food Guide 2007 www.hc-sc.gc.ca/fn-an/food-guide-aliment/context/index-eng.php	19+ years	5–15 μg (200–600 IU)	"Have 500 ml (two cups) of [fortified] milk every day for adequate vitamin D" and for "Men and women over 50 years the need for vitamin D increases after the age of 50. In addition to following Canada's Food Guide, everyone over the age of 50 should take a daily vitamin D supplement of 10 μg (400 IU)

Table 1
(continued)

Organization and date	Age group	Recommendation	Notes
Canadian Cancer Society 2007 http://www.cancer.ca	19+ years	25 µg (1,000 IU)	"Due to our northern latitude. . . we recommend that Canadian adults consider taking. . . 1,000 international units (IU) a day during fall and winter months. You're probably not getting enough vitamin D if you are elderly; have dark skin; don't go outside very much; wear clothing covering most of your skin. . . . talk to your doctor about whether you should take a vitamin D supplement of 1,000 IU every day, all year round"
National Osteoporosis Foundation 2008 http://www.nof.org	50+ years	20–25 µg (800–1,000 IU)	"Advice on adequate amounts. . . of vitamin D for individuals at risk of insufficiency"

Adequate Intake (AI) values, denoting the lack of scientific evidence needed to set a Recommended Daily Allowances (RDAs) in the DRI process *(25)*. Information on infants and pregnant and lactating women is covered in a different chapter. The 1997 AI is based on maintenance of serum 25(OH)D levels in the absence of sunlight (i.e., through the winter) at or above 27.5 nmol/l for most age groups *(26)*. It was acknowledged that a dietary intake should maintain serum 25(OH)D above the concentration below which vitamin D deficiency rickets or osteomalacia occurs in the absence of sun exposure. As there was not sufficient evidence to know what that dietary level should be, it was presumed that reported dietary intakes of a group of apparently healthy adults were sufficient, and this level was adjusted (multiplied by 2) for uncertainty *(18, 26)*. Maintenance of bone health (prevention of rickets and osteomalacia) is the only relationship considered for vitamin D by the current dietary guidelines used jointly by the United States and Canada *(26)*.

The first rendition of the DRIs for all nutrients is complete *(25)* and these are now slated for revision, beginning with vitamin D and calcium. In the meantime, other professional groups have made recommendations on vitamin D intakes (Table 1) in order to account for a decline in sun exposure due to sun safety messaging and to acknowledge recent evidence that higher doses are needed to achieve chronic disease prevention and other functions still under study *(2, 27)*. The groups listed in Table 1 recognized that the DRI recommendation did not take these new functions into account, nor was there recognition that both sun exposure and dietary intakes were inadequate in North Americans *(28, 29)*. For example, in 2005, the special needs for vitamin D were recognized in the Dietary Guidelines for Americans, for those in high-risk groups, defined as being unable under usual circumstances to make cholecalciferol in skin, were advised to consume an additional 25 μg (1,000 IU). Similarly, the Canadian Cancer Society made a recommendation for Canadians of 1,000 IU through the winter months, acknowledging the fact that Canada is primarily above 45°N, thus experiencing a 6-month "vitamin D winter." The most recent recommendation is in 2008 from the National Osteoporosis Foundation: men and women 50 years and older are advised to have 800–1,000 IU of vitamin for osteoporosis prevention and treatment.

The American Academy of Pediatrics (AAP) *(28)* in 2003 recommended that children up to age 18 years follow the DRI recommendations of 5 μg (200 IU) per day *(18)*. The AAP also recommends that the supplemental 5 μg (200 IU) is warranted when an infant or child is not ingesting fortified formula or milk. This recommendation for infants will prevent rickets but is sub-optimal for preventing risk of chronic disease development *(2)*. In Canada, new pediatric recommendations indicate that full-term breastfed infants should have 10 μg (400 IU); those in northern communities should have twice this much, which would also be true of other groups with little opportunity for skin synthesis of vitamin D by the infant and/or the mother *(30)*. Others concur that 400 IU is the preferred amount *(31, 32)*. Although prevention of rickets is the primary reason vitamin D supplements should be provided to breastfed infants, it is not the only reason. Vitamin D receptors are present in many organs, and vitamin D exerts anti-proliferative effects, which may explain why it protects against colorectal cancer; vitamin D supplementation during infancy may protect against type 1 diabetes mellitus *(33)* and has been shown to lead to higher bone mineral mass in pre-pubertal girls *(34)*.

4. THE TOLERABLE UPPER INTAKE LEVEL

As part of the 1997 DRIs, a daily tolerable upper intake level (UL) for vitamin D of 50 μg (2,000 IU) *(18)* was established for persons aged 1 year and older in order to discourage potentially dangerous self-medication. For infants 0–1 year of age, the UL was set at 25 μg (1,000 IU). The UL represents a safe intake (i.e., a zero risk of adverse effects in an otherwise healthy person), and when a patient is undergoing therapeutic treatment under a health professional's care, this amount can be exceeded. A recent risk assessment for vitamin D used new data (post-1997) to derive a "revised" UL *(35)*. Studies indicated that there was absence of any signs of toxicity when healthy adults were given over 250 μg (10,000 IU) daily. In situations of limited sun exposure, many researchers believe vitamin D intakes should approach 100 μg (4,000 IU) *(36, 37)*, a value greater than the UL. Therefore, it is important to consider whether the UL is too low to allow for ingestion of appropriate amounts of vitamin D to maintain optimal levels of 25(OH)D, particularly in the absence of sun exposure where this might be needed. In a dosing study, Heaney et al. *(38)* gave cholecalciferol to subjects in doses up to 250 μg (10,000 IU) and found no adverse effects in men treated for 5 months. The primary adverse effect that is expected at very high levels of vitamin D is hypercalcemia, which can lead, over time, to calcification of soft tissues such as arteries (arteriosclerosis) and kidney (nephrocalcinosis) *(39, 40)*. A less specific indicator is hypercalciuria, which may lead to increased risk of kidney stones *(2, 18)*. A recent risk assessment for vitamin D used new data (post-1997) to derive a revised UL *(2, 35)*. These newer studies indicated there was absence of any signs of toxicity when healthy adults were given over 250 μg (10,000 IU) daily. It should be noted that at this time, revised data for a UL have not considered children or infants.

5. ESTIMATED VITAMIN D INTAKES IN CANADA AND THE UNITED STATES

Relatively few studies take on the task of assessing vitamin D intake and when they do, largely all of them are inaccurate in their estimates of true vitamin D intake. This fact is attributed to the inaccuracies inherent in the methods used to determine food intake (e.g., Food Frequency Questionnaire, 24-hr recall, multiple-day food record) *(13)*, the large variability and incomplete nature of nutrient composition databases used to quantitate vitamin D intake *(41, 42)*, the variability in content of natural food sources due to seasonality *(13)*, the variation in dietary supplement use *(43, 44)*, or the inconsistency in food fortification practices *(43, 44)*. Vitamin D intake estimated from nationally representative surveys conducted in the United States, Canada, Ireland, and the United Kingdom vary with gender, age, and national fortification or supplementation practices (Table 2). Although flawed and somewhat outdated, these intake data alert us to important trends that target individuals in greater need. The trends evident in Table 2 include higher intakes in men than in women when food sources alone are considered and an increase in vitamin D intake with increasing age when supplements use is common. In a more recent NHANES 1999–2000 study, mean intake from food and mean intake

Table 2
Vitamin D Intake Varies with Gender, Age, and National Fortification/Supplementation Practice

Mean daily vitamin D intake, μg/d

Country	♀	♂	(age)	♀	♂	(age)	♀	(age)
Canada[a]			9–18 years			19–50 years		51–70 years
Food sources	5.4 (0.1)	5.1 (0.2)		5.8 (0.2)	5.1 (0.3)		7.0 (0.4)	
All sources	7.0	7.5		7.8	10.3		9.9	
United States[b]			9–18 years			19–50 years		51+ years
Food sources	4.4 (0.2)	4.2 (0.2)		5.4 (0.2)	4.7 (0.2)		5.3 (0.2)	
All sources	5.6 (0.3)	7.1 (0.3)		7.5 (0.3)	9.5 (0.4)		8.8 (0.3)	
United Kingdom[c]			19–24 years			35–49 years		50–64 years
Food sources	2.3 (1.6)	2.8 (2.1)		3.7 (2.3)	3.5 (2.4)		4.2 (2.4)	
All sources	2.9 (2.5)	3.5 (2.9)		4.2 (3.1)	5.1 (4.1)		4.9 (3.2)	
Ireland[d]			18–35 years			36–50 years		51–64 years
Food sources	2.4 (1.7)	2.9 (2.2)		3.9 (2.2)	3.4 (2.2)		4.0 (2.5)	
All sources	3.3 (3.1)	3.9 (3.4)		4.7 (3.3)	5.1 (4.9)		5.0 (4.5)	

[a]Canadian values are from the initial 24-h recall data from the Canadian Community Health Survey Cycle 2.2(CCHS 2.2, 2004), a nationally and provincially representative sample, n = 35,107; mean (SEM) or mean only are presented.

[b]United States values are from the 24-h recall data from the 1999–2000 National Health and Nutrition Examination Survey (NHANES 1999–2000), a nationally representative survey (n = 6,931). Values are presented in reference (39) as the mean (SEM).

[c]United Kingdom intake estimates are from the 2000/01 National Diet and Nutrition Survey: adults aged 19–64 years, Volume 3, 2003 reference (82). Values are represented as mean (SEM).

[d]Irish values are from a 2004 reference (83) and were estimated from 7-d food records from the North/South Ireland Food Consumption Survey of 18–64-year-old adults (n = 1,379), a nationally representative cross-sectional survey. Values are represented as mean (SD).

from food and supplements were reported *(39, 45)*. Intake from food, for all age/sex groups, averaged only 4–6 μg (160–240 IU), and most of the food-derived vitamin D was from fortified foods *(1)*. Supplement use contributed to vitamin D intake, adding 1–2 μg (40–80 IU) in each group except for women and men over 50 years who showed an increase of approximately 4–5 μg (160–200 IU). When considering total intakes, the mean intake of each age group met the current DRI recommendation except for adults over 50 years, where the AI of 10 μg (400 IU) and 15 μg (600 IU) were not met. Canadian intakes have recently been determined but for food intake alone *(40)*; these are difficult to compare with the US intakes given difference in fortification practices and nutrient compositional databases. Significant racial and ethnic differences in vitamin D nutritional status and intake can occur in a population as shown in Table 3. Black men and women in the United States have significantly lower serum 25(OH)D than whites and consume significantly lower vitamin D from milk, ready-to-eat cereals, and dietary supplements. Those in greatest need of dietary sources of vitamin D due to aging and/or darker skin have the lowest intakes of vitamin D, thus further contributing to their low circulating 25(OH)D. Similar low intakes and poor vitamin D status are observed in Canadian Aboriginal populations, especially among urban dwellers who no longer consume traditional diets, and among darker skinned Canadians of Asians ancestry who consume traditional vegetarian or vegan diets *(40, 46, 47)*.

Table 3
Racial Differences in Vitamin D Status, Intakes, Food Sources, and Supplements Use from the Third National Health and Nutrition Examination Survey, 1988–1994

	White adults	*Black adults*
n	6,456	4,316
Serum 25(OH)D, nmol/l	79.0 ± 0.95	48.2 ± 1.05*
Vitamin D intake		
All sources, μg/d	7.92 ± 0.15	6.20 ± 0.13*
Supplements only, μg/d	2.84 ± 0.14	2.0 ± 0.11*
Fluid milk and milk-based drinks, μg/d	1.99 ± 0.06	1.01 ± 0.44*
Plain fluid milk, μg/d	1.86 ± 0.06	0.92 ± 0.04*
Ready-to eat breakfast cereals, μg/d	0.39 ± 0.02	0.23 ± 0.04*

[a]Weighted mean ± SEM; adapted from reference *(43)*. *Significantly lower in black adults; $P < 0.0001$.
[b]Includes samples taken at all latitudes during both summer and winter, thus not controlled for differences in UV intensity.

6. FOOD PATTERNS AND DIETARY SOURCES OF VITAMIN D

When food alone is considered, mean intakes of vitamin D do not meet the current DRI recommendation for many, notably those in greatest need. Considering that current DRIs are considered too low to maintain 25(OH)D at levels necessary for all functions *(18)*, a need to improve dietary intakes is clear. Many factors can contribute to increases in vitamin D intake including change in nutrient composition databases, food availability, development of new analytical methods, changes in fortification, and in use and potency of dietary supplements. All of these factors are strategic actions for implementing improvements in vitamin D intake. Traditionally less attention has been given to improving the quantity or content of vitamin D in foods naturally rich in vitamin D_2 or D_3. In Canada and the United States, such foods are limited and not frequently consumed. Fatty fish represent the richest natural source of vitamin D commonly consumed in North America. Evidence from Japan and Norway illustrate how frequent consumption can be an effective way to maintain healthy 25(OH)D levels. Fish consumption contributes 1.5–1.8 μg/d to Norwegian daily intakes *(48)*, 3.21 μg/d to Belgian adolescents *(49)*, and 6.4 μg/d to older Japanese women *(50)* and provides 25% of the French RDA for vitamin D *(51)*.

The completeness of the food composition database used to estimate intake significantly influences national estimates of vitamin D intake. The United States is in need of updating and reanalyzing foods for vitamin D content in the US Department of Agriculture database (USDA Standard Reference) (SR) which is the authoritative source of food composition data for the United States *(41, 52)*. Release 20 of the USDA SR (SR20) contains vitamin D values for approximately 600 foods or only about 6% of the total 7,520 food items listed. Canada is further along in revising its nutrient file and has included items that are very poorly represented as sources of vitamin D in SR20, such as beef, lamb, veal, pork, and poultry products (Health Canada, personal communications). Applications of new analytical methods reveal measurable levels of vitamin D and its metabolite, 25(OH)D, in various meats shown in Table 4 *(53)*. Updated nutrient database and newly analyzed vitamin D food content values for natural and fortified foods are shown in Table 4. Given the differences in foods analyzed using very different nutrient content databases, it is difficult to accurately compare Canadian and American intake estimates for vitamin D. For example, the recent Canadian Community Health Survey, 2004, showed 2.3 μg out of a total intake of 6.72 μg for men 9 year and older and 1.74 μg out of a daily mean total of 5.42 μg for women were contributed by meat and meat alternatives. In Canada meat contribution to vitamin D intake was second only to the contributions from fortified milk products *(40)*. Strategies are being developed for the natural enhancement of vitamin D content of foods through changes in diet composition *(54)* and the pharmacologic supplementation (vitamin D_3 or 25(OH)D) of poultry and livestock feed to enhance meat and egg vitamin D content *(55–57)*. Postharvest light exposure techniques have been developed to significantly enhance the vitamin D content of fresh edible mushrooms (Table 4). A serving of light-exposed portabella mushrooms can supply 10 μg of vitamin D_2 (http://dolemushrooms.com/Vitamin%20D.htm). This represents an important natural food source of vitamin D for vegetarians and vegans *(44)*. The concept of irradiating ergosterol from yeast or fungi with light is not new. This was the major source of vitamin D used to fortify milk and to treat rickets spearheading the public health campaign that successfully eradicated rickets in North America by the 1930s, only to see it surface again with the new century *(58)*.

Table 4
Updated Nutrient Database and Newly Analyzed Vitamin D Food Content Values from Natural and Fortified Sources

Country of origin (reference)	Food	Vitamin D content (μg/ ~100 g)
Canada (40)	Margarine (fortified)	13.25
	Soy milk (fortified)	1.05
	Milk (fortified)	1.05
	Pork	0.59 μg D_3
		0.74 μg 25(OH)D[a]
		4.29 μg total D activity[a]
	Rainbow trout	0.80 μg D_3
		<0.02 μg 25(OH)D*
		<0.90 μg total D activity*
	Eggs, large	0.96 μg D_3
		0.38 μg 25(OH)D[a]
		2.86 μg total D activity[a]
United States (59)	Blue fish	7.0
	Cod	2.6
	Gray sole	1.4
	Farmed salmon	6.18
	Wild-caught salmon	24.7
	Farmed trout	9.7
	Ahi-yt tuna	10.1
(61)	Orange juice, calcium, and vitamin D fortified, 125 ml	1.25
(43)	Yogurt, selected fortified brands, 100 g	2.5
(USDA SR20)	White button mushrooms (84 g serving, fresh not light exposed)	0.13 μg D_2
(Mushroom Council)	White button mushrooms, light-exposed, fresh 84 g serving	10 μg D_2
(Mushroom Council)	Portabella mushrooms, light-exposed, fresh 84 g serving	10 μg D_2
(2)	Shitake mushrooms, sun-dried, 36 g	2.8 μg D_2
Ireland (83)	Low-fat spread	2.20

Table 4
(continued)

Country of origin (reference)	Food	Vitamin D content (μg/ ~100 g)
	70 % fat spread	1.98
	Cod fish, raw	1.49
	Premium ham	1.05
	Chicken breast, breaded, fried	1.11
	Turkey slices	2.17
(53)	Salmon, raw	5.4 μg D_3 0.11 μg 25(OH)D
(53)	Trout, raw	8.1 μg D_3 0.22 μg 25(OH)D
(53)	Chicken	0.25 μg 25(OH)D
(53)	Lean pork	<0.15 μg 25(OH)D
Holland *(84)*	Margarine (fortified)	7.2
	Butter (non-fortified)	1.2
	Full-fat milk (non-fortified)	0.15
	Reduced-fat milk (non-fortified)	0
	Mackerel, smoked	8.0
	Salmon, micro waved	8.7
France *(51)*	Eel	20
	Herring	17
	Salmon	15
	Sardine	11
	Mackerel	10
	Trout	8
	Oysters	8
	Anchovy	7
	Tuna	5
	Halibut	4.3

*Assume 5× activity factor.

Table 4 also illustrates how rich a vitamin D source fish are. The vitamin D content of fish varies with season as to be expected, but also varies with environmental conditions and diet. Farmed fish with lower access to vitamin D in their diets have significantly lower content of vitamin D than their wild-caught counterparts *(59)*. Given that wild fish are generally a good source of vitamin D, this information would be helpful if displayed on the label. In the United States and Canada, food labels are not required to list the vitamin D content of foods naturally rich in vitamin D. Only foods that have been fortified with vitamin D are required to list it on the label. In both countries the vitamin D content is listed as a percent of the Daily Value (%DV). The DVs were developed by the US Food and Drug Administration to help consumers compare the nutrient content of a product within the context of a 2,000 calorie diet. Canada uses a different value for the DV, 5 μg (200 IU), while a value of 10 μg (400 IU) is used on American labels. One possible solution to this source of confusion would be to adopt the method used by other countries where the vitamin D content per serving size is displayed in absolute amounts.

7. VITAMIN D FORTIFICATION OF FOOD

Canadians and Americans are largely dependent on fortified foods and dietary supplements to meet their vitamin D needs because foods that are naturally rich in vitamin D are less frequently consumed. While Canada and the United States use the same dietary guidelines (DRI) for vitamin D and calcium and the same upper limits of safe intake for these nutrients, they have very different regulatory approaches to the lawful addition of vitamin D and calcium to foods (43). Both countries recognize the potential for toxicity if vitamin D is consumed at very high doses, and therefore both countries carefully regulate vitamin D addition to food.

The Canadian approach is that of mandatory fortification of food staples through the Canadian Food and Drug Regulations (43). Fortification of staple foods ensures nutritional benefit to all Canadians. Milk and milk alternatives and margarine are required to be fortified in Canada. Fluid milk in Canada is labeled as providing 44% of the recommended daily intake (10 μg or 400 IU) per 250 ml serving (43). Other milk products that require vitamin D fortification include evaporated milk, powdered milk, goat's milk, and milks of plant origin (soy, rice, and other grains), which also must be fortified with calcium (43). All margarines in Canada are fortified with vitamin D at the level of 13.25 μg (530 IU) per 100 g (43). Other foods for which vitamin D addition is permitted are meal replacements, nutritional supplements, and formulated liquid diets. Addition of vitamin D to such foods are optional but may be no less than 2.5 μg (100 IU) and no more than 10 μg (400 IU) per 1,000 kcal, as long as the intended total energy intake is <2,500 kcal (43). Fortification of some egg products is also permitted at this time, but industrial milk used in baked goods and cheeses does not need to be fortified (43).

The approach to vitamin D fortification of foods in the United States is also very carefully regulated, but is largely optional for a number of food categories and is required only for fortified fluid milk and fortified evaporated milk (Table 5). The addition of vitamin D to foods as a nutrient supplement is in accordance with the US Code of Federal Regulations (CFR): 21 CFR 184.1 (b) (2) which imposes strict limitations with respect to the categories of foods, functional use, and level of use (43) as shown in Table 5. Such regulatory limitations provide a control mechanism that limit over-fortification with vitamin D. In accordance with 21 CFR 184.1 (b) (2), any addition of vitamin D to foods not in compliance with each of these established limitations requires a food additive regulation (43). More recently, vitamin D_3 may be added safely to foods as a nutrient supplement in accordance with 21 CFR 172.380 for those food uses presented in Table 5. In contrast to Canada, where fortification with vitamin D is mandatory for designated food staples, the lawful addition of vitamin D to eligible foods in the United States is voluntary in most cases, with the exception of fluid milk. Vitamin D fortification is required when the label declares that the milk is fortified. There is good evidence

Table 5
Lawful Additions of Vitamin D to Foods in the United States

Addition of Vitamins D_2 and D_3

Category of food	21 CFR citation	Fortification status	Maximum level allowed[a], µg vitamin D_2 or D_3
Cereal flours and related products			
Enriched farina	137.305	Optional	8.7 5 µg (350 IU)/100 g
Ready-to-eat breakfast cereals	137.305	Optional	8.75 µg (350 IU)/100 g
Enriched rice	137.350	Optional	2.25 µg (90 IU)/100 g
Enriched cornmeal products	137.260	Optional	2.25 µg (90 IU)/100 g
Enriched noodle products	139.155	Optional	2.25 µg (90 IU)/100 g
Enriched macaroni products	139.115	Optional	2.25 µg (90 IU)/100 g
Milk			
Fluid milk	131.110	Optional	1.05 µg (42 IU)/100 g
Acidified milk	131.11	Optional	1.05 µg (42 IU)/100 g
Cultured milk	131.112	Optional	1.05 µg (42 IU)/100 g
Concentrate milk	131.115	Optional	1.05 µg (42 IU)/100 g
Nonfat dry milk, A and D fortified	131.127	Required	1.05 µg (42 IU)/100 g
Evaporated milk, fortified	131.130	Required	1.05 µg (42 IU)/100 g
Dry whole milk	131.147	Optional	1.05 µg (42 IU)/100 g
Milk products			
Yogurt	131.200	Optional	2.22 µg (89 IU)/100 g
Low-fat yogurt	131.203	Optional	2.22 µg (89 IU)/100 g
Nonfat yogurt	131.206	Optional	2.22 µg (89 IU)/100 g
Cheese	133.165, 133.183	Optional	2.22 µg (89 IU)/100 g
Margarine	166.110	Optional	8.3 µg (331 IU)/100 g

Table 5
(continued)

Addition of Vitamins D_2 and D_3

Category of food	21 CFR citation	Fortification status	Maximum level allowed[a], μg vitamin D_2 or D_3
Lawful addition of vitamin D_3 to foods[b]			
Calcium-fortified 100% fruit juice[c]	172.380	Optional	2.5 μg (100 IU)/240 ml
Calcium-fortified fruit juice drinks[d]	172.380	Optional	2.5 μg (100 IU)/240 ml
Soy-protein meal-replacement beverage[e]	172.380	Optional	3.5 μg (140 IU)/240 ml
Meal-replacement and other type bars[f]	172.380	Optional	2.5 μg (100 IU)/40 g
Cheese and cheese products[g]	172.380	Optional	2.02 μg (81 IU)/30 g

[a]Maximal level of vitamin D that can be added in accordance with 21 CFR 184.1 (b) (2) for the category of food.

[b]Vitamin D_3 may be used safely in foods as a nutrient supplement in accordance with 21 CFR 172.380 at levels not to exceed those specified above.

[c]Vitamin D_3 may be added, at levels not to exceed 100 IU per 240 ml serving, to fruit juice drinks that are fortified with >10% of the RDI for calcium. (Excludes fruit juice drinks that are specially formulated or processed for infants.)

[d]Vitamin D_3 may be added at levels not to exceed 100 IU per 240 ml serving to fruit juice drinks that are fortified with >10% of the RDI for calcium. Excludes fruit juice drinks that are specially formulated or processed for infants.

[e]Vitamin D_3 may be added at levels not to exceed 140 IU per 240 ml of soy-protein-based meal-replacement beverage (powder or liquid) used for special dietary purposes, weight reduction, or maintenance at levels not to exceed 100 IU per 40 g.

[f]Vitamin D_3 may be added to meal-replacement and other bars used for special dietary purposes, weight reduction, or maintenance at levels not to exceed 100 IU per 40 g.

[g]Vitamin D_3 may be added to cheese and cheese products, excluding cottage cheese, ricotta cheese, and hard grinding cheese such as Parmesan and Romano, at levels not to exceed 81 IU per 30 g.

that vitamin D is bioavailable from natural food sources such as wild mushrooms or light-exposed mushrooms *(60)* or when added as a fortificant to calcium-fortified orange juice *(61)*, bread *(62, 63)*, cheese *(64)*, and calcium-fortified milk *(65, 66)*. Discussion of vitamin D additions to infant formulas in Canada or the United States is beyond the scope of this chapter.

Although many food categories are eligible for controlled levels of vitamin D fortification in the United States, there is a large discrepancy between the number of eligible foods and the number and variety of vitamin D-fortified foods currently in the US market place *(43)*. Fluid milk and ready-to-eat cereals are the major contributors to vitamin D intake in the United States, while milk and margarine are the major fortified foods contributing to Canadian vitamin D intake *(44)*. Significant racial/ethnic differences in vitamin D intake attributed to differences in the consumption of these fortified staples are shown in Tables 4 and 6. There are marked differences in these fortified food consumption patterns between whites and blacks in the United States which are attributed to higher prevalence of lactose intolerance and low milk consumption in black men and women. Selective fortification of only a few staple foods that are not commonly consumed by individuals at greatest risk of poor vitamin D status is a major barrier to optimizing vitamin D intake. By encouraging United States manufacturers to utilize these fortification options, the vitamin D intake of groups at risk could be significantly improved. Consideration could also be given to shifting optional fortification to mandatory for those eligible food staples commonly consumed by the entire US population. Cereal grain products (pasta, bread, and other baked goods) are frequently consumed by the general population and would serve as a good candidate for mandatory fortification with vitamin D. The inadequacy of vitamin D levels currently added to Canadian milk (1 μg/100 g fluid milk) to maintain healthy circulating levels of 25(OH)D during the winter has helped to persuade Health Canada to re-evaluate their food fortification regulations with respect to vitamin D fortification *(67, 68)*. It remains unclear whether or not Canada will allow more food categories to be fortified with vitamin D in the near future.

Table 6
Racial/Ethnic Differences in Prevalence of Hypovitaminosis D: Changes with Supplement Use[a]

Race/ethnic group	No supplement use		Supplement users	
	25(OH)D, nmol/l	Prevalence % below 70 nmol/l	25(OH)D, nmol/l	Prevalence, % below 70 nmol/l
Total	74.4	48	79.5	39
White	78.8	42	83.3	33
Black	48.0	86	54.6	77
Hispanic	62.7	66	66.2	60

[a]Modified from reference *(70)*. Values are presented as mean.

8. VITAMIN D SUPPLEMENTATION

Dietary supplement use is another option for improving vitamin D intake and status. Table 2 showing intakes from all sources illustrates how supplement use can significantly increase vitamin D intake in Europeans as well as North Americans by 1–2 μg. However, analyses of the NHANES III data demonstrate that the benefits gained from vitamin D containing supplement use are not uniform across all age, race/ethnic, and gender groups *(43, 69, 70)*. In black and Hispanic adults, supplement use shown in Table 6 had limited influence on reducing the prevalence of hypovitaminosis D defined as 25(OH)D levels <70 nmol/l *(70)*. The benefit of supplement use was gained more in those whose intakes were above the median intake and less gain was observed in individuals with low vitamin D intake from food *(43)*. Supplement use like fortified food consumption in the United States varies among race/ethnic groups (Table 3). Significant racial differences have been observed for the effects of daily intake of supplements containing >10 μg (400 IU) *(71, 72)*.The prevalence of vitamin D insufficiency, conservatively defined as <37.5 nmol/l, was shown in NHANES III to be markedly higher in black women compared to white women, even when both consume >10 μg (400 IU) vitamin D *(72)*. Intervention trials with higher levels of vitamin D intake present convincing evidence that vitamin D needs may be greater in some race/ethnic groups *(73–75)*. These newer findings underscore the need for race- or age-specific dietary supplements with higher vitamin D content that reflects the specific needs of the target population *(44)*.

Until recently non-prescription dietary supplements containing vitamin D were only available in multivitamin or single nutrient products containing 5–10 μg (200–400 IU) per tablet. New products have been introduced that contain 25–50 μg (1,000–2,000 IU) usually as cholecalciferol. The Dietary Supplements Labels Database of the US National Library of Medicine lists over 400 brands of dietary supplements containing cholecalciferol (vitamin D_3) or ergocalciferol (vitamin D_2). The database presents the amount of vitamin D in IU per manufacturer's recommended daily dose and the percent DV, but does not specify whether the product contains vitamin D as ergocalciferol or cholecalciferol. For the brands listed, the vitamin D content ranges from 17 IU (4 %DV) in a multiple vitamin and mineral product to 2,000 IU (500% DV) found in a single nutrient supplement.

The biological equivalency of ergocalciferol and cholecalciferol in humans is controversial *(17, 76, 77)*. Several studies have called into question whether or not ergocalciferol which is considered the plant-derived vitamin D_2 has the same potency as cholecalciferol or vitamin D_3 in raising circulating levels of 25(OH)D. This controversy stems from the faster clearance of vitamin D_2 when given as a single large dose *(76)*. When used on a daily basis or slightly less frequently, vitamin D_2 has been shown to effectively raise 25(OH)D levels in older women *(78, 79)*, adults *(80)*, and infants and toddlers *(81)*. The harmful publicity directed to ergocalciferol as an inferior dietary source of vitamin D is unfounded and consideration should be given to the fact that it is a potentially safer form of the vitamin to use in situations where very high doses given frequently over time are required in order to achieve rapid repletion in severe deficiency *(81)*.

9. CONSIDERATIONS FOR PUBLIC HEALTH ACTION

Vitamin D deficiency as measured by low circulating 25(OH)D and its association with increased risk of chronic disease and poor bone health is a significant global reality that is as strongly linked to poor dietary intakes of vitamin D as it is to poor sunlight exposure. Vitamin D intakes remain lower than the recommended dietary guidelines for the majority of North Americans, especially those at greatest risk of low 25(OH)D due to age, gender, or darker skin. A number of modifiable factors concerning the current Canadian and American food supply limit our ability to meet increased dietary needs when cutaneous synthesis of vitamin D is compromised. National fortification of staple foods can make a significant contribution to improving vitamin D intakes and nutritional status. Nutritional supplement use, especially in the United States, may be widespread and effective for many individuals, but it does not eliminate concern for at-risk populations. Consideration should be given to re-evaluating the current dietary guidelines in accordance with newer evidence to determine the intake levels needed to achieve good health for all organ systems and tissues identified with vitamin D receptors, not just bone, and for men and women of all ages and skin pigmentation. Dietary guidelines need to identify and emphasize consumption of food staples naturally rich or fortified with vitamin D among dark-skinned and older men and women through better labeling and manufacturing practices. Food and supplement manufacturers should be encouraged to lawfully add vitamin D to their products at the safest maximum levels permitted to provide the highest nutritional quality products.

REFERENCES

1. Whiting SJ, Green TJ, Calvo MS (2007) Vitamin D intakes in North America and Asia-Pacific countries are not sufficient to prevent vitamin D insufficiency. J Steroid Biochem Mol Biol 103:626–630
2. Holick MF (2007) Vitamin D Deficiency. N Engl J Med 357:266–281
3. Hollis BW, Wagner CL, Drezner MK, Binkley NC (2007) Circulating vitamin D3 and 25-hydroxyvitamin D in humans: an important tool to define adequate nutritional vitamin D status. J Steroid Biochem Mol Biol 103:631–634
4. Holick MF, Chen TC (2008) Vitamin D deficiency: a worldwide problem with health consequences. Am J Clin Nutr 87:1080S–1086S
5. Autier P, Gandini S (2007) Vitamin D supplementation and total mortality: a meta-analysis of randomized controlled trials. Arch Intern Med 167:1730–1737
6. Jia X, Aucott LS, McNeill G (2007) Nutritional status and subsequent all-cause mortality in men and women aged 75 years or over living in the community. Br J Nutr 98:593–599
7. Dobnig H, Pilz S, Scharnagl H et al (2008) Independent association of low serum 25-hydroxyvitamin D and 1,25-dihydroxyvitamin D levels with all-cause and cardiovascular mortality. Arch Intern Med 168:1340–1349
8. Melamed ML, Michos ED, Post W, Astor B (2008) 25-hydroxyvitamin D levels and the risk of mortality in the general population. Arch Intern Med 168:1629–1637
9. Richards JB, Valdes AM, Gardner JP et al (2007) Higher serum vitamin D concentrations are associated with longer leukocyte telomere length in women. Am J Clin Nutr 86:1420–1425
10. Cauley JA, LaCroix AZ, Wu L et al (2008) Serum 25-Hydroxyvitamin D concentrations and risk for hip fractures. Ann Intern Med 149:242–250
11. Looker AC, Mussolino ME (2008) Serum 25-hydroxyvitamin D and hip fracture risk in older U.S. white adults. J Bone Min Res 23:143–150

12. Lips P (2007) Vitamin D status and nutrition in Europe and Asia. J Steroid Biochem Mol Biol 103: 620–625
13. Calvo MS, Whiting SJ, Barton CN, Vitamin D (2005) Intake: a global perspective of current status. J Nutr 135:310–316
14. Prentice A, Goldberg GR, Schoenmakers I (2008) Vitamin D across the lifecycle: physiology and biomarkers. Am J Clin Nutr 88:500S–506S
15. Norman AW, Bouillon R, Whiting SJ, Vieth R, Lips P (2007) 13th workshop consensus for vitamin D nutritional guidelines. J Steroid Biochem Mol Biol 103:204–205
16. Dawson-Hughes B (2008) Serum 25-hydroxyvitamin D and functional outcomes in the elderly. Am J Clin Nutr 88:537S–540S
17. Norman AW (2008) From vitamin D to hormone D: fundamentals of the vitamin D endocrine system essential for good health. Am J Clin Nutr 88:491S–499S
18. Institute of Medicine (1997) Dietary Reference Intakes for calcium, phosphorus, magnesium, vitamin D and fluoride. National Academy Press, Washington, D.C.
19. Burns L, Ashwell M, Berry J et al (2003) UK Food Standards Agency Optimal Nutrition Status Workshop: environmental factors that affect bone health throughout life. Br J Nutr 89:835–840
20. Holick MF (2008) The vitamin D deficiency pandemic and consequences for nonskeletal health: mechanisms of action. Mol Aspects Med in press
21. Cranney A, Weiler HA, O'Donnell S, Puil L (2008) Summary of evidence-based review on vitamin D efficacy and safety in relation to bone health. Am J Clin Nutr 88:513S–519S
22. Brannon PM, Yetley EA, Bailey RL, Picciano MF (2008) Summary of roundtable discussion on vitamin D research needs. Am J Clin Nutr 88:587S–592S
23. Committee on Food and Nutrition, National Research Council (1941) Recommended allowances for the various dietary essentials. J Am Diet Assoc 17:565–567
24. Park EA (1940) The therapy of rickets. JAMA 115:370–379
25. Otten JJ, Hellwig PJ, Myers LD (2006) Dietary Reference Intakes: the Essential Guide to Nutrient Requirements. The National Academies Press, Washington, D.C.
26. Whiting SJ, Calvo MS (2005) Dietary recommendations for Vitamin D: a critical need for functional end points to establish an estimated average requirement. J Nutr 135:304–309
27. Schwalfenberg G (2007) Not enough vitamin D: health consequences for Canadians. Can Fam Physician 53:841–854
28. Gartner LM, Greer FR (2003) Prevention of rickets and vitamin D deficiency: new guidelines for vitamin D intake. Pediatrics 111:908–910
29. Looker AC, Dawson-Hughes B, Calvo MS, Gunter EW, Sahyoun NR (2002) Serum 25-hydroxyvitamin D status of adolescents and adults in two seasonal subpopulations from NHANES III. Bone 30:771–777
30. Canadian Pediatric Society (2007) Vitamin D supplementation: recommendations for Canadian mothers and infants. Paediatr Child Health 12:583–589
31. Ziegler EE, Hollis BW, Nelson SE, Jeter JM (2006) Vitamin D deficiency in breastfed infants in Iowa. Pediatrics 118:603–610
32. Lee JM, Smith JR, Philipp BL, Chen TC, Mathieu J, Holick MF (2007) Vitamin D deficiency in a healthy group of mothers and newborn infants. Clin Pediatr (Phila) 46:42–44
33. Mathieu C, Badenhoop K (2005) Vitamin D and type 1 diabetes mellitus: state of the art. Trends Endocrinol Metab 16:261–266
34. Cashman KD, Hill TR, Cotter AA et al (2008) Low vitamin D status adversely affects bone health parameters in adolescents. Am J Clin Nutr 87:1039–1044
35. Hathcock JN, Shao A, Vieth R, Heaney R (2007) Risk assessment for vitamin D. Am J Clin Nutr 85:6–18
36. Aloia JF, Patel M, DiMaano R et al (2008) Vitamin D intake to attain a desired serum 25-hydroxyvitamin D concentration. Am J Clin Nutr 87:1952–1958
37. Heaney RP (2007) The case for improving vitamin D status. J Steroid Biochem Mol Biol 103: 635–641

38. Heaney RP, Davies KM, Chen TC, Holick MF, Barger-Lux MJ (2003) Human serum 25-hydroxycholecalciferol response to extended oral dosing with cholecalciferol. Am J Clin Nutr 77: 204–210

39. Moore CE, Murphy MM, Holick MF (2005) Vitamin D Intakes by children and adults in the United States differ among ethnic groups. J Nutr 135:2478–2485

40. Health canada, Nutrient Intakes from Food, Provincial, Regional and National Summary Data Tables, Volume 1, 2008.

41. Holden JM, Lemar LE (2008) Assessing vitamin D contents in foods and supplements: challenges and needs. Am J Clin Nutr 88:551S–553S

42. Yetley EA (2008) Assessing the vitamin D status of the US population. Am J Clin Nutr 88:558S–564S

43. Calvo MS, Whiting SJ, Barton CN (2004) Vitamin D fortification in the United States and Canada: current status and data needs. Am J Clin Nutr 80:1710S–1716S

44. Calvo MS, Whiting SJ (2006) Public health strategies to overcome barriers to optimal vitamin D status in populations with special needs. J Nutr 136:1135–1139

45. Moore C, Murphy MM, Keast DR, Holick MF (2004) Vitamin D intake in the United States. J Am Diet Assoc 104:980–983

46. Gozdzik A, Barta JL, Wu H, Wagner D, Cole DE, Vieth R, Whiting S, Parra EJ (2008) Low winter-time vitamin D levels in a sample of healthy young adults of diverse ancestry living in the Toronto area: associations with vitamin D intake and skin pigmentation. BMC Public Health Sep 26;8:336. PMID:18817578 [Pub Med-indexed for MEDLINE]

47. Weiler HA, Leslie WD, Krahn J, Steiman PW, Metge CJ (2007) Canadian aboriginal women have a higher prevalence of vitamin D deficiency than non-aboriginal women despite similar dietary vitamin D intakes. J Nutr 137:461–465

48. Jorde R, Bonna KH (2000) Calcium from dairy products, vitamin D intake, and blood pressure: the Tromso Study. Am J Clin Nutr 71:1530–1535

49. Sioen I, Matthys C, De Backer G, Van Camp J, Henauw SD (2007) Importance of seafood as nutrient source in the diet of Belgian adolescents. J Hum Nutr Diet 20:580–589

50. Nakamura KNM, Okuda Y, Ota T, Yamamoto M (2002) Fish as a major source of vitamin D in the Japanese diet. Nutrition 18:415–416

51. Edouard Bourre J-M, Paquotte P (2008) Contributions (in 2005) of marine and fresh water products (finfish and shellfish, seafood, wild and farmed) to the French dietary intakes of vitamins D and B12, selenium, iodine and docosahexaenoic acid: impact on public health. Int J Food Sci Nutr 59:49–501

52. Holden JM, Lemar LE, Exler J (2008) Vitamin D in foods: development of the US Department of Agriculture database. Am J Clin Nutr 87:1092S–1096S

53. Ovesen L, Brot CJJ (2003) Food contents and biological activity of 25-hydroxyvitamin D: a vitamin D metabolite to be reckoned with?. Ann Nutr Metab 47:107–113

54. Mattila P, Valaja J, Rossow L, Venalainen E, Tupasela T (2004) Effect of vitamin D2- and D3-enriched diets on egg vitamin D content, production, and bird condition during an entire production period. Poult Sci 83:433–440

55. Carnagey KMH-LE, Lonergan SM, Trenkle A, Horst RL, Beitz DC (2008) Use of 25-hydroxyvitamin D-3 and dietary calcium to improve tenderness of beef from the round of beef cows. J An Sci 86: 1637–1648

56. Purchas RZM, Pearce P, Jackson F (2007) Concentrations of vitamin D3 and 25-hydroxyvitamin D3 in raw and cooked New Zealand beef and lamb. J Food Comp Anal 20:90–98

57. Jaobsen JMH, Bysted A, Sommer HM, Hels O (2007) 25-Hydroxyvitamin D3 affects vitamin D status similar to vitamin D3 in pigs – but meat produced has a lower content of vitamin D. Br J Nutr 98: 908–913

58. Holick MF (2006) Resurrection of vitamin D deficiency and rickets. J Clin Invest 116:2062–2072

59. Lu Z, Chen TC, Zhang A et al (2007) An evaluation of the vitamin D3 content in fish: is the vitamin D content adequate to satisfy the dietary requirement for vitamin D?. J Steroid Biochem Mol Biol 103:642–644

60. Outila TA, Mattila PH, Piironen VI, Lamberg-Allardt CJ (1999) Bioavailability of vitamin D from wild edible mushrooms (Cantharellus tubaeformis) as measured with a human bioassay. Am J Clin Nutr 69:95–98

61. Tangpricha V, Koutkia P, Rieke SM, Chen TC, Perez AA, Holick MF (2003) Fortification of orange juice with vitamin D: a novel approach for enhancing vitamin D nutritional health. Am J Clin Nutr 77:1478–1483

62. Tylavsky FA, Cheng S, Lyytikainen A, Viljakainen H, Lamberg-Allardt C (2006) Strategies to improve vitamin D status in Northern European children: exploring the merits of Vitamin D fortification and supplementation. J Nutr 136:1130–1134

63. Natri A-M SP, Vikstedt T, Palssa A, Huttunen M, Karkkainen MUM, Salovaara H, Piironen V, Jakobsen J, Lamberg-Allardt C (2006) Bread fortified with Cholecalciferol increases the serum 25-hydroxyvitamin D concentration in women as effectively as a cholecalciferol supplement. J Nutr 136:123–127

64. Wagner D, Sidhom G, Whiting SJ, Rousseau D, Vieth R (2008) The bioavailability of vitamin D from fortified cheeses and supplements is equivalent in adults. J Nutr 138:1365–1371

65. Daly R, Brown M, Bass S, Kukulijian S, Nowson C (2006) Calcium and vitamin D3-fortified milk reduces bone loss at clinically relevant skeletal sites in older men: a 2-year randomized controlled trial. J Bone Min Res 21:397–405

66. Daly R, Petrass N, Bass S, Nowson CA (2008) The skeletal benefits of calcium-and vitamin D3-fortified milk are sustained in older men after withdrawal of supplementation: an 18-mo follow-up study. Am J Clin Nutr 87:771–777

67. Vieth RCD, Hawker GA, Trang HM, Rubin LA (2001) Wintertime vitamin D insufficiency is common in young Canadian women, and their vitamin D intake does not prevent it. Eur J Clin Nutr 55:1091–1097

68. Roth DEMP, Yeo R, Prosser C, Bell M, Jones AB (2005) Are national vitamin D guidelines sufficient to maintain adequate blood levels in children?. Can J Pub Health 96:443–449

69. Zadshir ATN, Pan D, Norris K, Martins D (2005) The prevalence of hypovitaminosis D among US adults: data from the NHANES III. Ethn Dis 15:S5-97–S5-101

70. Taureen NMD, Zadshir A, Pan D, Norris K (2005) The impact of routine vitamin supplementation on serum levels of 25(OH)D3 among the general adult population and patients with chronic kidney disease. Ethn Dis 15:102–106

71. Kyriakidou-Himonas M, Aloia JF, Yeh JK (1999) Vitamin D Supplementation in Postmenopausal Black Women. J Clin Endocrinol Metab 84:3988–3990

72. Nesby-O'Dell SSK, Cogswell ME, Gillespie C, Hollis BW, Looker AC, Allen C, Dougherrtly C, Gunter EW, Bowman BA (2002) Hypovitaminosis D prevalence and determinants among African American and white women of reproductive age: third national health and nutrition examination survey, 1988–1994. Am J Clin Nutr 76:187–192

73. Aloia JF, Talwar SA, Pollack S, Yeh J (2005) A randomized controlled trial of vitamin D3 supplementation in African American women. Arch Intern Med 165:1618–1623

74. Aloia JF, Talwar SA, Pollack S, Feuerman M, Yeh JK (2006) Optimal vitamin D status and serum parathyroid hormone concentrations in African American women. Am J Clin Nutr 84:602–609

75. Aloia JF (2008) African Americans, 25-hydroxyvitamin D, and osteoporosis: a paradox. Am J Clin Nutr 88:545S–50

76. Armas LAG, Hollis BW, Heaney RP (2004) Vitamin D2 Is much less effective than vitamin D3 in humans. J Clin Endocrinol Metab 89:5387–5391

77. Houghton LA, Vieth R (2006) The case against ergocalciferol (vitamin D2) as a vitamin supplement. Am J Clin Nutr 84:694–697

78. Mastaglia SR, Mautalen CA, Parisi MS, Oliveri B (2006) Vitamin D2 dose required to rapidly increase 25OHD levels in osteoporotic women. Eur J Clin Nutr 60:681–687

79. Zhu K, Devine A, Dick IM, Wilson SG, Prince RL (2008) Effects of calcium and vitamin D supplementation on hip bone mineral density and calcium-related analytes in elderly ambulatory Australian women: a five-year randomized controlled trial. J Clin Endocrinol Metab 93:743–749

80. Holick MF, Biancuzzo RM, Chen TC et al (2008) Vitamin D2 is as effective as vitamin D3 in maintaining circulating concentrations of 25-hydroxyvitamin D. J Clin Endocrinol Metab 93:677–681

81. Gordon C, Williams AL, Feldman HA, May J, Sinclair L, Vasquez A, Cox J (2008) Treatment of hypovitaminosis D in infants and toddlers. J Clin Endocrinol Metab 93:2716–2721

82. Henderson LBC, Prentice A, Perks SG, Farron M. The national diet and nutrition survey: adults aged 19–64 years- vitamin and mineral intake and urinary analytes.http://www.foodstandards.gov.uk/multimedia/pdf [accessed April 15, 2004], 2003.

83. Hill TR, O'Brien MM, Cashman KD, Flynn A, Kiely M (2004) Vitamin D intakes in 18-64-y-old Irish adults. Eur J Clin Nutr 58:1509–1517

84. Van Dam RMSM, Dekker J, Stehouwer CDA, Bouter LM, Heine RJ, Lips P (2007) Potentially modifiable determinants of vitamin D status in an older population in the Netherlands: the Hoorn study. Am J Clin Nutr 85:755–761

19

25-Hydroxyvitamin D Assays and Their Clinical Utility

N. Binkley and G. Lensmeyer

Abstract   The vitamin D status of an individual is best assessed by measurement of circulating 25-hydroxyvitamin [25(OH)D]. Challenges to measurement of this analyte include the hydrophobic nature of vitamin D and the presence of two forms of vitamin D, ergocalciferol and cholecalciferol (vitamin D_2 and vitamin D_3, respectively). The absence of standard calibrators contributes to between-laboratory bias in 25(OH)D measurement. The current state of 25(OH)D measurement is reviewed; modest differences between methodologies persist. Until there is assay standardization, and subsequent evidence-based consensus, it is premature to recommend screening 25(OH)D measurement. Targeted measurement of 25(OH)D in those at increased risk for vitamin D deficiency, and those most likely to have a prompt positive response to supplementation, seems appropriate.

Key Words: 25-hydroxyvitamin D; 25-hydroxyvitamin D assay; radioimmunoassay; liquid chromatography; mass spectroscopy; 25-hydroxyvitamin D_2; 25-hydroxyvitamin D_3; CABP; 3-epi-25-hydroxyvitamin D_3; vitamin D deficiency

1. INTRODUCTION

Humans obtain vitamin D from either cutaneous production or ingestion. Upon skin exposure to ultraviolet B (UVB) radiation, 7-dehydrocholesterol is converted to vitamin D_3 (cholecalciferol). Additionally, dietary sources provide either vitamin D_3 or vitamin D_2 (ergocalciferol) *(1)*. Due to little dietary intake combined with sun avoidance, low vitamin D status is extremely common. Inadequate vitamin D status may have multiple adverse health consequences including calcium malabsorption and secondary hyperparathyroidism leading to skeletal fragility, muscle weakness, and, potentially, a multitude of non-skeletal morbidities.

2. VITAMIN D PHYSIOLOGY

As noted above, vitamin D exists in two forms: cholecalciferol (vitamin D_3) produced in the skin via a photochemical reaction with 7-dehydrocholesterol *(1, 2)* and ergocalciferol (vitamin D_2) derived from plants and used as a supplement.

From: *Nutrition and Health: Vitamin D*
Edited by: M.F. Holick, DOI 10.1007/978-1-60327-303-9_19,
© Springer Science+Business Media, LLC 2010

Vitamin D_2 or D_3 is hydroxylated to 25-hydroxyvitamin D [25(OH)D] in the liver and subsequently to 1,25-dihydroxyvitamin D [1,25(OH)$_2$D] in the kidneys *(3)*. As 1,25(OH)$_2$D is the active form of vitamin D, it may seem intuitive that measurement of this analyte would be the appropriate indicator of vitamin D status; however, this is not the case. Low vitamin D status induces parathyroid hormone (PTH) elevation, thereby enhancing renal 1α-hydroxylase activity which promotes conversion of available 25(OH)D (present in nanograms/ml amounts) to 1,25(OH)$_2$D (present in picograms/ml quantities). Thus, even in the setting of low vitamin D status, 1,25(OH)$_2$D may be maintained within the normal range. It is therefore appropriate that measurement of 25(OH)D, not 1,25(OH)$_2$D, is accepted as the appropriate indicator of an individual's vitamin D nutritional status by the Food and Nutrition Board of the Institute of Medicine *(4)*. As such, we will focus on the current state of 25(OH)D measurement in this chapter. A recent overview of 1,25(OH)$_2$D measurement is available elsewhere *(5)*.

When considering 25(OH)D measurement, it is necessary to appreciate that two chemically distinct forms of vitamin D exist; vitamin D_3 is a 27 carbon molecule whereas vitamin D_2 contains 28 carbons and differs from vitamin D_3 by containing an additional methyl group and a double bond between carbons 22 and 23. It has often been assumed that vitamins D_2 and D_3 had equal efficacy. However, some studies find vitamin D_3 to be more "potent" than vitamin D_2 in maintaining circulating 25(OH)D *(6, 7)*. This is currently an area of controversy as some recent work calls this into question *(8)*, though other recent work supports greater efficacy of vitamin D_3 *(9, 10)*. Given potentially greater potency of vitamin D_3, combined with the fact that this form is produced naturally by humans, it is logical to utilize cholecalciferol rather than ergocalciferol for supplementation. As such, whether vitamin D_2 or vitamin D_3 is more "potent," or whether D_2 or its metabolites impact 25(OH)D measurement, would be of little concern were it not for the fact that vitamin D_2 is the only high-dose preparation available by prescription in the USA. Given the clinical reality that clinicians will continue to prescribe vitamin D_2, potential differences in the ability of 25(OH)D assays to accurately detect 25(OH)D_2 and 25(OH)D_3 are, and will likely remain, of substantial clinical importance.

As the differing capability of vitamins D_2 and D_3 in maintaining 25(OH)D status implies, these compounds appear to be metabolized differently in humans. However, despite such differences, there are no data documenting differing physiologic effects, for example, on calcium absorption. Thus, at this time the clinical significance (if any) of differences between these two forms and their respective metabolites remains unknown. At this time there is no documented benefit in considering 25(OH)D_2 and 25(OH)D_3 separately; measurement of circulating total 25(OH)D is reported clinically as it has been designated the functional indicator of vitamin D nutritional status by the Food and Nutrition Board of the Institute of Medicine *(4)*. When 25(OH)D assay results are reported as 25(OH)D_2, 25(OH)D_3, and total, physicians should be cognizant that vitamin D_2 administration does reduce circulating 25(OH)D_3 concentration (Fig. 1). The physiologic impact of this reduction, if any, remains to be determined.

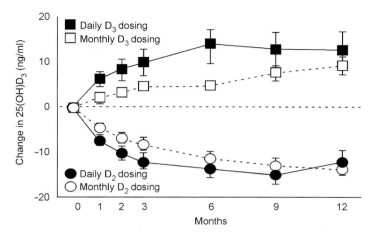

Fig. 1. Ergocalciferol supplementation reduces circulating 25(OH)D$_3$ concentration. In this study, ~14 older adults per treatment group received either ergocalciferol or cholecalciferol (1,600 IU daily or 50,000 IU monthly) *(9)*. Use of ergocalciferol supplementation promptly and persistently reduces circulating 25(OH)D$_3$. While the significance of this reduction (if any) remains unknown, clinicians should recognize this phenomenon and understand that a "low" 25(OH)D$_3$ value in the setting of ergocalciferol supplementation does not constitute vitamin D deficiency.

3. CURRENTLY AVAILABLE 25(OH)D ASSAYS

The clinician and laboratory scientist have a number of 25(OH)D assays to choose from. Current assays are displayed in Table 1 and are comprised of commercially available kits and procedures developed in-house. These assays can be placed into three general categories: competitive protein binding assays (CPBA) *(11, 12)*, immunochemical assays (IC) *(13–15)*, and chromatographic procedures which include gas chromatography/mass spectrometry (GC/MS) *(16, 17)*, high-performance liquid chromatography (HPLC) *(18–20)*, and liquid chromatography–tandem mass spectroscopy (LC-MS/MS) *(21–23)*. At present, no one method has been deemed the reference or "gold standard" assay; however, "physico-chemical" methods such as GC-MS or LC/MS/MS are likely candidates. The lack of standard 25(OH)D$_2$ and 25(OH)D$_3$ reference materials and the absence of consensus on a reference methodology have hindered the process to explain the significant variability in patient and proficiency testing results among this diverse group of methods. Additionally, it is important to recognize that challenges arise when measuring a compound such as 25(OH)D that is extremely hydrophobic and moreover exists in two forms, 25(OH)D$_2$ and 25(OH)D$_3$.

3.1. Chromatographic Assays

"Physico-chemical" methods include a variety of chromatographic-based assays that can separate and measure 25(OH)D$_2$ and 25(OH)D$_3$ individually. HPLC with spectrophotometric (UV) or electrochemical detection, GC-MS, and the highly selective LC-MS/MS comprise this group. Protein precipitation followed by solid-phase extraction or column-switching techniques are commonly used to isolate and purify the metabolites prior to instrumental analysis. In this group of assays, LC-MS/MS methods are an

Table 1
Currently Available 25(OH)D Assays

Assay type	Vendor	Technique	Separates 25(OH)D$_2$ and 25(OH)D$_3$	Detects 25(OH)D$_2$ fully	Notes
Chromatographic assays					
General-HPLC (High-performance liquid chromatography)	Developed and validated in-house (18–20)	SPE/HPLC with ultraviolet detection (SPE = solid-phase extraction)	Yes	Yes	Chemical separation of vitamin D metabolites on chromatography columns. Potential as a reference quality assay.
ESA-HPLC	ESA Biosciences, Chelmsford, MA	SPE/HPLC with electrochemical detection	Yes	Yes	Commercial assay that reports the sum of the two separated metabolites.
Chromsystems-HPLC	ChromSystems, Munich, Germany	SPE/HPLC with ultraviolet detection	Yes	Yes	Commercial assay
General-LC/MS/MS (Liquid chromatography–tandem mass spectrometry)	Developed and validated in-house ([21]>23)	SPE/HPLC with tandem mass spectrometric detection	Yes	Yes	Can be developed for high-throughput processing of samples. Potential as a reference quality assay
General-GC/MS (gas chromatography mass spectrometry)	Developed and validated in-house (16,17)	Extraction/GC with mass spectrometric detection	Yes	Yes	Limited use clinically. Potential as a reference quality assay.

Table 1
(continued)

Assay type	Vendor	Technique	Separates 25(OH)D₂ and 25(OH)D₃	Detects 25(OH)D₂ fully	Notes
Competitive protein binding assays (CPBA)					
General-CPBA	Developed and validated in-house (11, 12)	Protein binding assay with or without prior chromatography. ^3H-25(OH)D₃ tracer for detection	No	Uncertain	These assays are usually prone to matrix interferences and are not selective for 25(OH)D₃ and 25(OH)D₂.
Immunochemical assays (IC)					
Biosource-RIA	Biosource International, Camarillo, CA	IC-coated tube with labeled tracer detection	No	Uncertain	
Diasorin-RIA	Diasorin, Stillwater, MN	IC with detection of I^{125}-labeled tracer	No	"Co-specific" for the two forms (35)	
Diasorin Liaison 1	Diasorin, Stillwater, MN	IC antibody-coated magnetic particles chemiluminescence detection of isoluminol derivative	No	"Co-specific" for the two forms (35)	

Table 1
(continued)

Assay type	Vendor	Technique	Separates 25(OH)D$_2$ and 25(OH)D$_3$	Detects 25(OH)D$_2$ fully	Notes
Diasorin Liaison total	Diasorin, Stillwater, MN	IC with pre-extraction and chemiluminescence detection as with Liaison 1	No	"Co-specific" for the two forms (35)	Modified version of Liaison 1 to improve lower limit of detection.
IDS-EIA manual and automated formats	Immunodiagnostics, Bolden, UK	IC-ELISA biotinylated 25(OH)D tracer detected	No	No	Detects only a portion of the 25(OH)D$_2$ in a sample.
IDS-RIA	Immunodiagnostics, Bolden, UK	IC dual antibody with detection of I^{125} labeled tracer	No	No	Detects only a portion of the 25(OH)D$_2$ present in a sample.
Roche-ECI	Roche Diagnostics, Penzberg, Germany	IC with electro-chemiluminescence detection of ruthenium labeled 25(OH)D$_3$	No	No	Marketed to measure 25(OH)D$_3$ only.

attractive option enabling short run times with high throughput and as such are being utilized by some large reference laboratories.

Although most chromatographic assays separate the two metabolite forms, only a few can isolate the inactive isomer 3-epi 25(OH)D₃ found in infants, from 25(OH)D₃ *(24, 25)*. Moreover, assays that do not separate the 3-epi form or have significant cross-reactivity with the epimer will most likely report an erroneously high level for infants because the 3-epi constitutes a substantial fraction of total 25(OH)D. Appreciation of this assay limitation is appropriate for clinicians measuring vitamin D status in infants.

With chromatographic assays, metabolite concentrations are usually reported as separate results for 25(OH)D₂ and 25(OH)D₃ and as a total 25(OH)D value to minimize health care provider confusion when interpreted. The ideal way of reporting these results remains to be defined. However, as vitamins D₂ and D₃ are chemically different, metabolize differently, and show dissimilar affinity for vitamin D binding protein (VDBP) *(26)*, it is clear that these forms are unique and as such it is argued that they should be reported separately. As noted above, the question of whether the two compounds can be used interchangeably and/or their results interpreted in the same way is unclear.

3.2. Immunochemical and Competitive Protein Binding Assays (CPBA)

Manufacturers of immunochemical assays for 25(OH)D face the challenge of developing an antibody that is equally selective for 25(OH)D₂ and 25(OH)D₃ without cross-reactivity from one or several of the various vitamin D₂ and D₃ metabolites. Additionally, these assays must not be negatively influenced by sterol-like or other endogenous components that are normal constituents or by-products of disease that occur in blood. The above issues are not unique to 25(OH)D measurement; all immunoassays face similar challenges *(27, 28)*. Additionally, the hydrophobic nature of 25(OH)D renders it susceptible to matrix effects when measured by protein binding assays (PBA) or immunochemical assays (IC). Such matrix effects may change the ability of the antibody or binding protein to associate with 25(OH)D in the specimen, thereby reducing the accuracy of measurement. Based upon data from independent studies *(29, 30)* and proficiency testing, it appears that no commercial immunoassay, to date, has ideally met these challenges. Similarly, commercial CPBA have been fraught with problems resulting in limited acceptance clinically. Interestingly, some of the pioneering CPBA were very selective because they incorporated a chromatographic step before the ligand assay. Typically, these methods were labor intensive and used primarily in research laboratories. Commercial adaptation of these assays resulted in the abandonment of the chromatography purification step for the sake of speed which prompted the poor selectivity and matrix interferences reported *(5, 31)*. The now defunct Nichols Advantage methods represent the CPBA approach used clinically.

Currently available commercial IC and CPBA assays have the advantage of rapid turnaround time, high-throughput, and ease of automation. However, inconsistencies in 25(OH)D results reported for cohorts of patient samples have been reported *(29, 32)*. Moreover, and most importantly, some of the assays require the establishment of assay-specific decision limits *(30, 33)* as a compromise for limitations intrinsic to the assay.

3.3. Specimen Stability

Both 25(OH)D$_2$ and 25(OH)D$_3$ are exceptionally stable in serum and plasma even when samples are subjected to extreme conditions. A notable exception is when samples are exposed to prolonged unprotected direct sunlight (34, 35). This stability validates transport of light-protected samples by mail and other courier services to distant destinations for 25(OH)D testing such as is utilized by DEQAS (see below).

However, it should be noted that the assays used for the studies documenting stability were of reference quality and therefore less likely to suffer from potential matrix issues. Accordingly, one must be cautious when interpreting such stability data when using assays that are prone to matrix issues (IC, CPBA) in which degrading/ageing serum/plasma matrix components could adversely impact results (28). Consistent with this, when an RIA was used to test serum samples stored at 11°C for up to 14 days, the concentration of 25(OH)D increased ~7% (36).

3.4. Proficiency Testing and Assay Quality

DEQAS (Vitamin D External Quality Assessment Scheme) was founded in 1989 to compare the performance of assays for vitamin D measurement. It is currently the dominant proficiency testing scheme with more than 470 laboratories participating from over 30 countries. Recently, the College of American Pathologists (CAP) offered a proficiency program for 25(OH)D; however, problems associated with their adulterated serum have brought into question the usefulness of this approach (37).

The major aims of DEQAS are to ensure analytic reliability of 25(OH)D and 1,25(OH)$_2$D measurement and to establish a forum for exchange of information on vitamin D methodology. These goals are achieved through distribution of five unadulterated human serum challenge pools four times yearly. Results submitted by participants are analyzed, reports are generated, and a proficiency certificate is issued annually for those laboratories meeting the DEQAS advisory panel performance target. Additional information on DEQAS is available at www.deqas.org.

DEQAS currently uses the "All Laboratory Trimmed Mean" (ALTM) value to assess assay method performance. It is reassuring that the ALTM was demonstrated to be a good surrogate for the "true" value obtained by GC-MS in a study of a small number of samples (38). However, in the near future, DEQAS plans to employ a reference quality GC-MS assay to obtain a true target value for some of the challenge samples and thus allows improved assessment of accuracy rather than using a group consensus of all participant results (i.e., the ALTM) (38). Use of a true target value should avoid bias that might potentially occur if any one method dominates the current testing scheme.

Summary data of current methodology performance were recently reported by DEQAS (39). Briefly, cumulative proficiency data demonstrate some improvement in between-laboratory imprecision over time with CVs of 19–24% for 2006–2007 compared with values of >25% from 1994 to 1998. However, the current inter-laboratory CVs indicate that a high level of analytical imprecision persists. DEQAS additionally reported (Fig. 2) the current mean bias (%) from the ALTM for each method group. Here, substantial negative method bias was demonstrated for the Diasorin products whereas a positive method bias was observed for LC-MS/MS. Similar findings were

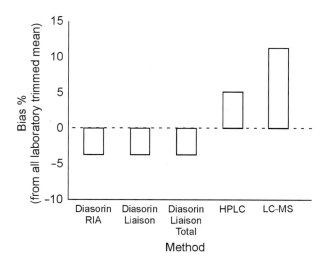

Fig. 2. Comparison of mean 25(OH)D results from selected current assays. DEQAS (*Vitamin D External Quality Assessment Scheme*) data demonstrating the mean % bias of selected methods compared to the all laboratory trimmed mean (ALTM) value. Data from DEQAS distributions from April 2007 through January 2008. Adapted with permission of Dr. G. Carter.

observed by Glendenning *(30)* and Roth *(29)* in their comparison studies with these products. The positive bias observed with the LC-MS/MS methods may be due to the lack of standard reference materials which would allow accurate calibration of the in-house-developed assays or alternatively, because the ALTM was used instead of a known target value for calculation of bias.

While DEQAS contributes substantially to enhancing 25(OH)D measurement reliability, it is important to recognize a limitation of this program in that the samples distributed rarely contain substantial amounts of 25(OH)D$_2$. In fact, of the last 55 serum specimens distributed by DEQAS, only one contained a measurable amount of 25(OH)D$_2$. Therefore, as ergocalciferol is frequently prescribed in the USA, and vitamin D$_2$ potentially complicates 25(OH)D measurement, the DEQAS distribution may not be representative of clinical reality, at least in the USA. A potential solution to this limitation would consist of spiking distributed specimens with 25(OH)D$_2$. Unfortunately, though chromatographic methods are not adversely impacted by such spiking, low recoveries of 25(OH)D$_2$ have been observed in spiked samples by some IC methods *(39)*. It is probable that adding exogenous 25(OH)D$_2$ disrupts the sensitive equilibrium of bound, unbound, and tracer, thereby leading to erroneous results. For this reason the DEQAS scheme remains limited to supplying challenge samples that contain only endogenous 25(OH)D$_2$. Additionally, the Roche method is marketed to measure only 25(OH)D$_3$. Figure. 3 demonstrates recovery of endogenous 25(OH)D$_2$ in one DEQAS sample distributed in 2008. Among the reported methods, the HPLC and LC-MS/MS methodologies displayed the highest recovery (66–73%) while, as expected, the Roche method recovered 25(OH)D$_2$ the least. These results beg for further assessment of the 25(OH)D$_2$ issue and the impact that the simultaneous presence of the two forms of

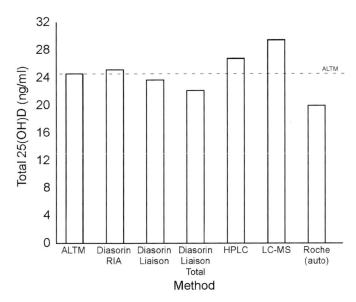

Fig. 3. Impact of the presence of 25(OH)D$_2$ on total 25(OH)D results. For the single specimen in recent DEQAS distributions that contained approximately one-third endogenous 25(OH)D$_2$ and two-thirds 25(OH)D$_3$, the mean values obtained by various methods are shown. Data from DEQAS; adapted with permission of Dr. G. Carter. ALTM, all laboratory trimmed mean.

25(OH)D has on the accuracy of individual testing methods, if any, and the certainty of diagnosis for individual patients.

3.5. Current Status of 25(OH)D Result Agreement

A recent evaluation finds fairly reasonable agreement of clinical 25(OH)D results when performed using HPLC, LC-MS/MS, and one RIA methodology *(40)*. In this small study, better agreement between different methods was observed at lower serum 25(OH)D concentrations *(20, 41)*. However, in a larger study, Roth et al *(29)* evaluated the accuracy of six versions of routinely available methodologies (chromatographic, IC, and CPBA) with ∼300 randomly selected patient samples. All methods tested in this study (IDS-RIA, IDS-EIA, Nichols Advantage, Diasorin Liaison 1, Diasorin Liaison 2 Total, and Roche-ECI on the Elecsys) with the exception of the HPLC assay showed a more or less considerable deviation of 25(OH)D$_3$ results compared to target values generated by a reference LC-MS/MS method. Bland–Altman plots (Fig. 4) from a sample of these data demonstrate substantial variability at a potential decision limit of 30 ng/ml such that results obtained from these assays could cause a clinician to incorrectly define an individual patient's vitamin D status. To further elucidate this variability, it is useful to consider a patient whose 25(OH)D result is "30–35 ng/ml" (noted by the rectangles in Fig. 4). Assuming that the LC-MS/MS methodology is "correct," very few individuals in the 30–35 ng/ml range would be classified as having insufficient vitamin D status (at a cutpoint of 30 ng/ml) using HPLC, whereas substantially more would be classified as low with either the Liaison or the IDS methodologies. Moreover, even at a value of

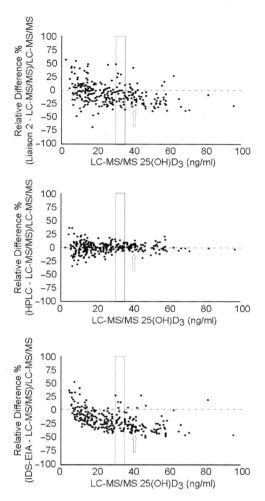

Fig. 4. Comparison of selected 25(OH)D assays with LC-MS/MS. Using the 25(OH)D result obtained by LC-MS/MS as a reference, substantial variability is observed with a chemiluminescent and an EIA assay. At a value of 30–35 ng/ml (*rectangles*) HPLC would classify few patients as vitamin D insufficient whereas a substantial number would be classified as low with the other two methodologies. Moreover, at a value of 40 ng/ml by LC-MS/MS (*arrows*) some patients would be classified as low with the chemiluminescent and EIA assays. Data adapted with permission from Roth et al. *(29)*.

"40 ng/ml," (arrows in Fig. 3) some patients would be classified as having low vitamin D status with the EIA and chemiluminescent methods. Such considerations are disturbing and highlight the importance of, and need for, 25(OH)D assay standardization.

One potential contributor to the differences between 25(OH)D measurement methodologies noted above is between-assay variation (bias) due to the absence of international standardized calibrators. It is anticipated that such calibrators will soon become available, thereby enhancing between-assay agreement. To this end, the National Institute of Standards and Technology (NIST) is developing standard reference material for 25(OH)D analysis *(42)*. These serum pools will contain differing concentrations of 25(OH)D$_2$, 25(OH)D$_3$, and C-3-epi-25(OH)D$_3$ and are designed to reproduce values encountered in clinical specimens.

3.6. 25(OH)D Variability

As noted above, varying degrees of disagreement in 25(OH)D measurement exists between laboratories and methods used. Additionally, it must be recognized that a single laboratory result is impacted by the analytical imprecision and inaccuracy present in all quantitative medical procedures due to method, human, and instrument limitations. Such variability is contributed to by analytical imperfection and systematic bias and is further confounded by biological variability.

An example of 25(OH)D assay variability is shown in Fig. 5 in which quality control data for a sample containing ~30 ng/ml of 25(OH)D$_2$ and ~30 ng/ml of 25(OH)D$_3$ is presented for 49 HPLC runs conducted in a research laboratory over approximately 1 year *(43)*. Modest assay variability is apparent such that the 95% confidence interval subsumes ~32 ± ~2 ng/ml. This assay variability is in a research environment using a single QA sera; it is possible that greater variability would exist in the clinical real world.

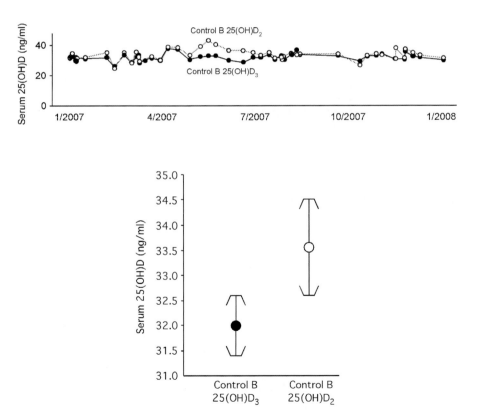

Fig. 5. Example of 25(OH)D analytical variability. Modest variability is observed for repeated measurement over time of a single specimen containing ~30 ng/ml of both 25(OH)D$_2$ and 25(OH)D$_3$ *(upper panel)* *(43)*. In this research HPLC system, the 95% confidence interval (whiskers in *lower panel*) surrounding the mean values, is approximately 2 ng/ml.

It is important that clinicians appreciate the above-noted variability that confounds use of any "cutpoint" diagnostic approach, e.g., a serum 25(OH)D of $\geq$30 ng/ml, to define optimal vitamin D status.

In addition to method and performance variability, within-individual biological variability in 25(OH)D measurement is worthy of consideration but has received only limited investigation. One study did find a 13–19% within-individual variability in 25(OH)D *(44)*, causing the authors to question the clinical utility of a single 25(OH)D measurement in an individual patient. However, it seems likely that this conclusion reflects the well-known seasonal 25(OH)D change and not day-to-day variability *(45)*. Given the relatively long half-life of 25(OH)D, day-to-day variability seems likely to have minimal effect on this measurement.

Finally, as intermittent dosing of relatively large amounts of vitamin D seems to be gaining clinical favor, for example, 50,000 IU once per week or once per month, it is appropriate that clinicians consider basic pharmacology when monitoring this approach. For example, following a 50,000 IU dose of vitamin D, existing data indicate that 25(OH)D levels peak within 1 week and subsequently decline, varying from peak to trough by ~6–10 ng/ml *(6)*. Given this, when utilizing intermittent high-dose vitamin D therapy, measurement of trough 25(OH)D seems appropriate.

3.7. Definition of Low Vitamin D Status/Reporting of Circulating 25(OH)D Results

Though there is agreement that measurement of circulating 25(OH)D is the functional measure of vitamin D status *(4)*, substantial controversy surrounds the definition of optimal vitamin D status and what cutpoint defines "low" vitamin D status *(46, 47)*. Some of this controversy is doubtlessly fueled by the variation in 25(OH)D methodology noted above. Perhaps in response to the ambiguity surrounding the definition of ideal vitamin D status, various reference laboratories report "optimal" 25(OH)D concentrations as being 30–80, 25–80, 20–100, and 32–100 ng/ml. Recognizing this controversy and accepting that a value selected at this time could potentially change in the future, there appears to be increasing agreement in the USA that circulating 25(OH)D values below 30–32 ng/ml are indicative of low vitamin D status *(46)*. Clearly, assay standardization followed by clinical consensus regarding the definition of optimal 25(OH)D status with subsequent standardization of result reporting is needed. Similar lack of standardization of potentially "toxic" levels confounds 25(OH)D result reporting in that some reference laboratories report possibly toxic levels as being >80 ng/ml with others include up to 100 ng/ml within the reference range. Such variability is not surprising as recent expert opinion suggests "The serum 25(OH)D concentration that is the threshold for vitamin D toxicity has not been established" *(48)*. However, a review of the published vitamin D toxicity cases finds all reports of hypercalcemia due to vitamin D intoxication to be associated with 25(OH)D concentrations >88 ng/ml *(49)*. Regarding reporting of "high" 25(OH)D values, it seems reasonable that highly sun-exposed individuals could be utilized to assist in the determination of "normal" vitamin D status *(50)*. When such individuals are evaluated, it appears that the highest attainable 25(OH)D values from cutaneous production are in the 70–80 ng/ml range *(41, 51)*. As such, the current approach of reporting 80–100 ng/ml as the upper limit of normal seems appropriate.

Two final caveats regarding 25(OH)D measurement are worthy of note. First, it is currently recommended that the total 25(OH)D concentration be utilized for clinical decision making. However, many laboratories utilizing LC-MS/MS or HPLC report not only the total 25(OH)D but also separate values for $25(OH)D_2$ and $25(OH)D_3$. This practice may allow assessment of compliance/persistence with prescribed ergocalciferol therapy. However, such reporting has the potential to confuse clinicians unfamiliar with these assays and the necessity of utilizing the total 25(OH)D to determine an individual's vitamin D status. In fact, in a recent small survey, 23% of health care providers indicated that a total serum 25(OH)D of 40 ng/ml (40 ng/ml $25(OH)D_2$, <5 ng/ml $25(OH)D_3$) is indicative of vitamin D deficiency *(52)*. Finally, some published work, and to our knowledge all clinical laboratories in the USA, report 25(OH)D values in ng/ml, whereas much of the scientific literature utilizes nmol/l. Simply dividing the 25(OH)D nmol/l value by 2.5 provides a close approximation of the ng/ml value.

3.8. When Should 25(OH)D Be Measured?

It has been suggested that measurement of circulating 25(OH)D should be performed in some groups of patients, e.g., those with cancer *(53)*, and even as routine screening *(54, 55)*. Such sweeping recommendations may well be appropriate given accumulating evidence that vitamin D deficiency contributes to a multitude of adverse health outcomes not limited to musculoskeletal disease but also including cancer, diabetes mellitus, and cardiovascular disease, among others. Despite the impressive evidence (ranging from in vitro through epidemiological studies) supporting these non-traditional roles of vitamin D one must recognize that large randomized controlled trials exist to document benefit only for musculoskeletal outcomes and perhaps cancer. While acknowledging potentially vast health care savings that might accompany optimization of vitamin D status, one must also recognize that 25(OH)D measurement of the entire population would produce a substantial health care cost in its own right. For perspective, if one assumes a cost of $100 per 25(OH)D measurement (recognizing that some measurements would cost more, others less), performing this on the US population (currently estimated at over 305 million) would lead to a direct expenditure of over $30 billion, approximately 7.5% of the total 2006 Medicare expenditure of $408 billion *(56)*. Despite this huge economic cost, it is plausible that such population-based screening is in fact cost-effective. However, prior to recommending such an approach, formal evidence-based assessment is needed. In this regard, a formal evaluation along the lines of that utilized by other groups, for example, the National Cholesterol Education Program (NCEP), seems warranted. Formal review of the literature by an expert panel with grading of the evidence regarding vitamin D measurement seems reasonable and indicated.

Until such an analysis occurs, and acknowledging that large and long-term studies to justify routine 25(OH)D screening have not been done, others have recently suggested that it is too early to recommend routine 25(OH) screening *(57)*. At this time, rather than advocating a population screening approach, it seems reasonable to measure 25(OH)D in those for whom a prompt musculoskeletal response to optimization of vitamin D status could be expected. Such groups of people include those with osteoporosis, a history of falls or high falls risk, malabsorption (e.g., celiac disease, radiation enteritis, bariatric surgery), individuals with liver disease, and those requiring medications known to alter vitamin D status, e.g., certain anticonvulsants.

4. CLINICAL UTILITY OF 25(OH)D MEASUREMENT

The definition of a target 25(OH)D concentration and clinical approaches to achieving this value are beyond the scope of this chapter. However, in brief, when the circulating 25(OH)D concentration is known, available data *(58)* indicate that, on average, for each additional 100 IU of vitamin D_3 ingested daily, the serum 25(OH)D should increase by $\sim$1 ng/ml. Thus, for example, addition of 2,000 IU daily could be expected to increase serum 25(OH)D by $\sim$20 ng/ml, an increment without toxicity.

5. CONCLUSIONS

Clinicians should be aware that substantial variability continues to exist in 25(OH)D assay methodology and that this will impact the results for any given individual patient. Efforts to enhance agreement are ongoing and should enhance between-assay agreement. When such international standards become available, methods that are inaccurate and/or lack robustness must be modified or abandoned. As clinical specimens (at least in the USA) seem likely to continue to contain ergocalciferol, it is essential that 25(OH)D assays are able to accurately measure 25(OH)D_2. Until there is assay standardization, and subsequent evidence-based consensus, it is premature to recommend screening 25(OH)D measurement. Targeted measurement in those at increased risk for vitamin D deficiency, and those most likely to have a prompt positive response to supplementation, is appropriate.

REFERENCES

1. Holick MF (1985) The photobiology of vitamin D and its consequences for humans. Ann N Y Acad Sci 453:1–13
2. Holick MF (1995) Environmental factors that influence the cutaneous production of vitamin D. Am J Clin Nutr 61:638S–645S
3. DeLuca HF (1988) The vitamin D story: a collaborative effort of basic science and clinical medicine. FASEB J 2:224–236
4. Standing Committee on the Scientific Evaluation of Dietary Reference Intakes Food and Nutrition Board, Institute of Medicine (1997) DRI Dietary Reference Intakes for calcium phosphorus, magnesium, vitamin D and fluoride. National Academy Press, Washington, DC
5. Hollis BW, Horst RL (2007) The assessment of circulating 25(OH)D and 1, 25 (OH)$_2$D: Where we are and where we are going. J Steroid Biochem Mol Biol 103:473–476
6. Armas LAG, Hollis BW, Heaney RP (2004) Vitamin D_2 is much less effective than vitamin D_3 in humans. J Clin Endocrinol Metab 89:5387–5391
7. Houghton LA, Vieth R (2006) The case against ergocalciferol (vitamin D_2) as a vitamin supplement. Am J Clin Nutr 84:694–697
8. Holick MF, Biancuzzo RM, Chen TC et al (2008) Vitamin D_2 is as effective as vitamin D_3 in maintaining circulating concentrations of 25-hydroxyvitamin D. J Clin Endocrinol Metab 93:677–681
9. Binkley N, Gemar D, Ramamurthy R et al (2007) Daily versus monthly oral vitamin D_2 and D_3: Effect on serum 25(OH)D concentration. J Bone Miner Res 22(suppl 1):S215

10. Glendenning P, Seymour H, Gillett MJ et al (2008) Ergocalciferol is not bioequivalent to cholecalciferol in vitamin D insufficient hip fracture cases. J Bone Miner Res 23(suppl):S81
11. Haddad JG, Chyu KJ (1971) Competitive protein-binding radioassay for 25-hydroxycholecalciferol. J Clin Endocrinol Metab 33:992–995
12. Belsey R, DeLuca H, Potts J (1971) Competitive binding assay for vitamin D and 25-OH vitamin D. J Clin Endocrinol Metab 33:554–557
13. Hollis BW, Napoli JL (1985) Improved radioimmunoassay for vitamin D and its use in assessing vitamin D status. Clin Chem 31:1815–1819
14. Anonymous (2005) IDA Product Insert: Gamma B 25-hydroxy Vitamin D RIA.
15. Leino A, Turpeinen U, Koskinen P (2008) Automated measurement of 25-OH vitamin D_3 on the Roche modular E170 analyzer. Clin Chem 54:1–4
16. Caldwell RD, Porteous CE, Trafford DJ et al (1987) Gas chromatography-mass spectrometry and the measurement of vitamin D metabolites in human serum or plasma. Steroids 49:155–196
17. Seamark DA, Trafford DJ, Makin HL (1980) The estimation of vitamin D and some metabolites in human plasma by mass fragmentography. Clin Chim Acta 106:51–62
18. Eisman JA, Shepard RM, DeLuca HF (1977) Determination of 25-hydroxyvitamin D_2 and of 25-hydroxyvitamin D_3 in plasma using high-pressure liquid chromatography. Anal Biochem 90:298–305
19. Turpeinen U, Hohenthal U, Stenman UH (2003) Determination of 25-hydroxyvitamin D in serum by HPLC and immunoassay. Clin Chem 49:1521–1524
20. Lensmeyer GL, Wiebe DA, Binkley N et al (2006) HPLC method for 25-hydroxyvitamin D measurement: Comparison with contemporary assays. Clin Chem 52:1120–1126
21. Maunsell Z, Wright DJ, Rainbow SJ (2005) Routine isotope-dilution liquid chromatography-tandem mass spectrometry assay for the simultaneous measurement of the 25-hydroxymetabolites of vitamins D_2 and D_3. Clin Chem 51:1683–1690
22. Vogeser M, Kyriatsoulis A, Huber E et al (2004) Candidate reference method for the quantitation of circulating 25-hydroxyvitamin D_3 by liquid chromatography tandem-mass spectrometry. Clin Chem 50:1415–1417
23. Chen H, McCoy LF, Schleichter RL et al (2008) Measurement of 25-hydroxyvitamin D_3 ($25OHD_3$) and 25-hydroxyvitamin D_2 ($25OHD_2$) in human serum using liquid chromatography-tandem mass spectrometry and its comparison to a radioimmunoassay method. Clin Chim Acta 391:6–12
24. Singh R, Taylor R, Reddy R et al (2006) C-3 epimers can account for a significant portion of total circulating 25-hydroxyvitamin D in infants, complicating accurate measurement and interpretation of vitamin D status. J Clin Endocrinol Metab 91:3055–3061
25. Lensmeyer GL, Wiebe DA, Binkley N et al (2006) Measurement of 25-hydroxyvitamin D revisited: Reply to letter. Clin Chem 52:2305–2306
26. Trang H, Cole D, Rubin L et al (1998) Evidence that vitamin D_3 increases serum 25-hydroxyvitamin D more efficiently than does vitamin D_2. Am J Clin Nutr 68:854–858
27. Stenman UH (2001) Immunoassay standardization: Is it possible, who is responsible, who is capable?. Clin Chem 47:815–820
28. Valdes RJ, Jortani S (2002) Unexpected suppression of immunoassay results by cross-reactivity: Now a demonstrated cause for concern. Clin Chem 48:405–406
29. Roth HJ, Schmidt-Gayk H, Weber H et al (2008) Accuracy and clinical implications of seven 25-hydroxyvitamin D methods compared with liquid chromatography-tandem mass spectrometry as a reference. Ann Clin Biochem 45:153–159
30. Glendenning P, Taranto M, Noble JM et al (2006) Current assays overestimate 25-hydroxyvitamin D_3 and underestimate 25-hydroxyvitamin D_2 compared with HPLC: need for assay-specific decision limits and metabolite-specific assays. Ann Clin Biochem 43:23–30
31. Ramakkrishnan K, Holick MF (2005) Underestimation of 25-OH vitamin D by Nichols Advantage Assay in patients receiving vitamin D replacement therapy. Clin Chem 51:1074
32. Cavalier E, Wallace AM, Knox S et al (2008) Serum vitamin D measurement may not reflect what you give to your patients. J Bone Miner Res 23:1864–1865
33. Lensmeyer GL, Binkley N, Drezner MK (2006) New horizons for assessment of vitamin D status in man, 2nd edn. Academic Press, San Diego

34. Lewis JG, Elder PA (2008) Serum 25-OH vitamin D_2 and D_3 are stable under exaggerated conditions. Clin Chem 54:1931–1932

35. Hollis BW (2008) Measuring 25-hydroxyvitamin D in a clinical environment: challenges and needs. Am J Clin Nutr 88(suppl):507S–510S

36. Drammeh BS, Schleicher RL, Pfeiffer CM et al (2008) Effects of delayed sample processing and freezing on serum concentrations of selected nutritional indicators. Clin Chem 54:1883–1891

37. Singh R (2008) Are clinical laboratories prepared for accurate testing of 25-hydroxy vitamin D?. Clin Chem 54:221–223

38. Carter G (2008) Personal Communication.

39. Carter G (2008) 2008 DEQAS Review Report for 25-hydroxyvitamin D and 1.25 dihydroxy-vitamin D.

40. Binkley N (2006) Vitamin D: Clinical Measurement and use. J Musculoskelet Neuronal Interact 6: 338–340

41. Binkley N, Novotny R, Krueger D et al (2007) Low vitamin D status despite abundant sun exposure. J Clin Endocrinol Metab 92:2130–2135

42. Phinney KW (2008) Development of a standard reference material for vitamin D in serum. Am J Clin Nutr 88(suppl):511S–512S

43. Binkley N, Krueger D, Engelke JA et al (2008) What is your patient's vitamin D status? Clinical consideration of variability in a 25(OH)D measurement. J Bone Miner Res 23(suppl 1):S351

44. Rejnmark L, Lauridsen AL, Brot C et al (2006) Vitamin D and its binding protein Gc: Long-term variability in peri- and postmenopausal women with and without hormone replacement therapy. Scand J Clin Lab Invest 66:227–238

45. Lester E, Skinner RK, Wills MR (1977) Seasonal variation in serum 25-hydroxyvitamin D in the elderly in Britain. Lancet 1:979–980

46. Dawson-Hughes B, Heaney RP, Holick MF et al (2005) Estimates of optimal vitamin D status. Osteoporos Int 16:713–716

47. Lips P (2001) Vitamin D deficiency and secondary hyperparathyroidism in the elderly: Consequences for bone loss and fractures and therapeutic implications. Endocr Rev 22:477–501

48. Vieth R (2007) Vitamin D toxicity, policy and science. J Bone Miner Res 22(suppl 2):V64-V68

49. Vieth R (1999) Vitamin D supplementation, 25-hydroxyvitamin D concentrations and safety. Am J Clin Nutr 69:842–856

50. Hollis BW (2005) Circulating 25-hydroxyvitamin D levels indicative of vitamin D sufficiency: Implications for establishing a new effective dietary intake recommendation for vitamin D. J Nutr 135: 317–322

51. Barger-Lux MJ, Heaney RP (2002) Effects of above average summer sun exposure on serum 25-hydroxyvitamin D and calcium absorption. J Clin Endocrinol Metab 87:4952–4956

52. Binkley N, Drezner MK, Hollis BW (2006) Laboratory reporting of 25-hydroxyvitamin D results: Potential for clinical misinterpretation. Clin Chem 52:2124–2125

53. Anonymous (2008) Vitamin D deficiency: information for cancer patients. The Bone and Cancer Foundation, New York

54. Holick MF (2002) Too little vitamin D in premenopausal women: Why should we care?. Am J Clin Nutr 76:3–4

55. Giovannucci E (2007) Can vitamin D reduce total mortality?. Arch Intern Med 167:1709–1710

56. Anonymous (2007) 2007 Annual report of the boards of trustees of the federal hospital insurance and federal supplementary medical insurance trust funds

57. Anonymous (2008) Vitamin D: Are you getting enough? UC Berkeley Wellness Letter.com, February 2008 ed. UC-Berkeley, Berkeley, CA

58. Heaney RP, Davies KM, Chen TC et al (2003) Human serum 25-hydroxycholecalciferol response to extended oral dosing with cholecalciferol. Am J Clin Nutr 77:204–210

20 Health Disparities and Vitamin D

Douglass Bibuld

Abstract   Research over the last two to three decades has slowly demonstrated that vitamin D, a long poorly understood and unappreciated hormone, is of profound importance to human health and survival. Vitamin D begins its synthesis in human skin with ultraviolet B radiation (UVB) from the sun. Melanin is a potent UVB blocker, protecting the skin from the high-intensity sunlight found on the tropical savannah into which humans evolved, but not impairing the skin's ability to synthesize generous quantities of vitamin D there. *Adaptation to environmental availability of ultraviolet B radiation from the sun appears to explain the variation in skin melanin content in indigenous human populations around the world.* Evidence shows populations in the United States have mean 25-hydroxyvitamin D levels that are associated with the relative amount of melanin in the skin in those populations, and that humans with insufficient levels of 25-hydroxyvitamin D suffer disproportionately from the diseases associated with health disparities. The differences in incidence and severity of cardiovascular diseases, the most common cancers, diabetes, tuberculosis, conditions associated with infant mortality, and total mortality in populations associated with health disparities in the United States are explored in this chapter. Indeed, the magnitude of the disparity in diseases associated with "health disparities," related to the effect of vitamin D on those diseases, is such that eliminating the differences in 25-hydroxyvitamin D levels between the populations would appear to virtually eliminate the health disparities between them.

Key Words: Vitamin D; black; health disparity; melanin; 25-hydroxyvitamin D; cancer; heart disease; diabetes

1. INTRODUCTION

The impetus to identify and address health disparities in the United States began with the recognition of a persistent, relatively fixed, disparity in the life expectancy for populations classified as white and black in the United States (Fig. 1).

The US National Institutes of Health (NIH) convened a working group in 1999 in response to a Clinton Administration directive to develop a strategic plan for reducing "health disparities." That group issued the following definition "Health disparities are differences in the incidence, prevalence, mortality and burden of diseases and other adverse health conditions that exist among specific population groups in the United States."*(1)* The Center for Disease Control (CDC) Office of Minority Health & Health Disparities (OHMD) takes the position that "Compelling evidence that race and ethnicity correlate with persistent, and often increasing, health disparities among U.S.

From: *Nutrition and Health: Vitamin D*
Edited by: M.F. Holick, DOI 10.1007/978-1-60327-303-9_20,
© Springer Science+Business Media, LLC 2010

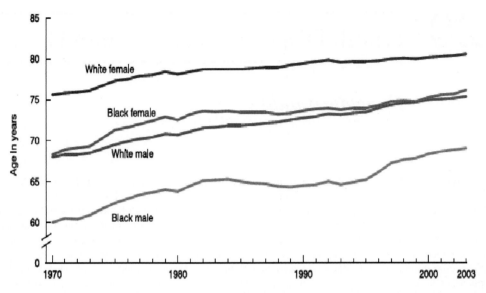

Fig. 1. Life expectancy at birth by race and sex: 1970–2003. From the CDC National Vital Statistics Reports, Volume 54 Number 14 United States Life Tables, 2003.

populations demands national attention." And that "Current information about the biologic and genetic characteristics of minority populations does not explain the health disparities experienced by these groups compared with the white, non-Hispanic population in the United States" *(1)*. As part of its Healthy People 2010 Initiative, the US Department of Health and Human Services (HHS) selected six focus areas in which racial and ethnic minorities experience serious disparities in health access and outcomes. HHS also identified additional diseases and conditions which disproportionately impact racial and ethnic minorities (Table 1) *(2)*.

This chapter will address the likelihood that vitamin D deficiency is a major factor in health disparities between populations in the United States, and particularly between the non-Hispanic white (NHW) and non-Hispanic black (NHB) populations between

Table 1
Diseases associated with health disparities – US department of HHS, ODPHP, Healthy People 2010

Healthy people 2010 focus areas	*Other diseases of health disparity*
Cancer screening and management	Mental health
	Hepatitis
Cardiovascular disease (CVD)	Syphilis
Diabetes	Tuberculosis (TB)
HIV infection/AIDS	
Immunizations	
Infant mortality	

whom there is most often the widest health disparities and about whom there is the greatest documentation. The Hispanic population usually falls in the midrange between those populations on measures of health disparities as its mean serum vitamin D levels fall midrange between the levels of those non-Hispanic populations.

Mattapan Community Health Center (MCHC), where the author of this chapter serves as Medical Director, became interested in the role of vitamin D deficiency as major factor affecting the health disparities noted in our community, which census figures identified as 92% black, compared with the city of Boston as a whole (in which our community is located). MCHC has been publishing an annual community health report card for several years as part of its yearly health outreach event "Health Care Revival" *(3)*. A chance reading of Moskilde's review article *(4)* led to a review of research into vitamin D, and subsequently prompted an investigation into the possibility that vitamin D deficiency might affect the incidence of low birth weight babies and infant mortality, a significant problem in Mattapan that MCHC had been founded in 1972 to address. The search for references on this relationship took place in May 2006, just a month after Mannion et al. (vide infra) had published their study on the correlation of vitamin D and milk intake with birth weight. Testing of 25(OH)D levels of random patients whose insurance would cover the cost, along with a cohort of prenatal registration patients, revealed a high prevalence of vitamin D deficiency in our population in the early months of 2007.

Vitamin D sufficiency is now accepted to be represented by serum levels of 25-hydroxyvitamin D [25(OH)D] of at least 30 ng/ml, with serum levels from 20 to 30 ng/ml considered to be insufficient and levels below 20 ng/ml deficient, by reference to levels consistent with maximal absorption of calcium from the gut and plateauing (at baseline) of parathyroid hormone (PTH) levels. However, optimal levels of vitamin D for health have been demonstrated to be even higher by measure of maximal rate of bone mineral deposition (36–40 ng/ml) *(5)*. Lower extremity muscle function, rate of fall prevention in the elderly, rate of fracture prevention, and depression of colorectal cancer incidence all appear to be maximal at 25(OH)D levels greater than 48 ng/ml, though by that level the rates of change in the measures are approaching minimal *(6)*.

The Third National Health and Nutrition Examination Survey (NHANESIII) found NHWs have a mean serum level of 25(OH)D of 32 ng/ml (Fig. 2). It found NHBs have a mean 25(OH)D of 19 ng/ml while those labeled Hispanic had a mean 25(OH)D of 25 ng/ml *(7)*. By these measures a substantial proportion of thepopulation in the United

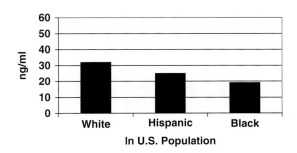

Fig. 2. 25(OH)D levels – NHANES III. Bibuld Data adapted from *(7)*.

States has suboptimal vitamin D levels. It is also quite evident that there is a large disparity in the levels of vitamin D sufficiency in the populations noted, and that to the extent that vitamin D is important to human health, this will be manifested in disparate health between the populations.

2. INFANT MORTALITY

Infant mortality rates in the United States vary by racially and ethnically categorized subpopulations. The 2003 US infant mortality rates from the CDC National Vital Statistics Reports (NVSR) Health of the United States are shown in Fig. 3. Similar to the overall mortality rates for the US populations, there has been a persistent infant mortality disparity between the black and white populations (Fig. 4).

Short gestation and low birth weight (SGLBW) were among the most important predictors of infant survival in the overall US population, second only to congenital

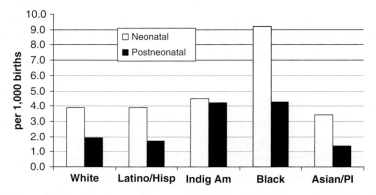

Fig. 3. Neonatal and postneonatal mortality rates. Bibuld Data assembled from Health, United States 2006 – Table 19 Linked Birth/Infant Death Data Set, National Center for Health Statistics, CDC.

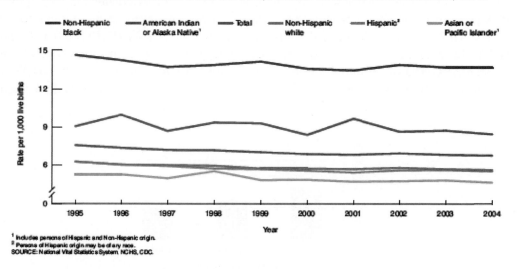

Fig. 4. Disparity in infant mortality rate. Adapted from Mathews and MacDorman *(8)*.

malformations, deformations, and chromosomal disorders (CMDCD) in 2004 *(8)*. Sudden Infant Death Syndrome (SIDS) ranks third as a cause of death.

In 2004 an analysis of the five leading causes of infant mortality showed significant differences in the proportions of death attributable to those causes in different ethnic populations. Specifically, while the number one cause of death was CMDCD in the non-Hispanic white (NHW) and Hispanic populations, the rate of CMDCD in the non-Hispanic black (NHB) population was only 56% that of disorders of short gestation and low birth weight (SGLBW) *(8)*. Accordingly, while death rates in the NHB population from CMDCD were 29% higher than the NHW population, the death rate from SGLBW was 285% higher or almost four times as much. Thus SGLBW was the leading cause of infant mortality for the NHB population *(8)*.

Gains in neonatal survival in the United States have been largely due to increases in survival of SGLBW babies. Currently, among very low birth weight babies and very premature births, the mortality rate does not reach 50% of babies born except for those born below 500 g, or at less than 24 weeks of gestation. For babies born at less than 3,500 g and less than 37 weeks gestation, neonatal mortality is lowest among NHB of the three ethnic groups *(9)*. It is highest in that group for NHWs.

Despite that, in 1995 through 1997 the overall mean birth weight for babies born to NHB resident mothers in the United States was 3,133 g; that for NHW babies was 3,413 g. The percentage of infant mortality in the NHB population attributable to birth weight less than 750 g (or VLBW) was 63.6%. This compared to 42.9% of infant mortality for Hispanic and 38.9% for white babies (Fig. 5). The incidence of VLBW babies born was 85 in NHB, 25 for Hispanic, and 20 for NHW per 10,000 births. Similarly, for infants born at less than 28 weeks gestation, the incidence among the NHB, Hispanic, and NHW populations were 139, 45, and 35 per 10,000 births, respectively (Fig. 6). The percent of mortality attributable to these very short gestation births was 44.6, 48.7, and 68.6% for NHW, Hispanic, and NHB populations, respectively *(9)*.

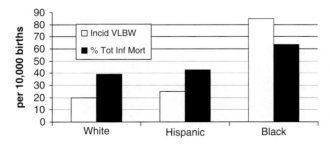

Fig. 5. Incidence of birth weight (<750g) and percent contribution to total infant mortality. Bibuld adapted from *(9)*.

Similarly, though less dramatically, there was disparity in incidence of overall prematurity and LBW. Babies born at less than 2,500 g were 1155, 538, and 491 per 10,000 and born at less than 37 weeks gestation were 1587, 987, and 803 per 10,000 births (Fig. 7) in the NBH, Hispanic, and NHW populations, respectively *(9)*. These figures show the incidence of SGLBW babies to be inversely proportional to serum levels of 25(OH)D in mothers and infants in these populations. The incidence of infants born at

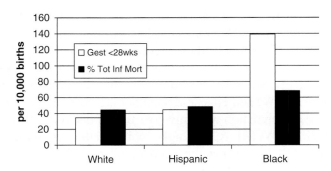

Fig. 6. Incidence of gestational age <28 weeks with comparison to percentage of total infant mortality due to VLBW. Bibuld adapted from *(9)*.

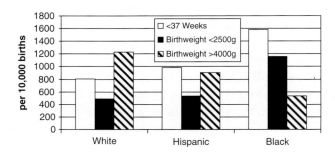

Fig. 7. Incidence premature, LBW and high BW babies. Bibuld adapted from *(9)*.

greater than 4,000 g shows the opposite skew to that seen in SGLBW, being 536, 905, and 1,233 per 10,000 births in the NHB, Hispanic, and NHW populations. The incidence of larger babies directly correlates with the levels of 25(OH)D in the respective populations (Fig. 2).

The death rate from SIDS in the NHB population was 105% higher than the NHW population *(8)*. SGLBW is a major risk factor for SIDS. Age of mother and maternal smoking history are the two major maternal risk factors for SIDS *(10)*. The incidence of cigarette smoking in the United States for NHB mothers was only 61% (at 8.4%) that of NHW mothers (13.8%) in 2004. The rate for Hispanic mothers was 2.6%; for AI/AN mothers the rate was 18.2% *(11)*. The rate of birth to mothers aged 15–19 years was 6.3% for NHB or 67% higher than for NHW mothers (3.8%) at the same age. This data suggest that the higher SGLBW rate in NHB neonates is the major factor in the higher rate of SIDS in this population.

The vitamin D status of the neonate and fetus appears to be totally dependent on vitamin D stores of the mother and specifically maternal 25(OH)D *(12–15)*. Various studies show fetal cord blood at birth containing 50–70% of maternal 25(OH)D levels *(12–16)*. They also show that when vitamin D supplementation is given to mothers, serum 25(OH)D in maternal and fetal cord increases in roughly the same proportion *(12, 14, 16)*. There is no significant correlation between fetal and maternal serum levels of 1,25(OH)$_2$D *(14)*.

Vitamin D deficiency has been well documented to be much more common in the NHB population in the United States than the NHW population. The Hispanic (or

Latino) population as a whole have serum levels of 25(OH)D roughly midway between the black and white populations (Fig. 2) *(7, 15, 17–21)*. This disparity in serum 25(OH)D is apparent throughout the age spectrum, and has been particularly well documented in women of child-bearing age and in pregnant women (Fig. 8) *(15, 17, 18, 20–22)*. 25(OH)D levels have been shown to be disparate in neonates as well *(15, 23)*.

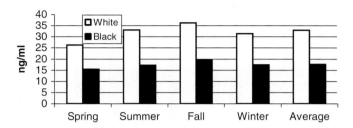

Fig. 8. Mean 25(OH)D in women 15–49 years in NHANES III. Bibuld adapted from *(21)*.

In data taken from the third National Health and Nutrition Examination Survey (NHANES III) *(21)* from 1988 to 1994, 1546 NHB and 1426 NHW non-pregnant women aged 15–49 were evaluated after exclusions for a number of variables. The NHB women had a mean serum 25(OH)D concentration of 17.7 ng/ml. The NHW women's serum 25(OH)D levels averaged 33 ng/ml.. Similarly, Bodnar et al. *(22)* studying 200 NHB and 200 NHW pregnant women in Pittsburgh found initial serum values (at 4–21 weeks of gestation) of 25(OH)D of 16.1 and 29.2 ng/ml in the respective cohorts. Women who did not carry to term were excluded from these values. Ninety percent of the women were taking prenatal supplements during the last 3 months of pregnancy. At birth (37–41 weeks) the NHB mothers showed a mean serum level of 19.8 ng/ml, and their cord blood 15.6 ng/ml. For NHW women the values were 32.2 ng/ml for serum and 26.9 ng/ml in cord blood at term *(22)*.

Vitamin D deficiency has been correlated with SGLBW babies *(12, 24–27)*. In 1980 in response to recognition of a high incidence of pregnancy-associated osteomalacia and decreased fetal size in association with vitamin D deficiency among Asian (primarily Indian) women in England, Brooke et al. *(26)* evaluated vitamin D supplementation in Asian women. That study included 59 pregnant women given 1,000 IU/day in their last trimester and a matched group of 67 women given placebo. They reported modest increase in birth weight of 123 g in the treatment group. The baseline maternal levels of 25(OH)D before supplementation (at 28 weeks) was 8.0 ng/ml. The term blood levels reported of 25(OH)D in the treated mothers (67.2 ng/ml) is highly inconsistent with expected increase in serum 25(OH)D levels at the reported level of supplementation (~18 ng/ml). However, incidence of SGA infants was 28.6% in the control group vs. 15.3% in the treated group. SGA was defined as weight less than the tenth percentile. Weight gain in the treated mothers increased significantly more (63 g/day) than in the control group (46 g/day). European women typically gain 71 g/day in their last trimester *(26)*. *The increase in birth weight of 123 g found in the treated group, who received up to 84,000 IU of vitamin D during the third trimester of pregnancy (or the equivalent of 300 IU per day for the entire pregnancy) was about 12 g per 30 IU of vitamin D per day over the course of the pregnancy. Significantly, this is similar to the weight gain associated with the differential in vitamin D intake found by Mannion et al. (24) as noted below.* In a follow-up study Brooke et al. *(28)* showed that the weight difference between the infants in the treated group increased modestly from birth through the first year and was 490 g greater at the end of first year. The length of infants from the treatment group was 3.3 cm (or 0.33 dm) greater at first year as well (Fig. 9).

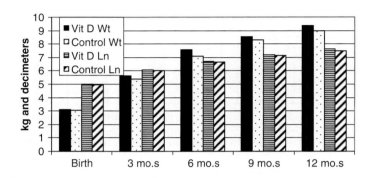

Fig. 9. Infant weight and length – maternal D3 supp vs control. Bibuld adapted from *(28)*.

Marya et al. *(27)* studied 25 women treated with 1,200 IU of vitamin D a day in their third trimester, 20 women treated with two doses of 600,000 IU in the seventh and eighth months of pregnancy, and 75 women who received no supplemental vitamin D in a study published in 1980. They reported a significantly greater increase in birth weight with either vitamin D supplementation, but greater increase with the 600,000 IU doses.

Brunvand et al. *(29)* evaluated 30 Pakistani women in Norway to look at the relation of elevated serum parathyroid hormone (PTH) levels, serum ionized calcium levels, and their relation to reduced fetal growth in vitamin D-deficient pregnant women. Women with complicated pregnancies were excluded. Twenty-nine of the 30 women had 25(OH)D levels below 12 ng/ml. Thirteen of them had high PTH levels. A positive correlation was found between the maternal serum ionized calcium level, and a negative correlation noted for serum PTH level, with crown–heel length of the fetus. PTH elevation and hypocalcemia are both complications of low vitamin D levels.

Specker *(30)*, in her review of studies of maternal and neonatal outcomes in vitamin D deficiency, reports on a study by Marya et al. *(31)*, published in 1988. One hundred women were given 600,000 IU 25(OH)D in both the seventh and eighth month of gestation and compared to 100 pregnant women who were not supplemented. Serum 25(OH)D levels were not reported, but greater birth weight and size were reported for infants born to the treated cohort.

Mannion et al. *(24)* evaluated the relationship of vitamin D intake through fortified milk in 279 pregnant women in Calgary, Alberta. Seventy-two of them reported intake of one cup of milk or less daily (restricted intake). Two hundred and seven drank more milk than one cup daily (unrestricted). It was calculated that the restricted group took in on average 316 IU vitamin D daily, compared to 524 IU per day in the unrestricted group. The birth weight of infants born to mothers in the restricted group was 3,410 and 3,530 g in the unrestricted cohort (Fig. 10). It was calculated that for each increase in daily intake of vitamin D by 40 IU, there was a corresponding increase in birth weight

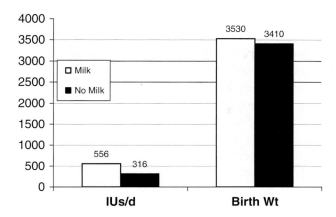

Fig. 10. Vitamin D intake from milk and supplements. Bibuld adapted from *(24)*.

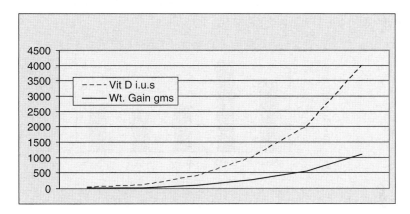

Fig. 11. Weight gain per increase in vitamin D intake. Figure adapted by Bibuld from *(24)*.

of 11 g. By extrapolation, this suggests a possible weight gain of 550 g with 2,000 IU vitamin D supplementation daily from the beginning of pregnancy (Fig. 11).

Vitamin D deficiency has also been associated with an increased incidence of pre-eclampsia, elevated BPs in pregnant women, hypocalcemia and hypocalcemic seizures in neonates, craniotabes, and other disorders of fetal development. However, documentation on the relationship of these disorders to overall infant mortality is lacking, and therefore they are not addressed here.

Infant mortality remains one of the most dramatic markers of health disparity in the United States. SGLBW babies are the most important component to that disparity, including their impact on the disparity in infant deaths due to SIDS. Abundant data tie the incidence of SGLBW to low maternal serum levels of 25(OH)D. Infant mortality and birth data demonstrate the inverse relationship of 25(OH)D levels with the incidence and severity of SGLBW in ethnically and racially defined populations. Evidence also shows that vitamin D supplementation increases birth weight and appears to reduce the incidence of SGLBW births.

Unfortunately, the number of studies and subjects in those studies of supplementation are limited and many questions remain unresolved. There is not enough data to suggest whether there is a threshold of circulating 25(OH)D for optimal outcomes, or whether improvement in weight gain and outcome is linearly related to 25(OH)D level. It does not appear that enough vitamin D has been given to mothers to raise their levels as high as optimal levels have been shown to be for dental attachment, maximal muscle strength, colon cancer prevention, and fracture prevention (levels above 48 ng/ml) *(5)*. Work that Hollis, Wagner et al. are doing in South Carolina compares supplementation of vitamin D of 400–6,000 IU per day beginning in early pregnancy should be helpful in answering some of these questions.

3. CARDIOVASCULAR DISEASE

Cardiovascular disease, including hypertension and cerebrovascular disease, is more prevalent and causes higher mortality in the black population in the United States than in the white population (Fig. 12) *(32, 33)*.

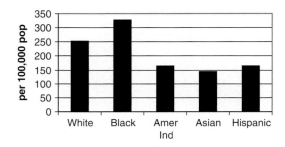

Fig. 12. Age adjusted heart disease mortality, United States. Bibuld adapted from *(32)*.

Risk factors traditionally associated with cardiovascular disease include family history, hyperlipidemia, smoking, diabetes mellitus, obesity, sedentary lifestyle, and the presence of hypertension itself. Recent research also strongly identifies hypovitaminosis D as a major risk factor for heart disease, and perhaps the most important single risk factor in the United States *(34, 35)*.

Evidence suggests several mechanisms by which vitamin D may protect against cardiovascular disease. These include inhibition of inflammatory cytokines and stimulation of production of anti-inflammatory cytokines (thereby inhibiting initiation of atherosclerosis), inhibition of smooth muscle proliferation (inhibiting vascular media thickening), inhibition of myocardial cell hypertrophy, inhibition of the renin–angiotensin system, and prevention of arterial calcification *(36–38)*.

Martins et al. *(7)* reported on the relationship between cardiovascular risk factors and serum levels of 25(OH)D. Their review of data from over 15,000 persons who participated in NHANES III made comparisons between those in the lowest quartile of serum 25(OH)D (less than 21 ng/ml) with those in the highest quartile (more than 37 ng/ml). It found that those in the lowest quartile had a 35% higher likelihood of hypertension, over twice as likely to have prediabetes or diabetes, over twice as likely to be obese,

38% more likely to have elevated triglycerides, and 26% more likely to have elevated cholesterol.

In May 2007, Forman et al. *(39)* looked at the likelihood of developing hypertension in 613 males participating in the Health Professionals' Follow-Up Study (HPFS) and in 1,198 women from the Nurses' Health Study. None of these professionals had hypertension when they entered the study. The two groups were compared by levels of serum 25(OH)D. Among the men, the chance of developing hypertension with a 25(OH)D level of less than 15 ng/ml was more than six times higher than men with a 25(OH)D level of more than 30 ng/ml after 4 years. Among the women, the chance of developing hypertension was almost three times higher for the low vitamin D group compared to the higher level (Fig. 13).

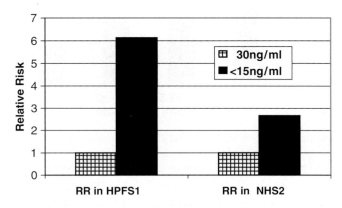

Fig. 13. Association between plasma 25(OH)D and risk of HTN. Bibuld adapted from *(39)*.

Wang et al. *(34)* reported on the impact of vitamin D levels on cardiovascular disease in 1,739 Framingham, MA patients with no evidence of cardiovascular disease (angina, heart attack, stroke, TIA, peripheral artery disease, or heart failure) at the start of the study. They were followed for the development of cardiovascular disease during the course of the study. The participants had an average age of 59 years, were 55% women, and all white (owing to the history of the participants followed in study which originated in the 1940s). The study compared patients with serum 25(OH)D levels <15 ng/ml D with those with levels ≥15 ng. It found that those with the lower levels of 25(OH)D had more than twice the incidence of cardiovascular disease than those with the higher levels after 5 years of follow-up. For patients with hypertension the risk of cardiovascular disease was two and a half times higher in the group with lower 25(OH)D. And while the risk of cardiovascular disease in patients with higher levels of 25(OH)D was 65% higher in those patients with hypertension than those without hypertension, it was over 300% higher (or four times as great) in those with lower vitamin D levels and hypertension than those with higher 25(OH)D and no hypertension. It was also reported that those with 25(OH)D <10 ng/ml had even greater risk than those who had a 25(OH)D of 10–15 ng/ml. Additionally, analysis showed that the probability of developing cardiovascular disease with normal blood pressure and lower 25(OH)D (less than 15 ng/ml) was greater than the chance of developing cardiovascular disease with high blood pressure and the higher level of 25(OH)D (at least 15 ng/ml) after 7 years of follow-up.

In June 2008 Giovanucci et al. *(35)* reported on a prospective, nested case-controlled study involving men free of cardiovascular disease in the 18,255 man HPFS who were followed for development of fatal and non-fatal MI and had blood samples collected for 25(OH)D. Four hundred and fifty-four men who had events were matched 2:1 with controls matched by age, date of blood collection, and smoking status. Compared with subjects whose 25(OH)D levels were ≥30 ng/ml, those whose levels were lower than 15 ng/ml had a relative risk (RR) for MI of 2.42 (Fig. 14). Subjects whose 25(OH)D levels were between 24 and 30 ng/ml had intermediate risk.

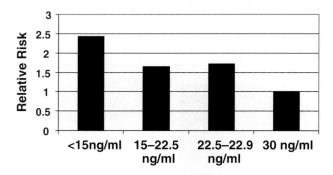

Fig. 14. 10 year risk of MI by 25(OH)D level. Bibuld adapted from *(35)*.

Dobrig et al. *(40)* reported on 3258 consecutive male and female patients undergoing coronary arteriography in Austria. They found twice the mortality for cardiovascular disease (RR of 2.22) for patients in the two lower quartiles of 25(OH)D (means of 7.6 and 13.3 ng/ml), than among patients in the highest quartile (median 28.4 ng/ml). The graphic data presented show a linear relationship between the quartile of 25(OH)D level and cardiovascular mortality risk (Fig. 15).

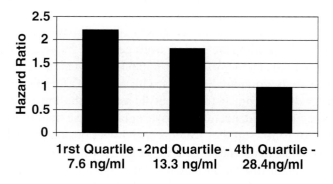

Fig. 15. Risk of CV death by 25(OH)D level. Bibuld adapted from *(40)*.

In August 2008 Melamed et al. *(41)* in an analysis of NHANES III data found an unadjusted increase of cardiovascular mortality of 70% (RR 1.70) among 13,331 participants aged 20 years or more in participants in the lowest quartile of 25(OH)D levels (<17.8 ng/ml) compared with those in the highest quartile (>32.1 ng/ml). Non-Hispanic blacks, Mexican Americans, and the elderly were oversampled in this study to allow for more precise estimates for those groups. When adjusted for age, race, sex, season, hypertension, CVD history, diabetes, smoking, HDL, total cholesterol, use of anti-lipids medication, GFR, albumin, albumin/creatinine ratio, C-reactive protein level, BMI, physical activity, use of vitamin D supplementation, and low-socioeconomic factor they still found a 20% increase in cardiovascular mortality between the quartiles. In their study they also reported on the vitamin D profiles of population groups in study and found that for non-Hispanic whites 43.5% were in the highest quartile of 25(OH)D level of the general population and 9.5% fell into the lowest quartile, while only 7.8% of non-Hispanic blacks fell into the highest quartile, and 50.3% were in the lowest quartile.

4. CANCER

Cancer is a heterogeneous group of diseases with multifactorial etiologies, many of which remain to be elucidated. The mechanisms of vitamin D's inhibition of development, growth, and spread of cancer are addressed elsewhere in this volume. An article by Garland et al. *(42)* published in February 2006 referred to over 1,000 laboratory and human population studies published concerning the association of vitamin D and cancer up until then.

There are major differences in the incidence of overall and individual cancers in different ethnically categorized subgroups in the United States, with NHB and NHW having higher incidence and mortality than other subgroups (Fig. 16) *(43)*. NHBs not only have a higher incidence of cancer compared to NHWs, but as significantly, the mortality rates of NHBs with cancer are higher than NHWs. Despite having a significantly lower incidence of cancer than those two population subgroups, AI/ANs have the highest ratio of mortality to incidence for cancer, while the rates of Hispanics and Asian/PIs are the lowest for all three measures. Some of the difference in mortality rates may be reflective of past differences in cigarette smoking (by comparison of mortality/incidence ratios to prevalence of cigarette smoking) *(44)*. Currently, tobacco use is less in NHBs than NHWs and least in Hispanics *(45)*.

Lung, breast, and colorectal cancers were responsible for 51% of cancer mortality during 1975–2005 in both NHW and NHB women *(43)*. Lung, prostate, and colorectal cancers were responsible for 51% of NHW male cancer death in that same period, but

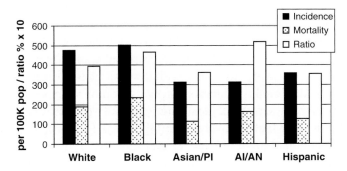

Fig. 16. Incidence and mortality of cancer – US populations (age-adjusted). Bibuld adapted from *(32)*.

59% in NHB males largely due to the greater impact of prostate cancer on mortality in that population. In all populations lung cancer was the greatest killer.

During the period 1992–1999 NHBs had an age-adjusted incidence of cancer that was 10% higher than NHWs. In females, the incidence of cancer was actually 5% higher in NHWs than in NHBs. In males NHBs had an incidence 24% higher. Mortality was 30% higher for NHBs than NHWs during the same period, being 19% higher for NHB females and 43% higher for NHB males than their NHW counterparts (Fig. 17) *(46)*.

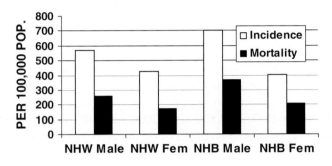

Fig. 17. Cancer incidence and mortality. Bibuld adapted from *(46)*.

The period 2001–2005 showed modest decreases in the incidence and more significant decreases in mortality in cancer for US populations, but significant disparities remained, especially in mortality *(43)*. The incidence of cancer in NHBs was 5% higher for the period compared to NHWs, but mortality was still 24% higher. In females, the incidence of cancer was 6% higher in NHWs than in NHBs. In males NHBs had an incidence 18% higher than NHWs. Mortality was 17% higher for NHB females and 36% higher for NHB males than NHW females and males, respectively.

Giovannucci et al. *(47)* developed a multiple linear regression model using multiple determinants of vitamin D exposure, including skin pigmentation, diet and vitamin D supplementation to compute the relative risk of cancer incidence and mortality based on serum 25(OH)D level in men from the HPFS. Their model found that an increase of 10 ng/ml of 25(OH)D was associated with a 17% reduction in cancer incidence and a 29% reduction in cancer mortality.

Looking at cancer incidence and mortality among black and white male health professionals from the same data (involving a total of 481 black and 43,468 white men over 16 years) Giovannucci et al. *(48)* concluded that blacks with low risk factors for hypovitaminosis D (average serum 25(OH)D; 21.6 ng/ml) had similar incidence and mortality (RR 0.95 and 1.55, respectively) to whites with low risk factors (average 25.8 ng/ml) for cancer. Blacks with additional risk factors for poor vitamin D status (average 17.2 ng/ml) had much higher risks for cancer incidence and mortality (RR 1.57 and 2.27) compared to whites with higher risk factors (average 25(OH)D; 23.1 ng/ml). Whites at higher risk for hypovitaminosis D had higher incidence and mortality than whites with lower risk, but lower mortality than blacks with higher risk. Giovannucci did not report analysis of cancer incidence and mortality for black and white subjects at equivalent serum levels of 25(OH)D.

Lappe et al. *(49)*, in a 4-year study of vitamin D and calcium supplementation in 1,179 women (all white) in eastern Nebraska, showed a 77% reduction in the incidence of cancer during the second to fourth years of the study with supplementation compared with placebo. Vitamin D supplementation was given at 1,100 IU per day. The decrease in cancer incidence was associated with an increase of serum 25(OH)D from a baseline of 28.7 to 38.4 ng/ml after 12 months. After 12 months the calcium supplement and placebo arms remained close to baseline at 28.4 ng/ml each. While there was a lesser reduction in cancer incidence with calcium supplementation alone, it did not reach a level of significance (Table 2).

Table 2
Cancer by Site and Treatment Arm (2–4 years). Bibuld adapted from *(49)*

	Placebo *(266)*	*Ca^{2+}* *(416)*	*Ca^{2+} + Vitamin D* *(403)*
Breast	7	6	4
Colon	2	0	0
Lung	3	2	1
Hematopoetic	4	4	2
Uterus	0	1	0
Other	2	2	1
Total (%)	18 (6.8)	15 (3.7)	8 (2.0)

Lung cancer incidence is 36% higher and mortality 31% higher in NHB males than NHW males. The incidence and mortality are slightly lower for NHB females than for NHW females (Fig. 18).

Lung cancer incidence appeared to be reduced by 15–20% in males associated with an increase of 10 ng/ml in serum 25(OH)D according to Giovannucci et al. *(47)*. Lappe et al. *(49)* found only one case of lung cancer in the vitamin D supplemented group (403 subjects) compared with 3 in the placebo group (266 subjects). Zhou et al. *(50)* found that patients with stage 1B–2B non-small cell lung cancer had a 55% better survival rate up to 12 years after diagnosis if serum levels of 25(OH)D were >21.5 ng/ml (highest quartile in their study) compared with <10.2 ng/ml (lowest quartile) (Table 3).

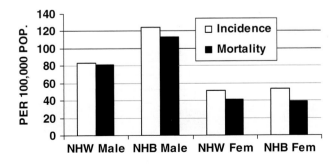

Fig. 18. Lung cancer and mortality. Bibuld adapted from *(39)*.

Table 3
Serum 25(OH)D and Mortality Among 447 Patients with Early-Stage NSCLC. Bibuld adapted from *(50)*

	Lowest quartile RR	Highest quartile RR	Higher level and intake by medians - RR
Overall survival	1	0.74	0.64
Stage IB–IIB	1	0.45	0.67

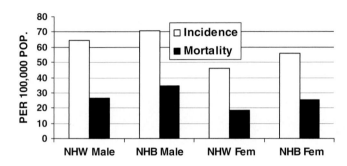

Fig. 19. Colorectal cancer and mortality. Bibuld adapted from *(35)*.

Colorectal cancer incidence was 22% higher in NHBs than NHWs in 2001–2005 *(43)*. The mortality was 42% higher in NHBs (Fig. 19). Numerous studies have shown significant decrease in incidence of colorectal cancer in the range of 50% with vitamin D supplementation and with adequate levels of serum 25(OH)D *(51–54)*.

The incidence of breast cancer in NHBs was only 90% that of NHWs in 2001–2005. Yet the mortality for NHBs was 37% higher for this disease (Fig. 20). Black women in the United States have a higher prevalence of mammography screening than do white women. Knight et al. *(55)* have presented evidence that higher levels of vitamin D intake and sun exposure, particularly during the period of adolescent breast development is associated with decreased later incidence of breast cancer. Lappe et al. *(49)* found a

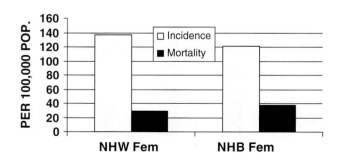

Fig. 20. Breast cancer and mortality. Bibuld adapted from *(40)*.

1.0% incidence of breast cancer in the group supplemented with vitamin D, compared to a 2.6% incidence in women receiving placebo from the second to fourth year of that study. Neuhouser et al. *(56)*, in a multiethnic study of breast cancer survivors, found that the stage of disease was independently associated with serum 25(OH)D levels with lower levels associated with more advanced disease.

Prostate cancer occurs 59% more frequently in NHB (275.3 per 100,000) than NHW males 172.9 per 100,000) in the United States. Its mortality rate is 2.4 times higher (75.1 vs. 32.9 per 100,000) in NHBs than NHWs (Fig. 21) *(43)*. In Trinidad the age-adjusted mortality rate from prostate cancer (32.3 per 100,000) was about the same as for NHWs in the United States (32.9 per 100,000) *(57)*(Fig. 22). Mauritius, a nation of very dark-skinned people located just north of Senegal at the western edge of the Sahara desert, had an age-adjusted prostate cancer mortality rate of 7.3 per 100,000.

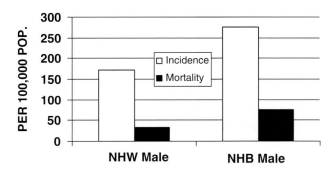

Fig. 21. Prostate cancer incidence and mortality. Bibuld adapted from *(43)*.

A nested control study in Finland *(58)* based on 13-year follow-up of 19,000 middle-aged men who were free of prostate cancer at baseline found a RR of 1.7 for men entering the study with serum 25(OH)D levels <16 ng/ml (the mean in that population) compared with those who had ≥16 ng/ml. However, the RR for males <52 years with serum 25(OH)D <16 ng/ml was 2.5, along with an RR of 6.3 for metastatic disease.

John et al. *(59)* found large reductions in the incidence of advanced prostate cancer with high recreational and occupational sun exposure in NHWs, as have others. Li et al. *(60)* found among US physicians that those whose median serum levels of 25(OH)D fell below 32 ng/ml for the summer and 25 ng/ml in the winter had significantly increased risk of total and aggressive prostate cancer. This was particularly the case if they had a relatively common vitamin D receptor polymorphism. However, among men with 25(OH)D levels above those means in summer and winter the genotype associated with that polymorphism was no longer associated with increased risk and was related to a 60–70% lower risk of total and aggressive prostate cancer compared to men with levels of 25(OH)D below those medians.

5. TOTAL MORTALITY

Figure 1 graphs the average life expectancy at birth by race and sex from 1970 to 2003. Despite the gradual increase in life expectancy over this period the gap between black and white life expectancy in the United States remains remarkably constant.

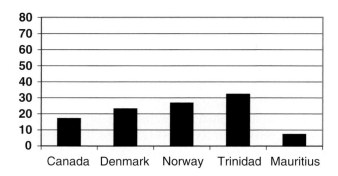

Fig. 22. Prostate cancer mortality rates in some other countries. Bibuld adapted from *(57)*.

Recent studies implicate the gap in serum 25(OH)D levels between the population groups as being a significant contributor to this disparity; sufficient perhaps to account for the entirety of it.

Autier and Gandini *(61)* conducted a meta-analysis of all available medical literature published up to November 2006 of randomized controlled trials in which vitamin D was given for any health condition and death was a reported outcome. A search of the PubMed, ISI Web of Science, EMBASE, and Cochrane databases revealed 18 such trials. The trials encompassed 57,311 participants and 4,777 deaths. The mean trial period was 5.7 years, with individual trials running from 6 to 84 months. The mean vitamin D supplementation dose was 528 IU daily, with a range of 300–2,000 IU daily. The analysis showed that patients receiving vitamin D supplementation in the trials had 7% less mortality than control subjects. In trials lasting at least 3 years vitamin D supplementation resulted in 8% fewer deaths. Where calcium supplementation was given with vitamin D, treated subjects had 7% less mortality, but in trials in which vitamin D was the sole supplement given subjects had 9% less mortality. In trials that were placebo controlled, vitamin D supplementation resulted in 8% less mortality.

Dobnig et al. *(40)* reported a prospective cohort study of 3,258 consecutive patients having coronary angiography, looking at the relationship of 25(OH)D and 1,25(OH)$_2$D levels to all-cause and cardiovascular mortality. Comparing patients by quartiles of 25(OH)D they found that patients in the lowest quartile (median serum 25(OH)D of 7.6 ng/ml) had a multivariate adjusted RR of 2.08 for all-cause mortality compared to the highest quartile (median 28.4 ng/ml). The RR of mortality for the next to lowest quartile (mean 25(OH)D of 13.3 ng/ml) was 1.53. Serum 1,25(OH)$_2$D levels were independently associated with all-cause (and cardiovascular) mortality, but the study found only weak correlation between serum 25(OH)D and 1,25(OH)$_2$D levels. The relative impact of quartile of serum 1,25(OH)$_2$D level on mortality did not approach that of 25(OH)D level above the lowest quartiles.

Melamed et al. *(41)* reviewed NHANES III data of 13,331 adults ≥20 years. 25(OH)D levels were collected from 1988 through 1994, and individuals were followed for mortality through 2000. Mortality rates were computed by quartile of 25(OH)D level. Those in the highest quartile had >32 ng/ml, those in the lowest <17.8 ng/ml. A total of 43.5% of NHWs fell into the highest quartile, while only 7.8% of NHBs were represented therein. A total of 9.5% of NHWs were found in the lowest quartile, along with 50.3% of blacks. Without adjustment those in the lowest quartile of vitamin D had a 78% increased risk of all-cause mortality than those in the highest quartile. After adjustment for age, sex, race, and season the mortality rate for the lowest quartile was 52% higher. After further adjustment for hypertension, CVD history, diabetes, smoking, HDL cholesterol, total cholesterol, use of anti-cholesterol medications, glomerular filtration rate, serum albumin, albumin creatinine ratio, C-reactive protein, body mass index, physical activity, vitamin D supplementation, and low socioeconomic status mortality remained 26% higher in the lowest quartile of serum 25(OH)D.

6. DIABETES

According to The Office of Minority Health (OMH) *(62)*, a department of HHS, "African Americans are 2.2 times as likely to have diabetes as Whites." As reported on its web site in November 2008 "Hispanics are 1.5 times as likely to have diabetes as Whites." They also noted that "American Indian and Alaskan Natives are 2.3 times as likely as non-Hispanic Whites of similar age to have diabetes." Age-adjusted death rates from diabetes mellitus have historically been highest in NHBs of the population groups mentioned. In 2005 the relative age-adjusted risk of mortality for diabetes for black, AI/AN, Hispanic, and whites were 2.2, 1.9, 1.6, and 1.0, respectively *(63)*.

The positive association of serum vitamin D levels with insulin sensitivity, insulin secretion from pancreatic β-islet cells, and glucose tolerance in both animals and humans has been well established *(64–66)*. Vitamin D deficiency has also been associated with glucose intolerance and Type 2 diabetes *(7, 64, 67, 68)*. Vitamin D supplementation has been shown to improve these measures in diabetic patients *(69–71)*.

These studies have not been detailed enough to determine the magnitude of impact of either serum 25(OH)D level or supplementation on the incidence or severity of this disease.

7. TUBERCULOSIS

Steele et al. *(72)* in a review for the CDC reports that "US-born blacks have consistently had tuberculosis rates eight times higher than US-born whites." In 2004 45% of US-born persons with tuberculosis were black.

Vitamin D has been shown to have a positive effect on macrophage function and to inhibit intracellular growth of tuberculosis and enhance killing of mycobacterium tuberculosis (MTB). 25(OH)D deficiency has been shown to be associated with active tuberculosis. It has been shown that sera from African-American individuals, low in 25(OH)D, were inefficient in producing cathelicidin, the antimicrobial peptide (AMP) responsible for enhancement of macrophage killing of MTB. It was also shown that restoring 25(OH)D in the sera of those same African-American individuals to physiologic range restored the killing ability of macrophages in that sera *(73–75)*.

8. HIV/AIDS

Continuing a historical pattern, the CDC-defined black population had the highest HIV/AIDS infection rate in 2004, with 69.3 cases per 100,000 population. Hispanics had the second highest rate at 26.6 cases, followed by American Indians, whites, and Asians with 10.2, 8.2, and 6.5 cases per 100,000 population, respectively (72).

While it is beyond the scope of this chapter to explore the curious history and epidemiology of HIV/AIDS infection in human beings, there are suggestions that vitamin D status may play an important role in defense against this virus. There has been little research to date in this area, but there is scientific basis for promise of positive results. Cannell (76) has pointed out that HIV has a lipoprotein coat, and viruses with lipoprotein coats are susceptible to attack by AMPs. 1,25(OH)D has been shown to induce production of AMPs such as cathelicidin and defensins (77). Defensins have been shown to inhibit HIV and suppress HIV infection (78–80).

Observation suggests that the spread of AIDs in subtropical southern Africa has been more of a problem than in tropical sub-Saharan Africa where it appears to have been stable for many years. In the United States the prevalence of HIV/AIDS has been historically much higher in the northeast compared to other areas of the country, and the prevalence of HIV/AIDS in the west well below the mean for United States despite the early loci of AIDS in Los Angeles and San Francisco and the higher black population rate in the South (75). Disparities in serum 25(OH)D levels could be a contributing factor to this distribution.

Similarly, despite the identification of Haitian natal origin as a major risk factor for HIV/AIDS early in the epidemic, Haiti, the poorest country in the western hemisphere, is not known for having a significant HIV problem, nor are other Caribbean countries afflicted in a major way. Observation at Mattapan Community Health Center (unpublished data, observed by author), which serves a proportionately large concentration of Haitian immigrants and transients, has shown vitamin D levels in transients and recent émigrés from Haiti and other Caribbean countries to be much higher than in residents of the United States. Mattapan, which has the largest Haitian community in Massachusetts, as well as being 83% black, has an HIV incidence rate that is 20% higher than Boston as a whole. Boston's black residents overall had an incidence rate of 39% higher than Boston's rate (81).

9. CONCLUSION

A lot of time, money, and research have been given to explain disparities in health outcome between so-called racial subpopulations in the United States. These disparities have been especially confusing given evidence that groups of immigrants moving to the United States from other parts of the world have appeared to assume the health characteristics of the traditional majority population of the United States within one to two generations.

It now appears that the bulk of these health disparities can be attributed to the disparity in vitamin D production related to the degree of melanin pigmentation and sun exposure in individual human beings. This evidence-based theory, though contrary to prevailing thought that unequal access to the technology of healthcare is at the root of health disparities, is entirely consistent with the history of public health science: environmental factors are the key effectors of human health.

REFERENCES

1. *From the Home Page of the CDC OMHD Website*, http://www.cdc.gov/omhd/AMH/AMH.html
2. *From the US Department of Health and Human Services, Office of Disease Prevention and Health Promotion, Healthy People 2010 Website,* http://www.healthypeople.gov
3. Lawson E, Young A (2002 Feb) Health care revival renews, rekindles, and revives. Am J Public Health 92:177–179
4. Mosekilde L(2005) Vitamin D in the elderly. Clin Endocrinol 62(3):265–281
5. Bischoff-Ferrari HA, Dietrich T, Orav EJ, Dawson-Hughes B (2004) Positive association between 25-hydroxy vitamin d levels and bone mineral density: a population-based study of younger and older adults. Am J Med 116:634—639
6. Bischoff-Ferrari HA, Giovannucci E, Willett WC, Dietrich T, Dawson-Hughes B (2006) Estimation of optimal serum concentrations of 25-hydroxyvitamin D for multiple health outcomes. Am J Clin Nutr 84:18–28
7. Martins D, Wolf M, Pan D et al (2007) Prevalence of cardiovascular risk factors and the serum levels of 25-Hydroxyvitamin D in the United States. Arch Int Med 167:1159–1165 Data from NHANES III
8. MathewsT J, MacDorman M. Infant Mortality Statistics from the 2004 Linked Birth/Infant Death Data Set, NVSR Vol 55, Number 14
9. Alexander G, Kogan M, Bader D, Carlo W, Allen M, Mor J(2003) US Birth weight/gestational age-specific neonatal mortality: 1995–1997 rates for whites, hispanics, and blacks. Pediatrics Jan 3(1): e61–e66
10. HoffmanH J, Damus K, Hillman L, Krongrad E (1988) Risk factors for SIDS. Results of the National Institute of Child Health and Human Development SIDS Cooperative Epidemiologic Study. Ann NY Acad Sci 533:13–30
11. Martin J, Hamilton B, Sutton P et al. Births: Final Data for 2004. NVSR Vol 55, Number 1. Hyattsville, MD: National Center for Health Statistics. 2006.
12. Hollis BW, Wagner CL (2004) Assessment of dietary vitamin D requirements during pregnancy and lactation. Am J Clin Nutr 79:717–726
13. Bouillon R, Van Baelen H, DeMoor D (1977) 25-Hydroxy-vitamin D and its binding protein in maternal and cord serum. J Clin Endocrinol Metab 45:679–684
14. Markestad T, Aksnes L, Ulstein M, Aarskog D (1984) 25-Hydroxyvitamin D and 1,25-dihydroxy vitamin D of D2 and D3 origin in maternal and umbilical cord serum after vitamin D2 supplementation in human pregnancy. Am J Clin Nutr 40:1057–1063
15. Hollis BW, Pittard WB (1984) Evaluation of the total fetomaternal vitamin D relationships at term: evidence for racial differences. J Clin Endocrinol Metab 59:652–657
16. Waiters B, Godel JC, Basu TK (1998) Perinatal vitamin D and calcium status of northern canadian mothers and their newborn infants. J Am Coll Nutr 18(1):122–126
17. Harris SS, Dawson-Hughes B (1998) Seasonal changes in plasma 25-hydroxyvitamin D concentrations of young American black and white women. Am J Clin Nutr 67:1232–1236
18. Looker AC, Dawson-Hughes B, Calvo MS et al (2002) Serum 25-hydroxyvitamin D status of adolescents and adults in two seasonal subpopulations from NHANES III. Bone 30:771–777
19. Harris SS, Soteriades E, Coolidge JA et al (2000) Vitamin D insufficiency and hyperparathyroidism in a low income, multiracial, elderly population. J Clin Endocrinol Metab 85:4125–4130
20. Harkness LS, Cromer BA (2005) Vitamin D deficiency in adolescent females. J Adolesc Health 37:75
21. Nesby-O'Dell S, Scanlon KS, Cogswell ME, Gillespie C, Hollis BW, Looker AC et al (2002) Hypovitaminosis D prevalence and determinants among African American and white women of reproductive age: third National Health and Nutrition Examination Survey, 1988–1994. Am J Clin Nutr 76: 187–192

22. Bodnar L, Simhan H, Powers R et al (2007) High prevalence of vitamin D insufficiency in black and white pregnant women residing in the Northern United States and their neonates. J Nutr 137: 447–452

23. Basile L, Taylor S, Wagner C, Quinones L, Hollis B(2007 Sep) Neonatal vitamin D status at birth at latitude 32°72': evidence of deficiency. J Perinatol 27(9):568–571

24. Mannion CA, Gray-Donald K et al(2006) Association of low intake of milk and vitamin D during pregnancy with decreased birth weight. CMAJ 174(9):1273–1277

25. Bibuld D, Johnson T, Young A. Is Vitamin D a Breakthrough for Addressing Health Disparities? Presented at the American Public Health Association Annual Meeting & Exposition in Washington, DC Nov 2007.

26. Brooke O, Brown I, Bone C et al(1980 Mar 15) Vitamin D supplements in pregnant Asian women: effects on calcium status and fetal growth. Br Med J 280(6216):751–754

27. Marya R, Rathee S, Lata V, Mudgil S(1981) Effects of vitamin D supplementation in Pregnancy. Gynecol Obstet Invest 12(3):155–161

28. Brooke O, Butters F, Wood C (1981 Oct 17) Intrauterine vitamin D nutrition and postnatal growth in Asian infants. Br Med J 283:1024

29. Brunvand L, Quigstad E, Urdal P, Haug E(1996 Jul 5) Vitamin D deficiency and fetal growth. Early Hum Dev 45(1–2):27–33

30. Specker B(2004) Vitamin D requirements during pregnancy. Am J Clin Nutr 80(suppl):1740S–7S

31. Marya R, Rathee S, Dua V, Sangwan K (1988) Effect of vitamin D supplementation during pregnancy on foetal growth. Indian J Med Res 88:488–492

32. National Center for Health Statistics, Health, United States, 2002 With Chartbook on Trends in the Health of Americans. (Hyattsville, MD: NCHS, 2002), accessed online at http://www.cdc.gov/nchs/products/pubs/pubd/hus/02tables.html

33. Morbidity and Mortality: 2007 Chart Book on Cardiovascular, Lung, and Blood Diseases Chart 3–14, National Heart Lung and Blood Institute

34. Wang T, Pencina M, Booth S et al (2008) Vitamin D deficiency and risk of cardiovascular disease. Circulation 117:1–9

35. Giovannucci E, Liu Y, Hollis B, Rimm E(2008) 25-hydroxyvitamin D and risk of myocardial infarction in men. Arch Intern Med 168(11):1174–1180

36. Zitterman A, Schleithoff S, Koerfer R (2005) Putting cardiovascular disease and vitamin D insufficiency into perspective. Br J Nutr 94:483–492

37. Targher G, Bertolini L, Paovani R et al (2006) Serum 25-hydroxyvitamin D3 concentrations and carotid artery intima-media thickness among type 2 diabetic patients. Clin Endocrinol 65:593–597

38. Melamed M, Muntner P, Michos E et al (2008) Serum 25-hydroxyvitamin D levels and the prevalence of peripheral arterial disease. Arterioscler Thromb Vasc Biol 28:1179–1185

39. Forman J, Giovannucci E, Holmes M et al (2007) Plasma 25-hydroxyvitamin D levels and risk of incident hypertension. Hypertension 49:1063–1069

40. Dobrig H, Pilz S, Schanargl H et al(2008) Independent association of low serum 25-hydroxyvitamin D and 1,25-dihydroxyvitamin D levels with all-cause and cardiovascular mortality. Arch Intern Med 168(12):1340–1349

41. Melamed M, Michos E, Post W, Astor B(2008) 25-Hydroxyvitamin D levels and the risk of mortality in the general population. Arch Intern Med 168(15):1629–1637

42. Garland C, Garland F, Gorham D et al (2006 Feb) The role of vitamin D in cancer prevention. Am J Pub Health 96(2):252–261

43. Ries LAG, Melbert D, Krapcho M et al. SEER Cancer Statistics Review, 1975–2005, National Cancer Institute. Bethesda, MD, http://seer.cancer.gov/csr/1975_2005/, based on November 2007 SEER data submission, posted to the SEER web site, 2008.

44. Percentage of Adults Who Used Cigars, Pipes, Chewing Tobacco, Snuff, or Any Form of Tobacco by Hispanic Origin, Race, and Sex-1987,1991, 1992 (combined) and 1998 and 2008 (combined). CDC Website @ http://www.cdc.gov/tobacco/data_statistics/tables/adult/table_9.htm

45. Pevalence of Current Smoking among Adults Aged 18 Years and Over: United States, 1997-September 2007. CDC @ http://www.cdc.gov/tobacco/data_statistics/tables/adult/table_9.htm

46. Incidence and Mortality Rates by Site, Race, and Ethnicity, US 1992–1999. From Cancer Facts & Figures, 2003, American Cancer Society.

47. Giovanucci E, Liu Y, Rimm E et al (2006) Prospective study of predictors of vitamin D status and cancer incidence and mortality in men. J Nat Cancer Inst 98(7):451–459

48. Giovannucci E, Liu Y, Willett W(2006) Cancer incidence and mortality and vitamin D in black and white male health professionals. Cancer Epidemiol Biomarkers Prev 15(12):2467–2472

49. Lapp J, Tavers-Gustafson D, Davies K et al (2007) Vitamin D and calcium supplementation reduces cancer risk: results of a randomized trial. Am Jell Clin Nutr 85:1586–1591

50. Zhou W, Heist R, Liu G et al (2007) Circulating 25-hydroxyvitamin D levels predict survival in early-stage non-small-cell lung cancer patients. J Clin Oncol 25:479–485

51. Garland C, Comstock G, Garland F et al (1989) Serum 25-hydroxyvitamin D and colon cancer: eight-year prospective study. Lancet 2:1176–1178

52. Gorham E, Garland C, Garland F et al(2007) Optimal vitamin D status for colorectal cancer prevention. A quantitative meta analysis. Am Jnl Prev Med 32(3):210–216

53. Sieg J, Sieg A, Dreyhaupt J et al (2007) Insufficient vitamin D supply as a possible co-factor in colorectal carcinogenesis. Anticancer Res 26:2729–2733

54. Feskanich D, Ma J, Fuchs C et al(2004) Plasma vitamin D metabolites and risk of colorectal cancer in women. Cancer Epidemiol Biomarkers Prev 13(9):1502–1508

55. Knight J, Lesosky M, Barnett H et al(2007) Vitamin D and reduced risk of breast cancer: A population-based case-control study. Cancer Epidemiol Biomarkers Prev 16(3):422–429

56. Neuhouser M, Sorensen B, Hollis B et al(2008) Vitamin D insufficiency in a multiethnic cohort of breast cancer survivors. Am Jnl Clin Nutr 88(1):133–139

57. Cancer Around the World, 2000, Death Rates per 100,000 Population for 45 Countries. From Cancer Facts & Figures, 2003, American Cancer Society.

58. Ahonen M, Tenkanen L, Teppo L et al(2000 Nov) Prostate cancer risk and prediagnostic serum 25-hydroxyvitamin D levels. Cancer Causes and Control 11(9):847–852

59. John E, Koo J, Schwartz G(2007) Sun exposure and prostate cancer risk: Evidence for a protective effect of early-life exposure. Cancer Epidemiol Biomarkers Prev 16(6):1283–1286

60. Li H, Stampfer M, Hollis B et al (2007 Mar) A prospective study of plasma vitamin D metabolites, vitamin D receptor polymorphisms, and prostate cancer. PLoS Medicine 4(3):0562–0571 @www.plosmedicine.org

61. Autier P, Gandine S (2007) Vitamin D supplementation and total mortality. Arch Int Med 17: 1730–1737

62. From the US Department of Health and Human Services, Office of Minority Health, Diabetes Data/Statistics, http://www.omhrc.gov/templates/browse.aspx?lvl=3&lvlid=62

63. Table 29, National Center for Health Statistics, Health, United States, 2007, Hyattsville, MD: 2007

64. Scragg R, Chonchol M (2007) 25-Hydroxyvitamind D, insulin resistance, and kidney function in the third national health and nutrition examination survey. Kidney Int 71:134–139

65. Chui K, Chu A, Go VLW, Saad M (2004) Hypovitaminosis D is associated with insulin resistance and β-cell dysfunction. Am J Clin Nutr 79:820–825

66. Boucher B, Mannan N, Noonan K et al (1995) Glucose intolerance and impairment of insulin secretion in relation to vitamin D deficiency in East London Asians. Diabetologia 38:1239–1245

67. Isaia G, Giorgino R, Adami S (2001 Aug) High prevalence of hypovitaminosis D in female type 2 diabetic population. Diabetes Care 24(8):1496

68. Di Cesar D, Ploutz-Snyder R, Weinstock R, Moses A(2006) Vitamin D deficiency is more common in type 2 than in type 1 diabetes. Diabetes Care 29(1):174

69. Borissova A, Tankova T, Kirilov G et al(2003) The effect of vitamin D3 on insulin secretion and peripheral insulin sensitivity in type 2 diabetic patients. Int J Clin Pract 57(4):258–261

70. Pittas A, Dawson-Hughes B, Li T et al(2006) Vitamin D and calcium intake in relation to type 2 diabetes in women. Diabetes Care 29(3):650–656

71. Schwalfenberg G (2008 Jun) Vitamin D and diabetes. Can Fam Physician 54:864–866

72. C. Brooke Steele, Lehida Meléndez-Morales, Richard Campoluci, Nickolas DeLuca, and Hazel D. Dean. Health Disparities in HIV/AIDS, Viral Hepatitis, Sexually Transmitted Diseases, and

Tuberculosis: Issues, Burden, and Response, A Retrospective Review, 2000–2004. Atlanta, GA: Department of Health and Human Services, Centers for Disease Control and Prevention, November 2007. Available at: http://www.cdc.gov/nchhstp/healthdisparities/

73. Chan TYK (2000) Vitamin D deficiency and susceptibility to tuberculosis. Calcif Tissue Int 66: 476–478

74. Liu PT, Stenger S, Li H et al (24 March 2006) Toll-like receptor triggering of a vitamin D-mediated human antimicrobial response. Science 311:1770–1773

75. Wilkinson R, Llewelyn M, Toosi Z et al(2000) Influence of vitamin D deficiency and vitamin D receptor polymorphisms on tuberculosis among Gujarati Asians in west London: a case-control study. Lancet 355(9204):618–621

76. Aloia J, Li-Ng M(2007) Correspondence: Epidemic influenza and vitamin D. Epidemiol Infect:1–4

77. Wang T, Nestal F, Bourdeau V et al (2004) 1,25-Dihydroxyvitamin D3 is a direct inducer of antimicrobial peptide gene expression. J Immunol 173:2909–2912

78. Sun L, Finnegan C, Kish-Catelone T et al (2005) Human b-defensins suppress human immunodeficiency virus infection: potential role in mucosal protection. J Virol 79:14318–14329

79. Chang TL, Francois F, Mosoian A, Klotman. ME (2003) CAF-mediated human immunodeficiency virus (HIV) type 1 transcriptional inhibition is distinct from alpha-defensin-1 HIV inhibition. J Virol 77:6777–6784

80. Chang TL, Vargas J Jr., DelPortillo A, Klotman ME (2005) Dual role of α-defensin-1 in anti-HIV-1 innate immunity. J Clin Investig 115:765–773

81. *The Health of Boston 2008*. Boston Public Health Commission Research Office. Boston, MA 2008

21 Vitamin D Deficiency in Canada

David A. Hanley

Abstract   Vitamin D deficiency is common in Canada, but the prevalence depends on how deficiency is defined. With the increasing use of serum 25-OH cholecalciferol measurements, the extent of the problem is being uncovered. If the lower limit of the optimal range for this vitamin D metabolite is regarded as the threshold for vitamin D deficiency, then the majority of Canadians are deficient for at least part of the year. Rickets is still a recognized problem for Canadian children, particularly in breast-fed infants of vitamin D-deficient mothers. Individuals at particularly high risk for vitamin D deficiency in Canada are aboriginal peoples and ambulant or institutionalized elderly individuals.

Key Words: Vitamin D deficiency; canada; serum 25-OH vitamin D

1. INTRODUCTION

Vitamin D deficiency is usually defined clinically by the diagnosis of rickets in children or osteomalacia in adults. However, vitamin D deficiency can be present for some time before the bone manifestations of the clinical deficiency state present to medical attention. Most definitions of vitamin D deficiency now utilize the measurement of serum 25-OH cholecalciferol [25(OH)D], which is the best available indicator of tissue vitamin D stores and adequacy of vitamin D metabolism (1). The various methods of measuring this metabolite are described elsewhere in this volume. Severe and symptomatic vitamin D deficiency is usually accompanied by a serum 25(OH)D below 20–25 nmol/l (8–10 ng/ml). However, absence of a serious deficiency state does not define optimal health or adequacy of vitamin D metabolism. With the recent interest in vitamin D "insufficiency" as a contributing factor to many medical disorders outside the realm of mineral metabolism (1), there has been major interest in re-defining the optimal levels of vitamin D metabolism. Depending on how the optimum 25(OH)D levels are defined, the incidence of mild to moderate vitamin D deficiency (often termed "insufficiency") is either very common or nearly universal in Canada (2).

Living above latitude 35°N, residents of Canada are not able to synthesize vitamin D from sunlight exposure of the skin for 4–5 months of the year (1). This means that without adequate dietary sources, an individual may be prone to having inadequate levels of vitamin D at some point during a year. In one small population-

From: *Nutrition and Health: Vitamin D*
Edited by: M.F. Holick, DOI 10.1007/978-1-60327-303-9_21,
© Springer Science+Business Media, LLC 2010

based study of western Canadians, vitamin D insufficiency was found to be very common *(2)*. It is also not surprising to find that rickets is still seen in Canada *(3)*.

This chapter will examine the small number of studies that have looked at the adequacy of vitamin D metabolism in Canadians. It will not address vitamin D deficiency accompanying medical conditions which cause vitamin D deficiency, such as malabsorption syndromes.

2. DIETARY STUDIES OF VITAMIN D IN CANADIANS

Official Canadian recommendations for total dietary and supplement intake of vitamin D are the same as those for the USA, as the expert panel convened by the Institute of Medicine (IOM) was designed to set standards for both countries *(4)*. The Adequate Intakes (AIs) for vitamin D are given in Table 1. As in the USA, the typical Canadian diet does not provide large quantities of vitamin D. Food fortification in Canada with vitamin D is restricted, mainly to milk *(5, 6)*, and the actual amount of vitamin D found in dairy products is variable *(7)*. In Canada, there are no published national survey data for estimation of total vitamin D intake (food plus supplements), but it is reasonable to assume that the average Canadian's dietary sources and use of vitamin D supplements are not very different from the USA. In the US Third National Health and Nutrition Examination Survey, 1988–1994, intakes of vitamin D from food alone and from food and supplements combined were estimated. Median intakes were generally below the AI for female subjects over 12 years of age and men over 50 years of age *(6)*. Recently (2009) the 2004 Canadian Community Health survey of over 30,000 Canadians was released by Health Canada, and includes data from 24-hour food frequency questionnaires. The vitamin D data are reported in volume 2, and median intake from food was approximately 5.3 and 4.4 µg/day for adult men and women, respectively (approximately 200 I.U.) *(8)*.

In the past 4 years, there has been pressure to move vitamin D intake recommendations upward from the IOM standards. This is primarily because measurements of serum

Table 1

Canada's Recommended Adequate Intakes (AIs) for Vitamin D *(4)*

Age	Children	Men	Women	Pregnancy	Lactation
Birth to 13 years	5 µg (200 IU)				
14–18 years		5 µg (200 IU)	5 µg (200 IU)	5 µg (200 IU)	5 µg (200 IU)
19–50 years		5 µg (200 IU)	5 µg (200 IU)	5 µg (200 IU)	5 µg (200 IU)
51–70 years		10 µg (400 IU)	10 µg (400 IU)		
71+ years		15 µg (600 IU)	15 µg (600 IU)		

25-OH cholecalciferol [25(OH)D] have become more available to researchers and clinicians, and recent studies, described elsewhere in this volume, have suggested that the optimal level of 25(OH)D is higher than the early estimates of the normal range. In most Canadian clinical laboratories, the lower limit of the optimal range for 25(OH)D is now regarded as 75–80 nmol/l. Current Dietary Reference Intakes (DRIs) or AIs are not likely to consistently achieve or maintain these levels, particularly with reliance upon diet and sunlight exposure for the only sources of vitamin D in a northern country like Canada, and as a result, many Canadian physicians are now advising much larger doses of vitamin D for their patients.

Health Canada and the Canadian Paediatric Society have recently recommended that, for the prevention of rickets, Canadian infants and children receive 10 μg (400 IU) of vitamin D per day, either through diet or through supplementation (9, 10). The Canadian Paediatric Society has now recommended that because of high rates of vitamin D deficiency in aboriginal peoples (First Nations and Inuit), particular attention to vitamin D supplementation be taken; and their general recommendations for children are as follows: premature infants and infants during the first year of life should receive 400 IU/day, and if living above the 55th parallel of latitude, children should be given 800 IU between October and April (11). They further recommended that pregnant and lactating women receive 2,000 IU daily. In recognition of the fact that food sources of vitamin D may not meet the AI, and that skin synthesis does not occur in winter in this country, the Canada Food Guide was recently revised to include a recommendation that all adult Canadians over the age of 50 years consume a vitamin D supplement of 400 IU daily (12).

Given the limitations of dietary estimates of vitamin D intake, and the availability of 25(OH)D assays to better assess adequacy of vitamin D nutrition or metabolism, only a few representative studies are noted in this chapter. A recent study of Quebec pre-school children indicated a level of vitamin D intake in winter of approximately 11 μg/day, by 3-day dietary recall. This would suggest that most children in the survey achieved an intake capable of preventing rickets (13). Canada's northern aboriginal population might be expected to be at even higher risk than other population groups in the country, because of their location at higher latitudes, and their shift from a traditional diet rich in vitamin D to a more typical diet of people living in the large North American urban centers. A recent survey of three groups of Canadian Arctic indigenous peoples found higher vitamin D intake in the Inuit, who were more likely to be consumers of hunted meats containing large amounts of vitamin D than the other two groups (14).

3. STUDIES OF CANADIANS UTILIZING MEASUREMENTS OF SERUM 25(OH)D

3.1. Healthy Ambulatory Adults

A number of studies have addressed vitamin D metabolism in Canadians using measurement of 25(OH)D in a variety of population groups. As noted above, unequivocal, severe vitamin D deficiency is usually associated with serum 25(OH)D levels <20–25 nmol/l, but the lower limit of the optimal level of 25(OH)D has been set in various laboratories at 40, 50, or 75–80 nmol/l (8, 10, or 30–32 ng/ml). Prior to 2007, most clinical laboratories in Canada used 40–50 nmol/l as the lower limit of the normal range, using the term vitamin D "insufficiency" or "inadequacy" to describe levels between frank deficiency (20–25 nmol/l) and 40–50 nmol/l. However, most Canadian clinical laboratories now list a serum 25(OH)D level of 75 or 80 nmol/l as representing the lower limit of optimal vitamin D metabolism/nutrition/body stores, with 25–75 or 80 nmol/l termed "moderate to mild deficiency," and <25 nmol/l representing "severe deficiency" *(15)*.

There has not been a systematic cross-Canada population-based assessment of vitamin D adequacy based upon serum 25(OH)D measurements. There is only one population-based study of Canadians which might be in a position to do this – the Canadian Multicentre Osteoporosis Study (CaMOS), a study of men and women over the age of 25 years. CaMOS subjects were recruited by random selection from telephone directories and included subjects living within 50 km of the center of nine urban centers across the country *(16)*. Unfortunately, CaMOS has not yet had the funding to collect blood for vitamin D testing in the whole cohort. However, a small but representative sample of the CaMOS cohort in Calgary, Alberta (latitude 51°N), was tested every 3 months for 1 year (1999) with serum 25(OH)D measurements and other parameters of calcium metabolism *(2)*. The subjects participating in the Calgary CaMOS vitamin D study (60 men, 128 women, ages 27–89 years) were selected randomly from the larger cohort, and, by all available criteria for comparison, they were considered a representative sample of the larger cohort. Three subjects found to have 25(OH)D levels in the severe deficiency range (<25 nmol/l) were excluded from the study. Participants were not taking more than 200 IU of vitamin D supplements at entry and were instructed not to increase vitamin D intake during the year of study. The expected seasonal changes in serum 25(OH)D were observed: 52.9 nmol/l ± 17.2 (standard deviation) in late fall/early winter and 71.6 nmol/l ± 23.6 in summer. The authors applied three definitions of vitamin D insufficiency to the results (Fig. 1): if vitamin D insufficiency was defined as a

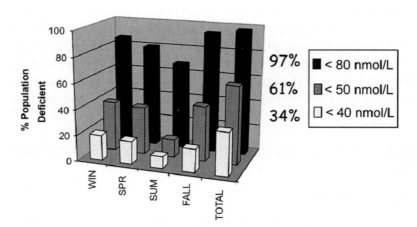

Fig. 1. Percentage of population from the Canadian Multicentre Osteoporosis Study Calgary cohort with vitamin D deficiency using three commonly proposed definitions of the lower limit of normal range of serum 25-OH cholecalciferol. Adapted from Rucker et al. *(2)* with permission from CMAJ.

serum 25(OH)D level below 40 nmol/l (16 ng/ml), approximately 1/3 of the population fell below that level, and if the definition of adequate vitamin D was set at 80 nmol/l (32 ng/ml), then, by inference, nearly 100% of the population of Calgary had a sub-optimal level of vitamin D in at least one 3-month period during the year. Although farther north than eastern Canadian major population centers, Calgary receives more days of sunshine than any other major Canadian city, and is at a much higher altitude (1,048 m above sea level), so this is a population that probably receives more ultraviolet light exposure than most Canadians (2).

Vieth et al. found that in a large sample of young women (aged 18–35 years) in Toronto (43°N latitude), there was a high prevalence of vitamin D "insufficiency" which was then defined as a serum 25(OH)D level below 40 nmol/l: 25.6% of 82 non-white, non-black women had vitamin D insufficiency vs. 14.8% of 702 white women tested ($p < 0.05$). Further, intakes of 10 μg (400 IU) vitamin D did not alter the vitamin D status, suggesting that to prevent vitamin D insufficiency, a higher dose of vitamin D is required (17).

The ambulant or institutionalized elderly are considered to be a population at very high risk for vitamin D deficiency. In Montreal, Quebec (45°N latitude), a study of blood samples, taken between 1994 and 1999, from 256 elderly (aged 65–94 years) apparently healthy, community-dwelling men and women, showed a very surprising 32% of women and 51% of men with 25(OH)D levels below 20 nmol/l (18). This is a level of deficiency that is likely to have an adverse effect on muscle function, bone, and fracture risk (1).

3.2. Children

Vitamin D deficiency in childhood is a concern because of effects on growth and bone health, as well as other medical problems as discussed elsewhere in this volume. A small study of 90 children and adolescents aged 2–16 years presenting to a pediatric emergency department in Edmonton, Alberta (53°N latitude), assessed both dietary intake and serum 25(OH)D levels at the end of winter. The mean 25(OH)D in the 68 children who provided blood samples was 47.2 nmol/l (95% CI 43.8–50.8 nmol/l), with 34% of participants <40 nmol/l and 6% in the severe deficiency range (<25 nmol/l). Boys and girls aged 9–16 years had a prevalence of "insufficiency" (defined as <40 nmol/l) of 69 and 35%, respectively, while boys and girls 2–8 years old had a prevalence of insufficiency of 22 and 8%, respectively. The authors found that dietary vitamin D intake (expressed as per kilogram body weight) was the most important independent determinant of 25(OH)D levels, and if they had an intake >0.45 μg/kg/day, no one was in the "insufficiency" range. The study was conducted in early spring, and the authors concluded this is a time of year when vitamin D deficiency (insufficiency) may be common among Canadian children and adolescents at the beginning of spring (19).

In contrast to the study of Quebec preschool children described above, a study of older children and adolescents in Quebec (45–48°N latitude) found a high incidence (93%) of vitamin D insufficiency, using 75 nmol/l as the lower limit of the optimal range (20). This is consistent with the Calgary data for adults (2). Even more disturbing in the Quebec study was the prevalence of 25(OH)D in the range classified as frank deficiency (the authors defined this as ≤27.5 nmol/l), and the incidence increased with age in both sexes ($p < 0.0001$). Vitamin D deficiency was found in 2, 3, and 13% in 9-, 13-, and 16-year-old boys and 2, 8, and 10% in 9-, 13-, and 16-year-old girls, respectively (20).

Rickets is still being diagnosed too frequently in Canada. From July 1, 2002, to June 30, 2004, 2325 Canadian pediatricians were surveyed monthly through the Canadian Paediatric Surveillance Program to determine the incidence, geographic distribution, and clinical profiles of confirmed cases of vitamin D deficiency rickets. They uncovered 104 confirmed cases, giving an incidence rate of 2.9 cases per 100,000 population of the same age group. Although incidence was highest in the northern territories (Yukon, Northwest Territories, and Nunavut), 68% of cases had lived in urban areas most of their lives. Darker or intermediate skin color was a major risk factor, as this classification fit 92% of the cases. Breast-feeding was a feature of 94% of cases, and in none of the cases had the children taken vitamin D supplements in the recommended range of 10 μg/day (400 IU). Maternal vitamin D deficiency was also a major risk factor for rickets in infants. The authors concluded that current intake recommendations for infants are capable of preventing rickets, but probably not enough for infants born to mothers with vitamin D deficiency (3). The findings of this study would tend to support the Canadian Paediatric Society's recommendations (referred to above) for children and pregnant and lactating women (11).

3.3. Aboriginal (First Nations and Inuit) Peoples

Canada's aboriginal population is generally believed to be at high risk for vitamin D deficiency. Between 1972 and 1984, 48 cases of rickets were documented at the Winnipeg Children's Hospital, and 40 of these cases were in aboriginal children (21). In the recent study by Ward et al. (3), a high proportion of the children identified with rickets were also aboriginal.

Vitamin D deficiency might be expected to be an even greater problem for those living in the far north. Von Westarp et al. conducted early study of serum 25(OH)D levels in Canada's northern aboriginal (Inuit) population residing in Nanisivik (63°N latitude) and Arctic Bay (73°N latitude). The serum samples were collected in the spring of 1980, when the subjects' own tissue stores of vitamin D would be expected to be at their lowest, along with samples from a small number of Caucasians who were also residing there (22). The Inuit subjects were consuming a diet rich in hunted meats, while the Caucasians had a standard southern Canadian diet with no hunted foods. The Caucasians' mean 25(OH)D level was 12.1 ng/ml (30 nmol/l), with half of the group falling in the severe deficiency range, while the Inuit adults living in Arctic Bay had a mean 25(OH)D of 25.9 ng/ml (65 nmol/l). The investigators also found that children taking a supplement of 400 IU vitamin D did not have higher 25(OH)D levels than children who were not taking supplements, suggesting that current recommendations were not effective in preventing vitamin D deficiency in a population receiving little vitamin D from skin synthesis.

More recently, Waiters et al. studied dietary vitamin D intake and 25(OH)D levels in mothers and newborns in a small number of Caucasians, Inuit, and Native Indians living in the Inuvik region of the Northwest Territories (68°N latitude) (23). Dietary intake of vitamin D correlated with levels of 25(OH)D, which were higher in the Caucasians.

Among the Inuit and Native patients, mean 25(OH)D levels were 50.1 ± 19.3 and 34.2 ± 13.1 nmol/l in mothers and infants, respectively. Although some vitamin D levels fell into the severe deficiency range, the authors state that none of the infants showed clinical signs of D deficiency.

Moving farther south, a study in a Manitoba First Nations community showed a very high prevalence of vitamin D deficiency and insufficiency in mothers and infants *(24)*. Diet assessment interviews were conducted with 80 mothers, each with a child at least 2 years old, and 25(OH)D levels measured. Ninety-one percent of mothers had initially breast-fed their children. The mean 25(OH)D level was 26.2 nmol/l for the children and 19.8 nmol/l for the mothers, reflecting a high incidence of vitamin D deficiency.

4. CONCLUSIONS

Vitamin D deficiency is common in Canada. With the increasing awareness of vitamin D's role in disorders outside the realm of calcium metabolism, this is becoming a major public health issue. Canada is a partner in the current IOM review of the DRIs for vitamin D, and guidelines for food fortification with vitamin D are under review. A number of specialty groups are now recommending higher intakes than the Institute of Medicine AIs recommended in 1997; e.g., Osteoporosis Canada's 2002 Clinical Practice Guidelines recommend a minimum intake of 20 μg (800 IU) vitamin D daily for individuals over age 50 years *(25)*.

Vitamin D deficiency is a particularly common problem for Canada's aboriginal peoples and the elderly. Rickets remains a problem and is more common in infants who are breast-fed by vitamin D-deficient mothers. Another high-risk group would be recent immigrants from countries with higher sunlight exposure, particularly if the individual has a darker skin color *(26)*.

Canada needs to revise its recommendations for vitamin D intake. Current recommendations, if followed, will prevent severe vitamin D deficiency, but may not achieve optimal vitamin D levels. With the current widespread avoidance of skin exposure to ultraviolet B solar or artificial irradiation, the only source of vitamin D that will allow Canadians to achieve optimal vitamin D metabolism has to be an increase in supplementation of diet and/or food fortification. For policy makers at Health Canada, this raises the risk of vitamin D toxicity. However, recent evidence would suggest that current estimates of the Tolerable Upper Level for daily vitamin D intake (50 μg [2000 IU]) are inappropriately low *(27)*, and higher intakes can be safely recommended.

REFERENCES

1. Holick MF (2007) Vitamin D deficiency. N Engl J Med 357:266–281
2. Rucker D, Allan JA, Fick GH, Hanley DA (2002) Vitamin D insufficiency in a population of healthy western Canadians. CMAJ 166:1517–1524
3. Ward LM, Gaboury I, Ladhani M, Zlotkin S (2007) Vitamin D-deficiency rickets among children in Canada. CMAJ 177:161–166
4. IOM. (Institute of Medicine) (1997) Dietary reference intakes for calcium, phosphorus, magnesium, vitamin D and fluoride. National Academy Press, Washington, DC

5. Institute of Medicine Food and Nutrition Board; Standing Committee on Use of Dietary Reference Intakes in Nutrition Labeling. (2003) Overview of food fortification in the United States and Canada. In: Dietary reference intakes: guiding principles for nutrition labeling and fortification. National Academies Press, Washington, DC

6. Calvo MS, Whiting SJ, Barton CN (2004) Vitamin D fortification in the United States and Canada: current status and data needs. Am J Clin Nutr 80:1710S–1716S

7. Faulkner H, Hussein A, Foran M, Szijarto L (2000) A survey of vitamin A and D contents of fortified fluid milk in Ontario. J Dairy Sci 83:1210–1216

8. Health Canada Canadian community health survey, cycle 2.2, Nutrition, 2004 - nutrient intakes from food, provincial, regional and national summary data tables. Volumes 1, 2 & 3. [computer file]. Ottawa, Ont.: Health Canada [producer and distributor], 2009. On-line access: http://www.hc-sc.gc.ca/fn-an/surveill/nutrition/commun/cchs_focus-volet_escc-eng.php#p1

9. Health Canada. (2004) Vitamin D supplementation for breastfed infants: 2004 Health Canada recommendation. Ottawa: Health Canada; Available: www.hcsc.gc.ca/fn-an/alt_formats/hpfb-dgpsa/pdf/nutrition/vita_d_supp_e.pdf (accessed 2007 May 25).

10. Statement of the Joint Working Group; Canadian Paediatric Society, Dieticians of Canada, Health Canada. (1998) Nutrition for healthy term infants. Ottawa: Minister of Public Works and Government Services Canada. Available: www.hc-sc.gc.ca/fn-an/alt_formats/hpfb-dgpsa/pdf/nutrition/infant-nourrisson_e.pdf (accessed 2007 May 25).

11. 10.First Nations, Inuit and Métis Health Committee, Canadian Paediatric Society (CPS). (2007) Vitamin D supplementation: recommendations for Canadian mothers and infants. Pediatr Child Health 12(7): 583–589

12. Anonymous (2007) Canada's Food Guide. Vitamin D for people over 50. http://www.hc-sc.gc.ca/fn-an/food-guide-aliment/context/evid-fond/vita_d-eng.php

13. Gagne D, Rhainds M, Galibois I (2004) Seasonal vitamin D intake in Quebec preschoolers. Can J Diet Pract Res 65:174–179

14. Kuhnlein HV, Receveur O, Soueida R, Berti PR (2008) Unique patterns of dietary adequacy in three cultures of Canadian Arctic indigenous peoples. Public Health Nutr 11:349–360

15. Hanley DA (2008) Informal survey of 25(OH)D reporting by clinical laboratories in British Columbia, Alberta, Manitoba, Ontario, Quebec, Nova Scotia and Newfoundland. Unpublished observations

16. Kreiger N, Tenenhouse A, Joseph L, Mackenzie T, Poliquin S, Brown JP, Prior JC, Rittmaster RS (1999) The Canadian multicentre osteoporosis study (CaMos): background, rationale, methods. Can J Aging 18:376–387

17. Vieth R, Cole DE, Hawker GA, Trang HM, Rubin LA (2001) Wintertime vitamin D insufficiency is common in young Canadian women, and their vitamin D intake does not prevent it.

18. Vecino-Vecino C, Gratton M, Kremer R, Rodriguez-Manas L, Duque G (2006) Seasonal variance in serum levels of vitamin D determines a compensatory response by parathyroid hormone: study in an ambulatory elderly population in Quebec. Gerontology 52:33–39

19. Roth DE, Martz P, Yeo R, Prosser C, Bell M, Jones AB (2005) Are national vitamin D guidelines sufficient to maintain adequate blood levels in children?. Can J Public Health 96:443–449

20. Mark S, Gray-Donald K, Delvin EE et al (2008) Low vitamin D status in a representative sample of youth from Quebec, Canada. Clin Chem 54:1283–1289

21. Haworth JC, Dilling LA (1986) Vitamin-D-deficient rickets in Manitoba, 1972-84. CMAJ 134:237–241

22. Von Westarp C, Outhet D, Eaton RDP (1981) Prevalence of vitamin D deficiency in two Canadian Arctic communities. In: Harvald B, Hart Hansen JP (eds) Circumpolar health 81. Proceedings of 5th international symposium on circumpolar health. Copenhagen, 9–13 August 1981. Stougard Jensen, Copenhagen, pp. 331–333

23. Waiters B, Godel JC, Basu TK (1999) Perinatal vitamin D and calcium status of northern Canadian mothers and their newborn infants. J Am Coll Nutr 18:122–126

24. Lebrun JB, Moffatt ME, Mundy RJ, Sangster RK, Postl BD, Dooley JP, Dilling LA, Godel JC, Haworth JC (1993) Vitamin D deficiency in a Manitoba community. Can J Public Health 84:394–396

25. Brown JP, Josse RG (2002) 2002 clinical practice guidelines for the diagnosis and management of osteoporosis in Canada. CMAJ 167:S1–34
26. Gibson RS, Bindra GS, Nizan P, Draper HH (1987) The vitamin D status of East Indian Punjabi immigrants to Canada. Br J Nutr 58:23–29
27. Hathcock JN, Shao A, Vieth R, Heaney R (2007) Risk assessment for vitamin D. Am J Clin Nutr 85:6–18

22 Vitamin D Deficiency and Its Health Consequences in Northern Europe

Leif Mosekilde

Abstract   Deficiency and insufficiency is common in the Northern, Western, and Central part of Europe. Around 43–92% of adults and 89–97% of teenage girls have plasma 25-hydroxyvitamin D [25(OH)D] levels < 50 nmol/l. The vitamin D deficiency is caused by low sun exposure especially during winter time and because of the high latitudes in the Northern regions. The lack of solar exposure is combined with a low dietary vitamin D intake especially in the Central and Western part of the continent. The high dietary intake of fatty fish, cod liver, and cod liver oil in Scandinavia and Iceland and the frequent use of supplements mitigate to some extent the effect of the reduced sun exposure. Across Europe, epidemiologic studies indicate a positive effect of latitude on average plasma levels of 25(OH)D. However, at the level of the individual country (e.g., France and the United Kingdom) with a more common fortification policy and supplementation tradition, latitude is inversely related to plasma 25(OH)D. At the individual level, European studies have confirmed that plasma 25(OH)D depends positively on dietary vitamin D, vitamin D supplementation, sun exposure, living in partnership, recent vacations to sunny regions, and sunbed use and inversely on use of sunscreen and covering clothes, smoking habits, and body mass index. Risk groups include breast-fed children, pregnant and lactating women, and older persons. Veiled and pigmented immigrant women with covering clothes and vegetarians constitute a special problem. Vitamin D status can be improved by encouraging safe sun exposure, improving dietary intake, e.g., by fish and cod liver oil, obligatory food fortification, and vitamin D supplements. Nutritional recommendations from governments and other regulatory institutions should be adapted to the present knowledge of the actual need of vitamin D.

Key Words: Vitamin D; 25-hydroxyvitamin D; fortification; dietary recommendation; Northern Europe

1. INTRODUCTION

Vitamin D is produced in the skin from 7-dehydrocholesterol in adequate quantities if sun exposure (ultraviolet B (UVB)) is sufficient and an appropriate part of the skin is exposed *(1, 2)*. The dermal production depends on skin pigmentation, both natural and induced by sun exposure, exposed area, exposure time, latitude and season, altitude, weather (cloudy or not), and pollution *(3)*. Vitamin D is also absorbed from the intestine depending on absorptive capacity, dietary fare, fortification, and supplementation

From: *Nutrition and Health: Vitamin D*
Edited by: M.F. Holick, DOI 10.1007/978-1-60327-303-9_22,
© Springer Science+Business Media, LLC 2010

(3, 4). Adequacy of vitamin D nutritional status is measured by the circulating levels of 25-hydroxyvitamin D [25(OH)D], which reflect the combined product of dermal synthesis from sun exposure and dietary sources *(4, 5)*. On average 80–90% of vitamin D is derived from dermal production with large variations between individuals and populations *(3–9)*.

At a global level mean plasma 25(OH)D in normal individuals is 54 ± 1.4 (SE) nmol/l, with borderline higher levels in females (56 ± 1.6 nmol/l) than in males (50 ± 2.6 nmol/l, $p = 0.05$) probably because of less pigmented skin *(9)*. Subjects <15 years of age have lower 25(OH)D levels (37 ± 5.7 nmol/l) than subjects between 15 and 65 years (57 ± 1.8 nmol/l, $p < 0.01$) and subjects >75 years have lower levels (47 ± 3.6 nmol/l) than subjects aged 66–75 years (57 ± 2.7 nmol/l, $p = 0.04$). Caucasians had on average 21.2 ± 5.1 nmol/l higher serum 25(OH)D levels than non-Caucasians (68 ± 3.2 versus 47 ± 4.0, $p < 0.01$) *(9)*.

2. DEFINITION OF VITAMIN D INSUFFICIENCY AND DEFICIENCY

In Europe vitamin D sufficiency is usually defined by plasma 25(OH)D levels >50 nmol/l *(10)* although other definitions are also used *(11)*. Based on the development of secondary hyperparathyroidism, increased bone turnover, and loss of bone mineral density, levels between 50 and 25 nmol/l are classified as vitamin D insufficiency, levels between 12 and 25 nmol/l as vitamin D deficiency, and levels below 12 nmol/l as severe vitamin D deficiency *(10)*. In the United States the cutoff level of 25(OH)D for sufficiency in relation to more general health outcomes is usually >75–80 nmol/l *(12–14)*. In the following different levels of 25(OH)D are used to define vitamin D sufficiency, insufficiency, deficiency, and severe deficiency (Tables 3 and 4).

3. GEOGRAPHY AND SOLAR UV RADIATION

United Nations describes Northern Europe as Scandinavia (Denmark, Norway, and Sweden), Finland, the Baltic countries (Estonia, Latvia, and Lithuania), the United Kingdom, Ireland, and Iceland *(15)*. In the present context, we will include part of Western and Central Europe (France, Germany, Holland, and Belgium) and part of Eastern Europe. The latitude of this geographic region varies from 81°N (Svalbard) to around 46°N (the Alps). Figure 1 illustrates the typical UV exposure pattern of the region in January and July *(16)*. In winter there is no UV irradiation north of Oslo and only between 0 and 60 mW/m² in the rest of the region compared with around 360 mW/m² in Australia. In summer UV irradiation is 60–120 mW/m² in the northern part and 120–160 mW/m² in the southern part of the region compared with 180–240 mW/m² in the Mediterranean region and 300–360 mW/m² in Sahara. It has been shown that at 52°N (Edmonton, Canada) there is no vitamin D production from October to March *(17)*. This period is longer with higher latitude. Therefore, the seasonal variation in UV exposure results in similar seasonal variations in average plasma 25(OH)D levels with a delay of about 6 weeks (Fig. 2), *(18, 19)*. However, even in Northern Norway (68–69°N), a positive relationship exists between hours of solar UV exposure per day

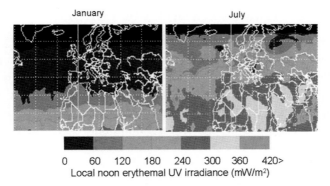

Fig. 1. Satellite-based global UV exposure measurements in January and July. Total Ozone Mapping Spectrometer. Erythemal UV exposure. http://toms.gsfc.nasa.gov/ery_uv/euv_v8.html (Accessed October 10, 2008.)

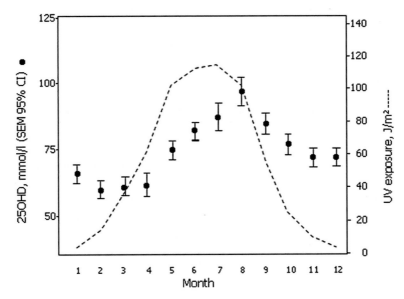

Fig. 2. Seasonal variation in UV exposure in Denmark related to variations in average plasma 25(OH)D levels measured by liquid chromatography-mass spectrometry. There is a delay of around 6 weeks from maximal exposure and zenith for plasma 25(OH)D.

and plasma 25(OH)D at UV-hours >3.5 per day and for those with low 25(OH)D already at UV-hours >1.5 per day *(20)*. Subjects with a low plasma 25(OH)D (< 30 nmol/l) seem to respond to UV radiation as early as in the beginning of March *(21)*. In Northern Europe also the cloudiness of the regions varies considerably with a tendency toward more clouds to the north and the west *(22)*.

4. OVERVIEW OF VITAMIN D STATUS IN EUROPE

In a review from 1992 McKenna *(23)* summarized the present knowledge from 1971–1990 on worldwide vitamin D status. He concluded that oral vitamin D intake was lower in Western and Central Europe (2–3 μg/day) than in both North America (5.5–7 μg/day) and Scandinavia (4–6 μg/day) due to a higher ingestion of vitamin D supplementation in Scandinavia and a higher intake of natural sources of vitamin D, such as fatty fish. Plasma 25(OH)D varied with season in both young adults and elderly and was lower during the winter and throughout the year in Central Europe (∼ 18 nmol/l) than in both North America (∼ 58 nmol/l) and Scandinavia (∼ 37 nmol/l). He also concluded that hypovitaminosis D and related skeletal abnormalities were most common in elderly residents in Europe, but were reported in all elderly populations.

In a European context van der Wielen et al. in 1995 *(24)* measured wintertime plasma 25(OH)D in 824 elderly people from 11 European countries. He concluded that free-living elderly Europeans, regardless of geographical location, were at substantial risk of inadequate vitamin D status during winter. Surprisingly the lowest mean 25(OH)D concentrations were seen in Southern Europe. This could largely be explained by attitudes toward sunlight exposure when users of vitamin D supplements and sunbeds were excluded. In 2005 Andersen et al. *(25)* stated that vitamin D status also was low in Northern Europe (Denmark, Finland, Ireland, and Poland) during winter. More than one-third of teenage girls had plasma 25(OH)D < 25 nmol/l and almost all were below 50 nmol/l. Two-thirds of the elderly community-dwelling women had vitamin D levels < 50 nmol/l (Table 2).

At the far North (65–72°N) dietary vitamin D intake is high due to a high intake of fatty fish, cod liver, and cod liver oil, and the prevalence of 25(OH)D levels < 50 nmol/l is relatively low (Table 2). At this latitude plasma 25(OH)D levels mainly depend not only on vitamin D intake but also, in spite of the restricted sunlight, on the estimated hours per day of exposure to solar UVB *(26, 27)*.

In an investigation of osteoporosis and osteopenia, Bruyère et al. in 2007 *(28)* measured plasma 25(OH)D in 8,532 European postmenopausal women with osteoporosis or osteopenia from France, Belgium, Denmark, Italy, Poland, Hungary, the United Kingdom, Spain, and Germany with a mean age of 74.2 (7.1) years. Mean plasma 25(OH)D was 61.0 (27.2) (SD) nmol/l. The lowest levels were found in France (51.5 (26.1) nmol/l) and the highest in Spain (85.2 (33.3) nmol/l). A total of 32% had levels < 50 nmol/l and 80% had levels < 80 nmol/l.

5. WHAT IS THE EFFECT OF LATITUDE ON VITAMIN D STATUS IN EUROPE?

UV exposure decreases markedly with increasing latitude (Fig. 1). However, a previous study in 824 elderly people in 11 European countries showed a positive relation between wintertime vitamin D status and latitude with lowest levels of 25(OH)D in the southern part of Europe *(24)*. A more recent cross-sectional ecological study *(29)* showed a highly significant positive correlation between latitude and plasma 25(OH)D in Europe ($r^2 = 0.42, p < 0.01$). The first study was a nutrition study (Euronut-SENECA) *(30)* and the second was based on baseline data from an osteoporosis intervention study *(29, 30)*. Participants in nutrition and osteoporosis studies may differ from the general population by being more health oriented with respect to outdoor life, sun exposure, diet, and vitamin D supplementation. This could have biased the outcomes of the studies. The finding that wintertime plasma 25(OH)D increased with latitude among the elderly in Europe also points toward differences between countries in dietary habits, fortification strategies, and supplementation. This interpretation is supported by a review that report higher plasma 25(OH)D levels among Scandinavians than in other parts of Europe especially during winter, but also a higher vitamin D intake from fish and fortified food, and a higher proportion taking supplements *(4)*. A recent meta-regression analysis of determinants of worldwide plasma 25(OH)D disclosed a significant overall decline in plasma 25(OH)D with latitude for Caucasians (-0.69 ± 0.30 nmol/l per degree, $p = 0.02$) *(9)*. However, subgroup analyses by continent showed a tendency toward a positive relation between latitude and plasma 25(OH)D in Europe (0.28 ± 0.16 nmol/l per degree, $p = 0.08$), in contrast to inverse relations in Asia, North America, and South America.

Opposite to these studies, an epidemiologic investigation of vitamin D status in the general urban population from 20 French cities between 43°N and 51°N reported a significant inverse effect of latitude ($r = -0.79, p = 0.01$) and a positive effect of average hours of sunshine per day ($r = 0.72, p = 0.03$) *(31)*. These findings suggest that within a more homogenous population with common fortification policies and supplementation traditions, the effects of latitude and thereby sun exposure on vitamin D status are more evident. This interpretation is supported by a recent study from Great Britain *(32)*, which showed that the multiple adjusted risk of having vitamin D insufficiency (25(OH)D < 40 nmol/l) was 1.13 (95% CI 1.0–1.3) in Midland and Wales compared with South England (reference), with a risk of 1.04 (0.9–1.3) in North England and 2.13 (1.7–2.6) in Scotland (p for trend < 0.0001).

6. NATIONAL NUTRITIONAL RECOMMENDATIONS AND DIET

Individual attitude to sun exposure, diet, and supplementation may depend on the national nutritional recommendations. Several countries, regions, and institutional bodies within Northern Europe have published recommendations for the dietary intake of vitamin D (Table 1) *(33)*. Most current recommendations specify different needs for subgroups of Caucasian whites, including special age groups (e.g., early childhood and elderly) and pregnant and lactating women, and they often supply some advice for high-risk groups. The dietary recommendations are typically defined in terms of target intakes that are given different names. If these target intakes are applied at an individual level, they are supposed to meet the needs of the majority of healthy individuals, including those with the highest requirements *(34)*. Most countries have no official specific recommendations for minorities, e.g., immigrants to Europe from Asia and Africa, and Inuit's, Laplanders, and other minorities in Northern Europe who are adapting to a westernized

Table 1
National and EU Dietary Vitamin D Recommendations for Northern, Western, and Central
Europe (33)

Recommendations	Year	Age group	μg/day
Nordic Nutrition Recommendations.	2004	2 weeks–2 years	10
Recommended intake (RI)[18]		2–60 years	7.5
		≥61 years	10 (+10)[a]
France. Apports Nutritionnels Conseillés (ANC)[16]	2001	<50 years	5
		>50 years	5/10–15[b]
Germany, Switzerland, Austria Referenzwerte[19]	2000	0–1 year	10
		1–64 years	5
		≥ 65 years	10
UK. Dietary reference values (DRV)	1998	0–6 months	8.5
Recommended nutrient intake (RNI)[20]		6 months–3 years	7
		4–64 years	0 (+10)[c]
		> 65 years	10
European community, population reference	1993	< 50 years	0–10[d]
Intake (PRI)[16]		> 50 years	0–10/10[d,e]

[a] Elderly people with little or no sun exposure should receive a supplement of 10 μg/day.
[b] Higher range in the elderly.
[c] 10 μg to risk groups.
[d] Higher end of interval for people with minimal dermal synthesis.
[e] Highest intake for age 65 years.

diet. In general, the recommended vitamin D intakes do not match the optimal 25(OH)D
levels described above. Furthermore, in most countries, population-based dietary vitamin D intakes are far lower than the recommended target levels (Table 2) (4).

7. FORTIFICATION POLICY IN NORTHERN EUROPE

Fortification may be mandatory, optional, or absent. In one review vitamin D intake
estimates from countries with the highest level of food fortification practices show
2–3 μg higher vitamin D intakes per day than those with optional or absent practices (4). Most European countries allow fortification of margarine with vitamin D
to the following amounts (data from 1995–1996 and 2003) (35, 36): Belgium (6.25–
7.5 μg/100 g), England (7–8.8 μg/100 g (obligatory)), Finland (5–10 μg/100 g (obligatory since 2003)), Holland (7.5 μg/100 g), Iceland (7.5 μg/100 g), Norway (8 μg/100 g),
Sweden (7.5–10 μg/100 g), and Germany and Austria (2.5 μg/100 g). In Denmark it is
allowed to add 7.5 μg/100 g but few products are fortified. Cooking oil can be fortified in Norway (8 μg/100 ml), Sweden (10 μg/100 ml), and Germany and Austria

Table 2
Plasma Concentrations of Plasma 25(OH)D in Population-Based Adult Individuals in Various Countries in Europe North of the Alps

Country	Season	Gender N	Age (years) SD (range)	Plasma 25(OH)D nmol/l SD (range)	<50 nmol/l (%) All	Summer	Winter	Dietary vitamin D µg/day SD (range)	Reference
Norway HPLC	All	♀♂ 300	52 ± 4	55 (8–143)	38	31	44	8.1 ± 7.0[d]	(26)
Finland HPLC	All	♂ 60 / ♂ 60	13 (0.4) / 72 (1.4)	29 (13–54) / 45 (15–90)	97 / 57			5.0 (1–12) / 5.0 (1–12)	(25)
Sweden EIA	Winter	♂ 116	61–86	69 ± 23				6.0 ± 1.8	(44)
Poland HPLC	All	♂ 61 / ♂ 65	13 (0.5) / 72 (1.4)	31 (15–62) / 33 (8–57)	87 / 92			3.1 (1–8) / 3.8 (1–22)	(25)
Scotland HPLC	All	♂ 3113	55 / 50–63	54 (10–190)	31[a] / 77[b]	25[a] / 73[b]	34[a] / 75[b]		(46)
England[e] EIA	All	♀ 3725 / ♂ 3712	45 / 45	S: 62, W: 41 / S: 59, W: 41		58[c] / 64[c]	89[c] / 85[c]		(32)
Ireland HPLC	All	♂ 95	60 / 51–69	Feb–Mar: 54 (28) / Aug–Sep: 75 (30)		17	53	3.2 (2.0) (diet) / 5.8 (5.6) (+suppl.)	(49)

Table 2
(continued)

Country	Season	Gender N	Age (years) SD (range)	Plasma 25(OH)D nmol/l SD (range)	<50 nmol/l (%)			Dietary vitamin D µg/day SD (range)	Reference
					All	Summer	Winter		
Denmark	All	♂ 19 ♂ 43	12 (0.8) 72 (1.5)	41 (19–59) 43 (17–89)	89 60			2.4 (1.2–7.5) 4.0 (1.4–13)	(25)
Holland RIA	ALL	♀ ♂ 1205	75 55–85	55	48			3.6 ± 0.25	(30, 56)
Belgium RIA	All	♂ 1196	77 (7.5) 50–100	53 (21) (13–175)	43				(54)
Germany RIA	All	♀ 1763 ♂ 2267	18–79	45 (31–69) 45 (31–72)	57 58	45 55	68 61	2.8 (1.9–4.4) 2.3 (1.5–3.6)	(41, 42)
France RIA	All	♀ 763 ♂ 804	45–60 35–60	62 ± 30 60 ± 30				4.0 ± 8.8 2.8 ± 6.4	(5)
Denmark HPLC	All	♂ 1097	51 17–87	65 ± 31 (7.5–226)	40			3.2 (0.1–27) (diet) 6.3 (0.2–29) (+suppl.)	(45, 59)

Table 2
(continued)

Country	Season	Gender N	Age (years) SD (range)	Plasma 25(OH)D nmol/l SD (range)	< 50 nmol/l (%) All	< 50 nmol/l (%) Summer	< 50 nmol/l (%) Winter	Dietary vitamin D μg/day SD (range)	Reference
Denmark	All	♂ 59	12 (0.5)	24 (9–57)	93			2.4 (0.9–4.9)	(25)
		♂ 53	72 (1.4)	48 (13–91)	55			3.4 (0.9–10.2)	
Hungary RIA	All	♂ 319	65 41–91	48 (12–135)					(57)

HPLC: high-pressure liquid chromatography. RIA: radioimmunoassay, EIA: enzyme immunoassay, suppl: supplementation

[a] < 16 ng/ml or 40 nmol/l.
[b] < 28 ng/ml or 70 nmol/l.
[c] < 75 nmol/l.
[d] including cod liver oil.
[e] from the 1958 British birth cohort.

Table 3

Plasma 25(OH)D in Free-Living Adult Females and Males in Denmark with Prevalence of Persons Having Values Below Different Cutoff Values for Vitamin D Deficiency or Insufficiency

Group	N	Age (years) Median (range)	Plasma 25(OH)D (nmol/l) Median (range)	<12	<25	<50	<75	Methods
Women								
Women	2624	51 (17–87)	72 (9–204)	0.7%	3.3%	22.3%	52.8%	LC-MS/MS All year
Women	395	64 (50–82)	53 (8–177)	0.8%	9.1%	46%	79%	RIA All year
Perimenopausal women	2006	50 (44–59)	58 (12–226)	0.5%	6.7%	39.9%	71.7%	HPLC All year)
Elderly women	67	73 (66–88)	36 (7–88)	11.9%	41.8%	82.1%	95.5%	RIA
Males								
Younger males	40	41 (29–51)	40 (15–91)	0.0%	20.0%	67.5%	95.0%	RIA
Elderly males	32	72 (66–88)	36 (9–80)	9.4%	25.0%	71.9%	96.9%	RIA

LC-MS/MS: Liquid chromatography-mass spectrometry; HPLC: High-pressure liquid chromatography. RIA: Radioimmunoassay. EIA: Enzyme immunoassay.

Table 4

Prevalence of Vitamin D Deficiency and Insufficiency in Adult Danes and Seasonal Variation in Plasma 25(OH)D According to Summer (April–September) and Winter Period (October–March)

Group	N	Age (years) Median (range)	Serum 25(OH)D (nmol/l) Median (range)	<12.5	<25	<50	<75	Methods
Blood donor Summer	203	43 (18–64)	85 (26–163)	0.0%	0.0%	4%	48%	EIA
Blood donor Winter	189	43 (18–64)	45 (13–128)	0.0%	18.0%	50.0%	91%	EIA
Perimenopausal women Summer	986	51 (45–58)	66 (12.5–226)	0.5%	4.1%	31.1%	63.0%	HPLC
Perimenopausal women Winter	1020	50 (44–59)	51 (12.5–208)	0.5%	9.2%	48.3%	80.2%	HPLC
Adult males and females Summer	223	64 (52–82)	60 (13–177)	0%	7.1%	34.1%	69.1%	RIA
Adult males and females Winter	172	63 (50–82)	45 (8–126)	1.7%	11.6%	61.6%	92.4%	RIA
Elderly males and females Summer	17	71 (66–83)	31 (12–74)	11.8%	35.3%	82.4%	100%	RIA
Elderly males and females Winter	82	72 (66–88)	32 (7–88)	11.0%	36.6%	78.0%	95.1%	RIA

HPLC: High-pressure liquid chromatography. RIA: Radioimmunoassay.
EIA: Enzyme immunoassay.

(2.5 μg/100 ml). Low fat milk is fortified in Finland (0.5 μg/100 ml (obligatory since 2003)), Iceland (0.38 μg/100 ml), and Sweden (0.38–0.5 μg/100 ml) and it is allowed to fortify milk (1 μg/100 ml) and other dietary products (1.25 μg/100 g) in France. Many European countries with no obligatory fortification have a much higher calcium intake but a lower vitamin D intake than in North America, where milk fortification is mandatory *(4)*. The effect of obligatory fortification has been evaluated in Finland with conflicting result. One study concluded that vitamin D fortification of milk and margarine products that started in February 2003 slightly but insufficiently improved winter vitamin D status in young men *(37)*. Other studies have showed that the fortification program substantially improved vitamin D status among young men *(38)* and 4-year-old children *(39)*.

8. SUPPLEMENTATION

It is difficult to obtain information on the intake of multivitamin in the different populations. Residents in North America and Scandinavia appear to be more apt to take vitamin D supplements than their counterparts in Western and Central Europe *(4)*. In Denmark multivitamins containing 5–10 μg of vitamin D are used by 40–68% of adult females, 29–47 of adult males, and 70% of children between 4 and 10 years *(36)*. Dietary supplements contributed to 42 to 49% of vitamin D intake in Norwegian men and women *(40)*. In Germany only 1.5% of males and 3.8% of females used vitamin D supplementation, whereas 12 and 19% used calcium supplementation, respectively *(41)*. In Norway and Iceland, cod liver and cod liver oil constitute a special form of supplementation. The highest intakes of vitamin D among adults have been observed in the northern part of Norway due to a high intake of cod liver and cod liver oil (8.1– 9.9 μg/day) *(26, 27)*. Among Icelandic children 40–68% used vitamin D supplementation or cod liver oil. The average vitamin D intake in these children was 10.4 μg/day compared with 2.7 μg/day in those not consuming any vitamin D supplements *(42)*. In a Swedish study plasma 25(OH)D increased by 25.5 nmol/l with two to three servings of fatty fish per week and by 11.0 nmol/l by regular use of vitamin D supplements compared with an increase of 6.2 nmol/l with a daily intake of 300 g vitamin D-fortified reduced-fat dairy products and 14.5 nmol/l with a sun vacation during winter *(43)*.

9. INDIVIDUAL LIFESTYLE FACTORS

Besides the influence of latitude, geographic location, climate, and governmental fortification policy, plasma 25(OH)D levels in European countries correlate positively to individual lifestyle factors like dietary vitamin D intake, vitamin D supplementation, sun bathing and use of sunbeds, recent holiday to sunny regions, and living in partnership and inversely to use of sunscreen and covering clothes, smoking habits, and body mass index *(11, 32, 41, 43–45)*. These facts impede a general description of the average vitamin D status and the occurrence of vitamin D deficiency within a certain geographical region such as the Northern Europe with its variation in climate, dietary habits, fortification policies, and lifestyles. Furthermore, recent year's immigration from

Palestine, India, Pakistan, Sri Lanka, Vietnam, and Africa has also resulted in more mixed populations with different risks of vitamin D deficiency (*41, 46–48*).

10. VITAMIN D INTAKES

It is only a few dietary ingestants that contain vitamin D. Fish, meat, and egg yolk are the most important. Vitamin D intakes vary with gender, age and national fortification, and supplementation policies *(4)* (Table 2). There is a trend toward higher intakes among males and with increasing age. In a French study the average intake was 2.8 μg/day among females and 4.0 μg/day among males *(5)*. In a study from the northern part of UK, dietary vitamin D intake was 4.2 ± 2.5 μg/day (range 0.1–24.7) among 2.598 postmenopausal women. If supplementation was included it rose to 5.8 ± 4.0 μg/day (range 0.2–33) *(45)*. In Ireland the intake is 3.4 μg/day from food and 5.1 μg/day from food and supplements *(49)*. Due to the high consumption of fish, vitamin D intakes for Norwegian men (6.8 μg/day) and women (5.9 μg/day) are higher than for British men (4.2 μg/day) and women (3.7 μg/day) even though England have mandatory fortification of some food items) *(4, 40)*. On average fish contributed to 1.5–1.8 μg/day to the Norwegians daily intake of vitamin D. A vegetarian diet is a well-known risk factor for vitamin D deficiency in United Kingdom and Finland *(50, 51)*. Asian immigrants to Northern Europe are at risk of developing vitamin D insufficiency not only because of lack of sunlight and covering clothes but also because of low intakes of vitamin D *(41, 46, 47, 48, 52, 53)*.

11. PREVALENCE OF VITAMIN D INSUFFICIENCY

The risk of being vitamin deficient is a result of the above-mentioned conditions. In spite of fortification policies and supplementation, several recent studies have identified a high prevalence of vitamin D deficiency and insufficiency (plasma 25(OH)D < 50 nmol/l) in Northern Europe (Table 2). The documentation is from Belgium *(54)*, Denmark *(25, 44)*, England *(32)*, Finland *(25, 55)*, France *(5)*, Holland *(56)*, Hungary *(57)*, Iceland *(42)*, Ireland *(11, 25, 49)*, Norway *(26, 27)*, Poland *(25)*, Scotland *(45)*, and Sweden *(43)*. The table also demonstrates that daily vitamin D intake is below 5 μg /day in most countries except in Norway, Sweden, and Finland.

12. VITAMIN D STATUS IN DENMARK

Vitamin D status is reported here in detail because more elaborate figures are available. Geographically, Denmark is situated in the middle of the region at 54–58°N. Bathing season is from June to August. The sky is often clouded also during summer period. Vitamin D status (plasma 25(OH)D) shows marked seasonal variations (Fig. 1). The recommended daily vitamin D intakes are 10 μg/day for < 2 years, 7.5 μg/day from 2 to 60 years, 10 μg/day for > 60 years plus extra 10 μg/day to people > 60 years with low sun exposure. An official program based on home nurse visits secures most infants <2 years a supplementation of 10 μg/day of vitamin D as droplets. Immigrants are advised an extra supplementation of 10 μg/day. The factual daily dietary intake in

the population is 2–4 μg, slightly increasing with age *(35)*. There is no obligatory fortification program or specific supplementation policy among children > 2 years or adults
(35, 36).

Frank rickets and osteomalacia (as diagnosed by bone histomorphometry) are very
rare (around 10 hospital admissions per year, population 5.5 million). In one study conducted during winter, 100% of puberty girls (average age of 12.5 years) had 25(OH)D
levels < 80 nmol/l, 93% had levels < 50 nmol/l, and 51% had levels < 25 nmol/l *(25)*.
Table 3gives the corresponding figures for adult Danes. Results vary with population
characteristics and the method applied for 25(OH)D estimation. Plasma 25(OH)D
< 75 nmol/l was seen in 52–97% of the population, levels < 50 nmol/l in 22–82%, lev-
els < 25 nmol/l in 3–42%, and levels <12 nmol/l in 0–12%. Low 25(OH)D levels were
more frequent during winter than summer time (Table 4). Dietary vitamin D intake,
vitamin supplementation, sunlight exposure, and use of sunbed were all significantly
related to 25(OH)D concentrations. Sun exposure seemed to contribute the most *(44)*.
Among 89 lactating women 67% had levels < 75 nmol/l in winter versus 27% in late
summer. Levels < 50 nmol/ were observed in 31 and 6%, respectively *(58)*. The vitamin D insufficiency was associated with secondary hyperparathyroidism. In 60 female
immigrants from Palestine, the average plasma 25(OH)D was substantially reduced (7.1
± 1.1 nmol/l) compared with 44 Danish controls (47.1 ± 4.6 nmol/l) *(46)*. Plasma PTH
also differed between the groups with secondary hyperparathyroidism in the immigrants
(15.6 ± 1.8 pmol/l versus 2.7 ± 0.3 pmol/l). Vitamin D deficiency was most pronounced
among those who wore the veil. Among Greenlanders living in Denmark on a westernized diet 98% had levels < 75 nmol/l, 79% had levels < 50 nmol/l, 27% had levels
< 25 nmol/l, and 1% had levels< 12 nmol/l *(8)*.

13. CONCLUSION

Randomized controlled trials have documented that vitamin D prevents fractures,
falls, myopathy, overall cancer risk, and overall mortality *(1–3, 18, 33)*. These results
should serve as incentives to implement public vitamin D fortification programs
(33). Furthermore, basic research and several epidemiological studies suggest that
vitamin D and its metabolites are important for preventing a number of frequent and
severe infections, diabetes, and autoimmune and cardiovascular diseases, as well as
the prevalence and course of several cancers *(1–3)*. In spite of these important biological and health-related effects of vitamin D, deficiency and insufficiency is common
in the Northern, Western, and Central parts of Europe. Around 43–92% of adults and
89–97% of pubertal females have plasma 25(OH)D levels <50 nmol/l. The vitamin D deficiency is caused by low sun exposure especially during winter time and
because of the high latitudes in the Northern regions *(16)*. The lack of solar exposure is combined with a low dietary vitamin D intake especially in the Central and
Western part of the continent *(5, 11, 25, 30, 41, 44)*. The high dietary intake of
fatty fish, cod liver, and cod liver oil in Scandinavia and Iceland and the frequent
use of supplements mitigate to some extent the effect of the reduced sun exposure
(25, 26, 42, 43). Air pollution and clouds may aggravate the lack of sun exposure *(22)*.
Across Europe epidemiologic studies indicate a positive effect of latitude on average

plasma levels of 25(OH)D *(24, 29, 30)*. However, at the level of the individual country (e.g., France and the United Kingdom) with a more uniform fortification policy and supplementation tradition, latitude is inversely related to plasma 25(OH)D *(31, 32)*. At the individual level European studies have confirmed that plasma 25(OH)D depends positively on dietary vitamin D, vitamin D supplementation, sun exposure, living in partnership, recent vacations to sunny regions, and sunbed use and inversely on use of sunscreen and covering clothes, smoking habits, and body mass index *(11, 32, 41, 43, 44)*. Risk groups include children, pregnant and lactating women, and older persons. Veiled and pigmented immigrant women with covering clothes and vegetarians constitute a special problem *(47, 48, 50, 51)*. Vitamin D status can be improved by encouraging safe sun exposure, improving dietary intake, e.g., by fish and cod liver oil, obligatory food fortification, and vitamin D supplements. Furthermore, nutritional recommendations from governments and other regulatory institutions should be adapted to the present knowledge of the actual need of vitamin D.

REFERENCES

1. Vieth R (2004) Why the optimal requirement for vitamin D_3 is probably much higher than what is officially recommended for adults. J Steroid Biochem Mol Biol 89–90:575–579
2. Holick MF, Vitamin D (2002) the underappreciated D-lightful hormone is important for skeletal and cellular health. Current Opin Endocrinol Diabetes 9:87–98
3. Holick MF (2003) Evolution and function of vitamin D. Recent Res Cancer Res 164:3–28
4. Calvo MS, Whiting SJ, Barton CN (2005) Vitamin D intake: a global perspective of current status. J Nutr 135:310–316
5. Ovesen L, Andersen R, Jakobsen J (2003) Geographical differences in vitamin D status with particular reference to European countries. Proc Nutr Soc 62:813–821
6. Iqbal SJ, Kaddam I, Wassif W et al (1994) Continuing clinically severe vitamin D deficiency in Asians in the UK (Leicester). Postgrad Med J 70:708–714
7. Nesby-O'Dell S, Scanlon KS, Cogswell ME et al (2002) Hypovitaminosis D prevalence and determinants among African American and white women of reproductive age: third National Health and Nutrition Examination Survey, 1988–1994. Am J Clin Nutr. 76:187–192
8. Rejnmark L, Jorgensen ME, Pedersen MB et al (2004) Vitamin D insufficiency in Greenlanders on a westernized fare. Ethnic differences in calcitropic hormones between Greenlanders and Danes. Calcif Tissue Int 74:255–263
9. Hagenau T, Vest R, Gissel TN et al (2009 May 6) Global vitamin D levels in relation to age, gender, skin pigmentation and latitude: an ecologic meta-regression analysis. Osteoporos Int. 20:133–140
10. Lips P (2001) Vitamin D deficiency and secondary hyperparathyroidism in the elderly: consequences for bone loss and fractures and therapeutic implications. Endocr Rev 22:477–501
11. Hill T, Collins A, O'Brien M et al (2005) Vitamin D intake and status in Irish postmenopausal women. Eur J Clin Nutr 59:404–410
12. Bischoff-Ferrari HA, Giovannucci E, Willett WC et al (2006) Estimation of optimal serum concentrations of 25-hydroxyvitamin D for multiple health outcomes. Am J Clin Nutr 84:18–28
13. Holick MF (2005) The vitamin D epidemic and its health consequences. J Nutr 135:2739S–2748S
14. Dawson-Hughes B, Heaney RP, Holick MF et al (2005) Estimates of optimal vitamin D status. Osteoporos Int 16:713–716
15. United Nations Statistic Division – Composition of macro geographical (continental) regions, geographical sub-regions, and selected economic and other groupings http://unstats.un.org/unsd/methods/m49/m49regin.htm#europe (Accessed October 10, 2008.)
16. Total Ozone Mapping Spectrometer. Erythemal UV exposure. http://toms.gsfc.nasa.gov/ery_uv/euv_v8.html (Accessed October 10, 2008)

17. Webb AR, Kline L, Holick MF (1998) Influence of season and latitude on the cutaneous synthesis of vitamin D_3: exposure to winter sunlight in Boston and Edmonton will not promote vitamin D_3 synthesis in human skin. J Clin Endocrinol Metab 67:373–378
18. Mosekilde L (2005) Vitamin D and the elderly. Clin Endocrinol (Oxf) 62:265–281
19. Moosgaard B, Vestergaard P, Heickendorff L et al (2005) Vitamin D status, seasonal variations, parathyroid adenoma weight and bone mineral density in primary hyperparathyroidism. Clin Endocrinol (Oxf) 63:506–513
20. Brustad M, Edwardsen K, Wilsgaard T et al (2007) Seasonality of UV-radiation and vitamin D status at 69 degrees north. Photochem Photobiol Sci 6:903–908
21. Edwardsen K, Brustad M, Engelsen O, Aksnes L (2007) The solar UV radiation level needed for cutaneous production of vitamin D_3 in the face. A study conducted among subjects living at a high latitude (68°N). Photochem Photobiol Sci 6:57–62
22. Goddard Space Flight Center. Monthly and Annual Mean Precipitation, Monthly Cloudiness, and Daily Weather Observations for Europe and Western Russia. http://gcmd.nasa.gov/records/GCMD_NCL00161_163_168_234.html. (Accessed October 10, 2008)
23. McKenna M (1992) Differences in vitamin D status between countries in young adults and the elderly. Am J Med 93:69–77
24. van der Wielen RP, Loewik MR, van den Berg H et al (1995) Serum vitamin D concentrations among elderly people in Europe. Lancet 346:207–210
25. Andersen R, Mølgaard C, Skovgaard LT et al (2005) Teenage girls and elderly women living in northern Europe have low winter vitamin D status. Eur J Clin Nutr 59:533–541
26. Brustad M, Alsaker E, Engelsen O et al (2004) Vitamin D status of middle-aged women at 65-71 degrees N in relation to dietary intake and exposure to ultraviolet radiation. Public Health Nutr 7:327–335
27. Brustad M, Sandager T, Aksnes L, Lund E (2004) Vitamin D status in a rural population of northern Norway with high fish liver consumption. Public Health Nutr 7:783–789
28. Bruyère O, Malaise O, Neuprez A et al (1944) Prevalence of vitamin D inadequacy in European postmenopausal women. Curr Med Res Opin 2007(23):1939–
29. Lips P, Duong T, Oleksik A et al (2001) A global study of vitamin D status and parathyroid function in postmenopausal women with osteoporosis: baseline data from the multiple outcomes of raloxifene evaluation clinical trial. J Clin Endocrinol Metab 86:1212–1221
30. Lips P (2007) Vitamin D status and nutrition in Europe and Asia. J Steroid Biochem Mol Biol 103: 620–625
31. Chapuy MC, Preziosi P, Maamer M et al (1997) Prevalence of vitamin D insufficiency in an adult normal population. Osteoporos Int 7:439–443
32. Hyppönen E, Power C: (2007) Hypovitaminosis D in British adults at age 45 y: nationwide cohort study of dietary and lifestyle predictors. Am J Clin Nutr 85:860–868
33. Mosekilde L (2008) Vitamin D requirement and setting recommendation levels: long- term perspectives. Nutr Rev 66(suppl 2):S182–S189
34. Prentice A (2002) What are the dietary requirements for calcium and vitamin D?. Calcif Tissue Int 70:83–88
35. Rasmussen L, Hansen GL, Hansen E et al (1998) Vitamin D. Should the supply in the Danish population be improved? (In Danish). Danish Institute for Food and Veterinary Research, publication number 246.
36. Danish Institute for Food and Veterinary Research. Vitamin D status in the Danish Population should be improved (in Danish with English summary), 2004. Available at: http://www.dfvf.dk/Default.aspx?ID=10082. (Accessed September 10, 2008)
37. Välimäki V-V, Löyttyniemi E, Välimäki MJ (2007) Vitamin D fortification of milk products does not resolve hypovitaminosis D in young Finnish men. Eur J Clin Nutr 61:493–497
38. Laaksi IT, Rouhola J-PS, Ylikomi TJ et al (2006) Vitamin D fortification as public health policy: significant improvement in vitamin D status in young men. Eur J Clin Nutr 60:1035–1038
39. Piirainen T, Laitinen K, Isolauri E (2007) Impact of fortification of fluid milks and margarine with vitamin D on dietary intake and serum 25-hydroxyvitamin D concentration in 4-year-old children. Eur Clin Nutr 61:123–128

40. Jorde R, Bønna KH (2000) Calcium form dairy products, vitamin D intake, and blood pressure: the Tromsø Study. Am. J. Clin. Nutr. 71:1530–1535
41. Hintzpeter B, Mensink GBM, Thierfelder W et al (2008) Vitamin D status and health correlates among German adults. Eur J Clin Nutr. 62:1079–1089
42. Thorsdôttir I, Gunnarsdôttir I (2005) Vitamin D nutrition in young Icelandic children. Laeknabladid 91: 581–586
43. Burgaz A, Åkesson A, Öster A et al (2007) Associations of diet, supplement use, and ultraviolet B radiation exposure with vitamin D status in Swedish women during winter. Am J Clin Nutr 86:1399–1404
44. Brot C, Vestergaard P, Kolthoff N et al (2001) Vitamin D status and its adequacy in healthy Danish perimenopausal women: relationships to dietary intake, sun exposure and serum parathyroid hormone. Br J Nutr 86:S97–S103
45. Macdonald HM, Mavroeidi A, Barr RJ et al (2008) Vitamin D status in postmenopausal women living at higher latitudes in the UK in relation to bone health, overweight, sunlight exposure and dietary vitamin D. Bone 42:996–1003
46. Glerup H, Mikkelsen K, Poulsen L et al (2000) Hypovitaminosis D myopathy without biochemical signs of osteomalacic bone involvement. Calcif Tissue Int 66:419–424
47. Holvik K, Meyer HE, Haug E, Brunvald L (2005) Prevalence and predictors of vitamin D deficiency in five immigrant groups living in Oslo, Norway: the Oslo Immigrant Health study. Eur J Clin Nutr.59: 57–63
48. Andersen R, Mølgaard C, Skovgaard LT et al Pakistani immigrant children and adults in Denmark have severely low vitamin D status (2008) Eur J Clin Nutr. 62:625–634
49. Hill TR, O'Brian MM, Lamberg-Allardt C et al (2006) Vitamin D status of 51-75 year old Irish Women: its determinants and impact on biochemical indices of bone turnover. Pubic Health Nutr. 9:225–233
50. Davey GK, Spencer EA, Appleby PN et al (2003) Lifestyle characteristics and nutrient intakes in a cohort of 33,883 meat-eaters and 31,546 non-meat-eaters in the United Kingdom. Public Health Nutr 6: 259–268
51. Outila TA, Karkkainen MUM, Seppanen RH, Lamberg-Allardt CJE (2000) Dietary intake of vitamin D in pre-menopausal, healthy vegans was insufficient to maintain concentrations of serum 25-hydroxyvitammin D and intact parathyroid hormone within normal ranges during the winter in Finland. J Am Diet Assoc 100:434–441
52. Shaw NJ, Pal BR (2002) Vitamin D deficiency in UK Asian families: activating a new concern. Arch Dis Child 86:147–149
53. Serhan E, Newton P, Ali HA, Walford S, Singh BM (1999) Prevalence of hypovitaminosis D in Indo-Asian patients attending a rheumatology clinic. Bone 25:609–611
54. Neuprez A, Bruyère O, Colette J et al (2007) Vitamin D inadequacy in Belgian postmenopausal osteoporotic women. BMC public Health 7:64
55. Kauppinen-Makelin R, Tahtela R, Loyttyniemi E et al (2001) A high prevalence of hypovitaminosis D in Finnish medical in- and outpatients. J Int Med 249:559–563
56. Sneijder MB, Lips P, Seidell JC et al (2007) Vitamin D status and parathyroid hormone levels in relation to blood pressure: A population-based study in older men and women. J Int Med 261:558–565
57. Bhattoa HP, Bettembuk P, Ganacharya S (2004) Prevalence and seasonal variation of hypovitaminosis D and its relationship to bone metabolism in community dwelling postmenopausal Hungarian women. Osteoporos Int 15:447–451
58. Møller UK, Ramlau-Hansen CH, Rejnmark L et al (2006) Postpartum vitamin D insufficiency and secondary hyperparathyroidism in healthy Danish women. Eur J Clin Nutr 60:1214–1221
59. Rejnmark L, Vestergaard P, Brot C, Mosekilde L (2008) Parathyroid response to vitamin D insufficiency: relations to bone, body composition and to lifestyle characteristics. Clin Endocrinol (Oxf) 69:29–35

23 Vitamin D Deficiency and Consequences for the Health of People in Mediterranean Countries

Jose Manuel Quesada-Gomez and Manuel Díaz-Curiel

Abstract The results of several recent cross-sectional observational studies conducted across Spain and other Mediterranean countries showed a very high prevalence of population with vitamin D deficiency, which produces skeletal and extra skeletal health consequences for the Mediterranean populations. According to these data it is important to highlight the need for improving patient's and physician's understanding of the optimization of vitamin D status irrespective of the hypothetical availability of hours of sun in Mediterranean countries. The medical community has an ethical responsibility to heighten efforts by individual health surveillance to ensure adequate vitamin D intake by patients and to teach the public about vitamin D.

Governmental policy and health policy makers will have to decide whether food fortification or supplements are the best way to achieve adequate vitamin D levels in Mediterranean populations.

KeyWords: Spain; mediterranean countries; 25-hydroxyvitamin D; vitamin D deficiency; North Africa; UV; sunlight; vitamin D; infants; clothing

1. INTRODUCTION

After the epidemic of rickets in the nineteenth century caused by vitamin D deficiency due to inadequate sun exposure, vitamin D inadequacy (deficiency or insufficiency) is nowadays once again recognized as a universal pandemic with severe consequences for human health *(1)*. Prolonged vitamin D deficiency causes rickets in children and osteomalacia in adults, while inadequacy of vitamin D is a main contributor to osteopenia, osteoporosis, loss of muscle mass, muscle weakness, falls, and fractures *(1–4)*. But in addition to these classic effects, vitamin D deficiency has been associated with an increased risk of chronic and degenerative diseases such as common cancers, autoimmune and infectious diseases, hypertension, and cardiovascular disease *(1, 5, 6)*.

From: *Nutrition and Health: Vitamin D*
Edited by: M.F. Holick, DOI 10.1007/978-1-60327-303-9_23,
© Springer Science+Business Media, LLC 2010

Vitamin D has a dual origin, both from skin synthesis under the influence of solar Ultraviolet B (UVB) radiation (wavelength, 290–315 nm) and from oral intake, comprised of the limited natural sources of vitamin D and fortified or supplemented foods. Vitamin "D" means the combination of vitamin D_2 and vitamin D_3. It was believed that vitamin D_2 was less efficient than vitamin D_3 in maintaining the levels of 25-hydroxyvitamin D [25(OH)D] due to its faster metabolism *(2)*. It has now been shown that both are equipotent for the maintenance of 25(OH)D serum levels *(7)*.

Vitamin D is metabolized in the liver to 25-hydroxyvitamin D, the major circulating metabolite of the vitamin D endocrine system, which has a long average life span (between 10 and 19 days), and it is commonly accepted as the clinical indicator of vitamin D status *(8)*, since it reflects levels either from dietary intake or from cutaneous synthesis.

The 25(OH)D status is critical for human health, because 25(OH)D is the substratum to form 1,25-dihydroxyvitamin D_3 [$1,25(OH)_2D$ or calcitriol] in the kidneys, where 1-α hydroxylation is strictly regulated by parathyroid hormone, calcium, and phosphorus serum levels and plays a key endocrine role in calcium and bone homeostasis. Renal-synthesized 1,25-dihydroxyvitamin D regulates gene transcription through nuclear high-affinity vitamin D receptor in the "classic" target organs: gut, bone, kidney, and parathyroid glands.

Besides, 25(OH)D is also the substratum to form 1,25-dihydroxyvitamin D in widespread organs and tissues such as muscle, heart, brain, breast, colon, pancreas, prostate, skin, and the immune system which possesses the activating 1α-hydroxylase (CYP27B1), inactivating (24-hydroxylase, CYP24A1) enzymes, and VDR, where $1,25(OH)_2D$ regulates about 3% of the human genome, regulates cell growth and maturation, inhibits renin production, stimulates insulin secretion, and modulates the function of activated T and B lymphocytes and macrophages and many other cellular functions in an autocrine–paracrine fashion *(9)*.

Vitamin D status quantified as 25(OH)D mainly depends on the season, the time of day of solar exposure, the duration of exposure, the use of sunscreens, skin pigmentation, and the latitude; in fact, vitamin D synthesis is extremely limited or impossible during winter months beyond 35°N and decreases substantially with aging. The dietary sources of vitamin D are minor; they include fatty fish, fish oils, and fortified dairy products (available only in some regions of the world) *(10)*.

Moreover, the measurement of 25(OH)D has been problematic and there is some concern about the reliability and consistency of serum 25(OH)D laboratory results *(10)*. Historically, measurements of 25(OH)D were performed in research centers using high-pressure liquid chromatography (HPLC) or competitive protein-binding methods (CBP). In the 1990s, validated radioimmunoassay (RIA) and other methods were developed, such as enzyme-linked immunosorbent assay (ELISA) or chemoluminiescence. The recent clinical availability of liquid chromatography tandem mass spectroscopy (LCMSMS) and HPLC technologies has improved 25(OH)D assay performance, leading to higher agreement between measurements obtained at different clinical laboratories, for example, DEQAS (vitamin D External Quality Assessment

Scheme), and although further standardization is desirable, recent data from 2006 demonstrate comparable values using RIA, HPLC, and LC-MS methodologies.

Despite the assay variability, and even though there is no universally established consensus on what levels of 25(OH)D constitute adequacy, there is a growing consensus that a serum 25(OH)D concentration above approximately 30 ng/ml constitutes optimal vitamin D status to assure bone health, and this is probably higher for other health outcomes (1, 5, 12). Therefore, it is considered that the 25(OH)D minimum desirable serum concentration should exceed 20 ng/ml in all individuals because this implies a population mean close to 30 ng/ml (13). Thus, there is a severe deficiency of 25(OH)D serum levels <10 ng/ml and a moderate deficiency (or insufficiency) from 10 to 20 ng/ml, and suboptimal 25(OH)D serum levels correspond to 20–30 ng/ml. A sufficient or adequate status of 25(OH)D serum level is above 30 nmol/L (14). The 25(OH)D concentration goal has not been clearly defined, but in highly sun-exposed populations, it does not appear possible to obtain a circulating 25(OH)D status above approximately 70 ng/ml (15). So attaining 25(OH)D levels between 30 and 70 ng/ml does seem physiologically sound.

Using such a definition, more than a half of the population worldwide has vitamin D deficiency or insufficiency, and globally this has been reported in healthy children, young adults, especially Afro-Americans, postmenopausal women, and middle-aged and elderly adults (1, 6, 12). Vitamin D inadequacy is particularly prevalent among osteoporotic patients, especially postmenopausal women and those with osteoporosis fractures (1).

Vitamin D status varies enormously between different countries in North America, Europe, the Middle East, and Asia, with seasonal variations in the countries below the 37th parallel (1, 2, 16). This is caused by varied exposure to sunshine, dietary intake of vitamin D, and the use of supplements.

The Mediterranean Basin, due to its location, weather, and its religious, social, cultural, and eating habits, constitutes an exceptional place in the world to assess the status of vitamin D and consider its implications, with regard to health consequences. The Mediterranean Sea is located in the Northern Hemisphere and it is a sea of the Atlantic Ocean, surrounded by the Mediterranean region and almost completely enclosed by land: on the north by Europe, on the south by Africa, and on the east by Asia. Nowadays, 21 modern states have a coastline on the Mediterranean Sea. The circle of latitude or parallel 40° runs more or less from Madrid (Spain) to Ankara (Turkey) and cuts the Mediterranean Basin into two parts (Fig. 1).

The combination of sharing a similar Mediterranean climate, with hot, dry summers and mild, rainy winters, and generally mild changes between summer and winter, rather than extreme heat or cold, and access to a common sea around which some of the most ancient human civilizations were organized, has led to numerous historical and cultural connections between ancient and modern societies around the Mediterranean, with traditional dietary patterns of countries which have inspired a modern nutritional recommendation, the so-called Mediterranean diet.

Fig. 1. Twenty-one countries have a coastline on the Mediterranean Sea. The parallel 40° runs from Madrid (Spain) to Ankara (Turkey) cut the Mediterranean Basin into two parts.

2. VITAMIN D STATUS IN MEDITERRANEAN POPULATIONS IN SPAIN AND OTHER EUROPEAN COUNTRIES

The Euronut SENECA study, a European Study developed in winter time, reported that half the elderly men and women in Spain, France, Italy, and Greece had vitamin D deficiency. The study showed a "paradoxical" relationship between serum 25(OH)D and northern latitude and reported higher levels in northern countries than in Mediterranean ones *(17)*. The baseline data of the MORE study, an intervention study in postmenopausal women with osteoporosis, confirmed this tendency *(18)*. Moreover, French and Spanish osteoporotic 67-year women outpatients in a multinational study also showed a high prevalence of vitamin D inadequacy, defined as serum 25(OH)D levels <30 ng/ml: 50 and 65% (French and Spaniards, respectively), with similar levels in osteoporosis-treated women *(19)*.

In the French EPIDOS study of good health, in aged women (80 years) living at home, the vitamin D status did not depend on the latitude but on the duration of individual exposure to sunshine, and the serum 25(OH)D level was lower than in young women in the OFELY study (34 years) (17 ± 10 vs. 25 ± 16 ng/ml) *(20)*.

However, in the French SUVIMAX study (latitude from 51° to 43°), a younger population (765 men and 804 women between 35 and 65 years) showed serum 25(OH)D levels varying from 17 ± 8 ng/ml with 1.06 h/day of sunshine in the north (29% hypovitaminosis D) to 37.5 ± 15.2 ng/ml (0% hypovitaminosis D) with 2 h/day of sunshine in the southwest. In this study, 25(OH)D serum levels correlated positively with sunshine exposure and negatively with latitude, as seems logical. But even in this younger healthy urban population in the Mediterranean Coast region, 7% of the subjects were vitamin D deficient (<12 ng/ml). The mean vitamin D intake was low: 3.4 ± 7.6 μg/day, much lower than the recommended intake of 10 μg/day *(21)*. Studies done in postmenopausal women in Italy showed 25(OH)D mean serum levels (18 ± 8 and 18.5 ± 11.ng/ml) rather unexpectedly, with prevalence of vitamin D deficiency of 28 and 32% *(22, 23)*, similar to 6,403 medical inpatients of both sexes studied in a town in northern Italy *(24)*.

A study in independent community-living elderly residents in Athens in Greece (latitude 38°N) *(25)* showed very low 25(OH)D serum levels during the winter and summer, 17.6 ± 8 ng/ml (combined women and men; aged 71), 20% of the elderly had vitamin D deficiency, with 25(OH)D below 10 ng/ml, and only 6.5% could be judged as vitamin D sufficient, with values above 32 ng/ml. The vitamin D status improved during the summer, with 25(OH)D serum levels of 36 ± 26 ng/ml, although 65% of the elderly subjects continued to have levels below 32 ng/ml, but no persons had values less than 10 ng/ml. Summertime vitamin D status was similar to young people (34 years) of the same study which however did not show 25(OH)D changes between seasons: 32 ± 11 vs. 34 ± 18 ng/ml. In teenagers (15–18 years) in the same country, the prevalence of vitamin D deficiency was 47% in winter, while the prevalence was lower (14%) at a younger age *(26)*.

The vitamin D status and the prevalence of vitamin D inadequacy in Spain in children, adults, community- or residence-living elderly, and osteoporotic-treated or osteoporotic-untreated postmenopausal women, representative of several studies *(18, 27–39)*, are given in Table 1.

Table 1
Vitamin D Status, Vitamin D Prevalence of Deficiency, and Other Data of the Study Populations in Spain According to Different Studies

Reference	Study population	Location	Season	Age (years)	Number	25(OH)D, mean ± SD (ng/ml)	Low serum 25(OH)D prevalence	Definition of low serum 25(OH)D (ng/ml)	Comments
(27)	Both sexes Living at home Living in a residence	Cordoba 37°6'	Spring	27–49 67–82 70–85	32 32 21	22.1 ± 11 14 ± 6 15 ± 10	32% 68% 100%	15	CBP
(28)	Both sexes Living at home	Cordoba 37°6'	Spring	20–59 60–79 > 8	81 31 17	38.0 ± 13 18 ± 14 9 ± 4.6			CBP
(28)	Blood donors Males Females	Cordoba 37°6'	Spring	18–65 18–64	116 9	18 ± 10–5 15 ± 9.2	14% 51% 65 %	10 20 30	HPLC p<0.02
(30)	Postmenopausal women	Granada 37°10'	Winter–spring	61 ± 7	161	19 ± 8	39%	15	RIA
(31)	Postmenopausal women	Madrid 40°26'	Winter–spring	47–66	171	13 ± 7	87% 64% 35%	20 15 10	RIA
(18)	Osteoporosis postmenopausal women	Spain 43–37°	Winter–summer	64 ± 7	132	24 ± 14	41.7% 10.6	20 10	RIA MORE study
(32)	Elderly, both sexes living in a residence	Sabadell 41°35'		61–96	100	10.2 ± 5.3.	87%	25	RIA

(33)	Elderly, both sexes living at home	Sabadell 41°35'	Winter–Spring	72 ± 5	239	17 ± 7.5	80%	25	RIA
(34)	Elderly outpatients, both sexes	Barcelona 41°23'	Winter–spring	75 ± 6	127		17%	10	RIA
							34.6%	10	RIA
(35)	Elderly living at home	Oviedo 43°22'	All year	68 ± 9	134	17 ± 8	72%	18	RIA
	Males		Winter–Spring	68 ± 9	134	17 ± 9	80%		
	Females			<65			72%		
				65–74					
				>65					
(36)	Elderly living at home	Murcia 37°59'	All year	77 ± 8	86	20 ± 13	58.2%	20	RIA
			Autumn–Winter			25 ± 15			
			Spring–summer			16 ± 9			
(37)	Children, elderly living at home	Cantabria 43°27'	Winter	8 ± 2	43	15 ± 5	31%	12	RIA
			Summer			29 ± 10	80%	20	p<0.001
(38)	Elderly, both sexes living at home	Valladolid 41°38'	All year	75 ± 85	197	15 ± 8	31	10	RIA
	Residence			83 ± 7	146	17 ± 7	79	20	
							32	10	
							91	20	
(39)	Osteoporotic postmenopausal women	Spain 43–28°	Late spring	71 ± 5	190	22 ± 10	11%	10	HPLC
	No treated			71 ± 5	146	27 ± 11	44%	20	
	Treated						76%	30	
							5%	10	
							29%	20	
							63%	30	

Even though the different methods employed or the inter-laboratory variation of 25(OH)D measurement makes these studies difficult to compare *(10, 11, 40–42)*, these reports highlight the fact that vitamin D insufficiency or even frank deficiency is just as common or even higher in southern European countries than in central or northern ones.

This paradox could be partially explained because vitamin D intake is generally much lower in southern countries (< 200 UI of mean) than in Scandinavian countries *(43)*, where it is as high as 400 UI due to the fatty fish intake and the contribution from vitamin D-fortified foods here *(1, 16–18)*. It is taken for granted by the European Mediterranean population that casual exposure to sunlight provides enough vitamin D, while skin synthesis of vitamin D may not compensate for the low nutritional intake because almost all the European Mediterranean countries are located at a high latitude from approximately 37°N, and photosynthesis of previtamin D$_3$ is scarce in winter and early spring *(1)* and people in southern Europe have a more pigmented skin, probably with less efficient vitamin D production; there is a poorer vitamin D status during the winter and the early spring, especially in elderly people. Nevertheless, the prevalence of low vitamin D status is the result of inadequate sun exposure, and because of the high temperatures in southern Mediterranean cities, as happens in Murcia or Cordoba during the summer (frequently 30–45°C), subjects avoid sunlight exposure and prefer to stay inside where the temperature is more comfortable. Moreover, older people especially are very concerned about the effect of direct sun exposure because of the risk of skin cancer. However, during the autumn and even winter months, elderly subjects frequently use light clothing and venture outdoors to benefit from the sunlight and generally warm weather (frequently 20–25°C).

The results of a recent cross-sectional observational study conducted from north to south across Spain showed that 63% of postmenopausal women receiving therapy for osteoporosis and 76% not receiving treatment had 25(OH)D levels of less than 30 ng/ml *(38)*, similar to other reports around the world *(18, 19, 44)* The high prevalence of vitamin D inadequacy in that study was consistent across all age groups and the geographic regions studied in Spain (Table 1).

Actually, the misconception among southern populations that sunlight provides enough vitamin D favors the low persistence of and adherence to vitamin D intake in osteoporosis and general treatment, and this is a cause for concern.

3. VITAMIN D STATUS IN MEDITERRANEAN POPULATIONS OF THE MIDDLE EAST AND NORTH AFRICAN COUNTRIES

Hypovitaminosis D is also highly prevalent in the Middle East and North African population, from Turkey to Morocco *(45–52)* (Table 2), latitude between 42° and 34°N (Fig. 1). Vitamin D intake is even lower than in European Mediterranean countries *(45, 46, 51, 53, 54)*, and its contribution to vitamin D status in those populations is also scarce. Moreover, while the warm weather and potentially high sunshine exposure facilitate vitamin D formation, as happens in southern European populations, other lifestyle factors determine vitamin D formation; in women in Mediterranean populations, excessive outdoor clothing, related to socio-cultural and religious factors, hinder cutaneous vitamin D production *(45, 46)*. In general, serum 25(OH)D was lower in women than

Table 2
Vitamin. D Status, Prevalence of Vitamin D Deficiency, and Vitamin D Intake in the Middle East and North Africa Mediterranean Countries According to Some Representative Studies

Reference	Study Population	Location	Season	Age (years)	Number	25(OH)D mean ± SD ng/mL	low serum 25(OH)D Prevalence	Definition of low serum 25(OH)D ng/mL	Coments
(47)	Women	Turkey Ankara 40°	Spring	14–44	48				
	Western clothing					22 ± 16			
	Traditional					13 ± 9.6			
	Completely veiled					3.6 ± 2.4			
(48)		Turkey Ankara	Autumn						
	Institutionalized women	40		75 ± 7	138	24 ± 30	40.7% 15.3%,		
	Men			76 ± 7	87	38 ± 29	<15 n		
	in own homes women			72 ± 5	171	41 ± 39	27.90%		
	Men			72 ± 5	24	63 ± 43	4.20%		
(49)	women	Turkey 37°3'	Summer						
	veiled			16–38	30	33 ± 16			
	non veiled			17–39	30	54 ± 27			
(50)	schoolchildren	Lebanon Beirut 33°5'	Spring Autumn	16-Oct	385				

Table 2
(continued)

Reference	Study Population	Location	Season	Age (years)	Number	25(OH)D mean ± SD ng/mL	low serum 25(OH)D Prevalence	Definition of low serum 25(OH)D ng/mL	Coments
	Spring								
	Boys					19 ± 7	9% / 55%	<10 / <20	
	Girls					15 ± 8	32% / 74%	<10 / <20	Low Status socioeconomic decreases Vitamin D status
	Autumn								
	Boys					24 ± 6	0% / 25%	<10 / <20	
	Girls					19 ± 7	7.50% / 53.50%	<10 / <20	
(51)		Lebanon Beirut 34°	Winter		316	10 ± 7	80.10%	intake	Vitamin D
	Women Total			30–50	217	7.6 ± 6	88.90%		100 ± 68 UI / 88 ± 58 UI
	Veiled						97%	<15	74 ± 50 UI
	Men			30–50	99	14 ± 7	50.50%		127 ± 80 UI

in men, and lower in urban areas than in rural ones. Moreover, as a rule, women wearing the traditional veil with face and hands uncovered (hijab) or covered (niqab) had poor vitamin D status and more prevalent vitamin D deficiency or insufficiency than did women with Western style clothing *(46)*.

There are consistent data that prove that vitamin D insufficiency and deficiency is common in Mediterranean populations throughout all lifetime periods, either in children, adolescents, or adults, and it is especially prevalent in postmenopausal women and elderly subjects, and this without doubt has consequences for the state of health in those populations *(1, 56)*.

Chronic severe vitamin D deficiency in infants and children is now frequently found with outbreaks of new cases of rickets in these countries in recent years *(57)*. Chronic vitamin D inadequacy in adults can result in secondary hyperparathyroidism, increased bone turnover, enhanced bone loss, increased muscle weakness, and falls with increased risk of fragility fracture. Some, but not all, observational studies have linked vitamin D inadequacy to an increased risk of hip and other non-vertebral fractures *(56)*. A supplement of calcium and vitamin D is recommended by all guidelines for osteoporosis treatment *(58)*. However, the results of several recent cross-sectional observational studies conducted across Spain *(19, 38)* and other Mediterranean countries *(19, 52)* showed a very high prevalence of postmenopausal women receiving therapy for osteoporosis who had 25(OH)D levels of less than 30 ng/ml, potentially reducing the effectiveness of the therapy, especially in patients with a low intake of calcium *(19, 38)*.

Moreover, on the basis of current evidence, vitamin D deficiency has extra skeletal health consequences for the Mediterranean populations. Increasingly, epidemiological, ecological, prospective, or retrospective studies indicate that vitamin D inadequacy is associated with increased risk of incident colon, prostate, and breast cancer, along with higher mortality from these cancers, and autoimmune diseases, such as type 1 diabetes mellitus, multiple sclerosis, rheumatoid arthritis, and inflammatory bowel disease. In addition, hypovitaminosis D also increases the risk of metabolic syndrome, hypertension, cardiovascular disease *(56)*, peripheral arterial disease, risk of myocardial infarction *(59)*, and mortality *(60, 61)*. Moreover, the intake of ordinary doses of vitamin D supplements seems to be associated with decreases in total mortality rates *(62)*.

According to these data it is important to highlight the need for improving patient's and physician's understanding of the optimization of vitamin D status, irrespective of the hypothetical availability of hours of sun in Mediterranean countries. The medical community has an ethical responsibility to heighten efforts by individual health surveillance to ensure adequate vitamin D intake by patients and to teach the public about vitamin D. Yet the public health message is complex. The safe dosage of sun exposure is unknown and may differ depending on skin pigmentation, and currently the scientific community paradoxically places greater emphasis on the risk of overexposure to ultraviolet (UV) light than to underexposure. We do know that certain populations, including breast-fed children, pregnant, postmenopausal women and older people, and especially women who cover most of their skin when outdoors, are at risk from vitamin D deficiency and sensitivity to cultural and religious practices is extremely important.

Governmental policy and health policy makers will have to decide whether food fortification or supplements are the best way to achieve adequate vitamin D levels in

Ref	Group	Location	Season	Age	N	Value	%	Threshold	Notes
(52)	Women	Lebanon	Winter				85%	<30	72% treatment
	Postmenopausal	34 N	Summer						67% vitamin D
	Christians			68 ± 6	151	22 ± 11			
	Muslims			67 ± 7	100	15 ± 7			
(53)	Geriatric patients	Israel	spring	78 ± 8	338	13.2 ± 8.8	supplements		vitamin D intake < 100 UI/day
(54)	Women	Tunis Tunesia 36°84'	winter	20–60	389				
	Veiled				14	48.90%	<15		vitamin D intake <76 UI/day
(55)	No veiled women	Rabat Morocco 32°2'	Summer	24–77	415	17	70.50%		
						18.4 ± 7.9.	91%	<30	
							43%	<15	
							4%	<5	
	Premenopausal			43 ± 6		19 ± 8			
	Postmenopausal			56 ± 7		18 ± 8			

Mediterranean populations. Awaiting these measures to become effective, and because of the very high prevalence of vitamin D insufficiency in these countries, vitamin D supplementation is imperative at least in all the above cited risk groups.

REFERENCES

1. Holick MF (2007) Vitamin D deficiency. N Engl J Med 357:266–281
2. Lips P (2001) Vitamin D deficiency and secondary hyperparathyroidism in the elderly: consequences for bone loss and fractures and therapeutic implications. Endocr Rev 22:477–501
3. Bischoff-Ferrari HA, Dawson-Hughes B, Willett WC et al (2004) Effect of Vitamin D on falls: a meta-analysis. JAMA 291:1999–2006
4. Quesada Gómez JM, Alonso J, Bouillon R (1996) Vitamin D insufficiency as a determinant of hip fractures. Osteoporos Int 6(Suppl 3):42–47
5. Bischoff-Ferrari HA, Giovannucci E, Willett WC (2006) Estimation of optimal serum concentrations of 25-hydroxyvitamin D for multiple health outcomes. Am J Clin Nutr 84:18–28
6. Bouillon R, Bischoff-Ferrari H, Willett W (2008) Vitamin D and health: perspectives from mice and man. J Bone Miner Res 23:974–979
7. Holick MF, Biancuzzo RM, Chen TC et al. (2008) Vitamin D_2 is as effective as vitamin D3 in maintaining circulating concentrations of 25-hydroxyvitamin D. J Clin Endocrinol Metab 93: 677–681
8. Institute of Medicine (U.S.). Standing Committee on the Scientific Evaluation of Dietary Reference Intakes (1997) Dietary reference intakes for calcium, phosphorus, magnesium, vitamin D, and fluoride/Standing Committee on the Scientific Evaluation of Dietary Reference Intakes, Food, and Nutrition Board, Institute of Medicine. National Academy Press, Washington, DC
9. Dusso AS, Brown AJ, Slatopolsky E (2005) Vitamin D. Am J Physiol Renal Physiol 289:F8–F28
10. Lips P, Chapuy MC, Dawson-Hughes B et al (1999) An international comparison of serum 25-hydroxyvitamin D measurements. Osteoporos Int 11:394–397
11. Lensmeyer GL, Wiebe DA, Binkley N et al (2006) HPLC method for 25-hydroxyvitamin D measurement: comparison with contemporary assays. Clin Chem 52:1120–1126
12. Dawson-Hughes B, Heaney RP, Holick MF et al (2005) Estimates of optimal vitamin D status. Osteoporos Int 16:713–716
13. Roux C, Bischoff-Ferrari HA, Papapoulos SE et al (2008) New insights into the role of vitamin D and calcium in osteoporosis management: an expert roundtable discussion. Curr Med Res Opin 24: 1363–1370
14. Mata-Granados JM, Luque de Castro MD, Quesada Gomez JM (2008) Inappropriate serum levels of retinol, alpha-tocopherol, 25 hydroxyvitamin D_3 and 24,25 dihydroxyvitamin D_3 levels in healthy Spanish adults: simultaneous assessment by HPLC. Clin Biochem 4:676–680
15. Barger-Lux MJ, Heaney RP (2002) Effects of above average summer sun exposure on serum 25-hydroxyvitamin D and calcium absorption. J Clin Endocrinol Metab 87:4952–4956
16. McKenna MJ (1992) Differences in vitamin D status between countries in young adults and the elderly. Am J Med 93:69–77
17. vd Wielen RPJ, Lowik MRH, vd Berg H et al (1995) Serum vitamin D concentrations among elderly people in Europe. Lancet 346:207–210
18. Lips P, Duong T, Oleksik AM et al for the MORE Study Group (2001) A global study of vitamin D status and parathyroid function in postmenopausal women with osteoporosis: baseline data from the multiple outcomes of raloxifene evaluation clinical trial. J Clin Endocrinol Metab 86: 1212–1221
19. Lips P, Hosking D, Lippuner K et al (2006) The prevalence of vitamin D inadequacy amongst women with osteoporosis: an international epidemiological investigation. J Intern Med 260:245–254
20. Chapuy MC, Schott AM, Garnero P et al (1996) Healthy elderly French women living at home have secondary hyperparathyroidism and high bone turnover in winter. J Clin Endocrinol Metab 81: 1129–1133

21. Chapuy MC, Preziosi P, Maamer M et al (1997) Prevalence of vitamin D insufficiency in an adult normal population. Osteoporos Int 7:439–443
22. Bettica P, Bevilacqua M, Vago T et al (1999) High prevalence of hypovitaminosis D among free-living postmenopausal women referred to an osteoporosis outpatient clinic in Northern Italy for initial screening. Osteoporos Int 9:226–229
23. Isaia G, Giorgino R, Rini GB et al (2003) Prevalence of hypovitaminosis D in elderly women in Italy: clinical consequences and risk factors. Osteoporos Int 14:577–582
24. Cigolini M, Miconi V, Soffiati G et al (2006) Hipovitaminosis D among unselected medical inpatients and outpatients in Northern Italy. Clin Endocrinol (Oxf) 64:475
25. Papapetrou PD, Triantaphyllopoulou M, Karga H et al (2007) Vitamin D deficiency in the elderly in Athens, Greece. J Bone Miner Metab 25:198–203
26. Lapatsanis D, Moules A, Cholevas V (2005) Vitamin D: a necessity for children and adolescents in Greece Calcif. Tissue Int 77:348–355
27. Quesada JM, Jans I, Benito P et al (1989) Vitamin D status of elderly people in Spain. Age Ageing 18:392–397
28. Quesada JM, Coopmans W, Ruiz P (1992) Influence of vitamin D on parathyroid function in the elderly. J Clin Endocrinol Metab 75:494–501
29. Mata-Granados JM, Luque de Castro MD, Quesada JM (2008) Inappropriate serum levels of retinol, α-tocopherol, 25 hydroxyvitamin D_3 and 24,25 dihydroxy vitamin D_3 levels in healthy Spanish adults: Simultaneous assessment by HPLC. Clin Biochem 41:676–680
30. Mezquita-Raya P, Muñoz-Torres M, Luna JD et al (2001) Relation between vitamin D insufficiency, bone density, and bone metabolism in healthy postmenopausal women. J Bone Miner Res 16:1408–1415
31. Aguado P, del Campo MT, Garces M et al (2000) Low vitamin D levels in outpatient postmenopausal women from a rheumatology clinic in Madrid, Spain: their relationship with bone mineral density. Osteoporos Int 11:739–744
32. Larrosa M, Gratacòs J, Vaqueiro M et al (2001) Prevalencia de hipovitaminosis D en una población anciana institucionalizada. Valoración del tratamiento sustitutivo. Med Clin (Barc) 117:611–614
33. Vaqueiro M, Baré ML, Anton E (2006) Valoración del umbral óptimo de vitamina D en la población mayor de 64 años. Med Clin (Barc). 127:648–650
34. González-Clemente JM, Martínez-Osaba MJ, Miñarro A et al (1999) Hipovitaminosis D: alta prevalencia en ancianos de Barcelona atendidos ambulatoriamente. Factores asociados. Med Clin (Barc) 113:641–645
35. Gómez-Alonso C, Naves-Díaz ML, Fernández-Martín JL et al (2003) Vitamin D status and secondary hyperparathyroidism: The importance of 25-hydroxyvitamin D cut-off levels. Kidney Int 63:S44–S48
36. Pérez-Llamas F, López-Contreras MJ, Blanco MJ et al (2008) Seemingly paradoxical seasonal influences on vitamin D status in nursing-home elderly people from a Mediterranean area. Nutrition 24:414–420
37. Docio S, Riancho JA, Pérez A et al (1998) Seasonal deficiency of vitamin D in children: A potential target for osteoporosis-preventing strategies?. J Bone Miner Res 13:544–548
38. Perez Castrillón JL, Niño Martin V (2008) Niveles de vitamina D en población mayor de 65 años. REEMO. 17:1–4
39. Quesada Gomez JM, Mata Granados JM, Delgadillo J et al (2007) Low calcium intake and insufficient serum vitamin D status in treated and non-treated postmenopausal osteoporotic women in Spain. J Bone Miner Metab 22:S309
40. Binkley N, Krueger D, Cowgill C et al (2003) Assay variation confounds hypovitaminosis D diagnosis: a call for standardization. J Clin Endocrinol Metab 89:3152–3157
41. Binkley N, Krueger D, Gemar D (2008) Correlation among 25-Hydroxy-Vitamin D Assays. J Clin Endocrinol Metab 89:3152–3157
42. Carter GD, Carter R, Jones JJB (2004) How accurate are assays for 25-hydroxyvitamin D? Data from the international vitamin D External Quality Assessment Scheme. Clin Chem 50:2195–2197

43. Brustad M, Sandanger T, Aksnes L (2004) Vitamin D status in a rural population of northern Norway with high fish liver consumption. Public Health Nutr 7:783–789

44. Holick MF, Siris ES, Binkley N et al. (2005) Prevalence of vitamin D inadequacy among post-menopausal North American women receiving osteoporosis therapy. J Clin Endocrinol Metab 90:3215–3224

45. Lips P (2007) Vitamin D status and nutrition in Europe and Asia. J Steroid Biochem Mol Biol 103: 620–625

46. Maalouf G, Gannagé-Yared MH, Ezzedine J et al (2007) Middle East and North Africa consensus on osteoporosis. J Musculoskelet Neuronal Interact 7:131–143

47. Alagol F, Shihadeh Y, Boztepe H (2000) Sunlight exposure and vitamin D deficiency in Turkish women. J Endocrinol Invest 23:173–177

48. Atli T, Gullu S, Uysal AR et al (2005) The prevalence of vitamin D deficiency and effects of ultraviolet light on vitamin D levels in elderly Turkish population, Arch. Gerontol Geriatr 40:53–60

49. Guzel R, Kozanoglu E, Guler-Uysal F et al (2001) Vitamin D status and bone mineral density of veiled and unveiled Turkish women. J Womens Health Gend Based Med 10:765–770

50. Gannage-Yared MH, Chemali R, Yaacoub N et al (2000) Hypovitaminosis D in a sunny country: relation to lifestyle and bone markers. J Bone Miner Res 15:1856–1862

51. El-Hajj Fuleihan G, Nabulsi M, Choucair M et al (2001) Hypovitaminosis D in healthy schoolchildren. Pediatrics 107:E53

52. Gannagé-Yared MH, Maalouf G, Khalife S et al (2008) Prevalence and predictors of vitamin D inade-quacy amongst Lebanese osteoporotic women. Br J Nutr 17:1–5

53. Goldray D, Mizrahi-Sasson E,, Merdler C et al (1989) Vitamin D deficiency in elderly patients in a general hospital. J Am Geriatr Soc 37:589–592

54. Meddeb N, Sahli H, Chahed M et al (2005) Vitamin D deficiency in Tunisia. Osteopor Int 16:180–183

55. Allali F, El Aichaoui S, Khazani H et al (2009) High Prevalence of Hypovitaminosis D in Morocco: Relationship to Lifestyle, Physical Performance, Bone Markers, and Bone Mineral Density. Semin Arthritis Rheum 38:444–451

56. Holick MF (2006) High prevalence of Vitamin D inadequacy and implications for health. Mayo Clin Proc 81:353–373

57. Yeste D, Carrascosa A on behalf of Grupo Interhospitalario para el Estudio del raquitismo Carencial en Cataluña (GIERCC) (2003) Raquitismo carencial en la infancia: análisis de 62 casos. Med Clin (Barc) 121:23–27

58. Kanis JA, Burlet N, Cooper C et al (2008) European guidance for the diagnosis and management of osteoporosis in postmenopausal women. Osteoporos Int 19:399–428

59. Giovannucci E, Liu Y, Hollis BW et al (2008) Independent Association of Low Serum 25-Hydroxyvitamin D and risk of myocardial infarction in men. Arch Intern Med 168:1174–1180

60. Dobnig H, Pilz S, Scharnagl H et al (2008) 25-Hydroxyvitamin D and 1,25-Dihydroxyvitamin D levels with all-cause and cardiovascular mortality. Arch Intern Med 168:1340–1349

61. Adit A, Ginde AA, Scragg R, et al (2009) Prospective study of serum 25-hydroxyvitamin D level, cardiovascular disease mortality, and all-cause mortality in older U.S. Adults. J Am Geriatr Soc 57: 1595–1603

62. Autier P, Gandini S (2007) Vitamin D supplementation and total mortality: a meta-analysis of random-ized controlled trials Arch Intern Med 167:1730–1737

24 Vitamin D Deficiency in the Middle East and Its Health Consequences

Ghada El-Hajj Fuleihan

Abstract   Despite its abundant sunshine the Middle East, a region spanning latitudes from 12° to 42°N allowing vitamin D synthesis year round, registers some of the lowest levels of vitamin D and the highest rates of hypovitaminosis D worldwide. This major public health problem affects individuals across all life stages including pregnant women, neonates, infants, children and adolescents, adults, and the elderly. Furthermore, while rickets is almost eradicated from developed countries, it is still reported in several countries in the Middle East. These observations can be explained by limited sun exposure due to cultural practices, dark skin color, and very hot climate in several countries in the gulf area, along with prolonged breast-feeding without vitamin D supplementation, decreased calcium content of diets and outdoor activity, obesity, and lack of government regulation for vitamin D fortification of food, in several if not all countries.

The lack of population-based studies renders estimates for the prevalence and incidence of rickets in the Middle East difficult, but several series from the region illustrate its dire consequences on growth and development. Furthermore, it is reported that 20–80% of apparently healthy individuals from several countries in this region have suboptimal vitamin D levels, depending on the cutoff used for defining hypovitaminosis D, the country, season, age group, and gender studied. Suboptimal levels have been associated with compromised skeletal health across age groups, and with poor muscular function and increased fall risk and osteoporotic fractures in the elderly. Studies detailing associations between low vitamin D levels and musculoskeletal health in the Middle East, and the impact of various treatment regimens are reviewed. Current recommendations for vitamin D derived from data in western subjects may not be sufficient for subjects from the Middle East; therefore, suggestions for vitamin D replacement doses based on evidence available to date are provided.

Hypovitaminosis D is a major public health problem across all life stages in the Middle East with deleterious immediate and latent manifestations. Long-term strategies to address this often silent disease should include public education, national health policies for screening and prevention through food fortification, and treatment through vitamin D supplementation.

Key Words: Rickets; hypovitaminosis D; Middle East; calcium; nutrition; neonates; infancy; adolescence; elderly; culture; socioeconomic; veiling

From: *Nutrition and Health: Vitamin D*
Edited by: M.F. Holick, DOI 10.1007/978-1-60327-303-9_24,
© Springer Science+Business Media, LLC 2010

1. INTRODUCTION

The Middle East region is a sub-continent that does not have precisely defined borders, and definitions vary from one source to the other. Traditionally the Middle East region includes western Asia and some parts of North Africa (Figure 1). A commonly used modern definition includes the following countries: Bahrain, Egypt with its Sinai Peninsula in Asia, Iran, Iraq, Israel, Jordan, Kuwait, Lebanon, Oman, Palestine, Qatar, Saudi Arabia, Syria, Turkey, the United Arab Emirates, and Yemen *(1, 2)*. The Middle East has a hot and arid climate and the latitudes it spans, from 12° to 42°N, allow vitamin D synthesis from ultraviolet B (UVB) rays for almost all months of the year, for more than 8 h a day *(3)*.

Fig. 1. Middle East in the context of major continents and regions. Definition of Middle Eastern countries varies; countries discussed in the chapter are labeled above. Countries considered in the chapter include: Bahrain, Egypt with its Sinai Peninsula in Asia, Iran, Iraq, Israel, Jordan, Kuwait, Lebanon, Oman, Palestine, Qatar, Saudi Arabia, Syria, Turkey, the United Arab Emirates, and Yemen.

While rickets is almost eradicated in western northern developed countries, excluding Asian immigrants, the Middle East is a region that registers some of the highest rates of rickets worldwide. This is in large part explained by limited sun exposure, cultural practices, and dark skin color and low calcium intake rather than vitamin D deficiency in several countries in Africa *(4)*. In addition, hypovitaminosis D, the latency disease of osteoporosis, constitutes a major public health problem in this region. Low vitamin D levels have also been associated with autoimmune disorders, certain cancers, and with the metabolic syndrome and type 2 diabetes, for which some of the highest rates reported arise from the Middle East. Despite the skin ability to synthesize vitamin D from exposure to UVB rays throughout the year, both rickets and hypovitaminosis D are more prevalent across all age groups in this region compared to the western populations.

Rickets, a symptomatic disease with overt clinical consequences, and hypovitaminosis D, a subclinical disorder with subtle and latent outcomes, will be discussed separately. Relevant articles were identified by performing a PubMed search entering the key terms rickets and Middle East, and a separate search was performed entering the key terms hypovitaminosis D and Middle East. Both searches had no restrictions on date, language, gender, or type of article. Another search was conducted by entering the individual names of the countries of interest detailed above with the term rickets or hypovitaminosis D. The abstracts of all identified articles were screened. Studies that could be retrieved and that discussed the presentation and manifestations of rickets, risk factors for its development, or the prevalence and presentations of hypovitaminosis D and its clinical consequences were included. Osteomalacia results from severe hypovitaminosis D in adults and presents with isolated or generalized skeletal pain. Case reports illustrate that this condition still exists in the Middle East, but the magnitude of the problem is unclear. Indeed, unless specifically searched for, it can be misdiagnosed as fibromyalgia, chronic fatigue syndrome, or depression, and therefore it is not discussed as a separate entity. The origin of many affected kindreds with vitamin D-dependant rickets type II, an autosomal recessive familial disorder with the unusual feature of alopecia, is concentrated in the Mediterranean region. However, studies reporting non-nutritional rickets, including vitamin D-dependant rickets type II, were excluded *(5)*. Additional articles were obtained from the author's library or from the reference list of the search identified articles.

2. NUTRITIONAL RICKETS

Nutritional rickets is a disease resulting from impaired bone mineralization due to insufficient calcium or phosphorus at the growth plate in growing children. It ranks as one of the five commonest diseases in children from developing countries and is still quite common in the Middle East *(4, 6)*. Although nutritional rickets in infants had been traditionally thought to be exclusively secondary to vitamin D deficiency, increasing evidence over the last several years points to a pivotal role of calcium deficiency in its pathogenesis as evidenced from studies conducted in older children from South Africa, Nigeria, and Egypt *(4, 7)*. Thus nutritional rickets can be the result of severe nutritional deficiencies in calcium or vitamin D, or even as likely, if not more likely, a combination of less severe deficiencies in both *(4, 8)*. However, the relative contribution of deficiency in either calcium or vitamin D to the disease has not been adequately investigated and is only reported in limited studies from certain countries.

2.1. Nutritional Rickets in the Middle East

The prevalence of nutritional rickets in the Middle East is not known due to the lack of population-based figures. The prevalence of rickets was estimated at 27% in a sample group of 197 children under 5 years of age attending a health center in a rural area in North Yemen *(9)*. Similarly, population-based incidence rates for rickets are lacking; however, annual incidence rates derived from selected studies report a rate of 10% in an older field survey conducted in a maternity and child welfare center of a rural community of 6000 inhabitants in Egypt *(10)*, of 1% between 1981 and 1986 in a study of Kuwaiti children less than 2 years of age *(11)*, of 0.5% between 1997 and 1999 in a study of Saudi infants less than 2 years *(12)*, and of less than 0.1% in 2007–2008 (down from 6%) in children 0–3 years from Turkey due to a national vitamin D campaign *(13)*. These figures, albeit based on selected samples of children, are nevertheless at least 100 folds higher than the reported annual incidence rates of 3–7.5 cases/100,000 children in western developed countries *(14)*. Furthermore, rickets accounts for a substantial number of pediatric hospital admissions in the region. A report of 500 cases of rickets admitted at the Suleimania hospital in Riyadh between 1986 and 1988 estimated that the diagnosis accounted for 1.8% of all hospital admissions for that period *(15)*, it was estimated at 1% in a hospital-based study conducted in Saudi Arabia *(16)*, and 6.5% in a similar study of newborns from Kuwait *(17)*. Rickets accounted for 10.6% of all admissions for infants presenting with an acute illness to a hospital in Jordan *(18)*. Limited sun exposure due to cultural practices, dress styles, limited time spent outdoors, and prolonged breast-feeding without vitamin D supplementation account in large part for the persistence of rickets in the Middle East despite its plentiful sunshine *(4, 6)*, as detailed below. This is probably more relevant in infants, whereas the pathophysiology is likely to be mixed in older children with variable contributions of concomitant low calcium intake. Finally, genetic factors such as possible resistance to vitamin D and polymorphisms in the vitamin D receptor (VDR) may modify the clinical manifestations of nutritional rickets *(7)*. In a recent study of 98 rachitic children from Egypt and Turkey, increased frequency of the F *(Fok1)* VDR allele was noted in patients from Turkey, and the BB *(Bsm)* polymorphism was associated with a lower vitamin D in patients and controls and with the severity of rickets *(7)*. Maternal vitamin D deficiency during pregnancy is an established risk factor for rickets presenting neonatally or in early infancy *(13, 17, 19, 20)*.

Turkey: Turkey, specially its eastern part, had been accepted as an endemic region for vitamin D deficiency rickets, with substantial improvement due to a national vitamin D campaign on most recent estimates *(13)*. A small retrospective study of 42 infants less than 3 months of age presenting with seizures, and a mean 25-hydroxyvitamin D [25(OH)D] of less than 15 ng/ml, identified exclusive breast-feeding, low maternal sun exposure, and concealed maternal clothing as risk factors for vitamin D deficiency in these infants *(21)*. They presented mostly in winter and spring and lacked the classical skeletal manifestations of rickets. Two additional studies revealed a similar risk profile. A survey conducted on 39,133 children ages 0–3 years presenting to outpatient clinics in Erzurum between March 2007 and Feb 2008, revealed that the 39 infants presenting with nutritional rickets, at a mean age of 10 months, and with a mean 25(OH)D level of 6 ng/ml, were more likely to come from larger families, of lower socioeconomic and educational level, and that both infants and their mothers spent less time outdoors and were less likely to take vitamin D supplementation compared to controls *(13)*. Similarly, a recent study of 68 rachitic children, ages 6 months–4 years, revealed that household crowding, lower parental socioeconomic status and educational level, suboptimal prenatal care, time spent outdoors during pregnancy, exclusive breast-feeding of the infant, and lower calcium intake to be risk factors for developing rickets *(7)*.

Iran: In a review of records of 30,000 admissions in a pediatric ward in Shiraz, Iran, between 1958 and 1967, 25 cases were diagnosed with vitamin D deficiency rickets *(22)*. The majority of cases (*N*=18) were below age 3, and had clinical findings consistent with rickets such as rachitic rosary, pigeon breast deformity, widening of the wrists, Harrison's groove, and lower extremity deformities. All patients had been exclusively breast-fed, had not received vitamin D supplementation, belonged to a group with a lower socioeconomic status, suffered from protein-calorie malnutrition, and were below the third percentile for height and weight and presented with an acute infection.

Iraq: In a study of 50 infants presenting with two or more clinical signs of rickets in a children's outpatient department in Mosul, mean age was 18 months, and all children were breast-fed, at least for the first 6 months of their life *(23)*. Clinical signs included delayed closure of the fontanelle, bossing of the head, delayed teething, beading of the ribs or rosary, widening of the wrists and bowing of the legs, and the majority of children suffered from growth failure, few were marasmic and hypoalbuminemic, invalidating the notion that rickets occurs in growing children who are not wasted *(23)*. Rickets was 2.5 folds more common in infants presenting with wheezy bronchitis compared to controls *(24)*.

Jordan: In a case–control study of 47 infants, mean age 8 months, admitted to a hospital with an acute illness and found to have rickets, child family ranking of second or more, large family size, breast-feeding, maternal clothing type (wearing head cover) were risk factors associated with rickets *(18)*.

Saudi Arabia: A total of 102 infants with vitamin D deficiency, half of whom were Saudis, and 30 control subjects were studied in a maternity and children's hospital in Makkah between 1993 and 1994 *(25)*. Exclusive breast-feeding and lower maternal educational status were more likely to be present in patients than controls, and families of patients tended to be larger than those of controls. Similarly, in a study of 61 infants with rickets, mean age of 16.5 months and mean 25(OH)D of 8 ng/ml, conducted between 2004 and 2005 in a major hospital in Ryadh, breast-feeding and limited sun exposure were more common in patients than controls *(26)*. Rickets was linked to vitamin D deficiency in 59% of adolescents, to poor calcium intake in 12%, and to genetic abnormalities in the remaining proportion (25-hydroxylase deficiency in 8.8%, vitamin D-dependant rickets type I in 6%, and hypophosphatemic rickets for 12%) *(27)*. Conversely, both limited sun exposure and poor calcium intake have been estimated to contribute to rickets in two other studies of children and adolescents, mean age 13 years, $N = 42$ and 21 subjects per study, respectively *(28, 29)*. In these studies, mean 25(OH)D levels were below 10 ng/ml, and mean calcium intake was below 500 mg in one study *(29)* and between 100 and 300 mg in the other *(28)*. In adolescents presenting symptoms were more likely to be exclusively musculoskeletal in nature, including aches and pains, tetany, muscle weakness, and limb deformities *(27, 29)*.

Kuwait: A study reported 75 neonates with rickets presenting with rachitic rosary at birth, thus pointing to the development of rickets due to maternal hypovitaminosis D during pregnancy *(17).* The mean 25(OH)D was less than 7 ng/ml, hypotonia, enlarged fontanelles and widened cranial sutures were other manifestations of the disease *(17).* Breast-feeding and poor sun exposure were major risk factors for the development of rickets in 250 infants less than 2 years of age *(11).* Similarly, exclusive breast-feeding, delayed weaning to solids, nutritional quality of semisolids, and low sun exposure were major risk factors in another study of 103 infants with rickets *(30).* At diagnosis these infants were shorter, lighter, and had lower 25(OH)D levels than controls, measured at 10 and 33 ng/ml, respectively.

United Arab Emirates: Thirty-eight Emirati children with rickets referred to pediatric clinics of two university hospitals and their mothers were compared to 50 non-rachitic children–mother controls. The patients' mean age was 13 months, serum calcium was 8.88 mg/dl, mean 25(OH)D was 3.2 ng/ml, and serum alkaline phosphatase was 834 IU/L, compared to mean 25(OH)D of 17.2 ng/ml in controls. The majority (92%) were breast-feeding, 8% were on supplements, and sunshine exposure was nil *(20).* However, in another retrospective study of 31 older children with rickets, mean age less than 3 years, only half had 25(OH)D levels below 10 ng/ml, while calcium deficiency contributed to the development of rickets in the other half *(31).*

Egypt: Fifty-four successive cases of clinically diagnosed rickets presenting at an outpatient clinic in Cairo, and subsequently confirmed radiologically, were compared to 28 controls *(32).* Half had signs of the disease in limbs, skull, and chest, and two-thirds could not stand. Children mostly presented because of failure to thrive before age 2, and they were lighter and shorter than controls. Whereas mean calcium intake was low, estimated at 300 mg/day, both in patients and controls, patients had a lower mean 25(OH)D level of 3.3 ng/ml compared to 10.1 ng/ml in controls. Similarly, in a more recent study of 30 children with rickets a mixed etiology, namely a low calcium intake <500 mg and suboptimal 25(OH)D levels in mid-teens, was felt to be responsible for the development of rickets *(7).*

2.2. Non-skeletal Manifestations of Nutritional Rickets

In addition to the classical skeletal manifestations of rickets outlined above, the proportion of patients presenting with convulsions varied from 4 to 79%, depending on the study *(11, 13, 15, 18, 21, 25, 33–36).* In addition, the studies detailed above illustrate an association between rickets and infectious conditions, in large part respiratory infections and diarrhea. The admitting diagnosis was recurrent chest infections in 66% of 500 cases of rickets in Saudi Arabia *(15);* broncho-pneumonia was present in 43% of 200 rachitic Iranian children *(33)* and 44% of 250 rachitic children from Kuwait *(11).* Similarly, an acute infection or respiratory diseases were the presenting manifestation in 20–60% of cases presenting with rickets in smaller studies from Turkey, Egypt, Jordan, and Saudi Arabia *(13, 18, 22, 25, 37, 38).* Gastroenteritis accounted for 8–56% of reasons for admission *(11, 18, 32, 33).*

Dilated cardiomyopathy secondary to hypocalcemia is an unusual manifestation of nutritional rickets *(39),* 16 such cases were reported from a children's hospital in

London, all were from dark-skinned ethnic minorities, 12/16 were exclusively breast-fed, and presented at the end of winter. Dilated cardiomyopathy, as a manifestation of vitamin D deficiency rickets, has been reported in three infants from the Middle East, one from Turkey and two from the United Arab Emirates *(40, 41)*. All three infants presented with heart failure had florid skeletal signs of rickets and hypocalcemia with 25(OH)D levels below 8 ng/ml. They all experienced rapid clinical improvement within a week of instituting therapy with inotropes, diuretics, in addition to calcium and vitamin D, and had resolution of their cardiomegaly and radiologic healing of rickets within few weeks of therapy initiation. The two infants from the United Arab Emirates were exclusively breast-fed, did not receive vitamin D supplementation, had little sun exposure, and both their mothers had 25(OH)D level of <8 ng/ml *(41)*.

2.3. Rickets in Immigrants from Middle East

Several studies have reported high rates of rickets in immigrants from the Middle East to western countries. In a study of 126 patients with rickets or vitamin D deficiency in Australia, 11% originated from the Middle East *(42)*. Similarly in a study of 41 adolescents diagnosed with rickets in France, 33/41 (80%) were immigrants, of those 15 were from South Africa, 10 from North Africa, 6 from Pakistan, and 2 from Turkey *(43)*.

3. HYPOVITAMINOSIS D

Hypovitaminosis D is a subclinical condition with latent manifestations. There is no consensus on optimal levels of 25(OH)D, but it has recently been re-defined by most experts by a cutoff of less than 20 ng/ml, for both children and adults *(14, 44–46)*. In recent publications it is often defined as vitamin D insufficiency when levels fall between 10 and 20 ng/ml, and vitamin D deficiency when levels fall below 10 ng/ml *(45)*. However, some of the studies detailed below had predated such definitions and have used variable cutoffs (Tables 1, 2, and 3).

3.1. Mothers–Neonates–Infants

Neonates born to mothers with low vitamin D levels have lower cord vitamin D levels, and may be at risk for rickets and other complications *(14, 47–52)*. Studies from the region reveal a high prevalence of hypovitaminosis D in pregnant mothers and their neonates (Table 1).

A study of 8 Beduin women and 41 Jewish Shepardi women from the Negev Desert revealed a mean 25(OHD) level of 7.8 ng/ml in the Beduin and of 3.2 ng/ml in the cord blood of their offsprings, compared with a mean 25(OH)D level of 25 ng/ml in the control mothers and of 12 ng/ml in the cord blood of their neonates *(52)*. Venous blood obtained from 100 consecutive and unselected Saudi Arabian mothers and their asymptomatic neonates, within 48 h of delivery, revealed that 59 mothers and 70 neonates had 25(OH)D levels below 10 ng/ml, along with asymptomatic hypocalcemia *(19)*. Similarly, a study of 90 Arab and South Asian mothers and their breast-fed infants at 6 weeks revealed that 61% of mothers and 80% of the offspring had 25(OH)D below 10 ng/ml *(48)*. The same authors studied a convenient sample of 50 infants admitted at 15 months

Table 1

Prevalence of Hypovitaminosis D by Country in Pregnant Women and Neonates in the Middle East

Author	Year	Country, city	Latitude	N	Gender	Age (years) mean ± SD or range	25(OH)D mean ± SD (ng/ml)	25(OH)D cutoff values (ng/ml)	Predictors	Comments
Serenius	1984	Saudi Arabia, Riyadh	24°N	119	Women at term and their neonates	–	*Median:* • All: 5.7	*Undetected levels:* • Mothers: 9% • Cord blood: 42% *% < 4:* • Mothers: 16% • Cord blood: 26%	High SES* Antenatal care Vitamin D supplement	Survey, 75% of subjects selected randomly from hospital
Bassir	2001	Iran, Tehran	35°N	50	Women at term and their neonates	16–40 years	• Mothers: 5.1±10.4 • New born: 2±3.8	% < 10–12: • Mothers: 80% • New born: 82%	Sun exposure	Pilot study, convenience sample from hospital
Molla	2005	Kuwait, Kuwait City	29°N	214	Women at term and their neonates	27.5±4.2	• Mothers: 14.6±10.7 • Newborn: 8.2±6.7	% < 10: • Mothers: 38–41% • New born: 60–70%	Maternal education	Mothers selected from two hospitals
Ainy	2006	Iran, Tehran	35°N	95	Women: 48 pregnant 47 control	26.2±5.0	• 1st term: 20.6±12 • 2nd term: 25.7±16.7 • 3rd term: 24.5±12.8 • Control: 23.0±12.9	% <10–12: • 1st term: 20% • 2nd term: 10% • 3rd term: 3% • Control: 15% % Between 10 and 20: • 1st term: 40% • 2nd term: 38% • 3rd term: 44% • Control: 25%	Pregnancy trimester	Cohort study, randomly selected from care centers

*SES: socioeconomic status

Table 2

Prevalence of Hypovitaminosis D by Country in Children and Adolescents in the Middle East

Author	Year	Country, city	Latitude	N	Gender	Age (years) mean ± SD or range	25(OH)D mean ± SD (ng/ml)	25(OH)D cutoff values (ng/ml)	Predictors	Comments
El-Hajj Fuleihan	2001	Lebanon, Beirut	33°N	346	*Spring:* 81 Boys 88 Girls	*Spring:* 13.3±1.6	*Spring:* • Boys: 19±7 • Girls: 15±8 • All: 17±8	% < 10–12: *Spring:* • Boys: 9% • Girls: 32% • All: 21%	Gender Season SES* Clothing	Children selected three schools of different SES*
					Fall: 83 Boys 94 Girls	*Fall:* 13.3±1.7	*Fall:* • Boys: 24±6 • Girls: 19±7 • All: 22±7	% Between 10 and 20: *Spring:* • Boys: 46% • Girls: 42% • All: 44%		
								Fall: • Boys: 0% • Girls: 8% • All: 4%		
								Fall: • Boys: 25% • Girls: 46% • All: 36%		
Bahijri	2001	Saudi Arabia, Jeddah	21°N	935	–	4–72 months	• 4–6 months: 26.2±14.1 • 6–12 months: 24.9±14.1 • 12–24 months: 24.6±14 • 24–36 months: 26.7±11.3 • 36–72 months: 24.2±11.5	% Between 5 and 10: • 4–6 months: 14% • 6–12 months: 13% • 12–24 months: 14% • 24–36 months: 4% • 36–72 months: 8%	Episodes of diarrhea Dietary intake of vitamin D Sun exposure	Random selection covering all districts and all SES*
Moussavi	2005	Iran, Isfahan	32°N	318	153 Boys 165 Girls	14–18	• Boys: 37.3±18.8 • Girls: 16.8±8.4	% < 20: • Boys: 18% • Girls: 72%	Gender Sun exposure	Cross-sectional, multistage random selection from schools

Table 2
(continued)

Author	Year	Country, city	Latitude	N	Gender	Age (years) mean ± SD or range	25(OH)D mean ± SD (ng/ml)	25(OH)D cutoff values (ng/ml)	Predictors	Comments
Dahifar	2006	Iran, Tehran	35°N	414	Girls	11–15	• All: 30	% < 10–12: • All: 3.6%	Ca intake Sun exposure	Cross-sectional, random selection from schools
El-Hajj Fuleihan	2006	Lebanon, Beirut	33°N	363	184 Boys 179 Girls	10–17	• All: 16±9	% < 10–12: • Boys: 12% • Girls: 33% % Between 10 and 20: • Boys: 66% • Girls: 51%	Gender Winter	Convenience sample, from four schools, balanced geographical and socioeconomic presentation
Siddiqui	2007	Saudi Arabia, Jeddah	21°N	433	Girls	12–15	–	% < 10–12: All: 81%	Family income Sun exposure Intake of dairy products	Randomly selected from different schools
Rabbani	2008	Iran, Tehran	35°N	963	424 Boys 539 Girls	7–18	• Boys: 21.6 • Girls: 18.4	% < 20: • Boys: 11% • Girls: 54% % <8: • Boys: 0.9% • Girls: 11%	Gender, age in girls	Random cluster sampling, puberty, private schools in winter, veiled

*SES: socioeconomic status

Table 3
Prevalence of Hypovitaminosis D by Country in Adults in the Middle East

Author	Year	Country, city	Latitude	N	Gender	Age (years) mean ± SD or range	25(OH)D mean ± SD (ng/ml)	25(OH)D cutoff values (ng/ml)	Predictors	Comments
El-Sonbaty	1996	Kuwait, Kuwait City	29°N	72	Women: 50 Veiled 22 Unveiled	14–45	• Veiled: 5.8±2 • Unveiled: 12±3.3	% <8: • Veiled: 86%	Veiling	Case–control study, convenience sampling
El-Hajj Fuleihan	1999	Lebanon, Beirut	33°N	465	Women	15–60	• All: 11±14	% < 10–12: • All: 60% % Between 10 and 20 • All: 35%	Veiling	Random sample from a village in central Lebanon
Ghannam	1999	Saudi Arabia, Riyadh	24°N	321	Women	10–50	• All: 10±8	% < 10–12: • All: 52%	Lactation Parity	Cohort, convenience sampling through advertisements

Table 3
(continued)

Author	Year	Country, city	Latitude	N	Gender	Age (years) mean ± SD or range	25(OH)D mean ± SD (ng/ml)	25(OH)D cutoff values (ng/ml)	Predictors	Comments
Alagol	2000	Turkey, Istanbul	41°N	48	Premenopausal	26±8	By dress style: • Western: 22±16 • Partial cover: 13±10 • Full cover: 4±2	% < 16: By dress style • Western: 44% • Partial cover: 60% • Full cover: 100%	Dress style	Convenience sampling
Gannage	2000	Lebanon, Beirut	33°N	310	99 Men 217 Women	30–50	• Men: 14.3±7.5 • Women 7.6±5.8 • All: 9.7±7.1	% <12: • Men: 48% • Women: 84% % <5: • Men: 41% • Veiled: 62%	Vitamin D intake Urban dwelling Veiling High parity	Convenience sampling, from different rural and urban centers
Mishal	2001	Jordan, Amman	31°N	146	22 Men 124 Women	18–45	*Summer:* • Men: 43.8±5.2 Women by dress style: • Western: 36.7±6.1 • Hijab: 28.3±4.5 • Niqab: 24.3±5.8 *Winter:* • Men: 34.7±4.2 Women by dress style: • Western: 30.9±4.6 • Hijab: 24.4±3.9 • Niqab: 22.7±3.0	% Between 5 and 12: *Summer:* • Men: 18% Women by dress style: • Western: 31% • Hijab: 55% • Niqab: 75% *Winter:* • Men: 46% Women by dress style: • Western: 75% • Hijab: 71% • Niqab: 82%	Dress style Winter	Convenience sampling

Table 3
(continued)

Author	Year	Country, city	Latitude	N	Gender	Age (years) mean ± SD or range	25(OH)D mean ± SD (ng/ml)	25(OH)D cutoff values (ng/ml)	Predictors	Comments
Mirsaeid	2004	Iran, Tehran	35°N	1172	682 Men 490 Women	3–69	• Men: 35±26 • Women: 21±22	% < 20: • Men: 35% • Women: 69%	Season	Cluster random sample. More variation than the women
Meddeb	2005	Tunisia, Tunis	36°N	389	128 Men 261 Women	20–60	• Veiled: 14 • Non-veiled: 17	% < 15: • All: 47% • Veiled: 70% • Non-veiled: 50%	Age Veiling Parity Menopause winter	Transverse descriptive inquiry
Atli	2005	Turkey, Ankara	40°N	420	Elderly: Men: 111 Women: 309	–	–	% < 15: Women in old-age home: 54% Women in own home: 18% Men old-age home: 18% Men own home: 4%	Age, UV* light index	Cross-sectional
Hashemipour	2006	Iran, Tehran	35°N	1210	495 Men 715 Women	20–69	• All: 13±16.5	% < 5: • All: 9% % 5–10: • All: 56%	Gender Season	Randomized clustered sampling from the Tehran population

Table 3
(continued)

Author	Year	Country, city	Latitude N	N	Gender	Age (years) mean ± SD or range	25(OH)D mean ± SD (ng/ml)	25(OH)D cutoff values (ng/ml)	Predictors	Comments
Saadi	2006	UAE, Al Ain	24°N	259	Women	20–85	• All: 10.1±4.3	% <8: • Premenopausal: 39%	Season (low in summer due to avoidance of heat)	Subjects recruited through advertisements. All had low vitamin D levels
Arabi	2006	Lebanon, Beirut	33°N	443	157 Men 286 Women	65–85	Median: • Men: 11.3 • Women: 9.6	% < 10–12: • Men: 37% • Women: 56% % Between 10 and 20: • Men: 57% • Women: 39%	Gender	Randomly recruited based on geographical maps
Hosseinpanah 2008		Iran, Tehran	35°N	245	Postmenopausal women	40–80	• All: 29.2±24.9	% < 10–12: • All: 5% % Between 10 and 20: • All: 38%	Menopause	Cross-sectional, random sampling

*UV: ultraviolet

to the hospital and their mothers and demonstrated that 50% of mothers and 22% of children had 25(OH)D levels below 10 ng/ml *(53)*. Over 2/3 were still breast-feeding and 40% did not receive vitamin D supplements. In a study of 119 pregnant woman and their newborns from Saudi Arabia, the median 25(OH)D concentration in mothers was 5.7 ng/ml, 50 cord samples had undetectable vitamin D levels, and it was below 4 ng/ml in another 81 cord specimens. Higher socioeconomic status, antenatal care, and vitamin D intake were associated with higher maternal vitamin D levels *(47)*. A cohort study of 48 Iranian pregnant women revealed that 20% had 25(OH)D levels below 10 ng/ml and 40% had levels below 20 ng/ml *(51)*. Fifty mothers and their term babies were studied at delivery in Tehran, 80% of the mothers had 25(OH)D below 10 ng/ml, mean cord serum levels were 2 ng/ml *(49)*; but there was no effect of maternal vitamin D levels on birth weight after adjusting for maternal height, age, and parity. Conversely, in a more recent study from Tehran, 449 apparently healthy women and their newborns were studied at the time of delivery *(54)*. Only one-third of women had adequate calcium and vitamin D intakes as determined by the Recommended Dietary Allowances *(55)*. Mean length at birth and 1-min Apgar scores in the neonates correlated with maternal calcium and vitamin D intake, and neonates of mothers with adequate calcium and vitamin D intake were 0.9 cm taller and had a higher Apgar at birth than those who consumed suboptimal levels *(54)*.

Several studies from the Middle East report that 10–60% of mothers and 40–80% of their apparently asymptomatic neonates have undetectable to low vitamin D levels (0–10 ng/ml) at delivery *(18, 47–52, 55)*. Socioeconomic status, educational level, antenatal care, and sun exposure were predictors of vitamin D levels.

3.2. Children and Adolescents

Hypovitaminosis D does not spare the pediatric age (Table 2) *(14, 46)*. A large proportion of adolescent girls, up to 70% in Iran *(56)* and 80% in Saudi Arabia *(57)* have 25(OH)D levels below 10 ng/ml. The proportions reported were lower, measured at 32% in Lebanese girls and between 9 and 12% in Lebanese adolescent boys *(58, 59)*. In a study of a random sample of 433 Saudi girls, ages 12–15 years, 81% had 25(OH)D levels below 10 ng/ml and 61% had no symptoms or rickets *(57)*. Similarly, a study of 318 adolescent Iranians reported that 25(OH)D levels were below 20 ng/ml in 46% of students, and in 72% in girls and 18% of boys *(56)*. The most recent cross-sectional, random-sampling survey of 963 students from Tehran, ages 7–18 years, revealed that the 25(OH)D level was below 20 ng/ml in 54% of girls (all of whom were veiled) and in 11% of boys (see Table 2 for mean levels) *(60)*. In contrast, the mean 25(OH)D level was higher, calculated at 30 ng/ml in another study of 414 Iranian girls *(61)*, only a subset of whom (N=15, 3.63%) had low 25(OH)D levels and hypocalcemia, unexpected findings in view of a reported mean vitamin D intake of 119 IU/day and sun exposure of 10 min/day *(61)*. A cross-sectional study of apparently healthy 458 Qatari children revealed a mean 25(OH)D of 13.4 ng/ml, vitamin D deficiency (defined as a 25(OH)D level less than 30 ng/ml) was reported in 315 of the children (68%); the highest proportions recorded were in adolescents, and the lowest were in children below age 5 years *(62)*. Fractures, delayed milestones, rickets, and gastroenteritis were more common in vitamin D-deficient children than control subjects *(62)*. The mean 25(OH)D (±SD) level in 346 adolescent Lebanese children was 17 (8) ng/ml at the end of winter and 22 (7) ng/ml at the end of summer *(58)*. Girls had lower sun exposure than boys, 9% of boys and 32% of girls had a level below 10 ng/ml, whereas 42% of girls and 46% of boys had levels between 10 and 20 ng/ml at the end of winter, with lower proportions at the end of summer *(58)*. The corresponding numbers for adolescents from the NHANES III study were 1–5% *(63)*.

Diarrhea and maternal vitamin D status were risk factors for low vitamin D levels in infants *(53, 64)*. Gender, physical activity, clothing style, season, and socioeconomic status were independent risk factors for low 25(OH)D levels in older children (Table 2) *(56 – 58, 61 – 62)*. Several of these risk factors were also predictors for vitamin D and calcium intake *(65)*. Indeed, although mean estimated calcium intake varied between studies, the mean intake was below the RDA recommendation for age in most studies when assessed.

3.3. Adults

In adults comparable results were observed (Table 3). One of the first studies conducted on apparently healthy adults from the region evaluated a small sample of university students and elderly from Saudi Arabia, and revealed a mean 25(OH)D level ranging between 4 and 12 ng/ml *(66)*. In a study of 316 Lebanese adult volunteers, mean 25(OH)D (SD) level was 10 *(7)* ng/ml, levels below 12 ng/ml were noted in 72% of the group, and the proportions were 84% in women and 48% in men *(67)*. In a sample of 465 adult Lebanese women of reproductive age, with conservative dress style, the mean 25(OH)D level in the summer was 11 ng/ml, 65% had a level below 10 ng/ml, 35% had levels between 10 and 20 ng/ml *(68)*; the mean vitamin D level was almost identical in a sample of 321 mostly premenopausal Saudi women *(69)*. The proportions for the same cutoff values in age-matched American women from the NHANES study were 12 and 82%, respectively *(70)*. In a study of 126 young adult Jordanians, 60% had 25(OH)D levels below 12 ng/ml, the proportions being 72% in the winter and 50% in the summer *(71)*. In a sample of 245 postmenopausal Iranian women, mean age 57 years, 5% had 25(OH)D levels below 10 ng/ml and 37% between 10 and 20 ng/ml *(72)*. However, the proportions were much higher in a random sample of 1200 Iranian men and women, 8% had 25(OH)D levels below 5 ng/ml and 60% had levels below 10 ng/ml *(73)*. The mean 25(OH)D level was 10 ng/ml in 259 Emirati women, vitamin D levels were lowest and PTH levels highest in the middle of summer, due to avoidance of going outdoors during the very hot summer season *(74)*. In a sample of 453 elderly Lebanese, aged 73 years, mean 25(OH)D level was 10 ng/ml, 37% of men and 56% of women had levels below 10 ng/ml; the proportions from a sample of similarly aged subjects from Turkey with a mean 25(OH)D below 15 ng/ml were between 28 and 54% for women and 4–18% for men, depending on their living environment (old age home versus own home) (Table 3) *(75 – 76)*. Similarly, it has been reported that up to 35% of elderly people in Israel have vitamin D deficiency or insufficiency; another subgroup of the population identified to be at higher risk were Ultra-Orthodox Jewish women, due to their conservative dress code *(77)*.

The mean 25(OH)D level was low averaging 10 ng/ml in studies conducted in Lebanese, Saudi, Emirati, and Iranian women *(67–69, 74)*, with a similar mean in elderly Lebanese *(75)*. The proportion of subjects with vitamin D levels below specific cut-offs varied between 60 and 65% for a vitamin D level <10 ng/ml in Lebanon, Jordan, and Iran *(68, 71 – 72)*. In the elderly Lebanese, 37% of men and 56% of women had vitamin D levels below 10 ng/ml *(75)*; the corresponding proportions were 8% for men and 14% for elderly subjects participating in the Longitudinal Aging Study Amsterdam *(78)*. Inadequate vitamin D intake, urban dwelling, female gender, conservative clothing style with a cover, season, age, and high parity were independent predictors of low vitamin D levels (Table 3) *(67, 71, 73 – 74, 79–81)*. Other high-risk subgroups are patients with specific diseases. In an international epidemiological investigation of women with osteoporosis, the highest proportion of hypovitaminosis D was noted in the Middle East *(82)*. Of 360 consecutive Qatari patients, mostly women, attending an outpatient rheumatology clinic, 56% had a mean 25(OH)D level below 20 ng/ml *(83)*. In a survey of 338 elderly admitted to a geriatric hospital in Tel Aviv Israel, with a mean 25(OH)D level of 13 ng/ml, 35% had a level below 10 ng/ml *(84)*, compared to 22% for the same cutoff in a group of 290 elderly patients admitted to a medical ward in Boston *(85)*. Finally, 80% of elderly with hip fractures in Israel had hypovitaminosis D *(77)*.

3.4. Immigrants from Middle East

Immigrants from Asian countries carry a high risk for severe vitamin D deficiency *(86 – 87)*. Serum 25(OH)D was reported to be below 10 ng/ml in 40% of non-western immigrants and to exceed 80% in pregnant women originating from Turkey and Morocco in the Netherlands *(88 – 89)*. In a cross-sectional population-based study of 1000 immigrants living in Oslo, natives of Turkey, Sri Lanka, Iran, Pakistan, and Vietnam, the median levels in Iranian women were 10.8 ng/ml, 45% had levels below 8 ng/ml; with comparable numbers in women from Turkey *(90)*.

3.5. Impact of Hypovitaminosis D on Musculoskeletal Outcomes

Although the two studies relating maternal or neonatal 25(OH)D levels to anthropometric measures of the neonates in the Middle East have yielded conflicting results, several studies worldwide suggest that neonate size or bone mass may be affected by maternal vitamin D status *(14, 91)*.

The deleterious impact of hypovitaminosis D on musculoskeletal health is unequivocal. It is illustrated in the inverse relationship between levels of 25(OH)D and parathyroid hormone, direct correlations between vitamin D and bone mass as well as muscle mass, and the efficacy of vitamin D in reducing falls and fractures *(14, 44, 92)*. Limited studies from the region yielded comparable results. The negative impact of low vitamin D on mineral metabolism is illustrated in the inverse relation between 25(OH)D and PTH levels noted in Lebanese adolescents, adults, and elderly, reminiscent of similar findings in other populations $(R = -0.19$ to $-0.22)$ *(58, 67, 75, 93)*. A similar negative relationship was noted in Saudi women $(R = -0.28)$ *(69)*, Emirati women $(R = -0.22)$ *(74)*, and Iranian women $(R = -0.25$ with Ln PTH) *(72)*. While no relationship between 25(OH)D level and BMD was detected in Saudi women, a positive relationship between

25(OH)D and spine, but not hip BMD Z-score was noted in postmenopausal Iranian women *(94)*. Similarly, a correlation was also noted between 25(OH)D level and spine, hip, and forearm BMD in elderly Lebanese ($R = 0.13$–0.3), but disappeared in analyses adjusting for age, height, lean mass, and PTH levels *(75)*; consistent with findings in Iranian women *(72)*. Vitamin D supplementation for 1 year increased lean mass, bone area, and bone mass in a randomized controlled trial in Lebanese adolescent girls *(59)*. In that trial, girls with the lowest levels had the highest increments in bone mass; and the effect was modulated by vitamin D receptor polymorphisms.

4. RECOMMENDATIONS FOR TREATMENT

Whereas there is little photosynthesis above and below latitudes of 40 degrees North and South during winter months, the latitudes of the Middle East allow the production of vitamin D throughout the year *(3)*. Barriers to sun exposure including fear of skin cancer and skin damage, cultural lifestyles including a conservative concealed clothing style, low outdoor activity and low intake of vitamin D-enriched foods, and the lack of governmental regulation for food fortification with vitamin D in several countries in the region, render supplementation the most efficacious mean to achieve desirable levels. A desirable 25(OH)D level of 30 ng/ml is recommended to optimize musculoskeletal health in adults *(3, 44, 92)*, and a similar target was suggested for the pediatric age group *(14, 46)*. Such a level was also demonstrated to optimize calcium transfer into the milk of a breast-feeding mother to satisfy her infant's needs *(44)*. Doses of vitamin D needed to reach the above desirable level are anticipated to exceed the recommended doses in subjects from western countries, due to the significantly lower baseline vitamin D levels in subjects from the Middle East. In a recent review of strategies to prevent and treat vitamin D deficiency, it was underscored that vitamin D_2 may be less effective than vitamin D_3 in maintaining serum 25(OH)D levels *(44)* and thus when available the use of D_3 would be preferable. Summarized below are results from interventional studies conducted in the Middle East. The type of vitamin D used in the various studies discussed is specified when reported. The evidence provided herein, and from other studies conducted worldwide, is used to derive treatment recommendations.

4.1. Rickets

Prevention of rickets: In a Cochrane review of interventions to prevent nutritional rickets in term infants, four studies qualified for inclusion, rickets did not occur in patients or controls in two of those studies, and only one study was conducted in the Middle East *(8)*. In the controlled trial from Turkey, 676 children aged 3–36 months, recruited from a rural community setting, received placebo or 400 IU of vitamin D daily over 12 months. The RR for developing rickets in the treated group was 0.04 [0–0.71] compared to the controls. In the second randomized study conducted in urban China, oral vitamin D_3 of 300 IU/day during the first 12 months administered with oral calcium supplementation from age 5 till 24 months, resulted in a RR for developing rickets of 0.76 [0.61–0.95]. The authors concluded that preventive measures using calcium and vitamin D in high-risk patients was indicated. The high-risk group specified included infants and toddlers living in Africa, Asia, and the Middle East *(8)*.

Treatment of rickets: Various regimens have been used to treat rickets in the Middle East, mostly conducted in non-randomized, non-blinded, interventional studies. In a study of 200 cases of rickets from Iran, subjects mostly below 3 years of age, classical rickets usually responded within 2–3 weeks of having received one injection of vitamin D at a dose of 600,000 IU *(33)*. Similarly, treatment of 47 Jordanian infants with nutritional rickets with one injection of 600,000 IU of vitamin D resulted in a substantial improvement in x-ray abnormalities, normalization of serum calcium and phosphate levels, and a significant decrement in alkaline phosphatase within 3 weeks of therapy *(18)*. In a dose-ranging study of 52 Turkish infants with nutritional rickets and a mean age of 10 months, subjects were randomized to receive one oral dose of vitamin D of 50,000, 300,000, or 600,000 IU and all received oral calcium for 1 week. On the 30th day there was no difference in the improvement between the three groups, and all patients had improved by the 60th day post-therapy; however, eight infants developed hypercalcemia, six of which were allocated to the high-dose group *(38)*. Treatment of 50 Iraqi infants with rickets and growth failure with a daily injection of 5,000 IU of vitamin D resulted in clinical and radiological improvement within 6 weeks of therapy (cumulative dose 210,000 IU) *(23)*. Treatment of 102 Saudi infants with daily vitamin D at doses of 2,500 IU/day for a month (75,000 IU total dose) followed by 400 IU/day of vitamin D resulted in normalization of the serum alkaline phosphatase from over 1,000 IU/L to below 500 IU/L in 90% of subjects 1 month post-therapy *(25)*. The efficacy of one intramuscular (IM) injection of 600,000 IU of vitamin D was compared with that of 2,000 IU/of oral vitamin D daily for 4 weeks in Kuwaiti infants with rickets, and both groups subsequently received 400 IU of vitamin D daily for 6 months *(11)*. Whereas a prompt response was noted in the group receiving IM therapy, no or a minimal response was noted in 40% of subjects assigned to the oral group. Conversely, comparison of a single oral dose of vitamin D_3 of 600,000 IU to an oral daily dose of 20,000 IU for 30 days in 20 Turkish infants with nutritional rickets resulted in comparable decrements in serum alkaline phosphatase, and similar increments in serum calcium, phosphorus and in bone mineral density *(95)*. In an interventional study conducted in 19 Emirati young children, with stunted growth and mean age 17 ± 6 months, treatment with ergocalciferol (2,000–5,000 IU/day) daily for 3 months followed by 400 IU/day for variable periods resulted in a mean increment in height Z score of 0.86 ± 0.95 *(31)*. Similarly, treatment of 46 Qatari infants with rickets, decreased height and growth velocity standard deviation scores, with one injection of vitamin D_3 of 300,000 IU resulted in significant increments in these variables, but not in normalization of stature *(96)*. A randomized trial compared calcium, vitamin D, and a combination of both in 42 Turkish infants with rickets randomized to calcium lactate 3 g/day, vitamin D 300, 000 IU IM once, or both *(97)*. Whereas all groups experienced increments in serum calcium and a decrease in serum alkaline phosphatase by the fourth week of therapy, the most substantial increment was in the combination group. Treatment of 34 Saudi adolescents with 1200 mg calcium and 4,000–5,000 IU of vitamin D_3/day or with α-calcidol 0.05–0.08 μg/kg resulted in normalization of serum alkaline phosphatase at a mean duration of 19 months (range 6–38 months) *(27)*.

In summary, treatment of rickets with intramuscular injections of vitamin D at doses of 150,000–600,000 IU/dose, with repeated dosing as needed, results in clinical, biochemical, and radiological improvement within weeks and may improve compliance. Conversely, daily oral supplementation with doses of 2,000–5,000 IU/day for several weeks may be as efficacious. Children with rickets from Egypt, Saudi Arabia, and the United Arab Emirates had a low calcium intake. Therefore, assessment of calcium nutrition and concomitant therapy with calcium is indicated and would expedite clinical recovery. Prevention of rickets in neonates and infants with vitamin D at a dose of 400 IU/day is justified based on evidence provided above.

4.2. Hypovitaminosis D

Pregnant/lactating mothers: The only interventional study in this age group from the region randomized 90 lactating and 88 nulliparous women to receive 2,000 IU D_2/day or 60,000 IU D_2/month over 3 months. The mean 25(OH)D levels in lactating women increased from 9.2 to 10.8 ng/ml to 15.2–16.8 ng/ml at 3 months, and the increments did not differ between the two regimens. The mean 25(OH)D levels in nulliparous women increased from 7.6 ng/ml to 15.6–16.8 ng/ml. The above doses were able to raise 25(OH)D levels to above 20 ng/ml only in 1/3 of subjects *(98)*. Although vitamin D nutrition during pregnancy and lactation seems to be important for offspring's skeletal health, and may contribute to genetic programming of the offspring for the development of chronic diseases in later life, trial-based evidence testing the safety and efficacy of vitamin D supplementation during these demanding reproductive years is currently lacking *(14, 91)*. It is, nevertheless, suggested that women should be replaced with vitamin D with monitoring of serum 25(OH)D levels to a target of 30 ng/ml *(44)*.

Infants/Children/Adolescents: Data pertaining to vitamin D nutrition during infancy and pre-pubertal years are also scarce. The most recent recommendations from the American Academy of Pediatrics recommend 400 IU/day of vitamin D in infants, children, and adolescents *(99)*. However, although such recommendations may be adequate for subjects with mild degrees of hypovitaminosis D, they would be suboptimal for subjects with more severe deficiencies as described for subjects from the Middle East, and at periods of peak bone mass accrual *(14, 46)*. A randomized double-blind placebo controlled trial conducted in Lebanese adolescent girls demonstrated the safety and efficacy of weekly vitamin D_3 administration at doses equivalent to 400 IU/day and 2,000 IU/day of vitamin D_3, in increasing bone mass, bone area, and muscle mass *(59, 100)*. Only the higher dose resulted in mean increments in 25(OH)D levels from the mid-teens (in ng/ml) to the mid-30s, whereas the mean increment with the 400 IU/day dose was 4 ng/ml, with a mean level remaining below 20 ng/ml *(59)*. There was a trend for further increments in musculoskeletal outcomes at the higher dose and in pre-pubertal girls *(59, 100)*.

Elderly: In a recent review of the evidence from several randomized controlled trials for the beneficial effect of vitamin D on musculoskeletal outcomes in the elderly, it was concluded that levels above 26 ng/ml would optimize bone health, reducing bone loss and fractures, and levels above 30 ng/ml would optimize muscle health, improving muscle performance and reducing fractures *(92)*. It has become increasingly recognized from data available in western countries that adequate intake for vitamin D of 600 IU/day for the elderly may be insufficient and that 800 IU/day may be more adequate *(44)*. However, this may not apply to elderly individuals from the Middle East, who have lower baseline 25(OH)D levels *(75 – 77, 101)*. A study comparing the efficacy of 3,000 IU/day of vitamin D_2 for 12 weeks, 25 μg/day of 25(OH)D for 1 week, and 0.5 μg/day of 1(OH)D for 8 weeks, in 30 elderly subjects from Beer Sheva, Israel, with a mean baseline 25(OH)D level of 15 ng/ml, demonstrated that supplementation with the D_3 and not the D_2 preparation significantly raised mean serum 25(OH)D levels *(102)*. More recently, a short-term randomized trial allocated 48 subjects with hip fractures from Haifa, mean age 81 years, mean 25(OH)D of 15 ng/ml, to vitamin D_3 1,5000 IU daily or 10,500 IU weekly or 45,000 IU once monthly. Such doses resulted in achieving a 25(OH)D level above 20 ng/ml in only half of the subjects *(101)*. Furthermore, although comparable mean 25(OH)D levels, varying between 29 and 37 ng/ml, were reached with the three dosing regimens, the once monthly dose may be particularly attractive because it may improve patient adherence. Furthermore, only the monthly dosing regimen substantially reduced circulating PTH levels *(101)*, an independent predictor of bone loss in the elderly *(103)*.

5. CONCLUSION

Vitamin D levels are quite low across all age groups in the Middle East. The persistence of rickets, and its sequela on morbidity and mortality in general, on growth and musculoskeletal outcomes in particular, is alarming. Whereas consistent predictors of rickets in infants were low maternal vitamin D levels and exclusive and prolonged breast-feeding; family crowding, lower socioeconomic status, lower parental educational level, and low calcium intake were additional risk factors in some studies. The skeletal manifestations of rickets were variable, and a common presentation was that of acute respiratory tract diseases and gastroenteritis, underscoring the importance of vitamin D as a potent immunomodulator. Various high-dose vitamin D regimens along with calcium supplementation resulted in the healing of clinical and radiographic features of rickets within weeks, but only partially corrected the growth retardation. Prevention of rickets with vitamin D, at a dose of 400 IU/day, is justified in neonates, infants, and toddlers.

Skin color is darker in some countries in the Gulf area, and obesity is a major public health problem in the Middle East across age groups, both conditions are known predictors of low vitamin D levels. Risk factors for hypovitaminosis D included female gender, multi-parity, season, conservative clothing style, low-socioeconomic status, and urban living. Overt clinical manifestations of insufficient vitamin D levels were almost nonexistent. However, the negative impact of low vitamin D levels on indices of mineral bone metabolism and the positive effect of vitamin D replacement on musculoskeletal outcomes support recommendations to optimize vitamin D status across all life stages. Whether vitamin D supplementation will also have a positive impact on chronic diseases associated with hypovitaminosis D, including the metabolic syndrome and type II diabetes, prevalent conditions in the Middle East, needs to be established.

Current recommendations for daily intake of vitamin D are likely to be suboptimal for subjects from the Middle East. The extensive body of evidence regarding the prevalence of hypovitaminosis D in the Middle East, information regarding the safety and efficacy of vitamin D at relatively high doses, and data from randomized dose-ranging studies in Lebanese adolescents and Israeli elderly justifies the following suggested doses. Daily vitamin D, at doses of 1,000–2,000 IU in infants and adolescents and 2,000–4,000 IU in elderly, with periodic monitoring coupled with adequate calcium nutrition is recommended. Intermittent dosing, using a cumulative amount of vitamin D proportionate to the daily dose (e.g., 7,000–14,000 IU/week instead of 1,000–2,000 IU/day), may enhance compliance. Further research is needed to conclusively define optimal vitamin D levels and determine the doses and regimens of vitamin D for pregnant and lactating women, infants and adolescents, and elderly subjects from the Middle East.

ACKNOWLEDGMENTS

This work was supported by institutional funds from the American University of Beirut, the Lebanese National Council for Scientific Research, the Nestle Foundation, and the World Health Organization Eastern Mediterranean Office. Special thanks to Ms Aida Farha for her help in the retrieval of selected articles and to Ms Rola El-Rassi and Mr Ghassan Baliki for their tireless assistance in performing PubMed searches, the retrieval of articles, and manuscript preparation.

REFERENCES

1. Koppes CR (1976) Captain Mahan, General Gordon and the origin of the term "Middle East". Middle East Stud 12:95–98
2. Beaumont P, Blake GH, Wagstaff JM (1998) The Middle East: a geographical study, 2nd ed. David Fulton Publishers, London, England
3. Holick M (2004) Vitamin D: importance in the prevention of cancers, type 1 diabetes, heart disease, and osteoporosis. Am J Clin Nutr 79:362–371
4. Pettifor JM (2004) Nutritional rickets: deficiency of vitamin D, calcium, or both?. Am J Clin Nutr 80(6 Suppl):1725S–9S
5. Liberman UA (2007) Vitamin D-resistant diseases. J Bon Miner Res Suppl 2:V105–V107
6. Thacher T, Fischer P, Strand M, Pettifor J (2006) Nutritional rickets around the world: causes and future directions. Ann Trop Paediatr 26:1–16
7. Baroncelli G, Bereket A, El Kholy M, Audì L, Cesur Y, Ozkan B, Rashad M, Fernández-Cancio M, Weisman Y, Saggese G, Hochberg Z (2008) Rickets in the Middle East: role of environment and genetic predisposition. J Clin Endocrinol Metab 93:1743–1750
8. Lerch C, Meissner T (2007) Interventions for the prevention of nutritional rickets in term born children. Cochrane Database Syst Rev 4:CD006164
9. Underwood P, Margetts B (1987) High levels of childhood rickets in rural North Yemen. Soc Sci Med 24:37–41
10. Awwaad S, Khalifa AS, Naga MA, Tolba KA, Fares R, Gaballa AS, Hayeg OE, Wahhab SA (1975) A field survey on child health in a rural community in Egypt. J Trop Med Hyg 78:20–25
11. Lubani MM, Al-Shab TS, Al-Saleh QA, Sharda DC, Quattawi SA, Ahmed SA, Moussa MA, Reavey PC (1989) Vitamin-D-deficiency rickets in Kuwait: the prevalence of a preventable problem. Ann Trop Paediatr 9:134–139
12. Fida NM (2003) Assessment of nutritional rickets in Western Saudi Arabia. Saudi Med J 24:337–340
13. Ozkan B, Doneray H, Karacan M, Vançelik S, Yıldırım ZK, Ozkan A, Kosan C, Aydın K (2009) Prevalence of vitamin D deficiency rickets in the eastern part of Turkey. Eur J Pediatr 168:95–100
14. Kimball S, El-Hajj Fuleihan G, Vieth R (2008) Vitamin D: a growing perspective. Crit Rev Clin Lab Sci 45:339–415

15. Abanamy A, Salman H, Cheriyan M, Shuja M, Siddrani S (1991) Vitamin D deficiency in Riyadh. Ann Saudi Med 11:35–39
16. Elidrissy ATH (1991) Vitamin D-deficiency rickets in Saudi Arabia. In: Glorieux EH (ed) Rickets. Nestle Nutrition Workshop Series, vol 21. Raven press, New York, 223–231
17. Ramavat LG (1999) Vitamin D deficiency rickets at birth in Kuwait. Indian J Pediatr 66:37–43
18. Najada AS, Mabashneh MS, Khader M (2004) The frequency of nutritional rickets among hospitalized infants and its relation to respiratory diseases. J Trop Pediatr 50:364–368
19. Taha SA, Dost SM, Sedrani SH (1984) 25-Hydroxyvitamin D and total calcium: extraordinarily low plasma concentrations in Saudi mothers and their neonates. Pediatr Res 18:739–741
20. Dawodu A, Agarwal M, Sankarankutty M, Hardy D, Kochiyil J, Badrinath P (2005) Higher prevalence of vitamin D deficiency in mothers of rachitic than nonrachitic children. J Pediatr 147:109–111
21. Hatun S, Ozkan B, Orbak Z, Doneray H, Cizmecioglu F, Toprak D, Calikoglu AS (2005) Vitamin D deficiency in early infancy. J Nutr 135:279–282
22. Amirhakimi GH (1973) Rickets in a developing country: Observations of general interest from Southern Iran. Clin Pediatr 12:88–92
23. Nagi NA (1972) Vitamin D deficiency rickets in malnourished children. J Trop Med Hyg 75:251–254
24. El Radhi AS, Majeed M, Mansor M, Ibrahim M (1982) High incidence of rickets in children with wheezy bronchitis in a developing country. J R Soc Med 75:884–887
25. Erfan A, Nafie O, Neyaz AA, Hassanein MA (1997) Vitamin D deficiency in maternity and children's hospital, Makkah, Saudi Arabia. Ann Saudi Med 17:371–373
26. Al-Mustafa ZH, Al-Madan M, Al-Majid HJ, Al-Muslem S, Al-Ateeq S, Al-Ali AK (2007) Vitamin D deficiency and rickets in the eastern province of Saudi Arabia. Ann Trop Paediatr 27: 63–67
27. Abdullah MA, Salhi HS, Bakry LA, Okamoto E, Abomelha AM, Stevens B, Mousa FM (2002) Adolescent rickets in Saudi Arabia: a rich and sunny country. J Pediatr Endocrinol Metab 15: 1017–1025
28. Al-Jurayyan NA, El-Desouki ME, Al-Herbish AS, Al-Mazyad AS, Al-Qhtani MM (2002) Nutritional rickets and osteomalacia in school children and adolescents. Saudi Med J 23:182–185
29. Narchi H, El Jamil M, Kulaylat N (2001) Symptomatic rickets in adolescents. Arch Dis Child 84: 501–503
30. Majid Molla A, Badawi MH, Al-Yaish S, Sharma P, el-Salam RS, Molla AM (2000) Risk factors for nutritional rickets among children in Kuwait. Pediatr Int 42:280–284
31. Rajah J, Jubeh J, Haq A, Shalash A, Parsons H (2008) Nutritional rickets and z scores for height in the United Arab Emirates: to D or not to D? Pediatr Int 50:424–428
32. Lawson DE, Cole TJ, Salem SI, Galal OM, el-Meligy R, Abdel-Azim S, Paul AA, el-Husseini S (1987) Etiology of rickets in Egyptian children. Hum Nutr Clin Nutr 41:199–208
33. Salimpour R (1975) Rickets in Tehran. Study of 200 cases. Arch Dis Child 50:63–66
34. Mathew PM, Imseeh GW (1992) Convulsions as a possible manifestation of vitamin D deficiency rickets in infants one to six months of age. Ann Saudi Med 12:34–37
35. Ahmed I, Atiq M, Iqbal J, Khurshid M, Whittaker P (1995) Vitamin D deficiency rickets in breast-fed infants presenting with hypocalcaemic seizures. Acta Paediatr 84:941–2
36. Erdeve O, Atasay B, Arsan S, Siklar Z, Ocal G, Berberoğlu M (2007) Hypocalcemic seizure due to congenital rickets in the first day of life. Turk J Pediatr 49:301–303
37. Karatekin G, Kaya A, Salihoğlu O, Balci H, Nuhoğlu A (2009) Association of subclinical vitamin D deficiency in newborns with acute lower respiratory infection and their mothers. Eur J Clin Nutr 63(4):473–477
38. Cesur Y, Caksen H, Gundem A, Kirimi E, Odabaş D (2003) Comparison of low and high dose of vitamin D treatment in nutritional vitamin D deficiency rickets. J Pediatr Endocrinol Metab 16: 1105–1109
39. Maiya S, Sullivan I, Allgrove J, Yates R, Malone M, Brain C, Archer N, Mok Q, Daubeney P, Tulloh R, Burch M (2008) Hypocalcaemia and vitamin D deficiency: an important, but preventable, cause of life-threatening infant heart failure. Heart 94:581–584
40. Kosecik M, Ertas T (2007) Dilated cardiomyopathy due to nutritional vitamin D deficiency rickets. Pediatr Int 49:397–399

41. Amirlak I, Al Dhaheri W, Narchi H (2008) Dilated cardiomyopathy secondary to nutritional rickets. Ann Trop Paediatr 28:227–230

42. Robinson PD, Hogler W, Craid ME, Verge CF, Walker JL, Piper AC, Woodhead HJ, Cowell CT, Ambler GR (2006) The re-emerging burden of rickets: a decade of experience from Sydney. Arch Dis Child 91:564–568

43. Mallet E, Gaudelus J, Reinert P, Le Luyer B, Lecointre C, Léger J, Loirat C, Quinet B, Bénichou JJ, Furioli J, Loeuille GA, Roussel B, Larchet M, Freycon F, Vidailhet M, Varet I (2004) Symptomatic rickets in adolescents. Arch Pediatr 11:871–878

44. Holick MF (2007) Vitamin D deficiency. N Engl J Med 357:266–281

45. Lips P (2007) Vitamin D status and nutrition in Europe and Asia. J Steroid Biochem Mol Biol 103:620–625

46. El-Hajj Fuleihan G, Vieth R (2007) Vitamin D insufficiency and musculoskeletal health in children and adolescents. International Congress Series, Elsevier 1297:91–108

47. Serenius F, Elidrissy AT, Dandona P (1984) Vitamin D nutrition in pregnant women at term and in newly born babies in Saudi Arabia. J Clin Pathol 37:444–447

48. Dawodu A, Agarwal M, Hossain M, Kochiyil J, Zayed R (2003) Hypovitaminosis D and vitamin D deficiency in exclusively breast-feeding infants and their mothers in summer; a justification for vitamin D supplementation of breast-feeding infants. J Pediatr 142:169–173

49. Bassir M, Laborie S, Lapillonne A, Claris O, Chappuis MC, Salle BL (2001) Vitamin D deficiency in Iranian mothers and their neonates: A pilot study. Acta Paediatrica 90:577–579

50. Molla AM, Al Badawi M, Hammoud MS, Molla AM, Shukkur M, Thalib L, Eliwa MS (2005) Vitamin D status of mothers and their neonates in Kuwait. Pediatr Int 47:649–652

51. Ainy E, Ghazi AA, Azizi F (2006) Changes in calcium, 25(OH) vitamin D3 and other biochemical factors during pregnancy. J Endocrinol Invest 29:303–307

52. Biale Y, Shany S, Levi M, Shainkin-Kestenbaum R, Berlyne GM (1979) 25 Hydroxy- cholecalciferol levels in Beduin women in labor and in cord blood of their infants. Am J Clin Nutr 32:2380–2382

53. Dawodu A, Dawson KP, Amirlak I, Kochiyil J, Agarwal M, Badrinath P (2001) Diet, clothing, sunshine exposure and micronutrient status of Arab infants and young children. Ann Trop Paediat 21:39–44

54. Sabour H, Hossein-Nezhad A, Maghbooli Z, Madani F, Mir E, Larijani B (2006) Relationship between pregnancy outcomes and maternal vitamin D and calcium intake: A cross-sectional study. Gynecol Endocrinol 22:585–589

55. Standing Committee on the Scientific Evaluation of Dietary Reference Intakes, Food and Nutrition Board, Institute of Medicine (1997) Dietary reference intakes for calcium, phosphorus, magnesium, vitamin D, and fluoride. http://books.nap.edu/openbook.php?record_id=5776&page=250. Accessed 21 December 2009

56. Moussavi M, Heidarpour R, Aminorroaya A, Pournaghshband Z, Amini M (2005) Prevalence of vitamin D deficiency in Isfahani high school students in 2004. Horm Res 64:144–148

57. Siddiqui AM, Kamfar HZ (2007) Prevalence of vitamin D deficiency rickets in adolescent school girls in western region, Saudi Arabia. Saudi Med J 28:441–444

58. El-Hajj Fuleihan G, Nabulsi M, Choucair M, Salamoun M, Hajj Shahine C, Kizirian A, Tannous R (2001) Hypovitaminosis D in healthy school children. Pediatrics 107:E53

59. El-Hajj Fuleihan G, Nabulsi M, Tamim H, Maalouf J, Salamoun M, Khalife H, Choucair M, Arabi A, Vieth R (2006) Effect of vitamin D replacement on musculoskeletal parameters in school children: A randomized controlled trial. J Clin Endocrinol Metab 91:405–412

60. Rabbani A, Alavian SM, Motlagh ME, Ashtiani MT, Ardalan G, Salavati A, Rabbani B, Rabbani A, Shams S, Parvaneh N (2008) Vitamin D insufficiency among children and adolescents living in Tehran, Iran. J Trop Pediatr [Epub ahead of print]

61. Dahifar H, Fraji A, Ghorbani A, Yassobi S (2006) Impact of dietary and lifestyle on vitamin D in healthy student girls aged 11–15 years. J Med Invest 53:204–208

62. Bener A, Al-Ali M, Hoffmann G (2008) Vitamin D deficiency in healthy children in a sunny country: associated factors. Int J Food Sci Nutr[Epub ahead of print]

63. Looker AC, Dawson-Hughes B, Calvo MS, Gunter EW, Sahyoun NR (2002) Serum 25-hydroxyvitamin D status of adolescents and adults in two seasonal subpopulations from NHANES III. Bone 30:771–777

64. Bahijri SM (2001) Serum 25-hydroxy cholecalciferol in infants and preschool children in the western region of Saudi Arabia: etiological factors. Saudi Med J 22:973–979

65. Salamoun MM, Kizirian AS, Tannous RI, Nabulsi MM, Choucair MK, Deeb ME, El-Hajj Fuleihan GA (2005) Low calcium and vitamin D intake in healthy children and adolescents and their correlates. Eur J Clin Nutr 59:177–184

66. Sedrani SH, Elidrissy AW, El Arabi KM (1983) Sunlight and vitamin D status in normal Saudi subjects. Am J Clin Nutr 38:129–132

67. Gannage-Yared MH, Chemali R, Yaacoub N, Halaby G (2000) Hypovitaminosis D in a sunny country: relation to lifestyle and bone markers. J Bone Miner Res 15:1856–1862

68. El-Hajj Fuleihan G, Deeb M (1999) Letter to the Editor. Hypovitaminosis in a Sunny Country. N Engl J Med 340:1840–41

69. Ghannam NN, Hammami MM, Bakheet SM, Khan BA (1999) Bone mineral density of the spine and femur in healthy Saudi females: Relation to vitamin D status, pregnancy, and lactation. Calcif Tissue Int 65:23–28

70. Looker AC, Gunter EW (1998) Hypovitaminosis D in medical patients. N Engl J Med 339:344–5

71. Mishal AA (2001) Effects of different dress styles on vitamin D levels in healthy young Jordanian women. Osteoporos Int 12:931–935

72. Hosseinpanah F, Rambod M, Hossein-Nejad A, Larijani B, Azizi F (2008) Association between vitamin D and bone mineral density in Iranian postmenopausal women. J Bone Miner Metab 26:86–92

73. Hashemipour S, Larijani B, Adibi H, Sedaghat M, Pajouhi M, Bastan-Hagh MH, Soltani A, Javadi E, Shafaei AR, Baradar-Jalili R, Hossein-Nezhad A (2006) The status of biochemical parameters in varying degrees of vitamin D deficiency. J Bone Miner Metab 24:213–218

74. Saadi HF, Nagelkerke N, Benedict S, Qazaq HS, Zilahi E, Mohamadiyeh MK, Al-Suhaili AI (2006) Predictors and relationships of serum 25 hydroxyvitamin D concentration with bone turnover markers, bone mineral density, and vitamin D receptor genotype in Emirati women. Bone 39:1136–1143

75. Arabi A, Baddoura R, Awada H, Salamoun M, Ayoub G, El-Hajj Fuleihan G (2006) Hypovitaminosis D osteopathy: Is it mediated through PTH, lean mass, or is it a direct effect? Bone 39:268–275

76. Atli T, Gullu S, Uysal AR, Erdogan G (2005) The prevalence of vitamin D deficiency and effects of ultraviolet light on vitamin D levels in elderly Turkish population. Arch Gerontol Geriat 40:53–60

77. Weisman Y (2003) Vitamin D deficiency rickets and osteomalacia in Israel. IMAJ 5:289–90

78. Snijder MB, van Dam RM, Visser M, Deeg DJ, Dekker JM, Bouter LM, Seidell JC, Lips P (2005) Adiposity in relation to vitamin D status and parathyroid hormone levels. A population-based study in older men and women. J Clin Endocrinol Metab 90:4119–4123

79. El-Sonbaty MR, Abdul-Ghaffar NU (1996) Vitamin D deficiency in veiled Kuwaiti women. Eur J Clin Nutr 50:315–318

80. Mirsaeid Ghazi AA, Rais Zadeh F, Pezeshk P, Azizi F (2004) Seasonal variation of serum 25 hydroxy D3 in residents of Tehran. J Endocrinol Invest 27:676–679

81. Alagol F, Shihadeh Y, Boztepe H, Tanakol R, Yarman S, Azizlerli H, Sandalci O (2000) Sunlight exposure and vitamin D deficiency in Turkish women. J Endocrinol Invest 7:783–789

82. Lips P, Hosking D, Lippuner K, Norquist JM, Wehren L, Maalouf G, Ragi-Eis S, Chandler J (2006) The prevalence of vitamin D inadequacy amongst women with osteoporosis: an international epidemiological investigation. J Int Med 260:245–254

83. Siam AR, Hammoudeh M, Khanjer I, Bener A, Sarakbi H, Mehdi S (2006) Vitamin D deficiency in Rheumatology clinic practice in Qatar. Qatar Med J 15:49–51

84. Goldray D, Mizrahi-Sasson E, Merdler C, Edelstein-Singer M, Algoetti A, Eisenberg Z, Jaccard N, Weisman Y (1989) Vitamin D deficiency in elderly patients in a general hospital. J Am Geriatr Soc 37:589–592

85. Thomas MK, Lloyd-Jones DM, Thadhani RI, Shaw AC, Deraska DJ, Kitch BT, Vamvakas EC, Dick IM, Prince RL, Finkelstein JS (1998) Hypovitaminosis D in medical inpatients. N Engl J Med 338:777–783

86. Preece MA, McIntosh WB, Tomlinson S, Ford JA, Dunnigan MG, O'Riordan JL (1973) Vitamin D deficiency among Asian immigrants to Britain. Lancet i:907–910

87. Torrente de la Jara G, Pecoud A, Farrat B (2004) Musculoskeletal pain in female asylum seekers and hypovitaminosis D3. BMJ 329:56–57

88. Van der Meer IM, Boeke AJ, Lips P, Grootjans-Geerts I, Wuister JD, Devillé WL, Wielders JP, Bouter LM, Middelkoop BJ (2008) Fatty fish and supplements are the greatest modifiable contributors to hydroxyvitamin D concentration in multi-ethnic population. Clin Endocrinol 68:466–472

89. Van der Meer IM, Karamali NS, Boeke AJ, Lips P, Middelkoop BJ, Verhoeven I, Wuister JD (2006) High prevalence of vitamin D deficiency in pregnant non-Western women in The Hague, Netherlands. Am J Clin Nutr 84:350–353

90. Holvik K, Meyer HE, Haug E, Brunvand L (2005) Prevalence and predictors of vitamin D deficiency in five immigrant groups living in Oslo, Norway, the Oslo Immigrant Health Study. Eur J Clin Nutr 59:57–63

91. Kovacs C, El-Hajj Fuleihan G (2006) Calcium and bone disorders during pregnancy and lactation. Endocrinol Metab Clin North Am 35:21–51

92. Dawson-Hughes B (2008) Serum 25-hydroxyvitamin D and functional outcomes in the elderly. Am J Clin Nutr (Suppl):537S–40S

93. Veith R, El-Hajj Fuleihan G (2005) There is no lower threshold for parathyroid hormone as 25-hydroxy vitamin D concentration increases. J Endocrinol Invest 28:183–186

94. Rassouli A, Milanian I, Moslemi-Zadeh M (2001) Determination of serum 25-hydroxyvitamin D (3) levels in early postmenopausal Iranian women: Relationship with bone mineral density. Bone 29:428–430

95. Akcam M, Yildiz M, Yilmaz A, Artan R (2006) Bone mineral density in response to two different regimes in rickets. Indian Pediatr 43:423–427

96. Soliman A, Al Khalaf F, AlHemaidi N, Al Ali M, Al Zyoud M, Yakoot K (2008) Linear growth in relation to the circulating concentrations of insulin-like growth factor I, parathyroid hormone, and 25-hydroxy vitamin D in children with nutritional rickets before and after treatment: endocrine adaptation to vitamin D deficiency. Metabolism 57:95–102

97. Kutluk G, Cetinkaya F, Basak M (2002) Comparisons of oral calcium, high dose vitamin D and a combination of these in the treatment of nutritional rickets in children. J Trop Pediatr 48:351–353

98. Saadi HF, Dawodu A, Afandi BO, Zayed R, Benedict S, Nagelkerke N (2007) Efficacy of daily and monthly high-dose calciferol in vitamin D-deficient nulliparous and lactating women. Am J Clin Nutr 85:1565–1571

99. Wagner CL, Greer FAmerican Academy of Pediatrics Section on Breastfeeding; American Academy of Pediatrics Committee on Nutrition (2008) Prevention of rickets and vitamin D deficiency in infants, children, and adolescents. Pediatrics 122:1142–1152

100. Maalouf J, Nabulsi M, Vieth R, Kimball S, El-Rassi R, Mahfoud Z, El-Hajj Fuleihan G (2008) Short- and long-term safety of weekly high-dose vitamin D_3 supplementation in school children. J Clin Endocrinol Metab 93:2693–2701

101. Ish Shalom S, Segal E, Salganik T, Raz B, Bromberg IL, Vieth R (2008) Comparison of daily, weekly, and monthly vitamin D_3 in ethanol dosing protocols for two months in elderly hip fracture patients. J Clin Endocrinol Metab 93:3430–3435

102. Shany S, Chaimovitz C, Yagev R, Bercovich M, Lowenthal MN (1988) Vitamin D-deficiency in the elderly: treatment with ergocalciferol and hydroxylated analogues of vitamin D_3. Israel J Med Sci 24:160–163

103. Arabi A, Baddoura R, El-Hajj Fuleihan G (2008) PTH and not vitamin D predicts age related bone loss in the elderly: a prospective population-based study. J Bone Miner Res 23:S417

25 Vitamin D Deficiency in the Middle East and Its Health Consequences for Adults

Samer El-Kaissi and Suphia Sherbeeni

Abstract   Information from limited studies suggests that vitamin D levels in Middle Eastern populations are among the lowest in the world. Vitamin D insufficiency or deficiency affects a large proportion of the population in most countries and almost everyone in some countries such as Saudi Arabia and United Arab Emirates. Particularly at risk are the young, women, and the elderly. The etiology is likely to be multifactorial, but is predominantly related to reduced sun exposure due to sociocultural and religious factors, in addition to extreme weather conditions in certain regions. Reduced consumption of vitamin D and calcium-rich foods and the absence of adequate food fortification programs are additional risk factors for vitamin D deficiency in the Middle East. Population-based studies are necessary to effectively address this problem and institute appropriate treatment and prevention strategies.

Key Words: Vitamin D deficiency; 25-hydroxyvitamin D; Lebanon; Middle East; rickets; calcium; veiling; infants; elderly; nutrition

1. INTRODUCTION

Despite the presence of year-round sunshine, vitamin D deficiency and insufficiency appear to be highly prevalent in the Middle East *(1)*. The task of estimating the magnitude of this problem is hampered by the lack of recent well-designed population-based studies in addition to the use of different cutoff levels to define what constitutes normal vitamin D status and poor inter-laboratory standardization of the 25-hydroxyvitamin D [25(OH)D] assay *(2)*. This chapter will review available evidence on the status of vitamin D in the Middle East and the potential causes of this emerging epidemic.

2. EPIDEMIOLOGY

An estimated 1 billion people around the world have vitamin D deficiency ($\leq$ 50 nmol/l (20 ng/ml)) or insufficiency ($\leq$75 nmol/l (30 ng/ml)) *(1)*. The Middle East is no exception despite an abundance of sunshine. In fact, in a multinational study

From: *Nutrition and Health: Vitamin D*
Edited by: M.F. Holick, DOI 10.1007/978-1-60327-303-9_25,
© Springer Science+Business Media, LLC 2010

of 2606 postmenopausal women with osteoporosis in 18 countries including Lebanon and Turkey, the lowest 25(OH)D mean of 51 nmol/l (20.4 ng/ml) was recorded in those two countries compared to 61.0 nmol/l (24.4 ng/ml) in Asia, 69.7 nmol/l (27.9 ng/ml) in Australia, 73.2 nmol/l (29.3 ng/ml) in Europe, and 73.3 nmol/l (29.3 ng/ml) in Latin America. In addition, the two Middle Eastern countries had the highest prevalence of vitamin D insufficiency (<75 nmol/l (30 ng/ml)) and deficiency (<22.5 nmol/l (9 ng/ml)) with 82 and 9% of women affected, respectively. The respective percentage of patients affected in the other continents were 71 and 2.6% in Asia, 60 and 1% in Australia, 58 and 1.6% in Europe, and 53 and 0.7% in Latin America (3).

Studies from Saudi Arabia suggest that up to 100% of the population have vitamin D deficiency or insufficiency. In 1983, Sedrani et al. (4) found unexpectedly low vitamin D levels in Saudi university students ($n = 59$) and in elderly subjects ($n = 24$), although the mean 25(OH)D levels were significantly lower in elderly subjects (9 nmol/l (3.6 ng/ml)) followed by young males (21 nmol/l (8.4 ng/ml)) and highest in young females (29 nmol/l (11.6 ng/ml)). Vitamin D levels were below 12.5 nmol/l (5.0 ng/ml) in 83% of the elderly subjects and below 25 nmol/l (10 ng/ml) in 73% of young males and in 33% of young females. In a subsequent study, up to 53% ($n = 175$) of Saudis and other Arab nationals living in the central region of Riyadh were found to have 25(OH)D levels below 25 nmol/l (10 ng/ml), with a mean of 32 nmol/l (12.8 ng/ml) (5). Similarly, Taha et al. (6) demonstrated that 59% of mothers and 70% of their neonates ($n = 100$) had 25(OH)D levels below 25 nmol/l (10 ng/ml) at birth, while Serenius et al. (7) detected severe vitamin D deficiency (25(OH)D less than 10 nmol/l (4 ng/ml)) in 25% of 119 Saudi mothers and 42% of their newborns.

In 1992, the first population-based study of the vitamin D status in Saudi Arabia was published (8). The mean 25(OH)D was 41 nmol/l (16.4 ng/ml), and vitamin D levels below 25 nmol/l (10 ng/ml) were documented in 26% of children and 18% of adults. The respective percentages of children and adults with 25(OH)D less than 12.5 nmol/l (5 ng/ml) were 4 and 6%. However, in studying the bone mineral density of healthy Saudi females, Ghannam et al. showed that 52% ($n = 321$) had 25(OH)D levels of 20 nmol/l (8 ng/ml) or less (9).

More recent studies from Saudi Arabia show that almost everyone has vitamin D deficiency or insufficiency. Of 360 Saudi patients with lower back pain of unknown etiology, 83% had vitamin D levels below 22.5 nmol/l (9 ng/ml) (10), while 81% of 433 school girls had levels below 25 nmol/l (10 ng/ml) and 40% were below 12.5 nmol/l (5 ng/ml) (11). Most alarming is a study of 71 ambulatory primary health-care patients (87% females) from the Eastern Province of Saudi Arabia where the mean 25(OH)D concentration was 19 nmol/l (7.6 ng/ml), with 100% of patients having levels below 75 nmol/l (30 ng/ml) and 97% below 50 nmol/l (20 ng/ml) (12).

In Turkey, 33% ($n = 420$) of elderly subjects had 25(OH)D levels below 37.5 nmolmL (15 ng/ml), with nursing home residents almost twice as likely to be affected as people living in their own homes (40 vs. 24%). This study also showed that females are at much higher risk of vitamin D deficiency whether they lived in nursing homes (54 vs. 18% for males) or in their own homes (28 vs. 4% for males) (13). Erkal et al. studied adult Turkish migrants in Germany and showed that more than 78% ($n = 893$) of all Turkish nationals had 25(OH)D levels below 50 nmol/l (20 ng/ml) whether they lived in Turkey or Germany, compared to 29% of Germans ($n = 101$). Severe vitamin D deficiency (<25 nmol/l (10 ng/ml)) was seen in 30 and 19% of Turkish females, compared to 8 and 6% of Turkish males living in Germany and Turkey, respectively *(14)*.

The vitamin D status in Iran seems somewhat better than other Middle Eastern countries. Hashemipour et al. *(15)* recruited 1082 adults from Tehran and found an average serum 25(OH)D of 32.5 nmol/l (13 ng/ml). Serum 25(OH)D levels were less than 12.5 nmol/l (5 ng/ml) in 9% of subjects, 12.5–35 nmol/l (5–14 ng/ml) in 70% but only 19% had levels greater than 35 nmol/l (14 ng/ml).

In a cross-sectional study of 318 Iranian high-school students, the mean 25(OH)D level at the end of winter was 93 nmol/l (37.2 ng/ml) in males and 42 nmol/l (16.8 ng/ml) in females. Vitamin D levels below 80, 50, and 20 nmol/l (32, 20, and 8 ng/ml) were detected in 73, 46, and 8% of the sample, respectively. Females were more likely than males to have vitamin D insufficiency (95 vs. 49%), mild deficiency (72 vs. 18%), and severe deficiency (15 vs. 0.7%) *(16)*.

In another study, 245 postmenopausal Iranian women had a mean 25(OH)D of 73 nmol/l (29.2 ng/ml). The vitamin D level was below 25 nmol/l (10 ng/ml) in 5% of women, between 25 and 50 nmol/l (10–20 ng/ml) in 38%, and above 80 nmol/l (32 ng/ml) in 25% *(17)*. Similarly, the mean 25(OH)D during winter was 75 nmol/l (30 ng/ml) in 414 Iranian girls aged 11–14 years, and less than 4% of girls had levels below 25 nmol/l (10 ng/ml) *(18)*. It is unclear why the mean vitamin D levels are higher in Iran despite similar cultural and socioeconomic factors as those of other Middle Eastern countries, but the effect of inter-laboratory variation in the vitamin D assay cannot be excluded *(2)*.

In the United Arab Emirates (UAE), Saadi et al. *(19)* compared 259 Emarati women to 7 Western women residing in the Emirates, and although all women in both groups had 25(OH)D levels below 80 nmol/l (32 ng/ml), the vitamin D means were significantly lower in Arabs (26 vs. 63 nmol/l (10.4 vs. 25.2 ng/ml)). It is interesting to note that the 25(OH)D means were much lower toward the end of the summer season than at the end of winter (18 vs. 29 nmol/l (7.2 vs. 11.6 ng/ml)), reflecting the lack of outdoor activities in the summer due to extreme heat.

In a similar study of 33 Emarati women of child-bearing age, 25 non-gulf Arabs and 17 Europeans, the mean serum 25(OH)D levels were 22, 32, and 161 nmol/l (8.8, 12.8 and 64.4 ng/ml), respectively. Furthermore, 24% of Emarati women and 12% of non-gulf Arabs had 25(OH)D levels below 12.5 nmol/l (5 ng/ml) *(20)*. The same authors demonstrated that 61% of breastfeeding women ($n = 90$) and 82% of their infants ($n = 78$) had 25(OH)D levels below 25 nmol/l (10 ng/ml) *(21)*, compared to 50 and 22%, respectively ($n = 51$), in children around 15 months old, possibly due to the introduction of food and vitamin D supplements *(22)*.

Studies from Lebanon also indicate that a substantial proportion of the community is affected by vitamin D deficiency or insufficiency. In a study of 316 adults (69% females), the mean 25(OH)D level was 24 nmol/l (9.6 ng/ml), with 73 and 31% of subjects having levels below 30 nmol/l (12 ng/ml) and 12.5 nmol/l (5 ng/ml), respectively *(23)*. A more recent study by the same group demonstrated that 85% of 251 postmenopausal Lebanese women with osteoporosis had 25(OH)D levels below 75 nmol/l (30 ng/ml)

(24). In both studies, veiled Muslim women had lower vitamin D levels compared to non-veiled women *(23, 24)*.

The 25(OH)D mean of Lebanese school children lies between 35 nmol/l (14 ng/ml) and 55 nmol/l (22 ng/ml), being lower in females and during the winter months *(25, 26)*. An estimated 44 and 36% of school children had 25(OH)D levels between 25 and 50 nmol/l (10–20 ng/ml) in spring and fall, respectively. In addition, 21% of students had vitamin D levels below 25 nmol/l (10 ng/ml) during spring, whereas only 4% were affected in fall. Girls were 3.5 times more likely to have severe vitamin D deficiency compared to boys *(26)*. The elderly seem prone to more severe vitamin D deficiency as up to 60% of the institutionalized and 49% of the ambulatory people over 70 years of age had 25(OH)D levels below 25 nmol/l (10 ng/ml) [*(27)* cited in *(23)*].

In the multiethnic society of Israel, Bedouins *(28, 29)* and Ethiopian migrants *(30)* are more likely than other ethnic groups to have vitamin D deficiency. In an earlier study, Shany et al. found that Bedouin men and women had 25(OH)D means up to 64 nmol/l (25.6 ng/ml) with slightly lower levels during pregnancy, but none of the Bedouins had levels below 25 nmol/l (10 ng/ml). In contrast, the mean vitamin D level in non-Bedouins in the city of Beersheba was much higher at 110 nmol/l (44 ng/ml) in non-pregnant females and up to 82 nmol/l (32.8 ng/ml) in males and in pregnant females *(28)*. More recently, the 25(OH)D mean of medical inpatients was demonstrated to be much lower at 53 nmol/l (21.2 ng/ml), and 26% of patients had vitamin D levels less than or equal to 37.5 nmol/l (15 ng/ml). The latter was seen in 47% of Bedouins, compared to 28% of Africans and Asians, and 20% of those of European or American origin, highlighting the importance of sociocultural and ethnic factors in the development of vitamin D deficiency *(29)*.

Limited information is available about the vitamin D status of other Middle Eastern countries. A study of 214 Kuwaiti mothers and their neonates at birth found mean 25(OH)D levels of 37 nmol/l (14.8 ng/ml) and 21 nmol/l (8.4 ng/ml), respectively. More importantly, up to 83% of mothers and 98% of their infants had 25(OH)D levels below 50 nmol/l (20 ng/ml) and up to 18% of mothers and 37% of infants were below 12.5 nmol/l (5 ng/ml) at birth. This study emphasized the role of maternal education and good antenatal care in reducing the risk of severe vitamin D deficiency in pregnancy *(31)*.

In Yemen, rickets was diagnosed clinically in up to 27% of children attending immunization clinics *(32)* and in 50% of children hospitalized with pneumonia *(33)*, while 11% of inpatient Jordanian children had clinical evidence of rickets, most of whom were admitted because of respiratory tract infections *(34)*. In North Africa, 48% of a mostly female Tunisian adult sample (*n* = 389) was found to have 25(OH)D levels below 37.5 nmol/l (15 ng/ml) *(35)*, while the mean vitamin D levels of 60 Egyptian females with lower back pain of unknown etiology and 20 normal controls were surprisingly high at 91 nmol/l (36.4 ng/ml) and 100 nmol/l (40 ng/ml), respectively *(36)*.

3. ETIOLOGY

There are multiple causes of vitamin D deficiency *(1)*. Of particular importance in the Middle East are reduced sun exposure, reduced dietary vitamin D and calcium intake, in addition to skin pigmentation in some ethnic groups which can reduce vitamin D synthesis by up to 99%, and increasing prevalence of malabsorptive disorders such as celiac disease *(37, 38)*.

Sun exposure is a major source of vitamin D in humans *(1)*. Apart from the extreme heat and frequent dust storms in certain regions of the Middle East , reduced sun exposure is primarily related to sociocultural and religious factors. Muslim women are required to wear a "hijab" which consists of a headscarf and full-length clothing so that only the face, hands, and feet are exposed. In some states like Saudi Arabia and Iran, women wear a full black cloak "abaya" and may also cover their face "niqab," hands, and feet. Men usually wear an ankle-length shirt "thawb" and apply a head cover "shemagh " so that only the face, hands, and feet are exposed. There are significant variations in the dress code between states, and western-style clothing is the norm in some states such as Turkey, Morocco, Tunisia, and Lebanon.

Vitamin D levels have been shown to be lower in veiled than non-veiled women in most studies *(14, 19, 20, 23, 24, 26, 35, 39–41)*, but not all *(42)*. In a study of Turkish migrants in Germany, females wearing a headscarf had a fivefold higher risk of vitamin D insufficiency *(14)*. Exposure of the face, hands, and feet as opposed to total body coverage in veiled Emarati women has not been shown to improve vitamin D levels, presumably because of skin pigmentation and limited outdoor activities *(19)*.

Reduced vitamin D levels in veiled women may not be entirely related to reduced sun exposure as suggested by the presence of additional risk factors for vitamin D deficiency such as a lower dietary intake of vitamin D compared to non-veiled women *(23)*, greater BMI *(23, 41)*, and higher parity *(41)*. The latter has been associated with a twofold increase in the risk of vitamin D insufficiency in Turkish women *(14)*.

Dietary vitamin D intake in the Middle East ranges from 55 IU/day in Saudi Arabia *(4)* to 150 IU/d in Lebanon *(26)*. Dietary calcium intake is also insufficient being as low as 100–300 mg/day in rachitic Saudi adolescents, and almost up to 800 mg/day in Lebanese students *(23, 26)*. Furthermore, a study from Lebanon shows that only 12% (*n* = 385) of students consume more than 200 IU of vitamin D and 1300 mg of dietary calcium per day *(43)*. In another Lebanese study, the vitamin D intake was greater in men than women (128 vs. 88 IU/d), in non-veiled than veiled women (100 vs. 74 IU/d), and in urban than rural inhabitants (130 vs. 70 IU/d) *(23)*.

Of particular concern is a study from Iran suggesting that more than two-thirds of pregnant women do not take any form of calcium or vitamin D supplementation during pregnancy *(44)*. Vitamin D fortification of dairy products is not mandatory in most parts of the Middle East , although the Saudi Ministry of Health mandates the fortification of liquid milk and has recently introduced measures to increase the amount of cholecalciferol in milk from 400 to 800 IU/l. Nevertheless, unpublished data from Saudi milk producers suggest that the average yearly consumption of milk in Saudi Arabia is less than 50 L per capita, possibly due to the relatively high incidence of lactose intolerance in the region *(45, 46)*. In addition to the low dietary vitamin D and calcium intake, vitamin D deficiency in the Middle East may be exacerbated by the consumption bread with a relatively high phytate content in certain regions such as Egypt *(47)*, rural Iran *(48)*, and Bedouins in Israel *(28)*. Phytate is a strong chelator and can reduce the

Table 3

absorption of calcium from the gut *(49)*, thereby exacerbating calcium deficiency which is associated with PTH-induced vitamin D metabolism *(1)*.

Low socioeconomic status has been associated with lower vitamin D levels in adolescent Turkish girls *(42)*, Lebanese high-school students *(26)*, and pregnant Kuwaiti and Saudi women *(7, 31)*. The latter has been attributed to the quality of antenatal care with insufficient maternal education and lower consumption of calcium and vitamin D supplements during pregnancy *(31)*.

Compared to males, females are at increased risk of vitamin D deficiency and insufficiency *(13, 14, 16, 23, 26, 35)*. Potential reasons for this finding include reduced sun exposure as a result of veiling *(14, 19, 20, 23, 24, 26, 41)* and limited outdoor activities, reduced consumption of dietary calcium and vitamin D *(23, 35)*, greater prevalence of obesity *(50–52)*, high parity *(14, 23, 35)*, and prolonged breastfeeding. The estimated duration of breastfeeding in Saudi women is around 2 years *(9)* and among breastfeeding Emarati mothers, only 40% consume vitamin D-fortified milk, despite which they did not meet the recommended daily allowance for vitamin D consumption *(21)*. Furthermore, prolonged and exclusive breastfeeding is an important contributor to vitamin D deficiency in breast-fed infants of mothers who are already vitamin D deficient, as has been demonstrated in rachitic children from Kuwait *(53)* and Jordan *(34)*.

4. CONCLUSIONS

Preliminary data suggest that vitamin D deficiency and insufficiency are widespread in the Middle East. Women, young children, and the elderly are particularly at risk of more severe vitamin D deficiency. Reduced sun exposure due to sociocultural and religious reasons appears to be the predominant factor in the etiology of vitamin D deficiency, although other causes such as poor dietary intake of calcium and vitamin D and prolonged breastfeeding cannot be discounted.

Given the potential public health implications of vitamin D deficiency *(1, 54–57)*, population-based studies using accepted definitions for optimal vitamin D levels are urgently needed to evaluate the magnitude of the problem and to institute adequate vitamin D replacement and food fortification programs.

REFERENCES

1. Holick MF (2007) Vitamin D deficiency. N Engl J Med 357:266–281
2. Lips P, Chapuy MC, Dawson-Hughes B et al. (1999) An international comparison of serum 25-hydroxyvitamin D measurements. Osteoporos Int 9:394–397
3. Lips P, Hosking D, Lippuner K et al. (2006) The prevalence of vitamin D inadequacy amongst women with osteoporosis: an international epidemiological investigation. J Intern Med 260:245–254
4. Sedrani SH, Elidrissy AW and El Arabi KM (1983) Sunlight and vitamin D status in normal Saudi subjects. Am J Clin Nutr 38:129–132
5. Sedrani SH (1984) Low 25-hydroxyvitamin D and normal serum calcium concentrations in Saudi Arabia: Riyadh region. Ann Nutr Metab 28:181–185
6. Taha SA, Dost SM and Sedrani SH (1984) 25-Hydroxyvitamin D and total calcium: extraordinarily low plasma concentrations in Saudi mothers and their neonates. Pediatr Res 18:739–741
7. Serenius F, Elidrissy AT and Dandona P (1984) Vitamin D nutrition in pregnant women at term and in newly born babies in Saudi Arabia. J Clin Pathol 37:444–447
8. Sedrani SH, Al-Arabi K, Abanmy A et al. (1992) Vitamin D status of Saudis: effect of age, sex and living accomodation. Saudi Med J 13:151–158
9. Ghannam NN, Hammami MM, Bakheet SM et al. (1999) Bone mineral density of the spine and femur in healthy Saudi females: relation to vitamin D status, pregnancy, and lactation. Calcif Tissue Int 65:23–28
10. Al Faraj S and Al Mutairi K (2003) Vitamin D deficiency and chronic low back pain in Saudi Arabia. Spine 28:177–179
11. Siddiqui AM and Kamfar HZ (2007) Prevalence of vitamin D deficiency rickets in adolescent school girls in Western region, Saudi Arabia. Saudi Med J 28:441–444

12. Eledrisi M, Alamoudi R, Alhaj B et al. (2007) Vitamin D: deficiency or no deficiency? South Med J 100:543–544
13. Atli T, Gullu S, Uysal AR et al. (2005) The prevalence of vitamin D deficiency and effects of ultraviolet light on vitamin D levels in elderly Turkish population. Arch Gerontol Geriatr 40:53–60
14. Erkal MZ, Wilde J, Bilgin Y et al. (2006) High prevalence of vitamin D deficiency, secondary hyperparathyroidism and generalized bone pain in Turkish immigrants in Germany: identification of risk factors. Osteoporos Int 17:1133–1140
15. Hashemipour S, Larijani B, Adibi H et al. (2006) The status of biochemical parameters in varying degrees of vitamin D deficiency. J Bone Miner Metab 24:213–218
16. Moussavi M, Heidarpour R, Aminorroaya A et al. (2005) Prevalence of vitamin D deficiency in Isfahani high school students in 2004. Horm Res 64:144–148
17. Hosseinpanah F, Rambod M, Hossein-nejad A et al. (2008) Association between vitamin D and bone mineral density in Iranian postmenopausal women. J Bone Miner Metab 26:86–92
18. Dahifar H, Faraji A, Ghorbani A et al. (2006) Impact of dietary and lifestyle on vitamin D in healthy student girls aged 11–15 years. J Med Invest 53:204–208
19. Saadi HF, Nagelkerke N, Benedict S et al. (2006) Predictors and relationships of serum 25 hydroxyvitamin D concentration with bone turnover markers, bone mineral density, and vitamin D receptor genotype in Emirati women. Bone 39:1136–1143
20. Dawodu A, Absood G, Patel M et al. (1998) Biosocial factors affecting vitamin D status of women of childbearing age in the United Arab Emirates. J Biosoc Sci 30:431–437
21. Dawodu A, Agarwal M, Hossain M et al. (2003) Hypovitaminosis D and vitamin D deficiency in exclusively breast-feeding infants and their mothers in summer: a justification for vitamin D supplementation of breast-feeding infants. J Pediatr 142:169–173
22. Dawodu A, Dawson KP, Amirlak I et al. (2001) Diet, clothing, sunshine exposure and micronutrient status of Arab infants and young children. Ann Trop Paediatr 21:39–44
23. Gannage-Yared MH, Chemali R, Yaacoub N et al. (2000) Hypovitaminosis D in a sunny country: relation to lifestyle and bone markers. J Bone Miner Res 15:1856–1862
24. Gannage-Yared MH, Maalouf G, Khalife S et al. (2008) Prevalence and predictors of vitamin D inadequacy amongst Lebanese osteoporotic women. Br J Nutr 1–5
25. El-Hajj Fuleihan G, Nabulsi M, Tamim H et al. (2006) Effect of vitamin D replacement on musculoskeletal parameters in school children: a randomized controlled trial. J Clin Endocrinol Metab 91:405–412
26. El-Hajj Fuleihan G, Nabulsi M, Choucair M et al. (2001) Hypovitaminosis D in healthy schoolchildren. Pediatrics 107:E53
27. Gannage-Yared MH, Brax H, Asmar A et al. (1998) Vitamin D status in aged subjects. Study of a Lebanese population. Presse Med 27:900–904
28. Shany S, Hirsh J and Berlyne GM (1976) 25-Hydroxycholecalciferol levels in bedouins in the Negev. Am J Clin Nutr 29:1104–1107
29. Hochwald O, Harman-Boehm I and Castel H (2004) Hypovitaminosis D among inpatients in a sunny country. Isr Med Assoc J 6:82–87
30. Fogelman Y, Rakover Y and Luboshitzky R (1995) High prevalence of vitamin D deficiency among Ethiopian women immigrants to Israel: exacerbation during pregnancy and lactation. Isr J Med Sci 31:221–224
31. Molla AM, Al Badawi M, Hammoud MS et al. (2005) Vitamin D status of mothers and their neonates in Kuwait. Pediatr Int 47:649–652
32. Underwood P and Margetts B (1987) High levels of childhood rickets in rural North Yemen. Soc Sci Med 24:37–41
33. Banajeh SM, al-Sunbali NN and al-Sanahani SH (1997) Clinical characteristics and outcome of children aged under 5 years hospitalized with severe pneumonia in Yemen. Ann Trop Paediatr 17:321–326
34. Najada AS, Habashneh MS and Khader M (2004) The frequency of nutritional rickets among hospitalized infants and its relation to respiratory diseases. J Trop Pediatr 50:364–368
35. Meddeb N, Sahli H, Chahed M et al. (2005) Vitamin D deficiency in Tunisia. Osteoporos Int 16:180–183

36. Lotfi A, Abdel-Nasser AM, Hamdy A et al. (2007) Hypovitaminosis D in female patients with chronic low back pain. Clin Rheumatol 26:1895–1901

37. Akbari MR, Mohammadkhani A, Fakheri H et al. (2006) Screening of the adult population in Iran for coeliac disease: comparison of the tissue-transglutaminase antibody and anti-endomysial antibody tests. Eur J Gastroenterol Hepatol 18:1181–1186

38. Cataldo F and Montalto G (2007) Celiac disease in the developing countries: a new and challenging public health problem. World J Gastroenterol 13:2153–2159

39. Alagol F, Shihadeh Y, Boztepe H et al. (2000) Sunlight exposure and vitamin D deficiency in Turkish women. J Endocrinol Invest 23:173–177

40. Mishal AA (2001) Effects of different dress styles on vitamin D levels in healthy young Jordanian women. Osteoporos Int 12:931–935

41. Guler T, Sivas F, Baskan BM et al. (2007) The effect of outfitting style on bone mineral density. Rheumatol Int 27:723–727

42. Olmez D, Bober E, Buyukgebiz A et al. (2006) The frequency of vitamin D insufficiency in healthy female adolescents. Acta Paediatr 95:1266–1269

43. Salamoun MM, Kizirian AS, Tannous RI et al. (2005) Low calcium and vitamin D intake in healthy children and adolescents and their correlates. Eur J Clin Nutr 59:177–184

44. Sabour H, Hossein-Nezhad A, Maghbooli Z et al. (2006) Relationship between pregnancy outcomes and maternal vitamin D and calcium intake: A cross-sectional study. Gynecol Endocrinol 22:585–589

45. Sadre M and Karbasi K (1979) Lactose intolerance in Iran. Am J Clin Nutr 32:1948–1954

46. Al-Sanae H, Saldanha W, Sugathan TN et al. (2003) Comparison of lactose intolerance in healthy Kuwaiti and Asian volunteers. Med Princ Pract 12:160–163

47. Abd-el-Fattah M, Gabrial GN, Shalaby SM et al. (1978) An epidemiological and biochemical study on osteomalacia among pregnant women in Egypt. Z Ernahrungswiss 17:140–144

48. Reinhold JG (1971) High phytate content of rural Iranian bread: a possible cause of human zinc deficiency. Am J Clin Nutr 24:1204–1206

49. Hurrell RF (2003) Influence of vegetable protein sources on trace element and mineral bioavailability. J Nutr 133:2973S–2977S

50. Fouad M, Rastam S, Ward K et al. (2006) Prevalence of obesity and its associated factors in Aleppo, Syria. Prev Control 2:85–94

51. Heidari B, Bijani K, Eissazadeh M et al. (2007) Exudative pleural effusion: effectiveness of pleural fluid analysis and pleural biopsy. East Mediterr Health J 13:765–773

52. Khader Y, Batieha A, Ajlouni H et al. (2008) Obesity in Jordan: prevalence, associated factors, comorbidities, and change in prevalence over ten years. Metab Syndr Relat Disord 6:113–120

53. Majid Molla A, Badawi MH, al-Yaish S et al. (2000) Risk factors for nutritional rickets among children in Kuwait. Pediatr Int 42:280–284

54. Holick MF (2004) Sunlight and vitamin D for bone health and prevention of autoimmune diseases, cancers, and cardiovascular disease. Am J Clin Nutr 80:1678S–1688S

55. Holick MF (2004) Vitamin D: importance in the prevention of cancers, type 1 diabetes, heart disease, and osteoporosis. Am J Clin Nutr 79:362–371

56. Holick MF (2006) High prevalence of vitamin D inadequacy and implications for health. Mayo Clin Proc 81:353–373

57. Holick MF and Chen TC (2008) Vitamin D deficiency: a worldwide problem with health consequences. Am J Clin Nutr 87:1080S–1086S

26 Vitamin D Deficiency and Its Health Consequences in Africa

Ann Prentice, Inez Schoenmakers, Kerry S. Jones, Landing M.A. Jarjou, and Gail R. Goldberg

Abstract   Rickets and osteomalacia are established consequences of clinical vitamin D deficiency, while emerging evidence suggests that vitamin D status, at plasma 25-hydroxyvitamin D[25(OH)D] concentrations above 25 nmol/l, may be linked to other health consequences such as diabetes, cancer and susceptibility to infections. Africa is a large continent, heterogeneous in its latitude, geography, climate, food availability, religious and cultural practices and skin pigmentation. It would be expected, therefore, that the prevalence of vitamin D deficiency would vary widely across and within African countries, in line with the known influences on skin exposure to UVB sunshine. Furthermore, the low calcium intakes and heavy burden of infectious disease common in many African countries may increase the utilisation and turnover of vitamin D. Published data from African studies suggest the presence of a spectrum of vitamin D status, as assessed using the plasma concentration of 25(OH)D, from those associated with clinical deficiency (<25 nmol/l) to values at the high end of the physiological range. This suggests that there are individuals in Africa at risk of clinical vitamin D deficiency and potentially of sub-optimal vitamin D status. However, these data are limited and there is a need for representative data to be collected in different countries across Africa, using comparable analytical techniques by accredited laboratories. Studies are also needed to determine relationships between vitamin D status and risk of infectious and chronic diseases that are relevant to the African context. Until more is known, public health measures aimed at securing vitamin D adequacy in Africa cannot encompass the whole continent and need to be developed on a local basis.

Key Words: Vitamin D deficiency; Africa; rickets; sunlight; skin pigment; UV radiation; black; malnutrition; calcium deficiency; fluorosis; vitamin D metabolism

1. INTRODUCTION

This chapter addresses the prevalence of vitamin D deficiency and issues that need to be considered when considering vitamin D status (supply, sufficiency, utilisation) and potential health consequences in Africa. The reader is referred to other chapters that discuss in detail some of the general background to vitamin D including: evolution, biologic functions and dietary requirements; photobiology; assays; pregnancy and lactation;

From: *Nutrition and Health: Vitamin D*
Edited by: M.F. Holick, DOI 10.1007/978-1-60327-303-9_26,
© Springer Science+Business Media, LLC 2010

calcium and vitamin D deficient rickets and children's bone health; dietary sources; sun deprivation and health consequences. Where appropriate these are also considered here because they are particularly relevant to, or may be different in, an African context.

Vitamin D status and supply is influenced by sunshine exposure, diet and underlying health conditions. People acquire vitamin D by cutaneous synthesis following exposure to sunlight, and from the diet. It is not possible to generalise about the proportions contributed by different sources because this varies considerably around the world, across population groups, and between individuals, mainly because of wide differences in skin exposure to ultraviolet B (UVB) radiation, the efficiency of cutaneous synthesis, dietary supplementation and food fortification practices (1). Reliance on dietary sources of vitamin D is greatest during the winter at latitudes outside the tropics (> 30°N and > 30°S) and among people with restricted skin sunshine exposure (1, 2). The endogenous supply of vitamin D3 (cholecalciferol) depends on the skin's exposure to UVB radiation at wavelengths of 290–315 nm, the efficiency of cutaneous vitamin D synthesis, and the extent to which vitamin D is degraded within the skin (3–5). The quantity of UVB radiation of the relevant wavelengths that reaches the skin depends on many factors (see Section 2.3) including time of day, and, at latitudes north and south from approximately 30°N and 30°S, on month of the year. In the summer months however, the amount of UVB containing wavelengths that can generate vitamin D synthesis is the same at all latitudes (6). In addition, even when there is abundant sunshine, the extent of UVB skin exposure depends on clothing, living and working environments and sunscreen use (7).

The status marker for vitamin D, plasma 25-hydroxyvitamin D [25(OH)D] concentration, depends on factors that determine the balance between supply and turnover (8). Supply depends on factors including host, dietary and environmental factors. Turnover of 25(OH)D and 1,25-dihydroxyvitamin D (1,25(OH)$_2$D) in renal and extra-renal tissues depends on many factors, including nutrient intake (of which calcium is the most important), protein–energy malnutrition, body composition, high disease burden and innate characteristics such as polymorphisms in the vitamin D receptor (VDR) and DBP genes (8–12).

2. THE CONTINENT OF AFRICA

2.1. Geography, Topography, Climate

Africa is the world's second-largest continent after Asia, occupying an area of about 30 million km^2 including adjacent islands. Thus Africa represents about 20% of the earth's total land area. It is the only continent that straddles the equator and has both northern and southern temperate zones. The distance between the most northerly (~37°N in Tunisia) and most southerly (~34°S in South Africa) points is ~8,000 km; from the westernmost (~17°W, Cape Verde) to the easternmost (~51°E in Somalia) points is ~7,400 km. Some of the larger African countries are distributed across more than one latitude band. The climate ranges from tropical to subarctic and the topography includes deserts, mountains, grassland plateaus, lakes and rivers. Much of northern Africa is predominantly desert or arid land, and central and southern areas contain deserts, savannah plains and dense rainforest. All these features affect the available water supply, food sources, population distribution and the sorts of clothing worn (e.g. for warmth, protection against the sun, wind, rain and sand). The weather in Africa is also very variable, and temperatures vary greatly throughout the continent. Many countries have wet and dry seasons which in turn affect food availability, farming activities and, therefore, levels of physical activity, weight and body composition changes, morbidity and mortality, including incidence of low birthweight infants, and the prevalence and incidence of malaria. About 74% of people in Africa live in areas that are highly endemic for malaria, 19% in epidemic-prone areas and only 7% in low-risk or malaria-free areas *(13)*.

2.2. Demography

Africa is the second most populated continent after Asia with more than 900 million people who account for ∼14% of the world's population. Africa comprises more than 50 countries ranging in landmass from 2.4 million km^2 (Sudan) to 455 km^2 (The Seychelles) (www.fao.org). Individual countries are not only very diverse with respect to size but also population, ranging from 145,000,000 (Nigeria) to 86,000 (The Seychelles) and the proportion of the population that live in urban areas (10% Burundi; 84% Gabon; average 38% in WHO Africa Region) *(14)*. Across the continent as a whole ∼40% of Africans are Christian, ∼40% are Muslim and ∼20% follow other religions (including indigenous African religions, Hinduism and Judaism). There is a great deal of variation within and between countries and population groups. Some religious beliefs and traditional practices can impact on customary dress and dietary habits.

2.3. Ultraviolet Radiation and Skin Pigmentation

The factors that affect the amount of ambient UVB radiation include stratospheric ozone levels, cloud cover, latitude, season and lower atmospheric pollution. The factors that affect an individual's UVB exposure and biological effects of UVB include sun-seeking and sun protective behaviours, skin pigmentation, and cultural dress and behaviour *(15)*. The concentration of atmospheric ozone that is able to absorb UVB, and the amount and spectral structure of radiation reaching the body is dependent on the angle at which the sun's rays pass through the atmosphere; at lower latitudes, closer to the equator there is more intense radiation. At higher altitudes, UVB radiation is increased because there is less mass of air for solar radiation to pass through. Time of day, season and presence of clouds, dust, haze and various organic compounds can alter the intensity of incident solar radiation. The UV Index, the maximum UV level on a given day during the 4-h period around solar noon, varies according to time and place, and with season and latitude. Levels of total annual UV radiation (UVR) vary approximately fourfold globally, but within any area there can be a tenfold difference in personal exposure related to behavioural and cultural factors. Individual exposure within population groups may vary from one-tenth to ten times the mean exposure in a particular location.

Different skin pigmentations are thought to have developed to protect against high ambient UVR whereby those inhabiting low latitudes with high UVR have darker pigmentation to protect against deleterious effects, whilst those inhabiting higher latitudes developed fair skin to maximise vitamin D production from much lower ambient UVR. However, in recent history migration of people of different skin types means that skin pigmentation is not necessarily so suited to the environment. For the purposes of assessing the effects of UVR on different skin pigmentations, a scale classifying skin types for UVR sensitivity is used which comprises six skin types *(16, 17)*. This is often broken down into three broader skin pigmentation groups *(15)*: lightly pigmented (Caucasian fair, medium and darker skin, skin types I–IV), intermediate pigmentation (Asian or Indian skin, skin type V) and deeply pigmented (Afro-Caribbean or black skin, skin type VI). Latitude provides a rough approximation to global variation in UVR, so in epidemiological studies of UVR, groups of different races or ethnicities can be categorised under these broad headings, but with recognition that there is great individual variation within racial and ethnic groups *(15)*. An illustration of the distribution of skin types at different latitudes in Africa is shown in Table 1

The efficiency of vitamin D cutaneous synthesis depends on skin pigmentation and age *(18, 19)*. Very little data from controlled interventional research are available on the differences in efficiency in vitamin D synthesis between different skin types, but a fourfold difference has been reported between type II and type V *(20)*. However, evidence from studies in African-American Blacks and Indians suggests that there is no difference in the total capacity of cutaneous vitamin D synthesis between individuals with different skin types *(21, 22)*. There are no studies in the literature investigating cutaneous synthesis in indigenous African people.

On the African continent, seasonality in cutaneous vitamin D synthesis would be expected in countries such as Morocco, Tunisia, Algeria, Libya, Egypt and South Africa which lie at latitudes > 30°N and > 30°S *(23)*. This is supported by studies in hip fracture patients in South Africa *(24)* and in pregnant women and their newborns in Algeria *(25)*. In countries close to the equator, seasonality in cutaneous vitamin D synthesis might be expected due to differences in cloud cover (http://nadir.nilu.no/,olaeng/fastrt/Vitamin D.html). In two studies, in The Gambia (14°N) and Guinea Bissau (12°N), no significant effect of season on plasma 25(OH)D concentrations was found *(26, 27)*. However, seasonal variations in plasma 25(OH)D concentration have been reported in older people in The Gambia where lower concentrations were observed in December–February *(1)*.

2.4. Diet and Foodstuffs

The African diet is very heterogeneous and varies from region to region. Historically, the climate, geography and history largely dictated where populations became pastoralists, agriculturalists and hunter-gatherers, and, therefore, local availability and ability to store fresh and salt-water fish, meat, poultry and milk and the kinds of crops that could be grown and thus traditionally consumed. Over the centuries, there have been many secular changes in diet and food composition. There are now also 'westernised' populations in most African countries; the proportions differ between and within countries

Table 1
WHO Sub-regions[a] Comprising African Countries, by Latitude, and % Distribution of Population Within a Region in That Latitude with Different Skin Pigmentation (Compiled From Information in (15))

Latitude	WHO sub-region, country (% of country in latitude band, if not 100%)	Lightly pigmented	Intermediate pigmentation	Deeply pigmented
0–10	Africa D: Cameroon, Equatorial Guinea, Gabon, Ghana, Liberia, Sao Tome and Principe, Seychelles, Sierra Leone, Angola (50%), Benin (90%), Nigeria (50%), Guinea (50%) Togo	<1	<1	41
	Africa E: Burundi, Central African Republic, Congo, Cote D'Ivoire, Kenya, Rwanda, Democratic Republic of Congo (95%), Ethiopia (50%), Tanzania	<<1	<1	59
	East Mediterranean D: Somalia, Sudan (20%)	<1	<1	3
10–20	Africa D: Cap Verde, Chad, Comoros, Gambia, Guinea-Bissau, Mauritius, Niger, Senegal, Angola (50%), Benin (10%), Burkina Faso, Guinea (50%), Madagascar (50%, Mauritania (50%), Nigeria (50%), Mali	<1	2.5	42
	Africa E: Eritrea, Malawi, Zimbabwe, Botswana (10%), Democratic Republic of Congo (5%), Mozambique (70%), Namibia (20%), Ethiopia (50%)	<1	<1	22
	East Mediterranean D: Sudan (79%)	8	1	4
20–30	Africa D: Algeria (10%), Madagascar 50%), Mauritania (50%)	<1	2	2
	Africa E: Lesotho, Swaziland, Zambia, Botswana (90%), Mozambique (30%), Namibia (80%), South Africa (80%)	1	1	13
	East Mediterranean B: Libya (50%), Morocco (10%)	22	2	<1
	East Mediterranean D: Sudan (1%), Egypt (50%)	<1	34	<1

Table 1
(continued)

Latitude	WHO sub-region, country (% of country in latitude band, if not 100%)	Lightly pigmented	Intermediate pigmentation	Deeply pigmented
30–40	*Africa D*: Algeria (90%)	<1	10	<1
	Africa E: South Africa (20%)	<1	<1	2
	East Mediterranean B: Libya (50%), Morocco (90%), Tunisia	70	2	<1
	East Mediterranean D: Egypt (50%)	14	34	<1

[a]Countries are listed according to both WHO Africa regions (D and E) and other countries on the African continent which come under WHO East Mediterranean regions B and D. All these WHO regions encompass the countries mentioned in this chapter.

and are changing as countries transit from less to more developed, and depending on natural and manmade environment. This transition affects dietary patterns, the sorts of foods consumed, nutrient intakes and sources of nutrients.

3. INDICATORS OF HEALTH AND BURDEN OF DISEASE IN AFRICA WHICH MAY AFFECT OR BE AFFECTED BY VITAMIN D STATUS

Africa is associated with a high prevalence of economic disadvantage and high burden of disease. This is still mostly infectious disease but increasingly also non-communicable disease. The various factors that may be especially important to consider in the context of Africa, and that may affect the interpretation of vitamin D status and deficiency are outlined below. In addition, vitamin D status has been implicated in the progression of disease, and this can add another layer of complexity in interpreting vitamin D status measurements. The emerging work on the possible relationship between vitamin D status and disease risk reduction (TB and HIV/AIDS) is discussed in Section 4.

3.1. Malnutrition

The percentage of infants born with low birthweight is ~14% in the WHO African region and in some African countries is >20%. There are many reasons for low birthweight but a major contributory factor in Africa is low maternal weight gain due to general shortage of foods or energy or specific nutrients. Low birthweight is the major risk factor for infant mortality and morbidity. The most recent estimate of infant mortality rate in the WHO Africa region is 94 per 1,000 live births. In some African countries the rate is >100. These figures are in the context of a global rate of 49. The under 5 mortality rate in Africa is 157 per 1,000 live births and in some African countries is >200 per 1,000 live births (global rate = 71). Recent estimates suggest that 43% of children in Africa are stunted and 23% are underweight for their age. In some African countries, the prevalence is >50% and >30% (14). Vitamin D or calcium-deficiency rickets are often found where more general malnutrition is prevalent.

3.2. Vitamin D Intake

There are very little data from Africa on dietary vitamin D intakes, although sources of other nutrients and food patterns indicate that, as observed elsewhere, food intake is unlikely to make a substantial contribution because there are few naturally occurring food sources that are rich in vitamin D (meat, offal, egg yolk, oily fish), and these are not eaten at all or only rarely (28–30). It is impossible to quantify or generalise about the possible contribution of vitamin D in foods supplied as food aid (some of which may be fortified) or in the form of multiple micronutrient supplements.

3.3. Low Calcium Intake

Where dietary calcium intakes have been assessed in African countries, the values from infants, children and adults are considerably below recommendations, and close to biological requirements. Calcium intakes of 200–300 mg/day have been reported from Egypt, Kenya, Nigeria, The Gambia and South Africa *(31)*. Dairy product intake is minimal in many African countries and typical diets also contain high amounts of phytates, oxalates and tannins that are likely to reduce the absorption of calcium. Children with bone deformities consistent with rickets but without overt vitamin D deficiency have been reported from Nigeria, The Gambia and South Africa. Dietary calcium deficiency has been identified as a possible contributing factor due to its effects on the metabolism of vitamin D *(32)*.

3.4. Fluorosis

Fluoride is present in all natural waters, but higher concentrations are found in groundwater from calcium-poor aquifers and in areas where fluoride-bearing minerals are common. In Africa the latter applies particularly to areas in the East African rift system, and many lakes in the rift valley (Sudan, Ethiopia, Uganda, Kenya and Tanzania) have very high levels of fluoride. High groundwater fluoride concentrations are associated with igneous and metamorphic rocks including those in western Africa (e.g. Nigeria, Senegal) and South Africa *(33)*. Endemic skeletal fluorosis has been associated with hypocalcaemia, phosphaturia and raised concentration of $1,25(OH)_2D$ *(34)*; all indices that are associated with various forms of rickets.

3.5. Tropical Enteropathy

Tropical enteropathy (tropical sprue) is common in many parts of Africa. It is characterised by intestinal permeability and poor absorption, abnormal villus structure and function in adults and children and thought to be caused by many factors including repeated episodes of infection and diahorrea *(35–38) (39)*. Poor nutrient absorption of calcium and other factors that may affect vitamin D metabolism may impact on vitamin D status, which in turn may affect gastrointestinal health. Thus tropical enteropathy may be a contributing factor in vitamin D deficiency and conversely may be exacerbated by poor vitamin D status. However, although these links are plausible, they have not yet been explored.

3.6. Malaria

Of the 350–500 million estimated cases of clinical malaria each year, nearly 60% of cases and 90% of deaths occur in Africa south of the Sahara with 75% of these in children under 5 *(13)*. Malaria affects everyone, but children under 5 years of age and pregnant women are the most vulnerable groups because of lower immunity. Malaria contributes to health problems and deaths in many ways, especially in younger children. These include frequent acute infections; anaemia as a result of repeated or chronic malaria infection; malaria in pregnancy resulting in low birthweight in infants; increased susceptibility to other diseases such as respiratory infections and diarrhoea *(40)*. The immunmodulatory properties of vitamin D discussed in Section 4 may also play a role in the progression of malaria and its co-morbidities.

3.7. HIV/AIDS

Sub-Saharan Africa is the region most affected by HIV/AIDS. In 2007 one in three people in the world with HIV lived in sub-Saharan Africa; a total of 22.5 million people. Cause-specific mortality rates per 100,000 population in 2006 were: 203 for HIV/AIDS (compared with a global rate of 34); 56 for TB among HIV negative people (global rate 22) and 26 for TB among HIV positive people (global rate 4). The mean prevalence in Africa (estimated in 2005) of HIV among people over 15 years of age is 4,459 per 100,000 population (global figure is 662) and in some African countries the figures are >20,000 *(14)*.

3.8. Tuberculosis

Globally tuberculosis (TB) is the second highest cause of death from infectious disease and is responsible for more deaths in those individuals infected with HIV than any other infectious disease *(41)*. Almost 30% of TB cases and 34% of TB-related deaths occur in Africa and it is estimated that between 1990 and 2005 the incidence of TB more than doubled in many African countries *(42)*. The prevalence of TB in Africa (estimated in 2006) is 547 per 100,000 population, and is >900 in some African countries (global prevalence is 219) *(14)*. Although HIV infection is a major cause, poverty, inadequate healthcare (in terms of diagnosis and treatment) and drug resistance have all contributed *(42)*.

3.9. Non-communicable Disease

In addition to a high burden of infectious disease in Africa, secular changes in diet and lifestyle particularly in North Africa and urban centres in sub-Saharan Africa are leading to increases in cardiovascular disease, cancer, and obesity and its co-morbidities including hypertension and type 2 diabetes *(43–47)*. Low vitamin D status has been implicated in diabetes and cancer (see Chapters 20, 49). Serum 25(OH)D has been found to be low in obese adults and may be due to sequestration of vitamin D in subcutaneous fat and its consequent reduced bioavailability *(48)*. Currently, cancer is estimated to cause only 4% of deaths in Africa, but because of large populations this equates to a large number of people, and furthermore, many cases are undiagnosed. The major cancers in African men are Kaposi's sarcoma (associated with HIV), liver, prostate, bladder and oesophageal cancer, and non-Hodgkins lymphoma. In African women the major cancers are Kaposi's sarcoma, cervical, breast, liver and ovarian cancer, and non-Hodgkins lymphoma *(49)*.

4. HEALTH CONSEQUENCES OF POOR VITAMIN D STATUS IN AFRICA

The classical effects of poor vitamin D status are rickets and osteomalacia, but there is also emerging evidence for a role of vitamin D in reduction in the progression or severity of TB and HIV/AIDS. The immunomodulatory effects of vitamin D means there are many potential health consequences of vitamin D deficiency in Africa where the infectious disease burden is high. Vitamin D deficiency may impact on the immune system by: decreasing innate immunity and immune surveillance; decreasing T lymphocyte number and function; and disrupting the Th1/Th2 balance (vitamin D normally inhibits Th1 profile). Vitamin D deficiency in an African setting may impact on the progression of infectious diseases such as TB (see below), HIV (see below) and schistosomiasis *(50)*.

4.1. Rickets and Osteomalacia

Well-established health consequences of vitamin D deficiency are rickets and osteomalacia (see Chapters 1, 34, 35). However, although vitamin D deficiency rickets is reported in Africa, there are very little data on the burden of disease this imposes, and on associations with malnutrition, poverty and other factors. It is therefore difficult to judge the extent to which primary vitamin D deficiency is a cause.

4.2. Tuberculosis

In Europe, as early as 1849 cod liver oil was recognised as beneficial in the treatment of TB and led in the early 1900s to the widespread use of vitamin D as a therapeutic treatment for both cutaneous and pulmonary TB *(51)*. The use of sanatoria was based on the belief that fresh air and sun exposure led to a positive outcome in the treatment of TB. Recently, a potential mechanism has been elucidated. In an in vitro study using serum from African-Americans and Caucasians, it was demonstrated that recognition of *Mycobacterium tuberculosis* particles by toll-like receptors on the surface of macrophages triggers the upregulation of VDR and CYP27B1 (1 alpha hydroxylase) expression. Intracellular $1,25(OH)_2D$ production and signalling upregulated production of the antimicrobial peptide, cathelicidin, that is able to destroy the intracellular mycobacterium. Importantly, the serum from African-Americans (~20 nmol/l [8 ng/ml] 25(OH)D versus ~80 nmol/l [32 ng/ml] in Caucasians) was less efficient in upregulating cathelicidin *(52)*.

Observational studies have demonstrated a link between low serum 25(OH)D concentration and susceptibility to TB *(53–58)*, and patients with active TB have lower serum 25(OH)D concentration than contacts (tuberculin skin-test positive) from similar ethnic and social backgrounds *(59)*. Despite the high incidence of TB in Africa, only one study, from Guinea-Bissau (latitude 12°N), has been published that compares patients with active TB and a control group. The latter consisted of both tuberculin skin-test positive and negative subjects *(27)*. There was a high prevalence of plasma 25(OH)D <75 nmol/l (30 ng/ml) in both groups but this was more common in patients than controls, although not after correction for socio-economic and demographic factors. However, mean serum 25(OH)D concentration remained lower in the TB group. A systematic review of studies conducted around the world, including in Kenyans, has recently concluded that low serum 25(OH)D is associated with greater risk of active TB *(60)*. However, although evidence suggests that lower vitamin D status is associated with increased risk of TB, none of the studies could disentangle cause and effect because of relatively small sample size.

Studies have investigated polymorphisms in the VDR gene in relation to TB and other diseases in Africa and elsewhere. The polymorphisms *Taq1* and *Fok1* affect transcription and expression of the VDR protein and may result in functional effects, including susceptibility to TB, response to anti-TB treatment and potentially vitamin D metabolism. Furthermore, because the VDR interacts with vitamin D metabolism, the effect of polymorphisms may depend on vitamin D status. Studies in West Africa *(61–63)*, South Africa *(64, 65)* and East Africa *(66, 67)* have investigated VDR polymorphisms and TB. However, results were inconsistent, and a systematic review and meta-analysis of VDR polymorphisms and TB concluded that results were inconclusive, studies included heterogeneous populations and were underpowered *(68)*.

A further area of potential study is the use of vitamin D as an adjuvant to TB drug treatment *(69)*. A study in Egypt *(70)* in which vitamin D was administered to children with TB concluded that clinical outcome in patients that received vitamin D plus drug treatment was better than those who received treatment alone, although differences were not significant. A double-blind, placebo-controlled vitamin D supplementation trial was performed in Indonesia in patients with moderately advanced pulmonary TB. Those receiving vitamin D had a significantly higher rate of sputum conversion and increased radiological improvement compared with the placebo group *(71)*. In vitro work has shown a positive effect of vitamin D; a single dose of 2.5 mg vitamin D was found to protect against reactivation of latent TB infection in whole blood from TB contacts *(72)*. A recent review *(51)* identified 14 reports of vitamin D administration to patients with pulmonary TB and concluded that major shortcomings in the studies prevented any clear conclusion from being drawn. Despite the strong mechanistic evidence for a possible beneficial role of vitamin D in the prevention and cure of TB, appropriately powered, randomised controlled trials are needed, especially in Africa where the disease burden is high. In the next few years, there should be more definitive evidence because a number of registered clinical trials (http://www.controlled-trials.com) are investigating the effects of adjuvant vitamin D therapy on recovery from TB. One of these is a randomised, placebo-controlled double-blind trial [ISRCTN352121232] to investigate the therapeutic effects of oral vitamin D_3 in 500 TB patients in Guinea-Bissau.

4.3. HIV/AIDS

The role of vitamin D in the progression of HIV infection is largely unknown, and studies have mostly been performed in Western settings. A recent review considers in detail the evidence for a role of vitamin D in HIV infection *(73)*. Decreased concentrations of 25(OH)D and 1,25(OH)$_2$D among HIV-infected people have been attributed to defects in renal hydroxylation and increased utilisation. An inverse association between 1,25(OH)$_2$D concentration and mortality has been reported from a small cohort of HIV-infected adults, and some cross-sectional studies have indicated positive correlations between 1,25(OH)$_2$D and CD4+ cell counts.

Poor vitamin D status may compound the negative effects of HIV infection and HIV anti-retrovirals on the skeleton (as evidenced by high incidence of osteoporosis and osteopenia in infected individuals). Drugs used in HIV treatment may inhibit or stimulate cytochrome P450 enzymes. Vitamin D and its metabolites are catabolised by P450 enzymes, so stimulation may lead to vitamin D deficiency *(62)*. More studies are needed to confirm the associations between vitamin D status and HIV disease progression.

5. STUDIES OF VITAMIN D STATUS IN AFRICA

Plasma 25(OH)D concentrations measured in studies conducted in Africa are summarised in Table 2. The data are from studies in healthy men and women, elderly people, pregnant and lactating women, healthy children and those with rickets, and clinical studies of TB and pneumonia. The subjects encompass people of different religions and in different settings. Full details of the studies can be found in the original references.

Table 2 highlights a number of important points. It demonstrates that there are very few nationally representative data; the studies are concentrated in The Gambia, Nigeria and South Africa and many studies compare control and 'unhealthy' groups. Table 2 also demonstrates that the 25(OH)D concentrations in the studies summarised should be compared and considered with caution; values were derived using 15 different methods. This leads to considerable variation in measured values for the same sample (74, 75), and the variation may even be larger when comparing older assays with modern methods. Only three papers reported the use of Quality Assurance materials and/or participation in a quality assurance scheme (e.g. the Vitamin D External Quality Assessment Scheme, www.deqas.org).

With these caveats, generally when compared to national data in the United Kingdom and the United States (1), Table 2 suggests that values in African countries are higher. Mean values range from 7 to 150 nmol/l (2.8–60 ng/ml) and the vast majority are above 25 nmol/l (10 ng/ml). Collectively, the prevalence of 25(OH)D <25 nmol/l (10 ng/ml) averages between 5–20% and 1–5% in most age groups in the United Kingdom and the United States. In North African countries there is a high prevalence of low vitamin D status, 25(OH)D is in the rachitic range. In tropical African countries the mean 25(OH)D values are high, but those with rickets have lower values than their peers in the same study, even though values are generally higher than 25 nmol/l (10 ng/ml). Similarly, individuals with TB or pneumonia have lower values than control groups within the same study. Veiled women or women in purdah have a lower vitamin D status than their peers within the same country. Finally, a geographical gradient is apparent across the African continent; 25(OH)D is lower in North African countries and in South Africa compared with tropical African countries (i.e. higher in the latter), and within a country (different regions of South Africa).

6. CONCLUSIONS

Studies in Africa that have assessed 25(OH)D suggest a range of status from deficiency (in the rachitic range) to relatively high values. Well-established health consequences of vitamin D deficiency in Africa include rickets and osteomalacia. Emerging evidence also suggests that consequences also include increased susceptibility to infectious disease. In some African countries, risk factors for vitamin D deficiency may also include very low calcium intake, and the burden of infectious disease whereby utilisation and turnover of vitamin D is increased. Thus a plasma 25(OH)D concentration considered to be sufficient in healthy people may not apply to those, for example, on very low calcium intakes, or who have, tropical enteropathies, HIV or TB.

Table 2

Studies that Include Assessment of Vitamin D Status Conducted in Africa

Study group	Age	Plasma 25(OH)D (nmol/l)[a] Mean±SD (if given), or median and range; n = number of subjects where specified
Healthy women at term delivery	Not specified	
Summer		11.8 ± 4.2[2] (n = 56)
Winter		9.0 ± 2.8 (n = 28)
Newborns	5 days	
Summer		9.0 ± 5.6 (n = 56)
Winter		7.6 ± 2.1 (n = 28)
Healthy children	0–3 years	25.3 ± 10.3 (n = 14)[3b]
Active rickets		9.3 ± 7.3 (n = 31)
Healthy men and non-pregnant women (24 M, 6F)	20–22 years	23.5 (range: 18–29)[7]
		25 (range: 17–46)
Healthy women at term delivery (n = 31)	22–28 years	24.9 ± 22.4[3a]
Children with active rickets (10 M, 6F)	0.25–2 years	34 (13–75, n = 19, n)[3a]
Healthy women at term delivery	26–45	20 (10–45, n = 14)
Cord blood	20–60 years	Veiled: 35.7[1]
Healthy veiled or non-veiled non-pregnant women		Non-veiled: 42.5
Healthy term infants (44 M, 35F) at 0,3,6 months	0–0.5 years	0m: 109.96 ± 42.5[2] (n = 27)
		3m: 148.0 ± 54.3 (n = 27)
		6m: 150.5 ± 64 (n = 28)

Table 2
(continued)

	Study group	Age	Plasma 25(OH)D (nmol/l)[a] Mean±SD (if given), or median and range n = number of subjects where specified
	Healthy children	8–12	95.0 ± 19.6 ($n = 44$)[4, 8]
	Children with non-active rickets	1–14 years	50.7 ± 12.8, ($n = 30$)
	Children with active rickets		42.4 ± 13.8 ($n = 13$)
The Gambia (26)	Lactating women 3 months postpartum	16–41 years	All year: 64.9 ± 18.5 ($n = 28$)[1]
			Jan-Mar: 64.1 ± 14.1 ($n = 9$)
			Apr-Jun: 66.1 ± 21.6 ($n = 9$)
			Jul-Sep: 54.7 ± 15.0 ($n = 5$)
			Oct-Dec: 73.9 ± 21.7 ($n = 5$)
The Gambia (84)	Healthy women	25–44	80.9 ± 22.8 ($n = 11$)[1]
		45–49	113.3 ± 26.5 ($n = 12$)
		50–54	95.7 ± 2.3 ($n = 14$)
		55–59	83.6 ± 19.4 ($n = 21$)
		60–64	97.7 ± 26.5 ($n = 27$)
		65–69	87.3 ± 25.1 ($n = 13$)
		70–74	87.0 ± 32.7 ($n = 7$)
		75+	72.3 ± 19.8 ($n = 8$)
The Gambia (85)	Healthy men and women	60–75 years	M: 64.3 ± 15.5 ($n = 15$)[4, 8]
			F: 72.8 ± 17.5 ($n = 15$)
Guinea Bissau (27)	Healthy men and women (239 M, 255F)	Not specified	85.3 ± 34.8[5, 8]
	TB patients (221 M, 142F)		78.3 ± 22.6

Table 2
(continued)

Study group	Age	Plasma 25(OH)D (nmol/l)[a] Mean±SD (if given), or median and range; n = number of subjects where specified
Kenya (55, 60)	Mean (sd)	Median and range
Healthy controls (8M, 7F)	33 ± 7.4y	65.5 (26.25–114.75, *n* = 15)[3e]
TB patients (9M, 6F)	35 ± 8.3y	39.75 (6.75–89.25, *n* = 15)
Nigeria (86, 87) Healthy age matched children (9M, 3F)	3–5 years	41 (29–50)[3c]
Active rickets (7M, 3F)		36 (22–84)
Nigeria (88) Healthy children	1–5 years	69 ± 22 (*n* = 20)[1]
Children with active rickets		36 ± 28 (*n* = 22)
Nigeria (89, 90) Healthy children	1–5 years	63 ± 17.8 (*n* = 47)[1]
Children with active rickets		43 ± 33.5 (*n* = 37)
Nigeria (91) Healthy children (matched for sex, age, religion, 19 M, 8F)	0.8–7 years	60.0 ± 18.8 (*n* = 27)[4]
Active rickets (12M, 4F)		35.3 ± 11.8 (*n* = 16)
Nigeria (92–94) Healthy children	Median 42 mo	51.3 ± 15.5[1]
Active rickets	Median 46 mo	34.8 ± 25.5
Nigeria (95, 96) Healthy children (6M, 4F)	1–8 years	52.3 ± 7.3[1] (*n* = 10)
Active rickets (3M, 7F)		24 ± 11.3 (*n* = 10)
Nigeria (97–99) Healthy children (6M, 9F)	2–8 years	37.5 ± 11.5[6] (*n* = 15)
Active rickets (6M, 9F)		37.5 ± 13.5 (*n* = 15)
Nigeria (100) Healthy children (5M, 4F)	0.5–5 years	130 ± 107 (*n* = 9)
Children with pneumonia (15M, 9F)		104 ± 59 (*n* = 18)

Table 2
(continued)

Study group	Age	Plasma 25(OH)D (nmol/l)[a] Mean±SD (if given), or median and range; n = number of subjects where specified
Nigeria (101) Children's survey (97 M, 121F)	6–35 mo	All: 64.3 ± 23.3[1] (n = 218) Healthy: 65 ± 24 (n = 198) Rickets: 56.5 ± 11.8 (n = 20)
Nigeria (102) Active rickets (9 M, 7F)	1.2–2.0 years	28.5 (range: 17–40)[4]
Nigeria (87, 103) Healthy women at term delivery (10 in purdah, 20 not in purdah) Cord bloods	Not given	Mean and range Purdah: 53 (37–64)[3d] Non-purdah: 90 (68–150) Purdah: 31 (24–59) Non-purdah: 58 (35–79)[3f]
Zaire (104) Healthy men (n = 33)	Mean 31y	65 ± 39[3f]
Southern Africa		
South Africa (105) Children with active rickets (2 M, 6F)	4–13 years	48.9 ± 10.1 (n = 8)[3 g]
South Africa (24) 232 patients with hip fracture	72.7 ± 13 years	All: 44.3 ± 23[3 g] Nov-June (s): 51 ± 26.9 June-Oct (w): 38.1 ± 17.0
South Africa (106) Survey of school children M/F rural M/F suburb M/F urban	7–12 years	72.3 ± 21.8 (n = 60)[3 g] 77.3 ± 23 (n = 60) 82.8 ± 18.8 (n = 60)
South Africa (107) Active rickets (2 M, 2F)	4–14 years	62.5 ± 21.5 (n = 4)[3 g]

Table 2
(continued)

	Study group	Age	Plasma 25(OH)D (nmol/l)[a] Mean±SD (if given), or median and range n = number of subjects where specified
South Africa (108)	Survey of pre-school children	3–5 years	85.5 ± 19 (n = 20)[3 h]
South Africa (109)	Healthy black children	6–18 years	6–9 years: 123.9 ± 12.0 (n = 17)[1]
			10–13 years: 115.6 ± 5.7 (n = 26)
			14–18 years: 90.2 ± 10.8 (n = 15)
	Healthy albino children		6–9 years: 102.9± 11.4 (n = 30)
			10–13 years: 86.1 ± 14.0 (n = 36)
			14–18 years: 90.7 ± 3.4 (n = 16)
South Africa (110)	Children:	1–12 years	
	Active vitamin D or Ca deficiency rickets		32.3 ± 18.8 (n = 21)[3 h]
	Hypophosphataemic rickets		52.5 ± 12.3 (n = 9)
	Healing rickets		61.8 ± 6.0 (n = 3)
South Africa (111)	Active rickets:	Median (range)	38.5 (15.3–111.5)[3i]
	Calcipenia	9.5 years (1.7–18.0)	26.0 (10.0–62.5)
	phosphopenia	5.7 years (0.3–16.0)	
South Africa (112)	Healthy women at term delivery	16–40 years	81.75 ± 28.8 (n = 43)[3j]
	Cord blood		170.97 ± 72.6 (n = 43)
South Africa (113)	Elderly female nursing home residents	80 ± 4 years	32 ± 11 (n = 60)[2]

Table 2
(continued)

	Study group	Age	Plasma 25(OH)D (nmol/l)[a] Mean±SD (if given), or median and range n = number of subjects where specified
South Africa (114)	Survey of women:	20–64 years	
	Premenopausal black		48.3 (17.0–114.0, n = 74)[3][g]
	Premenopausal white		65.8 (34.0–114.8, n = 105)
	Postmenopausal black		47.5 (15.8–80.8, n = 65)
	Postmenopausal white		64.5 (25.5–139.8, n = 50)

[a] To convert from nmol/l to ng/ml divide the value by 2.5.
[1] RIA Incstar.
[2] Not detailed
[3] In house assays.
[3a] Competitive binding assay after extraction and HPLC.
[3b] Competitive binding assay (115).
[3c] RIA after extraction (116).
[3d] Competitive binding assay (117).
[3e] RIA (118).
[3f] Competitive binding assay (119).
[3 g] Competitive binding assay (120).
[3 h] Competitive binding assay (121).
[3i] Competitive binding assay (no details given).
[3j] Competitive binding assay (122).
[4] RIA Diasorin.
[5] LC-MS/MS.
[6] RIA Nicols after ethanol extraction.
[7] HPLC.
[8] Participation in QA scheme DEQAS reported.

Underlying calcium nutrition must be considered when interpreting vitamin D status and considering the causes and health consequences of vitamin D deficiency in Africa. Vitamin D or calcium-deficiency rickets rather than being routinely screened for often presents as co-morbidities with infectious disease, for example, pneumonia *(76)*, and more work may give insights into the aetiology of various conditions where poor vitamin D status and/or low calcium intakes may be predisposing factors.

Studies are needed to confirm if emerging relationships between vitamin D status and health outcomes such as diabetes and cancer identified in Western countries can be replicated in African countries.

There is a need for the collection of nationally representative data of vitamin D status across Africa, with samples analysed using comparable techniques by accredited laboratories.

In summary, 'Africa' is not a homogenous entity with respect to geography, climate, water sources, food production and availability, or the religious and cultural practices, skin pigmentation, and burden of infectious and chronic disease of its people. All these factors are likely to affect and/or be affected by vitamin D status, and, therefore, how vitamin D deficiency is defined and its health consequences determined. It is therefore important to recognise that policy aspects of vitamin D cannot encompass the whole continent of Africa, but need to be considered in a more local context.

REFERENCES

1. Prentice A (2008) Vitamin D deficiency: a global perspective. Nut Revs 66(suppl 2):S153-S164
2. Scientific Advisory Committee on Nutrition (2007) Update on Vitamin D. The Stationery Office, London
3. Webb AR (2006) Who, what, where and when-influences on cutaneous vitamin D synthesis. Prog Biophys Mol Biol 92:17–25
4. Holick MF (2005) Photobiology of Vitamin D. In: Feldman D, Pike JW, Glorieux FH (ed) Vitamin D, 2nd ed. Elsevier Academic Press, Burlington, MA, USA
5. Norman A (2008) From vitamin D to hormone D: fundamentals of the vitamin D endocrine system essential for good health. Am J Clin Nutr 88:491S–499S
6. Kimlin MG, Olds WJ, Moore MR (2007) Location and vitamin D synthesis: is the hypothesis validated by geophysical data? J Photochem Photobio B, Biol 86:234–239
7. Schoenmakers I, Goldberg GR, Prentice A (2008) Abundant sunshine and vitamin D deficiency. Br J Nutr 99:1171–1173
8. Prentice A, Goldberg GR, Schoenmakers I (2008) Vitamin D across the lifecycle: physiology and biomarkers. Am J Clin Nutr 88:500S–506S
9. Thacher TD (2003) Calcium-deficiency rickets. Endocr Dev 6:105–125
10. Wondale Y, Shiferaw F, Lulseged S (2005) A systematic review of nutritional rickets in Ethiopia: status and prospects. Ethiop Med J 43:203–210
11. Lauridsen AL, Vestergaard P, Hermann AP et al (2005) Plasma concentrations of 25-hydroxy-vitamin D and 1,25-dihydroxy-vitamin D are related to the phenotype of Gc (vitamin D-binding protein): a cross-sectional study on 595 early postmenopausal women. Calcif Tiss Int 77:15–22
12. Zella LA, Shevde NK, Hollis BW, Cooke NE, Pike JW (2008) Vitamin D-binding protein influences total circulating levels of 1,25-dihydroxyvitamin D3 but does not directly modulate the bioactive levels of the hormone in vivo. Endocrinol 149:3656–3667
13. World Health Organisation Regional Offices for Africa and Eastern Mediterranean (2006) The Africa malaria report 2006. WHO, Geneva
14. World Health Organisation (2008) World health statistics 2008. WHO, Geneva

15. Lucas R, McMichael T, Smith W, Armstrong B (2006) Solar ultraviolet radiation: Global burden of disease from solar ultraviolet radiation. In: Pruss-Ustun A, Zeeb H, Mathers C, Repachoi M (eds) Environmental burden of disease series, No 13. World Health Organisation, Geneva

16. Quevedo WC, Fitzpatrick TB, Pathak MA, Jimbow K (1975) Role of light in human skin color variation. Am J Phys Anthropol 43:393–408

17. Cesarini JP, Roubin R, Fridman WH, Mouly R (1976) Characterization of the infiltration of human malignant melanoma according to the nature of the receiving surface. I. Study of fragments of tumors. Ann Dermatol Syphiligr (Paris) 103:312–315

18. Malvy DJ, Guinot C, Preziosi P et al (2000) Relationship between vitamin D status and skin phototype in general adult population. Photochem Photobiol 71:466–469

19. MacLaughlin J, Holick MF (1985) Aging decreases the capacity of human skin to produce vitamin D3. J Clin Invest 76:1536–1538

20. Holick MF (2004) Sunlight and vitamin D for bone health and prevention of autoimmune diseases, cancers, and cardiovascular disease. Am J Clin Nutr 80:1678S–1688S

21. Brazerol WF, McPhee AJ, Mimouni F, Specker BL, Tsang RC (1988) Serial ultraviolet B exposure and serum 25 hydroxyvitamin D response in young adult American blacks and whites: no racial differences. J Am Coll Nutr 7:111–118

22. Lo CW, Paris PW, Holick MF (1986) Indian and Pakistani immigrants have the same capacity as Caucasians to produce vitamin D in response to ultraviolet irradiation. Am J Clin Nutr 44:683–685

23. Jablonski NG, Chaplin G (2000) The evolution of human skin coloration. J Hum Evol 39:57–106

24. Pettifor JM, Ross FP, Solomon L (1978) Seasonal variation in serum 25-hydroxycholecalciferol concentrations in elderly South African patients with fractures of femoral neck. Br Med J 1:826–827

25. Garabedian M, Ben-Mekhbi H (1991) Is vitamin D-deficiency rickets a public health problem in France and Algeria?. Rickets 21:215–221

26. Prentice A, Yan L, Jarjou LM et al (1997) Vitamin D status does not influence the breast-milk calcium concentration of lactating mothers accustomed to a low calcium intake. Acta Paediatr 86:1006–1008

27. Wejse C, Olesen R, Rabna P et al (2007) Serum 25-hydroxyvitamin D in a West African population of tuberculosis patients and unmatched healthy controls. Am J Clin Nutr 86:1376–1383

28. MacKeown JM, Cleaton-Jones PE, Edwards AW, Turgeon-O'Brien H (1998) Energy, macro- and micronutrient intake of 5-year-old urban black South African children in 1984 and 1995. Ped Perinat Epidemiol 12:297–312

29. Murphy SP, Calloway DH, Beaton GH (1995) Schoolchildren have similar predicted prevalences of inadequate intakes as toddlers in village populations in Egypt, Kenya and Mexico. Eur J Clin Nutr 49:647–657

30. Bwibo NO, Neumann CG (2003) The need for animal source foods by Kenyan children. J Nutr 133(suppl 2):3936S–3940S

31. Thacher TD, Fischer PR, Strand MA, Pettifor JM (2006) Nutritional rickets around the world: causes and future directions. Ann Trop Paediatr 26:1–16

32. Prentice A, Ceesay M, Nigdikar S, Allen SJ, Pettifor JM (2008) FGF23 is elevated in Gambian children with rickets. Bone 42:788–797

33. Fawell J, Bailey K, Chilton J, Dahi E, Fewtrell L, Magara Y (2006) Fluoride in drinking-water. World Health Organisation and IWA Publishing, London

34. Pettifor JM, Schnitzler CM, Ross FP, Moodley GP (1989) Endemic skeletal fluorosis in children: hypocalcaemia and the presence of renal resistance to parathyroid hormone. Bone Miner 7:275–288

35. Anon (1972) The tropical intestine. Br Med J 1:2–3

36. Menzies IS, Zuckerman MJ, Nukajam WS et al (1999) Geography of intestinal permeability and absorption. Gut 44:483–489

37. Mayoral LG, Tripathy K, Garcia FT, Klahr S, Bolanos O, Ghitis J (1967) Malabsorption in the tropics: a second look. Am J Clin Nutr 20:866–883

38. Falaiye JM (1971) Present status of subclinical intestinal malabsorption in the tropics. Br Med J 4:454–458

39. Lunn PG (2002) Growth retardation and stunting of children in developing countries. Br J Nutr 88:109–110

40. Breman JG (2001) The ears of the hippomotamus: manifestations, determinants, and estimates of the malaria burden. Am J Trop Med Hyg 64:1–11
41. Young DB, Perkins MD,, Duncan K,, Barry CE 3rd (2008) Confronting the scientific obstacles to global control of tuberculosis. J Clin Invest 118:1255–1265
42. Chaisson RE, Martinson NA (2008) Tuberculosis in Africa – Combating an HIV-driven crisis. New Engl J Med 358:1089–1092
43. Amuna P, Zotor FB (2008) Epidemiological and nutrition transition in developing countries: impact on human health and development. Proc Nutr Soc 67:82–90
44. Boutayeb A (2006) The double burden of communicable and non-communicable diseases in developing countries. Trans Roy Soc Trop Med Hyg 100:191–199
45. Mufunda J, Chatora R, Ndambakuwa Y et al (2006) Emerging non-communicable disease epidemic in Africa: preventive measures from the WHO Regional Office for Africa. Ethn Dis 16:521–526
46. Vorster HH, Bourne LT, Venter CS, Oosthuizen W (1999) Contribution of nutrition to the health transition in developing countries: a framework for research and intervention. Nut Revs 57:341–349
47. Davis CD (2008) Vitamin D and cancer: current dilemmas and future research needs. Am J Clin Nutr 88:565S–569S
48. Rajakumar K, Fernstrom JD, Holick MF, Janosky JE, Greenspan SL (2008) Vitamin D status and response to Vitamin D(3) in obese vs. non-obese African American children. Obesity (Silver Spring) 16:90–95
49. Parkin DM, Sitas F, Chirenje M, Stein L, Abratt R, Wabinga H (2008) Part I. Cancer in Indigenous Africans – burden, distribution, and trends. Lancet Oncol 9:683–692
50. Snyman JR, de Sommers K, Steinmann MA, Lizamore DJ (1997) Effects of calcitriol on eosinophil activity and antibody responses in patients with schistosomiasis. Eur J Clin Pathol 52:277–280
51. Martineau AR, Honecker FU, Wilkinson RJ, Griffiths CJ (2007) Vitamin D in the treatment of pulmonary tuberculosis. J Steroid Biochem Mol Biol 103:793–798
52. Liu PT, Stenger S, Li H et al (2006) Toll-like receptor triggering of a vitamin D-mediated human antimicrobial response. Science 311:1770–1773
53. Sasidharan PK, Rajeev E, Vijayakumari V (2002) Tuberculosis and vitamin D deficiency. J Assoc Phys Ind 50:554–558
54. Grange JM, Davies PD, Brown RC, Woodhead JS, Kardjito T (1985) A study of vitamin D levels in Indonesian patients with untreated pulmonary tuberculosis. Tubercle 66:187–191
55. Davies PD, Church HA, Brown RC, Woodhead JS (1987) Raised serum calcium in tuberculosis patients in Africa. Eur J Resp Dis 71:341–344
56. Davies PD, Brown RC, Woodhead JS (1985) Serum concentrations of vitamin D metabolites in untreated tuberculosis. Thorax 40:187–190
57. Wilkinson RJ, Llewelyn M, Toossi Z et al (2000) Influence of vitamin D deficiency and vitamin D receptor polymorphisms on tuberculosis among Gujarati Asians in west London: a case-control study. Lancet Oncol 355:618–621
58. Ustianowski A, Shaffer R, Collin S, Wilkinson RJ, Davidson RN (2005) Prevalence and associations of vitamin D deficiency in foreign-born persons with tuberculosis in London. J Infect 50:432–437
59. Sita-Lumsden A, Lapthorn G, Swaminathan R, Milburn HJ (2007) Reactivation of tuberculosis and vitamin D deficiency: the contribution of diet and exposure to sunlight. Thorax 62:1003–1007
60. Nnoaham KE, Clarke A (2008) Low serum vitamin D levels and tuberculosis: a systematic review and meta-analysis. Int J Epidemiol 37:113–119
61. Bellamy R, Ruwende C, Corrah T et al (1999) Tuberculosis and chronic hepatitis B virus infection in Africans and variation in the vitamin D receptor gene. J Infect Dis 179:721–724
62. Bolland MA, Grey A, Horne A, Thomas M (2008) Osteomalacia in an HIV-infected man receiving rifabutin, a cytochrome P450 enzyme inducer: a case report. Ann Clin Microbiol Antimicrob 7:3
63. Olesen R, Wejse C, Velez DR et al (2007) DC-SIGN (CD209), pentraxin 3 and vitamin D receptor gene variants associate with pulmonary tuberculosis risk in West Africans. Genes Immun 8:456–467
64. Lombard Z, Dalton D-L, Venter PA, Williams RC, Bornman L (2006) Association of HLA-DR, -DQ, and vitamin D receptor alleles and haplotypes with tuberculosis in the Venda of South Africa. Hum Immunol 67:643–654

65. Babb C, van der Merwe L, Beyers N et al (2007) Vitamin D receptor gene polymorphisms and sputum conversion time in pulmonary tuberculosis patients. Tuberculosis (Edinb) 87:295–302
66. Fitness J, Floyd S, Warndorff DK et al (2004) Large-scale candidate gene study of tuberculosis susceptibility in the Karonga district of northern Malawi. Am J Trop Med Hyg 71:341–349
67. Soborg C, Andersen AB, Range N et al (2007) Influence of candidate susceptibility genes on tuberculosis in a high endemic region. Mol Immunol 44:2213–2220
68. Lewis SJ, Baker I, Davey Smith G (2005) Meta-analysis of vitamin D receptor polymorphisms and pulmonary tuberculosis risk. Int J Tuberc Lung Dis 9:1174–1177
69. Adams JS, Liu P, Chun R, Modlin RL, Hewison M (2007) Vitamin D in defense of the human immune response. Ann NY Acad Sci 1117:94–105
70. Morcos MM, Gabr AA, Samuel S et al (1998) Vitamin D administration to tuberculous children and its value. Bollettino Chimico Farmaceutico 137:157–164
71. Nursyam EW, Amin Z, Rumende CM (2006) The effect of vitamin D as supplementary treatment in patients with moderately advanced pulmonary tuberculous lesion. Acta Med Indones 38:3–5
72. Martineau ARR, Wilkinson J, Wilkinson KA et al (2007) A single dose of vitamin D enhances immunity to mycobacteria. Am J Respir Crit Care Med 176:208–213
73. Villamor E (2006) A potential role for vitamin D on HIV infection?. Nut Revs 64:226–233
74. Lensmeyer GL, Binkley N, Drezner MK (2006) New horizons for assesment of vitamin D status in man. In: Seibel MJ, Robins SP, Bilezikian JP (eds) Dynamics of bone and cartilage metabolism. Principles and clinical applications, 2nd ed. Academic Press, New York
75. Hollis BW (2008) Measuring 25-hydroxyvitamin D in a clinical environment: challenges and needs. Am J Clin Nutr 88:507S–510S
76. Muhe L, Lulseged S, Mason KE, Simoes EAF (1997) Case-control study of the role of nutritional rickets in the risk of developing pneumonia in Ethiopian children. Lancet 349:1801–1804
77. Lawson DE, Cole TJ, Salem SI et al (1987) Etiology of rickets in Egyptian children. Hum Nutr:Clin Nutr 41:199–208
78. Feleke Y, Abdulkadir J, Mshana R et al (1999) Low levels of serum calcidiol in an African population compared to a North European population. Eur J Endocrinol 141:358–360
79. Elzouki AY, Markestad T, Elgarrah M, Elhoni N, Aksnes L (1989) Serum concentrations of vitamin D metabolites in rachitic Libyan children. J Pediatr Gastroenterol Nutr 9:507–512
80. Markestad T, Elzouki A, Legnain M, Ulstein M, Aksnes L (1984) Serum concentrations of vitamin D metabolites in maternal and umbilical cord blood of Libyan and Norwegian women. Hum Nutr:Clin Nutr 38:55–62
81. Meddeb N, Sahli H, Chahed M et al (2005) Vitamin D deficiency in Tunisia. Osteoporos Int 16:180–183
82. Nguema-Asseko B, Ganga-Zandzou PS, Ovono F et al (2005) Vitamin D status in Gabonese children. Arch Pediatr 12:1587–1590
83. Dibba B, Prentice A, Ceesay M, Stirling DM, Cole TJ, Poskitt EME (2000) Effect of calcium supplementation on bone mineral accretion in Gambian children accustomed to a low-calcium diet. Am J Clin Nutr 71:544–549
84. Aspray TJ, Yan L, Prentice A (2005) Parathyroid hormone and rates of bone formation are raised in perimenopausal rural Gambian women. Bone 36:710–720
85. Yan L, Schoenmakers I, Zhou B et al (2009) Ethnic differences in parathyroid hormone secretion and mineral metabolism in response to oral phospahte administration. Bone 45:238–245
86. Okonofua F, Gill DS, Alabi ZO, Thomas M, Bell JL, Dandona P (1991) Rickets in Nigerian children: A consequence of calcium malnutrition. Metabolism 40:209–213
87. Okonofua FE (2002) Calcium and vitamin D nutrition in Nigerian women and children. In: Thacher TD (ed) Nutritional rickets in Nigerian children: the way forward. Nestle Nutrition, Vevey, Switzerland
88. Oginni LM, Worsfold M, Oyelami OA, Sharp CA, Powell DE, Davie MW (1996) Etiology of rickets in Nigerian children. J Pediatr 128:692–694
89. Oginni LM, Worsfold M, Sharp CA, Oyelami OA, Powell DE, Davie MW (1996) Plasma osteocalcin in healthy Nigerian children and in children with calcium-deficiency rickets. Calcif Tiss Int 59:424–427

90. Sharp CA, Oginni LM, Worsfold M et al (1997) Elevated collagen turnover in Nigerian children with calcium-deficiency rickets. Calcif Tiss Int 61:87–94

91. Walter EA, Scariano JK, Easington CR et al (1997) Rickets and protein malnutrition in northern Nigeria. J Trop Pediatr 43:98–102

92. Thacher TD, Fischer PR, Pettifor JM et al (1999) A comparison of calcium, vitamin D, or both for nutritional rickets in Nigerian children. New Engl J Med 341:563–568

93. Thacher TD, Fischer PR, Pettifor JM, Lawson JO, Isichei CO, Chan GM (2000) Case-control study of factors associated with nutritional rickets in Nigerian children. J Pediatr 137:367–373

94. Thacher TD (2002) Medical treatment of rickets in Nigerian children. In: Thacher TD (ed) Nutritional rickets in Nigerian children: the way forward. Nestle Nutrition, Vevey, Switzerland

95. Thacher T, Glew RH, Isichei C et al (1999) Rickets in Nigerian children: response to calcium supplementation. J Trop Pediatr 45:202–207

96. VanderJagt DJ, Peery B, Thacher T, Pastuszyn A, Hollis BW, Glew RH (1999) Aminoaciduria in calcium-deficiency rickets in northern Nigeria. J Trop Pediatr 45:258–264

97. Pam S (2002) Calcium absorption in Nigerian children with rickets. In: Thacher TD (ed) Nutritional rickets in Nigerian children: the way forward. Nestle Nutrition, Vevey, Switzerland

98. Graff M, Thacher TD, Fischer PR et al (2004) Calcium absorption in Nigerian children with rickets. Am J Clin Nutr 80:1415–1421

99. Oramasionwu GE, Thacher TD, Pam SD, Pettifor JM, Abrams SA (2008) Adaptation of calcium absorption during treatment of nutritional rickets in Nigerian children. Br J Nutr 100:387–392

100. Oduwole AO (2002) Vitamin D nutrition in Nigerian children with pneumonia. In: Thacher TD (ed) Nutritional rickets in Nigerian children: the way forward. Nestle Nutrition, Vevey, Switzerland

101. Pfitzner MA, Thacher TD, Pettifor JM et al (1998) Absence of vitamin D deficiency in young Nigerian children. J Pediatr 133:740–744

102. Thacher TD, Fischer PR, Isichei CO, Pettifor JM (2006) Early response to vitamin D2 in children with calcium deficiency rickets. J Pediatr 149:840–844

103. Okonofua F, Houlder S, Bell J, Dandona P (1986) Vitamin D nutrition in pregnant Nigerian women at term and their newborn infants. J Clin Pathol 39:650–653

104. Buyamba-Kabangu JRM, Fagard R, Lijnen P, Bouillon R, Lissens W, Amery A (1987) Calcium, vitamin D-endocrine system, and parathyroid hormone in black and white males. Calcif Tiss Int 41:70–74

105. Pettifor JM, Ross P, Wang J, Moodley G, Couper-Smith J (1978) Rickets in children of rural origin in South Africa: Is low dietary calcium a factor. J Pediatr 92:320–324

106. Pettifor JM, Ross P, Moodley G, Shuenyane E (1979) Calcium deficiency in rural black children in South Africa – a comparison between rural and urban communities. Am J Clin Nutr 32:2477–2483

107. Pettifor JM, Ross FP, Travers R, Glorieux FH, Deluca HF (1981) Dietary calcium deficiency: A syndrome associated with bone deformities and elevated serum 1,25-dihydroxyvitamin D concentrations. Metabol Bone Dis Rel Res 2:301–305

108. van der Westhuyzen J (1986) Biochemical evaluation of black preschool children in the northern Transvaal. S Afr Med J 70:146–148

109. Cornish DA, Maluleke V, Mhlanga T (2000) An investigation into a possible relationship between vitamin D, parathyroid hormone, calcium and magnesium in a normally pigmented and an albino rural black population in the Northern Province of South Africa. Biofactors 11:35–38

110. Bhimma R, Pettifor JM, Coovadia HM, Moodley M, Adhikari M (1995) Rickets in black children beyond infancy in Natal. S Afr Med J 85:668–672

111. Daniels ED, Pettifor JM, Moodley GP (2000) Serum osteocalcin has limited usefulness as a diagnostic marker for rickets. Eur J Pediatr 159:730–733

112. Fairney A, Sloan MA, Patel KV, Coumbe A (1987) Vitamin A and D status of black South African women and their babies. Hum Nutr:Clin Nutr 41:81–87

113. van Papendorp DH (1990) The vitamin D status of South African women living in old-age homes. S Afr Med J 78:556

114. Daniels ED, Pettifor JM, Schnitzler CM, Moodley GP, Zachen D (1997) Differences in mineral homeostasis, volumetric bone mass and femoral neck axis length in black and white South African women. Osteoporos Int 7:105–112

115. Lawson DE (1980) Competitive binding assay for 25-hydroxyvitamin D using specific binding proteins. Methods Enzymol 67:459–465

116. Hummer L, Tjellesen L, Rickers H, Christiansen C (1984) Measurement of 25-hydroxyvitamin D3 and 25-hydroxyvitamin D2 in clinical settings. Scand J Clin Lab Invest 44:595–601

117. Preece MA, O'Riordan JL, Lawson DE, Kodicek E (1974) A competitive protein-binding assay for 25-hydroxycholecalciferol and 25-hydroxyergocalciferol in serum. Clin Chem Acta 31:235–242

118. Clemens TL, Hendy GN, Graham RF, Baggiolini EG, Oskokovic MR, O'Riordan JL (1978) A radioimmunoassay for 1,25-dihydroxycholecalciferol. Clin Sci Mol Med 54:329–332

119. Bouillon R, Kerkhove PV, De Moor P (1976) Measurement of 25-hydroxyvitamin D3 in serum. Clin Chem 22:364–368

120. Pettifor JM, Ross FP, Wang J (1976) A competitive protein-binding assay for 25-hydroxyvitamin D. Clin Sci Mol Med Suppl 51:605–607

121. Haddad JG, Chyu KJ (1971) Competitive protein-binding radioassay for 25-hydroxycholecalciferol. J Clin Endocrinol Metab 33:992–995

122. Fairney A, Turner C, Hanson S, Zambon M (1979) A simple micromethod for 25-hydroxyvitamin D estimation. Ann Clin Biochem 16:106–110

27 Vitamin D Deficiency and Its Health Consequences in India

R.K. Marwaha and R. Goswami

Abstract   Studies on bone mineral health from different parts of India indicate wide prevalence of vitamin D deficiency (VDD) in all age groups including neonates, infants, school children, pregnant/lactating women, and adult males and females residing in rural and urban India. These have resulted due to poor sun exposure, dark skin complexion, vegetarian food habits, sedentary lifestyle, and lack of vitamin D food fortification programme. Supranormal serum parathyroid hormone values and low peak bone mass in apparently healthy children and adults reported in various studies could be linked to hypovitaminosis D in Indians. In such a scenario active intervention may be required in the form of a national policy for vitamin D fortification programme in our country.

Key Words: 25(OH)D; iPTH; BMD; rickets; osteomalacia

1. INTRODUCTION

Vitamin D status has a profound effect on the growth and development of children and has major implications for adult bone health. Optimal bone mineral health during childhood and adolescence leads to adequate peak bone mass which acts as a safeguard against osteoporosis and susceptibility to fractures at latter age. Considering importance of calcium and vitamin D nutrition on bone mass, industrialized countries have made fortification of milk and other food products with vitamin D a routine practice. In contrast, food fortification with vitamin D was never considered in countries like India due to widely held belief that adequate sunshine is available. However, the available literature shows evidence to the contrary. Vitamin D deficiency disorders like rickets, osteomalacia, and hypovitaminosis D osteopathy, which were prevalent in the 1900s, continue to exist in a significant proportion of Indian population *(1–4)*.

Recent evidence also suggests that vitamin D has important bioregulatory action and is not only necessary for the mere prevention of rickets/osteomalacia but also required for overall health due to its multitude effects. Suboptimal levels of vitamin D are associated with an elevated risk of a number of chronic diseases including malignancies, particularly of colon, breast, and prostrate gland; chronic inflammatory and autoimmune

From: *Nutrition and Health: Vitamin D*
Edited by: M.F. Holick, DOI 10.1007/978-1-60327-303-9_27,
© Springer Science+Business Media, LLC 2010

diseases like type I diabetes; inflammatory bowel disease; and multiple sclerosis *(5, 6)*.
A recent meta-analysis showed 29% reduction in the risk of type I diabetes in children receiving vitamin D supplementation *(7, 8)*. Similarly, in multiple sclerosis, every
50 nmol/l increase in serum 25(OH)D levels in healthy Caucasian population reduced
the risk of disease by 41% *(9)*. However, there are no systematic studies to date from
India, evaluating association between vitamin D deficiency and chronic diseases other
than bone mineral disorders.

2. HISTORICAL BACKGROUND

The earliest description of adolescent rickets and osteomalacia in India was in the
early 1990s *(10–12)*. However, Hodgkin et al. in 1973 reported that osteomalacia resulting from vitamin D deficiency was uncommon among Punjabis in India, in contrast
with Punjabis in Britain, and this disparity was explained by the difference in sunlight
exposure in the two populations. The conclusion of this study was based on estimation
of biochemical parameters like serum total calcium, inorganic phosphorus, and alkaline
phosphatase. Serum 25(OH)D, a currently used modern marker for vitamin D status,
was not assessed in the above study *(13)*. Assay for circulating 25(OH)D in human subjects has been available for nearly three decades. Systematic evaluation of 25(OH)D in
apparently healthy Indian population was first assessed in 2000 following a preliminary
report on low 25(OH)D in patients with primary hyperparathyroidism and a few healthy
controls from Delhi *(2, 14)*.

The definition of "normal" nutritional vitamin D status in human subjects was based
on a Gaussian distribution of circulating 25(OH)D levels in subjects who appear to be
free from disease. This method of establishing a normal range has limitations in view
of its not taking into consideration major factors affecting serum 25(OH)D levels like
lifestyle habits, race, sunscreen usage, age, and latitude. Using functional indicators
like intact parathyroid hormone, calcium absorption, and bone mineral density, several
studies have defined vitamin D deficiency/insufficiency as circulating levels of 25(OH)D
$\leq$ 80 nmol/l or 32.0 ng/ml *(15)*. Applicability of these values in Indians is yet to be
validated. No studies have been carried out to define normal vitamin D levels using all
the three functional indicators. Goswami et al. found the PTH rising in subjects with
vitamin D levels less than 25 nmol/l or 10.0 ng/ml *(16)*. Seth and Marwaha et al. found
25(OH)D value of 26.85 nmol/l (10.7 ng/ml) below which PTH rose beyond the upper
limit of normal in lactating mothers *(17)*.

3. VITAMIN D STATUS IN ADULTS

Vitamin D and calcium are critically important for growth and development of bones,
its mineralization and maintenance of calcium homeostasis, and the structural integrity
of the skeleton. Chronic vitamin D deficiency results in secondary hyperparathyroidism
causing minimal cortical bone loss and precipitate osteomalacia and osteoporosis in
predisposed individuals.

Vitamin D deficiency (VDD) may result from extrinsic factors like inadequate productions in the skin or low dietary intake of vitamin D and/or intrinsic factors like gastrointestinal disorders with malabsorption. In countries like India, where vitamin D deficiency is endemic due to cultural avoidance of skin exposure, increased skin pigmentation, crowded houses with limited sunlight exposure, frequent pregnancies coupled with calcium deficiency due to prolonged lactation, consumption of diet rich in fibre and phytate, and low calcium intake are important contributors in the pathogenesis of metabolic bone disease *(18, 19)*. Teotia et al. during the period 1963–2005, surveyed a population of 337.6 million from among the total population of 22 states of India. In this survey, he identified 4.1 lakh population with bone mineral metabolic disorders, of which 52% suffered from nutritional bone disease *(20)*.

A number of recent reports show a wide prevalence of hypovitaminosis D in adult Indians from north as well as south India. Goswami et al. assessed vitamin D status in groups of healthy subjects from Delhi, differing with respect to variables relevant to vitamin D and bone mineral status, such as direct sunlight exposure, season of measurement, skin pigmentations dietary calcium and phytate contents, and altered physiological states such as pregnancy and neonatal stage. Three groups consisting of soldiers, depigmented subjects, and physicians/nurses were studied in winter and other three groups comprising of pregnant mothers, newborns, and doctors/nurses group were studied in summer. The doctors/nurses group was evaluated again in summer to see the effect of seasonal variation on vitamin D status. Though all the groups had subnormal concentrations of 25(OH)D, the serum levels were highest in the soldiers (44 nmol/l or 17.0 ng/ml) followed by depigmented groups. The serum 25(OH)D levels were reduced to half in the winter months in physicians and nurses group. The study concluded that despite abundant sunlight, healthy persons in Delhi remain vitamin D deficient because of inadequate direct sun exposure and skin pigmentation *(2)*.

Population studies carried out in southern India on vitamin D status in rural and urban subjects residing in and around Tirupati in Andhra Pradesh (13.4°N and 99.2°E with a zenith angle of 9.92° in summers and 32.2° in winters) also showed wide prevalence of hypovitaminosis D despite adequate sunlight. This was explained by their dark skin colour *(21)*. The mean serum 25(OH)D levels were clearly seen to be higher in people from south than from north because of higher duration of cloud-free sunshine throughout the year (8–10 h/day) in comparison to just 3.1 h/day in the winters and 7 h/day in summers in Delhi (28.35°N and 77.12°E with zenith angle of 84.5° in peak summers and 38.5 in peak winters).

Similar occurrences of vitamin D deficiency were also noted among 164 postmenopausal women from Tirupati despite adequate exposure to sun for 4–6 h/day. Using commercial kit reference values (22.5–94 nmol/l or 9–37.6 ng/ml), 126 (77%) subjects had normal 25(OH)D levels and 38 (23%) had 25(OH)D deficiency. Using functional health-based reference values, 30 (18%) patients had normal 25(OH)D levels (>50 nmol/l or >20 ng/ml), 85 (52%) had mild 25(OH)D insufficiency (25–50 nmol/l or 10–20 ng/ml) and 49 (30%) had moderate to severe 25(OH)D insufficiency (<25 nmol/l or <10 ng/ml). The study concluded that population-based reference values underdiagnose vitamin D deficiency and overdiagnose normal vitamin D status *(22)*.

The most recent study from south India documenting dietary habits and concentrations of serum calcium, 25(OH)D, and parathyroid hormone in 943 healthy urban and 205 rural subjects showed that 25(OH)D concentration of rural subjects were significantly higher than the urban subjects ($p<0.001$) in both men and women. In the rural subjects, 25(OH)D deficiency (<50 nmol/l or <20 ng/ml), insufficiency (50–75 nmol/l or 20–30 ng/ml), and sufficiency (>75 nmol/l or >30 ng/ml) were noted in 44, 39.5, and 16.5% of men and 70, 29, and 1% of women, respectively, whereas in the urban subjects, vitamin D deficiency, insufficiency, and sufficiency states were observed in 62, 26, and 12% of men and 75, 19, and 6% of women. Mean 25(OH)D levels in urban and rural areas were the following: men, 46.35 ± 2.0 nmol/l or 18.54 ± 0.8 ng/ml and 59.32 ± 2.0 nmol/l or 23.73 ± 0.8 ng/ml and women, 38.75 ± 0.75 nmol/l or 15.5 ± 0.3 and 47.5 ± 2.22 nmol/l or 19 ± 0.89 ng/ml, respectively. The study concluded that low dietary calcium intake and 25(OH)D concentrations were associated with deleterious effect on bone mineral homeostasis (23, 24).

3.1. Vitamin D Deficiency and Bone Mineral Density

VDD in Asian Indians has been associated with higher serum PTH and lower serum calcium resulting from bone resorption leading to rickets, osteomalacia, and osteoporosis in predisposed individuals. While it is common to see lower T and Z scores in BMD reports from our country, there is paucity of population-based systematic data related to prevalence of osteoporosis in adult Indian population. Rarer are the adequately designed studies assessing association between serum 25(OH)D and BMD in Indian population. Shatrugna et al. studied prevalence of osteoporosis in 289 middle-aged women from Hyderabad. The prevalence of osteoporosis at the femoral neck was around 29%. The T scores in the BMD at all the skeletal sites were much lower than the values reported from the developed countries. BMD showed a decline after the age of 35 years at lumbar spine and femoral neck; on multiple regression analysis, calcium intake appeared as an important determinant of BMD. However, 25(OH)D status was not assessed in the study (25).

There have been only three studies which have directly assessed relationship between vitamin D status and BMD in apparently healthy Asian Indians. Arya et al. observed a positive correlation between serum 25(OH)D concentrations and hip BMD in a cohort of 92 healthy subjects from Lucknow (26). On the contrary, in a well-characterized group of subjects with near-normal vitamin D status and other nutritional and environmental factors known to influence bone mineral metabolism, no such correlation could be observed (27). Recently, Vupputuri et al. observed a significant relation between serum 25(OH)D concentrations and hip BMD but not with lumbar spine or forearm BMD in an urban Delhi cohort (28).

Differential effects of vitamin deficiency disorder (VDD) on BMD at various sites possibly involve factors other than hypovitaminosis D. In this context, genetic variation related to vitamin D receptor (VDR) polymorphism was studied by Vupputuri et al. Among the three VDR single nucleotide polymorphisms studied (*TaqI*, *FokI*, and *BsmI*), subjects with *TaqI* TT genotype had a significantly higher BMD at the forearm than did subjects with Tt and tt genotypes. The correlation of VDR gene polymorphism with forearm BMD among Indians is interesting and indicates that even in a severely hypovitaminosis D population, variations in BMD at forearm could be explained by genetic factors (28).

Overall results of various studies conducted to date in urban and rural India indicate that widely prevalent VDD is functionally relevant in view of associated increase in serum PTH and low BMD. However, there has been no systematic study which has assessed the association between osteoporosis-related fracture rates and serum 25(OH)D status among Indians.

4. VITAMIN D STATUS IN PREGNANCY AND LACTATION

Significant changes in maternal vitamin D and calcium metabolism occur during pregnancy to provide the calcium for foetal bone mineral accretion. Approximately 25–30 g of calcium is transferred to the foetal skeletal, most of which is transferred during the last trimester of pregnancy (29). Because of good correlation between maternal and foetal 25(OH)D levels, presence of vitamin D deficiency in pregnant mothers would result in foetus developing in the state of hypovitaminosis D. The latter is likely to adversely affect the offsprings in terms of foetal and childhood bone development and innate immune function (30).

Brooke et al. (31) first reported reduced incidence of low birth weight babies in vitamin D-supplemented Asian mothers. Vitamin D and calcium supplementations of the pregnant mothers were associated with improved anthropometry and bone mass in the offsprings (32). Marya et al. from Rohtak reported higher body weight, crown heel length, head circumference, and mid-arm circumference in mothers who received two doses of 60,000 IU of vitamin D3 during the third trimester pregnancy compared to those who did not receive vitamin D3 (33). Besides, newborns had a higher Apgar score at birth when mothers received calcium and vitamin D supplementations as opposed to mothers with inadequate calcium and vitamin D intakes (34).

Similarly, the vitamin D content of human milk is related to the lactating mother's vitamin D status (35, 36). Exclusively breast-fed infants born to mothers with vitamin D deficiency have a high prevalence of vitamin D deficiency based on serum 25(OH)D levels and are therefore at increased risk of developing vitamin D deficiency rickets (37, 38).

This problem of vitamin D deficiency in India is further compounded by the fact that very few women in India during pregnancy and lactation receive adequate vitamin D supplementation or dietary calcium intake. With severe maternal vitamin D deficiency, the foetus may develop rickets in utero or at birth, tetany, delayed growth, bone ossification, and abnormal enamel formation (39). Rickets during infancy has been associated with high prevalence of lower respiratory tract infections, which is a major cause of infant mortality in India.

Hypovitaminosis D and osteomalacia among pregnant south Asian women have been widely reported (40–48). However, most studies except a few (2, 40, 45, 46) were from temperate regions such as United Kingdom (41–44, 47) and Norway (48) where the already low availability of overhead sun is compounded for Asian women by poor outdoor activity, pigmented skin, and excessive clothing. Vitamin D deficiency has also been noted in pregnant muslim women from tropical countries, where practice of purdah might have played an important role (44, 45, 48–52).

Recent limited data on vitamin D status in pregnant and lactating Indian women from Delhi and Lucknow, India, reveal unexpectedly high prevalence of hypovitaminosis D (84–93%) *(17, 53)*. Goswami et al. also reported vitamin D deficiency in 29 pregnant women from Delhi studied in summers with a mean serum 25(OH)D concentration levels of 29.9 ± 10.7 nmol/l (8.6 ± 4.28 ng/ml) *(2)*. Recently Sachan et al. studied 207 urban and rural pregnant women belonging to low and middle socioeconomic strata at term and observed prevalence of hypovitaminosis D (<56.25 nmol/l or <22.5 ng/ml) in up to 84% of them with mean 25(OH)D of 35 ± 23.25 nmol/l or 14 ± 9.3 ng/ml. PTH rose above the normal range when 25(OH)D was ≤ 56.25 nmol/l or ≤ 22.5 ng/ml. Maternal serum 25(OH)D was ≤25 nmol/l or ≤ 10 ng/ml in 88 of them (42.5%) *(53)*.

In a recent study, we evaluated the vitamin D status of 180 healthy lactating mothers and its impact on their exclusively breast-fed infants *(17)*. The mean serum 25(OH)D value in lactating mothers was 27.18 ± 14.55 nmol/l (10.87 ± 5.82 ng/ml). Serum 25(OH)D levels of <25 nmol/l (10 ng/ml) were found in 47.8% mothers, and PTH levels exceeded the upper limit of normal (>66 pg/ml) when 25(OH)D levels fell below 26.85 nmol/l (10.74 ng/ml).

5. VITAMIN D STATUS IN NEONATES AND INFANTS

Rickets has often been thought of as a nineteenth century disease. Lack of exposure to sunlight was recognized as the major cause in the United Kingdom since high incidence of rickets was observed in children living in the large industrialized towns *(54, 55)*. In developing countries like India, there is rapid surge in the industrialization mainly in and around the metrocities. The adverse impact of atmospheric pollution on vitamin D status of infants and toddlers in Delhi was shown by Agarwal et al. *(56)*. Nutritional rickets due to vitamin D and/or calcium deficiency continues to exist as a major health problem in India.

Teotia et al. carried out several epidemiological surveys during 1963–2005 and reported 17,286 cases of rickets, 11,900 of infantile rickets due to lack of sunlight exposure, and 3 with congenital rickets born to mothers with vitamin D osteomalacia *(57)*.

While hypovitaminosis D has emerged as a significant health problem across all age groups, there is limited information on 25(OH)D status of newborns and breast-fed infants from India. Studies from India have shown significant correlation of 25(OH)D between mother and infant pairs *(2, 17, 52, 58)*. Goswami et al. showed mean cord blood 25(OH)D concentration of the newborn to be significantly lower than that of the pregnant group (16.72 ± 4.99 nmol/l (6.68 ± 1.96 ng/ml) vs. 21.9 ± 10.73 nmol/l (8.76 ± 4.29 ng/ml)). Correlative analysis of 25(OH)D values in pregnant women and their neonates showed a significant positive correlation between the two variables ($r = 0.7934$, $p<0.001$) *(2)*.

Similar observations were made by Sachan et al. in 207 rural and urban pregnant subjects, where mean serum 25(OH)D level in cord blood (21 ± 14.25 nmol/l or 8.4 ± 5.7 ng/ml) was significantly lower than the mean maternal serum 25(OH)D (35 ± 23.25 nmol/l or 14 ± 9.3 ng/ml) and that cord blood 25(OH)D strongly correlated with maternal values ($r = 0.79$, $p<0.001$) which is in keeping with the reports in literature *(41, 58–61)*.

Bhalala et al. from Mumbai determined vitamin D status of mother–newborn pairs at birth followed by repeat 25(OH)D estimation in the infants at 3 months while being breast-fed. Serum 25(OH)D values of <25 ng/ml or 10 nmol/l were observed in 50% mothers, 62% cord blood specimens, and 80% infants at 3 months. There was significant correlation noted between 25(OH)D values of the breast-fed infants and their mothers ($r = +0.552$, $p<0.001$) as well as with the cord blood ($r = +0.616$, $p<0.001$). The study concluded that subnormal maternal vitamin D status is associated with vitamin D deficiency in newborns and persists in exclusively breast-fed infants (62).

Anju and Marwaha et al. in a recent study evaluating the vitamin D status of 180 lactating mothers and their exclusively breast-fed infants in the age group between 2 and 24 weeks showed the presence of hypovitaminosis D in 91.1% infants using Lips criteria. Of these 47, 23, and 21% of them had mild (25–50 nmol/l or 10–20 ng/ml), moderate (12.5–25 nmol/l or 5–10 ng/ml), and severe (<12.5 nmol/l or <5 ng/ml) hypovitaminosis D, respectively. The mean 25(OH)D of the infants was 28.88 ± 20.75 nmol/l (11.55 ± 8.3 ng/ml), and there was a positive correlation between serum 25(OH)D levels of mother–infant pairs ($r = 0.324$, $p = 0.001$). Using logistic regression, infants born to mothers with 25(OH)D < 25 nmol/l (<10 ng/ml) had 3.79 times higher risk of developing moderate to severe hypovitaminosis D, as compared to those born to mothers with 25(OH)D ≥ 25 nmol/l (≥ 10 ng/ml; $p<0.001$) (17).

Another recent study evaluating the role of vitamin D deficiency in causation of neonatal hypocalcemic seizures revealed a strong positive correlation of 25(OH)D levels between mothers and their infants presenting with hypocalcemic seizures. The study suggested that infants born to vitamin D-deficient mothers are at a significantly higher risk to develop hypocalcemic seizures (63). A report by Balasubramanian et al. from Chennai, southern India, has also revealed similar observations. Serum 25(OH)D levels in mothers of all the 13 exclusively breast-fed infants with hypocalcemic seizures were in the vitamin D-deficient range, suggesting good mother–infant correlation in vitamin D deficiency (58). Thus, offsprings of mothers with suboptimal vitamin D status are at higher risk of hypovitaminosis D and hypocalcemic seizures.

6. VITAMIN D STATUS IN INDIAN CHILDREN

Approximately 40–50% of total skeletal mass is accumulated during childhood and adolescence (64). Studies show that 60–80% variability in bone mass is due to genetic factors, with nutrition, lifestyle, physical activity, and hormonal factors causing the rest (65–67). It is during this period calcium and vitamin D nutrition or non-pharmacologic strategies should be adopted to have the maximum impact on peak bone mass. Severe vitamin D deficiency, usually associated with 25(OH)D levels < 5.0 ng/ml or <12.5 nmol/l, results in rickets and osteomalacia. However, these clinically overt cases of vitamin D deficiency would represent only the "tip of an iceberg" of vitamin D insufficiency (68). Majority of children with biochemical hypovitaminosis D have adverse skeletal consequences secondary to raised PTH, increased bone turnover, enhanced bone loss, and fracture risk (69, 70).

Bone mass is the most important determinant of bone strength and is useful in the early detection of bone loss, monitoring of therapeutic responses, and prediction of fracture risk in osteoporosis. Prospective studies have noted an increased risk of fractures with decreasing bone density, and a two- to threefold increase of fracture risk for every reduction in the standard deviation of bone mass at the spine and hips (71). Significant reduction in fracture risk has been shown in elderly women by about 10–20% after supplementation with calcium and vitamin D (72).

Unlike adults, there is limited data on clinical and subclinical vitamin D deficiency status in Indian children. Our recent studies have filled the above lacuna and have provided detailed clinical and biochemical information on the prevalence vitamin D deficiency in Indian children and its impact on bone density.

6.1. Clinical Evidence of Rickets

Metabolic bone disorders secondary to vitamin D deficiency continues to be prevalent in Indian subcontinent as reported by hospital-based studies. The earliest description of adolescent rickets in India was given in 1925 (73). Our studies demonstrated clinical evidence of genu varum/genu valgum in 10.8–11.5% of apparently healthy children and adolescents in Delhi with no significant difference between the lower (LSES, 11.6%) and upper socioeconomic groups (USES, 9.7%) (4, 74). In children of Indian origin residing in South Africa, the prevalence of knock knees and bow legs with gaps of 2.5 cm or more was 6.1–19.4% (75). In Indian migrants in the United Kingdom, the prevalence of clinical vitamin D deficiency in children and adolescents was shown to be 5–30% (76–79), while in studies using biochemical and radiological variables, prevalence was 12.5–66% (67, 80, 81). Comparative prevalence of symptomatic rickets in 9.4% children has been reported from China (82).

7. SERUM 25(OH)D LEVELS

In two of our recent studies we have provided data on 25(OH)D levels of Indian children. Comparison of our data on serum vitamin D with other studies may not be entirely appropriate given the fact that different studies have been conducted in different seasons and using different assays. Nonethless, using the Lips classification (70), severe hypovitaminosis D (<5 ng/ml or <12.5 nmol/l) was seen in 4–8.6% of our study population in summer months (4, 74). Similar prevalence of hypovitaminosis D (6.7%) using Lips criteria was reported in Brazilian children (83). There are no data on 25(OH)D levels of Indian children during winter. However, study conducted by Du et al. has shown up to 45.2% of hypovitaminosis in Chinese adolescents during winter months (82). Studies from Finland has revealed that 13.5% of children demonstrate serum 25(OH)D between 8 and 10 ng/ml (84). However, in our studies 37 and 29.9% of children had serum 25(OH)D < 9 ng/ml (4, 74).

The mean serum concentrations of 25(OH)D reported in children and adolescents from urban northern India were 11.8 ± 7.2 and 13.84 ± 6.97 ng/ml, respectively (4, 74). These were significantly lower than that reported in children from southern India. The mean 25(OH)D concentrations in adolescents in urban and rural Andhrapradesh were 17 and 18 ng/ml, respectively (3).

Detailed evaluation of 25(OH)D status in higher and lower socioeconomic strata reveals no significant difference in the values when percentage of body surface area exposed to sunshine, duration of exposure, and use of UV protection sunscreen were taken into consideration *(74)*. The only other study from west, which has compared low and high socioeconomic groups, has also not revealed any difference in mean vitamin D concentration between the two groups *(83)*. An objective evaluation of association of nutrition and lifestyle clearly revealed significant correlation between serum 25(OH)D and estimated sun exposure ($r = 0.185$, $p = 0.001$) and percentage body surface area exposed ($r = 0.146$, $p = 0.004$) and not socioeconomic status suggesting that lifestyle-related factors contribute significantly to the vitamin D status of the apparently healthy school girls. In the absence of fortification, diet alone has insignificant role in determining serum 25(OH)D levels *(74)*.

The functional significance of low serum 25(OH)D in Indian children is reflected in their serum PTH values. A negative correlation between 25(OH)D and iPTH was observed in our studies ($r = -0.202$, $p<0.001$) *(4, 74)*. Up to 10.3–37.5% of subjects with vitamin D levels below 9 ng/dl demonstrate supranormal serum intact PTH concentrations. Variation in serum PTH levels despite comparable levels of 25(OH)D could possibily be due to variables such as serum 1,25(OH)D, prolonged exposure to low vitamin D status, and dietary calcium intake.

7.1. *Vitamin D Deficiency and Bone Mineral Density*

Various non-invasive techniques currently available to measure bone mineral density are central or peripheral dual-energy X-ray absorptiometry (cDXA and pDXA), quantitative computerized tomography (QCT), and quantitative ultrasound (QUS). Bone mass in the radius (non-weight bearing) and calcaneum (trabecular site) can be conveniently measured using portable pDXA machine with low precision errors, low radiation dose, and short data acquisition time. The values of bone density parameters obtained with pDXA correlate well with cDXA measurements and are now used to predict fracture risk in children and adults *(85–88)*. The role of pDXA in clinical practice in children is yet not determined because of lack of reference data and predictors of BMD at these sites.

While studies looking into predictors of BMD have mainly focused on lifestyle and anthropometry, studies assessing role of 25(OH)D and related biochemical parameters in determining variation in BMD are not available in children.

In the first study on BMD by pDXA in 555 north Indian children (225 boys and 330 girls), children from LSES had significantly lower BMD values at the forearm than those from USES ($p<0.01$) *(4)*. Subsequent BMD study carried out in girls from LSES ($n = 369$) and USES ($n = 295$) also showed significantly lower mean BMD at distal forearm and calcaneum in the latter (0.366 ± 0.075 vs. 0.337 ± 0.70 g/cm^2 and $0.464 + 0.093$ vs. $0.407 + 0.073$ g/cm^2, respectively) *(88)*. Comparison of BMD at calcaneum in Indian children *(86)* with data from healthy Caucasian children in the United Kingdom acquired with the same model of densitometer *(85)* showed that LSES subjects had lower values and USES subjects had higher values.

The lower bone density in LSES could be due to poor overall nutrition as evidenced by low BMI, poor dietary calcium intake *(89)*, and secondary hyperparathyroidism *(90)*. The important role played by nutrition in accrual of bone mass has been recently shown by Shatrugna et al., where bone density at neck of the femur was found to be significantly greater in children (6–16 years) supplemented with micronutrients for 14 months than the placebo group *(91)*.

On multiple regression analysis, after adjusting for the association of other factors, we did not find any significant association of either 25(OH)D or PTH on either site *(4, 88)*. Also, other studies available on adolescent subjects did not find any association of BMDdf with either 25(OH)D *(4, 84, 92)* or PTH *(4, 91)*. Interestingly, Qutilla et al. observed that subjects with 25(OH)D concentration ≤40 nmol/l had low BMD values at BMD at distal forearm compared to those with values ≥ 40 nmol/l *(84)*. The narrow range of 25(OH)D values in Indian children does not allow us to assess similar situation in Indian children.

In view of high prevalence of hypovitaminosis D in apparently healthy children because of lifestyle changes and cultural practices, awareness needs to be generated about benefits accrued by direct sunlight exposure. Significantly low BMD in LSES children when compared to those from USES suggests the need for interventional studies to evaluate the role of nutrition in improving BMD and peak bone mass.

7.2. Do We Have Protective Bioadaptation to Vitamin D Deficiency?

In the early stage of vitamin D deficiency, we adapt by increasing serum PTH levels to maintain calcium homeostasis. Though rise in serum PTH helps maintaining normal serum ionized calcium; in the long run, its effect is detrimental consequent to bone resorption and osteoporosis. Up to one-fourth of apparently normal Indians including physicians and other paramedical workers have supranormal serum PTH *(2, 4, 74)*.Our recent study shows lack of bio-adaptation in the intestinal calcium absorption in Indians with chronic hypovitaminosis D. The VDR mRNA expression level in the jejunal mucosa and peripheral blood mononuclear cells showed no significant difference in subjects with 25(OH)D < 10 ng/ml and > 10 ng/ml. The VDR mRNA levels of the peripheral mononuclear cell did not change after cholecalciferol supplementation *(28)*.

7.3. How to Tackle Vitamin D Deficiency

Awareness regarding vitamin D deficiency and its causes despite plenty of sunshine is the first step in the management of this deficiency at the community level. Since, serum 25(OH)D levels are low in majority of Indians, there is no point in carrying out costly assay in all individuals. Elevated serum alkaline phosphates and PTH are good markers of frank 25(OH)D deficiency. A good therapeutic regimen, which would successfully bring efficient serum 25(OH)D response in adults, is a sachet of cholecalciferol (60,000 IU) taken every week for 8 weeks and 1 g of elemental calcium per day (two tablespoons of 1,250 mg calcium carbonate). This regimen is practical and is not associated with any detrimental side effects such as hypercalciuria. After the initial 8 weeks, it would suffice to recommend 1 g of elemental calcium daily and 1,000–2,000 IU of vitamin D daily or 60,000 IU per month in the form of a single oral bolus dose *(93)*. Importance of direct sunshine exposure for at least half an hour daily should be reinforced, and 1 g elemental calcium intake in the diet needs to be promoted. In the long term, active intervention in the form of a National policy for vitamin D food fortification programme is required in our country.

8. CONCLUSION

There is a high prevalence of clinical and biochemical hypovitaminosis D in apparently healthy Indians of all age and sex groups despite adequate sunshine. This is primarily because of lifestyle changes and cultural practices and in principle is largely preventable. Hypovitaminosis D therefore is primarily an issue of public health rather than of individual patient care. Although it is unrealistic to expect Asian migrants to the United Kingdom or residents of India to change their habits or lifestyle, awareness needs to be generated about benefits accrued by sunlight exposure and alternate strategy like fortification of food articles is required to be adopted in Indian which will pay rich dividends in the long run.

REFERENCES

1. Bhattacharya AK (1992) Nutritional rickets in the tropics. World Rev Nutr Diet 67:140–197
2. Goswami R, Gupta N, Goswami D, Marwaha RK et al (2000) Prevalence and significance of low 25-hydroxyvitamin D concentrations in healthy subjects in Delhi. Am J Clin Nutr 72:472–475
3. Harinarayan CV, Ramalakssmi T, Prasad UV et al (2008) Vitamin D status in Andhra Pradesh: A population based study. Indian J Med Res 127:211–218
4. Marwaha RK, Tandon N, Reddy DHK et al (2005) Vitamin D and bone mineral density status of healthy schoolchildren in northern India. Am J Clin Nutr 82:477–482
5. Myron D (2005) Gross Vitamin D and Calcium in the prevention of prostate and Colon Cancer: New approaches for the identification of Needs American Society for nutritional Sciences:326–331.
6. Holick MF (2007) Vitamin D deficiency N. Engl J Med 357:266–281
7. Mohr SB, Garland CF, Gorham ED et al (2008) The association between ultraviolet B irradiance, vitamin D status and incidence rates of type 1 diabetes in 51 regions worldwide. Diabetologia (Epub)
8. Zipitis CS, Akobeng AK (2008) Vitamin D supplementation in early childhood and risk of type 1 diabetes: a systematic review and meta-analysis. Arch Dis Child 93:512–517
9. Munger KL, Levin LI, Hollis BW et al (2006) Serum 25- hydroxyvitamin D levels and risk of multiple sclerosis. JAMA 296:2832–2838
10. Stapleton G (1925) Late rickets and osteomalacia in Delhi. Lancet 2:1119–1123
11. Parfitt AM (1998) Osteomalacia and related disorders. In: Avioli LV, Krane SM, (eds) Metabolic bone disease and clinically related disorders, 3rd edn. Academic Press, San Diego
12. Teotia SPS, Teotia M (2002) Metabolic bone diseases: Changing perceptions. In: Thakur BB (ed) Post graduate medicine. The Association of Physicians of India, India
13. Hodgkin P, Kay GH, Hine PM et al (1973) Vitamin D deficiency in Asians at home and in Britain. Lancet 2:167–172
14. Harinarayan CV, Gupta N, Kochupillai N (1995) Vitamin D status in primary hyperparathyroidism. Clin Endocrinol (Oxf) 43:351–358
15. Hollis BW (2005) Circulating 25-Hydroxyvitamin D Levels Indicative of Vitamin D Sufficiency: Implications for Establishing a New Effective Dietary Intake recommendation for vitamin D. American Society for Nutritional Sciences; 317–322.
16. Goswami R, Mondal AM, Tomar N et al (2008) Presence of 25(OH)D deficiency and its effect on vitamin D receptor mRNA expression. Eur J Clin Nutr:1–4
17. Seth A, Marwaha RK, Singla B et al (2008) Vitamin D Nutritional status of exclusively breast fed infants and their mothers. Accepted in JPEM Manuscript in press
18. Scott AC (1916) A contribution to the study of osteomalacia in India. Ind J Med Res 4:140–182
19. Wilson DC (1932) Osteomalacia and rickets in Northern India. Lancet 18:951

20. Teotia SPS, Teotia M (2008) Nutritional and metabolic bone and stone disease an Asian perspective, 1st edn. CBS Publishers, New Delhi, pp. 274–289

21. Harinarayan CV, Ramalakshmi T, Venkataprasad U (2004) High prevalence of low dietary calcium and low vitamin D status in healthy south Indians. Asia Pac J Clin Nutr 13(4):359–364

22. Harinarayan CV (2005) Prevalence of vitamin D insufficiency in postmenopausal south Indian women. Osteoporos Int 16(4):397–402

23. Harinarayan CV, Ramalakshmi T, Prasad UV et al (2007) High prevalence of low dietary calcium, high phytate consumption, and vitamin D deficiency in healthy south Indians. Am J Clin Nutr 85: 1062–1067.

24. Harinarayan CV, Ramalakssmi T, Prasad UV et al (2008) Vitamin D status in Andhra Pradesh: A population based study. Indian J Med Res 127:211–218.

25. Shatrugna V, Kulkarni B, Kumar PA et al (2005) Bone status of Indian women from a low-income group and its relationship to the nutritional status. Osteoporos Int 16:1827–1835

26. Arya V, Bhambri R, Godbole MM et al (2004) Vitamin D status and its relationship with bone mineral density in healthy Asian Indians. Osteoporos Int 15:56–61

27. Tandon N, Marwaha RK, Kalra S et al (2003) Bone mineral parameters in healthy young Indian adults with optimal vitamin D availability. Nat Med J India 16:6

28. Vupputuri MR, Goswami R, Gupta N et al (2006) Prevalance and functional significance of 25-hydroxyvitamin D deficiency and vitamin D receptor gene polymorphisms in Asian Indians. Am J Clin Nutr 83:1411–1419

29. Secker B (2004) Vitamin D requirements during pregnancy. Am J Clin Nutr 80(suppl):1740S–1747S

30. Dawodu A, Wagner CL (2007) Mother-child vitamin D deficiency: an international perspective. Arch Dis Childhood 92:737–740

31. Brooke OG, Brown I, Bone CDM et al (1980) Vitamin D supplements in pregnant Asian women: effects on calcium status and fetal growth. BMJ 280:751–754

32. Raman L, Rajalakshmi K, Krishnamachari KAVR et al (1978) Effect of calcium supplementation to undernourished mothers during pregnancy on the bone density of the neonates. Am J Clin Nutr 31:466–469

33. Marya RK, Rathee S, Dua V et al (1988) Effect of vitamin D supplementation during pregnancy on fetal growth. Indian J Med Res 88:488–492

34. Sabour H, Hossein-Nezhad A, Maghbooli Z et al (2006) Relationship between pregnancy outcomes and maternal vitamin D and calcium intake: A cross-sectional study. Gynecol Endocrinol 22: 585–589

35. Hollis BW, Wagner CL (2004) Assessment of dietary vitamin D requirements during pregnancy and lactation. Am J Clin Nutr 79:717–726

36. Hollis BW, Wanger CL (2004) Vitamin D requirements during lactation: high dose maternal supplementation as therapy to prevent hypovitaminosis D in both mother and nursing infant. Am J Clin Nutr 80:1752S–1758S

37. Kreiter SR, Schwartz RP, Kirkman HN et al (2000) Nutritional rickets in African American breast-fed infants. J Pediatr 137:153–157

38. Holick MF (2006) Resurrection of vitamin D deficiency and rickets. J Clin Invest 116:2062–2072

39. Hollis BW, Wagner CL (2004) Assessment of dietary vitamin D requirements during pregnancy and lactation. Am J Clin Nutr 79:717–726

40. Marya RK, Rathee S, Dua V et al (1988) effect of vitamin D supplementation during pregnancy on fetal growth. Indian J Med Res 88:488–492

41. Heckmatt JZ, Pocock M, Davies AEJ et al (1979) Plasma 25-hydroxyvitamin D in pregnant asian women and their babies. Lancet 1:546–549

42. Brooke OG, Brown IRF, Cleeve HJW (1981) Sood A. Observations on the vitamin D state of pregnant Asian women in London. Br J Obstet Gynaecol 88:18–26

43. Howarth AT (1976) Biochemical indices of osteomalacia in pregnant Asian immigrants in Britain. J Clin Pathol 29:981–983

44. Dent CE, Gupta MM (1975) Plasma 25-hydroxyvitamin D levels during pregnancy in Caucasians and in vegetarian and non vegetarian Asians. Lancet 2:1057–1060

45. Rab SM, Baseer A (1976) Occult osteomalacia amongst healthy and pregnant women in Pakistan. Lancet 2:1211–1213
46. Atiq M, Suria A, Nizami SQ et al (1998) Maternal vitamin D deficiency in Pakistan. Acta Obstet Gynecol Scand 77:970–973
47. Datta S, Alfaham M, Davies DP et al (2002) Vitamin D deficiency in pregnant women from a non-European ethnic minority population: an interventional study. Br J Obstet Gynecol 109: 905–908
48. Henriksen C, Brunvand L, Stoltenberg C et al (1995) Diet and vitamin D status among pregnant Pakistani women in Oslo. Eur J Clin Nutr 49:211–218
49. Bassir M, Laborie S, Lapillone A et al (2001) Vitamin D deficiency in Iranian mothers and their neonates: a pilot study. Acta Paediatr 90:577–579
50. Brunvand L, Shah SS, Bergstrom S et al (1998) Vitamin D deficiency in pregnancy is not associated with obstructed labour: a study among Pakistani women in Karachi. Acta Obstet Gynecol Scand 77:303–306
51. Serenius F, Eldrissy A, Dandona P (1984) Vitamin D nutrition in pregnant women at term and in newly born babies in Saudi Arabia. J Clin Pathol 37:444–447
52. Taha SA, Dost SM, Sedrani SH (1984) 25-hydroxy D and total calcium: extraordinarily low plasma concentrations in Saudi mothers and their neonates. Paediatr Res 18:739–741
53. Sachan A, Gupta R, Das V et al (2005) High prevalence of vitamin D deficiency among pregnant women and their newborns in northern India. Am J Clin Nutr 81:1060–1064
54. Mozolowski W, Jedrzej S (1939) (1768–1883) on the cure of rickets. Nature 143:121
55. Palm TA (1890) The geographic distribution and etiology of rickets. Practitioner 45:321–342
56. Agarwal KS, Mughal MZ, Upadhyay P et al (2002) The impact of atmospheric pollution on vitamin D status of infants and toddlers in Delhi, India. Arch. Dis. Child 87:111–113
57. Teotia SPS, Teotia M (2002) Metabolic bone diseases: Changing perceptions. In: Thakur BB (ed) Post graduate medicine. The Association of Physicians of India, India
58. Balasubramanian S, Shivbalan S, Saravana KP (2006) Hypocalcemia due to vitamin D deficiency in exclusively breastfed infants. Indian Pediatrics 43
59. Okonofua F, Houlder S, Bell J et al (1986) Vitamin nutrition in pregnant Nigerian women at term and their newborn infants. J Clin Pathol 39:650–653
60. Zeghoud F, Vervel C, Guillozo H (1997) Subclinical vitamin D deficiency in neonates: definition and response to vitamin D supplements. Am J Clin Nutr 65:771–778
61. Hillman LS, Haddad JG (1974) Human perinatal vitamin D metabolism. I. 25-hydroxyvitamin D in maternal and cord blood. J Pediatr 84:742–749
62. Bhalala U, Desai M, Parekh P et al (2007) Subclinical hypovitaminosis D among exclusively breastfed young infants. Indian Pediatrics 44:897–901
63. Aneja S, Marwaha RK (2008) Hypovitaminosis D and Hypocalcemic seizures in Infancy. Accepted in EAP Nice France 2008.
64. Cadogan J, Blumsohn A, Barker ME et al (1998) A longitudinal study of bone gain in pubertal girls: anthropometric and biochemical correlates. J Bone Miner Res 13:1602–1612
65. Adams JS, Hollis BW (2002) Vitamin D synthesis, metabolism and clinical measurement. In: Coc FL, Favus MJ, (eds) Disorders of bone and mineral metabolism, 2nd edn. Lippincott, Williams & Wilkins, Philadelphia. PA, USA, 157–174
66. Guillemant J, Cabrol S, Allemandou A et al (1995) Vitamin D-dependant seasonal variation of PTH in growing male adolescents. Bone 17:513–516
67. Stryd RP, Gilbertson TJ, Brunden MN (1979) A seasonal variation study of 25-hydroxyvitamin D serum levels in normal humans. J Clin Endocrinol Metab 48:771–775
68. Ford JA, McIntosh WB, Butterfield R et al (1976) Clinical and sub clinical vitamin D deficiency in Bradford children. Arch Dis Child 51:939–943
69. Parfitt AM (1998) Osteomalacia and related disorders. In: Avioli LV, Krane SM (eds) Metabolic bone disease, 3rd edn. Academic press, San Diego, CA, USA
70. Lips P (2001) Vitamin D deficiency and secondary hyperparathyroidism in the elderly: consequences for bone loss and fractures and therapeutic implications. Endocr Rev 22:477–501

71. Marshall D, Iohnell O, Wedel H (1996) Metaanalysis of how well measures of bone mineral density predict occurrence of osteoporotic fractures. BMJ 312:1254–1259
72. Chapuy MC, Arlot ME, Dubocuf F et al (1992) Vitamin D and calcium to prevent hip fractures in elderly women. N Engl J Med 327:1637–1642
73. Stapleton G (1925) Late rickets and osteomalacia in Delhi. Lancet ii:1119–1123
74. Puri S, Marwaha RK, Agarwal N et al (2008) Vitamin D status of apparently healthy schoolgirls from two different socioeconomic strata in Delhi: relation to nutrition and lifestyle. Br J Nutr 99:876–882
75. Goel KM, Sweet EM, Logan RW et al (1976) Florid and subclinical rickets among immigrant children in Glasgow. Lancet I:1141–1145
76. Arncil GC (1975) Nutritional rickets in children in Glasgow. Proc Nutr Soc 34:101–109
77. Dunnigan MG, Paton JPJ, Haasc S et al (1962) Late rickets and osteomalacia in the Pakistani community in Glasgow. Scott Med J 7:159–167
78. Holmes AM, Enoch BA, Taylor JL et al (1973) Occult rickets and osteomalacia amongst the Asian immigrant population. Q J Med 42:125–149
79. Dunnigan MG, McIntosh WB, Sutherland GR et al (1981) Policy for prevention of Asian rickets in Britain: a preliminary assessment of the Glasgow rickets campaign. Br Med J 282:357–360
80. Ford JA, Colhoun EM, McIntosh WB (1972) Rickets and osteomalacia in the Glasgow Pakistani community, 1961–1971. BMJ 2:677–680
81. Moncrieff MW, Lunt HRW, Arthur LJH (1973) Nutritional rickets at puberty. Arch Dis Child 48:221–224
82. Du X, Greenfield H, Fraser DR, Ge K et al (2001) Vitamin D deficiency and associated factors in adolescent girls in Beijing. Am J Clin Nutr 74:494–500
83. Linhares ER, Jones DA, Round JM et al (1984) Effect of nutrition on vitamin D status: studies on healthy and poorly nourished Brazilian children. Am J Clin Nutr 39:625–630
84. Qutila TA, Karkkainen MUM, Lamberg-allardt CJE (2001) Vitamin D status affects serum parathyroid hormone concentrations during winter in female adolescents: associations with forearm bone mineral density. Am J Clin Nutr 74:206–210
85. Chinn DJ, Fordham JN, Kibirige MS et al (2005) Bone density at the os calcis: reference values reproducibility, and effects of fracture history of and physical activity. Arch Dis Child 90:30–35
86. Hernandez-Prado B, Lazcano-Ponee E, Cruz-Valdez A et al (2002) Validity of bone mineral density measurement in distal sites as an indicator of total bone mineral density in a group of pre-adolescent and adolescent women. Arch Med Res 33:33–39
87. IIer PD, Siris ES, Barrctt-Connor E et al (2002) Prediction of fracture risk in postmenopausal white women with peripheral bone densitometry: Evidence from the National Osteoporosis Risk Assessment. J Bone Miner Res 17:2222–2230
88. Marwaha RK, Tandon N, Reddy DHK et al (2006) Peripheral bone mineral density and its predictors in healthy schoolgirls from two different socioeconomic groups in Delhi. Osteoporos Int 2:23–27
89. Oliveri MB, Wittich A, Mautelen C et al (2000) Peripheral bone mass is not affected by winter vitamin D deficiency in children and young adults from Ushuaia. Calcify Tissue Int 67:220–224
90. Bonofiglio D, Maggiolini M, Catalano S et al (2000) Parathyroid hormone is elevated but bone markers and density are normal in young female subjects who consume inadequate dietary calcium. Br J Nutr; 84: 111–116.
91. Shatrugna V, Balakrishna N, Krishnaswamy K (2006) Effect of micronutrient supplement on health and nutritional status of schoolchildren: bone health and body composition. Nutrition 22:533–539
92. Kardinaal AF, Ando S, Charles P et al (1999) Dietary calcium and bone density in adolescent girls and young women in Europe. J bone Miner Res 14:583–592
93. Goswami R, Gupta N, Ray D et al (2008) Pattern of 25- hydroxyl vitamin D response at short (2 month) and long (1 year) interval after 8 weeks of oral supplementation with cholecalciferol in asian Indians with chronic hypovitaminosis D. Br J Nutr 1–4

28 Vitamin D Deficiency, Rickets, and Fluorosis in India

C. V. Harinarayan and Shashank R. Joshi

Abstract Data on the vitamin D status of the populations in a tropical country like India have seldom been documented. Vitamin D deficiency is presumed to be rare. Population studied by the author and others in the country has proved otherwise. Studies were carried out to document the dietary habits, serum calcium, 25-hydroxyvitamin D [25(OH)D], and parathyroid hormone levels of urban and rural population in a state in south India. The daily dietary calcium intake of both the urban and the rural population was low compared to that of recommended daily/dietary allowances (RDA) issued by Indian Council of Medical Research (ICMR). About 85% of the populations have varying degree of vitamin D status – either hypovitaminosis or vitamin D deficiency. The 25(OH)D levels of rural adult subjects were significantly higher ($P < 0.001$) than that of urban adult subjects in both male and female groups studied. The 25(OH)D levels of both the urban and the rural children were low.

Low dietary calcium intake and 25(OH)D levels are associated with deleterious effects on bone mineral homeostasis. The quality of diets has to be improved with enrichment/supplementation of calcium and vitamin D to suppress secondary hyperparathyroidism-induced bone loss and risk of fractures. Prospective longitudinal studies are required to assess the effect on bone mineral density, a surrogate marker for fracture risk and fracture rates. In the background of low dietary calcium and low vitamin D status an environmental toxin like fluoride has deleterious effects on the bone. The effect of fluoride and its concomitant renal tubular damage affects the bone mineral metabolism in a deleterious way.

Key Words: Vitamin D deficiency; rickets; fluorosis; calcium deficiency; India; sunlight; 25-hydroxyvitamin D

1. VITAMIN D DEFICIENCY IN INDIA

Vitamin D is a secosteroid which is converted into its active form via 1α-hydroxylase to $1,25(OH)_2D$. Besides its pivotal role in calcium homeostasis and bone mineral metabolism, vitamin D endocrine system is now recognized to subserve a wide range of fundamental biological functions in cell differentiation, inhibition of cell growth as well as immunomodulation. Traditionally known as antiricketic factor or sunshine

Partly adapted from the article previously published in The *Indian Journal of Medical Research (9)*. Published by the Indian Council of Medial Research, New Delhi, India.

From: *Nutrition and Health: Vitamin D*
Edited by: M.F. Holick, DOI 10.1007/978-1-60327-303-9_28,
© Springer Science+Business Media, LLC 2010

vitamin, it is a steroid that regulates complex system of genomic functions and has a role in prevention of neoplastic transformation. Recent evidences from genetic, nutritional, and epidemiological studies link vitamin D endocrine system with diseases like hypertension, myopathic disorders, proneness to infection, autoimmune disorders, and cancer.

Vitamin D deficiency in India has only recently received the attention it deserves. The FAO/WHO Expert Consultation (1) stated that in most locations in the world around the equator (between latitudes 42°N and 42°S) the most physiologically relevant and efficient way of acquiring vitamin D is to synthesize it endogenously from skin from 7-dehydrocholesterol present in the subcutaneous fat by 30 min of skin exposure (without sunscreen) of the arms and face to sun (ultraviolet spectrum of wavelength 290–310 nm). Skin synthesis of vitamin D is negatively influenced by latitude, season, aging process, skin pigmentation, clothing, and sunscreen. The skin pigmentation bestowed by the solar energy on the people of Indian subcontinent to protect them from cancerous insults deprive them the dermal ability to synthesize vitamin D. Vitamin D (vitamins D3 and D2) are not naturally present in the food consumed by the vast majority of the people in Indian subcontinent. There has been prevailing impressions among the medical professionals that abundance of sunlight prevents vitamin D deficiency in India. However, these facts are gradually being dispelled. Published evidences between 1995 and 2000 (2, 3) show that even doctors and nurses from northern latitudes of India have vitamin D deficiency. These observations are confirmed from the southern parts of the country (4).

Nutrition plays a vital role in bone homeostasis. Adequate calcium intake along with vitamin D helps to maintain bone mineral mass attained at the end of growth period (peak bone mass). During infancy, childhood, and adolescence, increasing dietary calcium intake favors bone mineral accrual (5). Adequate nutrition and sufficient activity provide mechanical impetus for bone development which may be critical in attaining bone growth potential. Increasing bone mineral content during periods of rapid growth (children and adolescence) increases "peak bone mass" and may effectively prevent osteoporosis.

Vitamin D deficiency and/or poor dietary calcium intake can together lead to defect in mineralization of bones (rickets in children and osteomalacia in adults). Globally there has been renewed interest and focus on the roles of vitamin D status and calcium intake in the prevention of rickets and osteoporosis. This renewed interest has come a major push, especially from researchers from United States to increase the generally accepted 25-hydroxyvitamin D [25(OH)D] reference values for categorizing the vitamin D status. The cutoff levels suggested are as follows (6): deficiency, serum 25(OH)D <20 ng/ml; insufficiency, serum 25(OH)D = 20–30 ng/ml; and insufficiency, serum 25(OH)D >30 ng/ml.

Previously we reported the prevalence of low 25(OH)D concentrations in India in a group of healthy subjects and in patients with primary hyperparathyroidism (2). Later, other reports ensued (3, 4, 7–11). It is surprising to find low concentrations of 25(OH)D in healthy subjects in a country with abundant sunshine. In a large population-based study (9) (a total of 1,148 adults and 146 children), we have documented the dietary habits and serum concentrations of calcium, 25(OH)D, and parathyroid hormone of the Indian population in subjects residing at Tirupati and the surrounding villages in southern Andhra Pradesh, south India, in the past 7 years. Tirupati is located at 13.4°N and 79.2°E. The rural population included men and women who were included after a demographic survey of seven villages. The rural subjects are agricultural workers starting their day from 8 am working outdoors till 5 pm with their face, chest, back, legs, and arms and forearms exposed to sunlight (Fig. 1). The urban population constituted medical and paramedical personnel and their relatives and postmenopausal women and their relatives. Urban subjects were fully dressed with only the face and forearms exposed to sunlight with a white collar job (working indoors between 10 am and 5 pm). Those of them not in a job are indoors most of the time. Urban children were those at home and as well as school going, fully dressed with only the face and forearms exposed to sunlight. Among the rural children, some were school going while the adolescent children (especially the boys) helped the family in agriculture (from 8 am working outdoors till 5 pm). Both urban and rural children (boys and girls) had their face, legs, and arms and forearms exposed to sunlight in their dress code. In addition the rural children (adolescent males) helping the family in agriculture had their torso and back exposed to sunlight. Patients with hepatic, renal, or dermatological disorders; alcoholics; and pregnant women were excluded from the study.

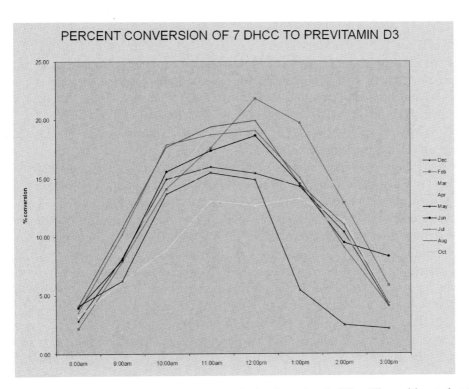

Fig. 1. Influence of time of day, and season on synthesis of previtamin D3 at Tirupati located at 13.4 N and 79.2 E. Biochemical analysis was done at Prof Holick Laboratory *(copyright Holick 2010 with permission).*

In the urban and rural locations, the average duration of cloud-free sunshine is 8–10 h/day throughout the year with the solar zenith angle of 9.92° in summer and 38.2° in winter. The UV index at the above-said latitude during those periods is 7–12. Winter is short with lowest temperature of 17°C (night) and 28°C (day) with scanty rainfall. Most often, there is a little seasonal variation of the peak intensity of sunlight.

The dietary assessment of total energy, calcium, phosphorus, and phytate was documented by recalling the diet consumed in the previous 5–7 days. The diet of urban subjects constituted 2,200 kcal/day. Carbohydrates contributed 55% of the total energy intake, proteins 10%, and fat 10%. Vegetables contributed 10% of the total energy intake and milk and milk products contributed 15%. The carbohydrate source was primarily from cereals with rice providing 50% of total carbohydrates, wheat 25%, and ragi (*Eleusine coracana*) 25%. Vegetable sources included amaranth leaves, cauliflower, carrots, lady's fingers, other seasonal vegetables, and tubers. Animal sources of protein were consumed once a week. The diet of rural subjects consisted of 1,700 kcal/day. Carbohydrates contributed 75% of the total energy intake, proteins 10%, fats 5%, vegetables 5%, and milk and milk products 5%. The carbohydrate source was from cereals (rice: 60%; ragi: 40%). Vegetable sources were drumstick leaves, brinjals, tomatoes, and so forth. Animal sources of protein were consumed once fortnightly. No other source of calcium or any other mineral was consumed in both groups.

Dietary calcium and phosphorus of adults and children were significantly lower ($P<0.0001$) in the rural subjects compared to that of the urban subjects. The dietary phytate/calcium ratio was significantly ($P<0.0001$) lower in urban subjects compared to that of rural subjects in both adults and children (Table 1). Dietary calcium intake correlated negatively with phytate/calcium ratio in adults – urban subjects ($r = -0.28$; $P<0.0001$) and rural subjects ($r = -0.43$; $P<0.0001$). The r values are significantly different from each other ($P=0.039$). In children the correlation in urban was $r = -0.31$; $P = 0.01$ and in rural was $r = -0.61$; $P<0.001$. The r value was significantly different between urban and rural locations ($P<0.02$).

The daily dietary calcium intake of both rural and urban adults and children were far below that of RDA of 400 mg/day for adults and 400–500 mg/day for children issued by the ICMR for the Indian population of both genders and age groups (Table 2) *(12, 13)*. The diet in rural subjects was high in phytate/calcium ratio thus retarding the absorption of already low dietary calcium. Intake of diet rich in phytate (inositol hexaphosphate) retards the absorption of calcium from the gut. Inositol hexaphosphate forms chelates with divalent calcium cations and reduces the absorption of calcium from the gut. It has been shown that the calculated values for all nutrients are significantly higher than the analytical values *(14)*. Hence, a patient with a calculated low intake of calcium with a background of diet containing foods high in phytates, as in the current study, may be more calcium deficient than calculated from dietary intake data. The quality of diet in the rural subjects is low in calcium and high in phytate/calcium ratio compared to that of urban diet. Hence the rural subjects are more affected. Though in rural subjects more body surface area is exposed to sunlight for longer duration by virtue of their occupation, the poor quality of diet impedes the bone homeostasis significantly.

Table 1
Comparison of Dietary Intake of Urban and Rural Groups – Adults and Children[a]

Parameter	Adults				Children			
	Males		Females		Males		Females	
	Urban (n = 32)	Rural (n = 109)	Urban (n = 476)	Rural (n = 96)	Urban (n = 28)	Rural (n = 34)	Urban (n = 43)	Rural (n = 36)
D.Ca (mg/day)	323 ± 8* (307–340)	271 ± 3 (263–280)	306 ± 2* (302–310)	262 ± 3 (253–271)	293 ± 6 (280–305)	277 ± 6 (265–289)	317 ± 9 (298–336)	270 ± 7* (254–285)
D.Pho (mg/day)	674 ± 17* (640–707)	493 ± 9 (475–511)	651 ± 9* (643–660)	481 ± 10 (462–501)	632 ± 18 (595–670)	483 ± 15* (451–514)	667 ± 15 (637–698)	489 ± 16* (458–521)
Phy/Ca ratio	0.5 ± 0.02* (0.47–0.54)	0.76 ± 0.01 (0.74–0.78)	0.51 ± 0.01* (0.50–0.52)	0.76 ± 0.01 (0.74–0.78)	0.49 ± 0.02 (0.45–0.53)	0.75 ± 0.02* (0.70–0.80)	0.52 ± 0.01 (0.49–0.55)	0.72 ± 0.03* (0.66–0.77)

[a]Values expressed as mean ± SEM (95% Confidence Intervals) (all such values). D.Ca – Dietary calcium; D.Phos–Dietary phosphorus; Phy/Ca – Phytate/calcium ratio; n = sample size. *$P < 0.0001$ compared to urban and rural.

Table 2
Recommended Dietary Allowances (RDA) of calcium in India and in
the United States

Category	India (19, 26, 27)	The United States (25)
Units (mg/day)		
Infants		
0–6 months	500	500
6–12 months	500	750
Children (boys and girls)		
1–9 years	400	800
10–15 years	500	1200–1300
16–18 years	500	1200–1300
Men	400	800–1000
Women	400	800–1000
Pregnant and Lactating mothers	1000	1200–1300

The serum calcium, phosphorous, and SAP levels of the urban and the rural adults and children were within the normal range (Table 3). Compared to urban children, the serum calcium was significantly high ($P<0.01$) and serum phosphorous and SAP low ($P<0.01$) in rural children. The 25(OH)D levels of rural subjects were significantly higher ($P<0.001$) than that of urban subjects in both male and female groups. The vitamin D-deficient, -insufficient, and -sufficient states of rural and urban adult subjects are depicted in Fig. 2a and rural and urban children in Fig. 2b. N-tact PTH levels negatively correlated with 25(OH)D in rural subjects ($r = -0.24$; $P<0.002$), and in urban subjects in adults ($r = -0.12$; $P<0.0001$) and in children ($r = -0.2$; $P<0.05$).

The calcium absorptive performance of the gut is a function of 25(OH)D status of an individual (15, 16). When the 25(OH)D concentrations are low, the effective calcium absorption from the gut is reduced (15, 16). In children where the calcium requirements are high, in the background of low dietary calcium and 25(OH)D levels, the peak bone mass achieved is low, which in turn leads to high risk of fractures in old age group at a later date. It has been shown that low dietary calcium converts the 25(OH)D to polar metabolites in the liver and leads to secondary 25(OH)D deficiency (17). The SHPT that ensues increases the risk of fractures, especially in postmenopausal women and elderly patients.

Also, low-calcium intake increases PTH which increases conversion of $25(OH)D_2$ to 1,25-dihydroxyvitamin D ($1,25(OH)_2D$) which in turn stimulates the intestinal calcium absorption. In addition, $1,25(OH)_2D$ induces its own destruction by increasing the 24-hydroxylase. This probably explains the low 25(OH)D levels in individuals on high-phytate diet or low-calcium diet.

In the present study, low prevalence of 25(OH)D deficiency is seen in rural male subjects compared to that of the urban male subjects. The same observation is made for females. This is probably due to occupation, dress code, and duration of exposure to

Table 3
Comparison of Biochemical and Hormonal Profile of Urban and Rural Groups – Adults and Children[a]

Parameter	Adults				Children			
	Males		Females		Males		Females	
	Urban	Rural	Urban	Rural	Urban	Rural	Urban	Rural
S.Ca (mg/dl)	9.74 ± 0.06 (9.63–9.85) (n = 100)	10.06 ± 0.06* (9.95–10.2) (n = 109)	9.68 ± 0.02 (9.64–9.73) (n = 678)	9.98 ± 0.06* (9.87–10.15) (n = 96)	9.67 ± 0.17 (9.33–10) (n = 28)	10.28 ± 0.09* (10.09–10.48) (n = 34)	9.78 ± 0.09 (9.59–9.97) (n = 36)	10.03 ± 0.11 (9.80–10.27) (n = 36)
S.Phos (mg/dl)	3.50 ± 0.07 (3.37–3.64) (n = 99)	2.84 ± 0.07* (2.27–2.97) (n = 109)	3.64 ± 0.03 (3.59–3.69) (n = 679)	2.74 ± 0.07* (2.79–3.09) (n = 96)	4.0 ± 0.22 (3.55–4.45) (n = 27)	3.15 ± 0.12* (2.92–3.39) (n = 33)	4.08 ± 0.16 (3.76–4.41) (n = 36)	3.31 ± 0.13* (3.05–3.56) (n = 35)
SAP (IU/l)	84.87 ± 3.87 (78.85–90.9) (n = 98)	55.67 ± 2.07* (49–61) (n = 109)	80.4 ± 3.07 (78–90.17) (n = 683)	62.7 ± 3.41* (56–69.4) (n = 96)	161.55 ± 17.5 (125.7–197) (n = 28)	90.7 ± 11.9* (66.5–115) (n = 34)	132 ± 16 (99–165) (n = 36)	87.6 ± 11.75** (63.8–111.5) (n = 36)
25(OH)D (ng/ml)	18.54 ± 0.8 (17–20) (n = 134)	23.73 ± 0.8* (22–25) (n = 109)	15.5 ± 0.3 (14.9–16) (n = 807)	19 ± 0.89* (17.54–21) (n = 96)	15.57 ± 1.21 (13–18) (n = 30)	17 ± 1.3 (14–20) (n = 34)	18.5 ± 1.66 (15–22) (n = 39)	19 ± 1.59 (16–22) (n = 36)
N-tact PTH (pg/ml)	27 ± 1.6 (23.9–30) (n = 135)	29.24 ± 1.6 (26–32.35) (n = 109)	28.35 ± 0.6 (27–29.5) (n = 803)	29.21 ± 1.7 (25.75–32.7) (n = 96)	36.7 ± 10 (16–58) (n = 28)	26.51 ± 1.6 (23–30) (n = 34)	21 ± 1.5 (18–24) (n = 41)	25.28 ± 1.8 (22–29) (n = 36)

[a]Values as mean ± SEM (95% Confidence Intervals) all such values. n = sample size. *$P < 0.0001$; **$P < 0.05$ compared to urban and rural. S.Ca – serum calcium; S.Phos – serum phosphorous; SAP – serum alkaline phosphatase; 25(OH)D – 25-hydroxyvitamin D; N-tact-PTH – immunoreactive parathyroid hormone. Conversion of 25(OH)D from ng/ml to nmol/l – multiply by factor 2.5.

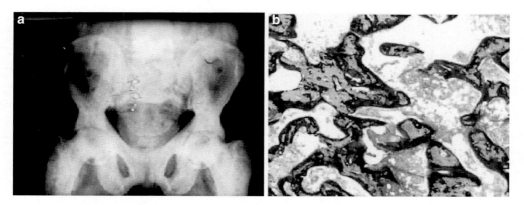

Fig. 2. (A) Osteosclerosis and ligamental calcification in fluorosis. (B) Undecalcified section of bone showing increased osteoid.

sunlight of the rural subjects, who are agricultural laborers working for about 8 h a day in sunlight. In the region where this study was conducted, season has little impact on cutaneous synthesis of vitamin D. There are reports from Indian subcontinent of very low dietary intakes of calcium (<300 mg/day) causing osteomalacia *(18, 19)*, low dietary calcium, and 25(OH)D status in postmenopausal women *(7, 8, 20)*, children *(21)*, and pregnant women and their offsprings *(3, 22)*.

Low dietary calcium intake results in the development of rickets among vitamin D-deficient baboons *(23, 24)*. In rats *(23, 24)* low-calcium diet or high-phytate diet resulted in increased catabolism of 25(OH)D concentrations leading to formation of inactive metabolites with resultant reduction in 25(OH)D concentrations. The pathogenesis of rickets in Asian community in the United Kingdom has been attributed to high cereal, low-calcium diet which induces mild hyperparathyroidism *(25)*. The role of low dietary calcium intake in the pathogenesis of 25(OH)D deficiency is probably greater than that is originally recognized.

Recently from a north Indian village around 100 km from Delhi despite of adequate sunlight Goswami et al. have reported that more than 70% of villagers were vitamin D deficient despite of more than daily 5 h of sunlight *(26)*. This finding and studies from Harinarayan from south India clearly tell us the role of melanin as a sunscreen in the Indian context *(9)*. Sahu et al. from Lucknow have reported vitamin D deficiency in rural girls and in pregnant women despite abundant sunshine in northern India *(27)*. They reported a higher prevalence of vitamin D deficiency (VDD) among pregnant women and adolescent girls from a rural Indian community with boys being relatively protected and a seasonal variation in serum 25(OH)D which was significant at latitude 26°N. In this context the current paper of Goswami and Kochupillai makes it one of the early studies in rural north Indian village which breaks the myth of the Indian sunshine *(26)*. Seventy percent of rural village residents in Bulandhar in Uttar Pradesh despite of plenty sunlight are VDD is indeed an eye-opener *(28)*. Multitude of factors apart from dark skin, pollution, and dietary malabsorption may play a role. The authors also make a strong case which is strongly supported by many of us for vitamin D fortification on a priority. This is a complex issue as unlike iodine where salt is an efficient vehicle it is difficult to get a vehicle to fortify for vitamin D except for milk and dairy products. With milk habits declining even if fortified, it makes the issue further difficult. This study thus is a classical vitamin D deficiency paradox despite plenty of sunshine and warns medical community about dangers of vitamin D deficiency epidemic which may swarm not only urban but also even rural Indian adults. But some urgent reprisal for vitamin D sufficiency and fortification is needed for vulnerable Indians. There is a need to evolve consensus guidelines for normal vitamin D levels and deficiency for India. It is time to screen vitamin D levels as well as treat them aggressively *(28)*.

2. NUTRITIONAL RICKETS IN INDIA

Classical vitamin D deficiency rickets in India has been prevalent for more than two centuries and the prevalence has not really come down despite of increase in the per capita milk consumption. Its omnipresence in the urban and rural India makes the policy makers a strong case for milk fortification of vitamin D *(28)*. The most distressing feature is the prevalence of neonatal hypocalcemia and rickets, which still persists, that are virtually eliminated from North America. Despite of the discovery of vitamin D and the role that ultraviolet light irradiation plays in vitamin D formation that rational and appropriate therapy became available and rickets was all but eradicated in a number of developed countries. Despite of the strong milk consumption and vitamin D fortification in North America and Europe, there has been a resurgence of the disease in many countries of Europe and the United States probably due to an increase in the prevalence of breast feeding, the immigration of dark-skinned families to countries of high latitude, and the avoidance of direct sunlight because of the risk of the development of skin cancers *(28, 29)*. The disease is still widely prevalent in many developing countries including southeast Asia especially India and Pakistan. The role and relative contribution of calcium and vitamin D deficiencies in development of rickets is still being studied but both have a substantial contribution. Studies have led to the realization that nutritional rickets may be caused by either vitamin D or calcium deficiency, but in the majority of situations variable combinations of both probably play a role. Although low dietary calcium intakes appear to be central to the pathogenesis of rickets in African countries like Nigeria, genetic and/or other environmental factors are likely to contribute *(29)*. But to date no single factor has been isolated as contribution significantly. The results of a recently conducted study suggest that in situation of low dietary calcium intakes vitamin D requirements may be higher than normal, possibly predisposing those children with vitamin D levels in the low normal range to rickets *(29)*. If this is so, it would indicate that the currently accepted normal range for vitamin D sufficiency would need to be adjusted depending on dietary calcium intakes. Yet we are still unclear as to the factors which predispose some children to the disease *(29)*.

In the south Asian subcontinent, rickets in infants, older children, and adolescents has been described from India, Nepal, Bhutan, Bangladesh, and Pakistan *(30)*. A number of factors have been indicated as being responsible for a high prevalence of vitamin D deficiency and rickets, including religious customs *(33)*, atmospheric pollution *(10)*, increased skin pigmentation, vegetarian diets *(31)*, and maternal vitamin D deficiency *(30, 22)*. Further, low dietary calcium and high phytate diets are also considered to play a major role in some communities. In one study it was suggested that low dietary calcium intakes were important as a factor in young children while in adolescent children vitamin D deficiency was mainly responsible *(32)*. Endemic fluorosis in children may also manifest with severe rachitic-like bone deformities and radiographic features of rickets at the growth plates *(33)*. Some 18 of the 33 states in India have problem with excessive fluoride concentration in ground water *(33)*. It is suggested that in the face of low dietary intakes, the increased calcium requirements associated with the increased mineralization of fluorotic bone may reduce 25(OH)D concentrations which together with the low calcium intakes induce rickets *(34)*.

Recently attention has been focused on the apparently high prevalence of rickets in children living in the southeastern coastal region of Bangladesh. Reports suggest that the disease has only been prevalent in the last two decades. Studies provide evidence that the pathogenesis is related to dietary calcium deficiency *(9, 26)*, possibly due to the introduction of irrigation which has been associated with an increase in the size of the annual rice harvest and a reduction in dietary variation. In other parts of Asia, such as the northern parts of China (including Tibet), Mongolia, and Afghanistan, rickets appears mainly due to vitamin D deficiency associated with the high latitude, cold winters, and limited skin exposure *(29)*.

The data on calcium intake in recently published Indian literature have been made available by Bhatia et al. *(35)*. Balasubramanian et al. *(32)* documented calcium deficiency rickets among toddlers attending an urban hospital in Lucknow. The mean daily calcium intake in this group of children belonging primarily to low and lower middle socioeconomic groups was 282 ± 114 mg. Mean serum 25-hydroxyvitamin D [25(OH)D] was normal, 20.0 ± 15.6 ng/ml, and rickets healed without any vitamin D supplementation in all children treated with calcium replenishment alone. In contrast to these toddlers, adolescent girls living in the same city, belonging to a similar socio economic group, presenting with rickets/osteomalacia, were documented to suffer from severe vitamin D deficiency (mean 25(OH)D 4.9 ± 2.7 ng/ml), with 20 out of 21 girls having values < 10 ng/ml. Vitamin D deficiency was thought to be contributed to by skin pigment, clothing, poor outdoor activity, and possibly their poor calcium intake (265 ± 199 mg/day). The reason for poor calcium intake was economic deprivation and the prohibitive cost of milk, in some; in others it was due to food fads. Unselected pregnant women attending a medical college hospital in northern India were documented to have a moderately good calcium intake of 842–459 mg/day, mainly due to supplements advised during antenatal care *(18)*. In contrast, rural, poor socioeconomic group pregnant women, studied in their home setting in Barabanki district, Uttar Pradesh, were documented to have a lower intake of 211 ± 158 mg/day. Reports from other regions include 556 healthy school children (adolescents 10–18 years of age) from New Delhi, who were documented to have an intake of 314 ± 194 mg/day among low socioeconomic group children and 713 ± 241 mg/day among high socioeconomic group children *(36)*. Rural and urban women from Tirupati had an intake of 264 ± 1.9 (mean $\pm$ SEM) and 356 ± 5.0 mg/day, respectively *(4, 9)*.

Vitamin D status of various studies from India along with dietary calcium intake is shown in Table 4. The table is arranged in such a way that the readers can have a view

Table 4
Indian studies on vitamin D status of population

Location	n	Study population	Age (years)	25(OH)D	Unit	Diet cal	Ref. No.
Kashmir	64	Men	28.8 4.9	37.7 30	nmol/l	230 63	39
Valley	28	Women	26.8 4.8	13.8 11	nmol/l	178 45	
Delhi	29	Pregnant women in summer	23 ± 3	21.9 + 10.73	nmol/l	345 ± 78	3
	29	New born in summer	new born	16.72 + 4.99	nmol/l	–	
Delhi	31	Soldier males in winter	21.2 ± 2	41.17 ± 11.73	nmol/l	1104 ± 666	
	19	Phys. and nurses in summer	23 ± 5	7.89 ± 3.49	nmol/l	879 ± 165	
	19	Phys. and nurses in winter	24 ± 4	17.97 ±7.98	nmol/l	880 ± 165	
	15	Depigmented persons in winter	43 + 16	18.2 ± 11.23	nmol/l	980 ± 300	
Delhi	26(5)	Toddlers (Mori gate)	16± 4 months	12.4 ± 7	ng/ml	–	10
	1	Infants (Gurgaon)	16 ± 4 months	28 ± 7	ng/ml	–	
Delhi	12	Controls	25–35	8.3 ± 2.5	µg/ml	~350	2
Delhi	40	Indian paramilitary forces Men	20–30	18.4 ± 5.3	ng/ml	~1100	37
	50	Indian paramilitary forces women	20–30	25.3 ± 7.4	ng/ml		
Delhi	32	Rural males	42.8 ± 16.6	44.2 ± 24.4	nmol/l	905± 409	26
	25	Rural females	43.4 ± 12.6	26.9 ± 15.9	nmol/l	595 ± 224	
Delhi slums	47	Sunder nagar, Jan 2001	9–30 months	96 ± 25.7	nmol/l	–	31
	49	Rajiv colony, Feb 2001	9–30 months	23.8 ± 27	nmol/l	–	
	48	Rajiv colony, Aug 2001	9–30 months	17.8 ± 22.4	nmol/l	–	
	52	Gurgoan, Aug 2001	9–30 months	19 ± 20	nmol/l	–	
Delhi	193	LSES school girls	12.4 ± 3.2	34.6 ± 17.43	nmol/l	454–187	38
	211	USES school girls	12.3 ± 3	29.4 ± 12.7	nmol/l	686–185	
	42	LSES school boys	10–12	12.4 ± 5.5	ng/ml	314 ± 194	

Table 4
(continued)

Location	n	Study population	Age (years)	25(OH)D	Unit	Diet cal	Ref. No.
	85		13–15	11.3 ± 5.8	ng/ml	314 ± 194	38
	40		16–18	11.3 ± 5.3	ng/ml	314 ± 194	
	33	USES school boys	10–12	19.3 ± 8.8	ng/ml	713 ± 241	
	70		13–15	13.1 ± 7	ng/ml	713 ± 241	
	55		16–18	13.5 ± 7	ng/ml	713 ± 241	
	78	LSES school girls	10–12	11 ± 6.5	ng/ml	314 ± 194	
	123		13–15	10 ± 6.2	ng/ml	314 ± 194	
	62		16–18	11 ± 5.7	ng/ml	314 ± 194	
	47	USES school girls	10–12	12.5 ± 8.9	ng/ml	713 ± 241	
	62		13–15	10.2 ± 5.7	ng/ml	713 ± 241	
	63		16–18	12.9 ± 10.5	ng/ml	713 ± 241	
Lucknow	92	Healthy volunteers	34.2 ± 6.7	12.3 ± 11	ng/ml	439 ± 123	11
Lucknow	140	Pregnant women (urban)	24 ± 4.1	14 ± 9.5	ng/ml	842 ± 459	22
	67	Pregnant women (rural)	24.7 ± 5.1	14 ± 9	ng/ml	549 ± 404	
	29	Cord blood (OSM)	–	12 ± 8	ng/ml	–	
	178	Cord blood (no OSM)	–	14.3 ± 9.5	ng/ml	–	
Lucknow	139	Pregnant women – summer	26.7 ± 4.1	55.5 ± 19.8	nmol/l	214 ± 150	27
	139	Pregnant women – winter		27.3 ± 12.3	nmol/l		
	28	Girls – winter	14.4 ± 2.7	31.3 ± 13.5	nmol/l	198 ± 159	
	34	Boys – winter	14 ± 3	67.5 ± 29	nmol/l	384 ± 600	
Lucknow	53	Controls		61 ± 36	nmol/l	404 ± 149	32
	40	Rickets/OSM		49 ± 38	nmol/l	285 ± 113	
Mumbai	42	Mothers suppl Ca 250–500 apart	20–35	23 ± 11	ng/ml	800–1500	40
	42	Cord blood	–	19.5 ± 9.6	ng/ml		
	35	Infants	3 months	18.2 ± 9.8	ng/ml		
Tirupati	134	Urban men[a]	47 ± 1.5	18.54 ± 0.8	ng/ml	323 ± 8	9

Table 4
(continued)

Location	n	Study population	Age (years)	25(OH)D	Unit	Diet cal	Ref. No.
	109	Rural men[a]	45 ± 1.4	23.7 ± 0.8	ng/ml	271 ± 3	9
	807	Urban women[a]	46 ± 0.4	15.5 ± 0.3	ng/ml	306 ± 2	
	96	Rural women[a]	41 ± 1.4	19 ± 0.9	ng/ml	262 ± 3	
	30	Urban children male[a]	11 ± 1	15.57 ± 1.2	ng/ml	293 ± 6	9
	34	Rural children male[a]	12 ± 0.7	17 ± 1.3	ng/ml	277 ± 6	
	39	Urban children female[a]	13.5 ± 0.6	18.5 ± 1.66	ng/ml	317 ± 9	
	36	Rural children female[a]	12.6 ± 0.5	19 ± 1.6	ng/ml	270 ± 7	
	164	Postmenopausal	54 ± 8	14.6 ± 7	ng/ml	322 ± 6	7
Vellore	150	Postmenopausal women	60.1 ± 5	20.85 ± 8.63	ng/ml	399 ± 190	41

[a]Mean ± SEM.

For conversion from nmol to ng – multiply by 0.4.

Latitude: Kashmir 34.6°N; Delhi 28.35°N; Lucknow 26. 55°N; Mumbai 18.56°N; Tirupati 13.4°N; and Vellore 12.55°N.

of vitamin D status studied from Kashmir valley in north India to Vellore in south India (Latitude: Kashmir 34.6°N to Vellore 12.55°N).

2.1. Skeletal Fluorosis in India: An Endemic Disease

Skeletal fluorosis is an endemic bone disease caused by excessive consumption of fluoride. In India, the most common cause of fluorosis is fluoride-laden water derived from bore wells dug deep into the earth, while fluorosis is most severe and widespread in the two largest countries – India and China – the UNICEF estimates that "fluorosis is endemic in at least 25 countries across the globe" (42). Of India's 35 states, 20 have been identified as "endemic" areas for fluorosis, with an estimated 20 million people impacted and another 66 million "at risk." Water exceeds 1 part per million (ppm) and has been found to occur in some communities with only 0.7 parts per million (48). To put this in perspective, an average of 1 ppm of fluoride is added to the water in artificial fluoridation programs in the United States and other fluoridating countries (e.g., Canada, England, Ireland, Israel, New Zealand, and Australia). While the elevated consumption of water in warm climates such as India along with the increased incidence of malnutrition makes direct comparisons of the Indian experience to the "West" difficult, it is striking to observe how narrow the margin is between the doses which cause advanced fluorosis in India and the doses that people are now regularly receiving in fluoridated communities. Fluorosis is an endemic disease prevalent in 20 states out of the 35 states and Union Territories of the Indian Republic. About 70–100% districts are affected in Andhra Pradesh, Gujarat, and Rajasthan. About 40–70% districts are affected in Bihar, National Capital Territory of Delhi, Haryana, Jharkhand, Karnataka, Maharashtra, Madhya Pradesh, Orissa, Tamil Nadu, and Uttar Pradesh. About 10–40% districts are affected in Assam, Jammu & Kashmir, Kerala, Chattisgarh, and West Bengal (43). While the endemicity for the rest of the states is not known but is even now reported from some parts where it is not known like western coastal belt by Bawaskar et al. (43). An elegant treatise on fluorosis is available from Susheela who now heads a India-specific fluorosis-dedicated "Fluorosis Research & Rural Development Foundation" in Delhi (44).

Fluoride can enter the body through drinking water, food, toothpastes, mouth rinses and other dental products, drugs, and fluoride dust and fumes from industries using fluoride-containing salt and or hydrofluoric acid. Fluorosis can occur as (a) water-borne fluorosis (hydro fluorosis), (b) food-borne fluorosis, (c) drug and cosmetic-induced fluorosis, and finally (d) industrial fluorosis. Fluorosis can affect young and old men and women alike. Clinically fluorosis occurs as (i) dental fluorosis, (ii) skeletal fluorosis, and (iii) non-skeletal fluorosis.

Teotia et al. have systematically surveyed several syndromes of bone disease and deformities consequent to disorders of nutrition, bone, and mineral metabolism for almost four decades from north India (45). They have surveyed 337.68 million population residing in 0.39 million villages in 22 states of India during the period 1963–2005; of the 4,11,744 patients identified with the disorders of bone and mineral metabolism, 2,13,760 (52%) had nutritional bone disease, 1,77,200 (43%) had endemic skeletal fluorosis, and 20,784 (5%) had metabolic bone disease and in 41 patients (0.19%) the bone disease was rare, mixed, or unidentified. Vitamin D deficiency osteomalacia and rickets caused by inadequate exposure to sunlight (290–315 nm), dietary calcium deficiency (<300 mg/day) and fluoride interaction syndromes, calcium deficiency-induced osteoporosis, and calcium and vitamin D deficiency-induced osteoporosis in the elderly were the commonest disorders responsible for bone disease and deformities, besides caused by endemic skeletal fluorosis as a single entity in endemic fluorosis villages. Only mothers with severely depleted bone mineral and vitamin D stores gave birth to their babies with congenital rickets. Vitamin D deficiency rickets in children and osteomalacia in the mothers are the commonest disorders prevalent in the rural population of India *(45)*. These disorders and the syndromes of calcium deficiency and fluoride interactions are largely responsible for the morbidity and mortality in the young and promising individuals, with economic consequences *(45)*.

3. PATHOGENESIS OF FLUOROTOXIC BONE DISEASE

In India two types of fluorotoxic bone disease are reported *(46)*. An endemic variety affecting the elderly, characterized by new bone formation, musculoskeletal dysfunction, arthritis with fixed flexion deformities, peripheral neuropathy, ankylosis of the spine with radiculopathy, and osteosclerosis with ligament calcification on radiology *(47–50)*, and another variety affecting predominantly children causing bone deformities such as genu valgum and bowing. Radiologically, their bones show a mixture of osteomalacia, osteosclerosis, and osteopenia affecting both the axial and the appendicular skeleton *(51)*. This type of skeletal fluorosis called as fluorotoxic metabolic bone disease (FMBD) is the commonest *(52–55)*. Different pathogenic mechanisms have been postulated for these two types of fluorosis *(56–58)*.

Recently a study by Harinarayan et al. *(42)* carried out a comprehensive set of investigations such as serum calcium, phosphorous, alkaline phosphatase (SAP), 25-hydroxyvitamin D [25(OH)D], 1,25-dihydroxyvitamin D [1,25(OH)$_2$D], and parathyroid hormone (PTH) levels as well as nephrologic parameters that assess renal handling of calcium and phosphorous along with skeletal dynamics at a microscopic level by bone histomorphometry on patients with FMBD.

Major clinical manifestations in this study group were bone pain (79%), dental mottling (38%), and tetany (12.5%). Radiological findings included osteosclerosis (96%), pseudofracture (33%) and ligamentous calcification (50%). These patients manifested with biochemical hypocalcemia and raised SAP with normal serum phosphorus. There was a positive correlation between phosphorous excretion index (PEI) and serum creatinine. There was a negative correlation between increasing renal loss of calcium and phosphorus and declining endogenous creatinine clearance (Cr Cl) as indicated by increased calcium to creatinine ratio and PEI. The serum 25(OH)D concentrations were in deficient range in 71% of them.

Figure 1a radiograph shows the typically increased bone density with osteosclerosis. Bone histomorphometry (Fig. 1b) revealed impairment of primary mineralization with hypomineralized lacunae, interstitial mineralization defects, and very thick and extended osteoid seams. This patient died of azotemia. Autopsy findings of this patient showed tubular atrophy with secondary glomerular changes. Available evidences in literature also support this observation of fluoride-induced renal damage. Jolly et al. *(59)* from north India have reported renal involvement in older patients with chronic endemic fluorosis *(33, 34)*. Reggabi et al. *(60, 61)* from Iran have reported tubular dysfunction in patients exposed to chronic fluoride intoxication.

Fluoride intoxication plays an important role in the pathogenesis of the unique osteo-renal syndrome *(61)*. Thus, acute fluoride toxicity following the use of fluorine-containing anesthetics has been reported to cause acute renal tubular necrosis and acute renal failure. There are possible mechanisms postulated for renal tubular damage in situations of excess fluoride ingestion in hot dry climates of tropics. Alkaline urine can promote effective and safe excretion of fluorides ingested *(35)*. But, in situations of excessive ingestion of acid-rich food and beverages, where the renal tubules have to handle excessive loads of hydrogen ions, high fluoride ingestion can result in generation of high concentration of hydrofluoric acid in the renal tubules. The flux of high concentration of H^+ and F^- ions across the tubular cells in the hot and dehydrating conditions of the tropics with high fluoride ingestion chronically may result in chronic renal damage.

In Indian subcontinent urinary crystalloids and ionic concentrations can reach several folds higher than what is common in temperate climates and fluoride ingestion through drinking water can reach a high level. Such mechanisms of predominant renal tubular damage seem plausible in hot dry tropical summer of north and south India.

4. CALCIUM –VITAMIN D–FLUORIDE INTERACTIONS

Gaster et al. *(62)* studied the interrelationships between dietary calcium, fluorine, and vitamin D in young rats. There was less weight gain and less ash content of the bones in the rats maintained on low-calcium diet. Supplementation of diet with fluoride decreased the growth and content of the bone ash without affecting the calcium and phosphorous content. Addition of fluoride to a low-calcium diet supplemented with vitamin D diminished the uptake of radioactive calcium. It was concluded that vitamin D added to a low-calcium diet does not exert a calcifying effect on bone, but rather increases calcium turnover. Fluoride, on the other hand, reduces the exchangeability of bone minerals. Thus fluoride supplement decreased uptake of radioactive calcium by bone. Studies by Teotia et al. *(63)* concluded that calcium deficiency per se does not cause rickets and the children reported with calcium deficiency rickets are in fact the "syndromes of calcium deficiency and fluoride interactions." It is suggested that each case of calcium deficiency presenting as rickets should be investigated for underexposure to sunlight (vitamin D deficiency) and overexposure to endemic fluoride (also fluoridated water) or some combinations *(64)*.

REFERENCES

1. Report of the Joint FAO/WHO Expert Consultation on vitamin and mineral requirement in human nutrition: Bangkok 1998. Second Edition FAO Rome, 2004. Available at http://whqlibdoc.who.int/publications/2004/9241546123.pdf.
2. Harinarayan CV, Gupta N, Kochupillai N (1995) Vitamin D status in primary hyperparathyroidism. Clin Endocrinol (oxf) 43:351–358. 28

3. Goswami R, Gupta N, Goswami D, Marwaha RK, Tandon N, Kochupillai N (2000) Prevalence and significance of low 25- hydroxyvitamin D concentrations in healthy subjects in Delhi. Am J Clin Nutr 72:472–475

4. Harinarayan CV, Ramalakshmi T, Prasad UV, Sudhakar D, Srinivasarao PVLN, Sarma KVS et al (2007) High prevalence of low-dietary calcium, high-phytate consumption, and vitamin D deficiency in healthy south Indians. Am J Clin Nutr 85:1062–1065

5. Parfitt AM, Gallagher JC, Heaney RP, Johnston CC, Neer R, Whedon GD, Vitamin D (1982) and bone health in the elderly. Am J Clin Nutr 35:1014–1031

6. Hollis BW (2005) Circulating 25 hydroxy D levels indicative of vitamin D insufficiency: implications of reestablishing a new effective dietary intake recommendation for vitamin D. J Nutr Biochem 135:317–322

7. Harinarayan CV (2005) Prevalence of vitamin D insufficiency in postmenopausal south Indian women. Osteoporos Int 16:397–402

8. Harinarayan CV, Ramalakshmi T, Venkataprasad V (2004) High prevalence of low dietary calcium and low vitamin D status in healthy south Indians. Asia Pac J Clin Nutr 13:359–365

9. Harinarayan CV, Ramalakshmi T, Prasad UV, Sudhakar D (2008) Vitamin D status in Andhra Pradesh: a population based study. Indian J Med Res 127(3):211–218

10. Agarwal KS, Mughal MZ, Upadhyay P, Berry JL, Marwer EB, Puliyel JM (2002) The impact of atmospheric pollution on vitamin D status of infants and toddlers in Delhi, India. Arch Dis Child 87:111–113

11. Arya V, Bhambari R, Godbole MM, Mithal A (2004) Vitamin D status and its relationship with bone mineral density in healthy Asian Indians. Osteoporos Int 1:56–61

12. Recommended Dietary Allowances. In: Essentials of Food and Nutrition. Fundamental aspects. Bappco Pub 1997; 1:50821.

13. Standing Committee on Scientific Evaluation of Dietary Reference Intakes (1999) Calcium. In: Dietary Reference Intakes for calcium. Food and Nutrition Board, National Academy Press, Washington, DC, pp. 71–145

14. Panwar B, Punia D (2000) Analysis of composite diets of rural pregnant women and comparison with calculated values. Nutr Health 14:217–223

15. Heaney RP, Dowell MS, Hale CA, Bendich A (2003) Calcium absorption varies within the reference range for serum 25-hydroxyvitamin D. J Am Coll Nutr 22:142–146

16. Heaney RP (2003) Vitamin D depletion and effective calcium absorption. J Bone Miner Res 18: 1342

17. Clements MR, Johnson L, Fraser DR (1987) A new mechanism for induced vitamin D deficiency in calcium deprivation. Nature 325:62–65

18. Rajeswari J, Balasubramanian K, Bhatia V, Sharma VP, Agarwal AK (2003) Aetiology and clinical profile of osteomalacia in adolescent girls in northern India. Natl Med J India 16:139–142

19. Mathew JT, Seshadri MS, Thomas K, Krishnaswami H, Cherian AM (1994) Osteomalacia – Fifty-five patients seen in a teaching institution over a 4-year period. J Assoc Phys India 42:692–694

20. Shatrugna V, Kulkarni B, Kumar PA, Rani KU, Balakrishna N (2005) Bone status of Indian women from low income group and its relationship to the nutritional status. Osteoporos Int 16: 1827–1835

21. Marwaha RK, Tandon N, Reddy DRHK, Aggarwal R, Singh R, Sawhney RC et al. (2005) Vitamin D and bone mineral density status of healthy school children in northern India. Am J Clin Nutr 82: 477–482

22. Sachan A, Gupta R, Das V, Agarwal A, Awasthi PK, Bhatia V (2005) High prevalence of vitamin D deficiency among pregnant women and their new borns in north India. Am J Clin Nutr 81: 1060–1064

23. Mellanby E (1919) An experimental investigation on rickets. Lancet 1:407–412

24. Sly MR, van der Walt WH, Du Bruyn D, Pettifor JM, Marie PJ (1984) Exacerbation of rickets and osteomalacia by maize: a study of bone histomorphometry and composition in young baboons. Calcif Tissue Int 36:370–379

25. Clements MR (1989) The problem of rickets in UK Asians. J Hum Nutr Diet 2:105–116

26. Goswami R, Goswami R, Kochupillai N, Gupta N, Goswami D, Singh N, Dudha A (2008) Presence of 25 (OH) D Deficiency in a Rural North Indian Village Despite Abundant Sunshine. J Assoc Phys India 56:755–757

27. Sahu Sahu M, Bhatia V, Aggarwal A, Rawat V, Saxena P, Pandey A, Das V (2008) Vitamin D deficiency in rural girls and pregnant women despite abundant sunshine in northern India. Clin Endocrinol (Oxf) 70(5):680–684

28. Joshi Shashank R (2008) Vitamin D paradox in plenty sunshine in rural India – an emerging threat. J Assoc Phys India 56:749–752

29. John M (March 2008) Pettifor. Vitamin D &/or calcium deficiency rickets in infants & children: a global perspective. Indian J Med Res 127:pp. 245–249

30. Bhattacharyya AK (1992) Nutritional rickets in the tropics. In: Simopoulos AP (ed) Nutritional triggers for health and in disease. Karger, Basel, pp. 140–197

31. Tiwari L, Puliyel JM (2004) Vitamin D level in slum children of Delhi. Indian Pediatr 41:1076–1077

32. Balasubramanian K, Rajeswari J, Gulab GYC, Agarwal AK, Kumar A et al (2003) Varying role of vitamin D deficiency in the etiology of rickets in young children vs. adolescents in northern India. J Trop Pediatr 49:201–206

33. Khandare AL, Harikumar R, Sivakumar B (2005) Severe bone deformities in young children from vitamin D deficiency and fluorosis in Bihar-India. Calcif Tissue Int 76:412–418

34. Pettifor JM, Schnitzler CM, Ross FP, Moodley GP (1989) Endemic skeletal fluorosis in children:hypocalcemia and the presence of renal resistance to parathyroid hormone. Bone Miner 7: 275–288

35. Marwaha RK, Tandon N, Reddy DR, Aggarwal R, Singh R, Sawhney RC et al. (2005) Vitamin D and bone mineral density status of healthy schoolchildren in northern India. Am J Clin Nutr 82:477–482

36. Bhatia V (2008) Dietary calcium intake – a critical reappraisal. Indian J Med Res 127:269–273

37. Tandon N, Marwaha RK, Kalra S, Gupta N, Dudha A, Kochupillai N (2003) Bone mineral parameters in healthy young Indian adults with optimal vitamin D availability. Natl Med J India 16(6): 298–302

38. Puri S, Marwaha RK, Agarwal N, Tandon N, Agarwal R, Grewal K, Reddy DH, Singh S (2008) Vitamin D status of apparently healthy schoolgirls from two different socioeconomic strata in Delhi: relation to nutrition and lifestyle. Br J Nutr 99(4):876–882

39. Zargar AH, Ahmad S, Masoodi SR, Wani AI, Bashir MI, Laway BA, Shah ZA (2007) Vitamin D status in apparently healthy adults in Kashmir Valley of Indian subcontinent. Postgrad Med J 83(985): 713–716

40. Bhalala U, Desai M, Parekh P, Mokal R, Chheda B (2007) Subclinical hypovitaminosis D among exclusively breastfed young infants. Indian Pediatr 44(12):897–901

41. Paul TV, Thomas N, Seshadri MS, Oommen R, Jose A, Mahendri NV (2008 Sep) Prevalence of osteoporosis in ambulatory postmenopausal women from a semiurban region in Southern India: relationship to calcium nutrition and vitamin D status. Endocr Pract 14(6):665–671

42. Harinarayan CV, Kochupillai N, Madhu V, Gupta N, Meunier PJ (2006) Endemic skeletal fluorosis in India. Fluorotoxic metabolic bone disease: an osteo-renal syndrome caused by excess fluoride ingestion in the tropics. Bone 39:907–914

43. Bawaskar HS, Bawaskar PH (2006) Endemic fluorosis in an Isolated Village in Western Maharashtra. India trop Doct 36(4):221–223

44. A.K. Susheela Treatise on Fluorosis: 3rd Revised Edition. Published by Fluorosis Research & Rural Development Foundation, Delhi, 2006

45. Teotia SPS, Teotia M (2008) Nutritional bone disease in Indian population. Indian J Med Res 127: 219–228

46. Krishnamachari KA (1986) Skeletal fluorosis in humans: a review of recent progress in the understanding of the disease. Prog Food Nutr Sci 10(3–4):279–314

47. Shortt HE, McRobert GR, Bernard TW, Nayar ASM (1937) Endemic fluorosis in Madras presidency. Indian J Med Res 25:553–568

48. Teotia SP, Teotia M (1984) Endemic fluorosis in India: a challenging national health problem. J Assoc Phys India 32(4):347–352

49. Siddiqui AH (1955) Fluorosis in Nalgonda District, Hyderabad-Deccan. Br J Med 4953:1408–1413
50. Singh A, Jolly SS, Bansal BC (1961) Skeletal fluorosis and its neurological complications. The Lancet I:197–200
51. Krishnamachari KA, Krishnaswamy K (1973) Genu valgum and osteoporosis in an area of endemic fluorosis. Lancet 2(7834):877–879
52. Christie DP (1980) The spectrum of radiographic bone changes in children with fluorosis. Radiology 136(1):85–90
53. Moudgil A, Srivastava RN, Vasudev A, Bagga A, Gupta A (1986) Fluorosis with crippling skeletal deformities. Indian Pediatr 23(10):767–773
54. Krishnamachari KA (1976) Further observations on the syndrome of endemic genu valgum of South India. Indian J Med Res 64(2):201–284
55. Teotia M, Teotia SP, Kunwar KB (1971) Endemic skeletal fluorosis. Arch Dis Child 46(249):686–691
56. Krishnamachari KA, Krishnaswamy K (1974 (Sep.)) An epidemiological study of the syndrome of genu valgum among residents of endemic areas for flurosis in Andhra Pradesh. Indian J Med Res 62(9):1415–1423
57. Narasinga Rao BS, Krishnamachari KA, Vijaya Sarathy C (1979 (Jan)) 47Ca turnover in endemic fluorosis and endemic genu valgum. Br J Nutr 41(1):7–14
58. Krishnaramachari KA (1982) Trace elements in serum and bone in endemic genu valgum – manifestation of fluorosis. Fluoride 15:25–31
59. Jolly SS, Singh BM, Mathur OC, Malhotra KC (1968 (Nov. 16)) Epidemiological, clinical, and biochemical study of endemic dental and skeletal fluorosis in Punjab. Br J Med 4(628):427–429
60. Reggabi M, Khelfat M, Tabet A, Azzoua M, Hamrour S, Alamir B et al (1984) Renal function in residents of an endemic fluorosis area in southern Algeria. Fluoride 17:35–41
61. Myers HM, (ed) (1989) Renal handling of fluoride (Chapter III). In: Metabolism and toxicity of fluoride. Monographs in oral science. Whitford GM (ed). Karger, Basel, pp. 51–66
62. Gaster D, Havivi. E, Guggenheim. K (1967) Interrelations of calcium, fluorine and vitamin D in bone metabolism. Br J Nutr 21:413–418
63. Teotia M, Teotia SPS, Singh KP (1998) Endemic chronic fluoride toxicity and dietary calcium deficiency interaction syndromes of bone disease and deformities in India-2000. Indian J Pediatr 65: 371–381
64. Teotia SPS, Teotia M (1983) Fluoride and calcium interactions: syndromes of bone disease and deformities (human studies).In: Frame B, Potts JT Jr (ed) Clinical disorders of bone and mineral metabolism. Excerpta Medica, Amsterdam, pp. 520–523

29 Vitamin D in Asia

Tim Green and Bernard Venn

Abstract   Suboptimal vitamin D status is common throughout Asia particularly in women and in infants. Vitamin D deficiency rickets is endemic in parts of northern Asia. Vitamin D inadequacy may also be contributing to the burden of osteoporosis and other diseases in the region. Vitamin D intakes are low in Asia and food fortification is uncommon. In the north cold temperatures and a lack of UV light prevent dermal vitamin D status in the winter months. In the south where there is sufficient UV year-round, cultural and religious practices that require women to wear clothing that conceals most of their skin when outdoors contribute to suboptimal vitamin D status. Further sun aversion for cosmetic reasons, avoidance of hot outdoor temperatures, and keeping infants indoors are also likely contributors. Strategies are required to improve the vitamin D status of Asian populations.

Key Words: Asia; vitamin D deficiency; 25-hydroxyvitamin D; China; rickets; Mongolia; sunlight; UV; Middle East; food fortification

1. VITAMIN D IN ASIA

Asia has a vast landmass extending from equatorial regions to the Arctic Circle. It comprises numerous countries inhabited by people who are culturally, ethnically, and religiously diverse. In this chapter we will assimilate much of the published evidence around the vitamin D status based on circulating concentrations of 25-hydroxyvitamin D [25(OH)D] of populations living within Asia. We will discuss vitamin D determinants and examine the literature for evidence of clinical manifestations of vitamin D deficiency within those populations. Finally, we will comment on strategies that might be used to increase vitamin D status. In general there are no nationally representative samples of vitamin D status in Asia. Thus, care should be taken when trying to extrapolate or generalize the results. Direct comparison among studies is also complicated by the use of different assays for 25(OH)D determination.

2. NORTHERN ASIA

The highest rates of vitamin D deficiency in the world have been reported in parts of northern Asia. This is due, in part, to the region's northerly latitude, which stretches from Hong Kong (22°N) in the south to the Arctic Circle (>66°N) in the north. However,

From: *Nutrition and Health: Vitamin D*
Edited by: M.F. Holick, DOI 10.1007/978-1-60327-303-9_29,
© Springer Science+Business Media, LLC 2010

low dietary intakes of vitamin D, greater skin pigmentation, and cultural practices also contribute to the high rate of deficiency in this region.

2.1. Vitamin D Status

Low 25(OH)D concentrations have been reported throughout northern Asia *(1–18)* (Table 1). Among Mongolian children 0.5–3 years the mean 25(OH)D was 9 ng/ml, and over 60% had a 25(OH)D less than 10 ng/ml *(11)*. At higher latitudes there are marked seasonal differences in 25(OH)D concentrations. In 1- to 2-year-old children living in Yuci, northern China (38°N), mean 25(OH)D was 51 ng/ml in September and 14 ng/ml in April, a greater than threefold difference *(1)*. In Beijing (39°N) girls mean 25(OH)D was 14 ng/ml in September and October compared with 5 ng/ml in January *(2)*. Even in southern areas of the region low 25(OH)D concentrations are common. In women of reproductive age living in sub-tropical Hong Kong (22°N), 18 and 92% of women had 25(OH)D concentrations indicative of deficiency (<10 ng/ml) and insufficiency (<20 ng/ml), respectively *(3)*. Circulating 25(OH)D concentrations in older women ranged from 9 ng/ml in those with hip fracture living in Yekaterinburg, Russia (59°N) to 32 ng/ml in healthy women living in Niigata, Japan *(9, 13)*. There are few studies of men; however, in two studies in Shenyang, China (42°N) men had lower 25(OH)D concentrations than women *(4, 6)*.

2.2. Consequences

Clinical signs of rickets are common in Mongolia and China, particularly in the northern provinces (Fig. 1). In a study of 250 children aged 12–24 months in the northern Shanxi province, 42% had clinical signs of rickets in the spring falling to 17% in the fall, and active rickets was present in 3.7% of the cohort *(19)*. In 1,248 girls aged 12–14 years living in Beijing, bone deformations indicative of rickets in early childhood were present in 25% of the girls although based on wrist radiography there were no cases of active rickets *(2)*. Vitamin D deficiency is undoubtedly a major cause of rickets in Mongolia and China but calcium intakes are also low at around 365 mg/day for Beijing girls (12–14 years) and 265 mg/day for Mongolian children (<5 years) *(2, 20)*. Using surveys carried out in the 1990s, estimates for the prevalence of rickets in Mongolian children aged 5 years or younger based on clinical signs range from 32 to 70% *(21, 22)*. A program to combat rickets in Mongolian children using vitamin D-fortified sprinkles showed only a 5% reduction in the prevalence; the authors concluded that the 10 μg dose of vitamin D may not have been sufficient *(23)*. Prevalence estimates for rickets are not available for other northern Asian countries although sporadic cases of rickets have been reported in Japan and there are unofficial reports of rickets from North Korea *(24, 25)*. In South Korean infants, a positive association between 25(OH)D and bone mineral content was found; also, babies born in winter had weight-adjusted total body bone mineral content 8% lower than summer-born babies *(14)*. In Kyoto, Japan, 22% of 1,120 neonates had craniotabes of whom nearly 40% had a 25(OH)D concentration less than 10 ng/ml; the incidence of craniotabes was highest in April and May and lowest in November *(26)*.

Table 1
25-Hydroxyvitamin D Concentrations and Prevalence of Vitamin D Deficiency and Insufficiency in north Asian Countries

	Location and latitude	Population (n)	Age (year range or $\bar{x}$)	Months	Method	25-Hydroxyvitamin D		
						Mean (ng/ml)	% Deficient (cutoff)	% Insufficient (cutoff)
China								
(1)	Yuci 38°N	Children (250)	1–2	Apr	Diasorin	14	75 (30)	84 (50)
		Children (176)	1–2	Sep	Diasorin	51	3 (30)	8 (50)
(2)	Beijing 39°N	Girls (254)	12–14	Sep–Oct	Diasorin	11	7 (12.5)	
		Girls (533)	12–14	Jan	Diasorin	5	45 (12.5)	
(3)	Beijing 39°N	Women (220)	20–35	Feb–May	Diasorin	12	40 (25)	94 (50)
	Hong Kong 22°N	Women (221)	20–35	Feb–May	Diasorin	14	18 (25)	92 (50)
(4)	Shenyang 42°N	Women (48)	25–35	Apr–May	Diasorin	16	29 (25)	
		Men (48)	25–35	Mar–May	Diasorin	12	13 (25)	
(5)	Hong Kong 22°N	Adults (382)	>50	Jan–Dec	Diasorin	28	1 (25)	23 (50)
(6)	Shenyang 42°N	Women (110)	62	Feb–Apr	Diasorin	12	39 (25)	
		Men (108)	68	Feb–Apr	Diasorin	11	52 (25)	
Japan								
(7)	Kobe 34°N	Pregnant term (24)		Jul–Sep	HPLC	33		
		Pregnant term (17)		Dec–Feb	HPLC	16		
		Neonate	5–6 days	Jul–Sep	HPLC	13		
		Neonate	5–6 days	Dec–Feb	HPLC	8		
(8)	Niigata 38°N	Women (77)	19–66	Dec–Feb	Diasorin	17	26 (30)	
(9)	Niigata 38°N	Women (122)	45–81	Feb	HPLC	24		
				Sep	HPLC	32		
(10)	Niigata 38°N	Women (600)	55–74	Nov	Diasorin	22	4 (25)	35 (50)

Table 1
(continued)

Location and latitude	Population (n)	Age (year range or x̄)	Months	Method	25-Hydroxyvitamin D		
					Mean (ng/ml)	% Deficient (cutoff)	% Insufficient (cutoff)
Mongolia							
(11) Several cities (~46°N)	Children (98)	0.5–3	Nov	Diasorin	9	61 (25)	90 (50)
(12) Ulaanbaatar (46°N)	Children (46)	9–11	Apr	Other	17	32 (37.5)	76 (50)
Russia							
(13) Yekaterinburg 59°N	Hip fracture (64)	70	Feb–Mar	Diasorin	9	65 (25)	100 (50)
	Controls (97)	70	Feb–Mar	Diasorin	11	47 (25)	98 (50)
South Korea							
(14) Seoul 37°N	Newborns (37)		Jul–Sep	Diasorin	30	47 (27.5)	
	Newborns (34)		Jan–Mar	Diasorin	11	97 (27.5)	
(15) Seoul 37°N	Women (179)	20–75	Sep	HPLC	23	17 (25)	
(16) Seoul 37°N	Osteoporotic women (101)	67	May–Apr	Nichols	18		~90 (75)
Taiwan							
(17) Taipei 25°N	Women (139)	40–72	Jun–Sep	Diasorin	13		
	Women (124)	40–72	Dec–Mar	Diasorin	12		
(18) Pingtung 22°N	Women and Men >65 (57)		–	Diasorin	36		

Fig. 1. Girl with rickets from Mongolia. Reproduced with permission from World Vision Canada (Photo by Philip Maher).

There are no published prevalence rates for osteomalacia in northern Asian countries but there is evidence that a lack of vitamin D may be contributing to osteoporosis in the region. Lower 25(OH)D concentrations were found in patients with hip fracture than in control subjects in Yekaterinburg, Russia (59°N), and in Hong Kong (22°N) *(13, 27)*. In a group of 276 postmenopausal women living in Seoul (37°N), women with 25(OH)D less than 12 ng/ml had lower bone mineral density compared with women whose 25(OH)D was more than 12 ng/ml *(28)*. In 600 ambulatory postmenopausal women living in Yokogshi, Japan, the prevalence of low bone mineral density for the femoral neck was 18–23% in those with a serum 25(OH)D less than 20 ng/ml compared to only 5% in those with a serum 25(OH)D greater than 28 ng/ml *(10)*. An inverse association between 25(OH)D and parathyroid hormone (PTH) and elevated PTH concentrations have been reported in several studies from the region (Fig. 2). Chronically elevated PTH (secondary hyperparathyroidism) accelerates bone turnover and is thought to be the key mechanism of bone loss related to vitamin D insufficiency *(29)*.

2.3. Causes of Vitamin D Deficiency

In northern Asia there is little vitamin D food fortification and the intake of vitamin D naturally present in foods, mainly sourced from oily fish, varies depending on proximity to coastal areas. In Japan where fish consumption is high mean vitamin D intake was 12 μg/day in postmenopausal women and was a determinant of 25(OH)D concentration,

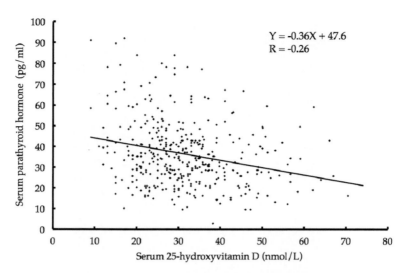

Fig. 2. Serum parathyroid hormone by 25-hydroxyvitamin D concentration in women (20–35 years; $n = 441$) from Hong Kong and Beijing. Figure depicts the slope of the fitted regression line. Reproduced from *(95)*.

whereas in Beijing adolescent girls vitamin D intake was only 1 μg/day *(2)*. Most of the north Asian population, therefore, is reliant on dermal vitamin D synthesis to meet their requirements. The UV index is a reasonable proxy for estimating the amount of UVB reaching the surface of the earth with a higher UV index indicating greater capacity for dermal synthesis. The UV indices at solar noon of several cities throughout Asia are shown for a typical December in Table 2. As indicated, northern cities have a low UV index through winter compromising dermal synthesis of pre-vitamin D. Aside from the low UV intensity winter temperatures as low as –45°C virtually preclude skin exposure. Maternal vitamin D deficiency probably contributes to rickets in these regions as does prolonged breastfeeding, with more than 80% of women in Mongolia still breastfeeding at 2 years *(23)*. Cultural practices such as infant swaddling also limit UV exposure throughout the year. In parts of rural China the custom of *zuo yuezi* involves confinement of mothers and their newborns indoors, in the case of infants for up to several months *(30)*. Strand and coworkers reported that in rural Shanxi Province children aged 12–24 months were taken outside for the first time at 132 days; on average, new mothers went outside for the first time at 62 days postpartum *(30)*.

Urbanization, high temperatures, and humidity in the summer months and prevention of skin tanning for cosmetic reasons may keep people indoors or covered up much of the day. Over 60% of women in a recent survey in Hong Kong ($n = 547$) indicated that they did not like going in the sun and many took measures to avoid sunlight such as use of parasols and sunscreen products *(31)*. Pollution, which is a serious problem in many northern Asian cities, filters UV light in addition to keeping people indoors. Time spent outdoors was positively associated with 25(OH)D in a number of studies from the region *(5, 15)*.

Table 2
UV Index in Various Cities in Asia

City	Index number	Exposure category
Jakarta	11	Extreme
Bangkok	9	Very high
Singapore	9	Very high
Hong Kong	6	High
Riyadh	5	Moderate
Delhi	4	Moderate
Shanghai	3	Moderate
Seoul	2	Low
Beijing	2	Low
Tehran	2	Low
Tokyo	2	Low
Istanbul	1–2	Low
Ulaanbaatar	0	Low

3. MIDDLE EAST

The Middle East is a subcontinent comprising a number of countries bounded by Africa to the west, India to the east, and Europe, European Russian, and Kazakhstan to the north. The land mass lies north of the equator spanning latitudes of 15–45°N. Yemen is the southernmost country of the Middle East grouping while Turkey and Uzbekistan are to the north. Much of the Middle East is low lying, although Yemen has mountains and highland plateaus and Iran, Iraq, and Pakistan contain mountainous regions.

3.1. Vitamin D Status

There is a wide range of 25(OH)D concentrations reported throughout the Middle East (Table 3) (32–60). Low 25(OH)D (<10 ng/ml) has been found in breast-fed infants from the United Arab Emirates and Turkey, Iranian and Kuwaiti neonates, and Turkish newborns with respiratory tract infections or rickets (41, 54, 55, 61–63). Other at-risk groups include socioeconomically deprived Afghani children and Lebanese schoolgirls with veiled mothers (32, 64). Wearing concealing clothing and veiling is frequently associated with vitamin D deficiency, including breastfeeding and pregnant women. Mean 25(OH)D concentrations less than 10 ng/ml have been found in women in the United Arab Emirates, Turkey, Lebanon, Iran, and Jordan (36, 41, 44, 49, 53, 55, 57, 58).

Mean 25(OH)D concentrations between 10 and 20 ng/ml have been found in veiled girls and women wearing concealing clothing in countries including Turkey (34, 38), the United Arab Emirates (58), and Jordan (53). Although veiling is a risk factor for low vitamin D status, there are accounts of women wearing different dress styles having mean 25(OH)D less than 20 ng/ml. These studies include women wearing western-style clothing in Jordan both in winter and in summer (53) and both orthodox and non-orthodox Jewish women in Israel (51). There are also indications that the vitamin D

Table 3

25-Hydroxyvitamin D Concentrations and Prevalence of Vitamin D Deficiency and Insufficiency in south Asian countries

Reference	Location and latitude	Population (n)	Age (year range or x̄)	Method	25-Hydroxyvitamin D		
					Mean (ng/ml)	% Deficient (cutoff)	% Insufficient (cutoff)
Bangladesh							
(85)	Dhaka 23°N	Garment factory workers (200)	18–36	IDS	15	15 (25)	~87 (50)
(86)	Dhaka 23°N	Non-veiled women (36)	22	Diasorin	12	39 (25)	78 (40)
		Veiled women (40)	48	Diasorin	13	30 (25)	83 (40)
India							
(87)	Mumbai 19°N	Infants (35)	3 months	Biosource	18	51(37.5)	80 (62.5)
(88)	Andhra Pradesh 13°N	Girls (75)	13	Diasorin	16		~68 (50)
		Boys (64)	13	Diasorin	14		~78 (50)
(89)	Delhi 28°N	Low SES girls (193)	12	Diasorin	14	25 (22)	90 (50)
		Upper SES girls (211)	12	Diasorin	13	34 (22)	92 (50)
(90)	Lucknow 27°N	Girls (121)	14	Diasorin	13		88 (50)
		Boys (34)	14	Diasorin	27		27 (50)
(91)	Delhi 28°N	Girls (435)	10–18	Diasorin	11	42 (22)	
		Boys (325)	10–18	Diasorin	13	27 (22)	
(92)	Kashmir 33°N	Women (28)	18–40	Diasorin	4		96 (50)
		Men (64)	18–40	Diasorin	15		76 (50)
(90)	Lucknow 27°N	Pregnant women (139)	27	Diasorin	15		74 (50)
(93)	Lucknow 27°N	Pregnant women (207)	24	Diasorin	9	43 (25)	
(88)	Andhra Pradesh 13°N	Women (899)	~45	Diasorin	16		~75 (50)
		Men (243)	~45	Diasorin	21		~55 (50)
(94)	Andhra Pradesh 13°N	Postmenopausal women (164)	60	Diasorin	15	49 (25)	82 (50)

Table 3
(continued)

Reference	Location and latitude	Population (n)	Age (year range or $\bar{x}$)	Method	25-Hydroxyvitamin D		
					Mean (ng/ml)	% Deficient (cutoff)	% Insufficient (cutoff)
Indonesia							
(95)	Jakarta 6°S	Women (126)	18–40	Diasorin	18	0 (17.5)	63 (50)
(96)	Jakarta 6°S	Women (100)	45–55	IDS	21		
Malaysia							
(95)	Kuala Lumpur	Malay women (133)	18–40	Diasorin	17	0 (17.5)	74 (50)
		Chinese women (123)	18–40	Diasorin	23	0 (17.5)	38 (50)
		Indian women (122)	18–40	Diasorin	18	1 (17.5)	68 (50)
(97)		Malay women (101)	50–65	Diasorin	18	2 (25)	73 (50)
		Chinese women (123)	50–65	Diasorin	27	0 (25)	12 (50)
Nepal							
(98)	Sarhali district 27°N	Pregnant women (1163)	24	Nichols	20	14 (25)	
Sri Lanka							
(99)	Kandy 7°N	Women (111)	30–60	Diasorin	25	0 (25)	34 (50)
		Men (85)	30–60	Diasorin	19	6 (25)	50 (50)
Thailand							
(100)	Khon Kaen 16°N	Urban women (65)	68	Diasorin	32		78 (87.5)
		Suburban women (40)	71	Diasorin	36		45 (87.5)

concentrations of other sections of the population may be within this range. These include men in Jordan in summer and winter and Lebanese men in Beirut and the Bekaa valley (49, 51). Groups of schoolchildren in Turkey, Iran, and Lebanon had mean concentrations equal to or less than 20 ng/ml (35, 42, 47). Lebanese school-age sons of veiled women had lower 25(OH)D than sons of unveiled mothers ($P \leq 0.05$), although both groups had mean concentrations less than 20 ng/ml (64). Low socioeconomic status was associated with low 25(OH)D in Turkish schoolgirls, a situation exacerbated in winter compared with summer (35).

Mean 25(OH)D concentrations in the range of 20–32 ng/ml have been found in groups of infants and children in the United Arab Emirates and Turkey (62, 65). Iranian men (66), women wearing western clothing in Turkey (36, 37) and the United Arab Emirates (58), and schoolgirls in Turkey (34, 35) and Beirut (47) have been found to have 25(OH)D within the 20–32 ng/ml range. Groups of middle-aged women are well represented in this range of 25(OH)D including women in Israel (52), Turkey (39, 67), and Iran (46).

Relatively high concentrations of 25(OH)D of 32 ng/ml or greater have been found in groups of schoolboys in Iran (42, 43), men in Iran (66), men in residential care, men and women aged over 65 years in Turkey, and women in western clothing (37, 67). An unusual finding is for a group of veiled women in Turkey that had a mean 25(OH)D concentration of 33 ng/ml. The women had worn the veil including black socks and gloves for a mean of 9 years, were not taking supplements, and two-thirds of them indicated that they were never exposed to direct sunlight. The 25(OH)D in this group is relatively high compared with the data from other studies that have included veiled women although as expected the concentration was lower compared with a group of unveiled women in the same study that had a higher mean concentration of 54 ng/ml.

3.2. Consequences

Over the last decade sporadic cases of vitamin D deficiency rickets have been reported throughout the region in Kuwait (68), Jordan (69), the United Arab Emirates (56), Pakistan (60), Qatar (70), Iran (71), and Saudi Arabia (59, 72), with at-risk children in rural Afghanistan (32). Although there are few good prevalence estimates, in 1998 the incidence of rickets in 0- to 3-year olds attributed to vitamin D deficiency in eastern Turkey was 6.1% based on clinical signs. Following a nationwide vitamin D campaign the prevalence of vitamin D-deficient rickets among 39,133 infants aged 0–3 years, sampled between March 2007 and February 2008 in an eastern region of Turkey, was 0.1% (73). There have been reports of osteomalacia, thought to be due largely to vitamin D deficiency, in Turkey (74), Saudi Arabia (75), Iran (76), and Pakistan affecting over 3% of female outpatients of reproductive age at a hospital in Hazara District, northwestern Pakistan (77).

Veiled women in the Middle East consistently have lower serum 25(OH)D concentrations compared with women wearing western-style clothing (34, 36–38, 58, 78). It might be expected that veiled women would have lower BMD than non-veiled women. However, BMD measurements carried out in several small groups of veiled or unveiled schoolgirls or premenopausal women generally do not show differences in mean BMD of the lumbar spine or the femoral neck despite differences in 25(OH)D between groups *(34, 36–38)*. In a larger group of women 24–72 years, a lower lumbar spine BMD was found in veiled ($n = 121$) compared with unveiled ($n = 207$) women *(78)*. Interestingly, 13-year-old schoolboys ($n = 50$) of veiled mothers were found to have lower BMD of the lumbar spine and femoral neck compared with the sons ($n = 111$) of unveiled women *(64)*. The authors suggested fetal programming as a possible explanation. However, familial customs may also have a bearing because the sons of veiled mothers tended to exercise less, had less sun exposure, and lower vitamin D status compared with the sons of unveiled mothers. An inverse relationship has been found between 25(OH)D and parathyroid hormone in an elderly Turkish group *(67)*. Quality of life assessed by questionnaire in Turkish women with osteoporosis has been positively correlated with 25(OH)D concentrations *(39)*.

3.3. Causes of Vitamin D Deficiency

As in other areas of Asia vitamin D intakes are generally low in the Middle East. Vitamin D intakes of pregnant urban- and rural-dwelling women in the north-west of Iran were 1.2 ± 0.6 and 2.1 ± 0.9 µg/day, respectively; the area is inland and none of the women reported eating fish *(79)*. The mean intakes of other groups within the region are generally in the 2–5 µg/day range including those of Lebanese and Iranian schoolchildren *(47, 80, 81)*, lactating Emirati women *(58)*, and urban adult residents of Beirut *(82)*. One notable exception is that the vitamin D intakes of pre- and post-menopausal Emirati women were 12.4 and 10.7 µg/day, respectively *(58)*. The mean fish intake of these women was 2.7 portions per week.

Many Middle Eastern countries are sunny and temperate. Most of the land mass lies within latitudes of 15–40°N, areas for which there is sufficient UV radiation for dermal vitamin D synthesis for all or most of the year *(83)*. However, northern countries of the region such as Iran, Turkey, Pakistan, and Afghanistan do not receive sufficient UV to make vitamin D in the winter months and seasonal fluctuations in 25(OH)D do occur *(35)*. Nevertheless, even the presence of the sun does not guarantee good sun exposure because of lifestyle, working conditions, pollution, excessive temperatures, sunscreens and cancer messages, dressing styles, and sun avoidance. Sun exposure has been assessed in a number of studies within the region.

Some Kuwaiti women keep neonates indoors for 6 months and do not expose them to sunlight over that time for fear of infection or darkening of the skin *(54)*. In nulliparous and lactating women living in the United Arab Emirates, mean sun exposure times of 0.4–6.6 min/day have been reported *(57)*, and of a group of veiled women in Adana, Turkey, two-thirds claimed that they were never exposed to direct sunlight *(37)*. In 414 schoolgirls aged 11–15 years in Tehran, the mean sunlight exposure in winter assessed by questionnaire was 10 ± 18 min/day *(81)*. Sun exposure times of 30 min or more have been reported in infants and young children in the United Arab Emirates *(65)* and of schoolchildren in Beirut in the spring *(47)*. Sun exposure in schoolchildren has been associated with socioeconomic status, with children of higher socioeconomic status receiving longer exposure *(47)*. Due to cultural norms and religious practices, the impact of clothing on the vitamin D status and bone mineral density (BMD) of women living in the Middle East has been investigated. The sun exposure of three population groups of women aged 19–44 years living in the United Arab Emirates was assessed through the summer as 17, 14, and 34 min/day for UAE nationals ($n = 33$), non-Gulf Arabs ($n = 25$), and Europeans ($n = 17$), respectively *(84)*. The duration of sunlight exposure and the extent of clothing were multiplied together to produce a UV score. The UV scores of UAE nationals, non-Gulf Arabs, and Europeans were 4.35, 7.56, and 47.7 and reflected the 25(OH)D concentrations being 9, 13, and 65 ng/ml in the three groups, respectively. The evidence overwhelmingly indicates the lower vitamin D status of women wearing clothing designed to minimize the amount of skin exposure, a dressing style that has the potential to adversely affect skeletal health.

4. SOUTHERN ASIA

South Asia includes the countries in the Indian subcontinent as well as Southeast Asia. Despite close proximity to the equator and sufficient sunlight 25(OH)D concentrations are surprisingly low in countries in south Asia, particularly India and Bangladesh.

4.1. Vitamin D Status

Although there are reports of relatively high 25(OH)D concentrations in southern Asia vitamin D insufficiency is still common (Table 4) *(85–100)*. In Kuala Lumpur, Malaysia (2°N), the mean 25(OH)D of women of reproductive age was 20 ng/ml, with women of Chinese ethnicity having lower rates of insufficiency (38% < 20 ng/ml) compared with Malay (74%) or Indian (68%) women *(95)*. Among Bangladeshi factory workers in Dhaka (23°N), most of the women had a 25(OH)D concentration less than 20 ng/ml with a mean of 15 ng/ml *(85)*. In Lucknow, the mean 25(OH)D concentration in boys was three times that of girls, 27 ng/ml compared with 9 ng/ml *(90)*. Interestingly in Sri Lanka women had higher 25(OH)D concentrations than men (25 vs 19 ng/ml) *(99)*. In India, there is a tendency for people living in the north to have lower 25(OH)D concentrations compared with those in the south *(88, 92, 101)*. For example, among Kashmiri (33°N) women in the north the mean 25(OH)D was 4 ng/ml compared with 18 ng/ml in women from Andhra Pradesh (13°N) in the south *(88, 92)*. The highest 25(OH)D concentrations in the region were reported among older northern Thai women living in suburban Khon Kaen (16°N) where the mean was 36 ng/ml *(100)*.

4.2. Consequences

In some areas of Bangladesh over 8% of children showed at least one clinical sign indicative of rickets such as pigeon chest or knocked knees *(102)*. To a large extent rickets in Bangladesh is thought to be as a result of calcium deficiency *(103)* and in a small study of children with active rickets only 14% had a 25(OH)D concentration less than 14 ng/ml *(104)*. Cases of rickets have been reported in India, particularly in the north, related to vitamin D deficiency; however, some of the rickets have been responsive to calcium supplementation without vitamin D *(105, 106)*. Sporadic cases of rickets have also been reported in Thailand *(107)*, Singapore *(108)*, and the Philippines *(109)*. There are no prevalence data for osteomalacia for this region although it has been reported

Table 4

25-Hydroxyvitamin D Concentrations and Prevalence of Vitamin D Deficiency and Insufficiency in Middle Eastern countries

Reference	Location and latitude	Population (n)	Age (year range or x̄)	Method	Months	25-Hydroxyvitamin D Mean (ng/ml)	% Deficient (cutoff)	% Insufficient (cutoff)
Afghanistan								
(32)	Kabul 34°N	Low SES children (107)	0.5–5	HPLC	Jan–May	md 5.0	73 (20)	
Turkey								
(33)	Erzurum and Van 40°N	Control infants and children (30)	0.5–4	Diasorin	Jan–Dec	26	17 (37.5)	
		Rachitic infants and children (68)	0.5–4	Diasorin	Jan–Dec	10	54 (37.5)	
(34)	Kocaeli 30–40°N	Suburban schoolgirls (29)	15	Biosource	Apr	23	1 (25)	48 (50)
		Urban schoolgirls (30)	15	Biosource	Apr	20	3 (25)	53 (50)
		Urban schoolgirls wearing concealing clothes (30)	15	Biosource	Apr	11	15 (25)	93 (50)
(35)	Izmar 38°N	Schoolgirls (32)	14–18	Biosource	Sep–Oct	20	25 (37.5)	
		Schoolgirls (32)	14–18	Biosource	Feb–Mar	14	59 (37.5)	
(36)	Ataturk 41°N	Women in western clothing (36)	20	Biosource	May	22		
		Women with only face and hands unclothed (31)	21	Biosource	May	6		
(37)	Adana 37°N	Women with only eyes exposed (30)	25	Biosource	Aug–Sep	33		
		Women in western-style clothing (30)	25	Biosource	Aug–Sep	54		

Table 4
(continued)

Reference	Location and latitude	Population (n)	Age (year range or x̄)	Method	Months	25-Hydroxyvitamin D		
						Mean (ng/ml)	% Deficient (cutoff)	% Insufficient (cutoff)
(38)	Ankara 40°N	Women in western clothing (40)	36	Biosource	Nov and Mar	26		
		Women with only face and hands unclothed (40)	35	Biosource	Nov and Mar	14		
(39)	Adana 37°N	Osteoporotic women (315)	61	HPLC	Apr–Sep	23	13.5 (30)	51 (50)
Iran								
(40)	Isfahan 32°N	Pregnant women (88)		Biosource	Aug–Sep			6 (50)
		Newborns (88)		Biosource	Aug–Sep		5 (31)	
(41)	Tehran 35°N	Pregnant women (552)		Biosource	Winter	11	67 (35)	
		Cord blood (552)		Biosource		7	86 (25)	
(42)	Isfahan 32°N	Schoolboys (153)	16	Biosource	Mar–Apr	37		18 (50)
		Schoolgirls (165)	16	Biosource	Mar–Apr	17		72 (50)
(43)	Tehran 35°N	Schoolboys (424)	7–18	Biosource	Winter	46		11 (50)
		Schoolgirls veiled outdoors (539)	7–18	Biosource	Winter	24		54 (50)
(44)	Tehran 35°N	Adults (1210)	20+	Biosource	Winter	md 8	67 (25)	
(45)	Tehran 35°N	Pregnant women (741)	26	IDS RIA	Jan–Dec	~9	71 (25)	
(46)	Tehran 35°N	Postmenopausal women (245)	58	DRG	Feb–Mar	29	5 (25)	43 (50)
(66)	Tehran 35°N	Men (37)		DRG	Nov	55		13 (50)
		Men (45)		DRG	Feb	24		51 (50)
		Women (38)		DRG	Nov	29		55 (50)
		Women (66)		DRG	Feb	14		83 (50)

Table 4
(continued)

Reference	Location and latitude	Population (n)	Age (year range or $\bar{x}$)	Method	Months	25-Hydroxyvitamin D Mean (ng/ml)	% Deficient (cutoff)	% Insufficient (cutoff)
Lebanon								
(47)	Beirut 34°N	Schoolchildren (169)	13	Diasorin	Mar–Apr	17	21 (25)	65 (50)
		Schoolchildren (177)	13	Diasorin	Nov–Dec	22	4 (25)	40 (50)
(48)	Nabi-Shit 34°N	Women (465)	15–59	Diasorin	Aug–Oct	11	60 (25)	95 (50)
(49)	Bekaa Valley and Beirut 34°N	Men (99)	41	Diasorin	Jan–Apr	14	49 (30)	
		Postmenopausal women (217)	39	Diasorin	Jan–Apr	8	84 (30)	
(50)	Beirut and Saida 34°N	Christian osteoporotic women (151)	68	Nichols	Jul–Aug	22		
		Muslim osteoporotic women (100)	67	Nichols	Feb–Mar	15		
Israel								
(51)	Tel Aviv 32°N	Non-orthodox Jewish mothers (185)	31	Other	Aug–Sep and Mar–May	18	14 (25)	
		Orthodox Jewish mothers (156)	27	Other	Aug–Sep Mar–May	14	33 (25)	
(52)	Beer Sheeva 31°N	Adult hospital admissions (293)	63	Other	Sep and Mar	21	15 (37.5)	

Table 4
(continued)

Reference	Location and latitude	Population (n)	Age (year range or x̄)	Method	Months	25-Hydroxyvitamin D		
						Mean (ng/ml)	% Deficient (cutoff)	% Insufficient (cutoff)
Jordan								
(53)	Amman 32°N	Women wearing western style (12)	~18–45	Diasorin	Jul–Sep	15	31 (30)	100 (50)
		Women wearing western style (8)	~18–45	Diasorin	Jan–Mar	12	75 (30)	99 (50)
		Women with face and hands (31)	~18–45	Diasorin	Jul–Sep	11	55 (30)	100 (50)
		Women with face and hands (49)	~18–45	Diasorin	Jan–Mar	10	78 (30)	100 (50)
		Women fully covered (11)	~18–45	Diasorin	Jul–Sep	10	83 (30)	100 (50)
		Women fully covered (12)	~18–45	Diasorin	Jan–Mar	9	82 (30)	100 (50)
		Men (11)	~18–45	Diasorin	Jul–Sep	18	18 (30)	96 (50)
		Men (11)	~18–45	Diasorin	Jan–Mar	14	45 (30)	99 (50)
Kuwait								
(54)	Al-Adan 29°N	New mothers (214)	~28	Diasorin	Sep–Jun	15	~40 (25)	~75 (50)
		Neonates (214)		Diasorin	Sep–Jun	8	>60 (25)	~95 (50)
United Arab Emirates								
(55)	Al Ain 24°N	Breastfeeding mothers (90)		HPLC	Apr–Oct	md 9	61 (25)	
		Breast-fed infants (78)	4–16 week	HPLC	Apr–Oct	md 5	82 (25)	
(56)	Al Ain 24°N	Young mothers (51)		HPLC		11	51 (25)	
		Non-rachitic hospitalized children (51)	15 months	HPLC		20	22 (25)	

Table 4
(continued)

Reference	Location and latitude	Population (n)	Age (year range or x̄)	Method	Months	25-Hydroxyvitamin D		
						Mean (ng/ml)	% Deficient (cutoff)	% Insufficient (cutoff)
(57)	Al Ain 24°N	Nulliparous women (88)	~29	Diasorin		~8		
		Lactating women (90)	~24	Diasorin		~10		
(58)	Al Ain 24°N	Women fully covered (n = 160)	46	Diasorin		10		
		Women with only eyes exposed (94)	41	Diasorin		10		
		Non-Arab women in western clothing (7)	42	Diasorin		25		
Saudi Arabia								
(59)	East 26°N	Infants (58)	17 months	Other	Jan–Dec		25 (20)	
		Rachitic infants (61)	15 months	Other	Jan–Dec		75 (20)	
(60)	Jeddah 22°N	Schoolgirls (433)	12–15		Oct–Feb			

in Indian girls living in Lucknow (27°N). Parathyroid hormone and/or alkaline phosphatase concentrations were inversely associated with 25(OH)D in a number of studies
from the region suggesting an adverse effect of low vitamin D status on bone health *(88,
91, 94, 97, 100, 110)*. In female Bangladeshi garment workers 25(OH)D ($\bar{x} = 15$ ng/ml)
was inversely associated with PTH and positively associated with BMD at the femoral
neck and spine. However, there was no association between 25(OH)D and BMD at the
distal forearm and heel in 664 girls (7–17 years) or 760 boys and girls (10–18 years) living in Delhi (28°N) despite low 25(OH)D concentrations ($\bar{x} = 13$ and 12 ng/ml, respectively) *(91, 110)*.

4.3. Causes of Vitamin D Deficiency

There are scant data on vitamin D intakes in this region although in areas where
fish consumption is low it is presumably minimal. In postmenopausal women in Kuala
Lumpur mean intake was 9 μg but daily fish consumption was high (70 g/day) *(97)*.
With the possible exception of Nepal and northern India there is abundant sunshine
throughout this region and the UV index remains high to extreme throughout the year
(98). Causes of vitamin D deficiency in south Asia are factors that limit sun exposure
and are similar to the Middle East. In addition, the highly pigmented skin of many
people from this region may limit endogenous vitamin D production *(111)*. Greater
time spent outdoors *(89, 92, 112, 113)* and rural suburban vs urban living *(100)* were
generally associated with better 25(OH)D indicating the importance of UV exposure.
Veiling probably explains the lower 25(OH)D concentrations reported in Malay women,
who are predominantly Muslim, vs Chinese women living in Kuala Lumpur *(95, 97)*. It
may be that veiling is a plausible explanation for low vitamin D status in India where
rickets has been found to be 3–4 times more common in the children of Muslims than
Hindus *(114)*. Among Bangladeshi women living in Dhaka no difference in 25(OH)D
concentrations was reported between veiled and non-veiled women, although over 80%
of women in the study indicated that they avoided direct sunlight. The poor vitamin D
status of female Bangladeshi garment factory workers, who number over 1.5 million,
probably relates to their indoor lifestyle including 14- to 16-h workdays *(85)*. The effect
of pollution on 25(OH)D concentrations was examined in a study of infants and toddlers
(16 months) living in Delhi; those living in the Mori Gate ($n = 26$), an area with high
atmospheric pollution, had 25(OH)D concentrations less than half that of a age-matched
control group living in Gurgaon ($n = 31$), a low pollution area (12 vs 27 ng/ml) *(115)*.

5. STRATEGIES TO IMPROVE VITAMIN D STATUS

The vitamin D status of many Asian populations needs improving. How is this best
accomplished? Given the large variation in the causes of poor vitamin D status in Asia
no single approach will work for all. Increasing sun exposure is one option especially
in sun-rich areas of the south. This approach was used as a strategy to improve the
biochemical indicators of rickets in Iranian girls whose treatment involved hand and face
exposure to sunlight for 1 h/day for 20 days *(81)*. Over the period alkaline phosphatase
and parathyroid hormone decreased and 25(OH)D increased from 7 to 14 ng/ml. Having

veiled women spend even 20 min a day outdoors, perhaps in a private courtyard, with their hands and faces exposed would certainly improve their vitamin D status. Designing fabrics that allow more ultraviolet light to pass through is another option *(116)*. When adequate UV light is received relatively high vitamin D status can be achieved in Asian population groups *(100)*.

In the absence of sufficient UV exposure from sunlight, additional vitamin D must be obtained from dietary or supplementary sources. Except where fish is readily available it will be impossible to improve dietary intake from natural sources of vitamin D. Food fortification is another option, as is done with milk in North America. At present few foods are fortified with vitamin D in Asia. Beijing girls given milk fortified with vitamin D_3 (average 3.3 μg/day) for 2 years had higher 25(OH)D, lower PTH, and greater bone mineral density than girls receiving milk without vitamin D or no milk *(117)*. Milk is not widely consumed in the region except in some places in the Middle East like Lebanon; however, fortification of other dairy products such as cheese, yoghurt, and ghee may help, especially in India. Other vehicles for vitamin D fortification such as wheat flour, rice, and oils could be considered. However, food fortification requires centralized processing or milling facilities, distribution networks, and strict quality control, which are not always available.

Vitamin D supplementation is the only viable option for much of the Asian population, especially they are to achieve 25(OH)D concentration thought to be necessary for optimal health *(118, 119)*. Pregnant women and breastfeeding infants need to be supplemented with vitamin D, especially in northern Asia to prevent rickets. The addition of fortified sprinkles to complementary foods is an attractive option when the effective dose is determined. Supplementation is not without cost, compliance, and safety issues *(120)*. Intermittent dosing may help overcome some of the cost and compliance problems *(57)*. Consideration of calcium supplementation is also necessary especially where intakes are low.

6. CONCLUSION

The vitamin D status of people living in Asia is highly variable. The expected seasonal variation in 25(OH)D, higher in summer and lower in winter, is evident in people living in north Asia. This obligatory seasonal variation in pre-vitamin D formation may be explained by reduced daylight hours, lower UV index, colder temperatures that necessitate extensive clothing, and an aversion to going outdoors in the winter compared with the summer months. Low vitamin D status caused by the environmentally restrictive conditions that do not allow for dermal synthesis of pre-vitamin D is compounded by diets lacking in vitamin D food sources. Low calcium intakes and vitamin D deficiency are manifested as overt rickets, a condition endemic in parts of north Asia. In contrast, there is ample sunshine for vitamin D synthesis to occur year-round in south Asia and much of the Middle East. The vitamin D status of infants and women in particular has been found to be low. Part of the explanation appears to be cultural and religious practices that require women to wear clothing that conceals most of their skin when outdoors. Sun aversion for cosmetic reasons, avoidance of extremely hot temperatures, and keeping infants indoors are also likely contributors.

The most well-described manifestations of vitamin D deficiencies in Asia are skeletal, principally rickets and osteomalacia, although other conditions associated with vitamin D deficiency that have been identified in the region include reduced quality of life in osteoporotic women *(39)* and severe acute respiratory infections in infants *(61)*. The vitamin D contribution to these burdens of disease may be preventable. Vitamin D status can be improved by sun exposure or by supplements in Asian people *(57, 81, 118)*. Improvement of vitamin D status using vitamin D supplements has been found to improve skeletal health *(48)*. Many investigators in Asian countries have called for programs aimed at improving the skeletal health of their young women and children through fortification or supplementation with vitamin D and/or calcium. To protect against long-term consequences, a consensus group for the prevention of osteoporosis in the region suggests vitamin D supplementation is imperative in all postmenopausal women *(121)*.

REFERENCES

1. Strand MA, Perry J, Zhao J et al (2007) Severe Vitamin D-deficiency and the health of North China children. Matern Child Health J 13(1):144–150
2. Du X, Greenfield H, Fraser DR et al (2001) Vitamin D deficiency and associated factors in adolescent girls in Beijing. Am J Clin Nutr 74:494–500
3. Woo J, Lam C, Leung J et al (2007) Very high rates of vitamin D insufficiency in women of childbearing age living in Beijing and Hong Kong. Brt J Nutr 99:1330–1334
4. Yan L, Prentice A, Zhang H et al (2000) Vitamin D status and parathyroid hormone concentrations in Chinese women and men from north-east of the People's Republic of China. Eur J Clin Nutr 54:68–72
5. Wat WZ, Leung JY, Tam S et al (2007) Prevalence and impact of vitamin D insufficiency in southern Chinese adults. Ann Nutr Metab 51:59–64
6. Yan L, Zhou B, Wang X et al (2003) Older people in China and the United Kingdom differ in the relationships among parathyroid hormone, vitamin D, and bone mineral status. Bone 33:620–627
7. Kuroda E, Okano T, Mizuno N et al (1981) Plasma levels of 25-hydroxyvitamin D2 and 25-hydroxyvitamin D3 in maternal, cord and neonatal blood. J Nutr Sci Vitaminol (Tokyo) 27:55–65
8. Nakamura K, Nashimoto M, Matsuyama S et al (2001) Low serum concentrations of 25-hydroxyvitamin D in young adult Japanese women: a cross sectional study. Nutrition 17: 921–925
9. Nakamura K, Nashimoto M, Yamamoto M (2000) Summer/winter differences in the serum 25-hydroxyvitamin D3 and parathyroid hormone levels of Japanese women. Int J Biometeorol 44: 186–189
10. Nakamura K, Tsugawa N, Saito T et al (2008) Vitamin D status, bone mass, and bone metabolism in home-dwelling postmenopausal Japanese women: yokogoshi Study. Bone 42:271–277
11. Lander RL, Enkhjargal T, Batjargal J et al (2008) Multiple micronutrient deficiencies persist during early childhood in Mongolia. Asia Pac J Clin Nutr 17:429–440
12. Ganmaa D, Tserendolgor U, Frazier L et al (2008) Effects of vitamin D fortified milk on vitamin D status in Mongolian school age children. Asia Pac J Clin Nutr 17:68–71
13. Bakhtiyarova S, Lesnyak O, Kyznesova N et al (2006) Vitamin D status among patients with hip fracture and elderly control subjects in Yekaterinburg, Russia. Osteoporos Int 17:441–446
14. Namgung R, Tsang RC, Lee C et al (1998) Low total body bone mineral content and high bone resorption in Korean winter-born versus summer-born newborn infants. J Pediatr 132: 421–425
15. Kim JH, Moon SJ (2000) Time spent outdoors and seasonal variation in serum concentrations of 25-hydroxyvitamin D in Korean women. Int J Food Sci Nutr 51:439–451
16. Rizzoli R, Eisman JA, Norquist J et al (2006) Risk factors for vitamin D inadequacy among women with osteoporosis: an international epidemiological study. Int J Clin Pract 60:1013–1019

17. Tsai KS, Hsu SH, Cheng JP et al (1997) Vitamin D stores of urban women in Taipei: effect on bone density and bone turnover, and seasonal variation. Bone 20:371–374

18. Lee WP, Lin LW, Yeh SH et al (2002) Correlations among serum calcium, vitamin D and parathyroid hormone levels in the elderly in southern Taiwan. J Nurs Res 10:65–72

19. Strand MA, Perry J, Jin M et al (2007) Diagnosis of rickets and reassessment of prevalence among rural children in northern China. Pediatr Int 49:202–209

20. Fraser DR (2004) Vitamin D-deficiency in Asia. J Steroid Biochem Mol Biol 89–90:491–495

21. Tserendolgor U, Mawson J, MacDonald A et al (1998) Prevalence of rickets in Mongolia. Asia Pac J Clin Nutr 7:325–328

22. Nutrition Research Center & UNICEF(2000) Report on the 1999 2nd National Child and Nutrition Survey. Ulaanbatar.

23. World Vision Mongolia(2005) Effectiveness of Home-Based Fortification of Complementary Foods with Sprinkles in an Integrated Nutrition Program to Address Rickets and Anemia. Ulaanbaatar.

24. Miyako K, Kinjo S, Kohno H (2005) Vitamin D deficiency rickets caused by improper lifestyle in Japanese children. Pediatr Int 47:142–146

25. Thacher TD, Fischer PR, Strand MA et al (2006) Nutritional rickets around the world: causes and future directions. Ann Trop Paediatri 26:1–16

26. Yorifuji J, Yorifuji T, Tachibana K et al (2008) Craniotabes in normal newborns: the earliest sign of subclinical vitamin D deficiency. J Clin Endocrinol Metab 93:1784–1788

27. Lau EM, Woo J, Swaminathan R et al (1989) Plasma 25-hydroxyvitamin D concentration in patients with hip fracture in Hong Kong. Gerontology 35:198–204

28. Lim SK, Kung AW, Sompongse S et al (2008) Vitamin D inadequacy in postmenopausal women in Eastern Asia. Curr Med Res Opin 24:99–106

29. Lips P (2001) Vitamin D deficiency and secondary hyperparathyroidism in the elderly: consequences for bone loss and fractures and therapeutic implications. Endocr Rev 22:477–501

30. Strand MA, Perry J, Guo J et al (2008) Doing the month: rickets and post-partum convalescence in rural China. Midwifery doi:10.1016/j.midw.2007.10.008

31. Kung AW, Lee KK (2006) Knowledge of vitamin D and perceptions and attitudes toward sunlight among Chinese middle-aged and elderly women: a population survey in Hong Kong. BMC Public Health 6:226

32. Manaseki-Holland S, Zulf Mughal M, Bhutta Z et al (2008) Vitamin D status of socio-economically deprived children in Kabul, Afghanistan. Int J Vitam Nutr Res 78:16–20

33. Baroncelli GI, Bertelloni S, Ceccarelli C et al (2000) Bone turnover in children with vitamin D deficiency rickets before and during treatment. Acta Paediatrca 89:513–518

34. Hatun S, Islam O, Cizmecioglu F et al (2005) Subclinical vitamin D deficiency is increased in adolescent girls who wear concealing clothing. J Nutr 135:218–222

35. Olmez D, Bober E, Buyukgebiz A et al (2006) The frequency of vitamin D insufficiency in healthy female adolescents. Acta Paediatrica 95:1266–1269

36. Budak N, Cicek B, Sahin H et al (2004) Bone mineral density and serum 25-hydroxyvitamin D level: is there any difference according to the dressing style of the female university students. Int J Food Sci Nutr 55:569–575

37. Guzel R, Kozanoglu E, Guler-Uysal F et al (2001) Vitamin D status and bone mineral density of veiled and unveiled Turkish women. J Womens Health Gend Based Med 10:765–770

38. Guler T, Sivas F, Baskan BM et al (2007) The effect of outfitting style on bone mineral density. Rheumatol Int 27:723–727

39. Basaran S, Guzel R, Coskun-Benlidayi I et al (2007) Vitamin D status: effects on quality of life in osteoporosis among Turkish women. Qual Life Res 16:1491–1499

40. Salek M, Hashemipour M, Aminorroaya A et al (2008) Vitamin D deficiency among pregnant women and their newborns in Isfahan, Iran. Exp Clin Endocrinol Diabetes 116:352–356

41. Maghbooli Z, Hossein-Nezhad A, Shafaei AR et al (2007) Vitamin D status in mothers and their newborns in Iran. BMC Pregnancy Childbirth 7:1

42. Moussavi M, Heidarpour R, Aminorroaya A et al (2005) Prevalence of vitamin D deficiency in Isfahani high school students in 2004. Horm Res 64:144–148

43. Rabbani A, Alavian SM, Motlagh ME et al (2009) Vitamin D insufficiency among children and adolescents living in Tehran, Iran. J Trop Pediatr 55:188–191

44. Hashemipour S, Larijani B, Adibi H et al (2004) Vitamin D deficiency and causative factors in the population of Tehran. BMC Public Health 4:38

45. Maghbooli Z, Hossein-nezhad A, Karimi F et al (2008) Correlation between vitamin D-3 deficiency and insulin resistance in pregnancy. Diabetes-Metab Res 24:27–32

46. Hosseinpanah F, Rambod M, Hossein-nejad A et al (2008) Association between vitamin D and bone mineral density in Iranian postmenopausal women. J Bone Miner Metab 26:86–92

47. Fuleihan GE, Nabulsi M, Choucair M et al (2001) Hypovitaminosis D in healthy schoolchildren. Pediatrics 107:E53

48. Fuleihan GE, Gundberg CM, Gleason R et al (1994) Racial differences in parathyroid hormone dynamics. J Clin Endocrinol Metab 79:1642–1647

49. Gannage-Yared MH, Chemali R, Yaacoub N et al (2000) Hypovitaminosis D in a sunny country: relation to lifestyle and bone markers. J Bone Miner Res 15:1856–1862

50. Gannage-Yared MH, Maalouf G, Khalife S et al (2009) Prevalence and predictors of vitamin D inadequacy amongst Lebanese osteoporotic women. Br J Nutr 101:487–491

51. Mukamel MN, Weisman Y, Somech R et al (2001) Vitamin D deficiency and insufficiency in Orthodox and non-Orthodox Jewish mothers in Israel. Isr Med Assoc J 3:419–421

52. Hochwald O (2004) Hypovitaminosis D among inpatients in sunny country. Isr Med Assoc J 6: 381

53. Mishal AA (2001) Effects of different dress styles on vitamin D levels in healthy young Jordanian women. Osteoporos Int 12:931–935

54. Molla AM, Al Badawi M, Hammoud MS et al (2005) Vitamin D status of mothers and their neonates in Kuwait. Pediatr Int 47:649–652

55. Dawodu A, Agarwal M, Hossain M et al (2003) Hypovitaminosis D and vitamin D deficiency in exclusively breast-feeding infants and their mothers in summer: a justification for vitamin D supplementation of breast-feeding infants. J Pediatr 142:169–173

56. Dawodu A, Agarwal M, Sankarankutty M et al (2005) Higher prevalence of vitamin D deficiency in mothers of rachitic than nonrachitic children. J Pediatr 147:109–111

57. Saadi HF, Dawodu A, Afandi BO et al (2007) Efficacy of daily and monthly high-dose calciferol in vitamin D-deficient nulliparous and lactating women. Am J Clin Nutr 85:1565–1571

58. Saadi HF, Nagelkerke N, Benedict S et al (2006) Predictors and relationships of serum 25 hydroxy-vitamin D concentration with bone turnover markers, bone mineral density, and vitamin D receptor genotype in Emirati women. Bone 39:1136–1143

59. Al-Mustafa ZH, Al-Madan M, Al-Majid HJ et al (2007) Vitamin D deficiency and rickets in the Eastern Province of Saudi Arabia. Ann Trop Paediatr 27:63–67

60. Siddiqui TS, Rai MI (2005) Presentation and predisposing factors of nutritional rickets in children of Hazara Division. J Ayub Med Coll Abbottabad 17:29–32

61. Karatekin G, Kaya A, Salihoglu O et al (2009) Association of subclinical vitamin D deficiency in newborns with acute lower respiratory infection and their mothers. Eur J Clin Nutr 63:373–477

62. Baroncelli GI, Bereket A, El Kholy M et al (2008) Rickets in the Middle East: role of environment and genetic predisposition. J Clin Endocrinol Metab 93:1743–1750

63. Andiran N, Yordam N, Ozon A (2002) Risk factors for vitamin D deficiency in breast-fed newborns and their mothers. Nutrition 18:47–50

64. Nabulsi M, Mahfoud Z, Maalouf J et al (2008) Impact of maternal veiling during pregnancy and socioeconomic status on offspring's musculoskeletal health. Osteoporos Int 19:295–302

65. Dawodu A, Dawson KP, Amirlak I et al (2001) Diet, clothing, sunshine exposure and micronutrient status of Arab infants and young children. Ann Trop Paediatr 21:39–44

66. Mirsaeid Ghazi AA, Rais Zadeh F, Pezeshk P et al (2004) Seasonal variation of serum 25 hydroxy D3 in residents of Tehran. J Endocrinol Invest 27:676–679

67. Atli T, Gullu S, Uysal AR et al (2005) The prevalence of Vitamin D deficiency and effects of ultra-violet light on Vitamin D levels in elderly Turkish population. Arch Gerontol Geriatr 40:53–60

68. Ramavat LG (1999) Vitamin D deficiency rickets at birth in Kuwait. Indian J Pediatr 66:37–43

69. Najada AS, Habashneh MS, Khader M (2004) The frequency of nutritional rickets among hospitalized infants and its relation to respiratory diseases. J Trop Pediatr 50:364–368

70. Soliman AT, Al Khalaf F, Alhemaidi N et al (2008) Linear growth in relation to the circulating concentrations of insulin-like growth factor I, parathyroid hormone, and 25-hydroxy vitamin D in children with nutritional rickets before and after treatment: endocrine adaptation to vitamin D deficiency. Metabolism 57:95–102

71. Dahifar H, Faraji A, Yassobi S et al (2007) Asymptomatic rickets in adolescent girls. Indian J Pediatr 74:571–575

72. Fida NM (2003) Assessment of nutritional rickets in Western Saudi Arabia. Saudi Med J 24:337–340

73. Ozkan B, Doneray H, Karacan M et al (2008) Prevalence of vitamin D deficiency rickets in the eastern part of Turkey. Eur J Pediatr

74. Gullu S, Erdogan MF, Uysal AR et al (1998) A potential risk for osteomalacia due to sociocultural lifestyle in Turkish women. Endocr J 45:675–678

75. El-Desouki MI, Othman SM, Fouda MA (2004) Bone mineral density and bone scintigraphy in adult Saudi female patients with osteomalacia. Saudi Med J 25:355–358

76. Sahibzada AS, Khan MS, Javed M (2004) Presentation of osteomalacia in Kohistani women. J Ayub Med Coll Abbottabad 16:63–65

77. Herm FB, Killguss H, Stewart AG (2005) Osteomalacia in Hazara District, Pakistan. Trop Doct 35: 8–10

78. Hayirlioglu DA, Gokaslan H, Cimsit C et al (2008) The impact of clothing style on bone mineral density among women in Turkey. Rheumatol Int 28:521–525

79. Esmaillzadeh A, Samareh S, Azadbakht L (2008) Dietary patterns among pregnant women in the west-north of Iran. Pak J Biol Sci 11:793–796

80. Salamoun MM, Kizirian AS, Tannous RI et al (2005) Low calcium and vitamin D intake in healthy children and adolescents and their correlates. Eur J Clin Nutr 59:177–184

81. Dahifar H, Faraji A, Ghorbani A et al (2006) Impact of dietary and lifestyle on vitamin D in healthy student girls aged 11–15 years. J Med Invest 53:204–208

82. Gannage-Yared MH, Chemali R, Sfeir C et al (2005) Dietary calcium and vitamin D intake in an adult Middle Eastern population: food sources and relation to lifestyle and PTH. Int J Vitam Nutr Res 75:281–289

83. Kimlin MG (2008) Geographic location and vitamin D synthesis. Mol Aspects Med 29(6):453–461

84. Dawodu A, Absood G, Patel M et al (1998) Biosocial factors affecting vitamin D status of women of childbearing age in the United Arab Emirates. J Biosoc Sci 30:431–437

85. Islam MZ, Shamim AA, Kemi V et al (2008) Vitamin D deficiency and low bone status in adult female garment factory workers in Bangladesh. Br J Nutr 99:1322–1329

86. Islam MZ, Akhtaruzzaman M, Lamberg-Allardt C (2006) Hypovitaminosis D is common in both veiled and nonveiled Bangladeshi women. Asia Pac J Clin Nutr 15:81–87

87. Bhalala U, Desai M, Parekh P et al (2007) Subclinical hypovitaminosis D among exclusively breast-fed young infants. Indian Pediatr 44:897–901

88. Harinarayan CV, Ramalakshmi T, Prasad UV et al (2008) Vitamin D status in Andhra Pradesh: a population based study. Indian J Med Res 127:211–218

89. Puri S, Marwaha RK, Agarwal N et al (2008) Vitamin D status of apparently healthy schoolgirls from two different socioeconomic strata in Delhi: relation to nutrition and lifestyle. Br J Nutr 99: 876–882

90. Sahu M, Bhatia V, Aggarwal A et al (2008) Vitamin D deficiency in rural girls and pregnant women despite abundant sunshine in northern India. Clin Endocrinol (Oxf) CEN3360 [pii] 10.1111/j.1365–2265.2008.03360.x

91. Marwaha RK, Tandon N, Reddy DR et al (2005) Vitamin D and bone mineral density status of healthy schoolchildren in northern India. Am J Clin Nutr 82:477–482

92. Zargar AH, Ahmad S, Masoodi SR et al (2007) Vitamin D status in apparently healthy adults in Kashmir Valley of Indian subcontinent. Postgrad Med J 83:713–716

93. Sachan A, Gupta R, Das V et al (2005) High prevalence of vitamin D deficiency among pregnant women and their newborns in northern India. Am J Clin Nutr 81:1060–1064

94. Harinarayan CV (2005) Prevalence of vitamin D insufficiency in postmenopausal south Indian women. Osteoporos Int 16:397–402

95. Green TJ, Skeaff CM, Rockell JE et al (2008) Vitamin D status and its association with parathyroid hormone concentrations in women of child-bearing age living in Jakarta and Kuala Lumpur. Eur J Clin Nutr 62:373–378

96. Oemardi M, Horowitz M, Wishart JM et al (2007) The effect of menopause on bone mineral density and bone-related biochemical variables in Indonesian women. Clin Endocrinol (Oxf) 67:93–100

97. Rahman SA, Chee WS, Yassin Z et al (2004) Vitamin D status among postmenopausal Malaysian women. Asia Pac J Clin Nutr 13:255–260

98. Jiang T, Christian P, Khatry SK et al (2005) Micronutrient deficiencies in early pregnancy are common, concurrent, and vary by season among rural Nepali pregnant women. J Nutr 135:1106–1112

99. Meyer HE, Holvik K, Lofthus CM et al (2008) Vitamin D status in Sri Lankans living in Sri Lanka and Norway. Br J Nutr 99:941–944

100. Soontrapa S, Chailurkit LO (2005) Difference in serum calcidiol and parathyroid hormone levels between elderly urban vs suburban women. J Med Assoc Thai 88(Suppl 5):S17–S20

101. Teotia SP, Teotia M (2008) Nutritional bone disease in Indian population. Indian J Med Res 127: 219–228

102. Kabir ML, Rahman M, Talukder K et al (2004) Rickets among children of a coastal area of Bangladesh. Mymensingh Med J 13:53–58

103. Craviari T, Pettifor JM, Thacher TD et al (2008) Rickets: an overview and future directions, with special reference to Bangladesh. A summary of the Rickets Convergence Group meeting, Dhaka, 26–27 January 2006. J Health Popul Nutr 26:112–121

104. Fischer PR, Rahman A, Cimma JP et al (1999) Nutritional rickets without vitamin D deficiency in Bangladesh. J Trop Pediatr 45:291–293

105. Girish M, Subramaniam G (2008) Rickets in exclusively breast fed babies. Indian J Pediatr 75: 641–643

106. Balasubramanian K, Rajeswari J, Gulab Y et al (2003) Varying role of vitamin D deficiency in the etiology of rickets in young children vs. adolescents in northern India. J Trop Pediatr 49:201–206

107. Unachak K, Visrutaratna P, Dejkamron P et al (2004) Infantile osteopetrosis in four Thai infants. J Pediatr Endocrinol Metab 17:1455–1459

108. Wong HB (1986) Rickets in Singapore infants and children. J Singapore Paediatr Soc 28:12–20

109. Stransky E, Dizon-Santos-Ocampo PO (1958) Clinical rickets in the Philippines; report of 22 cases in Manila. J Trop Pediatr 4:17–19

110. Marwaha RK, Tandon N, Reddy DH et al (2007) Peripheral bone mineral density and its predictors in healthy school girls from two different socioeconomic groups in Delhi. Osteoporos Int 18:375–383

111. Gozdzik A, Barta JL, Wu H et al (2008) Low wintertime vitamin D levels in a sample of healthy young adults of diverse ancestry living in the Toronto area: associations with vitamin D intake and skin pigmentation. BMC Public Health 8:336

112. Goswami R, Gupta N, Goswami D et al (2000) Prevalence and significance of low 25-hydroxy-vitamin D concentrations in healthy subjects in Delhi. Am J Clin Nutr 72:472–475

113. Harinarayan CV, Ramalakshmi T, Venkataprasad U (2004) High prevalence of low dietary calcium and low vitamin D status in healthy south Indians. Asia Pac J Clin Nutr 13:359–364

114. Bhattacharyya AK (1992) Nutritional rickets in the tropics. World Rev Nutr Diet 67:140–197

115. Agarwal KS, Mughal MZ, Upadhyay P et al (2002) The impact of atmospheric pollution on vitamin D status of infants and toddlers in Delhi, India. Arch Dis Child 87:111–113

116. Salih FM (2004) Effect of clothing varieties on solar photosynthesis of previtamin D3: an in vitro study. Photodermatol Photoimmunol Photomed 20:53–58

117. Du X, Zhu K, Trube A et al (2004) School-milk intervention trial enhances growth and bone mineral accretion in Chinese girls aged 10–12 years in Beijing. Br J Nutr 92:159–168

118. Maalouf J, Nabulsi M, Vieth R et al (2008) Short- and long-term safety of weekly high-dose vitamin D3 supplementation in school children. J Clin Endocrinol Metab 93:2693–2701

119. Holick MF (2007) Vitamin D deficiency. N Engl J Med 357:266–281
120. Li Tong GX-x (1995) Prevention of rickets and vitamin D intoxication in China. Acta Paediatrica 84:940–940
121. Maalouf G, Gannage-Yared MH, Ezzedine J et al (2007) Middle East and North Africa consensus on osteoporosis. J Musculoskelet Neuronal Interact 7:131–143

30 Vitamin D Deficiency and Its Health Consequences in New Zealand

Mark J. Bolland and Ian R. Reid

Abstract New Zealand is located in the southwestern Pacific Ocean and has warm summers with high UV exposure and some of the highest rates of skin cancers in the world. However, recent surveys have reported higher-than-expected rates of suboptimal vitamin D status. In healthy, community-dwelling, older women, 20–30% had serum 25-hydroxyvitamin D [25(OH)D] <50 nmol/L (<20 ng/ml) in summer and 50–75% in winter. In healthy, middle-aged and older men, the rates were much lower. In a national nutrition survey, 52% of adult women and 45% of adult men had serum 25(OH)D <50 nmol/l (<20 ng/ml). Similarly high rates of low vitamin D status were seen in children. In two cross-sectional studies, we found the major determinants of 25(OH)D in older men and women were surrogates of ultraviolet B exposure (skin pigmentation, month of blood sampling, exercise levels, age, and gender) and fat mass. Both seasonal variation of 25(OH)D and fat mass have a significant impact upon the diagnosis of vitamin D sufficiency. We found that summertime 25(OH)D levels approaching 70 nmol/l (28 ng/ml) in women and 90 nmol/l (36 ng/ml) in men were required to ensure year-round 25(OH)D > 50 nmol/l (>20 ng/ml). Therefore, clinicians should consider the month of sampling when interpreting the results of 25(OH)D measurements.

Key Words: New Zealand; 25-hydroxyvitamin D; sunlight, obesity; UV; vitamin D deficiency; season; summer; fat mass; skin pigment

1. INTRODUCTION

Studies performed in the Northern Hemisphere or in regions distant from the equator have suggested that there are significant seasonal variations in 25(OH)D levels, with corresponding seasonal fluctuations in PTH levels, markers of bone turnover, and BMD. Since sunlight exposure is the principal source of vitamin D, and this is affected by geographical location and lifestyle, studies in northern Europe and the United States are not likely to predict the prevalence of vitamin D deficiency in a subtropical climate, such as northern New Zealand. A number of cross-sectional studies have recently been performed which suggest that vitamin D status in New Zealand is suboptimal in many children and adults, despite warm summers with high summertime ultraviolet (UV) exposure.

From: *Nutrition and Health: Vitamin D*
Edited by: M.F. Holick, DOI 10.1007/978-1-60327-303-9_30,
© Springer Science+Business Media, LLC 2010

2. NEW ZEALAND LOCATION, CLIMATE, AND UV EXPOSURE

New Zealand is a country of 4 million people located in the southwestern Pacific Ocean. It comprises two larger islands ranging from approximately latitude 34–47°S and numerous other smaller, mainly uninhabited islands. In northern New Zealand, the climate is warm and subtropical, while further south the climate is cooler and temperate. Summertime UV levels over New Zealand are high – approximately 40% higher than levels in equivalent northern latitudes in Europe. These high UV levels are thought to result from lower atmospheric ozone concentrations over New Zealand because of the Antarctic ozone hole, from less polluted air over New Zealand, and because the sun is slightly closer to the Earth in December than in June *(1)*. As a consequence of these high UV levels and warm climate, New Zealand has among the highest rates of skin cancers in the world *(2)*.

3. PREVALENCE OF LOW 25(OH)D LEVELS

Recently, we carried out surveys of 25-hydroxyvitamin D (25(OH)D) levels in community-dwelling, healthy, older women, and in community-dwelling, healthy, middle-aged and older men living in Auckland, in northern New Zealand (latitude 37°S) *(3, 4)*. Participants were healthy volunteers recruited for studies of calcium supplementation and did not have any major medical conditions or take medications that might interfere with calcium metabolism. A total of 1,606 women aged >55 years (mean age 74 years) and 378 men aged >40 years (mean age 57 years) took part. 99.8% of the women and 92% of the men were of European descent. Serum 25(OH)D was measured by radioimmunoassay (DiaSorin, Stillwater, MN) in the women and in the first 252 men, then using a chemiluminescent assay (Nichols, San Juan Capistrano, CA) in the last 126 men. All data obtained using the Nichols assay were converted to predicted DiaSorin results using the derived equation DiaSorin = Nichols * 0.75 + 5.6 as previously described *(4)*.

Figure 1 shows the monthly variation in UV radiation and 25(OH)D. There were substantial circannual variations in UV dose with the highest levels in summer, as expected. The mean (SD) serum 25(OH)D in the women was 51 (19) nmol/l [20 (7.6) ng/ml] and in the men was 85 (31) nmol/l [34 (12) ng/ml]. The monthly mean 25(OH)D levels closely paralleled the seasonal fluctuation in UV levels for men and women, with a lag of 1–2 months, but, at each monthly time point, 25(OH)D levels were higher in men than in women.

Table 1 shows the prevalence of 25(OH)D <50 nmol/l (<20 ng/ml) and 25(OH)D <25 nmol/l (<10 ng/ml) by month of measurement. In the summer months (December–February), few men or women have 25(OH)D <25 nmol (<10 ng/ml), but 4–17% of men and 20–30% of women have 25(OH)D <50 nmol/l (<20 ng/ml). In winter, the prevalence of 25(OH)D <25 nmol/l (<10 ng/ml) and <50 nmol/l (<20 ng/ml) increased to 6–16% and 56–74%, respectively, for women, and 0–2% and 0–20%, respectively, for men.

Complementing these two surveys were data from a national nutrition survey *(5, 6)*. As part of this survey, 2,946 adults aged over 18 years had a measurement of 25(OH)D. The prevalence of 25(OH)D levels <17.5 nmol/l (<7 ng/ml), <50 nmol/l (<20 ng/ml),

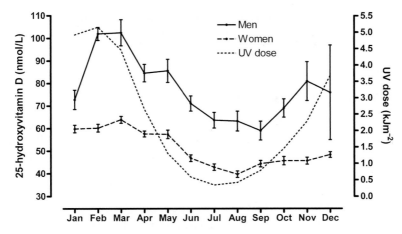

Fig. 1. Monthly variation of UV dose and 25(OH)D levels. UV dose is shown as the *dotted line*, 25(OH)D levels in men as the *solid line*, and 25(OH)D levels in women as the *dashed line*. Data are mean 25(OH)D levels for each month with error bars representing the SEM. To convert from nmol/l to ng/ml for 25(OH)D divide by 2.5. Reproduced from Bolland et al. *(4)* with kind permission from the American Society for Nutrition.

Table 1
Frequency of Low 25(OH)D Levels by Month of Sampling

Month	Women N	25(OH)D < 25 nmol/l (<10 ng/ml) (%)	25(OH)D < 50 nmol/l (<20 ng/ml) (%)	Men N	25(OH)D < 25 nmol/l (<10 ng/ml) (%)	25(OH)D < 50 nmol/l (<20 ng/ml) (%)
January	123	0.0	30.1	30	0.0	16.7
February	127	1.6	28.3	134	0.0	3.7
March	155	0.6	22.6	18	0.0	0.0
April	172	1.2	31.4	47	0.0	4.3
May	115	0.0	36.5	26	3.8	3.8
June	123	5.7	56.1	45	2.2	20.0
July	123	13.0	68.3	19	0.0	10.5
August	140	16.4	74.3	4	0.0	0.0
September	134	9.7	61.2	23	0.0	26.1
October	110	12.7	58.2	21	0.0	14.3
November	152	6.6	64.5	9	0.0	11.1
December	123	0.0	30.1	2	0.0	0.0

and <80 nmol/l (<30 ng/ml) were 4, 52, and 86%, respectively, in women and 2, 45, and 82%, respectively, in men *(6)*. In the corresponding survey of 1,585 children aged 5–14 years, the prevalence of 25(OH)D levels <17.5 nmol/l (<7 ng/ml) and <37.5 nmol/l (<15 ng/ml) were 4 and 36%, respectively, in girls, and 3 and 27%, respectively, in boys *(5)*.

These studies suggest that despite a warm and sunny climate with high UV exposure in the summer months, many adults and children in New Zealand have suboptimal vitamin D status.

4. DETERMINANTS OF 25(OH)D LEVELS

Next, we sought to define the major determinants of 25(OH)D levels in our two surveys and whether they differed between men and women. We used Pearson correlation analysis to identify variables with significant correlations with 25(OH)D (Table 2). For both men and women, correlations of similar strength and direction were observed between 25(OH)D and percentage body fat, fat mass, BMI, and physical activity. Lean mass was not correlated with 25(OH)D in either cohort. Differences between the cohorts occurred in the correlations between 25(OH)D and weight, and 25(OH)D and age. The inverse correlations between 25(OH)D and weight were of similar strength in both cohorts but the relationship was only significant in the much larger cohort of women. There was a significant negative correlation between 25(OH)D and age in the women but not in the men. The most likely explanation for this difference is age-related and gender-related differences in behavioral factors such as time spent outdoors, type and amount of physical activity, and other cultural factors such as concern regarding skin changes with aging, concerns about skin cancer, and use of sunscreen, rather than any gender-related biological differences.

On multivariate analysis, the month of blood sample, age, fat mass, and physical activity were significantly related to 25(OH)D in women (Table 3), while in men, the

Table 2
Pearson's Correlations Between 25(OH)D and Other Variables

	Men	*Women*
Age	−0.03	−0.15**
Height	0.11	0.04
Weight	−0.07	−0.11**
Body mass index	−0.14*	−0.14**
Lean mass	0.09	0.01
Fat mass	−0.21**	−0.15**
Percent fat	−0.23**	−0.15**
Dietary calcium	0.00	0.00
Physical activity	0.20**	0.09**
Gardening hours	0.14*	0.08**
Serum creatinine	0.07	0.02
Glucose	−0.06	−0.05*
Adjusted serum calcium	−0.13*	0.00
Alkaline phosphatase	−0.12*	−0.10**
Serum phosphate	0.03	−0.04
Albumin	0.08	0.04

* $P < 0.05$, ** $P < 0.01$.

Table 3
Predictors of 25(OH)D Levels in Multivariate Regression Models

Independent variable	Regression coefficient	95% CI	P	Partial r (2)
Model for women with 25(OH)D as dependent variable, for model P < 0.0001, r (2) = 0.21				
Month of blood sample			<0.0001	0.16
Age (years)	−0.72	−0.92 to −0.51	<0.0001	0.02
Physical activity (METS)	0.28	0.08−0.48	0.0056	0.01
Fat mass (kg)	−0.32	−0.42 to −0.23	<0.0001	0.03
Model for men with 25(OH)D as dependent variable, for model P < 0.0001, r (2) = 0.33				
Month of blood sample			<0.0001	0.25
Percent fat	−1.06	−1.52 to −0.60	<0.0001	0.05
Serum albumin	2.09	0.87−3.31	0.0009	0.02
Physical activity (METS)	0.92	0.31−1.53	0.0036	0.01
Model for combined datasets of men and women with 25(OH)D as dependent variable, for model P < 0.0001, r (2) = 0.42				
Gender (0 = male; 1 = female)	−13.3	−17.7 to −8.9	<0.0001	0.26
Month of blood sample			<0.0001	0.11
Percent fat	−0.43	−0.55 to −0.32	<0.0001	0.02
Age	−0.51	−0.67 to −0.35	<0.0001	0.02
Physical activity (METS)	0.39	0.19−0.58	0.0001	0.01
Serum albumin	0.53	0.13−0.94	0.01	0.002

month of the blood sample, percentage body fat, serum albumin, and physical activity were significantly related to 25(OH)D (Table 3). In the combined datasets of the men and women, gender, month of blood sampling, percentage body fat, age, physical activity, and serum albumin all were significantly related to 25(OH)D on multivariate analysis (Table 3). This six-variable model accounted for 42% of the variation in 25(OH)D levels.

There were few participants in these two surveys who were not of European descent (0.2% of women and 8% of men). In the men, the mean 25(OH)D level (SD) was higher in those of European descent, 87 (30) nmol/l [35 (12) ng/ml], compared with those of non-European descent, 66 (31) nmol/l [26 (12) ng/ml], $P < 0.001$. Similar findings were reported in the national nutrition surveys, in which people of Maori or Polynesian descent had mean 25(OH)D levels 10–20 nmol/l (4–8 ng/ml) lower than people of European descent *(5, 6)*.

Month of blood sampling, physical activity, and sunlight exposure are all indirect measures of exposure to UV-B. Similarly, the age and gender differences in 25(OH)D levels most likely reflect behavioral and cultural practices relating to sunlight exposure. The degree of skin pigmentation of an individual directly influences the amount of formation of vitamin D from solar UV-B. Taken together, these findings suggest that there are only two major biological determinants of 25(OH)D levels – UV-B exposure and fat mass – and that they do not differ between men and women. The relationship between 25(OH)D and fat mass is shown in Fig. 2.

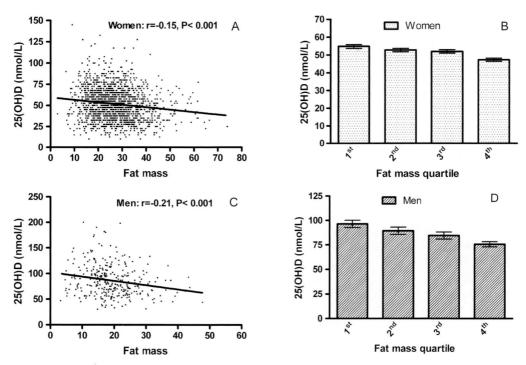

Fig. 2. The relationship between 25(OH)D and fat mass. Panel **A** shows a scatter plot of 25(OH)D versus fat mass with line of best fit in the women and panel **B** the mean 25(OH)D by quartile of fat mass in the women. Panel **C** shows a scatter plot of 25(OH)D versus fat mass with line of best fit in the men and panel **D** the mean 25(OH)D by quartile of fat mass in the men. The *error bars* represent the SEM. To convert from nmol/l to ng/ml for 25(OH)D divide by 2.5.

Individual exposure to UV-B radiation varies by season because the lower angle of incidence of incoming solar radiation during winter results in UV rays traveling a greater distance through the atmosphere, which in turn increases atmospheric absorption of UV radiation. Seasonal changes in cloud cover (with greater cloud cover in Auckland in winter) may also contribute to the increased atmospheric absorption of UV radiation.

In addition, exposure of the skin to UV-B is generally decreased during the colder winter months because more clothes are worn. The formation of vitamin D in the skin in response to UV-B radiation is affected by an individual's age, degree of skin pigmentation, and the intensity of sun exposure *(7)*. Seasonal changes in UV-B lead to seasonal changes in 25(OH)D with a lag of about 6–8 weeks between the maximum monthly UV dose and the peak 25(OH)D level. This lag period may represent the time taken to establish a new steady state since it corresponds to approximately 3–4 times the half-life of serum 25(OH)D.

The 25(OH)D levels are strongly related to measures of fat mass, moderately related to BMI or body weight, and not related to lean mass. The inverse relationship between 25(OH)D and fat mass has been attributed to increased sequestration of fat-soluble vitamin D in adipose tissue in obese individuals *(8, 9)*. Other possible explanations would be that overweight people have decreased exposure to sunlight because of their choice of clothing, or because of decreased exercise levels and mobility in heavier individuals. Variation of vitamin D-binding protein with body weight or fat mass does not appear to account for the relationship between 25(OH)D and fat mass *(10)*.

5. THE EFFECTS OF SEASONAL VARIATION OF 25(OH)D AND FAT MASS ON THE DIAGNOSIS OF VITAMIN D SUFFICIENCY

The serum level of 25(OH)D is considered to be the best estimate of body stores of vitamin D, but estimates of the level of serum 25(OH)D above which vitamin D stores are considered adequate vary widely, from 25 to 100 nmol/l (10–40 ng/ml) *(11)*. Since 25(OH)D levels vary by season, individuals could have 25(OH)D levels considered adequate in the summer and autumn months yet have suboptimal levels in winter and spring. Therefore, using the data from our two surveys, we set out to determine the effects of seasonal variation of 25(OH)D on a preselected threshold level for diagnosis of vitamin D sufficiency [50 nmol/l (20 ng/ml)] and whether fat mass or body weight modify these effects *(12)*.

25(OH)D levels were plotted against the day of the year the blood sample was taken in each of the cohorts and a sine curve fitted. It was assumed that 25(OH)D levels throughout the year for each participant would follow a similar sine curve to the population. Therefore the sine curve for any individual would be the sine curve for the population translated along the y-axis until it intersected with the known 25(OH)D level on the known day of the year for that individual. By solving the equation for the sine curve for each individual, we were able to predict the peak 25(OH)D level, the trough 25(OH)D level, and the number of days during which the 25(OH)D level was <50 nmol/l (<20 ng/ml). The equation for each sine curve was as follows: 25(OH)D = baseline + amplitude × sine (angular frequency × day of year + phase shift). The amplitude of the sine curve is the maximal deviation from the baseline [(peak value – trough value)/2]; the angular frequency is $2 \times \pi$/period ($2 \times \pi$/365); and the phase shift is the amount of translation along the x-axis.

Figure 3 shows the sine curve fitted for the cohort of women. The equation for the fitted sine curve is $y = 50.99 + 10.67 \times$ sine (frequency × x + 0.41). Figure 4 shows the sine curve fitted for the cohort of men. The equation for the fitted sine curve is

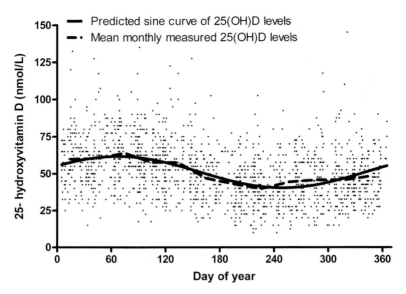

Fig. 3. Sine curve of best fit for 25(OH)D in the cohort of women (N = 1,606) with measured mean monthly 25(OH)D for comparison. To convert from nmol/l to ng/ml for 25(OH)D divide by 2.5.

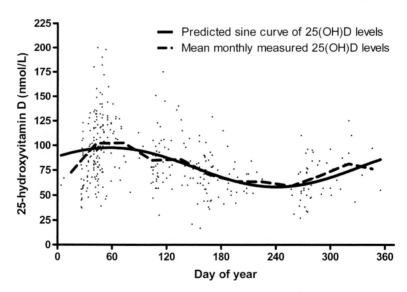

Fig. 4. Sine curve of best fit for 25(OH)D in the cohort of men ($N = 378$) with measured mean monthly 25(OH)D for comparison. To convert from nmol/l to ng/ml for 25(OH)D divide by 2.5.

$y = 77.89 + 19.67 \times$ sine (frequency $\times$ x + 0.57). There was excellent agreement between the fitted sine curves and the mean monthly 25(OH)D levels in both cohorts.

From the predicted sine curves, 73% of the women had a trough 25(OH)D <50 nmol/l (<20 ng/ml), whereas the observed prevalence of 25(OH)D <50 nmol/l (<20 ng/ml) was 49%, ranging from 23 to 74% depending on the month of blood sampling. Thirty-nine percent of the men had a predicted trough 25(OH)D <50 nmol/l (<20 ng/ml), in comparison to an observed prevalence of 25(OH)D <50 nmol/l (<20 ng/ml) of 9% (range 0–26%). Of the 73% of women predicted to have a trough 25(OH)D <50 nmol/l (<20 ng/ml), 28% were predicted to have 25(OH)D <50 nmol/l (<20 ng/ml) for the entire year, and the mean (SD) number of days with predicted 25(OH)D <50 nmol/l (<20 ng/ml) was 250 (108). Of the 39% of men predicted to have trough 25(OH)D <50 nmol/l (<20 ng/ml), 3% were predicted to have 25(OH)D <50 nmol/l (<20 ng/ml) for the whole year, and the mean (SD) number of days with predicted 25(OH)D <50 nmol/l (<20 ng/ml) was 165 (89). Table 4, shows the minimum 25(OH)D level for each month required to ensure that the 25(OH)D level was maintained >50 nmol/l (>20 ng/ml) throughout the year. During summer (December–March), this value approached 90 nmol/l (36 ng/ml) in men and 70 nmol/l (28 ng/ml) in women. The data in Table 2 can be readily applied to other thresholds for vitamin D sufficiency by adding the difference between any selected level for vitamin D sufficiency and 50 nmol/l (20 ng/ml) to the values in Table 2. For example, to maintain 25(OH)D levels >80 nmol/l (>30 ng/ml) year-round, men would need levels >100–120 nmol/l (>40–48 ng/ml) and women >90–100 nmol/l (>36–40 ng/ml) in the summer months.

Table 4
The Minimum 25(OH)D Level Required to Have a Predicted Trough
25(OH)D > 50 nmol/l (>20 ng/ml), by Month of Measurement

Month of year	Men		Women	
	nmol/l	ng/ml	nmol/l	ng/ml
January	81	32	65	26
February	87	35	69	28
March	87	35	71	28
April	79	32	67	27
May	69	28	62	25
June	59	24	57	23
July	52	21	52	21
August	50	20	50	20
September	50	20	50	20
October	53	21	51	20
November	61	24	54	22
December	71	28	60	24

Thus, seasonal variation in 25(OH)D levels does have a substantial impact upon the diagnosis of vitamin D insufficiency. Many people are predicted to have suboptimal 25(OH)D levels for a substantial proportion of the year despite having apparently adequate levels at the time of testing. Most commonly recommended thresholds for vitamin D sufficiency are between 50 and 80 nmol/l (20–30 ng/ml) (11), but the need to consider seasonal variation when interpreting these thresholds has not been widely discussed. Although not stated explicitly, it can be inferred that the definition of vitamin D sufficiency refers to the lowest 25(OH)D level during the year. Thus, in locations where seasonal variation of 25(OH)D occurs, thresholds for diagnosis of vitamin D sufficiency

Table 5
The Effect of Fat Mass on Seasonal Variation of 25-hydroxyvitamin D Levels

Quartile	First	Second	Third	Fourth	P_{gender}	$P_{quartile}$	$P_{gender \times quartile}$
Fat mass (kg)							
Men	10.4 (3.0)	16.3 (1.6)	20.7 (1.6)	29.3 (5.5)			
Women	16.0 (3.3)	23.2 (1.6)	29.1 (1.8)	39.6 (7.0)			
Peak 25(OH)D (nmol/l)							
Men	111 (102–120)	101 (91–110)	100 (91–110)	86 (78–93)	<0.01	<0.01	0.13
Women	67 (64–70)	63 (60–66)	61 (59–64)	55 (52–58)			
Trough 25(OH)D (nmol/l)							
Men	62 (46–78)	61 (44–78)	56 (42–70)	58 (47–69)	<0.01	0.72	0.75
Women	41 (37–44)	42 (39–45)	42 (39–45)	40 (37–43)			
Amplitude (nmol/l)							
Men	24 (14–35)	20 (9–31)	22 (13–32)	14 (6–22)	<0.01	<0.01	0.10
Women	13 (11–16)	10 (8–12)	10 (8–12)	7 (5–10)			

In each quartile there were 77 men or 374 women. Data are mean (95% confidence interval) except for fat mass which is mean (standard deviation). P are for the main effects in the models of peak 25-hydroxyvitamin D [25(OH)D], trough 25(OH)D], or amplitude versus gender, fat mass quartile, and gender × quartile using analysis of variance. To convert from nmol/l to ng/ml for 25(OH)D divide by 2.5. Reproduced from Bolland et al. (12) with kind permission from Springer Science + Business Media.

will also vary by season. These thresholds need to be individualized to the latitude and climate of a location because the amount of seasonal variation of 25(OH)D is likely to be determined by these factors, as discussed earlier.

To determine the effect of fat mass on thresholds for diagnosis of vitamin D sufficiency, we divided the cohorts into quartiles of fat mass and fitted a separate sine curve for each quartile of fat mass. The peak 25(OH)D level and amplitude of the sine curves decreased with increasing fat mass while the trough 25(OH)D levels were similar (Table 5). Thus, subjects in the highest quartile of fat mass had smaller seasonal excursions in 25(OH)D levels and lower peak 25(OH)D levels than subjects in the lowest quartile of fat mass (Fig. 5). In women, each 1 kg difference in fat mass was associated with the following changes: peak 25(OH)D of –0.52 nmol/l (–0.21 ng/ml), trough 25(OH)D of –0.05 nmol/l (–0.02 ng/ml), and amplitude of –0.23 nmol/l (–0.09 ng/ml). In men, each 1 kg difference in fat mass was associated with a change in peak 25(OH)D of –1.3 nmol/l (–0.52 ng/ml), trough 25(OH)D of –0.27 nmol/l (–0.11 ng/ml), and amplitude of –0.51 nmol/l (–0.20 ng/ml). Fat mass differences between genders could not fully explain the gender differences in the amount of seasonal variation of 25(OH)D.

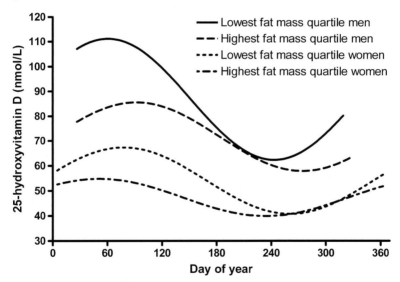

Fig. 5. The effect of fat mass on seasonal variation of 25(OH)D levels in men and women. Data are shown from the lowest and highest quartiles only for clarity. To convert from nmol/l to ng/ml for 25(OH)D divide by 2.5.

These findings that greater fat mass was associated with lower peak 25(OH)D levels and smaller seasonal variation of 25(OH)D, but no change in trough 25(OH)D levels, are consistent with an effect of both reduced sunlight exposure in heavier individuals and adipose tissue acting as a reservoir of vitamin D and its metabolites. Regardless of fat mass, reduced exposure to sunlight will lead to lower peak 25(OH)D levels. If adipocytes act as a reservoir of vitamin D and its metabolites and buffer against both higher 25(OH)D levels in summer and lower levels in winter, then increased adiposity would be associated with reduced peak 25(OH)D levels, reduced seasonal variation, and higher-than-expected trough 25(OH)D levels. A combination of these two hypotheses, both reduced sunlight exposure and an increased reservoir of vitamin D and its metabo-lites, would therefore explain our findings.

Fat mass does modify the effect of seasonal variation on thresholds for diagnosis of vitamin D sufficiency. People in the highest quartile of fat mass have lower peak 25(OH)D levels and smaller amounts of seasonal variation of 25(OH)D levels in com-parison to people in the lowest quartile of fat mass, while trough 25(OH)D levels are similar across fat mass quartiles. Thus, seasonally adjusted thresholds for diagnosis of vitamin D sufficiency decrease with increasing fat mass and increase with decreasing fat mass. However, this effect is small and may only be clinically relevant for people whose body weight or fat mass is greater than one standard deviation from the mean and who have 25(OH)D measurements taken in summer.

6. CONCLUSION

Despite a warm and sunny climate with high summertime UV levels, New Zealand has a high prevalence of low 25(OH)D levels in adults and children, particularly in the winter months. Healthy middle-aged and older men had higher 25(OH)D levels than healthy, older women, but the major determinants of 25(OH)D levels (surrogates of UV-B exposure and fat mass) did not differ between men and women. Because of seasonal variation of 25(OH)D, thresholds for diagnosis of vitamin D sufficiency also vary by season and need to be individualized for different climates and latitudes. Clinicians who interpret 25(OH)D levels should therefore take into account the season of the year when determining whether a patient is at risk of vitamin D insufficiency throughout the year. In New Zealand, to ensure year-round 25(OH)D levels >50 nmol/l (>20 ng/ml), sum-mertime 25(OH)D levels should exceed 70 nmol/l (28 ng/ml) in women and 90 nmol/l in men (36 ng/ml). Overall, our findings suggest that vitamin D supplementation, par-ticularly during winter, should become standard practice for elderly women. In contrast, routine vitamin D supplementation for healthy, older men is not warranted.

REFERENCES

1. Bodeker GE (2003) The WMO/UNEP 2002 ozone assessment:a New Zealand perspective. Water Atmos 11:1–4
2. Jones WO, Harman CR, Ng AK et al (1999) Incidence of malignant melanoma in Auckland, New Zealand: highest rates in the world. World J Surg 23:732–735
3. Lucas JA, Bolland MJ, Grey AB et al (2005) Determinants of vitamin D status in older women living in a subtropical climate. Osteoporos Int 16:1641–1648
4. Bolland MJ, Grey AB, Ames RW et al (2006) Determinants of vitamin D status in older men living in a subtropical climate. Osteoporos Int 17:1742–1748
5. Rockell JE, Green TJ, Skeaff CM et al (2005) Season and ethnicity are determinants of serum 25-hydroxyvitamin D concentrations in New Zealand children aged 5–14 year. J Nutr 135:2602–2608
6. Rockell JE, Skeaff CM, Williams SM et al (2006) Serum 25-hydroxyvitamin D concentrations of New Zealanders aged 15 years and older. Osteoporos Int 17:1382–1389
7. Lips P (2001) Vitamin D deficiency and secondary hyperparathyroidism in the elderly: consequences for bone loss and fractures and therapeutic implications. Endocr Rev 22:477–501
8. Need AG, Morris HA, Horowitz M et al (1993) Effects of skin thickness, age, body fat, and sunlight on serum 25-hydroxyvitamin D. Am J Clin Nutr 58:882–885

9. Wortsman J, Matsuoka LY, Chen TC et al (2000) Decreased bioavailability of vitamin D in obesity. Am J Clin Nutr 72:690–693

10. Bolland MJ, Grey AB, Ames RW et al (2007) Age-, gender-, and weight-related effects on levels of 25-hydroxyvitamin D are not mediated by vitamin D binding protein. Clin Endocrinol (Oxf) 67: 259–264

11. Dawson-Hughes B, Heaney RP, Holick MF et al (2005) Estimates of optimal vitamin D status. Osteoporos Int 16:713–716

12. Bolland MJ, Grey AB, Ames RW et al (2007) The effects of seasonal variation of 25-hydroxyvitamin D and fat mass on a diagnosis of vitamin D sufficiency. Am J Clin Nutr 86:959–964

31 Toxicity of Vitamin D

Reinhold Vieth

Abstract   There remains great concern by the medical community about the potential toxicity of vitamin D. Vitamin D intoxication is associated with hypercalciuria, hypercalcemia, and hyperphosphatemia which causes soft tissue calcification of the kidneys and blood vessels and increases the risk of kidney stones. The hallmark for vitamin D intoxication is a markedly elevated blood level of 25-hydroxyvitamin D that is usually greater than 200 ng/mL. The Institute of Medicine's Food and Nutrition Board proposed a tolerable upper level (UL) to be 2000 IU/day for vitamin D. However, dose ranging studies in healthy adults revealed that 10,000 IU/day of vitamin D for 5 months did not produce adverse changes in calcium metabolism. In a separate study, up to 40,000 IU (1 mg/d) vitamin D for 1 month did not affect calcium metabolism; however, urinary calcium and eventually plasma calcium will increase if such intake continues for several months. Therefore, it would be reasonable to increase the UL for vitamin D to 10,000 IU/day. This approximates the production of vitamin D with full-body skin exposure to sunlight and for which there is no concern about making too much of vitamin D from excessive exposure to sunlight. Therefore, vitamin D is safe and has a wide therapeutic index.

   Key Words: Vitamin D; vitamin D intoxication; toxicity; tanning; sunlight; 25-hydroxyvitamin D

Sunshine exposure of the full skin surface of an adult is thought to generate vitamin D in an amount equivalent to an oral dose of approximately 250 mcg/day (10,000 IU/day) of vitamin D3. Since humans evolved to acquire vitamin D through sunshine, it makes sense that 10,000 IU/day of intake is not generally regarded as harmful. However, like anything with a potent effect on the human body, vitamin D can be harmful if it is taken in excessive amounts (Fig. 1). The classic form of vitamin D toxicity is due to the hypercalcemia that results from excessive absorption of calcium from the gut, and excessive resorption of calcium from bone.

Vitamin D is a permitted component of rodenticides, where several milligrams taken at once by small animals cause death due to dehydration (1, 2). In combination with nicotine, extremely high vitamin D doses (several hundred thousand IUs per kg) are used to create an animal model of arteriosclerosis, a hardening of the arteries due to vascular calcification (3, 4).

From: *Nutrition and Health: Vitamin D*
Edited by: M.F. Holick, DOI 10.1007/978-1-60327-303-9_31,
© Springer Science+Business Media, LLC 2010

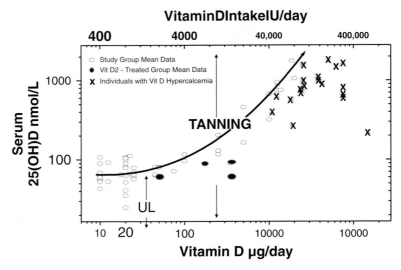

Fig. 1. Dose–response relationship between reported vitamin D consumption and serum 25(OH)D concentrations for groups of subjects (*circles*) and for case reports of vitamin D toxicity, as evidenced by hypercalcemia and the serum 25(OH)D concentration (X). *Open circles* represent mean values for groups dosed with vitamin D3, *black circles* represent mean values for groups dosed with vitamin D2. UL, upper level for vitamin D, according to Food and Nutrition Board. (Adapted from data in *(9, 53)*.)

1. SIGNS AND SYMPTOMS

The clinical signs and symptoms are the consequence of the hypercalcemia and the resulting dehydration (Table 1). Hypercalcemia impedes the action of antidiuretic hormone, causing diuresis and dehydration, and acute renal failure due to the lack of fluid. In one case, a patient was initially sent home from the emergency department because the symptoms were presumed to be due to gastroenteritis *(5)*.

2. TREATMENT

In severe cases, fluid replacement is essential to correct the dehydration and restore kidney function. Treatment of the hypercalcemia can involve removal of sources of dietary calcium, administration of glucocorticoid which appears to suppress 25(OH)D-1-hydroxylase, and use of bisphosphonate or calcitonin to impede bone resorption *(6, 7)*. After discontinuation of vitamin D, the decline in serum 25(OH)D proceeds with a half-life of about 2 months *(5)*, consistent with the long-established, in vivo half-life of radioactive vitamin D in humans *(8)*.

The therapeutic index for vitamin D – the ratio between the physiologically effective dose and the toxic level – is in the order of at least fivefold. This is similar to the safety margin of many other nutrients (including even water). Perhaps vitamin D has appeared to be toxic in the past it is potent in milligram amounts that far exceed physiologic levels of input. In contrast, milligram amounts of other nutrients are benign.

Table 1
Signs and Symptoms of Vitamin D Toxicity

Symptoms

- Extreme right-sided flank pain (due to kidney stone)
- Xonjunctivitis[a]
- Increased thirst[a]
- Vomiting
- In acute renal failure[a]
- Anorexia
- Fever, chills
- Constipation[a]
- Hporeflexia[a]

Laboratory evidence
- Hypercalcemia
- Hypercalciuria
- Secondary hypoparathyroidism
- Serum 25(OH)D well beyond 120 ng/ml (>280 nmol/l)

[a]A consequence of dehydration due to hypercalcemia.

3. MECHANISMS OF TOXICITY

The toxic limit for vitamin D is based largely upon saturation of circulating vitamin D-binding protein (DBP) by high levels of metabolites achievable only with oral intake. Excessive concentrations of vitamin D metabolites can exceed the capacity of vitamin D-binding protein, and displace the active metabolite, $1,25(OH)_2D$ from DBP, thus making active metabolite more accessible to tissues *(7, 9)*. At toxic doses, the freely circulating vitamin D, along with its metabolites accumulate in the usual storage tissues for vitamin D: adipose *(10)* and muscle *(11)*. The binding capacity of human vitamin D-binding protein is approximately 4,700 nmol/l *(12)*, and this is more than 20 times the physiologic total concentration of vitamin D metabolites normally present in the circulation.

Most cases of vitamin D toxicity have involved vitamin D2 *(6, 9)*, this may be due to the nature of vitamin D2, or because it is the most common prescription form. Toxicity with vitamin D3 has been less common, and cases have tended to be industrial accidents *(7, 13, 14)* or poisonings from an unknown source *(5)*. Blood levels of vitamin D and its metabolites measured by HPLC indicated that despite 25(OH)D concentrations ranging up to 2,400 nmol/l, there can be a large excess of unmetabolized vitamin D3 present (17,000 nmol/l), suggesting that the capacity of liver to hydroxylate vitamin D is limited *(5)*.

Vitamin D and its initial metabolite, 25(OH)D, are thought to be inactive within cells. It is very likely that the signaling molecule causing harm due to vitamin D is the active molecule $1,25(OH)_2D$. Ironically, circulating concentrations of $1,25(OH)_2D$ are not increased much by vitamin D intoxication. The unexpectedly low concentration in the serum of $1,25(OH)_2D$ in patients intoxicated with vitamin D reflects the high level of regulation of this hormone via both its synthesis and catabolism. Nonetheless, regulation of the $1,25(OH)_2D$ hormone is not capable of adequate suppression synthesis of circulating $1,25(OH)_2D$ when the substrate concentration is excessive. The circulating $1,25(OH)_2D$ hormone displaced from the vitamin D-binding protein (DBP) by the vast excess of other vitamin D metabolites that have higher affinity for the binding protein than does $1,25(OH)_2D$ *(15)*. This has been demonstrated by Pettifor and Bikle, who showed that patients intoxicated with vitamin D exhibited elevated "free" $1,25(OH)_2D$ concentrations in vitamin D, despite normal are only moderately elevated levels of total $1,25(OH)_2D$ *(7)*. It is also likely that at extremely high concentrations, $25(OH)D$ can bind nuclear vitamin D receptor to trigger a response as part of the mechanisms for toxicity *(16)*.

Another factor that makes vitamin D potentially toxic is the unique substrate relationship with its enzymes. In contrast to other parts of the endocrine system that involve enzymes not limited by substrate, the vitamin D system functions in an unsaturated state. In essence, the enzymes of the vitamin D system are functioning below their K_m values in vivo *(17)*. Until the enzymes can adapt, a doubling of the vitamin D input results in a doubling in the production $1,25(OH)_2D$. Adaptation to a higher $25(OH)D$ concentration involves both downregulation of the 1-hydroxylase enzyme, as well as upregulation of the catabolic pathway, via 24-hydroxylase *(17, 18)*. Another of the components of vitamin D toxicity is that the input of $25(OH)D$ exceeds the range within which its 1-hydroxylase and 24-hydroxylase can compensate.

3.1. Hypersensitivity to Vitamin D

Situations in which $25(OH)D$-1-hydroxylase cannot be properly downregulated result in apparent hypersensitivity to vitamin D. If supplementation or treatment with moderate doses of vitamin D (less than 40,000 IU/day) results in hypercalcemia, then there is aberrant control of $25(OH)D$-1-hydroxylase. This could occur secondary to primary hyperparathyroidism, where PTH continuously stimulates the enzyme in the kidney *(6, 19)*. Unregulated 1-hydroxylase can also occur as part of a granulomatous disease, where peripheral tissue loses ability to regulate the 1-hydroxylase that normally serves a autocrine/paracrine role *(9, 20)*.

In primary hyperparathyroidism, PTH both suppresses the breakdown enzyme, 24-hydroxylase, and it stimulates $25(OH)D$-1-hydroxylase. Together, both these effects raise the $1,25(OH)_2D$ level. Furthermore, this is an unusual form of vitamin D toxicity in that serum $25(OH)D$ declines more slowly than normal because of diminished catabolic enzyme activity *(6)*.

4. MANDATED SAFETY LEVELS

According to the Food and Nutrition Board, the "upper level" (UL) for vitamin D intake is 2,000 IU/day. This is the amount of vitamin D that adults in the general public can consume on a long-term basis with reasonable assurance that there will be no harm *(21)*. Europe has likewise also adopted the same upper level for vitamin D *(22)*. In the United Kingdom, the terminology is slightly different, but the "guidance level" limiting what might be regarded as to save intake is specified as 1,000 IU/day *(23)*. When considered on a per-kilogram-body-weight basis, these doses are remarkably conservative for adults, because they are a fraction of the per-kilogram dose of vitamin D recommended to prevent rickets in infants (24, 25). In the context of vitamin D safety, the reference to 2,000 IU/day has remained unchanged since the number was first mentioned in the 1968 Recommended Dietary Allowance book where the dose was implicated in infantile hypercalcemia (26).

The "no observed adverse effect level," is a formal term of the Food and Nutrition Board that specifies the highest dose shown to have no harmful effect. In a 1984 paper by Narang et al., 2,400 IU/day was the dose of vitamin D that statistically increased serum calcium, but not quite into the hypercalcemia range (27). The Food and Nutrition Board applied an uncertainty factor of 1.2 to return to a value for the UL that matched the older specification of 2,000 IU/day for vitamin D. When subsequent data were published indicating that 4,000 international unit/day was safe and failed to affect serum calcium, the safety margin was increased by The European Commission to a value of 2.0 so that again, the UL could remain at 2,000 IU/day (22). Eight weeks of supplementation with 275 mcg (12,500 IU)/day of vitamin D does not affect circulating calcium concentration (28). More recently, Kimball et al. reported a dose-escalation study design intended to define tolerability to vitamin D, i.e., to find the dose of vitamin D that would raise urine and/or serum calcium levels. However, doses ranging up to 40,000 IU/day for 1 month did not affect either of these calcium measures (29).

That is, the 40,000 IU/day dose is noncalcemic and safe by the criterion applied to drug studies of vitamin D analogs (30–32). Of course, there is no reason to advocate such a large daily dose of vitamin D; the point is simply that there is no strong evidence for harm through consumption of vitamin D up to that level.

With abundant sun exposure, the solar equivalent dose of vitamin D is easily equivalent to 100 mcg (4,000 IU)/day (33). With an additional oral dosage of 100 mcg/day of vitamin D, this would still be less than the doses of vitamin D shown to be safe in recent publications (29, 34). Since a long-term vitamin D consumption of at least 40,000 IU/day would be needed to cause hypercalcemia, there is a large margin of safety with 100 mcg (4,000 IU)/day.

The weight of published evidence on toxicity shows that the lowest dose of vitamin D proven based on serum 25(OH)D to cause hypercalcemia in some healthy adults is at least 1,000 mcg (40,000 IU)/day of the vitamin D2 form (9). This translates to 1,000 μg or 1 mg taken daily for several months. Hypercalciuria is a more sensitive indicator of vitamin D excess than is serum calcium (35). . Vitamin D excess will cause excessive absorption of calcium from the gut. Normally since adults are in equilibrium with calcium, net absorption of calcium from the intestine is compensated for by excretion of calcium into the urine – hence, hypercalciuria. It is only once the renal calcium-excretion capacity can no longer balance the input of calcium from the gut that serum calcium will increase.

In the early 1990s, a dairy in the Boston area, servicing 10,000 households, made prolonged, gross errors in fortifying milk with several milligrams per quart (13, 36). There were two deaths attributed to the incident. While hypercalcemia did occur, it was rare, with greatest susceptibility in people with diminished renal function. The average 25(OH)D concentration in serum of the confirmed cases of vitamin D toxicity was

560 nmol/l (214 ng/ml) *(13)*; in comparison, physiologically attained 25(OH)D concentrations, obtained through sunshine exposure, can reach 235 nmol/l safely, without hypercalcemia or hypercalciuria *(9, 37)*.

One concern sometimes expressed is that if adipose tissue were to break down, a sudden influx of vitamin D from adipose might be toxic *(38)*. In both rats and cattle, high doses of vitamin D are needed before vitamin D ends up as detectable in adipose tissue *(10, 11)*. Despite "significant" amounts of vitamin D in tissues, storage is not efficient. The proportion of vitamin D that enters the body via the skin or the diet and ends up being stored in adipose is a fraction of the total. In normal humans, adipose tissue content of vitamin D has been reported to be as high as 116 ng/g (approx. 5 IU/g adipose) *(39)*. In cattle intoxicated with 7.5 million IU vitamin D (to cause hypercalcemia, in an experimental process to activate proteases to make beef more tender after slaughter), muscle levels of vitamin D reached 91 ng/g tissue (4 IU/g). The highest tissue level reported in vitamin D-dosed cattle was in the liver, which contained vitamin D at 979 ng/g (39 IU/g) *(11)*. The point is that while there is "meaningful" storage of vitamin D in tissues, all the evidence indicates that only a fraction of any vitamin D dose ends up in tissues to be withdrawn at later times.

If there were a sudden breakdown of 1 kg of adipose tissue, or liver, this would release as much as 979 mcg (39,000 IU) of vitamin D into the body. A toxic excess of vitamin D would require the break down daily of 1 kg of adipose tissue that had been primed by prior vitamin D intoxication, with daily adipose catabolism to continue for several weeks. When toxic doses of vitamin D are administered, the effect will be manifest during the period of administration. There is no evidence that enough residual vitamin D can be stored in adipose tissue that vitamin D toxicity could possibly arise at some later time, because of weight loss. An interesting aside is that carnivorous animals such as dogs and cats synthesize cholesterol in their skin via a root other than 7-dehydrocholesterol. Dogs and cats do not produce vitamin D when exposed to sunshine. Instead, these animals require the meat of other animals as their natural source of vitamin D *(40)*.

5. THEORETICAL NONCALCEMIC RISKS OF VITAMIN D

It is of course impossible to prove the safety of anything. For vitamin D, the classic form of toxicity is easily detected, based upon serum and urine calcium levels. Even in the absence of the classic signs of vitamin D toxicity, unknown risks continue to linger. Some epidemiologic findings have suggested that higher serum 25(OH)D concentrations may be associated with higher risk of pancreatic cancer *(41)* and prostate cancer *(42)*. Such relationships have been the exception, because even for the cancers mentioned, most studies find benefit due to UV exposure or vitamin D *(43–45)*. The most distinctive feature common to the populations studied in the reports of Stolzenberg-Solomon and Tuohimaa are the extreme latitudes of Norway, Finland, and Sweden from which the study populations emanate. These countries are interesting, because despite their northern latitudes, serum 25(OH)D concentrations reported there are often higher than those of the more southerly countries in Europe *(46)*. Many forms of cancers, as well as multiple sclerosis and other conditions are more common with increasing latitude

(44, 47). Thus, it is possible that the effect of higher latitude on disease rates is not due to a simple diminution of population levels in serum 25(OH)D. Behavior plays a role in this, and people who live in the north tend to maximize their exposure to sunshine when it is available, whereas populations farther south often practice cultures that minimize sun exposure and maximize skin coverage, keeping 25(OH)D concentrations lower than would be expected based on latitude alone *(48, 49)*.

One mechanism that could explain why higher 25(OH)D concentrations in the most northerly countries may be associated with increased risk of some diseases is based on the unusual enzymology underlying the vitamin D endocrine system. Northern latitudes experience severe fluctuations in UVB light intensity, and progressively longer "vitamin D winter." If human populations originated in tropical climes, then surely UVB intensity would be steady throughout the year, with minimal fluctuations in serum 25(OH)D. This contrasts with high latitudes, where serum 25(OH)D concentrations rise and fall in annual cycles. This perpetual cycling in the serum 25(OH)D concentrations may itself be detrimental, because it forces the enzymes regulating $1,25(OH)_2D$ production to be perpetually in a non-steady state. In particular, it is during the declining phase of serum 25(OH)D that there would be a relative excess in the activity of the catabolic enzyme, 24-hydroxylase. This would result in a prolonged period during which $1,25(OH)_2D$ concentrations may not quite maintain optimal concentrations in peripheral tissues, where the hormone plays a paracrine role, regulating cellular behavior *(50)*. This hypothesis is partly supported by the evidence that in men being monitored for progressively rising levels of prostate-specific antigen (PSA), there is a summertime slowdown in the rate of rise of this tumor marker, the rate of rise in PSA increasing again toward autumn, despite similar average 25(OH)D concentrations *(51)*. Based on these considerations, it is theoretically possible that if vitamin D administration involves prolonged intermittent gaps between each dose, this could result in cycles of inappropriately higher tissue 24-hydroxylase activities. With a relative excess in this catabolic enzyme, there would be lower tissue $1,25(OH)_2D$ concentrations, and potentially greater risk of disease despite vitamin D treatment. While this is not proven, this mechanism might explain why, under certain circumstances there could be risk with higher doses of vitamin D that do not affect serum or urine calcium. This hypothesis could be tested in animal studies. The theory suggests that in clinical work, responses to vitamin D administration may vary with the dosage regimen. Until the mechanisms underlying the reports of Stolzenberg-Solomon et al. and of Tuohimaa et al *(41, 42)* are better understood, it would be prudent to approach with care any regimens involving high dosages given at intervals longer than the 2-month half-life of vitamin D in the body.

6. CONCLUSION

In the context of the amounts of vitamin D potentially available through unfortified foods, vitamin D is extremely safe, because the amounts generally present in foods are modest compared to even the amount of vitamin D derived from exposure of skin to sunshine. Exposure of human skin surface to sunshine is safe by definition, because such exposure is what human biology was designed for during our evolution. There is no known report of vitamin D toxicity due to sun exposure, because vitamin D production in the skin is self-limiting *(52)*. In the context of drugs, vitamin D is very safe, because it has a wide therapeutic index. Nonetheless, there are well-understood limits to the ability of the body to handle amounts of vitamin D that exceed by several fold the physiologic range of vitamin D. In relatively pure form, multiple milligrams of vitamin D can be taken easily, and if taken regularly they exceed the physiologic limit for vitamin D that is approximately equivalent to 10,000 IU daily (0.25 mg/day) *(53)*. It is important to mention that because of its long half-life, the "daily" dose of vitamin D is best thought of as an average rate of intake. The design of a dosing interval for a drug is in the context of the half-life *(54)*, hence, an occasional larger dose of vitamin D beyond 10,000 IU may be appropriate if it combines up to a month's worth of doses into one *(55, 56)*. The harm with vitamin D could come from the long-term, regular intake of amounts with an average-per-day intake beyond 40,000 IU.

REFERENCES

1. Murphy MJ (2002) Rodenticides. Vet Clin North Am Small Anim Pract 32:469–484, viii
2. Scheftel J, Setzer S, Walser M, Pertile T, Hegstad RL, Felice LJ, Murphy MJ (1991) Elevated 25-hydroxy and normal 1,25-dihydroxy cholecalciferol serum concentrations in a successfully-treated case of vitamin D3 toxicosis in a dog. Vet Hum Toxicol 33:345–348
3. Atkinson J, Poitevin P, Chillon J-M, Lartaud I, Levy B (1994) Vascular Ca overload produced by vitamin D3 plus nicotine diminishes arterial distensibility in rats. Am J Physiol 266:H540–H547
4. Marque V, Van EH, Struijker-Boudier HA, Atkinson J, Lartaud-Idjouadiene I (2001) Determination of aortic elastic modulus by pulse wave velocity and wall tracking in a rat model of aortic stiffness. J Vasc Res 38:546–550
5. Vieth R, Pinto T, Reen BS, Wong MM (2002) Vitamin D poisoning by table sugar. Lancet 359:672
6. Taskapan H, Vieth R, Oreopoulos DG (2008) Unusually prolonged vitamin D intoxication after discontinuation of vitamin D: possible role of primary hyperparathyroidism. Int Urol Nephrol 40:801–805
7. Pettifor JM, Bikle DD, Cavaleros M, Zachen D, Kamdar MC, Ross FP (1995) Serum levels of free 1,25-dihydroxyvitamin D in vitamin D toxicity. Ann Intern Med 122:511–513
8. Mawer EB, Backhouse J, Holman CA, Lumb GA, Stanbury SW (1972) The distribution and storage of vitamin D and its metabolites in human tissues. Clin Sci 43:413–431
9. Vieth R (1999) Vitamin D supplementation, 25-hydroxyvitamin D concentrations, and safety. Am J Clin Nutr 69:842–856
10. Brouwer DA, van Beek J, Ferwerda H, Brugman AM, van der Klis FR, van der Heiden HJ, Muskiet FA (1998) Rat adipose tissue rapidly accumulates and slowly releases an orally-administered high vitamin D dose. Br J Nutr 79:527–532
11. Montgomery JL, Parrish FC Jr., Beitz DC, Horst RL, Huff-Lonergan EJ, Trenkle AH (2000) The use of vitamin D3 to improve beef tenderness. J Anim Sci 78:2615–2621
12. Vieth R (1994) Simple method for determining specific binding capacity of vitamin D-binding protein and its use to calculate the concentration of "free" 1,25-dihydroxyvitamin D. Clin Chem 40:435–441
13. Blank S, Scanlon KS, Sinks TH, Lett S, Falk H (1995) An outbreak of hypervitaminosis D associated with the over fortification of milk from a home-delivery dairy. Am J Public Health 85:656–659
14. Koutkia P, Chen TC, Holick MF (2001) Vitamin D intoxication associated with an over-the-counter supplement. N Engl J Med 345:66–67
15. Vieth R (1990) The mechanisms of vitamin D toxicity. Bone Mineral 11:267–272
16. Lou YR, Murtola T, Tuohimaa P (2005) Regulation of aromatase and 5alpha-reductase by 25-hydroxyvitamin D(3), 1alpha,25-dihydroxyvitamin D(3), dexamethasone and progesterone in prostate cancer cells. J Steroid Biochem Mol Biol 94:151–157
17. Vieth R, McCarten K, Norwich KH (1990) Role of 25-hydroxyvitamin D3 dose in determining rat 1,25- dihydroxyvitamin D3 production. Am J Physiol 258:E780–E789

18. Vieth R, Fraser D, Kooh SW (1987) Low dietary calcium reduces 25-hydroxycholecalciferol in plasma of rats. J Nutr 117:914–918

19. Vieth R, Bayley TA, Walfish PG, Rosen IB, Pollard A (1991) Relevance of vitamin D metabolite concentrations in supporting the diagnosis of primary hyperparathyroidism. Surgery 110: 1043–1046; discussion 1046–1047

20. Bell NH (1998) Renal and nonrenal 25-hydroxyvitamin D-1alpha-hydroxylases and their clinical significance [In Process Citation]. J Bone Miner Res 13:350–353

21. Standing Committee on the Scientific Evaluation of Dietary Reference Intakes. (1997) Dietary reference intakes: calcium, phosphorus, magnesium, vitamin D, and fluoride. National Academy Press, Washington, D.C.

22. HEALTH & CONSUMER PROTECTION DIRECTORATE-GENERAL. Opinion of the Scientific Committee on Food on the Tolerable Upper Intake Level of Vitamin D. EUROPEAN COMMISSION. http://europa.eu.int/comm/food/fs/sc/scf/out157_en.pdf-Accessed August 11, 2003. 2002. Brussels, Belgium. 1-10-0030. (GENERIC)

23. SACN. Update on Vitamin D: position statement by the Scientific Advisory Committee on Nutrition. 2007. http://www.sacn.gov.uk/reports_position_statements/position_statements/update_on_vitamin_d__november_2007.html accessed December 3, 2008.

24. Godel JC, First Nations IAMHC (2007) Vitamin D supplementation: recommendations for Canadian mothers and infants. Paediatr Child Health 12:583–589

25. Misra M, Pacaud D, Petryk A, Collett-Solberg PF, Kappy M (2008) Vitamin D deficiency in children and its management: review of current knowledge and recommendations. Pediatrics 122:398–417

26. Anonymous (1968) Recommended dietary allowances. National Academy Press, Washington, D.C

27. Narang NK, Gupta RC, Jain MK, Aaronson K (1984) Role of vitamin D in pulmonary tuberculosis. J Assoc Physicians India 32:185–186

28. Heaney RP (2003) Quantifying human calcium absorption using pharmacokinetic methods. J Nutr 133:1224–1226

29. Kimball SM, Vieth R (2007) A comparison of automated methods for the quantitation of serum 25-hydroxyvitamin D and 1,25-dihydroxyvitamin D. Clin Biochem 40:1305–1310

30. Bouillon R, Verstuyf A, Verlinden L, Eelen G, Mathieu C (2003) Prospects for vitamin D receptor modulators as candidate drugs for cancer and (auto)immune diseases. Recent Results Cancer Res 164(353–356):353–356

31. Posner GH (2002) Low-calcemic vitamin d analogs (deltanoids) for human cancer prevention. J Nutr 132:3802S–3803S

32. van den Bemd GJ, Chang GT (2002) Vitamin D and vitamin D analogs in cancer treatment. Curr Drug Targets 3:85–94

33. Barger-Lux MJ, Heaney RP (2002) Effects of above average summer sun exposure on serum 25-hydroxyvitamin d and calcium absorption. J Clin Endocrinol Metab 87:4952–4956

34. Heaney RP, Davies KM, Chen TC, Holick MF, Barger-Lux MJ (2003) Human serum 25-hydroxycholecalciferol response to extended oral dosing with cholecalciferol. Am J Clin Nutr 77: 204–210

35. Kimball S, Vieth R (2008) Self-prescribed high-dose vitamin D3: effects on biochemical parameters in two men. Ann Clin Biochem 45:106–110

36. Jacobus CH, Holick MF, Shao Q, Chen TC, Holm IA, Kolodny JM, FuleihanG E, Seely EW (1992) Hypervitaminosis D associated with drinking milk [see comments]. N Engl J Med 326: 1173–1177

37. Kimball SM, Ursell MR, O'Connor P, Vieth R (2007) Safety of vitamin D3 in adults with multiple sclerosis. Am J Clin Nutr 86:645–651

38. Muskiet FA, Dijck-Brouwer DJ, van der Veer E, Schaafsma A (2001) Do we really need >/=100 &mgr; g vitamin D/d, and is it safe for all of us? Am J Clin Nutr 74:862–863

39. Lawson DE, Douglas J, Lean M, Sedrani S (1986) Estimation of vitamin D3 and 25-hydroxyvitamin D3 in muscle and adipose tissue of rats and man. Clin Chim Acta 157:175–181

40. How KL, Hazewinkel HA, Mol JA (1994) Dietary vitamin D dependence of cat and dog due to inadequate cutaneous synthesis of vitamin D. Gen Comp Endocrinol 96:12–18

41. Stolzenberg-Solomon RZ, Vieth R, Azad A, Pietinen P, Taylor PR, Virtamo J, Albanes D (2006) A prospective nested case-control study of vitamin D status and pancreatic cancer risk in male smokers. Cancer Res 66:10213–10219

42. Tuohimaa P, Tenkanen L, Ahonen M, Lumme S, Jellum E, Hallmans G, Stattin P, Harvei S, Hakulinen T, Luostarinen T, Dillner J, Lehtinen M, Hakama M (2004) Both high and low levels of blood vitamin D are associated with a higher prostate cancer risk: a longitudinal, nested case-control study in the Nordic countries. Int J Cancer 108:104–108

43. Giovannucci E, Liu Y, Rimm EB, Hollis BW, Fuchs CS, Stampfer MJ, Willett WC (2006) Prospective study of predictors of vitamin D status and cancer incidence and mortality in men. J Natl Cancer Inst 98:451–459

44. Schwartz GG (1992) Multiple sclerosis and prostate cancer: what do their similar geographies suggest? Neuroepidemiology 11:244–254

45. Schwartz GG, Eads D, Rao A, Cramer SD, Willingham MC, Chen TC, Jamieson DP, Wang L, Burnstein KL, Holick MF, Koumenis C (2004) Pancreatic cancer cells express 25-Hydroxyvitamin D-1{alpha}-hydroxylase and their proliferation is inhibited by the prohormone 25-hydroxyvitamin D3. Carcinogenesis 25:1015–1026

46. Lips P, Duong T, Oleksik A, Black D, Cummings S, Cox D, Nickelsen T (2001) A global study of vitamin D status and parathyroid function in postmenopausal women with osteoporosis: baseline data from the multiple outcomes of raloxifene evaluation clinical trial. J Clin Endocrinol Metab 86:1212–1221

47. Grant WB (2003) Ecologic studies of solar UV-B radiation and cancer mortality rates. Recent Results Cancer Res 164(371–377):371–377

48. Taha SA, Dost SM, Sedrani SH (1984) 25-Hydroxyvitamin D and total calcium: extraordinarily low plasma concentrations in Saudi mothers and their neonates. Pediatr Res 18:739–741

49. Al Faraj S, Al Mutairi K (2003) Vitamin d deficiency and chronic low back pain in Saudi Arabia. Spine 28:177–179

50. Vieth R (2004) Enzyme kinetics hypothesis to explain the U-shaped risk curve for prostate cancer vs. 25-hydroxyvitamin D in Nordic countries. Int J Cancer 111:468

51. Vieth R, Choo R, Deboer L, Danjoux C, Morton G, Klotz L (2006) Rise in Prostate-Specific Antigen in Men with Untreated Low-Grade Prostate Cancer Is Slower During Spring-Summer. Am J Ther 13:394–399

52. Holick MF, MacLaughlin JA, Doppelt SH (1981) Regulation of cutaneous previtamin D3 photosynthesis in man: skin pigment is not an essential regulator. Science 211:590–593

53. Vieth R (2007) Vitamin D toxicity, policy, and science. J Bone Miner Res 22(Suppl 2):V64–V68

54. Buxton ILO (2006) Pharmacokinetics and pharmacodynamics: the dynamics of drug absorption, distribution, action, and elimination. In: Brunton LL (ed) Goodman & Gilman's the pharmacological basis of therapeutics. McGraw-Hill, Medical Publishing Division, New York, pp. 1–41

55. Ilahi M, Armas LA, Heaney RP (2008) Pharmacokinetics of a single, large dose of cholecalciferol. Am J Clin Nutr 87:688–691

56. Ish-Shalom S, Segal E, Salganik T, Raz B, Bromberg IL, Vieth R (2008) Comparison of daily, weekly, and monthly vitamin d3 in ethanol dosing protocols for two months in elderly hip fracture patients. J Clin Endocrinol Metab 93:3430–3435

IV HEALTH CONSEQUENCES OF VITAMIN D DEFICIENCY AND RESISTANCE ON MUSCULOSKELETAL HEALTH

32 Vitamin D Deficiency in Pregnancy and Lactation and Health Consequences

Sarah N. Taylor, Carol L. Wagner, and Bruce W. Hollis

Abstract   Pregnancy and lactation represent time periods where health status affects two persons instead of one. With the monumental development and growth that occurs in fetal and in early life, optimizing health status is imperative. An important part of health status is achieving and sustaining vitamin D sufficiency with vitamin D's critical function in bone health, immune function, and cell proliferation. Investigation regarding the short-term and long-term effects of vitamin D status during these times is exponentially growing and, so far, demonstrates important roles for vitamin D in both maternal and infant bone health, glucose tolerance, and immune function. In addition, evidence points to a critical role for vitamin D in sustaining pregnancy and avoiding pregnancy-related diseases. As information grows concerning vitamin D function in these time periods, worldwide study reveals vitamin D deficiency in mothers and infants at epidemic proportions. However, of promise is ongoing investigation to further identify the maternal vitamin D supplementation that can safely support mother and infant during both pregnancy and lactation.

Key Words: Vitamin D; pregnancy; fetus; lactation; infant; breastfeeding

1. INTRODUCTION

The building of the human body during gestation and early infant life consists of countless cellular activities as the human structure develops and is programmed for successes and failures in health. Recently, knowledge of the importance of vitamin D not only in calcium metabolism and bone health *(1–4)* but also in immune function and cell proliferation has grown exponentially *(5–17)*. This information raises awareness that vitamin D status during early life has far-reaching consequences. As these consequences are identified, a persistent finding is the high prevalence of vitamin D insufficiency and deficiency in mother/infant populations. As the vitamin D supplementation recommended during pregnancy and lactation is found to have minimal effect on maternal vitamin D status, studies of vitamin D status in these women demonstrate similarities

From: *Nutrition and Health: Vitamin D*
Edited by: M.F. Holick, DOI 10.1007/978-1-60327-303-9_32,
© Springer Science+Business Media, LLC 2010

to other populations studied with high rates of deficiency with darker pigmentation, high latitudes, and winter season. However, adding to the deficient state of this population is the cultural female full body covering practiced in many areas with abundant sunshine.

An additional problem in this group is hesitancy to study these "high-risk" populations of pregnant and lactating women and their infants. The critical nature of these time periods not only should prompt caution in investigation but also should prompt urgency due to the potential far-reaching effects. This chapter details the results of investigations that have occurred and how these results affect clinical practice and guidelines during pregnancy and lactation.

2. MATERNAL/INFANT CALCIUM HOMEOSTASIS IN PREGNANCY

During pregnancy, a woman maintains her vitamin D requirements to support her own health, but adds the need to support the fetus and the placenta. In addition, pregnancy stimulates changes in calcium homeostasis which include important roles for vitamin D. These changes occur as the maternal calcium supply is responsible for the fetal skeleton and fetal growth. Approximately 25–30 g of calcium is transferred from mother to fetus during pregnancy (18, 19). To accumulate this quantity of calcium for the fetus, the mother's fractional calcium absorption and mother's concentration of the active vitamin D metabolite, 1,25-dihydroxyvitamin D [$1,25(OH)_2D$] increase nearly twofold by the last trimester (20–22). Increase in active calcium transport at the maternal intestinal tract is then followed by active transport of calcium to the fetus (23). The increase in maternal intestinal calcium absorption appears to be the primary source for calcium supply to the infant, but good evidence also demonstrates that calcium is mobilized from maternal bone during pregnancy (24). Calcium retention by the kidneys does not increase during pregnancy (19–21). The role that parathyroid hormone (PTH) plays in this increase in $1,25(OH)_2D$ production and intestinal calcium absorption and the decrease in bone mineralization is not clear (25–29). Studies present inconsistent results for changes of PTH during pregnancy (20, 21, 23, 25–27, 29–31). One reason for the conflicting findings may be that PTH is controlled by a mechanism other than pregnancy during this time. In a study of calcium homeostasis during pregnancy in Asian and Caucasian women, PTH was significantly inversely correlated with plasma calcium and 25-hydroxyvitamin D [25(OH)D] concentrations (31). Observed increases of PTH during pregnancy may be secondary to vitamin D deficiency (19, 26, 27).

3. CONTEMPORARY MATERNAL VITAMIN D STATUS IN PREGNANCY

The long-term unawareness of the health consequences of vitamin D insufficiency, defined by serum 25(OH)D < 32 ng/ml, has led to a high prevalence of this problem in most populations. The presence of vitamin D insufficiency raises concern not only for maternal health but also for infant health with the cord blood 25(OH)D commonly 50–70% of maternal concentrations (32–35). In a study of vitamin D status in Pittsburgh with 90% of the women taking prenatal vitamins, 54.1% of black women and 42.1% of white women demonstrated vitamin D insufficiency at the end of pregnancy. In the same study, vitamin D deficiency was defined as serum 25(OH)D <15 ng/ml and was seen in 29.2% of black women and 5% of white women at the end of pregnancy *(35)*.

The concerning presence of serum 25(OH)D < 15 ng/ml in pregnant black women in Pittsburgh has been echoed worldwide with alarming results. A representation of measured prevalence around the world is 8% in Western women in The Netherlands *(36)*, 18% in the United Kingdom *(37)*, 25% in the United Arab Emirates *(38)*, 42% in northern India *(39)*, 46% in Canada *(40)*, 50% in non-European ethnic minority women in South Wales *(41)*, 59–84% in women with darker pigmentation in The Netherlands *(36)*, 61% in New Zealand *(42)*, 71% in Pakistani women in Norway *(43)*, and 80% in Iran *(44)*. These studies and others show that high latitude, winter season, dark pigmentation, and full-body skin covering all increase the risk of vitamin D deficiency *(35, 36, 38, 39, 42, 44)*. These risk factors must be considered as appropriate oral vitamin D supplementation is defined.

4. CLINICAL TRIALS OF VITAMIN D SUPPLEMENTATION IN PREGNANCY

Study of the supplementation of vitamin D during the prenatal period is limited. A Cochrane report identified seven clinical trials of vitamin D supplementation with clinical outcomes reported in four *(45)*. The authors of the review concluded that there is no adequate evidence to date to evaluate the effects of vitamin D supplementation during pregnancy *(45)*.

Four trials have compared daily vitamin D supplementation at doses of 400 IU/day *(46)* and 1,000 IU/day *(47–49)* versus placebo. One limitation of these evaluations is the restricted effect of 400–1,000 IU/day on serum 25(OH)D concentrations. Table 1 shows serum 25(OH)D changes from baseline to the end of pregnancy of trials that reported maternal 25(OH)D. One trial, first published in 1981, exhibited a mean increase in serum 25(OH)D of 59.1 ng/ml with a dose of 1,000 IU/day vitamin D *(47)*. In light of later evaluation in both pregnant women *(48, 49)* and other populations *(50, 51)*, these results

Table 1

Clinical Trials of Maternal Vitamin D Supplementation During Pregnancy

Study	Vitamin D dose (IU)	Initial mean of 25(OH)D	At delivery, maternal mean of 25(OH)D	At delivery, cord mean of 25(OH)D	At delivery, control maternal mean of 25(OH)D
Cockburn et al. *(46)*.	400	15.6	17.1	11.2	13
Mallet et al. *(49)*.	1,000	No data	10.1	6.3	3.8
Brooke et al. *(47)*.	1,000	8.1	67.2	No data	6.5
Delvin et al. *(48)*.	1,000	No data	26	18	13

All serum 25(OH)D values are in ng/ml. Vitamin D doses are given daily. Control mothers received no supplemental vitamin D.

most likely reflect supplementation with a dose 10 times the dose documented or an invalid assay for 25(OH)D *(51, 52)*.

Results of evaluations of vitamin D supplementation effect on maternal health, serum markers of calcium homeostasis, bone health, and nutrition are presented in Table 2. Mean serum alkaline phosphatase values were consistently lower with vitamin D supplementation possibly signifying improved maternal bone health *(47, 53)*. No difference was seen in serum $1,25(OH)_2D$ or PTH *(48, 49)*. For serum calcium, no significant change was seen for doses of 400–1,200 IU/day *(46, 48, 53)*, but significant increase was seen with doses of 1,000 IU/day and a single oral dose of 600,000 IU in the seventh and eighth months of pregnancy *(47, 53)*. In evaluation of maternal nutrition, one study

Table 2
Maternal Calcium and Health Parameters with Vitamin D Supplementation During Pregnancy

Parameters	*Result for study group mean compared to control (0 vitamin D supplementation) mean*	*Vitamin D dose(s)*
Serum $1,25(OH)_2D$	No difference	1,000 IU/day *(48, 49)*; 200,000 IU/month *(49)*
Serum calcium	No difference	400 IU/day *(46)*; 1,000 IU/day *(48)*; 1,200 IU/day *(53)*
	Significantly higher	1,000 IU/day *(47)*; 600,000 IU/month *(53)*
Serum iPTH	No difference	1,000 IU/day *(48)*
Serum inorganic phosphorus	No difference	400 IU/day *(46)*; 1,000 IU/day *(47)*; 1,200 IU/day *(53)*
	Significantly higher	600,000 IU/month *(53)*
Serum alkaline phosphatase	Significantly lower	1,000 IU/day *(47)*; 1,200 IU/day *(53)*; 600,000 IU/month *(53)*
Serum magnesium	No difference	400 IU/day *(46)*
Serum protein	No difference	400 IU/day *(46)*; 1,200 IU/day *(53)*
	Higher retinol-binding protein, higher thyroid-binding prealbumin	1,000 IU/day *(54)*
Serum albumin	No difference	1,000 IU/day *(54)*
Maternal weight gain	No difference	600,000 IU/month *(87)*
	Significantly greater	1,000 IU/day *(54)*

of 1,000 IU/day dose demonstrated significantly higher maternal weight gain during pregnancy *(54)*. Study of protein status revealed inconsistent results *(46, 53, 54)*.

In the infants born to these clinical trial subjects, calcium homeostasis markers and growth parameters were studied with results provided in Table 3. The inconsistencies in these results mandate further evaluation to identify the effect of not only maternal vitamin D supplementation but also maternal vitamin D sufficiency during gestation. Of importance and as partially shown in Table 1, the vitamin D-supplemented pregnant

Table 3

Neonatal Calcium and Health Parameters with Maternal Vitamin D Supplementation During Pregnancy

Neonatal parameters	Results for supplemented group compared to control group	Vitamin D dose(s)
Serum $1,25(OH)_2D$	No difference	1,000 IU/day *(49)*
	Significantly lower	1,000 IU/day *(48)*
Serum calcium	No difference	400 IU/day *(46)*; 1,000 IU/day *(49)*; 1,000 IU/day *(48)*
	Significantly higher	600,000 IU/month *(53)*
Serum PTH	No difference	1,000 IU/day *(48)*
Serum inorganic phosphorus	No difference	400 IU/day *(46)*; 1,000 IU/day *(48)*
	Significantly higher	600,000 IU/month *(53)*
Serum alkaline phosphatase	Significantly lower	600,000 IU/month *(53)*
Serum magnesium	No difference	400 IU/day *(46)*
Birth weight	No difference	1,000 IU/day *(54)*; 1,000 IU/day *(49)*
	Significantly greater	1,200 IU/day *(53)*, 600,000 IU/month *(53, 87)*
Standard anthropometric measurements	No difference	1,000 IU/day *(54)*
Linear proportions	No difference	1,000 IU/day *(54)*
	Significantly greater	600,000 IU/month *(87)*
Head circumference	Significantly greater	600,000 IU/month *(87)*
Fontanelle area	Significantly less	1,000 IU/day *(47)*
Proportion of low birth weight infants	Significantly less	1,000 IU/day *(54)*
Growth in first year	Significantly greater	1,000 IU/day *(47)*
Dental enamel at 3 years	Significantly better	400 IU/day *(46)*

women in the majority of these trials did not achieve vitamin D sufficiency. This fact compels caution in interpreting the significance of these results.

5. VITAMIN D AND FETAL DEVELOPMENT

Maternal vitamin D status is integral to fetal development because the fetus receives all vitamin D support from the mother *(32–34)*. Maternal 25(OH)D readily crosses the placenta and is metabolized to $1,25(OH)_2D$ by the fetal kidneys for endocrine action and by other tissues for paracrine action *(17, 32–34, 55, 56)*. At birth the neonate's serum 25(OH)D status is 50–70% of maternal serum 25(OH)D concentrations *(32–35)*.

Observational studies have evaluated the role of vitamin D and calcium in bone development and fetal growth. The results of these studies demonstrate roles for both vitamin D and calcium *(37, 57)*, but the exact mechanism, as well as the role of iPTH concentration, remains to be determined. Failure of adequate intrauterine bone mineralization, secondary to both vitamin D and calcium deficiency, has led to cases of congenital rickets *(58–60)*. In a study of maternal vitamin D supplementation of 1,000 IU/day and neonatal bone mineralization, no significant difference was observed in infant bone mineral content measured by single photon absorptiometry *(61)*. Conversely, a longitudinal evaluation of 198 children demonstrated a positive association between maternal serum 25(OH)D in late pregnancy and child whole-body and lumbar-spine bone mineral content by dual-energy X-ray absorptiometry (DEXA) at 9 years of age *(37)*. A recent evaluation of neonatal bone mineralization by DEXA revealed higher whole-body bone mineral content not only associated with greater gestational age and higher weight at birth and higher infant 25(OH)D levels but also associated with lower maternal 25(OH)D concentrations *(40)*. On the other hand, another evaluation of maternal vitamin D deficiency and infant bone demonstrated an association of maternal vitamin D deficiency with delayed development of neonatal bone ossification centers at birth *(62)*. In growth, low maternal vitamin D status, often with low maternal calcium and elevated maternal iPTH, has been linked to decreased birth weight *(57)*, decreased neonatal length *(26, 57)*, and lower head circumference at 3 and 6 months *(63)*, but these results are not consistent.

6. FETAL VITAMIN D EXPOSURE AND IMMUNE FUNCTION

The role of vitamin D in immune function naturally raises the question of the role of vitamin D in development of this system. In a case–control study of newborns with acute lower respiratory infection, the cases and their mothers demonstrated significantly lower serum 25(OH)D concentrations than the healthy controls *(64)*. The newborns with infection exhibited a mean serum 25(OH)D concentration of 9.1 ng/ml, while the controls demonstrated a mean of 16.3 ng/ml *(64)*.

Another area of immune development with study of its relationship with the fetal vitamin D environment is allergic disease. In two birth cohorts, higher maternal vitamin D status in pregnancy was associated with decreased prevalence of wheezing at 3 and 5 years of age *(65, 66)*. In opposition, in a third birth cohort, higher maternal serum

25(OH)D status in late pregnancy was associated with increased risk of asthma at 9 years of age *(67)*.

Epidemiological study of other disease processes that raise the possibility for positive association with maternal vitamin D status in pregnancy include osteoporosis, multiple sclerosis, and schizophrenia *(68–71)*. With long-term programming in fetal development, identifying the effect of this compound which is important to bone health, immune function, and cell proliferation is critical *(72)*.

7. FETAL VITAMIN D EXPOSURE AND CARDIAC DEVELOPMENT

In the 1960s, vitamin D was considered the cause of supravalvular stenosis *(52, 56, 73, 74)*. The published hypothesis was that "toxic" amounts of vitamin D during pregnancy gave rise to a clinical condition titled "infantile hypercalcemia syndrome" *(74)*. Now this syndrome is known as Williams syndrome and is a genetic disease characterized by multiorgan involvement including supravalvular stenosis secondary to a defect in elastin formation. One in 7,500–20,000 people is born with this disorder which also exhibits an exaggerated response to oral vitamin D supplementation *(75)*. Abnormal vitamin D metabolism now is known to be a symptom of Williams syndrome and not a cause of supravalvular stenosis, but worldwide reductions in vitamin D supplementation policy that occurred in response to this concern have not been revised. Recommendations for infants to receive 4,000–5,000 IU/day in Finland in the 1960s are now reduced to current recommendation for 200–400 IU/day in the United States *(76, 77)*.

Ironically, vitamin D deficiency, not vitamin D toxicity, may lead to a severe and life-threatening cardiac disease – cardiomyopathy *(78–80)*. Despite unclear etiology as to whether the cardiac failure is due to hypocalcemia alone or with an intrinsic vitamin D effect on cardiac function, supplementation with vitamin D and calcium therapy prevented the need for cardiac transplantation *(79, 80)*. This relationship between vitamin D deficiency and cardiac function highlights the expanding knowledge of vitamin D physiology.

8. VITAMIN D AND PREGNANCY HEALTH

Achieving optimal maternal and infant health with adequate vitamin D status is not the only objective during pregnancy. The vitamin D status required to ensure the health of the pregnancy is vital. A pregnancy can develop severe disease processes that risk the health of both the mother and the infant. Preeclampsia and glucose intolerance characterize two of those processes. Preeclampsia is a pregnancy-specific disease process that affects 3–7% of pregnancies. Development of preeclampsia likely involves abnormal placental implantation, immune dysfunction, excessive inflammation, and defects in angiogenesis *(81–84)*. The placenta expresses 1α-hydroxylase, the enzyme required to convert 25(OH)D to the active metabolite $1,25(OH)_2D$ *(81)*. So far, this active vitamin D metabolite has been shown to regulate the transcription and function of genes associated with placental invasion, implantation, and angiogenesis *(81, 85)* and is being investigated for its role in early pregnancy loss, preterm birth, and preeclampsia *(81)*. In a case–control study, lower vitamin D status both in the early pregnancy and in the newborn was associated with the risk of preeclampsia during pregnancy. After adjustment for confounders, a deficit of 20 ng/ml was found to double the risk of preeclampsia *(85)*.

For another disease process in pregnancy, glucose intolerance, a cross-sectional study found women with gestational diabetes were more likely to exhibit vitamin D deficiency (<12.5 ng/ml) than women with normal glucose control during pregnancy *(86)*. With the known associations of both type I and type II diabetes with vitamin D status, this relationship is not surprising *(7–9, 76)*. Other maternal health markers, such as maternal weight gain in pregnancy, have demonstrated an inconsistent relationship with maternal vitamin D status *(54, 87)*. Further investigation to identify the vitamin D status required to sustain the health of the pregnancy is critical.

9. VITAMIN D REQUIREMENTS FOR MOTHER AND INFANT DURING LACTATION

The natural progression of human nutrition is for mother's milk to begin at infant delivery when placental supply ends. Human milk is considered the ideal infant nutrition with only one nutrient consistently in inadequate concentrations for the breastfeeding infant prior to 6 months of age – vitamin D *(77)*. Rickets is a disease process associated with exclusive breastfeeding and the low vitamin D content of human milk. With maternal supplementation of 400 IU/day, breast milk contains 33–68 IU/l of vitamin D activity which is far below the recommended daily vitamin D intake of 200–800 IU/day for infants to avoid rickets *(88–91)*. As new understanding of vitamin D sufficiency and vitamin D toxicity has led to acknowledgment of higher supplementation requirements for adults, the potential for maternal supplementation to produce human milk with adequate vitamin D activity to support an infant has evolved. Investigation to identify a safe vitamin D dose to support mother and infant health is being performed *(52)*. A review of the physiology and biochemistry of vitamin D activity in human milk and past evaluations of vitamin D supplementation during lactation is presented.

10. THE MECHANISM OF VITAMIN D IN HUMAN MILK

The usual role of the parent compound, vitamin D, is to produce circulating 25(OH)D which is then converted in an endocrine or paracrine fashion to $1,25(OH)_2D$ *(55, 92)*. For example, in pregnancy, maternal serum 25(OH)D crosses the placenta to provide the fetus a supply for conversion to $1,25(OH)_2D$ *(32–34)*. In contrast, human lactation is an unique circumstance where both maternal 25(OH)D and maternal vitamin D are supplied to support the vitamin D status of the nursing infant. Other metabolites including $1,25(OH)_2D$ are at insufficient concentrations to measure. The vitamin D_2, vitamin D_3, $25(OH)D_2$, and $25(OH)D_3$ compounds in human milk generate the vitamin D activity or antirachitic activity of the milk *(89, 93–95)*. Ligand-binding assays for the vitamin D content in human milk, developed in the 1980s, demonstrate that 30% of the maternal circulating vitamin D and 1% of the maternal circulating 25(OH)D are present *(88, 96, 97)*. Therefore, maternal vitamin D status plays an essential role in the vitamin D activity delivered to the infant. Due to the predominance of the parent compound, daily vitamin D supplementation of mother either through cutaneous production or through oral intake is required to optimize the vitamin D content of human milk *(88, 90, 91)*. Lactation is one circumstance where large weekly or monthly doses are not appropriate. Studies have demonstrated the ability to provide vitamin D activity in human milk with both maternal UVB exposure and maternal oral intake *(88, 90, 91, 98)*.

In 1984, Greer et al. *(88)* exposed lactating white women to UVB exposure equivalent to 30 min of sunshine at midday on a clear summer day at temperate latitudes. With this exposure, the vitamin D content of the milk significantly increased with a peak at 48 h and with a return to baseline at 7 days. At the same time, the women's circulating 25(OH)D concentrations increased from a mean of 13.9–20.5 ng/ml and remained significantly elevated for at least 14 days. However, despite this increase in circulating maternal 25(OH)D, the milk 25(OH)D did not change significantly. The predominance of vitamin D in promoting antirachitic activity in human milk and the short duration of effect of UVB-produced vitamin D underscore the significance of daily supplementation or production to sustain vitamin D activity in human milk. The next step required is identification of the appropriate oral supplementation or UVB exposure to be recommended for each lactating mother.

11. MATERNAL VITAMIN D SUPPLEMENTATION DURING LACTATION

Design of ligand-binding assays to measure the vitamin D content of human milk in the 1980s prompted the first evaluations of the effect of maternal supplementation *(93, 99, 100)*. Evaluation of the milk produced by a mother receiving 100,000 IU/day vitamin D_2 for treatment of hypoparathyroidism revealed an antirachitic activity of 7,600 IU/l. The infant receiving this milk had a serum 25(OH)D level of 251 ng/ml and no symptoms of hypercalcemia *(101)*.

Further evaluation of the efficacy and safety of maternal vitamin D supplementation to achieve vitamin D-sufficient human milk is limited secondary to the misconception that vitamin D toxicity occurs at a dose of 4,000 IU/day vitamin D *(102, 103)*. The first trials in the 1980s demonstrated no significant improvement in human milk vitamin D activity with maternal vitamin D intake of 500–1,000 IU/day vitamin D_2 *(98, 104–106)*. In 1986, a Finnish trial compared 15 weeks of maternal vitamin D_3 intake of 1,000 or 2,000 IU/day with infant intake of 400 IU/day and demonstrated healthy infant vitamin D status achieved in the 2,000 IU/day maternal intake group and the 400 IU/day infant intake group with mean serum 25(OH)D around 30 ng/ml. The infants of mothers receiving 1,000 IU/day vitamin D had serum 25(OH)D at a level approximately half of the other two groups *(98)*. However, due to misguided concerns that maternal supplementation of 2,000 IU/day was a dangerous supply of vitamin D, no further evaluation was performed for two decades.

Contemporary evaluations of maternal vitamin D supplementation to support infant vitamin D health (Table 4) confirm that maternal and infant vitamin D sufficiencies can be achieved with daily oral maternal vitamin D supplementation to the lactating women *(90, 91)*. With further study to identify the appropriate supplementation to achieve optimal efficacy and safety, a daily recommended dose for lactating mothers should be available.

Table 4
Contemporary Clinical Trials of Vitamin D Supplementation During Lactation *(From 90 and 91)*

Maternal vitamin D supplementation (IU/day)	Infant vitamin D supplementation (IU/day)	Maternal serum 25(OH)D (ng/ml)	Milk antirachitic activity (IU/l)	Infant serum 25(OH)D (ng/ml)
2,000	0	36.1	69.2	27.8
4,000	0	44.5	134.6	30.8
6,400	0	58.8	873	46
400	300	38.4	45.6–78.6	43

12. INFANT VITAMIN D SUPPLEMENTATION DURING LACTATION

Despite global promotion of guidelines recommending 400–1,000 IU/day for infants, the evidence supporting these doses is quite limited *(77, 107, 108)*. Clinical trials of direct infant supplementation during lactation have demonstrated that 400–1,426 IU/day vitamin D can be given daily to infants without toxic effect and to achieve and maintain vitamin D status above the serum 25(OH)D concentrations associated with the development of rickets *(109–112)*. These four trials occurred in the United States or in Europe with mostly lightly pigmented infants except for one study where 80% of the subjects were black *(111)*. Of note, the current National Academy of Sciences recommendation of 200 IU/day vitamin D to prevent vitamin D deficiency in normal children and adolescents *(103)* relied mainly on data from a prospective study from China *(62)*. In this study, vitamin D intake as low as 100 IU/day prevented overt rickets, but 30% of the infants exhibited serum 25(OH)D < 11 ng/ml, commonly accepted as vitamin D deficiency for infants *(62)*. Most infant formulas contain >400 IU/l vitamin D *(113)*.

13. CONSEQUENCES OF INFANT VITAMIN D STATUS

With current recognition of the effects of vitamin D status for bone health, immune function, and cellular proliferation, evaluation of the long-term consequence of infant vitamin D status is required. Studies in this area are limited, but interesting. In one retrospective cohort, vitamin D supplementation during infancy was associated with improved bone mineral mass at specific skeletal sites in prepubertal girls *(114)*. In a prospective cohort, infants who received at least 2,000 IU/day vitamin D in the first year of life exhibited an 80% decrease in risk for diabetes mellitus type I *(76)*. In the same prospective cohort, females who received regular vitamin D supplementation in the first year of life were 50% less likely to develop preeclampsia in their first pregnancy *(115)*. These long-term associations suggest a function of vitamin D in bone growth and immune programming in fetal and early life.

Prior to recognition of these potentially enduring effects, adequate vitamin D status in infants was defined as the serum 25(OH)D level required to avoid rickets and demonstrated a lower limit of 8–15 ng/ml depending on the reference study *(62, 103, 109–112, 116–123)*. To avoid rickets is to avoid a debilitating disease with bowing of the legs, curvature of the spine, and abnormalities especially at the wrist, ankles, and rib cage. In addition, delayed growth, muscle weakness, seizures, susceptibility to infection, and cardiomyopathy are associated *(78–80, 116–124)*. Concern for the high prevalence of vitamin D deficiency and the growing occurrence of rickets has been demonstrated worldwide *(116–123)*. Studies with measurement of both maternal and infant vitamin D status demonstrate that achieving adequate vitamin D status to avoid rickets is a continuum of the fetal and early infant vitamin D supply *(125, 126)*. Vitamin D status at birth is completely reliant on maternal vitamin D status *(32–34)*. Vitamin D status of exclusively breastfeeding infants is completely reliant on maternal vitamin D status if the infant is not directly provided oral supplementation or abundant sunlight exposure *(104, 105, 109, 110)*. The risk factors identified for breastfeeding infant vitamin D deficiency and rickets are identical to the risk factors for maternal vitamin D deficiency – high latitude, dark pigmentation, winter season, and maternal full-body skin covering *(60, 62, 77, 96, 107, 116–127)*.

Despite recommendations for infant vitamin D supplementation in many countries, recognition of rickets is increasing worldwide *(116–126)*. Between 2002 and 2004 in Canada, 104 confirmed cases of rickets were reported despite national recommendations for infant vitamin D supplementation. High latitude and dark pigmentation were appreciated risk factors *(107, 121)*. In the United States, a recent evaluation in Iowa identified 10% of breastfeeding infants at 9 months of age with serum 25(OH)D < 11 ng/ml. Most of these infants had measurements made in the winter, had dark skin, and had received no vitamin D supplements *(127)*.

In addition, vitamin D deficiency can occur at lower latitudes especially with darker pigmentation or maternal full-body clothing. In the United Arab Emirates in summer months, investigators identified serum 25(OH)D < 10 ng/ml in 61% of lactating mothers and in 82% of their children *(126)*. With the capacity for vitamin D supplementation to reverse the known health consequences for both mother and infant, identification of optimal supplementation and action to distribute both information and this nutritional support is critical at this time.

14. CONCLUSION

Although not identified as thoroughly as in other populations, current evidence during pregnancy and lactation points to both short-term and long-term effects of vitamin D deficiency in maternal and infant health consequences, which include bone development, immune function, and health of the pregnancy. Previous study was impeded by a general misunderstanding of vitamin D safety. With the realization that vitamin D doses upward of 10,000 IU/day are safe, this critical time of development deserves comprehensive investigation of vitamin D needs.

As current clinical trials evaluate the vitamin D needs in these populations, epidemiological study has revealed extensive worldwide vitamin D deficiency during pregnancy, lactation, and early infant life. The global risk factors for vitamin D deficiency are higher latitude, winter season, dark pigmentation, limited sun exposure, and full-body clothing, but, based on oral supplementation and sunlight exposure, no population exhibits zero risk. While evidence concerning the optimal supplementation for mothers and infants accumulates, emergent strategies to guarantee that pregnant women achieve adult vitamin D sufficiency defined as serum 25(OH)D > 32 ng/ml and that infants achieve infant D status to avoid rickets are imperative.

REFERENCES

1. Heaney RP, Dowell MS, Hale CA et al (2003) Calcium absorption varies within the reference range for serum 25-hydroxyvitamin D. J Am College Nutr 22:142–146
2. Bischoff-Ferrari H, Dietrich T, Orav E, Dawson-Hughes B (2004) Positive association between 25(OH)D levels and bone mineral density: a population-based study of younger and older adults. Amer J Med 116:634–639
3. Vieth R, Ladak Y, Walfish PG (2003) Age-related changes in the 25-hydroxyvitamin D versus parathyroid hormone relationship suggest a different reason why older adults require more vitamin D. J Clin Endocrinol Metab 88:185–191
4. Meier C, Woitge H, Witte K et al (2004) Supplementation with oral vitamin D3 and calcium during winter prevents seasonal bone loss: a randomized controlled open-label prospective trial. J Bone Miner Res 19:1221–1230
5. Harkness L, Cromer B (2005) Low levels of 25-hydroxy vitamin D are associated with elevated PTH in healthy adolescent females. Osteoporo Int 16:109–113
6. Liu PT, Stenger S, Li H et al (2006) Toll-like receptor triggering of a vitamin D-mediated human antimicrobial response. Science 311:1770–1773
7. Borissova AM, Tankova T, Kirilov G et al (2003) The effect of vitamin D3 on insulin secretion and peripheral insulin sensitivity in type 2 diabetic patients. Int J Clin Pract 57:258–261
8. Boucher BJ, Mannan N, Noonan K et al (1995) Glucose intolerance and impairment of insulin secretion in relation to vitamin D deficiency in east London Asians. Diabetologia 38:1239–1245
9. Chiu KC, Chu A, Go VLW et al (2004) Hypovitaminosis D is associated with insulin resistance and β cell dysfunction. Am J Clin Nutr 79:820–825
10. Merlino LA, Curtis J, Mikuls TR et al (2004) Vitamin D intake is inversely associated with rheumatoid arthritis. Arthritis Rheum 50:72–77
11. Munger K, Zhang S, O'Reilly E et al (2004) Vitamin D intake and incidence of multiple sclerosis. Neurology 62:60–65
12. Lappe JM, Travers-Gustafson D, Davies KM et al (2007) Vitamin D and calcium supplementation reduces cancer risk: results of a randomized trial. Am J Clin Nutr 85:1586–1591
13. Froicu M, Weaver V, Wynn T et al (2003) A crucial role for the vitamin D receptor in experimental inflammatory bowel diseases. Mol Endocrinol 17:2386–2392
14. Garland CF, Gorham ED, Mohr SB et al (2007) Vitamin D and prevention of breast cancer: pooled analysis. J Steroid Biochem Mol Biol 103:708–711
15. Gorham ED, Garland CF, Garland FC et al (2007) Optimal vitamin D status for colorectal cancer prevention: a quantitative meta analysis. Am J Prev Med 32:210–216
16. Holick MF (2004) Vitamin D: importance in the prevention of cancers, type 1 diabetes, heart disease, and osteoporosis. Am J Clin Nutr 79:362–371
17. Hollis BW (2005) Circulating 25-hydroxyvitamin D levels indicative of vitamin sufficiency: implications for establishing a new effective DRI for vitamin D. J Nutr 135:317–322
18. Widdowson EM (1981) Changes in body composition during growth. In: Davis JA, Dobbings J (eds) Scientific foundations of pediatrics. Wm Heinemann Medical Books, London
19. Specker B (2004) Vitamin D requirements during pregnancy. Am J Clin Nutr 80(suppl):1740S–1747S
20. Cross NA, Hillman LS, Allen SH et al (1995) Calcium homeostasis and bone metabolism during pregnancy, lactation, and post-weaning: a longitudinal study. Am J Clin Nutr 61:514–523
21. Ritchie LD, Fung EB, Halloran BP et al (1998) A longitudinal study of calcium homeostasis during human pregnancy and lactation and after resumption of menses. Am J Clin Nutr 67:693–701

22. Bikle DD, Gee E, Halloran B et al (1984) Free 1,25-dihydroxyvitamin D levels in serum from normal subjects, pregnant subjects, and subjects with liver disease. J Clin Invest 74:1966–1971

23. Pitkin RM (1985) Calcium metabolism in pregnancy and the perinatal period: a review. Am J Obstet Gynecol 151:99–101

24. Ensom MHH, Liu PY, Stephenson MD (2002) Effect of pregnancy on bone mineral density in healthy women. Obstet Gynecol 57:99–111

25. Davis OK, Hawkins DS, Rubin LP et al (1988) Serum parathyroid hormone in pregnant women determined by an immuno-radiometric assay for intact PTH. J Clin Endocrinol Metab 67:850–852

26. Brunvard L, Quigstad E, Urdal P et al (1996) Vitamin D deficiency and fetal growth. Early Hum Dev 45:27–33

27. Alfaham M, Woodhead S, Pask G et al (1995) Vitamin D deficiency: a concern in pregnant Asian women. Br J Nutr 73:881–887

28. Fleischman A, Rosen J, Cole J et al (1980) Maternal and fetal serum 1,25-dihydroxyvitamin D levels at term. J of Pediatr 97:640–642

29. Seki K, Makimura N, Mitsui C et al (1991) Calcium-regulating hormones and osteocalcin levels during pregnancy: a longitudinal study. Am J Obstet Gynecol 164:1248–1252

30. Bezerra FF, Laboissiere FP, King JC et al (2002) Pregnancy and lactation affect markers of calcium and bone metabolism differently in adolescent and adult women with low calcium intakes. J Nutr 132:2183–2187

31. Okonofua F, Menon RK, Houlder S et al (1987) Calcium, vitamin D, and parathyroid hormone relationships in pregnant Caucasian and Asian women and their neonates. Ann Clin Biochem 24:22–28

32. Markestad T, Aksnes L, Ulstein M, Aarskog D (1984) 25-Hydroxyvitamin D and 1,25-dihydroxy vitamin D of D_2 and D_3 origin in maternal and umbilical cord serum after vitamin D_2 supplementation in human pregnancy. Am J Clin Nutr 40:1057–1063

33. Bouillon R, Van Baelen H, DeMoor D (1977) 25-Hydroxy-vitamin D and its biding protein in maternal and cord serum. J Clin Endocrinol Metab 45:679–684

34. Hollis BW, Pittard WB (1984) Evaluation of the total fetomaternal vitamin D relationships at term: evidence for racial differences. J Clin Endocrinol Metab 59:652–657

35. Bodnar LM, Simhan HN, Powers RW et al (2007) High prevalence of vitamin D insufficiency in black and white pregnant women residing in the northern United States and their neonates. J Nutr 137:447–452

36. Van der Meer IM, Karamali NS, Boeke AJP et al (2006) High prevalence of vitamin D deficiency in pregnant non-western women in The Hague, Netherlands. Am J Clin Nutr 84:350–353

37. Javaid MK, Crozier SR, Harvey NC et al (2006) Maternal vitamin D status during pregnancy and childhood bone mass at age 9 years: a longitudinal study. Lancet 367:36–43

38. Dawodu A, Agarwal M, Patel M et al (1997) Serum 25-hydroxyvitamin D and calcium homeostasis in the United Arab Emirates mothers and neonates: a preliminary report. Middle East Paediatr 2: 9–12

39. Sachan A, Gupta R, Das V et al (2005) High prevalence of vitamin D deficiency among pregnant women and their newborns in northern India. Am J Clin Nutr 81:1060–1064

40. Weiler H, Fitzpatrick-Wong S, Veitch R et al (2005) Vitamin D deficiency and whole-body and femur bone mass relative to weight in healthy newborns. CMAJ 172:757–761

41. Datta S, Alfaham M, Davies DP et al (2002) Vitamin D deficiency in pregnant women from a non-European ethnic minority population – an interventional study. BJOG 109:905–908

42. Judkins A, Eagleton C (2006) Vitamin D deficiency in pregnant New Zealand women. N Z Med J 119:U2144

43. Brunvard L, Shah SS, Bergstroem S et al (1998) Vitamin D deficiency in pregnancy is not associated with obstructed labor. A study among Pakistani women in Karachi. Acta Obstet Gynecol Scand 77:303–306

44. Bassir M, Laborie S, Lapillonne A et al (2001) Vitamin D deficiency in Iranian mothers and their neonates: a pilot study. Acta Paediatr 90:577–579

45. Mahomed K, Gulmezoglu AM (2002) Vitamin D supplementation in pregnancy (Cochrane Review). Update Software, Oxford, UK.

46. Cockburn F, Belton NR, Purvis RJ et al (1980) Maternal vitamin D intake and mineral metabolism in mothers and their newborn infants. BMJ 5:11–14

47. Brooke OG, Brown IRF, Bone CDM et al (1980) Vitamin D supplements in pregnant Asian women: effects on calcium status and fetal growth. BMJ 1:751–754

48. Delvin EE, Salle BL, Glorieux FH et al (1986) Vitamin D supplementation during pregnancy: effect on neonatal calcium homeostasis. J Pediatr 109:328–334

49. Mallet E, Gugi B, Brunelle P (1986) Vitamin D supplementation in pregnancy: a controlled trial of two methods. Obstet Gynecol 68:300–304

50. Barger-Lux MJ, Heaney RP, Dowell S et al (1998) Vitamin D and its major metabolites: serum levels after graded oral dosing in healthy men. Osteoporos Int 8:222–230

51. Heaney RP, Davies LM, Chen TC et al (2003) Human serum 25-hydroxycholecalciferol response to extended oral dosing with cholecalciferol. Am J Clin Nutr 77:204–210

52. Hollis BW (2007) Vitamin D requirement during pregnancy and lactation. J Bone Miner Res 22: V39–V44

53. Marya RK, Rathee S, Lata V (1981) Effects of vitamin D supplementation in pregnancy. Gynecol obstet Invest 12:155–161

54. Maxwell JD, Ang L, Brooke OG et al (1981) Vitamin D supplements enhance weight gain and nutritional status in pregnant Asians. Br J Obstet Gynecol 88:987–991

55. Whiting SJ, Calvo MS (2005) Dietary recommendations to meet both endocrine and autocrine needs of vitamin D. J Steroid Biochem Mol Biol 97:7–12

56. Hollis BW, Wagner CL (2004) Assessment of dietary vitamin D requirements during pregnancy and lactation. Am J Clin Nutr 79:717–726

57. Sabour H, Hossein-Nezhad A, Maghbooli Z et al (2006) Relationship between pregnancy outcomes and maternal vitamin D and calcium intake: a cross-sectional study. Gynecol Endocrinol 22: 585–589

58. Russell JGB, Hill LF (1974) True fetal rickets. Br Radiol 47:732–734

59. Moncrief M, Fadahunsi TO (1974) Congenital rickets due to maternal vitamin D deficiency. Arch Dis Child 49:810–811

60. Zhou H (1991) Rickets in China. In: Glorieux FH (ed) Rickets. Raven Press, New York

61. Congdon P, Horsman A, Kirby PA et al (1983) Mineral content of the forearms of babies born to Asian and white mothers. Br Med J (Clin Res Ed) 286:1233–1235

62. Specker B, Ho M, Oestreich A et al (1992) Prospective study of vitamin D supplementation and rickets in China. J Pediatr 120:733–739

63. Pawley N, Bishop NJ (2004) Prenatal and infant predictors of bone health: the influence of vitamin D. Am J Clin Nutr 80(suppl):1748S–1751S

64. Karatekin G, Kaya A, Salihoğlu O et al (2007) Association of subclinical vitamin D deficiency in newborns with acute lower respiratory infection and their mothers. Eur J Clin Nutr [Epub ahead of print]

65. Devereux G, Litonjua AA, Turner SW et al (2007) Maternal vitamin D intake during pregnancy and early childhood wheezing. Am J Clin Nutr 85:853–859

66. Camargo CAJ, Rifas-Shiman SL, Litonjua AA et al (2007) Maternal intake of vitamin D during pregnancy and risk of recurrent wheeze in children at 3 years of age. Am J Clin Nutr 85:788–795

67. Gale CR, Robinson SM, Harvey NC et al (2008) Maternal vitamin D status during pregnancy and child outcomes. Eur J Clin Nutr 62:68–77

68. Dennison EM, Arden NK, Keen RW et al (2001) Birthweight, vitamin D receptor genotype and the programming of osteoporosis. Paediatr Perinat Epidemiol 15:211–219

69. Willer CJ, Dyment DA, Sadovnick AD et al (2005) Timing of birth and risk of multiple sclerosis: population based study. BMJ 330:120–124

70. McGrath J, Selten JP, Chant D (2002) Long-term trends in sunshine duration and its association with schizophrenia birth rates and age at first registration – data from Australia and the Netherlands. Schizophrenia Res 54:199–212

71. Kirkpatrick B, Tek C, Allardyce J (2002) Summer birth and deficit schizophrenia in Dumfries and Galloway, Southwestern Scotland. Am J Psychiatry 150:1382–1387

72. Moore SE (1998) Nutrition, immunity and the fetal and infant origins of disease hypothesis in developing countries. Proc Nutr Soc 57:241–247
73. Taussig HB (1966) Possible injury to the cardiovascular system from vitamin D. Ann Intern Med 65:1195–1200
74. Friedman WF (1967) Vitamin D as a cause of the supravalvular aortic stenosis syndrome. Am Heart J 73:718–720
75. Morris CA, Mervis CB (2000) William's syndrome and related disorders. Ann Rev Genomics Hum Genet 1:461–484
76. Hyppönen E, Läärä E, Reunanen A et al (2001) Intake of vitamin D and risk of type 1 diabetes: a birth-cohort study. Lancet 358:1500–1503
77. Gartner LM, Greer FR (2003) Prevention of rickets and vitamin D deficiency: new guidelines for vitamin D intake. Pediatrics 111:908–910
78. Maiya S, Sullivan I, Allgrove J et al (2008) Hypocalcaemia and vitamin D deficiency: an important, but preventable, cause of life-threatening infant heart failure. Heart 94:581–584
79. Abdullah M, Bigras JL, McCrindle BW et al (1999) Dilated cardiomyopathy as a first sign of nutritional vitamin D deficiency rickets in infancy. Can J Cardiol 15:699–701
80. Carlton-Conway D, Tulloh R, Wood L et al (2004) Vitamin D deficiency and cardiac failure in infancy. J R Soc Med 97:238–239
81. Evans KN, Bulmer JN, Kilby MD et al (2004) Vitamin D and placental-decidual function. J Soc Gynecol Investig 11:263–271
82. Cardus A, Parisi E, Gallego C et al (2006) 1,25-Dihydroxyvitamin D3 stimulates vascular smooth muscle cell proliferation through a VEGF-mediated pathway. Kidney Int 69:1377–1384
83. Hewison M (1992) Vitamin D and the immune system. J Endocrinol 132:173–175
84. Li YC, Kong J, Wei M et al (2002) 1,25-Dihydroxyvitamin D_3 is a negative endocrine regulator of the renin-angiotensin system. J Clin Invest 110:229–238
85. Bodnar LM, Catov JM, Simhan HN et al (2007) Maternal vitamin D deficiency increases the risk of preeclampsia. J Clin Endocrinol Metab 92:3517–3522
86. Maghbooli Z, Hossein-nezhad A, Karimi F et al (2008) Correlation between vitamin D_3 deficiency and insulin resistance in pregnancy. Diabetes Metab Res Rev 24:27–32
87. Marya RK, Rathee S, Dua V et al (1988) Effects of vitamin D supplementation during pregnancy on foetal growth. Indian J Med Res 88:488–492
88. Greer FR, Hollis BW, Cripps DJ et al (1984) Effects of maternal ultraviolet B irradiation on vitamin D content of human milk. J Pediatr 105:431–433
89. Hollis BW (1983) Individual quantitation of vitamin D_2, vitamin D_3, 25(OH)D_2, and 25(OH)D_3 in human milk. Anal Biochem 131:211–219
90. Hollis BW, Wagner CL (2004) Vitamin D requirements during lactation: high-dose maternal supplementation as therapy to prevent hypovitaminosis D for both the mother and the nursing infant. Am J Clin Nutr 80(suppl):1752S–1758S
91. Wagner CL, Hulsey TC, Fanning D et al (2006) High dose vitamin D_3 supplementation in a cohort of breastfeeding mothers and their infants: a six-month follow-up pilot study. Breastfeeding Med 2:59–70
92. Hewison M, Zehnder D, Chakraverty R (2004) Vitamin D and barrier function: a novel role for extra-renal 1 alpha-hydroxylase. Mol Cell Endocrinol 215:31–38
93. Hollis BW, Roos BA, Draper HH et al (1981) Vitamin D and its metabolites in human and bovine milk. J Nutr 111:1240
94. Reeve LE, Chesney RW, DeLuca HF (1982) Vitamin D of human milk: identification of biologically active forms. Am J Clin Nutr 36:122–126
95. Weisman Y, Bawnik JC, Eisenberg Z et al (1982) Vitamin D metabolites in human milk. J Pediatr 100:745–748
96. Specker BL, Tsang RC, Hollis BW (1985) Effect of race and diet on human-milk vitamin D and 25-hydroxyvitamin D. Am J Dis Child 139:1134–1137
97. Hollis BW, Pittard WB, Reinhardt TA (1986) Relationships among vitamin D, 25-hydroxyvitamin D, and vitamin D-binding protein concentrations in the plasma and milk of human subjects. J Clin Endocrinol Metab 62:41–44

98. Ala-Houhala M, Koskinen T, Terho A et al (1986) Maternal compared with infant vitamin D supplementation. Arch Dis Child 61:1159–1163

99. Hollis BW, Roos BA, Lambert PW (1981) Vitamin D in plasma: quantitation by a nonequilibrium ligand binding assay. Steroids 37:609–619

100. Horst RL, Reinhardt TA, Beitz DC et al (1981) A sensitive competitive protein binding assay for vitamin D in plasma. Steroids 37:581

101. Greer FR, Hollis BW, Napoli JL (1984) High concentrations of vitamin D_2 in human milk associated with pharmacologic doses of vitamin D_2. J Pediatr 105:61–64

102. Narang NK, Gupta RC, Jain MK (1984) Role of vitamin D in pulmonary tuberculosis. J Assoc Physicians India 32:185–188

103. Institute of Medicine, Food and Nutrition Board, Standing Committee on the Scientific Evaluation of Dietary Reference Intakes (1997) Vitamin D. In: Dietary reference intakes for calcium, phosphorus, magnesium, vitamin D, and fluoride. National Academy Press, Washington, DC

104. Ala-Houhala M (1985) 25-hydroxyvitamin D levels during breast-feeding with or without maternal or infantile supplementation of vitamin D. J Pediatr Gastroenterol Nutr 4:220–226

105. Rothberg AD, Pettifor JM, Cohen DF et al (1982) Maternal-infant vitamin D relationships during breast-feeding. J Pediatr 101:500–503

106. Takeuchi A, Okano T, Tsugawa N et al (1989) Effects of ergocalciferol supplementation on the concentration of vitamin D and its metabolites in human milk. J Nutr 119:1639–1646

107. First Nations, Inuit and Métis Health Committee, Canadian Paediatric Society (CPS) (2007) Vitamin D supplementation: recommendations for Canadian mothers and infants. Paediatr Child Health 12:583–589

108. ESPGAN Committee on Nutrition of the Preterm Infant (1987) Nutrition and feeding of the preterm infants. Acta Pediatr Scan supp l336:1–14

109. Greer FR, Marshall S (1989) Bone mineral content, serum vitamin D metabolite concentrations, and ultraviolet B light exposure in infants fed human milk with and without vitamin D_2 supplements. J Pediatr 114:204–212

110. Greer FR, Searcy JE, Levin RS et al (1982) Bone mineral content and serum 25-hydroxyvitamin D concentrations in breast-fed infants with and without supplemental vitamin D: one-year follow-up. J Pediatr 100:919–922

111. Pittard WB, Geddes KM, Hulsey TC et al (1991) How much vitamin D for neonates? Am J Dis Child 145:1147–1149

112. Zeghoud F, Vervel C, Guillozo H et al (1997) Subclinical vitamin D deficiency in neonates: definition and response to vitamin D supplements. Am J Clin Nutr 65:771–778

113. Atkinson SA, Tsang RC (2005) Calcium, magnesium, phosphorus, and vitamin D. In: Tsang RC, Uauy R, Koletzko B, Zlotkin SH (eds) Nutrition of the preterm infant: scientific basis and practical guidelines, 2nd edn. Digital Educational Publishing, Inc., Cincinnati, OH

114. Zamora SA, Rizzoli R, Belli DC et al (1999) Vitamin D supplementation during infancy is associated with higher bone mineral mass in prepubertal girls. J Clin Endocrinol Metab 84:4541–4544

115. Hyppönen E, Hartikainen AL, Sovio U et al (2007) Does vitamin D supplementation in infancy reduce the risk of pre-eclampsia? Eur J Clin Nutr 61:1136–1139

116. Dawodu A, Khadir A, Hardy DJ et al (2007) Nutritional rickets in the United Emirates: an unresolved cause of childhood morbidity. Middle East Paediatr 7:12–14

117. Greer FR (2004) Issues in establishing vitamin D recommendations for infants and children. Am J Clin Nutr 80:1759S–1762S

118. Kreiter SR, Schwartz RP, Kirkman HN et al (2000) Nutritional rickets in African American breast-fed infants. J Pediatr 137:153–157

119. Rajakumar K, Thomas SB (2005) Reemerging nutritional rickets: a historical perspective. Arch Pediatr Adolesc Med 159:335–341

120. Tomashek KM, Nesby S, Scanlon KS et al (2001) Nutritional rickets in Georgia. Pediatrics 107:e45

121. Ward LM, Gaboury I, Ladhani M et al (2007) Vitamin D-deficiency rickets among children in Canada. CMAJ 177:161–166

122. Weiler HA, Leslie WD, Krahn J et al (2007) Canadian Aboriginal women have a higher prevalence of vitamin D deficiency than non-Aboriginal women despite similar dietary vitamin D intakes. J Nutr 137:461–465
123. Weisburg P, Scanlon KS, Li R et al (2004) Nutritional rickets among children in the United States: review of cases reported between 1986 and 2003. Am J Clin Nutr 80(suppl):1697S–1705S
124. Hess AF (1929) The history of rickets. In: Hess AF (ed) Rickets including osteomalacia and tetany. Lea & Febiger, Philadelphia, PA
125. Dawodu A, Agarwal M, Sankarankutty M et al (2005) Higher prevalence of vitamin D deficiency in mothers of rachitic than nonrachitic children. J Pediatr 147:109–111
126. Dawodu A, Agarwal M, Hossain M et al (2003) Hypovitaminosis D and vitamin D deficiency in exclusively breast-feeding infants and their mothers in summer: a justification for vitamin D supplementation of breast-feeding infants. J Pediatr 142:169–173
127. Ziegler EE, Hollis BW, Nelson SE et al (2006) Vitamin D deficiency in breastfed infants in Iowa. Pediatrics 118:603–610

33 Vitamin D Deficiency in Children and Its Health Consequences

Amy D. DiVasta, Kristen K. van der Veen, and Catherine M. Gordon

Abstract   This chapter reviews data on the prevalence of vitamin D deficiency, risk factors that predispose children and adolescents to this problem, and current approaches to routine vitamin D supplementation and treatment of vitamin D deficiency. The increasing worldwide prevalence of vitamin D deficiency among otherwise healthy pediatric patients is explored. Risk factors for vitamin D deficiency have been identified and include, among others, northern geographical location, dark skin pigmentation, female gender, lack of supplementation, and diseases associated with malabsorption. While supplementation strategies to prevent vitamin D deficiency have not been standardized for infants, children, and adolescents, previous studies comparing different doses have been tested in both healthy youth and those with chronic disease. There is even less consensus on the appropriate treatment doses for young patients identified to have this deficiency. Health outcomes data are critically needed that provide guidance regarding the most appropriate supplementation and treatment regimens for the pediatric age group.

Key Words: Vitamin D deficiency; rickets; children; infants; supplementation; pediatrics

1. INTRODUCTION

Vitamin D deficiency in pediatric patients has serious implications for both childhood well-being and lifelong health. Several recent studies have demonstrated an alarming prevalence of medical conditions related to this nutritional problem, ranging from severe nutritional rickets to subclinical vitamin D deficiency, even within industrialized societies (1–4). However, calculation of precise prevalence rates among children and adolescents is difficult due to lack of consensus regarding the serum vitamin D concentration that defines vitamin D deficiency. An additional challenge is that neither the appropriate dosage for prophylactic vitamin D supplementation nor the optimal dose for repletion of low vitamin D stores has been established for pediatric patients.

In this chapter, we will review the prevalence of vitamin D deficiency in otherwise healthy pediatric populations and in children and adolescents with chronic diseases (including inflammatory bowel disease, cystic fibrosis, obesity, and anorexia nervosa)

From: *Nutrition and Health: Vitamin D*
Edited by: M.F. Holick, DOI 10.1007/978-1-60327-303-9_33,
© Springer Science+Business Media, LLC 2010

that place them at higher risk for nutritional deficits. Finally, we will assess what is known about the treatment of vitamin D deficiency in infants, children, and teenagers, including an analysis of safety and efficacy data drawn from studies published to date.

Both increasing the recognition of vitamin D deficiency as a common pediatric health problem and establishing evidence-based guidelines for treatment are important goals. Vitamin D is well known for its role as a mediator of calcium homeostasis and in the development and maintenance of the skeleton *(5)*. It is well established that vitamin D deficiency may be associated with rickets, low bone density, and fractures in children *(1, 6, 7)*. However, more recently, adult studies have examined the role of vitamin D in the maintenance of extraskeletal tissues and the function of bodily systems other than the skeleton. Hypovitaminosis D has been implicated in many diverse medical conditions, including cancers (e.g., breast, colon, and prostate), hypertension, multiple sclerosis, and diabetes mellitus *(5, 8–11)*. Preliminary pediatric data suggest the importance of vitamin D in the regulation of immune function. This relationship has been illustrated by studies linking suboptimal vitamin D concentrations to susceptibility to childhood diseases such as type 1 diabetes mellitus *(11–13)* and reactive airways disease *(14, 15)*. Vitamin D receptor allelic status and concentrations have also been associated with a predisposition to specific forms of malignancy *(6, 16–18)* and obesity *(19, 20)*. The role of vitamin D for these diverse health problems remains an area of fervent research. Future longitudinal studies will likely show whether vitamin D status during childhood confers protection against the development of many serious and common diseases in later life.

2. PREVALENCE DATA

An increasing number of studies from the United States *(21–24)*, the United Kingdom *(25)*, France *(26)*, Iran *(27)*, Greece *(28)*, Lebanon *(29)*, Turkey *(30)*, China *(31, 32)*, Finland *(33, 34)*, and Canada *(35)* highlight the fact that vitamin D deficiency is a global problem among the pediatric age group. In the United States alone, rickets has been reported in at least 17 states *(36)*. Vitamin D deficiency has been demonstrated in surprisingly high numbers in otherwise healthy infants *(22, 37–39)*, children *(23, 40)*, and adolescents *(21, 41)*. Vitamin D deficiency in otherwise healthy patients is often referred to as a "subclinical" deficiency, as within a typical clinical setting, it would be overlooked due to the absence of abnormal physical manifestations (e.g., genu varum in rickets).

This growing body of literature can be challenging to evaluate. Widely different thresholds for vitamin D "deficiency" are employed by various investigators; 25(OH)D concentrations ranging from 11 to 20 ng/mL (27.5–50.0 nmol/L) have been utilized *(23, 42)*. Varying and frequently confusing expressions such as vitamin D "insufficiency" and "inadequacy" are also used; there are also no standard definitions for these terms. Recent debate has centered around whether an "optimal" serum 25(OH)D status exists and suggests that a standard threshold of 25(OH)D > 30 ng/mL be established as a goal concentration, as favorable health outcomes occur when serum 25(OH)D ranges from 36 to 40 ng/mL *(6)*. A serum 25(OH)D concentration $\leq$ 20 ng/mL (50 nmol/L) is becoming widely accepted as a standard definition of vitamin D deficiency across the age spectrum *(43, 44)*.

2.1. Newborns, Infants, and Toddlers

The seasonally changing prevalence of infantile vitamin D deficiency and the relationship between neonatal and maternal vitamin D stores remain hot topics. High rates of vitamin D deficiency in neonates have been reported in many studies. One large Iranian study (*N*=552), which defined vitamin D deficiency as 25(OH)D < 14 ng/mL (35 nmol/L), found that 93.3% of neonatal cord blood samples and 66.8% of maternal serum samples were vitamin D deficient *(27)*. In northern latitudes in the United States, the rates of vitamin D deficiency in otherwise healthy newborns and infants ranged from 10 to 65%, depending on the population studied *(22, 38, 45)*.

Maternal vitamin D stores during pregnancy affect neonatal vitamin D status *(30, 38, 45, 46)*. Maternal and neonatal 25(OH)D concentrations consistently show a strong positive correlation *(38, 45, 46)*. Interestingly, pre-pregnancy BMI significantly effects both maternal and newborn 25(OH)D concentrations. Pre-gravid obese women (BMI $\geq$ 30 kg/m^2) had lower adjusted mean serum 25(OH)D prenatally (56.5 vs. 62.7 nmol/L, *p*<0.05) and a higher prevalence of vitamin D deficiency (61 vs. 36%, *p*<0.01) *(46)*. An increase in pre-pregnancy BMI from a normal to obese range (22–34 kg/m^2) was associated with a 2.1-fold increased odds of neonatal vitamin D deficiency *(46)*. Maternal vitamin D status affects infant 25(OH)D concentrations even outside of the neonatal period. One Turkish study examined mothers' dressing habits and distinguished between women who were covered and those who were not; the "covered" group had their arms, legs, head, and sometimes parts of their face covered in clothing *(30)*. These dressing habits, as well as the mothers' educational level, were significant predictors of maternal 25(OH)D concentration, which in turn predicted infants' 25(OH)D concentrations at 4 months of age *(30)*. These effects are most pronounced in mothers who exclusively breastfeed and in infants who do not receive vitamin D supplementation *(3, 36, 47–49)*. In two retrospective reviews of nutritional rickets in US infants, all identified cases were breast-fed and did not receive vitamin D supplementation; the vast majority were African American *(49, 50)*.

2.2. Older Children and Adolescents

Several large, recent studies highlight the surprisingly high rates of vitamin D deficiency among healthy children and adolescents *(21, 41, 51)*. Using data from the Third National Health and Nutrition Examination Survey (NHANES), Looker et al. examined rates of vitamin D deficiency [25(OH)D < 12.5 ng/mL (25 nmol/L)] and insufficiency [25(OH)D < 25 ng/mL (62.5 nmol/L)] in adolescents throughout the continental United States *(41)*. Rates of vitamin D deficiency of less than 1% were reported, while the prevalence of vitamin D insufficiency ranged between 21 and 58%. Interestingly, across all communities studied, females were two times more likely than males to have vitamin D insufficiency *(41)*. Similarly, a study of 12- and 15-year-old adolescents conducted in Ireland (54°–55°N latitude) found gender to be predictive of vitamin D status, with 55% of girls and only 38% of boys classified as vitamin D insufficient [25(OH)D levels < 20 ng/mL (50 nmol/L)] during the winter months *(51)*.

2.3. *Factors Influencing Vitamin D Status Throughout Childhood*

Both seasonal variation and skin pigmentation may significantly impact the vitamin D status of pediatric patients of all ages. A recent study examined the rate of craniotabes in neonates, a softening of the skull bones which can be associated with rickets *(52)*. In this investigation, craniotabes was most common during April and May, reaching a nadir in November. The authors concluded that the seasonal variation was most likely associated with in utero vitamin D status, perhaps resulting from low sun exposure of the mother *(52)*. During the winter months, rates of vitamin D deficiency in older children and adolescents are also increased. In the Irish study of Hill et al., 46% of all subjects were vitamin D insufficient during the winter months, while only 17% of subjects were insufficient during the summer months *(51)*. At an even higher northern latitude (44°N), 9- to 11-year-old girls in Maine were monitored for 3 years *(23)*. The girls were at highest risk for vitamin D insufficiency [25(OH)D levels < 20 ng/mL (50 nmol/L)] during the winter months. From September to March, a 28% mean decrease in 25(OH)D concentrations occurred *(23)*. Our group reported similar seasonal effects in a large (N=307) cross-sectional study of healthy adolescents in Boston *(21)*. The prevalence rate of vitamin D deficiency [25(OH)D ≤ 15 ng/mL (37.5 nmol/L)] in the winter was 39.4%, but only 12.1% in the summer months *(21)*. Several other recent reports support these findings *(26, 39, 53, 54)*, leading to the conclusion that vitamin D deficiency and insufficiency are most common during the winter months at northern latitudes.

While the seasonal variability of vitamin D deficiency is widely accepted, data are conflicting regarding the effects of skin pigmentation on vitamin D concentrations in infants. Prevalence rates of infantile vitamin D deficiency appear to be higher in darker-skinned populations *(38, 39)*. In one Pittsburgh study, 9.7% of white newborns were found to be vitamin D deficient [serum 25(OH)D < 15 ng/mL (37.5 nmol/L)], compared to 46.8% of black newborns *(45)*. However, in a Boston-based large study of healthy infants and toddlers, no association was found between skin pigmentation and serum vitamin D concentrations *(22)*. One possible explanation for this unexpected finding is that young infants and toddlers may be dressed more heavily at northern latitudes; therefore, sun exposure is minimized, negating the effect of skin pigmentation on cutaneous vitamin D synthesis.

Race also appears to play an important role in vitamin D deficiency among older children and adolescents. Prevalence rates in studies including only children of color are higher than those investigations which evaluate only Caucasian subjects *(24)*. Significant differences in rates of vitamin D deficiency between self-identified racial groups exist; our group reported that 35.9% of African-American adolescents were vitamin D deficient, compared to only 6.1% of white subjects *(21)*. Overall, African-American participants had mean 25(OH)D concentrations 40% lower than white participants *(21)*. Notably, the Hispanic participants had mean 25(OH)D concentrations 21.7% lower than the white participants.

While race seems to be associated with serum vitamin D concentrations during adolescence, it remains unclear if this relationship is due to cutaneous vitamin D synthesis, socioeconomic factors, or some other confounding variable. Vitamin D_3 synthesis is known to occur in the skin and is related to levels of epidermal melanin which produce skin pigmentation. Matsuoka et al. examined the cutaneous synthesis of vitamin D_3 across different racial groups and found that after UVB treatment, vitamin D_3 concentrations were significantly higher in white subjects than in black or Indian subjects *(55)*. However, the effect of race on post-UVB 25(OH)D concentrations was marginal. These findings suggest that additional variables likely contribute to racial differences in 25(OH)D concentrations, including variation in eating habits, availability of supplementation, sun exposure, access to healthcare, and educational level. Additional risk factors affect pediatric patients of all ages (Table 1). While the effects of skin pigmentation on vitamin D concentrations are still under debate, and may differ by age, this variable should be considered an additional risk factor for subclinical vitamin D deficiency.

Table 1
Risk Factors for Vitamin D Deficiency in Pediatric Patients

Female gender
African-American race
Extreme latitude (northern or southern)
Lack of vitamin D supplementation
Exclusive breastfeeding, without supplementation
Lack of sun exposure
Lack of milk consumption
Low socioeconomic status

Diseases
Obesity
Inflammatory bowel disease
Cystic fibrosis
Epilepsy
Celiac disease

Use of Specific Medications
Glucocorticoids
Anticonvulsants

3. PREVALENCE OF VITAMIN D DEFICIENCY IN AT-RISK GROUPS

3.1. Overweight

Recently, an inverse association between serum 25(OH)D concentration and total body fat has been described in several studies comparing normal-weight and overweight children and adolescents *(56–58)*. The inverse correlation between vitamin D and body fat may be due in part to augmented sequestration of vitamin D, a fat-soluble vitamin, in adipose tissue, resulting in lower levels of bioavailable 25(OH)D *(57)*. In fact, after 1 year of weight loss and a decrease in body fat, serum vitamin D levels increased in previously overweight children *(59)*. Interestingly, adolescent females with anorexia nervosa, a group with very low body fat, had a significantly *lower* prevalence of vitamin D deficiency than an age-matched group of healthy-weighted control subjects *(60)*.

Investigators who support the "adipose-storage theory" have suggested that the lower levels of bioavailable vitamin D associated with being overweight are linked to insulin resistance and metabolic syndrome. Vitamin D deficiency is associated with insulin resistance during pregnancy *(61)* and in healthy, glucose-tolerant adults. Results suggest that an increase in serum 25(OH)D concentration from 10 to 30 ng/mL could improve insulin sensitivity by 60% *(10)*. Alternatively, low vitamin D stores have been hypothesized to contribute to the causation of obesity, rather than to result from obesity. Low serum 25(OH)D concentrations have been associated with abnormal vitamin D signaling in beta cells; *(10, 20)* it is plausible that this abhorrent signaling could lead to an increased risk for obesity. Allelic variation in the vitamin D receptor (VDR) gene was associated with susceptibility to higher body fat and weight in adults *(19, 20)*, suggesting that vitamin D sensitivity may be a risk factor for obesity development. These two theories are not mutually exclusive. In fact, some small investigations have shown no relationship between weight status and vitamin D. In one small study ($N = 41$), African-American children were supplemented with vitamin D_3 400 IU daily for 1 month *(62)*. No significant effect of body fat on serum vitamin D levels was seen either before or after supplementation *(62)*. Thus, this area remains one of controversy and continued research.

3.2. Epilepsy and Other Neurological Disorders

Many children who are receiving chronic anticonvulsant therapy for epilepsy become not only vitamin D deficient, but also have a low bone mineral density *(63, 64)*. Nettekoven et al. found a significant difference in 25(OH)D concentrations between patients receiving first, second, and third generation antiepileptic drugs (AEDs) and a control group of healthy, geographically matched children. In this study, only 2.6% of subjects taking AEDs had an optimal level of vitamin D [25(OH)D > 30 ng/mL (75 nmol/L)] compared to 43.2% of the control group *(63)*. While the exact mechanism of the AED in altering vitamin D metabolism is not known, AEDs may activate the pregname X receptor, leading to an upregulation of the cytochrome P450 enzymatic system and inactivation of vitamin D metabolites *(65–67)*. Therefore, in addition to being at high risk for suboptimal vitamin D concentrations, patients treated with AEDs may require higher doses of vitamin D supplementation to prevent deficiency.

Low 25(OH)D concentrations have also been reported in patients with Duchenne muscular dystrophy (DMD) *(68)*. Traditionally, the observed low vitamin D concentrations were attributed to long-term use of glucocorticoids in patients with this disease, as has been reported in other medical conditions associated with chronic use of these agents *(68–70)*. However, Bianchi et al. found low serum 25(OH)D concentrations in a subgroup of DMD patients who were *not* being treated with glucocorticoids and who consumed normal diets and had exposure to sunlight *(68)*.

3.3. Inflammatory Bowel Disease, Celiac Disease, and Cystic Fibrosis

In patients with inflammatory bowel disease (IBD) and celiac disease, vitamin D absorption is compromised, placing patients at greater risk for vitamin D deficiency. One recent study examined a large group of patients with IBD, including subjects with both Crohn disease (n=94) and ulcerative colitis (n=36). Both cohorts had a relatively high rate of vitamin D deficiency, with 38% of subjects meeting criteria for deficiency [25(OH)D $\leq$ 15 ng/mL (37.5 nmol/L)] and 12.8% meeting criteria for severe deficiency [25(OH)D $\leq$8 ng/mL (20 nmol/L)] *(71)*. No between-group differences were noted. Similarly, hypovitaminosis D is prevalent in those with celiac disease, particularly in newly diagnosed patients or refractory patients who are unresponsive to dietary therapy *(72, 73)*; low baseline 25(OH)D concentrations appear to improve after dietary treatments are initiated in cases of newly diagnosed celiac disease. The identified low serum vitamin D levels have clinical correlates. In one study examining children and adolescents with Crohn's disease, vitamin D deficiency was associated with decreased bone mineralization; these effects were especially dramatic during the winter months *(74)*. Low concentrations of 25(OH)D have similarly been associated with symptomatic hypocalcemia *(75)* as well as low bone mineral density for age *(76)*.

Pediatric patients with cystic fibrosis (CF) are also at high risk for suboptimal vitamin D states due to dysfunction of the exocrine pancreas and poor absorption of fat-soluble vitamins *(77)*. Even with routine vitamin D supplementation, patients with CF frequently exhibit low 25(OH)D concentrations *(42)*. In one study of young adults with CF, approximately one-third of subjects were vitamin D deficient despite reported compliance with routine supplementation *(78)*. Other studies conducted in children and young adults have found a similar high prevalence of vitamin D deficiency in these patients *(79–82)*.

3.4. Hereditary Predisposition

Conflicting data exist regarding the relationship between vitamin D receptor (VDR) genotype, vitamin D deficiency, and nutritional rickets in children. The VDR is encoded by a single gene. Its several alleles are distinguished by the presence of different restriction enzyme recognition sequences, ranging in length from 4 to 6 nucleotides. Different alleles of the VDR gene have been associated with a wide range of disorders, including Graves' disease *(83)*, cancer *(84, 85)*, autoimmune thyroiditis *(83)*, and lower respiratory tract infections *(86)*. There is also a VDR allele that has been associated with 1,25(OH)$_2$D insensitive rickets *(87)*, an inherited form of rickets which is nonresponsive to vitamin D treatment. However, the relationships between other VDR alleles and nutritional rickets in children remain unclear.

To determine allelic status, restriction enzymes are used to cleave amplified genomic DNA. Using this technique, one can only examine mutations at highly localized regions within the introns and exons of a gene. In a study of Mongolian children (N=152), Kaneko et al. did not find a significant difference in allelic frequency between children with and without nutritional rickets *(88)*. However, only restriction sites in intron 8 and exon 9 were examined. Previous work that examined restriction sites in intron 8 and exons 2 and 9 found the only significant differences between cohorts (with and without rickets) existed in the Fok 1 restriction site of exon 2 *(89)*. Therefore, any significant contribution of allelic variation in the VDR may be linked solely to the Fok 1 restriction site.

4. ASSESSMENT OF VITAMIN D STATUS

The same principles and strategies that are used to evaluate the vitamin D status of adults are followed in the assessment of children and adolescents. Serum 25-hydroxyvitamin D [25(OH)D] concentrations provide the most reliable measurement of bioavailable vitamin D *(90)* and remain the most informative measure of circulating vitamin D concentrations and bodily stores. Serum 25(OH)D is the precursor of the active form of vitamin D, 1,25-dihydroxyvitamin D [1,25(OH)$_2$D], which is produced in the kidney. 1,25(OH)$_2$D is not an ideal marker for overall vitamin D stores in otherwise healthy patients because conversion to active vitamin D is tightly regulated by phosphorus, calcium, and parathyroid hormone (PTH) *(43)*. However, 1,25(OH)$_2$D measurements may offer helpful information in cases where renal conversion of 25(OH)D to active vitamin D may be impaired, including chronic renal disease *(90, 91)*. Additional serum measurements that aid in determining a child's vitamin D status include serum levels of calcium, phosphorus, and PTH. Total alkaline phosphatase concentrations reflect PTH secretion and can provide a marker of bone turnover.

Signs and symptoms of vitamin D deficiency may be overt or subtle (Table 2). The majority of patients have "subclinical deficiency," being completely asymptomatic at the time of presentation, which makes this problem a more widespread public health issue. Other pediatric patients will present with typical bony findings, such as genu varum when epiphyses are open, and radiographic changes, such as osteopenia and metaphyseal fraying (Fig. 1).

Table 2
Clinical Signs and Symptoms of Vitamin D Deficiency in Children

Craniotabes
Genu varum or valgum
Costchondral swellings ("rachitic rosary")
Growth delay or arrest (failure to thrive)
Muscle weakness
Harrison's grooves
Hair loss or alopecia
Delayed dentition
Refusal to walk
Fractures
Seizures
Tetany

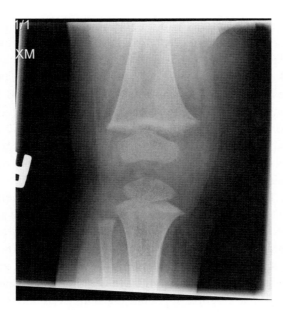

Fig. 1. Knee radiograph of an 11-month-old with vitamin D deficiency rickets, illustrating osteopenia and metaphyseal fraying. Reproduced from Williams et al. 2008 *(123)* with permission from Sage Publications.

5. SUPPLEMENTATION

In the past 50 years, debate has continued regarding the optimal daily intake and dosage of vitamin D supplementation needed to prevent vitamin D deficiency. In 1963, the American Academy of Pediatrics (AAP) Committee on Nutrition recommended a dose of vitamin D 400 IU daily as the standard of care for children *(92)*. However, by 1997, the National Academy of Sciences halved the recommended dose to 200 IU daily, citing this lower dose as the amount needed to prevent rickets *(93)*. Currently, these recommendations are under scrutiny; further studies are needed to examine whether a daily dose of 200 IU leads to vitamin D concentrations that provide adequate protection against other diseases in addition to rickets. The US Food and Drug Administration's (FDA) Daily Recommended Allowance of vitamin D is 400 IU *(44)*, consistent with the recommendations of the 1963 American Academy of Pediatrics (AAP) Committee on Nutrition *(92)*. In 2003, the AAP recommended supplementation with 200 IU daily across the pediatric age spectrum *(93)* and in 2008, increased the supplementation dose to 400 IU daily *(94)*. However, many clinicians believe that the requirements for adolescents are likely to be significantly higher than these current recommendations *(92, 95)*.

5.1. Supplementation in Infants

Several studies of newborns and infants have focused on the impact of breastfeeding without supplementation on vitamin D levels. Data consistently demonstrate that infants who are exclusively breast-fed are at higher risk to become or remain vitamin D deficient *(96, 97)* than supplemented or formula-fed infants. As mentioned, the AAP currently recommends supplementing infants with vitamin D 200 IU daily *(93)*. In contrast, the Canadian Paediatric Society (CPS) recently suggested that the current recommendations of 200 and 400 IU daily for breastfeeding infants and mothers, respectively, are too low, even though these doses may prevent vitamin D deficiency *(98)*. The CPS recommends that pregnant or breastfeeding mothers consume vitamin D 2,000 IU daily, especially during the winter months, to maintain optimal vitamin D stores in both the mother and the infant. This recommendation includes the caveat that periodic monitoring for side effects of supplementation should occur *(98)*. Vitamin D doses ranging between 200 and 800 IU daily are recommended for direct infant supplementation, depending on factors such as gestational age and BMI *(98)*.

One recent study found that prenatal maternal 25(OH)D concentrations were positively correlated with cord blood 25(OH)D measurements *(47)*. The infants who were exclusively breast-fed were followed longitudinally for 3 months; vitamin D deficiency persisted in these infants *(47)*. The etiology of the persistently low 25(OH)D concentrations is likely the sparse amount of vitamin D in human breast milk, which contains as little as 25–33 IU/L *(99, 100)* of vitamin D. Interestingly, data gathered from clinical practice indicate that a large fraction of pediatricians may erroneously believe that breast milk contains sufficient vitamin D for infant health. In one recent survey, only 44.6% of pediatricians ($N=383$) recommended vitamin D supplementation for all infants *(101)*. Of those who did not recommend infant supplementation, 83.1% believed that breast milk contained sufficient vitamin D levels to maintain adequate vitamin D stores *(101)*.

Fortunately, this is a preventable issue. Supplementation of breastfeeding mothers with vitamin D, 2,000 or 4,000 IU daily, conferred a protective effect against vitamin D deficiency in infants *(97)*. This strategy also had a positive effect in raising maternal 25(OH)D concentrations to optimal ranges. Importantly, no adverse effects related to treatment, such as hypercalciuria, were reported in infants or mothers. Direct infant supplementation has also been studied. Mawer et al. supplemented premature neonates (25–32 weeks gestational age) with oral vitamin D_2, either 1,000 or 3,000 IU daily *(95)*. No significant differences in 25(OH)D concentrations were found between the two groups; serum 25(OH)D plateaued at approximately 30 ng/mL (75 nmol/L) regardless of the supplementation dose received *(95)*.

Despite these data, infant formula should not be considered the optimal alternative. Researchers have analyzed the amount of vitamin D fortification in commercially available infant formula, realizing that many infants are not exclusively breast-fed. Of the 10 commercial infant formula samples evaluated, 70% contained more than 200% of the amount of vitamin D fortification claimed in the nutritional information label *(102)*.

Pediatricians should still encourage breast milk as the ideal infant food when paired with vitamin D supplementation to either the mother or the infant. High-dose treatment of the mother (vitamin D 2,000–4,000 IU daily) appears to provide adequate vitamin D supplementation for both mother and child with no associated adverse effects *(97)*. Doses as high as 6,400 IU daily have been administered to lactating mothers, leading to both an appreciable amount of vitamin D secreted in breast milk and a favorable safety profile *(103)*. Similarly, direct infant supplementation at doses of 200–800 IU daily may be safely utilized to prevent vitamin D deficiency and nutritional rickets. Future longitudinal studies will clarify the optimal dose range for use in this population.

5.2. Supplementation in Children and Adolescents

In children and adolescents, much research has been focused on food fortification with vitamin D rather than direct supplementation *(104–109)*. One Finnish longitudinal study compared 25(OH)D concentrations in adolescent females before and after the initiation of vitamin D fortification of fluid milk and margarines *(104)*. The cumulative prevalence of 25(OH)D concentrations < 75 nmol/L pre- and post-fortification was unchanged (93% at both time points), leading to the conclusion that fortification was inadequate to prevent vitamin D insufficiency. The authors postulated that this lack of effect was due to lack of consumption of the fortified products by this age group and that "cautious fortification" of more commonly eaten food products, such as bread, could make a more significant difference *(104)*. Similar data have been reported from the United States and Canada, leading many to conclude that existing official recommendations for vitamin D fortification are not adequate to prevent vitamin D deficiency. Several barriers may impede the efficacy of fortification as a means of preventing vitamin D deficiency, including low doses, low levels of milk consumption, and infrequent fortification of eligible milk products *(105)*. Both gender and racial differences also contribute to problems with food fortification. In the United States, female teenagers report the lowest intake of food sources of vitamin D when compared to the general population *(110)*. Significant racial differences exist among children aged 1–8 years; 52% of non-Hispanic blacks, 41% of non-Hispanic whites, and 31% of Mexican Americans do not achieve recommended daily intake of vitamin D through supplemented food *(111)* For these reasons, direct vitamin D supplementation may be a more efficacious way to maintain adequate serum vitamin D concentrations in both children and adolescents.

As previously noted, vitamin D deficiency is more common in the winter months than during summer *(21, 41, 54)*. Consistent supplementation would also help to address the significant seasonal variation of vitamin D deficiency that commonly occurs. In one study, higher 25(OH)D concentrations measured in September were associated with higher 25(OH)D concentrations in March, suggesting that building adequate vitamin D stores throughout the summer months could help stave off insufficiency during the winter *(23)*. Summer vitamin D concentrations were maintained through the winter months in adolescent boys supplemented with oral vitamin D 100,000 IU at three bimonthly time points (end of September, November, and January) *(26)*.

Given that socioeconomic status *(112)* and education level *(30)* have both been cited as predictors of rickets and 25(OH)D status, it is also important to take into consideration potential educational, economic, and cultural barriers when considering strategies to prevent vitamin D deficiency. In one recent study, immigrant parents presenting for primary medical care were randomized to receive either free vitamin D_2 drops and translated educational tools regarding the importance of vitamin D or usual medical care. The intervention group had higher serum 25(OH)D at follow-up than control subjects (93.5 vs. 72.7 nmol/L, $p = 0.03$) *(113)*. The implementation of similar innovative programs could help to bridge gaps in socioeconomic and educational status, and improve care for culturally diverse populations. More studies are needed to determine the optimal supplementation regimens to prevent vitamin D deficiency in pediatric patients, particularly in preparation for the winter months, and in identified high-risk groups, including females and dark-skinned individuals.

6. TREATMENT

The treatment of children who have been identified as vitamin D deficient is distinct from identifying individuals needing routine supplementation. In the United States, oral forms of vitamin D are the most easily obtained, and thus, the most commonly administered. Available preparations include ergocalciferol (vitamin D_2), cholecalciferol (vitamin D_3), and calcitriol. There are conflicting data as to whether significantly different responses to treatment with vitamin D_2 or vitamin D_3 exist. Some adult studies suggest that oral vitamin D_3 ingestion raises and maintains 25(OH)D concentrations better than vitamin D_2 intake *(114, 115)*. However, Holick et al. found no significant difference in 25(OH)D concentrations between participants ($n = 68$) treated with a regimen of either vitamin D_2 or vitamin D_3 in comparable dosages *(116)*.

No consensus exists as to whether optimal treatment for pediatric vitamin D deficiency involves a one-time dose, daily dose, or weekly dose of vitamin D; similarly, the optimal duration of therapy is unknown. Frequently recommended regimens include vitamin D_3 200,000 IU PO every 3 months, or 1,000–2,000 IU of either vitamin D_2 or vitamin D_3 PO daily for several weeks. In a Nigerian pediatric cohort, intramuscular therapy was successful when two injections of vitamin D_2 600,000 IU were administered 12 weeks apart and supplemented with daily calcium tablets *(117)*. A single daily oral dose of vitamin D 150,000 IU, 300,000 IU, or 600,000 IU has also been effective for the treatment of nutritional rickets *(118)*. While such single-dose therapy reduces problems with noncompliance, negative side effects of these high-dose regimens have been observed. For example, solitary dosing of vitamin D 300,000–600,000 IU (known as *stosstherapy*) may increase the risk of asymptomatic hypercalcemia *(112)*. Hypocalcemia, associated with the "hungry bone syndrome" can also result *(44)*. Therefore, high-dose *stosstherapy* is generally discouraged in an outpatient setting.

Recently, our group compared the effectiveness of three treatment regimens (vitamin D_2 2,000 IU daily, vitamin D_2 50,000 IU weekly, or vitamin D_3 2,000 IU daily) for the repletion of vitamin D stores in infants with known vitamin D deficiency [25(OH)D < 20 ng/mL (50 nmol/L)] *(22)*. No significant differences in efficacy or safety were found among these regimens. Neither symptomatic hypercalcemia nor other side effects of therapy were reported, in contrast to previous studies utilizing larger, single-dose protocols *(112)* Each regimen corrected the vitamin D deficiency with equal efficacy, implying that a clinician can tailor the treatment (weekly vs. daily therapy) to meet the needs of a given family.

In children with chronic diseases that are associated with malabsorption, prophylactic supplementation with higher doses of vitamin D can help prevent vitamin D deficiency and lead to long-term maintenance of optimal 25(OH)D concentrations. A study in adults with CF found that routine vitamin D supplementation significantly increased serum concentrations of 25(OH)D from baseline and maintained serum 25(OH)D >20 ng/mL *(119)*. Those individuals with the lowest baseline measures of 25(OH)D exhibited the largest gains in 25(OH)D *(119)*. Vitamin D supplementation in pediatric patients at risk for malabsorption, including children with CF, IBD, and celiac disease, is vitally important, because even with optimal sun exposure and adequate dietary intake, optimal levels of vitamin D are unlikely to be reached.

When treating infants and children with vitamin D deficiency, safety issues related to therapy should be considered. Overall, the occurrence of vitamin D toxicity is very rare *(120)* and the margin of safety wide. Generally, no signs of vitamin D toxicity occur until serum 25(OH)D concentrations $\geq$ 150 ng/mL (375 nmol/L) *(96)* are reached, or doses are given in excess of 10,000 IU daily *(121)*. Symptoms of vitamin D toxicity include weakness, headache, somnolence, metallic taste, nausea, constipation, and bone pain can occur, which may indicate hypercalcemia *(44)*. No evidence suggests that vitamin D consumption < 10,000 IU daily will result in ill effects *(122)*.

7. CONCLUSIONS

While recent research has led to a better appreciation of the increasing prevalence of vitamin D deficiency among infants, children, and adolescents, many gaps in knowledge still exist. Future research should focus on establishing optimal serum vitamin D concentrations for those within the pediatric age group and on determining whether childhood vitamin D deficiency impacts later health outcomes throughout the life span. Implementation of cost-effective educational and treatment programs to target high-risk groups will be crucial to alleviate the existing epidemic of vitamin D deficiency. In addition, the most effective vitamin D supplementation and treatment strategies must be confirmed: ones that are easy to follow, safe, inexpensive, and efficacious.

REFERENCES

1. Holick MF (2006) Resurrection of vitamin D deficiency and rickets. J Clin Invest 116:2062–2072
2. Robinson PD, Hogler W, Craig ME et al (2006) The re-emerging burden of rickets: a decade of experience from Sydney. Arch Dis Child 91:564–568
3. Cosgrove L, Dietrich A (1985) Nutritional rickets in breast-fed infants. J Fam Pract 21:205–209
4. Binet A, Kooh SW (1996) Persistence of vitamin D-deficiency rickets in Toronto in the 1990s. Can J Public Health 87:227–230
5. Cantorna MT (2006) Vitamin D and its role in immunology: multiple sclerosis, and inflammatory bowel disease. Prog Biophys Mol Biol 92:60–64
6. Bischoff-Ferrari HA, Giovannucci E, Willett WC, Dietrich T, Dawson-Hughes B (2006) Estimation of optimal serum concentrations of 25-hydroxyvitamin D for multiple health outcomes. Am J Clin Nutr 84:18–28
7. Cheng S, Tylavsky F, Kroger H et al (2003) Association of low 25-hydroxyvitamin D concentrations with elevated parathyroid hormone concentrations and low cortical bone density in early pubertal and prepubertal Finnish girls. Am J Clin Nutr 78:485–492
8. Munger KL, Levin LI, Hollis BW, Howard NS, Ascherio A (2006) Serum 25-hydroxyvitamin D levels and risk of multiple sclerosis. JAMA 296:2832–2838
9. Willer CJ, Dyment DA, Sadovnick AD, Rothwell PM, Murray TJ, Ebers GC (2005) Timing of birth and risk of multiple sclerosis: population based study. BMJ 330:120
10. Chiu KC, Chu A, Go VL, Saad MF (2004) Hypovitaminosis D is associated with insulin resistance and beta cell dysfunction. Am J Clin Nutr 79:820–825
11. Littorin B, Blom P, Scholin A et al (2006) Lower levels of plasma 25-hydroxyvitamin D among young adults at diagnosis of autoimmune type 1 diabetes compared with control subjects: results from the nationwide Diabetes Incidence Study in Sweden (DISS). Diabetologia 49:2847–2852
12. Hypponen E, Laara E, Reunanen A, Jarvelin MR, Virtanen SM (2001) Intake of vitamin D and risk of type 1 diabetes: a birth-cohort study. Lancet 358:1500–1503

13. Stene LC, Joner G (2003) Use of cod liver oil during the first year of life is associated with lower risk of childhood-onset type 1 diabetes: a large, population-based, case-control study. Am J Clin Nutr 78:1128–1134

14. Camargo CA, Jr, Rifas-Shiman SL, Litonjua AA et al (2007) Maternal intake of vitamin D during pregnancy and risk of recurrent wheeze in children at 3 y of age. Am J Clin Nutr 85:788–795

15. Devereux G, Litonjua AA, Turner SW et al (2007) Maternal vitamin D intake during pregnancy and early childhood wheezing. Am J Clin Nutr 85:853–859

16. Berwick M, Armstrong BK, Ben-Porat L et al (2005) Sun exposure and mortality from melanoma. J Natl Cancer Inst 97:195–199

17. Chang ET, Smedby KE, Hjalgrim H et al (2005) Family history of hematopoietic malignancy and risk of lymphoma. J Natl Cancer Inst 97:1466–1474

18. McGrath J (2001) Does 'imprinting' with low prenatal vitamin D contribute to the risk of various adult disorders? Med Hypotheses 56:367–371

19. Barger-Lux MJ, Heaney RP, Hayes J, DeLuca HF, Johnson ML, Gong G (1995) Vitamin D receptor gene polymorphism, bone mass, body size, and vitamin D receptor density. Calcif Tissue Int 57: 161–162

20. Martini LA, Wood RJ (2006) Vitamin D status and the metabolic syndrome. Nutr Rev 64:479–486

21. Gordon CM, DePeter KC, Feldman HA, Grace E, Emans SJ (2004) Prevalence of vitamin D deficiency among healthy adolescents. Arch Pediatr Adolesc Med 158:531–537

22. Gordon CM, Feldman HA, Sinclair L, Williams A, Cox J (2008) Prevalence of vitamin D deficiency among healthy infants and toddlers. Arch Pediatr Adolesc Med 162:505–512

23. Sullivan SS, Rosen CJ, Halteman WA, Chen TC, Holick MF (2005) Adolescent girls in Maine are at risk for vitamin D insufficiency. J Am Diet Assoc 105:971–974

24. Talwar SA, Swedler J, Yeh J, Pollack S, Aloia JF (2007) Vitamin-D nutrition and bone mass in adolescent black girls. J Natl Med Assoc 99:650–657

25. Lawson M, Thomas M (1999) Vitamin D concentrations in Asian children aged 2 years living in England: population survey. BMJ 318:28

26. Guillemant J, Le HT, Maria A, Allemandou A, Peres G, Guillemant S (2001) Wintertime vitamin D deficiency in male adolescents: effect on parathyroid function and response to vitamin D3 supplements. Osteoporos Int 12:875–879

27. Maghbooli Z, Hossein-Nezhad A, Shafaei AR, Karimi F, Madani FS, Larijani B (2007) Vitamin D status in mothers and their newborns in Iran. BMC Pregnancy Childbirth 7:1

28. Nicolaidou P, Hatzistamatiou Z, Papadopoulou A et al (2006) Low vitamin D status in mother-newborn pairs in Greece. Calcif Tissue Int 78:337–342

29. El-Hajj Fuleihan G, Nabulsi M, Choucair M et al (2001) Hypovitaminosis D in healthy schoolchildren. Pediatrics 107:E53

30. Pehlivan I, Hatun S, Aydogan M, Babaoglu K, Gokalp AS (2003) Maternal vitamin D deficiency and vitamin D supplementation in healthy infants. Turk J Pediatr 45:315–320

31. Du X, Greenfield H, Fraser DR, Ge K, Trube A, Wang Y (2001) Vitamin D deficiency and associated factors in adolescent girls in Beijing. Am J Clin Nutr 74:494–500

32. Specker BL, Ho ML, Oestreich A et al (1992) Prospective study of vitamin D supplementation and rickets in China. J Pediatr 120:733–739

33. Lehtonen-Veromaa M, Mottonen T, Irjala K et al (1999) Vitamin D intake is low and hypovitaminosis D common in healthy 9- to 15-year-old Finnish girls. Eur J Clin Nutr 53:746–751

34. Outila TA, Karkkainen MU, Lamberg-Allardt CJ (2001) Vitamin D status affects serum parathyroid hormone concentrations during winter in female adolescents: associations with forearm bone mineral density. Am J Clin Nutr 74:206–210

35. Ward LM, Gaboury I, Ladhani M, Zlotkin S (2007) Vitamin D-deficiency rickets among children in Canada. CMAJ 177:161–166

36. Weisberg P, Scanlon KS, Li R, Cogswell ME (2004) Nutritional rickets among children in the United States: review of cases reported between 1986 and 2003. Am J Clin Nutr 80:1697S–705S

37. Gessner BD, Plotnik J, Muth PT (2003) 25-hydroxyvitamin D levels among healthy children in Alaska. J Pediatr 143:434–437

38. Lee JM, Smith JR, Philipp BL, Chen TC, Mathieu J, Holick MF (2007) Vitamin D deficiency in a healthy group of mothers and newborn infants. Clin Pediatr (Phila) 46:42–44
39. Ziegler EE, Hollis BW, Nelson SE, Jeter JM (2006) Vitamin D deficiency in breastfed infants in Iowa. Pediatrics 118:603–610
40. Rajakumar K, Fernstrom JD, Janosky JE, Greenspan SL (2005) Vitamin D insufficiency in preadolescent African-American children. Clin Pediatr (Phila) 44:683–692
41. Looker AC, Dawson-Hughes B, Calvo MS, Gunter EW, Sahyoun NR (2002) Serum 25-hydroxyvitamin D status of adolescents and adults in two seasonal subpopulations from NHANES III. Bone 30:771–777
42. Rovner AJ, Stallings VA, Schall JI, Leonard MB, Zemel BS (2007) Vitamin D insufficiency in children, adolescents, and young adults with cystic fibrosis despite routine oral supplementation. Am J Clin Nutr 86:1694–1699
43. Holick MF (2007) Vitamin D deficiency. N Engl J Med 357:266–281
44. Huh SY, Gordon CM (2008) Vitamin D deficiency in children and adolescents: epidemiology, impact and treatment. Rev Endocr Metab Disord 9:161–170
45. Bodnar LM, Simhan HN, Powers RW, Frank MP, Cooperstein E, Roberts JM (2007) High prevalence of vitamin D insufficiency in black and white pregnant women residing in the northern United States and their neonates. J Nutr 137:447–452
46. Bodnar LM, Catov JM, Roberts JM, Simhan HN (2007) Prepregnancy obesity predicts poor vitamin D status in mothers and their neonates. J Nutr 137:2437–2442
47. Bhalala U, Desai M, Parekh P, Mokal R, Chheda B (2007) Subclinical hypovitaminosis D among exclusively breastfed young infants. Indian Pediatr 44:897–901
48. Chesney RW (2002) Rickets: the third wave. Clin Pediatr (Phila) 41:137–139
49. Kreiter SR, Schwartz RP, Kirkman HN, Jr, Charlton PA, Calikoglu AS, Davenport ML (2000) Nutritional rickets in African American breast-fed infants. J Pediatr 137:153–157
50. Mylott BM, Kump T, Bolton ML, Greenbaum LA (2004) Rickets in the Dairy State. WMJ 103:84–87
51. Hill TR, Cotter AA, Mitchell S et al (2008) Vitamin D status and its determinants in adolescents from the Northern Ireland Young Hearts 2000 cohort. Br J Nutr 99:1061–1067
52. Yorifuji J, Yorifuji T, Tachibana K et al (2008) Craniotabes in normal newborns; the earliest sign of subclinical vitamin D deficiency. J Clin Endocrinol Metab 93:1784–1788
53. McCarthy D, Collins A, O'Brien M et al (2006) Vitamin D intake and status in Irish elderly women and adolescent girls. Ir J Med Sci 175:14–20
54. Weng FL, Shults J, Leonard MB, Stallings VA, Zemel BS (2007) Risk factors for low serum 25-hydroxyvitamin D concentrations in otherwise healthy children and adolescents. Am J Clin Nutr 86:150–158
55. Matsuoka LY, Wortsman J, Haddad JG, Kolm P, Hollis BW (1991) Racial pigmentation and the cutaneous synthesis of vitamin D. Arch Dermatol 127:536–538
56. Alemzadeh R, Kichler J, Babar G, Calhoun M (2008) Hypovitaminosis D in obese children and adolescents: relationship with adiposity, insulin sensitivity, ethnicity, and season. Metabolism 57:183–191
57. Wortsman J, Matsuoka LY, Chen TC, Lu Z, Holick MF (2000) Decreased bioavailability of vitamin D in obesity. Am J Clin Nutr 72:690–693
58. Smotkin-Tangorra M, Purushothaman R, Gupta A, Nejati G, Anhalt H, Ten S (2007) Prevalence of vitamin D insufficiency in obese children and adolescents. J Pediatr Endocrinol Metab 20: 817–823
59. Reinehr T, de Sousa G, Alexy U, Kersting M, Andler W (2007) Vitamin D status and parathyroid hormone in obese children before and after weight loss. Eur J Endocrinol 157:225–232
60. Haagensen AL, Feldman HA, Ringelheim J, Gordon CM (2008) Low prevalence of vitamin D deficiency among adolescents with anorexia nervosa. Osteoporos Int 19:289–294
61. Maghbooli Z, Hossein-Nezhad A, Karimi F, Shafaei AR, Larijani B (2008) Correlation between vitamin D3 deficiency and insulin resistance in pregnancy. Diabetes Metab Res Rev 24:27–32
62. Rajakumar K, Fernstrom JD, Holick MF, Janosky JE, Greenspan SL (2008) Vitamin D status and response to Vitamin D(3) in obese vs.non-obese African American children. Obesity (Silver Spring) 16:90–95

63. Nettekoven S, Strohle A, Trunz B et al (2008) Effects of antiepileptic drug therapy on vitamin D status and biochemical markers of bone turnover in children with epilepsy. Eur J Pediatr 167:1369–1377

64. Samaniego EA, Sheth RD (2007) Bone consequences of epilepsy and antiepileptic medications. Semin Pediatr Neurol 14:196–200

65. Pascussi JM, Robert A, Nguyen M et al (2005) Possible involvement of pregnane X receptor-enhanced CYP24 expression in drug-induced osteomalacia. J Clin Invest 115:177–186

66. Valsamis HA, Arora SK, Labban B, McFarlane SI (2006) Antiepileptic drugs and bone metabolism. Nutr Metab (Lond) 3:36

67. Zhou C, Assem M, Tay JC et al (2006) Steroid and xenobiotic receptor and vitamin D receptor crosstalk mediates CYP24 expression and drug-induced osteomalacia. J Clin Invest 116:1703–1712

68. Bianchi ML, Mazzanti A, Galbiati E et al (2003) Bone mineral density and bone metabolism in Duchenne muscular dystrophy. Osteoporos Int 14:761–767

69. Bardare M, Bianchi ML, Furia M, Gandolini GG, Cohen E, Montesano A (1991) Bone mineral metabolism in juvenile chronic arthritis: the influence of steroids. Clin Exp Rheumatol 9(Suppl 6): 29–31

70. Klein RG, Arnaud SB, Gallagher JC, Deluca HF, Riggs BL (1977) Intestinal calcium absorption in exogenous hypercortisonism. Role of 25-hydroxyvitamin D and corticosteroid dose. J Clin Invest 60:253–259

71. Pappa HM, Gordon CM, Saslowsky TM et al (2006) Vitamin D status in children and young adults with inflammatory bowel disease. Pediatrics 118:1950–1961

72. Keaveny AP, Freaney R, McKenna MJ, Masterson J, O'Donoghue DP (1996) Bone remodeling indices and secondary hyperparathyroidism in celiac disease. Am J Gastroenterol 91:1226–1231

73. Corazza GR, Di Stefano M, Jorizzo RA, Cecchetti L, Minguzzi L, Gasbarrini G (1997) Propeptide of type I procollagen is predictive of posttreatment bone mass gain in adult celiac disease. Gastroenterology 113:67–71

74. Sentongo TA, Semaeo EJ, Stettler N, Piccoli DA, Stallings VA, Zemel BS (2002) Vitamin D status in children, adolescents, and young adults with Crohn disease. Am J Clin Nutr 76:1077–1081

75. Rakover Y, Hager H, Nussinson E, Luboshitzky R (1994) Celiac disease as a cause of transient hypocalcemia and hypovitaminosis D in a 13 year-old girl. J Pediatr Endocrinol 7:53–55

76. Kavak US, Yuce A, Kocak N et al (2003) Bone mineral density in children with untreated and treated celiac disease. J Pediatr Gastroenterol Nutr 37:434–436

77. Haaber AB, Rosenfalck AM, Hansen B, Hilsted J, Larsen S (2000) Bone mineral metabolism, bone mineral density, and body composition in patients with chronic pancreatitis and pancreatic exocrine insufficiency. Int J Pancreatol 27:21–27

78. Gordon CM, Anderson EJ, Herlyn K et al (2007) Nutrient status of adults with cystic fibrosis. J Am Diet Assoc 107:2114–2119

79. Chavasse RJ, Francis J, Balfour-Lynn I, Rosenthal M, Bush A (2004) Serum vitamin D levels in children with cystic fibrosis. Pediatr Pulmonol 38:119–122

80. Feranchak AP, Sontag MK, Wagener JS, Hammond KB, Accurso FJ, Sokol RJ (1999) Prospective, long-term study of fat-soluble vitamin status in children with cystic fibrosis identified by newborn screen. J Pediatr 135:601–610

81. Gordon CM, Binello E, LeBoff MS, Wohl ME, Rosen CJ, Colin AA (2006) Relationship between insulin-like growth factor I, dehydroepiandrosterone sulfate and proresorptive cytokines and bone density in cystic fibrosis. Osteoporos Int 17:783–790

82. Henderson RC, Lester G (1997) Vitamin D levels in children with cystic fibrosis. South Med J 90: 378–383

83. Stefanic M, Papic S, Suver M, Glavas-Obrovac L, Karner I (2008) Association of vitamin D receptor gene 3′-variants with Hashimoto's thyroiditis in the Croatian population. Int J Immunogenet 35: 125–131

84. Murtaugh MA, Sweeney C, Ma KN et al (2006) Vitamin D receptor gene polymorphisms, dietary promotion of insulin resistance, and colon and rectal cancer. Nutr Cancer 55:35–43

85. Slattery ML, Herrick J, Wolff RK, Caan BJ, Potter JD, Sweeney C (2007) CDX2 VDR polymorphism and colorectal cancer. Cancer Epidemiol Biomarkers Prev 16:2752–2755

86. Roth DE, Jones AB, Prosser C, Robinson JL, Vohra S (2008) Vitamin D Receptor Polymorphisms and the risk of acute lower respiratory tract infection in early childhood. J Infect Dis 197:676–680

87. Malloy PJ, Hochberg Z, Tiosano D, Pike JW, Hughes MR, Feldman D (1990) The molecular basis of hereditary 1,25-dihydroxyvitamin D3 resistant rickets in seven related families. J Clin Invest 86:2071–2079

88. Kaneko A, Urnaa V, Nakamura K et al (2007) Vitamin D receptor polymorphism among rickets children in Mongolia. J Epidemiol 17:25–29

89. Fischer PR, Thacher TD, Pettifor JM, Jorde LB, Eccleshall TR, Feldman D (2000) Vitamin D receptor polymorphisms and nutritional rickets in Nigerian children. J Bone Miner Res 15:2206–2210

90. Holick MF (1990) The use and interpretation of assays for vitamin D and its metabolites. J Nutr 120(Suppl 11):1464–1469

91. Clive DR, Sudhaker D, Giacherio D et al (2002) Analytical and clinical validation of a radioimmunoassay for the measurement of 1,25 dihydroxy vitamin D. Clin Biochem 35:517–521

92. Greer FR (2004) Issues in establishing vitamin D recommendations for infants and children. Am J Clin Nutr 80:1759S–1762S

93. Gartner LM, Greer FR (2003) Prevention of rickets and vitamin D deficiency: new guidelines for vitamin D intake. Pediatrics 111:908–910

94. Wagner CL, Greer FR (2008) Prevention of rickets and vitamin D deficiency in infants, children, and adolescents. Pediatrics 122(5):1142–1152

95. Mawer EB, Stanbury W, Robinson MJ, James J, Close C (1986) Vitamin D nutrition and vitamin D metabolism in the premature human neonate. Clin Endocrinol (Oxf) 25:641–649

96. Hollis BW, Wagner CL (2004) Assessment of dietary vitamin D requirements during pregnancy and lactation. Am J Clin Nutr 79:717–726

97. Hollis BW, Wagner CL (2004) Vitamin D requirements during lactation: high-dose maternal supplementation as therapy to prevent hypovitaminosis D for both the mother and the nursing infant. Am J Clin Nutr 80:1752S–1758S

98. Canadian Paediatric Society (CPS) (2007) Vitamin D supplementation: recommendations for Canadian mothers and infants. Paediatrics Child Health 12:583–589

99. Hollis BW, Roos BA, Draper HH, Lambert PW (1981) Vitamin D and its metabolites in human and bovine milk. J Nutr 111:1240–1248

100. Reeve LE, Chesney RW, DeLuca HF (1982) Vitamin D of human milk: identification of biologically active forms. Am J Clin Nutr 36:122–126

101. Davenport ML, Uckun A, Calikoglu AS (2004) Pediatrician patterns of prescribing vitamin supplementation for infants: do they contribute to rickets? Pediatrics 113:179–180

102. Holick MF, Shao Q, Liu WW, Chen TC (1992) The vitamin D content of fortified milk and infant formula. N Engl J Med 326:1178–1181

103. Wagner CL, Hulsey TC, Fanning D, Ebeling M, Hollis BW (2006) High-dose vitamin D3 supplementation in a cohort of breastfeeding mothers and their infants: a 6-month follow-up pilot study. Breastfeed Med 1:59–70

104. Lehtonen-Veromaa M, Mottonen T, Leino A, Heinonen OJ, Rautava E, Viikari J (2008) Prospective study on food fortification with vitamin D among adolescent females in Finland: minor effects. Br J Nutr 100:418–423

105. Calvo MS, Whiting SJ, Barton CN (2004) Vitamin D fortification in the United States and Canada: current status and data needs. Am J Clin Nutr 80:1710S–1716S

106. Hill TR, O'Brien MM, Cashman KD, Flynn A, Kiely M (2004) Vitamin D intakes in 18–64-y-old Irish adults. Eur J Clin Nutr 58:1509–1517

107. McKenna MJ, Freaney R, Byrne P et al (1995) Safety and efficacy of increasing wintertime vitamin D and calcium intake by milk fortification. QJM 88:895–898

108. Tylavsky FA, Cheng S, Lyytikainen A, Viljakainen H, Lamberg-Allardt C (2006) Strategies to improve vitamin D status in northern European children: exploring the merits of vitamin D fortification and supplementation. J Nutr 136:1130–1134

109. Valimaki VV, Loyttyniemi E, Valimaki MJ (2007) Vitamin D fortification of milk products does not resolve hypovitaminosis D in young Finnish men. Eur J Clin Nutr 61:493–497

110. Moore C, Murphy MM, Keast DR, Holick MF (2004) Vitamin D intake in the United States. J Am Diet Assoc 104:980–983

111. Moore CE, Murphy MM, Holick MF (2005) Vitamin D intakes by children and adults in the United States differ among ethnic groups. J Nutr 135:2478–2485

112. Cesur Y, Caksen H, Gundem A, Kirimi E, Odabas D (2003) Comparison of low and high dose of vitamin D treatment in nutritional vitamin D deficiency rickets. J Pediatr Endocrinol Metab 16:1105–1109

113. Madar AA, Klepp KI, Meyer HE (2008) Effect of free vitamin D(2) drops on serum 25-hydroxyvitamin D in infants with immigrant origin: a cluster randomized controlled trial. Eur J Clin Nutr 63:478–484

114. Armas LA, Hollis BW, Heaney RP (2004) Vitamin D2 is much less effective than vitamin D3 in humans. J Clin Endocrinol Metab 89:5387–5391

115. Trang HM, Cole DE, Rubin LA, Pierratos A, Siu S, Vieth R (1998) Evidence that vitamin D3 increases serum 25-hydroxyvitamin D more efficiently than does vitamin D2. Am J Clin Nutr 68:854–858

116. Holick MF, Biancuzzo RM, Chen TC et al (2008) Vitamin D2 is as effective as vitamin D3 in maintaining circulating concentrations of 25-hydroxyvitamin D. J Clin Endocrinol Metab 93:677–681

117. Thacher TD, Fischer PR, Pettifor JM et al (1999) A comparison of calcium, vitamin D, or both for nutritional rickets in Nigerian children. N Engl J Med 341:563–568

118. Shah BR, Finberg L (1994) Single-day therapy for nutritional vitamin D-deficiency rickets: a preferred method. J Pediatr 125:487–490

119. Stephenson A, Brotherwood M, Robert R, Atenafu E, Corey M, Tullis E (2007) Cholecalciferol significantly increases 25-hydroxyvitamin D concentrations in adults with cystic fibrosis. Am J Clin Nutr 85:1307–1311

120. Jacobus CH, Holick MF, Shao Q et al (1992) Hypervitaminosis D associated with drinking milk. N Engl J Med 326:1173–1177

121. Koutkia P, Chen TC, Holick MF (2001) Vitamin D intoxication associated with an over-the-counter supplement. N Engl J Med 345:66–67

122. Vieth R (1999) Vitamin D supplementation, 25-hydroxyvitamin D concentrations, and safety. Am J Clin Nutr 69:842–856

123. Williams AL, Cox J, Gordon CM (2008) Rickets in an otherwise healthy infant: a case report. Clin Pediatr (Phila) 47:409–412

34 Dietary Calcium Deficiency and Rickets

John M. Pettifor, Philip R. Fischer, and Tom D. Thacher

Abstract   The chapter reviews the role of low dietary calcium intakes in the pathogenesis of nutritional rickets in children. Calcium requirements in children are primarily determined by the skeletal demands for calcium during growth and the efficiency of intestinal calcium absorption to meet these demands. In large parts of the developing world, calcium intakes in children are considered to be low at approximately 300 mg/day. There is good evidence to indicate that children adapt to these low intakes by increasing fractional calcium absorption when the vitamin D status is adequate and by reducing renal calcium excretion. Nevertheless, despite these adaptive processes, calcium deficiency rickets has been described in children from a number of countries, such as South Africa, Nigeria, Bangladesh, and India. Calcium intakes in these children have been estimated to be approximately 200 mg/day. Biochemically, the hallmark is a marked elevation of serum 1,25-dihydroxyvitamin D [$1,25(OH)_2D$] concentrations in association with hypocalcemia and elevated alkaline phosphatase concentrations. 25-Hydroxyvitamin D [25(OH)D] concentrations are lower than in non-rachitic children but generally above levels usually associated with vitamin D deficiency rickets. A number of studies have documented healing of the disease by the supplementation of dietary calcium alone. There is increasing evidence to suggest that vitamin D requirements in children are influenced by their calcium requirements and the calcium content of the diet. Thus nutritional rickets has a spectrum of causes with dietary calcium deficiency being at one end of the spectrum and pure vitamin D deficiency at the other. In between a combination of relative vitamin D insufficiency and low dietary calcium intakes combine to exacerbate the development of rickets.

Key Words: Calcium; calcium deficiency; rickets; Africa; 25-hydroxyvitamin D; mineralization; bone growth; vitamin D; sunlight; Nigeria

1. INTRODUCTION

In the previous chapter, the causes and clinical manifestations of vitamin D deficiency rickets have been discussed. In this chapter, the pathogenesis of nutritional rickets is broadened to include the role played by low dietary calcium intakes either alone or in combination with other possible environmental or genetic factors.

From: *Nutrition and Health: Vitamin D*
Edited by: M.F. Holick, DOI 10.1007/978-1-60327-303-9_34,
© Springer Science+Business Media, LLC 2010

As our understanding of the physiology and metabolism of vitamin D has expanded over the last three decades, renewed interest has been focused on the pathogenesis of nutritional rickets not only in developed countries but also in developing countries. This interest has been aided and supported by the development of assays for the measurement of various vitamin D metabolites in serum (especially 25-hydroxyvitamin D [25(OH)D] and 1,25-dihydroxyvitamin D [1,25(OH)$_2$D]), which has helped to delineate the crucial role that dietary calcium intakes play in the pathogenesis of nutritional rickets in some communities and parts of the world.

2. CALCIUM HOMEOSTASIS

The major physiologic role that vitamin D plays in bone and mineral homeostasis during growth is in optimizing intestinal calcium absorption to meet the demands of the growing skeleton during childhood. Figure 1 depicts the calculated calcium accretion rate in the skeleton during growth *(1)*, while Fig. 2 documents the variability of the peak calcium accretion rate during the pubertal growth spurt in boys and girls *(2)*. Prior to the onset of the pubertal growth spurt calcium accretion remains fairly constant at between approximately 100 and 150 mg/day from the age of 2 years. During puberty calcium accretion rises rapidly to reach a peak of approximately 300 mg/day in girls and 340 mg/day in boys, with girls peaking approximately 2 years earlier than boys. These figures have been derived from children living in developed countries on typical western diets. There is little information from other communities or ethnic groups, although data from China suggest that calcium accretion rates during the pubertal growth spurt are very similar in Chinese girls to those of North American white girls whose dietary calcium intakes are 2–3 times those of the Chinese girls *(3)*. A recent study suggests that African-American adolescent girls have a calcium retention rate that is 185 mg/day

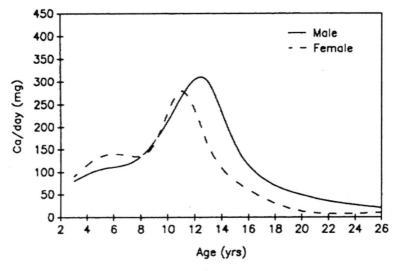

Fig. 1. Calculated daily calcium retention (mg/day) in boys and girls during childhood and adolescence (reproduced with permission from Peacock *(1)*).

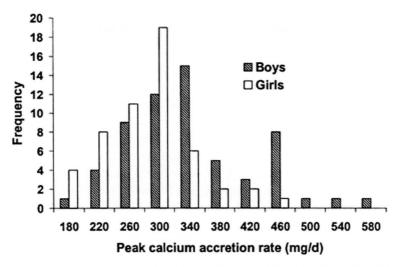

Fig. 2. Peak calcium accretion rates calculated from 60 boys and 53 girls. (Reproduced from *(2)* with permission of the American Society for Bone and Mineral Research.).

greater at any given calcium intake than that of their white peers *(4)*, but it is not known whether these findings are applicable to African black children having very different diets and growth patterns.

In order to provide intakes to meet these calcium demands during growth, dietary calcium requirements for children and adolescents have been set by a number of different bodies *(5–7)*; (Table 1). The data used to develop these recommendations have in general been derived from studies conducted on Caucasian subjects living in developed countries and consuming western-type diets. There is little or no information on dietary calcium requirements for children and adolescents of differing ethnic backgrounds and often living under very different circumstances in developing countries and resource poor communities. Heaney *(8)* has made calculations for African-American adults, which indicate that calcium requirements may be some 300 mg/day lower than those of whites and has suggested that African blacks may behave similarly, although

Table 1
Calcium Intake Recommendations (mg/day) from a Number of Countries (Adapted from *(5)*)

	Australia	United Kingdom	European Union	FAO/WHO	Canada and USA
Infancy	300	525	400	300–450	210–270
Childhood	530–800	350–550	400–550	500–700	500–800
Adolescence	1,000–1,200	1,000	1,000	1,000	1,300
–Males	800–1,000	800	800	1,000	1,300
–Females					

there is much less published information on which to base these conclusions. Heaney also concludes that there is no evidence to suggest that Oriental adults living in Asia have lower calcium requirements than those of Caucasians living western lifestyles *(8)*.

Calcium intakes in children vary widely across the world, although it is probably true to say that in general dietary calcium intakes are higher in children living in developed areas of the world than in developing areas as dairy products are more readily available and more affordable in industrialized countries. For example, in the USA early pubertal children have been estimated to have mean dietary calcium intakes of 1,022, 1,204, and 941 mg/day for white, Hispanic, and African-American children, respectively *(9)*, and in UK adolescents mean calcium intakes ranged from 700 to 1,100 mg/day *(7)*. In contrast, children consume 224 mg/day in Thailand *(10)*, 440 mg/day in China *(3)*, 360 mg/day by adolescent girls in Iran *(11)*, and 314 mg/day among adolescents of the lower socioeconomic group and 713 mg/day among those of the high socioeconomic group in India *(12)*. Thus it is clear that many children and adolescents throughout the world have calcium intakes which are less than 50% of recommended intakes, and yet good evidence of deleterious effects on bone health as a result of these poor intakes is not available in the majority of situations *(13)*. Nonetheless, some studies suggest an increased incidence of childhood fractures associated with low dietary calcium intakes *(14)*.

In order for children and adolescents to maintain adequate bone growth and mineralization in the face of habitually low dietary calcium intakes, their physiology has to adapt by increasing the fractional intestinal absorption of calcium, by reducing urinary and insensible calcium loss, or possibly by increasing the efficiency of bone mineralization. Fractional calcium absorption in vitamin D replete adults has been reported to be approximately 27% *(15, 16)*; with a range between 25 and 35% *(17)*. In children values tend to be higher. In neonates calcium retention is nearly 50% *(18)*. In 1–4 year olds fractional calcium absorption averaged 46% and was shown to correlate inversely with calcium intake so that absorption ranged from under 30% in children with the highest intakes to over 70% in those with the lowest intakes *(19)*. Similar differences in calcium absorption associated with differing habitual calcium intakes were found in 7-year-old Chinese children; those on low calcium intakes (mean 359 mg/day) had a fractional calcium absorption of 63% while those on a mean calcium intake of 862 mg/day had a fractional absorption of 55% *(20)*. In a group of US adolescents, aged between 10 and 14 years, intestinal calcium absorption averaged 36% (ranging from 10 to 75%) and was correlated with serum $1,25(OH)_2D$ but not $25(OH)D$ values *(21)*. Thus it appears that in children the intestine adapts to increased calcium demands or low dietary intakes by increasing its absorption efficiency so that a fractional absorption of over 70% can be obtained in certain situations.

Urine calcium excretion typically ranges from 2 to 5 mg/kg/day. In a study of adolescent boys and girls, Braun and coworkers *(22)* found that dietary calcium intake had little effect on urinary calcium excretion over a range of calcium intakes between 700 and 2,000 mg/day, with dietary calcium intake accounting for only 0.2% of the variance in urinary calcium excretion in boys and 6% in girls. At lower calcium intakes (about 380 mg/day) it appears that urine calcium excretion is reduced *(23)*, and in both South African and Nigerian studies, children on low dietary calcium intakes have very low urinary calcium excretion (<1 mg/day in some subjects). In several studies, African-American children have lower urinary calcium excretion than white children on similar diets *(23, 24)*.

Endogenous fecal calcium excretion is also influenced by calcium intakes *(23)*, although less information is available on this aspect of calcium homeostasis in children *(25)*. Much less is known about insensible calcium loss occurring from the skin *(25)* and it is unclear whether losses are influenced by dietary calcium intakes.

From the discussion above, it is apparent that children are able to adapt to poor dietary calcium intakes by increasing intestinal calcium absorption efficiency and reducing urinary calcium loss. This has led to discussion around what some researchers consider to be higher than necessary recommended dairy calcium intakes for children and adolescents *(26–28)*. Nevertheless, it is clear that at some point these adaptive responses will be overwhelmed if dietary calcium intakes are reduced sufficiently and for prolonged periods, resulting in perturbations of normal bone growth and development. It is these consequences that are discussed in the next section.

3. DIETARY CALCIUM DEFICIENCY RICKETS

Over 30 years ago, several case reports suggested that dietary lack of calcium was the possible underlying pathogenetic factor responsible for rickets in infants, who had chronic diarrhea for which they had been fed diets very low in calcium *(29, 30)*. These authors were probably the first to appreciate the fact that low dietary calcium intakes might cause rickets in the face of what was believed to be an adequate vitamin D status. Prior to that time, nutritional rickets was considered to be synonymous with vitamin D deficiency rickets, as all cases of nutritional rickets appeared to heal on vitamin D or vitamin D and calcium treatment. At that time, there was considerable opposition to the concept of dietary calcium lack per se causing or promoting rickets *(31)*. Since then, rickets, due primarily to a low dietary calcium intake, has been described by a number of different authors in children living in various communities *(32)*. Typically these children live in communities situated in the developing world where habitual calcium intakes are characteristically low due to the poor availability of dairy products and the lack of dietary diversity.

In 1978, Pettifor and colleagues described what they believed to be dietary calcium deficiency rickets in children living in a rural community in South Africa *(33, 34)*. They initially described nine such children between the ages of 4½ and 13 years, but since that time numerous other children with rickets have been diagnosed with calcium deficiency rickets from the same region *(35)* and other regions of the country *(36)*. The mean age of this larger group of children is similar to that of the initial 9 children, being 8.1 ± 3.6 years. Interestingly slightly more females than males (1.4:1) were diagnosed with active rickets. Typically the children presented with progressive lower limb deformities typical of rickets (bow legs, knock knees, and wind-swept deformities) (Fig. 3). Unlike classical vitamin D deficiency, bone pain and muscle weakness were not prominent symptoms and the majority of the children were physically quite active despite radiological active rickets and bony deformities. Biochemically, the features were similar to those found in vitamin D deficiency, with hypocalcemia, variable serum phosphate concentrations,

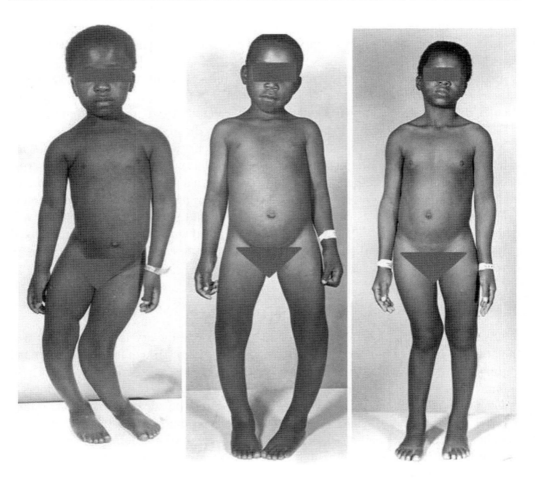

Fig. 3. Limb deformities associated with dietary calcium rickets in South African children.

and elevated alkaline phosphatase and parathyroid hormone values being characteristic. However, the biochemical features that differentiated these children with dietary calcium deficiency rickets from those with vitamin D deficiency rickets were the normal 25-hydroxyvitamin D [25(OH)D] (>10 ng/ml) and elevated 1,25-dihydroxyvitamin D [1,25(OH)$_2$D] concentrations *(34)*. Furthermore, fractional intestinal calcium absorption measured either isotopically or by balance studies was increased unlike the findings in vitamin D deficiency in the untreated state *(33)*. The histological features on bone biopsy were those of severe osteomalacia similar to that seen in vitamin D deficiency *(37)*. Calcium intakes in the children with biochemical features of rickets were significantly lower than those of age-matched controls living in the same community with calcium intakes averaging approximately 200 mg/day in affected subjects compared to over 300 mg/day in the controls *(38)*. The diet of affected children was typically devoid of dairy products and consisted mainly of a maize (corn)-based staple with the addition of seasonal vegetables for lunch and dinner and the occasional addition of chicken or beef. Children with active rickets responded well to calcium supplements (1,000 mg/day) alone *(33)*, while children in the community who had biochemical features of altered calcium homeostasis (elevated alkaline phosphatase concentrations and/or hypocalcemia) responded to a supplement of 500 mg/day *(39)*. Thus the evidence that low dietary calcium intakes were primarily responsible for the pathogenesis of rickets in these children seems fairly conclusive.

3.1. Dietary Calcium Intakes

Following the South African report a number of studies from Nigeria have postulated that dietary calcium deficiency was the cause of rickets in that country *(40–43)*, although vitamin D deficiency and an abnormality of vitamin D 25-hydroxylation have also been proposed *(44, 45)*. The rachitic Nigerian children at presentation were generally younger (average age of between 2 and 4 years) than the South African children, but still older than the usual age of presentation of vitamin D deficiency in young children in most previously reported communities. Nigerian calcium intakes were similar to those of the children in South Africa with mean intakes between 150 and 200 mg/day being recorded *(46, 40)*; however, in one large Nigerian study no difference in the dietary calcium intakes between affected and non-affected children was found, both groups having similarly low intakes *(46)*. The biochemical perturbations were similar to those of the South African children, with elevated 1,25-dihydroxyvitamin D levels in association with low-normal 25(OH)D concentrations being characteristic *(46, 41)*.

More recently, studies in the Chakaria region of Bangladesh suggest that dietary calcium deficiency rickets is widespread in that part of the country, occurring in children outside the infant age group and presenting with clinical and biochemical features similar to those described above *(47, 48)*. It has been estimated that some 9% of children between 1 and 15 years of age in southeastern Bangladesh have clinical features suggestive of rickets while 0.9% have radiographic evidence of active rickets *(49)*. Calcium intakes of affected children have been estimated to be approximately 150 mg/day *(47)*.

Investigations in northern India have indicated that low dietary calcium intakes might be responsible for rickets in young children *(50)* and perhaps somewhat surprisingly a study from the USA has also suggested that even some of the cases of nutritional rickets among weaned toddlers might be due to low dietary calcium intakes *(51)*. Finally, a few isolated cases of calcium deficiency rickets being diagnosed in children on macrobiotic or very restricted diets have been reported *(52, 53)*.

Although the studies conducted in South Africa clearly found lower calcium intakes in the children presenting with active rickets than in age-matched controls from the same community *(38)*, this is not true for several studies conducted in Nigeria *(46)*, even though the intakes were similar to those found in the South African children (approximately 200 mg/day). This raises the question as to whether or not other factors might play a role in the pathogenesis of the disease. The central role of low dietary calcium intakes in the pathogenesis of the disease has been established by the prompt response of both the biochemical and radiographic features of rickets to oral calcium supplements. In a randomized trial conducted in Nigeria to assess the response of children with active rickets to treatment with calcium (1,000 mg/day), vitamin D (600,000 IU/3 monthly), or both, Thacher and colleagues clearly showed the superiority of calcium alone to

Table 2
Response of Nigerian Children with Active Rickets to Calcium, Vitamin D, or a Combination of Both Treatments After 6 Months of Treatment (Adapted from *(54)*). Rickets was Considered to Be Healed if the Alkaline Phosphatase Concentration was <350 IU/l and the Radiographic Score of Active Rickets was ≤1.5

	Vitamin D$_2$ (600,000 IU/3 monthly) N = 37	*Calcium (1,000 mg/day) N = 34*	*Combination (Ca and Vit D) N = 38*
% children with healed rickets at 6 months	19%	61%	58%
Significance	*P<0.001* compared to Ca and combination groups	Not significantly different from combination group	Not significantly different from Ca group

vitamin D alone *(54)* Table 2). Nevertheless, it is likely that other environmental or genetic factors may predispose susceptible children to the development of rickets. In Fig. 4, a number of these possible factors are indicated; these include the presence of dietary constituents impairing intestinal calcium absorption, and the inability of affected children to be able to adapt appropriately to the stress of low dietary calcium intakes by increasing the fractional absorption of calcium, possibly through an inability to increase

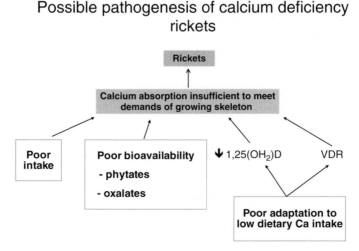

Fig. 4. Possible factors influence contributing to the development of rickets in children on low dietary calcium intakes.

1,25(OH)$_2$D concentrations adequately or through the intestinal mucosa not being able to respond appropriately to the high concentrations of 1,25(OH)$_2$D.

3.2. Intestinal Calcium Absorption

It is well established that certain dietary constituents such as high levels of oxalates or phytates may impair intestinal calcium absorption and predispose to the development of rickets (55, 56). The traditional diets of children in both Nigeria and South Africa are high in phytates due to the consumption of staples such as maize/corn, yet studies of the dietary patterns of affected and unaffected children in Nigeria have not found major differences in the dietary constituents between the two groups, thus making the likelihood of impaired absorption through differences in dietary patterns in the affected children less likely.

The most important factor controlling intestinal calcium absorption in situations of low dietary calcium content is the serum 1,25(OH)$_2$D concentration, which is dependent on renal 1α-hydroxlase activity (stimulated by increased parathyroid hormone concentrations and hypocalcemia) and substrate [25(OH)D] concentrations (a measure of the vitamin D status of the individual) (57). In dietary calcium deficiency rickets, elevated concentrations of 1,25(OH)$_2$D are consistently found in untreated children in an attempt to correct low serum calcium concentrations, despite 25(OH)D concentrations being lower than in aged-matched controls (Table 3) (46). In such situations, it would be expected that intestinal calcium absorption would be maximal in children with low dietary calcium intakes – a feature which would rule out vitamin D deficiency as the etiology of the disease. In the few studies in which intestinal absorption has been investigated by either balance studies or isotopic methods, calcium absorption has been found to be increased compared to values reported in children living in developed countries (58, 33), although in one study absorption (mean 61 ± 20%) was found to be no greater than that in non-rachitic control children (63 ± 13%), who had calcium intakes similar to those of affected subjects (≈200 mg/day) (58). Following calcium supplementation and healing of the bone disease, fractional calcium absorption remained elevated (81 ± 10%) and in fact was significantly higher than baseline values. This latter finding together with the fact that baseline absorption was not higher than that found in non-affected control children on low dietary calcium intakes might suggest that in affected untreated children intestinal calcium absorption had not been optimized, despite these rachitic children having higher 1,25(OH)$_2$D concentrations than controls.

Table 3
Serum 25(OH)D and 1,25(OH)$_2$D Concentrations in Children with Active Rickets and in Age-Matched Controls. Adapted from (54)

	Children with rickets	Age matched controls
25(OH)D (ng/ml)	13.9 ± 10.2	20.5 ± 6.2*
1,25(OH)$_2$D (pg/ml)	134 ± 40	116 ± 40*
	*$p<0.001$	

3.3. Vitamin D Requirements

Recent studies in Nigeria suggest that children with dietary calcium deficiency rickets may have relative vitamin D insufficiency which prevents them from optimizing $1,25(OH)_2D$ concentrations and maximizing intestinal calcium absorption *(59)*. The change in $1,25(OH)_2D$ concentrations in response to the administration of $25(OH)D$ has been used in population studies to determine the range of serum $25(OH)D$ concentrations needed to ensure an adequate vitamin D status. In a vitamin D replete state $1,25(OH)_2D$ levels should not rise after the administration of $25(OH)D$ or vitamin D, while they would be expected to do so, if vitamin D/$25(OH)D$ concentrations were insufficient *(60)*. This test was used in a group of untreated children with dietary calcium deficiency rickets; after administration of an oral bolus of 50,000 IU of vitamin D_2, serum concentrations of $25(OH)D$ and $1,25(OH)_2D$ were followed on days 1, 3, 7, and 14 (Fig. 5). Serum $1,25(OH)_2D$ concentrations rose from an elevated mean value of 184 pg/ml on day 0 to 349 pg/ml on day 3 and then declined to return to baseline values by day 14. The pattern of change in $1,25(OH)_2D$ values mirrored that of the $25(OH)D$ concentrations which reached a peak of 40 ng/ml on day 3 *(59)*. Concentrations of $1,25(OH)_2D$ correlated significantly with $25(OH)D$ values at days 0, 1, 3, and 7, even though mean $25(OH)D$ were well above the upper limit of vitamin D insufficiency (20 ng/ml in children *(61)*) on days 1, 3, and 7. This would suggest that renal 1α-hydroxylase is being driven by both hypocalcemia and secondary hyperparathyroidism as a consequence of dietary calcium deficiency in these children. These findings have implications for defining vitamin D requirements during childhood, as it has not been customary to consider dietary calcium content when assessing vitamin D needs, yet

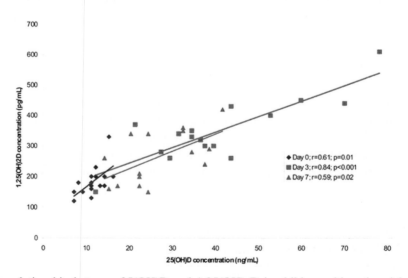

Fig. 5. The relationship between $25(OH)D$ and $1,25(OH)_2D$ in children with active rickets due to dietary calcium deficiency. Vitamin D_2 (50,000 IU) was administered orally as a bolus on day 0 and measurements were taken prior to administration and on days 1, 3, 7, and 14. Serum $1,25(OH)_2D$ concentrations were significantly correlated with serum $25(OH)D$ on days 0, 1, 3, and 7. (Reproduced from Thacher et al. *(59)* with permission from Elsevier.).

these studies in children with dietary calcium deficiency suggest that vitamin D requirements, as measured by circulating 25(OH)D concentrations, may be higher than generally considered adequate in calcium replete children.

The pathogenesis of the lower 25(OH)D concentrations in affected children than non-affected control children is of interest. We have been unable to find differences in sun exposure or skin coverage by clothing between the two groups *(46)*, and both South Africa and Nigeria have considerable hours of daily sunshine throughout the year, thus it is unlikely that skin synthesis of vitamin D is a limiting factor. Clements some years ago provided a rational explanation for the lower levels in the rachitic children *(62, 63)*. In rat experiments, he was able to show that low dietary calcium intakes or high $1,25(OH)_2D$ concentrations increase catabolism of 25(OH)D through the stimulation of 24-hydroxylase. In humans, a reduction in the half-life of 25(OH)D has been found in patients who were treated with $1,25(OH)_2D$ or those who had intestinal malabsorption *(64, 63)*. The effect of calcium supplementation on 25(OH)D concentrations in Nigerian children suggests that the low-normal levels in children with rickets might be due to increased catabolism of vitamin D due to hypocalcemia, secondary hyperparathyroidism, and elevated $1,25(OH)_2D$ concentrations (Figs. 6 and 7).

3.4. Possible Genetic Influences

Another interesting finding in the studies conducted in both South Africa and Nigeria has been the higher prevalence of rickets or bone deformities in first degree relatives of affected children compared to controls (14.6% compared to 3.1%) *(46)*. This raises the question of whether or not genetic factors might be playing a role, although it is

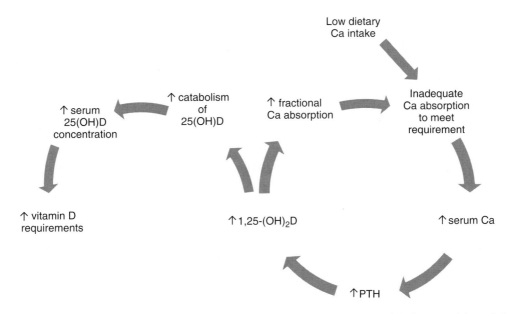

Fig. 6. The pathogenesis of low serum 25(OH)D concentrations in children with dietary calcium deficiency rickets.

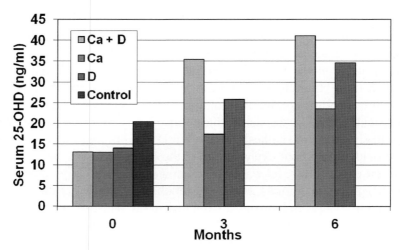

Fig. 7. Change in serum 25(OH)D concentrations in children with dietary calcium deficiency rickets treated on calcium, vitamin D, or both *(54)*. Note that children on calcium supplements alone increased their 25(OH)D concentrations significantly over the 6 months of treatment, suggesting that correction of the perturbations in calcium homeostasis reduced vitamin D catabolism. Supporting this is the finding that 25(OH)D concentrations improved more markedly in those children receiving calcium and vitamin D rather than vitamin D alone.

likely that the shared socioeconomic environment and dietary patterns in the families are major contributors. We considered the possibility that polymorphisms in the vitamin D receptor (VDR) gene might influence the ability of children to adapt adequately to low dietary calcium intakes, because of its central role in calcium homeostasis. In a group of Nigerian children with active rickets, the percentage of the *ff* genotype at the *Fok I* restriction site was less common, while the *FF* genotype was more common than in control children (*f* allele frequency 17% compared to 26%, respectively ($p = 0.03$)) *(65)*. It is of interest that a recent study of rachitic children (both vitamin D and calcium deficiency rickets) in Turkey and the Middle East found that rachitic children in Turkey, but not in Egypt, also had an increased frequency of the *F* allele compared to controls *(66)*. It is suggested that the *F* allele confers a transcriptionally more efficient VDR, but how this influences the development of rickets is unclear. It has been suggested that as the *F* allele is associated with greater bone mass in adults, children with the *F* allele might need higher calcium requirements to meet the demands of structurally larger bones.

Another intriguing piece of the jigsaw puzzle of the pathogenesis of rickets in Nigeria, that might suggest a genetic influence, is the finding of lower breast milk calcium concentrations in those lactating mothers who have had previous children with rickets than in those who have not (172 mg/l compared to 186 mg/l, respectively ($p = 0.034$)) *(67)*. In a number of studies, breast milk calcium concentrations in African women have been found to be considerably lower than in British women at all times during lactation and the calcium content appears to be unaltered by vitamin D status or dietary calcium intake *(68, 69)*, thus suggesting that genetic factors might influence the concentration. It is thus possible that mothers with children who develop rickets after weaning are genetically predisposed to produce breast milk with lower calcium content than mothers who do not have rachitic children. Lower breast milk calcium might be a disadvantage to the breastfed infant in that the calcium stores in the infant on weaning might be less able to withstand the stress of low dietary calcium diets which are characteristic of the weaning diet in many developing countries.

Recently, two families of children with severe rickets have been described in Nigeria, who were found to have a mutation in the 25-hydroxylase gene *(70)*. The children presented with low 25(OH)D concentrations and had a variable response to calcium supplements. There is no evidence at present to suggest that mutations or polymorphisms in the 25-hydroxylase gene are responsible for the high prevalence of dietary calcium deficiency rickets in Nigeria, although it is intriguing to postulate that polymorphisms in the *CYP2R1* gene might influence circulating levels of 25(OH)D and thus predispose children on low calcium intakes to rickets.

4. CONCLUSIONS

Over the past 30 years, a greater understanding has developed of the role that low dietary calcium intakes play in the pathogenesis of rickets in many developing countries. It is apparent that the pathogenesis of nutritional rickets should be seen as having a spectrum of causes *(71)* with pure vitamin D deficiency in, for example, breastfed infants at one end of the spectrum. At the opposite end of the spectrum is positioned dietary calcium deficiency with normal vitamin D status as probably occurs in South African rural children or Nigerian toddlers. However it is likely that the majority of nutritional rickets is caused by a combination of low vitamin D status and poor calcium intakes. In such situations a low dietary calcium intake results in secondary hyperparathyroidism and elevated $1,25(OH)_2D$ concentrations leading to an increased turnover of vitamin D, with the resultant development of vitamin D deficiency and the consequent reduction in intestinal calcium absorption and the development of rickets. Furthermore, we have evidence to suggest that low dietary calcium intakes might require higher circulating 25(OH)D concentrations to increase $1,25(OH)_2D$ values for optimized calcium homeostasis.

Even though nutritional rickets is a common disease and has been studied for many years, there remain a number of unanswered questions that need further investigation. For example, why are some children unable to adapt to low dietary calcium intakes and develop rickets while others tolerate similarly low calcium intakes with apparently no adverse effect? How are vitamin D requirements affected by dietary calcium intakes? What genetic factors predispose certain children to develop calcium deficiency rickets? What environmental factors, such as unknown dietary constituents, play a major role? What are the most effective strategies for preventing the disease in developing countries?

Despite our incomplete understanding of the mechanisms involved in the pathogenesis of dietary calcium deficiency rickets, the awareness of its frequency in a number of developing countries makes the use of calcium supplements in the treatment of rickets in children who have been weaned an important aspect of the management.

REFERENCES

 1. Peacock M (1991) Calcium absorption efficiency and calcium requirements in children and adolescents. Am J Clin Nutr 54(1):261S–265S
 2. Bailey DA, Martin AD, McKay HA, Whiting S, Mirwald R (2000) Calcium accretion in girls and boys during puberty: a longitudinal analysis. J Bone Miner Res 15:2245–2250
 3. Zhu K, Greenfield H, Zhang Q, Du X, Ma G, Foo LH et al (2008) Growth and bone mineral accretion during puberty in Chinese girls: a five-year longitudinal study. J Bone Miner Res 23(2):167–172
 4. Braun M, Palacios C, Wigertz K, Jackman LA, Bryant RJ, McCabe LD et al (2007) Racial differences in skeletal calcium retention in adolescent girls with varied controlled calcium intakes. Am J Clin Nutr 85(6):1657–1663
 5. Joint FAO/WHO expert consultation (2004) Calcium, vitamin and mineral requirements in human nutrition, 2nd edn. WHO/FAO, Geneva, 59–93
 6. Standing committee on the scientific evaluation of dietary reference intakes IoM (1997) Dietary reference intakes for calcium, phosphorus, magnesium, vitamin D, and fluoride. National Academy Press, Washington
 7. Department of health (1998) Nutrition and bone health: with particular reference to calcium and vitamin D. Her Majesty's Stationery Office, London
 8. Heaney RP (2002) Ethnicity, bone status, and the calcium requirement. Nutrition Res 22(1–2):153–178
 9. Weaver CM, McCabe LD, McCabe GP, Novotny R, Van Loan M, Going S et al (2007) Bone mineral and predictors of bone mass in White, Hispanic, and Asian early pubertal girls. Calcif Tissue Int 81(5):352–363
10. Gibson RS, Manger MS, Krittaphol W, Pongcharoen T, Gowachirapant S, Bailey KB et al (2007) Does zinc deficiency play a role in stunting among primary school children in NE Thailand? Br J Nutr 97(1):167–175
11. Dahifar H, Faraji A, Ghorbani A, Yassobi S (2006) Impact of dietary and lifestyle on vitamin D in healthy student girls aged 11–15 years. J Med Invest 53(3–4):204–208
12. Marwaha RK, Tandon N, Reddy DRH, Aggarwal R, Singh R, Sawhney RC et al (2005) Vitamin D and bone mineral density status of healthy schoolchildren in northern India. Am J Clin Nutr 82(2):477–482
13. Winzenberg T, Shaw K, Fryer J, Jones G (2006) Effects of calcium supplementation on bone density in healthy children: meta-analysis of randomised controlled trials. BMJ 333(7572):775–780
14. Greer FR, Krebs NF (2006) Committee on Nutrition. Optimizing bone health and calcium intakes of infants, children, and adolescents. Pediatrics 117(2):578–585
15. Kerstetter JE, O'Brien KO, Insogna KL (2003) Low protein intake: the impact on calcium and bone homeostasis in humans. J Nutr 133(3):855S–861
16. Hansen KE, Jones AN, Lindstrom MJ, Davis LA, Engelke JA, Shafer MM (2008) Vitamin D insufficiency: disease or no disease? J Bone Miner Res 23(7):1052–1060
17. Rafferty K, Heaney RP (2008) Nutrient effects on the calcium economy: emphasizing the potassium controversy. J Nutr 138(1):166S–171S
18. Bass JK, Chan GM (2006) Calcium nutrition and metabolism during infancy. Nutrition 22(10):1057–1066
19. Lynch MF, Griffin IJ, Hawthorne KM, Chen Z, Hamzo M, Abrams SA (2007) Calcium balance in 1–4-yr-old children. Am J Clin Nutr 85(3):750–754
20. Lee WT, Leung SS, Fairweather-Tait SJ, Leung DM, Tsang HS, Eagles J et al (1994) True fractional calcium absorption in Chinese children measured with stable isotopes (42Ca and 44Ca). Br J Nutr 72(6):883–897
21. Abrams SA, Griffin IJ, Hawthorne KM, Gunn SK, Gundberg CM, Carpenter TO (2005) Relationships among vitamin D levels, parathyroid hormone, and calcium absorption in young adolescents. J Clin Endocrinol Metab 90(10):5576–5581
22. Braun M, Martin BR, Kern M, McCabe GP, Peacock M, Jiang Z et al (2006) Calcium retention in adolescent boys on a range of controlled calcium intakes. Am J Clin Nutr 84(2):414–418

23. Abrams SA, Griffin IJ, Hicks PD, Gunn SK (2004) Pubertal girls only partially adapt to low dietary calcium intakes. J Bone Miner Res 19(5):759–763

24. Bell NH, Yergey AL, Vieira NE, Oexmann MJ, Shary JR (1993) Demonstration of a difference in urinary calcium, not calcium absorption, in black and white adolescents. J Bone Miner Res 8: 1111–1115

25. Weaver CM (1994) Age related calcium requirements due to changes in absorption and utilization. J Nutr 124(8_Suppl):1418S–1425S

26. Abrams SA, Strewler GJ (2007) Adolescence: how do we increase intestinal calcium absorption to allow for bone mineral mass accumulation? BoneKEy-Osteovision 4(5):147–157

27. Lanou AJ, Berkow SE, Barnard ND (2005) Calcium, dairy products, and bone health in children and young adults: a reevaluation of the evidence. Pediatrics 115(3):736–743

28. Winzenberg T, Jones G (2008) recommended calcium intakes in children: have we set the bar too high? IBMS BoneKEy 5(2):59–68

29. Maltz HE, Fish MB, Holliday MA (1970) Calcium deficiency rickets and the renal response to calcium infusion. Pediatrics 46:865–870

30. Kooh SW, Fraser D, Reilly BJ, Hamilton JR, Gall D, Bell L (1977) Rickets due to calcium deficiency. N Engl J Med 297:1264–1266

31. Walker ARP (1972) The human requirement of calcium: should low intakes be supplemented? Am J Clin Nutr 25:518–530

32. Thacher TD, Fischer PR, Strand MA, Pettifor JM (2006) Nutritional rickets around the world: causes and future directions. Ann Trop Paediatr 26(1):1–16

33. Pettifor JM, Ross P, Wang J, Moodley G, Couper-Smith J (1978) Rickets in children of rural origin in South Africa: is low dietary calcium a factor? J Pediatr 92:320–324

34. Pettifor JM, Ross FP, Travers R, Glorieux FH, DeLuca HF (1981) Dietary calcium deficiency: a syndrome associated with bone deformities and elevated serum 1,25-dihydroxyvitamin D concentrations. Metab Bone Rel Res 2:301–305

35. Pettifor JM (1991) Dietary calcium deficiency. In: Glorieux FH (ed) Rickets. Nestec, Vevey; Raven Press, New York, 123–143

36. Bhimma R, Pettifor JM, Coovadia HM, Moodley M, Adhikari M (1995) Rickets in black children beyond infancy in Natal. S Afr Med J 85:668–672

37. Marie PJ, Pettifor JM, Ross FP, Glorieux FH (1982) Histological osteomalacia due to dietary calcium deficiency in children. N Engl J Med 307:584–588

38. Eyberg C, Pettifor JM, Moodley G (1986) Dietary calcium intake in rural black South African children. The relationship between calcium intake and calcium nutritional status. Hum Nutr Clin Nutr 40C: 69–74

39. Pettifor JM, Ross FP, Moodley GP, Shuenyane E (1981) The effect of dietary calcium supplementation on serum calcium, phosphorus and alkaline phosphatase concentrations in a rural black population. Am J Clin Nutr 34:2187–2191

40. Okonofua F, Gill DS, Alabi ZO, Thomas M, Bell JL, Dandona P (1991) Rickets in Nigerian children: a consequence of calcium malnutrition. Metabolism 40:209–213

41. Oginni LM, Worsfold M, Oyelami OA, Sharp CA, Powell DE, Davie MW (1996) Etiology of rickets in Nigerian children. J Pediatr 128(5 Pt 1):692–694

42. Thacher TD, Ighogboja SI, Fischer PR (1997) Rickets without vitamin D deficiency in Nigerian children. Ambulatory Child Health 3:56–64

43. Akpede GO, Solomon EA, Jalo I, Addy EO, Banwo AI, Omotara BA (2001) Nutritional rickets in young Nigerian children in the Sahel savanna. East Afr Med J 78(11):568–575

44. Ekanem EE, Bassey DE, Eyong M (1995) Nutritional rickets in Calabar, Nigeria. Ann Trop Paediatr 15(4):303–306

45. Thacher T, Glew RH, Isichei C, Lawson JO, Scariano JK, Hollis BW et al (1999) Rickets in Nigerian children: response to calcium supplementation. J Trop Pediatr 45(4):202–207, [Erratum, J Trop Pediatr 2000; 46:62.]

46. Thacher TD, Fischer PR, Pettifor JM, Lawson JO, Isichei C, Chan GM (2000) Case-control study of factors associated with nutritional rickets in Nigerian children. J Pediatr 137:367–373

47. Fischer PR, Rahman A, Cimma JP, Kyaw-Myint TO, Kabir AR, Talukder K et al (1999) Nutritional rickets without vitamin D deficiency in Bangladesh. J Trop Pediatr 45(5):291–293

48. Arnaud J, Pettifor JM, Cimma JP, Fischer PR, Craviari T, Meisner C et al (2007) Clinical and radiographic improvement of rickets in Bangladeshi children as a result of nutritional advice. Ann Trop Paediatr 27(3):185–191

49. Kabir ML, Rahman M, Talukder K, Rahman A, Hossain Q, Mostafa G et al (2004) Rickets among children of a coastal area of Bangladesh. Mymensingh Med J 13(1):53–58

50. Balasubramanian K, Rajeswari J, Gulab GYC, Agarwal AK, Kumar A et al (2003) Varying role of vitamin D deficiency in the etiology of rickets in young children vs. adolescents in northern India. J Trop Pediatr 49(4):201–206

51. DeLucia MC, Mitnick ME, Carpenter TO (2003) Nutritional rickets with normal circulating 25-hydroxyvitamin D: a call for reexamining the role of dietary calcium intake in North American infants. J Clin Endocrinol Metab 88(8):3539–3545

52. Legius E, Proesmans W, Eggermont E, Vandamme-Lobaerts R, Bouillon R, Smet M (1989) Rickets due to dietary calcium deficiency. Eur J Pediatr 148(8):784–785

53. Dagnelie PC, Vergote FJVRA, van Staveren WA, van den Berg H, Dingjan PG, Hautvast JGAJ (1990) High prevalence of rickets in infants on macrobiotic diets. Am J Clin Nutr 51:202–208

54. Thacher TD, Fischer PR, Pettifor JM, Lawson JO, Isichei CO, Reading JC et al (1999) A comparison of calcium, vitamin D, or both for nutritional rickets in Nigerian children. N Engl J Med 341(8):563–568

55. Weaver CM, Proulx WR, Heaney R (1999) Choices for achieving adequate dietary calcium with a vegetarian diet. Am J Clin Nutr 70(suppl):543S–548S

56. Sly MR, van der Walt WH, Du Bruyn D, Pettifor JM, Marie PJ (1984) Exacerbation of rickets and osteomalacia by maize: a study of bone histomorphometry and composition in young baboons. Calcif Tissue Int 36:370–379

57. Weaver CM (2007) Vitamin D, calcium homeostasis, and skeleton accretion in children. J Bone Miner Res 22(s2):V45–V49

58. Graff M, Thacher TD, Fischer PR, Stadler D, Pam SD, Pettifor JM et al (2004) Calcium absorption in Nigerian children with rickets. Am J Clin Nutr 80:1415–1421

59. Thacher TD, Fischer PR, Isichei CO, Pettifor JM (2006) Early response to vitamin D2 in children with calcium deficiency rickets. J Pediatr 149(6):840–844

60. Docio S, Riancho JA, Perez A, Olmos JM, Amado JA, Gonzales-Macias J (1998) Seasonal deficiency of vitamin D in children: a potential target for osteoporosis-preventing strategies? J Bone Miner Res 13(4):544–548

61. Misra M, Pacaud D, Petryk A, Collett-Solberg PF, Kappy M (2008) On behalf of the drug and therapeutics committee of the Lawson Wilkins pediatric endocrine society. Vitamin D deficiency in children and its management: review of current knowledge and recommendations. Pediatrics 122(2):398–417

62. Clements MR, Johnson L, Fraser DR (1987) A new mechanism for induced vitamin D deficiency in calcium deprivation. Nature 325:62–65

63. Clements MR, Davies M, Hayes ME, Hickey CD, Lumb GA, Mawer EB et al (1992) The role of 1,25-dihydroxyvitamin D in the mechanism of acquired vitamin D deficiency. Clin Endocrinol 37:17–27

64. Batchelor AJ, Watson G, Compston JE (1982) Changes in plasma half-life and clearance of [^{3}H]-25-hydroxyvitamin D_3 in patients with intestinal malabsorption. Gut 23:1068–1071

65. Fischer PR, Thacher TD, Pettifor JM, Jorde LB, Eccleshall TR, Feldman D (2000) Vitamin D receptor polymorphisms and nutritional rickets in Nigerian children. J Bone Miner Res 15:2206–2210

66. Baroncelli GI, Bereket A, El Kholy M, Audi L, Cesur Y, Ozkan B et al (2008) Rickets in the Middle East: role of environment and genetic predisposition. J Clin Endocrinol Metab 93:1743–1750

67. Thacher TD, Pettifor JM, Fischer PR, Okolo SN, Prentice A (2006) Case-control study of breast milk calcium in mothers of children with and without nutritional rickets. Acta Paediatr 95(7):826–832

68. Prentice A, Jarjou LMA, Cole TJ, Stirling DM, Dibba B, Fairweather-Tait S (1995) Calcium requirements of lactating Gambian mothers: effects of a calcium supplement on breast-milk calcium concentration, maternal bone mineral content, and urinary calcium excretion. Am J Clin Nutr 62(1): 58–67

69. Prentice A, Yan L, Jarjou LM, Dibba B, Laskey MA, Stirling DM et al (1997) Vitamin D status does not influence the breast-milk calcium concentration of lactating mothers accustomed to a low calcium intake. Acta Paediatr 86(9):1006–1008

70. Levine MA, Dang A, Ding C, Fischer PR, Singh R, Thacher T (2007) Tropical rickets in Nigeria: Mutations of the CYP2R1 gene encoding vitamin D 25-hydroxylase as a cause of vitamin D dependent rickets. Bone 40(suppl):S22–S89

71. Pettifor JM (1994) Privational rickets: a modern perspective. J Roy Soc Med 87:723–725

35 Vitamin D in Fracture Prevention and Muscle Function and Fall Prevention

Heike Bischoff-Ferrari

Abstract   This chapter reviews the potential of vitamin D for the prevention of falls and fractures. Evidence from randomized-controlled trials will be reviewed for both endpoints, as well as epidemiological data that link higher 25-hydroxyvitamin D [25(OH)D] status to better bone and muscle health. The chapter addresses the evidence of fracture and fall prevention by dose of vitamin D, by type of dwelling, and by treatment duration. All data considered, this chapter summarizes the compelling dual benefit of vitamin D on fracture reduction by its bone and muscle target, a concept that is reviewed at the onset of the chapter.

Key Words: Vitamin D; falls; bone density; fractures; vitamin D deficiency; 25-hydroxyvitamin D; muscle function; sarcopenia; muscle weakness; meta-analysis; calcium

1. INTRODUCTION

Critical for the understanding and prevention of fractures at later age is their close relationship with muscle weakness *(1)* and falling *(2, 3)*. Over 90% of fractures occur after a fall and fall rates increase with age *(4)* and poor muscle strength or function (Fig. 1) *(4)*. Mechanistically, the circumstances *(5)* and the direction *(6)* of a fall determine the type of fracture, whereas bone density and factors that attenuate a fall, such as better strength or better padding, critically determine whether a fracture will take place when the faller lands on a certain bone *(7)*. Moreover, falling may affect bone density through increased immobility from self-restriction of activities *(8)*. It is well known that falls may lead to psychological trauma known as fear of falling *(9)*. After their first fall, about 30% of persons develop fear of falling *(8)* resulting in self-restriction of activities *(8)* and decreased quality of life. Figure 1 illustrates the fall–fracture construct that describes the complexity of osteoporosis prevention introduced by nonskeletal risk factors for fractures among older individuals.

Notably, anti-resorptive treatment alone may not reduce fractures among individuals 80 years and older in the presence of nonskeletal risk factors for fractures, such as muscle weakness and falling, despite an improvement in bone metabolism *(10)*.

From: *Nutrition and Health: Vitamin D*
Edited by: M.F. Holick, DOI 10.1007/978-1-60327-303-9_35,
© Springer Science+Business Media, LLC 2010

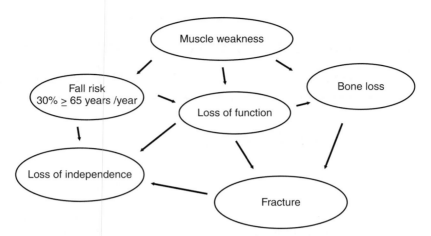

Fig. 1. The *graph* illustrates the central contribution of muscle weakness to the fall and fracture construct. *Supplementation with vitamin D* may address each component of the fall–fracture construct including muscle mass *(23)*, strength *(15)*, balance *(18)*, lower extremity function *(13, 14)*, falling *(17, 19)*, bone density *(32, 45)*, the risk of hip and non-vertebral fractures *(33)*, and the risk of nursing home admission *(46)*. Adapted from Bischoff-Ferrari et al.; Chapter Prevention of Falls in Treatment of Osteoporosis; 2008 Seventh Edition of The Primer on the Metabolic Bone Diseases and Disorders of Mineral Metabolism (American Society of Bone and Mineral Research).

2. VITAMIN D: ITS ROLE IN MUSCLE HEALTH

In humans, four lines of evidence support a role of vitamin D in muscle health and the prevention of sarcopenia. First, proximal muscle weakness is a prominent feature of the clinical syndrome of vitamin D deficiency. Second, VDR is expressed in human muscle tissue *(11)* and VDR activation may promote de novo protein synthesis in muscle *(12)*. Third, several observational studies suggest a positive association between 25-hydroxyvitamin D and muscle strength or lower extremity function in older persons *(13, 14)*. Finally, in several double-blind randomized-controlled trials, vitamin D supplementation increased muscle strength and balance *(15, 16)*, and reduced the risk of falling in community-dwelling individuals *(16–18)*, as well as in institutionalized individuals *(15, 19)*. Importantly, a study by Glerup and colleagues suggest that vitamin D deficiency may cause muscular impairment even before adverse effects on bone occur *(20)*.

Suggesting a role of vitamin D in muscle development, mice lacking the VDR show a skeletal muscle phenotype with smaller and variable muscle fibers and persistence of immature muscle gene expression during adult life *(21, 22)*. These abnormalities persist after correction of systemic calcium metabolism by a rescue diet *(22)*.

3. DESIRABLE 25-HYDROXYVITAMIN D STATUS FOR BETTER FUNCTION AND LOWER RISK OF SARCOPENIA

A dose–response relationship between vitamin D status and muscle health was examined in NHANES III including 4,100 ambulatory adults age 60 years and older. Muscle function measured as the 8-foot walk test and the repeated sit-to-stand test was poorest in subjects with the lowest 25-hydroxyvitamin D (below 20 nmol/l) levels. Similar results were found in a Dutch cohort of older individuals *(13)*. Notably, while from the smaller Dutch cohort a threshold of 50 nmol/l has been suggested for optimal function *(13)*, a threshold beyond which function would not further improve was not identified in the larger NHANES III survey, even beyond the upper end of the reference range (>100 nmol/l) *(14)*. In NHANES III, a similar benefit of higher 25-hydroxyvitamin D status was documented by gender, level of physical activity, and level of calcium intake. *With respect to sarcopenia*, appendicular skeletal muscle mass was measured in the Dutch cohort among 331 older adults using dual-energy X-ray absorptiometry *(23)*. After a 3-year follow-up, and after adjustment for physical activity level, season of data collection, serum creatinine concentration, chronic disease, smoking, and body mass index, persons with low (<25 nmol/l) baseline 25-hydroxyvitamin D levels were 2.14 (0.73–6.33, based on muscle mass) times more likely to experience sarcopenia, compared with those with higher (>50 nmol/l) levels. The associations were similar in men and women.

4. VITAMIN D AND MUSCLE WEAKNESS

Muscle weakness is a prominent feature of the clinical syndrome of vitamin D deficiency. Clinical findings in vitamin D deficiency myopathy include proximal muscle weakness, diffuse muscle pain, and gait impairments such as waddling way of walking *(24)*. The VDR is expressed in human muscle tissue *(11)*, and vitamin D bound to its nuclear receptor in muscle tissue may lead to de novo protein synthesis *(12, 25)*. One uncontrolled biopsy trial in postmenopausal women with osteoporosis documented a relative increase in the diameter and number of type II muscle fibers after a 3-month treatment with 1-alpha-calcidol *(12)*. These findings were confirmed by three recent double-blind RCTs with 800 IU vitamin D3 resulting in a 4–11% gain in lower extremity strength or function *(15, 16)*, and an up to 28% improvement in body sway *(16, 18)* in older adults age 65+ between 2 and 12 months of treatment.

5. VITAMIN D AND FALL PREVENTION

Summarizing five high-quality double-blind RCTs ($n = 1,237$), a meta-analysis published in 2004 found that vitamin D reduced the risk of falling by 22% (pooled corrected OR = 0.78; 95% CI [0.64, 0.92]) compared to calcium or placebo *(26)*. This risk reduction was independent of the type of vitamin D, duration of therapy, and gender. For the two trials with 259 subjects using 800 IU of cholecalciferol per day over 2–3 months *(15, 18)*, the corrected pooled OR was 0.65 (95% CI [0.40, 1.00]) *(26)*, while 400 IU was insufficient in reducing falls *(27)*. *The importance of dose of vitamin D in regard to anti-fall efficacy* was confirmed by one multi-dose double-blind RCT among 124 nursing home residents receiving 200, 400, 600, or 800 IU vitamin D compared to placebo over a 5-month period *(19)*. Participants in the 800 IU group had a 72% lower rate of falls than those taking placebo or a lower dose of vitamin D (rate ratio = 0.28; 95% confidence interval = 0.11–0.75) *(19)*.

Long-term supplementation over 3 years with 700 IU D3 plus 500 mg calcium among community-dwelling older women reduced the *odds of falling* by 46% (OR = 0.54; 95% confidence interval, 0.30–0.97) *(17)*. Similarly, long-term supplementation among institutionalized individuals with ergocalciferol for 2 years, initially 10,000 IU given once weekly and then 1,000 IU daily, reduced the incident rate ratio for falling by 26% (RR = 0.73, 95% CI, 0.57–0.95) *(28)*. A 49% fall reduction among older individuals with a history of a fall was demonstrated in a most recent 12-month trial with 1,000 IU ergocalciferol compared to placebo (OR = 0.61; 95% confidence interval [CI], 0.37–0.99) *(29)*.

The most up-to-date meta-analysis with a focus on anti-fall efficacy of vitamin D was published in October 2009 (30). For supplemental vitamin D, eight high-quality RCTs ($n = 2,426$) were identified and heterogeneity was observed for dose of vitamin D (low dose : <700 IU/day versus higher dose : 700–1,000 IU/day; p-value = 0.02) and achieved 25-hydroxyvitamin D level (<60 nmol/l versus $\geq$60 nmol/l; p-value = 0.005). *Higher dose supplemental vitamin D* reduced fall risk by 19% (pooled relative risk (RR) = 0.81; 95% CI, 0.71–0.92; $n = 1,921$ from seven trials). Falls were not reduced by *low-dose supplemental vitamin D* (pooled RR = 1.10, 95% CI, 0.89–1.35 from two trials) or by achieved serum 25-hydroxyvitamin D concentrations less than 60 nmol/l (pooled RR = 1.35, 95% CI, 0.98–1.84). At the higher dose of vitamin D, the meta-analysis documented a 38% reduction in the risk of falling with a treatment duration of 2–5 months and a sustained effect of 17% fall reduction with a treatment duration of 12–36 months. Indirect support for a rapid and sustained effect of vitamin D on falls comes from the large Chapuy fracture trial *(31)*, in which fracture risk reduction was already apparent at 6 months, which most likely resulted from lowered fall rate, although falls were not assessed. The 2009 meta-analysis also addressed anti-fall efficacy of supplemental versus 1-alpha-hydroxyvitamin D *(30)*. For 1-alpha-hydroxyvitamin D_3, two RCTs were identified ($n = 624$) and the pooled effect was 22% fall reduction (pooled RR = 0.78; 95% CI, 0.64–0.94). This effect was not statistically different from 700 to 1,000 IU supplemental vitamin D where the pooled effect from eight trials was 19%. Thus, for fall prevention the use of active D rather than 700–1,000 IU of supplemental vitamin D is not supported by these findings. On the other hand, a consistent benefit from 1-alpha-hydroxyvitamin D_3 adds to the evidence that improved vitamin D status will reduce the risk of falling in older individuals.

6. DESIRABLE 25-HYDROXYVITAMIN D STATUS FOR BETTER BONE HEALTH

A threshold for optimal 25(OH)D and hip BMD has been addressed among 13,432 individuals of NHANES III (The Third National Health and Nutrition Examination Survey) including both younger (20–49 years) and older (50+ years) individuals with different ethnic racial background *(32)*. Compared to the lowest quintile of 25(OH)D the highest quintile had higher mean BMD by 4.1% in younger whites (test for trend; $p < 0.0001$), by 4.8% in older whites ($p < 0.0001$), by 1.8% in younger Mexican Americans ($p = 0.004$), by 3.6% in older Mexican Americans ($p = 0.01$), by 1.2% in younger blacks ($p = 0.08$), and by 2.5% in older blacks ($p = 0.03$). In the regression plots higher

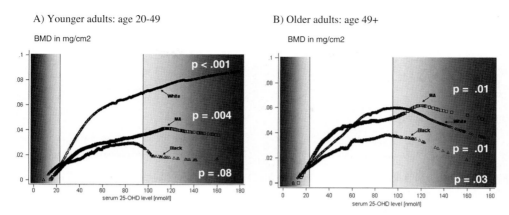

Fig. 2. Regression plot of difference in bone mineral density by 25(OH)D in younger (20–49 years, **a**) and older adults (50+ years, **b**). *Symbols* represent different ethnicities: *circles* are Caucasians, *squares* are Mexican Americans, and *triangles* are African American individuals. The intercept was set to "0" for all race/ethnicity groups to focus on the difference in BMD by 25(OH)D levels, as opposed to differences in BMD by race/ethnicity. The reference range of the 25(OH)D assay (22.5–94 nmol/l) is marked as *vertical lines*. The reference range of the Diasorin assay has been provided by the company and was established using 98 samples from apparently healthy normal volunteers collected in the southwestern United States (high latitude) in late autumn (www.fda.gov/cdrh/pdf3/k032844.pdf). Regression plots adjust for gender, age, body mass index, smoking, calcium intake, estrogen use, month, and poverty income ratio. Weighting accounts for NHANES III sampling weights, stratification, and clustering. Adapted from Bischoff-Ferrari HA et al. All Rights reserved *(32)*.

serum 25(OH)D levels were associated with higher BMD throughout the reference range of 22.5–94 nmol/l in all subgroups (Fig. 2a and b). In younger whites and younger Mexican Americans, higher 25(OH)D was associated with higher BMD even beyond 100 nmol/l.

Consistently, a 2005 meta-analysis of high-quality primary prevention RCTs of vitamin D and fracture risk found that anti-fracture efficacy of vitamin D increases with a higher achieved level of 25-hydroxyvitamin D in the treatment group *(33)*. The significant heterogeneity by achieved 25-hydroxyvitamin D level is illustrated in Fig. 3a and b for both hip fracture and any non-vertebral fracture prevention. Anti-fracture efficacy started at 25-hydroxyvitamin D levels of at least 75 nmol/l (30 ng/ml). This level was *reached only* in trials that gave 700–800 IU cholecalciferol (high-quality trials with oral ergocalciferol were not available at the time).

7. IMPORTANCE OF DOSE IN ANTI-FRACTURE EFFICACY WITH VITAMIN D

A dose–response relationship between vitamin D and fracture reduction is supported by epidemiologic data showing a significant positive trend between serum 25(OH)D concentrations and hip bone density *(32)* and lower extremity strength *(13, 14)*. A greater anti-fracture efficacy with higher achieved 25(OH)D levels was documented in the 2005 meta-analysis of high-quality primary prevention trials with supplemental

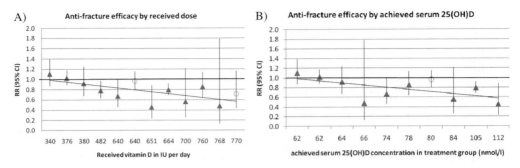

Fig. 3. (a) and **(b)** *Triangles* indicate trials with D3; *circles*, trials with D2. Line = Trendline. All 12 high-quality trials were included for the received dose meta-regression ($n = 42,279$ individuals). For achieved 25(OH)D levels two trials did not provide serum 25(OH)D levels measured in the study population during the trial period 2, 3. For any non-vertebral fractures, anti-fracture efficacy increased significantly with higher received dose (meta-regression: Beta = –0.0007; $p = 0.003$) and higher achieved 25-hydroxyvitamin D levels (meta-regression: Beta = –0.005; $p = 0.04$). Adapted from Bischoff-Ferrari et al. All rights reserved *(44)*.

vitamin D *(33)*. Summarizing five high-quality double-blind RCTs for hip fracture ($n = 9,294$) and seven RCTs for non-vertebral fracture risk ($n = 9,820$), variation between trials for each endpoint was larger than expected. 700–800 IU vitamin D per day reduced the relative risk (RR) of hip fracture by 26% (pooled RR = 0.74; 95% CI [0.61,0.88]) and any non-vertebral fracture by 23% (pooled RR = 0.77; 95% CI [0.68,0.87]) compared to calcium or placebo. No significant benefit was observed for RCTs with 400 IU vitamin D per day (pooled RR for hip fracture was 1.15; 95% CI [0.88,1.50] and for any non-vertebral fracture 1.03; 95% CI [0.86,1.24]).

Since the 2005 meta-analysis, anti-fracture benefits of vitamin D have been questioned by several more recent trials *(34–37)*. Two 2007 meta-analyses included most of these trials *(38, 39)* and concluded that vitamin D may not reduce fractures significantly or only in combination with calcium *(39)* and primarily among institutionalized older individuals *(38)*. A third 2007 meta-analysis concluded that calcium with or without vitamin D may reduce total fracture risk by 12% *(40)*, a result that was questioned by a most recent meta-analysis of high-quality trials of calcium supplementation alone where calcium had a neutral effect on non-vertebral fractures and a possible adverse effect on hip fracture risk *(41)*. Apart from the mixed data on calcium, the recent meta-analyses with vitamin D did not find heterogeneity by dose or achieved level of 25-hydroxyvitamin D. Factors that may have obscured a benefit of vitamin D or the importance of dose in the 2007 meta-analyses are the inclusion of trials with low adherence to treatment *(34–36)*, low dose of vitamin D *(36)*, or the use of less potent D2 *(42, 43)*. Furthermore, open design trials *(34)* also included in the primary analyses of the 2007 meta-analyses may have bias results toward the null because vitamin D is available over the counter.

The most up-to-date meta-analysis with a focus on anti-fracture efficacy of vitamin D was published in March 2009 (44)

Summarizing 12 double-blind RCTs for non-vertebral fractures ($n = 42,279$) and 8 RCTs for hip fractures ($n = 40,886$) comparing oral vitamin D with or without calcium, with calcium or placebo the pooled effect for both endpoints indicated larger variation between trials than expected. In order to incorporate adherence to treatment, we multiplied dose with percent adherence to estimate the mean received dose (dose × adherence) for each trial. Including all trials, anti-fracture efficacy increased significantly with higher dose and higher achieved blood 25-hydroxyvitamin D levels for both endpoints. Consistently, pooling trials with a higher received dose of more than 400 IU/day resolved heterogeneity. For the higher dose, the pooled RR was 0.80 (95% CI, 0.72–0.89; $n = 33,265$ from nine trials) for non-vertebral fractures and 0.82 (95% CI, 0.69–0.97; $n = 31,872$ from five trials) for hip fractures. The higher dose reduced non-vertebral fractures in community-dwelling (–29%) and institutionalized older individuals (–15%), and its effect was independent of additional calcium supplementation (–21% with additional calcium supplementation; –21% for the main effect of vitamin D).

Adding four open design trials in a sensitivity analysis, variation in results was seen between open design and double-blind trials suggesting that trial quality introduces heterogeneity.

8. SUMMARY

Based on evidence from randomized-controlled trials, vitamin D at a higher dose of 700 IU or more should reduce both falls by at least 19% and non-vertebral fractures by at least 20%, while a dose of 400 IU or less does not reduce falls or fractures. The benefit of vitamin D at the higher dose appears to be present in subgroups of community-dwelling and institutionalized older individuals and independent of additional calcium supplementation. For fall prevention, a minimum target 25-hydroxyvitamin D level of 60 nmol/l is required with further improvement thereafter and desirable levels may be at least 75 nmol/l. Consistently, for the anti-fracture benefit serum 25-hydroxyvitamin D levels no less than 75 nmol/l are desirable. The importance of dose of vitamin D is supported by epidemiologic data for lower extremity strength, risk of sarcopenia, and bone density. Notably, the effect of vitamin D on falling occurs early within months of treatment and is sustained over years. This may explain early anti-fracture efficacy of vitamin D observed in several larger trials.

REFERENCES

1. Cummings SR, Nevitt MC, Browner WS et al (1995) Risk factors for hip fracture in white women. Study of Osteoporotic Fractures Research Group. N Engl J Med 332(12):767–773
2. Centers for Disease Control and Prevention (CDC) (2006) Fatalities and injuries from falls among older adults – United States, 1993–2003 and 2001–2005. MMWR Morb Mortal Wkly Rep 55(45):1221–1224
3. Schwartz AV, Nevitt MC, Brown BW Jr, Kelsey JL (2005) Increased falling as a risk factor for fracture among older women: the study of osteoporotic fractures. Am J Epidemiol 161(2):180–185
4. Tinetti ME (1988) Risk factors for falls among elderly persons living in the community. N Engl J Med 319:1701–1707
5. Cummings SR, Nevitt MC (1994) Non-skeletal determinants of fractures: the potential importance of the mechanics of falls. Study of Osteoporotic Fractures Research Group. Osteoporos Int 4(Suppl 1):67–70

6. Nguyen ND, Frost SA, Center JR, Eisman JA, Nguyen TV (2007) Development of a nomogram for individualizing hip fracture risk in men and women. Osteoporos Int 17:17

7. Nevitt MC, Cummings SR (1993) Type of fall and risk of hip and wrist fractures: the study of osteoporotic fractures. The Study of Osteoporotic Fractures Research Group. J Am Geriatr Soc 41(11):1226–1234

8. Vellas BJ, Wayne SJ, Romero LJ, Baumgartner RN, Garry PJ (1997) Fear of falling and restriction of mobility in elderly fallers. Age Ageing 26(3):189–193

9. Arfken CL, Lach HW, Birge SJ, Miller JP (1994) The prevalence and correlates of fear of falling in elderly persons living in the community. Am J Public Health 84(4):565–570

10. McClung MR, Geusens P, Miller PD et al (2001) Effect of risedronate on the risk of hip fracture in elderly women. Hip Intervention Program Study Group. N Engl J Med 344(5):333–340

11. Bischoff-Ferrari HA, Borchers M, Gudat F, Durmuller U, Stahelin HB, Dick W (2004) Vitamin D receptor expression in human muscle tissue decreases with age. J Bone Miner Res 19(2):265–269

12. Sorensen OH, Lund B, Saltin B et al (1979) Myopathy in bone loss of ageing: improvement by treatment with 1 alpha-hydroxycholecalciferol and calcium. Clin Sci (Colch) 56(2):157–161

13. Wicherts IS, van Schoor NM, Boeke AJ et al (2007) Vitamin D status predicts physical performance and its decline in older persons. J Clin Endocrinol Metab 6:6

14. Bischoff-Ferrari HA, Dietrich T, Orav EJ et al (2004) Higher 25-hydroxyvitamin D concentrations are associated with better lower-extremity function in both active and inactive persons aged >=60 y. Am J Clin Nutr 80(3):752–758

15. Bischoff HA, Stahelin HB, Dick W et al (2003) Effects of vitamin D and calcium supplementation on falls: a randomized controlled trial. J Bone Miner Res 18(2):343–351

16. Pfeifer M, Begerow B, Minne HW, Suppan K, Fahrleitner-Pammer A, Dobnig H (2008) Effects of a long-term vitamin D and calcium supplementation on falls and parameters of muscle function in community-dwelling older individuals. Osteoporos Int 16:16

17. Bischoff-Ferrari HA, Orav EJ, Dawson-Hughes B (2006) Effect of cholecalciferol plus calcium on falling in ambulatory older men and women: a 3-year randomized controlled trial. Arch Intern Med 166(4):424–430

18. Pfeifer M, Begerow B, Minne HW, Abrams C, Nachtigall D, Hansen C (2000) Effects of a short-term vitamin D and calcium supplementation on body sway and secondary hyperparathyroidism in elderly women. J Bone Miner Res 15(6):1113–1118

19. Broe KE, Chen TC, Weinberg J, Bischoff-Ferrari HA, Holick MF, Kiel DP (2007) A higher dose of vitamin d reduces the risk of falls in nursing home residents: a randomized, multiple-dose study. J Am Geriatr Soc 55(2):234–239

20. Glerup H, Mikkelsen K, Poulsen L et al (2000) Hypovitaminosis D myopathy without biochemical signs of osteomalacic bone involvement. Calcif Tissue Int 66(6):419–424

21. Bouillon R, Bischoff-Ferrari H, Willett W (2008) Vitamin D and Health: perspectives from Mice and Man. J Bone Miner Res 28:28

22. Endo I, Inoue D, Mitsui T et al (2003) Deletion of vitamin D receptor gene in mice results in abnormal skeletal muscle development with deregulated expression of myoregulatory transcription factors. Endocrinology 144(12):5138–5144, E-pub 2003 Aug 13

23. Visser M, Deeg DJ, Lips P (2003) Low vitamin D and high parathyroid hormone levels as determinants of loss of muscle strength and muscle mass (sarcopenia): the Longitudinal Aging Study Amsterdam. J Clin Endocrinol Metab 88(12):5766–5772

24. Schott GD, Wills MR (1976) Muscle weakness in osteomalacia. Lancet 1(7960):626–629

25. Boland R (1986) Role of vitamin D in skeletal muscle function. Endocrine Rev 7:434–447

26. Bischoff-Ferrari HA, Dawson-Hughes B, Willett CW et al (2004) Effect of vitamin D on falls: a meta-analysis. JAMA 291(16):1999–2006

27. Graafmans WC, Ooms ME, Hofstee HM, Bezemer PD, Bouter LM, Lips P (1996) Falls in the elderly: a prospective study of risk factors and risk profiles. Am J Epidemiol 143(11):1129–1136

28. Flicker L, MacInnis RJ, Stein MS et al (2005) Should older people in residential care receive vitamin D to prevent falls? Results of a randomized trial. J Am Geriatr Soc 53(11):1881–1888

29. Prince RL, Austin N, Devine A, Dick IM, Bruce D, Zhu K (2008) Effects of ergocalciferol added to calcium on the risk of falls in elderly high-risk women. Arch Intern Med 168(1):103–108

30. Bischoff-Ferrari HA, Dawson-Hughes B, Staehelin HB et al (2009) Fall prevention with supplemental and alpha-hydroxylated vitamin D: a meta-analysis of randomized controlled trials. BMJ Oct 1;339:b3692. doi:10.1136/bmj.b3692

31. Chapuy MC, Arlot ME, Duboeuf F et al (1992) Vitamin D3 and calcium to prevent hip fractures in the elderly women. N Engl J Med 327(23):1637–1642

32. Bischoff-Ferrari HA, Dietrich T, Orav EJ, Dawson-Hughes B (2004) Positive association between 25-hydroxy vitamin d levels and bone mineral density: a population-based study of younger and older adults. Am J Med 116(9):634–639

33. Bischoff-Ferrari HA, Willett WC, Wong JB, Giovannucci E, Dietrich T, Dawson-Hughes B (2005) Fracture prevention with vitamin D supplementation: a meta-analysis of randomized controlled trials. JAMA 293(18):2257–2264

34. Porthouse J, Cockayne S, King C et al (2005) Randomised controlled trial of calcium and supplementation with cholecalciferol (vitamin D3) for prevention of fractures in primary care. BMJ 330(7498):1003

35. Grant AM, Avenell A, Campbell MK et al (2005) Oral vitamin D3 and calcium for secondary prevention of low-trauma fractures in elderly people (Randomised Evaluation of Calcium Or vitamin D, RECORD): a randomised placebo-controlled trial. Lancet 365(9471):1621–1628

36. Jackson RD, LaCroix AZ, Gass M et al (2006) Calcium plus vitamin D supplementation and the risk of fractures. N Engl J Med 354(7):669–683

37. Lyons RA, Johansen A, Brophy S et al (2007) Preventing fractures among older people living in institutional care: a pragmatic randomised double blind placebo controlled trial of vitamin D supplementation. Osteoporos Int 18(6):811–818

38. Cranny A, Horsley T, O'Donnell S et al: Effectiveness and safety of vitamin D in relation to bone health. http://www.ahrq.gov/clinic/tp/vitadtp.htm . 2007

39. Boonen S, Lips P, Bouillon R, Bischoff-Ferrari HA, Vanderschueren D, Haentjens P (2007) Need for additional calcium to reduce the risk of hip fracture with vitamin D supplementation: evidence from a comparative meta-analysis of randomized controlled trials. J Clin Endocrinol Metab 30:30

40. Tang BM, Eslick GD, Nowson C, Smith C, Bensoussan A (2007) Use of calcium or calcium in combination with vitamin D supplementation to prevent fractures and bone loss in people aged 50 years and older: a meta-analysis. Lancet 370(9588):657–666

41. Bischoff-Ferrari HA, Dawson-Hughes B, Baron JA et al (2007) Calcium intake and hip fracture risk in men and women: a meta-analysis of prospective cohort studies and randomized controlled trials. Am J Clin Nutr 86(6):1780–1790

42. Armas LA, Hollis BW, Heaney RP (2004) Vitamin D2 is much less effective than vitamin D3 in humans. J Clin Endocrinol Metab 89(11):5387–5391

43. Houghton LA, Vieth R (2006) The case against ergocalciferol (vitamin D2) as a vitamin supplement. Am J Clin Nutr 84(4):694–697

44. Bischoff-Ferrari HA, Willett WC, Wong JB et al (2009) Prevention of nonvertebral fractures with oral vitamin D and dose dependency: a meta-analysis of randomized controlled trials. Arch Intern Med 169:551–561

45. Dawson-Hughes B, Harris SS, Krall EA, Dallal GE (1997) Effect of calcium and vitamin D supplementation on bone density in men and women 65 years of age or older. N Engl J Med 337(10):670–676

46. Visser M, Deeg DJ, Puts MT, Seidell JC, Lips P (2006) Low serum concentrations of 25-hydroxyvitamin D in older persons and the risk of nursing home admission. Am J Clin Nutr 84(3):616–622, quiz 671–672

36 Inherited Defects of Vitamin D Metabolism

Marie B. Demay

Abstract   Inherited defects that interfere with vitamin D metabolism and action have contributed significantly to our understanding of the molecular and physiological actions of the vitamin D endocrine system. This chapter reviews the genetic syndromes that result from impaired hydroxylation of vitamin D, from impaired vitamin D receptor action, and the biological basis for the resultant phenotypes.

Key Words: Vitamin D resistance; rickets; vitamin D-resistant rickets; pseudovitamin D deficiency rickets; 25-hydroxyvitamin D; 1,25-dihydroxyvitamin D; 1-hydroxylase; vitamin D receptor

Abbreviations

PDDR   Pseudovitamin D deficiency Rickets
HVDRR   Hereditary Vitamin D-Resistant Rickets
PTH   Parathyroid hormone

1. INTRODUCTION

The absence of biologic effects of vitamin D, resulting from deficient synthesis, dietary intake, or lack of activation of vitamin D (or to resistance to the biologic effects of the active metabolite), presents primarily with signs and symptoms that reflect impaired intestinal calcium absorption. These include signs and symptoms of neuromuscular irritability, including tetany and seizures, which are a direct result of the hypocalcemia. Long-standing deficiency of, or resistance to, vitamin D metabolites leads to impaired bone mineralization as a result of calcium and phosphate deficiency. In the growing skeleton, growth plate abnormalities known as rickets are also observed. Secondary hyperparathyroidism ensues, due to the hypocalcemia and the lack of anti-proliferative and anti-transcriptional effects of 1,25-dihydroxyvitamin D ($1,25(OH)_2D$) on the parathyroid glands.

With the institution of solar radiation and fish oil therapy for the treatment of rickets, it became clear that there was a rare subset of individuals who were resistant to this treatment. Measurements of the circulating vitamin D metabolites in this group of patients revealed that they fell into two categories: those who had

From: *Nutrition and Health: Vitamin D*
Edited by: M.F. Holick, DOI 10.1007/978-1-60327-303-9_36,
© Springer Science+Business Media, LLC 2010

adequate levels of 25-hydroxyvitamin D (25(OH)D) but low or undetectable levels of 1,25-dihydroxyvitamin D and those in whom 1,25-dihydroxyvitamin D levels were elevated. These two groups were classified as having vitamin D-dependent rickets types I and II, respectively, now renamed pseudovitamin D deficiency rickets (PDDR) and hereditary vitamin D-resistant rickets (HVDRR). More recently, a patient with low circulating levels of 25-hydroxyvitamin D was found to have a homozygous mutation in the *CYP2R1* gene, suggesting that this is the biologically relevant vitamin D 25-hydroxylase *(1)*.

2. PSEUDOVITAMIN D DEFICIENCY RICKETS

The metabolic pathway by which vitamin D, derived from cutaneous production or dietary sources, is converted to the active hormone has been discussed extensively in Chapter 4. Following 25-hydroxylation in the liver, vitamin D is converted to its active metabolite, 1,25-dihydroxyvitamin D, in the proximal tubule of the kidney. While the 25-hydroxyvitamin D-1α-hydroxylase is expressed in numerous tissues, circulating 1,25-dihydroxyvitamin D levels are thought to reflect the activity of the renal enzyme in normal humans; however, 25-hydroxyvitamin D-lα-hydroxylase activity has also been detected in granulomata *(2)*, in the decidual cells of the placenta *(3)*, and in keratinocytes *(4)*. Pseudovitamin D deficiency rickets (PDDR) is a rare inborn error of metabolism, inherited in an autosomal recessive fashion. It is most commonly found in the French Canadian population. The cloning of the vitamin D-lα-hydroxylase cDNA permitted confirmation of the hypothesis that PDDR results from mutation of this gene *(4–7)*.

Since PDDR is an inborn error of metabolism, it presents within the first few months of life. Because the main effect of 1,25-dihydroxyvitamin D is to maintain mineral ion homeostasis by promoting intestinal calcium absorption, the first clinical signs are those of acute or chronic hypocalcemia. Infants may present with hypocalcemic seizures as early as 4 weeks of age or may present later (usually before the age of 2 years) with growth retardation, deformity, and bone pain secondary to rickets and osteomalacia. These abnormalities are clearly visible radiologically. In the in utero environment, the fetus' calcium homeostasis is thought to be relatively normal, due to maternal effects; therefore, teeth that calcify in utero are normal, but those that calcify postnatally often have marked enamel hypoplasia *(8)*. Because hypocalcemia elicits the normal physiologic response of secondary hyperparathyroidism, affected individuals may also present with aminoaciduria *(9)* in addition to hypocalcemia and hypophosphatemia. In patients with this clinical presentation, serum levels of vitamin D and its metabolites establish the diagnosis of PDDR. Levels of vitamin D and 25-hydroxyvitamin D are normal or elevated, whereas levels of the active metabolite, 1,25-dihydroxyvitamin D, are low *(10, 11)*. This observation was the basis for the hypothesis that impaired lα-hydroxylation is the underlying pathophysiologic defect. Although therapeutic responses have been observed using pharmacologic doses of vitamin D (40–54.5 mcg/kg/day) *(15)* and 25-hydroxyvitamin D (3–18 mcg/kg/day) *(9)*, these doses are two orders of magnitude higher than those required to cure classical rickets. Patients treated with these metabolites are often given insufficient doses or fail to respond to non-1α-hydroxylated metabolites, as evidenced by lack of remission of the

Table 1

Biochemical Data from Nine Patients Affected with PDDR Treated with Various Doses of Vitamin D.

Patient (age)	Ca mg/dl (9–10.5)	Pi mg/dl (3.8–5.0)	Alk Ptse IU/L (100–300)	iPTH mEq/L (20–150)	25(OH)D μg/L (18–36)	1,25(OH)₂D (ng/L) (33–35)	Dose of vitamin D₂
1 (4)	9.2	3.0	528	295	148	11	10,000 IU
2 (14)	6.1	3.6	1,160	292	108	9	5,000 IU
3 (14)	8.5	1.6	1,100	372	140	6	75,000 IU
4 (12)	6.3	4.2	595	258	135	6	60,000 IU
5 (15)	6.4	4.4	378	398	105	12	100,000 IU
6 (21)	8.6	2.6	395	375	262	12	50,000 IU
7 (1)	7.6	2.2	1,620	213	30	3	0
8 (1)	6.2	3.0	5,900	1,440	30	2	0
9 (10)	6.8	3.6	1,268	445	37	8	9,000–12,000 IU

Data in brackets represent normal ranges. Modified from *(10)*

clinical disorder (Table 1). Furthermore, although this treatment leads to markedly elevated levels of 25-hydroxyvitamin D, levels of 1,25-dihydroxyvitamin D remain low. The availability of 1α-hydroxylated metabolites of vitamin D has led to a complete remission of the clinical disorder (Fig. 1). Furthermore, the recommended doses of these 1α-hydroxylated metabolites are similar to those used to treat patients with secondary or acquired 1α-hydroxylation defects (such as hypoparathyroidism) and result in normal serum levels of 1,25-dihydroxyvitamin D. These observations support the finding that PDDR is a result of impaired or absent 25-hydroxyvitamin D-1α-hydroxylase activity.

The recommended treatment for this disorder is lifelong therapy with either 1α-hydroxyvitamin D (80–100 ng/kg/day) *(12)* or 1,25-dihydroxyvitamin D (8–400 ng/kg/day) *(10)*. This treatment results in a marked increase in intestinal calcium absorption (Fig. 2), accompanied by an increase in serum calcium within 24 h and radiologic healing of osteomalacic lesions within 2–3 months *(13)*. Histomorphometric documentation of healing of osteomalacic lesions has been obtained in two affected siblings after 9 and 10 months of treatment with 1,25-dihydroxyvitamin D *(10)*.

All patients receiving replacement therapy for vitamin D deficiency, including those with PDDR, require careful monitoring to insure adequate replacement and to minimize the risk of toxicity. The parameters to be monitored include serum calcium, which should normalize within 1 week of institution of adequate therapy. To suppress secondary hyperparathyroidism and promote healing of the osteomalacic lesions, adequate calcium intake should be insured. Serum levels of alkaline phosphatase may increase initially as the skeleton remineralizes, but should normalize within approximately 6 months. Secondary hyperparathyroidism regresses with adequate treatment; however, the time required for involution often reflects the duration and severity of untreated 1,25-dihydroxyvitamin D deficiency (Fig. 2). The main complications of therapy, hypercalcemia and nephrolithiasis, can be avoided by careful monitoring. Serum calcium

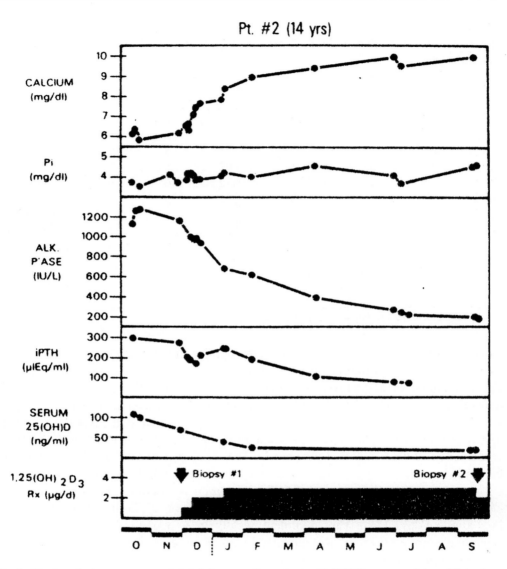

Fig. 1. Biochemical response to calcitriol therapy of a patient with PDDR serum calcium (Ca), phosphorus (Pi), alkaline phosphatase (AlkP'se), immunoreactive parathyroid hormone (iPTH), and 25-hydroxyvitamin D levels are shown, 1,25-dihydroxyvitamin D was increased from 1 to 3 mcg/day until a therapeutic response was observed. After the second bone biopsy revealed healing of osteomalacic lesions the 1,25-dihydroxyvitamin D dosage was decreased to 2 mcg/day (reprinted with permission from *(10)*).

levels should be maintained in the low-normal range and urinary calcium excretion below the lithogenic threshold (<4 mg/kg/day). Healing of the osteomalacic and rachitic lesions should be documented radiologically within 6–9 months of institution of therapy. In growing children, monitoring to insure adequate dosage and to avoid complications should be performed every 6–8 weeks. The required replacement doses of 1α-hydroxylated metabolites may change rapidly in those patients with a growing skeleton.

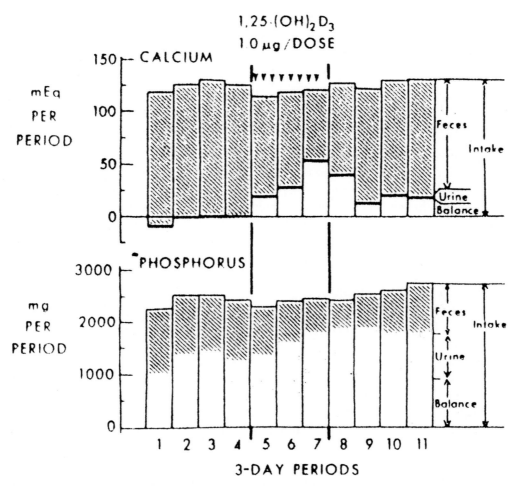

Fig. 2. Calcium balance studies in a patient with PDDR. The patient was treated with 1.0 mcg of 1,25-dihydroxyvitamin D daily in periods 5, 6, and 7. (Reprinted with permission from *(9)* copyright 1973 Massachusetts Medical Society. All rights reserved).

The growth rate of these children also needs to be followed carefully. In those patients with a mature skeleton, monitoring can be performed less frequently (every 3–6 months) since their 1,25-dihydroxyvitamin D requirements are more stable. However, any patient with intercurrent intestinal disease or pregnancy requires closer monitoring. Placental 1α-hydroxylase activity has been shown to be present in the decidua *(3, 13)* which is maternally derived; therefore, the physiologic increase in 1,25-dihydroxyvitamin D levels found in normal pregnancy is absent in affected individuals. During pregnancy, therefore, the dose of 1α-hydroxylated vitamin D metabolites may need to be increased by 50–100%, with careful monitoring of serum and urinary parameters.

Careful analysis of the biochemical parameters of affected individuals led to a therapy capable of reversing the clinical and metabolic abnormalities in affected individuals long before the molecular basis of PDDR was identified. The dramatic response to physiologic doses of 1α-hydroxylated vitamin D metabolites was the basis for the hypothesis that the disorder was caused by an inherited mutation in the 25-hydroxyvitamin D-1α-hydroxylase or in a factor required for its enzymatic activity. Early postnatal diagnosis of the disorder can prevent the development of skeletal and dental complications and help affected individuals achieve their normal growth potential.

3. HEREDITARY VITAMIN D-RESISTANT RICKETS

The therapeutic efficacy of 1α-hydroxylated vitamin D metabolites in the treatment of PDDR led to the identification of a clinical syndrome that did not respond to physiologic doses of these metabolites. Analysis of the biochemical parameters revealed that these patients, in fact, had normal or elevated levels of 1,25-dihydroxyvitamin D prior to treatment. This led to the hypothesis that affected individuals were resistant to the biologic effects of 1,25-dihydroxyvitamin D. The cloning of the vitamin D receptor (VDR) *(14)* enabled the identification of the molecular basis of this disease as mutations in the VDR (see Chapter 5). There are, however, reports of normal VDR cDNAs in kindreds with clinical and biochemical parameters suggestive of HVDRR *(15, 16)*.

Like PDDR, HVDRR is an inborn error of metabolism, inherited in an autosomal recessive fashion. HVDRR presents in infancy or childhood with rickets, osteomalacia, and signs of hypocalcemia including tetany and seizures *(17)*. Affected individuals may have enamel hypoplasia and oligodentia *(18)*. Interestingly, a few cases have not presented until late adolescence *(19)*. Like PDDR, affected patients have hypocalcemia, accompanied by secondary hyperparathyroidism, which results in hypophosphatemia and occasionally aminoaciduria. Normal levels of vitamin D and 25-hydroxyvitamin D are present, whereas levels of 1,25-dihydroxyvitamin D are elevated because of stimulation of the renal 1α-hydroxylase by parathyroid hormone (PTH) and decreased inactivation, due to the fact that expression of the vitamin D 24-hydroxylase is dependent upon the liganded VDR. One unique feature of HVDRR is the alopecia totalis observed in some kindreds (Fig. 3), which is thought to be associated with the finding that VDRs are normally found in the external root sheath of hair follicles *(20)*. It was suggested that patients presenting with alopecia had a more severe resistance to 1,25-dihydroxyvitamin D *(17, 21)*. However, spontaneous remissions have been described in patients with alopecia *(22)*, and severely affected individuals with normal hair growth have also been described *(23, 24)*. Recent investigations in a mouse model of HVDRR demonstrate that the ligand-dependent actions of the VDR are not required to prevent alopecia *(25)* and that the VDR is required for keratinocyte stem cell function *(26)*. Interestingly, the VDR cDNA in one patient with clinical and biochemical parameters suggestive of HVDRR, including alopecia, has been shown to be normal by sequence analysis and in vitro transcriptional activation studies *(16)*.

Histomorphometric analyses of bone biopsies from patients with HVDRR have demonstrated features consistent with osteomalacia *(19, 27–29)*. In patients with long-standing secondary hyperparathyroidism, marrow fibrosis may also be seen *(19, 28)*. Intestinal resistance to 1,25-dihydroxyvitamin D has been documented by marked impairment of intestinal calcium absorption and high fecal calcium losses in affected individuals, despite elevated serum levels of 1,25-dihydroxyvitamin D *(28–30)*;

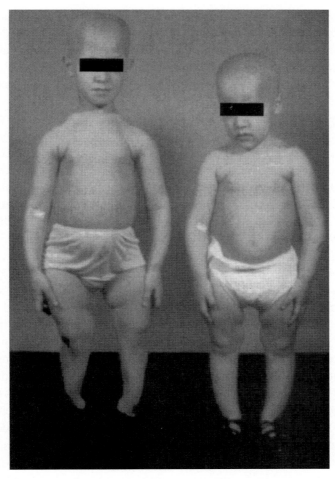

Fig. 3. Two sisters affected by HVDRR (ages 7 and 3) demonstrate short stature, bowing of the legs, and alopecia (reprinted with permission from *(29)*).

however, an improvement in intestinal calcium absorption has been observed concomitant with therapeutic responses to vitamin D metabolites.

Because the pathophysiologic abnormality underlying HVDRR is a receptor defect, no treatment has been uniformly successful. Affected individuals have been shown to have variable therapeutic responses to pharmacologic doses of vitamin D metabolites and oral calcium supplements (1–6 g/day). Although 1α-hydroxylated metabolites have been the favored treatment in recent years, the intact metabolic pathway of vitamin D activation suggests that vitamin D or 25-hydroxyvitamin D may be as effective. The recommended therapeutic doses vary from 0.1 to 50 mg/day of vitamin D, 0.05–1.5 mg/day of 25-hydroxyvitamin D, and 5–20 mcg/day of 1,25-dihydroxyvitamin D *(18)*, all of which result in markedly elevated levels of 1,25-dihydroxyvitamin D. In spite of these pharmacologic doses of vitamin D metabolites, in some severely affected individuals, hypocalcemia persists, as do the rachitic and osteomalacic lesions. Because the main physiologic effect of 1,25-dihydroxyvitamin D is thought to be the promotion of intestinal calcium absorption, some patients have been treated with parenteral calcium infusions in an attempt to circumvent the intestinal resistance to 1,25-dihydroxyvitamin D. This has led to near normalization of the associated biochemical abnormalities as well as clinical and radiologic remission of the osteomalacic lesions *(27, 31–33)*. In one case, bone biopsies prior to and after calcium infusion therapy documented healing of osteomalacia *(27)*.

In most cases of HVDRR, lifelong therapy is indicated. The severity of the disease varies markedly and has not been shown to be related to the domain of the VDR affected by the mutation. Occasional clinical remissions have occurred in individual patients who maintain normal or near normal biochemical and clinical parameters off therapy *(22, 23, 34–36)*. This phenomenon is poorly understood, since the underlying receptor mutation still persists. Perhaps these individuals reflect a subset of patients whose resistance is relatively mild and in whom normal calcium homeostasis can be maintained once the needs of a growing skeleton are met. One would expect, however, that clinical relapses may occur with pregnancy or intercurrent intestinal disease.

In patients who require long-term therapy with vitamin D metabolites, calcium levels should be measured frequently, every 4–6 weeks in those individuals with a growing skeleton, or every 3–6 months in adults. The efficacy of therapy should be assessed by monitoring serum PTH levels and alkaline phosphatase levels and documenting healing of rachitic and osteomalacic lesions. In those patients who fail to respond to vitamin D metabolites and oral calcium therapy, levels of 1,25-dihydroxyvitamin D should be measured to confirm therapeutic compliance and absorption of the medication. Unlike PDDR, in which normalization of serum 1,25-dihydroxyvitamin D levels has a predictive value for therapeutic response, in HVDRR, there is no correlation between these levels and clinical response. In fact, levels as high as 15,499 pg/ml have been documented in a patient who required parenteral calcium to normalize her biochemical parameters and cure her osteomalacic lesions *(27)*. As in this case, if administration of vitamin D metabolites does not lead to a clinical response, parenteral calcium infusions are indicated. Calcium infusions should be continued until clinical and biochemical remissions are documented. Once this end point has been achieved, close monitoring is required along with a repeat trial of conventional therapy *(37)*.

Like patients with PDDR, those with HVDDR are fertile *(18, 21)* but require increased doses of vitamin D metabolites to maintain normal mineral ion homeostasis during the second and third trimesters of pregnancy. Assessment of [^{3}H]-1,25-dihydroxyvitamin D binding and 1,25-dihydroxyvitamin D induction of 24-hydroxylase activity of amniotic cells has been used for antenatal diagnostic screening in the past *(38)*. Currently, characterization of the receptor defect in affected kindreds permits rapid prenatal diagnosis by polymerase chain reaction and sequencing of DNA obtained from amniotic cells.

4. 25-HYDROXYLASE DEFICIENCY

Casella et al. *(39)* described two siblings, the offspring of nonconsanguineous Nigerian parents, who presented with rickets at 2 and 7 years of age. Both patients had low normal serum calcium levels, increased alkaline phosphatase, and secondary hyperparathyroidism accompanied by hypophosphatemia. Levels of 25-hydroxyvitamin D were low, whereas those of 1,25-dihydroxyvitamin D were normal to high. Vitamin D absorption tests were normal. Supraphysiologic doses of vitamin D were required for normalization of the biochemical parameters, including 25-hydroxyvitamin D levels, and for healing of the bony lesions. Tapering of the vitamin D dosage resulted in a decrease in the 25-hydroxyvitamin D level, which returned to normal after institution of replacement doses of 25-hydroxyvitamin D. These clinical findings, which have also been described in a Turkish kindred *(40)*, are highly suggestive of impaired 25-hydroxylation of vitamin D. A homozygous mutation in the 25-hydroxylase *CYP2R1* gene in one affected individual *(1)* suggests that this enzyme is important for maintaining normal circulating levels of 25-hydroxyvitamin D.

REFERENCES

1. Cheng JB, Levine MA, Bell NH, Mangelsdorf DJ, Russell DW (2004) Genetic evidence that the human CYP2R1 enzyme is a key vitamin D 25-hydroxylase. Proc Natl Acad Sci USA 101(20): 7711–7715
2. Barbour GL, Coburn JW, Slatopolsky E, Norman AW, Horst RL (1981) Hypercalcemia in an anephric patient with sarcoidosis: evidence for extrarenal generation of 1,25-dihydroxyvitamin D. N Engl J Med 305:440–443
3. Weisman Y, Harell A, David SE, M ZS, Golander A (1979) 1α,25-Dihydroxyvitamin D$_3$ and 24,25-dihydroxyvitamin D$_3$ in vitro synthesis by human decidua and placenta. Nature 281: 317–319
4. Fu GK, Lin D, Zhang MY, Bikle DD, Shackleton CH, Miller WL, Portale AA (1997) Cloning of human 25-hydroxyvitamin D-1 alpha-hydroxylase and mutations causing vitamin D-dependent rickets type 1. Mol Endocrinol 11(13):1961–1970
5. Shinki T, Shimada H, Wakino S, Anazawa H, Hayashi M, Saruta T, DeLuca HF, Suda T (1997) Cloning and expression of rat 25-hydroxyvitamin D3 1-alpha-hydroxylase cDNA. Proc Nat Acad Sci USA 94:12920–12925
6. Takeyama K, Kitanaka S, Sato T, Kobori M, Yanagisawa J, Kato S (1997) 25-Hydroxyvitamin D3 1alpha-hydroxylase and vitamin D synthesis. Science 277(5333):1827–1830
7. St-Arnaud R, Messerlian S, Moir JM, Omdahl JL, Glorieux FH (1997) The 25-hydroxyvitamin D 1-alpha-hydroxylase gene maps to the pseudovitamin D-deficiency rickets [PDDR] disease locus. J Bone Miner Res 12(10):1552–1559
8. Arnaud C, Maijer R, Reade T, Scriver CR, Whelan DT (1970) Vitamin D dependency: an inherited postnatal syndrome with secondary hyperparathyroidism. Pediatrics 46:871–880
9. Fraser D, Kooh SW, Kind HP, Holick MF, Tanaka Y, DeLuca HF (1973) Pathogenesis of hereditary vitamin-D-dependent rickets: an inborn error of vitamin D metabolism involving defective conversion of 25-hydroxyvitamin D to 1α,25-dihydroxyvitamin D. N Engl J Med 289:817–822
10. Delvin EE, Glorieux FH, Marie PJ, Pettifor JM (1981) Vitamin D dependency: replacement therapy with calcitriol. J Pediatr 99:26–34
11. Scriver CR, Reade TM, DeLuca HF, Hamstra AJ (1978) Serum 1,25-dihydroxyvitamin D levels in normal subjects and in patients with hereditary rickets or bone disease. N Engl J Med 299:976–979
12. Reade TM, Scriver CR, Glorieux FH, Nogrady B, Delvin E, Poirier R, Holick MF, DeLuca HF (1975) Response to crystalline 1α-hydroxyvitamin D$_3$ in vitamin D dependency. Pediatr Res 9:593–599
13. Glorieux FH, Arabian A, Delvin EE (1995) Pseudo-vitamin D deficiency: absence of 25-hydroxyvitamin D 1α-hydroxylase activity in human placenta decidual cells. J Clin Endocrinol Metab 80:2255–2258
14. Baker AR, McDonnell DP, Hughes M, Crisp TM, Magelsdorf DJ, Haussler MR, Pike JW, Shine J, O'Malley BW (1988) Cloning and expression of full-length cDNA encoding human vitamin D receptor. Proc Natl Acad Sci USA 85:3294–3298

15. Giraldo A, Pino W, Garcia-Ramirez LF, Pineda M, Iglesias A (1995) Vitamin D dependent rickets type II and normal vitamin D receptor cDNA sequence. A cluster in a rural area of Cauca, Colombia, with more than 200 affected children. Clin Genet 48:57–65
16. Hewison M, Rut AR, Kristjansson K, Walker RE, Dillon MJ, Hughes MR, O'Riordan JLH (1993) Tissue resistance to 1,25-dihydroxyvitamin D without a mutation of the vitamin D receptor gene. Clin Endocrinol 39:663–670
17. Marx SJ, Liberman UA, Eil C, Gamblin GT, DeGrange DA, Balsan S (1984) Hereditary resistance to 1,25-dihydroxyvitamin D. Recent Prog Horm Res 40:589–615
18. Bell NH (1980) Vitamin D-dependent rickets type II. Calcif Tissue Int 31:89–91
19. Brooks MH, Bell NH, Love L, Stern PH, Orfei E, Queener SF, Hamstra AJ, DeLuca HF (1978) Vitamin D-dependent rickets type II: resistance of target organs to 1,25-Dihydroxyvitamin D. N Engl J Med 298:996–999
20. Stumpf WE, Sar M, Reid FA, Tanaka Y, DeLuca HF (1979) Target cells for 1,25-dihydroxyvitamin D3 in intestinal tract, stomach, kidney, skin, pituitary, and parathyroid. Science 206:1188–1190
21. Marx SJ, Bliziotes MM, Nanes M (1986) Analysis of the relation between alopecia and resistance to 1,25-dihydroxyvitamin D. Clin Endocrinol 25:373–381
22. Takeda E, Yokota I, Kawakami I, Hashimoto T, Kuroda Y, Arase S (1989) Two siblings with vitamin-D-dependent rickets type II: no recurrence of rickets for 14 years after cessation of therapy. Eur J Pediatr 149:54–57
23. Fraher LJ, Karmali R, Hinde FRJ, Hendy GN, Jani H, Nicholson L, Grant D, O'Riordan JLH (1986) Vitamin D-dependent rickets type II: extreme end organ resistance to 1,25-dihydroxyvitamin D3 in a patient without alopecia. Eur J Pediatr 145:389–395
24. Liberman UA, Eil C, Marx SJ (1986) Clinical features of hereditary resistance to 1,25-dihydroxyvitamin D: Hereditary hypocalcemic vitamin D resistant rickets type II. Adv Exp Med Biol 196:391–406
25. Skorija K, Cox M, Sisk JM, Dowd DR, MacDonald PN, Thompson CC, Demay MB (2005) Ligand-independent actions of the vitamin D receptor maintain hair follicle homeostasis. Mol Endocrinol 19(4):855–862
26. Cianferotti L, Cox M, Skorija K, Demay MB (2007) Vitamin D receptor is essential for normal keratinocyte stem cell function. Proc Natl Acad Sci USA 104(22):9428–9433
27. Balsan S, Garabedian M, Larchet M, Gorski A-M, Cournot G, Tau C, Bourdeau A, Silve C, Ricour C (1986) Long-term nocturnal calcium infusions can cure rickets and promote normal mineralization in hereditary resistance to 1,25-dihydroxyvitamin D. J Clin Invest 77:1661–1667
28. Beer S, Tieder M, Kohelet D, Liberman OA, Vure E, Bar-Joseph G, Gabizon D, Borochowitz ZU, Varon M, Modai D (1981) Vitamin D resistant rickets with alopecia: a form of end organ resistance to 1,25 dihydroxy vitamin D. Clin Endocrinol 14:395–402
29. Rosen JF, Fleischman AR, Finberg L, Hamstra A, DeLuca HF (1979) Rickets with alopecia: an in born error of vitamin D metabolism. J Pediatr 94:729–735
30. Tsuchiya Y, Matsuo N, Cho H, Kumagai M, Yasaka A, Suda T, Orimo H, Shiraki M (1980) An unusual form of vitamin D-dependent rickets in a child: alopecia and marked end-organ hyposensitivity to biologically active vitamin D. J Clin Endocrinol Metab 51:685–690
31. Bliziotes M, Yergey AL, Nanes MS, Muenzer J, Begley MG, Vieira NE, Kher KK, Brandi ML, Marx SJ (1988) Absent intestinal response to calciferols in hereditary resistance to 1,25-dihydroxyvitamin D: documentation and effective therapy with high dose intravenous calcium infusions. J Clin Endocrinol Metab 66:294–300
32. Walka MM, Däumling S, Hadorn HB, Kruse K, Belohradsky BH (1991) Vitamin D dependent rickets type II with myelofibrosis and immune dysfunction. Eur J Pediatr 150:665–668
33. Weisman Y, Bab I, Gazit D, Spirer Z, Jaffe M, Hochberg Z (1987) Long-term intracaval calcium infusion therapy in end-organ resistance to 1,25-dihydroxyvitamin D. Am J Med 83:984–990
34. Chen TL, Hirst MA, Cone CM, Hochberg Z, Tietze H-U, Feldman D (1984 1) 25-dihydroxyvitamin D resistance, rickets, and alopecia: analysis of receptors and bioresponse in cultured fibroblasts from patients and parents. J Clin Endocrinol Metab 59:383–388

35. Eil C, Liberman UA, Marx SJ (1986) The molecular basis for resistance to 1,25-dihydroxyvitamin D: studies in cells cultured from patients with hereditary hypocalcemic 1,25[OH]$_2$D$_3$-resistant rickets. Adv Exp Med Biol 196:407–422

36. Liberman UA, Halabe A, Samuel R, Kauli R, Edelstein S, Weisman Y, Papapoulos SE, Fraher LJ, Clemens TL, O'Riordan JLH (1980) End-organ resistance to 1,25-dihydroxycholecalciferol. Lancet 1:504–506

37. Al-Aqeel A, Ozand P, Sobki S, Sewairi W, Marx S (1993) The combined use of intravenous and oral calcium for the treatment of vitamin D dependent rickets type II [VDDRII]. Clin Endocrinol 39: 229–237

38. Weisman Y, Jaccard N, Legum C, Spirer Z, Yedwab G, Even L, Edelstein S, Kaye AM, Hochberg Z (1990) Prenatal diagnosis of vitamin D-dependent rickets, type II: response to 1,25-dihydroxyvitamin D in amniotic fluid cells and fetal tissues. J Clin Endocrinol Metab 71:937–943

39. Casella SJ, Reiner BJ, Chen TC, Holick MF, Harrison HE (1994) A possible genetic defect in 25-hydroxylation as a cause of rickets. J Pediatr 124:929–932

40. Nützenadel W, Mehls O, Klaus G (1995) A new defect in vitamin D metabolism. J Pediatrics 126: 676–677

37 Molecular Defects in the Vitamin D Receptor Associated with Hereditary 1,25-Dihydroxyvitamin D-Resistant Rickets (HVDRR)

Peter J. Malloy and David Feldman

Abstract   Vitamin D is important in skeletal development and in bone mineralization. The active form of vitamin D, $1\alpha,25$-dihydroxyvitamin D, $[1,25(OH)_2D]$, binds with high affinity to the vitamin D receptor (VDR), a member of the nuclear receptor family of transcription factors. Genetic mutations in the vitamin D receptor cause the rare genetic disease hereditary 1,25-dihydroxyvitamin D-resistant rickets (HVDRR). Children with HVDRR have rickets, hypocalcemia, hypophosphatemia, and secondary hyperparathyroidism. Some have total alopecia. A number of heterogeneous mutations have been identified in the VDR as the molecular cause of HVDRR. Mutations in DNA-binding domain inactivate the VDR by disrupting contact with VDREs in promoters of target genes. Mutations in the ligand-binding domain reduce the affinity of the VDR for $1,25(OH)_2D$, prevent the $1,25(OH)_2D$ from binding to the VDR, inhibit RXR heterodimerization, or abolish coactivator interactions. Other types of mutations have also been found including nonsense mutations, splice site mutations, deletions, insertions, and duplications. Children with HVDRR have been successfully treated with intravenous calcium that bypasses the intestinal defect in calcium transport due to the lack of $1,25(OH)_2D$ action on the mutant VDR.

Key Words: Hereditary vitamin D-resistant rickets; vitamin D receptor; rickets; mutation; hypophosphatemia; ligand binding; 1,25-dihydroxyvitamin D; 25-hydroxyvitamin D; DNA binding; splicing mutations

1. INTRODUCTION

Vitamin D, the primary regulator of calcium homeostasis in the body, is particularly important in skeletal development and in bone mineralization. The active form of vitamin D, $1\alpha,25$-dihydroxyvitamin D, $[1,25(OH)_2D$ or calcitriol], elicits its hormonal action by binding with high affinity to the vitamin D receptor (VDR) and regulating gene transcription by mechanisms described in other chapters in this volume.

Supported by NIH Grant DK 42482

From: *Nutrition and Health: Vitamin D*
Edited by: M.F. Holick, DOI 10.1007/978-1-60327-303-9_37,
© Springer Science+Business Media, LLC 2010

Genetic mutations in the *VDR* gene cause the rare genetic disease hereditary vitamin D-resistant rickets (HVDRR) also known as vitamin D-dependent rickets type II (VDDR-II, OMIM#277440). HVDRR results in a syndrome of generalized resistance to $1,25(OH)_2D$ caused by heterogeneous mutations in the *VDR* gene, which disrupt the function of the receptor ultimately leading to complete or partial resistance to $1,25(OH)_2D$.

Brooks et al. *(1)* and Marx et al. *(2)* described the first cases of HVDRR. The patients had hypocalcemia, hypophosphatemia, secondary hyperparathyroidism, and markedly elevated levels of serum $1,25(OH)_2D$ suggesting that the disease was due to end-organ resistance to $1,25(OH)_2D$. Since these initial studies there have been many reports of patients with apparent target organ resistance to $1,25(OH)_2D$ *(3)*. In this chapter, we will review the clinical features and the genetic basis underlying HVDRR and update our previous reviews of the subject *(3–6)*. An animal model of HVDRR has been developed by several laboratories in which the VDR has been knocked out *(7–10)*. Many of the attributes of clinical HVDRR are recapitulated in the mouse model.

2. HEREDITARY 1,25-DIHYDROXYVITAMIN D-RESISTANT RICKETS (HVDRR)

The major clinical features of HVDRR are hypocalcemia and rickets due to defective mineralization of newly formed bone. The hypocalcemia and rickets are usually severe and generally displayed early in life, usually within months of birth. Affected children are often growth retarded and suffer from bone pain, muscle weakness, and hypotonia. Convulsions due to hypocalcemia have sometimes occurred. In some cases children develop severe dental caries or exhibit enamel hypoplasia of the teeth *(11–17)*. Some infants have died from pneumonia as a result of poor respiratory movement due to severe rickets of the chest wall *(12, 15, 18)*. Many children have sparse body hair and some have total scalp and body alopecia including eyebrows and in some cases eyelashes. Children with alopecia may have skin lesions or papules. The alopecia and skin papules resemble the genetic disease atrichia with papular lesions (APL, OMIM#209500) that is caused by mutations in the *hairless* (*hr*) gene *(19, 20)*.

Typical serum biochemistry values found in HVDRR children include low serum concentrations of calcium and phosphate and elevated serum alkaline phosphatase activity. The patients have elevated parathyroid hormone (PTH) levels as the result of the hypocalcemia leading to secondary hyperparathyroidism. These findings are also common to patients with 1α-hydroxylase deficiency (OMIM#264700) caused by mutations in the *CYP27B1* (*1α-hydroxylase*) gene. In HVDRR, serum 25(OH)D values are usually normal or can be low but the $1,25(OH)_2D$ levels are very elevated. This singular feature distinguishes HVDRR from 1α-hydroxylase deficiency where the serum $1,25(OH)_2D$ values are severely depressed. Patients with 1α-hydroxylase deficiency can be treated with close to physiologic doses of calcitriol that bypass the 1α-hydroxylase deficiency and restore circulating $1,25(OH)_2D$ levels to normal. Patients with HVDRR, on the other hand, do not respond to physiologic doses of calcitriol and many are resistant to supraphysiologic doses of all forms of vitamin D therapy.

HVDRR is inherited as an autosomal recessive disease. The parents and siblings who are heterozygous carriers of the mutant *VDR* gene generally show no symptoms of the disease and have normal bone development. In most cases, consanguinity in the family lineage can be found and intermarriage is often associated with the disease. Males and females are equally affected *(21)*.

The primary biological process attributed to vitamin D is maintenance of calcium and bone homeostasis. $1,25(OH)_2D$ is essential for promoting the transport of calcium and phosphate across the small intestine and into the circulation. Adequate delivery of calcium and phosphate to the bone is essential for the normal mineralization of bone. The hypocalcemia and the resistance of the parathyroid gland to suppression by $1,25(OH)_2D$ because of defective VDR within the gland, in turn results in secondary hyperparathyroidism. The increase in circulating $1,25(OH)_2D$ levels is due to an increase in renal 1α-hydroxylase activity caused by both elevated PTH and hypophosphatemia that upregulate renal *CYP27B1* (*1α-hydroxylase*) gene expression as well as failure of elevated $1,25(OH)_2D$ to suppress 1α-hydroxylase. The hypophosphatemia results from the secondary hyperparathyroidism and the loss of a functional VDR in the kidney leading to renal loss as well as decreased intestinal absorption. The calcium and phosphate deficiencies compromise normal bone mineralization leading to rickets in children and osteomalacia in adults.

Alopecia totalis (sometimes called atrichia) is a clinical feature that is found in many patients with HVDRR. Patients may have sparse body hair or exhibit total scalp and body alopecia *(22–24)*. Children with extreme alopecia often lack eyebrows and in some cases eyelashes. Hair loss may be evident at birth or occurs during the first few months of life. Skin lesions or papules are often associated with alopecia. Skin biopsy has revealed apparently normal follicles with no hair. The lack of VDR action during a critical stage of hair follicle development is the suspected cause of alopecia. Patients with mutations in the *CYP27B1* (*1α-hydroxylase*) gene or other causes of vitamin D deficiency do not exhibit alopecia.

3. THE VITAMIN D RECEPTOR

The overall structure of the VDR protein is similar to the other members of the steroid-thyroid-retinoid receptor superfamily. Elucidation of the nature of naturally occurring mutations in children with HVDRR has helped to understand the functional domains of the VDR protein. At the N-terminus the VDR has a highly conserved DNA-binding domain (DBD) and in the C-terminal half of the protein a more variable ligand-binding domain (LBD). The DBD contains two finger-like structures of 12–13 amino acids each. Four cysteine residues bind one zinc atom to form each zinc finger structure *(25)*. Regions of the DBD are critical both for DNA binding and also serve as a dimerization interface for interaction with the retinoid X receptor (RXR) *(26, 27)*. The hinge region (amino acids residues 93–120) connects the DBD and LBD.

The VDR LBD is formed by amino acids 123–427. X-ray crystallography of the VDR showed that the LBD is composed of 12 α-helices (H1–H12) and 3 β-sheets (S1–S3) *(28)*. Helix H12 forms a retractable lid that traps and holds the ligand in position. Ligand binding causes a conformational change in the VDR that promotes heterodimerization with RXR. VDR heterodimerization with RXR involves residues in H9, H10, and an E1 domain that overlaps H4 and H5 within the LBD. An activating function domain 2 (AF-2 domain) residues 416–424 of helix H12 and the region between amino acids 232–272 encompassing H3 and H4 are essential for transactivation (28). The repositioning of helix H12 after ligand binding is critical for the formation of a hydrophobic groove that binds LxxLL motifs (where L is leucine and x is any amino acid) in the nuclear receptor interacting domains of coactivators. Other regions of the VDR also act to recruit coactivator proteins or facilitate contact with proteins associated with the core transcriptional machinery such as TFIIB or the TAFs $(29, 30)$. Mutations in the VDR LBD that cause HVDRR have been shown to completely abolish ligand binding, reduce its affinity for $1,25(OH)_2D$, or disrupt RXR heterodimerization or interactions with coactivators.

4. CELLULAR BASIS OF HVDRR

Feldman et al. (31) demonstrated that the VDR was present in fresh and cultured human foreskin as well as in cultured keratinocytes and dermal fibroblasts grown from adult skin biopsies. These findings triggered investigations into examining the VDR in patients with HVDRR. Eil et al. (32) showed in cultured skin fibroblasts that the cause of the cellular defect in patients with HVDRR was due to the defective nuclear uptake of $1,25(OH)_2D$. Feldman et al. (33) then demonstrated that cytosolic extracts of cultured fibroblasts from some HVDRR patients had undetectable levels of $[^3H]$-$1,25(OH)_2D$ binding. They also demonstrated that induction of 24-hydroxylase activity by $1,25(OH)_2D$ could serve as a marker of $1,25(OH)_2D$ responsiveness in the cultured fibroblasts. A number of other HVDRR cases were also examined using cultured skin fibroblasts $(15, 34–36)$ or cells derived from bone (37). Some patients' fibroblasts lacked specific $[^3H]$-$1,25(OH)_2D$ binding $(15, 34–37)$ and were unresponsive to $1,25(OH)_2D$ treatment while other cells exhibited normal $[^3H]$-$1,25(OH)_2D$ binding but were also unresponsive to $1,25(OH)_2D$ treatment $(15, 35, 37–40)$. The conclusion reached was that the HVDRR was caused by cellular resistance to $1,25(OH)_2D$ action and was due to at least two types of defects in the VDR, one that impaired ligand binding and one that retained ligand binding but caused resistance to $1,25(OH)_2D$ by a defect downstream of ligand binding (3).

As the number of reports on HVDRR increased, the heterogeneous nature of the defects in the VDR became more apparent. Hochberg et al. $(22, 23)$ reported the clinical findings in four patients from two unrelated families of Arab origin who exhibited HVDRR and alopecia. A follow-up study by Chen et al. (41) showed that fibroblasts from three of these patients and a patient from an unrelated family from Germany had no $[^3H]$-$1,25(OH)_2D$ binding and $1,25(OH)_2D$ treatment failed to induce 24-hydroxylase activity. Pike et al. (42) used a radioligand immunoassay (43) and a monoclonal antibody to the chick VDR $(44–46)$, to demonstrate the presence of an immunoreactive protein in cell extracts from fibroblasts of HVDRR patients that exhibited no ligand binding. The authors speculated that the lack of $[^3H]$-$1,25(OH)_2D$ binding in these patients was not due to defective synthesis of the VDR protein but was due to defects in the VDR LBD that prevented ligand binding (42). Liberman et al. (39), Gamblin et al. (47), and

Castells et al. *(48)* also used fibroblasts from affected patients to demonstrate various defects in VDR function causing $1,25(OH)_2D$ resistance.

Based on the concept that VDR binding to DNA is essential for activity, Hirst et al. *(38)* showed that defective DNA binding could be the cause of resistance in cases that had normal ligand binding. They showed that the fibroblasts from affected individuals had normal $[^3H]$-$1,25(OH)_2D$ binding but were resistant to $1,25(OH)_2D$ treatment. The authors demonstrated that the VDR from the affected fibroblasts exhibited a significant decrease in affinity for general DNA. The VDR from an unaffected sibling bound strongly to DNA whereas the patient's VDR bound weakly to DNA. Malloy et al. *(40)* identified a similar DNA-binding defect in the VDR from HVDRR patients who had normal $[^3H]$-$1,25(OH)_2D$ binding. The patient's VDR exhibited a low affinity for DNA. Furthermore, the cells from the parents had two forms of the VDR, one with a high affinity for DNA and the other with a low affinity for DNA. This was the first clear evidence of heterozygosity in parents of HVDRR children. It was suspected that the defects in these cases would likely be due to mutations in the DBD *(38, 40)*, which later proved to be correct *(49)*.

Other cell types have been used to study the VDR in HVDRR cases. These include peripheral mononuclear cells *(50)*, phytohemagglutinin (PHA)-stimulated lymphocytes *(51, 52)*, myeloid progenitor cells *(53)*, Epstein–Barr virus (EBV) immortalized B lymphoblasts *(21, 40, 49, 54)*, and HTLV-1 virus immortalized T lymphoblasts *(55)*. It is interesting to note that although EBV immortalized B lymphoblasts from normal subjects express wild-type VDR, they nevertheless fail to induce 24-hydroxylase activity or to show inhibition of cell growth in response to $1,25(OH)_2D$ *(21)*. On the other hand, PHA-stimulated lymphocytes and HTLV-1 immortalized T lymphoblasts from normal subjects do respond to $1,25(OH)_2D$ *(55, 56)*. Studies by Takeda et al. *(52)* showed that HVDRR can be diagnosed rapidly using lymphocytes by the failure of $1,25(OH)_2D$ to inhibit DNA synthesis or induce 24-hydroxylase activity *(51, 52)*. They also showed that lymphocytes from parents of children with HVDRR express intermediate levels of 24-hydroxylase in response to $1,25(OH)_2D$, whereas most other studies have failed to demonstrate abnormalities in the obligate heterozygotic parents or unaffected siblings or defects in their cultured cells.

5. MOLECULAR BASIS FOR HVDRR

5.1. *Mutations in the VDR DNA-Binding Domain (DBD)*

Investigations to determine the specific mutations in the VDR that cause HVDRR began shortly after the human *VDR* cDNA sequence was determined *(57)*. The initial studies were focused on deciphering the defects in the *VDR* gene from HVDRR patients whose fibroblasts displayed normal $[^3H]$-$1,25(OH)_2D$ binding but abnormal DNA binding. With the advent of the polymerase chain reaction (PCR), Hughes et al. *(49)* used this powerful technique to amplify the *VDR* gene from DNA isolated from two patients whose VDRs were defective in DNA binding *(38)*. In one family, the patient was shown to have a unique G to A single base change in exon 3 that replaced arginine with glutamine at amino acid residue 73 in the second zinc finger module (Arg73Gln) (Fig. 1). In the second family the patient was shown to have a unique G to A mutation in exon

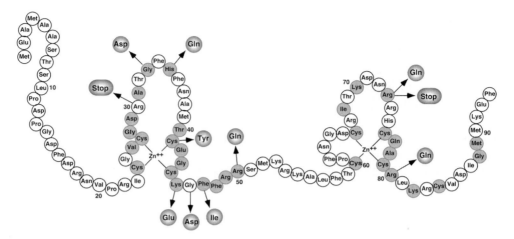

Fig. 1. Mutations in the VDR DBD that cause HVDRR. The DBD is composed of two zinc finger modules. Conserved amino acids are shown as *shaded circles*. Numbers refer to amino acid number.

2 that changed glycine to aspartic acid at amino acid residue 33 in the first zinc finger module (Gly33Asp) (Fig. 1). The parents had a normal and a mutant sequence demonstrating the genetic transmission and recessive nature of the disease. To prove that the missense mutations found in these families were the cause of $1,25(OH)_2D$ resistance seen in the HVDRR patients, the Arg73Gln and the Gly33Asp mutations were generated in the wild-type VDR cDNA by site-directed mutagenesis *(58)*. The recreated mutant VDRs exhibited normal $[^3H]$-$1,25(OH)_2D$ binding and weak binding to DNA similar to the VDR in the patient's fibroblasts. Furthermore, in transfected COS-1 cells the mutant VDRs failed to induce reporter gene activity when treated with $1,25(OH)_2D$ demonstrating that they were the cause of $1,25(OH)_2D$ resistance in the patients *(58)*.

Since the original study by Hughes et al. *(49)* a number of mutations in the VDR have been identified in patients with HVDRR. More than 100 cases of HVDRR have been recorded and a number of these have been analyzed at the biochemical and molecular level *(3, 4, 6)*. We have developed a comprehensive listing of all the mutations that have been identified as the basis of HVDRR. These are listed in Table 1 along with the type of defect and the reference to the work. A description of some of these mutations and the consequences of the abnormality in the VDR will be discussed below.

A number of mutations have now been identified in the VDR DBD as the cause of HVDRR and are shown in Fig. 1. These cases exhibit normal ligand binding but defective DNA binding *(3, 6)*. Sone et al. identified an Arg80Gln mutation in the DBD in a patient with HVDRR and alopecia *(59)*. The same mutation was also identified in two siblings with HVDRR and alopecia *(60)*. Both families with the Arg80Gln mutation had origins in North Africa *(59)*. Saijo et al. identified a missense mutation in the DBD that resulted in an Arg50Gln substitution in three HVDRR patients from two unrelated families of Japanese origin *(61)*. Skin fibroblasts from the patients showed normal $[^3H]$-$1,25(OH)_2D$ binding but the VDR exhibited abnormal nuclear binding *(51, 62, 63)*. Yagi et al. found a missense mutation in the DBD that caused a His35Gln substitution in three HVDRR patients *(64)*. Rut et al. identified two missense mutations in the DBD causing

Table 1
Properties of Mutant VDRs Causing HVDRR

Mutation	Ligand binding	DNA binding	RXR binding	Coactivator binding	Alopecia	Reference
Arg30stop	−	−	−	−	Yes	(72, 73)
Gly33Asp	+++	−	+++	+++	Yes	(49)
His35Gln	+++	−	+++	+++	Yes	(64)
Cys41Tyr	+++	−	+++	+++	Yes	(67)
Lys45Glu	+++	−	+++	+++	Yes	(65)
Gly46Asp	+++	−	+++	+++	Yes	(66)
Phe47Ile	+++	−	+++	+++	Yes	(65)
Phe48frameshift Splice site defect	−	−	−	−	Yes	(74)
Arg50Gln	+++	−	+++	+++	Yes	(61)
Arg73stop	−	−	−	−	Yes	(68, 69)
Arg73Gln	+++	−	+++	+++	Yes	(49)
Arg80Gln	+++	−	+++	+++	Yes	(59)
Leu141Trp142Ala143 Insertion/substitution	+	+	+	+	No	(90)
Gln152stop	−	−	−	−	Yes	(68, 70)
Glu92frameshift Splice site defect	−	−	−	−	No	(75)
Arg121frameshift 366delC	−	−	−	−	Yes	(87)
Leu233frameshift Splice site defect	−	−	−	−	Yes	(69)
ΔLys246	+++	−	−	−	Yes	(89)
Phe251Cys	+	+	+	+	Yes	(83)
Gln259Pro	+++	+	−	+++	Yes	(69)
Leu263Arg	−	−	−	−	Yes	(88)
Ile268Thr	++	+	+	+	No	(115)
Arg274Leu	+	+	+	+	No	(70)
Trp286Arg	−	−	−	−	No	(81)
Tyr295stop	−	−	−	−	Yes	(54)
His305Glu	++	++	++	++	No	(78)
Ile314Ser	+++	+	+	+++	No	(77)
Gln317stop	−	−	−	−	Yes	(71)
Glu329Lys	+++	−	−	−	Yes	(87)
Val346Met	+++	+	+	+	No	(84)
Arg391Cys	+++	+	+	+	Yes	(77)
Arg391Ser	+++	+	+	+	Yes	(88)
Tyr401stop Insertion/duplication	−	+	+	−	No	(91)
Glu420Lys	+++	+++	+++	−	No	(86)

(+) = present: graded + to +++; (−) = absent.

Lys45Glu and Phe47Ile substitutions in two patients with HVDRR and alopecia *(65)*. Lin et al. identified a DBD mutation resulting in a Gly46Asp substitution in a patient with HVDRR and alopecia *(66)*. A young Iranian boy with HVDRR and alopecia was found to have a missense mutation in the DBD resulting in a Cys41Tyr substitution *(67)*. An older brother with similar signs and symptoms died at the age of 2 years 8 months. The parents were related and both were heterozygous for the mutation and without symptoms as is the usual situation.

5.2. *Mutations Causing Premature Termination of the VDR*

5.2.1. PREMATURE STOP MUTATIONS

Mutations that introduce a premature stop signal cause early termination of the VDR protein. Depending upon how early in the mRNA sequence the mutation occurs will determine how much of the VDR protein is deleted. These cases usually exhibited absent ligand binding and often the entire VDR protein could not be detected. Ritchie et al. *(54)* described the first molecular analysis of three patients with HVDRR and alopecia that exhibited no [^{3}H]-1,25(OH)$_2$D binding. All three patients had a single base change in exon 8 that introduced a premature termination codon (Tyr295stop) (Fig. 2). The mutation deleted 132 amino acids of the carboxy terminus of the VDR that contains the LBD thus creating a ligand-binding negative phenotype. The recreated mutant VDR with a molecular size of 32,000 Da was unable to bind [^{3}H]-1,25(OH)$_2$D and failed to activate gene transcription demonstrating that this mutation was the cause of the hormone-resistant state.

The three patients described by Ritchie et al. *(54)* and patients from four additional families comprised a large kindred where consanguineous marriages were common.

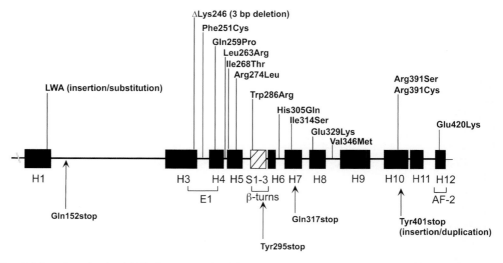

Fig. 2. Mutations in the LDB that cause HVDRR. The α-helices (H1–H12) of the VDR LBD are depicted as *shaded rectangles* and the β-turns are drawn as a *hatched rectangle*. The loops connecting the α-helices are drawn as *solid lines*. The E1 and AF-2 regions are shown below the α-helices. Missense mutations are shown *above* and stop mutations are shown *below* the protein structure.

Malloy et al. *(21)* analyzed eight children from this kindred that had HVDRR with alopecia. All of the affected children were homozygous for the Tyr295stop mutation and their parents were heterozygous *(21)*. In all but one case, the VDR mRNA was undetectable on northern blots using RNA isolated from the cultured fibroblasts or EBV-transformed lymphoblasts.

A number of other premature stop mutations have been identified in the VDR from patients with HVDRR (Figs. 1 and 2) *(3, 6)*. These include another family with the Tyr295stop mutation *(68)*, two patients with a premature stop codon at amino acid 73 (Arg73stop) *(68, 69)*, two patients with a premature stop codon at amino acid 152 (Gln152stop) *(68, 70)*, a patient with a premature stop codon at amino acid 317 (Gln317stop) *(71)*, and a premature stop codon at amino acid 30 (Arg30stop) in both a French–Canadian *(72)* and a Brazilian *(73)* family. Interestingly, in this last family, the parents had somewhat elevated serum $1,25(OH)_2D$ levels suggesting that the heterozygotic parents exhibited mild resistance to $1,25(OH)_2D$. All of the patients had alopecia.

5.2.2. SPLICING MUTATIONS

A number of splice site mutations have also been identified as the cause of HVDRR. These mutations usually cause a frameshift and eventually introduce a downstream premature stop signal resulting in a nonfunctional VDR. Hawa et al. *(74)* identified the first splice site mutation in the *VDR* gene in a young Greek girl with HVDRR and alopecia. No mutations were found in the coding sequence, however, a G to C base change was found in the 5′-donor splice site at the exon 3-intron E junction. This mutation changed the wild-type sequence from GTGAGT to GTGACT and altered the 5′-donor splice site (consensus sequence: GT(A/G)(A/T)G(T/A/C)). The loss of the 5′-donor splice site caused exon 3 to be skipped in the processing of the VDR transcript and resulted in a frameshift that introduced a premature stop codon in exon 4. The truncated VDR had no $[^3H]$-$1,25(OH)_2D$ binding and failed to induce 24-hydroxylase activity.

Two other splice site mutations were identified in two German patients with HVDRR. One patient with HVDRR and alopecia, had a single base change in exon 6 that introduced a cryptic 5′-donor splice site (GTCAGT to GTGAGT) that could be recognized by the spliceosome complex during RNA processing *(69)*. The mutation caused a 56 bp deletion that led to a frameshift in exon 7. A second patient with HVDRR without alopecia, had a single base substitution in the 5′-donor splice site at the exon 4-intron F junction *(75)*. The mutation caused exon 4 to be skipped and resulted in a frameshift that introduced a premature stop in exon 5.

5.3. *Mutations in the VDR Ligand-Binding Domain (LBD)*

5.3.1. MUTATIONS THAT AFFECT $1,25(OH)_2D$ BINDING

Rut et al. and Kristjansson et al. described the first missense mutation (Arg274Leu) in the VDR LBD in a patient with HVDRR (Fig. 2) *(70, 76)*. Unlike the patients with DBD and stop mutations, their patient did not have alopecia. The patient's fibroblasts expressed normal amounts of VDR but its affinity for $1,25(OH)_2D$ was significantly reduced compared to normal controls *(76)*. Arg274 is located in helix H5 and has been shown to be a contact point for the 1α-hydroxyl group of 1,25(OH)$_2$D *(28)*. The mutation alters the contact point and lowers the binding affinity of the VDR for 1,25(OH)$_2$D. Although the resistance caused by the defective VDR could be overcome by treating with high concentrations of 1,25(OH)$_2$D in vitro, the patient failed to respond to massive doses of the hormone and eventually died of pneumonia. The LBD mutations are depicted in Fig. 2.

Other missense mutations in the LBD include an Ile314Ser mutation in a girl who had HVDRR but without alopecia *(77)*. The patient's fibroblasts had normal 1,25(OH)$_2$D binding but had defective induction of 24-hydroxylase activity *(35)*. The mutation in helix H7 causes a subtle defect in heterodimerization with RXR and decreased response to 1,25(OH)$_2$D in transactivation assays. This patient showed a nearly complete cure when treated with pharmacological doses of 25-hydroxyvitamin D.

A His305Gln mutation was found in a Turkish boy with HVDRR without alopecia *(78)*. The patient also had two other disorders, congenital generalized lipoatrophic diabetes (Berardinelli–Seip syndrome) and persistent Müllerian duct syndrome *(79)*. The patient's fibroblasts required approximately a 5-fold higher concentration of 1,25(OH)$_2$D to induce 24-hydroxylase mRNA compared to control cells. Interestingly, [^{3}H]-1,25(OH)$_2$D-binding studies of the reconstructed mutant protein demonstrated an 8-fold lower affinity for 1,25(OH)$_2$D compared to the wild-type VDR when the assays were performed at 24°C (vs. 2-fold at 0°C). His305 is located in the interhelical loop between helix H6–H7 and is a contact point for the 25-hydroxyl group of 1,25(OH)$_2$D *(28)*. The His305Gln mutation therefore alters the contact point that lowers the binding affinity of the VDR for 1,25(OH)$_2$D. The child was treated with extremely high doses of calcitriol (Rocaltrol 12.5 μg/day) that eventually normalized his serum calcium and ultimately improved his rickets. However, the child died of apparently unrelated problems. The boy's sister, who also had HVDRR and the same mutation in the VDR, did not have any other disorders. No explanation was forthcoming for the presence of three genetic defects in a single individual. The congenital total lipodystrophy was later shown to be caused by a mutation in the *BSCL2* (*seipin*) gene *(80)*. A genetic cause for the persistent Müllerian duct syndrome in this patient has not been studied.

A Trp286Arg mutation was identified in an Algerian boy and his younger sister both with HVDRR without alopecia *(81)*. The patients' fibroblasts expressed a normal size VDR protein but no specific [^{3}H]-1,25(OH)$_2$D binding was observed and the cells were totally unresponsive to 1,25(OH)$_2$D treatment. Trp286 is located in the β1 sheet of the three-stranded β sheet between helices H5–H6. Trp286 makes contact with the α face of the C ring in 1,25(OH)$_2$D and is involved in forming the hydrophobic channel where the conjugated triene connecting the A and the C rings fits *(28)*. The Trp286Arg mutation therefore alters the contact point with the ligand and causes resistance to 1,25(OH)$_2$D *(81)*. An Ile268Thr mutation was found in a young Saudi Arabian girl with HVDRR but without alopecia *(5)*. The Ile268Thr mutant VDR exhibited an ~10-fold lower affinity for [^{3}H]-1,25(OH)$_2$D and a marked decrease in RXR binding compared to the WT VDR. Ile268 located in helix H5 directly interacts with 1,25(OH)$_2$D and is involved in the hydrophobic stabilization of helix H12. These cases demonstrate that disrupting a ligand contact point can be the basis for HVDRR.

5.3.2. MUTATIONS THAT AFFECT VDR–RXR HETERODIMERIZATION

The VDR requires heterodimerization with RXR for activity and disruption of this protein:protein interaction can cause 1,25(OH)$_2$D resistance. An Arg391Cys mutation was identified in a young girl who had HVDRR and alopecia *(77)*. The Arg391Cys mutation had no effect on 1,25(OH)$_2$D binding but required higher than normal concentrations of 1,25(OH)$_2$D for transactivation. However, the transactivation activity could be rescued and restored to normal when RXR levels were increased by co-transfection. The Arg391Cys mutant exhibited a lower capacity to form a complex with a VDRE than the wild-type VDR. Based on the crystallographic studies of RXRα, RAR-RXRα, and PPARγ-RXRα, the dimer interface is formed from helix H9 and helix H10 and the interhelical loops between H7–H8 and H8–H9 *(28)*. Arg391 is located in helix H10, therefore the Arg391Cys mutation would disrupt the dimer interface with RXR. This was the first report of a mutation in the VDR that interfered with RXR binding that caused HVDRR.

Several other patients with HVDRR have now been shown to have mutations in the VDR that affect RXR heterodimerization. A Gln259Pro mutation was identified in a brother and sister from India with HVDRR and alopecia *(69)*. The Gln259Pro had no effect on ligand binding but impaired VDR-RXR-VDRE formation. The Gln259Pro mutation is located in helix H4 in the E1 region (amino acids 244–263) that is important for RXR heterodimerization. The E1 region overlaps the C-terminal portion of helix H3, loop 3–4, and the N-terminal portion of helix H4 and is highly conserved throughout the nuclear receptor superfamily *(82)*.

A Phe251Cys mutation was found in a young Hmong boy with HVDRR and alopecia *(83)*. The Phe251Cys mutant VDR exhibited a lower affinity for 1,25(OH)$_2$D and exhibited a reduced capacity to bind RXR compared to the wild-type VDR. The Phe251Cys mutation is located in the E1 region in the interhelical loop between H3 and H4. This region of the VDR is not part of the dimer interface, however, the loop does appear to be positioned beneath the interhelical loop between H8 and H9 and helix H9. The mutation apparently disrupts the formation of the dimer interface of the VDR and results in a decreased ability to heterodimerize with RXR.

A Val346Met mutation was identified in a Bedouin boy with HVDRR and alopecia *(84)*. The patient also had papular lesions on the scalp and face. Two brothers and one sister were also affected. The affected children also had mild hearing loss and secondary speech abnormalities. Although hearing loss had not been noted before in other HVDRR patients, progressive hearing loss was found in VDR knockout mice *(85)*. V346 is located in the interhelical loop between helix H8–H9 and may be important in RXR heterodimerization.

5.3.3. MUTATIONS THAT AFFECT COACTIVATOR BINDING

The VDR also must recruit coactivators for transcriptional activity. Repositioning of helix H12 is a critical event that occurs as a consequence of ligand binding and is essential for transactivation *(28)*. The repositioning of helix 12 leads to the formation of the hydrophobic cleft critical for coactivator binding. Therefore, disruption of this surface interface may cause hormone resistance and HVDRR. A study by Malloy et al. *(86)*

examined the VDR in a young boy with HVDRR but without alopecia. The patient's fibroblasts exhibited normal 1,25(OH)$_2$D binding but the cells were totally resistant to 1,25(OH)$_2$D. A novel missense mutation was found in exon 9 of the *VDR* gene that caused a Glu420Lys mutation in helix H12. The recreated Glu420Lys mutant VDR showed no defects in VDR–RXR heterodimerization or binding to VDREs. However, the Glu420Lys mutation abolished binding by the coactivators SRC-1 and DRIP205. Co-transfection of SRC-1 failed to restore transactivation by the mutant VDR. This case was the first description of a naturally occurring mutation in the VDR that disrupted coactivator interactions and caused HVDRR. Interestingly, although the VDR was non-functional, the child did not have alopecia *(86)*. As discussed in Section 7, this finding provided further evidence that the presence of the VDR, even if it is nonfunctional, could prevent alopecia *(86)*.

5.4. Compound Heterozygous Mutations in the VDR

There are several reports of patients with HVDRR with compound heterozygous mutations in both alleles of the *VDR* gene. In these cases there is no consanguinity and the asymptomatic parents, each carrying a different heterozygous mutation in one allele of their *VDR* gene, unfortunately, both pass the defective allele to their offspring. One patient with HVDRR and alopecia had a heterozygous mutation in exon 8 that caused a Glu329Lys mutation and a second heterozygous mutation that deleted a single cytosine at nucleotide 366 (366delC) in exon 4 *(87)*. The deletion of the cytosine resulted in a frameshift in exon 4 and led to a premature termination signal in the VDR message that truncated most of the LBD. Although the effects of the Glu329Lys mutation on VDR function were not reported, the mutation occurs in helix H8 that is important in heterodimerization with RXR and would likely disrupt this activity. Thus, neither VDR alleles had normal function, resulting in HVDRR in this child.

Compound heterozygous mutations were identified in the *VDR* gene in a child with HVDRR, total alopecia, and early childhood-onset type 1 diabetes *(88)*. One mutation was found in exon 7 that caused a Leu263Arg mutation in helix H4. A second mutation was found in exon 9 that caused an Arg391Ser mutation in helix H10. As described above, the Arg391Cys mutation was previously shown to affect RXR heterodimerization *(77)*. The mutant VDRs in this case exhibited differential effects on 24-hydroxylase and RelB promoters. The 24-hydroxylase responses were abolished in the Leu263Arg mutant but only partially altered in the Arg391Ser mutant. On the other hand, RelB responses were normal for the Leu263Arg mutant but the Arg391Ser mutant was defective in this response *(88)*.

Compound heterozygous mutations were also found in the *VDR* gene in a young girl with HVDRR and alopecia *(89)*. One mutation changed the codon for arginine to a nonsense mutation at amino acid 30 (Arg30stop). The second mutation was a 3-bp deletion in exon 6. The 3-bp in-frame deletion removed the codon for lysine at amino acid 246 (△K246). The patient's fibroblasts expressed the VDR△K246 mutant protein but were unresponsive to 1,25(OH)$_2$D. The △K246 mutation abolished heterodimerization with RXR and binding to coactivators.

5.5. Other Mutations in the VDR

A young boy from Chile with HVDRR without alopecia was found to have a unique 5-bp deletion/8-bp insertion in exon 4 of the *VDR* gene *(90)*. The mutation in helix H1 of the LBD deleted His141 and Tyr142 and inserted three amino acids (Leu141, Trp142, and Ala143). The patient's fibroblasts had no demonstrable [^{3}H]-1,25(OH)$_2$D binding and were resistant to 1,25(OH)$_2$D treatment. When the effects of the three individual mutations were analyzed, only the insertion of A143 into the WT VDR disrupted VDR transactivation to the same extent of that observed with the natural mutation.

A young Jamaican boy with HVDRR and patchy alopecia was found to have a 102 bp insertion/duplication in the *VDR* gene that introduced a premature stop (Y401X) *(91)*. The mutation deleted helix H12. The patient's fibroblasts expressed the truncated VDR, but were resistant to 1,25(OH)$_2$D. The truncated VDR weakly bound [^{3}H]-1,25(OH)$_2$D but was able to heterodimerize with RXR, bind to DNA, and interact with the core-pressor hairless (HR). However, the truncated VDR failed to bind coactivators and was transactivation defective.

These is only a single reported case where investigators failed to detect a mutation in the VDR as the basis of HVDRR *(92)*. In this case the authors speculated that the resistance to the action of 1,25(OH)$_2$D was due to abnormal expression of hormone response element-binding proteins belonging to the hnRNP family that prevented the VDR–RXR complex from binding to vitamin D response elements in target genes *(93)*.

6. THERAPY OF HVDRR

6.1. General

Mutations in the VDR that cause HVDRR result in partial or total resistance to 1,25(OH)$_2$D. The patients become hypocalcemic predominately because of a lack of VDR action on the intestine to promote calcium absorption from the intestine to the circulation. The hypocalcemia leads to a decrease in bone mineralization and causes rickets. Prior to the advent of molecular analysis of the mutations causing HVDRR, therapy was administered on a trial and error basis. It was generally recognized that the patients had hypocalcemia, rickets, and elevated 1,25(OH)$_2$D levels indicating a vitamin D-resistant state. Physicians, usually pediatricians, reasonably would try to treat with various combinations of vitamin D metabolites and calcium in an attempt to raise the serum calcium levels. As discussed below, in some cases the children responded to this therapy but often they did not. Once the molecular diagnosis could be made, it became clearer whether the VDR contained a "fatal" flaw so that it was totally absent or unable to respond to even very high levels of calcitriol, or whether the defect could be overcome by elevating the levels of calcitriol to achieve adequate VDR binding to initiate transcription. This information could then inform the therapeutic decision. Our strategy for the management of HVDRR cases became simple, based on an analysis of the DNA and often the cultured fibroblasts from the patient. If the mutant VDR could respond to high concentrations of calcitriol, as was seen in some LBD mutations, we advised the physician to treat with calcitriol to raise the circulating concentration of 1,25(OH)$_2$D high enough to overcome the VDR-binding affinity defect. On the other hand, if the patient's fibroblasts harboring the mutant VDR, or recreated mutant cells, failed to respond even to 100 nM calcitriol, we advised treatment with calcium. For severe cases this usually meant intravenous (IV) calcium therapy. In some cases oral calcium in large doses did suffice, presumably because of VDR-independent intestinal calcium absorption.

6.2. Treatment with Vitamin D

Some patients with HVDRR have been reported to improve both clinically and radiologically when treated with pharmacological doses of vitamin D ranging from 5,000 to 40,000 IU/day (1–3, 6, 94); 20–200 μg/day of 25(OH)D; and 17–20 μg/day of 1,25(OH)$_2$D (2). One patient with the His305Gln mutation in the VDR LBD that is a contact point for the 25-hydroxyl group of 1,25(OH)$_2$D showed improvement with 12.5 μg/day calcitriol treatment (78, 79). The high dose of calcitriol used was enough to overcome the low affinity-binding defect of the mutant VDR. The patient with the Ile314Ser mutation was treated with 1 mg/day (40,000 IU) of vitamin D$_2$ from age 2 to 18 (94). At age 20 following an uneventful pregnancy the patient developed hypocalcemia and was treated successfully with 50 μg/day of 25(OH)D. On the other hand, the patient with the Arg274Leu mutation that is a contact point for the 1α-hydroxyl group of 1,25(OH)$_2$D was unresponsive to treatment with 600,000 IU vitamin D; up to 24 μg/day of 1,25(OH)$_2$D; or 12 μg/day 1(OH)D. The patient later died of pneumonia (18).

In general, HVDRR patients with alopecia are more resistant to treatment with vitamin D although a few reports indicate that patients have been treated successfully using vitamin D metabolites. The defects causing alopecia usually result in a VDR that is unresponsive to 1,25(OH)$_2$D, so an inability to respond to vitamin D therapy is expected. Also, these patients often raise their endogenous 1,25(OH)$_2$D to extraordinarily elevated levels to which they fail to respond. When patients with alopecia and unresponsive VDR in cultured cells show improvement on vitamin D therapy it is our opinion that this is likely due to the co-administration of calcium rather than a response to the vitamin D since the defective VDR is usually nonfunctional. It is possible of course that there are effects mediated by 1,25(OH)$_2$D that are not dependent on a functional VDR (95), however, the mechanism for this is not clear. Two patients with alopecia are reported to have shown improvement when treated with vitamin D or 1(OH)D (14, 96) and one patient responded to 25(OH)D as well as 1(OH)D (15). 1(OH)D and 1,25(OH)$_2$D have also been shown to be effective treatments in other cases (38, 48, 62, 97, 98) including patients with the Arg50Gln and Arg73Gln mutations (49, 61). Two siblings, with the Glu152stop mutation, showed no increase in serum calcium during high-dose vitamin D treatment despite raising their circulating 1,25(OH)$_2$D levels to more than 100-fold the mean normal range. However, notwithstanding their low serum calcium concentrations, healing of rickets and suppression of PTH was evident (99). In many cases when patients fail to respond to vitamin D or 1,25(OH)$_2$D, intensive calcium therapy is used as described below.

6.3. Calcium Therapy

Oral calcium can be absorbed in the intestine by both vitamin D-dependent and vitamin D-independent pathways. In children with nonfunctional VDR, the vitamin D-independent pathway becomes critical. When oral calcium therapy is successful, we believe the calcium levels have been raised high enough so that passive diffusion or other non-vitamin D-dependent absorption is adequate to maintain normocalcemia.

Balsan et al. *(100)* was the first to describe the use of IV calcium infusions to treat a child with HVDRR who had failed prior treatments with large doses of vitamin D derivatives and/or oral calcium *(15)*. Intravenous calcium therapy bypasses the calcium absorption defect in the intestine caused by the lack of action of the mutant VDR. Clinical improvement including relief of bone pain was observed within the first 2 weeks of therapy. Within 7 months, the child gained both weight and height. The serum calcium normalized, the secondary hyperparathyroidism was reversed, and the rickets ultimately resolved. However, when the IV infusions were discontinued the rickets returned. Several other groups have investigated the use of IV calcium infusions for treating HVDRR patients *(17, 101, 102)*. Two patients on IV calcium showed a decrease in their serum alkaline phosphatase activity and an increase in their serum calcium and phosphate over a 1-year period *(101)*. X-ray analysis showed resolution of the rickets with the appearance of normal mineralization of bone. In some cases, high-dose oral calcium therapy has been effective in maintaining normal serum calcium concentrations after radiological healing of the rickets has been achieved with IV calcium infusions *(102)*.

Oral calcium alone has also been used as a therapy for HVDRR patients. The patient with the Gly46Asp mutation in the VDR DBD who failed to respond to calciferols showed clinical improvement during 4 months of oral therapy with 3–4 g of elemental calcium per day *(103)*.

Spontaneous healing of rickets also has been observed in some HVDRR patients as they get older *(22, 38, 41)*. In our opinion, this may reflect a reduced demand for calcium once the skeleton is fully formed and an ability of oral calcium to satisfy the body's reduced calcium requirement. In one case, this was so striking that we reanalyzed the child's DNA to demonstrate that the VDR still contained the same defect that had caused the HVDRR in the first place *(38)*. However, in young children receiving IV calcium, the syndrome recurs slowly over time when the IV therapy is discontinued. There are many problems for families to maintain continuous IV therapy for years in small children. Unfortunately infection of the port is a recurrent problem. Because of these difficulties for the families, some children have been managed with intermittent IV regimens using oral calcium in the intervals. Once the rickets is healed and the children are doing well, some physicians have successfully initiated an intermittent IV schedule. By monitoring serum calcium, PTH, and alkaline phosphatase, physicians can detect the requirement to restart IV calcium in these children if necessary. Once the child is old enough that oral calcium suffices to maintain normocalcemia, the IV calcium regimen can be discontinued.

6.4. Lessons from the Therapy of HVDRR Cases

It was unexpected by us that raising the serum calcium to normal by IV calcium administration alone would reverse all aspects of HVDRR including hypocalcemia, hypophosphatemia, secondary hyperparathyroidism, rickets, elevated alkaline phosphatase, etc. except for alopecia (discussed below). The inescapable conclusion is that the most important actions of $1,25(OH)_2D$ on calcium and bone homeostasis occur in the intestine on calcium absorption and not in the bone. The ability of the rachitic bone abnormality to normalize in the absence of VDR-mediated vitamin D action was surprising. The data are incomplete in patients about whether the bones are entirely normal and Panda et al., using VDR knockout mice, have data suggesting subtle defects remain in the bones of VDR null mice whose serum calcium had been corrected by a rescue diet *(10)*. However, the reversal of all clinical aspects of HVDRR with IV calcium does indicate that healing of bone and reversal of secondary hyperparathyroidism can take place without normal VDR-mediated vitamin D action. There is no doubt that vitamin D has important actions on bone and parathyroid cells. However, these actions can apparently be compensated for in vivo if the calcium level is normalized.

In recent years there have been many new actions attributed to vitamin D that mediate important and widespread effects that are unrelated to calcium and bone homeostasis *(104, 105)*. These include actions to reduce the risk of cancer, autoimmune disease, infection, neurodegeneration, etc. At this time we have not detected a trend toward an increased risk for any of these potential problems in the children with HVDRR. However, there are very few cases of HVDRR and most of the cases are detected in infants so that it may be too early in their life to detect an increased tendency toward any of these potential health problems.

7. ALOPECIA

The molecular analysis of the VDR from HVDRR patients with and without alopecia has provided several clues to the functions of the VDR that are required to prevent alopecia. Patients with premature stop mutations have alopecia indicating that the intact VDR protein is critical for renewed hair growth after birth. VDR knockout mice also develop alopecia *(7, 8)* and targeted expression of the WT VDR in keratinocytes of VDR knockout mice prevented alopecia, a finding that further supports a role for the VDR in regulating hair growth *(106)*. Patients with DBD mutations have alopecia indicating that VDR binding to DNA is critical to prevent alopecia. HVDRR patients without alopecia all had mutations in the LBD. Some of these mutations abolished ligand binding suggesting that a ligand-independent action of the VDR is critical to regulate the normal hair cycle *(5, 86, 107, 108)*. This hypothesis is further supported by the fact that patients with 1α-hydroxylase deficiency and other forms of vitamin D deficiency do not have alopecia.

Patients with mutations that inhibit VDR heterodimerization with RXR but have no effect on ligand binding have alopecia, suggesting an essential role for VDR–RXR heterodimers in hair growth *(69, 77, 83)*. Also, inactivation of RXRα in keratinocytes in mice also caused alopecia clearly demonstrating a role for RXR in hair growth *(109)*.

The patient with the Glu420Lys mutation in helix H12 that abolished coactivator binding (but not ligand binding or RXR heterodimerization) did not have alopecia, indicating that VDR actions to regulate hair growth were independent of coactivator interactions *(81)*. Targeted expression of ligand-binding defective or coactivator-binding defective mutant VDRs to keratinocytes in VDR knockout mice that have alopecia also prevented alopecia *(108)*.

The alopecia associated with HVDRR is clinically and pathologically indistinguishable from the generalized atrichia with papules found in patients with mutations in the *hairless* (*hr*) gene *(87, 110, 111)*. Mutations in the *hr* gene do not cause hypocalcemia or rickets. The *hr* gene is expressed in many tissues especially in the skin and brain *(19)*. The *hr* gene product, HR acts as a corepressor and directly interacts with the VDR and suppresses $1,25(OH)_2D$-mediated transactivation *(107, 111, 112)*. It has been hypothesized that the role of the VDR in the hair cycle is to repress the expression of a gene(s) in a ligand-independent manner *(5, 86, 90, 107, 108, 111)*. The ligand-independent activity requires that the VDR heterodimerizes with RXR and binds to DNA even if it failed to transactivate *(5, 86)*. The corepressor actions of HR may also be required in order for the unliganded VDR to repress gene transcription during the hair cycle. Mutations in the VDR that disrupt the ability of the unliganded VDR to suppress gene transcription are hypothesized to lead to the derepression of a gene(s) whose product, when expressed inappropriately, disrupts the hair cycle that ultimately leads to alopecia *(5, 86, 90, 107, 108, 111)*. Inhibitors of the Wnt signaling pathway are possible candidates *(113, 114)*. Thus far, there have been no reports of mutations in the VDR that affect interactions with HR. The role of HR in regulating the unliganded action of the VDR during the hair cycle remains to be discovered.

8. CONCLUDING REMARKS

Since 1978, more than sixty families with HVDRR have been reported. Presently, 34 mutations have been identified in the *VDR* gene as the cause of HVDRR (Table 1). All of the DBD mutations prevent the VDR from binding to DNA causing total $1,25(OH)_2D$ resistance even though $1,25(OH)_2D$ binding is normal. Mutations in the LBD disrupt ligand binding, heterodimerization with RXR, or prevent coactivators from binding to the VDR and cause partial or total hormone resistance. Other mutations that cause total $1,25(OH)_2D$ resistance include premature termination codons, insertions/substitutions, insertions/duplications, deletions, and splice site mutations.

The biochemical and genetic analysis of the VDR in HVDRR patients has yielded important insights into the structure and function of the receptor in mediating $1,25(OH)_2D$ action. Similarly, studies of the affected children with HVDRR continue to provide further insight into the biological role of $1,25(OH)_2D$ in vivo. A concerted investigative approach to HVDRR at the clinical, cellular, and molecular level has proven exceedingly valuable in gaining understanding of the functions of the domains of the VDR and elucidating the detailed mechanism of action of $1,25(OH)_2D$. These studies have been essential to promote the well-being of the families with HVDRR and in improving the diagnostic and clinical management of this rare genetic disease.

REFERENCES

1. Brooks MH, Bell NH, Love L, Stern PH, Orfei E, Queener SF, Hamstra AJ, DeLuca HF (1978) Vitamin-D-dependent rickets type II. Resistance of target organs to 1,25-dihydroxyvitamin D. N Engl J Med 298:996–999

2. Marx SJ, Spiegel AM, Brown EM, Gardner DG, Downs RW Jr, Attie M, Hamstra AJ, DeLuca HF (1978) A familial syndrome of decrease in sensitivity to 1,25-dihydroxyvitamin D. J Clin Endocrinol Metab 47:1303–1310

3. Malloy PJ, Pike JW, Feldman D (1999) The vitamin D receptor and the syndrome of hereditary 1,25-dihydroxyvitamin D-resistant rickets. Endocr Rev 20:156–188

4. Malloy PJ, Feldman D (1998) Molecular defects in the vitamin D receptor associated with hereditary 1,25-dihydroxyvitamin D resistant rickets. In: Holick, MF (ed) Vitamin D: physiology, molecular biology, and clinical applications. Humana Press, Totowa.

5. Malloy PJ, Feldman D (2003) Hereditary 1,25-Dihydroxyvitamin D-resistant rickets. Endocr Dev 6:175–199

6. Malloy PJ, Pike JW, Feldman D (2005) Hereditary 1,25-dihydroxyvitamin D resistant rickets. In: Feldman, D, Glorieux, F, and Pike, JW (eds) Vitamin D, 2nd edn. Elsevier, San Diego.

7. Li YC, Pirro AE, Amling M, Delling G, Baron R, Bronson R, Demay MB (1997) Targeted ablation of the vitamin D receptor: an animal model of vitamin D-dependent rickets type II with alopecia. Proc Natl Acad Sci USA 94:9831–9835

8. Yoshizawa T, Handa Y, Uematsu Y, Takeda S, Sekine K, Yoshihara Y, Kawakami T, Arioka K, Sato H, Uchiyama Y, Masushige S, Fukamizu A, Matsumoto T, Kato S (1997) Mice lacking the vitamin D receptor exhibit impaired bone formation, uterine hypoplasia and growth retardation after weaning. Nat Genet 16:391–396

9. Bouillon R, Van Cromphaut S, Carmeliet G (2003) Intestinal calcium absorption: molecular vitamin D mediated mechanisms. J Cell Biochem 88:332–339

10. Panda DK, Miao D, Bolivar I, Li J, Huo R, Hendy GN, Goltzman D (2004) Inactivation of the 25-hydroxyvitamin D 1alpha-hydroxylase and vitamin D receptor demonstrates independent and interdependent effects of calcium and vitamin D on skeletal and mineral homeostasis. J Biol Chem 279:16754–16766

11. Rosen JF, Fleischman AR, Finberg L, Hamstra A, DeLuca HF (1979) Rickets with alopecia: an inborn error of vitamin D metabolism. J Pediatr 94:729–735

12. Liberman UA, Samuel R, Halabe A, Kauli R, Edelstein S, Weisman Y, Papapoulos SE, Clemens TL, Fraher LJ, O'Riordan JL (1980) End-organ resistance to 1,25-dihydroxycholecalciferol. Lancet 1:504–506

13. Sockalosky JJ, Ulstrom RA, DeLuca HF, Brown DM (1980) Vitamin D-resistant rickets: end-organ unresponsiveness to 1,25(OH)$_2$D$_3$. J Pediatr 96:701–703

14. Kudoh T, Kumagai T, Uetsuji N, Tsugawa S, Oyanagi K, Chiba Y, Minami R, Nakao T (1981) Vitamin D dependent rickets: decreased sensitivity to 1,25-dihydroxyvitamin D. Eur J Pediatr 137:307–311

15. Balsan S, Garabedian M, Liberman UA, Eil C, Bourdeau A, Guillozo H, Grimberg R, Le Deunff MJ, Lieberherr M, Guimbaud P, Broyer M, Marx SJ (1983) Rickets and alopecia with resistance to 1,25-dihydroxyvitamin D: two different clinical courses with two different cellular defects. J Clin Endocrinol Metab 57:803–811

16. Laufer D, Benderly A, Hochberg Z (1987) Dental pathology in calcitirol resistant rickets. J Oral Med 42:272–275

17. Bliziotes M, Yergey AL, Nanes MS, Muenzer J, Begley MG, Viera NE, Kher KK, Brandi ML, Marx SJ (1988) Absent intestinal response to calciferols in hereditary resistance to 1,25-dihydroxyvitamin D: documentation and effective therapy with high dose intravenous calcium infusions. J Clin Endocrinol Metab 66:294–300

18. Fraher LJ, Karmali R, Hinde FR, Hendy GN, Jani H, Nicholson L, Grant D, O'Riordan JL (1986) Vitamin D-dependent rickets type II: extreme end organ resistance to 1,25-dihydroxy vitamin D$_3$ in a patient without alopecia. Eur J Pediatr 145:389–395

19. Cichon S, Anker M, Vogt IR, Rohleder H, Putzstuck M, Hillmer A, Farooq SA, Al-Dhafri KS, Ahmad M, Haque S, Rietschel M, Propping P, Kruse R, Nothen MM (1998) Cloning, genomic organization, alternative transcripts and mutational analysis of the gene responsible for autosomal recessive universal congenital alopecia. Hum Mol Genet 7:1671–1679

20. Ahmad W, Zlotogorski A, Panteleyev AA, Lam H, Ahmad M, ul Haque MF, Abdallah HM, Dragan L, Christiano AM (1999) Genomic organization of the human hairless gene (HR) and identification of a mutation underlying congenital atrichia in an Arab Palestinian family. Genomics 56:141–148

21. Malloy PJ, Hochberg Z, Tiosano D, Pike JW, Hughes MR, Feldman D (1990) The molecular basis of hereditary 1,25-dihydroxyvitamin D_3 resistant rickets in seven related families. J Clin Invest 86:2071–2079

22. Hochberg Z, Benderli A, Levy J, Vardi P, Weisman Y, Chen T, Feldman D (1984) 1,25-Dihydroxyvitamin D resistance, rickets, and alopecia. Am J Med 77:805–811

23. Hochberg Z, Gilhar A, Haim S, Friedman-Birnbaum R, Levy J, Benderly A (1985) Calcitriol-resistant rickets with alopecia. Arch Dermatol 121:646–647

24. Marx SJ, Bliziotes MM, Nanes M (1986) Analysis of the relation between alopecia and resistance to 1,25-dihydroxyvitamin D. Clin Endocrinol 25:373–381

25. Berg JM (1988) Proposed structure for the zinc-binding domains from transcription factor IIIA and related proteins. Proc Natl Acad Sci USA 85:99–102

26. Mader S, Kumar V, de Verneuil H, Chambon P (1989) Three amino acids of the oestrogen receptor are essential to its ability to distinguish an oestrogen from a glucocorticoid-responsive element. Nature 338:271–274

27. Umesono K, Evans RM (1989) Determinants of target gene specificity for steroid/thyroid hormone receptors. Cell 57:1139–1146

28. Rochel N, Wurtz JM, Mitschler A, Klaholz B, Moras D (2000) The crystal structure of the nuclear receptor for vitamin D bound to its natural ligand. Mol Cell 5:173–179

29. Blanco JC, Wang IM, Tsai SY, Tsai MJ, O'Malley BW, Jurutka PW, Haussler MR, Ozato K (1995) Transcription factor TFIIB and the vitamin D receptor cooperatively activate ligand-dependent transcription. Proc Natl Acad Sci USA 92:1535–1539

30. MacDonald PN, Sherman DR, Dowd DR, Jefcoat SC Jr, DeLisle RK (1995) The vitamin D receptor interacts with general transcription factor IIB. J Biol Chem 270:4748–4752

31. Feldman D, Chen T, Hirst M, Colston K, Karasek M, Cone C (1980) Demonstration of 1,25-dihydroxyvitamin D_3 receptors in human skin biopsies. J Clin Endocrinol Metab 51:1463–1465

32. Eil C, Liberman UA, Rosen JF, Marx SJ (1981) A cellular defect in hereditary vitamin-D-dependent rickets type II: defective nuclear uptake of 1,25-dihydroxyvitamin D in cultured skin fibroblasts. N Engl J Med 304:1588–1591

33. Feldman D, Chen T, Cone C, Hirst M, Shani S, Benderli A, Hochberg Z (1982) Vitamin D resistant rickets with alopecia: cultured skin fibroblasts exhibit defective cytoplasmic receptors and unresponsiveness to $1,25(OH)_2D_3$. J Clin Endocrinol Metab 55:1020–1022

34. Clemens TL, Adams JS, Horiuchi N, Gilchrest BA, Cho H, Tsuchiya Y, Matsuo N, Suda T, Holick MF (1983) Interaction of 1,25-dihydroxyvitamin-D_3 with keratinocytes and fibroblasts from skin of normal subjects and a subject with vitamin-D-dependent rickets, type II: a model for study of the mode of action of 1,25-dihydroxyvitamin D_3. J Clin Endocrinol Metab 56:824–830

35. Griffin JE, Zerwekh JE (1983) Impaired stimulation of 25-hydroxyvitamin D-24-hydroxylase in fibroblasts from a patient with vitamin D-dependent rickets, type II. A form of receptor-positive resistance to 1,25-dihydroxyvitamin D_3. J Clin Invest 72:1190–1199

36. Liberman UA, Eil C, Marx SJ (1983) Resistance to $1,25(OH)_2D_3$: association with heterogeneous defects in cultured skin fibroblasts. J Clin Invest 71:192–200

37. Liberman UA, Eil C, Holst P, Rosen JF, Marx SJ (1983) Hereditary resistance to 1,25-dihydroxyvitamin D: defective function of receptors for 1,25-dihydroxyvitamin D in cells cultured from bone. J Clin Endocrinol Metab 57:958–962

38. Hirst MA, Hochman HI, Feldman D (1985) Vitamin D resistance and alopecia: a kindred with normal 1,25-dihydroxyvitamin D binding, but decreased receptor affinity for deoxyribonucleic acid. J Clin Endocrinol Metab 60:490–495

39. Liberman UA, Eil C, Marx SJ (1986) Receptor-positive hereditary resistance to 1,25-dihydroxyvitamin D: chromatography of receptor complexes on deoxyribonucleic acid-cellulose shows two classes of mutation. J Clin Endocrinol Metab 62:122–126

40. Malloy PJ, Hochberg Z, Pike JW, Feldman D (1989) Abnormal binding of vitamin D receptors to deoxyribonucleic acid in a kindred with vitamin D-dependent rickets, type II. J Clin Endocrinol Metab 68:263–269

41. Chen TL, Hirst MA, Cone CM, Hochberg Z, Tietze HU, Feldman D (1984) 1,25-dihydroxyvitamin D resistance, rickets, and alopecia: analysis of receptors and bioresponse in cultured fibroblasts from patients and parents. J Clin Endocrinol Metab 59:383–388

42. Pike JW, Dokoh S, Haussler MR, Liberman UA, Marx SJ, Eil C (1984) Vitamin D_3 – resistant fibroblasts have immunoassayable 1,25-dihydroxyvitamin D_3 receptors. Science 224: 879–881

43. Dokoh S, Haussler MR, Pike JW (1984) Development of a radioligand immunoassay for 1,25-dihydroxycholecalciferol receptors utilizing monoclonal antibody. Biochem J 221:129–136

44. Pike JW, Donaldson CA, Marion SL, Haussler MR (1982) Development of hybridomas secreting monoclonal antibodies to the chicken intestinal 1 alpha,25-dihydroxyvitamin D_3 receptor. Proc Natl Acad Sci USA 79:7719–7723

45. Pike JW, Marion SL, Donaldson CA, Haussler MR (1983) Serum and monoclonal antibodies against the chick intestinal receptor for 1,25-dihydroxyvitamin D_3. Generation by a preparation enriched in a 64,000-dalton protein. J Biol Chem 258:1289–1296

46. Pike JW (1984) Monoclonal antibodies to chick intestinal receptors for 1,25-dihydroxyvitamin D_3. Interaction and effects of binding on receptor function. J Biol Chem 259:1167–1173

47. Gamblin GT, Liberman UA, Eil C, Downs RWJ, Degrange DA, Marx SJ (1985) Vitamin D dependent rickets type II: Defective induction of 25-hydroxyvitamin D_{3-24}-hydroxylase by 1,25-dihydroxyvitamin D_3 in cultured skin fibroblasts. J Clin Invest 75:954–960

48. Castells S, Greig F, Fusi MA, Finberg L, Yasumura S, Liberman UA, Eil C, Marx SJ (1986) Severely deficient binding of 1,25-dihydroxyvitamin D to its receptors in a patient responsive to high doses of this hormone. J Clin Endocrinol Metab 63:252–256

49. Hughes MR, Malloy PJ, Kieback DG, Kesterson RA, Pike JW, Feldman D, O'Malley BW (1988) Point mutations in the human vitamin D receptor gene associated with hypocalcemic rickets. Science 242:1702–1705

50. Koren R, Ravid A, Liberman UA, Hochberg Z, Weisman Y, Novogrodsky A (1985) Defective binding and function of 1,25-dihydroxyvitamin D_3 receptors in peripheral mononuclear cells of patients with end-organ resistance to 1,25-dihydroxyvitamin D. J Clin Invest 76:2012–2015

51. Takeda E, Kuroda Y, Saijo T, Toshima K, Naito E, Kobashi H, Iwakuni Y, Miyao M (1986) Rapid diagnosis of vitamin D-dependent rickets type II by use of phytohemagglutinin-stimulated lymphocytes. Clin Chim Acta 155:245–250

52. Takeda E, Yokota I, Ito M, Kobashi H, Saijo T, Kuroda Y (1990) 25-Hydroxyvitamin D-24-hydroxylase in phytohemagglutinin-stimulated lymphocytes: intermediate bioresponse to 1,25-dihydroxyvitamin D_3 of cells from parents of patients with vitamin D-dependent rickets type II. J Clin Endocrinol Metab 70:1068–1074

53. Nagler A, Merchav S, Fabian I, Tatarsky I, Hochberg Z (1987) Myeloid progenitors from the bone marrow of patients with vitamin D resistant rickets (type II) fail to respond to 1,25(OH)$_2$D$_3$. Brit J Haematol 67:267–271

54. Ritchie HH, Hughes MR, Thompson ET, Malloy PJ, Hochberg Z, Feldman D, Pike JW, O'Malley BW (1989) An ochre mutation in the vitamin D receptor gene causes hereditary 1,25-dihydroxyvitamin D_3-resistant rickets in three families. Proc Natl Acad Sci USA 86:9783–9787

55. Koeffler HP, Bishop JE, Reichel H, Singer F, Nagler A, Tobler A, Walka M, Norman AW (1990) Lymphocyte cell lines from vitamin D-dependent rickets type II show functional defects in the 1 alpha,25-dihydroxyvitamin D_3 receptor. Mol Cell Endocrinol 70:1–11

56. Takeda E, Yokota I, Saijo T, Kawakami I, Ito M, Kuroda Y (1990) Effect of long-term treatment with massive doses of 1 alpha-hydroxyvitamin D_3 on calcium-phosphate balance in patients with vitamin D-dependent rickets type II. Acta Paediatr Jpn 32:39–43

57. Baker AR, McDonnell DP, Hughes M, Crisp TM, Mangelsdorf DJ, Haussler MR, Pike JW, Shine J, O'Malley BW (1988) Cloning and expression of full-length cDNA encoding human vitamin D receptor. Proc Natl Acad Sci USA 85:3294–3298

58. Sone T, Scott RA, Hughes MR, Malloy PJ, Feldman D, O'Malley BW, Pike JW (1989) Mutant vitamin D receptors which confer hereditary resistance to 1,25-dihydroxyvitamin D$_3$ in humans are transcriptionally inactive in vitro. J Biol Chem 264:20230–20234

59. Sone T, Marx SJ, Liberman UA, Pike JW (1990) A unique point mutation in the human vitamin D receptor chromosomal gene confers hereditary resistance to 1,25-dihydroxyvitamin D$_3$. Mol Endocrinol 4:623–631

60. Malloy PJ, Weisman Y, Feldman D (1994) Hereditary 1 alpha,25-dihydroxyvitamin D-resistant rickets resulting from a mutation in the vitamin D receptor deoxyribonucleic acid-binding domain. J Clin Endocrinol Metab 78:313–316

61. Saijo T, Ito M, Takeda E, Huq AH, Naito E, Yokota I, Sone T, Pike JW, Kuroda Y (1991) A unique mutation in the vitamin D receptor gene in three Japanese patients with vitamin D-dependent rickets type II: utility of single-strand conformation polymorphism analysis for heterozygous carrier detection. Am J Hum Genet 49:668–673

62. Takeda E, Yokota I, Kawakami I, Hashimoto T, Kuroda Y, Arase S (1989) Two siblings with vitamin-D-dependent rickets type II: no recurrence of rickets for 14 years after cessation of therapy. Eur J Pediatr 149:54–57

63. Yokota I, Takeda E, Ito M, Kobashi H, Saijo T, Kuroda Y (1991) Clinical and biochemical findings in parents of children with vitamin D-dependent rickets Type II. J Inherit Metab Dis 14:231–240

64. Yagi H, Ozono K, Miyake H, Nagashima K, Kuroume T, Pike JW (1993) A new point mutation in the deoxyribonucleic acid-binding domain of the vitamin D receptor in a kindred with hereditary 1,25-dihydroxyvitamin D-resistant rickets. J Clin Endocrinol Metab 76:509–512

65. Rut AR, Hewison M, Kristjansson K, Luisi B, Hughes MR, O'Riordan JL (1994) Two mutations causing vitamin D resistant rickets: modelling on the basis of steroid hormone receptor DNA-binding domain crystal structures. Clin Endocrinol 41:581–590

66. Lin NU-T, Malloy PJ, Sakati N, Al-Ashwal A, Feldman D (1996) A novel mutation in the deoxyribonucleic acid-binding domain of the vitamin D receptor gene causes hereditary 1,25-dihydroxyvitamin D resistant rickets. J Clin Endocrinol Metab 81:2564–2569

67. Shafeghati Y, Momenin N, Esfahani T, Reyniers E, Wuyts W (2008) Vitamin D-dependent rickets type II: report of a novel mutation in the vitamin D receptor gene. Arch Iran Med 11:330–334

68. Wiese RJ, Goto H, Prahl JM, Marx SJ, Thomas M, Aqeel A, DeLuca HF (1993) Vitamin D-dependency rickets type II: truncated vitamin D receptor in three kindreds. Mol Cell Endocrinol 90:197–201

69. Cockerill FJ, Hawa NS, Yousaf N, Hewison M, O'Riordan JL, Farrow SM (1997) Mutations in the vitamin D receptor gene in three kindreds associated with hereditary vitamin D resistant rickets. J Clin Endocrinol Metab 82:3156–3160

70. Kristjansson K, Rut AR, Hewison M, O'Riordan JL, Hughes MR (1993) Two mutations in the hormone binding domain of the vitamin D receptor cause tissue resistance to 1,25 dihydroxyvitamin D$_3$. J Clin Invest 92:12–16

71. Malloy PJ, Zhu W, Bouillon R, Feldman D (2002) A novel nonsense mutation in the ligand binding domain of the vitamin D receptor causes hereditary 1,25-dihydroxyvitamin D-resistant rickets. Mol Genet Metab 77:314–318

72. Zhu W, Malloy PJ, Delvin E, Chabot G, Feldman D (1998) Hereditary 1,25-dihydroxyvitamin D-resistant rickets due to an opal mutation causing premature termination of the vitamin D receptor. J Bone Miner Res 13:259–264

73. Mechica JB, Leite MO, Mendonca BB, Frazzatto ES, Borelli A, Latronico AC (1997) A novel nonsense mutation in the first zinc finger of the vitamin D receptor causing hereditary 1,25-dihydroxyvitamin D$_3$-resistant rickets. J Clin Endocrinol Metab 82:3892–3894

74. Hawa NS, Cockerill FJ, Vadher S, Hewison M, Rut AR, Pike JW, O'Riordan JL, Farrow SM (1996) Identification of a novel mutation in hereditary vitamin D resistant rickets causing exon skipping. Clin Endocrinol 45:85–92

75. Katavetin P, Wacharasindhu S, Shotelersuk V (2006) A girl with a novel splice site mutation in VDR supports the role of a ligand-independent VDR function on hair cycling. Horm Res 66: 273–276

76. Rut AR, Hewison M, Rowe P, Hughes M, Grant D, O'Riordan JLH (1991) A novel mutation in the steroid binding region of the vitamin D receptor (VDR) gene in hereditary vitamin D resistant rickets (HVDRR). In: Norman, AW, Bouillon, R, and Thomasset, M (eds) Vitamin D: gene regulation, structure-function analysis, and clinical application eighth workshop on vitamin D. Walter de Gruyter, New York

77. Whitfield GK, Selznick SH, Haussler CA, Hsieh JC, Galligan MA, Jurutka PW, Thompson PD, Lee SM, Zerwekh JE, Haussler MR (1996) Vitamin D receptors from patients with resistance to 1,25-dihydroxyvitamin D3: point mutations confer reduced transactivation in response to ligand and impaired interaction with the retinoid X receptor heterodimeric partner. Mol Endocrinol 10: 1617–1631

78. Malloy PJ, Eccleshall TR, Gross C, Van Maldergem L, Bouillon R, Feldman D (1997) Hereditary vitamin D resistant rickets caused by a novel mutation in the vitamin D receptor that results in decreased affinity for hormone and cellular hyporesponsiveness. J Clin Invest 99: 297–304

79. Van Maldergem L, Bachy A, Feldman D, Bouillon R, Maassen J, Dreyer M, Rey R, Holm C, Gillerot Y (1996) Syndrome of lipoatrophic diabetes, vitamin D resistant rickets, and persistent müllerian ducts in a Turkish boy born to consanguineous parents. Am J Med Genet 64:506–513

80. Magre J, Delepine M, Khallouf E, Gedde-Dahl T Jr, Van Maldergem L, Sobel E, Papp J, Meier M, Megarbane A, Bachy A, Verloes A, d'Abronzo FH, Seemanova E, Assan R, Baudic N, Bourut C, Czernichow P, Huet F, Grigorescu F, de Kerdanet M, Lacombe D, Labrune P, Lanza M, Loret H, Matsuda F, Navarro J, Nivelon-Chevalier A, Polak M, Robert JJ, Tric P, Tubiana-Rufi N, Vigouroux C, Weissenbach J, Savasta S, Maassen JA, Trygstad O, Bogalho P, Freitas P, Medina JL, Bonnicci F, Joffe BI, Loyson G, Panz VR, Raal FJ, O'Rahilly S, Stephenson T, Kahn CR, Lathrop M, Capeau J (2001) Identification of the gene altered in Berardinelli-Seip congenital lipodystrophy on chromosome 11q13. Nat Genet 28:365–370

81. Nguyen TM, Adiceam P, Kottler ML, Guillozo H, Rizk-Rabin M, Brouillard F, Lagier P, Palix C, Garnier JM, Garabedian M (2002) Tryptophan missense mutation in the ligand-binding domain of the vitamin D receptor causes severe resistance to 1,25-dihydroxyvitamin D. J Bone Miner Res 17: 1728–1737

82. Wurtz JM, Bourguet W, Renaud JP, Vivat V, Chambon P, Moras D, Gronemeyer H (1996) A canonical structure for the ligand-binding domain of nuclear receptors. Nat Struct Biol 3:87–94

83. Malloy PJ, Zhu W, Zhao XY, Pehling GB, Feldman D (2001) A novel inborn error in the ligand-binding domain of the vitamin D receptor causes hereditary vitamin D-resistant rickets. Mol Genet Metab 73:138–148

84. Arita K, Nanda A, Wessagowit V, Akiyama M, Alsaleh QA, McGrath JA (2008) A novel mutation in the VDR gene in hereditary vitamin D-resistant rickets. Br J Dermatol 158:168–171

85. Zou J, Minasyan A, Keisala T, Zhang Y, Wang JH, Lou YR, Kalueff A, Pyykko I, Tuohimaa P (2008) Progressive hearing loss in mice with a mutated vitamin D receptor gene. Audiol Neurootol 13: 219–230

86. Malloy PJ, Xu R, Peng L, Clark PA, Feldman D (2002) A novel mutation in helix 12 of the vitamin D receptor impairs coactivator interaction and causes hereditary 1,25-dihydroxyvitamin D-resistant rickets without alopecia. Mol Endocrinol 16:2538–2546

87. Miller J, Djabali K, Chen T, Liu Y, Ioffreda M, Lyle S, Christiano AM, Holick M, Cotsarelis G (2001) Atrichia caused by mutations in the vitamin D receptor gene is a phenocopy of generalized atrichia caused by mutations in the hairless gene. J Invest Dermatol 117:612–617

88. Nguyen M, d'Alesio A, Pascussi JM, Kumar R, Griffin MD, Dong X, Guillozo H, Rizk-Rabin M, Sinding C, Bougneres P, Jehan F, Garabedian M (2006) Vitamin D-resistant rickets and type 1 diabetes in a child with compound heterozygous mutations of the vitamin D receptor (L263R and R391S): dissociated responses of the CYP-24 and rel-B promoters to 1,25-dihydroxyvitamin D3. J Bone Miner Res 21:886–894

89. Zhou Y, Wang J, Malloy PJ, Dolezel Z, Feldman D (2009) Compound heterozygous mutations in the vitamin D receptor in a patient with hereditary 1,25-dihydroxyvitamin D-resistant rickets with alopecia. J Bone Miner Res 24:643–651

90. Malloy PJ, Xu R, Cattani A, Reyes L, Feldman D (2004) A unique insertion/substitution in helix H1 of the vitamin D receptor ligand binding domain in a patient with hereditary 1,25-dihydroxyvitamin D-resistant rickets. J Bone Miner Res 19:1018–1024

91. Malloy PJ, Wang J, Peng L, Nayak S, Sisk JM, Thompson CC, Feldman D (2007) A unique insertion/duplication in the VDR gene that truncates the VDR causing hereditary 1,25-dihydroxyvitamin D-resistant rickets without alopecia. Arch Biochem Biophys 460:285–292

92. Hewison M, Rut AR, Kristjansson K, Walker RE, Dillon MJ, Hughes MR, O'Riordan JL (1993) Tissue resistance to 1,25-dihydroxyvitamin D without a mutation of the vitamin D receptor gene. Clin Endocrinol 39:663–670

93. Chen H, Hewison M, Hu B, Adams JS (2003) Heterogeneous nuclear ribonucleoprotein (hnRNP) binding to hormone response elements: a cause of vitamin D resistance. Proc Natl Acad Sci USA 100:6109–6114

94. Zerwekh JE, Glass K, Jowsey J, Pak CY (1979) An unique form of osteomalacia associated with end organ refractoriness to 1,25-dihydroxyvitamin D and apparent defective synthesis of 25-hydroxyvitamin D. J Clin Endocrinol Metab 49:171–175

95. Goltzman D, Miao D, Panda DK, Hendy GN (2004) Effects of calcium and of the Vitamin D system on skeletal and calcium homeostasis: lessons from genetic models. J Steroid Biochem Mol Biol 89–90:485–489

96. Tsuchiya Y, Matsuo N, Cho H, Kumagai M, Yasaka A, Suda T, Orimo H, Shiraki M (1980) An unusual form of vitamin D-dependent rickets in a child: alopecia and marked end-organ hyposensitivity to biologically active vitamin D. J Clin Endocrinol Metab 51:685–690

97. Takeda E, Kuroda Y, Saijo T, Naito E, Kobashi H, Yokota I, Miyao M (1987) 1 alpha-hydroxyvitamin D$_3$ treatment of three patients with 1,25-dihydroxyvitamin D-receptor-defect rickets and alopecia. Pediatrics 80:97–101

98. Fujita T, Nomura M, Okajima S, Furuya H (1980) Adult-onset vitamin D-resistant osteomalacia with the unresponsiveness to parathyroid hormone. J Clin Endocrinol Metab 50:927–931

99. Kruse K, Feldmann E (1995) Healing of rickets during vitamin D therapy despite defective vitamin D receptors in two siblings with vitamin D-dependent rickets type II. J Pediatr 126:145–148

100. Balsan S, Garabedian M, Larchet M, Gorski AM, Cournot G, Tau C, Bourdeau A, Silve C, Ricour C (1986) Long-term nocturnal calcium infusions can cure rickets and promote normal mineralization in hereditary resistance to 1,25-dihydroxyvitamin D. J Clin Invest 77:1661–1667

101. Weisman Y, Bab I, Gazit D, Spirer Z, Jaffe M, Hochberg Z (1987) Long-term intracaval calcium infusion therapy in end-organ resistance to 1,25-dihydroxyvitamin D. Am J Med 83:984–990

102. Hochberg Z, Tiosano D, Even L (1992) Calcium therapy for calcitriol-resistant rickets. J Pediatr 121:803–808

103. Sakati N, Woodhouse NJY, Niles N, Harfi H, de Grange DA, Marx S (1986) Hereditary resistance to 1,25-dihydroxyvitamin D: clinical and radiological improvement during high-dose oral calcium therapy. Hormone Res 24:280–287

104. Feldman D, Glorieux FH, Pike JW (2005) Vitamin D. Elsevier Academic Press, San Diego

105. Feldman D, Malloy PJ, Krishnan AV, Balint E (2007) Vitamin D: biology, action, and clinical implications. In: Marcus, R, Feldman, D, Nelson, DA, and Rosen, CJ (eds) Osteoporosis, 3rd edn. Academic Press, San Diego

106. Chen CH, Sakai Y, Demay MB (2001) Targeting expression of the human vitamin D receptor to the keratinocytes of vitamin D receptor null mice prevents alopecia. Endocrinology 142:5386–5389

107. Hsieh JC, Sisk JM, Jurutka PW, Haussler CA, Slater SA, Haussler MR, Thompson CC (2003) Physical and functional interaction between the vitamin D receptor and hairless corepressor, two proteins required for hair cycling. J Biol Chem 278:38665–38674

108. Skorija K, Cox M, Sisk JM, Dowd DR, MacDonald PN, Thompson CC, Demay MB (2005) Ligand-independent actions of the vitamin D receptor maintain hair follicle homeostasis. Mol Endocrinol 19:855–862

109. Li M, Chiba H, Warot X, Messaddeq N, Gerard C, Chambon P, Metzger D (2001) RXR-alpha ablation in skin keratinocytes results in alopecia and epidermal alterations. Development 128:675–688
110. Ahmad W, Faiyaz ul Haque M, Brancolini V, Tsou HC, ul Haque S, Lam H, Aita VM, Owen J, Blaquiere M, Frank J, Cserhalmi-Friedman PB, Leask A, McGrath JA, Peacocke M, Ahmad M, Ott J, Christiano AM (1998) Alopecia universalis associated with a mutation in the human hairless gene. Science 279:720–724
111. Wang J, Malloy PJ, Feldman D (2007) Interactions of the vitamin D receptor with the corepressor hairless: analysis of hairless mutants in atrichia with papular lesions. J Biol Chem 282:25231–25239
112. Xie Z, Chang S, Oda Y, Bikle DD (2006) Hairless suppresses vitamin D receptor transactivation in human keratinocytes. Endocrinology 147:314–323
113. Beaudoin GM 3rd, Sisk JM, Coulombe PA, Thompson CC (2005) Hairless triggers reactivation of hair growth by promoting Wnt signaling. Proc Natl Acad Sci USA 102:14653–14658
114. Thompson CC, Sisk JM, Beaudoin GM 3rd (2006) Hairless and Wnt signaling: allies in epithelial stem cell differentiation. Cell Cycle 5:1913–1917
115. Malloy PJ, Xu R, Peng L, Peleg S, Al-Ashwal A, Feldman D (2004) Hereditary 1,25-dihydroxyvitamin D resistant rickets due to a mutation causing multiple defects in vitamin D receptor function. Endocrinology 145:5106–5114

38 Receptor-Independent Vitamin D Resistance in Subhuman and Human Primates

John S. Adams, Hong Chen, Thomas S. Lisse,
Rene F. Chun, and Martin Hewison

Abstract   *Learning from Nature.* Experiments of nature are crucial for informing scientific discovery. Twenty five years ago we began to investigate an outbreak of rachitic bone disease in adolescent, female New World primates residing at the Los Angles Zoo. Our investigation of this "experiment of nature" and that of an adolescent human female with a similar phenotype (Chen et al., *Proc Natl Acad Sci USA* 100:6109–6114, 2003) led us to the discovery of a novel means for relative resistance to vitamin D and estrogen in primates, including man. We coined these resistance-causing proteins as the vitamin D response element-binding protein (VDRE-BP) and estrogen response element-binding protein (ERE-BP) for their ability to compete *in trans* with the liganded vitamin D receptor and estrogen receptor α for their cognate response elements, identifying them as nucleic acid-binding proteins in the heterogeneous nuclear ribonu-cleoprotein C family (Chen et al., *J Biol Chem* 275:35557–35564, 2000; *J Clin Invest* 99:769–775, 1997). The purpose of this review is to examine the role of one of these proteins, the VDRE-BP, and other associated intracellular proteins that regulate the expression of vitamin D-controlled genes in nonhuman and human primates.

Key Words: Vitamin D resistance; subhuman; 1,25-dihydroxyvitamin D; monkeys; ribonu-clear protein; primate evolution; steroid hormone; new world monkeys; HIV; vitamin D responsive element

"Man with all his noble qualities…with his god-like intellect which has penetrated into the movements and constitution of the solar system…still bears in his bodily frame the indelible stamp of his lowly origin"

Charles Robert Darwin from the *Descent of Man*

1. EARLY PRIMATE EVOLUTION

In the Eocene period, 50–100 million years ago, the great southern hemispheric landmass, Pangea, broke apart. These tectonic events resulted in the American land mass and Madagascar moving away from Africa. This continental separation occurred early in the process of primate evolution, trapping primordial primates in South America, Africa,

From: *Nutrition and Health: Vitamin D*
Edited by: M.F. Holick, DOI 10.1007/978-1-60327-303-9_38,
© Springer Science+Business Media, LLC 2010

and Madagascar, respectively. As a consequence, the three major primate infraorders, platyrrhines or New World primates, catarrhines or Old World primates, and lemurs evolved independently of one another *(1)* (panel a, Fig. 1). Unlike Old World primates, including our own species, which have populated virtually every landmass on our planet, New World primates have remained confined to Central and South America for the last 50 million years. Compared to Old World primates, especially some of the terrestrial species like gorilla, New World primates are smaller in stature. This is a characteristic

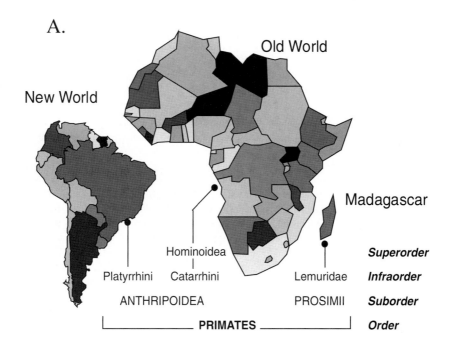

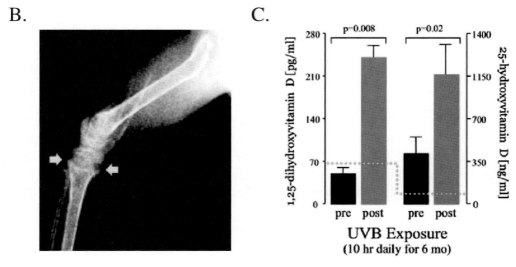

Fig. 1. (continued).

well suited to their lifestyle as plant-eating, arboreal sunbathers, residing in the canopy of the periequatorial rain forests of the Americas.

2. SIMIAN BONE DISEASE

The appearance of generalized metabolic bone disease in captive primates has been recognized for the last 150 years *(2)*. The disease, which has not been well studied from a histopathological standpoint, carries the clinical and radiological stigmata of rickets (panel b, Fig. 1); *(3)*. Compared to Old World primates reared in captivity, female New World primates are particularly susceptible to the disease. The disorder affects primarily young, growing animals and results in muscle weakness, skeletal fragility, and in many instances death of the affected individual primarily from complications of skeletal fracture and malnutrition owing to weakness of the muscles of mastication. Rachitic bone disease of this sort has long presented a problem to veterinarians caring for captive platyrrhines, particularly in North American and European zoos *(4)*, because death of pre-adolescent and adolescent primates, particularly females, prior to sexual maturity severely limits on-site breeding programs.

Because the disease was reported to be ameliorated by either the oral administration of vitamin D_3 in large doses or by ultraviolet B irradiation of affected primates, it was presumed to be caused by vitamin D deficiency *(4)*. The frequent occurrence of rickets and osteomalacia in New World primates was also ascribed to the relative inability of platyrrhines, compared to Old World primates including man, to effectively employ vitamin D_2 in their diet *(5)*; a similar observation had been made for chickens *(6)*. Using an assay technology that does not discriminate between 25-hydroxylated vitamin D_2 and vitamin D_3 metabolites, Marx and colleagues *(7)* determined that 25-hydroxyvitamin D [25(OH)D] levels were 2–3-fold higher when platyrrhines were dosed with supplemental vitamin D_3 than with vitamin D_2. These data suggested that 25-hydroxylation of

---

Fig. 1. (continued) New World primate evolution, rachitic bone disease, and vitamin D metabolites profiles of rachitic New World primates in response to ultraviolet B irradiation (UVB). **Panel A** describes in geographic terms the independent evolution of the three primate suborders, platyrrhini, catarrhini, and lemuridae, in South America (the New World), Africa (the Old World), and Madagascar, respectively. **Panel B** displays the characteristic "cupping" and "fraying" of the tibial metaphysis (*arrows*) in a New World primate resident of the Los Angeles Zoo with rickets. **Panel C** demonstrates biochemical indices of bone health in New World primate suffering from rickets compared to developmental age- and sex-matched nonrachitic Old World primates. The outstanding characteristic is a 1,25-dihydroxyvitamin D level two to three orders of magnitude greater than that observed in Old World primates, including man. Shown are the mean 25-hydroxyvitamin D (*left*) and 1,25-dihydroxyvitamin D levels (*right*) in seven different rachitic New World primates before (pre) and after (post) exposure to 6 months of artificial sunlight in their enclosures. The upper limits of the normal human Old World primate range is described by the *dotted line*. Both substrate and product rose significantly with UVB therapy and resulted in cure of rickets. These data are reprinted with permission of the authors from reference 27.

vitamin D substrate in New World primates was much more effective when vitamin D_3 was employed as substrate. However, in the same study two species of Old World primates demonstrated similar discrimination against vitamin D_2, in favor of vitamin D_3, to promote significantly more 25(OH)D produced. In summary, these results seemed to indicate that all subhuman primates, whether Old or New World, were relatively resistant to vitamin D2 in terms of its ability to engender an increase in serum levels of 25(OH)D. Finally, Hay and colleagues (8) suggested that New World primates may transport 25(OH)D in the serum by means and proteins that are dissimilar from those encountered in Old World primate species. This hypothesis was disproved by Bouillon et al. (9) who showed that the vitamin D-binding protein was the major carrier of 25(OH)D in their serum of both New and Old World primates.

3. STEROID HORMONE RESISTANCE IN NEW WORLD PRIMATES

The concept of generalized steroid hormone resistance in New World primates was first suggested by Brown et al. (10). These investigators discovered greatly elevated serum cortisol levels in platyrrhini compared to catarrhini. Despite biochemical evidence of resistance to glucocorticoids, platyrrhines affected with high cortisol levels showed no sign of glucocorticoid deficiency or toxicity at the level of the target organ; glucose homeostasis, electrolyte balance, blood pressure, and life expectancy were all similar to that observed in Old World primates (11). These data indicated that glucocorticoid resistance in New World primates was physiologically compensated by increased synthesis of the hormone. Increased production of the hormone was achieved by lack of feedback inhibition of pituitary ACTH production (12), adrenal (zona fasciculata) hypertrophy (13), and increased enzymatic synthesis and decreased catabolism of cortisol (14, 15). A relative increase in the availability of glucocorticoid to target tissues was also proposed to occur in New World primates (16) and participate in the response to cortisol resistance. What then is the proximate cause of glucocorticoid resistance in New World primates? Early studies from the group of Lipsett and Loriaux (11, 17, 18) suggested that resistance was caused by expression of a glucocorticoid receptor (GR) in New World primates with a lower affinity for cortisol. More recently, relative overexpression of FKBP51, the FK506-binding co-chaperone immunophilin, cyclophilin A, that normally interacts with the heat shock protein 90 (hsp90)–GR complex, was postulated by Scammel et al. (19) to be the cause of lowered affinity of the New World primate GR for its cognate ligand; in this context it is interesting to note that the same co-chaperone cyclophilin A also is a crucial inhibitor of retroviral, including HIV-1, infection in primate species (20). It appears that constitutive overexpression of the New World primate FKBP51 in Old World primate cells will squelch GR-directed transactivation (21). It has been subsequently theorized by Fuller et al. (22) that constitutive overexpression of FKBP51 in New World primates is associated with a consequent underexpression of its companion hsp90 co-chaperone FKBP52; FKBP52 normally co-interacts with outside face (i.e., not the ligand-binding pocket) of the GR (23) and hsp-90 to increase the affinity of the GR for glucocorticoid.

New World primates are also resistant to steroid hormones produced by the ovary *(24)*. Until recently, it was considered that estrogen and progesterone resistance resulted from a diminishment of the estrogen receptor (ER) and progesterone receptor (PR) population in target tissues. A similar mechanism was proposed for vitamin D resistance *(25)*. As will be discussed below, it now appears that the steroid hormone and vitamin D hormone receptor compliment of New World primates is not functionally distinct from that of their Old World primate counterparts. What is different is the relative overexpression in New World primate cells of at least three distinct families of intracellular proteins, the heterogeneous nuclear ribonucleoproteins (hnRNPs), heat shock proteins (hsps), and hsp-co-chaperones which conspire to legislate, by means outside the primary structure of the steroid hormone receptors themselves, the degree of steroid/sterol hormone resistance among the various New World primate species.

4. OUTBREAK OF RICKETS IN THE NEW WORLD PRIMATE COLONIES OF THE LOS ANGELES ZOO

In the early-to-mid 1980s we investigated an outbreak of deforming rachitic bone disease in the emperor tamarin (*Saguinus imperator*) colony at the Los Angeles Zoo *(26, 27)*. This disease was most prevalent in females during their adolescent growth spurts often resulting in death from fracture, inanition, and/or infection prior to complete sexual maturity and procreation. In order to investigate this rachitic syndrome, blood and urine were collected from involved monkeys as well as from control, nonrachitic New and Old World primates. Compared to Old World primates, New World primates, rachitic or not, expressed a biochemical phenotype that was most remarkable for an elevated serum 1,25-dihydroxyvitamin D [1,25(OH)$_2$D] level *(28)*. In fact, with the exception of nocturnal primates in the genus *Aotus*, New World primates in all other genera had vitamin D hormone levels ranging up to two orders of magnitude higher than that observed in Old World primates including man *(26, 29, 30)*.

In initial analyses New World primates affected with rickets were those with the lowest 1,25(OH)$_2$D levels, while their healthy counterparts were those with the highest serum 1,25(OH)$_2$D levels. These data were interpreted to mean that most New World primate genera were naturally resistant to the vitamin D hormone, and that the resistant state could be compensated by maintenance of high 1,25(OH)$_2$D levels. If this was true, then an increase in the serum 1,25(OH)$_2$D concentration in affected primates should result in biochemical compensation for the resistant state and resolution of their rachitic bone disease. In fact, when rachitic New World primates were exposed to 6 months of artificial sunlight in their enclosures, both substrate serum 25-hydroxyvitamin D [25(OH)D] and product 1,25(OH)$_2$D levels rose dramatically resulting in cure of rickets (panel c, Fig. 1); *(31)*. In summary, New World primates are periequatorial sunbathers for a reason. They require a lot of cutaneous vitamin D synthesis in order to push their 25(OH)D and 1,25(OH)$_2$D levels high enough to effectively interact with the vitamin D receptor (VDR). The question remained as to why these primates are resistant to all but the highest levels of the vitamin D hormone?

5. INVESTIGATING THE BIOCHEMICAL NATURE OF VITAMIN D RESISTANCE IN NEW WORLD PRIMATES

In order to answer the above question, cultured fibroblasts and virus-immortalized cell lines from both resistant and hormone-responsive New and Old World primates were used to track, step by step, the path taken by the vitamin D hormone from the serum vitamin D-binding protein (DBP) in the blood in route to the nucleus and transactivation of hormone-responsive genes (panel a, Fig. 2; *26, 27, 29–36*). It was determined that the movement of hormone from DBP across the cell membrane and through the cell cytoplasm and nuclear membrane in New World primate cells was indistinguishable from that observed in Old World primate cells. It was also determined that the ability of the New World primate VDR to bind 1,25(OH)$_2$D and induce receptor dimerization with the retinoid X receptor (RXR) was normal. In fact, when removed from the intranuclear environment and in distinction to previous reports *(25)*, the VDR in New World primates was similar to the Old World primate VDR in all biochemical and functional respects *(34)*. That which was not the same in New World primate cells was (i) the reduced ability of VDR–RXR complex to bind to its cognate *cis* vitamin D response element and transact hormone-responsive genes as well as (ii) the apparent build-up of hormone in the cytoplasm of the New World primate cell (panel b, Fig. 2).

In order to elucidate nuclear receptor events in New World primate cells, the nuclei of New World primate cells were isolated, extracted, and analyzed by electrophoretic

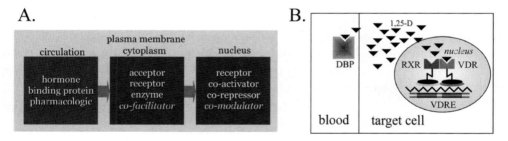

Fig. 2. Pathway of hormone 1,25-dihydroxyvitamin D [1,25(OH)$_2$D] from the blood to nucleus of the target cell in the vitamin D-resistant New World primate. **Panel A** shows the proteins and small molecules (pharmacologics like SERMs, *left panel*) that control access of the vitamin D hormone to its receptor and modifying enzymes (*center panel*) and subsequently legislate transcription of the hormone-controlled genes (*right panel*). Shown in *yellow* is the class of the hsp-related co-facilitator proteins, which include the intracellular vitamin D-binding protein (IDBP) and the hnRNP-related co-modulator proteins, like the VDRE-BP. In Old World primate cells 1,25(OH)$_2$D (*dark triangles*; **Panel B**) moves normally from the circulating vitamin D-binding protein (DBP), through the cell membrane and cytoplasm, and onto the vitamin D receptor (VDR) paired with the retinoid X receptor (RXR) forming a heterodimeric complex in the cell nucleus. This heterodimeric complex then interacts with its cognate *cis* element, the vitamin D response element (VDRE), and initiates transcription of vitamin D-regulated genes. **Panel B** demonstrates similar events as they occur in New World primate cells. The *jagged line* at the VDRE represents a relative inability of the heterodimeric receptor complex to engage the *cis* element owing to the overabundance of the VDRE-BP in New World primate cells. This and the accumulation of 1,25(OH)$_2$D in the cell cytoplasm owing to the overabundance of the IDBP there represent salient disparities in hormone handling and action in the New World and Old World primate cells.

mobility shift assay (EMSA) and by chromatin immunoprecipitation (ChIP) analysis. In addition to the VDR–RXR, it was determined that these extracts contained a second protein that was bound by the VDRE. This protein was coined the vitamin D response element-binding protein or VDRE-BP *(37)*. In EMSA using the VDRE as probe (panel a, Fig. 3) Old World primate cell extract contained only the VDR–RXR bound to the VDRE probe, while the New World primate extract contained two probe-reactive bands, one compatible with the VDR–RXR and a second, more pronounced VDRE-BP-VDRE band. This VDRE-BP-VDRE binding reaction was specific, as the VDRE-BP was competed away from VDRE probe by the addition of excess unlabelled VDRE. These data suggested that VDRE-BP might function as a dominant-negative inhibitor of receptor-response element binding by competing *in trans* with receptor, "knocking it off" the VDRE. That was the case. When recombinant human VDR and RXR were permitted to interact in EMSA with increasing amounts of nuclear extract from either vitamin D-resistant cells containing a VDRE-BP or from normal vitamin D-responsive cells, the addition of more control extract acted only to amplify the VDR–RXR-retarded

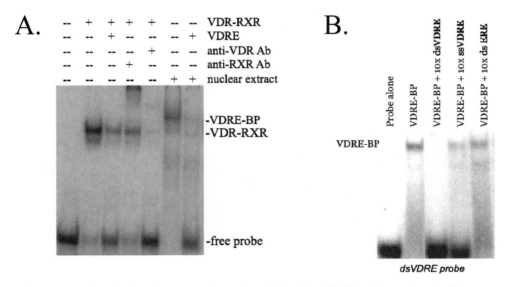

Fig. 3. Evidence for the dominant-negative action of the New World primate vitamin D response element-binding protein (VDRE-BP). **Panel A** is an electrophoretic mobility shift assay (EMSA) using consensus vitamin D response element as probe, showing the presence of a second *trans* binding protein, in addition to the vitamin D receptor (VDR)–retinoid X receptor (RXR), in nuclear extracts of vitamin D-resistant New World primate cells. Addition of excess unlabelled VDRE is shown in lanes 3 and 7. Addition of anti-RXRalpha antibody (lane 4) supershifts while addition of either anti-VDR antibody (lane 5) or New World primate nuclear extract containing the VDRE-BP (lane 6) competes away probe-VDR-RXR binding; these data are reprinted with permission of the authors from reference 34. **Panel B** shows the ability of the VDRE-BP to bind to the osteopontin VDRE and consensus ERE (AGGTCACAGTGACCTG) in either double-strand (ds) or single-strand (ss) format; affinity-purified human VDRE-BP and ERE-BP (150 pg of protein) were employed as a source of response element-binding protein and a 10-fold molar excess of response element-containing oligonucleotides, in either ds- or ss-format, as competitor DNA.

probe on the gel. By contrast, increasing amounts of the hormone-resistant extract competed away VDR–RXR-probe binding in favor VDRE-BP-probe binding.

6. VITAMIN D RESPONSE ELEMENT-BINDING PROTEIN

Purification, identification, and extensive functional characterization of the New World primate VDRE-BP and its Old World primate homolog have been accomplished. Once again, binding of the purified VDRE-BP to the VDRE was specific (right panel, Fig. 3); purified VDRE-BP was competed away from VDRE probe by the addition of excess unlabelled double-strand (ds) and single-strand (ss) VDRE but not by a radioinert ds ERE probe. The VDRE-BP is now known to be a distinct member of the hnRNP C super family, hnRNP C1/C2 *(38)*. Molecular cloning and overexpression of the VDRE-BP, previously considered to be only a ss mRNA-binding protein *(39)*, has confirmed it to be a dominant-negative inhibitor of 1,25(OH)$_2$D-VDR-directed transcription *(38)*. The VDRE-BP is also now recognized to be co-expressed with another category of intracellular chaperone proteins in the extended heat shock protein (hsp) family (panel a, Fig. 2 and below) that bind both (i) 25-hydroxylated vitamin D metabolites and (ii) other regulatory co-chaperone proteins. We have tagged the hsp-related binding protein as the intracellular vitamin D-binding protein (IDBP) and shown it to be identical to constitutively expressed hsc70. Results indicate that the VDRE-BP and IDBP act in concert to fine tune transcriptional control over 1,25(OH)$_2$D-driven genes in subhuman and human primates *(40, 41)*.

The ability of the purified VDRE-BP to compete *in trans* with the VDR for occupation of the VDRE in living primate cells is shown in panel a of Fig. 4. Shown is ChIP of

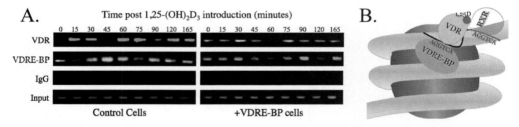

Fig. 4. Association of the VDR and VDRE-BP with the VDRE. Chromatin immunoprecipitation ChIP analysis of VDR, VDRE-BP, and IgG (negative control) on the native rat CYP24 VDRE was carried out in cell lysates isolated from control, vitamin D-responsive cells (*left panel*) and VDRE-BP overexpressing primate cells treated with or without 1,25(OH)$_2$D (10 nM) for periods from 0 to 165 minutes. The associated chromatin DNA fragments were amplified with primers to the 1,25(OH)$_2$D responsive VDRE region (–334 to –22 bp). The **Panel B** schematic shows a prototypical nucleosome bearing a specific VDRE-BP *cis* binding site, AGGTCA, on the exterior aspect of the double helix (*black line*) superimposed on dsDNA. It is proposed that initial VDRE-BP binding to the more distal, upstream VDRE half-site may open up (pioneer) that DNA *cis* element and provide access of the 1,25(OH)$_2$D-bound VDR–RXR dimer to compete with the VDRE-BP for binding to that *cis* element. Once the VDR–RXR has a grasp on the exterior-facing AGGTCA motif, this gives the VDR–RXR entree to the back side of the helix and the second AGGTCA half-site motif there. In this way the VDR–RXR would be able to straddle both sides of the helical VDRE, preparing that stretch of dsDNA for decondensation and subsequent transcription.

the VDR (top row) and VDRE-BP (second row) bound to the VDRE before (lane 1 in each blot) and time after the addition of hormone $1,25(OH)_2D$ to vitamin D-responsive (control cells, left panel) and vitamin D-resistant, VDRE-BP-overexpressing (+VDRE-BP cells, right panel) primate cells. In normal, vitamin D-responsive control cells in the absence of added hormone the VDRE is preferentially occupied by the VDRE-BP. After addition of a VDR-saturating concentration of $1,25(OH)_2D$ to control cells the ligand-bound VDR is recruited to the VDRE and the VDRE-BP competitively displaced from that *cis* element in a rhythmic, cyclic fashion. On the other hand, in the VDRE-BP-overexpressing primate cells the VDRE-BP remains bound to the VDRE diminishing and disrupting the rhythmic pattern of VDR interacting with the VDRE. The end result of this VDRE-BP-directed disruption is diminished transactivation of VDR-responsive genes, leading to vitamin D resistance in the host *(38, 42)*. In conclusion, organismal resistance to vitamin D is legislated, in part at least, by the ability of the VDRE-BP to compete with the VDR–RXR for occupancy of ds, *cis*-acting, VDREs. Because of the ability of VDRE-BP to specifically interact with ss as well as ds nucleic acid with an AGGTCA-like motif, we theorize that VDRE-BP is instrumental in marking the exposed, exterior surface of ds DNA as it wraps around the nucleosome (right panel, Fig. 4), readying that stretch of ds DNA and its underlying histone for chromatin remodeling and transcription (see below).

7. hnRNPC-RELATED PROTEINS AS MULTIFUNCTIONAL REGULATORS OF GENE EXPRESSION: BEYOND TRANSCRIPTION

Recent work *(43)* has led us to realize that the ability of the VDRE-BP to alter $1,25(OH)_2D$-directed gene expression is not limited to their dominant-negative actions at the level of the hormone response element (panel b, Fig. 5). By virtue of its capacity to interact with single-strand DNA (ssDNA), ssRNA as well as double-strand DNA (dsDNA; event 5, panel b, Fig. 5), the VDRE-BP has the potential to exert control over gene expression at multiple sites in the cell. As such, we now theorize that the VDRE-BP can act as a multisite participant in the synchronized expression of a *single* gene product by way of regulating, in succession, (i) chromatin remodeling (ssDNA binding), (ii) transcription (dsDNA binding), (iii) splicing (ssDNA and ssRNA binding), and (iv) microRNA (miRNA; ssRNA binding) (panel b, Fig. 5). The finding that the VDRE-BP occupies the VDRE in vivo in advance of these *cis* elements binding the VDR (see panel a, Fig. 4) and that protein attracting other elements of the chromatin remodeling machinery has led us to consider the possibility that interaction of the VDRE-BP with a 6-bp AGGTCA-like motif exposed on the outside of the DNA double helix (10–12 bp/turn) in heterochromatin (chromatin condensed about histones) would be an efficient means of starting the modification and opening of the chromatin structure for binding of transcriptionally active dimmers; the RXR–VDR requires both sides of dsDNA in a helical format and at least a 12–15 bp *cis* element for secure binding (see panel b, Fig. 4). In other words, we predict that, in and of itself, the VDRE-BP has the potential to regulate (1) the conversion of heterochromatin at that site in the genome to a more exposed, unwound euchromatic state as well as (2) the transcription that can subsequently take

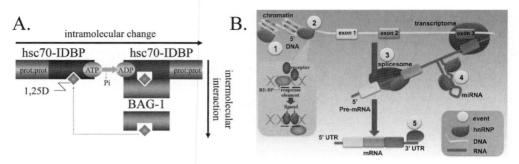

Fig. 5. Schematics for functions of the intracellular vitamin D-binding protein (IDBP; hsc70) and vitamin D response element-binding protein (RE-BP). **Panel A** displays a simplified view of the protein:protein interaction, vitamin D binding, and ATPase domains of the hsc70-related IDBP with the intramolecular (left to right) and intermolecular (top to bottom) changes in IDBP brought about by hormone binding and ATPase activity, respectively. The movement of bound hormone from the chaperone hsc70 to the co-chaperone, BAG-1 (Chun 2007) is shown. **Panel B** shows DNA and RNA *cis* elements that are known to be, or proposed to be (miRNA), bound by VDRE-BP hnRNP (see bottom right legend) during the process of mRNA generation. The numbered dots represent proposed, serial VDRE-BP-regulated events in the production and processing of a 1,25-dihydroxyvitamin D [1,25(OH)$_2$D]-induced mRNA. The *insert* at event 2 describes competition between the dominant negative-acting RE-BP and the dominant positive-acting liganded receptor for binding to the response element. The geographic proximity of these events within the cell leads us to propose a "hop-scotch" model of hnRNPs moving rapidly from one *cis* site to another to exert dynamic control over hormone-initiated gene expression.

place at that site in the genome. This would seem to provide a highly energy-efficient mode of co-control of both chromatin modification and transcription, as the VDRE-BP need not move from its *cis* element through the process of chromatin remodeling and initial stages of transcription (events 1 and 2, panel b, Fig. 5).

In similar fashion we predict that the same VDRE-BP need only "hop-scotch" a short distance from one *cis* site to another in the nuclear environment to exert control over splicing of the pre-mRNAs of the same 1,25(OH)$_2$D-directed genes (event 3, panel b, Fig. 5). Preliminary data *(43)* show that relative overexpression of the same VDRE-BP that regulates chromatin modification and transcription will bind to ssRNA and/or ssDNA (after strand separation) and alter splicing of the product of that transcriptional event. Finally, we hypothesize that by virtue of their ability to bind RNA, hnRNPs can effectively bind to noncoding (nc) RNA sequences, including microRNA (miRNA) sequence, transcribed from either a UTR or an intron (event 4, panel b, Fig. 5). These potential actions of the hnRNPs could effectively (1) limit the processing of mRNAs in the nucleus (e.g., through interaction with dsRNA) and/or (2) alter their hybridization with complimentary mRNA sequences in ss-format in the nucleus or cell cytoplasm in advance of their translation. In either proposed event, the action of VDRE-BP would be to take the ncRNA (e.g., miRNA) "out of play" and promote transcript translation to protein, conspiring with the effects rendered by the VDER-BP during chromatin remodeling, transcription, and splicing to formulate a composite functional response within the cell.

8. INTRACELLULAR VITAMIN D-BINDING PROTEIN

On the way to the discovery of the VDRE-BP in New World primate cells, it was also observed that these cells were extraordinarily efficient at accumulating 25-hydroxylated vitamin D metabolites in the cytoplasmic space (see right panel, Fig. 2). Accumulation here was the result of expression of another, distinct resistance-associated protein. The intracellular vitamin D-binding protein *(44–46)*, or IDBP as it has come to be called, exhibits both high capacity and high affinity for 25-hydroxylated vitamin D metabolites. In fact, among all of the vitamin D metabolites that have been tested, IDBP purified from vitamin D-resistant New World primate cells binds 25(OH)D and 25(OH)D best *(30, 45)*; in a competitive displacement assay using radioinert 25(OH)D as competitor and [^{3}H]-25(OH)D as labeled ligand, the concentration of metabolite required to achieve half-maximal displacement of labeled hormone (EC_{50}) is <1 nM. Although normally present in Old World primate including human cells, IDBP can be overexpressed some 50-fold in New World primate cells. The New World primate IDBP is highly homologous to constitutively expressed human heat shock protein-70 (hsc70; 44). The general domain structure of the IDBPs is shown in panel a of Fig. 5. They all contain an ATP-binding ATPase domain and a protein–protein interaction domain. Preliminary studies indicate that the vitamin D ligand-binding domain is in the middle of the molecule *(40)*.

Why is IDBP or hsc70 overexpressed in the hormone-resistant New World primate cell? Two countervailing, "sink" and "swim" hypotheses have been considered by us to explain the function of this protein. One hypothesis holds that IDBP is a "sink" molecule that works in cooperation with the VDRE-BP in the nucleus to exert vitamin D resistance by binding hormone and disallowing its access to the VDR and the nucleus of the cell. The opposing hypothesis holds that IDBP is a "swim" molecule that actually promotes the delivery of ligand to the VDR, improving the ability of the VDR to dimerize with the VDR and bind to DNA, antagonizing the actions of the VDRE-BP that was overexpressed in New World primate cells. In order to determine which of these hypotheses was correct, we stably overexpressed IDBP, in wild-type Old World primate cells and demonstrated that IDBP imparted pro-transactivating potential *(40)*; the endogenous transcriptional activity of three different 1,25(OH)$_2$D-responsive genes, the vitamin D-24-hydroxylase, osteopontin, and osteocalcin genes, in Old World primate wild-type cells was markedly enhanced. It was concluded from these studies, at least for the function of transactivation, that IDBP is a "swim" molecule for the vitamin D hormone, promoting delivery of ligand to the VDR.

Considering the fact that New World primates are required to maintain very high serum levels of 1,25(OH)$_2$D in order to avert rickets it was also hypothesized that the IDBPs, which are known to bind 25(OH)D even better than 1,25(OH)$_2$D, will also promote the synthesis of the active vitamin D metabolite via promotion of the 25(OH)D-CYP27B1-hydroxylase. When human kidney cells expressing the CYP27B1 gene were stably transfected with IDBP-1 and incubated with substrate 25(OH)D, 1,25(OH)$_2$D production went up 4–8-fold compared to untransfected wild-type cells *(41)*. This increase in specific 25(OH)D-1-hydroxylase activity occurred independent of a change in expression of the CYP27B1 gene *(42)*. If one overexpresses IDBP and incubates transfected IDBP-overexpressing primate cells with a fluorescently labeled

25-hydroxylated vitamin D metabolite, one will also observe a significant increase in the uptake of the labeled prohormone. These results indicate that the increase in hormone $1,25(OH)_2D$ production is the result of the ability of IDBP to promote the delivery of substrate 25(OH)D to the inner mitochondrial membrane and the 25(OH)D-CYP27B1-hydroxylase stabled there.

9. SUMMARY

The concept that a single *trans-acting* protein by way of its protein–DNA interaction capacity can influence the fate of a single hormone-regulated gene product in serial fashion at the level of chromatin modification, transcription, splicing as well as transcript handling in the nucleus and cytoplasm is an appealing prospect (panel b, Fig. 5). This is especially so when one considers that humans are getting by with far fewer structural genes (25–30,000) than either a mouse (35,000) or a plant (75,000). The complexity of our genome at the level of expression must, therefore, be invested in at least two powerful strategies: (1) the use of a single protein to function as a cog in more than a single cellular machine; and (2) the ability of the human cell to balance multiple levels of control to legislate a finely tuned bioresponse for the organism. Of additional intrigue and complexity is the fact that there exists another protein, the hsc70-related IDBP, that is present in normal human and subhuman primate cells and that is co-overexpressed with VDRE-BP in that state of vitamin D resistance characteristic of New World primates. The ability of the IDBP and its co-chaperone BAG-1 to (1) bind 25-hydroxylated metabolites (see panel a, Fig. 5), (2) promote cellular internalization of 25-hydroxylated vitamin D metabolites, and (3) localize the IDBP–BAG1 complex to the nucleus to potentiate VDR-mediated transactivation *(42, 46)* suggests that this chaperone-co-chaperone pair of molecules, like the VDRE-BP, may be multipurpose players in the cell, in this case trafficking vitamin Ds to specific intracellular destinations and to specific proteins like the VDR.

ACKNOWLEDGMENTS

This work was supported by NIH grants AR37399 and DK58891 to JSA.

REFERENCES

1. Pilbeam D (1984) The descent of hominoids and hominoids and hominoids. Sci Am 250:84–96
2. Bland Sutton JB (1884) Observation on rickets etc. in wild animals. J Anat 18:363–397
3. Krook L, Barrett RB (1962) Simian bone disease – A secondary hyperparathyroidism. Cornell Vet 52:459–492
4. Hershkovitz P (1977) Living New World monkeys (Platyrrhini): with an introduction to Primates. University of Chicago Press, Chicago, 1117 pp
5. Hunt RD, Garcia FG, Hegsted DM (1967) A comparison of vitamin D2 and D3 in New World primates. I. Production and regression of osteodystrophia fibrosa. Lab Anim Care 17:222–234
6. Steenbock H, Kletzein SWF, Halpin JG (1932) Reaction of chicken irradiated ergosterol and irradiated yeast as contrasted with natural vitamin D of fish liver oils. J Biol Chem 97:249–266
7. Marx SJ, Jones G, Weinstein RS, Chrousos GP, Renquist DM (1989) Differences in mineral metabolism among nonhuman primates receiving diets with only vitamin D3 or only vitamin D2. J Clin Endocrinol Metab 69:1282–1290

8. Hay AW (1975) The transport of 25-hydroxycholecalciferol in a New World monkey. Biochem J 151:193–196

9. Bouillon R, Van Baelen H, De Moor P (1976) The transport of vitamin D in the serum of primates. Biochem J 159:463–466

10. Brown GM, Grota LJ, Penney DP, Reichlin S (1970) Pituitary-adrenal function in the squirrel monkey. Endocrinology 86:519–529

11. Brandon DD, Markwick AJ, Chrousos GP, Loriaux DL (1989) Glucocorticoid resistance in humans and nonhuman primates. Cancer Res 49:2203–2213

12. Chrousos GP, Brandon D, Renquist DM, Tomita M, Johnson E, Loriaux DL, Lipsett MB (1984) Uterine estrogen and progesterone receptors in an estrogen and progesterone-resistant primates. J Clin Endocrinol Metab 58:516–520

13. Albertson BD, Maronian NC, Frederick KL, DiMattina M, Feuillan P, Dunn JF, Loriaux DL (1988) The effect of ketoconazole on steroidogenesis. II. Adrenocortical enzyme activity in vitro. Res Commun Chem Pathol Pharmacol 61:27–34

14. Albertson BD, Frederick KL, Maronian NC, Feuillan P, Schorer S, Dunn JF, Loriaux DL (1988) The effect of ketoconazole on steroidogenesis: I. Leydig cell enzyme activity in vitro. Res Commun Chem Pathol Pharmacol 61:17–26

15. Moore CC, Mellon SH, Murai J, Siiteri PK, Miller WL (1993) Structure and function of the hepatic form of 11 beta-hydroxysteroid dehydrogenase in the squirrel monkey, an animal model of glucocorticoid resistance. Endocrinology 133:368–375

16. Klosterman LL, Murai JT, Siiteri PK (1986) Cortisol levels, binding, and properties of corticosteroid-binding globulin in the serum of primates. Endocrinology 118:424–434

17. Chrousos GP, Renquist DM, Brandon D, Fowler D, Loriaux DL, Lipsett MB (1982) The squirrel monkey: receptor-mediated end-organ resistant to progesterone?. J Clin Endocrinol Metab 55: 364–368

18. Brandon DD, Markwick AJ, Flores M, Dixon K, Albertson BD, Loriaux DL (1991) Genetic variation of the glucocorticoid receptor from a steroid-resistant primate. Mol Endocrinol 7:89–96

19. Scammell JG, Denny WB, Valentine DL, Smith DF (2001) Overexpression of the FK506-binding immunophilin FKBP51 is the common cause of glucocorticoid resistance in three New World primates. Gen Comp Endocrinol 124:152–165

20. Newman RM, Hall L, Kirmaier A, Pozzi LA, Pery E, Farzan M, O'Neil SP, Johnson W (2008) Evolution of a TRIM5-CypA splice isoform in old world monkeys. PLoS Pathog 4:e1000003

21. Westberry JM, Sadosky PW, Hubler TR, Gross KL, Scammell JG (2006) Glucocorticoid resistance in squirrel monkeys results from a combination of a transcriptionally incompetent glucocorticoid receptor and overexpression of the glucocorticoid receptor co-chaperone FKBP51. J Steroid Biochem Mol Biol 100:34–41

22. Fuller PJ, Smith BJ, Rogerson FM (2004) Cortisol resistance in the New World revisited. Trends Endocrinol Metab 15:296–299

23. Riggs DL, Roberts PJ, Chirillo SC, Cheung-Flynn J, Prapapanich V, Ratajczak T, Gaber R, Picard D, Smith DF (2003) The Hsp90-binding peptidylprolyl isomerase FKBP52 potentiates glucocorticoid signaling in vivo. EMBO J 22:1158–1167

24. Chrousos GP, Brandon D, Renquist DM, Tomita M, Johnson E, Loriaux DL, Lipsett MB (1984) Uterine estrogen and progesterone receptors in an estrogen- and progesterone- "resistant" primate. J Clin Endocrinol Metab 58:516–520

25. Liberman UA, de Grange D, Marx SJ (1985) Low affinity of the receptor for 1 alpha,25-dihydroxyvitamin D3 in the marmoset, a New World monkey. FEBS Lett 182:385–388

26. Adams JS, Gacad MA, Baker AJ, Keuhn G, Rude RK (1985) Diminished internalization and action of 1,25-dihydroxyvitamin D in dermal fibroblasts cultured from New World primates. Endocrinology 116:2523–2527

27. Adams JS, Gacad MA (1988) Phenotypic diversity of the cellular 1,25-dihydroxyvitamin D-receptor interaction among different genera of New World primates. J Clin Endocrinol Metab 66:224–229

28. Adams JS, Gacad MA, Baker AJ, Gonzales B, Rude RK (1985) Serum concentrations of 1,25-dihydroxyvitamin D in Platyrrhini and Catarrhini: A phylogenetic appraisal. Am J Primatol 9:219–224

29. Adams JS, Gacad MA, Rude RK, Endres DB, Mallette LE (1987) Serum concentrations of immunore-active parathyroid hormone in Platyrrhini and Catarrhini: A comparative analysis with three different antisera. Am J Primatol 13:425–433
30. Gacad MA, Adams JS (1991) Evidence for endogenous blockage of cellular 1,25-dihydroxyvitamin D-receptor binding in New World primates. J Clin Invest 87:996–1001
31. Gacad MA, Adams JS (1992) Influence of ultraviolet B radiation on vitamin D3 metabolism in vitamin D3-resistant New World primates. Am J Primatol 28:263–270
32. Gacad MA, Adams JS (1992) Specificity of steroid binding in New World primate cells with a vitamin D-resistant phenotype. Endocrinology 131:2581–2587
33. Gacad MA, Adams JS (1993) Identification of a competitive binding component in vitamin D-resistant New World primate cells with a low affinity but high capacity for 1,25-dihydroxyvitmain D3. J Bone Miner Res 8:27–35
34. Chun RF, Chen H, Boldrick L, Sweet C, Adams JS (2001) Cloning, sequencing and functional characterization of the vitamin D receptor in vitamin D-resistant New World primates. Am J Primatol 54:107–118
35. Arbelle JE, Chen H, Gacad MA, Allegretto EA, Pike JW, Adams JS (1996) Inhibition of vitamin D receptor-retinoid X receptor-vitamin D response element complex formation by nuclear extracts of vitamin D-resistant New World primate cells. Endocrinology 137:786–789
36. Chen H, Arbelle JE, Gacad MA, Allegretto EA, Adams JS (1997) Vitamin D and gonadal steroid-resistant New World primate cells express an intracellular protein which competes with the estrogen receptor for binding to the estrogen response element. J Clin Invest 99:769–775
37. Chen H, Hu B, Allegretto EA, Adams JS (2000) The vitamin D response element binding proteins: novel dominant-negative-acting regulators of vitamin D-directed transactivation. J Biol Chem 275:35557–35564
38. Chen H, Hewison M, Adams JS (2006) Functional characterization of heterogeneous nuclear ribonuclear protein C1/C2 in vitamin D resistance: a novel response element-binding protein. J Biol Chem 281:39114–39120
39. Dreyfuss G, Matunis MJ, Pinol-Roma S, Burd CG (1993) hnRNP proteins and the biogenesis of mRNA. Annu Rev Biochem 62:289–321
40. Wu S, Ren S-Y, Gacad MA, Adams JS (2000) Intracellular vitamin D binding proteins: novel facilitators of vitamin D-directed transactivation. Mol Endocrinol 14:1387–1397
41. Wu S, Chun R, Ren S, Chen H, Adams JS (2002) Regulation of 1,25-dihydroxyvitamin D synthesis by intracellular vitamin D binding protien-1. Endocrinology 143:4135–4138
42. Adams JS, Chen H, Chun R, Ren S, Wu S, Gacad M, Nguyen L, Ride J, Liu P, Modlin R, Hewison M (2007) Substrate and enzyme trafficking as a means of regulating 1,25-dihydroxyvitamin D synthesis and action: the human innate immune response. J Bone Miner Res 22(Suppl 2):V20–V24
43. Chen H, Hewison M, Nanes M, Adams JS (2008) Evidence of a role for the human vitamin D response element binding protein, heterogeneous nuclear ribonucleoprotein (hnRNP) C1/C2 in chromatin remodeling and transcript splicing. J Bone Miner Res 23:S105
44. Gacad MA, LeBon TR, Chen H, Arbelle JE, Adams JS (1997) Functional characterization and purification of an intracellular vitamin D binding protein in vitamin D resistant New World primate cells: Amino acid sequence homology with proteins in the hsp-70 family. J Biol Chem 272:8433–8440
45. Gacad MA, Adams JS (1998) Proteins in the heat shock-70 family specifically bind 25-hydroxylated vitamin D metabolites and 17β-estradiol. J Clin Endocrinol Metab 83:1264–1267
46. Chun RF, Gacad M, Nguyen L, Hewison M, Adams JS (2007) Co-chaperone potentiation of vitamin D receptor-mediated transactivation: a role for Bcl2-associated anthanogene-1 as an intracellular-binding protein for 1,25-dihydroxyvitamin D3. J Mol Endocrinol 39:81–89

39 25-Hydroxyvitamin D-1α-Hydroxylase: Studies in Mouse Models and Implications for Human Disease

David Goltzman

Abstract   Genetic mouse models with targeted deletion ("knockout") of the 25-hydroxyvitamin D 1α-hydroxylase gene [$1\alpha(OH)ase^{-/-}$], as well as with targeted deletion of the vitamin D receptor (VDR) gene, when exposed to different dietary regimens, have provided considerable insight into the molecular regulation of skeletal physiology by the 1,25-dihydroxyvitamin D [$1,25(OH)_2D$]/VDR system. These regimens induced different phenotypic changes and demonstrated that parathyroid gland size and the development of the cartilaginous growth plate were each coordinately regulated by calcium and by $1,25(OH)_2D$, and that parathyroid hormone (PTH) secretion and mineralization of bone reflected ambient calcium (and phosphorus) levels rather than the direct actions of the $1,25(OH)_2D$/VDR system. In contrast, increased calcium absorption, optimal osteoblastogenesis, and baseline bone formation were observed to be modulated by $1,25(OH)_2D$/VDR signaling. These bone anabolic effects of endogenous $1,25(OH)_2D$ were evident in neonatal mice as well as in older animals, and exogenous $1,25(OH)_2D$ was also found to stimulate trabecular and cortical bone formation in neonatal double homozygous $1\alpha(OH)ase^{-/-}PTH^{-/-}$ mice. Furthermore, the anabolic effect of exogenously administered PTH appeared to be partly dependent on the stimulation of endogenous $1,25(OH)_2D$. Genetic mouse models have also been employed to study extra-skeletal actions modulated by the $1,25(OH)_2D$/VDR system. For example, increased blood pressure, activation of the renin/angiotensin system, myocardial hypertrophy, and cardiac dysfunction were observed in $1\alpha(OH)ase^{-/-}$ mice, and these alterations could be prevented by treatment with $1,25(OH)_2D$. These models allow controlled examination of the regulation of both skeletal and extra-skeletal pathophysiology associated with $1,25(OH)_2D$ deficiency, which appear to be relevant to humans and facilitate studies to prevent and treat these disorders by active vitamin D forms.

Key Words: 25-hydroxyvitamin D 1α-hydroxylase; vitamin D receptor; knockout mice; skeletal anabolic effects of vitamin D; vitamin D and renin

1. INTRODUCTION

The enzyme 25-hydroxyvitamin D_3-1α-hydroxylase [1α(OH)ase] is responsible for the introduction of a 1-hydroxyl group into the A ring of 25-hydroxyvitamin D[25(OH)D] and, as a result, is central to the production of the active form of

From: *Nutrition and Health: Vitamin D*
Edited by: M.F. Holick, DOI 10.1007/978-1-60327-303-9_39,
© Springer Science+Business Media, LLC 2010

vitamin D 1,25 dihydroxyvitamin D [1,25(OH)$_2$D] (1). The renal proximal tubular cell is the major source of the circulating pool of 1,25(OH)$_2$D, and the renal 1α(OH)ase, a mitochondrial enzyme, comprises a cytochrome P-450, a ferredoxin, and a ferredoxin reductase (2). The cytochrome P-450, which provides the specificity of the 25(OH)D for the 1α(OH)ase has been cloned from rat, mouse, and human (3–6).

PTH has been shown to be a central upregulator of the renal enzyme, acting through the PTH receptor by cyclic AMP signal transduction (7). Transcriptional upregulation of the gene encoding the cytochrome P450, CYP27B1, and an increase in protein synthesis result from the actions of PTH. 1,25(OH)$_2$D per se has been shown to downregulate the transcription of CYP27B1 through the direct interaction of its ligand-bound receptor with upstream elements in the promoter of the CYP27B1 gene, leading to a reduction in 1,25(OH)$_2$D synthesis. The CYP27B1 gene is also independently upregulated by low Ca^{2+} and PO$_4$$^{3-}$ signals (8), however, the signal transduction mechanisms that mediate these effects remain unclear. Recently, the osteocyte-derived peptide fibroblast growth factor 23 (FGF23) has been demonstrated to be a potent inhibitor of renal 1α(OH)ase (9), at least in part via transcriptional inhibition.

1,25(OH)$_2$D acts via its cognate receptor, the vitamin D receptor (VDR), a member of the nuclear receptor superfamily (10) and directly regulates gene expression at the transcriptional level of a wide variety of genes in vitamin D target cells (11). After being taken up by target cells and binding to intracellular shuttle proteins, the sterol binds to the VDR. In the nucleus of target cells, the ligand-activated VDR heterodimerizes with the retinoid X receptor and the dimer binds to response elements (VDREs) on target genes (12). Coregulators are recruited to link the dimer to the basal transcriptional machinery and thereby modulate gene transcription. Differences in tissue-specific, differentiation-stage-specific, and gene-specific coregulators present in vitamin D target cells modulate the wide array of genes that are regulated by vitamin D in each tissue at any given time. Furthermore, whether 1,25(OH)$_2$D causes upregulation or downregulation may be gene specific. Thus, various genes involved in calcium homeostasis (e.g., calbindins, Ca^{2+} channel proteins, osteocalcin, osteopontin, RANKL genes) are upregulated, whereas others (e.g., collagen and pre-pro-parathyroid hormone (PTH) genes) are downregulated by 1,25(OH)$_2$D. The broad distribution of the VDR, which is expressed in many cells throughout the body has lent support to the concept that vitamin D may have actions beyond the skeletal/calcium homeostatic system.

The 1α(OH)ase enzyme has also been shown to be expressed in various extra-renal sites (13), however, regulation of the extra-renal CYP27B1 appears to differ from the renal enzyme. For example cytokines, and not PTH, upregulate expression of CYP27B1 in the macrophage (14). By regulating the amount of ligand generated locally, and available for the vitamin D receptor (VDR), the extra-renal 1α(OH)ases may determine the extent of vitamin-D-dependent gene expression that occurs inside a variety of vitamin D target cells. Consequently, the extra-renal 1α(OH)ases may be part of an important intracrine, autocrine, or paracrine system for vitamin D action (15).

Despite the potential significance of the extra-renal actions of vitamin D, experiments of nature have confirmed the signal importance of vitamin D in skeletal and calcium homeostasis. Thus loss-of-function mutations in CYP27B1 in humans cause the autosomal recessive disorder vitamin D-dependent rickets type 1 (VDDR1), also known as pseudo-vitamin D-deficiency rickets (16). The disease is caused by homozygosity of a single abnormal *CYP27B1* gene or by compound heterozygosity of abnormal *CYP27B1* genes, leading to deficiency of 1α(OH)ase and is characterized by low serum calcium and phosphate, secondary hyperparathyroidism, and low circulating levels of $1,25(OH)_2D$. Many of the single-base substitutions in the mutated genes affect heme binding or ferredoxin docking, leading to a defective enzyme *(17)*. Loss-of-function mutations in the *VDR* in humans cause the autosomal recessive disorder vitamin D-dependent rickets type 2, also known as hereditary vitamin D-resistant rickets *(18)*. This syndrome generally presents with rachitic changes not responsive to vitamin D treatment and the circulating levels of $1,25(OH)_2D$ are elevated, differentiating it from VDDR1. Alopecia is seen in some families with vitamin D-dependent rickets type 2 and is usually associated with a more severe phenotype. Thus in human disorders characterized by either loss of the active vitamin D ligand or by loss of the capacity of vitamin D to function, skeletal and mineral abnormalities are the major phenotypic presentation pointing to the critical role for the vitamin D endocrine system in this system.

We and others have genetically engineered *CYP27B1 (19, 20)* and *VDR* knockout mice *(21–24)*, which have facilitated more controlled and extensive examination of the phenotypes than is possible by examination of the corresponding human disorders VDDR1 and VDDR2, respectively. These mouse mutants have shed interesting new light on both the skeletal and extra-skeletal actions of the vitamin D endocrine system.

2. SKELETAL AND MINERAL MODULATING ACTIONS OF 25-HYDROXYVITAMIN D_3-1α-HYDROXYLASE AS DETERMINED BY GENETIC MOUSE MODELS

We generated a *1α(OH)ase* null mouse by deleting exons VI and VII of CYP27B1 that encode the hormone-binding domain and exon VIII that encodes the heme-binding domain *(19)*. The null mutant mice [*1α(OH)ase$^{-/-}$*] appeared grossly normal from birth until weaning; however, after weaning at 3 weeks, they displayed marked growth retardation. Circulating concentrations of $1,25(OH)_2D$ were undetectable in the homozygous null mice and serum 25(OH)D concentrations were elevated in *1α(OH)ase$^{-/-}$* mice relative to wild-type controls. Both serum calcium and phosphate concentrations were reduced, serum PTH concentrations were markedly elevated, and urinary phosphate was increased. Examination of known $1,25(OH)_2D$ target genes revealed that the expression of the mRNA encoding the 25-hydroxyvitamin D-24-hydroxylase enzyme [24(OH)ase], which catalyzes the first step in the degradative pathway of vitamin D, was almost completely ablated in the null mutant mice, and the expression of intestinal calbindin $D_{9\,k}$, renal calbindin $D_{9\,k}$, and calbindin $D_{28\,k}$ was reduced. Typical features of advanced rickets were observed histologically in bone, including widening of the epiphyseal growth plates (predominantly because of an increase in the width of the hypertrophic zone, which was also disorganized); inadequate mineralization of cartilage, of the primary spongiosa, and of cortical bone; as well as an increase in osteoid in both trabecular and cortical bone. Osteoblasts lining bone surfaces were increased and the trabecular bone area in the primary spongiosa was greater in null mutant than in wild-type mice. This model therefore appeared to produce an accurate murine phenocopy of the human disorder VDDR1.

We next compared the phenotype of this mutant to that of the VDR null mouse (kindly provided by Dr Marie Demay of Mass. General Hospital, Boston, MA *(22)* in which a 5 kb fragment of genomic DNA encoding the second zinc finger of the receptor DNA-binding domain was deleted. We also examined $1\alpha(OH)ase^{-/-}VDR^{-/-}$ mice obtained by crossbreeding $1\alpha(OH)ase^{+/-}$ and $VDR^{+/-}$ mice to determine the consequences of $1\alpha(OH)$ase deficiency on the $VDR^{-/-}$ phenotype. We compared the phenotypes of these three genetically diverse mutants after exposure to different environments, i.e., diets with differing calcium intakes and after administering exogenous $1,25(OH)_2D$ *(25)*. All mice were maintained on a high calcium intake (without lactose) until weaning in order to facilitate reproduction. After weaning, mice received either a high calcium intake on which they remained hypocalcemic, a high calcium intake with injections of $1,25(OH)_2D$ intraperitoneally three times per week, or a "rescue" diet containing high calcium, high phosphorus, and 20% lactose. Animals were examined at 4 months of age.

2.1. Regulation of Calcium Absorption

The development of hypocalcemia after weaning, in mice with targeted deletion of the $1\alpha(OH)$ase and of the *VDR*, is well documented to occur on a normal calcium diet. It was also seen, although to a lesser extent, on the high-calcium diet we employed, in both $1\alpha(OH)ase^{-/-}$ and $VDR^{-/-}$ mice, reemphasizing the critical role of the vitamin D endocrine system in enhancing intestinal calcium absorption. We also found, as have others *(26)*, that the rescue diet containing lactose eliminated this hypocalcemia in $1\alpha(OH)ase^{-/-}$ and $VDR^{-/-}$ mice. This points to the important role of dietary lactose in increasing transport of calcium at least in the rodent intestine, independent of the $1,25(OH)_2D/VDR$ system and may account for the apparent absence of hypocalcemia in these mouse models which ingest lactose-containing maternal milk prior to weaning. The precise mechanism of the lactose effect remains to be determined but might involve enhanced passive intestinal transport of calcium via a paracellular route. The presence of hypocalcemia in both the $1\alpha(OH)ase^{-/-}$ mice with intact VDR but deficient $1,25(OH)_2D$ production and in the $VDR^{-/-}$ mice with elevated endogenous levels of $1,25(OH)_2D$ but deficient VDR suggests that both $1,25(OH)_2D$ and the VDR are necessary for optimal intestinal absorption of calcium. Treatment with exogenous $1,25(OH)_2D$ failed to normalize the serum calcium in $VDR^{-/-}$ mice. In contrast, $1\alpha(OH)ase^{-/-}$ mice on the same diet, treated with the same dose of $1,25(OH)_2D$ did normalize serum calcium both in our studies and others *(27)*. Intestinal calcium absorption, therefore, appears to require both $1,25(OH)_2D$ and the VDR.

2.2. Regulation of the 1α(OH)ase and of the 24(OH)ase Enzymes

Expression of renal $1\alpha(OH)$ase mRNA was elevated and $24(OH)$ase mRNA was suppressed in $VDR^{-/-}$ mice, resulting in high circulating $1,25(OH)_2D$ concentrations. Thus in the absence of VDR, $1\alpha(OH)$ase could not be suppressed or $24(OH)$ase stimulated by endogenous $1,25(OH)_2D$. Treatment with exogenous $1,25(OH)_2D$ of the $1\alpha(OH)ase^{-/-}$ mouse (which retains a normal VDR) enhanced $24(OH)$ase expression. This effect of $1,25(OH)_2D$ was not seen in the mutants lacking a VDR. These studies, therefore, suggest that both $1,25(OH)_2D$ and the VDR are required for regulation of gene expression of both the 1α(OH)ase and 24(OH)ase in vivo *(1, 11, 28)*. However, the exogenous administration of 1,25(OH)$_2$D to the *1α(OH)ase$^{-/-}$* mice also normalized serum calcium. Furthermore, elimination of hypocalcemia alone, using a rescue diet also normalized 1α(OH)ase levels and 24(OH)ase in both *1α(OH)ase$^{-/-}$* and *VDR$^{-/-}$* mice mutant mice This, therefore, demonstrates a calcium effect on gene expression of these enzymes independent of the 1,25(OH)$_2$D/VDR system. Whether this effect of calcium is entirely indirect, by suppressing ambient PTH concentrations *(1, 11, 28)*, or is partly direct remains to be determined.

2.3. Parathyroid Gland Function

Calcium is known to inhibit parathyroid hormone secretion via the calcium sensing receptor (CaSR) *(29)* and also to decrease parathyroid cell growth *(29)*. 1,25(OH)$_2$D has also been shown to inhibit PTH synthesis *(30)* and secretion *(31)* and has been reported to inhibit parathyroid cell growth in vitro *(32)*. On a normal or lactose-free high calcium diet, when hypocalcemia is present, increased circulating PTH concentrations and enlarged parathyroid glands have been described in the *1α(OH)ase$^{-/-}$* mice *(19, 20)* and the *VDR$^{-/-}$* *(22, 25)* mice, and we found similar increases in the *1α(OH)ase$^{-/-}$VDR$^{-/-}$* double mutants *(25)*. On a rescue diet, serum PTH concentrations fell in all mutants suggesting that raising the ambient calcium could alone normalize PTH secretion. Parathyroid gland size was smaller in the VDR deleted (*VDR$^{-/-}$*) mice with elevated endogenous 1,25(OH)$_2$D than in the *1α(OH)ase$^{-/-}$* and in the *1α(OH)ase$^{-/-}$VDR$^{-/-}$* mice *(25)*. The glands remained moderately enlarged in these latter mutants which lacked endogenous 1,25(OH)$_2$D even when hypocalcemia was rectified by the rescue diet. Treatment of *1α(OH)ase$^{-/-}$* mice with exogenous 1,25(OH)$_2$D normalized serum calcium and also normalized parathyroid gland size. Exogenous 1,25(OH)$_2$D could not, however, normalize parathyroid gland size in mutants deficient in VDR who remained hypocalcemic. Consequently both calcium and 1,25(OH)$_2$D appear to act cooperatively to diminish PTH production and parathyroid gland size (Fig. 1).

2.4. Development of the Cartilaginous Growth Plate

On a normal or high calcium lactose-free intake, all three hypocalcemic mutant mouse models, i.e., *1α(OH)ase$^{-/-}$*, *VDR$^{-/-}$*, and *1α(OH)ase$^{-/-}$VDR$^{-/-}$* mice, developed characteristic rachitic changes in long bones, i.e., enlarged and distorted cartilaginous growth plates with widened hypertrophic zones *(19–22, 33)*. These abnormalities appeared less severe in *VDR$^{-/-}$* mice with elevated endogenous 1,25(OH)$_2$D levels than in the other two mutants suggesting that 1,25(OH)$_2$D may modulate cartilage function independent of the VDR. Indeed non-genomic effects of 1,25(OH)$_2$D have been reported in cartilage cells *(34)*. Nevertheless, 1,25(OH)$_2$D per se could not normalize the growth plate if hypocalcemia was not normalized, i.e., in the *VDR$^{-/-}$* and *1α(OH)ase$^{-/-}$VDR$^{-/-}$* where intestinal calcium absorption was defective. However, elimination of hypocalcemia with the rescue diet did not completely normalize the growth plate in the mouse models that have deficient endogenous 1,25(OH)$_2$D, i.e., in the *1α(OH)ase$^{-/-}$* and *1α(OH)ase$^{-/-}$VDR$^{-/-}$* mice. Consequently both calcium

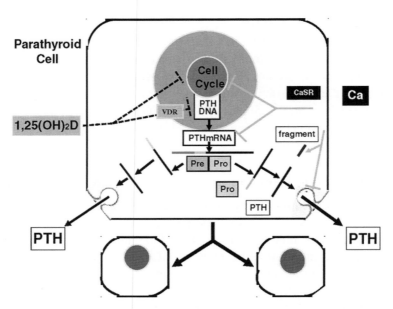

Fig. 1. Coordinate regulation of PTH production by calcium and 1,25(OH)$_2$D. Calcium (Ca), acting on the parathyroid cell via the calcium sensing receptor (CaSR), can inhibit PTH mRNA translation and reduce production of the biosynthetic precursor PrePrePTH, can increase intracellular proteolysis of PTH to fragments, and inhibit secretion of intact PTH. 1,25(OH)$_2$D can inhibit PTHrP gene transcription. Both calcium and 1,25(OH)$_2$D can inhibit cell cycle progression and reduce cell proliferation.

and 1,25(OH)$_2$D together appear necessary for normal development of the cartilaginous growth plate.

2.5. Bone and Cartilage Remodeling

Osteoblast numbers, bone formation, and bone volume were markedly increased in all hypocalcemic knockout models of the vitamin D/VDR system, i.e., $1\alpha(OH)ase^{-/-}$ mice, $VDR^{-/-}$ mice, and $1\alpha(OH)ase^{-/-}VDR^{-/-}$ double mutants, on either a lactose-free, normal calcium intake *(19, 20, 22, 25)* or a lactose-free, high calcium intake. This appears to be due to the "anabolic" effect of PTH, which is markedly elevated in association with the severe secondary hyperparathyroidism in these animals. Increased serum alkaline phosphatase reflected the increased osteoblastic stimulation by PTH and was normalized when PTH was normalized by eliminating the secondary hyperparathyroidism. The increased bone volume was largely due to increased unmineralized osteoid. Interestingly, a sustained elevation of PTH is generally associated with increased osteoclastic bone resorption as well as increased bone formation. Nevertheless, osteoclast number and resorbing surface were not generally elevated in these models compared to wild-type animals suggesting an "inappropriate" response to the increased PTH. This suggests, therefore, that there is uncoupling of bone turnover in the presence of a defective 1,25(OH)$_2$D/VDR system and the relatively low resorption may contribute, with increased osteoblast activity, to the increased bone volume. Therefore, although PTH and local modulators of bone resorption may sustain a "normal" level of osteoclastic resorption in these models, an intact 1,25(OH)$_2$D/VDR system appears to be required for an appropriate osteoclastic response to increased PTH. The precise molecular mechanism of this response remains to be determined.

In view of the fact that osteoclast/chondroclast production at the chondro–osseous junction may also be defective, diminished removal of hypertrophic chondrocytes may occur in this region leading to altered cartilage growth plate remodeling. Therefore the enlargement of the cartilaginous growth plate, and notably the hypertrophic zone, may also be in part due to reduced activity of the 1,25(OH)$_2$D/VDR system on the chondroclast/osteoclast system *(35)*.

2.6. Mineralization of Bone

On a rescue diet, mineralization of bone normalized, and osteoid accumulation returned to wild-type levels in all models studied, i.e., *1α(OH)ase$^{-/-}$* mice, *VDR$^{-/-}$* mice, and *1α(OH)ase$^{-/-}$VDR$^{-/-}$* double mutants *(25, 26, 35)*. Consequently mineralization of bone reflects ambient calcium and phosphate levels rather than the direct participation of the 1,25(OH)$_2$D/VDR system.

2.7. 1,25 Dihydroxyvitamin D as a Bone Anabolic Agent

2.7.1. 1ANABOLIC EFFECTS IN OLDER ANIMALS

In previous studies 1,25(OH)$_2$D has been shown to be a potent stimulator of osteoclastogenesis in vitro *(36)*, and we and others have demonstrated the capacity of high doses of 1,25(OH)$_2$D to exert an osteoclastogenic and bone resorbing effect in wild-type animals in vivo *(37)*. In our 4-month-old mutant models [*1α(OH)ase$^{-/-}$*, *VDR$^{-/-}$*, and *1α(OH)ase$^{-/-}$VDR$^{-/-}$*] however, when hypocalcemia and secondary hyperparathyroidism were prevented by the rescue diet, osteoblast numbers, mineral apposition rate, and bone volume were suppressed below levels seen in wild-type mice *(25)*. This suggested that the 1,25(OH)$_2$D/VDR system may exert a skeletal "anabolic" effect which is necessary to sustain basal bone forming activity and which is unmasked when the defective 1,25(OH)$_2$D/VDR system exists in the presence of normal PTH (Fig. 2). In view of the fact that this inhibition of bone formation was not previously observed in either *VDR$^{-/-}$* or mice on the rescue diet, we assumed that this may have reflected the older age of our mice at the time of analysis. Nevertheless, previous studies in other model systems had pointed to an anabolic effect of 1,25(OH)$_2$D *(38)*. Our studies therefore suggested that 1,25(OH)$_2$D may exert effects on bone turnover analogous to those of PTH.

2.7.2. ANABOLIC EFFECTS IN NEONATAL ANIMALS

We therefore next examined skeletal development in *1α(OH)ase$^{-/-}$* mice in the neonatal period and compared them to mice with targeted deletion of the PTH gene (*PTH$^{-/-}$* mice) and to double mutant *PTH$^{-/-}$1α(OH)ase$^{-/-}$* mice *(39)*. At 2 weeks of age, PTH$^{-/-}$ mice exhibited only minimal dysmorphic changes, whereas *1α(OH)ase$^{-/-}$* mice displayed epiphyseal dysgenesis which was most severe in the double mutants.

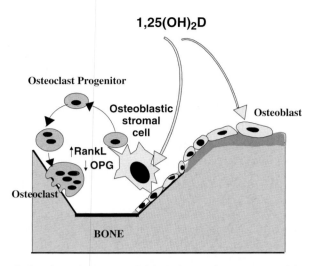

Fig. 2. Dual actions of vitamin D in bone. $1,25(OH)_2D$ can increase osteoblast differentiation and increase bone matrix production. $1,25(OH)_2D$ can also increase the release of RANKL and decrease the release of OPG from osteoblastic cells and stimulate osteoclastogenesis resulting in bone resorption.

Reduced osteoblastic bone formation was observed in both mutants. PTH deficiency caused a slight reduction in long bone length but a marked reduction in trabecular bone volume, whereas $1\alpha(OH)ase^{-/-}$ ablation caused a smaller reduction in trabecular bone volume but a significant decrease in bone length. These results therefore showed that $1\alpha(OH)ase^{-/-}$ mice are osteopenic as early as 2 weeks of age, and suggested that PTH plays a predominant role in appositional bone growth, whereas $1,25(OH)_2D$ acts predominantly, although not exclusively on endochondral bone formation.

2.7.3. ANABOLIC EFFECTS OF EXOGENOUS $1,25(OH)_2D$

To identify $1,25(OH)_2D$-mediated skeletal actions independent of PTH and to eliminate the potential confounding effects on the skeleton of endogenous PTH and $1,25(OH)_2D$, double homozygous $PTH^{-/-}1\alpha(OH)ase^{-/-}$ mice were treated with $1,25(OH)_2D$, from age 4 days to age 14 days and compared with vehicle-treated animals *(40)*. Parameters of endochondral bone formation, including long bone length, epiphyseal volume, chondrocyte proliferation and differentiation, and cartilage matrix mineralization, were all increased by $1,25(OH)_2D$. In addition exogenous $1,25(OH)_2D$ increased both trabecular and cortical bone; augmented both osteoblast number and type I collagen deposition in bone matrix; and upregulated expression levels of the osteoblastic genes alkaline phosphatase, type I collagen, and osteocalcin. Furthermore, in double mutants, osteoclastic bone resorption was not increased by $1,25(OH)_2D$ treatment. The results indicate that administered $1,25(OH)_2D$ can promote endochondral and appositional bone increases independent of endogenous PTH.

2.7.4. Interaction of Exogenous PTH and Endogenous 1,25(OH)$_2$D in Exerting a Bone Anabolic Effect

In view of the fact that PTH and 1,25(OH)$_2$D may each exert an anabolic effect on bone, we assessed the potential interaction of these agents in promoting skeletal anabolism by comparing the capacity of exogenous, intermittently injected PTH *(1–34)* to produce bone accrual in *1α(OH)ase$^{-/-}$* and in wild-type mice *(41)*. In initial studies, 3-month-old wild-type mice were either injected once daily or infused continuously with PTH *(1–34)* for up to 1 month. Infused PTH reduced bone mineral density (BMD), increased the bone resorption marker TRACP-5b, and raised serum calcium but did not increase serum 1,25(OH)$_2$D, presumably due to the suppressive effects of hypercalcemia on activity of the renal 1α(OH)ase. Injected PTH increased serum 1,25(OH)$_2$D and BMD, raised the bone formation marker osteocalcin more than did infused PTH, and did not produce sustained hypercalcemia as did PTH infusion. In subsequent studies, 3-month-old *1α(OH)ase$^{-/-}$* mice, raised on a rescue diet, and wild-type littermates were injected with PTH *(1–34)* either once daily or three times daily for 1 month. In *1α(OH)ase$^{-/-}$* mice on a rescue diet, baseline trabecular bone volume (BV/TV) and bone formation (BFR/BS) were lower than in wild-type mice as we had previously observed. PTH administered intermittently increased BV/TV and BFR/BS in a dose-dependent manner, but the increases were always less than in *1α(OH)ase$^{-/-}$* mice wild-type mice. These studies show that exogenous PTH administered continuously resorbs bone without raising endogenous 1,25(OH)$_2$D. Intermittently administered PTH can increase bone accrual in the absence of 1,25(OH)$_2$D, but 1,25(OH)$_2$D complements this action of PTH. An increase in endogenous 1,25(OH)$_2$D may therefore facilitate an optimal skeletal anabolic response to PTH and may be relevant to the development of improved therapeutics for enhancing skeletal anabolism.

These insights into the skeletal and mineral modulating actions of vitamin D based on analysis of carefully defined genetic mouse models may therefore provide a background for the development of newer vitamin D analogs which can minimize increases in intestinal calcium and phosphorus absorption and maximize beneficial effects on the skeleton for clinical use in metabolic bone disease.

3. ROLE OF 25-HYDROXYVITAMIN D$_3$-1α-HYDROXYLASE IN EXTRA-SKELETAL ACTIONS OF 1,25(OH)$_2$D AS DETERMINED BY GENETIC MOUSE MODELS

The widespread distribution of VDR action, determined from autoradiographic studies of receptor binding using radiolabeled calcitriol *(42)*, from the results of biochemical and immunohistochemical localization of VDR distribution *(43)*, as well as from the broad spectrum of vitamin D-dependent genes determined from gene chip arrays *(44)* all suggested that vitamin D might subserve important functions outside the realm of mineral and skeletal homeostasis. These functions include broad effects on differentiation and differentiated function in multiple organ systems. As a result, vitamin D action has been implicated in a number of critical processes including immune function, cancer development, and cardiovascular disease *(45)*. The identification of circulating

25(OH)D as a more valuable clinical biomarker than circulating 1,25(OH)$_2$D for the many and varied health-related effects in which vitamin D has been implicated *(45, 46)*, has lent support to the importance of extra-renal 1α(OH)ases *(14)* in the target cell production of 1,25(OH)$_2$D, as representing at least one mechanism to explain why circulating 25(OH)D, the substrate of these extra-renal 1α(OH)ases, better correlates with extra-skeletal health outcomes than does the circulating 1,25(OH)$_2$D, which is of renal origin.

Epidemiologic studies have supported and often advanced the cause for the broad importance of the extra-renal actions of vitamin D. One such study emphasized the decreased all-cause mortality (from 28.6 to 13.8% over 2 years) in hemodialysis patients supplemented with active vitamin D versus those not receiving vitamin D *(47)*. In view of the fact that the major cause of death in hemodialysis patients is due to cardiovascular death the study also showed a reduction in cardiovascular mortality from 14.6 to 7.6% over 2 years. End stage chronic kidney disease however represents an extreme case of deficiency of the active circulating metabolite of vitamin D. A less extreme scenario was evaluated in a meta-analysis of all-cause mortality in 18 randomized controlled trials in which vitamin D treatment was used, generally for osteoporosis and fracture prevention (in non-dialysis patients) *(48)*. Not unexpectedly this study demonstrated a more modest, but still significant mortality reduction of 7%.

The availability of *1α(OH)ase* and *VDR* gene knockout mice has facilitated examination in a carefully controlled fashion of the potential consequences of vitamin D deficiency on a number of extra-skeletal functions including immune function *(49)*, cancer development *(50)*, abnormal insulin secretion *(51)*, and blood pressure regulation *(52)*.

We examined the effect of the absence of 1,25(OH)$_2$D production on blood pressure regulation and cardiac structure and function in the *1α(OH)ase* null mouse and the consequences of repleting the animals with active vitamin D and other anti-hypertensives *(53)*. In the *1α(OH)ase*$^{-/-}$ mice on a normal diet, systolic blood pressure was increased, associated with an increase in both renal angiotensinogen and renal renin mRNA expression, as well as increased circulating renin, increased circulating angiotensin II, and increased aldosterone. Myocardial structure was abnormal with an increased heart to body ratio, increased myocyte diameter, increased interventricular septum thickness, increased left ventricular mass relative to body weight, and increased relative wall thickness. Myocardial function was reduced as evidenced by a reduction in percent fractional shortening and a reduction in the systolic ejection fraction. On a rescue diet, the hypocalcemia, hypophosphatemia, and elevated PTH levels, each of which could impact on the parameters measured and conceivably confound interpretation of the direct effects on vitamin D, were all normalized. In spite of this the elevated blood pressure, activation of the renin/angiotensin system, altered myocardial structure, and reduced myocardial function remained unchanged. The use of the *1α(OH)ase*$^{-/-}$ model facilitated treatment with 1,25(OH)$_2$D to assess effects on the cardiovascular phenotype. Such treatment was compared with the effects of the angiotensin-converting enzyme inhibitor captopril and with the angiotensin type 1 receptor antagonist losartan (Fig. 3). Each agent prevented the elevation in blood pressure, prevented the development of myocardial hypertrophy, and prevented a reduction in cardiac function. In this study cardiac renin and angiotensinogen were both elevated in untreated 1α(OH)ase$^{-/-}$ mice in addition to

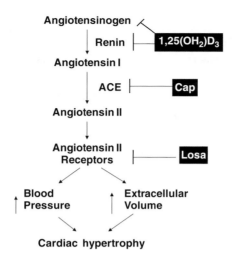

Fig. 3. Pharmacologic probing of the renin/angiotensin system. Different pharmacologic probes were used to influence the renin/angiotensin system in $1\alpha(OH)ase^{-/-}$ mice. These probes included $1,25(OH)_2D$, which can inhibit angiotensinogen gene expression and inhibit renin production, captopril (Cap) an inhibitor of angiotensin converting enzyme (ACE), and losartan (Losa), an antagonist of the type1 angiotensin II receptor.

the corresponding renal proteins. Furthermore VDR and $1\alpha(OH)$ase were both detected in whole heart of wild-type mice but the cardiac compartment in which these components of the vitamin D/VDR system was localized was not determined. As with many extra-skeletal functions of vitamin D it is uncertain if a local system of $1,25(OH)_2D$ production and action is the physiologically more relevant one, however in most cases, as in the example cited here, circulating $1,25(OH)_2D$ can access the tissue VDR and modify function.

In summary, although the actions of $1,25(OH)_2D$ on skeletal and mineral homeostasis are the most prominent, $1,25(OH)_2D$ actions may occur well beyond the skeletal system and may be pleiotropic and of broad importance. These pleiotropic effects of $1,25(OH)_2D$ may therefore be analogous to the pleiotropic effects of most nuclear receptor agonists.

4. ACKNOWLEDGMENTS

This work was supported by grants from the Canadian Institutes of Health Research to the author. The author thanks, Drs N Amizuka, D Panda, R Samadfam, Y Xue, GN Hendy, A Karaplis, and D Miao for their invaluable contributions to this work.

REFERENCES

1. Jones G, Strugnell SA, DeLuca HF (1998) Current understanding of the molecular actions of vitamin D. Physiol Rev 78:1193–1231
2. Ghazarian JG, Jefcoate CR, Knutson JC et al (1997) Mitochondrial cytochrome P-450. A component of chick kidney 25-hydroxycholecalciferol-1α–hydroxylase. J Biol Chem 249:3026–3033

3. St. Arnaud R, Messerlian S, Moir JM et al (1997) The 25-hydroxyvitamin D 1α-hydroxylase gene maps to the pseudovitamin D-deficiency rickets (PDDR) disease locus. J Bone Miner Res 12:1552–1559

4. Monkawa T, Yoshida T, Wakino S et al (1997) Molecular cloning of cDNA and genomic DNA for human 25-hydroxyvitamin D$_3$ 1α-hydroxylase. Biochem Biophys Res Commun 239:527–533

5. Takeyama K, Kitanaka S, Sato T et al (1997) 25-Hydroxyvitamin D$_3$ 1α-hydroxylase and vitamin D synthesis. Science 277:1827–1830

6. Panda DK, Al Kawas S, Seldin MF et al (2001) 25-hydroxyvitamin D 1alpha-hydroxylase: structure of the mouse gene, chromosomal assignment, and developmental expression. J Bone Miner Res 16:46–56

7. Brenza HL, DeLuca HF (2000) Regulation of 25-hydroxyvitamin D3 1alpha-hydroxylase gene expression by parathyroid hormone and 1,25-dihydroxyvitamin D3. Arch Biochem Biophys 381:143–152

8. Haussler MR, Whitfield GK, Haussler CA et al (1998) The nuclear vitamin D receptor: biological and molecular regulatory properties revealed. J Bone Miner Res 13:325–349

9. Bai X, Miao D, Li J et al (2004) Transgenic mice overexpressing human fibroblast growth factor 23 (R176Q) delineate a putative role for parathyroid hormone in renal phosphate wasting disorders. Endocrinology 145:5269–5279

10. Adams JS, Chen H, Chun R et al (2004) Response element binding proteins and intracellular vitamin D binding proteins: novel regulators of vitamin D trafficking, action and metabolism. J Steroid Biochem Mol Biol 89–90:461–465

11. Sutton AI, MacDonald PN (2003) Vitamin D: more than a "bone-a-fide" hormone. Molec Endocrinol 17:777–791

12. Jurutka PW, Whitfield GK, Hsieh J-C et al (2001) Molecular nature of the vitamin D receptor and its role in regulation of gene expression. Rev Endocr Metab Disord 2:203–216

13. Hewison M, Burke F, Evans KN et al (2007) Extra-renal 25-hydroxyvitamin D3–1 alpha-hydroxylase in human health and disease. J Steroid Biochem Mol Biol 103:316–321

14. Dusso AS, Kamimura S, Gallieni M et al (1997) γ gamma-Interferon-induced resistance to 1,25-(OH)2 D3 in human monocytes and macrophages: a mechanism for the hypercalcemia of various granulomatoses. J Clin Endocrinol Metab 82:2222–2232

15. Bikle DD, Chang S, Crumrine D et al (2004) 25 Hydroxyvitamin D 1 alpha-hydroxylase is required for optimal epidermal differentiation and permeability barrier homeostasis. J Invest Dermatol 122: 984–992

16. Glorieux FH, St-Arnaud R (1998) Molecular cloning of (25-OH D)-1 alpha-hydroxylase: an approach to the understanding of vitamin D pseudo-deficiency. Recent Prog Horm Res 53:341–349

17. Prosser DE, Jones G (2004) Enzymes involved in the activation and inactivation of vitamin D. Trends Biochem Sci 29:664–673

18. Koren R (2006) Vitamin D receptor defects: the story of hereditary resistance to vitamin D. Pediatr Endocrinol Rev 3 (Suppl 3):470–475, Review. Erratum in: (2007) Pediatr Endocrinol Rev 5:470

19. Panda DK, Miao D, Tremblay ML et al (2001) Targeted ablation of the 25-hydroxyvitamin D 1alpha-hydroxylase enzyme: evidence for skeletal, reproductive, and immune dysfunction. Proc Natl Acad Sci USA 98:7498–7503

20. Dardenne O, Prud'homme J, Arabian A et al (2001) Targeted inactivation of the 25-hydroxyvitamin D(3)-1(alpha)-hydroxylase gene (CYP27B1) creates an animal model of pseudovitamin D-deficiency rickets. Endocrinology 142:3135–3141

21. Yoshizawa T, Handa Y, Uematsu Y et al (1997) Mice lacking the vitamin D receptor exhibit impaired bone formation, uterine hypoplasia and growth retardation after weaning. Nat Genet 16:391–396

22. Li YC, Pirro AE, Amling M et al (1997) Targeted ablation of the vitamin D receptor: an animal model of vitamin D-dependent rickets type II with alopecia. Proc Natl Acad Sci USA 94:9831–9835

23. Van Cromphaut SJ, Dewerchin M, Hoenderop JG et al (2001) Duodenal calcium absorption in vitamin D receptor-knockout mice: functional and molecular aspects. Proc Natl Acad Sci USA 98:13324–13329

24. Erben RG, Soegiarto DW, Weber K et al (2002) Deletion of deoxyribonucleic acid binding domain of the vitamin D receptor abrogates genomic and nongenomic functions of vitamin D. Mol Endocrinol 16:1524–1537

25. Panda DK, Miao D, Bolivar I et al (2004) Inactivation of the 25-hydroxyvitamin D 1alpha-hydroxylase and vitamin D receptor demonstrates independent and interdependent effects of calcium and vitamin D on skeletal and mineral homeostasis. J Biol Chem 279:16754–16766

26. Dardenne O, Prud'homme J, Hacking SA et al (2003) Correction of the abnormal mineral ion homeostasis with a high-calcium, high-phosphorus, high-lactose diet rescues the PDDR phenotype of mice deficient for the 25-hydroxyvitamin D-1alpha-hydroxylase (CYP27B1.) Bone 32:332–340

27. Dardenne J, Prud'homme SA, Hacking FH et al (2003) Rescue of the pseudo-Vitamin D deficiency rickets phenotype of CYP27B1-deficient mice by treatment with 1,25-dihydroxyVitamin D3: biochemical, histomorphometric, and biomechanical analyses. J Bone Miner Res 18:637–643

28. Bouillon R, Okamura WH, Norman AW (1995) Structure – function relationships in the Vitamin D endocrine system. Endocr Rev 6:200–257

29. Brown EM, MacLeod RJ (2001) Extracellular calcium sensing and extracellular calcium signaling. Physiol Rev 81:239–297

30. Russell J, Silver J, Sherwood LM (1984) The effects of calcium and Vitamin D metabolites on cytoplasmic mRNA coding for pre-proparathyroid hormone in isolated parathyroid cells. Trans Assoc Am Physicians 97:296–303

31. Chan YL, McKay C, Dye E et al (1986) The effect of 1,25-dihydroxycholecalciferol on parathyroid hormone secretion by monolayer cultures of bovine parathyroid cells. Calcif Tissue Int 38:27–32

32. Kremer R, Bolivar I, Goltzman D et al (1989) Influence of calcium and 1,25-dihydroxycholecalciferol on proliferation and proto-oncogene expression in primary cultures of bovine parathyroid cells. Endocrinology 125:935–941

33. Yagishita N, Yamamoto Y, Yoshizawa T et al (2001) Aberrant growth plate development in VDR/RXR gamma double null mutant mice. Endocrinology 142:5332–5341

34. Boyan BD, Sylvia VL, Dean DD et al (1999) 1,25-(OH)2D3 modulates growth plate chondrocytes via membrane receptor-mediated protein kinase C by a mechanism that involves changes in phospholipid metabolism and the action of arachidonic acid and PGE2. Steroids 64:129–136

35. Amling M, Priemel M, Holzmann T et al (1999) Rescue of the skeletal phenotype of Vitamin D receptor-ablated mice in the setting of normal mineral ion homeostasis: formal histomorphometric and biomechanical analyses. Endocrinology 140:4982–4987

36. Takeda S, Yoshizawa T, T Y, Nagai Y et al (1999) Stimulation of osteoclast formation by 1,25-dihydroxyVitamin D requires its binding to Vitamin D receptor (VDR) in osteoblastic cells: studies using VDR knockout mice. Endocrinology 140:1005–1008

37. Lin R, Amizuka N, Sasaki T et al (2002) 1 Alpha 25-hydroxyvitamin D3 promotes vascular endothelial growth factor and matrix metalloproteinase 9. J Bone Miner Res 17:1604–1612

38. Van Leeuwen JP, van Driel M, van den Bemd GJ et al (2001) Vitamin D control of osteoblast function and bone extracellular matrix mineralization. Crit Rev Eukaryot Gene Expr 11:199–226

39. Xue Y, Karaplis AC, Hendy GN et al (2005) Genetic models show that parathyroid hormone and 1,25-dihydroxyvitamin D3 play distinct and synergistic roles in postnatal mineral ion homeostasis and skeletal development. Hum Mol Genet 14:1515–1528

40. Xue Y, Karaplis AC, Hendy GN et al (2006) Exogenous 1,25-dihydroxyvitamin D3 exerts a skeletal anabolic effect and improves mineral ion homeostasis in mice that are homozygous for both the 1alpha-hydroxylase and parathyroid hormone null alleles. Endocrinology 147:4801–4810

41. Samadfam R, Xia Q, Miao D et al (2008) Exogenous PTH and endogenous 1,25-dihydroxyvitamin D are complementary in inducing an anabolic effect on bone. J Bone Miner Res 23:1257–1266

42. Stumpf WE (1995) Vitamin D sites and mechanisms of action: a histochemical perspective. Reflections on the utility of autoradiography and cytopharmacology for drug targeting. Histochem Cell Biol 104:417–427

43. Pike JW (1991) Vitamin D_3 receptors: structure and function in transcription. Annu Rev Nutr 11:189–216

44. White JH (2004) Profiling 1,25-dihydroxyvitamin D_3-regulated gene expression by microarray analysis. J Steroid Biochem Mol Biol 89–90:239–244

45. Bouillon R, Bischoff-Ferrari H, Willett W (2008) Vitamin D and health: perspectives from mice and man. J Bone Miner Res 23:974–979

46. Jones G, Horst R, Carter G et al (2007) Contemporary diagnosis and treatment of vitamin D-related disorders. J Bone Miner Res 22(Suppl 2):V11–V47
47. Teng M, Wolf M, Ofsthun MN et al (2005) Activated injectable vitamin D and hemodialysis survival: a historical cohort study. J Am Soc Nephrol 16:1115–1125
48. Autier P, Gandini S (2007) Vitamin D supplementation and total mortality: a meta-analysis of randomized controlled trials. Arch Intern Med 167:1730–1737
49. Yu S, Cantorna MT (2008) The vitamin D receptor is required for iNKT cell development. Proc Natl Acad Sci U S A 105:5207–5212
50. Zinser GM, Suckow M, Welsh J (2005) Vitamin D receptor (VDR) ablation alters carcinogen-induced tumorigenesis in mammary gland, epidermis and lymphoid tissues. J Steroid Biochem Mol Biol 97:153–164
51. Zeitz U, Weber K, Soegiarto DW et al (2003) Impaired insulin secretory capacity in mice lacking a functional vitamin D receptor. FASEB J 17:509–511
52. Li YC, Kong J, Wei M et al (2002) 1,25-Dihydroxyvitamin D(3) is a negative endocrine regulator of the renin-angiotensin system. J Clin Invest 110:229–238
53. Zhou C, Lu F, Cao K et al (2008) Calcium-independent and 1,25(OH)2D3-dependent regulation of the renin-angiotensin system in 1alpha-hydroxylase knockout mice. Kidney Int 74:170–179

V SUNLIGHT, VITAMIN D AND CANCER

40 The Health Benefits of Solar Irradiance and Vitamin D and the Consequences of Their Deprivation

William B. Grant

Abstract Originally recognized for reducing the risk of rickets, the benefits of vitamin D were extended to bone health in general and, in the past three decades, to protection against many non-calcemic diseases. The primary source of vitamin D is production by solar ultraviolet-B (UVB) irradiance. Casual UVB irradiance in summer can generate 1,000–2,000 IU/day of vitamin D. Ecological studies of variations of disease end points as a function of geographic location or season of the year have often been the first studies to identify the effects of vitamin D. The advantages of ecological studies include large study populations, large variations in vitamin D production, and large variations in disease outcome. Other risk-modifying factors can generally be included in such studies. In this chapter, evidence that vitamin D reduces the risk of about 17 types of cancer, bacterial infections for diseases such as dental caries, periodontitis, septicemia, and tuberculosis, viral infections such as respiratory infections and Epstein–Barr virus, autoimmune diseases such as asthma and multiple sclerosis, cardiovascular diseases such as coronary heart disease, diabetes, and stroke, and dementia is reviewed. Several studies have also reported longer life expectancy with higher vitamin D supplement or serum 25-hydroxyvitamin D levels.

Key Words: Vitamin D; sunlight; UVB radiation; 25-hydroxyvitamin D; bone health; cancer; skin cancer; infectious disease; periodontal disease; autoimmune disease

1. INTRODUCTION

Life on Earth evolved in harmony with the Sun. By harnessing the Sun's energy, photosynthesis provides the energy-rich and structural molecules (as well as oxygen) on which most life on this planet depends *(1)*. For animals, the Sun provides light by which to see, radiation to heat the atmosphere and surface, and ultraviolet-B (UVB) radiation for production of vitamin D. Human skin pigmentation changed as man moved out of Africa to other solar UV climates *(2)*. For both plants and animals, the duration of

Disclosure. I receive funding from the UV Foundation (McLean, VA), the Vitamin D Society (Canada), the Sunlight Research Forum (Veldhoven), and Bio-Tech-Pharmacal (Fayetteville, AR).

From: *Nutrition and Health: Vitamin D*
Edited by: M.F. Holick, DOI 10.1007/978-1-60327-303-9_40,
© Springer Science+Business Media, LLC 2010

daylight and/or the spectral nature of sunlight provides information that helps regulate seasonal and diurnal activities. It would be hard to imagine a world without the Sun.

This chapter will address the scientific method as used to identify and quantify the health benefits of vitamin D as well as highlight some of the author's contributions to the field and thoughts about the directions the field is going.

2. SCIENTIFIC METHODS UTILIZED

In my view, the scientific methods are approaches that seek to systematize findings regarding relations between various factors, looking for relations or patterns that can be described, tested, and evaluated. In physics, it is easy to perform many types of experiments that can yield information leading to the development and testing of equations such as $F = ma$ (force = mass × acceleration) or $E = mc^2$ (energy = mass times the speed of light squared). In biology, it is often more difficult to do such experiments. Many of the important advances in biological science are, rather, made on the basis of observations. Charles Darwin, for example, did not perform a test tube experiment to develop the theory of evolution. Rather, he made careful observations of the flora and fauna during the voyage of the Beagle (3), leading, after much reflection, to *On the Origin of Species by Natural Selection (4)*. Much later the theory was confirmed and mutations in DNA found to be important in inducing changes.

In determining the health effects of vitamin D, the starting point is often with observational studies in the broad sense. I include ecological studies under this category along with case–control and cohort studies. Ecological studies were described by a colleague (Phil Jacklin, Saratoga, CA, private communication, 2007) as being triple-blind studies since neither the cases nor the controls know they are being studied. Use of observational studies involving solar UVB to determine the beneficial roles of vitamin D is facilitated by the fact that solar UVB has been and mostly continues to be the most important source of vitamin D for most people on Earth (5). An example of the use of the ecologic approach to deduce a beneficial effect of solar UVB in reducing the burden of disease is the paper by Cedric and Frank Garland (6) in which they hypothesized that the geographical pattern of colon cancer mortality rates in the United States, low in the Southwest, high in the Northeast, was due to vitamin D production by solar UVB. While ecological studies do not enjoy a good reputation among most health professionals, this low esteem is not deserved. One of the reasons for low esteem is that the classic paper (7) reporting that the most important risk factor for many types of cancers common in Western developed countries was animal fat. Unfortunately, case–control and cohort studies, such as the Harvard Nurses' Health Study, could not readily confirm the association between total fat and risk of breast cancer (8, 9). They realized that since diet in midlife did not seem to have much of an effect on breast cancer risk, diet in early life likely did (10). Recently, however, by using a younger cohort, they were able to make the confirmation of red meat as a risk factor (11). The primary risk factor for breast cancer is integrated lifetime estrogen exposure (12). Diets high in animal products increase the lifetime estrogen exposure through increased endogenous production (13).

Observational studies are often criticized for not being able to establish causality. In my opinion, that critique is not entirely justified. Koch (14) and Hill (15) developed the criteria that can be used to evaluate causality in the study of the environment and disease. Others have also discussed the application of these criteria *(16–18)*. The criteria, as laid down by Hill *(15)*, are as follows:

1. strength of association
2. consistency of findings
3. specificity between agent and effect
4. temporality (the cause must precede the effect)
5. biological gradient, preferably a linear dose–response relation
6. plausibility, based on the science of the day
7. coherence in terms of the natural history and biology of the disease
8. experiment, such as a randomized controlled trial (RCT)
9. analogy with other agent/diseases relationships

According to Hill, not all criteria need be satisfied, but the more that are, the more convincing the case is made. Several of the criteria do not apply to the study of UVB and vitamin D, such as specificity. The most difficult criterion to satisfy regarding UVB, vitamin D, and disease risk is that of experiment, although this problem should not be considered a major impediment to proceeding with revised vitamin D recommendations. There are a number of reasons for this difficulty. For one, vitamin D is so inexpensive ($10 for a year's supply at 3,000 IU/day; Bio-Tech Pharmacol, Inc., Fayetteville, AR) that there is little economic incentive to perform RCTs, especially when our disease-treatment system profits from treating, not preventing, disease. Such trials are expensive, due to the large number of people that have to be enrolled, and may take many years to conduct. Second, in earlier trials, doses of vitamin D used were too low to produce an observable effect *(19–21)*. This problem has been overcome in some recent trials that employed doses from 800 to 2,000 IU/day *(22–24)*. Third, the period of life when vitamin D is most effective may not be well known or there may not be studies of vitamin D supplementation on that age group. For example, solar UVB seems to reduce the risk of multiple sclerosis and prostate cancer early in life *(25, 26)*, and not many RCTs are conducted on children. Nonetheless, the health policy leaders want more RCTs to address a number of issues regarding increased vitamin D supplementation or food fortification at the population level *(27, 28)*.

Reasoning by analogy can be disadvantageous as well. One of the reasons given for the reluctance to accept vitamin D supplements as useful for preventing disease is the adverse effects for beta-carotene supplements with respect to risk of lung cancer among smokers *(29, 30)*.

So, what should one look for when trying to determine a possible link between vitamin D and disease risk reduction using observational studies? One is a variation of disease outcome (incidence, mortality, and/or prevalence rates and severity) with respect to an index of solar UVB dose *(6, 31)*. Of course confounding factors, such as dietary differences, should also be included in the analysis *(32, 33)*. A second is increased disease rates or severity in winter, such as done for epidemic influenza *(34)*. A third is differences in disease rates according to skin pigmentation for people living in the same region, such as done for tuberculosis *(35)*, cancer *(31, 36, 37)*, and respiratory syncytial virus *(38)*. A fourth is to look at serum 25-hydroxyvitamin D [25(OH)D] with respect to disease risk, as done recently for cardiovascular disease *(39–43)*. Interestingly, it is difficult to show using the ecologic approach that solar UVB is correlated with reduced risk of cardiovascular diseases due to the greater influences of diet and smoking on risk. Recent papers attempted to do that *(44, 45)*, but it does not appear that the effects of diet and smoking were properly considered. See, e.g., *(46, 47)* for some discussion of coronary heart disease rates in Europe and Grant *(32, 48, 49)* for ecologic studies that include national dietary supply as well as latitude as an index of solar UVB dose.

Vitamin D has been identified as beneficial in reducing the risk of many diseases. While there were a few reports of non-calcemic benefits of vitamin D in the first half of the twentieth century, such as for tuberculosis *(50)* and upper respiratory infections *(51)*, the investigation of the non-calcemic role of vitamin D did not start to receive serious attention until papers such as those by Cedric and Frank Garland on colon cancer *(6)* and on cardiovascular disease by Robert Scragg *(52)*. However, there was little progress on the vitamin D/non-calcemic disease paradigm until the 1990s, but in the first years of the twenty-first century interest has skyrocketed. Given the usual lag between scientific discovery and incorporation in health policy, it will probably be a few more years before there is widespread acceptance of the vitamin D/disease paradigm.

3. APPLICATION TO SPECIFIC HEALTH CONDITIONS

In the following sections, summaries of the understanding of the role of solar UVB and vitamin D in reducing the risk of various categories of diseases are presented.

3.1. Bone Health

It is assumed that all readers are familiar with the benefits of vitamin D in calcium absorption and metabolism related to bone health, so little will be said in this regard. However, I would like to draw readers' attention to a recent review of the history of vitamin D *(53)*. While the paper is largely in Portuguese, it has an interesting story of skull thickness among Persian and Egyptian soldiers written by Herodotus as well as other items of historical interest.

3.2. Cancer

The non-calcemic benefit of vitamin D that seems to be doing the most to drive the rapidly growing interest is reduction of cancer risk. First hypothesized by Garland and Garland *(6)*, the concept has now reached the stage of highly probable theory or nearly accepted principle. Several recent reviews have examined the evidence and concluded that the evidence, at least for several more common cancers in Western developed countries such as breast, colorectal, ovarian, and prostate cancer, is very strong *(54–59)*. The mechanisms are fairly well known and include effects on cells as well as on tumors such as anti-angiogenesis and anti-metastasis *(60)*.

There is good evidence that vitamin D reduces the risk of incidence and/or death of at least 17 types of cancer. The basis for this statement is that the mortality rate distributions of many types of cancer have similar patterns in the United States *(61)*. I showed that July solar UVB was inversely correlated with mortality rates for about 14 types of cancer in a linear regression analysis *(31)* and that the correlation remained for most of them in a multiple linear regression analysis including additional indices for alcohol consumption, ethnic background, smoking, socioeconomic status, and urban/rural residence *(33)*. Another multi-factorial regression analysis found that most of these cancers as well as a few others had both incidence and mortality rates inversely correlated with solar UVB *(62)*. The only back-tracking I found was that when an air pollution index was included in the analysis, one of the respiratory system cancers, laryngeal cancer, no longer had a statistically significant correlation with July UVB *(63)*.

Although indices of solar UVB dose are generally used in ecological studies of cancer incidence and mortality rates to provide evidence of a beneficial role of vitamin D, use of such indices has been challenged *(64)*. Such challenges seem to overlook the fact that summertime solar UVB in the United States, which has a much higher correlation with cancer rates than does latitude in general *(31, 33, 37)*, has an asymmetrical pattern *(65)*. The UVB doses in the Western United States in July are similar to those in the Eastern United States about 600–1,000 km to the south due to two reasons: the surface elevation is generally higher in the West and the stratospheric ozone layer is thinner due to the prevailing westerly winds crossing the Rocky Mountains and pushing the tropopause higher in the West.

Nonetheless, other UVB indices should be sought. One that has proven useful is development of non-melanoma skin cancer (NMSC), especially squamous cell carcinoma. The primary risk for squamous cell carcinoma is integrated lifetime solar UVB irradiance *(66)*, while for basal cell carcinoma, lifetime solar UVB and UVA and sunburning appear to be important. For melanoma, solar UVA is much more important than UVB *(67, 68)*. In a meta-analysis of second cancer after development of NMSC, I found significant reduced risk of several types of cancer *(69)*. I used lung cancer incidence rates in the study populations to reduce the effect of smoking on the relation, as smoking increases the risk of NMSC. Studies in The Netherlands found NMSC incidence rates inversely correlated with breast cancer *(70)* and prostate cancer *(71)*. A study of second cancers after diagnosis of NMSC in sunny countries found reduced risks for a number of cancers *(72)*. I pointed out that there was also information regarding smoking in that study *(73)*.

I also did an ecologic study of cancer mortality rates in Spain using NMSC mortality rates as the index of solar UVB irradiance *(74)*. Philippe Autier challenged the findings, pointing out that use of this index was not wise since NMSC deaths are uncommon. In checking the data in the *Atlas (75)*, I found that for males during the 15-year period, 1978–1992, there was a median value of about 60 deaths, while for females, the median value was about 50 deaths. For 60 cases, the 2-sigma uncertainty is 26%, while for 50, it is 28%.

Another problem with the analysis was that only single-factor regression results were presented. In an earlier work *(69)*, I had established that it was necessary to account for the effect of smoking since smoking increases the risk of both basal cell carcinoma and squamous cell carcinoma as well as many other types of cancer. Lung cancer incidence or mortality rates are convenient indices to use for the health effects of smoking *(33, 76)*.

Table 1
Multiple Linear Regression Results for Cancers in the Continental Provinces in Spain for
Males for Which Latitude and/or NMSC Had a Significant Correlation *(74)*

Cancer	Lung	Latitude × latitude	NMSC	Adjusted R^2, F, p
Brain	0.56,*	0.50,*	–	0.33, 12,*
Colon	0.53,*	–	−0.31, 0.01	0.34, 13,*
Esophageal	0.56,*	–	−0.31, 0.01	0.38, 15,*
Gallbladder	–	–	−0.31, 0.03	0.08, 4.9, 0.03
Gastric	–	0.41, 0.03	–	0.15, 9.5, 0.003
Lung	–	−0.36, 0.01	–	0.11, 5.9, 0.01
	–	–	0.02, 0.89	−0.02, 0.02, 0.89
NMSC	–	−0.50,*	–	0.24, 16,*
Pancreatic	–	0.55,*	–	0.28, 19,*
Pleural	0.38, 0.004	–	−0.40, 0.002	0.27, 9.5,*
Rectal	–	0.61,*	–	0.36, 28,*
Thyroid	0.38, 0.01	0.46, 0.002	–	0.19, 6.5, 0.003

* $p < 0.001$.

Thus, I decided to rerun the analysis using multiple linear regressions. The results are given in Tables 1 and 2. The effect of confounding was checked, and results that appeared to be strongly influenced by confounding were omitted. Only the best regression results for each cancer are presented in the tables. A summary of the findings is given in Table 1. A total of 21 cancers were found associated with the index for smoking. Fifteen cancers were found to be either inversely correlated with NMSC or correlated with increasing latitude. In Grant's *(69)* study, 17 cancers were reported inversely correlated with NMSC at a significant level. Bladder and uterine corpus are now removed from the list of UVB/vitamin D-sensitive cancers for Spain from this study.

It should be noted that use of NMSC incidence rates may be better than that of mortality rates if data for individuals are used. That was the basis of Grant's *(77)* and Tuohimaa et al.'s *(72)* study and discussed by Grant *(73)*.

There are about 17 types of cancer for which there is good evidence of a protective role by solar UVB and vitamin D: aerodigestive tract: esophageal, gastric, colon, rectal, gallbladder, pancreatic, and lung cancer; female cancers: breast, corpus uteri, and ovarian; urogenital tract: prostate, bladder, and renal cancer; thyroid cancer; and hematopoietic cancers: Hodgkin's lymphoma, non-Hodgkin's lymphoma, and multiple myeloma (see Table 1).

There are three cancers for which higher levels of serum 25(OH)D have been found correlated with increased risk of cancer: prostate cancer in Nordic countries *(78)* and aggressive prostate cancer in the United States *(79)*, pancreatic cancer in Finland *(80)*, and esophageal cancer in China *(81, 82)*. While it would take a separate paper to review these findings in detail, a few comments can be offered. As for prostate cancer, most studies of prediagnostic serum found no correlation with prostate cancer incidence. However, the ecological studies show a pronounced increase in prostate cancer mortality rates with increasing latitude *(31, 33, 83)*. My investigation of the geographic

Table 2
Multiple Linear Regression Results for Cancers in the Continental Provinces in Spain for
Females for Which Latitude and/or NMSC Had a Significant Correlation (74)

Cancer	Lung	Latitude × latitude	NMSC	Adjusted R^2, F, p
Brain	—	0.48,*	−0.30, 0.02	0.40, 16,*
Breast	0.30, 0.02	—	−0.49,*	0.39, 16,*
Gastric	—	0.46, 0.001	—	0.19, 12, 0.001
Hodgkin's lymphoma	0.30, 0.003	—	−0.38, 0.006	0.27, 9.6,*
Lung	—	—	−0.31, 0.03	0.08, 4.8, 0.03
	—	−0.05, 0.76	—	−0.02, 0.1, 0.76
Melanoma	—	—	−0.43, 0.002	0.17, 10, 0.002
Multiple myeloma	—	—	−0.45, 0.001	0.18, 12, 0.001
NHL	0.23, 0.06	—	−0.56,*	0.42, 18,*
NMSC	—	−0.33, 0.02	—	0.09, 5.6, 0.02
Ovarian	0.35, 0.008	—	−0.41, 0.002	0.35, 13,*
	0.48,*	0.23, 0.08	—	0.24, 8.5,*
Pancreatic	0.29, 0.03	0.42, 0.002	—	0.21, 7.4, 0.002
Pleural	—	—	−0.41, 0.005	0.15, 8.8, 0.005
Rectal	—	0.23, 0.10	−0.27, 0.06	0.14, 4.7, 0.01
Thyroid	—	0.48, 0.001	—	0.21, 14, 0.001

* $p < 0.001$.

variation of prostate cancer came up with the hypothesis that ethnic background plays an important role in the etiology of prostate cancer. The impetus for this hypothesis was seeing that the map of greatest ancestry by county in the United States for 2000 (84) had many features in common with prostate cancer mortality rates for the periods 1950-69 and 1970-94 (85). A factor linked to ancestry is prevalence of apolipoprotein E epsilon4 (ApoE4). In a multi-country ecological study, prevalence of ApoE4 was correlated with prostate cancer incidence and mortality rates (61). Support for this hypothesis is that ApoE4 increases cholesterol production, and cholesterol is a risk factor for prostate cancer (86). As for pancreatic cancer, it could be that a dietary source of vitamin D in Finland is a risk factor for pancreatic cancer. For esophageal cancer in China, I suspect viral infection and immunosuppression by solar UV, despite claims by the authors to the contrary (87).

3.3. Infectious Diseases

The role of vitamin D in strengthening the innate immune system to fight bacterial and viral infections is the subject of rapidly expanding interest. In the early twentieth century, those with tuberculosis were often sent to sanatoria where solar irradiance was a key part of the therapy (50). It is now known that 1,25-dihydroxyvitamin D induces production of human cathelicidin, LL-37, that effectively fights *Mycobacterium tuberculosis (88–90)*. "LL-37 has modest direct antimicrobial activity under physiological conditions, but has been demonstrated to have potent antiendotoxin activity in animal models, as well as the ability to resolve certain bacterial infections" *(91)*. Most of the studies of LL-37 and the innate immune system have looked at bacterial infections *(92, 93)*. Several bacterial and viral infections that LL-37 seems to combat are discussed here briefly.

My introduction to the role of vitamin D in the innate immune system came with participation on John Cannell's paper hypothesizing that epidemic influenza was seasonal due to the annual variation in solar UVB *(34)*. His hypothesis was quickly supported by a post-hoc analysis of a RCT of vitamin D supplementation for post-menopausal African-American women living on Long Island, NY. Those taking 800 IU of vitamin D/day had a 60% reduction in incidence of colds or influenza compared to those taking a placebo, while those taking 2,000 IU/day had a 90% reduction *(23)*.

Septicemia is an infection of the blood generally caused by *Streptococcus pneumoniae (94)*. LL-37 has been identified as an anti-septicemia agent *(91)*. The epidemiology of septicemia in the United States is strongly suggestive of an important role of vitamin D and LL-37 in reducing the risk. Septicemia rates are highest in winter *(95)*, African Americans have higher risk than white Americans *(96–98)*, rates are highest in the northeast, lowest in the West *(95)*, and the comorbid diseases are largely ones for which vitamin D reduces the risk: cancer of the breast, female reproductive system, gastrointestinal tract, prostate, urinary tract, and lymphoma (based on comparisons between African-Americans and whites), congestive heart failure, and pulmonary infection *(97)*. I recently published a paper on the topic *(99)*.

Bacterial pneumonia is another vitamin D-sensitive infection. There is some evidence for a role of vitamin D reducing the risk of pneumonia. In Ethiopia, there was still a 13-fold higher incidence of rickets among children with pneumonia than among controls [odds ratio 13.37 (95% CI 8.08–24.22), $p < 0.001$] *(100)*. In India, acute lower respiratory tract infection was linked to low serum 25(OH)D$_3$ for 25(OH)D > 22.5 nmol/l (odds ratio 0.09; 95% CI 0.03–0.24; $p < 0.001$) *(101)*.

It was recently reported that death during the influenza pandemic of 1918–1919 was attributed to the influenza permitting colonizing strains of bacteria to produce highly lethal pneumonias *(102)*. The same was true of the 1957–1958 influenza pandemic *(103)*. I found fatality rates and fraction of cases affected by pneumonia for 12 cities or regions in the US influenza pandemic of 1918–1919 *(104)*. In my ecologic study, I used both July 1992 UVB from the Total Ozone Mapping Spectrometer *(65)* and latitude, an index of wintertime UVB and vitamin D. The July index gave slightly stronger correlation for case fatality rates, while the two gave nearly identical results for influenza complicated by pneumonia.

Periodontal disease (PD) is the condition where bacteria form a biofilm on the teeth where their metabolic products including acids leech calcium from the teeth and alveolar bone, resulting in tooth attachment loss and, if left unchecked, tooth loss. PD is also important in many countries. There is good evidence that low vitamin D is an important risk factor for PD *(105, 106)*. PD has been associated with a number of diseases. The most prominent of these are cardiovascular diseases, diabetes mellitus, several types of cancer, and osteoporosis. While PD is often found in conjunction with these diseases, it is not clear that having PD is a risk factor for these diseases or that there are underlying risk-modifying factors that contribute to the development of PD and the associated diseases. There is also an interesting paper reporting an inverse correlation of dental caries with respect to hours of sunlight in the United States. In a study of male adolescents aged 12–14 years living in rural or semirural regions, the findings were 4.86 caries/adolescent for <2,200 h of sunshine/year; 3.74 for 2,200–2,600 h; 3.25 for 2,600–3,000 h; and 2.91 for >3,000 *(107)*.

3.4. Autoimmune Diseases

Multiple sclerosis (MS) is an autoimmune disease generally linked to infection and exhibiting an increase in prevalence with increasing latitude. The Epstein–Barr virus (EBV) appears to play a key role in the risk of developing MS *(86, 108)*. The strong latitudinal gradient has been observed in Europe *(109)*, the United States *(110)*, and Australia *(111)*. The variation in the United States is strongly linked to latitude, my index of wintertime solar UVB, rather than July UVB *(5, 65)*. I interpret this as vitamin D being able to reduce viral infections such as EBV *(25, 26)* (see below).

Type 1 diabetes mellitus (T1DM) is another autoimmune disease *(112)*. However, the risk appears to be related to problems with development of the islet beta cells in the pancreas, possibly related to infection. A study in New Zealand found that the month of birth for children with T1DM showed a statistically significant peak ($p < 0.01$) in summer, whereas the disease onset had a significant peak in winter ($p < 0.01$) *(113)*. This finding is consistent with low serum 25(OH)D levels in winter, either in utero or early in life. It appears that $1,25(OH)_2D$ is able to inactivate nuclear factor (NF)-kappa B and counteract cytokine-induced Fas expression in human islets *(114)*.

It was found in Sweden that risk factors included weaning infants early, and introducing them early to cow's milk formula and late to gluten, as well as a high consumption of milk at the age of 1 year *(115)*.

Based on the finding that MS prevalence in the United States has a latitudinal dependence rather than an asymmetrical distribution related to summertime solar UVB *(5)*, that respiratory and many other viral infections are more frequent in winter *(34, 116)*, and that vitamin D reduces the risk of such viral infections *(23)*, I hypothesized that the most important role of vitamin D in reducing the risk of MS was through reducing the risk of EBV infection *(25, 26)*. Since prostate cancer has a similar latitudinal gradient in the United States *(25, 33, 61, 85)*, I extended the hypothesis to prostate cancer and several other cancers such as Hodgkin's lymphoma and non-Hodgkin's lymphoma that have a significant correlation with latitude in a multiple linear regression analysis *(26)*. In support, it has been reported that solar UVB irradiance in youth reduces the risk of prostate cancer *(117, 118)*.

Asthma is an autoimmune disease *(119, 120)*. Asthma is associated with proinflammatory cytokines *(121, 122)* and vitamin D reduces their expression *(123)*. There is a growing body of literature reporting that respiratory viral infections early in life are important risk factors for incidence and exacerbation *(124–126)*. Since vitamin D reduces the risk of respiratory viral diseases such as the common cold and influenza *(23)* and RSV *(116)*, it would be expected that higher serum 25(OH)D levels at the susceptible period would reduce the risk of developing asthma. Maternal intake of vitamin D during pregnancy reduces risk *(127, 128)*. There is an increase of asthma prevalence in 2006 *(129)* with respect to latitude, an index of wintertime solar ultraviolet-B (UVB) doses and vitamin D *(25)*. There is a pronounced seasonal variation in asthma hospitalization, with peaks in winter and troughs in summer in South Carolina *(130)* and California *(131)*, corresponding to annual variations in solar UVB and vitamin D *(132)*. African-Americans have the highest rates of asthma *(133)* and the lowest serum 25-hydroxyvitamin D levels *(37, 134)*. Obesity increases risk, perhaps through decreased immunological tolerance as a consequence of immunological changes induced by adipokines (e.g., leptin and adiponectin) and cytokines [e.g., interleukin 6 (IL6) and tumor necrosis factor alpha (TNFα)] secreted by white adipose tissue *(135)*.

3.5. Metabolic Diseases

Type 2 diabetes mellitus (T2DM) is characterized by impaired insulin secretion, peripheral insulin resistance, and elevated serum glucose levels *(136)*. Recognized risk factors include diets high in simple carbohydrates and gylcemic index but low in fiber *(137, 138)* as well as obesity *(139)*.

A cross-sectional survey in New Zealand found an inverse correlation between serum 25(OH)D and newly diagnosed T2DM *(140)*. A similar study in the United States found an inverse relation across quartiles of 25(OH)D with respect to T2DM in a dose-dependent pattern [OR 0.25 (95% CI 0.11–0.60)] for non-Hispanic whites *(141)*. In the Nurses' Health Study, a combined daily intake of >1,200 mg calcium and >800 IU vitamin D was associated with a 33% lower risk of type 2 diabetes with RR of 0.67 (0.49–0.90) compared with an intake of <600 mg and 400 IU calcium and vitamin D, respectively *(142)*. However, vitamin D alone was not correlated with reduced risk. A second study with that cohort found that for those participants with impaired fasting glucose (FPG) at baseline, those who took combined calcium–vitamin D supplements had a lower rise in FPG at 3 years compared with those on placebo [0.02 mmol/l (0.4 mg/dl) vs. 0.34 mmol/l (6.1 mg/dl), respectively, $p = 0.042$] *(143)*. A meta-analysis of observational studies found a consistent pattern of lower incidence and prevalence of T2DM for those with higher serum 25(OH)D or vitamin D intake *(144)*.

A recent study in Finland based on stored sera and a 22-year follow-up period found the relative odds between the highest and the lowest quartiles was 0.28 (95% confidence interval 0.10–0.81) in men and 1.14 (0.60–2.17) in women after adjustment for smoking, body mass index, physical activity, and education *(136)*. Men had higher serum 25(OH)D levels based on the fourth (highest) quartile: 75.1 nmol/l vs. 62.5 nmol/l. Since vitamin D is used first for calcemic functions, the benefit for men but not women might be due to the higher 25(OH)D levels for men. Also, women may consume more added sugar than men *(145)*, a factor that seems to be associated with increased risk of coronary heart disease for women below the age of 65 years *(146)*. The authors, however, caution that their findings may not apply to those living in sunnier locations.

The mechanisms whereby vitamin D and calcium reduce the risk of T2DM include increased insulin sensitivity *(147, 148)* and benefit for beta-cell function *(147, 149)*.

3.6. Cardiovascular Diseases

There have been several papers published in 2008 reporting on observational studies finding a higher rate of cardiovascular disease incidence or mortality rate for those with lower serum 25(OH)D *(37–39, 41)*. There are also reviews of links between vitamin D and reduced risk of cardiovascular risk factors *(150–152)*. The mechanisms whereby vitamin D reduces the risk of cardiovascular diseases likely include lowering blood pressure *(153)*, increasing insulin sensitivity *(148)*, improving pancreas function and insulin secretion through increased plasma calcium levels, which regulate insulin synthesis and secretion, and a direct action on pancreatic beta-cell function *(154)*, reducing the risk of type 2 diabetes *(132, 138, 140)* and infection *(93, 155)*, blunting the deleterious impact of advanced glycation end products on endothelial cells *(156)*, and reducing inflammatory biomarkers. A good review is found in Zittermann et al. *(44)*.

3.7. Alzheimer's Disease

There is evidence that vitamin D can also reduce the risk of cognitive impairment and Alzheimer's disease. Both have vascular risk factors *(157, 158)*. The evidence falls into three categories: direct measures of serum 25(OH)D, correlations between serum 25(OH)D with risk factors for AD, and effects on diseases often found to precede cognitive impairment or AD. Two vitamin D receptors are correlated with risk of AD. Among risk factors, serum 25(OH)D inversely correlated with smoking, higher body fat, darker-skinned races in the United States, inflammation, infectious diseases, and vascular risk factors. Among the diseases found to precede AD, serum 25(OH)D correlated with osteoporosis, depression, tooth loss, diabetes mellitus, and infectious diseases. Most of this evidence supports a role of vitamin D in either reducing serum 25(OH)D levels or reducing the development of risk factors or diseases that lead to AD. At this stage, this evidence is suggestive of a causal role of vitamin D in reducing the risk of AD. It is hoped that this hypothesis will lead to further studies to evaluate it.

3.8. Life Expectancy

A pair of recent papers reported that higher vitamin D supplementation or serum 25(OH)D is associated with lower mortality rates. In one, a meta-analysis of vitamin D supplement trials with a trial size-adjusted mean of 5.7 years, a trial size-adjusted mean daily vitamin D dose was 528 IU and the summary relative risk for mortality from any cause was 0.93 (95% confidence interval 0.87–0.99) *(159)*. In the other, using data from the Third National Health and Nutrition Examination Survey (NHANES III), during a median 8.7 years of follow-up, compared with the highest quartile, being in the lowest quartile (25[OH]D levels < 17.8 ng/ml) was associated with a 26% increased rate of all-cause mortality (mortality rate ratio 1.26; 95% CI 1.08–1.46) *(42)*. Another related study found higher serum 25(OH)D concentrations associated with longer leukocyte telomere length in women *(160)*. Longer telomere length is associated with longer life expectancy *(161)*.

3.9. Historical Review

Table 3 presents my understanding of the history of early identifications of vitamin D-sensitive diseases along with more recent strong findings or reviews.

Table 3
Identification of Vitamin D-Sensitive Diseases

Disease	Early mention (year)	Early mention (reference)	Recent reports
Rickets	1840	(164)	(53)
Tuberculosis	1923	(50)	(89)
Upper respiratory infections	1932	(51)	(116)
Dental caries	1939	(107)*	–
Osteopenia, osteoporosis, and osteomalacia	1955	(165)	(166)
Pre-eclampsia	1969	(167)	(168)
Cancer	1937, 1980	(6, 169*, 170*)	(24, 33, 54, 55)
Cardiovascular disease	1978	(171)	(39–41)
Epidemic influenza	1981	(172)*	(23, 34)
Congestive heart failure	1984	(173)	(174, 175)
Type 2 diabetes mellitus	1985	(176)	(136, 142–144)
Multiple sclerosis	1986	(177)	(178)
Pneumonia	1997	(100)	(179)
Hypertension	1997	(180)	(150)
Chronic obstructive pulmonary disease	1998	(181)	(182)
Schizophrenia, risk in utero	1991	(183)	(184)
Type 1 diabetes mellitus	1999	(22, 185)	(186)
Periodontal disease	2004	(105, 106)	(187)
Asthma	2004	(188)	(128)
Septicemia	2005	(189)	(91, 99)
Depression	2006	(190)	(191)
Alzheimer's disease	2008	(192)	–

*Identified sunlight or solar UVB as a risk reduction factor, but did not identify vitamin D as the active factor.

4. SUMMARY AND CONCLUSION

The understanding of the health benefits of ultraviolet-B irradiance and vitamin D is increasing at a very rapid pace. It is as if the genie has been let out of Pandora's box. Vitamin D has been labeled the nutrient of the decade (162). The American Medical Association recently adopted Resolution 425: Appropriate Supplementation of Vitamin D (163). "RESOLVED, That our American Medical Association urge the Food and Nutrition Board of the Institute of Medicine to re-examine the Daily Reference Intake Values for Vitamin D in light of new scientific findings (Directive to Take Action)." It is anticipated that public health policies in the United States and other countries will change in the near future in recognition of the non-calcemic benefits of vitamin D.

There are several keys to a long and healthy life expectancy, don't smoke, eat good nutritious food, mainly from plants, exercise regularly, keep weight in a healthy range, and obtain adequate vitamin D from UVB irradiance or oral intake. Of these prescriptions, maintaining adequate vitamin D is the easiest to achieve.

REFERENCES

1. Demmig-Adams B, Adams WW 3rd (2000) Harvesting sunlight safely. Nature 403(371): 373–374
2. Jablonski NG, Chaplin G (2000) The evolution of human skin coloration. J Hum Evol 39: 57–106
3. Darwin C (1909) The Voyage of the Beagle. P. F. Collier & Son, New York
4. Darwin C (1959) On the Origin of Species by Natural Selection. Murray, London
5. Grant WB, Holick MF (2005) Benefits and requirements of vitamin D for optimal health: a review. Altern Med Rev 10:94–111
6. Garland CF, Garland FC (1980) Do sunlight and vitamin D reduce the likelihood of colon cancer?. Int J Epidemiol 9:227–231
7. Armstrong B, Doll R (1975) Environmental factors and cancer incidence and mortality in different countries, with special reference to dietary practices. Int J Cancer 15:617–631
8. Willett WC, Hunter DJ, Stampfer MJ et al (1992) Dietary fat and fiber in relation to risk of breast cancer. An 8-year follow-up. JAMA 268:2037–2044
9. Hunter DJ, Spiegelman D, Adami HO et al (1996) Cohort studies of fat intake and the risk of breast cancer – a pooled analysis. N Engl J Med 334:356–361
10. Hunter DJ, Willett WC (1996) Nutrition and breast cancer. Cancer Causes Control 7:56–68
11. Cho E, Chen WY, Hunter DJ et al (2006) Red meat intake and risk of breast cancer among premenopausal women. Arch Intern Med 166:2253–2259
12. Eliassen AH, Missmer SA, Tworoger SS et al (2006) Endogenous steroid hormone concentrations and risk of breast cancer among premenopausal women. J Natl Cancer Inst 98:1406–1415
13. Carruba G, Granata OM, Pala V et al (2006) A traditional Mediterranean diet decreases endogenous estrogens in healthy postmenopausal women. Nutr Cancer 56:253–259
14. Koch R (1982) Classics in infectious diseases. The etiology of tuberculosis: Robert Koch. Berlin, Germany 1882. Rev Infect Dis 4:1270–1274
15. Hill AB (1965) The Environment and Disease: Association or Causation?. Proc R Soc Med 58: 295–300
16. Potischman N, Weed DL (1999) Causal criteria in nutritional epidemiology. Am J Clin Nutr 69(1309):S–1314S
17. Weed DL, Hursting SD (1998) Biologic plausibility in causal inference: current method and practice. Am J Epidemiol 147:415–425
18. Weed DL (2000) Interpreting epidemiological evidence: how meta-analysis and causal inference methods are related. Int J Epidemiol 29:387–390
19. Grant WB, Garland CF (2004) A critical review of studies on vitamin D in relation to colorectal cancer. Nutr Cancer 48:115–123
20. Jackson RD, LaCroix AZ, Gass M et al (2006) Calcium plus vitamin D supplementation and the risk of fractures. N Engl J Med 354:669–683
21. Wactawski-Wende J, Kotchen JM, Anderson GL et al (2006) Calcium plus vitamin D supplementation and the risk of colorectal cancer. N Engl J Med 354:684–696
22. Hypponen E, Laara E, Reunanen A et al (2001) Intake of vitamin D and risk of type 1 diabetes: a birth-cohort study. Lancet 358:1500–1503
23. Aloia JF, Li-Ng M (2007) Re: epidemic influenza and vitamin D. Epidemiol Infect 135:1095–1096, author reply 1097–1098
24. Lappe JM, Travers-Gustafson D, Davies KM et al (2007) Vitamin D and calcium supplementation reduces cancer risk: results of a randomized trial. Am J Clin Nutr 85:1586–1591

25. Grant WB (2008) Hypothesis – ultraviolet-B irradiance and vitamin D reduce the risk of viral infections and thus their sequelae, including autoimmune diseases and some cancers. Photochem Photobiol 84:356–365

26. Grant WB (2008) Response to Comments by Norval and Woods to my Hypothesis Regarding Vitamin D Viral Infections and their Sequelae. Photochem Photobiol 84:806–808

27. Brannon PM, Yetley EA, Bailey RL et al (2008) Summary of roundtable discussion on vitamin D research needs. Am J Clin Nutr 88:587S–592S

28. Davis CD (2008) Vitamin D and cancer: current dilemmas and future research needs. Am J Clin Nutr 88:565S–569S

29. Omenn GS, Goodman GE, Thornquist MD et al (1996) Effects of a combination of beta carotene and vitamin A on lung cancer and cardiovascular disease. N Engl J Med 334:1150–1155

30. Virtamo J, Pietinen P, Huttunen JK et al (2003) Incidence of cancer and mortality following alpha-tocopherol and beta-carotene supplementation: a postintervention follow-up. JAMA 290:476–485

31. Grant WB (2002) An estimate of premature cancer mortality in the U.S. due to inadequate doses of solar ultraviolet-B radiation. Cancer 94:1867–1875

32. Grant WB (2003) Ecologic studies of solar UV-B radiation and cancer mortality rates. Recent Results Cancer Res 164:371–377

33. Grant WB, Garland CF (2006) The association of solar ultraviolet B (UVB) with reducing risk of cancer: multifactorial ecologic analysis of geographic variation in age-adjusted cancer mortality rates. Anticancer Res 26:2687–2699

34. Cannell JJ, Vieth R, Umhau JC et al (2006) Epidemic influenza and vitamin D. Epidemiol Infect 134:1129–1140

35. Bakhshi SS, Hawker J, Ali S (1997) The epidemiology of tuberculosis by ethnic group in Birmingham and its implications for future trends in tuberculosis in the UK. Ethn Health 2:147–153

36. Giovannucci E, Liu Y, Willett WC (2006) Cancer incidence and mortality and vitamin D in black and white male health professionals. Cancer Epidemiol Biomarkers Prev 15:2467–2472

37. Grant WB (2006) Lower vitamin-D production from solar ultraviolet-B irradiance may explain some differences in cancer survival rates. J Natl Med Assoc 98:357–364

38. Grant WB (2008) Variations in vitamin D production could possibly explain the seasonality of childhood respiratory infections in Hawaii. Pediatr Infect Dis J 27:853

39. Wang TJ, Pencina MJ, Booth SL et al (2008) Vitamin D deficiency and risk of cardiovascular disease. Circulation 117:503–511

40. Dobnig H, Pilz S, Scharnagl H et al (2008) Independent association of low serum 25-hydroxyvitamin d and 1,25-dihydroxyvitamin d levels with all-cause and cardiovascular mortality. Arch Intern Med 168:1340–1349

41. Giovannucci E, Liu Y, Hollis BW et al (2008) 25-hydroxyvitamin D and risk of myocardial infarction in men: a prospective study. Arch Intern Med 168:1174–1180

42. Melamed ML, Michos ED, Post W et al (2008) 25-hydroxyvitamin D levels and the risk of mortality in the general population. Arch Intern Med 168:1629–1637

43. Pilz S, Dobnig H, Winklhofer-Roob B et al (2008) Low serum levels of 25-hydroxyvitamin D predict fatal cancer in patients referred to coronary angiography. Cancer Epidemiol Biomarkers Prev 17:1228–1233

44. Zittermann A, Schleithoff SS, Koerfer R (2005) Putting cardiovascular disease and vitamin D insufficiency into perspective. Br J Nutr 94:483–492

45. Wong A (2008) Incident solar radiation and coronary heart disease mortality rates in Europe. Eur J Epidemiol 23:609–614

46. Muller-Nordhorn J, Binting S, Roll S et al (2008) An update on regional variation in cardiovascular mortality within Europe. Eur Heart J 29:1316–1326

47. Zatonski W, Campos H, Willett W (2008) Rapid declines in coronary heart disease mortality in Eastern Europe are associated with increased consumption of oils rich in alpha-linolenic acid. Eur J Epidemiol 23:3–10

48. Grant WB (2002) An ecologic study of dietary and solar ultraviolet-B links to breast carcinoma mortality rates. Cancer 94:272–281

49. Grant WB (2006) The likely role of vitamin D from solar ultraviolet-B irradiance in increasing cancer survival. Anticancer Res 26:2605–2614

50. Rollier A (1923) Heliotherapy – with Special Consideration of Surgical Tuberculosis. Oxford Medical Publications, London

51. Hess AF (1932) Diet, nutrition, and infection. Acta Paediatrica 13:206–224

52. Scragg R (1981) Seasonality of cardiovascular disease mortality and the possible protective effect of ultra-violet radiation. Int J Epidemiol 10:337–341

53. Martins e Silva J (2007) Brief history of rickets and of the discovery of vitamin D. Acta Reumatol Port 32:205–229

54. Garland CF, Garland FC, Gorham ED et al (2006) The role of vitamin D in cancer prevention. Am J Public Health 96:252–261

55. Giovannucci E, Liu Y, Rimm EB et al (2006) Prospective study of predictors of vitamin D status and cancer incidence and mortality in men. J Natl Cancer Inst 98:451–459

56. Grant WB (2006) Epidemiology of disease risks in relation to vitamin D insufficiency. Prog Biophys Mol Biol 92:65–79

57. Kricker A, Armstrong B (2006) Does sunlight have a beneficial influence on certain cancers?. Prog Biophys Mol Biol 92:132–139

58. Garland CF, Grant WB, Mohr SB et al (2007) What is the dose-response relationship between vitamin D and cancer risk?. Nutr Rev 65:S91–S95

59. Giovannucci E (2007) Epidemiological evidence for vitamin D and colorectal cancer. J Bone Miner Res 22(Suppl 2):V81–V85

60. Ingraham BA, Bragdon B, Nohe A (2008) Molecular basis of the potential of vitamin D to prevent cancer. Curr Med Res Opin 24:139–149

61. Devesa SS, Grauman DJ, Blot WJ et al. Atlas of Cancer Mortality in the United States, 1950–1994. NIH Publication No 99-4564: National Institute of Health, 1999.

62. Boscoe FP, Schymura MJ (2006) Solar ultraviolet-B exposure and cancer incidence and mortality in the United States, 1993–2002. BMC Cancer 6:264

63. Grant WB (2009) Air pollution in relation to U.S. cancer mortality rates: an ecological study; likely role of carbonaceous aerosols and polycyclic aromatic hydrocarbons. Anticancer Res, 29(9): 3537–3545

64. Kimlin MG, Olds WJ, Moore MR (2007) Location and vitamin D synthesis: is the hypothesis validated by geophysical data?. J Photochem Photobiol B 86:234–239

65. Leffell DJ, Brash DE (1996) Sunlight and skin cancer. Sci Am 275(52–53):56–59

66. Armstrong BK, Kricker A (2001) The epidemiology of UV induced skin cancer. J Photochem Photobiol B 63:8–18

67. Garland CF, Garland FC, Gorham ED (1993) Rising trends in melanoma. An hypothesis concerning sunscreen effectiveness. Ann Epidemiol 3:103–110

68. Moan J, Porojnicu AC, Dahlback A et al (2008) Addressing the health benefits and risks, involving vitamin D or skin cancer, of increased sun exposure. Proc Natl Acad Sci U S A 105:668–673

69. Grant WB (2007) A meta-analysis of second cancers after a diagnosis of nonmelanoma skin cancer: additional evidence that solar ultraviolet-B irradiance reduces the risk of internal cancers. J Steroid Biochem Mol Biol 103:668–674

70. Soerjomataram I, Louwman WJ, Lemmens VE et al (2008) Are patients with skin cancer at lower risk of developing colorectal or breast cancer?. Am J Epidemiol 167:1421–1429

71. de Vries E, Soerjomataram I, Houterman S et al (2007) Decreased risk of prostate cancer after skin cancer diagnosis: a protective role of ultraviolet radiation?. Am J Epidemiol 165:966–972

72. Tuohimaa P, Pukkala E, Scelo G et al (2007) Does solar exposure, as indicated by the non-melanoma skin cancers, protect from solid cancers: vitamin D as a possible explanation. Eur J Cancer 43: 1701–1712

73. Grant WB (2008) The effect of solar UVB doses and vitamin D production, skin cancer action spectra, and smoking in explaining links between skin cancers and solid tumours. Eur J Cancer 44:12–15

74. Grant WB (2007) An ecologic study of cancer mortality rates in Spain with respect to indices of solar UVB irradiance and smoking. Int J Cancer 120:1123–1128

75. (1996) Atlas of Cancer Mortality and Other Causes of Death in Spain 1978–1992. Fundación Científica de la Asociación Española Contra el Cáncer Madrid. http://www2.uca.es/hospital/atlas92/www/Atlas92.html. Accessed 2 August 2006.
76. Leistikow B (2004) Lung cancer rates as an index of tobacco smoke exposures: validation against black male approximate non-lung cancer death rates, 1969–2000. Prev Med 38:511–515
77. Grant WB (2007) Does solar ultraviolet irradiation affect cancer mortality rates in China?. Asian Pac J Cancer Prev 8:236–242
78. Tuohimaa P, Tenkanen L, Ahonen M et al (2004) Both high and low levels of blood vitamin D are associated with a higher prostate cancer risk: a longitudinal, nested case-control study in the Nordic countries. Int J Cancer 108:104–108
79. Ahn J, Peters U, Albanes D et al (2008) Serum vitamin D concentration and prostate cancer risk: a nested case-control study. J Natl Cancer Inst 100:796–804
80. Stolzenberg-Solomon RZ, Vieth R, Azad A et al (2006) A prospective nested case-control study of vitamin D status and pancreatic cancer risk in male smokers. Cancer Res 66:10213–10219
81. Abnet CC, Chen W, Dawsey SM et al (2007) Serum 25(OH)-vitamin D concentration and risk of esophageal squamous dysplasia. Cancer Epidemiol Biomarkers Prev 16:1889–1893
82. Chen W, Dawsey SM, Qiao YL et al (2007) Prospective study of serum 25(OH)-vitamin D concentration and risk of oesophageal and gastric cancers. Br J Cancer 97:123–128
83. Hanchette CL, Schwartz GG (1992) Geographic patterns of prostate cancer mortality. Evidence for a protective effect of ultraviolet radiation. Cancer 70:2861–2869
84. Brittingham A, de la Cruz GP (2004) Ancestry 2000. Census 2000 Brief CK2BR-35. In: U. S. Dept. of Commerce CB, Editor, Washington, DC.
85. Grant WB (in press) An ecological study of dietary and genetic risk factors for prostate cancer: cholesterol as an important factor. Anticancer Res.
86. Iso H, Ikeda A, Inoue M, Sato S, Tsugane S (2009) Serum cholesterol levels in relation to the incidence of cancer: the JPHC Study Cohorts. Int J Cancer 125:2679–2686
87. Gao GF, Roth MJ, Wei WQ et al (2006) No association between HPV infection and the neoplastic progression of esophageal squamous cell carcinoma: result from a cross-sectional study in a high-risk region of China. Int J Cancer 119:1354–1359
88. Liu PT, Stenger S, Li H et al (2006) Toll-like receptor triggering of a vitamin D-mediated human antimicrobial response. Science 311:1770–1773
89. Liu PT, Stenger S, Tang DH et al (2007) Cutting edge: vitamin D-mediated human antimicrobial activity against Mycobacterium tuberculosis is dependent on the induction of cathelicidin. J Immunol 179:2060–2063
90. Liu PT, Modlin RL (2008) Human macrophage host defense against Mycobacterium tuberculosis. Curr Opin Immunol 20:371–376
91. Mookherjee N, Rehaume LM, Hancock RE (2007) Cathelicidins and functional analogues as anti-sepsis molecules. Expert Opin Ther Targets 11:993–1004
92. Bikle DD (2008) Vitamin D and the immune system: role in protection against bacterial infection. Curr Opin Nephrol Hypertens 17:348–352
93. White JH (2008) Vitamin d signaling, infectious diseases, and regulation of innate immunity. Infect Immun 76:3837–3843
94. Hsieh YC, Lee WS, Shao PL et al (2008) The transforming Streptococcus pneumoniae in the 21st century. Chang Gung Med J 31:117–124
95. Danai PA, Sinha S, Moss M et al (2007) Seasonal variation in the epidemiology of sepsis. Crit Care Med 35:410–415
96. Martin GS, Mannino DM, Eaton S et al (2003) The epidemiology of sepsis in the United States from 1979 to 2000. N Engl J Med 348:1546–1554
97. Danai PA, Moss M, Mannino DM et al (2006) The epidemiology of sepsis in patients with malignancy. Chest 129:1432–1440
98. Esper AM, Moss M, Lewis CA et al (2006) The role of infection and comorbidity: factors that influence disparities in sepsis. Crit Care Med 34:2576–2582
99. Grant WB (2009, in press) Solar ultraviolet-B irradiance and vitamin D may reduce the risk of septicemia. Dermato-Endocrinology 1:37–42

100. Muhe L, Lulseged S, Mason KE et al (1997) Case-control study of the role of nutritional rickets in the risk of developing pneumonia in Ethiopian children. Lancet 349:1801–1804

101. Wayse V, Yousafzai A, Mogale K et al (2004) Association of subclinical vitamin D deficiency with severe acute lower respiratory infection in Indian children under 5 year. Eur J Clin Nutr 58: 563–567

102. Brundage JF, Shanks GD (2008) Deaths from bacterial pneumonia during 1918–1919 influenza pandemic. Emerg Infect Dis 14:1193–1199

103. McDonald JC (1958) Asian influenza in great Britain 1957–9558. Proc R Soc Med 51:1016–1018

104. Britten RH (1932) The incidence of epidemic influenza, 1918–1919. Pub Health Rep 47: 303–339

105. Dietrich T, Joshipura KJ, Dawson-Hughes B et al (2004) Association between serum concentrations of 25-hydroxyvitamin D3 and periodontal disease in the US population. Am J Clin Nutr 80: 108–113

106. Dietrich T, Nunn M, Dawson-Hughes B et al (2005) Association between serum concentrations of 25-hydroxyvitamin D and gingival inflammation. Am J Clin Nutr 82:575–580

107. East BR (1939) Mean Annual Hours of Sunshine and the Incidence of Dental Caries. Am J Public Health Nations Health 29:777–780

108. DeLorenze GN, Munger KL, Lennette ET et al (2006) Epstein-Barr virus and multiple sclerosis: evidence of association from a prospective study with long-term follow-up. Arch Neurol 63:839–844

109. Kurtzke JF (1980) Geographic distribution of multiple sclerosis: An update with special reference to Europe and the Mediterranean region. Acta Neurol Scand 62:65–80

110. Wallin MT, Page WF, Kurtzke JF (2004) Multiple sclerosis in US veterans of the Vietnam era and later military service: race, sex, and geography. Ann Neurol 55:65–71

111. van der Mei IA, Ponsonby AL, Blizzard L et al (2001) Regional variation in multiple sclerosis prevalence in Australia and its association with ambient ultraviolet radiation. Neuroepidemiology 20: 168–174

112. Taplin CE, Barker JM (2008) Autoantibodies in type 1 diabetes. Autoimmunity 41:11–18

113. Willis JA, Scott RS, Darlow BA et al (2002) Seasonality of birth and onset of clinical disease in children and adolescents (0–19 years) with type 1 diabetes mellitus in Canterbury, New Zealand. J Pediatr Endocrinol Metab 15:645–647

114. Riachy R, Vandewalle B, Moerman E et al (2006) 1,25-Dihydroxyvitamin D3 protects human pancreatic islets against cytokine-induced apoptosis via down-regulation of the Fas receptor. Apoptosis 11:151–159

115. Wahlberg J, Vaarala O, Ludvigsson J (2006) Dietary risk factors for the emergence of type 1 diabetes-related autoantibodies in 21/2 year-old Swedish children. Br J Nutr 95:603–608

116. Yusuf S, Piedimonte G, Auais A et al (2007) The relationship of meteorological conditions to the epidemic activity of respiratory syncytial virus. Epidemiol Infect 135:1077–1090

117. Luscombe CJ, Fryer AA, French ME et al (2001) Exposure to ultraviolet radiation: association with susceptibility and age at presentation with prostate cancer. Lancet 358:641–642

118. John EM, Koo J, Schwartz GG (2007) Sun exposure and prostate cancer risk: evidence for a protective effect of early-life exposure. Cancer Epidemiol Biomarkers Prev 16:1283–1286

119. Shurin MR, Smolkin YS (2007) Immune-mediated diseases: where do we stand?. Adv Exp Med Biol 601:3–12

120. Becker A, Chan-Yeung M (2008) Primary asthma prevention: is it possible?. Curr Allergy Asthma Rep 8:255–261

121. Commins S, Steinke JW, Borish L (2008) The extended IL-10 superfamily: IL-10, IL-19, IL-20, IL-22, IL-24, IL-26, IL-28, and IL-29. J Allergy Clin Immunol 121:1108–1111

122. Hanania NA (2008) Targeting airway inflammation in asthma: current and future therapies. Chest 133:989–998

123. Banerjee A, Damera G, Bhandare R et al (2008) Vitamin D and glucocorticoids differentially modulate chemokine expression in human airway smooth muscle cells. Br J Pharmacol 155:84–92

124. Friedlander SL, Jackson DJ, Gangnon RE et al (2005) Viral infections, cytokine dysregulation and the origins of childhood asthma and allergic diseases. Pediatr Infect Dis J 24:S170–S176, discussion S174–175

125. Schleithoff SS, Zittermann A, Tenderich G et al (2006) Vitamin D supplementation improves cytokine profiles in patients with congestive heart failure: a double-blind, randomized, placebo-controlled trial. Am J Clin Nutr 83:754–759

126. Lemanske RF Jr., Jackson DJ, Gangnon RE et al (2005) Rhinovirus illnesses during infancy predict subsequent childhood wheezing. J Allergy Clin Immunol 116:571–577

127. Camargo CA Jr., Rifas-Shiman SL, Litonjua AA et al (2007) Maternal intake of vitamin D during pregnancy and risk of recurrent wheeze in children at 3 year of age. Am J Clin Nutr 85:788–795

128. Devereux G, Litonjua AA, Turner SW et al (2007) Maternal vitamin D intake during pregnancy and early childhood wheezing. Am J Clin Nutr 85:853–859

129. Kilmer G, Roberts H, Hughes E et al (2008) Surveillance of certain health behaviors and conditions among states and selected local areas – Behavioral Risk Factor Surveillance System (BRFSS), United States, 2006. MMWR Surveill Summ 57:1–188

130. Chen CH, Xirasagar S, Lin HC (2006) Seasonality in adult asthma admissions, air pollutant levels, and climate: a population-based study. J Asthma 43:287–292

131. Milet M, Tran S, Eatherton M et al (2007) The Burden of Asthma in California, A Surveillance Report. In: CDoH Services (ed). Environmental Health Investigations Branch, Richmond, CA

132. Hypponen E, Power C (2007) Hypovitaminosis D in British adults at age 45 year: nationwide cohort study of dietary and lifestyle predictors. Am J Clin Nutr 85:860–868

133. Meng YY, Babey SH, Hastert TA et al (2007) California's racial and ethnic minorities more adversely affected by asthma. Policy Brief UCLA Cent Health Policy Res PB2007-3:1–7

134. Nesby-O'Dell S, Scanlon KS, Cogswell ME et al (2002) Hypovitaminosis D prevalence and determinants among African American and white women of reproductive age: third National Health and Nutrition Examination Survey, 1988–1994. Am J Clin Nutr 76:187–192

135. Hersoug LG, Linneberg A (2007) The link between the epidemics of obesity and allergic diseases: does obesity induce decreased immune tolerance?. Allergy 62:1205–1213

136. Knekt P, Laaksonen M, Mattila C et al (2008) Serum vitamin D and subsequent occurrence of type 2 diabetes. Epidemiology 19:666–671

137. Schulze MB, Liu S, Rimm EB et al (2004) Glycemic index, glycemic load, and dietary fiber intake and incidence of type 2 diabetes in younger and middle-aged women. Am J Clin Nutr 80: 348–356

138. Montonen J, Jarvinen R, Knekt P et al (2007) Consumption of sweetened beverages and intakes of fructose and glucose predict type 2 diabetes occurrence. J Nutr 137:1447–1454

139. Freemantle N, Holmes J, Hockey A et al (2008) How strong is the association between abdominal obesity and the incidence of type 2 diabetes? Int J Clin Pract 62:1391–1396

140. Scragg R, Holdaway I, Singh V et al (1995) Serum 25-hydroxyvitamin D3 levels decreased in impaired glucose tolerance and diabetes mellitus. Diabetes Res Clin Pract 27:181–188

141. Scragg R, Sowers M, Bell C (2004) Serum 25-hydroxyvitamin D, diabetes, and ethnicity in the Third National Health and Nutrition Examination Survey. Diabetes Care 27:2813–2818

142. Pittas AG, Dawson-Hughes B, Li T et al (2006) Vitamin D and calcium intake in relation to type 2 diabetes in women. Diabetes Care 29:650–656

143. Pittas AG, Harris SS, Stark PC et al (2007) The effects of calcium and vitamin D supplementation on blood glucose and markers of inflammation in nondiabetic adults. Diabetes Care 30:980–986

144. Pittas AG, Lau J, Hu FB et al (2007) The role of vitamin D and calcium in type 2 diabetes. A systematic review and meta-analysis. J Clin Endocrinol Metab 92:2017–2029

145. Barker ME, Thompson KA, McClean SI (1996) Do type as eat differently? A comparison of men and women. Appetite 26:277–285

146. Grant WB (1998) Reassessing the role of sugar in the etiology of heart disease. J Orthomolec Med 13:95–104

147. Chiu KC, Chu A, Go VL et al (2004) Hypovitaminosis D is associated with insulin resistance and beta cell dysfunction. Am J Clin Nutr 79:820–825

148. Forouhi NG, Luan J, Cooper A et al (2008) Baseline serum 25-hydroxy vitamin d is predictive of future glycemic status and insulin resistance: the Medical Research Council Ely Prospective Study 1990–2000. Diabetes 57:2619–2625

149. Christakos S, Dhawan P, Benn B et al (2007) Vitamin D: molecular mechanism of action. Ann N Y Acad Sci 1116:340–348

150. Forman JP, Giovannucci E, Holmes MD et al (2007) Plasma 25-hydroxyvitamin D levels and risk of incident hypertension. Hypertension 49:1063–1069

151. Martins D, Wolf M, Pan D et al (2007) Prevalence of cardiovascular risk factors and the serum levels of 25-hydroxyvitamin D in the United States: data from the Third National Health and Nutrition Examination Survey. Arch Intern Med 167:1159–1165

152. Michos ED, Melamed ML (2008) Vitamin D and cardiovascular disease risk. Curr Opin Clin Nutr Metab Care 11:7–12

153. Li YC, Qiao G, Uskokovic M et al (2004) Vitamin D: a negative endocrine regulator of the renin-angiotensin system and blood pressure. J Steroid Biochem Mol Biol 89–90:387–392

154. Palomer X, Gonzalez-Clemente JM, Blanco-Vaca F et al (2008) Role of vitamin D in the pathogenesis of type 2 diabetes mellitus. Diabetes Obes Metab 10:185–197

155. Adams JS, Hewison M (2008) Unexpected actions of vitamin D: new perspectives on the regulation of innate and adaptive immunity. Nat Clin Pract Endocrinol Metab 4:80–90

156. Talmor Y, Golan E, Benchetrit S et al (2008) Calcitriol blunts the deleterious impact of advanced glycation end products on endothelial cells. Am J Physiol Renal Physiol 294:F1059–F1064

157. de la Torre JC (2002) Vascular basis of Alzheimer's pathogenesis. Ann N Y Acad Sci 977: 196–215

158. Cechetto DF, Hachinski V, Whitehead SN (2008) Vascular risk factors and Alzheimer's disease. Expert Rev Neurother 8:743–750

159. Autier P, Gandini S (2007) Vitamin D supplementation and total mortality: a meta-analysis of randomized controlled trials. Arch Intern Med 167:1730–1737

160. Richards JB, Valdes AM, Gardner JP et al (2007) Higher serum vitamin D concentrations are associated with longer leukocyte telomere length in women. Am J Clin Nutr 86:1420–1425

161. Farzaneh-Far R, Cawthon RM, Na B et al (2008) Prognostic value of leukocyte telomere length in patients with stable coronary artery disease: data from the Heart and Soul Study. Arterioscler Thromb Vasc Biol 28:1379–1384

162. Brody JE. An oldie vies for nutrient of the decade. New York Times. New York Feb. 19, 2008.

163. Pennsylvania Delegation American Medical Association House of Delegates. (2008) Resolution: 425 – Appropriate Supplementation of Vitamin D. http://www.ama-assn.org/ama1/pub/upload/mm/471/425.doc. Accessed 15 December 2008.

164. Sniadecki J, Sniadecki J (1909) On the cure for rickets (1840). Cited by Mozolowski W. Nature 143:121

165. De Seze S, Lichtwitz A, Hioco D et al (1955) [Modifications of calcium in urine caused by vitamin D in osteomalacia of supply and osteoporosis.]. Bull Mem Soc Med Hop Paris 71:961–966

166. Holick MF (2006) The role of vitamin D for bone health and fracture prevention. Curr Osteoporos Rep 4:96–102

167. Glatzel H (1969) [Nutrition of the pregnant and nursing mother]. Z Allgemeinmed 45:333–352

168. Bodnar LM, Catov JM, Simhan HN et al (2007) Maternal vitamin D deficiency increases the risk of preeclampsia. J Clin Endocrinol Metab 92:3517–3522

169. Peller S, Stephenson CS (1937) Skin irritation and cancer in the United states Navy. Am J Med Sci 194:326–333

170. Apperly FL (1941) The relation of solar radiation to cancer mortality in North America. Cancer Res 1:191–195

171. Lund B, Badskjaer J, Soerensen OH (1978) Vitamin D and ischaemic heart disease. Horm Metab Res 10:553–556

172. Hope-Simpson RE (1981) The role of season in the epidemiology of influenza. J Hyg (Lond 86: 35–47

173. McGonigle RJ, Fowler MB, Timmis AB et al (1984) Uremic cardiomyopathy: potential role of vitamin D and parathyroid hormone. Nephron 36:94–100

174. Zittermann A, Schleithoff SS, Gotting C et al (2008) Poor outcome in end-stage heart failure patients with low circulating calcitriol levels. Eur J Heart Fail 10:321–327

175. Pilz S, Marz W, Wellnitz B et al (2008) Association of vitamin D deficiency with heart failure and sudden cardiac death in a large cross-sectional study of patients referred for coronary angiography. J Clin Endocrinol Metab 93:3927–3935

176. Ishida H, Seino Y, Matsukura S et al (1985) Diabetic osteopenia and circulating levels of vitamin D metabolites in type 2 (noninsulin-dependent) diabetes. Metabolism 34:797–801

177. Goldberg P, Fleming MC, Picard EH (1986) Multiple sclerosis: decreased relapse rate through dietary supplementation with calcium, magnesium and vitamin D. Med Hypotheses 21:193–200

178. Munger KL, Levin LI, Hollis BW et al (2006) Serum 25-hydroxyvitamin D levels and risk of multiple sclerosis. JAMA 296:2832–2838

179. Roth DE, Jones AB, Prosser C et al (2008) Vitamin D receptor polymorphisms and the risk of acute lower respiratory tract infection in early childhood. J Infect Dis 197:676–680

180. Rostand SG (1997) Ultraviolet light may contribute to geographic and racial blood pressure differences. Hypertension 30:150–156

181. Schellenberg D, Pare PD, Weir TD et al (1998) Vitamin D binding protein variants and the risk of COPD. Am J Respir Crit Care Med 157:957–961

182. Black PN, Scragg R (2005) Relationship between serum 25-hydroxyvitamin d and pulmonary function in the third national health and nutrition examination survey. Chest 128:3792–3798

183. McGrath J (1999) Hypothesis: is low prenatal vitamin D a risk-modifying factor for schizophrenia?. Schizophr Res 40:173–177

184. Edwards MJ (2007) Hyperthermia in utero due to maternal influenza is an environmental risk factor for schizophrenia. Congenit Anom (Kyoto) 47:84–89

185. The EURODIAB Substudy 2 Study Group (1999) Vitamin D supplement in early childhood and risk for Type I (insulin-dependent) diabetes mellitus. Diabetologia 42:51–54

186. Zipitis CS, AkobengAK (2008) Vitamin D supplementation in early childhood and risk of type 1 diabetes: a systematic review and meta-analysis. Arch Dis Child 93:512–517

187. Park KS, Nam JH, Choi J (2006) The short vitamin D receptor is associated with increased risk for generalized aggressive periodontitis. J Clin Periodontol 33:524–528

188. Poon AH, Laprise C, Lemire M et al (2004) Association of vitamin D receptor genetic variants with susceptibility to asthma and atopy. Am J Respir Crit Care Med 170:967–973

189. Fukumoto K, Nagaoka I, Yamataka A et al (2005) Effect of antibacterial cathelicidin peptide CAP18/LL-37 on sepsis in neonatal rats. Pediatr Surg Int 21:20–24

190. Wilkins CH, Sheline YI, Roe CM et al (2006) Vitamin D deficiency is associated with low mood and worse cognitive performance in older adults. Am J Geriatr Psychiatry 14:1032–1040

191. Hoogendijk WJ, Lips P, Dik MG et al (2008) Depression is associated with decreased 25-hydroxyvitamin D and increased parathyroid hormone levels in older adults. Arch Gen Psychiatry 65:508–512

192. Grant WB (2009) Does vitamin D reduce the risk of dementia? J Alz Dis 17:151–159

41 Vitamin D Status, Solar Radiation and Cancer Prognosis

Johan Moan, Øyvind Sverre Bruland, Arne Dahlback, Asta Juzeniene, and Alina Carmen Porojnicu

Abstract   Vitamin D plays an important role in cancer prevention and prognosis. A good vitamin D status at the time of diagnosis and therapy start seems to improve survival. Due to the sun, the serum concentration of 25-hydroxyvitamin D [25(OH)D] is higher in summer than in winter. The vitamin D status in Norway and its impact on cancer prognosis was reviewed: 137,706 cases of breast, colon and prostate cancer and Hodgkin's lymphoma were analysed with respect to survival 3 years after diagnosis. The survival rates for these cancer forms were about 20% higher for summer diagnosis than for winter diagnosis. The effect was largest for the youngest patients. The concentration of 25(OH)D in serum was 49 ± 2 nmol/l in the winter and 66 ± 5 nmol/l in the summer, while the level of 1,25-dihydroxyvitamin D [$1,25(OH)_2D$] remained almost constant during the year, except for persons with high body mass index, for whom there was a similar seasonal variation as for 25(OH)D. It seems that 25(OH)D, rather than $1,25(OH)_2D$, may be the main metabolite influencing cancer survival. Vitamin D supplementation or, alternatively, exposure to ultraviolet radiation may be considered as adjuvants in cancer therapy.

Key Words: Sun; vitamin D; cancer prognosis

1. INTRODUCTION

The health effects of solar ultraviolet (UV) radiation are being debated worldwide. UV has both negative health effects, namely induction of skin cancer *(1)*, and beneficial health effects through generation of vitamin D *(2, 3)*. Solar UV is a recognized carcinogen, inducing more than half of all skin cancers and causing about 250 deaths per year in Norway *(4)*. Large sun-safety campaigns have been initiated, which seem to have had a significant impact, since the increasing trend of skin cancer incidence rates observed from 1960 is reversed for young persons after about 1990 *(5)*. However, in the same time period, i.e. after 1990, vitamin D deficiency became prevalent in many populations *(3)*. Since solar UV is a main source of vitamin D, this deficiency may, at least partly, be due to reduced sun exposure.

From: *Nutrition and Health: Vitamin D*
Edited by: M.F. Holick, DOI 10.1007/978-1-60327-303-9_41,
© Springer Science+Business Media, LLC 2010

One of the first epidemiological investigations on the association between vitamin D deficiency and cancer mortality was published by Garland et al. in 1980 *(6)*. The authors found a clear positive association between latitude and mortality from colon cancer in the United States. They hypothesized that this might be related to sun-induced vitamin D. One year later, the anti-cancer effect of vitamin D was demonstrated in vitro in a melanoma cell line *(7)*. After these early reports, the interest in the potential use of vitamin D as an anti-cancer agent has grown rapidly. A search in PubMed using the words "vitamin D" and "cancer" yielded over 4,500 hits describing various aspects of the vitamin D–cancer relationship, ranging from epidemiological studies to genomics and proteomics. These investigations have shown that vitamin D can protect against cancer induction and/or improve cancer prognosis *(8–12)*. In several countries the incidence rates and/or the mortality rates of breast cancer, colon cancer and prostate cancer are higher in the north than in the south *(13–20)*. It has been suggested that this is due to differences in concentrations of vitamin D or of one of its active metabolites in serum between populations in south and in north, mainly due to the effect of sun-induced production of vitamin D in skin.

There is yet no consensus on the optimal levels of vitamin D associated with reduced cancer risk. From evolutionary, epidemiological and experimental perspectives, the optimal level of vitamin D may be defined as the amount equivalent to what an adult can acquire through exposing the whole skin surface to summer sunshine *(21)*. A number of recent meta-analyses have evaluated the threshold for reducing cancer risk and point to levels above 80 nmol/l *(22–24)*. Based on this, a physiological intake of vitamin D for an adult might range up to 250 µg/day (i.e. 30 times the daily recommended dose in Norway) *(25)*. Alternatively, controlled summer sun or winter sunbed exposures in suberythemal doses may be used as efficient sources of vitamin D *(26–29)*.

In this chapter, we sum up our investigations where we have analysed the impact of vitamin D status at the time of diagnosis on survival of the three major internal cancer forms (prostate, breast and colon cancer) and on Hodgkin's lymphoma in Norway. We observed large seasonal variations in prognosis for all these cancer types investigated, despite the relatively constant incidence rates throughout the year. We further discuss the possible contribution of vitamin D from sun exposure and from food in explaining these findings.

2. SKIN SYNTHESIS AND ACTIVATION OF VITAMIN D

The vitamin D precursor 7-dehydrocholesterol (provitamin D) is present in skin and can be converted by UVB photons (ultraviolet B, 280–320 nm) to previtamin D, a thermodynamically unstable compound. Vitamin D is formed through thermal reactions from previtamin D in the skin *(30, 31)*. It is then transported in the blood, bound to vitamin D-binding protein and carried to the liver, where it is hydroxylated to 25-hydroxyvitamin D [25(OH)D]. 25(OH)D is transported to the kidneys and to other tissues where another hydroxyl group is added, whereby the hormone, 1,25 dihydroxyvitamin D [1,25(OH)$_2$D], is formed *(32)*. Even moderate exposures to sun or sunbeds contribute substantially to the vitamin D status. In fact, these contributions may under certain conditions be more important than the contributions from food and supplements *(29, 33)*.

Clinicians use 25(OH)D as the most reliable marker of vitamin D status since it is present in highest concentration in human serum and its synthesis is not tightly regulated. Thus, its concentration can be used to estimate the relative contribution of sun exposure and of vitamin D intake to the vitamin D status.

It is believed that photosynthesis of vitamin D in skin is responsible for up to 90% of the circulating levels of 25(OH)D *(34)*. Therefore, a seasonal variation of 25(OH)D in human serum is expected and has been found in most investigations *(35–48)*. Moreover, since Norway spans over a long distance from north to south, the annual UV dose from the sun being 50% higher in the south than in the north *(49)*, a latitude gradient in serum vitamin D is expected. We have in our work been looking for both seasonal and latitudinal variations.

3. CALCULATION OF SEASONAL AND LATITUDINAL VARIATIONS IN VITAMIN D STATUS

There are few studies comparing the vitamin D status of people living at different places in Norway. However, we believe that physical calculations for estimation of seasonal and latitudinal gradients are good alternatives.

The action spectrum for formation of previtamin D *(50)* and that for erythema, the so-called CIE reference spectrum *(51)*, are similar. Furthermore, they closely resemble the erythema spectrum measured in human skin by use of UV lasers *(52)*. The spectrum of melanoma induction measured in fish is shifted far into the UVA region compared with the two mentioned spectra *(53)*. We have earlier given a number of reasons for believing that the action spectrum for melanoma in humans also has a strong UVA component *(54)* and in the following we will use the action spectrum for melanoma measured in *Xiphophorus* by Setlow et al. *(53)*. The relevant part of the sun spectrum in Norway is well known. The same is true for its variation with time, latitude and ozone concentrations. Using these spectra, we were able to calculate daily, as well as latitudinal, variations of the exposures of erythema-forming, vitamin D-forming and melanoma-inducing UV radiation. Figure 1 shows the fluence rate of vitamin D-forming and melanoma-inducing UV radiation during a midsummer day in Oslo. In panel a of the figure, we have used a horizontal flat surface to represent the human skin (as done by practically all researchers until now) and in panel b, we have used a rotating, vertical cylinder. We have earlier argued for the latter choice *(5)* and, therefore, used it to determine the latitudinal variation of the annual fluences of vitamin D generating UVB and of melanoma-inducing UVA in Fig. 2.

The annual fluence of solar UV for vitamin D production is roughly 60% higher in Oslo (56°N) than in Tromsø (70°N). Knowledge of seasonal variations of the sun spectrum, including ozone variations, can be applied to calculate seasonal variations of vitamin D-producing solar radiation as we have done earlier *(49)*. At our latitudes there is no sun induction of vitamin D in skin in the winter (late October–early March). This is in agreement with our measurements of 25(OH)D in serum (see below).

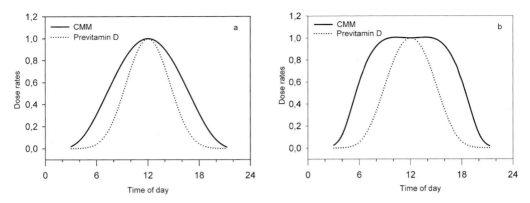

Fig. 1. The variation of UVB and UVA fluence rates over a midsummer day (June 21) in Oslo. In panel **a**, a horizontal planar surface was used as representation of human skin, while in panel **b** a vertical, rotating cylinder (excluding the top and the bottom circular disk) was used. UVB was calculated using the action spectrum of previtamin D formation *(81)* and UVA was calculated according to the Setlow melanoma spectrum *(53)*.

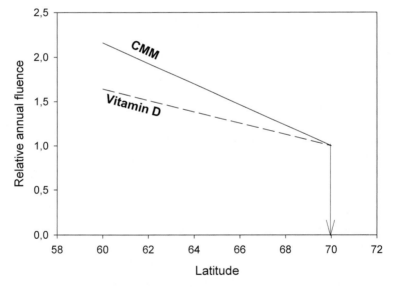

Fig. 2. Latitude variations of annual doses of vitamin D generating UV radiation (*dashed line*) and those of melanoma induction (*full line*).

4. SERUM MEASUREMENTS OF VITAMIN D METABOLITES

In the Nordic countries several investigations have been performed to elucidate the vitamin D status (Table 1) *(35–48)*. Such investigations can be used only for a rough comparison, and they seem to indicate that there is no large latitudinal gradient of the vitamin D status in the Nordic countries. However, practically all published investigations report seasonal variations in the range 10–100%. Such a wide range of seasonal amplitudes may be explained by differences in characteristics of the populations studied

Table 1

Seasonal Variation of Average Values of Serum 25(OH)D and 1,25(OH)$_2$D Measured in a Number of Nordic Investigations

Country (reference number)	Latitude (°N)	25(OH)D (nmol/l)		1,25(OH)$_2$D (pmol/l)	
		Winter	Summer	Winter	Summer
Norway (our data)	59	47	53	96	95
Norway (48)	70	53	81	–	–
Norway (36)	68	49.5	61	–	–
Norway (41)	61	42.7	62.2	–	–
Norway (45)	59	54.2	70.5	–	–
Finland (37)	63	54	89	–	–
Finland (47)	62	24	44	–	–
Finland (44)	62	36.3	82.5	–	–
Finland (39)	60	34	63	–	–
Sweden (40)	59	57.5	70	–	–
Sweden (38)	57	69.1	114	77	66.2
Denmark (35)	56	45.1	92.5	–	–
Denmark (42)	56	75	95	70	75
Denmark (43)	55	38	51	71	58
Denmark (46)	55	55.4	83	66.4	64.1

(age, gender, body mass impact on the vitamin D levels (55–57)) and by differences in the methods used for quantification of 25(OH)D.

The 1,25(OH)$_2$D level is almost constant throughout the year (Table 1), which is in agreement with previously published work (38, 43, 46, 58). However, in a recent work we have found a significant seasonal variation of 1,25(OH)$_2$D for people with a high BMI, almost as large as that of 25(OH)D.

The seasonal variation of 25(OH)D is explained by the photoproduction of vitamin D in skin exposed to solar UVB radiation. The efficiency spectrum of sun-induced vitamin D is obtained by convoluting the action spectrum of vitamin D photosynthesis (50) with the solar spectrum and lies in the wavelength region 300–310 nm. This spectrum is similar to the erythema efficiency spectrum as well as to the melanogenesis efficiency spectrum (59), which implies that sunscreens, properly applied, will almost eliminate the sun-induced vitamin D production, and that tanning beds are efficient sources of vitamin D. Both statements are supported by experimental data, also from our group (29, 60–62).

Whole-body single exposure to one minimal erythema dose (the UVB dose that induces a slight redness of the skin) gives a marked increase in the vitamin D concentration, but only a small increase in the 25(OH)D concentration in serum and an even smaller increase in the 1,25(OH)$_2$D concentration (32). This is due to regulatory mechanisms. Therefore, one may expect that the level of active vitamin D metabolites has a smaller seasonal variation than the level of 25(OH)D as measured in a number of investigations, including ours, while the level of vitamin D itself might have a larger seasonal variation. Furthermore, both previtamin D and vitamin D stay for some time, up to days, in the skin, and both are prone to UV photodegradation *(32)*. Therefore, the supply of 25(OH)D from UV exposure is likely to be delayed compared with the actual time of maximal UV exposure. This is supported by clinical studies that show a peak of the 25(OH)D levels in late summer/autumn *(49)*.

If we suppose, in a rough simplification, that there is linearity between the annual UV doses and their 25(OH)D contribution, and that there is a real seasonal variation of 25(OH)D equal to the average of the results reported in the Nordic measurements (Table 1), we can estimate that the annual average of 25(OH)D would be 10% larger in south Norway than in north Norway. The difference is in reality smaller for three reasons: (1) the 25(OH)D regulation plays a role (see above); (2) UV-induced breakdown of previtamin D and vitamin D may take place; and (3) a larger food intake of vitamin D from fish in north than in south may partly compensate for the low levels of solar radiation in the north. The Norwegian National Dietary Survey conducted in 1997 *(63)* shows that the vitamin D intake is 13% larger in north than in south Norway. Hence, our conclusion is that there is no significant north–south gradient in the level of vitamin D metabolites in serum in Norway.

5. EPIDEMIOLOGICAL INVESTIGATIONS OF RELATIVE RISK OF DEATH FROM BREAST, COLON AND PROSTATE CANCER

Norway was divided into three regions, based on latitude and incidence rates of skin cancer, as previously described *(64–67)*. Registration of the cancer population was done using the ID numbers (11 digits) which were introduced by Statistical Central Bureau in 1960 and brought into use in 1964. By help of this type of identification one has access to information regarding year of birth, living place, vital status, number of child births, occupation and education. The Norwegian Cancer Registry has listed all cancer diagnoses in Norway since 1953. In our work, 137,706 patients born in Norway during the period 1900–1966 and with the diagnosis of breast (49,821), colon (33,525), prostate cancer (46,205) and Hodgkin's lymphoma (3,139) during the period 1964–1992 were included. All deaths from cancer were registered *(68)*. Living region is listed as the place where the person lived at the time of cancer diagnosis.

The time points of diagnosis were grouped in seasons: winter (1st of December–February), spring (1st of March–May), summer (1st of June–August) and autumn (1st of September–November). The values for winter/spring were grouped together, as were values for summer/autumn, yielding then two large groups. The risk for death, RR, after a given cancer diagnosis is presented in relative units, and the value for winter/spring is set to unity (Table 2); 95% confidence intervals are given. Several factors that might influence the results have been taken into consideration. Such factors include stadium, age, birth cohort, period of diagnosis, sex, education, number of child birth and occupational exposure to solar radiation.

We found no significant seasonal variation in the number of new cases (data not shown), with number of diagnosed cases being $25 \pm 1.5\%$ per season. In contrast, the death rate was about 20% lower for the summer/autumn diagnosis than for the

Table 2

Seasonal Variation in 36 Months Survival from Breast, Colon and Prostate Cancer. Relative Risk (RR) of Death for Cases Diagnosed During Spring/Winter Was Set to 1.

Cancer type	Risk of death (RR) for summer and autumn/ winter and spring diagnosis	
	RR	95% CI
Breast cancer	0.75	0.72–0.79
Colon cancer	0.8	0.76–0.82
Prostate cancer	0.76	0.73–0.79
Hodgkin's lymphoma	0.84	0.73–0.97
Hodgkin's lymphoma (<30 years old)	0.52	0.42–0.95
Hodgkin's lymphoma (≥30 years old)	0.85	0.72–1.05

winter/spring diagnosis. The seasonal variation was largest for the youngest individuals, as exemplified here by Hodgkin's patients with ages below and above 30 years. The results are shown in Table 2, where the relative risk of death within 36 months after diagnosis is given. North–south gradients in cancer survival have been documented in several investigations, and have been related to sun-induced vitamin D production *(13–20)*. We found no such gradients, and possible reasons for this are discussed above.

We found a significant seasonal variation of the level of 25(OH)D. In agreement with this, we found significantly better survival for summer and autumn diagnosis than for winter and spring diagnosis (Table 2). Thus, our investigation supports the hypothesis that a good vitamin D status might improve cancer prognosis. We suggest that vitamin D and/or one of its metabolites, acts positively together with conventional treatments *(69–71)*. Furthermore, our work indicates that nontoxic vitamin D levels are large enough to give positive effects. Vitamin D seems to act in the early phase of treatment after diagnosis.

6. ANTI-NEOPLASTIC MECHANISMS OF ACTION

The inhibiting effect of vitamin D derivatives on replication of tumour cells in culture and on progression of tumours in vivo has been documented in many investigations *(72–74)*. Two mechanisms of action have been described: a genomic pathway by which derivatives of vitamin D influence the expression of cancer-related genes and a rapid pathway mediated by binding of the derivatives to cell membranes *(75, 76)*. However, until recently *(69)* vitamin D derivatives have not been brought into extensive use in cancer therapy because of the risk for hypercalcaemia *(76)*.

It has earlier been assumed that $1,25(OH)_2D$, and not $25(OH)D$, acts anti-carcinogenic, since it has been shown that $1,25(OH)_2D$ has the highest affinity for receptors *(77)*. However, the 500–1,000 times higher serum concentration of $25(OH)D$ *(78)* suggests that it may be active under certain conditions and in certain biochemical reactions. Moreover, while the level of $1,25(OH)_2D$ is strictly regulated by parathyroid hormone (PTH), the level of 25(OH)D is not regulated in this way. Therefore, changes of its level (like after solar exposure or vitamin D intake) can activate the vitamin D receptor.

7. SUMMARY

In conclusion, our work indicates that 25(OH)D, rather than $1,25(OH)_2D$, may influence cancer progression. This is in agreement with other investigations *(10, 79, 80)* and suggests that the action of 25(OH)D may have been underestimated so far. Moreover, our data indicate that relatively high levels of 25(OH)D, typical for summer/autumn, at the time of diagnosis and treatment start may improve the 3 years cancer survival by about 20% overall and even more for the youngest patients, in agreement with their higher efficiency of vitamin D photosynthesis.

Given the high probability of benefit for at least four major cancer types, clinical trials aimed at increasing 25(OH)D seem warranted. Investigators have various options to follow: testing of vitamin D compounds with low hypercalcaemic activity or evaluating the effect of moderate, controlled UV exposures. It should be remarked that sun lotions with high erythema protection factors may reduce vitamin D production *(60)* and, therefore, act negatively with respect to cancer survival, while moderate use of tanning beds may act positively.

Finally, it should be noted that noon is the optimal time to get vitamin D from the sun at a minimal melanoma risk. This is due to the difference between the action spectrum of vitamin D generation (mainly in UVB) and that of melanoma induction (shifted into UVA), and to the difference in atmospheric scattering and absorption between UVB and UVA. The statement is even more firmly founded if one assumes that a large fraction of human skin is vertically oriented when exposed to the sun. The health recommendation to avoid midday sun should be reevaluated.

REFERENCES

1. Armstrong BK, Kricker A, English DR (1997) Sun exposure and skin cancer. Australas J Dermatol 38(Suppl 1:S1–S6):S1–S6
2. Holick MF (2004) Vitamin D: importance in the prevention of cancers, type 1 diabetes, heart disease, and osteoporosis. Am J Clin Nutr 79:362–371
3. Holick MF (2007) Vitamin D deficiency. N Engl J Med 357:266–281
4. Cancer in Norway 2005 (2005) www.kreftregisteret.no . Accessed June 2008
5. Moan J, Porojnicu AC, Dahlback A et al (2008) Addressing the health benefits and risks, involving vitamin D or skin cancer, of increased sun exposure. Proc Natl Acad Sci USA 105:668–673
6. Garland CF, Garland FC (1980) Do sunlight and vitamin D reduce the likelihood of colon cancer? Int J Epidemiol 9:227–231
7. Colston K, Colston MJ, Feldman D (1981) 1,25-Dihydroxyvitamin D3 and malignant melanoma: the presence of receptors and inhibition of cell growth in culture. Endocrinology 108:1083–1086
8. Christakos S, Dhawan P, Shen Q et al (2006) New insights into the mechanisms involved in the pleiotropic actions of 1,25-dihydroxyvitamin D3. Ann N Y Acad Sci 1068(194–203):194–203
9. Garland CF, Garland FC, Gorham ED et al (2006) The role of vitamin D in cancer prevention. Am J Public Health 96:252–261
10. Giovannucci E (2005) The epidemiology of vitamin D and cancer incidence and mortality: a review (United States). Cancer Causes Control 16:83–95

11. Giovannucci E, Liu Y, Rimm EB et al (2006) Prospective study of predictors of vitamin D status and cancer incidence and mortality in men. J Natl Cancer Inst 98:451–459

12. Holick MF (2006) Vitamin D: its role in cancer prevention and treatment. Prog Biophys Mol Biol 92:49–59

13. Freedman DM, Dosemeci M, McGlynn K (2002) Sunlight and mortality from breast, ovarian, colon, prostate, and non-melanoma skin cancer: a composite death certificate based case–control study. Occup Environ Med 59:257–262

14. Garland FC, Garland CF, Gorham ED et al (1990) Geographic variation in breast cancer mortality in the United States: a hypothesis involving exposure to solar radiation. Prev Med 19:614–622

15. Gorham ED, Garland FC, Garland CF (1990) Sunlight and breast cancer incidence in the USSR. Int J Epidemiol 19:820–824

16. Grant WB (2002) An estimate of premature cancer mortality in the U.S. due to inadequate doses of solar ultraviolet-B radiation. Cancer 94:1867–1875

17. Grant WB (2002) An ecologic study of dietary and solar ultraviolet-B links to breast carcinoma mortality rates. Cancer 94:272–281

18. Hanchette CL, Schwartz GG (1992) Geographic patterns of prostate cancer mortality. Evidence for a protective effect of ultraviolet radiation. Cancer 70:2861–2869

19. John EM, Schwartz GG, Dreon DM et al (1999) Vitamin D and breast cancer risk: the NHANES I Epidemiologic follow-up study, 1971–1975 to 1992. National Health and Nutrition Examination Survey. Cancer Epidemiol Biomarkers Prev 8:399–406

20. Schwartz GG (2005) Vitamin D and the epidemiology of prostate cancer. Semin Dial 18:276–289

21. Vieth R (2004) Why the optimal requirement for Vitamin D3 is probably much higher than what is officially recommended for adults. J Steroid Biochem Mol Biol 89–90:575–579

22. Freedman DM, Looker AC, Chang SC et al (2007) Prospective study of serum vitamin D and cancer mortality in the United States. J Natl Cancer Inst 99:1594–1602

23. Garland CF, Gorham ED, Mohr SB et al (2007) Vitamin D and prevention of breast cancer: pooled analysis. J Steroid Biochem Mol Biol 103:708–711

24. Gorham ED, Garland CF, Garland FC et al (2007) Optimal vitamin D status for colorectal cancer prevention: a quantitative meta analysis. Am J Prev Med 32:210–216

25. Vieth R (2006) Critique of the considerations for establishing the tolerable upper intake level for vitamin D: critical need for revision upwards. J Nutr 136:1117–1122

26. Chandra P, Wolfenden LL, Ziegler TR et al (2007) Treatment of vitamin D deficiency with UV light in patients with malabsorption syndromes: a case series. Photodermatol Photoimmunol Photomed 23:179–185

27. Chel VG, Ooms ME, Popp-Snijders C et al (1998) Ultraviolet irradiation corrects vitamin D deficiency and suppresses secondary hyperparathyroidism in the elderly. J Bone Miner Res 13: 1238–1242

28. Davie MW, Lawson DE, Emberson C et al (1982) Vitamin D from skin: contribution to vitamin D status compared with oral vitamin D in normal and anticonvulsant-treated subjects. Clin Sci (Lond) 63:461–472

29. Tangpricha V, Turner A, Spina C et al (2004) Tanning is associated with optimal vitamin D status (serum 25-hydroxyvitamin D concentration) and higher bone mineral density. Am J Clin Nutr 80:1645–1649

30. Holick MF, MacLaughlin JA, Clark MB et al (1980) Photosynthesis of previtamin D3 in human skin and the physiologic consequences. Science 210:203–205

31. Holick MF, MacLaughlin JA, Doppelt SH (1981) Regulation of cutaneous previtamin D3 photosynthesis in man: skin pigment is not an essential regulator. Science 211:590–593

32. Holick MF, Vitamin D (1994) Photobiology, metabolism and clinical application. In: Arias IM, Boyer JL, Fausto N, Jakoby WB, Schachter D, Shafritz DA (eds) The Liver: Biology and Photobiology. Raven Press, New York, pp. 543–562

33. Vieth R (1999) Vitamin D supplementation, 25-hydroxyvitamin D concentrations, and safety. Am J Clin Nutr 69:842–856

34. Holick MF (2003) Vitamin D: a millenium perspective. J Cell Biochem 88:296–307

35. Brot C, Vestergaard P, Kolthoff N et al (2001) Vitamin D status and its adequacy in healthy Danish perimenopausal women: relationships to dietary intake, sun exposure and serum parathyroid hormone. Br J Nutr 86(Suppl 1):S97–S103
36. Brustad M, Alsaker E, Engelsen O et al (2004) Vitamin D status of middle-aged women at 65–71 degrees N in relation to dietary intake and exposure to ultraviolet radiation. Public Health Nutr 7: 327–335
37. Lamberg-Allardt C (1984) Vitamin D intake, sunlight exposure and 25-hydroxyvitamin D levels in the elderly during one year. Ann Nutr Metab 28:144–150
38. Landin-Wilhelmsen K, Wilhelmsen L, Wilske J et al (1995) Sunlight increases serum 25(OH) vitamin D concentration whereas 1,25(OH)2D3 is unaffected. Results from a general population study in Goteborg, Sweden (The WHO MONICA Project. Eur J Clin Nutr 49:400–407
39. Lehtonen-Veromaa M, Mottonen T, Irjala K et al (1999) Vitamin D intake is low and hypovitaminosis D common in healthy 9- to 15-year-old Finnish girls. Eur J Clin Nutr 53:746–751
40. Melin A, Wilske J, Ringertz H et al (2001) Seasonal variations in serum levels of 25-hydroxyvitamin D and parathyroid hormone but no detectable change in femoral neck bone density in an older population with regular outdoor exposure. J Am Geriatr Soc 49:1190–1196
41. Mowe M, Bohmer T, Haug E (1998) Vitamin D deficiency among hospitalized and home-bound elderly. Tidsskr Nor Laegeforen 118:3929–3931
42. Overgaard K, Nilas L, Johansen JS et al (1988) Lack of seasonal variation in bone mass and biochemical estimates of bone turnover. Bone 9:285–288
43. Rejnmark L, Jorgensen ME, Pedersen MB et al (2004) Vitamin D insufficiency in Greenlanders on a westernized fare: ethnic differences in calcitropic hormones between Greenlanders and Danes. Calcif Tissue Int 74:255–263
44. Savolainen K, Maenpaa PH, Alhava EM et al (1980) A seasonal difference in serum 25-hydroxyvitamin D3 in a Finnish population. Med Biol 58:49–52
45. Sem SW, Sjoen RJ, Trygg K et al (1987) Vitamin D status of two groups of elderly in Oslo: living in old people's homes and living in own homes. Compr Gerontol 1:126–130
46. Tjellesen L, Christiansen C (1983) Vitamin D metabolites in normal subjects during one year. A longitudinal study. Scand J Clin Lab Invest 43:85–89
47. Valimaki VV, Alfthan H, Lehmuskallio E et al (2004) Vitamin D status as a determinant of peak bone mass in young Finnish men. J Clin Endocrinol Metab 89:76–80
48. Vik T, Try K, Stromme JH (1980) The vitamin D status of man at 70 degrees north. Scand J Clin Lab Invest 40:227–232
49. Moan J, Porojnicu AC, Robsahm TE et al (2005) Solar radiation, vitamin D and survival rate of colon cancer in Norway. J Photochem Photobiol B 78:189–193
50. MacLaughlin JA, Anderson RR, Holick MF (1982) Spectral character of sunlight modulates photosynthesis of previtamin D3 and its photoisomers in human skin. Science 216: 1001–1003
51. McKinlay AF, Diffey BL (1987) A reference action spectrum for ultraviolet-induced erythema in human skin. In: Paschier DL, Bosnajakovic BF (eds) Human Exposure to Ultraviolet Radiation: Risks and Regulations. Elsevier, Amsterdam, pp. 83–87
52. Anders A, Altheide HJ, Knalmann M et al (1995) Action spectrum for erythema in humans investigated with dye lasers. Photochem Photobiol 61:200–205
53. Setlow RB, Grist E, Thompson K et al (1993) Wavelengths effective in induction of malignant melanoma. Proc Natl Acad Sci USA 90:6666–6670
54. Moan J, Dahlback A, Porojnicu AC (2008) At what time should one go out in the sun? Adv Exp Med Biol 624(86–88):86–88
55. MacLaughlin J, Holick MF (1985) Aging decreases the capacity of human skin to produce vitamin D3. J Clin Invest 76:1536–1538
56. Panidis D, Balaris C, Farmakiotis D et al (2005) Serum parathyroid hormone concentrations are increased in women with polycystic ovary syndrome. Clin Chem 51:1691–1697
57. Wortsman J, Matsuoka LY, Chen TC et al (2000) Decreased bioavailability of vitamin D in obesity. Am J Clin Nutr 72:690–693

58. Hine TJ, Roberts NB (1994) Seasonal variation in serum 25-hydroxy vitamin D3 does not affect 1,25-dihydroxy vitamin D. Ann Clin Biochem 31:31–34
59. Parrish JA, Jaenicke KF, Anderson RR (1982) Erythema and melanogenesis action spectra of normal human skin. Photochem Photobiol 36:187–191
60. Matsuoka LY, Ide L, Wortsman J et al (1987) Sunscreens suppress cutaneous vitamin D3 synthesis. J Clin Endocrinol Metab 64:1165–1168
61. Matsuoka LY, Wortsman J, Hanifan N et al (1988) Chronic sunscreen use decreases circulating concentrations of 25-hydroxyvitamin D. A preliminary study. Arch Dermatol 124:1802–1804
62. Porojnicu AC, Bruland OS, Aksnes L et al (2008) Sun beds and cod liver oil as vitamin D sources. J Photochem Photobiol B 29:125–131
63. Johansson L, Solvoll K, Norkost 1997 Norwegian National Dietary Survey, Statens råd for ernæring of fysisk aktivitet, Oslo, 1999, p. 45.
64. Lagunova Z, Porojnicu AC, Dahlback A et al (2007) Prostate cancer survival is dependent on season of diagnosis. Prostate 67:1362–1370
65. Porojnicu AC, Robsahm TE, Hansen Ree A et al (2005) Season of diagnosis is a prognostic factor in Hodgkin lymphoma. A possible role of sun-induced vitamin D. Br J Cancer 93:571–574
66. Porojnicu AC, Robsahm TE, Dahlback A et al (2007) Seasonal and geographical variations in lung cancer prognosis in Norway. Does vitamin D from the sun play a role? Lung Cancer 55:263–270
67. Porojnicu AC, Lagunova Z, Robsahm TE et al (2007) Changes in risk of death from breast cancer with season and latitude: sun exposure and breast cancer survival in Norway. Breast Cancer Res Treat 102:323–328
68. Robsahm TE, Tretli S, Dahlback A et al (2004) Vitamin D3 from sunlight may improve the prognosis of breast-, colon- and prostate cancer (Norway). Cancer Causes Control 15:149–158
69. Beer TM (2005) ASCENT: the androgen-independent prostate cancer study of calcitriol enhancing taxotere. BJU Int 96:508–513
70. Beer TM, Javle M, Lam GN et al (2005) Pharmacokinetics and tolerability of a single dose of DN-101, a new formulation of calcitriol, in patients with cancer. Clin Cancer Res 11:7794–7799
71. Pelczynska M, Switalska M, Maciejewska M et al (2006) Antiproliferative activity of vitamin D compounds in combination with cytostatics. Anticancer Res 26:2701–2705
72. Welsh J (2004) Vitamin D and breast cancer: insights from animal models. Am J Clin Nutr 80:1721S–1724S
73. Trump DL, Hershberger PA, Bernardi RJ et al (2004) Anti-tumor activity of calcitriol: pre-clinical and clinical studies. J Steroid Biochem Mol Biol 89–90:519–526
74. Lou YR, Qiao S, Talonpoika R et al (2004) The role of Vitamin D3 metabolism in prostate cancer. J Steroid Biochem Mol Biol 92:317–325
75. Lin R, White JH (2004) The pleiotropic actions of vitamin D. Bioessays 26:21–28
76. Mehta RG, Mehta RR (2002) Vitamin D and cancer. J Nutr Biochem 13:252–264
77. Colodro IH, Brickman AS, Coburn JW et al (1978) Effect of 25-hydroxy-vitamin D3 on intestinal absorption of calcium in normal man and patients with renal failure. Metabolism 27:745–753
78. DeLuca HF (1979) The vitamin D system in the regulation of calcium and phosphorus metabolism. Nutr Rev 37:161–193
79. Colston KW, Lowe LC, Mansi JL et al (2006) Vitamin D status and breast cancer risk. Anticancer Res 26:2573–2580
80. Tuohimaa P, Tenkanen L, Ahonen M et al (2004) Both high and low levels of blood vitamin D are associated with a higher prostate cancer risk: a longitudinal, nested case–control study in the Nordic countries. Int J Cancer 108:104–108
81. Galkin ON, Terenetskaya IP (1999) "Vitamin D" biodosimeter: basic characteristics and potential applications. J Photochem Photobiol B 53:12–19

42 The Epidemiology of Vitamin D and Cancer Risk

Edward Giovannucci

Abstract   The relation between vitamin D status and cancer risk has been investigated in a number of epidemiologic studies, while data from interventional studies remain scarce. The approaches to estimate vitamin D status have been varied and include direct measures of circulating 25-hydroxyvitamin D [25(OH)D] levels, surrogates or determinants of 25(OH)D, including region of residence, intake, and sun exposure estimates. In terms of cancer sites, the body of evidence is most extensive for colorectal cancer, for which support comes from studies of 25(OH)D, vitamin D intake, multiple predictors of 25(OH)D, and region of residence in a sunny climate. The evidence for breast cancer is also intriguing but prospective studies of 25(OH)D are sparse and somewhat conflicting. In one case–control study, retrospectively reported sun exposure during ages 10–19 was most strongly associated with reduced risk of breast cancer. For prostate cancer, the data on circulating 25(OH)D have been equivocal, suggesting no association or a weak inverse association, but studies tend to support a benefit of sun exposure on prostate cancer risk. It is plausible that for prostate cancer, vitamin D level much longer before the time of diagnosis is the most relevant exposure. Most of the epidemiologic studies to date have examined vitamin D status in relation to risk of cancer, but emerging evidence suggests that vitamin D may be an important factor for cancer progression and mortality, independently of any effects on incidence. Further study is needed to establish the precise role of vitamin D on carcinogenesis, especially in terms of when in the life span and on what stages of carcinogenesis vitamin D is relevant, the precise intakes and levels required, the magnitude of the association, and which cancer sites are most affected.

Key Words: Cancer; epidemiology; cancer risk; UVB radiation; solar; sunlight; colon cancer; 25-hydroxyvitamin D; dietary intake; prostate cancer

1. INTRODUCTION

In 1937, Peller hypothesized that sunlight exposure, by inducing skin cancer, could confer some secondary immunity on internal cancers *(1)*. Shortly thereafter in 1941, an association between latitude and cancer mortality was demonstrated by Apperly, who proposed a direct benefit of sunlight independent of any effect on skin cancer *(2)*. This potential anticancer property of sun exposure was largely ignored for about four decades, until the early 1980s when Garland and Garland hypothesized that poor vitamin D status from lower solar UVB radiation exposure explained the association

From: *Nutrition and Health: Vitamin D*
Edited by: M.F. Holick, DOI 10.1007/978-1-60327-303-9_42,
© Springer Science+Business Media, LLC 2010

between higher latitudes and increased mortality of a number of cancers, including colon cancer *(3)*, breast cancer *(4)*, and ovarian cancer *(5)*. Subsequently, a benefit of vitamin D was extended to prostate cancer *(6, 7)* and to other neoplasms *(8)*. Over the past several decades, laboratory studies have documented a number of anticarcinogenic properties of vitamin D, including inducing differentiation and inhibiting proliferation, invasiveness, angiogenesis, and metastatic potential.

This chapter will review the epidemiologic evidence from cohort and case–control studies of the association between vitamin D status and cancer risk. Various study designs have been utilized to assess individual exposure to vitamin D and to relate estimated vitamin D exposure to risk of a specific cancer or total cancer. These include studies directly measuring circulating levels of 25-hydroxyvitamin D [25(OH)D], the presumed relevant metabolite of vitamin D status, and surrogates or determinants of 25(OH)D level. The cancers that have been most thoroughly studied include colorectal and prostate cancer, while breast, pancreatic, and ovarian cancers have received some study. In addition, risk of total cancer has been assessed in several studies. The major strengths and limitations of the various approaches used are summarized in Table 1. Data on genetic polymorphisms on the vitamin D pathway (for example, the vitamin D receptor) may prove to be informative *(9)*, but because the data on functionality on known polymorphisms are yet unclear and these studies do not inform directly on vitamin D status, these will not be reviewed in this chapter.

2. COLORECTAL CANCER

2.1. 25(OH)D Level

Colorectal cancer was the first cancer type hypothesized to be related to vitamin D status *(3)*. Of note, in ecologic studies examining regional UVB exposure and mortality from various cancers, the magnitude of the association appeared to be strongest for this cancer *(8)*. Studies that have examined circulating 25(OH)D levels prospectively in relation to risk of colorectal cancer or adenoma, which is a precursor to presumably the majority of colorectal cancers, have generally supported an inverse association *(10–18)*. In a recent meta-analysis of the colorectal cancer studies, based on 535 cases, individuals with ≥82 nmol/l (33 ng/ml) serum 25(OH) level had 50% lower incidence of colorectal cancer (*P*<0.01) compared to those with relatively low levels of less than 30 nmol/l (12 ng/ml) *(19)*. The two largest studies included in the meta-analysis were from the Nurses' Health Study and the Women's Health Initiative. The Nurses' Health Study *(12)* was based on 193 incident cases of colorectal cancer. Two controls were matched per case on year of birth and month of blood draw. After adjusting for age, body mass index, physical activity, smoking, family history, use of hormone replacement therapy, aspirin use, and dietary intakes, the relative risk (RR) decreased monotonically across quintiles of plasma 25(OH)D concentration, with a RR of 0.53 (95% confidence interval (CI) = 0.27–1.04) for quintile 5 vs. 1. The median 25(OH)D concentration in quintile 5 was approximately 50 nmol/l (20 ng/ml) higher than that in quintile 1.

In the Women's Health Initiative, 322 total cases of colorectal cancer were documented. A similar inverse association was observed between baseline 25(OH)D level and colorectal cancer risk, though detailed analyses on potential confounders were not

Table 1

Summary of Major Strengths and Limitations in Overview of Study Designs in Studies of Vitamin D and Cancer

Circulating 25(OH)vitamin D

Strengths

- Direct measure of biologically relevant biomarker
- Accounts for skin exposure to UVB radiation, vitamin D intake, and for factors such as skin pigmentation
- Relatively long half-life ($t_{1/2}$) in the circulation of about 2–3 weeks

Limitations

- Typically only one measurement is made in studies
- The time may not reflect the etiologically relevant time period
- Levels fluctuate seasonally throughout the year due to variances in sun exposure

Vitamin D intake

Strengths

- Feasible in many studies, including retrospective studies
- In some populations, especially in winter months in regions at high latitude, an important contributor to vitamin D status
- Some studies have multiple updated measures, which improve assessment of long-term exposure

Limitations

- Intakes are relatively low in general because of the scarcity of vitamin D in natural foods
- In a specific population, intake of vitamin D may be predominantly from one or a few sources, such as fatty fish, fortified milk, or supplements, making high correlations with other dietary factors (e.g., omega-3 fatty acids in fish, calcium in milk, and other vitamins and minerals in supplement)

Predicted 25(OH)D

Strengths

- Feasible in some studies
- By incorporating multiple determinants of vitamin D status (for e.g., vitamin D intake, region of residence, outdoor activity level, skin color, body mass index), greater variability of 25(OH)D levels can be explained than for just one of these factors

Limitations

- May not incorporate actual sun exposure behaviors or differences in metabolism of vitamin D

Studies of sun exposure

Strengths

- Self-reported sun exposure or surrogates such as region of residence, number of sunburns can be feasibly used in epidemiologic studies, including retrospective studies

Table 1
(continued)

• Confounding may be better controlled than in ecologic studies
• Sun exposure is a major determinant of vitamin D status for most
  individuals
• Objective methods, such as the use of reflectometry, may be
  particularly useful
• May be useful in estimating vitamin D status at points earlier in life
Limitations
• Some surrogates that have been used (such as sunburns) may
  represent acute short-term exposures to sun rather than chronic
  exposures
• Measurement error and perhaps recall bias in case–control studies in
  assessing past exposures

Randomized trials (RCTs)
Strengths
• Double-blinded, placebo-controlled RCTs are the "gold standard" in
  establishing a causal association because confounding can be largely
  limited
• If a significant association is found, RCTs provide the strongest
  evidence of a causal association
Limitations
• Practical limitations include high expense, selection of the effective
  dose, varying baseline levels of the exposure of interest, poor
  compliance, contamination of the placebo group, and unknown
  induction time for the disease
• One or more of the limitations mentioned above could produce a null
  association

---

done. The Women's Health Initiative was primarily a randomized placebo-controlled trial of 400 IU vitamin D plus 1,000 mg a day of elemental calcium in 36,282 postmenopausal women. As discussed below, the interventional component of this study did not support a protective role of vitamin D intake *(18)*.

Since this meta-analysis was completed, three additional studies on colorectal cancer have been published. A similar reduced risk of colorectal cancer has been observed in the Health Professionals Follow-Up Study *(20)*. That study showed a nonstatistically significant inverse association between higher plasma 25(OH)D concentration and risk of colorectal cancer and a statistically significant inverse association for colon cancer (highest vs. lowest quintile: RR = 0.46, 95% CI = 0.24–0.89; P(trend) =0.005). After pooling the results from the Health Professionals Follow-Up Study and the Nurses' Health Study, higher plasma 25(OH)D levels were associated with decreased risks of both colorectal cancer (RR = 0.66, 95% CI = 0.42–1.05; P(trend) =0.01) and colon cancer (RR = 0.54, 95% CI = 0.34–0.86; P(trend) = 0.002). Inverse associations with plasma 25(OH)D concentration did not differ by location of colon cancer (proximal vs. distal), but the number of patients was small. The results for rectal cancer were inconsistent, though the number of cases was small.

The association between plasma 25(OH)D and the subsequent colorectal cancer incidence risk was examined in The Japan Public Health Center-based Prospective Study *(21)*. This nested case–control study covered 375 incident cases of colorectal cancer from 38,373 study subjects during 11.5 years of follow-up after blood collection. Two controls were matched per case on sex, age, study area, date of blood draw, and fasting time. In the multivariate analysis with matched pairs further adjusted for smoking, alcohol consumption, body mass index, physical exercise, vitamin supplement use, and family history of colorectal cancer, plasma 25(OH)D was not significantly associated with colorectal cancer. However, the lowest category of plasma 25(OH)D was associated with an elevated risk of rectal cancer in both men (RR = 4.6, 95% CI = 1.0–20) and women (RR = 2.7, 95% CI = 0.94–7.6), compared with the combined category of the other quartiles.

The association between 25(OH)D and colorectal cancer was examined in 16,818 participants followed from 1988–1994 to 2000 in the Third National Health and Nutrition Examination Survey *(22)*. Colorectal cancer mortality ($n = 66$ cases) was inversely related to serum 25(OH)D level, measured at baseline, with levels 80 nmol/l (32 ng/ml) or higher associated with a 72% risk reduction (95% CI = 32–89%) compared with levels <50 nmol/l (20 ng/ml), P(trend) =0.02. Several studies have examined circulating 25(OH)D levels and risk of colorectal adenoma, which are well-accepted precursors to colorectal cancer. On the whole, these studies suggest an inverse association with 25(OH)D and possibly 1,25(OH)$_2$D *(13–16, 23)*, particularly for advanced adenomas *(16)*.

Thus, based on multiple studies of circulating 25(OH)D and colorectal cancer, individuals in the high quartile or quintile of 25(OH)D had about a 40–50% risk reduction of colorectal cancer relative to those in the lowest group. The dose–response appears linear up to a 25(OH)D level of at least approximately 90 nmol/l (36 ng/ml), with no obvious threshold or nonlinear relationship, and controlling for multiple covariates has had little influence on the findings. The results are somewhat inconsistent in distinguishing whether the association is stronger for colon cancer or for rectal cancer, possibly due to small numbers, but in general the association has been observed for both anatomic sites. Similar results are observed for colorectal adenomas.

2.2. Predicted 25(OH)D Level

An approach to estimate predicted 25(OH)D and then relating this score to risk of colorectal cancer was used in the Health Professionals Follow-Up Study *(24)*. For this analysis, first, in a sample of 1,095 men who had plasma 25(OH)D levels measured, multiple linear regression was used to develop a predicted 25(OH)D score based on geographical region, skin pigmentation, dietary intake, supplement intake, body mass index, and leisure-time physical activity (a surrogate of potential exposure to sunlight UVB) as the independent variables *(24)*. Then, the score, after being validated, was calculated for each of approximately 47,000 cohort members and examined in relation to subsequent risk of cancer. For colorectal cancer, based on 691 cases diagnosed from

1986 to 2000, a 25 nmol/l (10 ng/ml) increment in 25(OH)D was associated with a reduced risk (multivariate RR=0.63, 95% CI = 0.48–0.83). This association persisted after controlling for body mass index or physical activity, which are contributors to the 25(OH)D score.

2.3. Dietary Intake

The association between colorectal cancer risk and dietary or supplementary vitamin D has been investigated in cohort studies of men (25, 26) and women (27–29) or both sexes (30, 31) and in case–control studies (32–39). The majority of these studies suggested inverse associations for colon or rectal cancer, or both (25–28, 31, 33, 35, 37, 38, 40). Most importantly, the studies of colorectal cancer that generally took into account supplementary vitamin D reported an inverse association. In these studies, the cutpoint for the top category was from approximately 500 to 600 IU/day, with an average intake of approximately 700–800 IU/day in this category. The risk reduction in the top vs. bottom category was as follows: 46% (27), 34% (26), 58% (28), 24% (29), 30% (38), 29% male, 0% female (31), 50% males, 40% females (39), and 28% male, 11% female (40).

2.4. Sun Exposure

A large death certificate-based case–control study examined mortality from female breast, ovarian, colon, and prostate cancers to examine associations with residential and occupational exposure to sunlight; nonmelanoma skin cancer served as a positive "control" (41). The cases consisted of all deaths from these cancers between 1984 and 1995 in 24 states of the United States. The controls were age frequency matched to a series of cases, and deaths from cancer and certain neurological diseases were excluded because of possible relationships with sun exposure. For colon cancer, based on 153,511 cases, those with a high compared to low exposure to sun based on residence were at decreased risk (RR = 0.73, 95% CI = 0.71–0.74). Further, those with outdoor occupations (RR = 0.90, 95% CI = 0.86–0.94) and occupations that required more physical activity (RR = 0.89, 95% CI = 0.86–0.92) were at lower risk. The inverse association with outdoor occupation was strongest among those living in the highest sunlight region. The multivariate analyses controlled for age, sex, race, mutual adjustment for residence, occupation (outdoor vs. indoor), occupational physical activity levels, and socioeconomic status.

2.5. Randomized Controlled Trial

The Women's Health Initiative was a randomized placebo-controlled trial of 400 IU vitamin D plus 1,000 mg/day of elemental calcium in 36,282 postmenopausal women (18). This study, based on 332 colorectal cancer cases, did not support a protective role of calcium and vitamin D on colorectal cancer risk over a period of 7 years. However, this trial had some important limitations. First, the vitamin D dose of 400 IU/day was probably inadequate to yield a substantial contrast between the treated and the control groups. Specifically, the expected increase of serum 25(OH)D level following an increment of 400 IU/day would be approximately 7.5 nmol/l (3 ng/ml). In comparison, in the epidemiologic studies of 25(OH)D, the contrast between the high and low quintiles was generally at least 50 nmol/l (20 ng/ml), a wide range likely due primarily to differences in sun exposure. Further, the adherence was suboptimal and a high percentage of women took nonstudy supplements, so the actual contrast of 25(OH)D tested between the treated and the placebo group in the intent-to-treat analysis was further reduced. Additionally, whether 7 years is sufficiently long to show an effect is unknown. The epidemiologic data on duration, although limited, suggest that any influence of calcium and vitamin D intakes may require at least 10 years for a risk reduction to emerge for colorectal cancer (28). Finally, the study design was a factorial with hormonal replacement use, and a post hoc analysis suggested an interaction whereby women on hormones did not benefit from the vitamin D and calcium, but women not taking hormones may have benefited (42).

3. PROSTATE CANCER

3.1. 25(OH)Vitamin D

Ecologic studies of regional UVB exposure support an inverse association with prostate cancer mortality, yet this association is weaker than that for colorectal or breast cancer (8) and perhaps limited to countries north of 40°N latitude, where vitamin D synthesis is limited to non-winter months (43). Circulating 25(OH)D level has been relatively well studied in relation to prostate cancer risk. Most studies do not show clear associations with risk for prostate cancer, although some of the studies suggest weak inverse associations (17, 44–48). Only two studies (49, 50), conducted in Nordic countries, supported an inverse association for 25(OH)D. However, one of these studies also found an increased risk in men with the highest 25(OH)D values, suggestive of a U-shaped relationship (50). Although $1,25(OH)_2D$ that is produced intracellularly is believed to be more important than circulating $1,25(OH)_2D$, several studies found supportive (44) or suggestive (45) inverse associations for circulating $1,25(OH)_2D$ and aggressive prostate cancer. In the Physicians' Health Study, men with both low 25(OH)D and $1,25(OH)_2D$ were at higher risk of aggressive prostate cancer (RR = 1.9) (51). In the Health Professionals Follow-Up Study, both lower 25(OH)D and $1,25(OH)_2D$ appeared to be associated with lower (mostly early-stage) prostate cancer risk (47), but possibly with higher risk of advanced prostate cancer, although numbers of advanced cases were limited ($n=60$) (47). Analysis of the Prostate, Lung, Colorectal and Ovarian (PLCO) Cancer Screening Trial based on 749 cases and 781 controls found no association and perhaps even a suggestion of an increased risk of aggressive prostate cancer among men with higher 25(OH)D levels (52). Thus, overall the studies of circulating 25(OH)D have been equivocal for prostate cancer and in general have not supported an association.

3.2. Predicted 25(OH)D Level

An approach to estimate predicted 25(OH)D used in the Health Professionals Follow-Up Study was described above (24). In the same population, prostate cancer risk was examined. The analysis was limited to more advanced cases of prostate cancer. Based on 461 cases of advanced prostate cancer, an approximately 20% reduction in risk was observed with a 25 nmol/l (10 ng/ml) increment in predicted 25(OH)D, though this was not statistically significant.

3.3. Vitamin D Intake

Studies of vitamin D intake and prostate cancer risk have generally not supported an association with prostate cancer incidence (53–56). Although studies of vitamin D intake have limitations (see Table 1), particularly since intake is generally not the main source of vitamin D for most populations, similar studies of colorectal cancer (often in the same study population) tend to support an inverse association between vitamin D intake and colorectal cancer risk, as described above.

3.4. Sun Exposure

Several case–control and cohort studies have assessed surrogates of sun exposure in relation to cancer risk. Prostate cancer appears to be the most thoroughly studied cancer using this method. In a cohort study of 3,414 white men based on NHANES I data, 153 developed prostate cancer over follow-up. Residence in the South at baseline (RR = 0.68), state of longest residence in the South (RR = 0.62), and high solar radiation in the state of birth (RR = 0.49) were associated with significant reductions in prostate cancer risk (57). This result suggests early-life exposure to sun is most critical for prostate cancer prevention.

One case–control study was based on use of a reflectometer to measure constitutive skin pigmentation on the upper underarm (a sun-protected site) and facultative pigmentation on the forehead (a sun-exposed site) to calculate a sun exposure index (58). The difference between facultative skin pigmentation and constitutive pigmentation is a function of overall sun exposure, at least on the forehead. This measurement of sun exposure predicted risk of advanced prostate cancer in this study (58). Specifically, a reduced risk of advanced prostate cancer was associated with high sun exposure determined by reflectometry (RR = 0.51) and high occupational outdoor activity level (RR = 0.73).

Other investigators have examined factors such as childhood sunburns, holidays in a hot climate, and skin type in case–control studies in relation to prostate cancer risk. In a study in the United Kingdom, subgroups stratified by childhood sunburns, holidays in a hot climate and skin type demonstrated a 13-fold higher risk of prostate cancer in those with combinations of high sun exposure/light skin compared to low sun exposure/darker skin type (59, 60). Further, self-reported UV exposure parameters and skin type in 553 men with prostate cancer were studied in association with stage, Gleason score, and survival after starting hormone manipulation therapy (61). UV exposures 10, 20, and 30 years before diagnosis were inversely associated with stage, and the RR for UV exposure 10 years before diagnosis was lowest (RR = 0.69, 95% CI = 0.56–0.86). RRs were lower in men with skin types I/II than III/IV. Additionally, skin types I/II were associated with longer survival after commencing hormone therapy (RR = 0.62, 95% CI = 0.40–0.95).

In a large death certificate-based case–control study of mortality from five cancers, as described above (for colon cancer), residential exposure to sunlight was inversely associated with mortality from prostate cancer (RR = 0.90, 95% CI = 0.86–0.91) *(41)*. However, occupation exposure to sunlight was not associated with prostate cancer risk (RR = 1.00, 95% CI = 0.96–1.05). The association with prostate cancer was weaker than that observed for other cancer sites (e.g., colorectal, breast, and ovarian cancers).

4. BREAST CANCER

4.1. 25(OH)Vitamin D

An inverse correlation has been observed between regionally estimated UVB and breast cancer mortality *(8)*. Relatively few studies of 25(OH)D level and risk of breast cancer have been reported. One nested case–control study based on 96 breast cancer cases found no association between prediagnostic $1,25(OH)_2D$ concentration and risk of breast cancer; circulating 25(OH)D, the more relevant metabolite, was not examined. Two studies examined postdiagnostic 25(OH)D levels in breast cancer cases and controls. In one, breast cancer cases had lower 25(OH)D levels than did controls *(62)*. In the other study, serum level of 25(OH)D was significantly higher in patients with early-stage breast cancer than in women with locally advanced or metastatic disease *(63)*. In these studies, the possibility of reverse causation cannot be ruled out because 25(OH)D levels were assessed in women who already had breast cancer.

Two large prospective studies have reported on 25(OH)D levels in relation to breast cancer risk. In the Nurses' Health Study, stored plasma samples were assessed in 701 breast cancer cases and 724 controls *(64)*. Women in the highest quintile of 25(OH)D had a RR of 0.73 (P trend=0.06) compared with those in the lowest quintile. The association was stronger in women aged 60 years and older, suggesting that vitamin D may be more important for postmenopausal breast cancer. The other prospective study of 25(OH)D level and breast cancer risk was based on the PLCO study *(65)*. In this cohort, 1,005 incident cases of breast cancer were frequency matched with 1,005 controls, over follow-up from 1993 to 2005 (mean time between blood draw and diagnosis was 3.9 years). No association was found comparing high to low quintiles (RR = 1.04, 95% CI = 0.75–1.45; P(trend)=0.81). Risk did not differ substantially in younger vs. older women or after excluding the first 2 years of follow-up. The range of 25(OH)D was 13.6–39.2 ng/ml (bottom vs. top quintile means). In a small nested case–control study of only 28 cases, a nonsignificant inverse association was found for breast cancer *(22)*.

4.2. Vitamin D Intake

A meta-analysis of studies of vitamin D intake and risk of breast cancer for studies conducted up to June 2007 identified six relevant studies *(66)*. No association was found between vitamin D intake and risk of breast cancer (summary RR = 0.98, 95% CI = 0.93–1.03), but significant heterogeneity was noted (P<0.01). Vitamin D intake was very low in some studies, with most individuals having intakes below 400 IU. When the studies were stratified into those with intakes higher and lower than 400 IU, a modest association was observed in those with intakes ≥400 IU (RR = 0.92, 95% CI=0.87–0.97; P(heterogeneity)=0.14, n=3 studies), but not in those with intakes <400 IU. One of the studies, the Nurses' Health Study, is of particular interest because vitamin D intake was assessed multiple times (every 2–4 years), presumably allowing a better estimate of long-term intake *(67)*. In that study, based on 3,482 incident cases of breast cancer, total vitamin D intake (dietary plus supplements) was associated with a lower risk of breast cancer (RR = 0.72, 95% CI = 0.55–0.94) for >500 vs. ≤150 IU/day of vitamin D. Although similar inverse associations were observed with other components of dairy foods (lactose, calcium), total vitamin D intake was more strongly associated with lower risk than was dietary or supplemental vitamin D intake individually, suggestive of an independent association with vitamin D.

4.3. Sun Exposure

An inverse association between regional sunlight exposure and breast cancer mortality has been observed in several ecologic analyses *(4, 8, 41)*. However, an analysis within the Nurses' Health Study did not find the expected geographic gradient for breast cancer incidence *(68)*. In data from NHANES I, based on 190 women with incident breast cancer from a cohort of 5,009 women, several measures of sunlight exposure and dietary vitamin D intake were associated with a moderate reduction in breast cancer risk *(69)*. Also, in a population-based case–control study of 1,788 newly diagnosed cases and 2,129 controls among Hispanic, African-American, and non-Hispanic white women aged 35–79 years from California (1995–2003), high sun exposure index based on reflectometry was associated with reduced risk of advanced breast cancer among women with light constitutive skin pigmentation (OR = 0.53, 95% CI = 0.31, 0.91) *(70)*. However, no associations were found for women with medium or dark pigmentation. The authors speculated that these measures better reflect vitamin D status in more lightly pigmented women. In another population-based case–control study conducted in Canada of 972 cases and 1,135 controls, reduced breast cancer risks were associated with increasing sun exposure from ages 10 to 19 (RR = 0.65, 95% CI = 0.50–0.85 for the highest quartile of outdoor activities vs. the lowest; *P* for trend = 0.0006) *(71)*. Weaker associations from ages 20 to 29 years were observed and no evidence for ages 45–54 years. These results suggest that exposure earlier in life, particularly during breast development, may be most relevant for breast cancer prevention.

In a large death certificate-based case–control study of cancer mortality to examine associations with residential and occupational exposure to sunlight, as described above, residential (RR = 0.74, 95% CI = 0.72–0.76) and occupational (RR = 0.82, 95% CI = 0.70–0.97) exposure to sunlight was inversely associated with mortality from female breast cancer *(41)*. The relationship between outdoor employment and female breast cancer mortality was strongest in the region of greatest residential sunlight (OR = 0.75, 95% CI = 0.55–1.03).

5. PANCREATIC CANCER

5.1. 25(OH)D Level

Serum 25(OH)D was examined in relation to pancreatic cancer risk in the Alpha-Tocopherol, Beta-Carotene Cancer Prevention cohort of male Finnish smokers *(72)*. This study was based on 200 cases of pancreatic adenocarcinoma and 400 controls. Contrary to expectation, this study found a significant positive association between higher

25(OH)D levels and increased risk of pancreatic cancer. Higher vitamin D concentrations were associated with a threefold increased risk for pancreatic cancer (highest vs. lowest quintile, >65.5 vs. <32.0 nmol/l: RR = 2.92, 95% CI = 1.56–5.48; P(trend) = 0.001). This association persisted in multivariate analysis and after excluding cases early in follow-up (to avoid reverse causation). Of note, all the men in this study were smokers, and smoking is a strong risk factor for pancreatic cancer.

5.2. Predicted 25(OH)D

As described above, one analysis based on the Health Professionals Follow-Up Study used a surrogate of 25(OH)D to examine risk of cancer (73). Based on 170 cases of pancreatic cancer, a 25 nmol/l (10 ng/ml) increment in predicted 25(OH)D was associated with a 51% reduction in pancreatic cancer incidence (multivariate RR = 0.49, 95% CI = 0.28–0.86).

5.3. Vitamin D Intake

One report, which combined data from the Nurses' Health Study and the Health Professionals Follow-Up Study, examined total vitamin D intake (from diet and supplements) in relation to pancreatic cancer risk based on 365 incident cases over 16 years of follow-up (74). This study found a linear inverse association, with a significant 41% reduction in risk comparing high (≥600 IU/day) to low (<150 IU/day) total vitamin D intake. Controlling for various other dietary factors did not alter this association.

6. ESOPHAGEAL AND GASTRIC CANCERS

6.1. 25(OH)D Level

One study in China, nested in randomized trial of micronutrients, examined pre-trial 25(OH)D levels in relation to esophageal and gastric cancers over 5.25 years of follow-up (75). The analysis included 545 esophageal squamous cell carcinomas, 353 gastric cardia adenocarcinomas, and 81 gastric noncardia adenocarcinomas, in an area with an extremely high rate of these malignancies. For esophageal squamous cell carcinomas, when comparing men in the fourth quartile of serum 25(OH)D concentrations to those in the first, a positive association was found (RR = 1.77, 95% CI = 1.16–2.70; P(trend)=0.0033); however, no association was found in women (RR = 1.06, 95% CI = 0.71–1.59; P(trend)=0.70). Also, no associations were found for gastric cardia or noncardia adenocarcinoma. The cut-point for the top quartile was relatively low (48.7 nmol/l or 19.5 ng/ml).

In another study from Linxian, China, the same authors measured serum 25(OH)D levels in a cross-sectional analysis of 720 subjects who underwent endoscopy and biopsy and were categorized by the presence or absence of histologic esophageal squamous dysplasia (76). In total, 230 of 720 subjects, about one-third, had evidence of squamous dysplasia. In multivariate analyses, those in the highest compared with the lowest quartile of 25(OH)D were at a significantly increased risk of squamous dysplasia (total: RR = 1.86, 95% CI = 1.35–2.62; men: RR = 1.74, 95% CI = 1.08–2.93; women: RR = 1.96, 95% CI = 1.28–3.18). The mean level of 25(OH)D was low in this population (35 nmol/l or 13.9 ng/ml). The findings from these two studies are interesting and unexpected, but somewhat difficult to generalize to other populations. The risk of esophageal cancer is extraordinarily high in this population, and the etiology is unclear and likely to be different from that in most other populations. Almost all of the 25(OH)D in this population results from sun light exposure, so relatively high 25(OH)D level could represent some other consequence of high sun exposure, such as an immunosuppressant effect.

7. OVARIAN CANCER

7.1. 25(OH)D

A nested case–control study of plasma 25(OH)D in relation to risk of epithelial ovarian cancer was conducted using data from three prospective cohorts: the Nurses' Health Study, the Nurses' Health Study II, and the Women's Health Study (77). The analysis, based on 224 cases and 603 controls, showed no significant association between 25(OH)D and ovarian cancer risk (top vs. bottom quartile: RR = 0.83, 95% CI = 0.49–1.39; P(trend) = 0.57). When the first 2 years of follow-up were excluded, a suggestive inverse association (RR = 0.67, 95% CI = 0.43–1.05) was noted for adequate 25(OH)D ($\geq$32 ng/ml). In addition, there was a significant inverse association among overweight and obese women for 25(OH)D levels (RR = 0.39, 95% CI = 0.16–0.93; P(trend) = 0.04), and women with adequate vs. inadequate 25(OH)D levels had a modestly decreased risk of serous ovarian cancer (RR = 0.64, 95% CI = 0.39–1.05). Other studies are needed to confirm these subgroup findings.

7.2. Sun Exposure

In a large death certificate-based case–control study of ovarian cancer mortality in association with residential and occupational exposure to sunlight (described above) (41), the investigators found that residential (RR = 0.84, 95% CI = 0.81–0.88) but not occupational exposure to sunlight was inversely associated with mortality from ovarian cancer.

8. NON-HODGKIN'S LYMPHOMA (NHL)

8.1. Sun Exposure

The relationship between sun exposure and NHL risk was examined in a pooled analysis of 10 case–control studies, which included 5 published studies, participating in the International Lymphoma Epidemiology Consortium (InterLymph) (78). The studies covered 8,243 cases and 9,697 controls in the United States, Europe, and Australia, and the participants were of European origin. Four kinds of measures of self-reported personal sun exposure were assessed at interview: outdoors and not in the shade in warmer months or summer; in the sun in leisure activities, in sunlight, sun bathing in summer. The risk of NHL fell significantly with the composite measure of increasing recreational sun exposure; the multivariate (adjusting for smoking and alcohol) pooled RR = 0.76 (95% CI = 0.63–0.91) for the highest exposure category (P for trend 0.005). An inverse but not significant trend in risk was observed with increasing total sun exposure,

OR = 0.87 (95% CI = 0.71–1.05; P = 0.08). The inverse association between recreational sun exposure and NHL risk was statistically significant at 18–40 years of age and in the 10 years before diagnosis. The results also were statistically significant for B-cell, but not T-cell, lymphomas, though the numbers for the T-cell lymphomas were small.

A large case–control study based on death certificates of residential and occupational sun exposure and NHL mortality was conducted in 24 states in the United States *(79)*. The study was based on over 33,000 fatal cases of NHL. The adjusted RR for residing in states with the highest sunlight exposure was 0.83 (95% CI = 0.81–0.86) and it was 0.44 (95% CI = 0.28–0.67) for those under 45 years of age. NHL mortality risk was also reduced with higher occupational sunlight exposure (RR = 0.88, 95% CI = 0.81–0.96).

9. TOTAL CANCER

9.1. Circulating 25(OH)D

One study *(22)* examined cancer risk in 16,818 participants 17 years or older at enrollment in the Third National Health and Nutrition Examination Survey who were followed from 1988 to 1994. Through 2000, 536 cancer deaths were identified. Total cancer mortality was unrelated to baseline vitamin D status, although a nonsignificant inverse trend (P =0.12) was observed in women only. Among specific cancer sites, colorectal cancer mortality was inversely related to serum 25(OH)D level (discussed above), and a nonsignificant inverse association was observed for breast cancer, though based on only 28 cases.

25(OH)D was measured in 3,299 patients who provided a blood sample in the morning before coronary angiography from the Ludwigshafen Risk and Cardiovascular Health study *(80)*. A total of 95 patients died due to cancer during a median follow-up period of 7.75 years. After adjustment for age, sex, body mass index, smoking, retinol, exercise, alcohol and diabetes history, the RR for the fourth quartile of 25(OH)D was 0.45 (95% CI = 0.22–0.93) compared with the first quartile, and the RR per increase of 25 nmol/l (10 ng/ml) in serum 25(OH)D concentrations was 0.66 (95% CI = 0.49–0.89).

Pretransplant 25(OH)D levels were examined in 363 renal transplant recipients at Saint-Jacques University hospital at Besancon, France *(81)*. Mean 25(OH)D was low at pretransplant (17.6 ng/ml) and further reduced posttransplant (posttransplant patients are advised to avoid sun exposure). Thirty-two cancers were diagnosed over 5 years of follow-up. 25(OH)D levels were lower in patients who developed cancer after transplantation (13.7 ± 6 vs. 18.3 ± 17.8 ng/ml, P=0.022). The risk of total cancer increased by 12% for each unit increment in 25(OH)D (RR = 1.12, 95% CI = 1.04–1.23; P= 0.021).

9.2. Predicted 25(OH)D

In the Health Professionals Follow-Up Study cohort, predicted 25(OH)D levels (see above) were examined in relation to risk of total cancer in men. From 1986 to January 31, 2000, 4,286 incident cancers (excluding organ-confined prostate cancer and nonmelanoma skin cancer) and 2,025 deaths from cancer were documented in the cohort. An increment of 25 nmol/l (10 ng/ml) in predicted 25(OH)D level was associated with a 17% reduction in total cancer incidence (multivariate RR = 0.83, 95% CI = 0.74–0.92) and a 29% reduction in total cancer mortality (RR = 0.71, 95% CI = 0.60–0.83). The reduction was largely confined to cancers of the digestive tract system, including esophagus, stomach, pancreas, colon and rectum, showing a 45% reduction in mortality (RR = 0.55, 95% CI = 0.41–0.74). A reduced risk was also observed for oral cancer and leukemias. Results were similar when controlled further for body mass index or physical activity level.

9.3. Randomized Trials (RCT)

A RCT of 2037 men and 649 women aged 65–85 years living in the general community in the United Kingdom was conducted using 100,000 IU oral vitamin D (cholecalciferol) supplementation or placebo every 4 months over 5 years (82). After treatment, 25(OH)D level was 74.3 nmol/l in the vitamin D group and 53.4 nmol/l in the placebo group. There were 188 incident cancer cases in the vitamin D group and 173 in the placebo group, and no overall reduction was observed for cancer risk (RR = 1.09, 95% CI = 86–1.36). There was a slight, nonsignificant suggestion of a reduction in risk of cancer mortality (RR = 0.86, 95% CI = 0.61–1.20).

A 4-year, population-based, double-blind, placebo-controlled RCT of vitamin D and calcium was conducted with the primary outcome being fracture incidence and the principal secondary outcome being cancer incidence (83). The community-based study population comprised of 1,179 Nebraskan (US) women aged >55 years. These women were randomly assigned to receive 1,400–1,500 mg supplemental calcium/day alone (Ca-only), supplemental calcium plus 1,100 IU vitamin D/day (Ca + D), or placebo. After treatment, the achieved 25(OH)D level was 96 nmol/l in the vitamin D group and 71 nmol/l in the non-vitamin D groups. When analyzed by intention to treat, cancer incidence was lower in the Ca + D women than in the placebo control subjects (P < 0.03). The RR of incident cancer was 0.40 (P = 0.01) in the Ca + D group and 0.53 (P = 0.06) in the Ca-only group. When analysis was confined to cancers diagnosed after the first 12 months, the RR for the Ca + D group was 0.23 (95% CI = 0.09–0.60; P < 0.005), but no significant risk reduction was observed for the Ca-only group. In multivariate models, both vitamin D treatment and 25(OH)D levels were significant, independent predictors of cancer risk. A limitation of the study was the relatively small number of total cancers (50 in total; 13 in the first year and 37 thereafter).

10. SYNTHESIS AND SUMMARY

The relation between vitamin D status and cancer risk has been investigated in a number of epidemiologic studies, while data from interventional studies remain scarce. The approaches to estimate vitamin D status have been varied. These include direct measures of circulating 25(OH)D levels, surrogates or determinants of 25(OH)D, including region of residence, intake, and sun exposure estimates. Arguably, the optimal assessment for epidemiologic studies is a direct measure of circulating 25(OH)D level, which is the biologically relevant metabolite. However, some additional insights could be provided by other assessments, as most studies of 25(OH)D have relied on a single assessment and measures such as sun exposure early in life could provide clues when vitamin D could potentially be most etiologically relevant.

In terms of cancer sites, the body of evidence is most extensive for colorectal cancer, for which support comes from studies of 25(OH)D, vitamin D intake, multiple predictors of 25(OH)D, and region of residence in a sunny climate. Evidence also suggests that vitamin D status is inversely associated with risk of adenoma. These epidemiologic data converge with ecologic data, which demonstrate the strongest inverse association with colorectal cancer. Also intriguing is that ecologic data tend to show cancers of the gastrointestinal tract most strongly associated with vitamin D status, a pattern also observed in the one study of (predicted) 25(OH)D that had sufficient power to examine digestive tract cancers. Data on cancers of the digestive tract other than colorectal cancer are sparse, and studies of these cancers should be a priority. A few studies have indicated potentially positive associations between 25(OH)D and esophageal and pancreatic cancers in special high-risk populations of these malignancies, but predicted vitamin D was inversely associated with these in a US population. Further study is clearly needed for these cancers.

The evidence for breast cancer is also intriguing but somewhat conflicting. There have been only two large studies of circulating 25(OH)D levels, with inconsistent results. Studies of sun exposure are generally supportive of a benefit, and studies of diet are modestly supportive though hindered by the generally low intakes of vitamin D in most populations. A potentially important finding from one case–control study is that retrospectively reported sun exposure during ages 10–19 was most strongly associated with reduced risk of breast cancer. While this finding could be a result of recall bias, many lines of evidence suggest that breast cancer risk is strongly influenced by exposures during breast development in girls. Replicating this result in a prospective setting to exclude the possibility of recall bias is important.

For prostate cancer, the data on circulating 25(OH)D have been equivocal, suggesting no association or at least an association of a much weaker magnitude than has been observed for colorectal cancer. Studies tend to support a benefit of sun exposure on prostate cancer risk. It is plausible that for prostate cancer, vitamin D level much longer before the time of diagnosis is most relevant, consistent with the notion that the process of prostate carcinogenesis encompasses a very long time period. Prostate cancer cells appear to lose 1-alpha-hydroxylase activity early in carcinogenesis (84, 85), so it is plausible that exposure to vitamin D early in life (during very early stages of carcinogenesis) is most relevant. In addition, determinants of prostate cancer incidence may differ from prostate cancer progression and ultimately mortality, and most of the available data have assessed incident prostate cancer, as opposed to aggressive or fatal prostate cancer. Future studies need to focus on exposures earlier in life and on progression or mortality rather than incidence.

Most of the epidemiologic studies to date have examined vitamin D status in relation to risk of cancer. However, several lines of evidence suggest that vitamin D may be an important factor for cancer progression and mortality, independently of any effects on incidence. First, associations with vitamin D have often been stronger for mortality or advanced cancers than for cancer incidence. Second, in some studies, those diagnosed/treated during the summer months, when vitamin D status is higher, have a better prognosis than those diagnosed and/or treated during the winter months *(86)*. Finally, direct measures of vitamin D status prediagnostically or at the time of diagnosis/treatment have been associated with better survival *(87, 88)*. These data have the same limitations regarding establishing a causal association as those on studies of incidence, but nonetheless suggest the intriguing possibility that improving vitamin D status at the time of and following diagnosis could potentially prolong survival. Because the time frame to demonstrate a benefit would be shorter and the numbers of subjects requiring randomization would be smaller than in primary prevention studies and because the vitamin D status of cancer patients is typically poor, intervention studies of treatment may be feasible and should be a priority.

In conclusion, a large body of epidemiologic evidence supports the hypothesis that vitamin D status is inversely associated with cancer incidence and/or progression. The evidence to date is strongest for colorectal cancer, for which the magnitude of the effect might be stronger than for other cancers. Thus far, randomized intervention studies are limited in scope and inconsistent. Further study is clearly needed to establish the precise role of vitamin D on carcinogenesis, especially in terms of when in the life span and on what stages of carcinogenesis vitamin D is relevant, the precise intakes and levels required, the magnitude of the association, and which cancer sites are most affected.

REFERENCES

1. Peller S, Stephenson CS (1937) Skin irritation and cancer in the United States Navy. Am J Med Sci 194:326–333
2. Apperly FL (1941) The relation of solar radiation to cancer mortality in North America. Cancer Res 1:191–195
3. Garland CF, Garland FC (1980) Do sunlight and vitamin D reduce the likelihood of colon cancer? Int J Epidemiol 9:227–231
4. Garland FC, Garland CF, Gorham ED et al (1990) Geographic variation in breast cancer mortality in the United States: a hypothesis involving exposure to solar radiation. Prev Med 19:614–622
5. Lefkowitz ES, Garland CF (1994) Sunlight, vitamin D, and ovarian cancer mortality rates in US women. Int J Epidemiol 23:1133–1136
6. Schwartz GG, Hulka BS (1990) Is vitamin D deficiency a risk factor for prostate cancer? (Hypothesis). Anticancer Res 10:1307–1311
7. Hanchette CL, Schwartz GG (1992) Geographic patterns of prostate cancer mortality. Cancer 70: 2861–2869
8. Grant WB (2002) An estimate of premature cancer mortality in the U.S. due to inadequate doses of solar ultraviolet-B radiation. Cancer 94:1867–1875
9. Jacobs ET, Haussler MR, Martinez ME (2005) Vitamin D activity and colorectal neoplasia: a pathway approach to epidemiologic studies. Cancer Epidemiol Biomarkers Prev 14:2061–2063
10. Garland CF, Comstock GW, Garland FC et al (1989) Serum 25-hydroxyvitamin D and colon cancer: eight-year prospective study. Lancet 2:1176–1178
11. Tangrea J, Helzlsouer K, Pietinen P et al (1997) Serum levels of vitamin D metabolites and the subsequent risk of colon and rectal cancer in Finnish men. Cancer Causes Control 8:615–625
12. Feskanich D, Ma J, Fuchs CS et al (2004) Plasma vitamin D metabolites and risk of colorectal cancer in women. Cancer Epidemiol Biomarkers Prev 13:1502–1508
13. Levine AJ, Harper JM, Ervin CM et al (2001) Serum 25-hydroxyvitamin D, dietary calcium in take, and distal colorectal adenoma risk. Nutr Cancer 39:35–41
14. Peters U, McGlynn KA, Chatterjee N et al (2001) Vitamin D, calcium, and vitamin D receptor polymorphism in colorectal adenomas. Cancer Epidemiol Biomarkers Prev 10:1267–1274
15. Platz EA, Hankinson SE, Hollis BW et al (2000) Plasma 1,25-dihydroxy- and 25-hydroxyvitamin D and adenomatous polyps of the distal colorectum. Cancer Epidemiol Biomarkers Prev 9:1059–1065

16. Grau MV, Baron JA, Sandler RS et al (2003) Vitamin D, calcium supplementation, and colorectal adenomas: results of a randomized trail. J Natl Cancer Inst 95:1765–1771

17. Braun MM, Helzlsouer KJ, Hollis BW et al (1995) Prostate cancer and prediagnostic levels of serum vitamin D metabolites (Maryland, United States). Cancer Causes Control 6:235–239

18. Wactawski-Wende J, Kotchen JM, Anderson GL et al (2006) Calcium plus vitamin D supplementation and the risk of colorectal cancer. N Engl J Med 354:684–696

19. Gorham ED, Garland CF, Garland FC et al (2007) Optimal vitamin D status for colorectal cancer prevention. A quantitative meta analysis. Am J Prev Med 32:210–216

20. Wu K, Feskanich D, Fuchs CS et al (2007) A nested case–control study of plasma 25-hydroxyvitamin D concentrations and risk of colorectal cancer. J Natl Cancer Inst 99:1120–1129

21. Otani T, Iwasaki M, Sasazuki S et al (2007) Plasma vitamin D and risk of colorectal cancer: the Japan Public Health Center-Based Prospective Study. Br J Cancer 97:446–451

22. Freedman DM, Looker AC, Chang SC et al (2007) Prospective study of serum vitamin D and cancer mortality in the United States. J Natl Cancer Inst 99:1594–1602

23. Peters U, Hayes RB, Chatterjee N et al (2004) Circulating vitamin D metabolites, polymorphism in vitamin D receptor, and colorectal adenoma risk. The Prostate, Lung, Colorectal and Ovarian Cancer Screening Project Team. Cancer Epidemiol Biomarkers Prev 13:546–552

24. Giovannucci E, Liu Y, Rimm EB, et al (2007) Prospective study of predictors of vitamin D status and cancer incidence and mortality in men. J Natl Cancer Inst 2006 98:451–459

25. Garland C, Shekelle RB, Barrett-Conner E et al (1985) Dietary vitamin D and calcium and risk of colorectal cancer: a 19-year prospective study in men. Lancet 1:307–309

26. Kearney J, Giovannucci E, Rimm EB et al (1996) Calcium, vitamin D and dairy foods and the occurrence of colon cancer in men. Am J Epidemiol 143:907–917

27. Bostick RM, Potter JD, Sellers TA et al (1993) Relation of calcium, vitamin D, and dairy food intake to incidence of colon cancer in older women. Am J Epidemiol 137:1302–1317

28. Martinez ME, Giovannucci EL, Colditz GA et al (1996) Calcium, vitamin D, and the occurrence of colorectal cancer among women. J Natl Cancer Inst 88:1375–1382

29. Zheng W, Anderson KE, Kushi LH et al (1998) A prospective cohort study of intake of calcium, vitamin D, and other micronutrients in relation to incidence of rectal cancer among postmenopausal women. Cancer Epidemiol Biomarkers Prev 7:221–225

30. Jarvinen R, Knekt P, Hakulinen T et al (2001) Prospective study on milk products, calcium and cancers of the colon and rectum. Eur J Clin Nutr 55:1000–1007

31. McCullough ML, Robertson AS, Rodriguez C et al (2003) Calcium, vitamin D, dairy products, and risk of colorectal cancer in the cancer prevention study II nutrition cohort (United States). Cancer Causes Control 14:1–12

32. Heilbrun LK, Nomura A, Hankin JH et al (1985) Dietary vitamin D and calcium and risk of colorectal cancer (letter). Lancet 1:925

33. Benito E, Stiggelbout A, Bosch FX et al (1991) Nutritional factors in colorectal cancer risk: a case–control study in Majorca. Int J Cancer 49:161–167

34. Peters RK, Pike MC, Garabrandt D et al (1992) Diet and colon cancer in Los Angeles County, California. Cancer Causes Control 3:457–473

35. Ferraroni M, La Vecchia C, D'Avanzo B et al (1994) Selected micronutrient intake and the risk of colorectal cancer. Br J Cancer 70:1150–1155

36. Boutron MC, Faivre J, Marteau P et al (1996) Calcium, phosphorus, vitamin D, dairy products and colorectal carcinogenesis: a French case–control study. Br J Cancer 74:145–151

37. Pritchard RS, Baron JA, Gerhardsson de Verdier M (1996) Dietary calcium, vitamin D, and the risk of colorectal cancer in Stockholm, Sweden. Cancer Epidemiol Biomarkers Prev 5: 897–900

38. Marcus PM, Newcomb PA (1998) The association of calcium and vitamin D, and colon and rectal cancer in Wisconsin women. Int J Epidemiol 27:788–793

39. Kampman E, Slattery ML, Caan B et al (2000) Calcium, vitamin D, sunshine exposures, dairy products and colon cancer risk (United States). Cancer Causes Control 11:459–466

40. Park SY, Murphy SP, Wilkens LR et al (2007) Calcium, vitamin D, and dairy product intake and prostate cancer risk: the Multiethnic Cohort Study. Am J Epidemiol 166:1259–1269

41. Freedman DM, Dosemeci M, McGlynn K (2002) Sunlight and mortality from breast, ovarian, colon, prostate, and non-melanoma skin cancer: a composite death certificate based case–control study. Occup Environ Med 59:257–262

42. Ding EL, Mehta S, Fawzi WW et al (2008) Interaction of estrogen therapy with calcium and vitamin D supplementation on colorectal cancer risk: reanalysis of Women's Health Initiative randomized trial. Int J Cancer 122:1690–1694

43. Schwartz GG, Hanchette CL (2006) UV latitude, and spatial trends in prostate cancer mortality: all sunlight is not the same (United States). Cancer Causes Control 17:1091–1101

44. Corder EH, Guess HA, Hulka BS et al (1993) Vitamin D and prostate cancer: a prediagnostic study with stored sera. Cancer Epidemiol Biomarkers Prev 2:467–472

45. Gann PH, Ma J, Hennekens CH et al (1996) Circulating vitamin D metabolites in relation to subsequent development of prostate cancer. Cancer Epidemiol Biomarkers Prev 5:121–126

46. Nomura AM, Stemmermann GN, Lee J et al (1998) Serum vitamin D metabolite levels and the subsequent development of prostate cancer. Cancer Causes Control 9:425–432

47. Platz EA, Leitzmann MF, Hollis BW et al (2004) Plasma 1,25-dihydroxy- and 25-hydroxyvitamin D and subsequent risk of prostate cancer. Cancer Causes Control 15:255–265

48. Jacobs ET, Giuliano AR, Martinez ME et al (2004) Plasma levels of 25-hydroxyvitamin D, 1,25-dihydroxyvitamin D and the risk of prostate cancer. J Steroid Biochem Mol Biol 89–90:533–537

49. Ahonen MH, Tenkanen L, Teppo L et al (2000) Prostate cancer risk and prediagnostic serum 25-hydroxyvitamin D levels (Finland). Cancer Causes Control 11:847–852

50. Tuohimaa P, Tenkanen L, Ahonen M et al (2004) Both high and low levels of blood vitamin D are associated with a higher prostate cancer risk: a longitudinal, nested case–control study in the Nordic countries. Int J Cancer 108:104–108

51. Li H, Stampfer MJ, Hollis BW et al (2007) A prospective study of plasma vitamin D metabolites, vitamin D receptor polymorphisms, and prostate cancer. PLoS Med 4:e103

52. Ahn J, Peters U, Albanes D et al (2008) Serum vitamin D concentration and prostate cancer risk: a nested case–control study. J Natl Cancer Inst 100:796–804

53. Giovannucci E, Rimm EB, Wolk A et al (1998) Calcium and fructose intake in relation to risk of prostate cancer. Cancer Res 58:442–447

54. Chan JM, Giovannucci E, Andersson SO et al (1998) Dairy products, calcium, phosphorous, vitamin D, and risk of prostate cancer. Cancer Causes Control 9:559–566

55. Chan JM, Pietinen P, Virtanen M et al (2000) Diet and prostate cancer risk in a cohort of smokers, with a specific focus on calcium and phosphorus (Finland). Cancer Causes Control 11:859–867

56. Kristal AR, Cohen JH, Qu P et al (2002) Associations of energy, fat, calcium, and vitamin D with prostate cancer risk. Cancer Epidemiol Biomarkers Prev 11:719–725

57. John EM, Dreon DM, Koo J et al (2004) Residential sunlight exposure is associated with a decreased risk of prostate cancer. J Steroid Biochem Mol Biol 89–90:549–552

58. John EM, Schwartz GG, Koo J et al (2005) Sun exposure, vitamin D receptor gene polymorphisms, and risk of advanced prostate cancer. Cancer Res 65:5470–5479

59. Luscombe CJ, Fryer AA, French ME et al (2001) Exposure to ultraviolet radiation: association with susceptibility and age at presentation with prostate cancer. Lancet 358:641–642

60. Bodiwala D, Luscombe CJ, French ME et al (2003) Associations between prostate cancer susceptibility and parameters of exposure to ultraviolet radiation. Cancer Lett 200:141–148

61. Rukin N, Blagojevic M, Luscombe CJ et al (2007) Associations between timing of exposure to ultraviolet radiation, T-stage and survival in prostate cancer. Cancer Detect Prev 31(6):443–449

62. Colston KW, Lowe LC, Mansi JL et al (2006) Vitamin D status and breast cancer risk. Anticancer Res 26:2573–2580

63. Palmieri C, MacGregor T, Girgis S et al (2006) Serum 25-hydroxyvitamin D levels in early and advanced breast cancer. J Clin Pathol 59:1334–1336

64. Bertone-Johnson E, Chen WY, Holick MF et al (2005) Plasma 25-hydroxyvitamin D and 1,25-dihydroxyvitamin D and risk of breast cancer. Cancer Epidemiol Biomarkers Prev 14:1991–1997

65. Freedman DM, Chang SC, Falk RT et al (2008) Serum levels of vitamin D metabolites and breast cancer risk in the prostate, lung, colorectal, and ovarian cancer screening trial. Cancer Epidemiol Biomarkers Prev 17:889–894

66. Gissel T, Rejnmark L, Mosekilde L et al (2008) Intake of vitamin D and risk of breast cancer: a meta-analysis. J Steroid Biochem Mol Biol Jun 11 (Epub ahead of print)

67. Shin MH, Holmes MD, Hankinson SE et al (2002) Intake of dairy products, calcium, and vitamin D and risk of breast cancer. J Natl Cancer Inst 94:1301–1310

68. Laden F, Spiegelman D, Neas LM et al (1997) Geographic variation in breast cancer incidence rates in a cohort of U.S. women. J Natl Cancer Inst 89:1373–1378

69. John EM, Schwartz GG, Dreon DM et al (1999) Vitamin D and breast cancer risk: the NHANES I Epidemiologic follow-up study, 1971–1975 to 1992. National Health and Nutrition Examination Survey. Cancer Epidemiol Biomarkers Prev 8:399–406

70. John EM, Schwartz GG, Koo J et al (2007) Sun exposure, vitamin D receptor gene polymorphisms, and breast cancer risk in a multiethnic population. Am J Epidemiol 166:1409–1419

71. Knight JA, Lesosky M, Barnett H et al (2007) Vitamin D and reduced risk of breast cancer: a population-based case–control study. Cancer Epidemiol Biomarkers Prev 16:422–429

72. Stolzenberg-Solomon RZ, Vieth R, Azad A et al (2006) A prospective nested case–control study of vitamin D status and pancreatic cancer risk in male smokers. Cancer Res 66:10213–10219

73. Giovannucci E, Liu Y, Rimm EB et al (2006) Prospective study of predictors of vitamin D status and cancer incidence and mortality in men. J Natl Cancer Inst 98:451–459

74. Skinner HG, Michaud DS, Giovannucci E et al (2006) Vitamin D intake and the risk for pancreatic cancer in two cohort studies. Cancer Epidemiol Biomarkers Prev 15:1688–1695

75. Chen W, Dawsey SM, Qiao YL et al (2007) Prospective study of serum 25(OH)-vitamin D concentration and risk of oesophageal and gastric cancers. Br J Cancer 97:123–128

76. Abnet CC, Chen W, Dawsey SM et al (2007) Serum 25(OH)-vitamin D concentration and risk of esophageal squamous dysplasia. Cancer Epidemiol Biomarkers Prev 16:1889–1893

77. Tworoger SS, Lee IM, Buring JE et al (2007) Plasma 25-hydroxyvitamin D and 1,25-dihydroxyvitamin D and risk of incident ovarian cancer. Cancer Epidemiol Biomarkers Prev 16:783–788

78. Kricker A, Armstrong BK, Hughes AM et al (2008) Personal sun exposure and risk of non Hodgkin lymphoma: a pooled analysis from the Interlymph Consortium. Int J Cancer 122:144–154

79. Freedman DM, Zahm SH, Dosemeci M (1997) Residential and occupational exposure to sunlight and mortality from non-Hodgkin's lymphoma: composite (threefold) case–control study. BMJ 314: 1451–1455

80. Pilz S, Dobnig H, Winklhofer-Roob B et al (2008) Low serum levels of 25-hydroxyvitamin D predict fatal cancer in patients referred to coronary angiography. Cancer Epidemiol Biomarkers Prev 17: 1228–1233

81. Ducloux D, Courivaud C, Bamoulid J et al (2008) Pretransplant serum vitamin D levels and risk of cancer after renal transplantation. Transplantation 85:1755–1759

82. Trivedi DP, Doll R, Khaw KT (2003) Effect of four monthly oral vitamin D3 (cholecalciferol) supplementation on fractures and mortality in men and women living in the community: randomized double blind controlled trial. BMJ 326:469–475

83. Lappe JM, Travers-Gustafson D, Davies KM et al (2007) Vitamin D and calcium supplementation reduces cancer risk: results of a randomized trial. Am J Clin Nutr 85:1586–1591

84. Whitlatch LW, Young MV, Schwartz GG et al. (2002) 25-Hydroxyvitamin D-1alpha-hydroxylase activity is diminished in human prostate cancer cells and is enhanced by gene transfer. J Steroid Biochem Mol Biol 81:135–140

85. Chen TC, Wang L, Whitlatch LW et al (2003) Prostatic 25-hydroxyvitamin D-1alpha-hydroxylase and its implication in prostate cancer. J Cell Biochem 88:315–322

86. Porojnicu A, Robsahm TE, Berg JP et al (2007) Season of diagnosis is a predictor of cancer survival. Sun-induced vitamin D may be involved: a possible role of sun-induced Vitamin D. J Steroid Biochem Mol Biol 103:675–678

87. Ng K, Meyerhardt JA, Wu K et al (2008) Circulating 25-hydroxyvitamin D levels and survival in patients with colorectal cancer. J Clin Oncol 26:2984–1991

88. Zhou W, Heist RS, Liu G et al (2007) Circulating 25-hydroxyvitamin D levels predict survival in early-stage non-small-cell lung cancer patients. J Clin Oncol 25:479–485

43 Vitamin D Deficiency and the Epidemiology of Prostate Cancer

Gary G. Schwartz

Abstract The hypothesis that vitamin D deficiency increases risk the for clinical prostate cancer has stimulated a large body of epidemiologic research including observational studies of sunlight exposure, serum vitamin D metabolites, and polymorphisms in the vitamin D receptor. The many studies on sunlight exposure strongly support a protective role for sunlight exposure. Conversely, the data from serological studies are mixed. Paradoxically, two studies suggest that high vitamin D levels may be associated with increased risk. Our recent demonstration that serum levels of calcium are strongly and positively related to prostate cancer risk may help resolve this discrepancy as high levels of vitamin D may be associated with normal, but relatively high values for serum calcium. These data indicate that elements of the vitamin D endocrine system whose investigation previously has attracted little attention, i.e., serum calcium and parathyroid hormone, are likely to exert important effects on the natural history of prostate cancer.

Key Words: Vitamin D; prostate cancer; 25-hydroxyvitamin D; epidemiology; 1,25-dihydroxyvitamin D; 1-hydroxylase; vitamin D receptor; polymorphism; sunlight; calcium

1. INTRODUCTION

The idea that prostate cancer is causally related to vitamin D deficiency, first proposed 20 years ago, has come of age. Once considered to be on the fringes of science, research on the roles of vitamin D in prostate cancer now features prominently not only in prostate cancer etiology, but in prostate cancer therapy, palliation, and survival *(1–3)*. This chapter summarizes the current status of vitamin D deficiency in prostate cancer etiology. It may be helpful to first summarize some clinical aspects of prostate cancer and vitamin D synthesis.

2. PROSTATE CANCER, CLINICAL CONSIDERATIONS

Prostate cancer is the commonest (nonskin) cancer among men in the Western world. Worldwide, age-adjusted mortality rates for prostate cancer vary more than 20-fold and are highest among African Americans and northern Europeans *(4)*. Unlike mortality rates, subclinical prostate cancer (also known as "latent" or "autopsy" prostate cancer)

From: *Nutrition and Health: Vitamin D*
Edited by: M.F. Holick, DOI 10.1007/978-1-60327-303-9_43,
© Springer Science+Business Media, LLC 2010

is highly prevalent among older men regardless of their race or geographic location
(5). This suggests that the occurrence of clinical prostate cancer depends upon factors
that govern the growth of subclinical prostate cancers. The widespread use of prostate-
specific antigen (PSA) as a screening test has markedly increased the detection of subclinical prostate tumors but has not been proven to reduce prostate cancer mortality rates
(6).

Prostate cancers that are contained within the prostate gland are potentially curable
(via surgical removal of the prostate gland, prostatectomy, or via radiation therapy).
Prostate cancers that have spread beyond the prostate gland are treatable with ionizing
radiation and androgen deprivation ("hormonal" therapy). Most prostate cells depend
upon androgen for growth and androgen deprivation, via medical or surgical castra-
tion, results in their death. Androgen deprivation is an effective palliative therapy but
the duration of response for men with metastatic disease typically is generally less
than 2 years, after which virtually all prostate cancers become androgen – independent
(7). Docetaxel-based therapies provide a small survival benefit for men with androgen-
independent metastatic prostate cancer, the median survival for which is 19 months *(8).*

3. SYNTHESIS OF VITAMIN D METABOLITES

As detailed in earlier chapters of this book, the synthesis of $1,25(OH)_2D$ (calcitriol)
begins with the production of vitamin D_3 (cholecalciferol) after 7-dehydrocholesterol
in the skin is exposed to UV-B radiation or after vitamin D is ingested from the diet.
Approximately 90% of vitamin D is sunlight derived *(9).* Vitamin D is hydroxylated
first in the liver at the 25th carbon, forming the prohormone, 25-hydroxyvitamin D
[25(OH)D], and again at the 1-α position, forming $1,25(OH)_2D$, the active, hormonal
form of vitamin D *(10).* Classically, the hydroxylation of 25(OH)D at the 1α position
was presumed to occur exclusively in the kidney and the function of $1,25(OH)_2D$ was
thought to be in the control of serum calcium and phosphorus *(11).* However, we now
understand that local synthesis of $1,25(OH)_2D$ occurs in an autocrine or paracrine fashion in prostate (and other) cells, where $1,25(OH)_2D$ controls key processes involving
cell differentiation, proliferation and invasion *(12, 13).*

4. PROSTATE CANCER AND THE VITAMIN D HYPOTHESIS

In 1990, we noted that the major risk factors for prostate cancer, older age, Black
race, and residence at northern latitudes, are all associated with a decreased synthesis
of vitamin D. We proposed that vitamin D maintained the normal phenotype of prostatic cells and that vitamin D deficiency promoted the development of clinical prostate
cancer from its preclinical precursors *(14).* This idea was supported by geographic analyses published in 1992, in which we showed that U.S. county-wide mortality rates for
prostate cancer among Caucasian men were inversely correlated with the availability
of ultraviolet radiation, the major source of vitamin D (Figs. 1 and 2) *(15, 16).* The
same year, Miller and colleagues demonstrated that prostate cancer cells possessed
high-affinity receptors for $1,25(OH)_2D$ (vitamin D receptors, VDR) *(17).* Pleiotropic
anticancer effects of $1,25(OH)_2D$ on normal and cancerous prostate cells later were

described by numerous laboratories. The mechanisms for these effects in the prostate are not completely understood but include inhibition of: cell proliferation (via cell cycle arrest) *(18)*; invasion through the basement membrane (via inhibition of matrix metallo-proteinases) *(19)*; migration *(20)*; metastasis *(21, 22)*; and angiogenesis *(23)*. In general, the inhibitory roles of vitamin D in prostate carcinogenesis are believed to occur in the various stages of cancer promotion/progression, not in tumor initiation (Table 1).

Table 1
Risk Factors for Prostate Cancer and Their Interpretation by the Vitamin D Deficiency Hypothesis

Risk factor	Explanation by vitamin D deficiency hypothesis
Age	The prevalence of vitamin D deficiency increases with age
Race	
Black	Melanin inhibits synthesis of vitamin D
Asian	Traditional diet high in vitamin D (fish oil) protects against clinical cancer. Protection wanes as migrants adopt a Western diet
Geography	U.S. mortality rates from prostate cancer are inversely correlated with ultraviolet radiation

Adapted from Schwartz and Hulka *(14)*.

The inverse relationship between UV radiation and prostate cancer mortality in the United States that we showed in the early 1990 s has been replicated by different authors using different analytic techniques and over different periods of time *(24–27)*. However, all of these studies are ecologic, i.e., they based on data at the level of the group (e.g., states or counties). Findings made at the group cannot guarantee that these relationships exist for individuals *(28, 29)*. Hypothesis testing about individuals requires epidemiologic studies of individuals, i.e., cohort and case–control studies (reviewed Sections 7–11, below).

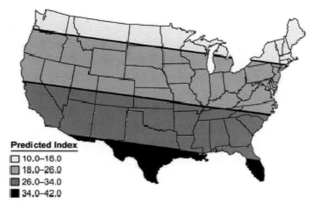

Predicted Index
☐ 10.0–18.0
▨ 18.0–26.0
▩ 26.0–34.0
■ 34.0–42.0

Fig. 1. Linear trend surface map of ultraviolet radiation in 3073 counties of the contiguous United States. (Redrawn from *(15)*).

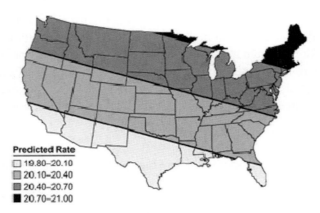

Fig. 2. Linear trend surface analysis of age-adjusted prostate cancer mortality by county among White men, 1970–1979, in the contiguous United States. (Redrawn from *(15)*).

5. 1,25(OH)$_2$D IS SYNTHESIZED BY NORMAL PROSTATE CELLS

In 1990, most endocrine textbooks were emphatic that, with the exception of rare disorders like sarcoidosis, the kidney was the sole organ to hydroxylate the vitamin D prohormone into the active hormone. This view was based on the observations that circulating levels of 1,25(OH)$_2$D are basically undetectable in anephric individuals. We now recognize that this generalization is untrue; many organs have the capacity to synthesize 1,25(OH)$_2$D locally (i.e., as an autocrine hormone) yet do not release the hormone into the circulation as an endocrine hormone. The clue to the discovery of the autocrine synthesis of 1,25(OH)$_2$D by prostate cells came from the descriptive features of prostate cancer epidemiology. The north–south gradient in prostate cancer mortality and higher rates among Blacks suggested a deficiency in 25(OH)D, whose serum levels are known to be lower at higher latitudes and among persons with dark pigmentation. However, the active vitamin D hormone is 1,25(OH)$_2$D, not 25(OH)D. Serum levels of 1,25(OH)$_2$D are tightly regulated, are not lower in Blacks than Whites, and (in normal individuals) are not correlated with serum levels of 25(OH)D *(30)*. Thus, it was difficult to understand how the north–south gradient in prostate cancer mortality rates and the racial difference could be related to vitamin D deficiency. We reasoned that this paradox would be solved if prostate cells synthesized 1,25(OH)$_2$D intra-prostatically from circulating serum levels of 25(OH)D. In 1998, we demonstrated that normal human prostate cells possess 25-hydroxyvitamin D$_3$-1α-hydroxylase (1α(OH)ase) and indeed synthesize 1,25(OH)$_2$D from 25(OH)D *(31)*. Moreover, we and others showed that 25(OH)D inhibits the proliferation of prostate cells that possess 1α-OHase *(32)*. Thus, the autocrine synthesis of 1,25(OH)$_2$D in normal prostate cells provides a biochemical mechanism by which exposure to sunlight or vitamin D might prevent prostate cancer since normal prostate cells manufacture a tumor inhibitor from the body's exposure to sunlight *(33)* (Fig. 3).

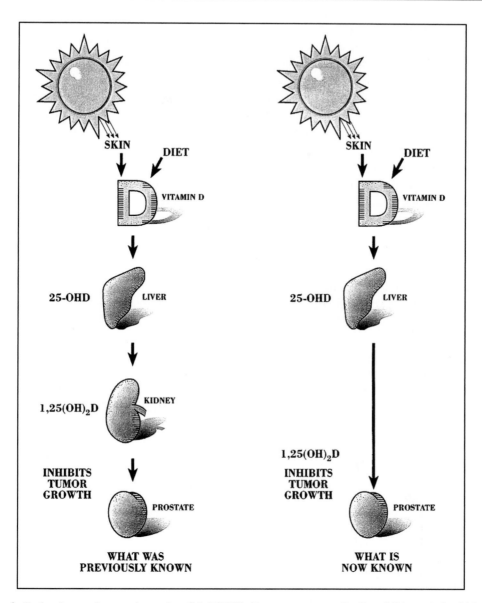

Fig. 3. Endocrine and autocrine role of $1,25(OH)_2D$ on prostate cells. In addition to the kidney (endocrine system), the prostate synthesizes its own $1,25(OH)_2D$ (autocrine system) that controls local growth and proliferation. (From Schwartz Semin Dialysis 2005; IS:276–289 *(43)*).

6. PROSTATE CANCER CELLS EXPRESS VDR BUT LOSE EXPRESSION OF 1α-OHase

The biological activity of $1,25(OH)_2D$ in tissues requires the presence of VDR, a ligand-dependent transcription factor that is a member of the steroid nuclear receptor superfamily *(34–36)*. The existence of VDR in human prostate cancer cells, first identified by Miller et al. in 1992, has been confirmed repeatedly by numerous investigators *(37–39)*. Krill et al. examined the expression patterns of VDR using immuno-histochemistry in 27 clinical samples of normal human prostates that were free from adenocarcinoma or suspected carcinoma. They showed that VDR are widely expressed in human prostate cells and were expressed more abundantly in the peripheral zone of the prostate (the site of origin of most prostate cancers) than in the central zone.

Although normal prostate cancer cells express high levels of 1α-OHase, 1α-OHase expression appears to be diminished in prostate cancer epithelial cells. Hsu and colleagues compared 1α-hydroxylase activity in samples from normal prostate epithelial cells, cancer-derived prostate epithelial cells, prostate cancer cell lines, and samples of benign prostatic hyperplasia (BPH). Expression of 1α-OHase was significantly reduced in BPH cells and was reduced further in the cancer-derived cells and cell lines *(40)*. Decreased expression of 1α-OHase was correlated with a decrease in growth inhibition in response to 25(OH)D. A loss of 1α-OHase expression in prostate cancers vs. benign and noncancerous prostates was also shown by Whitlatch et al., who demonstrated that transfection of the cDNA for 1α-OHase into prostate cancer cells that did not express 1α-OHase (LNCaP cells) and were not growth inhibited by 25(OH)D conferred growth inhibition by 25(OH)D in these cells *(41)*. These findings have important implications for prostate cancer therapy and prevention: Because normal prostate epithelial cells possess 1α-OHase, they could, in theory, be treated preventively with vitamin D. However, because many prostate cancers have reduced expression of 1α-OHase, they are unlikely to respond to vitamin D or 25(OH)D. However, because prostate cancer cells still express VDR, they could be treated with $1,25(OH)_2D$ and its analogs *(42)* (Fig. 4).

In summary, laboratory investigations of the vitamin D hypothesis have provided strong experimental support for broad, anticancer effects of vitamin D. These studies

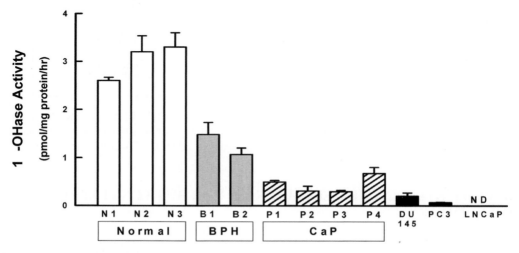

Fig. 4. 1α-OHase activity in primary cultures of normal, BPH, and prostate cancer (CaP) and in human prostate cancer cell lines, DU 145, PC-3, and LNCaP cells. Bars are standard deviations of three determinations. (Reprinted from Whitlatch et al. *(41)*, with permission).

have stimulated epidemiologic investigations of vitamin D exposures in individual men. Epidemiologic studies include studies of the VDR, of serum vitamin D concentrations, and of exposure to sunlight *(43)*.

7. VDR POLYMORPHISMS

The VDR protein is polymorphic (in the 3' end, the middle [the *FokI* polymorphism], and 5' upstream regulatory region of the gene) and numerous studies have evaluated prostate cancer risk in association with these polymorphisms. Two early studies found three- to fourfold increased risks of prostate cancer associated with polymorphisms in the 3' end of the VDR gene *(44, 45)*. A recent meta-analysis of 17 studies that assessed the 3' polymorphisms *TaqI*, *BsmI*, and *poly-A* repeat polymorphisms as well as the *FokI* polymorphism concluded that none of these variants was a major determinant of prostate cancer risk *(46)*. This conclusion is not surprising since it is likely that the effect of functional VDR polymorphisms would be observable only in the presence of low levels of vitamin D (i.e., VDR are likely to be risk modifiers). It is noteworthy in this regard that John et al. found reduced risks associated with both *TaqI* *tt* and *BglI* *BB* genotypes but only in the presence of high sun exposure *(47)*. Similarly, Ma et al. *(48)* reported reduced risk associated with the *TaqI* *tt* genotype but only among men with low serum 25(OH)D. Similar findings recently were reported by Li et al. *(49)*.

There is some evidence that *VDR* polymorphisms may be more strongly associated with advanced than with local disease *(50, 51)*. This may reflect the fact that in the post-PSA era, many localized cancers are screen-detected cancers with little malignant potential. Because most previous studies have included a mix of localized and advanced cases, the inclusion of localized cases would dilute risk estimates, which may explain some of the inconsistent findings in this area.

8. SEROLOGICAL STUDIES OF THE VITAMIN D DEFICIENCY HYPOTHESIS

Prediagnostic serum levels of 25(OH)D have been assessed in several prospective studies. A threefold increased risk was observed in Finnish men with low 25(OH)D (<40 nmol/l [<16 ng/ml] *(52)*. In addition to an increase in risk associated with low 25(OH)D, risk in a second Finnish study also was elevated in men with high 25(OH)D levels *(53)*. Several U.S. studies conducted in northern California *(54)*, Hawaii *(55)*, Maryland *(56)*, the Southeast *(57)*, and among physicians *(58)* and health professionals *(59)* did not observe an association with serum 25(OH)D levels (reviewed in John et al., 2007). However, the cut points for low 25(OH)D in these studies ranged from <21.4 ng/ml to <24.1 ng/ml and <34 ng/ml, approximately twice those in the Finnish study (<12 ng/ml). It is possible that increased risks are observed only when the low exposure category reflects vitamin D deficiency (i.e., below 20 ng/ml).

Recently, Li et al. identified 1,066 men with incident prostate cancer from 14,916 men initially free of cancer from the Physician's Health Study. They reported that compared to men with plasma 25(OH)D levels above the median and with the *FokI* FF or Ff genotype, men with 25(OH)D levels below the median and the less functional

*Fok*I ff genotype had increased risks of total (OR = 1.9, 95% CI = 1.1–3.3) and aggressive prostate cancer (OR = 2.5, CI = 1.1–5.8; aggressive was defined as stage C or D, Gleason grade 7–10, and fatal prostate cancer). Conversely, in a case–control study nested within the prostate, lung, colon, and ovarian (PLCO) cancer screening trial, in which patients were diagnosed from 1 to 8 years after blood draw, Ahn et al. reported no effect for season-adjusted 25(OH)D and overall prostate cancer risk. However, they reported that serum concentrations greater than the lowest quintile were associated with increased risk of aggressive cancer (defined as Gleason sum ≥7 or clinical stage III or IV disease) *(60)*.

In summary, the serologic evidence in favor of the vitamin D hypothesis is mixed. The evidence appears strongest for studies conducted at high latitudes, e.g., England and Scandinavia, where the prevalence of vitamin D deficiency and insufficiency is higher than in the United States. In the original Finnish study, half the men had 25(OH)D levels below 40 nmol/l (16 ng/ml), which is near a common clinical indicator of vitamin D deficiency (<15 ng/ml). In the U.S. studies, considerably smaller proportions of men had deficient 25(OH)D levels ranging from 5 to 13.3%. In the Hawaiian study, none of the men had 25(OH)D levels below 21 ng/ml. Paradoxically, data from one Finnish study and the U.S. PLCO trial raise the possibility that high levels of 25(OH)D might be associated with increased prostate cancer risk.

9. EPIDEMIOLOGIC STUDIES OF SOLAR EXPOSURE

At least nine epidemiologic studies have examined the risk of prostate cancer in relation to exposure to ultraviolet (UV) radiation, and their results are consistent with a protective effect of sunlight exposure. For example, Luscombe and colleagues compared 210 men with prostate cancer in the United Kingdom to 155 men with benign prostatic hypertrophy (noncancer controls) on various measures of lifetime sunlight exposure *(61)*. They found that a high sunbathing score was significantly protective (OR = 0.83, 95% CI = 0.76–0.89). Conversely, a low exposure to ultraviolet radiation was associated with a significantly increased risk for prostate cancer (OR = 3.03, 95% CI = 1.59–5.78). Multiple sunburns during childhood were significantly inversely associated with prostate cancer risk (OR = 0.18, 95% CI = 0.08–0.38).

An important issue in retrospective studies such as Luscombe's is recall bias. Due to the popularization of the sunlight/prostate cancer story in the popular press (e.g., *(62)*), some men with prostate cancer may underestimate their actual exposures to UV radiation, leading to a bias in support of the hypothesis.

Four studies that circumvent this problem are those of Friedman et al., Weinrich et al., and two studies by John et al. Freedman et al. conducted a death certificate-based case–control study of mortality from prostate cancer in association with exposure to sunlight *(63)*. Cases were deaths from cancer between 1984 and 1995 in 24 states. Controls were deaths from causes other than cancer and other diseases thought to involve sunlight exposure. Residential exposure to sunlight was classified by state of residence at birth and at death. High residential exposure to sunlight was associated with a significantly decreased risk of fatal prostate cancer (OR = 0.90, 95% CI = 0.87–0.93). Because exposure to sunlight was classified from information on the death certificate, this study is unlikely to be influenced by recall bias.

Weinrich et al. measured the association between self-reported solar exposure and an abnormal serum PSA among men attending a prostate cancer-screening program. After adjustment for age, education, and income, frequent sun exposure (three or more times per week) was associated with a 55% reduction in the odds of an abnormal PSA ($R = 0.45$, 95% CI = 0.21–0.97) *(64)*.

John et al. analyzed data from the first National Health and Nutrition Examination Survey (NHANES I) Epidemiologic Follow-up Study in order to test the hypothesis that sunlight exposure reduces the risk of developing prostate cancer *(65)*. One hundred and fifty-three men with incident prostate cancer were identified from a cohort of 3,414 White men who completed the dermatologic examination. State of longest residence in the South (RR = 0.58, 95% CI = 0.38–0.88, $p<0.01$), and high solar radiation in the state of birth (RR = 0.48, 95% CI = 0.30–0.76, $p<0.01$) were associated with substantial and significant reductions in the risk of prostate cancer. The data were adjusted for the confounding effects of education, income, body mass index (BMI), height, alcohol consumption, smoking, physical activity, energy intake, and intake of fat and calcium. The prospective design of this study precludes the possibility of recall bias.

In a subsequent study, John et al. conducted a population-based case–control study of advanced prostate cancer among men from the San Francisco Bay area *(66)*. Interview data were collected on lifetime sun exposure for 450 Caucasian cases and 455 Caucasian controls. Using a reflectometer, skin pigmentation was measured at two sites: the upper underarm (a measure of sun-unexposed skin) and the forehead (a measure of sun-exposed skin). A halving of the risk of prostate cancer was seen for men with high sun exposure as determined by reflectometry (OR = 0.51, 95% CI = 0.33–0.80). Reductions in risk were also observed among men with high occupational outdoor activity. The use of skin reflectometry adds objective evidence for the effects of solar radiation that cannot be due to reporting bias.

Using estimates of vitamin D status derived from the Health Professionals Follow-up Study, Giovannucci et al. reported inverse, nonsignificant protective effects of vitamin D status on prostate cancer incidence and mortality *(67)*. Using cancer registry data from Norway, a country with a very large variation in seasonal levels of 25(OH)D, Robsahm et al. reported a significantly longer cancer survival from prostate cancer for cases diagnosed in the summer and autumn, when 25(OH)D levels are at their highest *(68)*.

In an ingeniously-designed study, de Vries et al. addressed the role of ultraviolet radiation in prostate cancer risk using nonmelanoma skin cancer as a surrogate measure for chronic sunlight exposure. Among Caucasians, nonmelanoma skin cancers, particularly those occurring in the head and neck region, are widely believed to reflect chronic sunlight exposure. de Vries and colleagues examined the risk of subsequent prostate cancer among elderly men with nonmelanoma skin cancer, who were part of a population-based cancer registry in the Netherlands. Men with nonmelanoma skin cancer subsequently experienced a significantly decreased incidence of prostate cancer (Standardized Incidence Ratio = 0.73, 95% CI = 0.56–0.94) *(69)*. These findings are in agreement with those of Luscombe et al. and are not susceptible to problems of recall bias. The decreased risk of prostate cancer among men with nonmelanoma skin cancer is difficult to explain on the basis of detection bias since patients with a diagnosis of cancer tend to be examined more intensely for the occurrence of subsequent tumors.

In summary, in contrast to the mixed result from serological studies, epidemiologic studies of sunlight exposure demonstrate consistently protective effects of sunlight exposure on prostate cancer risk. Why the contrast between the two types of studies? It is important to note that the serologic studies have important limitations. The most important of these concerns the problem of the timing of the exposure measurement. Athough 25(OH)D is the best serological measure of vitamin D status, its half-life in blood is only about 3 weeks. Because we do not know when during life vitamin D exposures are most important, it is possible that, for example, studies of middle-aged men a few years prior to diagnosis may have missed the window of opportunity for vitamin D to exert a preventive effect *(70)*. This interpretation is supported by the study by John et al., which found that early life exposure is more important than exposure later in life. In this regard, studies of habitual sunlight exposure, including those with objective measurements of skin pigmentation (e.g., John et al.) and diagnoses of nonmalignant skin cancer (e.g., de Vries et al.), are likely to be more sensitive to typical vitamin D status than is a single measurement in serum at a time that may not be relevant to the natural history of prostate cancer.

10. VITAMIN D DEFICIENCY AND PROSTATE CANCER: POTENTIAL MECHANISMS

Reduced intraprostatic conversion of 25(OH)D to 1,25(OH)$_2$D has generally been considered an adequate explanation for the geographic distribution of prostate cancer and the increased risk observed among African Americans. However, the local conversion of 25(OH)D to 1,25(OH)$_2$D is only one of several possible mechanisms by which vitamin D deficiency could increase prostate cancer risk. For example, in addition to VDR and 1α-OHase, prostate cancer cells express receptors for other vitamin D-related hormones, e.g., parathyroid hormone (PTH-Type I receptors) and for calcium (calcium-sensing receptors). At physiological levels, PTH promotes the growth and metastases of prostate cancer cells in tissue culture (see *(71)* for a review). Because serum levels of PTH rise as a consequence of vitamin D deficiency, vitamin D deficiency could increase prostate cancer risk indirectly, via its effects on serum PTH. This possibility is important because this effect would occur independently of the prostatic VDR and prostatic 1α-OHase. This observation also could help to explain how vitamin D deficiency could increase prostate cancer risk in the presence of reduced levels of prostatic 1α-OHase.

Additionally several epidemiologic studies have reported increased risk of prostate cancer with high intakes of dietary calcium *(72)*. In the Health Professionals Study, Giovannucci and colleagues reported a significantly increased risk of advanced and fatal cancer at calcium intakes of 1,500–1,999 mg/day that was greatest at intakes >2,000 mg/day (the present recommended allowance for men aged 51 and older is 1,200 mg/day) *(73)*. Other prospective studies, especially those investigating more moderate calcium intake, e.g. *(74)*, have not found evidence of increased risk. Giovannucci et al. speculated that the risk of elevation is caused by calcium's inhibition of the renal hydroxylation of 25(OH)D to 1,25(OH)$_2$D *(75)*. However, this mechanism seems unsatisfactory because prostate cancer risk appears more closely associated with serum 25(OH)D, rather than 1,25(OH)$_2$D.

11. SERUM CALCIUM AND PROSTATE CANCER RISK

We recently examined the association between calcium levels *in serum* and the risk for prostate cancer using a prospective cohort, the National Health and Nutrition Examination Survey (NHANES) and the NHANES Epidemiologic Follow-up Study (NHEFS). Eighty-five incident cases of prostate cancer and 25 prostate cancer deaths occurred over 46,188 person-years of follow-up. Serum calcium was determined an average of 9.9 years prior to the diagnosis of prostate cancer. Comparing men in the top to men in the bottom tertile of serum calcium, the multivariable adjusted relative hazard for fatal prostate cancers was 2.68 (95% CI = 1.02–6.99; P_{trend} = 0.04). For incident prostate cancer, the relative risk for the same comparison was 1.31 (95% CI = 0.77–2.20; P_{trend}=0.34). These results support the hypothesis that high serum calcium or a factor strongly associated with it, e.g., high serum parathyroid hormone, increases risk for fatal prostate cancer *(76)* (Fig. 5). These findings have recently been confirmed in a second, independent cohort *(77)*.

These recent findings may shed light on two unresolved issues in the literature on the epidemiology of prostate cancer and vitamin D: conflicting data on the effects of dietary calcium and on the apparently paradoxical effects of high levels of serum vitamin D. Although, in general, there is a poor correlation between calcium in diet and calcium levels in serum, individuals who consume large quantities of calcium, e.g., via calcium supplements, may increase their serum calcium levels *(78)*. This could help explain the null effects of several studies that examined calcium intakes within the normal range.

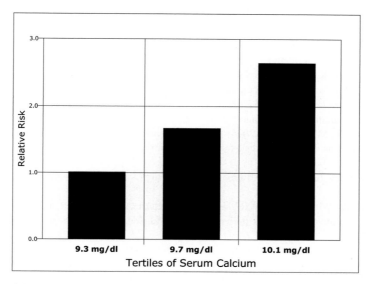

Fig. 5. Relative risk of fatal prostate cancer according to tertile of prediagnostic serum calcium. Data drawn from Skinner and Schwartz *(76)*.

Second, calcium absorption is known to vary within the normal reference range for 25(OH)D. Heaney and colleagues showed that absorption rates for calcium were 65% higher at serum 25(OH)D levels of 86.5 nmol/l compared to <50 nmol/l *(79)*. Both of these vitamin D levels are within the normal reference range. These data suggest that despite acting as substrate for the prostatic conversion of 25(OH)D to 1,25(OH)$_2$D, by acting to increase serum calcium levels, high levels of 25(OH)D, in the presence of high dietary calcium, might act to increase prostate cancer risk. It is important to note that the increases in serum 25(OH)D need not be high enough to induce hypercalcemia (often defined as calcium > normal reference range of 10.5 mg/dl), but would only need to move an individual's serum calcium into the high end of the normal range.

These recent developments indicate that the role of the vitamin D endocrine system in prostate cancer is more complex than originally envisioned. Our earliest model of prostate cancer risk involved vitamin D deficiency as the sole actor *(14)*. However, recent data indicate that serum calcium and PTH contribute to vitamin D-dependent and vitamin D-independent effects *(71, 76, 77)*. If our recent findings for serum calcium predicting risk for fatal prostate cancer are confirmed, it is conceivable that optimum levels of vitamin D for prostate cancer prevention may be those which maximize intra-prostatic synthesis of 1,25(OH)$_2$D and suppress serum levels of PTH, but which are associated with serum levels of calcium in the lower normal range. A complete understanding of the role of vitamin D deficiency in the etiology of prostate cancer will require a comprehensive understanding of these related endocrine and autocrine effects.

REFERENCES

1. Schwartz GG (2009) Vitamin D and intervention trials in prostate cancer: from theory to therapy. Ann Epidemiol 19:96–102
2. Beer TM, Myrthue A (2006) Calcitriol in the treatment of prostate cancer. Anticancer Res 26: 2647–2651
3. Lagunova Z, Porojnicu AC, Dahlback A, Berg JP, Beer TM, Moan J (2007) Prostate cancer survival is dependent on season of diagnosis. Prostate 67:1362–1370
4. Kurihara M, Aoki K, Hismamichi S (eds) (1989) Cancer Mortality Statistics in the World, 1950–1985. University of Nagoya Press, Nagoya, Japan
5. Yatani R, Chigusa I, Akazaki K, Stermmerman G, Welsh R, Correa P (1984) Geographic pathology of latent prostate carcinoma. Int J Cancer 29:611–616
6. Lin K, Lipsitz R, Miller T et al (2008) Benefits and harms of prostate-specific antigen screening for prostate cancer: an evidence update for the U.S. Preventive Services Task Force. Ann Intern Med 149:192–199
7. Sharifi N, Gulley JL, Dahut WI (2005) Androgen deprivation therapy for prostate cancer. JAMA 294:238–244
8. Tannock IF, de Wit R, Berry WR, Horti J, Pluzanska A, Chi KN et al (2004) Docetaxel plus prednisone or mitoxantrone plus prednisone for advanced prostate cancer. New Engl J Med 351: 1502–1512
9. Haddad JG Jr, Hahn TJ (1973) Natural and synthetic sources of circulating 25-hydroxyvitamin D in man. Nature 244:525–527
10. Holick MF (1997) Photobiology of vitamin D. In: Feldman D, Gloriexux FH, Pike JW, (eds) Vitamin D, 1st ed. Academic Press, San Diego, CA, 33–39
11. Holick MF (2007) Vitamin D deficiency. N Engl J Med 19:266–281

12. Hansen CM, Binderup L, Hamberg KJ, Carlberg C (2001) Vitamin D and cancer: effects of $1,25(OH)_2D_3$ and its analogs on growth control and tumorigenesis. Front Biosc 6:D820–D848

13. Hewison M, Burke F, Evans KN, Lammas DA, Sansum DM, Liu P et al (2007) Extra-renal 25-hydroxyvitamin $D_{3-1}\alpha$-hydroxylase in human health and disease. J Steroid Biochem Mol Biol 103:316–321

14. Schwartz GG, Hulka BS (1990) Is vitamin D deficiency a risk factor for prostate cancer (hypothesis)?. Anticancer Res 10:1307–1311

15. Hanchette CL, Schwartz GG (1992) Geographic patterns of prostate cancer mortality: evidence for a protective effect of ultraviolet radiation. Cancer 70:2861–2869

16. Schwartz GG, Hanchette CL (2006) UV, latitude, and prostate cancer mortality: all sunlight is not the same (United States). Cancer Causes Control 17:1091–1101

17. Miller GJ, Stapleton GE, Ferrara JA, Lucia MS, Pfister S, Hedlund TE et al (1992) The human prostatic carcinoma cell line LNCaP expresses biologically active, specific receptors for 1α, 25-dihydroxyvitamin D_3. Cancer Res 52:515–520

18. Moreno J, Krishnan AV, Feldeman D (2005) Molecular mechanisms mediating the anti-proliferative effects of vitamin D in prostate cancer. J Steroid Biochem Bol Biol 97:31–36

19. Schwartz GG, Wang M-H, Zhang M, Singh RK, Siegal GP (1997) 1α, 25-Dihydroxyvitamin D (calcitriol) inhibits the invasiveness of human prostate cancer cells. Cancer Epidem Biomark Prev 6:727–732

20. Sung V, Feldman D (2000) 1,25-Dihydroxyvitamin D_3 decreases human prostate cancer cell adhesion and migration. Mol Cell Endocrinol 164:133–143

21. Lokeshwar BL, Schwartz GG, Selzer MG, Burnstein KL, Zhuang S-H, Block NL, Binderup L (1999) Inhibition of prostate cancer metastasis in vivo: a comparison of 1,25-dihydroxyvitamin D (calcitriol) and EB1089. Cancer Epidem Biomarkers Preven 8:241–248

22. Bao BY, Yeh SD, Lee YF (2006) 1α,25-Dihydroxyvitamin D_3 inhibits prostate cancer cell invasion via modulation of selective proteases. Carcinogenesis 27:32–42

23. Bao Y-B, Yao J, Lee Y-F (2007) 1α-Dihydroxyvitamin D_3 suppresses interleukin-8-mediated prostate cancer cell angiogenesis. Carcinogenesis 27:1883–1889

24. Kafadar K (1997) Geographic trends in prostate cancer mortality: an application of spatial smoothers and the need for adjustment. Ann Epidemiol 1997(7):35–45

25. Grant WB (2002) An estimate of premature cancer mortality in the U.S. due to inadequate doses of solar ultraviolet-B radiation. Cancer 94:1867–1875

26. Bosco FP, Schymura MJ (2006) Solar ultraviolet-B exposure and cancer incidence and mortality in the United States, 1993–2002. BMC Cancer 10:264–

27. Schwartz GG, Hanchette CLUV (2006) Latitude, and prostate cancer mortality: all sunlight is not the same (United States). Cancer Causes Control 17:1091–1101

28. Robinson WS (1950) Ecological correlations and the behavior of individuals. Am Sociol Rev 15:351–357

29. Schwartz GG, Porta M (2007) Vitamin D ecologic studies and endometrial cancer. Prev Med 45:323–324

30. Chesney RW, Rosen JF, Hanstra AJ, Smith C, Mahaffey K, DeLuca HF (1981) Absence of seasonal variations in serum concentrations of 1,25-dihydroxyvitamin D despite a rise in 25-hydroxyvitamin D in summer. J Clin Endocrinol Metab 53:139–143

31. Schwartz GG, Whitlatch LW, Chen TC, Lokeshwar BL, Holick MF (1998) Human prostate cells synthesize 1,25-dihydroxyvitamin D_3 from 25-hydroxyvitamin D_3. Cancer Epidemiol Biomark Prev 7:391–395

32. Barreto A, Schwartz GG, Woodrugg R, Cramer SD (2000) 25-Hydroxyvitamin D_3, the prohormonal form of 1,25-dihydroxyvitamin D_3, inhibits the proliferation of primary prostatic epithelial cells. Cancer Epidemiol Biomarkers Prev 9:265–270

33. Young MV, Schwartz GG, Wang L, Jamieson DP, Whitlatch LW, Flanagan JN, Lokeshwar BL, Holick MF, Chen TC (2004) The prostate 25-hydroxyvitamin D-1α-hydroxylase is not influenced by parathyroid hormone and calcium: implications for prostate cancer chemoprevention by vitamin D. Carcinogenesis 25:967–971

34. Haussler MR, Jurutka PW, Hseih JC, Thompson PD, Selznick SH, Hausller CA et al (1995) New understanding of the molecular mechanism of receptor-mediated genomic actions of vitamin D hormone. Bone 17:33S–38S

35. Zhuang S-H, Schwartz GG, Cameron D, Burnstein KL (1997) Vitamin D receptor content and transcriptional activity do not fully predict antiproliferative effects of vitamin D in human prostate cancer cells. Mol Cell Endocrinol 126:83–90

36. Miller GJ, Stapleton GE, Hedlund TE, Moffatt KA (1995) Vitamin D receptor expression, 24-hydroxylase activity, and inhibition of growth by $1\alpha,25$-dihydroxyvitamin D_3 in seven human prostatic carcinoma cell lines. Clin Cancer Res 1:997–1003

37. Kivineva M, Bläuer M, Syvälä H, Tammela T (1998) Tuohimaa. Localization of 1,25-dihydroxyvitamin D_3 receptor (VDR) expression in human prostate. J Steroid Biochem Mol Biol 66:121–127

38. Blutt SE, Weigel NL (1999) Vitamin D and prostate cancer. Proc Soc Exp Biol Med 221:89–98

39. Miller GJ (1998) Vitamin D and prostate cancer: biological interactions and clinical potentials. Cancer Metastasis Rev 17:353–360

40. Hsu JY, Feldman D, McNeal JE, Peehl DM (2001) Reduced 1alpha-hydroxylase activity in human prostate cancer cells correlates with decreased susceptibility to 25-hydroxyvitamin D_3-induced growth inhibition. Cancer Res 61:2852–2856

41. Whitlatch LW, Young MV, Schwartz GG, Flanagan JN, Burnstein KL, Lokeshwar BL, Rich ES, Holick MF, Chen TC (2002) 25-Hydroxyvitamin D-1-alpha-hydroxylase activity is diminished in human prostate cancer cells and is enhanced by gene transfer. J Steroid Biochem Mol Biol 81: 135–140

42. Chen TC, Schwartz GG, Burnstein KL, Lokeshwar BL, Holick MF (2000) The use of 25-hydroxyvitamin D_3 and 19-nor-1,25-dihydroxyvitamin D_2 as therapeutic agents for prostate cancer. Clin Cancer Res 6:901–908

43. Schwartz GG (2005) Vitamin D and the epidemiology of prostate cancer. Semin Dial 19: 276–289

44. Taylor JA, Hirvonen A, Watson M et al (1996) Association of prostate cancer with vitamin D receptor gene polymorphism. Cancer Res 56:4108–4110

45. Ingles SA, Ross RK, Yu MC et al (1997) Association of prostate cancer with genetic polymorphisms in vitamin D receptor and androgen receptor. J Natl Cancer Inst 89:166–170

46. Ntais C, Polycarpou A, Ioannidid JP (2003) Vitamin D receptor gene polymorphisms and risk of prostate cancer: a meta-analysis. Cancer Epidemiol Biomarkers Prev 12:1395–1402

47. John EM, Schwartz GG, Koo J, Van Den Berg D (2005) Ingles SA exposure, vitamin D gene polymorphisms and risk of advanced prostate cancer. Cancer Res 65:5470–5479

48. Ma J, Stampfer MJ, Gann PH et al (1998) Vitamin D receptor polymorphisms, circulating vitamin D metabolites, and risk of prostate cancer in United States physicians. Cancer Epidemiol Biomarkers Prev 7:385–390

49. Li H, Stampfer MJ, Hollis JB et al (2007) A prospective study of plasma vitamin D metabolites, vitamin D receptor polymorphisms, and prostate cancer. PLoS Med 4:e103

50. Ingles SA, Coetzee GA, Ross RK et al (1998) Association of prostate cancer with vitamin D receptor haplotypes in African-Americans. Cancer Res 58:1620–1623

51. Hamasaki T, Inatomi H, Katoh T et al (2002) Significance of vitamin D receptor gene polymorphism for risk and disease severity of prostate cancer and benign prostatic hyperplasia in Japanese. Urol Int 68:226–231

52. Ahonen MH, Tenkanen I, Teppo L et al (2000) Prostate cancer risk and prediagnostic serum 25-hydroxyvitamin D levels (Finland). Cancer Causes Control 11:847–852

53. Tuohimaa P, Tenkanen L, Ahonen M et al (2004) Both high and low levels of blood vitamin D are associated with a higher prostate cancer risk: a longitudinal, nested case–control study in the Nordic countries. Int J Cancer 108:104–108

54. Corder EH, Guess HA, Hulka BS et al (1993) Vitamin D and prostate cancer: a prediagnostic study with stored sera. Cancer Epidemiol Biomarkers Prev 2:467–472

55. Nomura AM, Stemmermann GN, Lee J et al (1998) Serum vitamin D metabolite levels and the subsequent development of prostate cancer (Hawaii, United States). Cancer Causes Control 9:425–432

56. Braun MM, Helzlsouer KJ, Hollis BW et al (1995) Prostate cancer and prediagnostic levels of serum vitamin D metabolites (Maryland, United States). Cancer Cause Control 9:425–432

57. Jacobs ET, Giuliano AB, Martinez ME et al (2004) Plasma levels of 25-hydroxyvitamin D, 1,25-Dihydroxyvitamin D and the risk of prostate cancer. J Steroid Biochem Mol Biol 89–90:533–537

58. Gann PH, Ma J, Hennekens CH et al (1996) Circulating vitamin D metabolites in relation to subsequent development of prostate cancer. Cancer Epidemiol Biomarkers Prev 5:121–126

59. Platz EA, Leitzmann MF, Hollis BW (2004) Plasma 1,25-dihydroxy- and 25-hydroxyvitain D and subsequent risk of prostate cancer. Cancer Causes Control 15:255–265

60. Ahn J, Peters U, Albanes D et al (2008) Serum vitamin D concentrations and prostate cancer risk: a nested case–control study. J Natl Cancer Inst 100:796–804

61. Luscombe CJ, French ME, Siu S, Saxby MR, Jones PW, Fryer AA et al (2001) Prostate cancer risk: associations with ultraviolet radiation, tyrosinase and melanocortin-1 receptor genotypes. Br J Cancer 85:1504–1509

62. Walsh PC, Worthington JF (1995) *The Prostate: A Guide for Men.* Johns Hopkins University' Press, Baltimore, MD

63. Freedman DM, Dosemeci M, McGlynn K (2002) Sunlight and mortality from breast, ovarian, colon, prostate, and non-melanoma skin cancer: a composite death certificate based case–control study. Occup Environ Med 59:257–262

64. Weinrich S, Ellison G, Weinrich M, Ross K, Reis-Starr C (2001) Low sun exposure and elevated serum prostate specific antigen in African American and Caucasian men. Am J Health Studies 17:148–150

65. John EM, Dreon DM, Koo J, Schwartz GG (2004) Residential sunlight exposure is associated with a decreased risk of prostate cancer. J Steroid Biochem Mol Biol 89–90:549–552

66. John EM, Schwartz GG, Koo J, Van Den Berg D, Ingles SA (2005) Sun exposure, vitamin D gene polymorphisms and risk of advanced prostate cancer. Cancer Res 65:5470–5479

67. Giovannucci E, Liu Y, Rimm EB et al (2006) Prospective study of predictors of vitamin D status and cancer incidence and mortality in men. J Natl Cancer Inst 98:451–459

68. Robsahm TE, Tretli S, Dahlback A, Moan J (2004) Vitamin D3 from sunlight may improve the prognosis of breast-, colon- and prostate cancer (Norway). Cancer Causes Control 15:149–158

69. De Vries D, Soerjomataram I, Houterman S et al (2007) Decreased risk of prostate cancer after skin cancer diagnosis: a protective role of ultraviolet radiation? Am J Epidemiol 165:966–972

70. Schwartz GG (2007) The "Cocaine Blues" and other problems in epidemiologic studies of vitamin D and cancer. Nutr Rev 65:S

71. Schwartz GG (2008) Prostate cancer, serum parathyroid hormone and the progression of skeletal metastases. Cancer Epidem Biomark Prev 17:478–483

72. Gao X, LaValley MP, Tucker KL (2005) Prospective studies of dairy product and calcium intakes and prostate cancer risk: a meta-analysis. J Natl Cancer Inst 97:1768–1777

73. Straub DS (2007) Calcium supplementation in clinical practice: a review of forms, doses, and indications. Nutr Clin Pract 22:286–296

74. Berndt SI, Carter HB, Landis PK et al (2002) Calcium intake and prostate cancer risk in a long-term aging study: the Baltimore Longitudinal Study of Aging. Urology 60:1118–1123

75. Giovannucci E, Lui Y, Stampfer MJ et al (2006) A prospective study of calcium intake and incidence of fatal prostate cancer. Cancer Epidemiol Biomarkers Prev 15:203–210

76. Skinner HG, Schwartz GG (2008) Serum calcium and incident and fatal prostate cancer in the National Health and Nutrition Examination Survey. Cancer Epidem Biomark Prev 17:2302–2305

77. Skinner HG, Schwartz GG (2009) A prospective study of total and ionized serum calcium and fatal prostate cancer. Cancer Epidem Biomark Prev 18:573–578

78. Muldowney WP, Mazbar SA (1996) Rolaids-yogurt syndrome: a 1990s version of milk-alkali syndrome. Am J Kidney Dis 27:270–272

79. Heaney RP, Dowell MS, Hale CA et al (2003) Calcium absorption varies within the reference range for serum 25-hydroxyvitamin D. J Am Coll Nutr 22:142–146

44 Vitamin D for Cancer Prevention and Survival

Edward D. Gorham, Sharif B. Mohr,
Frank C. Garland, and Cedric F. Garland

Abstract   Higher levels of the principal circulating form of vitamin D, 25-hydroxyvitamin D [25(OH)D], are associated with substantially lower incidence and death rates from colon, breast, and ovarian cancer, with a linear dose–response gradient. The accumulated evidence from observational studies and a randomized trial reveal that population serum levels of 25(OH)D in the range of 40–60 ng/ml will markedly reduce incidence and mortality rates of several cancers including those of the breast, colon, and ovary. There is an immediate clinical need for cancer care providers worldwide to assure that a serum 25(OH)D level > 40 ng/ml is achieved as soon as feasible after diagnosis of patients with breast and colon cancer, unless specifically contraindicated by preexisting hypercalcemia. This serum target could be revisited after further rigorous studies are performed, but the evidence that has accumulated during the past 30 years is sufficiently strong now to adopt the above dosages and serum targets for professional and public health action. Such prompt action is likely to cut mortality from these cancers by half within approximately 5 years. Research on a wider range of cancer types with higher serum 25(OH)D levels ($\geq$50 ng/ml or 125 nmol/l) is needed. In the meantime, vitamin D_3 intake by everyone in the continental United States and Canada aged 1 year and older should be no less than 2,000 IU/day of vitamin D_3 and 1,000 IU/day for infants.

   Key Words: Vitamin D; 25-hydroxyvitamin D; breast neoplasms; colonic neoplasms; ovarian neoplasms; dose–response; pathogenesis; prevention

1. INTRODUCTION

Studies linking vitamin D deficiency with higher incidence of cancer have created great scientific interest in public health and medicine *(1, 2)*. A total of 2,750 research studies have been published investigating this association, including 80 epidemiological studies. The epidemiological studies generally have found that higher serum 25(OH)D levels, or an environment typically associated with higher UVB irradiance, are associated with lower rates of cancer of the breast *(3–12)*, colon *(6, 8, 9, 13–19)*, ovary *(6, 20–22)*, and many other anatomic sites *(1, 6, 10–12, 15, 17, 18, 22–24)*, with similar findings *(14, 25–31)* or trends *(32)* for oral intake of vitamin D. Higher solar irradiance

From: *Nutrition and Health: Vitamin D*
Edited by: M.F. Holick, DOI 10.1007/978-1-60327-303-9_44,
© Springer Science+Business Media, LLC 2010

levels in childhood and adolescence are associated with substantially lower incidence of prostate cancer later in life *(33, 34)*.

The present rate of publication of epidemiological studies of the relationship between vitamin D and cancer is approximately 50 new articles per year. The vast majority of these studies support this relationship. They include cohort and case–control studies, natural experiments, and a randomized clinical trial using 1,100 IU/day of vitamin D_3 with calcium *(35)*. These positive studies have been accompanied by supportive laboratory research that has provided a new and unexpected mechanism for preventive action of vitamin D, in conjunction with calcium, on cancer incidence and survival. This mechanism is the vitamin D-deficiency-driven Disjunction–Natural Selection–Invasion (DNI) model *(36)*.

2. SERUM 25-HYDROXYVITAMIN D AND RISK OF BREAST CANCER

There was an approximately linear inverse relationship between serum 25(OH)D and estimated relative risk of breast cancer in a case–control study by Lowe and colleagues (Fig. 1) *(10)*, a nested case–control study of the Nurses' Health Cohort study by Bertone-Johnson et al. (Fig. 2) *(11)*, and two large case–control studies by Abbas et al. (Figs. 3–4) *(12, 48)*. The studies by Abbas et al. analyzed premenopausal and postmenopausal women separately and demonstrated that the inverse association of serum 25(OH)D deficiency with risk of breast cancer was similar in direction and magnitude in pre- and postmenopausal women *(12, 48)*.

A pooled analysis of published data from studies of breast cancer that reported odds ratios by quantiles *(10–12)* revealed that a median serum 25(OH)D level > 38 ng/ml

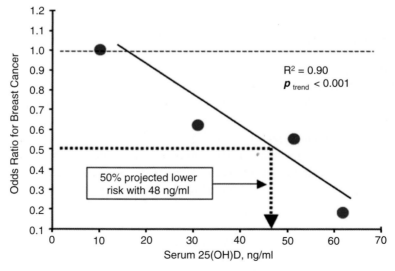

Fig. 1. Dose–response gradient of risk of breast cancer according to serum 25-hydroxyvitamin D concentration, Saint George's Hospital, London. Data from Lowe et al. *(10)*.

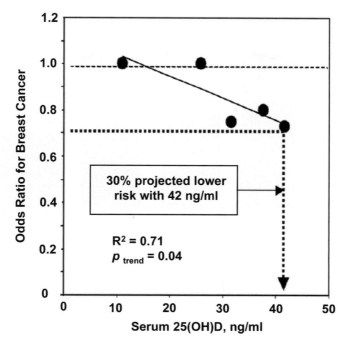

Fig. 2. Dose–response gradient of risk of breast cancer according to prediagnostic serum 25-hydroxyvitamin D concentration, Harvard Nurses' Health Study. Data from Bertone-Johnson et al. *(11).*

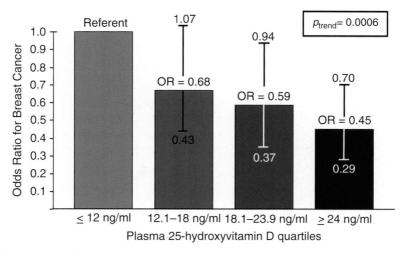

Fig. 3. Multivariate-adjusted odds ratios for *pre-menopausal* breast cancer by plasma 25-hydroxyvitamin D concentration by quartiles, 289 cases, 595 matched controls, Heidelberg, Germany. Source: Abbas et al. *(48).*

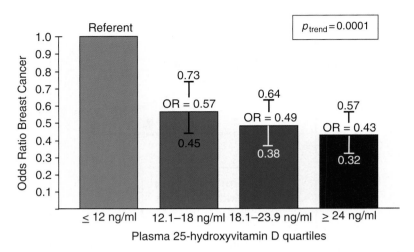

Fig. 4. Multivariate-adjusted odds ratios for *post-menopausal* breast cancer by plasma 25-hydroxyvitamin D concentration by quartiles, 1,394 cases, 1,365 matched controls, Heidelberg, Germany. Source: Abbas et al. *(48)*.

(95 nmol/l) (top quintile) was associated with an odds ratio of 0.42, corresponding to 58% lower risk of breast cancer in the top quintile compared to those in the lowest (<16 ng/ml) (Fig. 5) *(37)*. In this analysis, most of the variation in estimated relative risk of breast cancer was accounted for by the serum 25(OH)D level. The results of this analysis strongly suggest that raising the serum 25(OH)D level to 52 ng/ml is virtually certain to eradicate half of all new breast cancer cases occurring in North America and Europe, based on the combined results of observational studies and consideration of

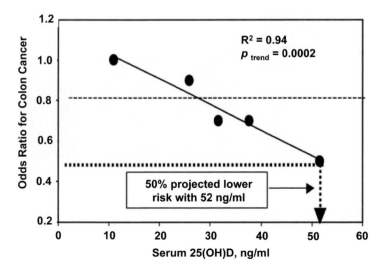

Fig. 5. Dose–response gradient of risk of breast cancer according to serum 25-hydroxyvitamin D concentration, pooled analysis. Source: Garland et al. *(37)*.

the standard A. B. Hill criteria for assessing causality in observational studies *(38)*, including dose–response relationship *(10–12, 36)*, strength of association *(10–12, 36)*, consistency among studies *(10–12, 36)*, coherence with biological knowledge *(1, 36, 39–42)*, and temporal sequence *(11, 43)*.

Freedman and associates of the National Cancer Institute *(43)* recently reported that women in the National Health and Nutrition Examination Survey III (NHANES III) cohort with serum 25(OH)D $\geq$ 25 ng/ml (62 nmol/l) had a relative risk of dying of breast cancer of only 0.28 (95% CI 0.08–0.93), compared to those with < 25 ng/ml (62 nmol/l) at baseline, corresponding to 72% reduction in mortality in women in the top half of the distribution of serum 25(OH)D compared to the bottom half (Fig. 6).

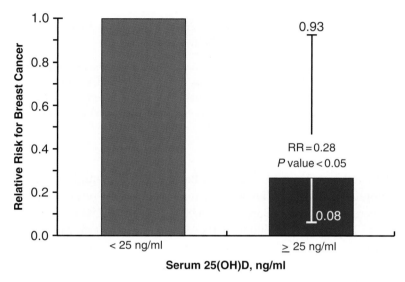

Fig. 6. Relative risk of breast cancer mortality, by baseline serum 25-hydroxyvitamin D concentration, divided at the median, *NHANES III* cohort, 1988–2000. Source: Freedman et al. *(43)*.

A nested case–control study of a subset of individuals enrolled in the Women's Health Initiative (WHI) found that women in the highest quintile of serum 25(OH)D (> 27 ng/ml) at the time of enrollment had 20% lower incidence of breast cancer than those in the lowest quintile (<13 ng/ml) ($p < 0.04$) *(44)* (Fig. 7). These differences could be reduced slightly by adjusting for body mass index and physical activity, but such adjustments are arguably not appropriate for obesity. Serum 25(OH)D is in the causal sequence leading from obesity to breast cancer, since obese individuals have lower than average serum levels of 25(OH)D *(45, 46)*. Physical activity performed outdoors generally raises 25(OH)D due to solar UVB exposure *(47)*. Therefore, an adjustment for physical activity could needlessly overmatch and possibly mask an inverse association of 25(OH)D with breast cancer.

A nested case–control study in the Nurses' Health cohort found that the statistically significant inverse association between serum 25(OH)D and breast cancer persisted after adjustment for physical activity and body mass index *(11)* and that neither physical activity nor body mass index was independently associated with incidence of breast cancer *(11)*.

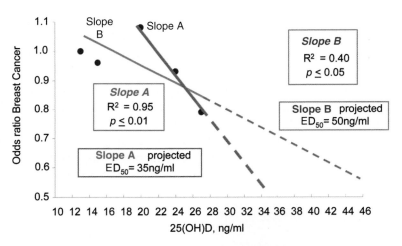

Fig. 7. Breast cancer odds ratios by serum 25-hydroxyvitamin D concentration, Women's Health Initiative Nested Case–Control Study with two slopes. Slope A excludes values below the 20 ng/ml threshold for prevention. Slope B includes all values. Slope A is the most accurate, since it is not influenced by random variation below the threshold. Source: Chlebowski et al. *(44)*.

The WHI nested case–control study *(44)* had a result that might have been expected on pharmacological grounds, by an analogy to classical dose–response curves. Two different slopes were evident for the association between serum 25(OH)D and risk of breast cancer (Fig. 7). The most prominent of these was a downward, diagonal linear slope (Slope A), starting at 20 ng/ml and continuing monotonically downward. The value of the slope was –0.04, and the p-value for trend for the slope was $p < 0.04$. There was no association for the range from 12 to 18 ng/ml, suggesting that there is a low-end asymptotic segment that is below a possible threshold for breast cancer risk reduction. Slope B includes values below the apparant threshold, and is somewhat more shallow (Fig. 7). A possible threshold was also discernible at 25 ng/ml in the Nurses Health cohort (Fig. 2) *(11)* and a case–control study of premenopausal women *(48)*. However, no threshold was evident in the study by Lowe et al. (Fig. 1) *(10)*. Some observational studies have failed to detect an association between serum 25(OH)D and risk of breast cancer *(49– 51)*. One of these studies identified an inverse association of serum 1,25(OH)$_2$D with risk of breast cancer *(50)*. Occasional failures to detect the associations that have been found in more recent studies, or those of different design, may have occurred due to analysis of serum specimens that were not collected with vitamin D research as the objective *(49)*, and possibly not fully protected from light during serum collection and processing. It is essential that serum collected for vitamin D research be collected without exposure to light and stored at –70°C or lower. Exposure to light or ultraviolet irradiance can seriously affect the results through unintended in vitro biosynthesis of vitamin D. Another reason that may explain an occasional failure to find the association is use of controls who are not closely enough matched on the date the specimen was collected. A case–control study of prediagnostic sera *(51)* may have missed the association because it did not individually match controls to cases on month and year that the sample was drawn. Climatological conditions vary substantially from year to year, as well as from month to month *(52)*.

3. SERUM 25-HYDROXYVITAMIN D AND RISK OF COLON CANCER

There are five studies of the association of serum 25(OH)D and risk of colorectal cancer (Fig. 8) that reported incidence rates by quintiles of serum 25(OH)D *(15, 18, 19, 53)*. A pooled analysis of published results of these studies yielded an inverse linear dose–response gradient between 25(OH)D and odds ratios for colon cancer *(36, 54)* (Fig. 9). A serum 25(OH)D level > 34 ng/ml (85 nmol/l) (top quintile) was associated with an odds ratio of 0.50, corresponding to 50% lower risk of colorectal cancer compared to the risk in subjects with 25(OH)D < 16 ng/ml (bottom quintile) *(54)*.

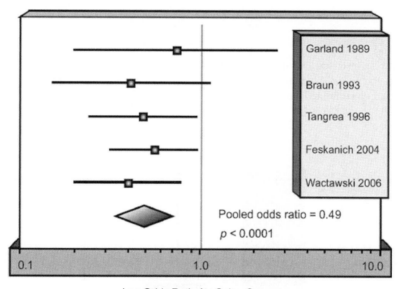

Fig. 8. Forest plot of all studies of serum 25(OH) vitamin D and risk of colorectal cancer. The upper and lower 95% confidence limits on the odds ratio are denoted by horizontal lines for each study, and the 95% confidence limits for the combined estimate for all studies are denoted by the points of the diamond. The odds ratios compare the highest quintile to the lowest. Source: Gorham et al. *(54)*.

A similarly inverse linear association with serum 25(OH)D was found for colorectal cancer mortality in the US national NHANES III cohort (Fig. 10) *(43)*. Individuals with serum 25(OH)D ≥ 32 ng/ml (80 nmol/l) had a relative risk of colorectal cancer death of only 0.28 (95% CI 0.11–0.68) compared with those with <20 ng/ml (50 nmol/l) serum 25(OH)D at baseline. This corresponds to 72% reduction in mortality from colorectal cancer in individuals in the top third of the serum 25(OH)D distribution.

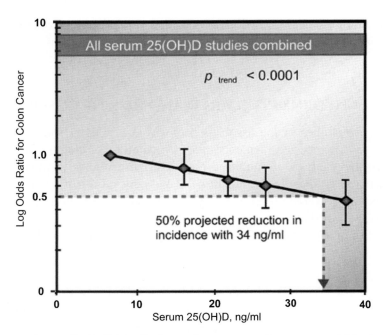

Fig. 9. Dose–response gradient of risk of colorectal cancer according to serum 25-hydroxyvitamin D concentration for all five studies combined. The five points are the odds ratios for each quintile of 25(OH)D based on combined data for the five studies. Source: Gorham et al. *(54)*.

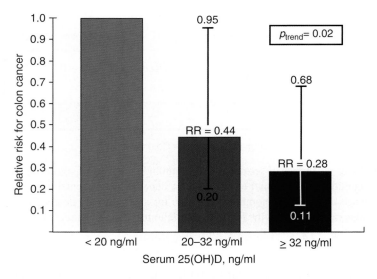

Fig. 10. Relative risk of colon cancer mortality, by baseline serum 25-hydroxyvitamin D concentration in tertiles, NHANES III cohort, 1988–2000. Source: Freedman et al. *(43)*.

4. LATITUDE, ULTRAVIOLET B, SERUM 25-HYDROXYVITAMIN D AND RISK OF CANCER OF THE OVARY

There have been similar findings for other cancer sites. Ecological studies first identified higher ovarian cancer mortality rates at higher latitudes and lower levels of solar irradiance *(6, 20, 24, 55)*. The latitudinal gradient was most evident in perimenopausal women. These studies have been supported by a study revealing that lower serum 25(OH)D is associated with higher risk of ovarian cancer in overweight women, although not in thinner women *(22)* (Fig. 11). Higher oral intake of vitamin D is also associated with substantially lower overall risk of ovarian cancer regardless of body weight *(29)*.

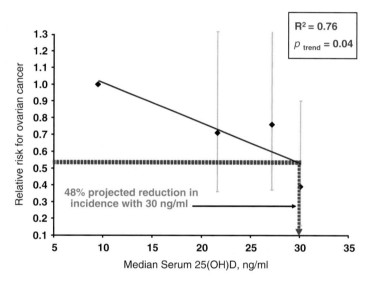

Fig. 11. Dose–response gradient of ovarian cancer according to prediagnostic serum 25-hydroxyvitamin concentration, subgroup of women with body mass index > 25. Source: Tworoger et al. *(22)*.

5. VITAMIN D AND CANCER OF OTHER SITES

Ecological studies have identified inverse associations of solar irradiance or UVB with 20 other sites of cancer, including renal *(6, 56)*, endometrial *(56)*, gastric *(6)*, and pancreatic *(6)*. A case–control study of oral vitamin D intake *(30)* and ecological studies of solar UVB *(23, 56)* found inverse associations with risk of endometrial cancer.

Vitamin D status is inversely associated with reduced incidence of most common cancers. The associations found in ecological studies of latitude or UVB and cancer risk have generally persisted when multiple adjustments were included in regression models, including per capita health care expenditures, and are therefore unlikely to be due to confounding *(23, 56–58)*. These associations should be examined in observational studies, but not at the expense of delaying action on the nearly universal need for vitamin D$_3$ repletion for cancer prevention in North America and Europe, and anywhere that serum 25(OH)D levels are low (<40 ng/ml, 120 nmol/l) due to habits such as having dark skin pigmentation, living a predominantly urban, indoor lifestyle or wearing clothing that limits access of sunlight to the skin.

There have been a few exceptions to the inverse association of markers of vitamin D with risk of other cancers, including a weak positive association with pancreatic cancer in Finnish male heavy smokers, mainly during winter months *(59)*, and of squamous dysplasia and malignancy of the esophagus in impoverished, agrarian, Chinese men *(60)*. The association in Finnish male heavy smokers was most evident during months when no vitamin D photosynthesis is possible in Finland. This is likely to be an indirect association, probably resulting from consumption of preserved and processed oily fish such as pickled herring, that are popular in winter months in the Nordic region. These contain substantial amounts of vitamin D but are toxic due to nitrosamines resulting from decomposition of protein in the presence of nitrate and nitrite salts *(61)*. The study did not address specific types of fish that were consumed, such as oily fish contaminated with carcinogenic compounds, nor specifically address intake of fish preserved with nitrates and nitrites, and therefore probably contaminated with nitrosamines *(59)*.

The finding that vitamin D status was positively related to squamous dysplasia or malignancy of the esophagus could also be indirect. Freshwater and marine fish are preserved with sodium nitrate and nitrite in some parts of China. These preserved fish contain nitrosamines, such as NDMA, that are highly carcinogenic to the pancreas, as well as containing vitamin D, so an indirect association analogous to that possibly present for preserved oily fish in Finland may exist for preserved fish in rural Chinese men *(62)*. While 25(OH)D and vitamin D$_3$ are almost exclusively found in food of animal origin, they also occur in a few species of toxic plants, such as *Solanum glaucophyllum* and related species that grow in abundance in disturbed soil of many rural agrarian areas worldwide and are eaten by some poor rural men due to a cultural belief that their consumption increases strength and masculinity *(63)*. This could create an indirect, artifactual association confined to men, since women would not normally engage in this practice. Although it is probable that these plants were present in the environment, it has not been confirmed that they grow in the region of China where the study was performed.

6. RANDOMIZED TRIALS

Three randomized clinical trials have reported results of assignment to vitamin D regarding reduction of cancer incidence. The most recently begun trial was a randomized, placebo-controlled clinical trial performed by Joan Lappe and her colleagues *(35)*. It found that supplementation of postmenopausal women with 1,100 IU/day of vitamin D$_3$ and 1,450 mg/day of calcium resulted in a 60% reduction in incidence of all invasive cancers combined ($p \leq 0.01$) (Fig. 12). There was a 77% reduction in incidence when cases diagnosed during the first year of follow-up were excluded ($p \leq 0.01$) (Fig. 13) *(35)*. This unexpectedly profound reduction in incidence of all invasive cancers combined occurred surprisingly promptly.

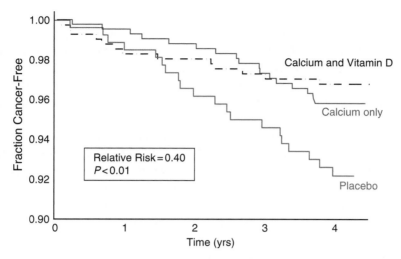

Fig. 12. Results of a randomized controlled trial of vitamin D and calcium after 4 years of follow-up among 1,179 healthy women in Omaha, NE, with a mean age 66.7 ± 7.3 years. Baseline serum 25(OH)D was 29 ± 8 ng/ml. The three treatment groups were vitamin D_3 (1,100 IU/day) and calcium (1450 mg/day); calcium (1,450 mg/day); and placebo. Serum 25(OH)D level was 38±9 ng/ml in the vitamin D_3 and calcium group. The outcome was all cancers (mainly breast, lung, and colon). Source: Lappe et al. *(35)*.

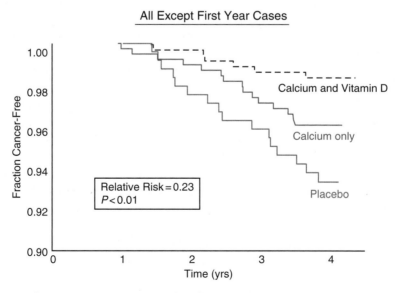

Fig. 13. Results of a randomized controlled trial of vitamin D and calcium after 4 years of follow-up among 1,179 healthy women in Omaha, NE, excluding first year cases. Baseline serum 25(OH)D was 29 ± 8 ng/ml. The three treatment groups were vitamin D_3 (1,100 IU/day) and calcium (1450 mg/day); calcium (1,450 mg/day); and placebo. Serum 25(OH)D level was 38±9 ng/ml in the vitamin D_3 and calcium group. The outcome was all cancers (mainly breast, lung, and colon). Source: Lappe et al. *(35)*.

On the other hand, the Women's Health Initiative (WHI) randomized controlled trial, which used a low dose of vitamin D (400 IU/day) and 1,000 mg/day of calcium, was unable to identify a substantial difference in risk of colorectal *(19)* or breast *(44)* cancer in the intervention group versus the comparison group. It is likely that this was because the 400 IU vitamin D intervention had only a minor effect (a predicted increase of 4 ng/ml or 10 nmol/l) on the serum level of 25(OH)D between intervention and comparison subjects.

A substantial number of subjects in the WHI study took vitamin D supplements on their own, or as directed by their personal physician, since this was allowed for ethical reasons in a population of women who were at typically high risk of osteoporotic fractures. After taking into consideration the inevitable contamination and crossover that occurred in WHI, the difference between treated and control subjects may have been in the range of 2.5–3.0 ng/ml, and the effect may have been lost in the noise. Questions have also been raised regarding the validity of the analysis, since the study had a factorial design and there were unexpected interactions among factors *(64)*. After accounting for interactions, a favorable effect of the vitamin D–calcium intervention was apparent, although it was of borderline statistical significance *(64)*.

7. FUTURE POTENTIAL FOR CANCER PREVENTION

It would be worthwhile to determine with precision whether additional reduction in risk of cancer at higher serum levels of 25(OH)D beyond the range that was present in most existing observational studies and clinical trial. An exception is the study of breast cancer reported by Lowe et al., that found that the inverse association of serum 25(OH)D persisted through at least 60 ng/ml (150 nmol/l) *(10)*. Classical dose–response curves for disease-preventing micronutrients typically have an early plateau at very low intakes, below the lowest dose threshold for efficacy, followed by a downward linear slope to the intake level at which the disease becomes virtually nonexistent *(65)*.

The inverse dose–response relationship for breast cancer is linear and downward beginning at 20 ng/ml (50 nmol/l) and continuing downward through at least 60 ng/ml (150 nmol/l), where risk was 83% lower than in the lowest quartile ($p < 0.001$) (Fig. 1) *(10, 36)*. Cohort studies of colorectal cancer have revealed a downward slope, similar to that of breast cancer *(36, 54)*. The tendency for linear extrapolations of these curves to approach nearly zero incidence within the physiological range of 25(OH)D is consistent with ecological studies that have examined incidence of breast *(21)* and colon cancer according to latitude and solar UVB irradiance *(66)*. The projected near-zero intercept is 90 ng/ml (225 nmol/l) for breast cancer and 60 ng/ml (150 nmol/l) for colon cancer. However, modeling that includes the most recent data *(12, 48)* suggests the near-zero intercept for breast cancer actually could be closer to 65 ng/ml (162 nmol/l).

Bertone-Johnson *(11)* has drawn attention to the possibility that there may be substantial subpopulations of individuals who have unrecognized vitamin D resistance, possibly due to previously unrecognized polymorphisms of the nuclear vitamin D receptor (VDR), that could influence the dose–response gradient. Further nested case–control studies would be desirable to determine the dose–response gradient in regions of the world where serum levels of 25(OH)D $\geq$ 50 ng/ml are commonly encountered.

8. SERUM 25-HYDROXYVITAMIN D AND CANCER SURVIVAL

Breast cancer patients with serum 25(OH)D > 32 ng/ml (80 nmol/l) at diagnosis had a 42% lower overall age-adjusted mortality rate than those with <20 ng/ml according to a follow-up study by Goodwin and her colleagues (67). Similarly, an analysis of survival of colorectal cancer patients at the Dana Farber Cancer Center by Ng and associates found that those with serum 25(OH)D ≥ 32 ng/ml (80 nmol/l) at diagnosis had only half the death rate as those with <20 ng/ml (50 nmol/l) (68). The difference persisted after adjustment for age. These very important follow-up studies confirmed earlier research that found substantially lower case-fatality rates in patients with breast (69) and colon (70) cancer who were diagnosed in months when serum 25(OH)D levels were highest (71).

9. VITAMIN D AND GLOBAL CANCER PREVENTION

According to calculations based on pooled analyses of relative risk of breast and colorectal cancer from observational studies (36, 54, 72), incidence estimates from the IARC GLOBOCAN database (73), estimates of ultraviolet B irradiance based on geophysical characteristics (74), and data on cloud cover from the NASA International Satellite Cloud Climatology Project (75), approximately 200,000 cases of breast (Fig. 14) and 250,000 cases of colorectal (Fig. 15) cancer annually worldwide would be prevented by raising serum 25(OH)D concentrations to 40–60 ng/ml. This serum 25(OH)D level is, in general, associated with oral intake of 2,000 IU of vitamin D3/day.

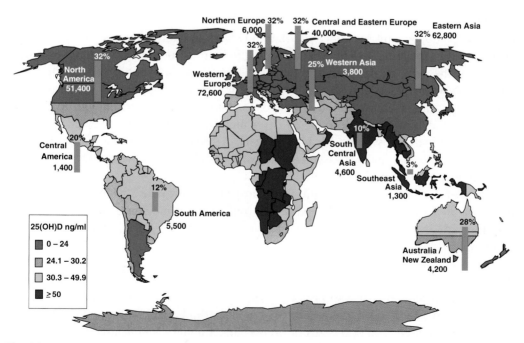

Fig. 14. Estimated 25(OH)D serum levels (see *legend*) and projected percentage prevention of *colon cancer* cases (bars) with 2,000 IU/day of vitamin D3 and 3–10 min daily of noon sunlight seasonally, when weather permits.

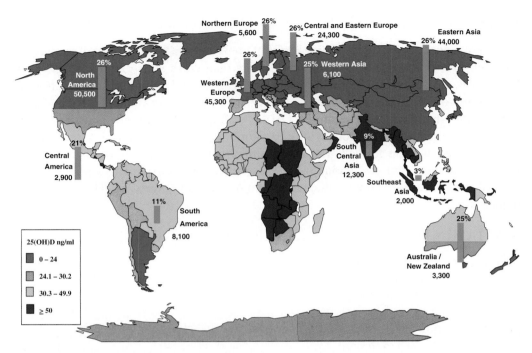

Fig. 15. Estimated 25(OH)D serum levels (see *legend*) and projected percentage prevention of *breast cancer* cases (*bars*) with 2,000 IU/day of vitamin D₃ and 3–10 min daily of noon sunlight seasonally, when weather permits.

Intake of 2,000 IU/day is the National Academy of Sciences Institute of Medicine (NAS-IOM) upper limit for intake of vitamin D *(76)*. In many climates, oral intake may be prudently supplemented by very modest (5–10 min) noontime sunlight exposure, to 40% of the skin, on days when weather allows. This corresponds to wearing shorts with no shirt for males, and shorts and a top for females that bares the shoulders and upper back, without sunscreen. Such exposure must be kept short enough to produce no visible reddening of the skin; longer exposures are counterproductive and would not substantially increase vitamin D₃ biosynthesis. A broad-brimmed hat should be worn whenever in the sun, and photosensitive persons should take a vitamin D₃ supplement in preference to solar exposure.

Intake of 2,000 IU/day of vitamin D₃ is projected to provide 26% reduction in incidence of breast cancer (Fig. 14) and a 32% reduction in incidence of colorectal cancer (Fig. 15) in the United States and Canada. The number of cases and deaths that would be prevented was projected based on these values, annual incidence rates *(73)* and case–fatality data *(77)*. The analysis projected that approximately 50,000 cases of breast cancer would be prevented in the United States and Canada each year and 12,000 deaths from it annually. It also projected that approximately 47,000 cases of colorectal cancer and 17,000 deaths would be prevented in the United States and Canada each year. Approximately 100,000 cases of breast cancer and 24,000 deaths per year would be prevented with intake of 4,000 IV/day of vitamin D₃. This intake also would prevent 94,000 cases of colorectal cancer and 34,000 deaths from it annually.

10. DISJUNCTION–NATURAL SELECTION–INVASION (DNI) MODEL

An understanding of the mechanisms of vitamin D action in cancer prevention suggests new directions for research and action. There are, of course, gaps in the research on mechanisms that will need to be filled. The level of understanding is moving forward, but is still rudimentary and emerging, so the summary that follows is necessarily partly speculative. The suggested model presented here is not intended to replace the current initiation–promotion model, but rather to place it into a context that accommodates features of cancer pathogenesis that are not predicted by the current models.

Vitamin D is not an antioxidant and does not prevent reactive oxygen species from attacking DNA. Its activity in preventing and reducing death rates from cancer could not be predicted from few-hit (78, 79) or many-hit deterministic sequential carcinogenesis models (80). Existing deterministic initiation–promotion models mainly address accumulation of defects in (or hypermethylation of) DNA, advancing toward malignancy. They did not accommodate the possibility of reversal of the time sequence from malignancy to normal tissue function.

A revised model for cancers of epithelial origin that accommodates the actions of vitamin D includes a phase that begins before classical initiation and continues through metastasis (36). This model may be termed the Disjunction–Natural Selection–Invasion (DNI) model. Its first phase is *disjunction,* consisting of the loss of intercellular adherence of epithelial cells within a tissue compartment, such as a breast terminal ductal lobular unit or a colonic epithelial crypt (Fig. 16) (36, 81, 82). Disjunction is a hallmark of malignancy that has been observed in time-lapse microphotography of mammary epithelium (83).

Most cancer cells lack intercellular junctions, except for some highly differentiated cells in locations distant from the primary malignancy. The capability of human epithelial cells for disjunction, and autonomous proliferation, is not surprising, since life earth existed only in unicellular form for approximately 85% of its 4.25 billion years on (84). Some of the DNA code present in contemporary human cells was very likely in place by the time of evolution of the first multicellular organisms approximately 650 million years ago (84).

10.1. Natural Selection: Basis of Pathogenesis

When disjunction of cells allows autonomy and a degree of mobility of cells in a tissue compartment, a population dynamic is created that allows the next phase, *natural selection.* This phase consists of competitive selection of the fastest-reproducing, most aggressive cells, a central process of evolution. Some of these cells appear to express the rapidly mitotic characteristics of their progenitor stem cells (85). Natural selection of rapidly mitotic epithelial cells does not ordinarily occur in well-differentiated epithelial cells that have intact intercellular junctions, but rather in clusters of cells that are not growth inhibited by tight intercellular contact with adjacent cells (86). An early step in carcinogenesis is natural selection of the fastest reproducing, disjunct cells, which may include epithelial cells that have reverted to acting as stem cells (87).

The driver of evolution is the gene (88), and its medium is any autonomous unit that reproduces the gene. When cells are bound to one another to constitute a multicellular

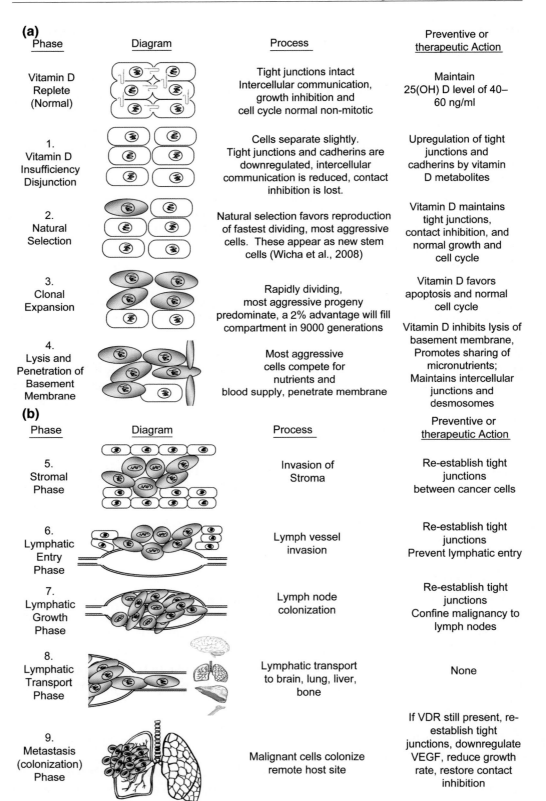

Fig. 16. (a, b) Disjunction-Natural Selection-Invasion (DNI) theory of cancer.

organism, evolution occurs at the level of the organism. When cells are no longer tightly bound, the cell itself is the vehicle of evolution. A cell having acquired a polymorphism associated with faster reproduction or aggression against other cells for limited resources, will eventually be over-represented in the tissue compartment (36). In settings where contact inhibition is absent, the progeny of a cell that has a 2% advantage in mitotic rate will occupy 99% of the tissue compartment in 9,000 generations, corresponding to approximately 20 years in most populations of actively mitotic epithelial cells (36).

Competing rapidly reproducing, aggressive cells that have arisen within a tissue compartment by natural selection have no way of sensing that their behavior will ultimately lead to the death of the individual. Of course, natural selection at the level of the organism would reduce this tendency, but its influence virtually stops at the end of the reproductive period. This has previously been suggested as a possible reason that incidence rates of cancer are usually low below age 50 years, after which they rise exponentially.

Leukemia, central nervous system tumors, and Wilm's tumors in children are exceptions. These malignancies might result from natural selection of very fast-reproducing aggressive stem cells during embryogenesis, when many stem cells are migratory, dissociated, and are reproducing rapidly. Natural selection is accelerated by agents that alter DNA, increasing genetic variation (89, 90). It can be accelerated by classical growth factors, such as insulin-like growth factors and 17-β-estradiol, that increase the number of generations per unit in cells having intact receptors (91).

10.2. Invasion

The next phase is *invasion* of the epithelial basement membrane. The basement membrane is composed of collagens and amino acids (92). A possible reason for lysis of the basement membrane could be erosive action on it by peripheral cells of a rapidly reproducing malignant clone that has become starved, due to hyperproliferation and crowding within the tissue compartment, for essential amino acids. Human cells cannot synthesize these compounds, but they are present in abundance as components of the acellular basement membrane. Aggressive cells near the basement membrane may begin to dissolve it enzymatically and by changing extracellular pH, to obtain the essential amino acids it contains. If the erosion of the membrane occurs at a faster rate than the repair rate, an unintended result may be weakening or localized lysis of the membrane, eventually allowing penetration by aggressive cells, which are then designated as malignant.

Invasion of the basement membrane may be followed by more advanced phases, from stromal invasion to metastasis (Fig. 16 a,b). Intercellular binding and communication of malignant cells with functional VDR receptors and response elements theoretically might be restored, allowing arrest of some tumors in more advanced stages of growth, such as through contact inhibition and differentiation. The theory that there may be beneficial action of vitamin D metabolites late in the natural history of malignancy is mainly based on observational data.

Invasion of distal tissues by malignant cells may be facilitated by lesions of the host tissue, including some that reduce its physical barrier function. These could include

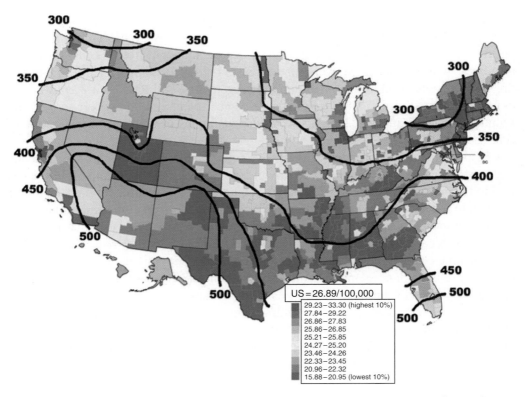

Fig. 17. Age-adjusted breast cancer mortality rates, by county area, and contours of annual mean daily solar irradiance in Langleys, in calories/cm^2, women, United States, 1970–1994. Source: Garland et al. *(1)*. Developed through use of National Cancer Institute and National Oceanic and Atmospheric Administration data sources (available at http://www3.cancer.gov/atlasplus/charts.html and http://www.noaa.gov).

loss of intercellular adherence due to weak or absent intercellular junctions, allowing metastatic cells access to the host tissue by the more open paracellular route.

Clinical studies of the effect of vitamin D$_3$ repletion in humans in all stages of malignancy would be the logical next step, allowing a comparison with historical survival data from local hospital and regional cancer registries and the Survival, Epidemiology and End Results Program (SEER) of the National Cancer Institute. Randomized trials may also be of some help, although waiting for their completion should not delay taking action on vitamin D for cancer patients, since the adverse effects of vitamin D$_3$ at the ordinary doses needed for repletion are well known and a benefit/risk calculation based on existing knowledge would be highly favorable for cancer patients.

10.3. Role of Calcium and Dietary Factors

Extracellular calcium ions are absolutely required for intercellular adherence *(93, 94)*. Intercellular junctions endocytose in response to very low concentrations of calcium in the extracellular fluid and exocytose upon its restoration *(94)*. Some exogenous agents also produce endocytosis of intercellular junctions, including linolenic acid (C18:3n-6),

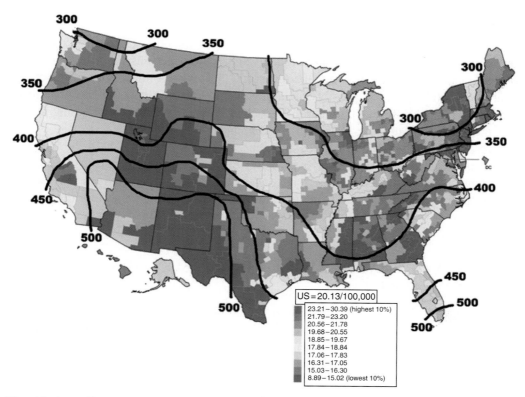

Fig. 18. Age-adjusted colon cancer mortality rates, by county area, and contours of annual mean daily solar irradiance in Langleys, in calories/cm², United States, 1970–1994. Source: Garland et al. *(1)*. Developed through use of National Cancer Institute and National Oceanic and Atmospheric Administration data sources (available at http://www3.cancer.gov/atlasplus/charts.html and http://www.noaa.gov).

and its precursor, linoleic acid (C18:2n-6) *(95)*. Unfortunately, n-6 linoleic acid is among the most common polyunsaturated fatty acids consumed in the Western diet (the median intake is 15 g/day). Many popular vegetable oils contain substantial amounts of n-6 linoleic and linolenic acid *(95)*, but their content is considerably lower in olive and canola oils than in other oils. Chenodeoxycholic acid and some other human bile acids also produce disjunction *(96, 97)*, and high lumenal concentrations predispose to colon cancer in humans *(97)* who do not consume sufficient calcium, consistent with a principle advanced and articulated by Newmark *(98)*.

Ideally, calcium intake should be adequate to allow vitamin D repletion to prevent malignancies, and intake of certain fats that induce disjunction should not be excessive. In some individuals disjunction may be due mainly to severe calcium deficiency and might be preventable with calcium repletion. Disjunction has its closest historical analogy in rickets, where deficiency of either vitamin D or calcium will cause the disease to occur *(99)* and repletion of both has the most substantial benefits *(100)*. Similar to their role in rickets, vitamin D and calcium play a joint and inextricable role in prevention of breast, colon, and ovarian cancer *(101)*.

10.4. Implications for Cancer Treatment

Recent follow-up observational studies that support the concept that endocytosis of intercellular junctions, the first and arguably the primary lesions preceding evolution of malignancy within a tissue, may be reversible throughout the life span of the malignant clone and its progeny, possibly including some metastatic cells that have colonized the host's remote tissues. The vitamin D nuclear receptor is robust and it may remain functional throughout the evolution of a malignancy *(102, 103)*.

Extreme exposures to powerful carcinogens, such as tobacco smoke or high intakes of ethanol and mycotoxins, may overwhelm the beneficial influence of vitamin D, and they do so for cancer of the lung and bronchus in smokers. Such carcinogen-driven cancers can be prevented only by eliminating exposure to tobacco or the relevant carcinogen.

11. SAFETY OF VITAMIN D FOR CANCER PREVENTION AND AS AN ADJUNCT TO TREATMENT

There have been 748 randomized controlled clinical trials (RCTs) that have randomly assigned vitamin D supplements to study participants. Most were performed to examine the effect of vitamin D on bone disease and to determine whether there were any side effects or toxicity. Dosages of vitamin D used in RCTs have included 400 *(19)*, 800 *(104)*, 1,100 *(35)*, 2,000 *(105–107)*, and 4,000 IU/day *(108, 109)*. A pediatric clinical trial used 800 IU/kg body weight for premature infants, corresponding to 2,000 IU for a 4.5 kg infant *(110)*. One study used a bolus dose of 300,000 IU given intramuscularly *(100)* for treatment of rickets. Reports of toxicity from use of vitamin D in the ordinary preventive and therapeutic range have been uncommon and virtually all minor *(111, 112)*, probably because most intake is below the normal human rate of excretion of vitamin D metabolites of 3,000–5,000 IU/day *(113)*.

Adoption of the new national guidelines for adequate intake (AI) of vitamin D_3 would substantially reduce the incidence of breast, colon, and ovarian cancer. A literature review found no reliable adverse effect of vitamin D_3 intake at or below 4,000 IU/day that would justify a lower AI *(35, 111, 113–116)*. Rare exceptions to the AI, such as for persons with sarcoidosis or certain other granulomatous diseases, should be mentioned in the recommendations. The evidence also indicates that it would be wise to raise the upper limit (UL) to at least 5,000 IU/day or higher, based on expected benefits compared to anticipated minor risks; some vitamin D scientists have recommended a UL of 10,000 IU/day *(111)*. Vitamin D_3 (cholecalciferol) should replace vitamin D_2 (ergocalciferol). Virtually all evidence reported to date on the efficacy of vitamin D for cancer and chronic disease prevention has been based on vitamin D_3. Vitamin D_3 is the normal product of photosynthesis of vitamin D in humans and other animals, and the most common form of vitamin D in the US diet. Vitamin D_3 is more effective in raising serum 25(OH)D in humans, at least in larger doses *(117)*, although vitamin D_2 and vitamin D_3 produce similar serum levels of $25(OH)D_2$ or $25(OH)D_3$ when taken at lower doses *(118)*. Vieth *(112)* has described the extensive literature supporting the low likelihood of toxicity of vitamin D_3 at intakes of 2,000–4,000 IU/day and concluded that a serum 25(OH)D target of at least 60 ng/ml (150 nmol/l) is optimal for most adults.

Restoring vitamin D and calcium to levels needed for lifelong bone health and minimizing the risk of cancer and other chronic diseases due to vitamin D deficiency may have some adverse health consequences, especially when applied to a large and diverse population. Principal among these is a possible minor increase in incidence of kidney stones in predisposed individuals or, possibly, at a rare extreme, renal calcinosis and renal failure. It is presently impossible to identify all persons predisposed to kidney stones on a population basis. With regard to vitamin D and calcium, restriction of intake is not generally recommended for reducing incidence of kidney stones *(119)*.

Most kidney stones in the United States and Canada are due to dehydration *(120)*, combined with underlying conditions such as excessive salt intake and metabolic disorders such as hypocitraturia *(121, 122)*, hyperoxaluria *(123)*, hyperuricosuria *(124)*, and an excessively acid or alkaline urinary pH *(120)*. Risk is particularly high in desert environments, apparently due to dehydration *(125)*. Excellent reviews on risk factors are available *(121, 122)*. Restriction of intake of calcium may paradoxically increase the risk of kidney stones, since oxalate bound in the intestine to calcium is not absorbed and does not appear in the urine *(126)*.

The preponderance of important preventive effects of higher vitamin D$_3$ intake for cancer and other chronic diseases has led 16 vitamin D scientists and physicians in the United States and Canada to join in a Call to Action recommending universal daily intake of 2,000 IU of vitamin D$_3$ at age 1 year through adulthood. This intake is the same as the current upper limit that NAS-IOM previously found, upon review and extensive analysis of scientific research, to have no adverse health effects *(76)*. This call to action arose from the cancer prevention research reviewed in this chapter and in part from studies that have identified inverse associations of vitamin D with risk of myocardial infarction *(127, 128)*, type 1 diabetes *(129, 130)*, multiple sclerosis *(131, 132)*, and falls *(104)*. These studies have moved the scientific debate past a threshold where the risk of inadequate vitamin D status greatly exceeds any credible minor known risk that might be associated with intake of 2,000 IU/day of vitamin D$_3$.

12. GEOGRAPHIC DISTRIBUTION OF CANCER AND ENVIRONMENTAL FACTORS

The findings from observational and laboratory studies and the clinical trial of 1,100 IU of vitamin D$_3$ and adequate calcium provide an explanation for the unusual geographic distribution of death rates from breast (Fig. 17) and colon (Fig. 18) cancer in the United States and help solve a mystery of epidemiology. In the northeast, turbid and carbonaceous sulfate air pollution from burning high-sulfur coal for electricity generation *(133)*, a disproportionately thick ozone layer *(4)*, and a climate influenced in winter by cold Canadian air masses *(4, 133)* eliminate any opportunity for photosynthesis of vitamin D during half of each year *(134)*, creating a severe epidemic of winter vitamin D deficiency *(135, 136)*. This increases the risk of breast, colorectal, and several other common malignancies.

13. SUMMARY

Mounting scientific evidence highlights the importance of preventing vitamin D deficiency to reduce cancer risk and using vitamin D repletion, with adequate calcium, to enhance the efficacy of existing cancer treatments. Universally preventing serum 25(OH)D deficiency will prevent a wide range of cancers, except those due to well-recognized carcinogens, such as tobacco smoke and certain viruses.

Further research should be performed to determine the extent to which the benefits of vitamin D repletion may apply to a wider range of cancers and to describe dose–response relationships with cancer incidence and mortality at higher serum 25(OH)D levels, such as 50–80 ng/ml (125–200 nmol/l) and oral intakes of vitamin D_3 above 2,000 IU/day.

If the DNI model is correct, there is no longer a pressing need for searching for a strong carcinogen to account for each cancer. Many cancers may arise in the absence of carcinogens, solely as a result of disjunction, natural selection, and invasion due to infidelity in reproduction of DNA or epigenetic factors. If disjunction is prevented by vitamin D repletion with adequate intake of calcium, autonomy and adverse natural selection leading to cancer will not occur. Restoration of contact inhibition and the capacity for apoptosis by vitamin D and calcium repletion may prevent uncontrolled proliferation of tumor clones, opening new avenues for enhancing treatment and improving survival.

Populations living at or above 30° latitude in either the northern or southern hemisphere, or whose members have a mainly indoor lifestyle, should be considered at high risk of breast, colon, and ovarian cancers as a result of highly prevalent vitamin D deficiency. The studies described in this chapter provide the scientific basis for a new era of public health action using 2,000 IU/day of vitamin D_3, adequate calcium, and a serum 25(OH)D target of 40–60 ng/ml (100–150 nmol/l) to substantially reduce incidence and mortality rates from cancer and markedly increase survival of patients with breast and colon cancer. The time to act is now, before more new cases occur and more lives are needlessly lost to these readily preventable diseases.

ACKNOWLEDGMENTS

This research was partially supported by a Special Interest Congressional Project allocation to the Penn State Cancer Institute of the Milton S. Hersey Medical Center, Hershey, PA, through the Department of the Navy, Bureau of Medicine and Surgery, and the Naval Health Research Center in San Diego, CA, Work Unit No. 60126. The views expressed in this chapter are those of the authors and do not represent an official position of the Department of the Navy, Department of Defense, or the US Government.

REFERENCES

1. Garland CF, Garland FC, Gorham ED, Lipkin M, Newmark H, Mohr SB et al (2006) The role of vitamin D in cancer prevention. Am J Public Health 96:252–261
2. Holick MF (2009) Vitamin D status: measurement, interpretation, and clinical application. Ann Epidemiol 19:73–78
3. Gorham E, Garland C, Garland F (1989) Acid haze air pollution and breast and colon cancer in 20 Canadian cities. Can J Publ Health 80:96–100

 4. Garland F, Garland C, Gorham E, Young J Jr (1990) Geographic variation in breast cancer mortality in the United States: a hypothesis involving exposure to solar radiation. Prev Med 19: 614–622

 5. Gorham ED, Garland FC, Garland CF (1990) Sunlight and breast cancer incidence in the USSR. Int J Epidemiol 19:820–824

 6. Grant WB (2002) An estimate of premature cancer mortality in the U.S. due to inadequate doses of solar ultraviolet-B radiation. Cancer 94:1867–1875

 7. Grant WB (2002) An ecologic study of dietary and solar ultraviolet-B links to breast carcinoma mortality rates. Cancer 94:272–281

 8. Freedman D, Dosemeci M, McGlynn K (2002) Sunlight and mortality from breast, ovarian, colon, prostate, and non-melanoma skin cancer: a composite death certificate based case–control study. Occup Environ Med 59:257–262

 9. Boscoe FP, Schymura MJ (2006) Solar ultraviolet-B exposure and cancer incidence and mortality in the United States, 1993–2002. BMC Cancer 6:264

10. Lowe LC, Guy M, Mansi JL, Peckitt C, Bliss J, Wilson RG et al (2005) Plasma 25-hydroxy vitamin D concentrations, vitamin D receptor genotype and breast cancer risk in a UK Caucasian population. Eur J Cancer 41:1164–1169

11. Bertone-Johnson ER, Chen WY, Holick MF, Hollis BW, Colditz GA, Willett WC et al (2005) Plasma 25-hydroxyvitamin D and 1,25-dihydroxyvitamin D and risk of breast cancer. Cancer Epidemiol Biomarkers Prev 14:1991–1997

12. Abbas S, Linseisen J, Slanger T, Kropp S, Mutschelknauss EJ, Flesch-Janys D et al (2008) Serum 25-hydroxyvitamin D and risk of post-menopausal breast cancer – results of a large case–control study. Carcinogenesis 29:93–99

13. Garland C, Garland F (1980) Do sunlight and vitamin D reduce the likelihood of colon cancer? Int J Epidemiol 9:227–231

14. Garland C, Shekelle RB, Barrett-Connor E, Criqui MH, Rossof AH, Paul O (1985) Dietary vitamin D and calcium and risk of colorectal cancer: a 19-year prospective study in men. Lancet 1: 307–309

15. Garland C, Comstock G, Garland F, Helsing K, Shaw E, Gorham E (1989) Serum 25-hydroxyvitamin D and colon cancer: eight-year prospective study. Lancet 2:1176–1178

16. Mizoue T (2004) Ecological studies of solar radiation and cancer mortality in Japan. Health Phys 87:532–538

17. Feskanich D, Ma J, Fuchs CS, Kirkner GJ, Hankinson SE, Hollis BW et al (2004) Plasma vitamin D metabolites and risk of colorectal cancer in women. Cancer Epidemiol Biomarkers Prev 13: 1502–1508

18. Tangrea J, Helzlsouer K, Pietinen P, Taylor P, Hollis B, Virtamo J et al (1997) Serum levels of vitamin D metabolites and the subsequent risk of colon and rectal cancer in Finnish men. Cancer Causes Control 8:615–625

19. Wactawski-Wende J, Kotchen JM, Anderson GL, Assaf AR, Brunner RL, O'Sullivan MJ et al (2006) Calcium plus vitamin D supplementation and the risk of colorectal cancer. N Engl J Med 354: 684–696

20. Lefkowitz ES, Garland CF (1994) Sunlight, vitamin D, and ovarian cancer mortality rates in US women. Int J Epidemiol 23:1133–1136

21. Mohr SB, Garland CF, Gorham ED, Grant WB, Garland FC (2008) Relationship between low ultraviolet B irradiance and higher breast cancer risk in 107 countries. Breast J 14:255–260

22. Tworoger SS, Lee IM, Buring JE, Rosner B, Hollis BW, Hankinson SE (2007) Plasma 25-hydroxyvitamin D and 1,25-dihydroxyvitamin D and risk of incident ovarian cancer. Cancer Epidemiol Biomarkers Prev 16:783–788

23. Grant W, Garland C (2006) The association of solar ultraviolet B with reducing risk of cancer: multifactorial ecological analysis of geographic variation in age-adjusted cancer mortality rates. Anticancer Res 26:2687–2699

24. Grant WB (2003) Ecologic studies of solar UV-B radiation and cancer mortality rates. Recent Results Cancer Res 164:371–377

25. Martinez ME, Giovannucci EL, Colditz GA, Stampfer MJ, Hunter DJ, Speizer FE et al (1996) Calcium, vitamin D, and the occurrence of colorectal cancer among women. J Natl Cancer Inst 88: 1375–1382

26. Kearney J, Giovannucci E, Rimm EB, Ascherio A, Stampfer MJ, Colditz GA et al (1996) Calcium, vitamin D, and dairy foods and the occurrence of colon cancer in men. Am J Epidemiol 143:907–917

27. Pritchard RS, Baron JA, Gerhardsson de Verdier M (1996) Dietary calcium, vitamin D, and the risk of colorectal cancer in Stockholm, Sweden. Cancer Epidemiol Biomarkers Prev 5:897–900

28. LaVecchia C, Braga C, Negri E, Francesci S, Russo A et al (1997) Intake of selected micronutrients and risk of colorectal cancer. Int J Cancer 73:525–530

29. Salazar-Martinez E, Lazcano-Ponce EC, Gonzalez Lira-Lira G, Escudero-De los Rios P, Hernandez-Avila M (2002) Nutritional determinants of epithelial ovarian cancer risk: a case–control study in Mexico. Oncology 63:151–157

30. Salazar-Martinez E, Lazcano-Ponce E, Sanchez-Zamorano LM, Gonzalez-Lira G, Escudero DELRP, Hernandez-Avila M (2005) Dietary factors and endometrial cancer risk. Results of a case–control study in Mexico. Int J Gynecol Cancer 15:938–945

31. Lin J, Manson JE, Lee IM, Cook NR, Buring JE, Zhang SM (2007) Intakes of calcium and vitamin D and breast cancer risk in women. Arch Intern Med 167:1050–1059

32. John E, Schwartz G, Dreon D, Koo J (1999) Vitamin D and breast cancer risk: The NHANES I epidemiologic follow-up study, 1971–1975 to 1992. Cancer Epidemiol Biomark Prev 8: 399–406

33. John EM, Dreon DM, Koo J, Schwartz GG (2004) Residential sunlight exposure is associated with a decreased risk of prostate cancer. J Steroid Biochem Mol Biol 89–90:549–552

34. John EM, Koo J, Schwartz GG (2007) Sun exposure and prostate cancer risk: evidence for a protective effect of early-life exposure. Cancer Epidemiol Biomarkers Prev 16:1283–1286

35. Lappe JM, Travers-Gustafson D, Davies KM, Recker RR, Heaney RP (2007) Vitamin D and calcium supplementation reduces cancer risk: results of a randomized trial. Am J Clin Nutr 85:1586–1591

36. Garland C, Grant W, Mohr S, Gorham E, Garland F (2007) What is the dose–response relationship between vitamin D and cancer risk?. Nutr Rev 65:S91–S95

37. Garland CF, Gorham ED, Mohr SB, Grant WB, Giovannucci EL, Lipkin M et al (2007) Vitamin D and prevention of breast cancer: pooled analysis. J Steroid Biochem Mol Biol 103:708–711

38. Hill AB (1965) The environment and disease: association or causation? Proc R Soc Med 58:295–300

39. Lipkin M, Newmark HL (1999) Vitamin D, calcium and prevention of breast cancer: a review. J Am Coll Nutr 18:392S–397S

40. Brenner B, Russell N, Albrecht S, Davies R (1998) The effect of dietary vitamin D3 on the intracellular calcium gradient in mammalian colonic crypts. Cancer Lett 12:43–53

41. Tangpricha V, Flanagan JN, Whitlatch LW, Tseng CC, Chen TC, Holt PR et al (2001) 25-Hydroxyvitamin D-1alpha-hydroxylase in normal and malignant colon tissue. Lancet 357:1673–1674

42. Cross HS, Lipkin M, Kallay E (2006) Nutrients regulate the colonic vitamin D system in mice: relevance for human colon malignancy. J Nutr 136:561–564

43. Freedman DM, Looker AC, Chang SC, Graubard BI (2007) Prospective study of serum vitamin D and cancer mortality in the United States. J Natl Cancer Inst 99:1594–1602

44. Chlebowski R, Johnson K, Kooperberg C et al (2008) Calcium plus vitamin D supplementation and the risk of breast cancer. J Natl Cancer Inst 100:1581–1591

45. Liel Y, Ulmer E, Shary J, Hollis BW, Bell NH (1988) Low circulating vitamin D in obesity. Calcif Tissue Int 43:199–201

46. Snijder MB, van Dam RM, Visser M, Deeg DJ, Dekker JM, Bouter LM et al (2005) Adiposity in relation to vitamin D status and parathyroid hormone levels: a population-based study in older men and women. J Clin Endocrinol Metab 90:4119–4123

47. Scragg R, Camargo CA Jr. (2008) Frequency of leisure-time physical activity and serum 25-hydroxyvitamin D levels in the US population: results from the Third National Health and Nutrition Examination Survey. Am J Epidemiol 168:577–586

48. Abbas S, Chang-Claude J, Linseisen J (2009) Plasma 25-hydroxyvitamin D and premenopausal breast cancer risk in a German case–control study. Int J Cancer 124:250–255

49. Hiatt R, Krieger N, Lobaugh B, Drezner M, Vogelman J, Orentreich N (1998) Prediagnostic serum vitamin D and breast cancer. J Natl Cancer Inst 90:461–463

50. Janowsky EC, Lester GE, Weinberg CR, Millikan RC, Schildkraut JM, Garrett PA et al (1999) Association between low levels of 1,25-dihydroxyvitamin D and breast cancer risk. Public Health Nutr 2:283–291

51. Freedman DM, Chang SC, Falk RT, Purdue MP, Huang WY, McCarty CA et al (2008) Serum levels of vitamin D metabolites and breast cancer risk in the prostate, lung, colorectal, and ovarian cancer screening trial. Cancer Epidemiol Biomarkers Prev 17:889–894

52. Garland CF, Gorham ED, Baggerly CA, Garland FC (2008) Re: Prospective study of vitamin D and cancer mortality in the United States. J Natl Cancer Inst 100:826–827

53. Braun MM, Helzlsouer KJ, Hollis BW, Comstock GW (1995) Colon cancer and serum vitamin D metabolite levels 10–17 years prior to diagnosis. Am J Epidemiol 142:608–611

54. Gorham ED, Garland CF, Garland FC, Grant WB, Mohr SB, Lipkin M et al (2007) Optimal vitamin D status for colorectal cancer prevention: a quantitative meta analysis. Am J Prev Med 32: 210–216

55. Garland CF, Mohr SB, Gorham ED, Grant WB, Garland FC (2006) Role of ultraviolet B irradiance and vitamin D in prevention of ovarian cancer. Am J Prev Med 31:512–514

56. Mohr SB, Garland CF, Gorham ED, Grant WB, Garland FC (2007) Is ultraviolet B irradiance inversely associated with incidence rates of endometrial cancer? an ecological study of 107 countries. Prev Med 45:327–331

57. Mohr SB, Gorham ED, Garland CF, Grant WB, Garland FC (2006) Are low ultraviolet B and high animal protein intake associated with risk of renal cancer? Int J Cancer 119: 2705–2709

58. Mohr S, Gorham E, Garland C, Grant W, Garland F (2006) Are low UVB and high animal energy intake associated UVB and vitamin D higher incidence rates of renal cancer? Int J Cancer 119:2705–2709

59. Stolzenberg-Solomon RZ, Vieth R, Azad A, Pietinen P, Taylor PR, Virtamo J et al (2006) A prospective nested case–control study of vitamin D status and pancreatic cancer risk in male smokers. Cancer Res 66:10213–10219

60. Abnet CC, Chen W, Dawsey SM, Wei WQ, Roth MJ, Liu B et al (2007) Serum 25(OH)-vitamin D concentration and risk of esophageal squamous dysplasia. Cancer Epidemiol Biomarkers Prev 16:1889–1893

61. Knekt P, Jarvinen R, Dich J, Hakulinen T (1999) Risk of colorectal and other gastro-intestinal cancers after exposure to nitrate, nitrite and N-nitroso compounds: a follow-up study. Int J Cancer 80:852–856

62. Zou XN, Lu SH, Liu B (1994) Volatile *N*-nitrosamines and their precursors in Chinese salted fish – a possible etiological factor for NPC in China. Int J Cancer 59:155–158

63. Dallorso ME, Gil S, Pawlak E, Lema F, Marquez A (2001) 1,25(OH)2 vitamin D concentration in the plasma of *Solanum glaucophyllum* intoxicated rabbits. Aust Vet J 79:419–423

64. Ding EL, Mehta S, Fawzi WW, Giovannucci EL (2008) Interaction of estrogen therapy with calcium and vitamin D supplementation on colorectal cancer risk: reanalysis of Women's Health Initiative randomized trial. Int J Cancer 122:1690–1694

65. Pratt W, Taylor P, Goldstein A (1990) Principles of Drug Action: The Basis of Pharmacoloy. Churchill Livingstone, New York

66. Mohr S, Garland C, Gorham E, Grant W, Highfill R, Garland F (2005) Mapping vitamin D deficiency, breast cancer and colorectal cancer. Proceedings of the ESRI International User Conference, San Diego, CA: 1778

67. Goodwin P, Ennis M, Pritchard K, Koo J, Hood N, Lunenfeld S et al(2008) Vitamin D deficiency is common at breast cancer diagnosis and is associated with a significantly higher risk of distant recurrence and death in a prospective cohort study of T1–3, N0–1, M0 BC (abstract). American Society for Clinical Oncology Annual Meeting, Chicago, IL, May 17, 2008

68. Ng K, Meyerhardt JA, Wu K, Feskanich D, Hollis BW, Giovannucci EL et al (2008) Circulating 25-hydroxyvitamin d levels and survival in patients with colorectal cancer. J Clin Oncol 26: 2984–2991

69. Porojnicu AC, Dahlback A, Moan J (2008) Sun exposure and cancer survival in Norway: changes in the risk of death with season of diagnosis and latitude. Adv Exp Med Biol 624:43–54

70. Robsahm TE, Tretli S, Dahlback A, Moan J (2004) Vitamin D_3 from sunlight may improve the prognosis of breast-, colon- and prostate cancer (Norway. Cancer Causes Control 15:149–158

71. Stryd RP, Gilbertson TJ, Brunden MN (1979) A seasonal variation study of 25-hydroxyvitamin D_3 serum levels in normal humans. J Clin Endocrinol Metab 48:771–775

72. Garland C, Gorham MS, Grant W, Garland F (2008) Pooled analysis of vitamin D and breast cancer. American Association for Cancer Research Annual Meeting; San Diego CA.

73. Ferlay J, Bray F, Pisani P, Parkin D (2008) GLOBOCAN 2002: Cancer Incidence, Mortality and Prevalence Worldwide. IARC CancerBase No. 5. version 2.0. http://www-dep.iarc.fr /. Accessed 10 July 2008

74. Columbia U (2008) Center for International Earth Science Information Network. Available from: http://sedac.ciesin.columbia.edu/ozone/rtm/mval.html , accessed 10 July 2008

75. National A, Space A International Satellite Cloud Climatology Project database. http://isccp.giss.nasa.gov/products/isccpDsets.html Accessed 31 December 2008.

76. National Academy of Sciences-Institute of Medicine-Food and Nutrition Board (1997) Dietary Reference Intakes for Calcium, Phosphorus, Magnesium, Vitamin D, and Fluoride. National Academy Press, Washington, DC.

77. Coleman MP, Quaresma M, Berrino F, Lutz JM, De Angelis R, Capocaccia R et al (2008) Cancer survival in five continents: a worldwide population-based study (CONCORD. Lancet Oncol 9: 730–756

78. Vilenchik MM, Knudson AG (2006) Radiation dose-rate effects, endogenous DNA damage, and signaling resonance. Proc Natl Acad Sci USA 103:17874–17879

79. Cairns J (2002) Somatic stem cells and the kinetics of mutagenesis and carcinogenesis. Proc Natl Acad Sci USA 99:10567–10570

80. Jones S, Chen WD, Parmigiani G, Diehl F, Beerenwinkel N, Antal T et al (2008) Comparative lesion sequencing provides insights into tumor evolution. Proc Natl Acad Sci USA 105:4283–4288

81. Gumbiner BM (1996) Cell adhesion: the molecular basis of tissue architecture and morphogenesis. Cell 84:345–357

82. Hay ED (1995) An overview of epithelio-mesenchymal transformation. Acta Anat 154:8–20

83. Pearson GW, Hunter T (2007) Real-time imaging reveals that noninvasive mammary epithelial acini can contain motile cells. J Cell Biol 179:1555–1567

84. Gradstein F, Ogg J, Smith A (2004) A Geologic Time Scale. Cambridge University Press, Cambridge

85. Wicha MS (2006) Identification of murine mammary stem cells: implications for studies of mammary development and carcinogenesis. Breast Cancer Res 8:109

86. Shimkin M (1979) Contrary to nature. U. S. Department of Health, Education and Welfare, Bethesda, MD

87. Liu S, Dontu G, Wicha MS (2005) Mammary stem cells, self-renewal pathways, and carcinogenesis. Breast Cancer Res 7:86–95

88. Dawkins R (2006) The Selfish Gene. Oxford University Press, New York

89. Zheng B, Mills AA, Bradley A (2001) Introducing defined chromosomal rearrangements into the mouse genome. Methods 24:81–94

90. Okamura M, Yasuno N, Ohtsuka M, Tanaka A, Shikazono N, Hase Y (2003) Wide variety of color- and shape mutants regenerated from leaf cultures irradiated with ion beams. Nucl Instrum Methods Phys Res B 206:574–578

91. Eigeliene N, Harkonen P, Erkkola R (2006) Effects of estradiol and medroxyprogesterone acetate on morphology, proliferation and apoptosis of human breast tissue in organ cultures. BMC Cancer 6: 246

92. Ruotsalainen H, Sipila L, Vapola M, Sormunen R, Salo AM, Uitto L et al (2006) Glycosylation catalyzed by lysyl hydroxylase 3 is essential for basement membranes. J Cell Sci 119:625–635

93. Pitelka DR, Taggart BN, Hamamoto ST (1983) Effects of extracellular calcium depletion on membrane topography and occluding junctions of mammary epithelial cells in culture. J Cell Biol 96: 613–624

94. Contreras RG, Miller JH, Zamora M, Gonzalez-Mariscal L, Cereijido M (1992) Interaction of calcium with plasma membrane of epithelial (MDCK) cells during junction formation. Am J Physiol 263:C313–C318

95. Whelan J, McEntee MF (2004) Dietary (n-6) PUFA and intestinal tumorigenesis. J Nutr 134: 3421S–3426S

96. Raimondi F, Santoro P, Barone MV, Pappacoda S, Barretta ML, Nanayakkara M et al (2008) Bile acids modulate tight junction structure and barrier function of Caco-2 monolayers via EGFR activation. Am J Physiol Gastrointest Liver Physiol 294:G906–G913

97. Debruyne PR, Bruyneel EA, Li X, Zimber A, Gespach C, Mareel MM (2001) The role of bile acids in carcinogenesis. Mutat Res 480–481:359–369

98. Newmark H, Wargovich M, Bruce W (1984) Colon cancer, and dietary fat, phosphate, and calcium: a hypothesis. J Natl Cancer Inst 72:1321–1325

99. Pettifor JM (2008) Vitamin D &/or calcium deficiency rickets in infants & children: a global perspective. Indian J Med Res 127:245–249

100. Kutluk G, Cetinkaya F, Basak M (2002) Comparisons of oral calcium, high dose vitamin D and a combination of these in the treatment of nutritional rickets in children. J Trop Pediatr 48: 351–353

101. Garland C, Garland F, Gorham E (1991) Colon cancer parallels rickets. In: Lipkin M, Newmark H, Kelloff G, (eds) Calcium, Vitamin D, and Prevention of Colon Cancer. National Cancer Institute and CRC Press, Bethesda, MD and Boca, Raton, FL, pp. 81–109

102. Murillo G, Matusiak D, Benya RV, Mehta RG (2007) Chemopreventive efficacy of 25-hydroxyvitamin D3 in colon cancer. J Steroid Biochem Mol Biol 103:763–767

103. Vantieghem K, Overbergh L, Carmeliet G, De Haes P, Bouillon R, Segaert S (2006) UVB-induced 1,25(OH)2D3 production and vitamin D activity in intestinal CaCo-2 cells and in THP-1 macrophages pretreated with a sterol Delta7-reductase inhibitor. J Cell Biochem 99:229–240

104. Broe KE, Chen TC, Weinberg J, Bischoff-Ferrari HA, Holick MF, Kiel DP (2007) A higher dose of vitamin D reduces the risk of falls in nursing home residents: a randomized, multiple-dose study. J Am Geriatr Soc 55:234–239

105. Talwar SA, Aloia JF, Pollack S, Yeh JK (2007) Dose response to vitamin D supplementation among postmenopausal African American women. Am J Clin Nutr 86:1657–1662

106. Aloia JF, Talwar SA, Pollack S, Yeh J (2005) A randomized controlled trial of vitamin D3 supplementation in African American women. Arch Intern Med 165:1618–1623

107. Wicklow BA, Taback SP (2006) Feasibility of a type 1 diabetes primary prevention trial using 2,000 IU vitamin D3 in infants from the general population with increased HLA-associated risk. Ann NY Acad Sci 1079:310–312

108. Saadi HF, Dawodu A, Afandi BO, Zayed R, Benedict S, Nagelkerke N (2007) Efficacy of daily and monthly high-dose calciferol in vitamin D-deficient nulliparous and lactating women. Am J Clin Nutr 85:1565–1571

109. Basile LA, Taylor SN, Wagner CL, Horst RL, Hollis BW (2006) The effect of high-dose vitamin D supplementation on serum vitamin D levels and milk calcium concentration in lactating women and their infants. Breastfeed Med 1:27–35

110. Kislal FM, Dilmen U (2008) Effect of different doses of vitamin D on osteocalcin and deoxypyridinoline in preterm infants. Pediatr Int 50:204–207

111. Hathcock JN, Shao A, Vieth R, Heaney R (2007) Risk assessment for vitamin D. Am J Clin Nutr 85:6–18

112. Vieth R (2009) The risk of additional vitamin D. Ann Epidemiol 19:441–445

113. Heaney RP, Davies KM, Chen TC, Holick MF, Barger-Lux MJ (2003) Human serum 25-hydroxycholecalciferol response to extended oral dosing with cholecalciferol. Am J Clin Nutr 77:204–210

114. Vieth R (1999) Vitamin D supplementation, 25-hydroxyvitamin D concentrations, and safety. Am J Clin Nutr 69:842–856

115. Vieth R, Chan PC, MacFarlane GD (2001) Efficacy and safety of vitamin D3 intake exceeding the lowest observed adverse effect level. Am J Clin Nutr 73:288–294

116. Vieth R (2006) Critique of the consideration for establishing the vitamin D tolerable intake: critical need for revision. J Nutr 136:1117–1122

117. Armas LA, Hollis BW, Heaney RP (2004) Vitamin D_2 is much less effective than vitamin D_3 in humans. J Clin Endocrinol Metab 89:5387–5391

118. Holick MF, Biancuzzo RM, Chen TC, Klein EK, Young A, Bibuld D et al (2008) Vitamin D_2 is as effective as vitamin D_3 in maintaining circulating concentrations of 25-hydroxyvitamin D. J Clin Endocrinol Metab 93:677–681

119. Hall PM (2002) Preventing kidney stones: calcium restriction not warranted. Cleve Clin J Med 69:885–888

120. Meschi T, Schianchi T, Ridolo E, Adorni G, Allegri F, Guerra A et al (2004) Body weight, diet and water intake in preventing stone disease. Urol Int 72(Suppl 1):29–33

121. Pak CY (2008) Medical stone management: 35 years of advances. J Urol 180:813–819

122. Spivacow FR, Negri AL, del Valle EE, Calvino I, Fradinger E, Zanchetta JR (2008) Metabolic risk factors in children with kidney stone disease. Pediatr Nephrol 23:1129–1133

123. Massey LK (2003) Dietary influences on urinary oxalate and risk of kidney stones. Front Biosci 8:s584–s594

124. Kramer HJ, Choi HK, Atkinson K, Stampfer M, Curhan GC (2003) The association between gout and nephrolithiasis in men: The Health Professionals' Follow-Up Study. Kidney Int 64:1022–1026

125. Cramer JS, Forrest K (2006) Renal lithiasis: addressing the risks of austere desert deployments. Aviat Space Environ Med 77:649–653

126. Curhan GC (1997) Dietary calcium, dietary protein, and kidney stone formation. Miner Electrolyte Metab 23:261–264

127. Scragg R, Jackson R, Holdaway I, Lim T, Beaglehole R (1990) Myocardial infarction is inversely associated with plasma 25-hydroxyvitamin D_3 levels: a community-based study. Int J Epidemiol 19:559–563

128. Giovannucci E, Liu Y, Hollis BW, Rimm EB (2008) 25-Hydroxyvitamin D and risk of myocardial infarction in men: a prospective study. Arch Intern Med 168:1174–1180

129. Hypponen E, Laara E, Reunanen A, Jarvelin MR, Virtanen SM (2001) Intake of vitamin D and risk of type 1 diabetes: a birth-cohort study. Lancet 358:1500–1503

130. Mohr SB, Garland CF, Gorham ED, Garland FC (2008) The association between ultraviolet B irradiance, vitamin D status and incidence rates of type 1 diabetes in 51 regions worldwide. Diabetologia 51:1391–1398

131. Goldberg P, Fleming MC, Picard EH (1986) Multiple sclerosis: decreased relapse rate through dietary supplementation with calcium, magnesium and vitamin D. Med Hypotheses 21:193–200

132. Munger KL, Levin LI, Hollis BW, Howard NS, Ascherio A (2006) Serum 25-hydroxyvitamin D levels and risk of multiple sclerosis. JAMA 296:2832–2838

133. Gorham ED, Garland CF, Garland FC (1989) Acid haze air pollution and breast and colon cancer in 20 Canadian cities. Can J Publ Health 80:96–100

134. Webb AR, Kline L, Holick MF (1988) Influence of season and latitude on the cutaneous synthesis of vitamin D_3: exposure to winter sunlight in Boston and Edmonton will not promote vitamin D3 synthesis in human skin. J Clin Endocrinol Metab 67:373–378

135. Vieth R, Bischoff-Ferrari H, Boucher BJ, Dawson-Hughes B, Garland CF, Heaney RP et al (2007) The urgent need to recommend an intake of vitamin D that is effective. Am J Clin Nutr 85:649–650

136. Kakarala RR, Chandana SR, Harris SS, Kocharla LP, Dvorin E (2007) Prevalence of vitamin D deficiency in uninsured women. J Gen Intern Med 22:1180–1183

45 The Anti-cancer Effect of Vitamin D: What Do the Randomized Trials Show?

Joan M. Lappe and Robert P. Heaney

Abstract One of the most exciting roles of vitamin D is its potential for preventing or slowing the development of cancer. The inverse association between sunlight exposure and malignancy has been recognized for decades, and numerous epidemiological studies support the inverse association of cancer and vitamin D. In vitro and in vivo studies have provided understanding of an underlying mechanism for the vitamin D effect. However, recently there has been increased intensity in vitamin D/cancer research after publication of a randomized trial that found that vitamin D_3 and calcium reduced the risk of all-type cancer by 60–77%. The only other randomized trial had conflicting results. In this chapter, we evaluate the strength of those randomized trials and the associated hypothesis that adequate vitamin D status reduces incident cancer rate. We also describe present estimates of serum 25-hydroxyvitamin D [25(OH)D] needed to prevent cancer and propose doses of vitamin D_3 supplementation needed to maintain those 25(OH)D levels.

Key Words: Anti-cancer; 25-hydroxyvitamin D; vitamin D; postmenopausal women; calcium; cancer; breast cancer; colon cancer; vitamin D deficiency; bone health; osteoporosis

1. INTRODUCTION

The importance of vitamin D, the sunshine vitamin, has long been recognized, particularly for its role in bone health. However, recently the significance of vitamin D for promotion of overall health and prevention of a multitude of diseases has become a major focus of scientists, the lay community, and policy makers. One of the most exciting roles of vitamin D is its potential for preventing or slowing the development of cancer.

The inverse association between sunlight exposure and malignancy has been recognized for decades. In the 1930s, Peller and Stephenson noted that US Navy personnel with abundant sunlight exposure had higher rates of skin cancer, but lower rates of other malignancies (1). Several years afterward, Apperly reported an inverse association between latitude and cancer mortality rates (2). These observations led to the hypothesis that skin cancer somehow conferred immunity for other cancers. Surprisingly, no

From: *Nutrition and Health: Vitamin D*
Edited by: M.F. Holick, DOI 10.1007/978-1-60327-303-9_45,
© Springer Science+Business Media, LLC 2010

841

one seems to have explored the mechanism of this effect until 1980 when Garland and Garland proposed that the apparent benefit of sunlight exposure was mediated by vitamin D *(3)*. Since then a massive research effort, across a broad array of disciplines, has been devoted to elucidating the role of vitamin D in cancer development and prevention. In vitro and in vivo studies have provided understanding, albeit not all of the details, of an underlying mechanism for the vitamin D effect. Numerous epidemiological studies have evaluated the association of cancer and vitamin D in the form of sunlight exposure, dietary intake, and serum 25-hydroxyvitamin D 25(OH)D. These are discussed thoroughly in Chapters 42–44. Two randomized trials evaluating the effect of vitamin D and calcium supplementation on cancer incidence have been reported. In this chapter, we evaluate the strength of those studies and the associated hypothesis that adequate vitamin D status reduces incident cancer rate. We also present estimates of serum 25(OH)D needed to prevent cancer and propose doses of vitamin D_3 supplementation needed to maintain those 25(OH)D levels.

2. POTENTIAL MECHANISM OF VITAMIN D ANTI-CANCER EFFECTS

Vitamin D metabolism has been addressed in previous chapters. The mechanism for vitamin D's effect on cancer works through its autocrine mode of action (see Fig. 1) Through the autocrine pathway, in various cells of the immune system as well as in many epithelial cell types (breast, colon, lung, skin, and prostate), tissue-level 1-α-

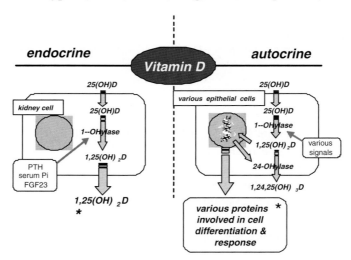

Fig. 1. Schematic depiction of the two faces of vitamin D function. The principal outputs of each are designated by the symbol ★. In the case of the endocrine side, the stimulus to expression of the 1-α-hydroxylase is typically PTH and FGF23, and the principal output is calcitriol (1,25(OH)₂D). On the autocrine side the stimulus to expression of the 1-α-hydroxylase will vary from tissue to tissue, and the principal output of the process (but not measurably released from the cell) will be the proteins signaling, e.g., cell differentiation and apoptosis. For the autocrine function the principal input variable will be serum 25(OH)D, as circulating calcitriol levels are not usually high enough to elicit the full autocrine response. Instead, each tissue controls its own autocrine activation independently of other tissues, but is dependent on an adequate circulating level of 25(OH)D (Copyright 2006, Robert P. Heaney, M.D. used with permission).

hydroxylases convert 25(OH)D to 1,25 dihyroxyvitamin D [1,25(OH)$_2$D] intracellularly *(4–11)*. Then this freshly synthesized 1,25(OH)$_2$D binds to the vitamin D receptor (VDR), and in combination with tissue- and stimulus-specific proteins binds to one of more than 3,000 vitamin D response elements (VDREs) on the chromosomes, inducing transcription of the corresponding proteins. These include proteins responsible for cell proliferation, differentiation, adhesion, and apoptosis, activities that are necessary for initiation and promotion of cancer *(9, 12)*. In addition, 1,25(OH)$_2$D induces transcription of the vitamin D 24-hydroxylase that degrades the 1,25(OH)$_2$D, also intracellularly. As the tissue level 1-α-hydroxylase operates well below its k_M, this local conversion of 25(OH)D to 1,25(OH)$_2$D is dependent on circulating 25(OH)D levels. Adequate serum 25(OH)D is necessary for the presumed anti-cancer effect of vitamin D$_3$.

A likely mechanism of vitamin D's action was detailed in a fascinating study reported by Liu et al. *(13)*, which is described in more detail in the chapter by Modlin. Their study addressed the role of vitamin D in innate immunity, but vitamin D may play a similar role in cancer prevention. Liu et al. found that human macrophages exposed to a mycobacterial antigen express first the VDR and the CYP-27B (1-α-hydroxylase) genes, but without vitamin D in the medium go no further in counteracting the antigen. With 25(OH)D in the culture medium, 1,25(OH)$_2$D is produced intracellularly, complexes with the VDR–antigen–receptor complex, and binds to the vitamin D response element in the chromosome that encodes for a bactericidal peptide (cathelicidin). The same effect was produced by adding 1,25(OH)$_2$D itself to the medium, but the concentration required was higher than would be observed in normal serum (whereas medium 25(OH)D concentration fell within the range found in healthy humans). That serum 25(OH)D is the key precursor is also indicated by the fact that the first gene expressed in the macrophage is the 1-α-hydroxylase, a step that would make sense only if the cell were expecting to utilize circulating 25(OH)D as the enzyme substrate. In at least one mammary tumor model, a key defect is loss of the 1-α-hydroxylase *(14)*. Adding 1,25(OH)$_2$D to the culture medium in that model results in a reversion of the cell phenotype toward normal; however, high medium 1,25(OH)$_2$D levels are required. Something similar may underlie human myelodysplasia, which reverts to quasi-normal on high doses of 1,25(OH)$_2$D *(15)*.

Liu et al. *(13)* further found that, in addition to the gene for cathelicidin, the macrophage also produces the vitamin D-24 hydroxylase (CYP 24), resulting in a rapid degradation of the 1,25(OH)$_2$D synthesized within the cell. Thus, vitamin D serves as a quick on–off switch, necessary for expression of certain cellular actions but also limiting their duration and extent. This developed model illustrates the key role of vitamin D in mediating certain cellular responses to external signals. Consistent with this general mode of action is the observation that mammary epithelium in vitamin D receptor knockout mice over-responds to normal physiological stimuli, such as ovarian estrogen and progesterone production *(14)*.

Garland et al. *(16)* recently synthesized the published studies suggesting anti-cancer mechanisms of vitamin D. Reports indicate that vitamin D may act to decrease the incidence of a variety of cancers by inhibiting tumor angiogenesis *(17, 18)*, increasing cell adherence (thereby decreasing metastatic spread) *(19)*, and facilitating intercellular communication through gap junctions (thereby strengthening contact inhibition) *(20)*.

3. VITAMIN D AND CALCIUM IN PREVENTION OF CANCER

Prevention of cancer may also be due partially to the effects of vitamin D on calcium. It is widely accepted that calcium is a pivotal regulator of numerous cell functions and modulates cell properties *(21)*. Of particular importance are its role in cell division and its regulation of cell proliferation and differentiation *(21)*. It has been shown that low levels of intracellular ionized calcium play a role in cell proliferation and that increasing the calcium concentration in cell and organ culture media decreases cell proliferation and stimulates cell differentiation in numerous types of cells *(22–24)*. In this model system, persistent calcium malnutrition leads to a decrease in calcium concentration in extracellular fluid compartments, which is translated into modulation of cell functions that can lead to initiation of cancer. Vitamin D is necessary for absorption of calcium.

4. RANDOMIZED TRIALS OF VITAMIN D SUPPLEMENTATION AND CANCER

Two randomized trials have reported the effects of vitamin D supplementation on incidence of cancer. The first was a 4-year population-based randomized, placebo-controlled trial of vitamin D_3 and calcium with the primary outcome being fracture incidence and the principal secondary outcome, cancer incidence *(25)*. The study participants were 1,179 community-dwelling women randomly selected from the population of healthy postmenopausal women aged 55 and over in a nine-county rural area of Nebraska. Participants were randomly assigned to receive 1,400–1,500 mg/day supplemental calcium alone (Ca-only), supplemental calcium plus 1,100 IU/day vitamin D_3 (Ca + D), or double placebo.

The mean serum 25(OH)D at baseline in the three treatment groups was 72 nmol/l. Vitamin D_3 produced a statistically significant elevation in serum 25(OH)D in the Ca + D group of 23.9 ± 17.8 nmol/l, whereas the placebo and Ca-only groups had no significant change. By intention-to-treat analysis, the Ca + D group had significantly fewer incident cancers of all types (RR 0.40–0.23, $P<0.03$). Furthermore, baseline and treatment-induced serum 25(OH)D concentrations themselves were strong predictors of cancer risk.

The Ca-only group had a risk of cancer that was intermediate between that of the Ca + D group and the placebo group, but its relative risk was not statistically significantly different from that of the placebo groups. While the supplemental calcium may have contributed to the aggregate effect found in the Ca + D group, the study power was not sufficient to evaluate the difference possibly produced by calcium itself.

The principal weakness of this study is that cancer was a secondary outcome variable. The study has also been criticized for its small sample size and the small number of cancer cases. However, the power, and hence the sample size, was obviously adequate as there was a statistically significant effect of supplementation.

The other randomized trial was the Women's Health Initiative (WHI) in which 36,282 postmenopausal women were randomly assigned to 1,000 mg calcium and 400 IU vitamin D per day or placebo pills for both. The incidence of colorectal cancer and the incidence of breast cancer were both secondary outcomes (26, 27). The primary analyses by treatment group found no effect on colorectal cancer incidence (26). Although there were fewer cases of breast cancer in the supplemented ($N = 528$) versus the placebo group ($N = 546$), the difference was not statistically significant (27). The lack of an evident anti-cancer effect is not surprising since the dose of 400 IU (and factoring in compliance, a mean dose of only ~200 IU) was inadequate to raise blood levels of 25(OH)D to an optimal level (above 75–80 nmol/l). There were several other limitations to the study: (1) personal use of calcium and vitamin D was allowed, and 15% of placebo subjects crossed into, (i.e., "dropped in") to the active group; (2) serum 25(OH)D levels were not obtained during the study to ascertain the effects of supplement on raising serum 25(OH)D levels, but from other studies it is now known that the mean ingested dose in WHI would be expected to raise serum 25(OH)D by no more than 5 nmol/l (28); (3) baseline calcium intake was high at 1,150 mg/day (29) so supplemental calcium may have provided no additional benefit to cancer prevention; and (4) 58% of the participants in the supplement study were also assigned to hormone replacement therapy (HRT). In fact, the study was stopped early because in an interim analysis the combination of estrogen and progesterone was found to increase the risk of breast cancer (http://www.nhibi.nih.gov/whi/pr-02-7-9).

Although in WHI the primary analysis found no association between supplementation and colorectal cancer, a nested case–control study within WHI found a highly statistically significant inverse relationship between baseline 25(OH)D and incident colorectal cancer risk. The risk of colorectal cancer in the lowest quartile of serum 25(OH)D (<31.0 nmol/l) was more than two times higher than in the highest quartile ($\geq$58.4 nmol/l) (RR 2.53, 95% CI 1.49–4.32) (26).

A second nested case–control study from within WHI included 895 breast cancer cases and 898 matched controls that were breast cancer free. The mean baseline level of 25(OH)D in the cases was 50.0 ± 21.0 nmol/l, while the level in the controls was 52.0 ± 21.1 nmol/l (NS). Thus, both groups were, on average, vitamin D insufficient. However, in a multivariate analysis that included hormone therapy as one of several independent variables, higher baseline 25(OH)D was significantly associated with lower risk of breast cancer (P=0.04). Furthermore, among women in the lowest baseline quartile of vitamin D intake (from diet and supplement), fewer cancers occurred in the supplemented group (HR = 0.79, 95% CI 0.65–0.97, $P_{interaction}$ = 0.003). Unfortunately, these findings from the two nested case-control studies within WHI are not widely known.

A key difference between the two randomized trials (Lappe and WHI) is dosage and the corresponding serum 25(OH)D level produced by treatment. When WHI was designed, 400 IU/day was considered an adequate dose, and no evidence existed with respect to the serum 25(OH)D level that reflected sufficiency. The study by Lappe et al. (25) used nearly three times the dose of vitamin D (which, factoring in compliance, resulted in a dose five times greater than WHI) and produced a serum level at least twice that achieved in WHI. The findings of the Lappe et al. trial are consistent with and strongly support the findings of numerous ecologic and epidemiologic studies showing the efficacy of vitamin D_3 in prevention of cancer.

5. LEVELS OF 25(OH)D FOR CANCER PREVENTION

The indicator of vitamin D status is serum 25(OH)D *(30)*. Until recently, the index disease for vitamin D deficiency in adults has been osteomalacia, which is associated with serum 25(OH)D concentrations <20 nmol/l (8 ng/ml) *(31)*. However, it is now recognized that serum 25(OH)D above the osteomalacia threshold may also produce skeletal disease and a variety of other disorders *(32–35)* including cancer *(31, 36–39)*. There is currently a growing consensus that the lower bound of the range for 25(OH)D values lies above 75–80 nmol/l (30–32 ng/ml) for most populations. This is based on studies of the inverse relationship between serum parathyroid hormone (PTH) and 25(OH)D, showing that PTH concentrations plateau at serum 25(OH)D levels of 70–100 nmol/l (28–40 ng/ml) *(40–44)*, the relationship between 25(OH)D and both BMD and lower extremity neuromuscular function in NHANES-III *(32, 45)*, and the serum level of 25(OH)D needed to optimize calcium absorption efficacy *(46)*.

Although fully optimal levels of 25(OH)D have not been established, several studies show that 25(OH)D levels above 75–80 nmol/l (30–32 ng/ml) are associated with reduced incidence of colorectal adenomas and cancer *(25, 37, 47, 48)*. For example, in the Lappe et al. study *(25)* the mean achieved 25(OH)D level in the group with the lowest risk of cancer (RR 0.40, CI 0.20–0.82) was 96.0 ± 21.4 nmol/l (39 ± 8.6 ng/ml).

Garland et al. *(49)* recently described the dose–response gradient between serum 25(OH)D and breast and colon cancer. They combined data from observational studies to estimate the dose–response gradients. Then they confirmed the gradients with an analysis of modeled and reported 25(OH)D levels and estimated age-standardized cancer incidence rates for 177 countries from the International Agency for Research on Cancer (IARC) GLOBOCAN database. They found that the first apparent increase in prevention of colorectal cancer is seen at serum 25(OH)D levels ≥22 ng/ml (55 nmol/l), and the first apparent increase in prevention of breast cancer is at ≥32 ng/ml (80 nmol/l). They concluded that differences in serum 25(OH)D below those levels would most likely not affect cancer risk.

Based on known data points, Garland et al. further estimated that maintaining serum 25(OH)D levels ≥34 ng/ml in the US population could prevent 50% of colon cancer incidence *(49)*. Maintaining serum 25(OH)D levels ≥42 ng/ml would prevent about 30% of breast cancer cases. Using linear extrapolation of the known data points, an estimated 50% of breast cancer could be prevented with serum 25(OH)D levels ≥52 ng/ml.

6. DOSES OF VITAMIN D SUPPLEMENTATION FOR CANCER PREVENTION

Heaney established the 25(OH)D response to various doses of cholecalciferol (vitamin D_3) up to 10,000 IU *(28, 50)*. In brief, the serum 25(OH)D rises by about 0.7 nmol/l (0.28 ng/ml) for every microgram of additional D3 *(28)*. Applying this information to the NHANES III national distribution data for serum 25(OH)D, Heaney showed explicitly what the distribution would be if everyone in the US population received an additional 2,000 IU vitamin D_3/day (51). The mean would rise by about 14 ng/ml, and about 80–85% of the population would have 25(OH)D value above 32 ng/ml (80 nmol/l).

It is important to stress that both the currently recommended intakes of vitamin D and the generally accepted "normal" range for serum 25(OH)D are probably inadequate for prevention of cancer. In fact, 400 IU vitamin D_3 per day, the currently recommended dose for ages 50–70 will raise serum 25(OH)D by no more than 7–10 nmol/l (3–4 ng/ml) (28).

According to Garland et al. (49), a 50% reduction in colorectal cancer risk would require a population intake of 2,000 IU of vitamin D_3/day. However, a 50% reduction in breast cancer would require a higher dose, 3,500 IU/day. The safety of higher doses of vitamin D supplementation has been discussed in Chapter 31. The National Academy of Sciences Institute of Medicine (IOM) had designated 2,000 IU/day as the TUIL (tolerable upper intake level) of vitamin D intake in 1997 but did not relate that dose to serum 25(OH)D concentrations (30). Barger-Lux and Heaney (52) found that healthy men who completed a summer season of outdoor work had a mean serum 25(OH)D of 122 nmol/l (45 ng/ml), with some men exceeding 200 nmol/l (80 ng/ml). This mean sun exposure response was equivalent in dosing to 70 µg/day (2,800 IU/day). This means that natural sunlight exposure provides healthy young people with considerably higher doses of vitamin D than currently considered safe for oral dosing. In fact, there is no evidence of adverse effects of vitamin D_3 intake at or below 10,000 IU/day (53). Today scientists and members of the lay community are calling for higher recommended levels of vitamin D_3.

7. SUMMARY

As indicated in previous chapters, vitamin D deficiency is epidemic in persons across the life span and around the world (Part III). Numerous ecologic and epidemiologic studies consistently and strongly support the efficacy of vitamin D_3 in prevention of cancer. In addition, one rigorous randomized trial with cancer as a secondary outcome showed a remarkable decrease in incidence of all-type cancer. It is true that a definitive randomized trial of vitamin D_3 and calcium with cancer as a primary outcome has not been reported, and such studies need to be done. However, in light of evidence reported over more than 70 years, it seems prudent to urge increasing vitamin D at population levels. Although sunlight exposure is a natural source of vitamin D, the levels of exposure needed to maintain adequate serum 25(OH)D levels in persons of varied ages and skin tone have not been determined. Concern exists about the risk of skin cancers and photoaging with sunlight exposure, although evidence to date suggests that only short exposure is needed for vitamin D dosing. For example, Holick has shown that only about 15 min outdoors on a sunny day within an hour of noon with ≥40% of skin area exposed can enhance vitamin D status (9). It should be noted that sunscreen obliterates almost all conversion of vitamin D in the skin. Also, at latitudes above 38° no vitamin D conversion in the skin occurs during the winter months because of the low solar angle. Thus, for many persons vitamin D_3 supplementation, which is safe and inexpensive, is an appropriate option.

The American Cancer Society estimated 1,437,180 new cases of cancer in 2008 (http://seer.cancer.gov/statfacts/html/all.html). Preventing even half of these cancers would have a dramatic impact on quality of life and health-care costs to the individual and to society. Promotion of vitamin D sufficiency for all populations is critical.

REFERENCES

1. Peller S, Stephenson C (1937) Skin irritation and cancer in the United States Navy. Am J Med 194: 326–333
2. Apperly F (1941) The relation of solar radiation to cancer mortality in North America. Cancer Res 1:191–195
3. Garland C, Garland F (1980) Do sunlight and Vitamin D reduce the likelihood of colon cancer?. Int J Epidemiol 9:227–231
4. Schwartz G, Whitlatch L, Chen T, Lokeshwar B, Holick M (1998) Human prostate cells synthesize 1, 25-dihydroxyvitamin D3 from 25-hydroxyvitamin D3. Cancer Epidemiol Biomarkers Prev 7: 391–395
5. Mawer E, Hayes M, Heys S (1994) Constitutive synthesis of 1, 25-dihydroxyvitamin D3 by a human small cell lung cancer cell line. J Clin Endocrinol Metab 79:554–560
6. Cross H, Bareis P, Hofer H (2001) 25-hydroxyvitamin D3–1-alpha-hydroxlyase and vitamin D receptor gene expresion in human colonic mucosa is elevated during early cancerogenesis. Steroids 66:287–292
7. Tangpricha V, Flanagan J, Whitlatch L (2001) 25-hydroxyvitamin D-1-alpha-hydroxylase in normal and malignant colon tissue. Lancet 357:1673–4
8. Holick MF (2006) High prevalance of vitamin D inadequacy and implications for health. Mayo Clin Proc 81:353–373
9. Holick M (2003) Vitamin D a millennium perspective. J Cell Biochem 88:296–307
10. DeLuca H (2004) Overview of general physiologic features and functions of vitamin D. Am J Clin Nutr 80(supp):1689S–96S
11. Holick MF, Vitamin D (2006) Its role in cancer prevention and treatment. Prog Biophys Mol Biol 92:49–59
12. Rachez C, Freedman LP (2000) Mechanisms of gene regulation by vitamin D(3) receptor: a network of coactivator interactions. Gene 246:9–21
13. Liu PT, Stenger S, Li H, Wenzel L, Tan B (2006) Toll-like receptor triggering of a vitamin D-mediated human antimicrobial response. Science 311:1770–1773
14. Welsh J (2004) Vitamin D and breast cancer: insights from animal models. Am J Clin Nutr 80 (suppl):1721S–4S
15. Mellibovsky L, Diez A, Perez-Vila E, Serrano S, Nacher M, Aubia J et al (1998) Vitamin D treatment in myelodysplastic syndromes. Br J Haematol 100:516–520
16. Garland C, Garland F, Gorham E, Lipkin M, Newmark H, Mohr S et al (2006) The role of vitamin D in Cancer Prevention. Am J Public Health 96:252–261
17. Iseki K, Tatsuta M, Uehara H (1999) Inhibition of angiogenesis as a mechanism for inhibition by 1 alpha-hydroxyvitamin D3 and 1,25-dihydroxyvitamin D3 of colon carcinogenesis induced by azoxymethane in Wistar rats. Int J Cancer 81:730–733
18. Mantell D, Owens P, Bundred N, Mawer E, Canfield A (2000) 1 alpha,25-dihydroxyvitamin D3 inhibits angiogenesis in vitro and in vivo. Circ Res 87:214–220
19. Palmer H, Gonzalez-Sancho J, Espada J (2001) Vitamin D(3) promotes the differentiation of colon carcinoma cells by the induction of E-cadherin and the inhibition of beta-catenin signaling. J Cell Biol 154:369–387
20. Fujioka T, Suzuki Y, Okamoto T, Mastuschita N, Hasegawa M, Omori S (2000) Prevention of renal cell carcinoma by active vitamin D3. World J Surg 24:1205–1210
21. Rasmussen H (1986) The calcium messenger system (2). New England J Med 314:1164–1170
22. Hennings H, Michael D, Cheng C, Steinert P, Holbrook K, Yuspa SH (1980) Calcium regulation of growth and differentiation of mouse epidermal cells in culture. Cell 19:245–254

23. McGrath CM, Soule HD (1984) Calcium regulation of normal human mammary epithelial cell growth in culture. In Vitro 20:652–662

24. Babcock MS, Marino MR, Gunning WT III, Stoner GD (1983) Clonal growth and serial propagation of rat esophageal epithelial cells. In Vitro 19:403–415

25. Lappe JM, Travers Gustafson D, Davies KM, Recker RR, Heaney RP (2007) Vitamin D and calcium supplementation reduces cancer risk: results of a randomized trial. Am J Clin Nutr 85:1586–1591

26. Wactawski Wende J, Kotchen JM, Anderson GL, Assaf AR, Brunner RL, O'Sullivan MJ et al (2006) Calcium plus vitamin D supplementation and the risk of colorectal cancer. New Engl J Med 354:684–696

27. Chlebowski R, Johnson K, Kooperberg C, Pettinger M, Wactawski-Wende J, Rohan T et al (2008) Calcium plus vitamin D supplementation and the risk of breast cancer. J Natil Cancer Inst 100:1581–1591

28. Heaney RP, Davies KM, Chen T, Holick M, Barger-Lux J (2003) Human serum 25-hydroxycholecalciferol response to extended oral dosing with cholecalciferol. Am J Clin Nutr 77:204–210

29. Jackson RD, LaCroix AZ, Gass M, Wallace RB, Robbins J, Lewis CE et al (2006) Calcium plus vitamin D supplementation and the risk of fractures. N.Engl.J.Med 354:669–683

30. National Academy of Science (1999) Dietary reference intakes for calcium, magnesium, phosphorus, vitamin D, and fluoride. Food and Nutrition Board, Institute of Medicine. National Academy Press, Washington, D.C.

31. Heaney RP (2004) Functional indices of vitamin D status and ramifications of vitamin D deficiency. Am J Clin Nutr 80(suppl):1706S–9S

32. Bischoff-Ferrari H, Dietrich T, Orav E (2004) Higher 25-hydroxy-vitamin D concentrations are associated with better lower-extremity function in both active and inactive persons aged over 60 year. Am J Clin Nutr 80:752–758

33. Scragg R, Sowers MF, Bell C (2004) Serum 25-hydroxy vitamin D, diabetes, and ethnicity in the third national health and nutrition examination survey. Diabetes Cure 27:2813–2818

34. Jorde R, Bonaa K (2000) Calcium from dairy products, vitamin D intake, and blood pressure; the Tromso study. Am J Clin Nutr 71:1530–1535

35. Rostand S (1997) Ultraviolet light may contribute to geographic and racial blood pressure differences. Hypertension 30:150–156

36. Bertone-Johnson E, Chen W, Holick M, Hollis B, Colditz G, Willett W et al (2005) Plasma 25-hydroxyvitamin D and 1,25-dihydroxyvitamin D and risk of breast cancer. Cancer Epidemiol Biomarkers Prev 14:1991–1997

37. Grau M, Baron J, Sandler R, Haile R, Beach M, Church T et al (2003) Vitamin D, calcium supplementation, and colorectal adenomas: results of a randomized trial. J Nat Cancer Inst 95:1765–1771

38. Skinner H, Michaud DS, Giovannucci E, Willett W, Colditz G, Fuchs CS (2006) Vitamin D intake and the risk for pancreatic cancer in two cohort studies. Cancer Epidemiol Biomarkers Prev 15(9):1688–1695

39. Ahonen M, Tenkanen L, Teppo L (2000) Prostate cancer risk and prediagnostic serum 25-hydroxyvitamin D levels (Finland). Cancer Causes Control. 11:847–852

40. Thomas M, Lloyd-Jones D, Thadhani R, Shaw A, Deraska D, Kitch B et al (1998) Hypovitaminosis D in medical inpatients. New Engl J Med 338:777–783

41. Chapuy M, Prezoisi P, Maamer M (1997) Prevalence of vitamin D insufficiency in an adult normal population. Osteoporos Int 7:439–443

42. Need A, O'Loughlin P, Morris H, Horowitz M, Nordin B (2004) The effects of age and other variables on serum parathyroid hormone in postmenopausal women attending an osteoporosis center. J Clin Endocrinol Metab 89:1646–1649

43. Holick M, Siris E, Binkley N, Beard M, Khan A, Katzer J et al (2005) Prevalence of vitamin D inadequacy among postmenopausal North American women receiving osteoporosis therapy. J Clin Endocrinol Metab 90:3215–3224

44. Lappe J, Travers-Gustafson D, Davies M, Recker R, Heaney R (2006) Vitamin D status in a rural postmenopausal female population. JACN 25:395–402

45. Bischoff-Ferrari H, Dietrich T, Orav EJ, Dawson-Hughes B (2004) Positive Association between 25-hydroxy vitamin D levels and bone mineral density; a population-based study of younger and older adults. Am J Med 116:634–639

46. Heaney RP, Dowell Ms, Hale CA, Bendich A (2003) Calcium absorption varies within the reference range for serum 25-hydroxyvitamin D. J Am Coll Nutr 22(2):142–146

47. Feskanich D, Ma J, Fuchs CS, Kirkner GJ, Hankinson SE, Hollis BW (2004) Plasma vitamin D metabolites and risk of colorectal cancer in women. Cancer Epidemiol Biomarkers Prev 13:1501–1508

48. Platz E, Hankinson S, Hollis B (2000) Plasma 1,25-dihydroxyvitamin D and adenomatous polyps of the distal colorectum. Cancer Epidemiol Biomarkers Prev 9:1059–1065

49. Garland CF, Grant WB, Mohr SB, Gorham ED, Garland FC (2007) What is the dose-response relationship between vitamin D and cancer risk? . Nutr Rev 65:S91–S95

50. Heaney RP, Barger-Lux J, Dowell MS, Chen T, Holick MF (1997) Calcium absorptive effects of vitamin D and its major metabolites. J Clin Endocrinol Metab 82:4111–4116

51. Heaney R (2005) The vitamin D requirement in health and disease. J Steroid Biochem Mol Biol 97:13–19

52. Barger-Lux J, Heaney R (2002) Effects of above average summer sun exposure on serum 25-hydroxyvitamin D and calcium absorption fraction. J Clin Endocrinol Metab 87:4952–4956

53. Hathcock J, Shao A, Vieth R, Heaney R (2007) Risk assessment for vitamin D. Am J Clin Nutr 85:6–18

46 Sunlight, Skin Cancer, and Vitamin D

Jörg Reichrath

Abstract   Both in the public and in the scientific community, there is an ongoing debate about the relevance of beneficial and adverse effects of UV exposure. There is no doubt that solar UV exposure is the most important environmental risk factor for the development of non-melanoma skin cancer. Consequently, sun protection has been advocated as a key principle of skin cancer prevention campaigns. However, 90% of all vitamin D needed by the human body has to be formed in the skin through the action of the sun – a serious conflict, for an association of vitamin D deficiency and multiple independent diseases including various types of cancer, bone diseases, autoimmune diseases, infectious diseases, hypertension, and cardiovascular disease has now been convincingly demonstrated in epidemiologic and laboratory studies. An important link that advanced our understanding of these new findings was the discovery that the biologically active vitamin D metabolite 1,25-dihydroxyvitamin D [$1,25(OH)_2$ D] is not exclusively produced in the kidney, but in many other tissues including prostate, colon, skin, breast, and osteoblasts. Extrarenally produced $1,25(OH)_2D$ is now considered to represent an autocrine or paracrine hormone that regulates various cellular functions including cell growth. A paradigm shift occurred with the discovery that plasma levels of 25-hydroxyvitamin D [25(OH)D, calcidiol] are physiologically important for the extrarenal production of $1,25(OH)_2D$. It is now evident that 25(OH)D plasma levels represent a more accurate measure of the vitamin D status of an individual than the more tightly regulated plasma levels of $1,25(OH)_2D$. This development has revitalized the field of vitamin D research and led to the conclusion that the minimally desired 25(OH)D plasma levels should rise from 30 to 75 nmol/l. We and others have demonstrated that strict sun protection causes vitamin D deficiency in risk groups. In the context of new scientific findings that convincingly demonstrate an association of vitamin D deficiency with a variety of severe diseases including various types of cancer, the detection and treatment of vitamin D deficiency in sun-deprived risk groups is very important. It should be accentuated that vitamin D status should be monitored in groups that are at high risk of developing vitamin D deficiency (e.g., nursing-home residents or patients under immunosuppressive therapy). Vitamin D deficiency should be treated, e.g., by giving vitamin D orally. Dermatologists and other clinicians have to recognize that there is convincing evidence that the protective effect of less intense solar UV radiation outweighs its mutagenic effects. Although further work is necessary to define an adequate vitamin D status and adequate guidelines for solar UV exposure, it is at present mandatory that public health campaigns and recommendations of dermatologists on sun protection consider these facts. Well-balanced recommendations on sun protection have to ensure an adequate vitamin D status, thereby protecting us against adverse effects of strict sun protection without significantly increasing the risk to develop UV-induced skin cancer.

Key Words: Sunlight; skin cancer; vitamin D; solar UV exposure; cancer

From: *Nutrition and Health: Vitamin D*
Edited by: M.F. Holick, DOI 10.1007/978-1-60327-303-9_46,
© Springer Science+Business Media, LLC 2010

1. THE MOST IMPORTANT NEGATIVE HEALTH EFFECT OF SOLAR UV EXPOSURE: INDUCTION OF SKIN CANCER

1.1. Non-melanoma Skin Cancer, Malignant Melanoma, and Solar UV Exposure

Historically, the connection between solar UV exposure and epithelial skin cancer was first described by Unna and Dubreuilh in the last decade of the nineteenth century (1, 2). They observed actinic keratoses and squamous cell carcinomas in chronically sun-exposed skin from sailors and vineyard workers. Today, there is no doubt that solar UV exposure is the most important environmental risk factor for the development of non-melanoma skin cancer (3–8). Three main types of skin cancer can be distinguished: squamous cell carcinoma (9, 10), basal cell carcinoma (10), and malignant melanoma (11). Actinic keratoses are now considered to represent cutaneous squamous cell carcinomas in situ (10). Actinic keratoses are more frequent in men, in sun-sensitive subjects exposed to chronic sun, and in individuals who have a history of sunburn (12). During the last decades, epidemiological data have demonstrated that painful sunburns are implicated in the pathogenesis of squamous cell carcinoma (13), basal cell carcinoma (4), and malignant melanoma (5). Chronic sun exposure is the most important cause of squamous cell carcinoma (14), but may be less important for the development of basal cell carcinoma (4). In addition, various reports analyzing sun exposure parameters have consistently demonstrated an association between the development of malignant melanoma and short-term intense ultraviolet (UV) exposure, particularly burning in childhood (15). Many studies have shown that the incidence of malignant melanoma increases with decreasing latitude toward the equator (16). However, in contrast to short-term intense exposure, more chronic less intense exposure has not been found to be a risk factor for the development of malignant melanoma and in fact has been found in some studies to be protective (5, 17, 18). It may be speculated whether these connections may be an explanation for the finding of an increased risk to develop melanoma after sunscreen use that was reported (19). Recently, a large European case–control study investigated the association between sunbed use and cutaneous melanoma in an adult population aged between 18 and 49 years (20). Between 1999 and 2001 sun and sunbed exposure was recorded and analyzed in 597 newly diagnosed melanoma cases and 622 controls in Belgium, France, The Netherlands, Sweden, and the United Kingdom. In this study, 53% of cases and 57% of controls ever used sunbeds. There was a South-to-North gradient with high prevalence of sunbed exposure in Northern Europe and lower prevalence in the South (prevalence of use in France 20% compared to 83% in Sweden). The authors found that dose and lag time between first exposure to sunbeds and time of study were not associated with melanoma risk, neither were sunbathing and sunburns (20).

1.2. Photocarcinogenesis of Skin Cancer

The solar UV spectrum consists of UV-C (wavelength below 280 nm), UV-B (280–315 nm), and UV-A (315–400 nm) bands (Fig. 1) (9). The predominant part of the shortwave, high-energy, and destructive UV spectrum cannot reach the earth's surface,

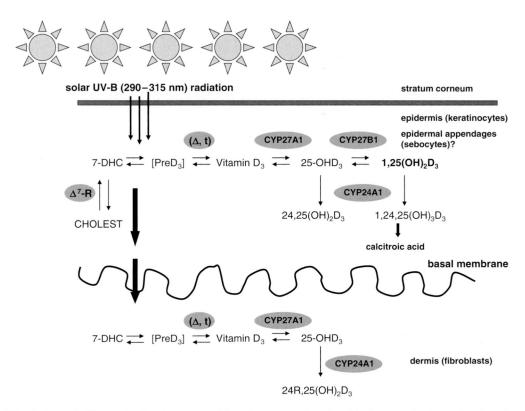

Fig. 1. Vitamin D metabolism in human skin. Please note that the skin is the only tissue that has the capacity for the complete synthesis of 1,25-dihydroxyvitamin D from 7-DHC.

for the ozone layer of the outer earth atmosphere absorbs the shorter wavelength up to approximately 310 nm (UV-C and part of UV-B radiations) (9). In human skin, UV light is absorbed by the different layers in a wavelength-dependent manner. UV-B is almost completely absorbed by the epidermis, only 20% of UV-B energy reach the epidermal basal cell layer or the dermal stratum papillare (9). UV-A penetrates deeper into the dermis and deposits 30–50% of its energy in the dermal stratum papillare. These absorption characteristics explain at least in part why UV-B effects (skin cancer development) have to be expected predominantly in the epidermis and UV-A effects (solar elastosis, skin aging) in the dermis (9). DNA is a major epidermal chromophore with an absorption maximum of 260 nm. Both UV-A and UV-B can induce structural damage to DNA. In 1928 Gates described for the first time that the bactericidal effect of UV radiation is connected with its absorption by prokaryotic DNA (21). Thirteen years later, Hollaender and Emmons observed that the occurrence of mutations in eukaryotic (Fungi) DNA due to UV radiation is wavelength dependent (22). Thus, the indispensable importance of DNA concerning cell survival and cell transformation on the one hand and the relevance of UV light to induce potentially lethal genomic disruptions on the other was already known in the first half of the last century (for review, see (23–25). In the 1960 s, the essential molecular characteristics of DNA alterations caused directly by UV-C and UV-B radiation were uncovered: Beukers and Berends (26), as well as Setlow and Carrier (27)

found covalent interactions between two adjacent pyrimidine bases forming cyclobutane pyrimidine dimers (CPD: thymine dimers, cytosine dimers); later on Varghese and
Patrick *(28)* described another typical dimer formation at dipyrimidine sites, the 6-4 photoproducts (6-4 PP: thymine–cytosine dimers) (Fig. 2). Today, it is well accepted that
UV-B induces molecular rearrangements of the DNA with a characteristic formation of
specific photoproducts (typically cyclobutane pyrimidine dimers or 6-4 photoproducts),
which are known to be mutagenic. The genotoxic potential of UV-A is predominantly
due to indirect mechanisms that include oxidative damage. Gene mutations that have
been shown to be of importance for the pathogenesis of skin cancer include mutations in
the p53 gene (actinic keratoses, squamous cell carcinomas) and mutations in the patched
(PTCH)/sonic hedgehog pathway (basal cell carcinomas). The UV-induced development of skin carcinomas has been analyzed using multiple model systems. Mutation-
associated inactivation of p53 tumor suppressor gene plays a critical role both for stages
of initiation and progression of SCC (review in *(29)*). Analysis of data on gene mutations in human premalignant actinic keratosis (AK) lesions and data from UV-induced
carcinogenesis experiments in mice have suggested that the first step involves acquisition of UV-induced mutations in the p53 gene by epidermal keratinocytes (review
in *(29)*). This defect diminishes sunburn cell formation and enhances cell survival

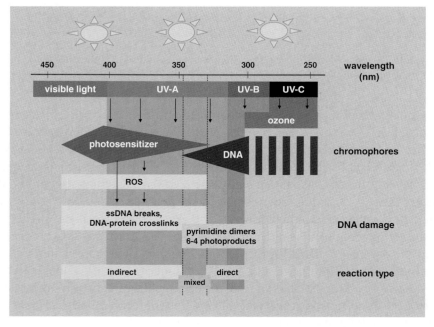

Fig. 2. UV-induced DNA damage. The reaction type of DNA damage is wavelength dependent: direct
induction of DNA photoproducts (CPD, 6-4 PP) by UV-B/UV-C and shortwave UV-A spectrum (315–
327 nm); indirect induction of oxidative DNA disruptions (ssDNA breaks, DNA–protein cross-links)
by longwave UV-A spectrum (347–400 nm); mixed reaction type with direct and indirect effects
by UV-A wavelength 327–347 nm. Oxidative DNA damage is mediated by photosensitizers (Type I
reaction) or by photosensitizer-induced reactive oxygen species (ROS, Type II reaction).

allowing retention of initiated, precancerous keratinocytes (review in *(29)*). Second, chronic exposures to solar UV result in the accumulation of p53 mutations in skin, which confer a selective growth advantage to initiated keratinocytes and allow their clonal expansion, leading to formation of AK (review in *(29)*). The expanded cell death-defective clones represent a larger target for additional UV-induced p53 mutations or mutations in other genes, thus enabling progression to carcinomas. Concerning the pathogenesis of basal cell carcinomas, the importance of PTCH, SMOH, and TP53 mutations has been demonstrated *(30)*. The importance of the DNA repair machinery for skin cancer pathogenesis is impressively demonstrated in xeroderma pigmentosum, a disease that is characterized by defective DNA repair that is associated with the occurrence of malignant skin tumors early in life (Fig. 3) (review in *(9)*). Suppression of the skin's immune system has been shown to be another one of the mechanisms by which solar ultraviolet (UV) radiation induces and promotes skin cancer growth, even at suberythemogenic doses (review in *(31)*). Immunosuppressive effects that are induced by short-time intense solar UV exposure (sunburns) may also be of importance for the pathogenesis of malignant melanoma (Fig. 4) *(32, 33)*. Immunosuppressive properties have been demonstrated for both UV-B and UV-A *(32, 33)*. It has been speculated whether UV-B-induced production of vitamin D may be involved in UV-B-induced immunosuppression *(34)*.

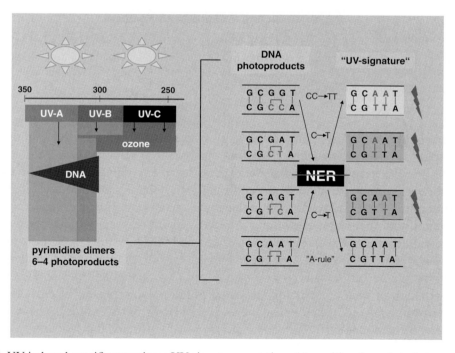

Fig. 3. UV-induced specific mutations: UV-signature mutations (.) resulting from 6-4 photoproducts can be identified as C→T transitions at dipyrimidine sites (cytosine–thymine, thymine–cytosine); cytosine cyclobutane dimers result in CC→TT tandem transitions. Thymine cyclobutane dimers do not yield a mutation, because non-informative bases on the template DNA strand were substituted by an adenine on the opposite DNA strand ("A-rule"). NER = nucleotide excision repair.

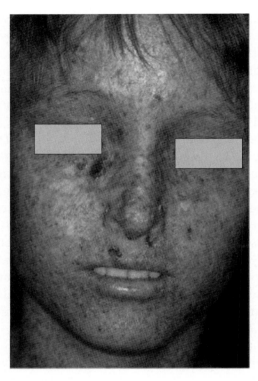

Fig. 4. A 9-year-old boy with xeroderma pigmentosum. Note that as a result of defective DNA repair, this patient presents with typical poikiloderma in UV-exposed skin areas, including multiple actinic keratoses, actinic cheilitis, squamous cell carcinoma (bridge of nose), and basal cell carcinomas (right nose-canthus region, upper lip).

Interestingly, a contribution of the skin vitamin D system to the pathogenesis and prognosis of malignancies including malignant melanoma has been demonstrated *(35)*. We have characterized the expression of key components of the vitamin D endocrine system (vitamin D receptor [VDR], vitamin D-25OHase, 25(OH)D-1αOHase, 1,25(OH)$_2$D-24OHase) in cutaneous squamous cell carcinomas, basal cell carcinomas, and malignant melanoma *(36–39)*. Our findings provide supportive evidence for the concept that endogenous synthesis and metabolism of vitamin D metabolites as well as VDR expression may regulate growth characteristics of basal cell carcinomas, cutaneous squamous cell carcinomas, and malignant melanoma *(36–39)*. An association of Fok 1 restriction fragment length polymorphisms of the VDR with occurrence and outcome of malignant melanoma, as predicted by tumor (Breslow) thickness, has been reported *(40)*. The same laboratory demonstrated that a polymorphism in the promotor region of VDR (A-1012G, adenine–guanine substitution -1012 bp relative to the exon 1a transcription start site) is related in melanoma patients to thicker Breslow thickness groups and to the development of metastasis *(41)*. The authors concluded that polymorphisms of the VDR gene, which can be expected to result in impaired function of biologically active vitamin D metabolites, are associated with susceptibility and prognosis in malignant melanoma. Using array CGH, amplification of the

1,25(OH)$_2$D-metabolizing enzyme 1,25(OH)$_2$D-24-hydroxylase was recently detected as a likely target oncogene of the amplification unit 20q13.2 in breast cancer cell lines and tumors *(42)*. It has been speculated that over-expression of 24-hydroxylase due to gene amplification may abrogate 1,25(OH)$_2$D-mediated growth control. Additionally, amplification of the 25(OH)D-1α-hydroxylase gene has been reported in human malignant glioma *(43)*. The significance of these findings remains to be investigated. We have analyzed metastases of malignant melanomas and found no evidence of amplification of 1α-hydroxylase or 24-hydroxylase genes using Southern analysis *(38)*. However, we detected various splicing variants of the 25(OH)D-1α-hydroxylase gene in cutaneous malignancies *(43)*. The clinical significance of this finding remains to be elucidated. Additionally, we have demonstrated that serum 25(OH)D levels are not significantly reduced in stage I melanoma patients *(44)*.

2. RECOMMENDATIONS FOR SUN PROTECTION IN SKIN CANCER PREVENTION CAMPAIGNS

During the last decades, public health campaigns have improved our knowledge regarding risk of UV radiation for skin cancers. However, it can be speculated that positive effects of UV light were not adequately considered in most of these campaigns that in general proposed a strict "no sun policy." Strict sun protection recommendations still represent a fundamental part of public health campaigns and prevention programs aimed at reducing UV radiation-induced skin damage and skin cancer. These sun protection recommendations include use of sunscreens, protective clothing, and avoidance of sunlight. Clothing is extremely effective in absorbing all UV-B radiation thereby preventing any UV-B photons from reaching the skin *(45, 46)*. Most sunscreen products combine chemical UV-absorbing sunscreens and physical inorganic sunscreens, which reflect UV, to provide broad spectrum protection. Nowadays, most sunscreen products protect against both UV-B and UV-A light.

3. A CHALLENGING PERSPECTIVE: UNDERSTANDING AND FIGHTING VITAMIN D DEFICIENCY

3.1. Vitamin D Deficiency – A Serious and Underappreciated Health Problem

Approximately 90% of all requisite vitamin D has to be formed in the skin through the action of the sun – a serious problem, for a connection between vitamin-D deficiency and various types of cancer (e.g., colon, prostate, and breast cancer) has been confirmed in a large number of studies *(47–50)*. The idea that sunlight and vitamin D inhibit the growth of human cancers is not new. When Peller noticed an apparent deficit of non-skin cancer among US Navy personnel, who experienced an excess of skin cancer, he concluded in 1936 that skin cancers induce a relative immunity to other types of cancer *(51)*. Consequently, he advocated the deliberate induction of non-melanoma skin cancers, which were easy to detect and to treat, as a form of vaccination against more life-threatening and less treatable cancers. It was in 1941 when the pathologist

Frank Apperly published geographic data that demonstrated for the first time an inverse correlation between levels of UV radiation in North America and mortality rates from non-skin cancers *(52)*. Apperly concluded that "the presence of skin cancer is really only an occasional accompaniment of a relative cancer immunity in some way related to exposure to ultraviolet radiation." "A closer study of the action of solar radiation on the body," he reasoned, "might well reveal the nature of cancer immunity." Since the time of Apperly's first report, an association between increased risk of dying of various internal malignancies (e.g., breast, colon, prostate, and ovarian cancer) and decreasing latitude toward the equator has now been confirmed *(50)*. A correlation of latitudinal association with sun exposure and decreased vitamin D serum levels has been demonstrated *(48, 50)*. Interestingly, black men, who have an increased risk to develop vitamin D deficiency, have also an increased risk of prostate cancer and develop a more aggressive form of this disease. Moreover, it has been reported that sun exposure is associated with a relatively favorable prognosis and increased survival rate in various other malignancies, including malignant melanoma *(53)*. It has been speculated that these findings were related to UV exposure-induced relatively high serum levels of vitamin D. Berwick et al. have evaluated the association between measures of skin screening and death from cutaneous melanoma in case subjects ($n = 528$) from a population-based study of cutaneous melanoma that were followed for an average of more than 5 years *(53)*. They found that sunburn, high intermittent sun exposure, and solar elastosis were statistically significantly inversely associated with death from melanoma and concluded that sun exposure is associated with increased survival from melanoma *(53)*. Cell and animal experiments reported in the literature, as well as epidemiological data from some countries, relate survival of various malignancies including colon cancer with sun exposure, latitude, and vitamin D_3 synthesis in skin (Fig. 5) *(54)*.

It can be summarized that the evolution of our understanding of the role of vitamin D in cancer parallels our understanding of the importance of vitamin D for rickets *(45)*. In both diseases, epidemiologic observations about consequences of sun exposure preceded experimental observations and were subsequently validated by them. Apperly's insightful observations on sunlight exposure and cancer, like those of Theobald Palm on the protective effects of UV radiation on rickets a half century earlier *(55)*, passed virtually unnoticed for many years, only to be re-discovered by epidemiologists decades later. During recent years, great progress has been made in laboratory investigations that searched for the "missing link" between the vitamin D and cancer connection. Of high importance was the discovery that in contrast to earlier assumptions, skin, prostate, colon, breast, and many other tissues express the enzyme to convert 25(OH)D to its active form, 1,25(OH)$_2$D *(56, 57)*. Therefore, 1,25(OH)$_2$D is now not exclusively considered as a calcium and bone metabolism regulating hormone but also as a hormone that is locally produced in a broad variety of human tissues where it is a potent regulator of cell growth and other cellular functions (review in *(57)*). Figure 1 shows our present understanding of the cutaneous metabolism of vitamin D compounds.

In conclusion, the lack of sunlight exposure results in more than bone disease and an increased risk for cancer – there are multiple other added benefits that include controlling insulin secretion, blood pressure, and cholesterol. Cholesterol is required as a substrate for conversion into vitamin D precursors. With adequate sun exposure, vitamin D

HGF/SF Transgenic Mice	UV-Treatment	Melanoma Outcome
neonate	single erythemal exposure	melanomas with junctional activity arise with significant reduction in latency
juvenile	dual erythemal exposure	no change from single exposure exept for increased multiplicity
adult	chronic sub-erythemal exposure	no change from untreated mice
adult	none	dermal melanomas arise in aged mice

Fig. 5. UV-induced melanomagenesis in the HGF/SF transgenic mouse *(33, 32)*. Dermal melanomas arise in untreated mice with a mean onset age of approximately 21 months, a latency that is not overtly altered in response to chronic suberythemal or skin non-reddening UV irradiation *(33, 32)*. In contrast, a single erythemal dose to 3.5-day-old-neonatal HGF/SF mice induces cutaneous melanoma with significantly reduced latency *(33, 32)*. Note that the UV-induced murine melanomas frequently resemble their human counterparts with respect to histopathological appearance and graded progression. Exposure of HGF/SF transgenic neonates to a second erythemal dose of UV irradiation does not accelerate melanomagenesis; however, the dual exposure significantly increases the number of melanocytic lesions arising per mouse *(33, 32)*.

precursors turn to vitamin D, a mechanism that stimulates the production of vitamin D precursors from cholesterol. It has been speculated that the increased concentration of blood cholesterol during winter months and the fact that outdoor activity (gardening) is associated with lower circulating cholesterol levels in the summer, but not in winter, may explain geographical differences in coronary heart disease incidence *(58)*.

3.2. Consequent Sun Protection Increases the Risk of Vitamin D Deficiency

We have analyzed whether patients that need to protect themselves for medical reasons from sun exposure are at risk to develop vitamin D deficiency. Plasma 25(OH)D levels were analyzed in renal transplant patients with adequate renal function and in an age- and gender-matched control group at the end of wintertime (February/March). All renal transplant patients had practised solar UV protection after transplantation due to the increased risk to develop skin cancer that is associated with their immunosuppressive medication. Plasma 25(OH)D levels were significantly lower in renal transplant patients as compared to controls *(59)*. In another pilot study, we analyzed basal 25(OH)D plasma levels in a small group of patients with xeroderma pigmentosum (XP, $n = 3$) and basal cell nevus syndrome (BCNS, $n = 1$) at the end of wintertime (February/March). 25(OH)D levels in all four patients were markedly decreased with a mean value of 9.5 ng/ml (normal range: 15.0–90.0 ng/ml). In conclusion, we could demonstrate reduced plasma 25(OH)D levels in sunlight-deprived risk groups *(59)*.

3.3. How Much Vitamin D Do We Need?

How much vitamin D do we need to guarantee a protecting effect against malignancies and other diseases? The US Recommended Dietary Allowance (RDA) of vitamin D from 1989 is 200 IU *(60)*. However, several investigations have convincingly shown that 200 IU/day has no effect on bone status *(61)*. It has been recommended that adults may need, at a minimum, five times the RDA, or 1,000 IU, to adequately prevent bone fractures, protect against some malignancies, and derive other broad-ranging health benefits *(60)*. In conclusion, the 1989 RDA of 200 IU is antiquated, and the newer 600 IU daily reference intake (DRI) dose for adults older than 70 is still not sufficient *(60)*. It has been suggested that even the 2,000 IU upper tolerable intake, the official safety limit, does not deliver the amounts of vitamin D that may be optimal *(60)*. On a sunny summer day, total body sun exposure produces approximately 10,000–40,000 IU vitamin D per day. As a result, concerns about toxic overdose with dietary supplements that exceed 800 IU are poorly founded. It has been speculated that a healthy person would have to consume almost 67 times more vitamin D than the current 600 IU recommended intake for older adults to experience symptoms of overdosage *(60)*. Vieth believes people need 4,000–10,000 IU vitamin D daily and that toxic side effects are not a concern until a 40,000 IU/day dose *(60)*.

Other researchers agree with these findings. They suggest that older adults, sick adults, and "perhaps all adults" need 800–1,000 IU daily. They indicate that daily doses of 2,400 IU – four times the recommended intake – can be consumed safely *(60)*. In a recent meta-analysis, Gorham and coworkers concluded that daily intake of 1,000–2,000 IU vitamin D could reduce the incidence of colorectal cancer with minimal risk *(62)*.

4. CONCLUSIONS

Which consequences do we draw from these findings, most importantly the demonstration of an association between vitamin D deficiency and the occurrence of various types of malignancies and other diseases? The most important take-home message, especially for dermatologists, is that strict sun protection procedures that are still recommended in skin cancer prevention campaigns increase the severe health risk to develop vitamin D deficiency. There is no doubt that UV radiation is mutagenic and is the main reason for the development of non-melanoma skin cancer. Therefore, excessive sun exposure has to be avoided, particularly burning in childhood. To reach this goal, the use of sunscreens as well as the wearing of protective clothes and glasses is absolutely important. Additionally, sun exposure around midday should be avoided during the summer in most latitudes. However, the dermatological community has to recognize that there is convincing evidence that the protective effect of less intense solar radiation outweighs its mutagenic effect. In consequence, it has been speculated that many lives could be prolonged through careful exposure to sunlight or more safely, vitamin D supplementation, especially in non-summer months. It has been suggested that, in a population with a high prevalence of vitamin D deficiency (e.g., Europe, the United States), a moderate increase in UV exposure may most likely result in more beneficial as negative health effects *(63, 64)*. Moreover, the economic burden in the United States due to insufficient solar UV irradiance has been estimated for 2004 to be approximately 40–56 billion USD *(65)*. In contrast, the economic burden due to exaggerated solar UV irradiance in the same time period has been estimated to be approximately only 6–7 billion USD *(65)*. The authors concluded that, at least in the United States, increased vitamin D levels (UV-B, nutrition) will result in reduced economic burden *(65)*. Therefore, recommendations of dermatologists on sun protection should be moderated. As Michael Holick reported *(66)*, we have learned that at most latitudes such as Boston, USA, very short and limited solar exposure is sufficient to achieve "adequate" vitamin D levels. Exposure of the body in a bathing suit to one minimal erythemal dose (MED) of sunlight is equivalent to ingesting about 10,000–40,000 IU of vitamin D and it has been reported that exposure of less than 18% of the body surface (hands, arms, and face) two to three times a week to a third to a half of an MED (about 5 min for skin type 2 adult in Boston at noon in July) in the spring, summer, and autumn is more than adequate. Anyone intending to stay exposed to sunlight longer than recommended above should apply a sunscreen with a sufficient sun protection factor to prevent sunburn and the damaging effects of excessive exposure to sunlight. A recent report of the International Agency for Research on Cancer *(67)* that relativizes the connection of vitamin D and cancer has been heavily criticized *(68)*. In response to this IARC report, Dr. Grant wrote a thoughtful and critical review of the report noting not only many deficiencies in the interpretation of the data that the committee members used to base their recommendations on but also that the ratio of expertise on the committee favoring UV-B association with cancer was fourfold higher than expertise and publication record in the field of vitamin D. Grant's thoughtful and critical review was appreciated in a editorial of Dr. Holick in the same issue of *Dermato-Endocrinology (69)*. Although further work is necessary to define the influence of vitamin D deficiency on the occurrence of melanoma and non-melanoma skin cancer, it is at present mandatory that especially dermatologists strengthen the importance of an adequate vitamin D status if sun exposure is seriously curtailed. It has to be emphasized that in groups that are at high risk of developing vitamin D deficiency (e.g., nursing-home residents; patients with skin type I, or patients under immunosuppressive therapy that must be protected from the sun exposure), vitamin D status should be monitored subsequently. Vitamin D deficiency should be treated, e.g., by giving vitamin D orally as recommended previously *(60, 66)*. It has been shown that a single dose of 50,000 IU vitamin D once a week for 8 weeks is efficient and safe to treat vitamin D deficiency *(60)*. Another means of guaranteeing vitamin D sufficiency, especially in nursing-home residents, is to give 50,000 IU of vitamin D once a month. An alternative to prevent vitamin D deficiency would be the use of vitamin D containing ointments. However, it should be noted that vitamin D containing ointments are, at least in Europe, not allowed as cosmetics. These antiquated laws are the result of the fear of vitamin D intoxication that was evident in Europe in the 1950s *(70)* and should be re-evaluated, for they do not reflect our present scientific knowledge. If we follow the guidelines discussed above carefully, they will ensure an adequate vitamin D status, thereby protecting us against adverse effects of strict sun protection recommendations. Most importantly, these measures will protect us sufficiently against the influence of vitamin D deficiency on the occurrence of various malignancies without increasing our risk to develop UV-induced skin cancer. To reach this goal it is of high importance that this information is transferred to every clinician, especially to dermatologists. Otherwise dermatologists will not be prepared for the moderation of sun protection recommendations that are necessary to protect us against vitamin D deficiency, cancer, and other diseases.

REFERENCES

1. Unna PG (1894) Histopathologie der Hautkrankheiten. August Hirschwald Verlag, Berlin
2. Dubreuilh W (1896) Des hyperkeratoses circumscriptes. Ann Derm et Syph 7:1158–1204
3. Preston DS, Stern RS (1992) Nonmelanoma cancers of the skin. N Engl J Med 327:1649–1662
4. Kricker A, Armstrong BK, English DR (1994) Sun exposure and non-melanocytic skin cancer. Cancer Causes Control 5:367–392
5. Elwood JM, Jopson J (1997) Melanoma and sun exposure: An overview of published studies. Int J Cancer 73:198–203
6. Armstrong BK, Kricker A (2001) The epidemiology of UV induced skin cancer. J Photochem Photobiol 63:8–18
7. Tilgen W, Rass K, Reichrath J (2005) 30 Jahre dermatologische Onkologie. Akt. Dermatol 31:79–88
8. Wang H, Diepgen TL (2006) The Epidemiology of basal cell and squamous cell carcinoma. In: Reichrath J (ed) Molecular mechanisms of basal cell and squamous cell carcinomas. Landes Bioscience, Georgetown, USA and Springer Science and Business Media, New York, USA, p. 1–9
9. Rass K (2006) UV-damage and DNA-repair in basal and squamous cell carcinomas. In: Reichrath J (ed) Molecular mechanisms of basal cell and squamous cell carcinomas. Landes Bioscience, Georgetown, USA and Springer Science and Business Media, New York, USA, p. 18–30
10. Reichrath J, Querings K (2006) Histology of epithelial skin tumors. In: Reichrath J (ed) Molecular mechanisms of basal cell and squamous cell carcinomas. Landes Bioscience, Georgetown, USA and Springer Science and Business Media, New York, USA, p. 10–17
11. Gilchrest BA, Eller MS, Geller AC, Yaar M (1999) The pathogenesis of melanoma induced by ultraviolet radiation. N Engl J Med 340:1341–1348
12. Frost CA, Green AC (1994) Epidemiology of solar keratoses. Br J Dermatol 131:455–464
13. Green A, Battistutta D (1990) Incidence and determinants of skin cancer in a high-risk Australian population. Int J Cancer 46:356–361
14. Alam M, Ratner D (2001) Cutaneous squamous-cell carcinoma. N Engl J Med 344:975–983
15. Osterlind A, Tucker MA, Stone BJ, Jensen OM (1988) The Danish case-control study of cutaneous malignant melanoma. II importance of UV-light exposure. Int J Cancer 42:319–324
16. Green A, Siskind V (1983) Geographical distribution of cutaneous melanoma in Queensland. Med J Aust 1:407–410
17. Elwood JM, Gallagher RP, Hill GB, Pearson JC (1985) Cutaneous melanoma in relation to intermittent and constant sun exposure – the Western Canada Melanoma Study. Int J Cancer 35:427–433
18. Kennedy C, Bajdik CD, Willemze R, de Gruijl FR, Bouwes Bavinck JN (2003) The influence of painful sunburns and lifetime sun exposure on the risk of actinic keratoses, seborrheic warts, melanocytic nevi, atypical nevi, and skin cancer. J Invest Dermatol 120(6):1087–1093
19. Westerdahl J, Olsson H, Måsbäck A, Ingvar C, Jonsson N (1995) Is the use of sunscreens a risk factor for melanoma?. Melanoma Res 5:59–65
20. Bataille V, Boniel M, De Vries E, Severi G, Brandberg Y, Sasieni P, Cuzick J, Eggermont A, Ringberg U, Grivegnee AR, Coebergh JW, Chignol MC, Dore JF, Autier P (2005) A multicentre epidemiological study on sunbed use and cutaneous melanoma in Europe. Eur J Cancer 41(14):2141–2149
21. Gates FL (1928) On nuclear derivatives and lethal action of ultraviolet light. Science 68:479–480
22. Hollaender A, Emmons CW (1941) Wavelength dependence of mutation production in the ultraviolet with special emphasis on fungi. Cold Spring Harbor Symp Quand Biol 9:179–186
23. Coohill TP (1997) Historical aspects of ultraviolet action spectroscopy. Photochem Photobiol 65S:123S–128S

24. De Grujil FR (1999) Skin cancer and solar uv radiation. Eur J Cancer 35:2003–2009
25. Setlow RB (1997) DNA damage and repair: a photobiological odyssey. Photochem Photobiol 65S:119S–122S
26. Beukers R, Berends W (1960) Isolation and identification of the irradiation product of thymine. Biochim Biophys Acta 41:550–551
27. Setlow RB, Carrier WL (1966) Pyrimidine dimers in ultraviolet-irradiated DNA's. J Mol Biol 17: 237–254
28. Varghese AJ, Patrick MH (1969) Cytosine derived heteroadduct formation in ultraviolet-irradiated DNA. Nature 223:299–300
29. Melnikova VO, Ananthaswamy HN (2006) p53 Protein and non-melanoma skin cancer. In: Reichrath J (ed) Molecular mechanisms of basal cell and squamous cell carcinomas. Landes Bioscience, Georgetown, USA and Springer Science and Business Media, New York, USA 2006, pp. 66–79
30. Reifenberger J, Wolter M, Knobbe CB, Kohler B, Schonicke A, Scharwachter C, Kumar K, Blaschke B, Ruzicka T, Reifenberger G (2005) Somatic mutations in the PTCH, SMOH, SUFUH and TP53 genes in sporadic basal cell carcinomas. Br J Dermatol. 152(1):43–51
31. Baron ED (2006) The immune system and nonmelanoma skin cancers. In: Reichrath J (ed), Molecular mechanisms of basal cell and squamous cell carcinomas. Landes Bioscience, Georgetown, USA and Springer Science and Business Media, New York, USA, p. 43–48.
32. Noonan FP, Recio JA, Takayama H, Duray P, Anver MR, Rush WL, De Fabo EC, Merlino G (2001) Neonatal sunburn and melanoma in mice. Nature 413(6853):271–272
33. Noonan FP, Otsuka T, Bang S, Anver MR, Merlino G (2000) Accelerated ultraviolet radiation-induced carcinogenesis in hepatocyte growth factor/scatter factor transgenic mice. Cancer Res 60(14): 3738–3743
34. Reichrath J, Rappl G (2003) Ultraviolet (UV)-induced immunosuppression: is vitamin D the missing link?. J Cell Biochem 89(1):6–8
35. Osborne JE, Hutchinson PE (2002) Vitamin D and systemic cancer: is this relevant to malignant melanoma?. Br J Dermatol 147:197–213
36. Reichrath J, Kamradt J, Zhu XH, Kong Xf, Tilgen W, Holick MF (1999) Analysis of 1,25-dihydroxyvitamin D_3 receptors in basal cell carcinomas. Am J Pathol 155(2):583–589
37. Reichrath J, Rafi L, Rech M, Mitschele T, Meineke V, Gärtner BC, Tilgen W, Holick MF (2004) Analysis of the vitamin D system in cutaneous squamous cell carcinomas (SCC). J Cut Pathol 31(3):224–231
38. Reichrath J, Rafi L, Rech M, Meineke V, Tilgen W, Seifert M (2004) No evidence for amplification of 25-hydroxyvitamin D-1α-OHase (1α-OHase) or 1,25-dihydroxyvitamin D-24-OHase (24-OHase) genes in malignant melanoma (MM). J Steroid Biochem Mol Biol 89–90:163–166
39. Seifert M, Rech M, Meineke V, Tilgen W, Reichrath J (2004) Differential biological effects of 1,25-dihydroxyvitamin D_3 on melanoma cell lines in vitro. J Steroid Biochem Mol Biol 89–90:375–379
40. Hutchinson PE, Osborne JE, Lear JT, Smith AG, Bowers PW, Morris PN, Jones PW, York C, Strange RC, Fryer AA (2000) Vitamin D receptor polymorphisms are associated with altered prognosis in patients with malignant melanoma. Clin Cancer Res 6(2):498–504
41. Halsall JA, Osborne JE, Potter L, Pringle JH, Hutchinson PE (2004) A novel polymorphism in the 1A promoter region of the vitamin D receptor is associated with altered susceptibility and prognosis in malignant melanoma. Br J Cancer 91(4):765–770
42. Albertson DG, Ylstra B, Segraves R, Collins C, Dairkee SH, Kowbel D, Kuo WL, Gray JW, Pinkel D (2000) Quantitative mapping of amplicon structure by array CGH identifies CYP24 as a candidate oncogene. Nat Genet. 25(2):144–146
43. Diesel B, Radermacher J, Bureik M, Bernhardt R, Seifert M, Reichrath J, Fischer U, Meese E (2005) Vitamin D_3 metabolism in human glioblastoma multiforme: functionality of CYP27B1 splice variants, metabolism of calcidiol, and effect of calcitriol. Clin Cancer Res 11(15):5370–5380
44. Reichrath J, Querings K (2004) No evidence for reduced 25-hydroxyvitamin D serum levels in melanoma patients. Cancer Causes Control 15:97–98
45. Holick MF (2003) Evolution and function of vitamin D. Recent Results Cancer Res. 164:3–28
46. Matsuoka LY, Wortsman J, Dannenberg MJ, Hollis BW, Lu Z, Holick MF (1992) Clothing prevents ultraviolet-B radiation-dependent photosynthesis of vitamin D_3. J Clin Endocrinol Metab 75(4): 1099–1103

47. Gorham ED, Garland FC, Garland CF (1990) Sunlight and breast cancer incidence in the USSR. Int J Epidemiol 19:614–622
48. Garland CF, Comstock GW, Garland FC, Helsing KJ, Shaw EK, Gorham ED (1989) Serum 25-hydroxyvitamin D and colon cancer: eight year prospective study. Lancet 2(8673):1176–1178
49. Garland CF, Garland FC, Gorham ED (1991) Can colon cancer incidence and death rates be reduced with calcium and vitamin D?. Am J Clin Nutr 54:193S–201S
50. Grant WB (2002) An estimate of premature cancer mortality in the U.S. due to inadequate doses of solar ultraviolet-B radiation. Cancer 94:1867–1875
51. Peller S (1936) Carcinogenesis as a means of reducing cancer mortality. Lancet 2:552–556
52. Apperly FL (1941) The relation of solar radiation to cancer mortality in North America. Cancer Res 1:191–195
53. Berwick M, Armstrong BK, Ben-Porat L, Fine J, Kricker A, Eberle C, Barnhill R (2005) Sun exposure and mortality from melanoma. J Natl Cancer Inst 97(3):195–199
54. Moan J, Porojnicu AC, Robsahm TE, Dahlback A, Juzeniene A, Tretli S, Grant W (2005) Solar radiation, vitamin D and survival rate of colon cancer in Norway. J Photochem Photobiol B 78(3):189–193
55. Palm TA (1890) The geographical distribution and etiology of rickets. Practitioner 45:270–279
56. Schwartz GG, Whitlatch LW, Chen TC, Lokeshwar BL, Holick MF (1998) Human prostate cells synthesize 1,25-dihydroxyvitamin D_3 from 25-hydroxyvitamin D_3. Cancer Epidemiol Biomarkers Prev 7:391–395
57. Lehmann B, Querings K, Reichrath J (2004) Vitamin D and skin: new aspects for dermatology. Exp Dermatol 13(s4):11–15
58. Grimes DS, Hindle E, Dyer T (1996) Sunlight, cholesterol and coronary heart disease. Q J Med 89:579–589
59. Querings K, Reichrath J (2004) A plea for detection and treatment of vitamin D deficiency in patients under photoprotection, including patients with xeroderma pigmentosum and basal cell nevus syndrome. Cancer Causes Control 15(2):219
60. Vieth R (1999) Vitamin D supplementation, 25-hydroxyvitamin D concentrations, and safety. Am J Clin Nutr 69:842–856
61. Dawson-Hughes B, Harris SS, Krall EA, Dallal GE, Falconer G, Green CL (1995) Rates of bone loss in post-menopausal women randomly assigned to one of two dosages of vitamin D. Am J Clin Nutr 61:1140–1145
62. Gorham ED, Garland CF, Garland FC, Grant WB, Mohr SB, Lipkin M, Newmark HL, Giovannucci E, Wei M, Holick MF (2007) Optimal vitamin D status for colorectal cancer prevention: a quantitative meta analysis. Am J Prev Med 32(3):210–216
63. Giovannucci E, Liu Y, Rimm EB, Hollis BW, Fuchs CS, Stampfer MJ, Willett WC (2006) Prospective study of predictors of vitamin D status and cancer incidence and mortality in men. J Natl Cancer Inst 98:451–459
64. Moan J, Porojnicu AC, Dahlback A, Setlow RB (2008) Addressing the health benefits and risks, involving vitamin D or skin cancer, of increased sun exposure. PNAS 105(2):668–673
65. Grant WB, Garland FC, Holick MF (2005) Comparison of estimated economic burden due to insufficient solar ultraviolet irradiance and vitamin D and excess solar UV irradiance for the united states. Photochem Photobiol 81(6):1276–1286
66. Holick MF (2001) Sunlight "D"ilemma: risk of skin cancer or bone disease and muscle weakness. Lancet 357:961
67. IARC. IARC report on cancer. IARC working group reports, Lyon, France: International Agency for Research on Cancer 2008. http://www.iarc.fr/en/Media-Centre/IARC-News/Vitamin-D-and-Cancer
68. Grant WB (2009) A critical review of vitamin D and cancer: a report of the IARC working group on vitamin D. Dermato-Endocrinology 1:25–33
69. Holick MF (2009) Shining light on the vitamin D-cancer connection IARC report. Dermato-Endocrinology 1:4–6
70. British Pediatric Association (1956) Hypercalcemia in infants and vitamin D. BMJ 2:149

VI   VITAMIN D
DEFICIENCY AND
CHRONIC DISEASE

47 Vitamin D and the Risk of Type 1 Diabetes

Elina Hyppönen

Abstract   Type 1 diabetes is an autoimmune disease resulting from a progressive destruction of the insulin-secreting β-cells. There is no cure or preventive treatment for type 1 diabetes. However, increased vitamin D intake is currently considered as one of the most promising candidates for prevention. The biological mechanisms for a role of vitamin D in diabetes development are plausible and as reviewed in this chapter, evidence to support an association has been obtained from various lines of investigation.

Key Words: Type 1 diabetes; vitamin D; epidemiology; animal experiments; genetic studies

Type 1 diabetes is a chronic disease which results from progressive autoimmune destruction of the insulin-secreting β-cells *(1)*. Latency period leading to disease onset is commonly long, and pancreatic β-cell destruction may already start several years before the diagnosis, in some cases before birth *(2)*. Type 1 diabetes is a multifactorial disease, and both genetic predisposition and environmental factors are believed to be important in the development of the disease. However, the etiology of type 1 diabetes is still not fully understood. The steeply increasing incidence of type 1 diabetes *(3)*, studies on migrant populations *(4)*, and the low concordance rate in monozygotic twins *(5)* indicate that environmental factors must have a role in the pathogenic process. It is likely that the effect of an environmental factor may differ at different stages of life, it may act at different stages of the pathogenic process, and multiple hits may often be needed. Disease etiology may further be complicated by interactions between different environmental exposures as well as with genetic factors *(6)*. There is no cure or preventive treatment for type 1 diabetes *(7)*. However, increased vitamin D intake is currently considered as one of the most promising candidates for prevention *(8)*. The biological mechanisms for a role of vitamin D in diabetes development are plausible and as reviewed in this chapter, evidence to support an association has been obtained from various lines of investigation.

1. MECHANISMS

Type 1 diabetes (also called diabetes Type 1A) is an autoimmune disease where the destruction of insulin-secreting β-cells occurs in a T-cell-dependent process *(6)*. Chronic

From: *Nutrition and Health: Vitamin D*
Edited by: M.F. Holick, DOI 10.1007/978-1-60327-303-9_47,
© Springer Science+Business Media, LLC 2010

867

inflammatory infiltrate (i.e., insulitis) consisting of CD8+ and CD4+ T cells, B lympho-cytes, macrophages, and natural killer cells is found in the islets of Langerhans near the time of onset of type 1 diabetes and in patients with long-standing disease *(10)*. Disease onset is typically preceded by an asymptomatic inflammation of the insulin-secreting β-cells, during which autoantibodies to diabetes-specific antigens can often be detected *(12)*. As in other autoimmune diseases, both T-cell and autoantibody responses are typi-cally directed against multiple self-antigens. With the progression of autoimmunity from initial activation to a chronic state, the number of islet autoantigens targeted by T cells and autoantibodies is often increased.

Disruption in the polarization between T helper type 1 (Th1) and Th2 toward Th1 upregulation is believed to be central to the pathogenesis of type 1 diabetes *(8)*. In vitro studies and animal experiments have shown that the active form of vitamin D 1,25-dihydroxyvitamin D [1,25(OH)$_2$D or calcitriol] affects T-cell activity, which could potentially influence various stages of the autoimmune process leading to type 1 dia-betes. In vitro, calcitriol inhibits T-cell proliferation and decreases the production of Th1 cytokines IL-2 and INF-γ *(14)*. Furthermore, there is evidence to show that cal-citriol (or its analogues) may suppress the activation of Th1 cells via direct influences on the differentiation and maturation of dendritic cells or by enhancing the presence or function of suppressor T cells *(8)*.

Other proposed mechanisms include influences through reductions in infections pre-disposing to type 1 diabetes (rather than influences on the autoimmune process per se) *(15)* and reduction in diabetes risk through direct effects of calcitriol on insulin secretion in the β-cell *(16)*. Furthermore, it has also been suggested that after the autoimmune acti-vation, free radicals produced by macrophages and T cells will contribute to the β-cell dysfunction and death *(17)* and that calcitriol may have beneficial effects through acting as a free radical scavenger *(18)*. However, this mechanism remains largely hypotheti-cal, especially as there are very little data to show an effect of other dietary antioxidant vitamins on the risk of type 1 diabetes *(19)*.

2. ECOLOGICAL CORRELATIONS

Given the strong influence of time of year and latitude on vitamin D production in the skin, seasonal and geographical variations are commonly evaluated to obtain evidence for a possible influence of vitamin D on disease development.

Geographical variations It has been proposed that geographical variations in the incidence of type 1 diabetes are explained by lower average temperature and reduced number of sunshine hours, which have been independently associated with diabetes incidence *(20)*. There are wide variations in the incidence of type 1 diabetes between populations, and the rates are 400 times higher in Finland with the highest incidence (40/100,000) compared to China or Venezuela where the reported incidence is the low-est (0.1/100,000) *(21)*. As seen in Fig. 1, the incidence of type 1 diabetes increases with increasing distance from the equator *(22, 23)*, which may reflect variations in ultraviolet radiation-induced vitamin D synthesis in the skin. Support for a north–south gradient in type 1 diabetes incidence has been obtained from comparisons of incidence rates between countries but patterns have also been observed within a single country *(23, 24)*.

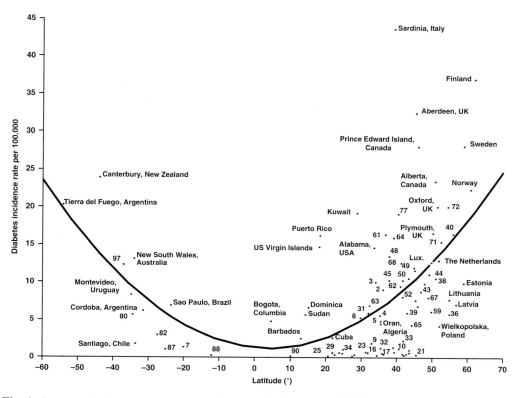

Fig. 1. Age-standardized incidence rates of type 1 diabetes per 100,000 boys <14 years of age, by latitude, in 51 regions worldwide, 2002. Data points are shown by *dots*; names shown adjacent to the *dots* denote location, where space allows. Where space was limited, numerical codes (*below*) designate location. Source: data from WHO DiaMond. Lux., Luxembourg. Numerical codes for areas: *2.* Beja, Tunisia; *3.* Gafsa, Tunisia; *4.* Kairoan, Tunisia; *5.* Monastir, Tunisia; *7.* Mauritius; *8.* Wuhan, China; *9.* Sichuan, China; *10.* Huhehot, China; *16.* Nanjing, China; *17.* Jinan, China; *21.* Harbin, China; *23.* Changsha, China; *25.* Hainan, China; *29.* Hong Kong, China; *31.* Israel; *32.* Chiba, Japan; *33.* Hokkaido, Japan; *34.* Okinawa, Japan; *36.* Novosibirsk, Russia; *38.* Antwerp, Belgium; *39.* Varna, Bulgaria; *40.* Denmark; *43.* France; *44.* Baden, Germany; *45.* Attica, Greece; *48.* Sicily, Italy; *49.* Pavia, Italy; *50.* Marche, Italy; *52.* Lazio, Italy; *59.* Krakow, Poland; *61.* Algarve, Portugal; *62.* Coimbra, Portugal; *63.* Madeira Island, Portugal; *64.* Portalegre, Portugal; *65.* Bucharest, Romania; *67.* Slovakia; *68.* Catalonia, Spain; *71.* Leicestershire, UK; *72.* Northern Ireland, UK; *77.* Allegheny, PA, USA; *80.* Avellaneda, Argentina; *82.* Corrientes, Argentina; *87.* Paraguay; *88.* Lima, Peru; *90.* Caracas, Venezuela; *97.* Auckland, New Zealand. Data points not labeled because of space constraints (latitude in degrees, rate per 100,000): *11.* Dalian, China (39, 1.1); *12.* Guilin, China (24, 0.6); *13.* Beijing, China (40, 0.7); *14.* Shanghai, China (32, 0.7); *15.* Chang Chun, China (44, 0.6); *18.* Jilin, China (43, 0.4); *19.* Shenyang, China (42, 0.4); *20.* Lanzhou, China (36, 0.5); *22.* Nanning, China (23, 0.3); *24.* Zhengzhou, China (35, 0.2); *26.* Tie Ling, China (42, 0.2); *27.* Zunyi, China (28, 0.1); *28.* Wulumuqi, China (44, 0.9); *35.* Karachi, Pakistan (25, 0.5); *37.* Austria (48, 9.8); *46.* Hungary (47, 8.7); *51.* Turin, Italy (45, 11.9); *53.* Lombardia, Italy (46, 7.6); *66.* Slovenia (46, 6.8); *79.* Chicago, IL, USA (42, 10.2). $R^2 = 0.25$, $p < 0.001$. Reproduced with kind permission from Springer Science+Business Media: Mohr et al. (*22*).

There are some exceptions to the global north-to-south pattern, some of which can be explained by differences in genetic susceptibility. For example, in Europe, Finland and the Mediterranean island of Sardinia (3,000 km south of Finland) share equally high incidence of type 1 diabetes. Studies on children of Sardinian heritage who live and have been born in Lazio region in mainland Italy (across the sea from Sardinia) have shown that the incidence of type 1 diabetes was fourfold for children with Sardinian heritage compared to those with parents from mainland Italy *(25)*. This suggests that the very high incidence of type 1 diabetes seen in Sardinia (in contrast to its geographical location/rest of Italy) is likely to be due to genetic factors rather than environmental influences.

Seasonal variation Onset of type 1 diabetes demonstrates seasonal patterns, and most typically highest rates are seen during winter and the lowest during the summer months *(21, 26–28)*. These observed seasonal patterns may suggest a presence of season-dependent precipitating factor, such as low vitamin D status or viral infection, or a combination of the two. The evidence for the influence of time of birth on diabetes incidence is not consistent: significant variations by month of birth are seen in some studies *(29, 30)* but the observed patterns are not consistent nor is variation seen in all populations *(31, 32)*. However, it is problematic to make inferences about the possible importance of prenatal/early life vitamin D status based on month of birth only, given the multifactorial and complex nature of type 1 diabetes, variations in vitamin D supplementation recommendations in pregnancy, and lack of information on the relevant time window for the possible influence of low vitamin D status on future diabetes risk *(33)*.

Temporal changes Some indication for a possible association between vitamin D intake and type 1 diabetes can also be obtained from temporal changes in the incidence rate and corresponding changes in vitamin D intake at the population level. The incidence of type 1 diabetes has been increasing in many industrialized countries including the United Kingdom *(3, 34, 35)*, where since World War II there has been a gradual decrease in the vitamin D fortification of foods and a breakdown of systematic vitamin D/cod liver oil supplementation strategy *(36)*. In Finland, vitamin D supplementation to infants has been reduced to one-tenth of the recommended dose in the 1940s–1950s *(37)*, and the incidence of type 1 diabetes has increased fourfold since the first nation-wide surveys in the early 1950s *(38)*. The recommended dose of vitamin D was decreased in steps, and in 1964 the dose was recommended to be reduced from the dose interval of 4,000–5,000 IU down to 2,000 IU/day *(37)*, and in 1975 it was further reduced to 1,000 IU/day. In Finland, the transfer to the current recommendation of 400 IU/day took place in 1992. Given the observed decrease in the compliance to supplementation recommendations *(39, 40)*, it could be speculated that the reduced vitamin D intake in infancy could to some extent have contributed to increase in the incidence of type 1 diabetes in Finland during the past decades.

3. ANIMAL EXPERIMENTS

Observations for the beneficial effect of calcitriol administration in other autoimmune diseases (e.g., experimental encephalitis and thyroiditis) *(41–43)* stimulated the first animal studies on type 1 diabetes in the early 1990s. The effect of calcitriol on type 1 diabetes was first tested using nonobese diabetic (NOD) mice, an animal model which spontaneously develops autoimmune diabetes. In these early experiments, administration of calcitriol to NOD mice (5 μg/kg i.p. every other day from weaning) strongly reduced the proportion of animals developing insulitis or diabetes; insulitis was seen in 41% of the treated group compared to 75% of the controls (day 100) (44), and from the treated mice only 8% progressed to overt disease by day 200 compared to 56% in the control group (45). Subsequent experiments with the NOD mouse suggest that using sufficiently large doses of calcitriol, a complete protection from progression to diabetes may be achieved (46). Zella et al. (46) gave NOD mice 50 ng of calcitriol orally every day from weaning and reported that none of the treated mice had developed diabetes by day 200. Animals in the control group were also used in a further experiment on diabetes risk in which the effect of adequate vitamin D intake (ensured by housing animals under fluorescent light) was compared to deprivation (no intake either from diet or UVB radiation). In this experiment, adequate vitamin D intake completely protected the male (0% developed diabetes in sufficient vs. 44% in deficient group) but not the female animals (44% vs. 88%, respectively) (46).

Typically, diabetes prevention in experiments using pharmacological doses of calcitriol has led to increases in serum calcium levels in the treated animals. Given the potentially very serious consequences of hypercalcemia, there is great research interest in determining whether corresponding effects on diabetes risk can be achieved by non-hypercalcemic structural analogues of calcitriol. There are over 2,000 vitamin D analogues (47). Findings are promising, and some vitamin D analogues have been found to be as effective as calcitriol in preventing diabetes in the NOD mouse, although complete protection against disease progression has not been achieved. For example, diabetes incidence was reduced equally by administration of KH1060 (48) or TX527 (49) as compared to calcitriol in the earlier studies (45). The effect of analogues was observed to be dependent on the dosage used, and diabetes protection was achieved without increases in serum calcium levels.

Some animal experiments have attempted to evaluate the critical period when calcitriol/analogue administration may exert the influences on diabetes risk. In one experiment, NOD mice that were vitamin D deficient in utero and early life (up to day 100) had earlier disease onset and higher incidence of diabetes compared to controls (50). The more aggressive disease pattern in the vitamin D-deficient group continued after restoration of normal vitamin D status and metabolism, suggesting that early life influences may contribute to the disease progression in the NOD mouse. However, it should be noted that in the NOD mouse, supplementation with vitamin D per se is ineffective (51) and in all studies any benefits on diabetes risk have been seen with the use of active calcitriol or related analogues (52).

Animal experiments suggest that it is possible for the progression to diabetes to be halted by vitamin D analogues even after the initiation of autoimmune attack. In an experiment using MC1288 after established insulitis with or without cyclosporin A (a known immunosuppressant), MC1288 alone was not effective, whereas the combination treatment led to a reduction in diabetes incidence (53). Treatment was well tolerated with no signs of generalized immunosuppression or major side effects on calcium metabolism. Further support was obtained from a study using vitamin D analogue

RO-26-219, where analogue administration to adult NOD mice reduced the progression to diabetes, arresting Th1 infiltration and insulitis, with the longest treatment giving the best results *(54)*.

Interestingly, despite suggestions for beneficial effects of calcitriol administration in experimental models, support for beneficial effects of vitamin D on diabetes development was not obtained in an experiment on vitamin D receptor (VDR) knockout NOD mice *(55)*. Neither onset of insulitis nor progression to diabetes was altered in the animals lacking vitamin D receptor, despite observation for intensified defects in both innate and adaptive immunity *(55)*. Although this could be used to question the role of vitamin D in diabetes development in the NOD mouse, the authors provided several examples where the lack of receptor has produced differing effects compared to the situation where receptor is present but ligand is missing (as in the case of vitamin D deficiency) and concluded that their results cannot be used to discount a possible influence of vitamin D deficiency for the development of diabetes either in mice or in humans *(55)*.

4. STUDIES IN HUMANS

Despite increasing interest in the topic, to date there are no published clinical trials investigating the role of vitamin D in the prevention of type 1 diabetes. The available observational studies have largely been restricted to looking at either very early exposure (i.e., vitamin D supplementation in uterus/during infancy) or vitamin D status and supplementation at or after diabetes diagnosis. The first study on the association between vitamin D supplementation in infancy and diabetes risk came from the EURODIAB project reporting data from seven European countries. This multinational case–control study suggested a 33% reduction (OR 0.67) in the subsequent risk of developing type 1 diabetes for vitamin D supplementation during the first year of life *(56)*. This association was supported by the findings from a recent meta-analysis, combining information on all four case–control studies published to date *(57)*. The pooled effect estimate for infant vitamin D supplementation (at any dosage) and the risk of type 1 diabetes was OR 0.71 (95% CI 0.60–0.84). In our own prospective study on the 1966 Northern Finland Birth cohort, there was a remarkably consistent association between several indicators of vitamin D intake and status with risk of type 1 diabetes, which both showed evidence for a dose–response effect and was robust to adjustment for a wide range of neonatal, anthropometrical, and social indicators *(39)*. Infants who had received vitamin D supplementation regularly during the first year had 88% lower risk of type 1 diabetes by age 31 compared to participants receiving no supplementation. Furthermore, in the subgroup analyses among children who had received vitamin D supplementation regularly ($n = 9087$), a further 86% risk reduction was seen if the dose given to the infant had been at least 2,000 IU (50 μg) per day, the level of the contemporary recommendation *(37, 39)*. Infants who were suspected by the healthcare personnel of having had rickets during the first year of life had a threefold risk of developing diabetes compared to others.

In some countries (including Norway) vitamin D supplementation is provided in the form of cod liver oil which, in addition to vitamin D, is also rich in omega-3 fatty acids.

In two Norwegian case–control studies, associations with diabetes risk were reported for cod liver oil consumption but not vitamin D per se, which commonly was obtained from multivitamin supplements. In the first study, offspring to mothers who had used cod liver oil supplements during pregnancy had a reduced risk of type 1 diabetes with inconclusive findings on the effect of vitamin D supplementation during pregnancy or in infancy (estimated effect ranged from a nearly 90% reduction to a twofold increase in diabetes risk) *(58)*. In the second, a somewhat larger case–control study, infants who had received cod liver oil had a reduced risk of type 1 diabetes, while vitamin D supplementations in infancy or maternal vitamin D supplements or cod liver oil were not significantly associated with diabetes risk *(59)*. Further support for beneficial effects of higher vitamin D intakes during pregnancy on offspring diabetes risk has been obtained from recent studies showing reduced occurrence of diabetes-specific autoantibodies in offspring of mothers with comparatively high vitamin D intakes *(60, 61)*. Nevertheless, it has been recently suggested that omega-3 fatty acids may have an independent beneficial influence on the progression of type 1 diabetes related autoimmunity. In the longitudinal Diabetes Autoimmunity Study in the Young (DAISY), omega-3 fatty acid intake during childhood was significantly associated with reduced development of islet cell autoantibodies *(62)*. According to the authors (data not shown), vitamin D intake had no independent association with islet autoimmunity nor did adjustment for vitamin D intake affect the estimates for risk reduction observed for omega-3 fatty acids. In their earlier report, DAISY investigators had reported reduced seroconversion rates for offspring to mothers with higher compared to lower vitamin D intakes *(61)*, and it is possible that the apparent discrepancy between these findings reflects either an early critical window for the influence of vitamin D on diabetes-related autoimmunity (pregnancy/infancy, not childhood) or instability in the effect estimation caused by small numbers. In the Northern Finland 1966 cohort, it was clearly vitamin D supplementation and not cod liver oil which accounted for the observed benefits on type 1 diabetes risk *(39)*. Incidence of diabetes in the relatively small group of children receiving cod liver oil ($n = 86$) was higher compared to those who received the supplementation as vitamin D droplets (78.15/per 100,000 vs. 24.66/100,000 person-years at risk), and when we compared to infants receiving cod liver oil regularly with those who were given the recommended dose of vitamin D (2,000 IU/day), the data were suggestive of an increased risk of type 1 diabetes for the cod liver oil users (HR 3.9, 95% CI 0. 96–16, $p = 0.058$) (Hyppönen and Järvelin, unpublished). These data are likely to reflect a higher concentration of vitamin D in droplets compared to cod liver oil (rather than harmful effects of cod liver oil). However, no support is clearly provided for a superior influence on type 1 diabetes risk for supplements containing omega-3 fatty acids compared to supplements containing vitamin D only.

There is evidence to suggest that nutritional vitamin D status (measured by the circulating concentrations of 25-hydroxyvitamin D) *(63, 64)* and in some studies also calcitriol concentrations *(65, 66)* are lower in newly diagnosed type 1 diabetes patients compared to population controls. In a study investigating the effect of administration of calcitriol ($1,25(OH)_2D$) (compared to nicotinamide) to children with newly diagnosed diabetes, there appeared to be some short-term benefits with calcitriol treatment (0.25 μg every other day) compared to nicotinamide on the required insulin dose at

3 and 6 months *(67)*. However, C-peptide or HbA1c concentrations (reflecting residual insulin secretion and glucose homeostasis) did not differ between the calcitriol and nicotinamide groups at 1 year, suggesting that long-term benefits for calcitriol supplementation with treatment initiated after the diagnosis of diabetes are likely to be limited.

5. GENETIC STUDIES

Vitamin D metabolism and status in the body is known to be strongly influenced by genetic variations, and hence, there is great interest in examining the possible effect of vitamin D-related polymorphisms and the risk of type 1 diabetes. The hormonal actions of vitamin D are mediated by vitamin D receptors (VDRs), and VDR polymorphisms have been related to insulin secretion capacity in humans *(68)*. Several studies have investigated the association between VDR polymorphisms and type 1 diabetes, however, with conflicting results. A meta-analysis including information on all association and family studies published from 1997 to 2005 for the four most common variations (*Apa*I, *Bsm*I, *Fok*I, and *Taq*I) did not find support for an association between any of the sites and type 1 diabetes *(69)*. Accordingly, another meta-analysis on the same VDR polymorphisms and bone mineral density also failed to identify significant associations *(70)*. Given that variations in vitamin D status are well known to influence bone mineral density and the risk of osteoporosis *(71)*, these data may suggest that these common variations may not be helpful in detecting variations in VDR activity or function. Indeed, of the four major polymorphic sites, *Fok*I has been shown to result in an alternative transcription initiation site *(72)*, while *Taq*I, *Bsm*I, and *Apa*I sites are presented at the noncoding regions.

There are some studies investigating associations of polymorphisms in other vitamin D-related genes and type 1 diabetes [including variations in 25-hydroxylase (CYP2R1), 1-α hydroxylase (CYP27B1), and vitamin D-binding protein (Gc-globulin)], and again for none of the sites is there currently robust evidence for an association with replication. Variations in CYP27B1 have been the most extensively studied. In a British study, C-allele of CYP27B1 -1260 C/A polymorphism was both associated with increased risk of type 1 diabetes and preferentially transmitted to children with diabetes in an independent family collection *(73)*. Findings from earlier studies of CYP27B1 genotype and type 1 diabetes had been less consistent *(74, 75)* which, however, is likely to be due to past errors in genotyping and data analysis *(73)*. Evidence for a role of CYP27B1 in type 1 diabetes has also been obtained from a recent gene expression study, where CYP27B1 mRNA expression was lower in diabetic patients compared to others *(76)*. CYP27B1 -1260 polymorphisms were associated both with altered mRNA expression and circulating concentrations of hormonally active calcitriol. Results from this study were used to suggest that CYP27B1 could play a functional role in the pathogenesis of type 1 diabetes, through modulation of its mRNA expression and influence of serum calcitriol concentrations *(76)*. There is one study which shows both an association between CYP2R1 genotype (rs10741657, G allele) and type 1 diabetes in a case–control series and a preferential transmission to affected offspring *(77)*, while there was no difference in gene expression between cases and controls *(76)*. Genetic variations in vitamin D-binding protein (Group Specific Component, Gc) have not been consistently associated with type 1 diabetes, and suggestive differences have been seen in some *(78, 79)* but not in all studies *(80, 81)*.

Taken together, there is some evidence for vitamin D-related genetic susceptibility of type 1 diabetes; however, replication is urgently needed as for most polymorphisms there are only a handful of studies. There is also a need to identify better functional markers of vitamin D metabolism, as many of the markers used to date are known to be presented in noncoding regions of the genome.

6. CONCLUSIONS

Type 1 diabetes is a complex multifactorial disease and it is unlikely that vitamin D deficiency would be the only or perhaps even the main cause. However, there is abundant evidence to suggest that it may have a strong influence on the underlying susceptibility to develop type 1 diabetes and that it may act as a factor accelerating or promoting progression to disease. It is possible that by using large-dose vitamin D supplementation in infancy the pattern of immune regulation can be influenced in a manner that makes subsequent progression to type 1 diabetes less likely, even in an individual who is genetically predisposed to develop the disease. It is also possible that by ensuring adequate vitamin D intakes later in childhood, it is easier to defend from the influence of other environmental insults that might otherwise lead to the progression of the overt disease. More studies, including clinical trials, are clearly required to establish the role played by vitamin D in type 1 diabetes and whether by using supplementation either with vitamin D or with vitamin D analogues some of this serious disease could be safely prevented.

REFERENCES

1. Atkinson MA, Maclaren NK (1994) The pathogenesis of insulin-dependent diabetes mellitus. N Engl J Med 331(21):1428–1436
2. Lindberg B, Ivarsson SA, Landin-Olsson M, Sundkvist G, Svanberg L, Lernmark A (1999) Islet autoantibodies in cord blood from children who developed type I (insulin-dependent) diabetes mellitus before 15 years of age. Diabetologia 42(2):181–187
3. Gale EA (2002) The rise of childhood type 1 diabetes in the 20th century. Diabetes 51(12):3353–3361
4. Bodansky HJ, Staines A, Stephenson C, Haigh D, Cartwright R (1992) Evidence for an environmental effect in the aetiology of insulin dependent diabetes in a transmigratory population. BMJ 304(6833):1020–1022
5. Kaprio J, Tuomilehto J, Koskenvuo M, Romanov K, Reunanen A, Eriksson J et al (1992) Concordance for type 1 (insulin-dependent) and type 2 (non-insulin-dependent) diabetes mellitus in a population-based cohort of twins in Finland. Diabetologia 35(11):1060–1067
6. Eisenbarth GS (2007) Update in type 1 diabetes. J Clin Endocrinol Metab 92(7):2403–2407
7. Haller MJ, Gottlieb PA, Schatz DA (2007) Type 1 diabetes intervention trials 2007: where are we and where are we going? Curr Opin Endocrinol Diabetes Obes 14(4):283–287
8. Zella JB, DeLuca HF, Vitamin D (2003) Autoimmune diabetes. J Cell Biochem 88(2):216–222
9. Roep BO (2003) The role of T-cells in the pathogenesis of Type 1 diabetes: from cause to cure. Diabetologia 46(3):305–321
10. Foulis AK, McGill M, Farquharson MA (1991) Insulitis in type 1 (insulin-dependent) diabetes mellitus in man – macrophages, lymphocytes, and interferon-gamma containing cells. J Pathol 165(2):97–103
11. Pietropaolo M, Barinas-Mitchell E, Kuller LH (2007) The heterogeneity of diabetes: unraveling a dispute: is systemic inflammation related to islet autoimmunity? Diabetes 56(5):1189–1197
12. Taplin CE, Barker JM (2008) Autoantibodies in type 1 diabetes. Autoimmunity 41(1):11–18

13. Katz JD, Benoist C, Mathis D (1995) T helper cell subsets in insulin-dependent diabetes. Science 268(5214):1185–1188
14. Lemire J (2000) 1,25-Dihydroxyvitamin D3 – a hormone with immunomodulatory properties. Z Rheumatol 59(Suppl 1):24–27
15. Grant WB (2008) Hypothesis – ultraviolet-B irradiance and vitamin D reduce the risk of viral infections and thus their sequelae, including autoimmune diseases and some cancers. Photochem Photobiol 84(2):356–365
16. Norman AW, Frankel JB, Heldt AM, Grodsky GM (1980) Vitamin D deficiency inhibits pancreatic secretion of insulin. Science 209(4458):823–825
17. Eizirik DL, Darville MI (2001) Beta-cell apoptosis and defense mechanisms: lessons from type 1 diabetes. Diabetes 50(Suppl 1):S64–S69
18. Bao BY, Ting HJ, Hsu JW, Lee YF (2008) Protective role of 1 alpha, 25-dihydroxyvitamin D3 against oxidative stress in nonmalignant human prostate epithelial cells. Int J Cancer 122(12): 2699–2706
19. Hyppönen E (2004) Micronutrients and the risk of Type 1 diabetes: vitamin D, vitamin E and nicotinamide. Nutr Rev 62(9):340–347
20. Dahlquist G, Mustonen L (1994) Childhood onset diabetes – time trends and climatological factors. Int J Epidemiol 23(6):1234–1241
21. Soltesz G, Patterson CC, Dahlquist G (2007) Worldwide childhood type 1 diabetes incidence – what can we learn from epidemiology? Pediatr Diabetes 8(Suppl 6):6–14
22. Mohr SB, Garland CF, Gorham ED, Garland FC (2008) The association between ultraviolet B irradiance, vitamin D status and incidence rates of type 1 diabetes in 51 regions worldwide. Diabetologia 51(8):1391–1398
23. The EURODIAB ACE Study Group (2000) Variation and trends in incidence of childhood diabetes in Europe. Lancet 355(9207):873–876.
24. Nystrom L, Dahlquist G, Ostman J, Wall S, Arnqvist H, Blohme G et al (1992) Risk of developing insulin-dependent diabetes mellitus (IDDM) before 35 years of age: indications of climatological determinants for age at onset. Int J Epidemiol 21(2):352–358
25. Muntoni S, Fonte MT, Stoduto S, Marietti G, Bizzarri C, Crino A et al (1997) Incidence of insulin-dependent diabetes mellitus among Sardinian-heritage children born in Lazio region, Italy. Lancet 349(9046):160–162
26. Levy-Marchal C, Patterson C, Green A (1995) Variation by age group and seasonality at diagnosis of childhood IDDM in Europe. The EURODIAB ACE Study Group. Diabetologia 38(7): 823–830
27. Mooney JA, Helms PJ, Jolliffe IT, Smail P (2004) Seasonality of type 1 diabetes mellitus in children and its modification by weekends and holidays: retrospective observational study. Arch Dis Child 89(10):970–973
28. Siemiatycki J, Colle E, Aubert D, Campbell S, Belmonte MM (1986) The distribution of type I (insulin-dependent) diabetes mellitus by age, sex, secular trend, seasonality, time clusters, and space-time clusters: evidence from Montreal, 1971–1983. Am J Epidemiol 124(4):545–560
29. Lewy H, Hampe CS, Kordonouri O, Haberland H, Landin-Olsson M, Torn C et al (2008) Seasonality of month of birth differs between type 1 diabetes patients with pronounced beta-cell autoimmunity and individuals with lesser or no beta-cell autoimmunity. Pediatr Diabetes 9(1):46–52
30. Vaiserman AM, Carstensen B, Voitenko VP, Tronko MD, Kravchenko VI, Khalangot MD et al (2007) Seasonality of birth in children and young adults (0–29 years) with type 1 diabetes in Ukraine. Diabetologia 50(1):32–35
31. Rothwell PM, Gutnikov SA, McKinney PA, Schober E, Ionescu-Tirgoviste C, Neu A (1999) Seasonality of birth in children with diabetes in Europe: multicentre cohort study. European Diabetes Study Group. BMJ 319(7214):887–888
32. Muntoni S, Karvonen M, Muntoni S, Tuomilehto J (2002) Seasonality of birth in patients with type 1 diabetes. Lancet 359(9313):1246–1248
33. Hyppönen E, Jarvelin MR, Virtanen SM (2008) Seasonality of birth in patients with type 1 diabetes. Lancet 359:1247–1248, Authors reply

34. Gillespie KM, Bain SC, Barnett AH, Bingley PJ, Christie MR, Gill GV et al (2004) The rising incidence of childhood type 1 diabetes and reduced contribution of high-risk HLA haplotypes. Lancet 364(9446):1699–1700

35. Zhao HX, Stenhouse E, Soper C, Hughes P, Sanderson E, Baumer JH et al (1999) Incidence of childhood-onset Type 1 diabetes mellitus in Devon and Cornwall, England, 1975–1996. Diabet Med 16(12):1030–1035

36. Bivins R (2007) "The English disease" or "Asian rickets"? Medical responses to postcolonial immigration. Bull Hist Med 81(3):533–568

37. Hallman N, Hultin H, Visakorpi J (1964) Riisitaudin ennakkotorjunnasta. [Prevention of rickets, in Finnish]. Duodecim 80:185–189

38. Somersalo O (1955) Studies of childhood diabetes. 1. Incidence in Finland. Ann Paediatr Fenn 1: 239–249.

39. Hyppönen E, Läärä E, Reunanen A, Järvelin M, Virtanen SM (2001) Intake of vitamin D and risk of type 1 diabetes: a birth-cohort study. Lancet 358(9292):1500–1503

40. Sihvola S (1994) Lapsen terveys ja lapsiperheiden hyvinvointi. Sosiaalipediatrinen tutkimus suomalaisesta lapsesta. [Child health, and the well-being of families with children. Social-pediatric study of Finnish Child, in Finnish]. Mannerheimin Lastensuojeluliitto, Helsinki

41. Koizumi T, Nakao Y, Matsui T, Nakagawa T, Matsuda S, Komoriya K et al (1985) Effects of corticosteroid and 1,24R-dihydroxy-vitamin D3 administration on lymphoproliferation and autoimmune disease in MRL/MP-lpr/lpr mice. Int Arch Allergy Appl Immunol 77(4):396–404

42. Fournier C, Gepner P, Sadouk M, Charreire J (1990) In vivo beneficial effects of cyclosporin A and 1,25-dihydroxyvitamin D3 on the induction of experimental autoimmune thyroiditis. Clin Immunol Immunopathol 54(1):53–63

43. Lemire JM, Archer DC (1991) 1,25-Dihydroxyvitamin D3 prevents the in vivo induction of murine experimental autoimmune encephalomyelitis. J Clin Invest 87(3):1103–1107

44. Mathieu C, Laureys J, Sobis H, Vandeputte M, Waer M, Bouillon R (1992) 1,25-Dihydroxyvitamin D3 prevents insulitis in NOD mice. Diabetes 41(11):1491–1495

45. Mathieu C, Waer M, Laureys J, Rutgeerts O, Bouillon R (1994) Prevention of autoimmune diabetes in NOD mice by 1,25 dihydroxyvitamin D3. Diabetologia 37(6):552–558

46. Zella JB, McCary LC, DeLuca HF (2003) Oral administration of 1,25-dihydroxyvitamin D(3) completely protects NOD mice from insulin-dependent diabetes mellitus. Arch Biochem Biophys 417(1):77–80

47. Guyton KZ, Kensler TW, Posner GH (2003) Vitamin D and vitamin D analogs as cancer chemopreventive agents. Nutr Rev 61(7):227–238

48. Mathieu C, Waer M, Casteels K, Laureys J, Bouillon R (1995) Prevention of type I diabetes in NOD mice by nonhypercalcemic doses of a new structural analog of 1,25-dihydroxyvitamin D3, KH1060. Endocrinology 136(3):866–872

49. Van Etten E, Decallonne B, Verlinden L, Verstuyf A, Bouillon R, Mathieu C (2003) Analogs of 1alpha,25-dihydroxyvitamin D3 as pluripotent immunomodulators. J Cell Biochem 88(2):223–226

50. Giulietti A, Gysemans C, Stoffels K, Van Etten E, Decallonne B, Overbergh L et al (2004) Vitamin D deficiency in early life accelerates Type 1 diabetes in non-obese diabetic mice. Diabetologia 47: 451–462

51. Hawa MI, Valorani MG, Buckley LR, Beales PE, Afeltra A, Cacciapaglia F et al (2004) Lack of effect of vitamin D administration during pregnancy and early life on diabetes incidence in the non-obese diabetic mouse. Horm Metab Res 36(9):620–624

52. Mathieu C, Van Etten E, Gysemans C, Decallonne B, Bouillon R (2002) Seasonality of birth in patients with type 1 diabetes. Lancet 359(9313):1248

53. Casteels K, Waer M, Bouillon R, Allewaert K, Laureys J, Mathieu C (1996) Prevention of type I diabetes by late intervention with nonhypercalcemic analogues of vitamin D3 in combination with cyclosporin A. Transplant Proc 28(6):3095

54. Gregori S, Giarratana N, Smiroldo S, Uskokovic M, Adorini LA (2002) 1Alpha,25-dihydroxyvitamin D(3) analog enhances regulatory T-cells and arrests autoimmune diabetes in NOD mice. Diabetes 51(5):1367–1374

55. Gysemans C, Van Etten E, Overbergh L, Giulietti A, Eelen G, Waer M et al (2008) Unaltered diabetes presentation in NOD mice lacking the vitamin D receptor. Diabetes 57(1):269–275
56. The EURODIAB Substudy 2 Study Group (1999) Vitamin D supplement in early childhood and risk for Type I (insulin-dependent) diabetes mellitus. Diabetologia 42(1):51–54
57. Zipitis CS, Akobeng AK (2008) Vitamin D supplementation in early childhood and risk of type 1 diabetes: a systematic review and meta-analysis. Arch Dis Child 93:512–517
58. Stene LC, Ulriksen J, Magnus P, Joner G (2000) Use of cod liver oil during pregnancy associated with lower risk of Type I diabetes in the offspring. Diabetologia 43(9):1093–1098
59. Stene LC, Joner G (2003) Use of cod liver oil during the first year of life is associated with lower risk of childhood-onset type 1 diabetes: a large, population-based, case–control study. Am J Clin Nutr 78(6):1128–1134
60. Brekke HK, Ludvigsson J (2007) Vitamin D supplementation and diabetes-related autoimmunity in the ABIS study. Pediatr Diabetes 8(1):11–14
61. Fronczak CM, Baron AE, Chase HP, Ross C, Brady HL, Hoffman M et al (2003) In utero dietary exposures and risk of islet autoimmunity in children. Diabetes Care 26(12):3237–3242
62. Norris JM, Yin X, Lamb MM, Barriga K, Seifert J Hoffman M et al (2007) Omega-3 polyunsaturated fatty acid intake and islet autoimmunity in children at increased risk for type 1 diabetes. JAMA 298(12):1420–1428
63. Greer RM, Rogers MA, Bowling FG, Buntain HM, Harris M, Leong GM et al (2007) Australian children and adolescents with type 1 diabetes have low vitamin D levels. Med J Aust 187(1):59–60
64. Littorin B, Blom P, Scholin A, Arnqvist HJ, Blohme G, Bolinder J et al (2006) Lower levels of plasma 25-hydroxyvitamin D among young adults at diagnosis of autoimmune type 1 diabetes compared with control subjects: results from the nationwide Diabetes Incidence Study in Sweden (DISS). Diabetologia 49(12):2847–2852
65. Pozzilli P, Cherubini V, Pinelli A, Suraci C, IMIDIAB GROUP (2002) Low levels of 1,25-dihydroxyvitamin D3 in patients with recent onset Type 1 diabetes living in a Mediterranean country. Diabetes 51(Suppl. 2):A289
66. Baumgartl HJ, Standl E, Schmidt-Gayk H, Kolb HJ, Janka HU, Ziegler AG (1991) Changes of vitamin D3 serum concentrations at the onset of immune-mediated type 1 (insulin-dependent) diabetes mellitus. Diabetes Res 16(3):145–148
67. Pitocco D, Crino A, Di Stasio E, Manfrini S, Guglielmi C, Spera S et al (2006) The effects of calcitriol and nicotinamide on residual pancreatic beta-cell function in patients with recent-onset Type 1 diabetes (IMIDIAB XI). Diabet Med 23(8):920–923
68. Ogunkolade BW, Boucher BJ, Prahl JM, Bustin SA, Burrin JM, Noonan K et al (2002) Vitamin D receptor (VDR) mRNA and VDR protein levels in relation to vitamin D status, insulin secretory capacity, and VDR genotype in Bangladeshi Asians. Diabetes 51(7):2294–2300
69. Guo SW, Magnuson VL, Schiller JJ, Wang X, Wu Y, Ghosh S (2006) Meta-analysis of vitamin D receptor polymorphisms and type 1 diabetes: a HuGE review of genetic association studies. Am J Epidemiol 164(8):711–724
70. Uitterlinden AG, Ralston SH, Brandi ML, Carey AH, Grinberg D, Langdahl BL et al (2006) The association between common vitamin D receptor gene variations and osteoporosis: a participant-level meta-analysis. Ann Intern Med 145(4):255–264
71. Holick MF (2007) Vitamin D deficiency. N Engl J Med 357(3):266–281
72. Gross C, Krishnan AV, Malloy PJ, Eccleshall TR, Zhao XY, Feldman D (1998) The vitamin D receptor gene start codon polymorphism: a functional analysis of FokI variants. J Bone Miner Res 13(11):1691–1699
73. Bailey R, Cooper J, Zeitel L, Smyth DJ, Yang J, Walker NM et al (2007) Genetic evidence that vitamin D metabolism is an etiological factor in type 1 diabetes. Diabetes 56(10):2616–2621
74. Pani MA, Regulla K, Segni M, Krause M, Hofmann S, Hufner M et al (2002) Vitamin D 1alpha-hydroxylase (CYP1alpha) polymorphism in Graves' disease, Hashimoto's thyroiditis and type 1 diabetes mellitus. Eur J Endocrinol 146(6):777–781

75. Lopez ER, Regulla K, Pani MA, Krause M, Usadel KH, Badenhoop K (2004) CYP27B1 polymorphisms variants are associated with type 1 diabetes mellitus in Germans. J Steroid Biochem Mol Biol 89–90(1–5):155–157

76. Ramos-Lopez E, Bruck P, Jansen T, Pfeilschifter JM, Radeke HH, Badenhoop K (2007) CYP2R1-, CYP27B1- and CYP24-mRNA expression in German type 1 diabetes patients. J Steroid Biochem Mol Biol 103(3–5):807–810

77. Ramos-Lopez E, Bruck P, Jansen T, Herwig J, Badenhoop K (2007) CYP2R1 (vitamin D 25-hydroxylase) gene is associated with susceptibility to type 1 diabetes and vitamin D levels in Germans. Diabetes Metab Res Rev 23(8):631–636

78. Ongagna JC, Kaltenbacher MC, Sapin R, Pinget M, Belcourt A (2001) The HLA-DQB alleles and amino acid variants of the vitamin D-binding protein in diabetic patients in Alsace. Clin Biochem 34(1):59–63

79. Ongagna JC, Pinget M, Belcourt A (2005) Vitamin D-binding protein gene polymorphism association with IA-2 autoantibodies in type 1 diabetes. Clin Biochem 38(5):415–419

80. Pani MA, Donner H, Herwig J, Usadel KH, Badenhoop K (1999) Vitamin D binding protein alleles and susceptibility for type 1 diabetes in Germans. Autoimmunity 31(1):67–72

81. Klupa T, Malecki M, Hanna L, Sieradzka J, Frey J, Warram JH et al (1999) Amino acid variants of the vitamin D-binding protein and risk of diabetes in white Americans of European origin. Eur J Endocrinol 141(5):490–493

48 Vitamin D and Multiple Sclerosis

Alberto Ascherio and Kassandra L. Munger

Abstract   The etiology of multiple sclerosis (MS) is unknown, but there is growing evidence that vitamin D deficiency may contribute to MS development. In this chapter we critically review the current evidence from epidemiological studies. Results of case–control studies of childhood sun exposure and risk of MS are mixed, but largely support a protective effect with less sun exposure reported by individuals with MS. However, selection and recall biases cannot be ruled out as explanations of these results. Stronger evidence for a role for vitamin D comes from two longitudinal studies. In these studies, dietary intake of vitamin D in healthy adult women and elevated levels of serum 25-hydroxyvitamin D [25(OH)D] in young adults were both associated with a reduced risk of MS. Randomized clinical trials of vitamin D supplementation are needed to determine whether vitamin D can effectively be used in MS prevention or as a therapeutic agent among individuals with MS.

   Key Words: Multiple sclerosis; risk factor; epidemiology

1. INTRODUCTION

   Multiple sclerosis (MS) is a chronic inflammatory disease of the central nervous system characterized by large multiple focal areas of demyelination called lesions, or plaques, which can be distributed from the cerebrum through the spinal cord. The symptoms and signs of MS and the rate of disease progression are variable, reflecting both the location and severity of the demyelinating lesions. MS typically begins with relapsing–remitting neurological deficits, but later transitions to a more steadily progressive course with the accumulation of irreversible neurological deficit and disability (1). The prevailing theory is that MS is an autoimmune disorder whereby an unknown self-antigen triggers a T-cell-mediated inflammatory attack (2), but a role of an exogenous infection, such as Epstein–Barr virus, has not been excluded (3, 4, 5). There is also evidence that non-infectious environmental factors contribute to the etiology of MS, including vitamin D deficiency. As discussed in more detail below, the observations that MS is more common in northern areas and less common in populations with high fish consumption first led to the suggestion that vitamin D may protect against MS (6), and there is a growing body of evidence supporting this hypothesis.

From: *Nutrition and Health: Vitamin D*
Edited by: M.F. Holick, DOI 10.1007/978-1-60327-303-9_48,
© Springer Science+Business Media, LLC 2010

2. MS EPIDEMIOLOGY

MS is rare in childhood. Incidence rates increase in late adolescence, reach a peak in the late 1920 s and early 1930 s, and then gradually decline *(7)*. Women are about twice as likely to develop MS as compared to men *(8, 9)*, but there is some evidence that this ratio has increased over time, with women three times more likely to develop MS in younger birth cohorts *(10)*. Lifetime risk among White women in the United States and Denmark, which are areas of relatively high incidence, is about 1 in 200 *(11, 12)*.

The most notable feature of MS epidemiology is a worldwide latitude gradient (Fig. 1), with high prevalence and incidence in northern Europe *(9)*, Canada *(13, 14)*, the

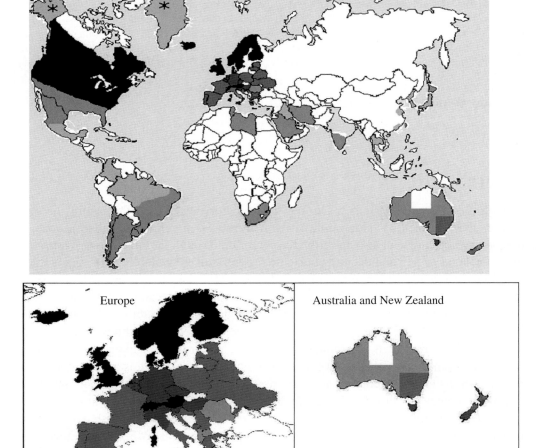

Fig. 1. Worldwide prevalence estimates for MS. *Black*: >90/100,000; *dark gray*: 60–89/100,000; *medium-dark gray*: 30–59/100,000; *medium-light gray*: 5–29/100,000; *light gray*: <5/100,000; *White*: no or insufficient data. Data for this region/country are based on old data and should be interpreted cautiously. From Ascherio and Munger *(72)*. Reproduced with permission by Elsevier.

northern United States *(8, 15, 16, 12)*, and southern Australia *(17)* and decreasing prevalence and incidence in regions closer to the equator *(18)*. There are exceptions to the latitude gradient, however, which include a lower than expected prevalence in Japan *(19)* and higher than expected prevalence and incidence in Sardinia and Sicily *(9)*. Some have argued that this gradient may be explained by genetic differences between populations *(20, 21)*, but the observations that the gradient is attenuating over time (MS incidence was threefold higher in the northern United States (above 41–42°N latitude) than in the southern United States (<37°N) for women born before 1946, but did not change at all with latitude among women born in 1947 or later *(12)*; and similarly, in Vietnam/Gulf War US veterans the latitude gradient in MS incidence was weaker than among World War II US veterans *(22)*) and that migrants or their offspring may acquire the risk of their new location *(23)* support a role for an environmental factor in determining the latitude gradient.

One of the strongest correlates of latitude is the duration and intensity of sunlight. As described in other sections of this book, exposure to sunlight is the main source of vitamin D for most people *(24)*. Therefore, average levels of vitamin D also display a strong latitude gradient, and an increased MS risk among people who have low vitamin D could contribute to explain the latitude gradient *(6)*. It is also noteworthy that among people living at high latitudes, where sunlight during most of the year does not reach the intensity required for vitamin D synthesis, a lower prevalence of MS has been found in fishing communities with high consumption of fatty fish, which is a major source of dietary vitamin D *(6)*. Interest in a possible protective effect of vitamin D increased with the discovery that vitamin D is a potent immunomodulator *(25)* and has preventive and therapeutic effects in experimental autoimmune encephalomyelitis (EAE), an animal model of MS *(26, 27, 28)*.

3. SUN EXPOSURE, VITAMIN D, AND MS RISK

The results of ecological studies *(29, 30, 31, 32)* support the hypothesis that high vitamin D levels may reduce MS risk, as strong inverse correlations have been found between average sunshine and average UV radiation of a geographic location and frequency of MS. However, evidence from these studies is rather weak because they compare MS frequency between groups of individuals who share some common characteristics, such as latitude of place of residence, and they cannot rule out confounding by other factors as an explanation of the results. Therefore, several investigations have been conducted to examine whether, within the same population, individuals with higher levels of vitamin D have a lower risk of developing MS, as reviewed below. An important factor to consider in interpreting these studies is age at exposure. Among migrants from high- to low-risk areas, MS risk seems to decline only or more markedly when migration occurs in childhood, suggesting that there is an age of susceptibility for MS during which vitamin D levels are critically important. For this reason, some investigators attempted to estimate the amount of sun exposure in childhood, as a proxy for levels of vitamin D, rather than in the years immediately preceding MS onset.

3.1. Database Analyses

One study conducted in the United States accessed death certificates to compare the type of occupation and sunlight intensity of residential area between individuals who died from MS and those from other causes *(33)*. After adjusting for age, sex, and socioeconomic status, outdoor work was associated with a 26% reduction in risk of MS mortality (odds ratio [OR] = 0.74, 95% confidence interval [CI]: 0.61–0.89) and residence in a high sunlight area with a 47% reduced risk (OR = 0.53, 95% CI: 0.48–0.57), consistent with a protective effect of sunlight exposure. Further, compared with working indoors and living in an area of low sunlight, working outdoors was associated with a significantly lower MS mortality in areas of high (OR = 0.24, 95% CI: 0.15–0.38), but not low (OR = 0.89, 95% CI: 0.64–1.22), sunlight. In this same study, skin cancer mortality, a surrogate for lifetime exposure to sunlight, followed an opposite pattern, with outdoor workers in high sunlight areas having the highest risk, indirectly supporting the validity of using occupation and residence as a proxy for sunlight exposure. Because these are proxies for sunlight exposure during adult life rather than in childhood (although a positive correlation is of course possible), the results of this study support a protective effect of sunlight or vitamin D during adulthood. Major limitations of this study include reliance on death certificates, which may misreport cause of death and occupation, and the possibility of "reverse causation," i.e., individuals with MS could preferentially choose an indoor rather than an outdoor occupation. Reverse causation is also a concern in a UK study comparing the observed rates of skin cancer among individuals with MS to that expected according to sex- and age-adjusted population rates in the same districts *(34)*. Among individuals with MS, skin cancer rates were about 50% lower than expected ($p = 0.03$), which is consistent with a lower level of sun exposure among people with MS. Thus, whether decreased sun exposure preceded the onset of MS cannot be determined.

3.2. Case–Control Studies

In traditional case–control studies, the investigators recruit individuals who have the disease (cases) and compare their history of exposure to the factors of interest with that of a control group, typically comprising healthy individuals of the same age and sex. Although case–control studies have played an important role in the epidemiology of chronic diseases, they are prone to bias because of the difficulty of identifying and recruiting an appropriate control group (selection bias) and of collecting comparable information from cases and controls (recall bias). Selection bias is exacerbated by the low participation rates, particularly among controls, in many studies; typically, there is little evidence to support the claim that the distribution of the exposure of interest is similar among those controls who agreed to participate and those who did not, a claim that is critical to the validity of the study. Recall bias is particularly a problem with exposures that are difficult to measure accurately and change over time, such as time spent outdoors in the summer or dietary vitamin D intake. A further problem is that the case series often includes individuals who have had the disease for many years or even decades ("prevalent cases"); as such, factors that prolong survival in individuals with the disease will be overrepresented among the cases, which can lead to erroneous inference about the associations with the disease.

Considering these limitations, it is perhaps not surprising that the results of case–control studies comparing history of sun exposure between MS cases and controls have been mixed. The results of the first of these studies, conducted in Israel and including 241 MS patients, suggested sun exposure was associated with an increased risk of MS, contrary to a protective effect of vitamin D (35). In a second study, conducted in Poland and including 300 cases, no association was found between sun exposure in childhood and MS risk (36). Both studies included prevalent cases of MS, and a spurious result would have been obtained if, for example, individuals with MS who reported high exposure to sunlight in childhood had a healthier lifestyle and lived longer than those with low sun exposure.

In contrast, results consistent with a protective effect of sun exposure were reported in three recent case–control studies: a study in Tasmania (37), in which information on time spent in the sun was complemented by measuring skin actinic damage, a biomarker of UV light exposure; a study in Norway (38), which included questions on both sunlight exposure and dietary intake of vitamin D; and a study of childhood sun exposure in monozygotic twins in the United States (39). The Tasmanian study (37) included 136 individuals with MS and 272 age- and sex-matched controls chosen from a comprehensive listing of registered electors. Exposure to sunlight at different ages and seasons was assessed using an interviewer-administered questionnaire. Prior to the interview, participants were asked to recall periods of relatively constant sun exposure during each year of their life by completing a "lifetime calendar" of sun exposure. These measures were complemented by use of silicon casts of the skin surface to assess the degree of actinic damage, a marker of cumulative sun exposure. The strongest association with MS risk was found for sunlight exposure between the ages of 6 and 10 years – individuals who reported at least 1 h/day in the sun during the winter months had a 53% reduced risk of MS (OR = 0.47, 95% CI: 0.26–0.84) compared to individuals reporting less than 1 h of sun exposure/day, and individuals who reported at least 2 h/day in the sun during summer months had a 50% reduced risk (OR = 0.50, 95% CI: 0.24–1.02) as compared to individuals reporting less than 2 h/day, although the overall trends were not statistically significant. Somewhat stronger associations and significant trends were found using information derived from the lifetime calendar of sun exposure. An inverse association was also found between the level of actinic damage and MS risk, but because exposure to sunlight after MS onset contributes to actinic damage, the latter association may be a consequence of sun avoidance among MS patients. The persistence of the inverse association between actinic damage and MS risk after adjusting for self-reported sun exposure after the onset of MS symptoms and in analyses restricted to MS cases of more recent onset, however (37), provide some evidence against reverse causation. The Norwegian study (38) conducted in a population living above the Arctic Circle, where the contribution of vitamin D from diet and supplements is high because of the low intensity of sunlight during most of the year, collected information on summer outdoor activities at different ages using a mailed questionnaire, with response rates of 83% among cases and 65% among controls. Analyses included 152 MS patients and 402 matched controls. As in the Tasmanian study, individuals who reported spending more time outdoors in the summer had a lower risk of MS, but exposure between the ages of 16 and 20 years was the strongest predictor of MS risk (RR = 0.55, 95% CI: 0.39–0.78), rather than exposure in childhood (6–10 or 11–15 years). There were also trends toward a lower MS risk among individuals with higher fish intake, whereas results on use of cod liver oil supplements, an important source of vitamin D, were inconclusive. Finally, in a study of 79 monozygotic twin pairs discordant for MS, twins with MS reported lower levels of sun exposure during childhood on average *(39)*. Although this study is interesting in that results cannot be explained by individuals at high risk of MS having a genetically determined lower propensity to sun exposure, recall bias due to differential reporting of sun exposure between affected and non-affected twins remains a concern.

3.3. Longitudinal Studies

Because of the limitations of traditional case–control studies discussed above, stronger evidence relating vitamin D to MS risk is provided by two longitudinal studies, one based on assessment of dietary vitamin D intake *(40)* and one on serum levels of 25(OH)D *(41)*. In these studies, individuals without MS reported dietary vitamin D intake *(40)* or provided a blood sample for 25(OH)D measurement *(41)* and were followed for the occurrence of MS. The reported vitamin D intake, or serum 25(OH)D levels, in those who developed MS were then compared to those who did not. This type of study design is preferred over the traditional case–control design because it not only eliminates recall bias but also limits selection bias as the controls are representative of the population that gave rise to the cases.

The relation between dietary vitamin D intake and MS risk has been studied in over 200,000 women in the Nurses Health Study (NHS) and the Nurses Health Study II (NHS-II) cohorts *(40)*. Newly diagnosed cases of MS were identified by self-report on biennial questionnaires sent to all participants and confirmed by asking the treating neurologist to complete a questionnaire on the certainty of the diagnosis (definite, probable, possible, not MS), clinical history (including date of MS diagnosis and date of the first symptoms of MS), and supporting laboratory tests. Vitamin D intake from foods and supplements was assessed from comprehensive and previously validated semi-quantitative food frequency questionnaires administered every 4 years during the follow-up of the cohorts *(42, 43)*. The validity of vitamin D intake was assessed by comparing it with the plasma concentrations of 25(OH)D among 323 healthy NHS women *(44)*. The mean 25(OH)D was 55 nmol/l (22.0 ng/ml) among women in the bottom quintile of vitamin D intake and 75.3 nmol/l (30.1 ng/ml) among women in the top quintile; for plasma collected in winter (January–April), the corresponding values were 39.8 nmol/l (15.9 ng/ml) and 69.8 nmol/l (27.9 ng/ml). Validity of estimated vitamin D intake in the NHS is further supported by its inverse association with risk of hip fractures *(44)*.

During the follow-up (18 years in NHS and 8 years in NHS-II), 173 incident cases of MS were documented. Total vitamin D intake was inversely associated with risk of MS – the age-adjusted relative risk (RR) comparing the highest with the lowest quintile of consumption was 0.67 (95% CI: 0.40–1.12; *p* for trend = 0.03) (Fig. 2a). Intake of 400 IU/day of vitamin D from supplements only was associated with a 40% lower MS

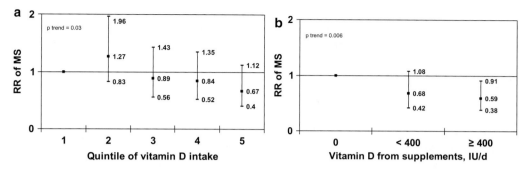

Fig. 2. (**a**) Relative risk (RR) of multiple sclerosis (MS) according to vitamin D intake, *p* for trend = 0.03. (**b**) Relative risk of MS according to use of vitamin D supplements, *p* for trend = 0.006. Data from Munger and colleagues *(40)*. From Ascherio and Munger *(72)*. Reproduced with permission by John Wiley & Sons, Inc.

risk (RR = 0.59, 95% CI: 0.38–0.91) (Fig. 2b). These RRs did not materially change after further adjustment for pack-years of smoking and latitude at birth. Confounding by other micronutrients cannot be excluded, however, because supplemental vitamin D was mostly from multivitamins rather than from vitamin D-specific supplements, but there is no convincing evidence that other components of multivitamins would reduce MS risk. It is also noteworthy that the youngest women in the Nurses Health Studies were 25 years old at the time of recruitment – it seems therefore that vitamin D exposure during adult life and not only during childhood or adolescence could be important for MS prevention.

Serum levels of 25(OH)D are an integrated measure of vitamin D from both diet and sun exposure and a marker of bioavailability; therefore, if vitamin D is protective, high serum levels of 25(OH)D would be expected to predict a lower risk of MS in healthy individuals. This question was addressed in a collaborative, prospective case–control study using the Department of Defense Serum Repository (DoDSR) *(41)*. The study population included over 7 million active duty US military personnel who had at least one serum sample stored in DoDSR. Personnel generally provide one sample at entry into the military for HIV testing and, on average, a sample every 2 years thereafter *(45)*. Potential cases of MS were identified by searching the physical disability databases of the US Army and Navy/Marines for active duty personnel discharged with a diagnosis of MS between 1992 and 2004. Medical records were reviewed to confirm the diagnosis of MS according to standard criteria. Two controls for each case were randomly selected, matched by year of birth, sex, race/ethnicity (non-Hispanic White, non-Hispanic Black, Hispanic, others), and dates of sample collection. Serum levels of 25(OH)D were measured using a radioimmunoassay *(46)*. Although this study involves identifying cases and selecting controls, as a traditional case–control study does, the major strength here is that selection bias is minimized because all cases and controls arise from the same cohort (individuals in the military); therefore, the exposure distribution in the controls is representative of the population that gave rise to the cases, and using previously collected serum samples ensures 100% "participation." Further, as noted above, serum

Table 1
Selected Characteristics of US Military Personnel in Study of 25(OH)D and MS Risk

Characteristic	Cases n = 257	Controls n = 514
Male, n (%)	174 (68)	348 (68)
Race, n (%)		
White, non-Hispanic	148 (57.6)	296 (57.6)
Black, non-Hispanic	77 (30)	154 (30)
Hispanic/others	32 (12.5)	64 (12.5)
Age first sample collected, y mean	23.3	23.3
Range	16–40	17–41
Age MS onset, y mean	28.5	NA
Range	18–48	
Time between first sample and MS onset, y mean	5.3 <1–13	NA
Range		
Relapsing–remitting MS (%)	72	NA

Adapted from Munger and colleagues (41)

samples were collected prior to the onset of MS, so reverse causation cannot explain these results.

Analyses included 257 military personnel with confirmed MS and at least two serum samples collected before the onset of MS symptoms and their matched controls (Table 1) and were stratified by race/ethnicity, because 25(OH)D levels were, as expected, much lower in Blacks (45.5 nmol/l) than in Whites (75.2 nmol/l). For each subject, the average of season-adjusted 25(OH)D levels from all the available samples collected before the onset of MS (or before the onset of MS in the matched case for controls) was used to obtain an integrated measure of long-term pre-clinical 25(OH)D level. The analyses were adjusted for latitude of residence at time of entry into active military duty, thus removing possible confounding due to the correlation of sunlight exposure to place of residence. Among Whites, there was a statistically significant linear trend ($p = 0.03$), with a predicted 41% decrease in MS risk for every 50 nmol/l increase in 25(OH)D (RR = 0.59, 95% CI: 0.36–0.95). Due to smaller sample sizes, the results among Blacks and Hispanic/other race were more unstable and the tests of linear trend were not significant; the predicted RR for a 50 nmol/l increase in 25(OH)D was 0.62 (95% CI: 0.23–1.70, $p = 0.36$) in Blacks and 0.88 (95% CI: 0.25–3.04, $p = 0.84$) in Hispanics/other race. In analyses based on categorical levels of 25(OH)D, risk of MS was 51% lower among White individuals with 25(OH)D ≥100 nmol/l as compared to those with levels below 75 nmol/l (Fig. 3a). Although in secondary analyses we found that the inverse association between 25(OH)D levels and MS risk tended to be stronger for samples collected at younger ages (aged 17–19) (Fig. 3b), the mean age at time of collection of the first blood sample was 23 years, and these results are thus consistent with those of the Nurses' cohorts in suggesting that vitamin D levels during adult life are an important predictor of MS risk. It is important to note, however, that while these results support a direct role for vitamin D in protecting against MS in Whites, other explanations, such as

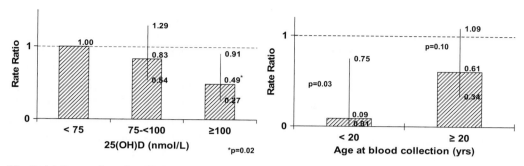

Fig. 3. (a) Rate ratios of multiple sclerosis (MS) by category of serum 25(OH)D level for *Whites*. *p = 0.02. (b) Rate ratios of MS comparing 25(OH)D ≥ 100 versus <100 nmol/l by age at blood collection for Whites. Data from Munger and colleagues *(41)*. From Ascherio and Munger *(73)*. Reproduced with permission by John Wiley & Sons, Inc.

a genetic predisposition to both low 25(OH)D levels and MS, or non-vitamin D-related effects of sun exposure, cannot be excluded.

4. SEASONAL VARIATIONS IN BIRTH PATTERNS

Several studies have suggested that individuals with MS are more likely to be born in the spring months (when mothers are pregnant for little, if any, time during the summer)*(47, 48, 49)*. Vitamin D exposure in utero has been proposed as a possible explanation for the peak in MS incidence (10% over the expected) among individuals born in May in the northern hemisphere (when mothers are likely to be vitamin D deficient for most of their pregnancy) and the dip among those born in November (9% below the expected), according to data from Canada and northern Europe *(49)*. This interpretation, however, is speculative, and it is unclear how vitamin D levels, which change gradually with season, could explain the rather abrupt changes in risk with month of birth.

5. VITAMIN D AND MS PROGRESSION

Individuals with MS tend to have low vitamin D levels *(50, 51, 52, 53)*, probably at least in part, because of the changes in lifestyle imposed by the disease and the fact that heat, and thus direct sunlight, may exacerbate symptoms. It is unclear, however, whether vitamin D levels can influence the progression of MS. 25(OH)D levels have been shown to be lower during MS relapses than remissions *(52, 54)* and the more severe course of MS in African-American patients *(55, 56)* is consistent with an adverse effect of vitamin D deficiency, but could also be due to genetic or other factors. The results of some *(57, 58, 59)* but not all *(60, 61)* studies suggest some seasonality in the occurrence of active lesions and relapses, and the former seems to parallel the seasonal fluctuations in 25(OH)D levels *(62)*. Vitamin D levels, however, reach their lowest point when respiratory infections are at their peak, and because infections can trigger MS relapses the association between vitamin D and the clinical course of MS could be spurious. Further, vitamin D itself could reduce the risk of respiratory infections *(63)*, and thus have an indirect effect on MS. Because of the difficulty of determining causality in this complex setting, the question of whether vitamin D supplementation can favorably affect MS progression is better pursued in randomized trials. In a year-long pilot tolerability study of oral $1,25(OH)_2D_3$ (2.5 µg/day), exacerbation rates were 27% less than baseline rates *(64)*. Another safety and tolerability trial examined the effect of high doses of oral vitamin D_3 (28,000–280,000 IU/week)*(65)*. Over the 28-week study, the mean number of active lesions dropped from 1.75 at baseline to 0.83. However, having an active lesion was required for inclusion in the study and resolution of some active lesions over time would be expected. There was no effect of vitamin D reported on other disease parameters such as EDSS. Large trials to determine long-term effects of vitamin D in individuals with MS should thus be feasible and are being pursued.

6. SUMMARY

Overall, the results of epidemiological studies strongly support a protective effect of vitamin D on MS risk and suggest that vitamin D levels are important not only in childhood but also during adolescence and early adult life. If vitamin D truly reduced the risk of MS, the public health implications would be enormous. Considering that almost half of Whites and two-thirds of Black adults in the United States have 25(OH)D levels below 70 nmol/l *(66)*, increasing the vitamin D levels of all adolescents and young adults to an average above 100 nmol/l would be expected to cause a substantial reduction in MS incidence. If we assume, based on the study among military personnel described above, that 80% of individuals have 25(OH)D levels less than 100 nmol/l, and that this level during adolescence is associated with a 10-fold decreased risk of MS (as shown in Fig. 3b), then 72% of MS cases could be prevented by increasing 25(OH)D levels to 100 nmol/l or greater during adolescence and young adulthood. This level could be achieved in most individuals by taking between 1,000 and 4,000 IU/day of supplemental vitamin D *(67, 68, 69, 70)*. Although much higher than the current Institute of Medicine's adequate intake of vitamin D for adults younger than 50 (200 IU/day) *(71)* and the typical amount in multivitamins (400 IU), these amounts do not cause hypercalcemia or other known adverse side effects and would result in 25(OH)D levels still below the average among healthy individuals with regular high exposure to sunlight *(67)*. A high priority should thus be given to the conduction of a large randomized trial to assess whether vitamin D supplementation in the general population prevents MS. The trial would have to be very large, however, as MS is a relatively rare disease. An attractive option would be a national or multinational study based on randomization of school districts or other suitable units and including multiple outcomes that may be affected by vitamin D, such as obesity, diabetes, respiratory infections, and asthma. Observational data on the effect of vitamin D on MS progression are more sparse and less persuasive, and whether vitamin D has any therapeutic value could be more effectively addressed in randomized trials in patients with MS.

REFERENCES

1. Ebers G (1998) Natural history of multiple sclerosis. In: Compston, A (ed) McAlpine's multiple sclerosis. Churchill Livingstone, New York, pp 191–221
2. Noseworthy JH et al. (2000) Multiple sclerosis. N Engl J Med 343:938–952

3. Gilden DH (2005) Infectious causes of multiple sclerosis. Lancet Neurol 4:195–202
4. Ascherio A, Munger KL (2007) Environmental risk factors for multiple sclerosis. Part I: the role of infection. Ann Neurol 61:288–299
5. Serafini B et al (2007) Dysregulated Epstein-Barr virus infection in the multiple sclerosis brain. J Exp Med 204:2899–2912
6. Goldberg P (1974) Multiple sclerosis: vitamin D and calcium as environmental determinants of prevalence (A viewpoint) Part 1: sunlight, dietary factors and epidemiology. Intern J Environ Stud 6:19–27
7. Koch-Henriksen N (1999) The Danish Multiple Sclerosis Registry: a 50-year follow-up. Mult Scler 5:293–296
8. Kurtzke JF et al (1979) Epidemiology of multiple sclerosis in U.S. veterans: 1. Race, sex, and geographic distribution. Neurology 29:1228–1235
9. Pugliatti M et al (2006) The epidemiology of multiple sclerosis in Europe. Eur J Neurol 13:700–722
10. Orton SM et al (2006) Sex ratio of multiple sclerosis in Canada: a longitudinal study. Lancet Neurol 5:932–936
11. Koch-Henriksen N, Hyllested K (1988) Epidemiology of multiple sclerosis: incidence and prevalence rates in Denmark 1948–1964 based on the Danish Multiple Sclerosis Registry. Acta Neurol Scand 78:369–380
12. Hernán MA et al (1999) Geographic variation of MS incidence in two prospective studies of US women. Neurology 53:1711–1718
13. Kurland LT (1952) The frequency and geographic distribution of multiple sclerosis as indicated by mortality statistics and morbidity surveys in the United States and Canada. Am J Hyg 55: 457–476
14. Stazio A et al (1967) Multiple sclerosis in New Orleans, Louisiana, and Winnipeg, Manitoba, Canada: follow-up of a previous survey in New Orleans, and comparison between the patient populations in the two communities. J Chron Dis 20:311–332
15. Visscher BR et al (1977) Latitude, Migration, and the prevalence of multiple sclerosis. Am J Epidemiol 106:470–475
16. Baum HM, Rothschild BB (1981) The incidence and prevalence of reported multiple sclerosis. Ann Neurol 10:420–428
17. Miller DH et al (1990) Multiple sclerosis in Australia and New Zealand: are the determinants genetic or environmental? J Neurol Neurosurg Psychiatry 53:903–905
18. Kurtzke JF (1995) MS epidemiology world wide. One view of current status. Acta Neurol Scand 161(Suppl):23–33
19. Kira J (2003) Multiple sclerosis in the Japanese population. Lancet Neurol 2:117–127
20. Bulman DE, Ebers GC (1992) The geography of MS reflects genetic susceptibility. J Trop Geographic Neurol 2:66–72
21. Page WF et al (1993) Epidemiology of multiple sclerosis in US veterans: V. Ancestry and the risk of multiple sclerosis. Ann Neurol 33:632–639
22. Wallin MT et al (2004) Multiple sclerosis in US veterans of the Vietnam era and later military service: race, sex, and geography. Ann Neurol 55:65–71
23. Gale CR, Martyn CN (1995) Migrant studies in multiple sclerosis. Prog Neurobiol 47:425–448
24. Holick MF (2004) Sunlight and vitamin D for bone health and prevention of autoimmune diseases, cancers, and cardiovascular disease. Am J Clin Nutr 80:1678S–88S
25. Hayes CE et al. (2003) The immonological functions of the vitamin D endocrine system. Cell Mol Biol 49:277–300
26. Lemire JM, Archer DC (1991) 1,25-dihydroxyvitamin D3 prevents the in vivo induction of murine experimental autoimmune encephalomyelitis. J Clin Invest 87:1103–1107
27. Cantorna MT et al (1996) 1,25-dihydroxyvitamin D3 reversibly blocks the progression of relapsing encephalomyelitis, a model of multiple sclerosis. Proc Natl Acad Sci USA 93:7861–7864
28. Garcion E et al (2003) Treatment of experimental autoimmune encephalomyelitis in rat by 1,25-dihydroxyvitamin D(3) leads to early effects within the central nervous system. Acta Neuropathol (Berl) 105:438–448
29. Acheson ED et al (1960) Some comments on the relationship of the distribution of multiple sclerosis to latitude, solar radiation, and other variables. Acta Psychiatr Scand 147:132–147

30. Sutherland JM et al (1962) The prevalence of multiple sclerosis in Australia. Brain 85:146–164
31. Leibowitz U et al (1967) Geographical considerations in multiple sclerosis. Brain 90:871–886
32. van der Mei IA et al (2001) Regional variation in multiple sclerosis prevalence in Australia and its association with ambient ultraviolet radiation. Neuroepidemiology 20:168–174
33. Freedman DM et al (2000) Mortality from multiple sclerosis and exposure to residential and occupational solar radiation: a case-control study based on death certificates. Occup Environ Med 57:418–421
34. Goldacre MJ et al (2004) Skin cancer in people with multiple sclerosis: a record linkage study. J Epidemiol Community Health 58:142–144
35. Antonovsky A et al (1965) Epidemiologic study of multiple sclerosis in Israel. Arch Neurol 13: 183–193
36. Cendrowski W et al (1969) Epidemiological study of multiple sclerosis in Western Poland. Eur Neurol 2:90–108
37. van der Mei IAF et al (2003) Past exposure to sun, skin phenotype and risk of multiple sclerosis: a case-control study. BMJ 327:316–321
38. Kampman MT et al (2007) Outdoor activities and diet in childhood and adolescence relate to MS risk above the Arctic Circle. J Neurol 254:471–477
39. Islam T et al (2007) Childhood sun exposure influences risk of multiple sclerosis in monozygotic twins. Neurology 69:381–388
40. Munger KL et al (2004) Vitamin D intake and incidence of multiple sclerosis. Neurology 62:60–65
41. Munger KL et al (2006) Serum 25-hydroxyvitamin D levels and risk of multiple sclerosis. Jama 296:2832–2838
42. Willett WC et al (1988) The use of a self-administered questionnaire to assess diet four years in the past. Am J Epidemiol 127:188–199
43. Salvini S et al (1989) Food-based validation of a dietary questionnaire: the effects of week-to-week variation in food consumption. Int J Epidemiol 18:858–867
44. Feskanich D et al (2003) Calcium, vitamin D, milk consumption, and hip fractures: a prospective study among postmenopausal women. Am J Clin Nutr 77:504–511
45. Rubertone MV, Brundage JF (2002) The Defense Medical Surveillance System and the Department of Defense serum repository: glimpses of the future of public health surveillance. Am J Public Health 92:1900–1904
46. Hollis BW et al (1993) Determination of vitamin D status by radioimmunoassay with an 125I-labeled tracer. Clin Chem 39:529–533
47. Templer DI et al (1992) Season of birth in multiple sclerosis. Acta Neurol Scand 85:107–109
48. Sadovnick AD, Yee IM (1994) Season of birth in multiple sclerosis. Acta Neurol Scand 89:190–191
49. Willer CJ et al (2005) Timing of birth and risk of multiple sclerosis: population based study. BMJ 330:120
50. Nieves J et al (1994) High prevalence of vitamin D deficiency and reduced bone mass in multiple sclerosis. Neurology 44:1687–1692
51. Ozgocmen S et al (2005) Vitamin D deficiency and reduced bone mineral density in multiple sclerosis: effect of ambulatory status and functional capacity. J Bone Miner Metab 23:309–313
52. Soilu-Hanninen M et al (2005) 25-Hydroxyvitamin D levels in serum at the onset of multiple sclerosis. Mult Scler 11:266–271
53. van der Mei IA et al (2007) Vitamin D levels in people with multiple sclerosis and community controls in Tasmania, Australia. J Neurol 254:581–590
54. Soilu-Hanninen M et al (2008) A longitudinal study of serum 25-hydroxyvitamin D and intact PTH levels indicate the importance of vitamin D and calcium homeostasis regulation in multiple sclerosis. J Neurol Neurosurg Psychiatry 79:152–157
55. Cree BA et al (2004) Clinical characteristics of African Americans vs. Caucasian Americans with multiple sclerosis. Neurology 63:2039–2045
56. Marrie RA et al (2006) Does multiple sclerosis-associated disability differ between races? Neurology 66:1235–1240
57. Wuthrich R, Rieder HP (1970) The seasonal incidence of multiple sclerosis in Switzerland. Eur Neurol 3:257–264

58. Bamford CR et al (1983) Seasonal variation of multiple sclerosis exacerbations in Arizona. Neurology 33:697–701

59. Auer DP et al (2000) Seasonal fluctuations of gadolinium-enhancing magnetic resonance imaging lesions in multiple sclerosis. Ann Neurol 47:276–277

60. Rovaris M et al (2001) Effects of seasons on magnetic resonance imaging–measured disease activity in patients with multiple sclerosis. Ann Neurol 49:415–416

61. Killestein J et al (2002) Seasonal variation in immune measurements and MRI markers of disease activity in MS. Neurology 58:1077–1080

62. Embry AF et al (2000) Vitamin D and seasonal fluctuations of gadolinium-enhancing magnetic resonance imaging lesions in multiple sclerosis. Ann Neurol 48:271–272

63. Cannell JJ et al (2006) Epidemic influenza and vitamin D. Epidemiol Infect 134:1129–1140

64. Wingerchuk DM et al (2005) A pilot study of oral calcitriol (1,25-dihydroxyvitamin D3) for relapsing-remitting multiple sclerosis. J Neurol Neurosurg Psychiatry 76:1294–1296

65. Kimball SM et al (2007) Safety of vitamin D3 in adults with multiple sclerosis. Am J Clin Nutr 86:645–651

66. Zadshir A et al (2005) The prevalence of hypovitaminosis D among US adults: data from the NHANES III. Ethn Dis 15:S5–S97

67. Vieth R (1999) Vitamin D supplementation, 25-hydroxyvitamin D concentrations and safety. Am J Clin Nutr 69:842–856

68. Heaney RP et al (2003) Human serum 25-hydroxycholecalciferol response to extended oral dosing with cholecalciferol. Am J Clin Nutr 77:204–210

69. Hollis BW (2005) Circulating 25-hydroxyvitamin D levels indicative of vitamin D sufficiency: implications for establishing a new effective dietary intake recommendation for vitamin D. J Nutr 135:317–322

70. Dawson-Hughes B et al (2005) Estimates of optimal vitamin D status. Osteoporos Int 16:713–716

71. Standing Committee on the Scientific Evaluation of Dietary Reference Intakes FaNB, Institute of Medicine. (1997) Dietary reference intakes for calcium, phosphorus, magnesium, vitamin D, and fluoride. National Academy Press, Washington, DC

72. Ascherio A, Munger KL (2010) Epidemiology of multiple sclerosis environmental factors. In: Lucchinetti CF, Hohlfeld R, eds. Multiple sclerosis 3. 1st ed. Saunders/Elsevier, Philadelphia pp. 57–82

73. Ascherio A, Munger KL (2007) Environmental risk factors for multiple sclerosis: Part II. Noninfectious factors. Ann Neurol 61:504–513

49 Vitamin D and Type 2 Diabetes

Myrto Eliades and Anastassios G. Pittas

Abstract Vitamin D has been reported to have a variety of non-skeletal actions, including on glucose metabolism. There has been increasing evidence from animal and human studies, as reviewed in this chapter, to suggest that vitamin D may be important in modifying risk of type 2 diabetes. Vitamin D is thought to have both direct (through activation of the vitamin D receptor) and indirect (via regulation of calcium homeostasis) effects on various mechanisms related to the pathophysiology of type 2 diabetes, including pancreatic beta-cell dysfunction, impaired insulin action, and systemic inflammation. The evidence from human studies comes primarily from cross-sectional and a few prospective observational studies showing an inverse association between vitamin D status and prevalence or incidence of type 2 diabetes. Because there is paucity of trials that have specifically examined the role of vitamin D in prevention or treatment of type 2 diabetes, it remains to be seen whether improving vitamin D status plays a role in the prevention or treatment of type 2 diabetes.

Key Words: Vitamin D; type II diabetes; 25-hydroxyvitamin D; epidemiology; latitude; glucose; insulin; inflammation; blacks; calcium

1. EPIDEMIOLOGY AND BURDEN OF TYPE 2 DIABETES

Type 2 diabetes is rapidly emerging as a significant global health-care problem. In the United States alone, total prevalence is expected to more than double in the next few decades from 6% of the population in 2005 (16 million) to 12% in 2050 (48 million) (1). The total number of people with diabetes worldwide is expected to rise from 171 million in 2000 to 366 million by 2030 (2). Type 2 diabetes, traditionally thought as a disease of adulthood, is now diagnosed with increasing frequency among those younger than 20 years, especially in developed countries (3). Type 2 diabetes is associated with serious morbidity and increased mortality. It is the leading cause of blindness, kidney disease, heart disease, and stroke. In the United States alone, each year up to 25,000 individuals lose their sight and as many as 28,000 initiate treatment for chronic kidney failure because of diabetes. People with diabetes are four times more likely to have heart disease or suffer a stroke as compared to those without diabetes. Diabetes confers an equivalent risk of aging of 15 years (4). Beyond its devastating human toll, type 2 diabetes is also associated with increasing costs. In the United States, the total

From: *Nutrition and Health: Vitamin D*
Edited by: M.F. Holick, DOI 10.1007/978-1-60327-303-9_49,
© Springer Science+Business Media, LLC 2010

(direct and indirect) costs of diabetes were estimated at $132 billion in 2002 and are expected to increase to $192 billion by 2020 *(5)*.

Although therapies for type 2 diabetes and its complications have improved over the last few decades, the rapidly increasing incidence and prevalence of type 2 diabetes highlight the need for innovative approaches for the management and prevention of the disease. Epidemiologic data suggest that 9 out of 10 cases of type 2 diabetes could be attributed to modifiable habits and lifestyle *(6)*. Although lifestyle interventions, most notably weight loss through caloric restriction and physical activity, have been shown to be successful in delaying the progression to type 2 diabetes among those with pre-diabetes *(7)*, it is difficult to achieve and maintain long term. Therefore, identification of easily modifiable environmental risk factors for prevention of type 2 diabetes is urgently needed. Recently, there has been increasing evidence from animal and human studies, as reviewed in this chapter, to suggest that vitamin D may be important in modifying risk of type 2 diabetes *(8)*.

2. VITAMIN D METABOLISM

Vitamin D metabolism is described in Chapter 3 and will not be reviewed here. The well-recognized role of vitamin D on skeletal homeostasis is mediated by the binding of vitamin D to its receptor (VDR), a member of the nuclear receptor superfamily. However, the recent recognition that the VDR has a wide distribution, essentially in all tissues and cells in the body, extended the potential role of vitamin D on a variety of non-skeletal functions. These include modulation of the immune response, regulation of cellular proliferation and differentiation, as well as regulation of insulin secretion and action *(9)*. Vitamin D may also act through transcription-independent mechanisms, specifically through signal transduction pathways initiated by binding to membrane receptors *(10)*. Whether through transcription or through membrane-initiated signal transduction, the action of vitamin D appears to be dependent on adequate vitamin D status.

3. POTENTIAL MECHANISMS OF ACTION OF VITAMIN D ON GLUCOSE METABOLISM

For glucose intolerance and type 2 diabetes to develop, impaired pancreatic beta-cell function, insulin resistance, and systemic inflammation are often present. There is evidence to support that vitamin D influences these mechanisms, as summarized next.

3.1. *Vitamin D and Pancreatic Beta-Cell Function/Insulin Secretion*

There are several lines of evidence supporting a beneficial role for vitamin D in pancreatic beta-cell function (Fig. 1). In in vitro and in vivo studies, vitamin D deficiency impairs glucose-mediated insulin secretion from rat beta cells *(11–14)*, while vitamin D supplementation restores insulin secretion *(11, 13–16)*. Vitamin D may have a direct effect on beta-cell function mediated by binding of the circulating active form, 1,25(OH)$_2$D, to VDR, which is expressed in beta cells *(17)*. Furthermore, mice

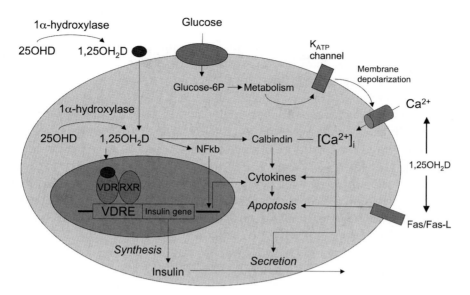

Fig. 1. Vitamin D and pancreatic beta-cell function. Vitamin D can promote pancreatic beta-cell function in several ways. The active form of vitamin D, (1,25(OH)$_2$D), enters the beta cell from the circulation and interacts with the vitamin D receptor-retinoic acid x-receptor complex (VDR-RXR), which binds to the vitamin D response element (VDRE) found in the human insulin gene promoter, to enhance the transcriptional activation of the insulin gene and increase the synthesis of insulin. Vitamin D may promote beta-cell survival by modulating the generation (through inactivation of nuclear factor-κB [NF-κb]) and effects of cytokines. The anti-apoptotic effect of vitamin D may also be mediated by downregulating the Fas-related pathways (Fas/Fas-L). Activation of vitamin D also occurs intracellularly by 1α-hydroxylase, which is expressed in pancreatic beta cells. The effects of vitamin D may be mediated indirectly via its important and well-recognized role in regulating extracellular calcium (Ca^{2+}), calcium flux through the beta cell, and intracellular calcium (Ca^{2+})$_i$. Alterations in calcium flux can directly influence insulin secretion, which is a calcium-dependent process. Vitamin D also regulates calbindin, a cytosolic calcium-binding protein found in beta cells, which acts as a modulator of depolarization-stimulated insulin release via regulation of intracellular calcium. Calbindin may also protect against apoptotic cell death via its ability to buffer intracellular calcium.

lacking functional VDR show impaired insulin secretory response *(18)*. The presence of the vitamin D response element in the human insulin gene promoter *(19)* and transcriptional activation of the human insulin gene caused by 1,25(OH)$_2$D *(20)* further support a direct effect of vitamin D on insulin synthesis and secretion. Alternatively, activation of vitamin D may occur within the beta cell by the 1α-hydroxylase enzyme, which is expressed in beta cells *(21)*. An indirect effect of vitamin D on the beta cell may be mediated via its regulation of extracellular calcium concentration and calcium flux through the beta cell *(22)*. Insulin secretion is a calcium-dependent process *(23)*; therefore, alterations in calcium flux can affect insulin secretion. Vitamin D also regulates calbindin, a cytosolic calcium-binding protein found in many tissues including beta cells *(17, 24)*. Calbindin is a modulator of depolarization-stimulated insulin release via regulation of intracellular calcium *(25)*.

3.2. Vitamin D and Insulin Sensitivity

In peripheral insulin target tissues, vitamin D may directly enhance insulin sensitivity by stimulating the expression of insulin receptors *(26)* and/or by activating peroxisome proliferator-activated receptor (PPAR-δ), a transcription factor implicated in the regulation of fatty acid metabolism in skeletal muscle and adipose tissue *(27)* (Fig. 2). An indirect effect of vitamin D on insulin sensitivity might also be exerted via its role in regulating extracellular calcium concentration and flux through cell membranes. Calcium is known to be essential for insulin-mediated intracellular processes in insulin-responsive tissues such as skeletal muscle and adipose tissue *(28, 29)* with a very narrow range of intracellular calcium needed for optimal insulin-mediated functions *(30)*. Changes in intracellular calcium in primary insulin target tissues may contribute to peripheral insulin resistance *(30–37)* via impaired insulin signal transduction *(37, 38)* leading to decreased glucose transporter activity *(37–39)*.

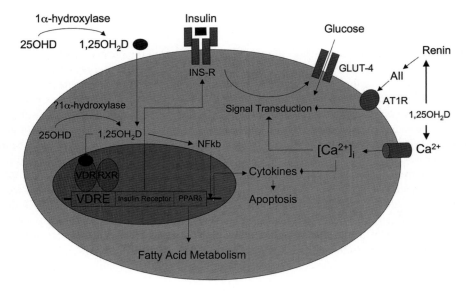

Fig. 2. Vitamin D and insulin action. In peripheral insulin-target cells, vitamin D may directly enhance insulin sensitivity by stimulating the expression of insulin receptors (INS-R) and/or by activating peroxisome proliferator-activated receptor (PPAR-δ), a transcription factor implicated in the regulation of fatty acid metabolism in skeletal muscle and adipose tissue. The effects of vitamin D may be mediated indirectly via its important and well-recognized role in regulating extracellular calcium (Ca^{2+}), calcium flux through the cell, and intracellular calcium (Ca^{2+})$_i$. Vitamin D may promote beta-cell survival by modulating the generation (through inactivation of nuclear factor-κB [NF-κb]) and effects of cytokines. Vitamin D may also affect insulin resistance indirectly through the renin–angiotensin (AII)–aldosterone system.

Vitamin D may also affect insulin resistance indirectly through the renin–angiotensin–aldosterone system (RAAS). Angiotensin II is thought to contribute to the development of insulin resistance by inhibiting the action of insulin in vascular and skeletal muscle tissue, leading to impaired glucose uptake *(40, 41)*. Recently, it was reported that renin expression and angiotensin II production were increased several fold in VDR-null mice, while administration of 1,25(OH)$_2$D suppressed rennin *(42, 43)*. Vitamin D, therefore, appears to be a negative endocrine regulation of RAAS *(43)*, providing another potential mechanism linking vitamin D to a decreased risk of type 2 diabetes.

3.3. Vitamin D and Systemic Inflammation

It is currently recognized that systemic inflammation plays an important role in the pathogenesis of type 2 diabetes, mostly via increasing insulin resistance but beta-cell function may also be affected via cytokine-induced apoptosis *(44–47)*. Vitamin D may improve insulin sensitivity and protect against beta-cell cytokine-induced apoptosis by directly modulating the expression and activity of cytokines *(48, 49)*. One such pathway that may, at least in part, mediate the anti-apoptotic effect of vitamin D on beta cell is through counteracting cytokine-induced Fas expression *(50)*. Several other immunomodulating effects of vitamin D have also been described and include blockade of dendritic cell differentiation, inhibition of lymphocyte proliferation, enhanced regulatory T-lymphocyte development *(49)*, and downregulation of cytokine expression from abnormally stimulated monocytes derived from patients with type 2 diabetes *(51)*. The immune-modulatory function of vitamin D may provide an additional protective effect against inflammation-induced type 2 diabetes.

4. EVIDENCE FROM OBSERVATIONAL HUMAN STUDIES

4.1. Seasonal Variation and Type 2 Diabetes

A potential role of vitamin D in type 2 diabetes is suggested by a reported seasonal variation in glycemic control in patients with type 2 diabetes, being worse in the winter *(52, 53)* when hypovitaminosis D is highly prevalent.

4.2. Case–Control and Cross-Sectional Studies of Vitamin D and Type 2 Diabetes

Additional evidence comes from case–control studies with small number of participants where most *(54–63)* but not all *(60, 64, 65)* reported that patients with type 2 diabetes or glucose intolerance have lower serum 25-hydroxyvitamin D [25(OH)D] concentration compared to controls without diabetes.

Over the last few years, several cross-sectional studies have examined the association between vitamin D status (assessed by serum 25(OH)D concentration or self-reported vitamin D intake) and prevalence of glucose intolerance or type 2 diabetes. Most have reported an inverse association between vitamin D status and glucose intolerance *(58, 66–72)* or the metabolic syndrome *(67, 68, 73, 74)*, while others failed to show such an association *(66, 67, 71, 72, 75–79)*.

In the largest cross-sectional study to date, using data from the National Health Nutrition Examination Survey (NHANES), serum 25(OH)D concentration (after multivariate adjustment, including BMI) was inversely associated with prevalence of diabetes in a dose-dependent pattern in non-Hispanic whites and Mexican Americans *(72)* (Table 1). There was no association in non-Hispanic blacks despite lower levels of 25(OH)D found

Table 1

Major Cross-Sectional Studies Reporting an Association Between Vitamin D Status, Calcium Intake, and Prevalence of Type 2 Diabetes/Metabolic Syndrome in Non-pregnant Adults

Study First author, year	Sex M/F	Age (mean or range) years	Cohort	Outcome (assessment)	Predictor, range, or category	Main study results	Adjustments	Comments and other outcomes
Vitamin D status (25(OH)D concentration or vitamin D intake)								
Scragg et al. (2004) (72)	M/F	>20	NHANES, N = 2,766 Non-Hispanic whites	T2DM (FPG)	25(OH)D, <18 ng/ml >32 ng/ml	Odds ratio 1.00 0.25 (0.11–0.60)	Age, sex, race, BMI, exercise, season	25(OH)D inversely associated with IR (HOMA)
	M/F	>20	NHANES, N = 1,726 Mexican Americans	T2DM (FPG)	25(OH)D, <18 ng/ml >32 ng/ml	Odds ratio 1.00 0.17 (0.08–0.37)	Age, sex, race, BMI, exercise, season	25(OH)D inversely associated with IR (HOMA)
	M/F	>20	NHANES, N = 1,726 Non-Hispanic blacks	T2DM (FPG)	25(OH)D, <18 ng/ml >32 ng/ml	Odds ratio 1.00 3.40 (1.07–10.86)	Age, sex, race, BMI, exercise, season	
Ford et al. (2005) (68)	M/F	>20	NHANES, N = 8,241	T2DM (FPG)	25(OH)D, <19 ng/ml >39 ng/ml	Odds ratio 1.00 0.54 (0.35–0.84)	Age, sex, abdominal obesity, race, exercise, smoking, alcohol, diet, vitamin use, cholesterol, CRP, education, season	

Study	Sex	Age	Population/N	Outcome	25(OH)D	Odds ratio	Adjusted for	Notes
Snijder et al. (2006) (78)	M/F	75	N = 1,235	T2DM (self-report)	25(OH)D <10 ng/ml ≥30 ng/ml	Odds ratio 1.0 1.23 (0.50–3.02)	Age, sex, WHR exercise, smoking, alcohol, region, season	Association pronounced among obese
Hypponen and Power (2006) (58)	M/F	45	Caucasians, N = 7,198	Hemoglobin A1c (%)	25(OH)D <10 ng/ml ≥30 ng/ml	Hemoglobin A1c concentration 5.4% 5.1%	Sex, season	
Martins et al. (2007) (69)	M/F	≥20	NHANES III (1988–1994) N = 7,186 M N = 7,902 F 25(OH)D, 30 ng/ml	Diabetes (history and/or FPG ≥126)	25(OH)D ≥37 ng/ml <21 ng/ml	OR 1.00 1.98 (1.57–2.74)	Age, sex, ethnicity	
Reis et al. (2007) (71)	M	44–96	Rancho Bernardo Study, N = 410	Hyperglycemia (FPG >110 mg/dl)	25(OH)D <35 ng/ml >50 ng/ml	Odds ratio 1.00 0.43 (0.20–0.95)	Age, season, smoking, alcohol, exercise	
	F	44–96	Rancho Bernardo Study, N = 660	Hyperglycemia (FPG >110 mg/dl)	25(OH)D <31 ng/ml >48 ng/ml	Odds ratio 1.00 0.62 (0.26–1.46)	Age, season, smoking, alcohol, exercise	

Table 1
(continued)

Study First author, year	Sex M/F	Age (mean or range) years	Cohort	Outcome (assessment)	Predictor, range, or category	Main study results	Adjustments	Comments and other outcomes
Hypponen et al. (2008) (74)	M/F	44–46	1958 British Birth Cohort Study, N =6,810	High A1c	25(OH)D <18 ng/ml >24 ng/ml	Odds ratio 1.00 0.30 (0.16–0.57)	Age, sex, BMI, WC, month, exercise, smoking, alcohol, hour of measurement, IGF-1, social class	
Ford et al. (2005) (68)	M/F	>20	NHANES, N = 8,241	Metabolic syndrome	25(OH)D, <19 ng/ml >38 ng/ml	Odds ratio 1.00 0.46 (0.32–0.67)	Age, sex, race, exercise, smoking, alcohol, diet, vitamin use, cholesterol, CRP, education, season	
Liu et al. (2005) (75)	F	>45	Women's Health Study, N = 10,066	Metabolic syndrome	Vitamin D intake, ≤159 IU/day ≥511 IU/day	Odds ratio 1.00 1.05 (0.84–1.32)	Age, exercise, smoking, alcohol, vitamin use, history of myocardial infarction, calcium intake	
Hypponen et al. (2008) (74)	M/F	44–46	1958 British Birth Cohort Study, N = 6,810	MetS	25(OH)D <18 ng/ml >24 ng/ml	Odds ratio 1.00 0.31	Age, sex, month, exercise, smoking, alcohol, hour of measurement, IGF-1, social class	

Study	Sex	Age	Cohort	Outcome	25(OH)D	Odds ratio	Adjustments
Reis et al. (2007) (71)	M	44–96	Rancho Bernardo Study, N = 410	MetS (ATP III)	25(OH)D <35 ng/ml >50 ng/ml	Odds ratio 1.00 0.57 (0.26–1.25)	Age, season, smoking, alcohol, exercise
	F	44–96	Rancho Bernardo Study, N = 660	MetS (ATP III)	25(OH)D <31 ng/ml >48 ng/ml	Odds ratio 1.00 0.88 (0.43–1.80)	Age, season, smoking, alcohol, exercise
Martins et al. (2007) (69)	M/F	≥20	NHANES III (1988–1994) N = 7,186 M N = 7,902 F 25(OH)D, 30 ng/ml	Hypertension (history and/or BP ≥140/90)	25(OH)D ≥37 ng/ml <21 ng/ml	OR 1.00 1.30 (1.13–1.49)	Age, sex, ethnicity

BMI, body mass index; NR, not reported; NGT, normal glucose tolerance (based on FPG or 2hPG); IGT, impaired glucose tolerance (based on FPG or 2hPG); T2DM, type 2 diabetes mellitus (based on FPG, 2hPG, or self-report); FPG, fasting plasma glucose; 1hPG, plasma glucose 1 h after 75 g glucose load; 2hPG, plasma glucose 2 h after 75 g glucose load; GLU$_{AUC}$, glucose area under the curve after 75 g glucose load; IR, insulin resistance; HOMA, homeostasis model assessment; CRP, C-reactive protein; WHR, waist–hip ratio; 25(OH)D: 25-hydroxyvitamin D; NHANES, National Health and Nutrition Examination Survey; BWHS, Black Women's Health Study; CARDIA, Coronary Artery Risk Development in Young Adults study; HPFS, Health Professionals Follow-Up Study.
↓ Decreased (statistically significant), ↑ increased (statistically significant), ↔ no difference (no statistical significance).
To convert 25(OH)D concentration to SI units, multiply by 2.459.

in this group, which may be explained by the observation that non-whites exhibit a different vitamin D, calcium, and PTH homeostasis compared to whites *(80)*. In the same study, serum 25(OH)D also correlated with measures of insulin resistance but did not correlate with beta-cell function *(72)*. More recent studies using NHANES data have confirmed the inverse association between 25(OH)D and fasting glycemia *(68, 69)*. Two other large cross-sectional cohorts have reported inverse associations between 25(OH)D and hyperglycemia. In the 1958 British Birth Cohort study, 25(OH)D was inversely associated with prevalent elevated hemoglobin A1c *(74)*. In the Rancho Bernardo study, a community-based cohort of older adults in southern California, 25(OH)D was inversely associated with prevalent hyperglycemia only among men *(71)*.

Vitamin D status has also been inversely associated with the prevalence of the metabolic syndrome in some *(67, 73)* but not all studies *(76)*. The metabolic syndrome is defined as a constellation of physical signs and laboratory findings (central adiposity, hyperglycemia, hypertriglyceridemia, low HDL cholesterol, and hypertension) and is closely associated with both type 2 diabetes and cardiovascular disease. In NHANES, serum 25(OH)D concentration (after multivariate adjustment) was inversely associated with having the metabolic syndrome among both genders and all three major racial or ethnic groups *(68)*. The inverse association was also reported in the 1958 British Birth Cohort *(74)*. The components of the metabolic syndrome that have consistently been independently associated with low 25OHD are abdominal obesity and hyperglycemia; therefore, the results of the cross-sectional studies may simply reflect the inverse association between serum 25(OH)D and body weight or fatness *(72, 81, 82)*. In a cross-sectional analysis of the Women's Health Study, a large randomized trial designed to evaluate the effects of low-dose aspirin and vitamin E in cardiovascular disease, the inverse association between vitamin D intake and prevalence of metabolic syndrome was dissipated after adjustment for calcium intake *(75)*.

4.3. Prospective Studies of Vitamin D and Type 2 Diabetes

There is relative paucity of prospective observational studies on the association of vitamin D status with incident type 2 diabetes (Table 2). In the Women's Health Study, an intake of 511 IU/day or more of vitamin D was associated with lower risk of incident type 2 diabetes compared with an intake of 159 IU/day or less (2.7 vs. 5.6% of the cohort developed type 2 diabetes, respectively) *(75)*. However, this analysis did not adjust for any risk factors of type 2 diabetes other than age. In the Nurses Health Study – the largest prospective study to date – women with the highest vitamin D intake (>800 IU/day) had a 33% lower risk of incident type 2 diabetes compared to women with the lowest vitamin D intake (<200 IU/day) after adjustment for multiple risk factors for type 2 diabetes *(83)*. These two studies, however, defined vitamin D status using self-reported vitamin D intake. There are two prospective studies that have examined the association between 25(OH)D concentration and incident type 2 diabetes. In the first study from Finland, the inverse association between serum 25(OH)D and incident type 2 diabetes was attenuated after adjustments for BMI, physical activity, smoking, and education *(84)*. The second study, also from Finland, pooled data from two nested case–control studies, collected by the Finnish Mobile Clinic in 1973–1980 *(85)*. Men had higher serum

Table 2
Major Prospective Studies (Including Case–Control Studies Nested Within a Prospective Cohort) Reporting an Association Between Vitamin D Status, Calcium Intake, and Incidence of Type 2 Diabetes/Metabolic Syndrome in Non-pregnant Adults

Study First author, Year	Sex M/F	Age at baseline (mean or range) year	Cohort Total N/no. of cases	Outcome (assessment)	Predictor Lowest and highest category	Main study results	Adjustments	Comments
Vitamin D status (25(OH)D concentration or vitamin D intake)								
Liu et al. (2005) (75)	F	>45	Women's Health Study, 10,066/NR	T2DM (validated self-report)	Vitamin D intake ≤ 159 IU/day ≥ 511 IU/day	Percentage of cohort with T2DM 5.6 2.7	Age	
Pittas et al. (2006) (83)	F	46	Nurses Health Study, 83,779/4,843	T2DM (validated self-report)	Vitamin D intake < 200 IU/day > 800 IU/day	Relative risk 1.00 0.87 (0.69–1.09)	Age, BMI, exercise, diabetes family history, smoking, alcohol, coffee diet, hypertension, calcium intake	
Mattila et al. (2007) (84)	M/F	40–69	Mini-Finland Health Survey 4,097/187 Baseline 25(OH)D 17 ng/ml	T2DM (nationwide registry of diabetes medications)	25(OH)D <30 nmol/l >55 nmol/l	Relative risk 1.00 0.70 (0.42–1.16)	Age, sex, month of collecting blood samples, BMI, smoking education	

Table 2
(continued)

Study First author, Year	Sex M/F	Age at baseline (mean or range) year	Cohort Total N/no. of cases	Outcome (assessment)	Predictor Lowest and highest category	Main study results	Adjustments	Comments
Knekt et al. (2008) (85)	M	40–74	Finnish Mobile Clinic 626/185	T2DM (medication-treated, registry-based)	25(OH)D	OR 1.00 0.28 (0.10–0.81)	BMI, smoking, exercise, education	Nested case control
	F	40–74	Finnish Mobile Clinic 748/220	T2DM (medication-treated, registry-based)	25(OH)D	OR 1.00 1.14 (0.60–2.17)	BMI, smoking, exercise, education	Nested case control
Calcium Intake								
Liu et al. (2005) (75)	F	>45	Women's Health Study, 10,066/NR	T2DM (validated self-report)	Calcium intake ≤ 610 mg/day ≥1,284 mg/day	Percentage of cohort with T2DM 5.6 2.7	Age	
Pittas et al. (2006) (83)	F	46	Nurses Health Study, 83,779/4,843	T2DM (validated self-report)	Calcium intake ≤ 600 mg/day >1,200 mg/day	Relative risk 1.00 0.79 (0.70–0.90)	Age, BMI, exercise, diabetes family history, smoking, alcohol, coffee diet, hypertension, calcium intake	

Study	Sex	Age	Cohort, N	Outcome	Calcium intake	Relative risk	Adjustments	Notes
van Dam et al. (2006) (86)	F	39	BWHS, 41,186/1,964	T2DM (validated self-report)	219 mg/day 661 mg/day	1.00 0.86 (0.74–1.00)	Age, BMI, exercise, diabetes family history, smoking, alcohol, coffee, diet, education	Association dissipated after adjustment for magnesium intake
Pereira et al. (2002) (87)	M/F	18–30	CARDIA, 3,157	MetS (ATP-3 criteria)	<600 mg/day >1,200 mg/day	1.00 0.79 (0.61–1.03) Among overweight (BMI >25) only	Age, sex, BMI, exercise, smoking, diet, vitamin use, energy intake	Association dissipated after adjusting for dairy intake

Table 2
(continued)

Study First author; Year	Sex M/F	Age at baseline (mean or range) year	Cohort Total N/no. of cases	Outcome (assessment)	Predictor Lowest and highest category	Main study results	Adjustments	Comments
Combined vitamin D and calcium intake								
Pittas et al. (2006) (83)	F	46	Nurses Health Study, 83,779/4,843	T2DM (validated self-report)	Vitamin D and calcium ≤ 400 IU/day and ≤ 600 mg/day > 800 IU/day and >1,200 mg/day	Relative risk 1.00 0.67 (0.49–0.90)	Age, BMI, exercise, diabetes family history, smoking, alcohol, coffee diet, hypertension	

BMI, body mass index; NR, not reported; NGT, normal glucose tolerance (based on FPG or 2hPG); IGT, impaired glucose tolerance (based on FPG or 2hPG); T2DM, type 2 diabetes mellitus (based on FPG, 2hPG, or self-report); FPG, fasting plasma glucose; 1hPG, plasma glucose 1 h after 75 g glucose load; 2hPG, plasma glucose 2 h after 75 g glucose load; GLU$_{AUC}$, glucose area under the curve after 75 g glucose load; IR, insulin resistance; HOMA, homeostasis model assessment; CRP, C-reactive protein; WHR, waist–hip ratio; 25(OH)D: 25-hydroxyvitamin D; NHANES, National Health and Nutrition Examination Survey; BWHS, Black Women's Health Study; CARDIA, Coronary Artery Risk Development in Young Adults study; HPFS, Health Professionals Follow-Up Study.
↓ Decreased (statistically significant), ↑ increased (statistically significant), ↔ no difference (no statistical significance).
To convert 25(OH)D concentration to SI units, multiply by 2.459.

vitamin D concentrations than women and showed a reduced risk of type 2 diabetes in the highest vitamin D quartile after multivariate adjustment.

4.4. The Role of Calcium Intake and Type 2 Diabetes or the Metabolic Syndrome

A potentially important role for calcium status in the development of type 2 diabetes is suggested by case–control studies in which calcium intake was found to be lower in patients with diabetes compared with controls *(59)*. In a cross-sectional analysis from the Women's Health Study, calcium intake (after adjustment for vitamin D intake) was inversely associated with the prevalence of metabolic syndrome *(75)*. In prospective studies (Table 2), low calcium intake was consistently found to be inversely associated with type 2 diabetes *(75, 83, 86)* or the metabolic syndrome *(87)*.

In an effort to clarify the individual contribution of each nutrient to future type 2 diabetes risk, in the Nurses Health Study, our group examined the combined effects of total (food + supplements) vitamin D and calcium intake on risk of incident t2DM (Fig. 3). We observed that, after multivariate adjustment, women with the highest calcium (>1,200 mg/day) *and* vitamin D (>800 IU/day) intakes (1.3% of the cohort) had a 33% lower risk of type 2 diabetes compared to women with the lowest calcium (<600 mg/day) and vitamin D (<400 IU/day) intakes. The lower risk seen with the combined intake was more than that seen with the highest intake of each nutrient separately,

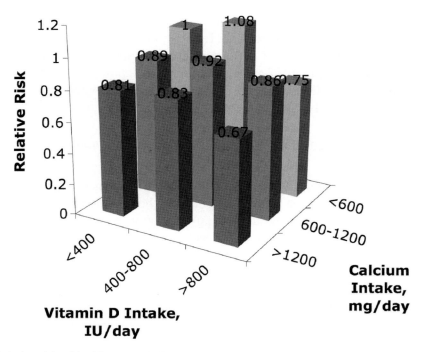

Fig. 3. Relative risk of incident type 2 diabetes by self-reported intake of vitamin D and calcium in the Nurses Health Study, a large observation cohort.

which suggests a synergy and highlights the importance of both nutrients as potential risk modifiers.

4.5. Summary of Evidence from Human Observational Studies

Overall, the evidence from observational studies suggests an association between low vitamin D status and risk of type 2 diabetes or metabolic syndrome. However, definite conclusions cannot be drawn for a variety of reasons: (1) In cross-sectional or case–control studies, vitamin D was measured in patients with glucose tolerance or established diabetes; therefore, these measures may not reflect vitamin D status prior to diagnosis and, as a result the causative nature of the reported associations cannot be established. (2) There is considerable variability among the various cohorts [normal glucose tolerance vs. diabetes (newly diagnosed vs. established), age, ethnicity, latitude, mean vitamin D status, etc.], which makes it difficult to compare, contrast, and combine results. (3) In most studies, there is lack of adjustment for important confounders (adiposity, physical activity, calcium status), although most of the recent studies have accounted for these variables. (4) Finally, vitamin D status is an excellent marker of overall health status; therefore, residual confounding may explain some of the inverse association with type 2 diabetes.

5. EVIDENCE FROM INTERVENTION HUMAN STUDIES

5.1. Effect of Vitamin D Supplementation on Type 2 Diabetes

There are four small-scale short-term trials and one long-term controlled trial that have examined the effect of supplementation with a variety of formulations of vitamin D on type 2 diabetes parameters (Table 3). In the small-scale short-term (less than 12 weeks) studies, the effect of vitamin D supplementation on fasting or postprandial glycemia was neutral *(77, 88–90)*. Similarly, in a post hoc analysis of a 2-year trial designed for bone-related outcomes, supplementation with vitamin D_3 or $1(OH)D_3$ had no effect on fasting glycemia in postmenopausal non-diabetic women *(91)*.

5.2. Effect of Combined Vitamin D and Calcium Supplementation on Type 2 Diabetes

There are two trials that have reported the effect of combined vitamin D_3 and calcium supplementation on type 2 diabetes, both post hoc. In a trial designed to assess bone-related outcomes, combined daily supplementation with 700 IU of vitamin D_3 and 500 mg of calcium prevented the expected rise in glycemia and insulin resistance over a 3-year period among participants with impaired fasting glucose at baseline, but had no effect on glycemia or insulin resistance among subjects with normal glucose tolerance at baseline. Importantly, the effect size of combined vitamin D and calcium supplementation in this high-risk group on fasting glycemia was similar to the effect size seen in the Diabetes Prevention Program with intensive lifestyle or metformin (0.2 mg/dl in the lifestyle and 0.2 mg/dl in the metformin arm vs. 5.5 mg/dl in placebo) *(7)*. In contrast, combined vitamin D_3 (400 IU/day) and calcium supplementation (1,000 mg/day) in the Women's Health Initiative failed to reduce the risk of

Table 3
Major Randomized Controlled Trials of the Effect of Vitamin D and/or Calcium Supplementation on Glucose Tolerance

Study First author, year	Sex M/F	Age (mean or range) years	Study participants	25(OH)D concentration and calcium intake at baseline	Intervention Type and dose	Duration	Main outcome (glycemia)	Comment and other outcomes
Vitamin D alone								
Nilas et al. (1984) (91)	F	45–54	Non-diabetic; N = 151	NR	Vitamin D₃ 2,000 IU/day [N = 25] vs. 1(OH)D 0.25 mcg/day [N = 23] vs. placebo [N = 103] All received 500 mg/day calcium Post hoc analysis	104 weeks	↔FPG (change from baseline [mg/dl]: +2.2 vs. –0.33 vs. +0.1269)	
Inomata et al. (1986) (89)	M/F	36–80	T2DM; N = 14	NR	1(OH)D 2 mcg/day [N = 7] vs. placebo [N = 7]	3 weeks	↔GLU$_{AUC}$ (change from baseline [mg/2 h/dl]: –21.2 vs. –.3)	↑ INS$_{AUC}$

Table 3
(continued)

Study First author, year	Sex M/F	Age (mean or range) years	Study participants	25(OH)D concentration and calcium intake at baseline	Intervention Type and dose	Duration	Main outcome (glycemia)	Comment and other outcomes
Ljunghall et al. (1987) (90)	M	61–65	IGT/mild t2DM; $N = 65$	25(OH)D 38 ng/ml	1(OH)D 0.75 mcg/day [$N = 33$] vs. placebo [$N = 32$]	12 weeks	↔FPG (baseline to end-of-study [mg/dl]: 117–117 vs. 115–117) ↔A1c (baseline to end [%]: 6.46–5.90 vs. 6.28–5.70)	↔ IR_{IVGTT}
Orwoll et al. (1994) (77)	M/F	40–70	Non-insulin-treated T2DM; $N = 20$	25(OH)D 14 ng/ml	1,25(OH)D 1 mcg/day vs. placebo [cross-over trial, $N = 20$]	4 days	↔FPG (baseline to end of study [mg/dl]: 214–209 vs. 214–198) ↔ Meal-stimulated PG (data NR)	↔ IR_{FI}, ↔ INS_{AUC} ↑ INS_{AUC} if diabetes is of short duration

Study	Sex	Age	Population	25(OH)D status	Intervention	Duration	Results
Fliser et al. (1997) (88)	M	26	Healthy, non-diabetic; $N = 18$	NR	1,25(OH)$_2$D 1.5 mcg/day [$N = 9$] vs. placebo [$N = 9$]	1 weeks	↔FPG (baseline to end of study [mg/dl]: 84–86 vs. 86–88) ↔IR$_M$

Combined vitamin D plus calcium supplementation

Study	Sex	Age	Population	25(OH)D status	Intervention	Duration	Results
Pittas et al. (2006) (83)	M/F	71	Normal fasting glucose; $N = 222$	25(OH)D, 30 ng/ml; calcium intake, 750 mg/day	D$_3$ 700 IU/day + calcium citrate 500 mg/day [$N = 108$] vs. placebo [$N = 114$] Post hoc analysis	3 year	↔FPG (change from baseline [mg/dl]: 2.7 vs. 2.2) ↔IR$_{HOMA}$
	M/F	71	Impaired fasting glucose; $N = 92$	25(OH)D, 30 ng/ml; calcium intake, 680 mg/day	D$_3$ 700 IU/day + calcium citrate 500 mg/day [$N = 45$] vs. placebo [$N = 47$] Post hoc analysis	3 year	↓FPG (change from baseline [mg/dl]: 0.4 vs. 6.1) ↓IR$_{HOMA}$

Table 3
(continued)

Study First author, year	Sex M/F	Age (mean or range) years	Study participants	25(OH)D concentration and calcium intake at baseline	Intervention Type and dose	Duration	Main outcome (glycemia)	Comment and other outcomes
De Boer et al. (2008) (92)	F	50–79	WHI No diabetes (self-report) $N = 33{,}951$	25(OH)D <32 ng/ml for 89% of participants	D_3 400 IU/day + calcium carbonate 1,000 mg/day [$N = 16{,}999$] vs. placebo [$N = 16{,}952$] Post hoc analysis	7 year	↔Incident diabetes (self-report)	↔ IR_{HOMA}

NR, not reported; IGT, impaired glucose tolerance (based on FPG or 2hPG); T2DM, type 2 diabetes mellitus (based on FPG, 2hPG, or self-report); FPG, fasting plasma glucose; 2hPG, plasma glucose 2 h after 75 g glucose load; GLU_{AUC}, glucose area under the curve after 75 g glucose load; INS_{AUC}, insulin area under the curve after 75 g glucose load; INS_{120}, insulin value at 120' after glucose load is given; IR, insulin resistance; 25(OH)D: 25-hydroxyvitamin D; IR_{FI}, insulin resistance by fasting insulin; IR_{HOMA}, insulin resistance by homeostasis model assessment; IR_M, insulin resistance after euglycemic hyperinsulinemic clamp; IR_{IVGTT}, insulin resistance after intravenous glucose tolerance test.

↓ Decreased (statistically significant), ↑increased (statistically significant), ↔ no difference (no statistical significance).

To convert 25(OH)D concentration to SI units, multiply by 2.459; to convert FPG to SI units, multiply by 0.0555.

developing self-reported diabetes over a 7-year period *(92)*. In that trial, there was also no significant effect of treatment on simple indices of insulin resistance. This null result in the Women's Health Initiative study may be attributed to insufficient doses of vitamin D given to the active treatment group.

5.3. Summary of Evidence from Intervention Studies

It is difficult to draw definitive conclusions from the results of the above trials, primarily because none of the three largest trials were specifically designed to examine the effect of vitamin D on type 2 diabetes. However, one can conclude that vitamin D taken at the currently recommended intakes has no benefit in lowering risk of type 2 diabetes among the general population. Whether a higher intake of vitamin D, alone or in combination with calcium, has a role in the progression to type 2 diabetes among high-risk patients (i.e., those with glucose intolerance) remains to be seen.

6. LIMITATIONS IN THE STUDY OF VITAMIN D

As described above, some of the effects of vitamin D on glucose homeostasis may be mediated by its role in regulating calcium balance. Calcium may also have effects on glucose metabolism independent of vitamin D status. Therefore, when one studies the role of vitamin D in glucose homeostasis and type 2 diabetes, it is also important to examine the contribution of calcium status including any potential interaction between them on the outcomes of interest. Additional issues related to food synergy or residual confounding further complicate the study of the association between vitamin D and type 2 diabetes *(93)*. For example, a higher vitamin D status is often accompanied by a "prudent diet" (including higher consumption of dairy) and favorable lifestyle (including increased leisure-time physical activity). Although analyses can adjust for such variables, unmeasured synergy between associated behaviors or residual confounding may bias the results toward or away from null.

7. OPTIMAL INTAKE OF VITAMIN D AND CALCIUM IN RELATION TO TYPE 2 DIABETES

Currently recommended intakes for calcium are 1,200 mg/day for adults aged >50 years and for vitamin D are 400 IU/day for those aged 51–70 years and 600 IU/day for those aged >70 years *(94)* (see also Chapter 1. However, there is growing consensus that vitamin D intakes above the current recommendations may be associated with better health outcomes. For a variety of skeletal and non-skeletal outcomes, the optimal 25(OH)D level appears to be 30–40 ng/ml *(95)*. In relation to type 2 diabetes, it is difficult to draw definitive conclusion but the data suggest that serum 25(OH)D concentration above 20 ng/ml are desirable, while a level above 40 ng/ml may be optimal. To achieve such 25(OH)D concentration, an intake of approximately 1,000 IU/day of vitamin D is needed *(95, 96)*. In relation to calcium intake, the evidence suggests that intakes above 600 mg/day are desirable but intakes above 1,200 mg may be optimal in relation to type 2 diabetes.

8. CONCLUSIONS AND FUTURE DIRECTIONS

An inverse association between vitamin D status and type 2 diabetes is suggested based on the available evidence to date. However, the lack of large prospective observational studies that have measured 25(OH)D as the exposure variable and the lack of randomized trials specifically designed to test the effects of vitamin D limit any firm conclusions. Type 2 diabetes is a multifactorial disease and it is unlikely that vitamin D deficiency would be a major cause of the disease or a major therapeutic target. However, vitamin D sufficiency may have a role in preventing progression from glucose intolerance to clinical diabetes in high-risk populations. To better define the clinical role of vitamin D and calcium as potential interventions for prevention and management of type 2 diabetes, randomized trials specifically designed to test such a hypothesis are needed.

REFERENCES

1. Narayan KM et al (2006) Impact of recent increase in incidence on future diabetes burden: US, 2005–2050. Diabetes Care 29(9):2114–2116
2. Wild S et al (2004) Global prevalence of diabetes: estimates for the year 2000 and projections for 2030. Diabetes Care 27(5):1047–1053
3. Dabelea D et al (2007) Incidence of diabetes in youth in the United States. JAMA 297(24):2716–2724
4. Booth GL et al (2006) Relation between age and cardiovascular disease in men and women with diabetes compared with non-diabetic people: a population-based retrospective cohort study. Lancet 368(9529):29–36
5. Hogan P, Dall T, Nikolov P (2003) Economic costs of diabetes in the US in 2002. Diabetes Care 26(3):917–932
6. Hu FB et al (2001) Diet, lifestyle, and the risk of type 2 diabetes mellitus in women. N Engl J Med 345(11):790–797
7. Knowler WC et al (2002) Reduction in the incidence of type 2 diabetes with lifestyle intervention or metformin. N Engl J Med 346(6):393–403
8. Pittas AG et al (2007) The role of vitamin D and calcium in type 2 diabetes. A systematic review and meta-analysis. J Clin Endocrinol Metab 92(6):2017–2029
9. Holick MF (2006) Resurrection of vitamin D deficiency and rickets. J Clin Invest 116(8): 2062–2072
10. Fleet JC (2004) Rapid, membrane-initiated actions of 1,25 dihydroxyvitamin D: what are they and what do they mean? J Nutr 134(12):3215–3218
11. Norman AW et al (1980) Vitamin D deficiency inhibits pancreatic secretion of insulin. Science 209(4458):823–825
12. Kadowaki S, Norman AW (1984) Dietary vitamin D is essential for normal insulin secretion from the perfused rat pancreas. J Clin Invest 73(3):759–766
13. Tanaka Y et al (1984) Effect of vitamin D3 on the pancreatic secretion of insulin and somatostatin. Acta Endocrinol (Copenh) 105(4):528–533
14. Cade C, Norman AW (1986) Vitamin D3 improves impaired glucose tolerance and insulin secretion in the vitamin D-deficient rat in vivo. Endocrinology 119(1):84–90
15. Bourlon PM, Faure-Dussert A, Billaudel B (1999) The de novo synthesis of numerous proteins is decreased during vitamin D3 deficiency and is gradually restored by 1, 25-dihydroxyvitamin D3 repletion in the islets of langerhans of rats. J Endocrinol 162(1):101–109
16. Clark SA, Stumpf WE, Sar M (1981) Effect of 1,25 dihydroxyvitamin D3 on insulin secretion. Diabetes 30(5):382–386
17. Johnson JA et al (1994) Immunohistochemical localization of the 1,25(OH)2D3 receptor and calbindin D28k in human and rat pancreas. Am J Physiol 267(3 Pt 1):E356–E360

18. Zeitz U et al (2003) Impaired insulin secretory capacity in mice lacking a functional vitamin D receptor. Faseb J 17(3):509–511

19. Maestro B et al (2003) Identification of a Vitamin D response element in the human insulin receptor gene promoter. J Steroid Biochem Mol Biol 84(2–3):223–230

20. Maestro B et al (2002) Transcriptional activation of the human insulin receptor gene by 1,25-dihydroxyvitamin D(3). Cell Biochem Funct 20(3):227–232

21. Bland R et al (2004) Expression of 25-hydroxyvitamin D3-1alpha-hydroxylase in pancreatic islets. J Steroid Biochem Mol Biol 89–90(1–5):121–125

22. Sergeev IN, Rhoten WB (1995) 1,25-Dihydroxyvitamin D3 evokes oscillations of intracellular calcium in a pancreatic beta-cell line. Endocrinology 136(7):2852–2861

23. Milner RD, Hales CN (1967) The role of calcium and magnesium in insulin secretion from rabbit pancreas studied in vitro. Diabetologia 3(1):47–49

24. Kadowaki S, Norman AW (1984) Pancreatic vitamin D-dependent calcium binding protein: biochemical properties and response to vitamin D. Arch Biochem Biophys 233(1):228–236

25. Sooy K et al (1999) Calbindin-D(28 k) controls [Ca(2+)](i) and insulin release. Evidence obtained from calbindin-d(28 k) knockout mice and beta cell lines. J Biol Chem 274(48): 34343–34349

26. Maestro B et al (2000) Stimulation by 1,25-dihydroxyvitamin D3 of insulin receptor expression and insulin responsiveness for glucose transport in U-937 human promonocytic cells. Endocr J 47(4): 383–391

27. Dunlop TW et al (2005) The human peroxisome proliferator-activated receptor delta gene is a primary target of 1alpha,25-dihydroxyvitamin D3 and its nuclear receptor. J Mol Biol 349(2):248–260

28. Ojuka EO (2004) Role of calcium and AMP kinase in the regulation of mitochondrial biogenesis and GLUT4 levels in muscle. Proc Nutr Soc 63(2):275–278

29. Wright DC et al (2004) Ca2+ and AMPK both mediate stimulation of glucose transport by muscle contractions. Diabetes 53(2):330–335

30. Draznin B et al (1987) The existence of an optimal range of cytosolic free calcium for insulin-stimulated glucose transport in rat adipocytes. J Biol Chem 262(30):14385–14388

31. Byyny RL et al (1992) Cytosolic calcium and insulin resistance in elderly patients with essential hypertension. Am J Hypertens 5(7):459–464

32. Draznin B et al (1989) Mechanism of insulin resistance induced by sustained levels of cytosolic free calcium in rat adipocytes. Endocrinology 125(5):2341–2349

33. Draznin B et al (1988) Possible role of cytosolic free calcium concentrations in mediating insulin resistance of obesity and hyperinsulinemia. J Clin Invest 82(6):1848–1852

34. Draznin B et al (1988) Relationship between cytosolic free calcium concentration and 2-deoxyglucose uptake in adipocytes isolated from 2- and 12-month-old rats. Endocrinology 122(6): 2578–2583

35. Ohno Y et al (1993) Impaired insulin sensitivity in young, lean normotensive offspring of essential hypertensives: possible role of disturbed calcium metabolism. J Hypertens 11(4):421–426

36. Segal S et al (1990) Postprandial changes in cytosolic free calcium and glucose uptake in adipocytes in obesity and non-insulin-dependent diabetes mellitus. Horm Res 34(1):39–44

37. Zemel MB (1998) Nutritional and endocrine modulation of intracellular calcium: implications in obesity, insulin resistance and hypertension. Mol Cell Biochem 188(1–2):129–136

38. Williams PF et al (1990) High affinity insulin binding and insulin receptor-effector coupling: modulation by Ca^{2+}. Cell Calcium 11(8):547–556

39. Reusch JE et al (1991) Regulation of GLUT-4 phosphorylation by intracellular calcium in adipocytes. Endocrinology 129(6):3269–3273

40. Sowers JR (2004) Insulin resistance and hypertension. Am J Physiol Heart Circ Physiol 286(5):H1597–H1602

41. Wei Y et al (2008) Angiotensin II-induced skeletal muscle insulin resistance mediated by NF-kappaB activation via NADPH oxidase. Am J Physiol Endocrinol Metab 294(2):E345–E351

42. Li YC et al (2002) 1,25-Dihydroxyvitamin D(3) is a negative endocrine regulator of the renin-angiotensin system. J Clin Invest 110(2):229–238

43. Yuan W et al (2007) 1,25-dihydroxyvitamin D3 suppresses renin gene transcription by blocking the activity of the cyclic AMP response element in the renin gene promoter. J Biol Chem 282(41): 29821–29830
44. Duncan BB et al (2003) Low-grade systemic inflammation and the development of type 2 diabetes: the atherosclerosis risk in communities study. Diabetes 52(7):1799–1805
45. Hu FB et al (2004) Inflammatory markers and risk of developing type 2 diabetes in women. Diabetes 53(3):693–700
46. Pittas AG, Joseph NA, Greenberg AS (2004) Adipocytokines and insulin resistance. J Clin Endocrinol Metab 89(2):447–452
47. Pradhan AD et al (2001) C-reactive protein, interleukin 6, and risk of developing type 2 diabetes mellitus. Jama 286(3):327–334
48. Riachy R et al (2002) 1,25-dihydroxyvitamin D3 protects RINm5F and human islet cells against cytokine-induced apoptosis: implication of the antiapoptotic protein A20. Endocrinology 143(12):4809–4819
49. van Etten E, Mathieu C (2005) Immunoregulation by 1,25-dihydroxyvitamin D3: basic concepts. J Steroid Biochem Mol Biol 97(1–2):93–101
50. Riachy R et al (2006) 1,25-Dihydroxyvitamin D3 protects human pancreatic islets against cytokine-induced apoptosis via down-regulation of the Fas receptor. Apoptosis 11(2):151–159
51. Giulietti A et al (2007) Monocytes from type 2 diabetic patients have a pro-inflammatory profile. 1,25-Dihydroxyvitamin D(3) works as anti-inflammatory. Diabetes Res Clin Pract 77(1):47–57
52. Behall KM et al (1984) Seasonal variation in plasma glucose and hormone levels in adult men and women. Am J Clin Nutr 40(6 Suppl):1352–1356
53. Campbell IT, Jarrett RJ, Keen H (1975) Diurnal and seasonal variation in oral glucose tolerance: studies in the Antarctic. Diabetologia 11(2):139–145
54. Aksoy H et al (2000) Serum 1,25 dihydroxy vitamin D (1,25(OH)2D3), 25 hydroxy vitamin D (25(OH)D) and parathormone levels in diabetic retinopathy. Clin Biochem 33(1):47–51
55. Boucher BJ et al (1995) Glucose intolerance and impairment of insulin secretion in relation to vitamin D deficiency in east London Asians. Diabetologia 38(10):1239–1245
56. Christiansen C et al (1982) Vitamin D metabolites in diabetic patients: decreased serum concentration of 24,25-dihydroxyvitamin D. Scand J Clin Lab Invest 42(6):487–491
57. Cigolini M et al (2006) Serum 25-hydroxyvitamin D3 concentrations and prevalence of cardiovascular disease among type 2 diabetic patients. Diabetes Care 29(3):722–724
58. Hypponen E, Power C (2006) Vitamin D status and glucose homeostasis in the 1958 British birth cohort: the role of obesity. Diabetes Care 29(10):2244–2246
59. Isaia G, Giorgino R, Adami S (2001) High prevalence of hypovitaminosis D in female type 2 diabetic population. Diabetes Care 24(8):1496
60. Nyomba BL et al (1986) Vitamin D metabolites and their binding protein in adult diabetic patients. Diabetes 35(8):911–915
61. Pietschmann P, Schernthaner G, Woloszczuk W (1988) Serum osteocalcin levels in diabetes mellitus: analysis of the type of diabetes and microvascular complications. Diabetologia 31(12): 892–895
62. Scragg R et al (1995) Serum 25-hydroxyvitamin D3 levels decreased in impaired glucose tolerance and diabetes mellitus. Diabetes Res Clin Pract 27(3):181–188
63. Stepan J et al (1982) Plasma 25-hydroxycholecalciferol in oral sulfonylurea treated diabetes mellitus. Horm Metab Res 14(2):98–100
64. Heath H 3rd et al (1979) Calcium homeostasis in diabetes mellitus. J Clin Endocrinol Metab 49(3):462–466
65. Ishida H et al (1985) Diabetic osteopenia and circulating levels of vitamin D metabolites in type 2 (noninsulin-dependent) diabetes. Metabolism 34(9):797–801
66. Baynes KC et al (1997) Vitamin D, glucose tolerance and insulinaemia in elderly men. Diabetologia 40(3):344–347
67. Chiu KC et al (2004) Hypovitaminosis D is associated with insulin resistance and beta cell dysfunction. Am J Clin Nutr 79(5):820–825

68. Ford ES et al (2005) Concentrations of serum vitamin D and the metabolic syndrome among US adults. Diabetes Care 28(5):1228–1230

69. Martins D et al (2007) Prevalence of cardiovascular risk factors and the serum levels of 25-hydroxyvitamin D in the United States: data from the Third National Health and Nutrition Examination Survey. Arch Intern Med 167(11):1159–1165

70. Need AG et al (2005) Relationship between fasting serum glucose, age, body mass index and serum 25 hydroxyvitamin D in postmenopausal women. Clin Endocrinol (Oxf) 62(6):738–741

71. Reis JP et al (2007) Vitamin D, parathyroid hormone levels, and the prevalence of metabolic syndrome in community-dwelling older adults. Diabetes Care 30(6):1549–1555

72. Scragg R, Sowers M, Bell C (2004) Serum 25-hydroxyvitamin D, diabetes, and ethnicity in the Third National Health and Nutrition Examination Survey. Diabetes Care 27(12):2813–2818

73. Botella-Carretero JI et al (2007) Vitamin D deficiency is associated with the metabolic syndrome in morbid obesity. Clin Nutr 26(5):573–580

74. Hypponen E et al (2008) 25-hydroxyvitamin D, IGF-1, and metabolic syndrome at 45 years of age: a cross-sectional study in the 1958 British Birth Cohort. Diabetes 57(2):298–305

75. Liu S et al (2005) Dietary calcium, vitamin D, and the prevalence of metabolic syndrome in middle-aged and older US women. Diabetes Care 28(12):2926–2932

76. McGill AT et al (2008) Relationships of low serum vitamin D3 with anthropometry and markers of the metabolic syndrome and diabetes in overweight and obesity. Nutr J 7:4

77. Orwoll E, Riddle M, Prince M (1994) Effects of vitamin D on insulin and glucagon secretion in non-insulin-dependent diabetes mellitus. Am J Clin Nutr 59(5):1083–1087

78. Snijder M et al To: Mathieu C, Gysemans C, Giulietti A, Bouillon R (2005) Vitamin D and diabetes. Diabetologia 48:1247–1257. Diabetologia 49(1):217–218

79. Wareham NJ et al (1997) Glucose intolerance is associated with altered calcium homeostasis: a possible link between increased serum calcium concentration and cardiovascular disease mortality. Metabolism 46(10):1171–1177

80. Bell NH et al (1985) Evidence for alteration of the vitamin D-endocrine system in blacks. J Clin Invest 76(2):470–473

81. Parikh SJ et al (2004) The relationship between obesity and serum 1,25-dihydroxy vitamin D concentrations in healthy adults. J Clin Endocrinol Metab 89(3):1196–1199

82. Wortsman J et al (2000) Decreased bioavailability of vitamin D in obesity. Am J Clin Nutr 72(3):690–693

83. Pittas AG et al (2006) Vitamin D and calcium intake in relation to type 2 diabetes in women. Diabetes Care 29(3):650–656

84. Mattila C et al (2007) Serum 25-hydroxyvitamin D concentration and subsequent risk of type 2 diabetes. Diabetes Care 30(10):2569–2570

85. Knekt P et al (2008) Serum Vitamin D and subsequent occurrence of type 2 diabetes. Epidemiology 19:666–671

86. van Dam RM et al (2006) Dietary calcium and magnesium, major food sources, and risk of type 2 diabetes in US black women. Diabetes Care 29(10):2238–2243

87. Pereira MA et al (2002) Dairy consumption, obesity, and the insulin resistance syndrome in young adults: the CARDIA Study. JAMA 287(16):2081–2089

88. Fliser D et al (1997) No effect of calcitriol on insulin-mediated glucose uptake in healthy subjects. Eur J Clin Invest 27(7):629–633

89. Inomata S et al (1986) Effect of 1 alpha (OH)-vitamin D3 on insulin secretion in diabetes mellitus. Bone Miner 1(3):187–192

90. Ljunghall S et al (1987) Treatment with one-alpha-hydroxycholecalciferol in middle-aged men with impaired glucose tolerance – a prospective randomized double-blind study. Acta Med Scand 222(4):361–367

91. Nilas L, Christiansen C (1984) Treatment with vitamin D or its analogues does not change body weight or blood glucose level in postmenopausal women. Int J Obes 8(5):407–411

92. de Boer IH et al (2008) Calcium plus vitamin D supplementation and the risk of incident diabetes in the Women's Health Initiative. Diabetes Care 31(4):701–707

93. Jacobs DR Jr, Steffen LM (2003) Nutrients, foods, and dietary patterns as exposures in research: a framework for food synergy. Am J Clin Nutr 78(3 Suppl):508S–513S

94. Food and Nutrient Board I.O.M. (2003) Dietary reference intakes for calcium, phosphorus, magnesium, vitamin D and fluoride. National Academy Press, Washington, DC

95. Bischoff-Ferrari HA et al (2006) Estimation of optimal serum concentrations of 25-hydroxyvitamin D for multiple health outcomes. Am J Clin Nutr 84(1):18–28

96. Hollis BW (2005) Circulating 25-hydroxyvitamin D levels indicative of vitamin D sufficiency: implications for establishing a new effective dietary intake recommendation for vitamin D. J Nutr 135(2):317–322

50 Role of Vitamin D for Cardiovascular Health

Robert Scragg

Abstract   The prevailing medical opinion up to the 1960s held that vitamin D was a risk factor for cardiovascular disease, arising from research showing that megadoses of vitamin D caused arteriosclerosis in animals. However, the development of competitive protein-binding assays for 25-hydroxyvitamin D [25(OH)D] in the 1970s showed that only a small proportion of body vitamin D comes from diet, with more than 80% synthesized from sun exposure. A 1981 review of the epidemiology, which found that cardiovascular disease rates increased in winter with increasing latitude and were reduced at high altitudes, led to the hypothesis that ultraviolet radiation, by increasing vitamin D levels, protects against cardiovascular disease. Since then, observational epidemiological studies have reported inverse associations between serum 25-hydroxyvitamin D levels and risk of cardiovascular disease. However, randomized clinical trials are required to confirm that vitamin D protects against cardiovascular disease. Several mechanisms have been proposed for the putative protective effect of vitamin D against cardiovascular disease. These include beneficial changes in cardiac function, blood pressure (including endothelial function), insulin resistance, and inflammatory processes.

   Key Words: Blood pressure; cardiac function; cardiovascular disease; coronary heart disease; immune function; seasons; stroke; vitamin D; ultraviolet radiation

1. HISTORICAL REVIEW

1.1. Adverse Cardiovascular Effects from Very High Intake of vitamin D

   Shortly after the discovery of vitamin D in the 1920s by Mellanby in England and McCollum's group in Baltimore *(1)*, experimental studies in a range of animals found that extremely high doses of vitamin D – up to 10 mg (400,000 IU)/day – resulted in widespread calcification of arteries, heart, kidneys, and other organs *(2)*. Cases of infantile hypercalcemia and congenital supravalvular aortic stenosis were attributed to a susceptibility to excess maternal vitamin D intake during pregnancy *(2, 3)*; although some US physicians reported neither of the above conditions in the offspring of women who took large doses of vitamin D during pregnancy *(4, 5)*. Reports by the American Academy of Pediatrics were more cautious in linking vitamin D with these diseases in 1963 *(6)* and in 1967 concluded the hypothesis that vitamin D caused infantile hypercalcemia was unproven *(7)*.

From: *Nutrition and Health: Vitamin D*
Edited by: M.F. Holick, DOI 10.1007/978-1-60327-303-9_50,
© Springer Science+Business Media, LLC 2010

The development of animal models of arteriosclerosis caused by hypervitaminosis D, studies in which megadoses of 5,000–10,000 IU/kg/day were given *(8, 9)*– equivalent to daily doses of 350,000–700,000 IU for a 70 kg adult human – supported medical opinion in the 1960s/1970s that increased intake of vitamin D was a risk factor for vascular damage and cardiovascular disease *(10–12)*. Of interest, there were isolated reports that vitamin Dcould prevent myocardial calcification in rats *(13)* and assist with the treatment of vascular calcification, heart failure, and cardiac arrhythmias in humans *(14–17)*, while exposure to ultraviolet radiation (wavelength not specified) could attenuate the adverse effects of a high cholesterol diet on the arteries of rats *(18)*.

1.2. *Early Epidemiological Studies*

The prevailing opinion in the early 1970s that vitamin D was a cause of cardiovascular disease influenced early epidemiological studies. An ecological study of regions within England and Wales showed positive correlations between vitamin D intake measured in national surveys and standardized mortality ratios for ischemic heart disease ($r = 0.58$) and cerebrovascular disease ($r = 0.49$) *(19)*. A case–control study from Tromso (Norway) reported that myocardial infarction cases had significantly higher mean dietary intake of vitamin D compared to age- and sex-matched controls *(20)*. The Tromso study also reported a significant positive association ($p = 0.0013$) between dietary vitamin D intake and serum cholesterol in a random cross-sectional sample of men aged 20–50 years *(21)*.

The development in the 1970s of competitive protein-binding assays for 25-hydroxyvitamin D [25(OH)D] *(22)*, the main marker of vitamin D status, revealed that diet contributed only a small proportion of vitamin D, with more than 80% synthesized from sun exposure *(23, 24)*. This new method of measuring vitamin D status showed in studies from Germany and Denmark, surprisingly at the time, that cases of coronary heart disease had similar, not elevated, levels of serum 25(OH)D compared with controls *(25, 26)*. Importantly, the Tromso Heart Study, which had previously reported higher vitamin D intakes in cases *(20)* also found in a nested case–control comparison that myocardial infarction cases had slightly lower serum 25(OH)D levels compared with controls matched for age and time of year, after correcting for vitamin D-binding protein ($p = 0.024$) *(27)*. Further, two of these studies found no association between serum 25(OH)D and serum cholesterol *(26, 27)*, a result that has been confirmed by subsequent research *(28–31)*.

2. ECOLOGICAL STUDIES

2.1. *Hypothesis That Sunlight and vitamin D Protect Against Cardiovascular Disease*

Drawing on ecological studies of variations in cardiovascular disease by season, latitude, and altitude, in 1981 the author published a hypothesis that sunlight and vitamin D may protect against cardiovascular disease *(32)*. This paper, which was subsequently expanded *(33)*, proposed that the winter excess in cardiovascular mortality; the positive relationship between latitude and cardiovascular mortality; the inverse association between altitude and cardiovascular mortality; plus the increased cardiovascular rates in older people and those with increased skin pigmentation, such as African Americans, who have lower body levels of vitamin D due to decreased skin synthesis *(34, 35)* were all consistent with an inverse association between vitamin D status and cardiovascular disease.

Most vitamin D in humans is synthesized in the skin through exposure to the ultraviolet (UV) B radiation (wavelengths 280–315 nm) *(23, 24, 36)*. The intensity of UV radiation on the surface of the earth varies with season being lowest in winter, decreases with increasing latitude from the equator, and increases with altitude by about 15% per 1,000 m *(33)*. A substantial body of research on the seasonality of cardiovascular disease has been published since the earlier review *(33)* which is summarized below.

2.2. Winter Excess in Cardiovascular Disease

Studies in both the northern and southern hemispheres have shown consistent seasonal variations in vitamin D status, with low levels in winter. For example, the Third National Health and Nutrition Examination Survey (NHANES III), a representative sample of the US civilian population carried out during 1988–1994, shows a pronounced seasonal variation in serum 25(OH)D levels, varying from a peak of 82 nmol/l (33 ng/ml) during July–August down to 68 nmol/l (27 ng/ml) from January to April *(37)*. The actual seasonal variation in serum 25(OH)D among the US population is probably greater than this since NHANES III, for practical reasons, surveyed the south in winter and the north in summer; with the consequence that people with very low winter levels in the north during winter, and very high summer levels in the south, were not included. A national survey in New Zealand showed a similar seasonal variation – varying from 36 to 45 nmol/l (14–18 ng/ml) in winter and spring, up to 67–70 nmol/l (27–28 ng/ml) in summer *(38)*.

The following studies have reported a winter excess in cardiovascular disease. Data from the first and second National Register of Myocardial Infarction, which recruited 83,541 acute myocardial infarction (MI) cases from throughout the United States during 1990–1994 *(39)* and 259,891 cases during 1994–1996 *(40)*, showed 10 and 53% more MI cases, respectively, in winter than in summer for all geographic regions. Mortality from coronary heart disease and stroke was 18.6 and 19.9%, respectively, higher in January than in September in Canada during 1980–1982 and 1990–1992 *(41)*. A seasonal variation in the incidence of ischemic stroke (but not hemorrhagic), with a peak in mid-May, has been observed in US Veterans *(42)*. Data from an international collaboration of neurosurgical centers in 14 countries showed a seasonal variation in the incidence of subarachnoid hemorrhage, with a peak in February in the Northern Hemisphere *(43)*.

Other components of the cardiovascular disease spectrum, besides coronary artery disease, exhibit a winter excess in disease events. The incidence of emergency department visits for congestive heart failure at seven New Jersey hospitals during 1988–1998 was 14.3% higher in December, and 15.5% lower in August, compared with the average of all months *(44)*. Hospital admissions for heart failure in Scotland during 1990–1996 were 14% more common in women, and 16% more common in men, in December compared with July, while mortality from heart failure was 21 and 25% higher in December than in July for women and men, respectively *(45)*. In this Scottish study, approximately one-fifth of the winter excess in admissions for heart failure was attributable to respiratory disease including pneumonia *(45)*. Total mortality was significantly increased during winter among patients with ventricular tachyarrhythmias in a US registry *(46)*. The risk of endocarditis in US patients referred for echocardiography was twice as high in fall/winter (6.4%) than spring/summer (3.0%, $p = 0.004$) *(47)*. US health statistics data show mortality from pulmonary embolism is 13% higher during January–March compared with April–September *(48)*.

The winter excess in CV disease is attributed frequently to the cold temperatures of winter *(49)*. However, temperature may be a proxy measure for UV radiation since both are highly correlated (e.g., $r = 0.94$ in Queensland, Australia *(50)*). Of particular importance, the winter excess in cardiovascular disease has been observed in warm climates, with 33% more coronary artery deaths during winter than summer in Los Angeles *(51)* and 22% more during March than August in Hawaii, despite a small seasonal variation in temperature between 22.8 and 27.8°C *(52)*, in addition to the winter excess observed in tropical northern Queensland at latitudes 17–19°S *(33)*. It does not seem plausible that the mild-winter temperatures experienced in the above locations are a major factor in the winter excess from CV disease.

3. RECENT EPIDEMIOLOGICAL STUDIES

3.1. *Observational Studies*

Since the publication of the vitamin D hypothesis *(32)*, a range of epidemiological study designs have been used to determine if vitamin D is associated with reduced risk of cardiovascular disease. A population-based case–control study from New Zealand, restricted to incident cases providing blood samples within 12 h of onset of symptoms, observed an inverse association between plasma 25(OH)D and risk of myocardial infarction, with the odds ratio for those in the highest 25(OH)D quartile being 0.30 (95% CI: 0.15, 0.61) compared with the lowest quartile *(53)*. In contrast, a hospital-based case–control study from India, which recruited prevalent cases of coronary artery disease whose vitamin D status may not have reflected that at the time of disease onset, reported a significantly ($p < 0.001$) higher proportion of cases (59.4%) than controls (22.1%) had serum 25(OH)D levels above 222.5 nmol/l (89 ng/ml) *(54)*. A case–control study from Cambridge (UK) found that mean Z score of 25(OH)D for incident stroke cases measured within 30 days of disease onset was significantly below that expected for a sample of healthy controls (–1.4, 95% CI: –1.7, –1.1; $p < 0.0001$) *(55)*.

Recent cohort studies have reported inverse associations between vitamin D and risk of cardiovascular disease. Participants in the Framingham Offspring Study cohort with baseline serum 25(OH)D levels < 10 ng/ml (25 nmol/l) had a hazard ratio of 1.80 (95% CI: 1.05, 3.08) for cardiovascular disease during the 5-year follow-up period, compared with those > 15 ng/ml (37.5 nmol/l), adjusting for major cardiovascular risk factors including age, sex, systolic blood pressure, antihypertensive treatment, diabetes, total-to-high lipoprotein ratio, cigarette smoking, and body mass index, while further adjustment for C-reactive protein, physical activity, and vitamin use did not affect the findings *(56)*. The effect was evident in participants with hypertension but not in normotensive people, suggesting that hypertension could magnify the adverse effects of vitamin D on the cardiovascular system.

Other cohort studies of healthy populations include the US Health Professionals Follow-up Study which found, after 10 years follow-up, that men with baseline plasma 25(OH)D levels ≤ 15 ng/ml (37.4 nmol/l) had an adjusted relative risk of 2.09 (95% CI: 1.24, 3.54) for myocardial infarction (fatal plus non-fatal) compared to those with 25(OH)D > 30 ng/ml (74.9 nmol/l) *(57)* and follow-up of the NHANES III cohort which found that participants in the lowest quartile of baseline serum 25(OH)D < 17.8 ng/ml (44.4 nmol/l) had a 26% (95% CI: 8, 46) increased risk of all-cause mortality during a median 8.7 years follow-up, compared to the highest 25(OH)D quartile *(58)*.

The remaining cohort studies are of mostly hemodialysis patients, which have all observed decreased cardiovascular mortality in those supplemented with active vitamin D (vitamin D receptor agonists). A study of Japanese dialysis patients reported an adjusted hazard ratio = 0.38 (95% CI: 0.25, 0.58) of cardiovascular mortality over 5 years for users of 1α-calcidol compared with non-users *(59)*. A study of 51,000 US hemodialysis patients found a cardiovascular disease incidence rate of 7.6 per 100 person-years in the vitamin D-treated group compared with 14.6 per 100 person-years in the non-vitamin D group ($p < 0.001$) *(60)*. Recently, a cohort study of incident hemodialysis patients, using the nested case–control design, observed increased cardiovascular mortality after 90 days follow-up in those with low baseline 25(OH)D levels in patients not on vitamin D therapy, while no association with baseline 25(OH)D was observed in those on vitamin D *(61)*. Lastly, a German study of 3,258 patients referred for coronary angiography found that those with baseline serum 25(OH)D levels in the bottom quartile had a significantly increased adjusted risk of all-cause mortality (hazard ratio = 2.08; 95% CI: 1.60, 2.70) and cardiovascular mortality (hazard ratio = 2.22; 95% CI: 1.57, 3.13) compared with those in the highest baseline 25(OH)D quartile, after a median follow-up period of 7.7 years *(62)*. This study also reported that baseline serum 1,25(OH)$_2$D was inversely associated with follow-up risk of all-cause and cardiovascular mortality, independent of 25(OH)D.

3.2. Experimental Studies

The only clinical trial to date of vitamin D in the general population, the Women's Health Initiative trial, failed to detect any effect of vitamin D and calcium supplementation on cardiovascular mortality and morbidity *(63)*. In this study, 36,282 postmenopausal women aged 50–79 years at 40 clinical sites in the United States were randomized to take calcium carbonate 500 mg with vitamin D 200 IU twice daily or placebo. After 7 years of follow-up, the adjusted hazard ratios in the treated group versus control were 1.04 (95% CI: 0.92–1.18) for coronary heart disease and 0.95 (95% CI: 0.82–1.10) for stroke. However, this study is not a proper test of the hypothesis that vitamin D protects against cardiovascular disease since the dose of vitamin D was only 400 IU/day, the control group was able to continue taking vitamin D supplements so that contamination occurred, and compliance was low as only 59% of participants took ≥80% of the study medication *(63, 64)*.

A recent meta-analysis of 18 randomized clinical trials (RCTs) published from 1992 to 2006, which included data from the Women's Health Initiative trial *(65)*, found that vitamin D supplementation reduces total mortality by 7% *(66)*. This further strengthens the evidence that vitamin D protects against cardiovascular disease since the latter is the main cause of death in developed countries. The RCTs were mainly carried out in Europe ($n = 15$) reflecting the greater concern there about vitamin D deficiency, with two studies from Australia and New Zealand and one – the Women's Health Initiative (WHI) – although the largest, from the United States *(65)*. The results of this meta-analysis are consistent with cohort studies showing that dialysis patients prescribed active vitamin D have lower all-cause mortality than patients not on vitamin D *(60, 61, 67–72)*.

The weighted vitamin D dose of 528 IU/day for all studies in the meta-analysis is likely to have only increased blood 25(OH)D levels by 10–15 nmol/l (4–6 ng/ml), based on a carefully controlled study which compared change in 25(OH)D with dose of vitamin D *(73)*. This is much lower than the daily dose of 1,700 IU currently recommended by international experts to maintain serum 25(OH)D above 75–80 nmol/l (30–32 ng/ml) at levels required for optimum health *(74)*. Several reports from NHANES III have shown that risk measures for cardiovascular, bone, and lung disease are lowest in people with 25(OH)D levels above 75–80 nmol/l (30–32 ng/ml) *(37, 75–77)*. Thus, the potential beneficial effect of vitamin D supplementation on all-cause mortality may be higher than 7% if vitamin D doses are given which increase blood 25(OH)D levels up to these levels. Most of the prevented deaths in the treated group are likely to have been from cardiovascular and infectious diseases, since the weighted mean follow-up period was 5.7 years *(66)*, too short to detect any benefit in preventing cancer deaths *(78)*.

4. POSSIBLE CARDIO-PREVENTIVE MECHANISMS OF VITAMIN D

Several mechanisms have been proposed for the putative protective effect of vitamin D against cardiovascular disease. These mechanisms, reviewed below, involve beneficial changes in cardiac function, blood pressure (including endothelial function), insulin resistance, and inflammatory processes (Fig. 1).

4.1. Cardiac Function

There is accumulating evidence that low vitamin D status adversely affects cardiac function. A receptor to the active metabolite 1,25-dihyroxyvitamin D_3 has been identified in the rat heart *(79)*. Vitamin D deficiency results in increased cardiac contractility, hypertrophy, and fibrosis in rats *(80–82)*. Matrix metalloproteinases (MMPs) may be involved in the pathophysiology arising from vitamin D deficiency *(83, 84)*, since vitamin D supplementation lowers blood MMP-9 and MMP-2 *(85)*, while plasma levels of MMP-9 were raised in men from the Framingham Study who had increased left ventricular end-diastolic dimensions and wall thickness *(86)*; and left ventricular hypertrophy increases the risk of cardiovascular disease mortality and morbidity *(87)*.

Cardiomegaly, reversed by calcium and vitamin D supplementation, has been described in children with rickets *(88–91)* and in an adult with congestive heart failure *(92)*. Observational studies of patients with congestive heart failure, who typically have

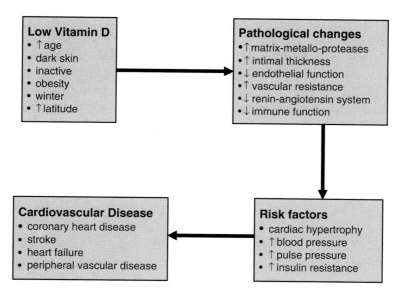

Fig. 1. Mechanisms by which low vitamin D status may increase the risk of cardiovascular disease.

enlarged hearts, have shown they have low serum 25(OH)D levels *(93, 94)*, while low serum 1,25-dihydroxyvitamin D (but not 25(OH)D) predicted increased risk of death or need for heart transplant in patients with end-stage congestive heart failure *(95)*. In contrast, a recent German RCT of 93 patients with congestive heart failure failed to show an effect of vitamin D supplementation on measures of cardiac function with echocardiography *(96)*. RCTs have shown that vitamin D supplementation reduces radial pulse rates, an albeit crude measure of cardiac function *(30, 97)*, although of potential importance since heart rate is related with increased coronary heart disease and mortality *(98)*.

4.2. Blood Pressure

A receptor to 1,25-dihydroxyvitamin D has been described in smooth muscle tissue, supporting a potential role for vitamin D in the regulation of smooth muscle contraction and blood pressure *(99)*.

Observational studies of dietary vitamin D have produced conflicting results. While an earlier study in Iowa women reported an inverse association between oral vitamin D intake and systolic blood pressure *(100)*, results recently published from three large US cohorts did not show an association between dietary vitamin D and incident hypertension *(101)*. A likely explanation of the failure to find an association in these cohort studies is that dietary sources of vitamin D contribute only a small proportion of the total vitamin D entering the body each day, which is mainly derived from sun exposure *(24)*. When two of these cohort studies were re-analyzed using plasma 25(OH)D, which measures vitamin D from all sources, both measured 25(OH)D and estimated 25(OH)D were inversely associated with risk of incident hypertension in both men and women *(102)*. This finding is supported by a recent publication from NHANES III which found that serum 25(OH)D was inversely associated with both systolic blood pressure and pulse pressure *(37)*.

In experimental studies, supplementation with a vitamin D analogue (α-calcidol) has been shown to lower blood pressure in vitamin D-replete patients with hypercalcemia or impaired glucose tolerance in Sweden *(103, 104)*, while in German studies, vitamin D supplements reduced both systolic blood pressure and pulse rate *(97)* and exposure to ultraviolet (UV) B radiation, which increases vitamin D, lowered blood pressure by 6 mmHg in the treated group compared with the control ($p < 0.05$) *(105)*. In contrast, supplementation with vitamin D_3 did not lower blood pressure in a small sample ($n = 65$) of normotensive white men in Oregon *(106)* nor in elderly men and women in England *(30)*.

Several mechanisms have been proposed for the antihypertensive effect of vitamin D. vitamin D may lower blood pressure by direct suppression of the renin–angiotensin system *(107, 108)* or through increased intimal thickening of blood vessels caused by matrix metalloproteinases (MMPs) *(109)*, which are suppressed by vitamin D *(85)* and which may decrease arterial compliance (i.e., increase arterial stiffness) *(37)*. Arterial stiffness is caused by impaired endothelial function arising from reduced nitric oxide synthesis by the endothelium *(110)* and is an independent predictor of cardiovascular disease *(111)*.

Several studies have reported significant associations between vitamin D and arterial function. Serum 25(OH)D levels were found in an earlier study of hypertension patients, after 3 min of arterial occlusion of the calf, to be associated positively with blood flow ($r = 0.72$) and negatively with vascular resistance ($r = -0.78$) *(112)*. In patients with end-stage renal disease, serum 25(OH)D was correlated positively with brachial artery distensibility and flow-mediated dilatation, after adjustment for age and blood pressure *(113)*. A significant inverse association between serum 25(OH)D and carotid artery intimal medial thickening has been observed in type 2 diabetes patients *(114)*. Recently, vitamin D supplementation has been shown to increase flow-mediated brachial artery dilatation in type 2 diabetes patients with low vitamin D levels (25(OH)D < 50 nmol/l or 20 ng/ml) *(115)*. These studies are consistent with a Japanese report of an inverse association between serum 25(OH)D and microvascular complications in patients with type 2 diabetes *(116)* and with recent analyses of NHANES data (for 2001–2004) showing that the prevalence ratio of peripheral arterial disease increased by 1.35 (95% CI: 1.15, 1.59) for each 20 ng/ml (25 nmol/l) decrease in serum 25(OH)D *(117)*. The above studies showing inverse associations between serum 25(OH)D and flow-mediated dilatation suggest that vitamin D may improve nitric oxide-mediated endothelial function to reduce risk of coronary heart disease *(118, 119)*.

4.3. Inflammatory Factors

It is now well established that inflammatory factors are centrally involved in the process of atherosclerosis and plaque rupture *(120, 121)*. Blood levels of inflammatory markers, such as C-reactive protein (CRP) and the cytokine interleukin-6 (IL-6), predict subsequent risk of cardiovascular disease *(122, 123)*. Positive associations have been reported between IL-6 and insulin resistance *(124–126)*. The latter is a risk factor for type 2 diabetes, which is itself inversely related to vitamin D status *(127)*.

There is increasing laboratory research showing that vitamin D influences immune response. In vitro studies show that 1,25-dihydroxyvitamin D_3 decreases production of pro-inflammatory cytokines such as IL-6 and tumor necrosis factor α (TNFα) by macrophages and lymphocytes *(128–130)* and up-regulates synthesis of IL-10 *(131)*. However, there is conflicting clinical information of the effect of vitamin D supplementation on immune response in humans. Oral or intervenous supplementation with the active metabolite 1,25-dihydroxyvitamin D_3 decreased blood levels of IL-1 and IL-6 in hemodialysis patients *(124)*. An RCT of German patients with congestive heart failure (reported above) found that vitamin D (2,000 IU/day for 9 months) decreased TNFα and increased IL-10, with no effect on CRP *(96)*. In contrast, studies which gave lower doses of vitamin D did not find any effect from it on cytokines. A study of 47 healthy postmenopausal Lebanese women, which had no control group and gave only 800 IU/day of vitamin D_3 for 12 weeks, found no change in IL-6 and TNFα *(132)*. The likely reason for this is the small daily dose of vitamin D which only increased serum 25(OH)D levels from 10.57 ng/ml (26.4 nmol/l) to 25.84 ng/ml (64.5 nmol/l), well below the optimum levels above 30–40 ng/ml (75–100 nmol/l). A US study found that vitamin D supplementation over 3 years did not change serum levels of CRP and IL-6, again possibly due to the low dose of 700 IU/day given to women who were relatively vitamin D replete *(133)*.

Recently, vitamin D has been shown to have an important role in the innate immune system by stimulating the synthesis of the antimicrobial peptide cathelicidin *(134)*. This new evidence provides an explanation for the historical link between sun exposure, vitamin D, and tuberculosis (TB) which has recently been reviewed *(135)*. The association between rickets and infection has been known since the 1960s *(136)*; low vitamin D status also has been linked to respiratory infection in people without rickets *(137, 138)*, while a clinical trial to prevent bone loss found that women who received vitamin D supplements reported fewer respiratory symptoms than the placebo group *(139)*.

4.4. Summary of Possible Mechanisms

Vitamin D may protect against cardiovascular disease through a number of mechanisms. First, these may involve remodeling of connective tissue resulting in hypertrophy of cardiac and smooth muscle caused by elevated levels of matrix metalloproteinases (e.g., MMP-9) *(83, 84)*. Second, nitric oxide-mediated endothelial dysfunction, which has an etiological role in coronary heart disease *(118, 119)*, may also be involved, resulting in increased arterial stiffness and pulse pressure. Lastly, low vitamin D levels may alter both the innate and acquired immune systems to increase the inflammatory process involved with atherosclerosis, endothelial dysfunction, and insulin resistance to increase risk of cardiovascular disease and associated components such as hypertension and insulin resistance *(120, 140, 141)*.

REFERENCES

1. Chick H (1975) The discovery of vitamins. Prog Food Nutr Sci 1:1–20
2. Seelig MS (1969) vitamin D and cardiovascular, renal, and brain damage in infancy and childhood. Ann NY Acad Sci 147:539–582

3. (1966) Congenital supravalvular aortic stenosis, idiopathic hypercalcemia, and vitamin D. Nutr Rev 24:311–313
4. Forbes GB, Cafarelli C, Manning J (1968) vitamin D and infantile hypercalcemia. Pediatrics 42: 203–204
5. Goodenday LS, Gordon GS (1971) No risk from vitamin D in pregnancy. Ann Intern Med 75: 807–808
6. American Academy of Pediatrics, Committee on Nutrition (1963) The prophylactic requirement and toxicity of vitamin D. Pediatrics 31:512–525
7. Fraser D (1967) The relation between infantile hypercalcemia and vitamin D – public health implications in North America. Pediatrics 40:1050–1061
8. Eisenstein R, Zeruolis L (1964) vitamin D-induced aortic calcification. Arch Pathol 77:27–35
9. Schenk EA, Penn I, Schwartz S (1965) Experimental atherosclerosis in the dog: a morphologic evaluation. Arch Pathol 80:102–109
10. Taussig HB (1966) Possible injury to the cardiovascular system from vitamin D. Ann Intern Med 65:1195–1200
11. Taylor CB, Hass GM, Ho KJ, Liu LB (1972) Risk factors in the pathogenesis of atherosclerotic heart disease and generalized atherosclerosis. Ann Clin Lab Sci 2:239–243
12. Kummerow FA (1979) Nutrition imbalance and angiotoxins as dietary risk factors in coronary heart disease. Am J Clin Nutr 32:58–83
13. Bajusz E, Jasmin G (1963) Action of parathyroidectomy and dihydrotachysterol on myocardial calcification due to coronary vein obstruction. Proc Soc Exp Biol Med 112:752–755
14. Verberckmoes R, Bouillon R, Krempien B (1975) Disappearance of vascular calcifications during treatment of renal osteodystrophy. Two patients treated with high doses of vitamin D and aluminum hydroxide. Ann Intern Med 82:529–533
15. Aryanpur I, Farhoudi A, Zangeneh F (1974) Congestive heart failure secondary to idiopathic hypoparathyroidism. Am J Dis Child 127:738–739
16. Falko JM, Bush CA, Tzagournis M, Thomas FB (1976) Case report. Congestive heart failure complicating the hungry bone syndrome. Am J Med Sci 271:85–89
17. Johnson JD, Jennings R (1968) Hypocalcemia and cardiac arrhythmias. Am J Dis Child 115:373–376
18. Altschul R (1953) Inhibition of experimental cholesterol arteriosclerosis by ultraviolet irradiation. N Engl J Med 249:96–99
19. Knox EG (1973) Ischaemic-heart-disease mortality and dietary intake of calcium. Lancet 1: 1465–1467
20. Linden V (1974) vitamin D and myocardial infarction. Br Med J 3:647–650
21. Linden V (1975) vitamin D and serum cholesterol. Scand J Soc Med 3:83–85
22. Hollis BW, Horst RL (2007) The assessment of circulating 25(OH)D and 1,25(OH)2D: where we are and where we are going. J Steroid Biochem Mol Biol 103:473–476
23. Haddad JG Jr, Hahn TJ (1973) Natural and synthetic sources of circulating 25-hydroxyvitamin D in man. Nature 244:515–517
24. Poskitt EM, Cole TJ, Lawson DE (1979) Diet, sunlight, and 25-hydroxy vitamin D in healthy children and adults. Br Med J 1:221–223
25. Schmidt-Gayk H, Goossen J, Lendle F, Seidel D (1977) Serum 25-hydroxycalciferol in myocardial infarction. Atherosclerosis 26:55–58
26. Lund B, Badskjaer J, Lund B, Soerensen OH (1978) vitamin D and ischaemic heart disease. Horm Metab Res 10:553–556
27. Vik B, Try K, Thelle DS, Forde OH (1979) Tromso Heart Study: vitamin D metabolism and myocardial infarction. Br Med J 2:176
28. Scragg R, Holdaway I, Jackson R, Lim T (1992) Plasma 25-hydroxyvitamin D3 and its relation to physical activity and other heart disease risk factors in the general population. Ann Epidemiol 2: 697–703
29. Scragg R, Holdaway I, Singh V, Metcalf P, Baker J, Dryson E (1995) Serum 25-hydroxyvitamin D3 is related to physical activity and ethnicity but not obesity in a multicultural workforce. Aust N Z J Med 25:218–223

30. Scragg R, Khaw KT, Murphy S (1995) Effect of winter oral vitamin D3 supplementation on cardio-vascular risk factors in elderly adults. Eur J Clin Nutr 49:640–646
31. Martins D, Wolf M, Pan D, Zadshir A, Tareen N, Thadhani R, Felsenfeld A, Levine B, Mehrotra R, Norris K (2007) Prevalence of cardiovascular risk factors and the serum levels of 25-hydroxyvitamin D in the United States: data from the Third National Health and Nutrition Examination Survey. Arch Intern Med 167:1159–1165
32. Scragg R (1981) Seasonality of cardiovascular disease mortality and the possible protective effect of ultra-violet radiation. Int J Epidemiol 10:337–341
33. Scragg R (1995) Sunlight, vitamin D and cardiovascular disease. In: Crass MF, Avioloi LV (eds) Calcium-regulating hormones and cardiovascular function. CRC Press, Boca Raton, pp 213–237
34. MacLaughlin J, Holick MF (1985) Aging decreases the capacity of human skin to produce vitamin D3. J Clin Invest 76:1536–1538
35. Clemens TL, Adams JS, Henderson SL, Holick MF (1982) Increased skin pigment reduces the capacity of skin to synthesise vitamin D3. Lancet 1:74–76
36. Holick MF, MacLaughlin JA, Doppelt SH (1981) Regulation of cutaneous previtamin D3 photosynthesis in man: skin pigment is not an essential regulator. Science 211:590–593
37. Scragg R, Sowers M, Bell C (2007) Serum 25-hydroxyvitamin D, ethnicity, and blood pressure in the Third National Health and Nutrition Examination Survey. Am J Hypertens 20:713–719
38. Rockell JE, Skeaff CM, Williams SM, Green TJ (2006) Serum 25-hydroxyvitamin D concentrations of New Zealanders aged 15 years and older. Osteoporos Int 17:1382–1389
39. Ornato JP, Peberdy MA, Chandra NC, Bush DE (1996) Seasonal pattern of acute myocardial infarction in the National Registry of Myocardial Infarction. J Am Coll Cardiol 28:1684–1688
40. Spencer FA, Goldberg RJ, Becker RC, Gore JM (1998) Seasonal distribution of acute myocardial infarction in the second National Registry of Myocardial Infarction. J Am Coll Cardiol 31:1226–1233
41. Sheth T, Nair C, Muller J, Yusuf S (1999) Increased winter mortality from acute myocardial infarction and stroke: the effect of age. J Am Coll Cardiol 33:1916–1919
42. Oberg AL, Ferguson JA, McIntyre LM, Horner RD (2000) Incidence of stroke and season of the year: evidence of an association. Am J Epidemiol 152:558–564
43. Vinall PE, Maislin G, Michele JJ, Deitch C, Simeone FA (1994) Seasonal and latitudinal occurrence of cerebral vasospasm and subarachnoid hemorrhage in the northern hemisphere. Epidemiology 5:302–308
44. Allegra JR, Cochrane DG, Biglow R (2001) Monthly, weekly, and daily patterns in the incidence of congestive heart failure. Acad Emerg Med 8:682–685
45. Stewart S, McIntyre K, Capewell S, McMurray JJ (2002) Heart failure in a cold climate. Seasonal variation in heart failure-related morbidity and mortality. J Am Coll Cardiol 39:760–766
46. Page RL, Zipes DP, Powell JL, Luceri RM, Gold MR, Peters R, Russo AM, Bigger JT Jr, Sung RJ, McBurnie MA (2004) Seasonal variation of mortality in the Antiarrhythmics Versus Implantable Defibrillators (AVID) study registry. Heart Rhythm 1:435–440
47. Finkelhor RS, Cater G, Qureshi A, Einstadter D, Hecker MT, Bosich G (2005) Seasonal diagnosis of echocardiographically demonstrated endocarditis. Chest 128:2588–2592
48. Stein PD, Kayali F, Beemath A, Skaf E, Alnas M, Alesh I, Olson RE (2005) Mortality from acute pulmonary embolism according to season. Chest 128:3156–3158
49. Kvaloy JT, Skogvoll E (2007) Modelling seasonal and weather dependency of cardiac arrests using the covariate order method. Stat Med 26:3315–3329
50. Scragg R (1982) Seasonal variation of mortality in Queensland. Community Health Stud 6:120–129
51. Kloner RA, Poole WK, Perritt RL (1999) When throughout the year is coronary death most likely to occur? A 12-year population-based analysis of more than 220,000 cases. Circulation 100:1630–1634
52. Seto TB, Mittleman MA, Davis RB, Taira DA, Kawachi I (1998) Seasonal variation in coronary artery disease mortality in Hawaii: observational study. BMJ 316:1946–1947
53. Scragg R, Jackson R, Holdaway IM, Lim T, Beaglehole R (1990) Myocardial infarction is inversely associated with plasma 25-hydroxyvitamin D3 levels: a community-based study. Int J Epidemiol 19:559–563

54. Rajasree S, Rajpal K, Kartha CC, Sarma PS, Kutty VR, Iyer CS, Girija G (2001) Serum 25-hydroxyvitamin D3 levels are elevated in South Indian patients with ischemic heart disease. Eur J Epidemiol 17:567–571

55. Poole KE, Loveridge N, Barker PJ, Halsall DJ, Rose C, Reeve J, Warburton EA (2006) Reduced vitamin D in acute stroke. Stroke 37:243–245

56. Wang TJ, Pencina MJ, Booth SL, Jacques PF, Ingelsson E, Lanier K, Benjamin EJ, D'Agostino RB, Wolf M, Vasan RS (2008) vitamin D deficiency and risk of cardiovascular disease. Circulation 117:503–511

57. Giovannucci E, Liu Y, Hollis BW, Rimm EB (2008) 25-hydroxyvitamin D and risk of myocardial infarction in men: a prospective study. Arch Intern Med 168:1174–1180

58. Melamed ML, Michos ED, Post W, Astor B (2008) 25-hydroxyvitamin D levels and the risk of mortality in the general population. Arch Intern Med 168:1629–1637

59. Shoji T, Shinohara K, Kimoto E, Emoto M, Tahara H, Koyama H, Inaba M, Fukumoto S, Ishimura E, Miki T et al (2004) Lower risk for cardiovascular mortality in oral 1alpha-hydroxy vitamin D3 users in a haemodialysis population. Nephrol Dial Transplant 19:179–184

60. Teng M, Wolf M, Ofsthun MN, Lazarus JM, Hernan MA, Camargo CA Jr, Thadhani R (2005) Activated injectable vitamin D and hemodialysis survival: a historical cohort study. J Am Soc Nephrol 16:1115–1125

61. Wolf M, Shah A, Gutierrez O, Ankers E, Monroy M, Tamez H, Steele D, Chang Y, Camargo CA Jr, Tonelli M et al (2007) vitamin D levels and early mortality among incident hemodialysis patients. Kidney Int 72:1004–1013

62. Dobnig H, Pilz S, Scharnagl H, Renner W, Seelhorst U, Wellnitz B, Kinkeldei J, Boehm BO, Weihrauch G, Maerz W (2008) Independent association of low serum 25-hydroxyvitamin D and 1,25-dihydroxyvitamin D levels with all-cause and cardiovascular mortality. Arch Intern Med 168: 1340–1349

63. Hsia J, Heiss G, Ren H, Allison M, Dolan NC, Greenland P, Heckbert SR, Johnson KC, Manson JE, Sidney S et al (2007) Calcium/vitamin D supplementation and cardiovascular events. Circulation 115:846–854

64. Newmark HL, Heaney RP (2006) Calcium, vitamin D, and risk reduction of colorectal cancer. Nutr Cancer 56:1–2

65. Jackson RD, LaCroix AZ, Gass M, Wallace RB, Robbins J, Lewis CE, Bassford T, Beresford SA, Black HR, Blanchette P et al (2006) Calcium plus vitamin D supplementation and the risk of fractures. N Engl J Med 354:669–683

66. Autier P, Gandini S (2007) vitamin D supplementation and total mortality: a meta-analysis of randomized controlled trials. Arch Intern Med 167:1730–1737

67. Teng M, Wolf M, Lowrie E, Ofsthun N, Lazarus JM, Thadhani R (2003) Survival of patients undergoing hemodialysis with paricalcitol or calcitriol therapy. N Engl J Med 349:446–456

68. Kalantar-Zadeh K, Kuwae N, Regidor DL, Kovesdy CP, Kilpatrick RD, Shinaberger CS, McAllister CJ, Budoff MJ, Salusky IB, Kopple JD (2006) Survival predictability of time-varying indicators of bone disease in maintenance hemodialysis patients. Kidney Int 70:771–780

69. Melamed ML, Eustace JA, Plantinga L, Jaar BG, Fink NE, Coresh J, Klag MJ, Powe NR (2006) Changes in serum calcium, phosphate, and PTH and the risk of death in incident dialysis patients: a longitudinal study. Kidney Int 70:351–357

70. Tentori F, Hunt WC, Stidley CA, Rohrscheib MR, Bedrick EJ, Meyer KB, Johnson HK, Zager PG (2006) Mortality risk among hemodialysis patients receiving different vitamin D analogs. Kidney Int 70:1858–1865

71. Kovesdy CP, Ahmadzadeh S, Anderson JE, Kalantar-Zadeh K (2008) Association of activated vitamin D treatment and mortality in chronic kidney disease. Arch Intern Med 168:397–403

72. Wolf M, Betancourt J, Chang Y, Shah A, Teng M, Tamez H, Gutierrez O, Camargo CA Jr, Melamed M, Norris K et al (2008) Impact of activated vitamin D and race on survival among hemodialysis patients. J Am Soc Nephrol 19:1379–1388

73. Barger-Lux MJ, Heaney RP, Dowell S, Chen TC, Holick MF (1998) vitamin D and its major metabolites: serum levels after graded oral dosing in healthy men. Osteoporos Int 8:222–230

74. Vieth R, Bischoff-Ferrari H, Boucher BJ, Dawson-Hughes B, Garland CF, Heaney RP, Holick MF, Hollis BW, Lamberg-Allardt C, McGrath JJ et al (2007) The urgent need to recommend an intake of vitamin D that is effective. Am J Clin Nutr 85:649–650

75. Bischoff-Ferrari HA, Giovannucci E, Willett WC, Dietrich T, Dawson-Hughes B (2006) Estimation of optimal serum concentrations of 25-hydroxyvitamin D for multiple health outcomes. Am J Clin Nutr 84:18–28

76. Scragg R, Sowers M, Bell C (2004) Serum 25-hydroxyvitamin D, diabetes, and ethnicity in the Third National Health and Nutrition Examination Survey. Diabetes Care 27:2813–2818

77. Black PN, Scragg R (2005) Relationship between serum 25-hydroxyvitamin D and pulmonary function in the third national health and nutrition examination survey. Chest 128:3792–3798

78. Giovannucci E (2007) Can vitamin D reduce total mortality? Arch Intern Med 167:1709–1710

79. Simpson RU (1983) Evidence for a specific 1,25-dihydroxyvitamin D3 receptor in rat heart (abstract). Circulation 68:239

80. Weishaar RE, Simpson RU (1989) The involvement of the endocrine system in regulating cardiovascular function: emphasis on vitamin D3. Endocr Rev 10:351–365

81. Weishaar RE, Kim SN, Saunders DE, Simpson RU (1990) Involvement of vitamin D3 with cardiovascular function. III. Effects on physical and morphological properties. Am J Physiol 258:E134–E142

82. Simpson RU, Hershey SH, Nibbelink KA (2007) Characterization of heart size and blood pressure in the vitamin D receptor knockout mouse. J Steroid Biochem Mol Biol 103:521–524

83. Loftus IM, Thompson MM (2002) The role of matrix metalloproteinases in vascular disease. Vasc Med 7:117–133

84. Perlstein TS, Lee RT (2006) Smoking, metalloproteinases, and vascular disease. Arterioscler Thromb Vasc Biol 26:250–256

85. Timms PM, Mannan N, Hitman GA, Noonan K, Mills PG, Syndercombe-Court D, Aganna E, Price CP, Boucher BJ (2002) Circulating MMP9, vitamin D and variation in the TIMP-1 response with VDR genotype: mechanisms for inflammatory damage in chronic disorders? QJM 95: 787–796

86. Sundstrom J, Evans JC, Benjamin EJ, Levy D, Larson MG, Sawyer DB, Siwik DA, Colucci WS, Sutherland P, Wilson PW et al (2004) Relations of plasma matrix metalloproteinase-9 to clinical cardiovascular risk factors and echocardiographic left ventricular measures: the Framingham Heart Study. Circulation 109:2850–2856

87. Lorell BH, Carabello BA (2000) Left ventricular hypertrophy: pathogenesis, detection, and prognosis. Circulation 102:470–479

88. Gillor A, Groneck P, Kaiser J, Schmitz-Stolbrink A (1989) Congestive heart failure in rickets caused by vitamin D deficiency. Monatsschr Kinderheilkd 137:108–110

89. Memmi I, Brauner R, Sidi D, Sauvion S, Souberbielle JC, Garabedian M (1993) Neonatal cardiac failure secondary to hypocalcemia caused by maternal vitamin D deficiency. Arch Fr Pediatr 50: 787–791

90. Brunvand L, Haga P, Tangsrud SE, Haug E (1995) Congestive heart failure caused by vitamin D deficiency? Acta Paediatr 84:106–108

91. Uysal S, Kalayci AG, Baysal K (1999) Cardiac functions in children with vitamin D deficiency rickets. Pediatr Cardiol 20:283–286

92. Connor TB, Rosen BL, Blaustein MP, Applefeld MM, Doyle LA (1982) Hypocalcemia precipitating congestive heart failure. N Engl J Med 307:869–872

93. Shane E, Mancini D, Aaronson K, Silverberg SJ, Seibel MJ, Addesso V, McMahon DJ (1997) Bone mass, vitamin D deficiency, and hyperparathyroidism in congestive heart failure. Am J Med 103: 197–207

94. Zittermann A, Schleithoff SS, Tenderich G, Berthold HK, Korfer R, Stehle P (2003) Low vitamin D status: a contributing factor in the pathogenesis of congestive heart failure? J Am Coll Cardiol 41:105–112

95. Zittermann A, Schleithoff SS, Gotting C, Dronow O, Fuchs U, Kuhn J, Kleesiek K, Tenderich G, Koerfer R (2008) Poor outcome in end-stage heart failure patients with low circulating calcitriol levels. Eur J Heart Fail 10:321–327

96. Schleithoff SS, Zittermann A, Tenderich G, Berthold HK, Stehle P, Koerfer R (2006) vitamin D supplementation improves cytokine profiles in patients with congestive heart failure: a double-blind, randomized, placebo-controlled trial. Am J Clin Nutr 83:754–759

97. Pfeifer M, Begerow B, Minne HW, Nachtigall D, Hansen C (2001) Effects of a short-term vitamin D(3) and calcium supplementation on blood pressure and parathyroid hormone levels in elderly women. J Clin Endocrinol Metab 86:1633–1637

98. Dyer AR, Persky V, Stamler J, Paul O, Shekelle RB, Berkson DM, Lepper M, Schoenberger JA, Lindberg HA (1980) Heart rate as a prognostic factor for coronary heart disease and mortality: findings in three Chicago epidemiologic studies. Am J Epidemiol 112:736–749

99. Kawashima H (1987) Receptor for 1,25-dihydroxyvitamin D in a vascular smooth muscle cell line derived from rat aorta. Biochem Biophys Res Commun 146:1–6

100. Sowers MR, Wallace RB, Lemke JH (1985) The association of intakes of vitamin D and calcium with blood pressure among women. Am J Clin Nutr 42:135–142

101. Forman JP, Bischoff-Ferrari HA, Willett WC, Stampfer MJ, Curhan GC (2005) vitamin D intake and risk of incident hypertension: results from three large prospective cohort studies. Hypertension 46:676–682

102. Forman JP, Giovannucci E, Holmes MD, Bischoff-Ferrari HA, Tworoger SS, Willett WC, Curhan GC (2007) Plasma 25-hydroxyvitamin D levels and risk of incident hypertension. Hypertension 49:1063–1069

103. Lind L, Wengle B, Ljunghall S (1987) Blood pressure is lowered by vitamin D (alphacalcidol) during long-term treatment of patients with intermittent hypercalcaemia. A double-blind, placebo-controlled study. Acta Med Scand 222:423–427

104. Lind L, Lithell H, Skarfors E, Wide L, Ljunghall S (1988) Reduction of blood pressure by treatment with alphacalcidol. A double-blind, placebo-controlled study in subjects with impaired glucose tolerance. Acta Med Scand 223:211–217

105. Krause R, Buhring M, Hopfenmuller W, Holick MF, Sharma AM (1998) Ultraviolet B and blood pressure. Lancet 352:709–710

106. Orwoll ES, Oviatt S (1990) Relationship of mineral metabolism and long-term calcium and cholecalciferol supplementation to blood pressure in normotensive men. Am J Clin Nutr 52:717–721

107. Resnick LM, Muller FB, Laragh JH (1986) Calcium-regulating hormones in essential hypertension. Relation to plasma renin activity and sodium metabolism. Ann Intern Med 105:649–654

108. Li YC, Kong J, Wei M, Chen ZF, Liu SQ, Cao LP (2002) 1,25-Dihydroxyvitamin D(3) is a negative endocrine regulator of the renin-angiotensin system. J Clin Invest 110:229–238

109. Newby AC (2005) Dual role of matrix metalloproteinases (matrixins) in intimal thickening and atherosclerotic plaque rupture. Physiol Rev 85:1–31

110. Zieman SJ, Melenovsky V, Kass DA (2005) Mechanisms, pathophysiology, and therapy of arterial stiffness. Arterioscler Thromb Vasc Biol 25:932–943

111. Khoshdel AR, Carney SL, Nair BR, Gillies A (2007) Better management of cardiovascular diseases by pulse wave velocity: combining clinical practice with clinical research using evidence-based medicine. Clin Med Res 5:45–52

112. Duprez D, De Buyzere M, De Backer T, Clement D (1993) Relationship between vitamin D and the regional blood flow and vascular resistance in moderate arterial hypertension. J Hypertens Suppl 11:S304–S305

113. London GM, Guerin AP, Verbeke FH, Pannier B, Boutouyrie P, Marchais SJ, Metivier F (2007) Mineral metabolism and arterial functions in end-stage renal disease: potential role of 25-hydroxyvitamin D deficiency. J Am Soc Nephrol 18:613–620

114. Targher G, Bertolini L, Padovani R, Zenari L, Scala L, Cigolini M, Arcaro G (2006) Serum 25-hydroxyvitamin D3 concentrations and carotid artery intima-media thickness among type 2 diabetic patients. Clin Endocrinol (Oxf) 65:593–597

115. Sugden JA, Davies JI, Witham MD, Morris AD, Struthers AD (2008) vitamin D improves endothelial function in patients with Type 2 diabetes mellitus and low vitamin D levels. Diabet Med 25:320–325

116. Suzuki A, Kotake M, Ono Y, Kato T, Oda N, Hayakawa N, Hashimoto S, Itoh M (2006) Hypovitaminosis D in type 2 diabetes mellitus: association with microvascular complications and type of treatment. Endocr J 53:503–510

117. Melamed ML, Muntner P, Michos ED, Uribarri J, Weber C, Sharma J, Raggi P (2008) Serum 25-hydroxyvitamin D levels and the prevalence of peripheral arterial disease. Results from NHANES 2001 to 2004. Arterioscler Thromb Vasc Biol 28:1179–1185

118. Abrams J (1997) Role of endothelial dysfunction in coronary artery disease. Am J Cardiol 79:2–9

119. Monnink SH, Tio RA, van Boven AJ, van Gilst WH, van Veldhuisen DJ (2004) The role of coronary endothelial function testing in patients suspected for angina pectoris. Int J Cardiol 96:123–129

120. Ross R (1999) Atherosclerosis – an inflammatory disease. N Engl J Med 340:115–126

121. Hansson GK, Libby P, Schonbeck U, Yan ZQ (2002) Innate and adaptive immunity in the pathogenesis of atherosclerosis. Circ Res 91:281–291

122. Rao M, Jaber BL, Balakrishnan VS (2006) Inflammatory biomarkers and cardiovascular risk: association or cause and effect? Semin Dial 19:129–135

123. Tousoulis D, Antoniades C, Koumallos N, Stefanadis C (2006) Pro-inflammatory cytokines in acute coronary syndromes: from bench to bedside. Cytokine Growth Factor Rev 17:225–233

124. Turk S, Akbulut M, Yildiz A, Gurbilek M, Gonen S, Tombul Z, Yeksan M (2002) Comparative effect of oral pulse and intravenous calcitriol treatment in hemodialysis patients: the effect on serum IL-1 and IL-6 levels and bone mineral density. Nephron 90:188–194

125. Mohamed-Ali V, Goodrick S, Rawesh A, Katz DR, Miles JM, Yudkin JS, Klein S, Coppack SW (1997) Subcutaneous adipose tissue releases interleukin-6, but not tumor necrosis factor-alpha, in vivo. J Clin Endocrinol Metab 82:4196–4200

126. Bastard JP, Jardel C, Bruckert E, Blondy P, Capeau J, Laville M, Vidal H, Hainque B (2000) Elevated levels of interleukin 6 are reduced in serum and subcutaneous adipose tissue of obese women after weight loss. J Clin Endocrinol Metab 85:3338–3342

127. Pittas AG, Lau J, Hu FB, Dawson-Hughes B (2007) The role of vitamin D and calcium in type 2 diabetes. A systematic review and meta-analysis. J Clin Endocrinol Metab 92:2017–2029

128. Muller K, Diamant M, Bendtzen K (1991) Inhibition of production and function of interleukin-6 by 1,25-dihydroxyvitamin D3. Immunol Lett 28:115–120

129. Willheim M, Thien R, Schrattbauer K, Bajna E, Holub M, Gruber R, Baier K, Pietschmann P, Reinisch W, Scheiner O et al (1999) Regulatory effects of 1alpha,25-dihydroxyvitamin D3 on the cytokine production of human peripheral blood lymphocytes. J Clin Endocrinol Metab 84:3739–3744

130. Zhu Y, Mahon BD, Froicu M, Cantorna MT (2005) Calcium and 1 alpha,25-dihydroxyvitamin D3 target the TNF-alpha pathway to suppress experimental inflammatory bowel disease. Eur J Immunol 35:217–224

131. Canning MO, Grotenhuis K, de Wit H, Ruwhof C, Drexhage HA (2001) 1-alpha,25-Dihydroxyvitamin D3 (1,25(OH)(2)D(3)) hampers the maturation of fully active immature dendritic cells from monocytes. Eur J Endocrinol 145:351–357

132. Gannage-Yared MH, Azoury M, Mansour I, Baddoura R, Halaby G, Naaman R (2003) Effects of a short-term calcium and vitamin D treatment on serum cytokines, bone markers, insulin and lipid concentrations in healthy post-menopausal women. J Endocrinol Invest 26:748–753

133. Pittas AG, Harris SS, Stark PC, Dawson-Hughes B (2007) The effects of calcium and vitamin D supplementation on blood glucose and markers of inflammation in nondiabetic adults. Diabetes Care 30:980–986

134. Liu PT, Stenger S, Li H, Wenzel L, Tan BH, Krutzik SR, Ochoa MT, Schauber J, Wu K, Meinken C et al (2006) Toll-like receptor triggering of a vitamin D-mediated human antimicrobial response. Science 311:1770–1773

135. Martineau AR, Honecker FU, Wilkinson RJ, Griffiths CJ (2007) vitamin D in the treatment of pulmonary tuberculosis. J Steroid Biochem Mol Biol 103:793–798

136. Stroder J, Kasal P (1970) Phagocytosis in vitamin D deficient rickets. Klin Wochenschr 48:383–384

137. Karatekin G, Kaya A, Salihoglu O, Balci H, Nuhoglu A (2009) Association of subclinical vitamin D deficiency in newborns with acute lower respiratory infection and their mothers. Eur J Clin Nutr 63:473–477

138. Laaksi I, Ruohola JP, Tuohimaa P, Auvinen A, Haataja R, Pihlajamaki H, Ylikomi T (2007) An association of serum vitamin D concentrations <40 nmol/l with acute respiratory tract infection in young Finnish men. Am J Clin Nutr 86:714–717

139. Aloia JF, Li-Ng M (2007) Re: epidemic influenza and vitamin D. Epidemiol Infect 135:1095–1096; author reply 1097–1098

140. Daxecker H, Raab M, Markovic S, Karimi A, Griesmacher A, Mueller MM (2002) Endothelial adhesion molecule expression in an in vitro model of inflammation. Clin Chim Acta 325:171–175

141. Visser M, Bouter LM, McQuillan GM, Wener MH, Harris TB (1999) Elevated C-reactive protein levels in overweight and obese adults. JAMA 282:2131–2135

51 Vitamin D, Renin, and Blood Pressure

Yan Chun Li

Abstract   Hypertension, or high blood pressure, is one of the most prevalent health problems in the world. Hypertension increases the risk of heart attack, heart failure, stroke, progressive atherosclerosis, and kidney disease. Accumulating body of evidence from epidemiological and clinical studies has demonstrated a connection between vitamin D deficiency or insufficiency and high blood pressure and cardiovascular problems in humans; however, until recently the mechanism underlying this link was unknown. The finding that vitamin D negatively regulates the renin–angiotensin system, a central regulator of blood pressure and cardiovascular functions, provides a mechanistic insight into this observed connection. Recent genetic, physiological, biochemical, and molecular data obtained from cell and animal studies have demonstrated an inhibitory role of 1,25-dihydroxyvitamin D [1,25(OH) D] in renin gene expression, and this is consistent with the inverse correlation between circulating 1,25-dihydroxyvitamin D levels and plasma renin activity seen in humans. The finding of 1,25-dihydroxyvitamin D as a negative regulator of renin expression also provides a molecular basis to explore the use of vitamin D analogs as novel renin inhibitors for therapeutic purposes. Recent data have demonstrated that vitamin D analogs used in combination with the classic renin–angiotensin inhibitors can block the unwanted compensatory renin increase and thus markedly increase the therapeutic efficacy. This combination strategy can be applied to a number of renal and cardiovascular diseases.

Key Words: Vitamin D; rennin; blood pressure; hypertension; mouse model; angiotensin; cardiac function; 25-hydroxyvitamin D; 1,25-dihydroxyvitamin D; renal failure

Abbreviations

VDR	vitamin D receptor
RAS	renin–angiotensin system
AGT	angiotensinogen
Ang	angiotensin
JG	juxtaglomerular
ACE	angiotensin-converting enzyme
$1,25(OH)_2D$	1,25-dihydroxyvitamin D

1. INTRODUCTION

Hypertension, or high blood pressure (systolic pressure $\geq$ 140 mmHg, diastolic pressure $\geq$ 90 mmHg), is one of the most prevalent health problems in the world. Hypertension increases the risk of heart attack, heart failure, stroke, progressive atherosclerosis,

From: *Nutrition and Health: Vitamin D*
Edited by: M.F. Holick, DOI 10.1007/978-1-60327-303-9_51,
© Springer Science+Business Media, LLC 2010

and kidney disease. Data from the National Health and Nutrition Examination Surveys (NHANES) show that in the United States, 29% of adults have hypertension in 2005–2006. Despite intense prevention and intervention efforts, there is still no change in the prevalence of hypertension during 1999–2006 *(1)*. Thus, prevention and therapeutic intervention of hypertension remain a major medical challenge.

The renin–angiotensin system (RAS) is one of the most important systems involved in the regulation of cardiovascular functions. This system plays a central role in the regulation of blood pressure, extracellular volume, and electrolyte homeostasis. Blood pressure is mainly determined by cardiac output and total systemic vascular resistance, both of which are regulated by the RAS, and over-activation of the RAS is a main cause for the development of hypertension. In fact, drugs that target the RAS have been the most widely prescribed anti-hypertensive drugs.

In the last decades, data from epidemiological and clinical studies have suggested a connection between vitamin D deficiency and increased risks of high blood pressure and other cardiovascular problems; however, until recently, the molecular mechanism underlying this link was unclear. The discovery of the vitamin D hormone as a negative endocrine regulator of renin biosynthesis provides part of the molecular explanation for the observed connection. Thus, the main focus of this chapter is to review the relationship between the vitamin D endocrine system and blood pressure and cardiovascular functions from the perspective of the RAS.

2. VITAMIN D, BLOOD PRESSURE, AND CARDIAC FUNCTIONS: EVIDENCE FOR A CONNECTION

2.1. Epidemiological Evidence

Evidence accumulated in the last decades has suggested a link between sunlight exposure, vitamin D, and blood pressure. As ultraviolet (UV) irradiation is essential for the cutaneous synthesis of vitamin D, circulating 25-hydroxyvitamin D [25(OH) D] levels are greatly influenced by geographic locations, seasonal changes, and skin pigmentations. UV irradiation decreases with the increase in latitude, and the data from the INTERSALT study show that the increase of latitude is correlated with the rise of blood pressure and the prevalence of hypertension in the general population around the globe *(2)*. Similarly, an increasing gradient of hypertension prevalence and stroke incidents from south to north has also been reported in big countries such as China *(3)*. Seasonal variations in blood pressure are seen in temperate climates, with blood pressure higher in the winter (low UV irradiation) than in the summer (high UV irradiation) *(4, 5)*. Winter season is also associated with high incident of myocardial infarction *(6)*. Dark skin pigmentation in the black population, which affects an efficient UV light penetration *(7)*, has been associated with higher blood pressure *(8, 9)*. In fact, a small clinical study has shown that direct UVB irradiation is able to lower blood pressure in patients with mild essential hypertension *(10)*.

Vitamin D deficiency is a global health problem. Recent epidemiological studies have established a correlation between vitamin D deficiency or insufficiency and high blood pressure and cardiovascular risk factors. As reported by Scragg et al., examination of the data from the Third US National Health and Nutrition Examination Survey (NHANES III), a cross-sectional survey representative of the US civilian population during 1988–1994, revealed an inverse relationship between serum 25-(OH) D and blood pressure *(11)*. In this non-hypertensive population ($n > 2{,}500$), systolic and diastolic blood pressures (adjusted for sex, age, ethnicity and physical activity) are 3 and 1.6 mmHg higher, respectively, in the lowest quintile of serum 25-(OH) D compared to highest quintile. This difference, although modest, has public health significance, as 2–3 mmHg decrease in blood pressure could account for 10–15% decline in cardiovascular mortality on a population basis *(12)*. Based on the same NHANES III data from 1988 to 1992, another study found that serum 25-(OH) D levels are inversely correlated with systolic blood pressure in non-hypertensive whites in the United States *(13)*. Furthermore, based on the NHANES III database Martins et al. reported that among more than 15,000 adults serum 25-(OH) D levels are also inversely associated with increased prevalence of cardiovascular risk factors including hypertension, diabetes, obesity, and hyperlipidemia *(14)*, and Melamed et al. showed that low-serum 25-(OH) D levels are associated with increased rate of all-cause mortality during a median 8.7 years of follow-up *(15)*.

Data from other large epidemiological studies also confirm the association between vitamin D deficiency and hypertension. In a prospective study, Forman et al. examined two prospective cohort studies including 613 men from the Health Professionals' Follow-Up Study (HPFS) and 1,198 women from the Nurses' Health Study (NHS) without baseline hypertension and found that serum 25-(OH) D levels are inversely associated with risk of incident hypertension during 4 years of follow-up *(16)*. A nested case–control prospective study conducted by Giovannucci et al. in more than 18,000 men in the Health Professionals' Follow-Up Study also demonstrated an association of low-serum 25-(OH) D levels with higher risk of myocardial infarction, even after adjusting for factors known to be associated with coronary artery disease *(17)*. Wang et al. studied 1,739 Framingham Offspring Study participants without prior cardiovascular disease and concluded that during the mean follow-up of 5.4 years low-serum 25-(OH) D (<10–15 ng/ml) is associated with incident cardiovascular disease. This association is not affected with adjustment for C-reactive protein, physical activity, or vitamin use *(18)*.

A similar inverse association is also found between serum levels of 1,25-dihydroxyvitamin D [$1{,}25(OH)_2D_3$], the active hormone, and blood pressure. In a cross-sectional, multivariate study involving 100 normotensive male industrial employees, Kristal-Boneh et al. showed an inverse and statistically significant association between serum $1{,}25(OH)_2D$ levels and systolic blood pressure independent of serum parathyroid hormone and calcium levels *(19)*. In another population-based study with 34 middle-aged men, serum levels of $1{,}25(OH)_2D$ were also found to be inversely correlated to the blood pressure *(20)*.

More interestingly, such an inverse relationship has also been reported between circulating $1{,}25(OH)_2D$ levels and plasma renin activity. Resnick et al. first demonstrated an inverse correlation between serum levels of $1{,}25(OH)_2D$ and plasma renin activity in a study with 51 patients with essential hypertension *(21)*. Subsequently, in another smaller study involving 10 subjects with high renin hypertension, the inverse correlation between the change in circulating $1{,}25(OH)_2D$ and the change in plasma renin activity was confirmed by Burgess et al. *(22)*. Given the critical role of the RAS in blood pressure control, the relationship between vitamin D and plasma renin activity is likely part of the mechanism underlying the relationship between vitamin D and blood pressure.

2.2. Clinical Evidence

Numerous clinical studies have demonstrated clear cardiovascular benefits of vitamin D supplementation or therapy. For instance, in a double-blinded, placebo-controlled clinical trials with 39 hypertensive patients, blood pressure was significantly reduced in the patients by long-term treatment (4 months) with 1α-hydroxyvitamin D_3 (23). In a clinical trial involving 148 elderly women, 8 weeks of vitamin D_3 and calcium supplement was reported to significantly reduce blood pressure in these subjects (24). Wang et al. recently reported a large prospective study with 28,886 middle-aged women in the United States, in which incident cases of hypertension were identified from annual follow-up questionnaires during 10 years of follow-up. They found that dietary intake of dairy products, calcium, and vitamin D is each inversely associated with risk of hypertension (25). In addition, numerous recent clinical observational studies with large cohorts of hemodialysis patients have demonstrated multiple beneficial effects of vitamin D therapy, including significant reduction in the risk of cardiovascular death (26–28). Thus, it appears that vitamin D therapy offers the patients significant survival advantage.

However, not all reported studies support a role of vitamin D in blood pressure control. For instance, a small clinical study involving 16 hypertensive and 15 normotensive men did not find clear association between vitamin D and blood pressure (29). In a clinical trial involving 189 elderly subjects, a single oral dose of 2.5 mg cholecalciferol in winter failed to change blood pressure (30). A possible explanation is that a single dose of vitamin D supplement in the winter is not sufficient to raise the circulating vitamin D level to the blood pressure-affecting threshold. However, results from three large prospective cohort studies including Nurse Health Study I and II and Health Professionals' Follow-up Study also failed to find an association between high intake of vitamin D and low risk of incident hypertension (31). These conflicting reports call for more rigorous and well-controlled investigations into the effects of vitamin D on blood pressure control.

Consistent with the observational data concerning the inverse correlation between 1,25(OH)$_2$D and plasma renin activity (21, 22), data from clinical studies also support a connection between vitamin D and renin. In double-blinded, placebo-controlled clinical trial (32 subjects), 16 weeks of daily oral calcium supplementation, which suppresses plasma 1,25(OH)$_2$D levels, resulted in a significant elevation of plasma renin activity, suggesting a suppressive role of vitamin D in renin regulation (32). In a clinical study involving 25 hypertensive patients with end-stage renal disease, 15 weeks of calcitriol (1,25(OH)$_2$D) treatment reduced myocardial hypertrophy, with concomitant reduction in plasma renin activity, Ang II, and ANP levels (33). In another clinical case, 1,25(OH)$_2$D treatment was reported to reduce blood pressure and plasma renin activity in a patient with pseudohyperparathyroidism and high plasma renin activity (34).

Therefore, the beneficial effects of vitamin D on cardiovascular system reported in the epidemiological and clinical studies likely include regulation of the RAS *(35, 36)*.

3. THE RENIN–ANGIOTENSIN SYSTEM: A CENTRAL REGULATOR OF BLOOD PRESSURE

3.1. *The RAS Cascade and Its Functions*

The RAS is a systemic endocrine regulatory cascade consisting of multiple components (Fig. 1). The substrate of this cascade is angiotensinogen (AGT), produced predominantly in the liver. AGT is cleaved to angiotensin (Ang) I in the circulation by renin, and this is the rate-limiting step of the RAS cascade. Ang I, an inactive 10 amino acid peptide, is converted to Ang II, an 8 amino acid peptide, by the angiotensin-converting enzyme (ACE), which primarily resides in the endothelial cells in blood vessels. Further processing of Ang II by aminopeptidase A and N produces Ang III and Ang IV *(37)*. ACE2, an ACE homolog, can convert Ang II to Ang(1–7) *(38, 39)*, and this enzyme is thought to play an essential role in heart functions *(40)*.

Ang II is the central biological effector of the RAS and plays a central role in blood pressure regulation (Fig. 1). Ang II is the most potent vasoconstrictor; it acts on smooth muscle cells in the vasculature to increase vasoconstriction and thus peripheral resistance. Ang II stimulates the synthesis and secretion of aldosterone from the adrenal cortex, a hormone that promotes sodium reabsorption in the kidney. Ang II stimulates the release of anti-diuretic hormone (also called vasopressin) from the pituitary/hypothalamus, which increases water retention in the kidney; it also stimulates the sensation of thirst in the central nerve system and thus increases water intake. Therefore, activation of the renin–angiotensin cascade leads to an increase in extracellular volume and peripheral resistance. As blood pressure is determined by the combination of cardiac output and total systemic vascular resistance, over-activation of the systemic RAS thus results in hypertension. In addition to blood pressure control, Ang II has other diverse

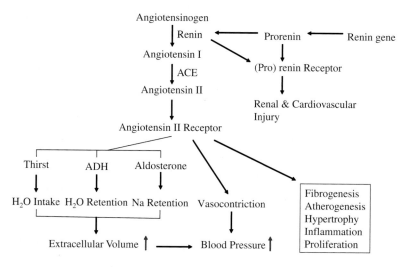

Fig. 1. The renin–angiotensin system.

physiological and pathological activities, being involved in such pathological processes as fibrogenesis, atherogenesis, inflammation, cell proliferation, and hypertrophy. Thus excessive activation of the RAS usually poses detrimental effects.

In addition to the systemic RAS, components of the RAS have also been found inside many tissues such as the brain, heart, vasculature, kidney, and reproductive system *(41)*. The tissue-specific RAS may function in a paracrine fashion and in some cases is involved in tissue damages. For example, the RAS within the brain is involved in the control of water drinking and blood pressure *(42, 43)*, the RAS within the heart may be involved in adaptive response to myocardial stress, and the RAS in the vasculature may be involved in vascular tone and endothelial functions *(41)*. The intrarenal RAS is well known to play a key role in hyperglycemia-induced renal injury in diabetes mellitus *(44)*.

The wide spectrum of activities of Ang II is mediated by several G protein-coupled receptors widely distributed in tissues *(45)*. Among these receptors, the type 1 receptor (AT1) mediates most of the activities involved in vasoconstriction, hypertrophy, and sodium retention, whereas the type 2 receptor (AT2) is involved in vasodilation, natriuresis, and growth inhibition. Hypertension and hypertension-related organ damage resulting from excessive activation of the RAS are mostly mediated by the AT1 receptor *(45)*.

3.2. Renin, the Rate-Limiting Enzyme of the RAS Cascade

Renin is a highly specific aspartic peptidase with AGT as the sole known substrate. Renin is also species-specific, in that human renin is not able to cleave murine AGT and vice versa *(46–48)*. The structure of renin is composed of two β-sheet domains with the enzymatic active site residing in a cleft between these two domains *(49)*. Renin and its inactive precursor, prorenin, are synthesized and secreted from the juxtaglomerular (JG) cells, highly granulated smooth muscle cells located in the media of the afferent arteriole at the vascular pole of the glomerulus. Renin is stored in secretory granules and secreted from these granules through exocytosis upon stimulation. In the plasma the prorenin concentration is usually much higher (10–100 times) than renin concentration. Prorenin is 43 amino acids longer than mature renin at the NH_2 terminus, and this NH_2-terminal prosegment is thought to block the enzymatic active site located in the cleft, thus preventing the interaction of the active site with the substrate AGT. Although in vitro studies have shown that prorenin can be activated by endopeptidase such as trypsin and cathepsin B or by low pH, the mechanism involved in proteolytic prorenin activation in vivo and its physiological role remain not fully understood. High renin activity can result in inappropriate activation of RAS, leading to hypertension and end-organ damage. Increased plasma renin activity is associated with hypertension *(50)*, left ventricular hypertrophy *(51)*, and renal dysfunction *(52)*.

The (pro)renin receptor, a single transmembrane receptor initially identified in mesangial cells and vascular smooth muscle cells *(53)*, binds to both renin and prorenin with high affinity. Renin bound to the receptor exhibits increased catalytic activity, and prorenin bound to this receptor exhibits full enzymatic activity comparable to that of mature renin. In addition, binding of renin or prorenin to this receptor triggers intracellular signaling and phosphorylation of MAP kinase independent of Ang II generation. For example, stimulation of (pro)renin receptor in mesangial cells with puri-fied renin or prorenin promotes the synthesis of TGFβ *(54)*, a profibrotic factor involved in the development of diabetic nephropathy. Thus, increased renin/prorenin can also cause tissue injury through the (pro)renin receptor independent of Ang II (Fig. 1). For instance, transgenic rats overexpressing prorenin in the liver with a high level of circu-lating prorenin develop severe vascular damage and diabetic renal complications in the absence of high blood pressure *(55)*.

Because of its central role in the renin–angiotensin cascade, the biosynthesis and secretion of renin is tightly regulated. The most common physiological factors that influ-ence renin secretion include renal perfusion pressure, renal sympathetic nerve activity, and tubular sodium chloride load *(56)*. The perfusion pressure in the renal artery is the most profound parameter to influence renin secretion: when the renal perfusion pressure falls, renin secretion rises, and when the renal perfusion pressure rises, renin secretion falls. This effect is mediated by a baroreceptor or stretch receptor mechanism in the JG cells *(57)*. The JG apparatus has rich sympathetic nerve endings, and stimulation of renin synthesis and release by sympathetic nerve activity are mediated by β-adrenergic receptors and intracellular cyclic AMP *(58)*. This pathway may exert a tonic stimulatory influence on renin production *(59)*. Renin secretion is also tightly regulated by the tubu-lar sodium chloride load *(60)*. There is an inverse relationship between dietary sodium chloride intake and renin secretion. Tubular control of renin release is mediated by the macula densa, which is part of the distal tubule and anatomically in close association with the JG cells. The macula densa senses the sodium chloride load and transduces the signal, possibly via adenosine and ATP, to the JG cells to influence renin production and secretion *(61)*. At the local level, renin synthesis and release are influenced by a variety of bioactive molecules. For instance, prostaglandins, nitric oxide, and adrenomedullin are known to stimulate renin secretion, whereas Ang II, endothelin, vasopressin, and adenosine are inhibitors of renin production *(56, 61)*. Ang II is a feedback inhibitor of renin production and secretion *(62)*, and this effect is mediated by the AT1 receptor *(63, 64)*.

At the molecular level, renin gene expression is regulated by a complex network of transcriptional factors *(65)*. In the renin gene promoter and enhancer regions, mul-tiple transcription factor-binding sites have been identified, which are responsive to various signal transduction pathways including cAMP, retinoic acid, endothelin-1, and cytokines, to alter renin gene transcription. Thus, the production of renin is determined by interplay of multiple transcriptional regulators available or activated under a particu-lar physiological condition.

3.3. Pharmacological Inhibition of the Renin–Angiotensin System

Because of its critical role in the regulation of blood pressure, the RAS has been a major therapeutic target for intervention of hypertension. Small molecules designed to target the RAS include ACE inhibitors (ACEI), Ang II type 1 receptor blockers (ARB), and renin inhibitors. These drugs inhibit each major step of the renin–angiotensin cas-cade, respectively (Fig. 2). ACEIs and ARBs are probably the most successful and most widely prescribed anti-hypertensive drugs *(66, 67)*. Aliskiren is the first FDA-approved

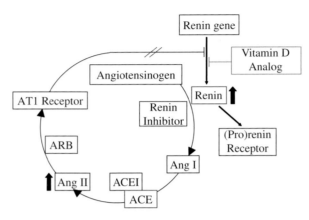

Fig. 2. Mechanism underlying the compensatory renin increase and its inhibition by vitamin D analogs. The three classes of RAS inhibitors including renin inhibitors, ACE inhibitors, and AT1 receptor blockers all disrupt the feedback inhibition of renin biosynthesis, leading to an increase in renin expression and accumulation of plasma renin concentration, which ultimately stimulates Ang II production. Increased Ang II levels reduce the efficacy of RAS inhibition by the inhibitors. Vitamin D analogs can suppress the compensatory renin increase by blocking the renin gene transcription when used in combination with the RAS inhibitors.

renin inhibitor that specifically inhibits the enzymatic activity of renin and lowers blood pressure in hypertensive subjects *(68, 69)*.

There are a number of issues associated with the current ACEIs and ARBs that may reduce the efficacy of RAS blockade and cause undesired side effects. For example, Ang II conversion from Ang I can be catalyzed by other enzymes such as chymase *(70)* and thus bypass the ACE *(71)*, which reduces the efficacy of ACEIs. Moreover, in addition to Ang I, ACE also recognizes other substrates such as bradykinin *(72)*; thus inhibition of ACE may also alter bradykinin metabolism and evoke undesirable side effects. Because Ang II has multiple receptors (e.g., AT1, AT2) with different functions *(45)*, blocking the AT1 receptor with ABRs may increase the availability of Ang II to the AT2 receptor, leading to enhancement of unwanted AT2 activity. Ang II increase due to AT1 receptor blockade can lead to elevation of various Ang II metabolites, such as Ang (1–7), Ang III, and Ang IV *(37)*, which are bioactive and may cause a variety of unwanted effects.

The major problem associated with all current RAS inhibitors is the compensatory increase of renin concentration *(73)*. Patients receiving chronic dosing of ACEIs initially have lower plasma Ang II levels; however, Ang II, as well as aldosterone, often rises to the earlier baseline levels *(74, 75)*. This phenomenon, often termed "ACE escape," is caused by the disruption of the feedback inhibition loop in renin biosynthesis, leading to increased renin production and secretion. The problem of plasma renin increase also exists in the case of ARBs and renin inhibitor aliskiren *(76)*. The huge increase in renin activity in the plasma stimulates the conversion of Ang I, which ultimately leads to the increase in Ang I, Ang II, and other angiotensin metabolites in the body, through ACE-dependent and -independent pathways (Fig. 2). This reactive renin increase compromises the efficacy of RAS inhibition and explains why the current RAS inhibitors are only suboptimal. Therefore, it is expected that agents that block the compensatory renin increase will enhance the efficacy of the RAS inhibitors.

4. VITAMIN D REGULATION OF THE RENIN–ANGIOTENSIN SYSTEM: MECHANISM FOR THE CONNECTION?

4.1. Vitamin D: Negative Endocrine Regulator of the Renin–Angiotensin System

The mechanism underlying the inverse relationship between circulating 1,25(OH)$_2$D levels and plasma renin activity reported about two decades ago *(21, 22)* was unclear until recently when Li et al. demonstrated that 1,25(OH)$_2$D functions as a negative endocrine regulator of renin gene expression in vivo *(77)*. This group found that vitamin D receptor (VDR)-null mutant mice, which lack VDR-mediated vitamin D signaling, developed hyperreninemia due to dramatic up-regulation of renin expression in the kidney. Renin increase was detected at both mRNA and protein levels, leading to marked increase in plasma renin activity and plasma Ang II levels. As a consequence of aberrant RAS overstimulation, VDR knockout mice developed high blood pressure and cardiac hypertrophy, which were corrected by treatment with ACEI or ARB, confirming that RAS activation is responsible for these abnormalities *(77, 78)*. As expected, plasma and urinary aldosterone levels are also markedly elevated in VDR knockout mice *(79)*. Moreover, renin expression in the brain is also up-regulated in VDR knockout mice, leading to activation of the local RAS in the brain, which is partly responsible for the overdrinking and polyuric phenotypes seen in the mutant mice *(80)*. In wild-type mice receiving several doses of 1,25(OH)$_2$D injection, renin expression is significantly suppressed *(77)*. These findings strongly suggest that 1,25(OH)$_2$D suppresses renin expression in vivo.

The critical role of vitamin D in regulation of the RAS in vivo is confirmed in another genetic mutant mouse model of vitamin D deficiency. These mutant mice lack Cyp27b1, the rate-limiting 1α-hydroxylase required for the biosynthesis of 1,25(OH)$_2$D in the kidney. Zhou et al. showed that, like VDR knockout mice, Cyp27b1 knockout mice also developed hyperreninemia, hypertension, and cardiac hypertrophy, as a result of renin up-regulation. Importantly, in this model these abnormalities can be corrected not only by ACEI or ARB but also by treating the mice with exogenous 1,25(OH)$_2$D *(81)*.

The inhibitory role of vitamin D in renin biosynthesis is further confirmed by a transgenic approach. Kong et al. recently produced transgenic mice that overexpress the human VDR in the JG cells. In these transgenic mice, renal renin mRNA levels and plasma renin activity were significantly suppressed without changes in serum calcium or parathyroid hormone levels. When the human VDR transgene was bred into VDR knockout mice to generate genetic mutant mice that express VDR only in the JG cells, renin up-regulation was markedly reduced in these mice compared to VDR knockout mice *(82)*. These data demonstrate that 1,25(OH)$_2$D suppresses renin expression in vivo independent of parathyroid hormone and calcium.

The finding of vitamin D regulation of renin expression explains the relationship between 1,25(OH)$_2$D and plasma renin activity reported in humans, and at least in part, underlies the epidemiological and clinical observations concerning vitamin D deficiency and high blood pressure. Based on this finding, it is predicted that long-term vitamin D deficiency will increase the risk of hypertension, whereas vitamin D supplement or therapy may be beneficial to the cardiovascular system.

4.2. Mechanism of Renin Suppression

As a ligand-activated transcription factor, VDR is involved in both positive and negative transcriptional regulations. While most positive regulations are mediated by a vitamin D response element (VDRE) in the vitamin D target gene promoter, the mechanism of negative regulation is diverse. Yuan et al. demonstrated that $1,25(OH)_2D$ suppresses renin gene expression by targeting the cAMP signaling pathway *(83)*, a major regulatory pathway involved in renin biosynthesis.

It is well established that cAMP signals through cyclic AMP response element (CRE), which interacts with members of the ATF/CREB/CREM bZIP transcription factor family in homodimeric or heterodimeric forms. Intracellular cAMP is converted from ATP by adenylate cyclase activated by membrane receptors; cAMP binds to the regulatory subunit of protein kinase A (PKA) to free the catalytic subunit, which enters the nucleus and phosphorylates CREB at serine-133 or CREM at serine-117, resulting in the recruitment of ubiquitous co-activators CBP/p300 to promote gene transcription *(84–86)*.

Through systematic deletion analysis of the mouse renin gene promoter, Yuan et al. found that a CRE at –2,688 is required to mediate the suppression of renin gene transcription by $1,25(OH)_2D$. This CRE has been shown to play a critical role in the basal expression of renin *(87)*. Experimental data obtained from EMSA, ChIP assays, GST pull-down assays, and cell transfection experiments indicate that $1,25(OH)_2D$ disrupts the formation of the DNA–protein complex on the CRE site, which contains CREB/CREM and CBP/p300, and liganded VDR blocks CREB binding to this CRE by directly interacting with CREB *(83)* (Fig. 3). This latter finding is not surprising because liganded VDR has previously been shown to physically interact with a variety of regulatory proteins including Smad3, β-catenin, and p65 NF-κB *(88–91)*. These data establish that $1,25(OH)_2D$ suppresses renin gene transcription, at least in part, by direct inhibition of CRE-mediated transcriptional activity. In the basal state CREB, CREM, and CBP/p300 are recruited to the CRE to drive renin gene expression; in the presence of $1,25(OH)_2D$, liganded VDR binds to CREB, thus blocking CREB binding to the CRE and disrupting the formation of CRE–CREB–CBP/p300 complex, leading to reduction in renin gene expression (Fig. 3).

Cyclic AMP is long known to be a major intracellular signal that stimulates renin production in the JG cells. Intracellular cAMP is thought to be critically involved in the stimulation of renin expression by sympathetic nerve activity (mediated by

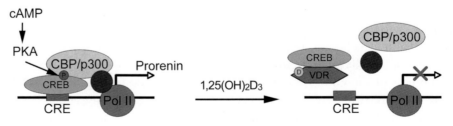

Fig. 3. Schematic model of $1,25(OH)_2D$-induced transrepression of renin gene transcription. PKA, protein kinase A; D, $1,25(OH)_2D$; Pol II, RNA polymerase II, CRE, cyclic AMP response element; CREB, CAMP response element binding protein; CBP, CREB-binding protein.

β-adrenergic receptor) or by low tubular sodium chloride concentration (mediated by the macula densa cells) *(61)*. It is hypothesized that, by targeting the cAMP signaling pathway, $1,25(OH)_2D$ may function as a gatekeeper to counterbalance the other renin-stimulating factors and prevent the detrimental over-production of renin.

4.3. Vitamin D Analogs as Novel Renin Inhibitors

The finding that $1,25(OH)_2D$ suppresses renin biosynthesis provides a molecular basis to explore the use of vitamin D analogs as renin synthesis inhibitors for therapeutic purposes. In the past decades, a large number of vitamin D analogs have been synthesized with a wide range of potency and calcemic index. A few vitamin D analogs have been approved for clinical use *(92, 93)*. Therefore, the concept of vitamin D analogs as therapeutic drugs is already sound, and issues such as safety and pharmacokinetic properties can be easily dealt with. Low calcemic vitamin D analogs that have potent renin-inhibiting activity are particularly valuable. The vitamin D analog-based renin inhibitors, which inhibit renin gene expression, are different from another class of aliskiren-like renin inhibitors, which inhibit renin enzymatic activity. There are advantages to have two classes of renin inhibitors. For example, a combination of these two classes of drugs can simultaneously inhibit renin at both the biosynthetic and enzymatic levels and increase therapeutic efficacy.

Numerous studies have demonstrated renin-inhibiting activity of vitamin D analogs in a variety of models. Through in vitro cell culture screening, Qiao et al. identified a group of vitamin D analog compounds that inhibit renin expression in vitro and in vivo without invoking severe hypercalcemia in mice *(94)*. Some analogs are much more potent than $1,25(OH)_2D$ in terms of renin suppression. Paricalcitol (19-nor-1,25-dihydroxyvitamin D_2), an activated vitamin D analog, has been shown to suppress renin expression in mice with the same potency as $1,25(OH)_2D$ but without induction of hypercalcemia *(95)*. Paricalcitol can suppress renin expression in kidney mesangial cells *(96)*. Freundlich et al. treated rat remnant kidney model of chronic renal failure with paricalcitol and showed that this analog can significantly lower blood pressure in these rats and suppress the RAS in the remnant kidney. This study suggests that the beneficial effects of the vitamin D analog in experimental chronic renal failure are due, at least in part, to down-regulation of the RAS *(97)*. Bodyak et al. showed that paricalcitol markedly attenuated left ventricular hypertrophy in Dahl salt-sensitive hypertensive rats and suppressed cardiac renin expression; however, paricalcitol did not significantly lower blood pressure in this model *(98)*. The effect of vitamin D analogs on the RAS in humans has not been reported; however, $1,25(OH)_2D$ has been reported to suppress plasma renin and Ang II in patients with kidney disease *(33)*.

As discussed above, the dramatic compensatory increase in plasma renin concentration that is associated with the current RAS inhibitors compromises the efficacy of these drugs (Fig. 2). One important application of vitamin D analogs is to block the compensatory renin increase at the transcriptional level when used together with the classic RAS inhibitors, and the combination is expected to enhance the efficacy of inhibition of the RAS *(99)*. Zhang et al. have used a combination therapy with an ARB and an activated vitamin D analog to test this notion in a model of diabetic nephropathy *(100)*. It is well known that activation of the intrarenal RAS by hyperglycemia plays a key role in development of diabetic nephropathy *(44)*, a main course of renal failure. The combination of losartan and paricalcitol produced synergistic therapeutic effects in this model, leading to complete prevention of albuminuria and marked amelioration of glomerulosclerosis, two main pathogenic hallmarks of diabetic nephropathy. The molecular basis underlying the synergistic effect is the blockade of the compensatory renin increase and Ang II accumulation in the kidney by this combination *(100)*. Given its excellent therapeutic efficacy, this combination strategy can be applied to other renal and cardiovascular diseases in which activation of the RAS is a main cause.

5. CONCLUSION

Accumulating body of evidence from a large number of epidemiological and clinical studies have demonstrated a connection between vitamin D deficiency or insufficiency and high blood pressure and a number of cardiovascular problems in humans; however, until recently the mechanism underlying this link was unknown. The finding that vitamin D negatively regulates the RAS, a central regulator of blood pressure and cardiovascular functions, provides a mechanistic insight into this observed connection. Recent genetic, physiological, biochemical, and molecular data obtained from cell and animal studies have demonstrated an inhibitory role of $1,25(OH)_2D$ in renin gene expression, and this is consistent with the inverse correlation between circulating $1,25(OH)_2D$ levels and plasma renin activity reported in humans. The molecular basis of this negative regulation is that liganded VDR disrupts the cAMP signal pathway, a physiologically important stimulatory pathway involved in renin gene expression. The finding of $1,25(OH)_2D$ as a negative regulator of renin expression also provides a molecular basis to explore the use of vitamin D analogs as novel renin inhibitors for therapeutic purposes. Particularly, recent data have demonstrated that vitamin D analogs used in combination with the classic renin–angiotensin inhibitors can block the unwanted compensatory renin increase and thus increase the therapeutic efficacy. This combination strategy may potentially be applied to many renal and cardiovascular diseases in which inhibition of the RAS is used as a treatment method. Overall, the physiological and clinical implications of the knowledge gained from these studies are profound.

6. ACKNOWLEDGMENTS

This work was supported in part by NIH grant R01HL085793.

REFERENCES

1. Ostchega Y, Yoon SS, Hughes J, Louis T (2008) Hypertension awareness, treatment, and control – continued disparities in adults: United States, 2005–2006. NCHS Data Brief No. 3. National Center for Health Statistics, Hyattsville, Maryland
2. Rostand SG (1997) Ultraviolet light may contribute to geographic and racial blood pressure differences. Hypertension 30:150–156
3. He J, Klag MJ, Wu Z, Whelton PK (1995) Stroke in the people's republic of China. I. Geographic variations in incidence and risk factors. Stroke 26:2222–2227

4. Kunes J, Tremblay J, Bellavance F, Hamet P (1991) Influence of environmental temperature on the blood pressure of hypertensive patients in Montreal. Am J Hypertens 4:422–426

5. Woodhouse PR, Khaw KT, Plummer M (1993) Seasonal variation of blood pressure and its relationship to ambient temperature in an elderly population. J Hypertens 11:1267–1274

6. Spencer FA, Goldberg RJ, Becker RC, Gore JM (1998) Seasonal distribution of acute myocardial infarction in the second National Registry of Myocardial Infarction. J Am Coll Cardiol 31:1226–1233

7. Holick MF (1987) Photosynthesis of vitamin D in the skin: effect of environmental and life-style variables. Fed Proc 46:1876–1882

8. Harburg E, Gleibermann L, Roeper P, Schork MA, Schull WJ (1978) Skin color, ethnicity, and blood pressure I: detroit blacks. Am J Public Health 68:1177–1183

9. Klag MJ, Whelton PK, Coresh J, Grim CE, Kuller LH (1991) The association of skin color with blood pressure in US blacks with low socioeconomic status. JAMA 265:599–602

10. Krause R, Buhring M, Hopfenmuller W, Holick MF, Sharma AM (1998) Ultraviolet B and blood pressure. Lancet 352:709–710

11. Scragg R, Sowers M, Bell C (2007) Serum 25-hydroxyvitamin D, ethnicity, and blood pressure in the Third National Health and Nutrition Examination Survey. Am J Hypertens 20:713–719

12. Lewington S, Clarke R, Qizilbash N, Peto R, Collins R (2002) Age-specific relevance of usual blood pressure to vascular mortality: a meta-analysis of individual data for one million adults in 61 prospective studies. Lancet 360:1903–1913

13. Judd SE, Nanes MS, Ziegler TR, Wilson PW, Tangpricha V (2008) Optimal vitamin D status attenuates the age-associated increase in systolic blood pressure in white Americans: results from the third National Health and Nutrition Examination Survey. Am J Clin Nutr 87:136–141

14. Martins D, Wolf M, Pan D et al (2007) Prevalence of cardiovascular risk factors and the serum levels of 25-hydroxyvitamin D in the United States: data from the Third National Health and Nutrition Examination Survey. Arch Intern Med 167:1159–1165

15. Melamed ML, Michos ED, Post W, Astor B (2008) 25-hydroxyvitamin D levels and the risk of mortality in the general population. Arch Intern Med 168:1629–1637

16. Forman JP, Giovannucci E, Holmes MD et al (2007) Plasma 25-hydroxyvitamin D levels and risk of incident hypertension. Hypertension 49:1063–1069

17. Giovannucci E, Liu Y, Hollis BW, Rimm EB (2008) 25-hydroxyvitamin D and risk of myocardial infarction in men: a prospective study. Arch Intern Med 168:1174–1180

18. Wang TJ, Pencina MJ, Booth SL et al (2008) Vitamin D deficiency and risk of cardiovascular disease. Circulation 117:503–511

19. Kristal-Boneh E, Froom P, Harari G, Ribak J (1997) Association of calcitriol and blood pressure in normotensive men. Hypertension 30:1289–1294

20. Lind L, Hanni A, Lithell H, Hvarfner A, Sorensen OH, Ljunghall S (1995) Vitamin D is related to blood pressure and other cardiovascular risk factors in middle-aged men. Am J Hypertens 8:894–901

21. Resnick LM, Muller FB, Laragh JH (1986) Calcium-regulating hormones in essential hypertension. Relation to plasma renin activity and sodium metabolism. Ann Intern Med 105:649–654

22. Burgess ED, Hawkins RG, Watanabe M (1990) Interaction of 1,25-dihydroxyvitamin D and plasma renin activity in high renin essential hypertension. Am J Hypertens 3:903–905

23. Lind L, Wengle B, Wide L, Ljunghall S (1989) Reduction of blood pressure during long-term treatment with active vitamin D (alphacalcidol) is dependent on plasma renin activity and calcium status. A double-blind, placebo-controlled study. Am J Hypertens 2:20–25

24. Pfeifer M, Begerow B, Minne HW, Nachtigall D, Hansen C (2001) Effects of a short-term vitamin D(3) and calcium supplementation on blood pressure and parathyroid hormone levels in elderly women. J Clin Endocrinol Metab 86:1633–1637

25. Wang L, Manson JE, Buring JE, Lee IM, Sesso HD (2008) Dietary intake of dairy products, calcium, and vitamin D and the risk of hypertension in middle-aged and older women. Hypertension 51:1073–1079

26. Teng M, Wolf M, Lowrie E, Ofsthun N, Lazarus JM, Thadhani R (2003) Survival of patients undergoing hemodialysis with paricalcitol or calcitriol therapy. N Engl J Med 349:446–456

27. Shoji T, Shinohara K, Kimoto E et al (2004) Lower risk for cardiovascular mortality in oral 1alpha-hydroxy vitamin D3 users in a haemodialysis population. Nephrol Dial Transplant 19:179–184

28. Teng M, Wolf M, Ofsthun MN et al (2005) Activated injectable vitamin D and hemodialysis survival: a historical cohort study. J Am Soc Nephrol 16:1115–1125

29. Young EW, Morris CD, Holcomb S, McMillan G, McCarron DA (1995) Regulation of parathyroid hormone and vitamin D in essential hypertension. Am J Hypertens 8:957–964

30. Scragg R, Khaw KT, Murphy S (1995) Effect of winter oral vitamin D3 supplementation on cardiovascular risk factors in elderly adults. Eur J Clin Nutr 49:640–646

31. Forman JP, Bischoff-Ferrari HA, Willett WC, Stampfer MJ, Curhan GC (2005) Vitamin D intake and risk of incident hypertension: results from three large prospective cohort studies. Hypertension 46:676–682

32. Petrov V, Lijnen P (1999) Modification of intracellular calcium and plasma renin by dietary calcium in men. Am J Hypertens 12:1217–1224

33. Park CW, Oh YS, Shin YS et al (1999) Intravenous calcitriol regresses myocardial hypertrophy in hemodialysis patients with secondary hyperparathyroidism. Am J Kidney Dis 33:73–81

34. Kimura Y, Kawamura M, Owada M et al (1999) Effectiveness of 1,25-dihydroxyvitamin D supplementation on blood pressure reduction in a pseudohypoparathyroidism patient with high renin activity. Intern Med 38:31–35

35. Levin A, Li YC (2005) Vitamin D and its analogues: do they protect against cardiovascular disease in patients with kidney disease? Kidney Int 68:1973–1981

36. Achinger SG, Ayus JC (2005) The role of vitamin D in left ventricular hypertrophy and cardiac function. Kidney Int Suppl 95:S37–S42

37. Reudelhuber TL (2005) The renin-angiotensin system: peptides and enzymes beyond angiotensin II. Curr Opin Nephrol Hypertens 14:155–159

38. Donoghue M, Hsieh F, Baronas E et al (2000) A novel angiotensin-converting enzyme-related carboxypeptidase (ACE2) converts angiotensin I to angiotensin 1–9. Circ Res 87:E1–E9

39. Tipnis SR, Hooper NM, Hyde R, Karran E, Christie G, Turner AJ (2000) A human homolog of angiotensin-converting enzyme. Cloning and functional expression as a captopril-insensitive carboxypeptidase. J Biol Chem 275:33238–33243

40. Crackower MA, Sarao R, Oudit GY et al (2002) Angiotensin-converting enzyme 2 is an essential regulator of heart function. Nature 417:822–828

41. Paul M, Poyan Mehr A, Kreutz R (2006) Physiology of local renin-angiotensin systems. Physiol Rev 86:747–803

42. Davisson RL, Oliverio MI, Coffman TM, Sigmund CD (2000) Divergent functions of angiotensin II receptor isoforms in the brain. J Clin Invest 106:103–106

43. Sakai K, Agassandian K, Morimoto S et al (2007) Local production of angiotensin II in the subfornical organ causes elevated drinking. J Clin Invest 117:1088–1095

44. Carey RM, Siragy HM (2003) The intrarenal renin-angiotensin system and diabetic nephropathy. Trends Endocrinol Metab 14:274–281

45. Berry C, Touyz R, Dominiczak AF, Webb RC, Johns DG (2001) Angiotensin receptors: signaling, vascular pathophysiology, and interactions with ceramide. Am J Physiol Heart Circ Physiol 281:H2337–H2365

46. Burton J, Quinn T (1988) The amino-acid residues on the C-terminal side of the cleavage site of angiotensinogen influence the species specificity of reaction with renin. Biochim Biophys Acta 952:8–12

47. Hatae T, Takimoto E, Murakami K, Fukamizu A (1994) Comparative studies on species-specific reactivity between renin and angiotensinogen. Mol Cell Biochem 131:43–47

48. Yang G, Merrill DC, Thompson MW, Robillard JE, Sigmund CD (1994) Functional expression of the human angiotensinogen gene in transgenic mice. J Biol Chem 269:32497–32502

49. Danser AH, Deinum J (2005) Renin, prorenin and the putative (pro)renin receptor. Hypertension 46:1069–1076

50. Mullins JJ, Peters J, Ganten D (1990) Fulminant hypertension in transgenic rats harbouring the mouse Ren-2 gene. Nature 344:541–544

51. Malmqvist K, Ohman KP, Lind L, Nystrom F, Kahan T (2002) Relationships between left ventricular mass and the renin-angiotensin system, catecholamines, insulin and leptin. J Intern Med 252:430–439

52. Kehoe B, Keeton GR, Hill C (1986) Elevated plasma renin activity associated with renal dysfunction. Nephron 44:51–57

53. Nguyen G, Delarue F, Burckle C, Bouzhir L, Giller T, Sraer JD (2002) Pivotal role of the renin/prorenin receptor in angiotensin II production and cellular responses to renin. J Clin Invest 109:1417–1427

54. Huang Y, Wongamorntham S, Kasting J et al (2006) Renin increases mesangial cell transforming growth factor-beta1 and matrix proteins through receptor-mediated, angiotensin II-independent mechanisms. Kidney Int 69:105–113

55. Veniant M, Menard J, Bruneval P, Morley S, Gonzales MF, Mullins J (1996) Vascular damage without hypertension in transgenic rats expressing prorenin exclusively in the liver. J Clin Invest 98:1966–1970

56. Hackenthal E, Paul M, Ganten D, Taugner R (1990) Morphology, physiology, and molecular biology of renin secretion. Physiol Rev 70:1067–1116

57. Carey RM, McGrath HE, Pentz ES, Gomez RA, Barrett PQ (1997) Biomechanical coupling in renin-releasing cells. J Clin Invest 100:1566–1574

58. Holmer SR, Kaissling B, Putnik K et al (1997) Beta-adrenergic stimulation of renin expression in vivo. J Hypertens 15:1471–1479

59. Wagner C, Hinder M, Kramer BK, Kurtz A (1999) Role of renal nerves in the stimulation of the renin system by reduced renal arterial pressure. Hypertension 34:1101–1105

60. Skott O, Briggs JP (1987) Direct demonstration of macula densa-mediated renin secretion. Science 237:1618–1620

61. Bader M, Ganten D (2000) Regulation of renin: new evidence from cultured cells and genetically modified mice [In Process Citation]. J Mol Med 78:130–139

62. Vander AJ, Geelhoed GW (1965) Inhibition of renin secretion by angiotensin. II. Proc Soc Exp Biol Med 120:399–403

63. Sugaya T, Nishimatsu S, Tanimoto K et al (1995) Angiotensin II type 1a receptor-deficient mice with hypotension and hyperreninemia. J Biol Chem 270:18719–18722

64. Shricker K, Holmer S, Kramer BK, Riegger GA, Kurtz A (1997) The role of angiotensin II in the feedback control of renin gene expression. Pflugers Arch 434:166–172

65. Pan L, Gross KW (2005) Transcriptional regulation of renin: an update. Hypertension 45:3–8

66. August P (2003) Initial treatment of hypertension. N Engl J Med 348:610–617

67. Cheung BM (2002) Blockade of the renin-angiotensin system. Hong Kong Med J 8:185–191

68. Wood JM, Schnell CR, Cumin F, Menard J, Webb RL (2005) Aliskiren, a novel, orally effective renin inhibitor, lowers blood pressure in marmosets and spontaneously hypertensive rats. J Hypertens 23:417–426

69. Nussberger J, Wuerzner G, Jensen C, Brunner HR (2002) Angiotensin II suppression in humans by the orally active renin inhibitor Aliskiren (SPP100): comparison with enalapril. Hypertension 39:E1–E8

70. Urata H, Kinoshita A, Misono KS, Bumpus FM, Husain A (1990) Identification of a highly specific chymase as the major angiotensin II-forming enzyme in the human heart. J Biol Chem 265:22348–22357

71. Padmanabhan N, Jardine AG, McGrath JC, Connell JM (1999) Angiotensin-converting enzyme-independent contraction to angiotensin I in human resistance arteries. Circulation 99:2914–2920

72. Tom B, Dendorfer A, Danser AH (2003) Bradykinin, angiotensin-(1–7), and ACE inhibitors: how do they interact? Int J Biochem Cell Biol 35:792–801

73. Muller DN, Luft FC (2006) Direct renin inhibition with aliskiren in hypertension and target organ damage. Clin J Am Soc Nephrol 1:221–228

74. Borghi C, Boschi S, Ambrosioni E, Melandri G, Branzi A, Magnani B (1993) Evidence of a partial escape of renin-angiotensin-aldosterone blockade in patients with acute myocardial infarction treated with ACE inhibitors. J Clin Pharmacol 33:40–45

Part VI / Vitamin D Deficiency and Chronic Disease

75. Roig E, Perez-Villa F, Morales M et al (2000) Clinical implications of increased plasma angiotensin II despite ACE inhibitor therapy in patients with congestive heart failure. Eur Heart J 21:53–57

76. Azizi M, Menard J, Bissery A et al (2004) Pharmacologic demonstration of the synergistic effects of a combination of the renin inhibitor aliskiren and the AT1 receptor antagonist valsartan on the angiotensin II-renin feedback interruption. J Am Soc Nephrol 15:3126–3133

77. Li YC, Kong J, Wei M, Chen ZF, Liu SQ, Cao LP (2002) 1,25-Dihydroxyvitamin D(3) is a negative endocrine regulator of the renin-angiotensin system. J Clin Invest 110:229–238

78. Xiang W, Kong J, Chen S et al (2005) Cardiac hypertrophy in vitamin D receptor knockout mice: role of the systemic and cardiac renin-angiotensin systems. Am J Physiol Endocrinol Metab 288: E125–E132

79. Kong J, Li YC (2003) Effect of angiotensin II type I receptor antagonist and angiotensin-converting enzyme inhibitor on vitamin D receptor null mice. Am J Physiol Regul Integr Comp Physiol 285:R255–R261

80. Kong J, Zhang Z, Li D et al (2008) Loss of vitamin D receptor produces polyuria by increasing thirst. J Am Soc Nephrol 19:2396–2405

81. Zhou C, Lu F, Cao K, Xu D, Goltzman D, Miao D (2008) Calcium-independent and 1,25(OH)2D3-dependent regulation of the renin-angiotensin system in 1alpha-hydroxylase knockout mice. Kidney Int 74:170–179

82. Kong J, Qiao G, Zhang Z, Liu SQ, Li YC (2008) Targeted vitamin D receptor expression in juxtaglomerular cells suppresses renin expression independent of parathyroid hormone and calcium. Kidney Int 74:1577–1581

83. Yuan W, Pan W, Kong J et al (2007) 1,25-Dihydroxyvitamin D3 suppresses renin gene transcription by blocking the activity of the cyclic AMP response element in the renin gene promoter. J Biol Chem 282:29821–29830

84. De Cesare D, Fimia GM, Sassone-Corsi P (1999) Signaling routes to CREM and CREB: plasticity in transcriptional activation. Trends Biochem Sci 24:281–285

85. Montminy M (1997) Transcriptional regulation by cyclic AMP. Annu Rev Biochem 66:807–822

86. Andrisani OM (1999) CREB-mediated transcriptional control. Crit Rev Eukaryot Gene Expr 9:19–32

87. Pan L, Black TA, Shi Q et al (2001) Critical roles of a cyclic AMP responsive element and an E-box in regulation of mouse renin gene expression. J Biol Chem 276:45530–45538

88. Yanagisawa J, Yanagi Y, Masuhiro Y et al (1999) Convergence of transforming growth factor-beta and vitamin D signaling pathways on SMAD transcriptional coactivators. Science 283: 1317–1321

89. Palmer HG, Gonzalez-Sancho JM, Espada J et al (2001) Vitamin D(3) promotes the differentiation of colon carcinoma cells by the induction of E-cadherin and the inhibition of beta-catenin signaling. J Cell Biol 154:369–387

90. Lu X, Farmer P, Rubin J, Nanes MS (2004) Integration of the NfkappaB p65 subunit into the vitamin D receptor transcriptional complex: identification of p65 domains that inhibit 1,25-dihydroxyvitamin D3-stimulated transcription. J Cell Biochem 92:833–848

91. Sun J, Kong J, Duan Y et al (2006) Increased NF-{kappa}B activity in fibroblasts lacking the vitamin D receptor. Am J Physiol Endocrinol Metab 291:E315–E322

92. Brown AJ, Dusso AS, Slatopolsky E (2002) Vitamin D analogues for secondary hyperparathyroidism. Nephrol Dial Transplant 17(Suppl 10):10–19

93. Malluche HH, Mawad H, Koszewski NJ (2002) Update on vitamin D and its newer analogues: actions and rationale for treatment in chronic renal failure. Kidney Int 62:367–374

94. Qiao G, Kong J, Uskokovic M, Li YC (2005) Analogs of 1alpha,25-dihydroxyvitamin D3 as novel inhibitors of renin biosynthesis. J Steroid Biochem Mol Biol 96:59–66

95. Fryer RM, Rakestraw PA, Nakane M et al (2007) Differential inhibition of renin mRNA expression by paricalcitol and calcitriol in C57/BL6 mice. Nephron Physiol 106:p76–p81

96. Zhang Z, Sun L, Wang Y et al (2008) Renoprotective role of the vitamin D receptor in diabetic nephropathy. Kidney Int 73:163–171

97. Freundlich M, Quiroz Y, Zhang Z et al (2008) Suppression of renin-angiotensin gene expression in the kidney by paricalcitol. Kidney Int 74:1394–1402

98. Bodyak N, Ayus JC, Achinger S et al (2007) Activated vitamin D attenuates left ventricular abnormalities induced by dietary sodium in Dahl salt-sensitive animals. Proc Natl Acad Sci USA 104: 16810–16815

99. Li YC (2007) Inhibition of renin: an updated review of the development of renin inhibitors. Curr Opin Investig Drugs 8:750–757

100. Zhang Z, Zhang Y, Ning G, Deb DK, Kong J, Li YC (2008) Combination therapy with AT1 blocker and vitamin D analog markedly ameliorates diabetic nephropathy: blockade of compensatory renin increase. Proc Natl Acad Sci USA 105:15896–15901

52 Role of Vitamin D and Vitamin D Analogs for Bone Health and Survival in Chronic Kidney Disease

Ishir Bhan, Hector Tamez, and Ravi Thadhani

Abstract   Chronic kidney disease (CKD) is accompanied by reduced conversion of 25-hydroxy-vitamin D to 1,25-dihydroxyvitamin D and is often also associated with a deficiency of 25-hydroxyvitamin D. These and other metabolic changes can lead to secondary hyperparathyroidism and a range of disorders of bone turnover, mineralization, and volume, the most prominent of which is osteitis fibrosa cystica. Attempts to prevent this disorder have led to widespread use of vitamin D receptor agonists, but may increase the risk of adynamic bone disease. Beyond its use in regulating bone metabolism, several studies suggest that vitamin D may influence survival in advanced CKD. Potential mechanisms for this effect are likely to influence cardiovascular disease, the leading cause of death in CKD. Alterations in left ventricular function and the renin–angiotensin–aldosterone axis appear to be modulated by vitamin D. We discuss these and other aspects of the role vitamin D may play in the CKD population as well as the current state of evidence supporting therapeutic use.

Key Words: Chronic kidney disease; vitamin D; 25-hydroxyvitamin D; kidney disease; 1,25-dihydroxyvitamin D; 1α-hydroxylase; bone disease; osteomalacia; vitamin D deficiency; parathyroid hormone

1. VITAMIN D AND METABOLIC CHANGES OF CHRONIC KIDNEY DISEASE

Chronic kidney disease (CKD) is characterized by a progressive fall in glomerular filtration rate (GFR) that is associated with a host of metabolic changes, including significant effects on the normal physiology of vitamin D. As renal function worsens, renal conversion of 25-hydroxyvitamin D to 1,25-dihydroxyvitamin D declines *(1)*. As a result of declining levels of 1,25-dihydroxyvitamin D, calcium absorption is impaired and blood levels begin to fall *(2)*. In contrast, phosphorous excretion becomes progressively impaired with declining glomerular filtration. Hyperphosphatemia itself then further suppresses renal 1,25-dihydroxyvitamin D production *(3)*. This process is mediated,

From: *Nutrition and Health: Vitamin D*
Edited by: M.F. Holick, DOI 10.1007/978-1-60327-303-9_52,
© Springer Science+Business Media, LLC 2010

at least in part, by the phosphatonin fibroblast growth factor 23 (FGF-23) which inhibits 1α-hydroxylase activity, reducing 1,25-dihydroxyvitamin D production; levels of FGF-23 rise steadily as CKD progresses and may precede overt hyperphosphatemia (4).

Production of 1,25-dihydroxyvitamin D may be further restricted by deficiency of its precursor, 25-hydroxyvitamin, which is common in CKD (5–7). One potential mechanism for this may be the loss of the vitamin D-binding protein (DBP) and DBP-bound 25-hydroxyvitamin D in proteinuric individuals (8). Impaired skin production of precursors is another reason why circulating 25-hydroxyvitamin D levels are low in subjects with renal failure (9).

As 1,25-dihydroxyvitamin D can suppress parathyroid hormone (PTH) gene transcription, its deficiency results in secondary hyperparathyroidism (10). This is further compounded by decreased vitamin D receptor (VDR), decreased binding of 1,25-dihydroxyvitamin to the VDR, and altered binding of the VDR to vitamin D response elements (VDRE) in DNA (11–13). The fall in serum calcium that accompanies the decline in vitamin D action may be an even more important contributor to stimulating PTH production (14).

In CKD, secondary hyperparathyroidism has been linked to the development of bone disease. Changes in calcium and phosphate homeostasis as well as bone disease have served as a stimulus for replacement of 1,25-dihydroxyvitamin D using calcitriol and other analogs. Clinical guidelines have used PTH levels to guide the decision on whether or not treatment with these medications is warranted (15). More recently, it has become apparent that alterations in PTH, calcium, and phosphate homeostasis might be related to overall survival (16). The results of several hypothesis-generating observational studies in humans and animal studies combined with the recognition of the broad distribution of the VDR have emphasized a potential role for vitamin D in CKD survival. Judicious administration of vitamin D analogs thus appears to be important in preserving bone health in CKD and may also be important in improving longevity in the CKD population (see below).

2. BONE DISEASE AND VITAMIN D IN CKD

The wide range of changes in bone morphology seen in CKD have been collectively referred to as renal osteodystrophy. In 2006, the Kidney Disease: Improving Global Outcomes (KDIGO) group classified osteodystrophy according to changes in three variables: turnover, mineralization, and volume (TMV) (17). Turnover refers to the rate of skeletal remodeling, a function of both bone loss and formation. Mineralization refers to the presence of normal or abnormal calcification of collagen. Volume is a measure of bone volume relative to total tissue volume. Three principal forms of osteodystrophy are described, each with a distinct relationship to vitamin D physiology. Mixed lesions, with characteristics of more than one disorder, can also occur.

3. OSTEITIS FIBROSA CYSTICA

Osteitis fibrosa cystica is characterized primarily by increased rates of bone turnover with preserved mineralization and a variable bone volume; marrow fibrosis often accompanies this disorder *(17)*. Osteitis fibrosa is felt primarily to be the result of hyper-parathyroidism (milder hyperparathyroid-related bone disease can also occur), though there may also be a direct effect of 1,25-dihydroxyvitamin D deficiency. This high-turnover state results in altered bony structure, with the loss of lamellar bone and increased predominance of the weaker woven bone *(18)*. As a result, patients with osteitis fibrosa are predisposed to fractures out of proportion to measurements of bone density.

Vitamin D therapy has been central to the treatment of this disease. Early studies in humans with end-stage renal disease (ESRD, kidney disease severe enough to require dialysis or transplantation) demonstrated histological improvement of biopsy-proven osteitis fibrosa after treatment with 25-hydroxyvitamin D, with decreased numbers of osteoclasts and a reduction in marrow fibrosis *(19)*. More recently, treatment of dialysis patients with intravenous calcitriol resulted in an increase in osteoblastic osteoid and a reduction in marrow fibrosis *(20)*. As expected, bone turnover was reduced. Regression of radiographic findings associated with osteitis fibrosa has also been reported follow-ing treatment with oral alfacalcidol *(21)*. Additional regimens, including doxercalciferol and paricalcitol, have also been used to target suppression of PTH in the prevention of osteitis fibrosa *(22, 23)*.

4. ADYNAMIC BONE DISEASE

In contrast to the high-turnover state of osteitis fibrosa, adynamic bone disease is characterized by abnormally low bone turnover, while mineralization is preserved. Bone volume is typically reduced. The disease is associated with increased skeletal fragility and risk of fracture, though the precise risk of fracture is unknown *(24, 25)*. In contrast to osteitis fibrosa, the prevalence of adynamic bone disease may be increasing over time *(26)*. Though it is not clear if increased treatment with vitamin D analogs is responsible for his change, studies of biopsy-proven osteitis fibrosa (or milder lesions suggestive of secondary hyperparathyroidism) showed the development of adynamic bone disease in dialysis patients following treatment with calcitriol *(27)*. Similar effects have been seen with alfacalcidol *(28)*. The development of adynamic bone disease appears to correlate with a reduction in PTH and over-suppression of PTH, and interventions such as vitamin D analogs and calcimimetics may be the major culprits in the emergence of this disorder *(29)*.

PTH levels are typically used to guide therapy with active vitamin D analogs in CKD, but the relationship between low PTH levels and adynamic bone disease is complicated by reduced skeletal responsiveness to the actions of PTH in CKD *(30)*. As a result, the level of PTH required for normal bone may be considerably higher than that of the non-CKD population *(31)*. Furthermore, the assumption that a low or normal PTH is necessarily accompanied by adynamic bone disease is uncertain given variable degree of PTH resistance. Elevated cytokine levels and reduced activity of insulin-like growth factor, along with advanced age and malnutrition, may increase the risk of adynamic bone disease in individuals independently of PTH *(25)*.

There is now some belief that, although increased PTH suppression may be a risk factor for the development of adynamic bone disease, vitamin D analogs may have additional effects on this disorder. Vitamin D receptor activation (VDRA) appears to stimulate bone formation in CKD, and thus has the potential to address one of the deficits in this disorder *(25)*. Varying actions of different vitamin D analogs on bone formation and resorption remain a topic of active investigation and may be critical to balancing the beneficial effects of this therapy with the risk of adynamic bone disease. As osteoblasts have intrinsic 1α-hydroxylase activity, therapy with 25-hydroxyvitamin may play a role in the management as well *(32)*.

5. OSTEOMALACIA

Osteomalacia, like adynamic bone disease, is characterized by low bone turnover in the setting of normal to reduced bone volume. While bone mineralization is preserved in adynamic bone disease, it is abnormal in osteomalacia. This disease is characterized by unmineralized osteoid, a feature not seen in the other two major forms of renal osteodystrophy. Unlike adynamic bone disease, the major cause of osteomalacia is felt to be the deposition of aluminum and other heavy metals in bone, which impairs mineralization and promotes accumulation of bony matrix *(18)*. With declining use of aluminum as a phosphate binder, the prevalence of aluminum-associated osteomalacia appears to be decreasing *(33, 34)*. Early descriptions of renal osteomalacia suggested that it was not affected by treatment with active vitamin D, but that 25-hydroxyvitamin D may play a beneficial role in therapy *(18, 35, 36)*.

6. VITAMIN D AND SURVIVAL

Early studies of activated vitamin D focused on the effects of this therapy on PTH, calcium and phosphorus as surrogates of adverse outcomes *(37)*. In the ensuing years, it became apparent that vitamin's actions extended beyond the traditional mineral axis. Animal studies revealed important roles in the renin–angiotensin axis as well as in the vascular endothelium *(38, 39)*. As newer analogs of vitamin D emerged, questions emerged regarding the relative efficacy of different vitamin D receptor activators. In 2003, Teng et al. performed an observational study comparing paricalcitol with calcitriol in a cohort of over 60,000 incident hemodialysis patients *(40)*. There was a 16% relative mortality rate reduction with paricalcitol treatment, which held in multivariate and stratified analyses even after adjustment for calcium, phosphorus, and PTH. Furthermore, subjects who were switched from calcitriol to paricalcitol had a survival pattern similar to the paricalcitol group. These findings suggested an effect of VDRA on mortality above and beyond the traditional effects on mineral metabolism.

A follow-up study by the same group compared subjects who received *any* form of active vitamin D replacement to those who received none (Fig. 1). Activated vitamin D was associated with a 26% 2-year reduction in mortality compared with those who did not receive this therapy, with the cardiovascular mortality nearly halved *(41)*. Levels of 25-hydroxyvitamin D also appear to influence mortality in this population, particularly among those not treated VDRAs (Fig. 2).

Other groups have echoed these findings in different populations. Shoji et al. reported that the administration of alfacalcidol was associated with a survival benefit in a cohort

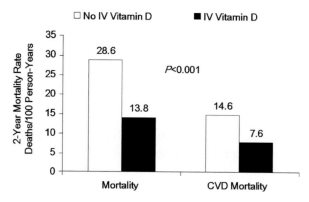

Fig. 1. Two-year all-cause and cardiovascular disease (CVD) mortality rate (deaths/100 person-years) for subjects who did or did not receive IV vitamin D on dialysis. Subjects receiving IV vitamin D were significantly less likely to die within 2 years of initiating dialysis ($p < 0.0001$) for both comparisons. Adapted from Teng et al. *(41)*.

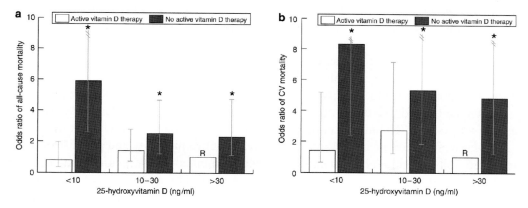

Fig. 2. Multivariate-adjusted odds ratios of 90-day all-cause and cardiovascular (CV) mortality on hemodialysis according to 25-hydroxyvitamin D levels and whether or not patients received active vitamin D. For both all-cause (**a**) and cardiovascular (**b**) mortality, there was a monotonic increase in odds of death for untreated patients as 25-hydroxyvitamin D levels declined. Adapted from Wolf et al. *(7)*.

of 242 Japanese patients with ESRD *(42)*. Melamed et al. noted a 26% mortality reduction associated with injectable vitamin D as part of the Choices for Health Outcomes in Caring for ESRD (CHOICE) study *(16)*. In yet another study with almost 15,000 patients on hemodialysis, there was a 17% reduction in the hazard comparing treatment with either calcitriol, paricalcitol, or doxercalciferol with no active vitamin D analog therapy *(43)*. Kalantar-Zadeh et al., in a large cohort with over 58,000 hemodialysis patients in the United States, found that paricalcitol administration was associated with improved survival compared with those who did not receive vitamin D *(44)*.

The potential role for active vitamin D in the mortality profile of the pre-ESRD CKD population not yet requiring dialysis has not been as thoroughly explored. Kovesdy et al. examined 512 patients with CKD stages 3–5 and reported a 65% reduction in the risk of death comparing treatment with calcitriol versus no treatment *(45)*. In a study of 1,418 patients with CKD stages 3 and 4 and hyperparathyroidism, oral calcitriol was linked to a 26% reduction in the risk of death *(46)*. These findings suggest that the potential survival benefits of activated vitamin D treatment may extend to the non-dialysis population.

Black patients with chronic kidney disease have been previously observed to survive longer than their Caucasian counterparts. Wolf et al. reported that black patients are more likely than Caucasians to be treated with an activated vitamin D analog *(47)*. After controlling for therapy with activated vitamin D, the association of race with survival disappeared, suggesting that disparities in vitamin D treatment may account for at least some of racial differences in outcome. More importantly, however, among those untreated, blacks had significantly worse survival compared to Caucasians. Activated vitamin D therapy may be one factor explaining the differences in dialysis outcomes according to race.

The potential for vitamin D, a widely available and relatively inexpensive therapy, to have an impact on survival in CKD has generated considerable interest. The support for a survival effect of vitamin D in CKD comes exclusively from observational studies. These studies are useful for generating hypotheses, but ultimately are prone to bias due to confounding factors that are either unmeasured or not included in multivariate analyses. Therefore, these studies cannot prove a causative effect of vitamin D treatment on survival. Randomized controlled trials are needed to determine if vitamin D treatment itself is responsible for the observed effects.

Given the limited trial data available for vitamin D, Palmer and colleagues conducted a meta-analysis of 76 studies and failed to find an effect of activated vitamin D therapy *(48)*. Many of these were not designed or powered with a mortality end point. Outcomes from pre-dialysis and dialysis populations were combined, as were adult and pediatric studies. Underlying heterogeneity in effects of vitamin D could mask potential effects when such disparate studies are combined. Despite this, this study underscored the importance of well-controlled trials in this area that directly examine outcomes rather than surrogates.

7. POTENTIAL MECHANISMS OF A CARDIOVASCULAR SURVIVAL BENEFIT

Cardiovascular mortality is 10–20 times higher than that observed in the general population and it is the leading cause of death in CKD *(49, 50)*. In addition to the relationship with overall mortality described above, observational studies have noted a reduction in cardiovascular-related mortality associated with vitamin D administration *(40, 41)*. There is growing evidence to suggest that beneficial cardiovascular outcomes may be mediated by the effects of vitamin D, either directly or indirectly, on cardiac structure and function. Notably, humans with rickets who are profoundly deficient in vitamin D may present with severe heart failure *(51)*. In animals, the VDR-knockout mouse model demonstrates increased cardiac renin expression and marked cardiomyocyte hypertrophy, while 1,25-hydroxyvitamin D attenuates cardiomyocyte proliferation and hypertrophy in vitro *(52–54)*. As cardiomyocytes express the vitamin D receptor, these actions may be due to a direct effect of vitamin D on the myocardium *(55)*.

Bodyak et al. reported that paricalcitol treatment attenuated the development of left ventricular hypertrophy in Dahl salt-sensitive rats *(56)*. These animals when fed with high-salt diet develop left ventricular hypertrophy (LVH) and heart failure. Because of renal losses of DBP, they also become vitamin D deficient. Paricalcitol-treated rats had improved cardiac contraction by M-mode echocardiography, lower end-diastolic pressure by cardiac catheterization, and smaller heart dimensions compared to those rats with untreated control animals. Gene expression significantly differed between groups, suggesting a direct effect of vitamin D in cardiomyocytes. Continuous invasive monitoring did not reveal any blood pressure differences between the two groups. In a retrospective study of 21 human dialysis patients by this group, increased E/A ratios, higher ejection fractions, and lower posterior wall thickness by echocardiogram were seen in paricalcitol-treated individuals compared with those who received no such treatment. These results suggest that the effects observed in the Dahl salt-sensitive rats may extend to humans and certainly provide sufficient support for pursuing related end points in human trials.

Zhou et al. studied the effects of 1,25-dihydroxyvitamin D deficiency and replacement in 1α-hydroxylase knockout mice, which are unable to produce this form of vitamin D *(57)*. These animals developed an array of cardiac complications including hypertension, cardiac hypertrophy, and impaired cardiac function, which did not improve despite modified diet that preserved normal calcium and phosphorous levels. Calcitriol administration, apparently acting via suppression of the renin–angiotensin axis, however, was able to normalize blood pressure as well as cardiac structure and function. Calcitriol's effect was similar to that of an angiotensin-converting enzyme inhibitor and an angiotensin receptor blocker. These drugs are commonly used in the management of hypertension and heart failure and are associated with improved clinical outcomes in a wide range of cardiovascular disease *(58)*.

While large observational studies laid the groundwork for the notion that vitamin D treatment may directly influence survival on dialysis, experimental animal data have further substantiated this by demonstrating an influence of vitamin D on the development of several cardiac abnormalities, particularly LVH. LVH is present in over 70% of patients initiating dialysis and is closely linked to cardiac death in this population *(50, 59)*. Given these findings, Thadhani et al. initiated the *Paricalcitol Benefits in Renal failure Induced Cardiac Morbidity* (PRIMO) study in subjects with chronic kidney disease. PRIMO is a multi-national, double-blind randomized controlled clinical trial in both CKD stages 3b and 4 (PRIMO I) and ESRD (PRIMO II) comparing the cardiovascular effects of paricalcitol versus placebo (clinicaltrials.gov NCT00497146 and NCT00616902). The primary outcome in these trials is left ventricular hypertrophy progression/regression assessed by cardiac magnetic resonance during a 48-week follow-up. These and other randomized controlled trials will look directly at physiologic changes and mortality effects of therapy with vitamin D.

8. INFECTIOUS DISEASE SURVIVAL AND VITAMIN D

Although cardiovascular disease is the leading cause of death in ESRD, infectious mortality follows close behind. Infection occurs at rates several fold above the population without kidney disease *(50)*. Some of this risk has been attributed to risk factors related to dialysis access, such as the use of indwelling catheters. However, patients on dialysis are also at increased risk of non-access-related infections *(60)*. Human cationic antimicrobial peptide 18 (hCAP18, sometimes referred to by the name of its active fragment, LL-37) is the only human member of a class of proteins called cathelicidins *(61)*. These proteins form a key component of innate immunity in most multicellular organisms and are capable of rapidly killing both gram-positive and gram-negative bacteria (see Chapter 58) *(62)*. Preliminary data suggests that levels low levels of hCAP18 predict an increased risk of infectious mortality in the dialysis population *(63)*. Importantly, hCAP18 appears to be regulated at a genetic level by 1,25-dihydroxyvitamin D *(64)*. Modulation of hCAP18 may thus be another mechanism by which vitamin D administration could influence outcomes in the CKD population.

9. SUMMARY

Chronic kidney disease is accompanied by a host of metabolic changes that lead to a reduction in the production of and response to 1,25-dihydroxyvitamin D. The consequent hyperparathyroidism results in excessive bone turnover that culminates in osteitis fibrosa cystica. Therapy with active vitamin D analogs can suppress PTH and reduce the risk of high-turnover bone disease, though it may also increase the risk of the low-turnover adynamic bone disease. An increasing number of observational studies now suggest that vitamin D's effects may extend well beyond its role in management of osteodystrophy. Several observational studies now suggest a significant overall and cardiovascular-specific survival benefit. These studies alone cannot prove that activated vitamin D itself is the cause of improved outcomes, but have served to generate exciting hypotheses regarding the potential therapeutic benefits of vitamin D. Animal data demonstrate that important cardiovascular changes may result directly from vitamin D action on the myocardium and renin–angiotensin axis. Vitamin D, via influence on innate immunity, may also prove important in sustaining a response against bacterial infection in this population as well. The fascinating biology revealed in these animal studies combined with observational data in humans is encouraging. Clinical trials are currently underway to determine the relevance of these exciting findings to use of vitamin D in humans with CKD.

REFERENCES

1. Martinez I, Saracho R, Montenegro J et al (1997) The importance of dietary calcium and phosphorous in the secondary hyperparathyroidism of patients with early renal failure. Am J Kidney Dis 29:496–502
2. Silver J, Russell J, Sherwood LM (1985) Regulation by vitamin D metabolites of messenger ribonucleic acid for preproparathyroid hormone in isolated bovine parathyroid cells. Proc Natl Acad Sci USA 82:4270–4273
3. Tanaka Y, DeLuca HF (1973) The control of 25-hydroxyvitamin D metabolism by inorganic phosphorus. Arch Biochem Biophys 154:566–574

4. Gutierrez O, Isakova T, Rhee E et al (2005) Fibroblast growth factor-23 mitigates hyperphosphatemia but accentuates calcitriol deficiency in chronic kidney disease. J Am Soc Nephrol 16: 2205–2215

5. Ishimura E, Nishizawa Y, Inaba M et al (1999) Serum levels of 1,25-dihydroxyvitamin D, 24,25-dihydroxyvitamin D, and 25-hydroxyvitamin D in nondialyzed patients with chronic renal failure. Kidney Int 55:1019–1027

6. LaClair RE, Hellman RN, Karp SL et al (2005) Prevalence of calcidiol deficiency in CKD: a cross-sectional study across latitudes in the United States. Am J Kidney Dis 45:1026–1033

7. Wolf M, Shah A, Gutierrez O et al (2007) Vitamin D levels and early mortality among incident hemodialysis patients. Kidney Int 72:1004–1013

8. González EA, Sachdeva A, Oliver DA et al (2004) Vitamin D insufficiency and deficiency in chronic kidney disease. A single center observational study. Am J Nephrol 24:503–510

9. Jacob AI, Sallman A, Santiz Z et al (1984) Defective photoproduction of cholecalciferol in normal and uremic humans. J Nutr 114:1313–1319

10. Russell J, Lettieri D, Sherwood LM (1986) Suppression by 1,25(OH)2D3 of transcription of the pre-proparathyroid hormone gene. Endocrinology 119:2864–2866

11. Korkor AB (1987) Reduced binding of [^{3}H]-1,25-dihydroxyvitamin D3 in the parathyroid glands of patients with renal failure. N Engl J Med 316:1573–1577

12. Merke J, Hügel U, Zlotkowski A et al (1987) Diminished parathyroid 1,25(OH)2D3 receptors in experimental uremia. Kidney Int 32:350–353

13. Patel SR, Ke HQ, Vanholder R et al (1995) Inhibition of calcitriol receptor binding to vitamin D response elements by uremic toxins. J Clin Invest 96:50–59

14. Li YC, Amling M, Pirro AE et al (1998) Normalization of mineral ion homeostasis by dietary means prevents hyperparathyroidism, rickets, and osteomalacia, but not alopecia in vitamin D receptor-ablated mice. Endocrinology 139:4391–4396

15. Foundation NK (2003) K/DOQI clinical practice guidelines for bone metabolism and disease in chronic kidney disease. Am J Kidney Dis 42:S1–S201

16. Melamed ML, Eustace JA, Plantinga L et al (2006) Changes in serum calcium, phosphate, and PTH and the risk of death in incident dialysis patients: a longitudinal study. Kidney Int 70:351–357

17. Moe S, Drüeke T, Cunningham J et al (2006) Definition, evaluation, and classification of renal osteodystrophy: a position statement from Kidney Disease: Improving Global Outcomes (KDIGO). Kidney Int 69:1945–1953

18. Hruska KA, Teitelbaum SL (1995) Renal osteodystrophy. N Engl J Med 333:166–174

19. Teitelbaum SL, Bone JM, Stein PM et al (1976) Calcifediol in chronic renal insufficiency. Skeletal response. JAMA 235:164–167

20. Andress D, Norris KC, Coburn JW et al (1989) Intravenous calcitriol in the treatment of refractory osteitis fibrosa of chronic renal failure. N Engl J Med 321:274–279

21. Arabi A (2006) Regression of skeletal manifestations of hyperparathyroidism with oral vitamin D. J Clin Endocrinol Metab 91:2480–2483

22. Coburn JW, Maung HM, Elangovan L et al (2004) Doxercalciferol safely suppresses PTH levels in patients with secondary hyperparathyroidism associated with chronic kidney disease stages 3 and 4. Am J Kidney Dis 43:877–890

23. Coyne D, Acharya M, Qiu P et al (2006) Paricalcitol capsule for the treatment of secondary hyperparathyroidism in stages 3 and 4 CKD. Am J Kidney Dis 47:263–276

24. Hruska KA, Saab G, Mathew S et al (2007) Renal osteodystrophy, phosphate homeostasis, and vascular calcification. Semin Dial 20:309–315

25. Andress DL (2008) Adynamic bone in patients with chronic kidney disease. Kidney Int 69:33–43

26. Malluche HH, Monier-Faugere MC (1992) Risk of adynamic bone disease in dialyzed patients. Kidney Int Suppl 38:S62–S67

27. Goodman WG, Ramirez JA, Belin TR et al (1994) Development of adynamic bone in patients with secondary hyperparathyroidism after intermittent calcitriol therapy. Kidney Int 46:1160–1166

28. Hamdy NA, Kanis JA, Beneton MN et al (1995) Effect of alfacalcidol on natural course of renal bone disease in mild to moderate renal failure. BMJ 310:358–363

29. Hercz G, Pei Y, Greenwood C et al (1993) Aplastic osteodystrophy without aluminum: the role of "suppressed" parathyroid function. Kidney Int 44:860–866

30. Massry SG, Coburn JW, Lee DB et al (1973) Skeletal resistance to parathyroid hormone in renal failure. Studies in 105 human subjects. Ann Intern Med 78:357–364

31. Quarles LD, Lobaugh B, Murphy G (1992) Intact parathyroid hormone overestimates the presence and severity of parathyroid-mediated osseous abnormalities in uremia. J Clin Endocrinol Metab 75: 145–150

32. Atkins GJ, Anderson PH, Findlay DM et al (2007) Metabolism of vitamin D3 in human osteoblasts: evidence for autocrine and paracrine activities of 1 alpha,25-dihydroxyvitamin D3. Bone 40: 1517–1528

33. Morinière P, Cohen-Solal M, Belbrik S et al (1989) Disappearance of aluminic bone disease in a long term asymptomatic dialysis population restricting Al(OH)3 intake: emergence of an idiopathic adynamic bone disease not related to aluminum. Nephron 53:93–101

34. Gonzalez EA, Martin KJ (1992) Aluminum and renal osteodystrophy A diminishing clinical problem. Trends Endocrinol Metab 3:371–375

35. Coburn JW, Sherrard DJ, Brickman AS et al (1980) A skeletal mineralizing defect in dialysis patients: a syndrome resembling osteomalacia but unrelated to viatamin D. Contrib Nephrol 18:172–183

36. Eastwood JB, Stamp TC, De Wardener HE et al (1977) The effect of 25-hydroxy vitamin D3 in the osteomalacia of chronic renal failure. Clin Sci Mol Med 52:499–508

37. Blacher J, Guerin AP, Pannier B et al (1999) Impact of aortic stiffness on survival in end-stage renal disease. Circulation 99:2434–2439

38. Li YC, Kong J, Wei M et al (2002) 1,25-Dihydroxyvitamin D(3) is a negative endocrine regulator of the renin-angiotensin system. J Clin Invest 110:229–238

39. Yamamoto T, Kozawa O, Tanabe K et al (2002) 1,25-dihydroxyvitamin D3 stimulates vascular endothelial growth factor release in aortic smooth muscle cells: role of p38 mitogen-activated protein kinase. Arch Biochem Biophys 398:1–6

40. Teng M, Wolf M, Lowrie E et al (2003) Survival of patients undergoing hemodialysis with paricalcitol or calcitriol therapy. N Engl J Med 349:446–456

41. Teng M, Wolf M, Ofsthun MN et al (2005) Activated injectable vitamin D and hemodialysis survival: a historical cohort study. J Am Soc Nephrol 16:1115–1125

42. Shoji T, Shinohara K, Kimoto E et al (2004) Lower risk for cardiovascular mortality in oral 1alpha-hydroxy vitamin D3 users in a haemodialysis population. Nephrol Dial Transplant 19: 179–184

43. Tentori F, Hunt WC, Stidley CA et al (2006) Mortality risk among hemodialysis patients receiving different vitamin D analogs. Kidney Int 70:1858–1865

44. Kalantar-Zadeh K, Kuwae N, Regidor DL et al (2006) Survival predictability of time-varying indicators of bone disease in maintenance hemodialysis patients. Kidney Int 70:771–780

45. Kovesdy CP, Ahmadzadeh S, Anderson JE et al (2008) Association of activated vitamin D treatment and mortality in chronic kidney disease. Arch Intern Med 168:397–403

46. Shoben AB, Rudser KD, De Boer IH et al (2008) Association of oral calcitriol with improved survival in nondialyzed CKD. J Am Soc Nephrol 7

47. Wolf M, Betancourt J, Chang Y et al (2008) Impact of activated vitamin d and race on survival among hemodialysis patients. J Am Soc Nephrol 19:1379–1388

48. Palmer SC, McGregor DO, Macaskill P et al (2007) Meta-analysis: vitamin D compounds in chronic kidney disease. Ann Intern Med 147:840–853

49. Foley RN, Parfrey PS, Sarnak MJ (1998) Epidemiology of cardiovascular disease in chronic renal disease. J Am Soc Nephrol 9:S16–S23

50. System, USRD (2006) USRDS 2006 Annual Data Report.

51. Maiya S, Sullivan I, Allgrove J et al (2008) Hypocalcaemia and vitamin D deficiency: an important, but preventable, cause of life-threatening infant heart failure. Heart 94:581–584

52. Nibbelink KA, Tishkoff DX, Hershey SD et al (2007) 1,25(OH)2-vitamin D3 actions on cell proliferation, size, gene expression, and receptor localization, in the HL-1 cardiac myocyte. J Steroid Biochem Mol Biol 103:533–537

53. Wu J, Garami M, Cheng T et al (1996) 1,25(OH)2 vitamin D3, and retinoic acid antagonize endothelin-stimulated hypertrophy of neonatal rat cardiac myocytes. J Clin Invest 97:1577–1588

54. Xiang W, Kong J, Chen S et al (2005) Cardiac hypertrophy in vitamin D receptor knockout mice: role of the systemic and cardiac renin-angiotensin systems. Am J Physiol Endocrinol Metab 288: E125–E132

55. Tishkoff DX, Nibbelink KA, Holmberg KH et al (2008) Functional vitamin D receptor (VDR) in the t-tubules of cardiac myocytes: VDR knockout cardiomyocyte contractility. Endocrinology 149: 558–564

56. Bodyak N, Ayus JC, Achinger S et al (2007) Activated vitamin D attenuates left ventricular abnormalities induced by dietary sodium in Dahl salt-sensitive animals. Proc Natl Acad Sci USA 104: 16810–16815

57. Zhou C, Lu F, Cao K et al (2008) Calcium-independent and 1,25(OH)(2)D(3)-dependent regulation of the renin-angiotensin system in 1alpha-hydroxylase knockout mice. Kidney Int 74:141–143

58. Volpe M, Savoia C, De Paolis P et al (2002) The renin-angiotensin system as a risk factor and therapeutic target for cardiovascular and renal disease. J Am Soc Nephrol 13(Suppl 3):S173–S1788

59. Foley RN, Parfrey PS, Harnett JD et al (1995) The prognostic importance of left ventricular geometry in uremic cardiomyopathy. J Am Soc Nephrol 5:2024–2031

60. Edfeldt K, Agerberth B, Rottenberg ME et al (2006) Involvement of the antimicrobial peptide LL-37 in human atherosclerosis. Arterioscler Thromb Vasc Biol 26:1551–1557

61. Sorensen O, Cowland JB, Askaa J et al (1997) An ELISA for hCAP-18, the cathelicidin present in human neutrophils and plasma. J Immunol Methods 206:53–59

62. Zasloff M (2006) Inducing endogenous antimicrobial peptides to battle infections. Proc Natl Acad Sci U S A 103:8913–8914

63. Gombart A, Bhan I, Borregaard N et al (2009) Low plasma level of cathelicidin antimicrobial peptide (hCAP18) predicts increased infectious disease mortality in patients undergoing hemodialysis. Clin Infect Dis 48:418–424

64. Liu PT, Stenger S, Li H et al (2006) Toll-like receptor triggering of a vitamin D-mediated human antimicrobial response. Science 311:1770–1773

53 Role of Vitamin D and Ultraviolet Radiation in Chronic Kidney Disease

Rolfdieter Krause

Abstract   The decrease of kidney function includes exocrine and endocrine capacity. Our group developed a schedule of serial UV irradiation in patients with end-stage kidney disease on dialysis. We used sun-simulating UV lamps with two different procedures. Whole-body irradiation was done in a standing position in a cabin. Partial-body irradiation should integrate UV radiation into the measure of routine haemodialysis treatment; we irradiated only the ventral part of the legs (approx. 15% of skin surface). Using a thrice-weekly exposure with both procedures, maximal increases of circulating vitamin D were found after 8–14 weeks. Between 14 and 26 weeks parathyroid harmone levels lowered and bone mineral density increased. Physical capacity and muscle strength improved as well as cardiocirculatory adaptation and parasympathetic tone with a reduction of the disturbances of cardiocirculatory neuroregulation. No serious events and no melanoma skin cancer were registered over a total observation period of 15 years. In our experience the increase and normalization of all circulating vitamin D metabolites, and the skeletal and the pleiotropic effects seem to be superior to enteral or parenteral application without major side effects.

Key Words: Vitamin D and chronic kidney disease; UV irradiation in haemodialysis patients; Vitamin D metabolites after UV irradiation; vitamin D metabolism; renal osteodystrophy; bone mineral density; physical capacity; blood pressure; erythropoiesis; heart rate variability; UV exposure in prevention and treatment of CKD patients

1. VITAMIN D STATUS IN CHRONIC KIDNEY DISEASE

Chronic kidney disease (CKD) is a slow ongoing decrease of kidney function that includes exocrine and endocrine capacity. The role of the kidney as an endocrine organ is still underestimated but important: expression of erythropoietin, of renin and regulation of the RAS, and conversion to the hormonal effective vitamin D metabolite calcitriol. Independent of the cause of renal failure, endocrine function impairment also has different time courses *(1, 2)*.

The use of renal replacement therapy (RRT) has been established for 50 years, and the main focus was and still is to keep patients alive. At the moment, people with RRT can live up to 30 years and longer. To date the majority of patients are more than 65 years

From: *Nutrition and Health: Vitamin D*
Edited by: M.F. Holick, DOI 10.1007/978-1-60327-303-9_53,
© Springer Science+Business Media, LLC 2010

old, and the main causes of RRT are diabetes mellitus type II and arterial hypertension *(3)*. Therefore, the major challenge is to establish therapeutic measures that include all aspects of quality of life.

Historically, the focus of the role of vitamin D has been renal bone disease and secondary hyperparathyroidism. The therapy focused on suppression of the elevated levels of parathyroid hormone (PTH) and on improvement of renal osteodystrophy (ROD). However, down-regulation of the vitamin D receptor at the parathyroid gland often needs high doses of calcitriol. Regular side effects due to the physiological action of calcitriol are hypercalcemia and hyperphosphatemia.

For more than two decades it has been known that the skin has the capacity to synthesize not only cholecalciferol but also the other vitamin D metabolites. Therefore our group developed a schedule of sun-simulating UV irradiation in patients with end-stage kidney disease on dialysis to use this natural pathway in the skin *(4–6)*.

The historical background includes reports from ancient Egypt. More than 4500 years ago sunshine was used to improve general health and resistance, and to improve weakness of muscle and joints. The best established ultraviolet therapy was the cure of tuberculosis in the nineteenth century. Also, more than 100 years later, a few years ago the antibacterial substance cathelicidin was found which is expressed via a nuclear vitamin D receptor pathway in the monocyte cells *(6)*.

2. VITAMIN D SUPPLEMENTATION IN CHRONIC KIDNEY DISEASE

Medication of vitamin D metabolites has continued for nearly as long as chronic dialysis therapy has been established. Due to the main focuses of suppressing elevated PTH and high bone turnover, medication with calcitriol was the option.

Within the last 20 years it became evident that many organs can synthesize calcitriol by an autocrine mode via local vitamin D receptor activation by circulating 25-hydroxyvitamin D [25(OH)D] *(7–17)*.

During the last decade the recommendations changed, and a primary goal has also been to normalize vitamin D status in people with end-stage kidney disease. Therefore the actual guidelines recommend: if the serum level of 25(OH)D is <30 ng/ml, supplementation with vitamin D_2 (ergocalciferol) should be initiated *(2)*.

3. VITAMIN D AND UV RADIATION IN CHRONIC KIDNEY DISEASE PATIENTS ON HAEMODIALYSIS (OWN EXPERIENCES AND DATA)

The time course of chronic kidney disease typically runs for many years up to decades from early stage 1 to end-stage 5. The disturbances of the different exocrine and endocrine functions of the kidney mostly start without clinical signs and have different time courses. With the decrease of the glomerular filtration rate (GFR) the capacity to convert 1,25-dihydroxyvitamin D_3 [1,25(OH)$_2$D] is also reduced due to the damage of nephrons and by the diminished 1α-hydroxylase. It is evident that a reduction of GFR <60 ml/min decreases circulating 1,25-dihydroxyvitamin D_3. Moreover, the majority of patients with chronic kidney disease and especially with ESKD on dialysis often have a marked deficiency of 25(OH)D *(18–20)*.

Therefore, our conception was to use the ability of the skin not only to convert provitamin D via previtamin D to native vitamin D (cholecalciferol), but also to improve 25(OH)D and 1α-hyloxylase as well as the vitamin D receptor activity of the skin to synthesize calcidiol and calcitriol in the skin. Probably, under pathological conditions both 25-hydroxyvitamin D and 1,25-dihydroxyvitamin D will be given into the circulation. On the other hand, activation of the whole vitamin D metabolism due to ultraviolet radiation will improve the residual capacity of the endocrine function of the kidney by a higher serum level of 25(OH)D.

4. UV TRIALS WITH PATIENTS ON HAEMODIALYSIS

To mimic the sun's capacity we used sun-simulating UV lamps with two different procedures. Both clinical trials were done with the approval of the Ethical Commitee of the University Hospital Benjamin Franklin of the Free University of Berlin (now the Benjamin Franklin Campus of the Charité – University Medical Center Berlin).

We had enrolled a total of 22 patients on intermittent haemodialysis in both trials. The trials ran over a period of 26 weeks (whole-body radiation) and 21 weeks (partial-body radiation). The irradiation took place three times weekly before the start of or during the routine haemodialysis sessions.

4.1. Whole-Body Irradiation

The irradiation was done in a standing position in a cabin with three metal halide lamps (MHL): UVB irradiance 0.37 mW/cm^2; UVA irradiance 6.4 mW/cm^2; UV$_{tot}$ (250–400 nm) 39 W/m^2; E$_{UVB}$ (280–315 nm) 4.2 W/m^2; E$_{vitD}$ (250–400 nm) 2.3 W/m^2; E$_{erythema}$ (250–400 nm) 2.0 W/m^2. (Measurements were done by the Institute of Lighting Technics of the Technical University of Berlin; the model of an artificial sun-simulator was due to Doctors Yeni and Kaase *(21)*) (Fig. 1).

The medians of vitamin D effective irradiance (E$_{VitD}$) for the total group of 22 patients were 3.2, 8.9, and 24.5 kJ/m^2 after 7, 14, and 26 weeks, respectively, and the corresponding medians of the minimal erythema doses were 12.5, 35, and 96 MED$_{II}$.

There is a gap of 14% between E$_{VitD}$ and E$_{erythema}$ for safe suberythemal activation of the vitamin D metabolism in the skin, the "therapeutic window" *(23, 21)* (Fig. 2).

4.2. Partial-Body Irradiation

The aim was to integrate the procedure of serial UV radiation into the measure of routine haemodialysis treatment (Fig. 3). Therefore we used a tripod-leg-based UV-device for irradiating only the ventral part of the legs (e.g. approximately 15% of body surface) with fluorescent lamps (FL) and an UVB output of 3.5% *(22)* (Fig. 4).

The median of the vitamin D effective irradiance was 6.2 kJ (=24.6 MED$_{II}$). The mean increase of irradiance over time was 2.5% per session, with an absolute higher increase in the first half and a lower one in the second half.

Fig. 1. Sun-simulating UV irradiation cabin *(21)*.

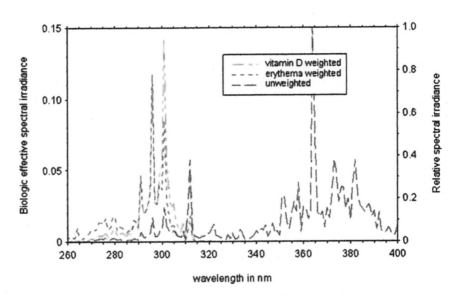

Fig. 2. Spectral output of sun-simulating metal halide lamps.

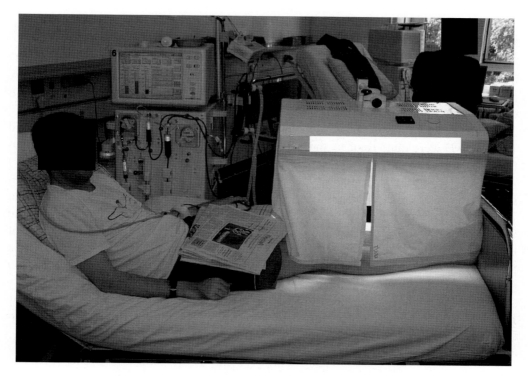

Fig. 3. Partial-body UV irradiation during a routine haemodialysis session.

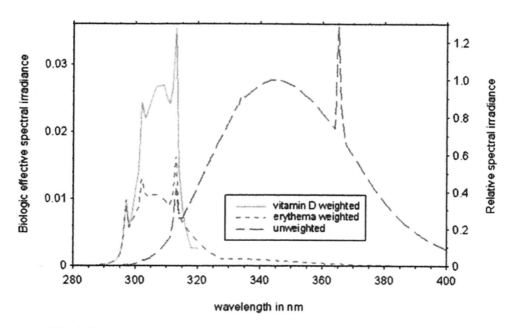

Fig. 4. Spectral output of the fluorescent lamp used for partial-body irradiation.

4.3. Vitamin D Metabolism During UVB Irradiation

During whole-body irradiation controls of serum samples of all major vitamin D metabolites (cholecalciferol, 25(OH)D, 1,25(OH)$_2$D) were done prior to and after 7, 14, and 26 weeks of the whole-body irradiation, as well as during a follow-up of another 6 months after 33, 40, and 52 weeks, respectively *(24–26)*.

During partial-body irradiation controls of serum samples of 25(OH)D and 1,25(OH)$_2$D were done prior to and after 2, 4, 6, 8, 10, 14, 17, and 21 weeks of irradiation, and after 7 weeks of follow-up, respectively *(22)*.

For a subgroup of 7 patients of the whole-body trial an UV-dose–response relationship of the vitamin D synthesis was calculated. The 7 patients received 21, 58, and 78 UV exposures with a cumulative vitamin D weighted irradiance of 4.5, 12.3, and 29.1 kJ/m^2, respectively, after 7, 14, and 26 weeks. Interestingly, the time courses of the three vitamin D metabolites were different. Cholecalciferol (native vitamin D) has a peak after 14 weeks, 25(OH)D rises constantly, and 1,25(OH)$_2$D passes into a plateau *(27, 28)* (Fig. 5).

For all 22 patients of the whole-body trial the three vitamin D metabolites show a nearly parallel course with the 7-patient subgroup. Cholecalciferol rises immedi-

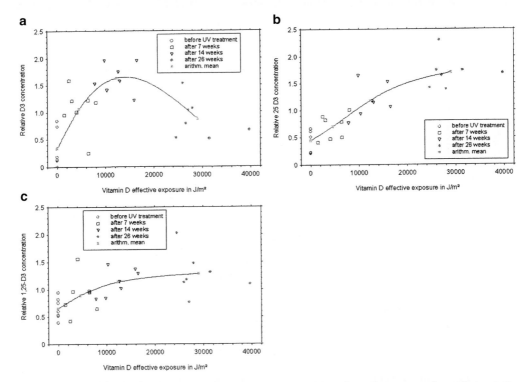

Fig. 5. Relationship of vitamin D effective UV exposure and the adjusted concentrations of circulating (**a**) vitamin D$_3$ (cholecalciferol), (**b**) 25-hydroxyvitamin D$_3$, (**c**) 1,25-dihydroxyvitamin D$_3$ over 6 months of whole-body UV irradiation.

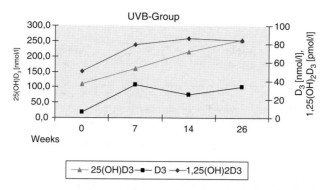

Fig. 6. Time course of the circulating blood levels of cholecalciferol (D), calcidiol (25(OH)D), and calcitriol (1,25(OH)$_2$D) during 6 months of whole-body UV irradiation.

ately, 25(OH)D increases continuously over time, and 1,25(OH)$_2$D initially follows the increase of 25(OH)D with a plateau level later on *(29, 26)* (Fig. 6).

A group of 8 patients was observed additionally over a follow-up period of another 6 months. Cholecalciferol and 1,25-dihydroxyvitamin D levels went down immediately after cessation of irradiation, and 7 weeks later both metabolites again reached the low preirradiation level. Only 25-hydroxyvitamin D remained in a normal upper level *(30, 26)* (Fig. 7).

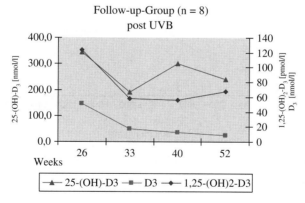

Fig. 7. Time course of the circulating blood levels of cholecalciferol (D), calcidiol (25(OH)D), and calcitriol (1,25(OH)$_2$D) over 6 months post-UVB irradiation.

Correspondingly, during partial-body irradiation, the increase of 25(OH)D had its maximum of 35% after 8 weeks from the middle normal range (51 µg/l) to the upper recommended range (77.5 µg/l). And following a physiological delay, circulating 1,25-dihydroxyvitamin D$_3$ reached its maximal increase of 48% 4 weeks later *(31, 22)* (Fig. 8).

For the vitamin D weighted irradiance dose and the serum level of 25-hydroxyvitamin D we calculated a nonlinear correlation *(32, 6)* (Fig. 9).

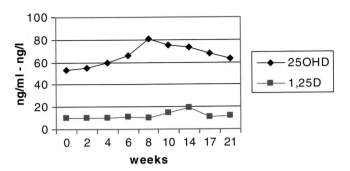

Fig. 8. Time course of the circulating blood levels of calcidiol (25(OH)D) and calcitriol (1,25(OH)$_2$D) during 5 months of partial-body irradiation of the forefront of the legs.

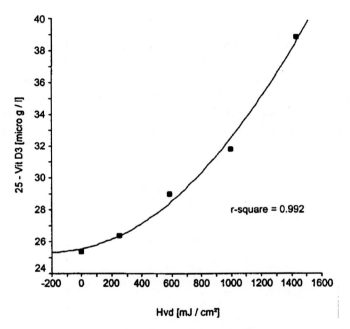

Fig. 9. Nonlinear correlation between vitamin D effective UV dose and calcidiol [25(OH)D] during partial-body irradiation.

Therefore, as well documented in healthy people, regular partial-body UVB exposure of approximately 15% of the body surface is sufficient to guarantee a normal status of vitamin D, and it is comparable with whole-body exposure *(33–36, 31)*.

In summary, from our experience using sun-simulating whole-body irradiation (with an UV output of 13%) or partial-body irradiation of approximately 15% of the skin (with an UV output of 3.5%) with thrice weekly exposure with both procedures the maximal increases of circulating cholecalciferol were found between 8 and 14 weeks. With this course our data mimic the action of the natural sun storing previtamin D in the skin, of accumulating 25-hydroxyvitamin D in the circulation, and of levelling-off 1,25-dihydroxyvitamin D$_3$ by a feed-back control.

The accumulation of each suberythemal radiation over the period of 2–3.5 months leads to a total of 15–35 MED$_{II}$, which is between one-sixth and one-third of the recommended minimal erythema dose per person for 1 year.

And, in contrast to the oral administration of active vitamin D compounds, as widely usual in the treatment of dialysis patients, the serum levels of calcium (2.3–2.4 mmol/l) and of inorganic phosphate (2.1–2.2 mmol/l) remain constant during the whole-body irradiation period of 6 months. Moreover, during the partial-body irradiation period of 21 weeks the serum levels of calcium and inorganic phosphate continuously decrease by 7 and 13%, respectively (37, 32, 22) (Fig. 10).

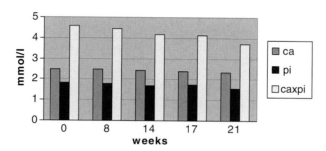

Fig. 10. Time course of blood levels of total calcium, inorganic phosphate, and calcium–phosphate product during 5 months of partial body irradiation.

4.4. Renal Osteodystrophy, Bone Mineral Density, and Bone Turnover

In general, in dialysis patients the bone mineral density (BMD) was found to be reduced by 16 to 22% in comparison with age-matched healthy people (38, 5). In our small control group of 7 patients without vitamin D supplementation over 6 months we observed a BMD reduction of 11% (median).

In 16 of the 22 patients of the whole-body trial prior to and after 26 weeks the trabecular BMD was measured by the DE-QCT technique (Somaton DRH, Siemens Erlangen). After 6 months of whole-body UVB irradiation we found a decrease of BMD of only 1.8% or 2.6%, resp. ($n = 8/7$); following 6 months after stopping the UV irradiation a further loss of only 5.6% ($n = 8$) was found(24, 39–44) (Fig. 11).

In addition, 22 patients with partial-body UV irradiation were measured by ultrasound (Lunar Corp.) before and after 21 weeks. No changes of the Z-Score or of the stiffness score were found (37).

This is in agreement with the only reference in the literature to a decrease of BMD by 2.9% also over 6 months (45).

For observing the course of PTH blood levels during the whole-body irradiation period we divided the entire group of 22 patients into a low-PTH- subgroup and a high-PTH- subgroup (both $n = 12$), with the cut-off of 120 pg/l (median of iPTH of 120 [range 280–1050 pg/l]). The course of PTH was analysed at the time when 1,25-dihydroxyvitamin D$_3$ had reached its individual maximum. Here we found a decrease of 49% in the low-PTH subgroup and of 36% in the high-PTH subgroup, respectively (26).

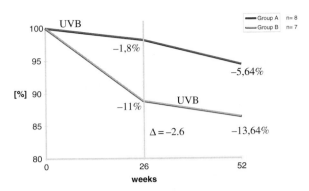

Fig. 11. Time course and decrease of bone mineral density (DE-QCT) after 6 months of whole-body UV irradiation and after following 6 months post-UVB irradiation, and over 6 months without intervention (control phase).

As markers of bone turnover we controlled osteocalcin and bone-alkaline-phosphatase. For osteocalcin we found no changes in the low-PTH subgroup, but a decrease of 31% in the high-PTH group. Correspondingly, for bone-alkaline-phosphatase no change was seen in the low-PTH subgroup, but a decrease of 40% was seen in the high-PTH subgroup.

In summary, it seems that vitamin D actions are different at the vitamin D receptor on the parathyroid gland and on the bone. With a higher level of PTH the action of vitamin D at the parathyroid VDR seems to be reduced. But, the improvement of (serum) markers of bone turnover as well as in bone mineral density was remarkable. Therefore, the guidelines of prevention and treatment of renal osteodystrophy have to recommend a vitamin D status within the (upper) normal range as well as a level of PTH within the upper normal and upward of a twofold normal range.

4.5. Physical Work Capacity and Blood Pressure

It is evident that sun exposure increases physical and endurance capacity as well as muscle power (46–50). In a subgroup of 19 patients (mean age 55 years) after three months of serial UVB irradiation we found a significant increase of workload of 6% in the bicycle ergometer test, with a significant increase in oxygen uptake of 11%, and a reduction of the accumulation of lactic acid of 9% (39, 51–58).

Additionally, we found a decrease of resting heart rate of 3%, and a reduction of systolic (significant) and diastolic blood pressure of 8 and 4%, respectively (59, 60) (Fig. 12).

This could be confirmed in 12 of the 22 patients during partial-body irradiation. Using 24-h-ABPM a significant decrease of systolic (7/12) and diastolic (10/12, $p < 0.05$) blood pressure without traces of antihypertensive medication was found (61–63, 65–67).

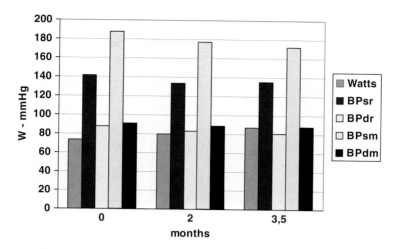

Fig. 12. Systolic and diastolic blood pressure (BP) at rest and at maximal ergometric workload after 2 and 3.5 months of whole-body UV irradiation.

4.5.1. ERYTHROPOIESIS AND UV RADIATION IN ESKD PATIENTS

Moreover, in a subgroup of eight patients the need for erythropoietin fell by 17%, and in three quarters of the patients haematocrit rose. And there was a significant increase in 2,3-diphosphoglycerate (2,3-DPG) of 17% *(25, 64)* (Figs. 13 and 14).

Also in ESKD patients the well-known increases of overall physical capacity, and of muscular and cardiocirculatory adaptation could be confirmed *(4, 23, 48, 64)*. In addition to the vitamin D mediated effects, UVB irradiation itself may have an additional effect on haematopoesis, on the intraerythrocytic oxygen affinity, and the antioxidative capacity *(68, 72, 73)*.

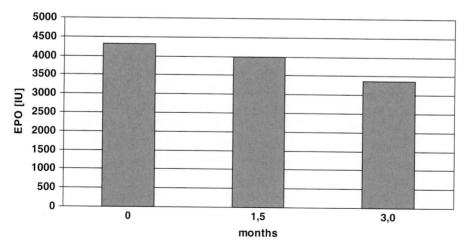

Fig. 13. Time course of close of erythropoietin administration after 1.5 and 3 months of whole-body UV irradiation.

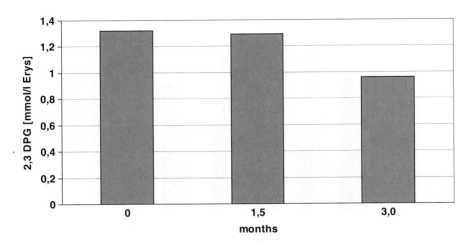

Fig. 14. Time course of the serum level of 2,3-diphosphoglycerate (2,3-DPG) after 1.5 and 3 months of whole-body UV- irradiation.

4.6. Heart Rate Variability and UV Radiation in ESKD Patients

In dialysis patients the rate of cardiac death is more than double in comparison with the general population. One cause may be the uremic neuropathy with an increase of the sympathetic tone. Heart rate variability (HRV) is one marker of the sympathetic–parasympathetic balance, and of the risk of sudden death.

In 18 patients (9 female; median age 55.5 [range 37-74] years) controls of the heart rate variability were done. We analysed the mean RR intervals and their standard deviation as well as the beat-to-beat intervals and their standard deviation.

During the whole-body irradiation intervention period we found an increase of the medians of the RR interval of 20%, of the beat-to-beat difference of 25% and of their standard deviation of 5%. The improvement of these parameters of HRV was observed at the time when the serum level of $1,25(OH)_2D$ reached the maximum *(69–71)* (Fig. 15).

In correlation with the decrease of blood pressure at rest and also during submaximal and maximal ergometric workload these data confirm previous findings from our group and from the literature in healthy people and in cardiac patients that UV exposure improves cardiocirculatory adaptation in an equal manner to that of endurance training. It also seems that serial sun(-like) UV exposure can improve parasympathetic tone and reduce the cardiocirculatory neuroregulation disturbance.

4.7. UVB Radiation in Clinical Routine of a Dialysis Unit

Over the last 15 years our dialysis unit had implemented serial UVB radiation for the treatment of renal vitamin D deficiency experience instead of oral or intravenous application of native vitamin D or active vitamin D compounds.

Therefore, a number of well-rehabilitated patients took whole-body irradiation regularly two or three times before starting the routine haemodialysis session. With this

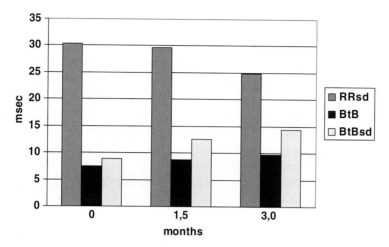

Fig. 15. Heart rate variability and beat-to-beat differences at individual maximum of blood level of calcitriol (1,25(OH)$_2$D max) after 1.5 and 3 months of whole-body UV irradiation.

measure as means, the patients had a normal level of circulating 25-hydroxyvitamin D (mean over 4–10 years: 45.3 µg/l) and near normal level of circulating 1,25-dihydroxyvitamin D$_3$ (mean over 4–10 years: 30.4 ng/l) (Fig. 16).

Moreover, in some patients the elevated blood pressure was lowered or even normalized, and/or the number of antihypertensive remedies was decreased.

No serious adverse events (e.g. severe erythema or sunburn) were registered, and no melanoma skin cancer was observed over the 15-year period, also after kidney transplantation and immunosuppressive therapy.

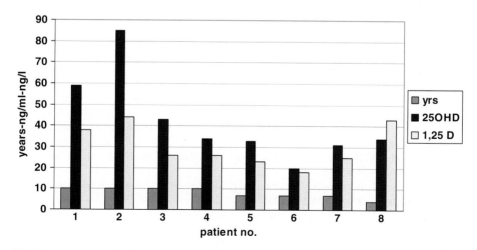

Fig. 16. Time course of the circulating blood levels of 25-hydroxyvitamin D$_3$ and 1,25-dihydroxyvitamin D$_3$ in 8 haemodialysis patients over 6–10 years of routine whole-body UV irradiation 2–3 times weekly [unpubl.].

5. VITAMIN D AND UV EXPOSURE IN PREVENTION AND TREATMENT OF CHRONIC KIDNEY DISEASE

The vitamin D deficiency in CKD patients, and especially in patients on dialysis is an independent risk factor for the function of many organs and/or metabolic and enzymatic pathways, as well as for overall and cardiovascular mortality *(9, 72, 73, 68)*.

On the other hand, approximately 50% of the (elderly) incipient patients manifest a need for dialysis treatment due to the primary diagnoses of arterial hypertension (approx. 20%) and diabetes mellitus type II (approx. 30%). For both diseases vitamin D deficiency is also an independent risk factor.

In addition to the impairment of the vitamin D metabolism, with the course of CKD and due to a reduction of physical capacity and multiple comorbidities, people with renal insufficiency often are sun deprived.

As documented in the K/DOQI guidelines vitamin D as the measure of vitamin D status and the basis for extrarenal autocrine hormonal action of vitamin D has to hold within or supplement into the normal range *(74, 13, 1, 2)*.

As demonstrated, the health benefits of UV exposure in patients on intermittent haemodialysis are in accordance with the experiences within the literature for all other groups of the population. Moreover, we had found no (major) side effects. Therefore regular outdoor sun exposure during summertime (April to September/October in latitudes >45° N/S) and/or serial suberythemal sun-simulating UVB irradiation should also be recommended as measures of prevention and treatment in people of all stages of chronic kidney disease.

REFERENCES

1. National Kidney Foundation/KDOQI (2002) Clinical practice guidelines for chronic kidney disease. Evaluation, classification, and stratification. Am J Kidney Dis 39(Suppl.1):S1–S266
2. National Kidney Foundation/KDOQI (2003) Clinical practice guidelines for bone metabolism and disease in chronic kidney disease. Am J Kidney Dis 42(Suppl.3):S1–S201
3. Frei U, Schober-Halstenberg HJ (2008) Nierenersatztherapie in Deutschland 2006/2007. QuaSiNiere gGmbH
4. Bühring M, Bocionek P, Schulz-Amling W et al (1982) Unterschiedliche Effekte einer Bestrahlung mit UVA und mit UVB. Kreislauffunktionswerte und Vigilanz nach einmaliger und nach serieller Exposition. Strahlenttherapie 158:490–497
5. Chel VGM, Ooms ME, Popp-Snijders C et al (1998) Ultraviolet irradiation corrects vitamin D deficiency and suppresses secondary hyperparathyroidism in the elderly. JBMR 3:1238–1242
6. Krause R (2010) Heliotherapie. In: Kraft K, Stange, R (ed) Lehrbuch der Naturheilverfahren. Hippokrates, Stuttgart 5:388–403
7. Bischoff-Ferrari HA (2008) Optimal serum 25-hydroxyvitamin D levels for multiple health outcomes. Adv Exp Med Biol 624:55–71
8. Bischoff-Ferrari HA, Borchers M, Gudat F et al (2004) Vitamin D receptor expression in human muscle tissue decrease with age. JBMR 19:265
9. Diffey BL (1991) Solar ultraviolet radiation effects on biological systems. Phys Med Biol 36:299–328
10. Holick MF (1981) The cutaneous photosynthesis of previtamin D3: A unique photoendocrine system. J Invest Dermatol 77:51–58
11. Holick MF (1987) Photosynthesis of vitamin D in the skin: Effect of enviromental and life-style variables. Fed Proc 46:1876–1882

12. Holick MF (1999) Noncalcemic Actions of 1,25-Dihydroxyvitamin D_3 and clinical Implications. In: Holick MF (ed) Vitamin D-physiology, molecular biology, and clinical applications. Humana Press, Totowa, NJ

13. Holick MF (2007) Vitamin D-Deficiency. NEJM 357:266–281

14. Holick MF, Uskokovis M, Henley JW et al (1980) The photoproductions of $1\alpha,25$-dihydroxyvitamin D in skin. NEJM 303:349–354

15. Lambert PW, Stern PH, Avoili RC et al (1982) Evidence for extrarenal production of $1\alpha25$-dihydroxyvitamin D in man. J Clin Invest 69:722–725

16. Lehmann B, Genehr T, Knuschke P et al (2001) UVB-induced conversion of 7-dihydrocholesterol to 1α, 25-dihydrocholesterol D3 in an in vitro human skin equivalent model. J Invest Dermatol 117: 1179–1185

17. Zehnder D, Bland R, Williams MC et al (2001) Extrarenal expression of 25-hydroxyvitamin D3-1α-Hydroxylase. J Clin Endocrinol Metab 86:888–894

18. Krause R, Roth HJ, Haas K et al (2008a) Vitamin-D-Status bei terminal Nierenkranken in Deutschland. Nieren Hochdruckkr 37:463

19. Krause R, Edenharter G, Haas K et al (2008b) Vitamin-D-Mangel als Mortalitätsrisiko bei terminal Nierenkranken. Nieren Hochdruckkr 37:464

20. Krause R, Roth HJ, Edenharter G et al (2008c) Vitamin D deficiency and cardiac mortality in patients on renal replacement therapy (RRT). JASN 19:74A

21. Yeni M (2004) Optimierte Technik für Photo- und Lichttherapie [Thesis]. Technische Universität, Berlin

22. Winter P (2007) Teilkörperbestrahlung mit einem sonnenähnlichen Wellenspektrum (UVB + UVA) während der Hämodialyse und dessen Einfluss auf den Vitamin-D-Metabolismus [MD-Thesis]. Charité, Berlin

23. Dobberke J (2009) Vergleichende Untersuchungen zur Wirkung von serieller UV-Bestrahlung und oraler Substitution von Cholecalciferol bei leichter essentieller Hypertonie [MD-Thesis]. Charité, Berlin

24. Albrecht C (1998) Wirkungen serieller UV-B-Bestrahlung auf den Knochenmineralgehalt, die Knochenmorphometrie und die Vitamin D_3-Synthese bei ambulanten chronischen Hämodialysepatienten [MD-Thesis]. Freie Universität, Berlin

25. Klamroth R (1996) Der Einfluß serieller UVB-Exposition auf die körperliche Leistungsfähigkeit ambulanter Hämodialyse-Patienten unter besonderer Berücksichtigung des Vitamin D-Stoffwechsels [MD-Thesis]. Freie Universität, Berlin

26. Matulla B (1998) Serielle UV-Therapie bei sekundärem Hyperparathyreoidismus und renaler Osteopathie (klinische Studie und Grundlagenforschung bei Hämodialysepatienten mit Vitamin D-Mangel) [MD-Thesis]. Freie Universität, Berlin

27. Krause R, Klamroth R, Bennhold I et al (1995a) Welchen Einfluß hat eine serielle UV-Bestrahlung auf den Vitamin D-Pool und die Vitamin D-Mangelfolgen. Nieren Hochdruckkr 24:511

28. Matulla B, Grothmann K et al (1998) Vorteile einer dosierten Heliotherapie gegenüber einer Substitution mit einem Vitamin D-Monopräparat. Phys Rehab Kur Med 8:140–141

29. Krause R, Matulla B, Bühring M et al (1994a) Extrarenale Calcitriol-Produktion unter serieller UV-Bestrahlung bei chronisch Nierenkranken. Phys Rehab Kur Med 4:153

30. Krause R, Matulla B, Chen TC et al (1996a) Regular UV(B)-irradiation. An effective way to normalize calcitriol deficiency in hemodialysis patients. Clin Lab 42:305–307

31. Krause R, Matulla B, Albrecht C et al (1999a) UVB can normalize Vitamin D, 25 (OH) D3 and calcitriol in ESRD patients without side effects: result of extrarenal synthesis of 1,25 (OH2) D3. J Am Soc Nephrol 10:621A

32. Krause R, Winter P, Matulla-Nolte B et al (2006) Vitamin-D-Kinetik unter UV-Exposition. Med Klin 101:95(A)

33. Krause R, Albrecht C, Bühring M et al (1996b) Vitamin-D-Haushalt und Knochendichte bei Dialysepatienten nach serieller UV-Bestrahlung. Phys Rehab Kur Med 6:155

34. Krause R, Bennhold I, Albrecht C et al (1996c) Bone mineral density, vitamin D metabolism and vitamin D receptor after UV(B) irradiation in hemodialysis patients. In: Holick MF, Jung EG (eds) Biologic effects of light 1995. de Gruyter, Berlin, New York

35. Krause R, Bennhold I, Matulla B et al (1996d) Regular UV(B) irradiation in prevention and therapy of vitamin D deficiency in hemodialysis patients. In: Holick MF, Jung EG (eds) Biologic effects of light 1995. de Gruyter, Berlin, New York
36. Krause R, Albrecht C, Matulla-Nolte B et al (1999b) Exposure to a suberythemal dose of ultraviolet irradiation prevents bone loss in hemodialysis patients. In: Holick MF, Jung EG (eds) Biologic effects of light. Kluwer Academic, Boston
37. Krause R, Winter P, Dobberke J et al (2001) Partialbody UV(B) irradiation on dialysis. Photoderm Photoimmunol Photomed 17:143
38. Chan TM, Pun KK, Cheng IKP (1992) Total and regional bone densities in dialysis patients. Nephrol Dial Transplant 7:835–839
39. Krause R, Bennhold I, Bühring M (1994b) Heliotherapie bei Dialysepatienten – eineAlternative in der Therapie der renalen Osteopathie. Nieren Hochdruckkr 23:450
40. Krause R, Albrecht C, Felsenberg D et al (1995b) Knochenmineralgehalt und Vitamin D-Haushalt unter UV(B)-Bestrahlung bei Dialysepatienten. Nieren- Hochdruckkr 24:511
41. Krause R, Bühring M, Bennhold I et al (1995c) Renale Osteopathie und serielle SUP-D3-Bestrahlung bei chronischen Hämodialysepatienten. Phys Rehab Kur Med 5:4–45
42. Krause R, Albrecht C, Felsenberg D et al (1996e) Bone mineral density and vertebral morphometry after UV-therapy in hemodialysis patients. Nephrol Dial Transplant 1:1212
43. Krause R, Albrecht C, Gowin W et al (1996f) Knochenmineralgehalt, Vitamin-D- und Parathormon-Spiegel während und nach serieller UV-Bestrahlung bei Dialysepatienten. Nieren Hochdruckkr 25:467
44. Krause R, Gowin W, Bennhold I et al (1996g) UV(B) irradiation stops loss of bone mass in hemodialysis patients. J Am Soc Nephrol 7:1792
45. Karantanas AH, Kalef-Ezra JA, Sferopoulos G et al (1996) Quantitative computed tomography for spine bone mineral measurements in chronic renal failure. Br J Radiol 69:132–136
46. Klein W, Weis V (1953) Statistische Untersuchungen über die Steigerung der Leistungsfähigkeit durch Ultraviolettbestrahlung. Arbeitsphys 15:85–92
47. Rummel M, Falkenbach R, Föhrenbach R et al (1991) Die körperliche Leistungsfähigkeit nach einer seriellen Bestrahlung mit UVB. In: Bernett P, Jeschke D (ed) Sport und medizin – Pro und contra. Zuckschwerdt, München
48. Schuh A, Kneist W, Schmidt HJ (1993) Steigerung der Ausdauerleistungsfähigkeit von durchschnittlich trainierten Personen durch natürliche Sonnenstrahlung (Heliotherapie). Phys Rehab Kur Med 3:95–99
49. Scragg R, Jackson R, Holdaway IM et al (1990) Myocardial infarction is inversely associated with plasma 25-hydroxyvitamin D3 levels: A community-based study. Int J Epidemiol 19:559–563
50. Spellerberg BAE (1952) Sportliche Leistungssteigerung durch systematische UV-Bestrahlung. Strahlentherapie 88:567–570
51. Bühring M (1986) Kreislauf- und metabolische Effekte serieller UV-Expositionen. Z Phys Med Baln Med Klim 15:170–172
52. Bühring M, Britzke K, Krause R et al (1996h) Serielle UV-Exposition mit einem natürlichen Strahlenspektrum (UVA und UVB) verbessert die Kreislaufregulation und die aerobe Kapazität (Laktatstoffwechsel) bei Patienten mit koronarer Herzerkrankung. Phys Rehab Kur Med 6:16–18
53. Glerup H, Eriksen E (2001) Hypovitaminosis D Myopathy. In: Holick MF (ed) Biological effects of light. Kluwer Academic Publishers, Boston
54. Klamroth R, Krause R, Bühring M et al (1996) Ausdauertrainingsähnliche Kreislauf- und Stoffwechseleffekte unter serieller UV-Bestrahlung bei Dialysepatienten. Phys Rehab Kur Med 6:155
55. Krause R, Bennhold I, Bühring M (1994c) Trainingsähnliche Kreislauf- und Stoffwechseleffekte bei Hämodialyse-Patienten durch serielle Heliotherapie. Nieren Hochdruckkr 11:450
56. Krause R, Matulla B, Klamroth R et al (1994d) Bestrahlungsdosis, Vitamin D-Spiegel und Kreislaufwirkungen während serieller Heliotherapie bei Hämodialysepatienten. Phys Rehab Kur Med 4:153
57. Krause R, Klamroth R, Bennhold I et al (1995d) Positive Beeinflussung von Blutdruckregulation und Myopathie bei Dialysepatienten durch serielle UV(B)-Bestrahlung. Nieren Hochdruckkr 24:510
58. Krause R, Fuhrmann I, Weber F et al (2004) Improvement of physical capacity by exercising and UV-irradiation during hemodialysis. JASN 15:412

59. Krause R, Klamroth R, Bennhold I et al (1996i) UV(B) irradiation reduces cardiac risk in hemodialysis patients. J Am Soc Nephrol 7:1452

60. Krause R, Klamroth R, Bennhold I et al (1996k) Welchen Stellenwert haben Vitamin D und Parathormon in der Blutdruckregulation bei Dialysepatienten. Nieren Hochdruckkr 25:470

61. Krause R, Klamroth R, Bennhold I et al (1996l) UV(B) irradiation reduces cardiac risk in hemodialysis (HD) patients. J Am Soc Nephrol 7:1452–1453

62. Krause R, Bühring M, Hopfenmüller W et al (1998a) Ultraviolet B and blood pressure. Lancet 352:709–710

63. Krause R, Klamroth R, Matulla-Nolte B et al (1998b) Kardioprotektive Wirkungen des UV-Lichtes. Phys Rehab Kur Med 8:140

64. Krause R, Klamroth R, Matulla-Nolte B et al (1999b) Are UV-irradiation effects similar to erythropoietin? Case study in hemodialysis patients during serial UVB-irradiation. In: Holick MF, Jung EG (eds) Biologic effects of light. Kluwer Academic, Boston

65. Krause R, Bennhold I, Britzke K et al (1995e) Reduction of cardiac risk factors in coronary and hemodialysis patients after UVB therapy. In: Holick MF , Jung EG(eds) Biologic effects of light. de Gruyter, Berlin

66. Schnaufer H (2003) Der Einfluß serieller UV-Therapie auf arteriellen Blutdruck, Herzfrequenz und Kalziumstoffwechsel bei Patienten mit milder essentieller Hypertonie [MD-Thesis]. Charité, Berlin

67. Wallis DE, Penckofer S, Sizemore GW (2008) The "Sunshine Deficit" and cardiovascular disease. Circulation 118:1476–1485

68. Röckl T (2003) Die Auswirkung von Ganzkörperbestrahlung mit verschiedenen UV-Spektren auf Antioxidantien im intravasalem Kompartiment [MD-Thesis]. Freie Universität, Berlin

69. Abel HH, Krause R, Klein B et al (1995e) Beeinflußt UV(B)-Bestrahlung die autonome Neuropathie bei Hämodialysepatienten. Nieren Hochdruckkr 24:483–484

70. Abel HH, Krause R, Klein B et al (1995f) Herzfrequenzvariabilität und Behandlung mit UV-Licht. Phys Rehab Kur Med 5:45–46

71. Claßen B (2007) Der Einfluß serieller UVB-Expositionen auf Parameter der Herzfrequenzvariabilität als Ausdruck der autonomen Neuropathie von chronischen Hämodialysepatienten unter Berücksichtigung des Vitamin-D-Stoffwechsels [MD-Thesis]. Charité, Berlin

72. Deuse U (2002) Der Einfluß serieller UV-Exposition auf die antioxidative Homöostase im menschlichen Blutplasma [MD-Thesis]. Freie Universität, Berlin

73. Howest S (2007) Verhalten der antioxidativen Eigenschaften der Blutplasmas nach UV-Ganzkörperbestrahlung [MD-Thesis]. Charité, Berlin

74. Holick MF (2005) The vitamin D epidemic and its health consequences. J Nutr 135:2739–2748

54 Role of Vitamin D in Rheumatoid Arthritis

Linda A. Merlino

Abstract Rheumatoid arthritis is an autoimmune disease that causes pain, swelling, and deformity of the joints and may result in other systemic effects. In animal models, vitamin D has been shown to suppress autoimmunity; however, in humans the role of vitamin D is less clear as to the effects on rheumatoid arthritis. There do appear to be suppressive effects of specific disease mechanisms; however, epidemiologic results on the effect of vitamin D on the development of the disease are contradictory. A better understanding of the rheumatoid arthritis immune system and the effects of vitamin D is needed before conclusions can be drawn.

Key Words: Rheumatoid arthritis; vitamin D

1. INTRODUCTION

Rheumatoid arthritis is an autoimmune disease, a disease in which the body identifies "self" as being foreign and initiates an attack on itself. Rheumatoid arthritis is characterized by chronic, painful symmetrical joint inflammation, joint stiffness, synovial swelling, and deformed joints due to bone and cartilage destruction. The disease may affect other organs and tissues in the body, such as the eyes and lungs. Rheumatoid arthritis is a disease which has existed for thousands of years [1]. The prevalence of rheumatoid arthritis among the general population is slightly less than 1% [2]. It is a complex disorder in which both genetic and environmental factors contribute. Rheumatoid arthritis is more common in women, most often develops between the ages of 30–50, and increases in prevalence with increasing age in both men and women. Potential factors that have been identified as contributing to rheumatoid arthritis include smoking, hormones (oral contraception use, pregnancy, lactation, hormone replacement therapy), and viruses [3–5]. The disease is characterized by periods of remission and flares. Additionally, rheumatoid arthritis has been associated with an increased mortality [6].

From: *Nutrition and Health: Vitamin D*
Edited by: M.F. Holick, DOI 10.1007/978-1-60327-303-9_54,
© Springer Science+Business Media, LLC 2010

2. RHEUMATOID ARTHRITIS DISEASE MECHANISM

The normal immune response is regulated so as not to perpetuate a reaction once a foreign pathogen has been contained, as well as to have a "memory" for a pathogen and a recognition of "self." However, in rheumatoid arthritis, as with other autoimmune diseases, the immune system is dysfunctional and continues the assault on the body's tissues. In the case of rheumatoid arthritis, the assault is concentrated in the synovium of joints. What initiates and perpetuates the inflammation in rheumatoid arthritis and other autoimmune diseases is very complex and not completely understood. However, studying other autoimmune and inflammatory diseases as well as the constituents of synovial tissue in established disease has provided insight into the disease mechanism of rheumatoid arthritis.

2.1. Genetic Involvement

As a disease that has both genetic and environmental contributions, one genetic area that has been identified as contributing to rheumatoid arthritis is the major histocompatibility complex (MHC) on chromosome 6. The MHC contains the human leukocyte antigen (HLA) genes that have a role in antigen presentation in rheumatoid arthritis. These genes have a function in the immune system for the recognition of "self" *(7)*. One of the antigen-presenting cells (APCs) expressing the MHC molecule is the dendritic cell. Maturation of the dendritic cell has been shown to be inhibited by 1,25-dihydroxyvitamin $D_3[1,25(OH)_2D]$ *(8, 9)*. The genes for the vitamin D receptor (VDR), which is found in the cells of the immune system, are located on chromosome 12 and have been associated with rheumatoid arthritis and bone loss *(10)*, osteoporosis *(11)*, and disease severity *(12)*. The VDR is activated by $1,25(OH)_2D$ and acts with other factors to alter the rate of transcription in genes *(8)*.

2.2. Rheumatoid Arthritis Immune System Involvement

Rheumatoid arthritis involves a very complex mechanism that is not completely understood, although it has been theorized that innate immune mechanisms may initiate the disease which then proceeds to acquire adaptive immunity mechanisms that may differ at each phase of the disease process *(7)*. Promonocytes are cells made in bone marrow, released into the blood as monocytes, and differentiated into macrophages or dendritic cells and are associated with innate immunity. Macrophages can be concentrated in synovial tissue and osteoclasts and can mature into dendritic cells. Lymphocyte (B and T cells) cells are produced in bone marrow. B cells produce antibodies and cells that differentiate along this line are considered a TH2 (T helper type 2) immune response subtype. The T cells, on the other hand, get sensitized in the thymus and upon presentation of antigens differentiate into helper, regulating, or suppressor cells, depending on the antigen. These are non-antibody producing and are associated with acquired immunity. Cells that differentiate along this line are considered a TH1 (T helper type 1) immune response subtype. Rheumatoid arthritis is a cell-mediated disease because of the predominating effects of the TH1 immune response.

Once the disease has been initiated in response to a pathogen, macrophages and lymphocytes infiltrate the synovial tissue and produce pro-inflammatory cytokines and chemokines which regulate the disease process that ultimately results in joint damage *(13)*. There are many regulatory factors and mediators involved in the disease mechanism that may activate or suppress multiple cellular structures *(14)*. The inflammatory response and cytokine milieu has been an area of intense interest that has resulted in treatments for the disease and helped lead to a better understanding of the disease process.

T-cell immune responses have classically been described as differentiated into the TH1 immune response subtype (cell-mediated immunity) or the TH2 subtype (antibody-mediated immunity). Differentiation into the TH1 subtype, generally associated with rheumatoid arthritis, is regulated by the interaction of naïve CD4$^+$ T cells and antigen-presenting cells (dendritic or macrophage) that express class II MHC molecules in the presence of co-stimulatory molecules that stimulate macrophages, synovial fibroblasts, and monocytes to produce cytokines *(15)*. The TH1 subtype results in the production of the cytokines, IL-2, tumor necrosis factor α (TNFα), and interferon-γ (INF-γ) *(16)*. IL-12 produced by activated antigen-presenting cells is a necessary cytokine for the differentiation into the TH1 subtype, whereas IL-4 is necessary for differentiation into the TH2 subtype *(17)*. It has been theorized that TH1 cytokines prevent naïve T cells from differentiating into TH2 cells and thus prevent the release of cytokines that suppress the inflammatory response (such as IL-4, IL-6, IL-10) *(7)*.

More recently, another immune response subtype, classified as TH17, has been identified which produces inflammatory responses and cytokines more consistent with that seen in rheumatoid arthritis *(18)*. This subtype also has considerable overlap with the TH1 subtype *(19)*. The cytokines IL-1β, IL-6, and IL-23 promote T-cell differentiation into the TH17 subtype response. This new line of immune subtype response is important because it has been found to be involved in the differentiation of cells and thus cytokines that promote excessive osteoclast activity and helps explain the lack of INF-γ and IL-2 in synovial joints *(20)*. Osteoclast activity is inhibited by INF-γ and enhanced by IL-17, which is dependent on TH17 helper cells producing IL-17. Since INF-γ is produced with the TH1 subtype, the inhibition of osteoclast activity is not consistent with that subtype *(21)*. Osteoclastogenesis is a normal function of bone resorption. However, the excessive osteoclastogenesis associated with rheumatoid arthritis results in the bone destruction, which is a hallmark of the disease. In addition to the above subtypes, naïve T cells can also differentiate into regulatory T cells (Treg) that regulate the helper T cells and can have a role in immune tolerance *(16)*.

3. VITAMIN D AND RHEUMATOID ARTHRITIS

3.1. Background

The effects vitamin D might exert in rheumatoid arthritis disease pathogenesis were recognized in the early 1940s when it was noticed that patients with rheumatoid arthritis were prone to an increased susceptibility to bone fractures *(22)*. Vitamin D was also used as a treatment for rheumatoid arthritis *(23)*. In 1974, a vitamin D deficiency in postmenopausal women with rheumatoid arthritis who had suffered fractures compared with postmenopausal women with rheumatoid arthritis who had not suffered fractures was reported *(24)*. The disparity was attributed to a deficiency in the dietary consumption of vitamin D as well as a lack of exposure to sun since the patients were housebound due to their disease. Other studies showing similar effects were also reported *(25)*. The use of corticosteroids in the treatment of rheumatoid arthritis was also known to contribute to corticosteroid-induced osteoporosis *(26)*, with vitamin D often prescribed to prevent this side effect. The foundation for the relationship of vitamin D with rheumatoid arthritis was the recognition that many people with rheumatoid arthritis had problems with their bone mineralization. In addition to these studies associated with rheumatoid arthritis, studies of the effects of vitamin D on other diseases affected by immune system problems such as tuberculosis, sarcoidosis, multiple sclerosis, psoriasis, and cancer were being reported *(27, 28)*.

3.1.1. IMMUNE SYSTEM EFFECTS OF VITAMIN D AND RHEUMATOID ARTHRITIS

Rheumatoid arthritis is a disease in which the immune system is aberrant. The finding that rheumatoid arthritis patients had more receptors for $1,25(OH_2)D$ in circulating lymphocytes compared to controls established a link between the disease and vitamin D *(29)*. The discovery of receptors for $1,25(OH_2)D$ in thymus and lymphocyte tissue *(30)* and in activated lymphocytes *(31)* provided further support for the idea that the immune system had a role in the production of $1,25(OH_2)D$ and that vitamin D was not just a regulator of bone mineralization and calcium homeostasis.

Using the knowledge that vitamin D had a function with the immune system and that rheumatoid arthritis resulted in bone and cartilage destruction, researchers looked for vitamin D metabolites in the synovium in order to try and understand the function of vitamin D in the pathogenesis of the disease. The fact that measureable amounts of the vitamin D metabolites $24(OH)D$ and $24,25-(OH)_2D$ were obtainable from synovial fluid and were also found in serum from patients with rheumatoid arthritis as well as other rheumatic conditions *(32)* provided a starting point for assessing the vitamin D–synovium relationship.

The synthesis of the active form, $1,25(OH_2)D$, in the synovial fluid of subjects with inflammatory arthritis compared to those with non-inflammatory arthritis suggested that macrophages appeared to be the source of production and thus played a role in the inflammatory arthritis *(33, 34)*. Since macrophages are found in abundance in synovial tissue, these findings were significant in helping to try and determine the role of vitamin D in the pathogenesis of rheumatoid arthritis. The finding that the enzyme $25-OHD_3$-1α-hydroxylase was expressed in a diverse distribution of tissue implied a role for peripheral synthesis of vitamin D and likely modulation of the immune system *(35)*.

Sophisticated tools were developed that allowed researchers to advance their understanding of the immune system at the molecular level. Additionally, concurrent research conducted in other specialized fields provided insight into the mechanisms that might be involved in the immune system and the actions of vitamin D. Some of this research showed that an infiltration of T cells and macrophages, commonly found in synovial tissue of rheumatoid arthritis patients, produced cytokines that were involved in the disease process. The association of vitamin D levels in synovial tissue with cytokine production has also been studied with the report of significantly increased levels of vitamin D metabolites in rheumatoid arthritis patients compared to osteoarthritis patients. This finding strongly suggested an extra-renal synthesis of vitamin D in synovial tissues which was stimulated by the pro-inflammatory cytokines IL-1 and/or IL-2 (36). IL-1 is also one of the macrophage-produced cytokines involved in the rheumatoid arthritis disease process.

In the immune system, vitamin D stimulates macrophages and suppresses T lymphocytes (37). Vitamin D also downregulates the expression of MCH class II groups, thus proliferation of T cells is decreased (38–42). In addition to vitamin D effects on several cytokines, one action of T-cell suppression occurs due to the inhibitory effects on IL-12 and IL-2 and INF production which suppress the TH1 subtype (43, 44). Vitamin D, through effects on naïve T cells, also promotes TH2 development (i.e., the subtype that suppresses the rheumatoid arthritis inflammatory process) (45–53).

3.2. Joint Destruction in Rheumatoid Arthritis

With the immune actions of vitamin D in mind, studying the effects of vitamin D, cytokines, and bone metabolism associated with rheumatoid arthritis was undertaken. Postmenopausal women were divided into varying levels of disease severity. The results showed that in women with high disease activity, low serum levels of $1,25(OH_2)D$ were inversely related to markers of T-cell activation, but this same marker had a positive correlation with disease activity. Thus the researchers speculated that the increased binding of $1,25(OH)_2D$ by the vitamin D receptor on activated T cells reduced serum levels of $1,25(OH)_2D$. This study also found that the pro-inflammatory cytokine IL-6 was the main determinant of increased bone resorption in postmenopausal women, thus contributing to osteoporosis (54). IL-6 was also the cytokine cited as being involved in synovial joint destruction (55).

Bones undergo a remodeling process in which osteoblasts form new bone tissue and osteoclasts resorp old bone tissue. Bone remodeling usually occurs in a balanced manner. The destruction of bone and cartilage in the joints of patients with rheumatoid arthritis has been the focus of much research in trying to understand the disease and its process. Macrophages in the synovial joints may differentiate into dendritic cells or osteoclasts. Excessive osteoclast activity (osteoclastogenesis) results in the bone destruction, which is a hallmark of rheumatoid arthritis. Osteoclast activity is dependent on TH17 helper cells producing IL17 (21). In the rheumatoid synovial joint, in the presence of excessive levels of cytokines, osteoclast precursors differentiate into mature osteoclasts, while they are also activated to produce cytokines that amplify the inflammatory response (56). The cytokine receptor activator necessary for osteoclast differentiation (RANKL) is expressed on activated T cells. The pro-inflammatory cytokines associated with rheumatoid arthritis can induce RANKL, thus initiating osteoclastogenesis (57). $1,25(OH_2)D$ along with the vitamin D receptor is also a critical component for osteoclastogenesis by stimulating RANKL which promotes osteoblast maturation to osteoclasts and inhibits a decoy receptor for RANKL – osteopoteregin – which suppresses osteoclastogenesis (45). It is not too ill conceived, therefore, that an action of

$1,25(OH_2)D$ is a possible modulator of the disease mechanism involved in the TH17 immune response subtype in rheumatoid arthritis. However, future research will have to elucidate that role.

3.3. Clinical, Animal, and Epidemiologic Studies of Vitamin D and Rheumatoid Arthritis

The remissions and flares associated with rheumatoid arthritis have also been studied in relation to further the understanding of the disease process. Women with rheumatoid arthritis have reported remission of their symptoms during pregnancy which often flare postpartum. An analysis of various cytokines and modulators of the immune system was performed and found that cytokines and substances that suppress immune function (IL-12, cortisol and 1,25-dihydroxyvitamin D_3) were higher in the third trimester of pregnancy compared to postpartum and TNFα (an immune function stimulator) was lower in the third trimester and rebounded postpartum (58).

Since studies of other autoimmune conditions have shown a latitude effect with prevalence, and the fact that vitamin D levels fluctuate with seasons, researchers looked at these issues as well as the relationship with rheumatoid arthritis severity. Latitude and seasonal effects were observed in both winter and summer with the northern latitude having lower serum levels of 25(OH)D in rheumatoid arthritis patients compared to controls. Additionally, both patients and controls had lower levels in the winter compared to the summer. However, mixed effects with regard to disease activity as measured by a rating instrument were observed in this same study (59). A prospective cohort study that followed over 83,000 nurses for 28 years reported on the geographic variation and migration patterns of subjects with new on-set rheumatoid arthritis. Place of residence at birth, age 15, and age 30 was assessed. Subjects living in the Western United States at all time periods had the lowest risk for rheumatoid arthritis and those in the Midwest and Northeast had the highest. One possible explanation was UV exposure, which stimulates the production of vitamin D (60).

In a study using c-reactive protein levels as the measure of disease severity, the c-reactive protein levels were worse for subjects with low levels of $1,25(OH_2)D$ compared to those with higher levels, whereas no relationship was seen for 25(OH)D. This was true irrespective of whether subjects were taking corticosteroids or were not taking corticosteroids (61). Controlling for disease treatment and duration, similar results were obtained in another study looking at vitamin D metabolism. This same study found a seasonal difference in serum vitamin D levels (62). Patients with recently diagnosed rheumatoid arthritis were also assessed for serum levels of 25(OH)D and $1,25(OH)_2D$ and early disease activity. Mean levels of both vitamin D metabolites were lower for patients satisfying American College of Rheumatology criteria for rheumatoid arthritis. Patients deficient in 25(OH)D at baseline had worse disease activity and severity at baseline and 1 year follow-up. 25(OH)D showed a stronger relationship than did $1,25(OH)_2D$ for disease activity and severity (63).

Rodent models have produced supporting evidence showing the suppressive inflammatory effects of vitamin D in relation to rheumatoid arthritis. Cantorna, who had studied the effects of the administration of vitamin D analogs in mice with experimental autoimmune encephalomyelitis (a multiple sclerosis-like disease), extended her studies to mice with collagen-induced arthritis (an inflammatory disease with similarities to rheumatoid arthritis). The incidence of the arthritis in the mice treated with vitamin D was reduced by 50% and those who did get arthritis had milder symptoms *(64)*. Similarly, in a study by Larsson, Mattsson et al. *(65)* they administered a vitamin D analog at various time points to rats that were immunized to induce collagen arthritis. Collagen-induced arthritis was inhibited in rats that were injected prior to or around the time of immunization and in those that were administered the vitamin D analog after immunization had a reduction in the severity of their disease. This study suggested that the timing of exposure of vitamin D relative to the pathogen exposure was important on the outcome. The animal studies further confirmed the suppressive effects of vitamin D on the immune system.

Recognizing the benefits of vitamin D not only for bone mineralization but also for suppression of immune activity, vitamin D has also been studied as a treatment for rheumatoid arthritis. Administration of alpha calcidiol for 3 months resulted in an improvement of disease activity in 89% of patients *(66)*. Another assessment of the nutritional status of patients with "general rheumatology" compared to patients with osteoarthritis showed lower serum levels of 25(OH)D, suggesting that new guidelines and greater attention to the assessment of nutritional status of rheumatology patients with regard to vitamin D need to be enacted *(67)*. Based on the lack of rheumatoid arthritis prevalence in sub-Saharan Africans, McCarty advocated a vegan diet rich in omega-3-rich fish, along with supplementation with vitamin D as a prevention strategy for rheumatoid arthritis *(68)*. Similar to the suggestions of the previous findings, a review article on the topic suggested that a more widely advocated assessment of vitamin D status and treatment in the clinical setting was warranted, given the weight of evidence of low vitamin D status in rheumatoid arthritis patients and the potential benefits of suppressing the immune response *(69)*.

There are very few epidemiologic studies that have assessed the association between vitamin D and rheumatoid arthritis. Two prospective cohort studies that have been published have produced contradictory results. Both of these studies were based on self-report of dietary intake of vitamin D. The Iowa Women's Health Study had 11 years of follow-up of 29, 368 postmenopausal women who were over the age of 55 at baseline and developed documented rheumatoid arthritis after baseline. This study found a protective association with increasing levels of dietary/supplemental vitamin D intake but did not include any measures of sun exposure *(70)*. The Nurses' Health Study had 22 years of follow-up of 186,389 women who were aged 30–55 at baseline and developed documented rheumatoid arthritis after baseline. This study did not find any effect of dietary/supplemental vitamin D intake and did include a proxy measure for sun exposure early in life *(71)*. A third study from Amsterdam that measured serum levels of 25(OH)D in blood samples of 79 patients and matched controls collected prior to onset of rheumatoid arthritis looked at the association of serum levels of vitamin D and newly diagnosed cases of rheumatoid arthritis. They looked at three timeframes prior to onset and found no association with the development of rheumatoid arthritis when compared to matched controls *(72)*. Studies on the relationship of vitamin D in the etiology of rheumatoid arthritis are contradictory and need further investigation.

Vitamin D has a role in specific mechanisms of the immune system of rheumatoid arthritis patients. However, a comprehensive understanding of the individual mechanisms on the complete activation of the immune system is not clear. There are multiple possible areas of action in which vitamin D may impact on the pathogenesis of rheumatoid arthritis, especially given the extensive effects of this hormone. Vitamin D has been shown to suppress T-cell activity through several actions and to modulate cytokines that can alter the course of the disease. Whether the effects of vitamin D are limited to disease severity/progression or have a role in the etiology is not completely clear. Ongoing research is leading to a better understanding of the immune system with regard to rheumatoid arthritis. The knowledge gained will provide better information, in the future, on the role of vitamin D in the pathogenesis of rheumatoid arthritis.

REFERENCES

1. Rothschild B, Woods R, Rothschild C et al (1992) Geographic distribution of rheumatoid arthritis in ancient North America: implications for pathogenesis. Semin Arthritis Rheum 22:181–187
2. Helmick C, Felson D, Lawrence R et al (2008) Estimates of the prevalence of arthritis and other rheumatic conditions in the United States. Part I. Arthritis Rheum 58:15–25
3. Hernandez-Avila M, Liang M, Willett W et al (1990) Reproductive factors, smoking, and the risk for rheumatoid arthritis. Epidemiol 1:285–291
4. Silman A, Newman J, MacGregor A (1996) Cigarette smoking increases the risk of rheumatoid arthritis. Results from a nationwide study of disease-discordant twins [see comments]. Arthritis Rheum 39:732–735
5. Criswell L, Merlino L, Cerhan J et al (2002) Cigarette smoking and the risk of rheumatoid arthritis among postmenopausal women: results from the Iowa Women's Health Study. Am J Med 112:465–471
6. Mikuls T, Saag K, Criswell L et al (2002) Mortality risk associated with rheumatoid arthritis in a prospective cohort of older women: results from the Iowa Women's Health Study. Ann Rheum Dis 61:994–999
7. Harris E, Budd R, Genovese M, Firestein G, Sargent J, Ruddy S, Sledge C (2005) Kelley's textbook of rheumatology, 7th edn. Saunders, Philadelphia
8. Griffin M, Xing N, Kumar R (2003) Vitamin D and its analogs as regulators of immune activation and antigen presentation. Annu Rev Nutr 23:117–145
9. Piemonti L, Monti P, Sironi M et al (2000) Vitamin D3 affects differentiation, maturation, and function of human monocyte-derived dendritic cells. J Immunol 164:4443–4451
10. Gough A, Sambrook P, Devlin J et al (1998) Effect of vitamin D receptor gene alleles on bone loss in early rheumatoid arthritis. J Rheumatol 25:864–868
11. Rass P, Pakozdi A, Lakatos P et al (2006) Vitamin D receptor gene polymorphism in rheumatoid arthritis and associated osteoporosis. Rheumatol Int 26:964–971
12. Gomez-Vaquero C, Fiter J, Enjuanes A et al (2007) Influence of the BsmI polymorphism of the vitamin D receptor gene on rheumatoid arthritis clinical activity. J Rheumatol 34:1823–1826
13. Firestein G (2007) Evolving concepts of rheumatoid arthritis. Nature 423:356–361
14. Choy E, Panayi G (2001) Cytokine pathways and joint inflammation in rheumatoid arthritis. N Engl J Med 344:907–916
15. Feldmann M, Maini S (2008) Role of cytokines in rheumatoid arthritis: an education in pathophysiology and therapeutics. Immunol Rev 223:7–19
16. Boissier M, Assier E, Falgarone G et al (2008) Shifting the imbalance from Th1/Th2 to Th17/treg: the changing rheumatoid arthritis paradigm. Joint Bone Spine 75:373–375
17. Park H, Li Z, Yang XO et al (2005) A distinct lineage of CD4 T cells regulates tissue inflammation by producing interleukin 17. Nat Immunol 6:1133–1141
18. McGeachy M, Cua D (2008) Th17 cell differentiation: the long and winding road. Immunity 28: 445–453

19. Harrington L, Hatton R, Mangan P et al (2005) Interleukin 17-producing CD4+ effector T cells develop via a lineage distinct from the T helper type 1 and 2 lineages. Nat Immunol 6:1123–1132
20. Sato K (2008) Th17 cells and rheumatoid arthritis -from the standpoint of osteoclast differentiation. Allergol Int 57:109–114
21. Nakashima T, Takayanagi H (2008) The dynamic interplay between osteoclasts and the immune system. Arch Biochem Biophys 473:166–171
22. Baer GJ (1941) Fractures in chronic arthritis. Ann Rheum Dis 2:269–273
23. Addis H, Currie R (1950) Hypercalcaemia during vitamin D treatment of rheumatoid arthritis. Br Med J 1:877–879
24. Maddison P, Bacon P (1974) Vitamin D deficiency, spontaneous fractures, and osteopenia in rheumatoid arthritis. Br Med J 4:433–435
25. O'Driscoll S, O'Driscoll M (1980) Osteomalacia in rheumatoid arthritis. Ann Rheum Dis 39:1–6
26. Demartini F, Grokoest A, Ragan C (1952) Pathological fractures in patients with rheumatoid arthritis treated with cortisone. J Am Med Assoc 149:750–752
27. Hayes C, Cantorna M, DeLuca H (1997) Vitamin D and multiple sclerosis. Proc Soc Exp Biol Med 216:21–27
28. Garland C, Garland F, Gorham E (1999) Calcium and vitamin D. Their potential roles in colon and breast cancer prevention. Ann NY Acad Sci 889:107–119
29. Manolagas S, Werntz D, Tsoukas C et al (1986) 1,25-Dihydroxyvitamin D3 receptors in lymphocytes from patients with rheumatoid arthritis. J Lab Clin Med 108:596–600
30. Provvedini D, Rulot C, Sobol R et al (1987) 1 alpha,25-Dihydroxyvitamin D3 receptors in human thymic and tonsillar lymphocytes. J Bone Miner Res 2:239–247
31. Bhalla A, Amento E, Serog B et al (1984) 1,25-Dihydroxyvitamin D3 inhibits antigen-induced T cell activation. J Immunol 133:1748–1754
32. Fairney A, Straffen A, May C et al (1987) Vitamin D metabolites in synovial fluid. Ann Rheum Dis 46:370–374
33. Hayes M, Denton J, Freemont A et al (1989) Synthesis of the active metabolite of vitamin D, $1,25(OH)_2D_3$, by synovial fluid macrophages in arthritic diseases. Ann Rheum Dis 48:723–729
34. Smith S, Hayes M, Selby P et al (1999) Autocrine control of vitamin D metabolism in synovial cells from arthritic patients. Ann Rheum Dis 58:372–378
35. Zehnder D, Bland R, Williams M et al (2001) Extrarenal expression of 25-hydroxyvitamin d(3)-1 alpha-hydroxylase. J Clin Endocrinol Metab 86:888–894
36. Inaba M, Yukioka K, Furumitsu Y et al (1997) Positive correlation between levels of IL-1 or IL-2 and $1,25(OH)_2D/25$-OH-D ratio in synovial fluid of patients with rheumatoid arthritis. Life Sci 61: 977–985
37 Lemire J (1992) Immunomodulatory role of 1,25-dihydroxyvitamin D3. J Cell Biochem 49:26–31
38. Tsoukas C, Watry D, Escobar S et al (1989) Inhibition of interleukin-1 production by 1,25-dihydroxyvitamin D3. J Clin Endocrinol Metab 69:127–133
39. Tokuda N, Mizuki N (1992) 1,25 Dihydroxyvitamin D3 down regulation of HLA-DR on human peripheral blood monocytes. Immunology 75:349–354
40. Rigby W, Waugh M (1992) Decreased accessory cell function and co-stimulatory activity by 1,25 dihydroxyvitamin D3 treated monocytes. Arthritis Rheum 35:110–119
41. Tokuda N, Kano M, Meiri H et al (2000) Calcitriol therapy modulates the cellular immune responses in hemodialysis patients. Am Nephrol 20:129–137
42. Clavreul A, D'Hellencourt C, Montero-Menei C et al (1998) Vitamin D differentially regulates B7.1 and B7.2 expression on human peripheral blood monocytes. Immunology 95:272–277
43 Manetti R, Parronchi P, Guidizi M (1993) Natural killer cell stimulations factor [IL-12] induces T helper type 1 (TH1) specific immune responses and inhibits the development of IL-4 producing Th cells. J Exp Med 177:1199–1204
44. Mattner F, Smiroldo S, Galbiati F et al (2000) Inhibition of Th1 development and treatment of chronic-relapsing experimental allergic encephalomyelitis by a non-hypercalcemic analogue of 1,25-dihydroxyvitamin D(3). Eur J Immunol 30:498–508
45. Dusso A, Brown A, Slatopolsky E (2005) Vitamin D. Am J Physiol Renal Physiol 289:F8–F28

46 Reichel H, Koeffler H, Tobler A (1987) 1a,25 dihydroxyvitamin D3 inhibits gamma interferon synthesis by normal human peripheral blood lymphocytes. Proc Natl Acad Sci USA 84:3385–3389

47 Gepner P, Amor B, Fournier C (1989) 1,25-dihydroxyvitamin D3 potentiates the in vitro inhibitory effects of cyclosporin A on T cells from rheumatoid arthritis patients. Arthritis Rheum 32:31–36

48. Rigby W, Hamilton B, Waugh M (1990) 1,25 Dihydroxyvitamin D3 modulates the effects of interleukin 2 independent of IL2 receptor binding. Cell Immunol 125:396–414

49. al Janaki M, al-Balla S, al-Dalaan A, et al (1993) Cytokine profile in systemic lupus erythematosus, rheumatoid arthritis, and other rheumatic diseases. J Clin Immunol 13:58–66

50. Muller K, Odun N, Bendtzen K (1993) 1,25 Dihydroxyvitamin D3 selectively reduces interleukin 2 levels and proliferation of human T cell lines in vitro. Immunology 35(Lett.):177–178

51. Spiegelberg J, Beck L, Stevenson D (1994) Recognition of T cell epitopes and lymphokine secretion by rye grass allergen - specific human T cell clones. J Immunol 152:4706–4711

52. Jirapongsananuruk O, Melamed I, Leung DY (2000) Additive immunosuppressive effects of 1,25-dihydroxyvitamin D3 and corticosteroids on TH1, but not TH2, responses. J Allergy Clin Immunol 106:981–985

53. Rausch-Fan X, Leutmezer F, Willheim M et al (2002) Regulation of cytokine production in human peripheral blood mononuclear cells and allergen-specific Th cell clones by 1alpha,25-dihydroxyvitamin D3. Int Arch Allergy Immunol 128:33–41

54. Oelzner P, Franke S, Muller A et al (1999) Relationship between soluble markers of immune activation and bone turnover in post-menopausal women with rheumatoid arthritis. Rheumatology 38:841–847

55. van Leeuwen M, Westra J, Limburg P et al (1995) Interleukin-6 in relation to other proinflammatory cytokines, chemotactic activity and neutrophil activation in rheumatoid synovial fluid. Ann Rheum Dis 54:33–38

56. Boyce B, Schwarz E, Xing L (2006) Osteoclast precursors: cytokine-stimulated immunomodulators of inflammatory bone disease. Curr Opin Rheumatol 18:427–432

57. Sato K, Takayanagi H (2006) Osteoclasts, rheumatoid arthritis and osteoimmunology. Curr Opin Rheumatol 18:419–426

58. Elenkov I, Wilder R, Bakalov V et al (2001) IL-12, TNF-alpha, and hormonal changes during late pregnancy and early postpartum: implications for autoimmune disease activity during these times. J Clin Endocrinol Metab 86:4933–4938

59. Cutolo M, Otsa K, Laas K et al (2006) Circannual vitamin d serum levels and disease activity in rheumatoid arthritis: Northern versus Southern Europe. Clin Exp Rheumatol 24:702–704

60. Costenbader K, Chang S, Laden F et al (2008) Geographic variation in rheumatoid arthritis incidence among women in the United States. Arch Intern Med 168:1664–1670

61. Oelzner P, Muller A, Deschner F et al (1998) Relationship between disease activity and serum levels of vitamin D metabolites and PTH in rheumatoid arthritis. Calcif Tissue Int 62:193–198

62. Kroger H, Penttila I, Alhava E (1993) Low serum vitamin D metabolites in women with rheumatoid arthritis. Scan J Rheumatol 22:172–177

63. Patel S, Farragher T, Berry J et al (2007) Association between serum vitamin D metabolite levels and disease activity in patients with early inflammatory polyarthritis. Arthritis Rheum 56:2143–2149

64. Cantorna M, Munsick C, Bermiss C et al (2000) 1,25 Dihydroxycholecalciferol prevents and ameliorates symptoms of experimental murine inflammatory bowel disease. J Nutr 130:2648–2652

65. Larsson P, Mattsson L, Klareskog L et al (1998) A vitamin D analogue (MC 1288) has immunomodulatory properties and suppresses collagen-induced arthritis (CIA) without causing hypercalcaemia. Clin Exp Immunol 114:277–283

66. Andjelkovic Z, Vojinovic J, Pejnovic N et al (1999) Disease modifying and immunomodulatory effects of high dose 1 alpha (OH) D3 in rheumatoid arthritis patients. Clin Exp Rheum 17:453–456

67. Mouyis M, Ostor A, Crisp A et al (2008) Hypovitaminosis D among rheumatology outpatients in clinical practice. Rheumatology 47:1348–1351

68. McCarty M (2001) Upregulation of lymphocyte apoptosis as a strategy for preventing and treating autoimmune disorders: a role for whole-food vegan diets, fish oil and dopamine agonists. Med Hypotheses 57:258–275

69. Leventis P, Patel S (2008) Clinical aspects of vitamin D in the management of rheumatoid arthritis. Rheumatology 47:1617–1621
70. Merlino L, Curtis J, Mikuls T et al (2004) Vitamin D intake is inversely associated with rheumatoid arthritis: results from the Iowa Women's Health Study. Arthritis Rheum 50:72–77
71. Costenbader K, Feskanich D, Holmes M et al (2008) Vitamin D intake and risks of systemic lupus erythematosus and rheumatoid arthritis in women. Ann Rheum Dis 67:530–553
72. Nielen M, van Schaardenburg D, Lems W et al (2006) Vitamin D deficiency does not increase the risk of rheumatoid arthritis: comment on the article by Merlino et al. Arthritis Rheum 54:3719–3720

55 Vitamin D, Respiratory Infections, and Obstructive Airway Diseases

Carlos A. Camargo Jr, Adit A. Ginde, and Jonathan M. Mansbach

Abstract   Over the past decade, interest has grown in the effects of vitamin D on respiratory infections and obstructive airway diseases (OADs), such as asthma and chronic obstructive pulmonary disease (COPD). Studies suggest that low vitamin D levels increase the risk of acute respiratory infections, which may contribute to incident wheezing illness and cause asthma exacerbations. Although unproven, the increased risk of susceptible hosts to specific respiratory pathogens may contribute to some cases of incident asthma. Likewise, the effect of vitamin D on COPD, while intriguing, is largely unknown. Emerging evidence provides biological plausibility for some of these respiratory findings. For example, vitamin D-mediated innate immunity, particularly through enhanced expression of the human cathelicidin antimicrobial peptide, is important in host defenses against respiratory pathogens. Vitamin D also modulates regulatory T-cell function and interleukin-10 production, which may increase the therapeutic response to corticosteroids in corticosteroid-resistant asthma. Finally, low vitamin D levels may have a role in the pathogenesis of allergies, including anaphylaxis. Further studies, especially randomized controlled trials, are needed to better understand vitamin D's effects on respiratory infections and OADs.

Key Words: Respiratory infections; bronchiolitis; asthma; chronic obstructive pulmonary disease; obstructive airway disease; hCAP-18; interleukin-10; atopic dermatitis; anaphylaxis

1. INTRODUCTION

Research on the effects of vitamin D on the respiratory system could be said to have begun more than a century ago, with work on the beneficial effects of sunlight on tuberculosis (1). This research led to important changes in clinical practice, including the development of health facilities in the mountains to provide tuberculosis patients with better access to sunlight, fresh air, and other putative health benefits. In subsequent years, however, major advances in anti-tuberculosis medications, along with the discovery of vitamin D per se (and its beneficial effects on rickets and overall bone health), led to a dramatically decreased interest in the role of sunlight (or vitamin D) on lung health.

Over the last decade, researchers have again taken interest in the possible antimicrobial effects of vitamin D (2, 3). In addition to studies on the respiratory infection

From: *Nutrition and Health: Vitamin D*
Edited by: M.F. Holick, DOI 10.1007/978-1-60327-303-9_55,
© Springer Science+Business Media, LLC 2010

risk of nutritionally deficient, often rachitic, children in developing nations *(4, 5)*, more recent studies have examined the vitamin D–respiratory infection association in otherwise well-nourished individuals in developed nations *(6, 7)*. In 1999, emerging data on the immunologic effects of vitamin D led Wjst and Dold to propose that vitamin D supplementation might be the cause of global increases in asthma and allergies *(8)*. Genetic association studies were undertaken to examine this putative harm [e.g., by examining polymorphisms of the vitamin D receptor (VDR) gene] but they provided conflicting results *(9–13)*.

It was in this context that Camargo and colleagues proposed the *opposite* hypothesis at the 2006 meeting of the American Academy of Allergy, Asthma, and Immunology *(14)* – i.e., widespread vitamin D *insufficiency* (not excess) might explain recent increases in childhood wheezing and asthma. The investigators presented original evidence for this hypothesis from a birth cohort study involving almost 1,200 Boston mother–child pairs. In brief, they found that maternal intake of vitamin D during pregnancy had a striking inverse association with recurrent wheezing in children (Fig. 1). Because most wheezing illness of childhood represents uncomplicated respiratory infection, as opposed to incident asthma *(15)*, the authors cautioned that further research would need to disentangle these two overlapping outcomes (i.e., respiratory infections and asthma).

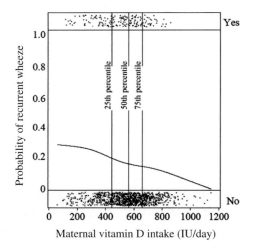

Fig. 1. Association between maternal vitamin D intake during pregnancy and risk of recurrent wheeze in offspring (*n* = 1,194 mother–child pairs in Boston area). The *top band* of *dots* (Yes) represents children who developed recurrent wheeze, and the *bottom band* of *dots* (No) represents children who did not. The *vertical lines* represent percentiles of maternal vitamin D intake during pregnancy (adapted from Ref. *(56)*).

The relation of vitamin D status to respiratory infections and asthma has become a focus of research teams around the world. In this chapter, we will examine these two outcomes, as well as the potential role of vitamin D on chronic obstructive pulmonary disease (COPD). We will also touch briefly on biological mechanisms that may underlie the complex relationship between vitamin D, respiratory infections, and obstructive airway diseases (OADs).

2. RESPIRATORY INFECTIONS

Respiratory infections are a major cause of morbidity and mortality *(16)*. Although they are often discussed collectively, they are a surprisingly diverse and complex group of infections. In addition to basic issues of disease chronicity (i.e., acute versus chronic infection), respiratory infections are usually categorized by the anatomical areas involved (i.e., upper versus lower tract disease). Acute *upper* respiratory infections include a variety of diagnoses, such as acute nasopharyngitis (e.g., the "common cold"), acute sinusitis, acute pharyngitis, and acute laryngitis and tracheitis (e.g., croup); some researchers also include acute otitis media. By contrast, acute *lower* respiratory infections include many other important and common diagnoses, including bronchiolitis, bronchitis, and pneumonia. Although this may sound quite orderly, it is important to recognize that respiratory infections may (and often do) involve multiple locations.

Another layer of complexity is introduced by the categorization of respiratory infections according to pathogen. For example, acute respiratory infections may be caused by the so-called "cold" viruses [e.g., rhinovirus, adenovirus, respiratory syncytial virus (RSV), parainfluenza] or common bacterial pathogens (e.g., *Streptococcus pneumoniae, Haemophilus influenzae*). The catch-all phrase "respiratory infection" also includes distinctive diseases due to specific pathogens of major public health importance. One example is influenza, the virus responsible for the 1918 flu pandemic that killed more than 50 million people worldwide *(17)*. Another important example is *Mycobacterium tuberculosis*, the cause of tuberculosis.

Bronchiolitis provides an excellent example of the complexity of research on acute respiratory infections *(18)*. Although bronchiolitis is a leading cause of infant hospitalization, with an annual cost of $543 million in the United States alone *(19)*, it remains a clinical diagnosis with rather vague diagnostic criteria *(20–22)*. For example, in 2006, the American Academy of Pediatrics (AAP) position paper described the typical child with bronchiolitis as being of age <2 years and having "rhinitis, tachypnea, wheezing, cough, crackles, use of accessory muscles, and/or nasal flaring" *(23)*. In other countries, however, we know of strong opinions that bronchiolitis only should be diagnosed among children of age <1 year or that the presence of specific exam findings (e.g., crackles) should be mandatory.

One might think that a better understanding of the microbiologic causes of bronchiolitis would help with disease classification. Indeed, when discussing the clinical course of bronchiolitis, clinicians rely heavily on descriptions of bronchiolitis caused by RSV, the most common cause *(24, 25)*. While almost all children are infected with RSV by age 2 years *(26, 27)*, most children do not present to an acute care setting with clinical bronchiolitis *(26, 28)*. For example, we estimate that <10% of US children infected with RSV will present to the emergency department with bronchiolitis and that only 2–3% are hospitalized *(29, 30)*. Clearly, there is a spectrum of disease severity that extends beyond current understanding. To further complicate matters, many other viruses have been linked to bronchiolitis (Table 1), including rhinovirus *(31, 32)* and a continually

Table 1

Frequency of Common Viruses Linked to Bronchiolitis in Three Care Settings, As Identified by Polymerase Chain Reaction (Adapted from *(18)*)

	Outpatient (%)[a]	*Emergency (%)*[b]	*Inpatient (%)*[c]
Respiratory syncytial virus	11–27	64	69
Parainfluenza	5–13	na	3
Influenza	1–5	4	2
Rhinovirus	46–49	16	21
Metapneumovirus	2	7	5
Combination infections	10–17	14	16

Abbreviation: na denotes not available.

[a]Based on upper and lower respiratory infections during infants' first winter *(136)* and first year *(137)*.

[b]Based on children presenting to 14 US emergency departments *(138)*.

[c]Based on children hospitalized to 15 US hospitals (Camargo et al., unpublished data).

growing number of new viruses and strains. An unknown percentage of cases may be caused by bacteria such as *Mycoplasma pneumoniae (33)*. Thus, increased appreciation of the actual microbiologic cause of bronchiolitis, and its short-term and long-term clinical implications, would further complicate the nosology of this common childhood disease.

Given the complexity and diversity of "respiratory infections," one might expect that it would be difficult to link an individual factor like vitamin D to risk of developing this composite outcome. To be more specific, vitamin D might have an important association with specific types of respiratory infections, as defined by specific anatomical areas and by specific pathogenic organisms. Vitamin D might also influence disease severity. If these hypotheses were confirmed, the common practice of collapsing all of these diverse infections into one composite outcome would tend to obscure real associations – i.e., one would conclude that there was no association when, in fact, there was one. The scientific literature on vitamin D and respiratory diseases needs to be interpreted with this important caveat in mind.

3. OBSTRUCTIVE AIRWAY DISEASES

3.1. Asthma

Asthma is another common medical condition that is associated with high morbidity and health-care utilization *(34, 35)*. Although the definition of asthma has challenged clinicians for literally centuries, there is fairly widespread acceptance today of the 1987 definition from the American Thoracic Society *(36)*: a chronic lung disease characterized by (1) airway narrowing that is reversible (though not always completely), either spontaneously or with treatment; (2) airway inflammation; and (3) bronchial hyper-responsiveness (BHR) to a variety of stimuli.

Although this definition has proven helpful for some types of researchers, it is very difficult to apply in epidemiologic studies, where subjects may be dispersed over large geographic areas and where spirometry and hyper-responsiveness testing usually are not feasible. For that reason, asthma epidemiologists have relied on much simpler definitions of asthma, such as an affirmative response to the question: "Do you have doctor-diagnosed asthma?" Although the limitations of this epidemiologic approach are self-evident, this approach actually performs with adequate accuracy (37) to allow epidemiologists to perform asthma surveillance and to begin to examine risk factors of the disease. Similar yes/no questions have led to widespread recognition and concern about the dramatic rise in "asthma" over the past few decades (34).

While respiratory infections occur commonly across the lifespan, and viral infections are common triggers of asthma exacerbations (38, 39), the vast majority of asthma begins in early childhood. Estimates vary but approximately 80–90% of asthma begins before the age of 6 years, with 70% of asthmatic children having asthma symptoms before 3 years of age (40, 41). The etiology of asthma has proven elusive, but we can infer from these clinical observations that the major risk factors must be present in early life. Atopy is one of the strongest risk factors of asthma (42), yet many asthmatic individuals are non-atopic (43). Indeed, there is growing appreciation that the term "asthma" is a syndrome composed of heterogeneous diseases (44).

3.2. Chronic Obstructive Pulmonary Disease

COPD is another common medical condition that is associated with high morbidity and mortality (45). The most widely accepted definition for COPD comes from the Global Initiative for Chronic Obstructive Lung Disease (GOLD): "a preventable and treatable disease with some significant extrapulmonary effects that may contribute to the severity in individual patients. Its pulmonary component is characterized by airflow limitation that is not fully reversible. The airflow limitation is usually progressive and associated with an abnormal inflammatory response of the lung to noxious particles or gases" (45).

As with asthma, there have been significant disagreements about the definition of COPD over the past few decades. Classical descriptions of COPD have emphasized two major types: "chronic bronchitis" and "emphysema." Although these COPD types continue to be taught in medical schools, there is now widespread recognition of their considerable overlap. Uncertainty regarding COPD diagnosis has been compounded by the fact that chronic bronchitis was, for many years, a clinical diagnosis (e.g., requiring that the patient has a productive cough for a specified number of months over a specific number of years), while emphysema was an anatomic diagnosis requiring actual parenchymal destruction. The GOLD definition now acknowledges that the chronic airflow limitation characteristic of COPD is caused by a mixture of small airway disease (obstructive bronchiolitis) and parenchymal destruction (emphysema), with the relative contributions varying from person to person (45).

4. CHILDHOOD WHEEZING

While experts may differ on the precise definitions of specific respiratory infections, asthma, and COPD, everyone agrees on the heterogeneity of early childhood wheezing. After all, wheezes are just musical sounds caused by the passage of air through narrowed respiratory tract airways, a condition with many different possible causes. Across the life span, this airway narrowing is most frequently caused by respiratory infections (edema) and/or OADs (mixture of infection, inflammation, and bronchocon- striction). Wheezing can also be caused by many, less common conditions in children (e.g., airway foreign body) and in adults (e.g., decompensated heart failure with the so-called "cardiac wheezing").

Cohort studies have provided important information about the natural history of child- hood wheezing. For example, the landmark studies of Martinez and colleagues clearly demonstrate that many children who wheeze in early childhood have transient episodes during respiratory infections and do not go on to develop asthma *(15)*. Researchers try, with varying levels of success, to divide wheezing children into three differ- ent clinical groups: transient early wheezers, non-atopic wheezers, and IgE-associated wheeze/asthma (Fig. 2). The figure nicely demonstrates the clinical adage that "All that wheezes is not asthma." To further underscore the difference between childhood wheez- ing and asthma, many young children with asthma often present with recurrent nocturnal cough and no wheezing whatsoever *(46)*. For all of these reasons, one should be cau- tious about generalizing from findings on even *recurrent* childhood wheezing to actual asthma.

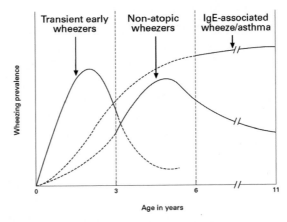

Fig. 2. Hypothetical prevalence of three different wheezing phenotypes in childhood. For each age interval, the overall wheezing prevalence is the sum of the areas under each *curve*. The *dashed lines* emphasize the possibility of different curve shapes due to many factors, including overlap between the three groups (adapted from Ref. *(135)*).

5. VITAMIN D AND RESPIRATORY INFECTION

5.1. Tuberculosis

As noted earlier, the first links between vitamin D and respiratory disease stretch back to more than a century. Niels Ryberg Finsen was awarded the 1903 Nobel Prize in Phys- iology and Medicine *(1)* in recognition of his innovative work showing that concentrated light radiation could effectively treat lupus vulgaris [skin tuberculosis (TB)]. Although he did not know the mechanism of this therapeutic benefit, it is likely that at least part of it was due to increased levels of vitamin D. Likewise, for a large part of the twentieth century, sunlight exposure was used as treatment for TB.

In recent years, many investigators have linked vitamin D more directly to TB (47, 48). For example, a hospital-based case–control study in London found that vitamin D deficiency was associated with an odds ratio (OR) of 2.9 [95% confidence interval (CI) 1.3–6.5] for having active TB (49). Susceptibility to TB has also been linked to vitamin D receptor polymorphisms, with the presence of FokI F allele protecting against TB infection, and the TaqI t allele protecting against active disease but not infection (50). In another recent study, Martineau and colleagues (51) reported that a single dose of 2.5 mg (100,000 IU) of vitamin D (ergocalciferol) enhanced immunity to M. tuberculosis. The relationship between vitamin D status and TB is an active area of research.

5.2. *Epidemiologic Studies on Respiratory Infections*

In addition to TB, vitamin D may have antimicrobial activity against a broad range of respiratory pathogens. Many studies have reported that children with rickets commonly present to hospitals with respiratory infections (52). Although 25-hydroxyvitamin D [25(OH)D] levels of at least 10 ng/ml (25 nmol/l) are known to prevent rickets, the relationship between higher 25(OH)D thresholds and respiratory infection is of greater public health interest, particularly in developed nations. Recent clinical studies have demonstrated a fairly consistent association between lower, but non-rachitic 25(OH)D levels, and increased risk of respiratory infections (6, 7, 53, 54).

For example, a case–control study in India demonstrated that children 2–60-months old with serum 25(OH)D levels <20 ng/ml had 12-fold higher odds of acquiring a severe acute lower respiratory infection (53). Moreover, a Turkish case–control study found that serum 25(OH)D levels were lower in neonatal cases of acute lower respiratory infection (9.1 ng/ml) than in age-matched controls (16.3 ng/ml) (54).

In contrast, a recent Canadian case–control study of children 1–25-months old (55) found no difference in mean serum 25(OH)D levels between patients with acute lower respiratory tract infection (30.8 ng/ml) and hospital controls (30.9 ng/ml). Of note, the average vitamin D status of these children was >30 ng/ml because "virtually all of the infants ... consumed vitamin D" through fortified infant formula or supplements. We believe that inadequate exposure variation did not allow for proper testing of the study hypothesis.

One of the first cohort studies on this topic was reported in 2007 by Laaksi and colleagues in Finland (6). They reported that young male soldiers with serum 25(OH)D levels <16 ng/ml at baseline had a 63% increased risk of absence from duty due to respiratory infection over the following 6 months, as compared to soldiers with levels of ≥16 ng/ml ($P = 0.004$).

Camargo and colleagues established an important relationship between vitamin D status and incident wheezing in two separate birth cohorts at higher latitudes: one in Massachusetts (56) and one in New Zealand (7). In the Massachusetts cohort, Project Viva, the investigators found that lower maternal intake of vitamin D during pregnancy was associated with significantly increased risk of recurrent wheezing (56), as shown in Fig. 1. In these 1,194 mother–child pairs, the mean (SD) total vitamin D intake during pregnancy was 548 (167) IU/day. By the age of 3 years, 186 children (16%) had recurrent wheeze. Compared with mothers in the lowest quartile of daily intake (median 356 IU), those in the highest quartile (724 IU) had a lower risk of having a child with recurrent wheeze (OR 0.39; 95% CI 0.25–0.62; P for trend <0.001). A 100-IU increase in vitamin D intake was associated with lower risk (OR 0.81; 95% CI 0.74–0.89), regardless of whether vitamin D was from the diet (OR 0.81) or supplements (OR 0.82). Adjustment for 12 potential confounders, including maternal intake of other dietary factors, did not change the results.

This Boston finding was replicated in an independent birth cohort by Devereux and colleagues in Scotland (57). In the 1,212 mother–child pairs, the median (interquartile range) total vitamin D intake during pregnancy was 128 (103–165) IU/day. Compared to mothers in the lowest quintile of daily intake (median 77 IU), those in the highest quintile (275 IU) had lower risk for ever wheeze (OR 0.48; 95% CI 0.25–0.91), wheeze in the previous year (OR 0.35; 95% CI 0.15–0.83), and persistent wheeze (OR 0.33; 95% CI 0.11–0.98) in their 5–year-old children. By contrast, there was no association between maternal vitamin D intake during pregnancy and doctor-diagnosed asthma at the age of 5 years ($P = 0.98$).

In a New Zealand birth cohort, Camargo and colleagues (7) were the first to examine the association between low cord blood levels of 25(OH)D and subsequent risk of respiratory infections and childhood wheezing. In this seemingly healthy cohort of >900 children, 20% of children had a cord blood 25(OH)D level <10 ng/ml, and 73% (cumulative) had a 25(OH)D level <30 ng/ml. Cord blood 25(OH)D level had an inverse association with the risk of wheezing illness at ages of 15 months, 3 years, and 5 years (all $P < 0.05$). Moreover, cord blood 25(OH)D level was inversely associated with the risk of respiratory infection by the age of 3 months: children with 25(OH)D levels <10 ng/ml were at a twofold higher risk than those with levels ≥30 ng/ml. These results were independent of season and other potential confounders.

To further explore this issue, we analyzed nationally representative data from Third National Health and Nutrition Examination Survey (NHANES-III) to examine the association between vitamin D status and upper respiratory infection (58). In brief, we found that lower serum 25(OH)D levels were associated with an increased adjusted OR of recent upper respiratory infection (compared with ≥30 ng/ml: OR 1.36 for <10 ng/ml and 1.24 for 10–29 ng/ml groups) (58). As shown in Fig. 3, the inverse association was present in all seasons. We also found that the association between serum 25(OH)D <10 ng/ml and upper respiratory infection was stronger among individuals with asthma (OR 5.67) versus those without asthma (OR 1.24; P for interaction $= 0.007$).

5.3. Preliminary Evidence from Interventional Trials

Although there have been few interventional trials to date, the preliminary data are promising. For example, two interventional cohort studies with 600–700IU of vitamin D intake daily from cod liver oil/multivitamin supplementation (59) and 60,000 IU weekly from a vitamin D/calcium supplement (60) noted a decrease in respiratory infections in children receiving supplementation. One randomized controlled trial of bone loss in postmenopausal black women found that 7.7% of women randomized to

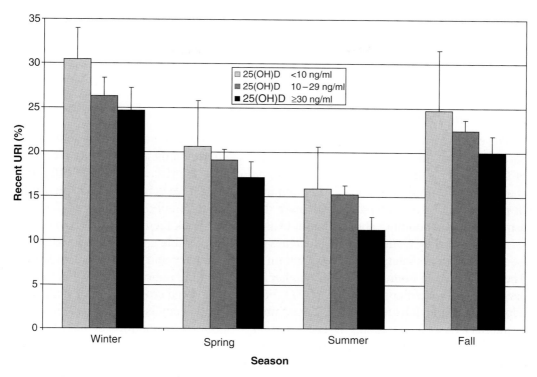

Fig. 3. Association between serum 25-hydroxyvitamin D level and recent upper respiratory infection (URI), by season ($n = 18,883$ participants in Third National Health and Nutrition Examination Survey). Error bars represent standard errors of the estimates (adapted from Ref. *(58)*).

vitamin D, 800–2,000 IU/day, and reported respiratory symptoms over the 3-year follow-up, compared with 25.0% in the control group *(61)*. Another substudy of a randomized controlled trial for fracture prevention *(62)* found a possible reduction in wintertime infection in participants randomized to vitamin D, 800 IU/day (adjusted OR 0.90; 95% CI 0.76–1.07), though the result was not statistically significant.

The first interventional study specifically designed to test the vitamin D–respiratory infection hypothesis has been reported in abstract form *(63)*. The investigators randomized 162 adults in New York to vitamin D3 2,000 IU/day versus placebo for 12 weeks between December and March. The mean 25(OH)D level at baseline was similar between the two groups (vitamin D 25.6 ng/ml versus control 25.2 ng/ml) and increased to 35.5 ng/ml in the vitamin D group while staying "virtually constant" among controls. Despite this increase, the investigators found no benefit on incidence or severity of viral upper respiratory tract infections in this relatively small and healthy sample of adults.

Taken together, the interventional studies provide support for an inverse association between vitamin D supplementation and respiratory infection – despite some recent null studies. The positive studies were not designed to test the study hypothesis, and they were limited by the relatively low dose of vitamin D supplementation used; the impact of elevating 25(OH)D levels to >30 ng/ml (or even >40 ng/ml) merits investigation. Such trials would be particularly attractive in high-risk populations, such as children in day-care centers, residents of nursing homes, hospital personnel, or individuals with OAD.

6. VITAMIN D AND ASTHMA

6.1. Asthma Pathogenesis

The exact role of vitamin D in the pathogenesis of asthma remains unclear. Given the early age of asthma onset (64, 65), and growing evidence on the developmental origins of health and disease (66), the role of maternal diet on the risk of asthma in offspring is a particularly intriguing topic. Although investigators have debated the potential role of nutrient intake on asthma risk for more than a decade, multiple reviews of diet and asthma have said nothing about vitamin D as a potential risk factor for asthma (67–70).

As discussed by Camargo and colleagues (56), considerable epidemiologic data suggest a possible association between vitamin D insufficiency and the asthma epidemic. The prevalence of both conditions is higher in racial/ethnic minorities, obese individuals, and westernized populations (34, 71). Furthermore, large cross-sectional studies of adolescents and adults have found that vitamin D insufficiency is correlated with lower pulmonary function, including forced expiratory volume in 1 s and forced vital capacity (72, 73). Of course, the causal nature of these cross-sectional associations is uncertain. Although it is possible that low levels of vitamin D cause asthma, it is also possible that individuals with asthma exercise outdoors less often and that simple lifestyle differences could create vitamin D insufficiency. Likewise, some children with asthma historically have avoided milk intake (an important dietary source of vitamin D in US children) (56), and this food avoidance could also create a spurious inverse association between vitamin D intake and asthma. Prospective studies (especially randomized trials) could help investigators to address these important methodological concerns.

As noted earlier, the hypothesis that low vitamin D may cause asthma – and that sunlight exposure and supplements may prevent asthma – is disputed by Wjst, who has described vitamin D supplementation as an important cause of asthma (8, 74). The strongest support for this putative harm comes from a birth cohort in northern Finland (75). In this study, Hypponen, Wjst, and colleagues reported that regular vitamin D supplementation (≥2,000 IU/day) in the first year of life increased the risks of developing atopy, allergic rhinitis, and asthma by the age of 31 years. However, this finding may be related to the very high dose of vitamin D supplementation (i.e., a dose-specific effect). The Finnish study was also limited by the absence of data on maternal intake of vitamin D and the inability to control for major confounders. Furthermore, recall bias may have affected the ascertainment of early life asthma and allergies. Another recent study from the United Kingdom also raised the possibility that higher levels of vitamin D might cause atopy and asthma (76). This study also has important methodological limitations (e.g., small sample and only 30% follow-up at the age of 9 years). Nevertheless, these two European studies raise possible concerns.

Other evidence, however, suggests that vitamin D has no apparent association with incident asthma or may even have a modest protective effect. For example, preliminary evidence from two family-based studies demonstrated that gene polymorphisms on the vitamin D receptor were associated with childhood and adult asthma (9, 10), but these results were not confirmed (11, 13). Also, as described in Section 5, birth cohorts in Boston (56) and Scotland (57) found an inverse association between maternal intake of vitamin D during pregnancy and risk of childhood wheezing. The Scottish investigators reported that lower maternal vitamin D intake was associated with a borderline significant decrease in bronchodilator response ($P = 0.04$), but there was no association with doctor-diagnosed asthma or other asthma-related outcomes, such as spirometry or exhaled nitric oxide concentration (57). Both findings were limited, however, by the measurement of vitamin D status from diet and supplements alone. Thus, the unique contribution of the New Zealand birth cohort is the concurrent demonstration that low serum 25(OH)D levels in cord blood were associated with increased risk of respiratory infections and childhood wheezing but had no association with the risk of current asthma at the age of 5 years (7). The null finding applied to both atopic asthma and non-atopic asthma, with atopy defined by skin-prick testing for common allergens at the age of 15 months.

Taken together, it seems unlikely that vitamin D status in pregnancy is a major contributor to the current asthma epidemic. We hypothesize that if a modest association exists between vitamin D insufficiency and incident asthma, then it would be mediated through increased risk of respiratory infection in early life. In a recent study from the Childhood Origins of Asthma (COAST) birth cohort in Wisconsin, Jackson and colleagues (77) found that wheezing rhinovirus illness during infancy predicted development of asthma at 6 years of age. Linking these rhinovirus data and the aforementioned birth cohort data, it is possible that vitamin D insufficiency may contribute to the seasonal nature of the infectious bronchiolitis in infants (78). Early viral respiratory infections in a genetically susceptible host may induce subsequent asthma development, and if vitamin D mediates some part of this pathway, there might be a new strategy for preventing some cases of asthma. Likewise, chronic infection with atypical bacteria may participate in asthma pathogenesis for a subset of patients, in addition to playing a role in lung inflammation and corticosteroid resistance in asthma (79). For these susceptible individuals, correction of vitamin D deficiency not only would provide overall health benefits (71) but might contribute to asthma prevention.

6.2. Asthma Control

Most people with asthma respond well to inhaled corticosteroids, the "preferred" long-term control medication for asthma according to the 2007 NIH asthma guidelines (80). Nevertheless, asthma control on even "optimal" controller therapy leaves room for improvement (81). This is particularly true in a subset of asthmatic patients with known corticosteroid resistance (82). Xystrakis and colleagues in London recently studied the effect of vitamin D in this relatively small asthma subgroup (83). They administered vitamin D to a small group of healthy individuals and steroid-resistant asthmatic patients and found that the intervention enhanced subsequent responsiveness to dexamethasone due to induction of interleukin (IL)-10. The authors concluded that vitamin D could potentially increase the therapeutic response to corticosteroids in steroid-resistant asthma patients. These experimental results await replication. Moreover, the implications of this finding to the larger population of asthmatic patients with suboptimal control (despite the use of inhaled corticosteroids) are not known.

6.3. *Asthma Exacerbation*

Viral respiratory infections, particularly rhinovirus, are associated with 50–85% of asthma exacerbations *(38, 39)*. Most adults experience 2–4 upper respiratory infections per year, while children experience 6–10 *(84)*. Individuals with asthma may be more susceptible to respiratory infection and have increased frequency of lower respiratory tract symptoms of higher severity and duration *(85)*. The emerging role of vitamin D in innate immune responses may explain predisposition to infection and asthma exacerbation in certain populations.

As presented earlier in this review, there is growing evidence of a "protective" association between vitamin D and respiratory infection in the general population *(6, 7, 53, 54, 59, 61)*. However, there are only sparse data on the association in asthmatic individuals. In our analysis of the NHANES-III data *(58)*, we found that the association between serum 25(OH)D levels <10 ng/ml and upper respiratory infection was stronger among individuals with asthma (OR 5.67) compared with those without asthma (OR 1.24; P for interaction = 0.007). Moreover, Litonjua and colleagues *(86)* recently examined the association between serum 25(OH)D levels and risk of severe asthma exacerbations (defined as an asthma-related emergency department visit or hospitalization). In 1,022 asthmatic children in the Childhood Asthma Management Program (CAMP), they found that children with low baseline 25(OH)D levels (<30 ng/ml) were more likely to have a severe asthma exacerbation over a 4-month period (OR 1.50; 95% CI 1.13–1.98).

7. VITAMIN D AND COPD

7.1. *COPD Pathogenesis*

Although vitamin D insufficiency may also play a role in COPD pathogenesis, data are sparse. Indirect evidence for an association comes from numerous studies reporting an increased association between osteoporosis and COPD *(87–92)*, usually independent of corticosteroid use. Although serum 25(OH)D levels have not been directly linked to COPD, Black and Scragg *(72)* reported a strong dose–response correlation in NHANES between serum 25(OH)D levels and pulmonary function, including forced expiratory volume in the first second (FEV_1), which is a key marker for increased susceptibility and diagnosis of COPD. They found that the adjusted FEV_1 was 126 ml lower in the lowest 25(OH)D quintile compared to the highest quartile. Moreover, this difference appeared higher in participants with doctor diagnosis of chronic bronchitis (248 ml) or emphysema (344 ml), although the test for interaction was underpowered and not statistically significant. Proposed mechanisms were explored in the accompanying editorial *(93)* and include the role of 1,25(OH)D in modulating the formation of matrix metalloproteinases, fibroblast proliferation, and collagen synthesis *(94, 95)*. Additionally, vitamin D-mediated immunomodulation could affect airway inflammation, a central process in the pathogenesis of COPD *(96)*.

Genetic studies have also suggested a possible association between vitamin D and COPD pathogenesis *(12, 97–99)*. Specifically, the vitamin D-binding protein [also known as the group-specific component of serum globulin (Gc-globulin)] is the major carrier protein of 25(OH)D in blood. In a Japanese population, Ito and colleagues found a strong association between Gc-globulin polymorphism and increased susceptibility to COPD *(99)*. While these data appear promising, the association between vitamin D-binding protein and COPD progression, as measured by the rate of decline in pulmonary function, has yielded mixed results *(99, 100)*. Large prospective studies are needed to address the potential role of vitamin D in COPD pathogenesis.

7.2. Acute Exacerbations of COPD

Similar to asthma, many acute exacerbations of COPD (AECOPD) are caused by common respiratory viral pathogens, such as rhinovirus, influenza, parainfluenza, and respiratory syncytial virus. While no direct data link vitamin D and AECOPD, vitamin D-mediated immune mechanisms may play a role in the prevention of respiratory infection and associated AECOPD. In our analysis of the NHANES-III data *(58)*, we found that the association between serum 25(OH)D levels <10 ng/ml and upper respiratory infection might be stronger among individuals with COPD (OR 2.26) compared with those without COPD (OR 1.27); statistical power was limited, however, and the difference was not statistically significant (*P* for interaction = 0.30). Further studies of the role of vitamin D in AECOPD are warranted.

8. POTENTIAL MECHANISMS

8.1. Vitamin D and Innate Immunity

Emerging evidence indicates that vitamin D has an important role in the innate immune system, which helps to prevent infection without the need for immunologic memory from previous exposure to the pathogen *(52)*. Innate immunity includes the production of antimicrobial peptides that are important in host defenses against respiratory pathogens, such as viruses, bacteria, and fungi *(101–105)*. These peptides include β-defensins and cathelicidins (e.g., hCAP-18 or LL-37), and are produced on epithelial surfaces and within circulating leukocytes *(52)*.

The only human cathelicidin antimicrobial peptide (hCAP-18) provides a particularly attractive explanation for the apparent inverse association between serum 25(OH)D levels and risk of respiratory infection. hCAP-18 enhances microbial killing in phagocytic vacuoles, acts as a chemoattractant for neutrophils and monocytes, and has a defined vitamin D-dependent mechanism *(105)*. In a landmark study, Liu and colleagues *(102)* reported that in *M. tuberculosis*-infected macrophages, there was a 30-fold increased cathelicidin expression in $1,25(OH)_2D$-treated cells compared with controls, which corresponded to a 50% reduction in *M. tuberculosis* viability at 3 days. The individuals with serum 25(OH)D levels less than approximately 10 ng/ml had the least efficient cathelicidin expression, and those with serum 25(OH)D levels above approximately 30 ng/ml had the highest induction of cathelicidin mRNA. Furthermore, black individuals, known to have increased susceptibility to TB infection, had low serum 25(OH)D levels and inefficient cathelicidin mRNA induction, but supplementation of 25(OH)D to normal range enhanced cathelicidin induction fivefold, to similar levels as the white patients.

Liu and colleagues extended these findings to provide further evidence that cathelicidin is the mechanism that enhances vitamin D-mediated antimicrobial activity against *M. tuberculosis (103)*. In these experiments, a short, interfering RNA was used specifically to block cathelicidin mRNA and protein expression, which eliminated vitamin D-mediated enhanced intracellular killing of *M. tuberculosis* that was observed in controls.

Thus, vitamin D-mediated increases in cathelicidin provide an excellent explanation for the extensively described link between sun exposure, vitamin D, and TB *(47, 48)* – and presumably other respiratory infections. In brief, pathogenic antigens interact with Toll-like receptors on macrophages to upregulate the expression of genes that code for the vitamin D receptor and for the 1α-hydroxylase enzyme that converts 25(OH)D into the biologically active 1,25-$(OH)_2$D *(102)*. In turn, 1,25-$(OH)_2$D interacts with the promoter on the cathelicidin gene and enhances hCAP-18 production in myeloid cells *(101)*, bronchial epithelial cells *(104)*, and keratinocytes *(106)*. Furthermore, Weber and colleagues found that 25(OH)D could induce intracellular hCAP-18 through the autocrine induction of the 1α-hydroxylase enzyme *(106)*. These diverse pathways would be of particular importance for neonates *(107)*, a group with a significant correlation ($r = 0.23$; $P = 0.002$) between cord blood levels of 25(OH)D and hCAP-18 (Camargo et al., unpublished data).

8.2. Vitamin D and Adaptive Immunity

Another proposed mechanism for vitamin D-mediated effects on respiratory infections and OADs includes modulation of antigen-presenting cells such as macrophages *(108, 109)*. Vitamin D also affects the generation of regulatory T cells (Treg) *(110, 111)* that express potentially inhibitory cytokines (IL-10 and TGFβ) and the ability to potently inhibit antigen-specific T-cell activation *(112)*. A murine model of pulmonary eosinophilic inflammation demonstrated that vitamin D supplementation of adult mice led to changes in cytokines, IgE levels, and airway eosinophilia during allergen sensitization *(113)*.

Although many laboratory studies suggest that vitamin D induces a shift in the balance between Th1- and Th2-type cytokines toward Th2 dominance *(52, 114)*, Pichler and colleagues *(115)* found that in $CD4^+$ and $CD8^+$ human *cord* blood cells, vitamin D not only inhibits IL-12-generated interferon-gamma production (Th1 type) but also suppresses IL-4 and IL-4-induced expression of IL-13 (Th2 type). In theory, this balanced Th1–Th2 regulation may modulate asthma and other allergic diseases. The role of regulatory T cells and IL-10 in the balance of the T-helper type 1 (Th1)-type and Th2-type cytokines and asthma phenotype was recently reviewed *(116)*. Thus, the differences between the studies on the Th1–Th2 dominance may lie in the timing of exposure of the cells to vitamin D (i.e., prenatal versus postnatal); the response of naïve T cells to vitamin D exposure may differ from that of mature cells when exposed to vitamin D *(117)*. Another possibility is that the association depends on the vitamin D status of the individual. In other words, lower vitamin D intakes (e.g., to correct a deficiency state) may have different consequences than relatively high-dose supplementation, where an excess of vitamin D may indeed have adverse effects. These hypotheses merit further investigation.

8.3. Vitamin D, Atopy, and Allergies

Although this chapter focuses on the respiratory system, the allergic foundations of asthma support a brief discussion of how vitamin D may affect atopy and allergy risk. *Atopy* is defined as IgE sensitization to a variety of allergens, such as different foods (e.g., peanut) and insect venoms. Atopy is different from an actual *allergic reaction*, which requires actual clinical manifestations (e.g., atopic dermatitis, allergic rhinitis, and allergic asthma). Finally, *anaphylaxis* is the most severe form of an acute allergic reaction and requires involvement of multiple organ systems *(118)*. The biological mechanisms of how vitamin D might influence atopy and allergies are presented briefly above (section 8.2) and discussed in more detail elsewhere *(52, 108, 114)*. Although the clinical and epidemiological data are limited, there are a few studies of potential relevance to the present discussion that we will briefly summarize here.

An allergic condition of particular interest in the study of asthma is *atopic dermatitis*, an early step in the classic "allergic march" *(119)*. The nosology of this and related conditions (e.g., eczema) is controversial but we consider eczema, the more inclusive skin condition, of which atopic eczema (atopic dermatitis) is one example. Based on vitamin D-mediated effects on innate immunity, and the anticipated reductions in bacterial colonization of the skin, we hypothesize that vitamin D might have beneficial effects on both the incidence and the severity of atopic dermatitis. Recent evidence that vitamin D supplementation increases skin production of cathelicidin in patients with atopic dermatitis *(120)* supports this hypothesis. Indeed, ecologic studies suggest higher prevalence of eczema at higher latitudes *(121, 122)*. Moreover, a recent case–control study demonstrated that atopic dermatitis was more common among obese individuals with low 25(OH)D levels (<10 ng/ml), as compared to obese individuals with 25(OH)D levels of ≥30 ng/ml *(123)*. Another epidemiologic study, however, suggested a potential *increase* in eczema risk *(76)*, and two birth cohorts with crude outcome data found no association at all *(56, 57)*. In terms of disease modification, investigators in Boston performed the first randomized, double-blind, placebo-controlled trial in a small sample of children with winter-related atopic dermatitis and found evidence of skin improvement with vitamin D supplementation *(124)*. Larger clinical trials are underway.

There are sparse data on the relation of vitamin D with allergic rhinitis. The Finnish study by Hypponen and colleagues suggested that supplementation of infants with high daily doses (≥2,000 IU/day) increased risk *(75)*. A cross-sectional study by Wjst and Hypponen also reported a positive association between serum 25(OH)D levels and prevalence of allergic rhinitis *(125)*, consistent with their hypothesis that higher levels of vitamin D cause allergic disease. Although the authors acknowledge that these results could be due to confounding, we believe that a more likely explanation is reverse causality (e.g., individuals with allergic rhinitis are less likely to go outside and, consequently, have lower levels of 25(OH)D). Prospective cohort studies are needed to better define the relation of vitamin D with allergic rhinitis, which is a strong risk factor for childhood asthma *(42)*.

More evidence for a link between vitamin D and allergic conditions comes from recent ecologic studies of EpiPen prescribing *(126–128)*, a surrogate marker for anaphylaxis *(129)*. Camargo and colleagues *(126)* observed a strong north–south gradient for the prescription of EpiPens in the United States, with the highest rates found in New England (Fig. 4). Population-adjusted rates were positively associated with several factors (e.g., number of health-care providers); however, a multivariate analysis controlling for all of these factors suggested that they did not mediate the strong north–south gradient. The investigators hypothesized that vitamin D may play a role in the etiology of anaphylaxis *(126)*. They recently confirmed this north–south gradient using a similar study design but restricted to urban areas in order to better control for this important determinant of health-care utilization *(127)*. In a third ecologic analysis, Mullin and Camargo recently demonstrated a south–north gradient for prescription of self-injectable epinephrine in Australia *(128)*, consistent with the distribution of UVB in the Southern hemisphere.

Although the atopic dermatitis findings can be explained by vitamin D-mediated benefits on innate immunity *(130)*, the anaphylaxis findings *(126–128)* present a conundrum: How is it that vitamin D lowers the risk of childhood wheezing (respiratory

EpiPen Prescriptions per 1,000 State Residents

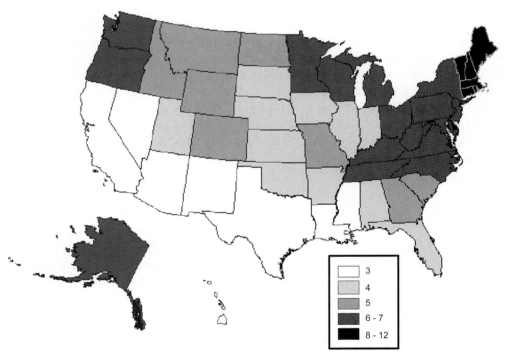

Fig. 4. Regional differences in EpiPen prescriptions (per 1,000 persons), a surrogate measure of anaphylaxis prevalence (adapted from Ref. *(126)*).

infections) and possibly anaphylaxis but does not lower the risk of actual asthma? The answer undoubtedly involves a number of host and environmental factors, as well the heterogeneity of the respiratory/allergy outcomes.

9. FUTURE RESEARCH ON VITAMIN D, RESPIRATORY INFECTIONS, AND OADs

Although the preliminary data in this chapter are promising, many scientific gaps remain. Figure 5 provides a summary of what is known in early 2009. Low levels of vitamin D appear to increase the risk of infections, with an unproven (but probably small) role on incident asthma. By contrast, it seems likely that increased risk of respiratory infections among asthmatic individuals would worsen asthma severity and control. There are sparse data on the effect of vitamin D on the two major components of asthma (inflammation and BHR) but vitamin D status in early life has no apparent association with incident childhood asthma. Finally, the relation of low vitamin D status to atopy and allergic conditions is uncertain, with theoretical reasons suggesting why allergy risk should be higher with supplementation but ecologic studies suggesting the opposite (i.e., areas of highest UVB exposure have the lowest rates of EpiPen prescribing).

Low vitamin D levels

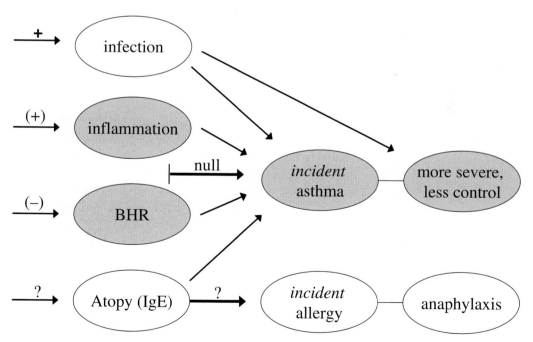

Fig. 5. Summary of current understanding of the associations between low vitamin D status and major respiratory/allergic conditions. BHR denotes bronchial hyper-responsiveness. Associations are summarized as follows: + denotes positive; (+), maybe positive; null, no apparent association; (−), maybe negative; and ?, mixed results or unknown.

Although there are many possible "next steps," an initial one would be to examine the association between vitamin D status at birth (e.g., using cord blood 25(OH)D levels) with respiratory infections in early childhood. It is important for other groups to confirm the New Zealand birth cohort findings *(7)* in diverse populations of children. Another important step would be to further describe the association between vitamin D and respiratory infection as it relates to the development of asthma, chronic asthma control, and risk for asthma exacerbations. These issues also are critically important for COPD, a field that remains almost entirely unexplored.

In all of these epidemiologic studies, the issue of reverse causation should be carefully addressed, particularly in cross-sectional studies. Because prevalent OAD, decreased OAD control, and increased OAD exacerbations probably contribute to decreased time outdoors (and, therefore, decreased sunlight exposure), the presence of vitamin D insufficiency may actually follow, rather than cause, the outcomes measured. The use of longitudinal cohort designs will help to address this very important methodological issue. Future epidemiologic studies also will need to carefully account for the likelihood of confounding from the many health-related factors associated with frequent sun exposure, such as outdoor physical activity *(131)*.

Additionally, measurement of immune markers, including hCAP-18, regulatory T cells, IL-10, and other markers of the Th1–Th2 balance, may help to elucidate the mechanisms of the observed associations and provide additional face validity. Animal models in which these relevant pathways can be manipulated or knocked out will provide additional support and rationale for the associations between vitamin D and OAD. However, animal models should be selected with care. For example, it appears that only primates have the vitamin D response element on the promoter of the cathelicidin gene. Accordingly, mouse, rat, and dog cell lines do not appear to require vitamin D for cathelicidin expression and thus have been unsuccessful in evaluating this vitamin D-mediated pathway *(101)*. Thus, many animal models may be of limited utility in testing the hypothesized association between vitamin D and respiratory infections.

Ultimately, large randomized controlled trials of vitamin D supplementation will be needed to confirm the ability to reverse the suboptimal outcomes associated with vitamin D insufficiency. In these studies, higher doses of supplementation (at least 1,000–2,000 IU/day and probably more in high-risk populations during winter months) may be required to maximize potential benefit. Current IOM recommendations for vitamin D supplementation (e.g., 200–600 IU/day, depending on age) are unlikely to achieve the serum 25(OH)D levels (i.e., at least 30–40 ng/ml) that appear necessary for good general health, including prevention of infections. Many experts already argue that current recommended doses of vitamin D supplementation are woefully inadequate to meet the need for higher serum 25(OH)D levels *(71, 132)*. For example, raising serum 25(OH)D levels from 20 to 32 ng/ml requires an additional 1,700 IU of vitamin D per day *(133)*.

Consideration to adequate dosage in future trials and measurement of pre- and postsupplementation serum 25(OH)D levels will help to optimize trial conditions. Additionally, as the relative importance of vitamin D in different age and racial/ethnic groups is unknown, diverse study participants or multiple studies of different demographic subgroups will help us to understand this potential interaction.

Progress in these endeavors will undoubtedly suggest new research avenues. For example, there is a growing research on the importance of vitamin D supplementation in *cystic fibrosis* patients *(134)*. Confirmation of a vitamin D benefit for respiratory infections and OADs suggests that vitamin D may be of particular importance for diseases such as *bronchiectasis*, which represents a combination of recurrent infections and obstructive airway disease. One could present a similar argument for *chronic sinusitis* and other conditions that we have not directly addressed in this review. The demonstration of a vitamin D-related benefit would add to the relatively few effective treatment options available for these conditions today.

10. CONCLUSIONS

Over the past decade, interest has grown in the effects of vitamin D on respiratory infections and obstructive airway diseases (OADs), such as asthma and chronic obstructive pulmonary disease (COPD). Observational studies suggest that low vitamin D levels increase the risk of acute respiratory infections from diverse respiratory pathogens in both lower and upper airways. Post-hoc analyses of randomized trials on vitamin D and bone health support this possibility. This increased risk of infection may contribute to incident wheezing illness in children and adults and cause asthma exacerbations. Although unproven, the increased risk of susceptible hosts to specific respiratory pathogens may contribute to some cases of incident asthma. Likewise, the effect of vitamin D on COPD, while intriguing, is largely unknown. Emerging evidence provides biological plausibility for some of these respiratory findings. For example, vitamin D-mediated innate immunity, particularly through enhanced expression of the human cathelicidin antimicrobial peptide (hCAP-18), is important in host defenses against respiratory pathogens. Vitamin D also modulates regulatory T-cell function and interleukin-10 production, which may increase the therapeutic response to corticosteroids in corticosteroid-resistant asthma. Finally, low vitamin D levels may have a role in the pathogenesis of allergies, including anaphylaxis. Further studies are needed to better understand vitamin D's effects on respiratory infections and OADs. Randomized controlled trials of higher dose vitamin D supplementation (e.g., at least 1,000 IU/day), particularly in the winter season at higher latitudes, will help clarify if vitamin D can prevent or reduce the severity of respiratory infections. In turn, these insights may prove helpful in preventing OAD exacerbations and the achievement of disease control.

ACKNOWLEDGMENTS

Dr. Camargo was supported by the Massachusetts General Hospital Center for D-receptor Activation Research (Boston, MA). All authors were supported by the National Institutes of Health (Bethesda, MD): Dr Camargo by grants R01 HL-84401 and R01 HL-64925; Dr Ginde by grant K12 RR-25779; and Dr Mansbach by grant K23 AI-77801.

REFERENCES

1. Lawrence G (1784) Tools of the trade: the Finsen Light. Lancet 2002:359
2. Cannell JJ, Vieth R, Umhau JC et al (2006) Epidemic influenza and vitamin D. Epidemiol Infect 134:1129–1140
3. Zasloff M (2006) Fighting infections with vitamin D. Nat Med 12:388–390
4. Muhe L, Lulseged S, Mason KE, Simoes EA (1997) Case–control study of the role of nutritional rickets in the risk of developing pneumonia in Ethiopian children. Lancet 349:1801–1804
5. Najada AS, Habashneh MS, Khader M (2004) The frequency of nutritional rickets among hospitalized infants and its relation to respiratory diseases. J Trop Pediatr 50:364–368
6. Laaksi I, Ruohola JP, Tuohimaa P et al (2007) An association of serum vitamin D concentrations <40 nmol/l with acute respiratory tract infection in young Finnish men. Am J Clin Nutr 86: 714–717
7. Camargo CA Jr, Ingham T, Wickens K et al (2008) Cord blood 25-hydroxyvitamin D levels and risk of childhood wheeze in New Zealand [abstract]. Am J Respir Crit Care Med 177(suppl):A993
8. Wjst M, Dold S (1999) Genes, factor X, and allergens: what causes allergic diseases? Allergy 54: 757–759
9. Poon AH, Laprise C, Lemire M et al (2004) Association of vitamin D receptor genetic variants with susceptibility to asthma and atopy. Am J Respir Crit Care Med 170:967–973
10. Raby BA, Lazarus R, Silverman EK et al (2004) Association of vitamin D receptor gene polymorphisms with childhood and adult asthma. Am J Respir Crit Care Med 170:1057–1065
11. Vollmert C, Illig T, Altmuller J et al (2004) Single nucleotide polymorphism screening and association analysis – exclusion of integrin beta 7 and vitamin D receptor (chromosome 12q) as candidate genes for asthma. Clin Exp Allergy 34:1841–1850
12. Laufs J, Andrason H, Sigvaldason A et al (2004) Association of vitamin D binding protein variants with chronic mucus hypersecretion in Iceland. Am J Pharmacogenomics 4:63–68
13. Wjst M (2005) Variants in the vitamin D receptor gene and asthma. BMC Genet 6:2
14. Camargo CA Jr, Rifas-Shiman SL, Litonjua AA et al (2006) Prospective study of maternal intake of vitamin D during pregnancy and risk of wheezing illnesses in children at age 2 years [abstract]. J Allergy Clin Immunol 117:721–722
15. Martinez FD, Wright AL, Taussig LM, Holberg CJ, Halonen M, Morgan WJ (1995) Asthma and wheezing in the first six years of life. The Group Health Medical Associates. N Engl J Med 332: 133–138
16. Graham NM (1990) The epidemiology of acute respiratory infections in children and adults: a global perspective. Epidemiol Rev 12:149–178
17. Taubenberger JK, Morens DM (1918) Influenza: the mother of all pandemics. Emerg Infect Dis 2006(12):15–22
18. Mansbach JM, Camargo CA Jr (2009) Update on bronchiolitis. Emerg Med Crit Care Rev 29(4): 741–755
19. Pelletier AJ, Mansbach JM, Camargo CA Jr (2006) Direct medical costs of bronchiolitis hospitalizations in the United States. Pediatrics 118:2418–2423
20. Fleisher GR (2000) Infectious disease emergencies. In: Fleisher GR, Ludwig S (eds) Textbook of pediatric emergency medicine, 4th edn. Philadelphia, Lippincott William & Wilkins, pp. 754–755
21. Hanson IC, Shearer WT (1999) Bronchiolitis. In: McMillan JA, DeAngelis CD, Feigin RD, Warshaw JB (eds) Oski's pediatrics: principles and practice, 3rd edn. Lippincott Williams & Wilkins, Philadelphia, pp. 1214–1216
22. Welliver R (2004) Bronchiolitis and infectious asthma. In: Feigin R, Cherry J, Demmler G, Kaplan S (eds) Textbook of pediatric infectious diseases, 5th edn. Philadelphia, Saunders, pp. 273–285
23. American Academy of Pediatrics Subcommittee on Diagnosis and Management of Bronchiolitis (2006) Diagnosis and management of bronchiolitis. Pediatrics 118:1774–1793.
24. Glezen WP, Loda FA, Clyde WA Jr et al (1971) Epidemiologic patterns of acute lower respiratory disease of children in a pediatric group practice. J Pediatr 78:397–406

25. Hall CB, Walsh EE, Schnabel KC et al (1990) Occurrence of groups A and B of respiratory syncytial virus over 15 years: associated epidemiologic and clinical characteristics in hospitalized and ambulatory children. J Infect Dis 162:1283–1290

26. Glezen WP, Taber LH, Frank AL, Kasel JA (1986) Risk of primary infection and reinfection with respiratory syncytial virus. Am J Dis Child 140:543–546

27. Ukkonen P, Hovi T, von Bonsdorff CH, Saikku P, Penttinen K (1984) Age-specific prevalence of complement-fixing antibodies to sixteen viral antigens: a computer analysis of 58,500 patients covering a period of eight years. J Med Virol 13:131–148

28. Levine DA, Platt SL, Dayan PS et al (2004) Risk of serious bacterial infection in young febrile infants with respiratory syncytial virus infections. Pediatrics 113:1728–1734

29. Mansbach JM, Emond JA, Camargo CA Jr (2005) Bronchiolitis in US emergency departments 1992–2000: epidemiology and practice variation. Pediatr Emerg Care 21:242–247

30. Mansbach JM, Clark S, Christopher NC et al (2008) Prospective multicenter study of bronchiolitis: predicting safe discharges from the emergency department. Pediatrics 121:680–688

31. Papadopoulos NG, Bates PJ, Bardin PG et al (2000) Rhinoviruses infect the lower airways. J Infect Dis 181:1875–1884

32. Korppi M, Kotaniemi-Syrjanen A, Waris M, Vainionpaa R, Reijonen TM (2004) Rhinovirus-associated wheezing in infancy: comparison with respiratory syncytial virus bronchiolitis. Pediatr Infect Dis J 23:995–999

33. Henderson FW, Clyde WA Jr, Collier AM et al (1979) The etiologic and epidemiologic spectrum of bronchiolitis in pediatric practice. J Pediatr 95:183–190

34. Moorman JE, Rudd RA, Johnson CA et al (2007) National surveillance for asthma – United States, 1980–2004. MMWR Surveill Summ 56:1–54

35. Ginde AA, Espinola JA, Camargo CA Jr (2008) Improved overall trends but persistent racial disparities in emergency department visits for acute asthma, 1993–2005. J Allergy Clin Immunol 122:313–318

36. Standards for the diagnosis and care of patients with chronic obstructive pulmonary disease (COPD) and asthma (1987). This official statement of the American Thoracic Society was adopted by the ATS Board of Directors, November 1986. Am Rev Respir Dis 136:225–244.

37. Toren K, Brisman J, Jarvholm B (1993) Asthma and asthma-like symptoms in adults assessed by questionnaires. A literature review. Chest 104:600–608

38. Nicholson KG, Kent J, Ireland DC (1993) Respiratory viruses and exacerbations of asthma in adults. BMJ 307:982–986

39. Johnston SL, Pattemore PK, Sanderson G et al (1995) Community study of role of viral infections in exacerbations of asthma in 9–11 year old children. BMJ 310:1225–1229

40. Yunginger JW, Reed CE, O'Connell EJ, Melton LJ 3rd, O'Fallon WM, Silverstein MD (1992) A community-based study of the epidemiology of asthma. Incidence rates, 1964–1983. Am Rev Respir Dis 146:888–894

41. Wainwright C, Isles AF, Francis PW (1997) Asthma in children. Med J Aust 167:218–223

42. Guilbert TW, Morgan WJ, Zeiger RS et al (2004) Atopic characteristics of children with recurrent wheezing at high risk for the development of childhood asthma. J Allergy Clin Immunol 114:1282–1287

43. Arbes SJ Jr, Gergen PJ, Vaughn B, Zeldin DC (2007) Asthma cases attributable to atopy: results from the Third National Health and Nutrition Examination Survey. J Allergy Clin Immunol 120:1139–1145

44. Borish L, Culp JA (2008) Asthma: a syndrome composed of heterogeneous diseases. Ann Allergy Asthma Immunol 101:1–8; quiz 8–11, 50

45. Global strategy for the diagnosis, management, and prevention of COPD. Available from: www.goldcopd.org . Last accessed 1 Dec 2008 (Accessed 1 Dec 2008, at www.goldcopd.org .)

46. Abouzgheib W, Pratter MR, Bartter T (2007) Cough and asthma. Curr Opin Pulm Med 13:44–48

47. Nnoaham KE, Clarke A (2008) Low serum vitamin D levels and tuberculosis: a systematic review and meta-analysis. Int J Epidemiol 37:113–119

48. Martineau AR, Honecker FU, Wilkinson RJ, Griffiths CJ (2007) Vitamin D in the treatment of pulmonary tuberculosis. J Steroid Biochem Mol Biol 103:793–798

49. Wilkinson RJ, Llewelyn M, Toossi Z et al (2000) Influence of vitamin D deficiency and vitamin D receptor polymorphisms on tuberculosis among Gujarati Asians in west London: a case–control study. Lancet 355:618–621

50. Wilbur AK, Kubatko LS, Hurtado AM, Hill KR, Stone AC (2007) Vitamin D receptor gene polymorphisms and susceptibility M. tuberculosis in native Paraguayans. Tuberculosis (Edinb) 87:329–337

51. Martineau AR, Wilkinson RJ, Wilkinson KA et al (2007) A single dose of vitamin D enhances immunity to mycobacteria. Am J Respir Crit Care Med 176:208–213

52. Adams JS, Hewison M (2008) Unexpected actions of vitamin D: new perspectives on the regulation of innate and adaptive immunity. Nat Clin Pract Endocrinol Metab 4:80–90

53. Wayse V, Yousafzai A, Mogale K, Filteau S (2004) Association of subclinical vitamin D deficiency with severe acute lower respiratory infection in Indian children under 5 years. Eur J Clin Nutr 58: 563–567

54. Karatekin G, Kaya A, Salihoğlu O, Balci H, Nuhoğlu A (2009) Association of subclinical vitamin D deficiency in newborns with acute lower respiratory infection and their mothers. Eur J Clin Nutr 63:473–477

55. Roth DE, Jones AB, Prosser C, Robinson JL, Vohra S (2009) Vitamin D status is not associated with the risk of hospitalization for acute bronchiolitis in early childhood. Eur J Clin Nutr 63:297–299

56. Camargo CA Jr, Rifas-Shiman SL, Litonjua AA et al (2007) Maternal intake of vitamin D during pregnancy and risk of recurrent wheeze in children at 3 years of age. Am J Clin Nutr 85:788–795

57. Devereux G, Litonjua AA, Turner SW et al (2007) Maternal vitamin D intake during pregnancy and early childhood wheezing. Am J Clin Nutr 85:853–859

58. Ginde AA, Mansbach JM, Camargo CA Jr (2009) Association between serum 25-hydroxyvitamin D level and upper respiratory tract infections in the Third National Health and Nutrition Examination Survey. Arch Intern Med 169:384–390

59. Linday LA, Shindledecker RD, Tapia-Mendoza J, Dolitsky JN (2004) Effect of daily cod liver oil and a multivitamin-mineral supplement with selenium on upper respiratory tract pediatric visits by young, inner-city, Latino children: randomized pediatric sites. Ann Otol Rhinol Laryngol 113:891–901

60. Rehman PK (1994) Sub-clinical rickets and recurrent infection. J Trop Pediatr 40:58

61. Aloia JF, Li-Ng M (2007) Re: epidemic influenza and vitamin D. Epidemiol Infect 135:1095–1096; author reply 7–8

62. Avenell A, Cook JA, Maclennan GS, Macpherson GC (2007) Vitamin D supplementation to prevent infections: a sub-study of a randomised placebo-controlled trial in older people (RECORD trial, ISRCTN 51647438). Age Ageing 36:574–577

63. Mikhail M, Aloia JF, Pollack S et al (2008) A randomized controlled trial of vitamin D3 supplementation for the prevention of viral upper respiratory tract infections [abstract]. In: 30th American Society for Bone and Mineral Research Annual Meeting. Montreal, Canada

64. Gern JE, Lemanske RF Jr, Busse WW (1999) Early life origins of asthma. J Clin Invest 104: 837–843

65. Warner JA, Jones CA, Jones AC, Warner JO (2000) Prenatal origins of allergic disease. J Allergy Clin Immunol 105:S493–S498

66. Gillman MW (2005) Developmental origins of health and disease. N Engl J Med 353:1848–1850

67. Weiss ST (1997) Diet as a risk factor for asthma. Ciba Found Symp 206:244–257; discussion 53–57

68. Romieu I, Trenga C (2001) Diet and obstructive lung diseases. Epidemiol Rev 23:268–287

69. Spector SL, Surette ME (2003) Diet and asthma: has the role of dietary lipids been overlooked in the management of asthma? Ann Allergy Asthma Immunol 90:371–377; quiz 7–8, 421

70. Devereux G, Seaton A (2005) Diet as a risk factor for atopy and asthma. J Allergy Clin Immunol 115:1109–1117; quiz 18

71. Holick MF, Vitamin D (2007) Deficiency. N Engl J Med 357:266–281

72. Black PN, Scragg R (2005) Relationship between serum 25-hydroxyvitamin d and pulmonary function in the third national health and nutrition examination survey. Chest 128:3792–3798

73. Burns JS, Dockery DW, Neas LM et al (2007) Low dietary nutrient intakes and respiratory health in adolescents. Chest 132:238–245

74. Wjst M (2006) The vitamin D slant on allergy. Pediatr Allergy Immunol 17:477–483

75. Hypponen E, Sovio U, Wjst M et al (2004) Infant vitamin d supplementation and allergic conditions in adulthood: northern Finland birth cohort 1966. Ann NY Acad Sci 1037:84–95
76. Gale CR, Robinson SM, Harvey NC et al (2008) Maternal vitamin D status during pregnancy and child outcomes. Eur J Clin Nutr 62:68–77
77. Jackson DJ, Gangnon RE, Evans MD et al (2008) Wheezing rhinovirus illnesses in early life predict asthma development in high-risk children. Am J Respir Crit Care Med 178:667–672
78. Mansbach JM, Camargo CA Jr (2008) Bronchiolitis: lingering questions about its definition and the potential role of vitamin D. Pediatrics 122:177–179
79. Sutherland ER, Martin RJ (2007) Asthma and atypical bacterial infection. Chest 132:1962–1966
80. National Asthma Education and Prevention Program (2007) Expert Panel report 3: guidelines for the diagnosis and management of asthma: full report. In. Washington DC: US Government Printing Office; 2007:417 pp. NIH Publication No. 07–4051.
81. Bateman ED, Boushey HA, Bousquet J et al (2004) Can guideline-defined asthma control be achieved? The Gaining Optimal Asthma ControL study. Am J Respir Crit Care Med 170:836–844
82. Ito K, Chung KF, Adcock IM (2006) Update on glucocorticoid action and resistance. J Allergy Clin Immunol 117:522–543
83. Xystrakis E, Kusumakar S, Boswell S et al (2006) Reversing the defective induction of IL-10-secreting regulatory T cells in glucocorticoid-resistant asthma patients. J Clin Invest 116:146–155
84. Heikkinen T, Jarvinen A (2003) The common cold. Lancet 361:51–59
85. van Elden LJ, Sachs AP, van Loon AM et al (2008) Enhanced severity of virus associated lower respiratory tract disease in asthma patients may not be associated with delayed viral clearance and increased viral load in the upper respiratory tract. J Clin Virol 41:116–121
86. Litonjua AA, Hollis BW, Schuemann B et al (2008) Low serum vitamin D levels are associated with greater risks for severe exacerbations in childhood asthmatics [abstract]. Am J Respir Crit Care Med 177(suppl):A993
87. Leech JA, Dulberg C, Kellie S, Pattee L, Gay J (1990) Relationship of lung function to severity of osteoporosis in women. Am Rev Respir Dis 141:68–71
88. Biskobing DM (2002) COPD and osteoporosis. Chest 121:609–620
89. Katsura H, Kida K (2002) A comparison of bone mineral density in elderly female patients with COPD and bronchial asthma. Chest 122:1949–1955
90. Jorgensen NR, Schwarz P, Holme I, Henriksen BM, Petersen LJ, Backer V (2007) The prevalence of osteoporosis in patients with chronic obstructive pulmonary disease: a cross sectional study. Respir Med 101:177–185
91. Kjensli A, Mowinckel P, Ryg MS, Falch JA (2007) Low bone mineral density is related to severity of chronic obstructive pulmonary disease. Bone 40:493–497
92. Ohara T, Hirai T, Muro S et al (2008) Relationship between pulmonary emphysema and osteoporosis assessed by CT in patients with COPD. Chest 134:1244–1249
93. Wright RJ (2005) Make no bones about it: increasing epidemiologic evidence links vitamin D to pulmonary function and COPD. Chest 128:3781–3783
94. Dobak J, Grzybowski J, Liu FT, Landon B, Dobke M (1994) 1,25-Dihydroxyvitamin D3 increases collagen production in dermal fibroblasts. J Dermatol Sci 8:18–24
95. Koli K, Keski-Oja J (2000) 1alpha,25-dihydroxyvitamin D3 and its analogues down-regulate cell invasion-associated proteases in cultured malignant cells. Cell Growth Differ 11:221–229
96. Spurzem JR, Rennard SI (2005) Pathogenesis of COPD. Semin Respir Crit Care Med 26:142–153
97. Schellenberg D, Pare PD, Weir TD, Spinelli JJ, Walker BA, Sandford AJ (1998) Vitamin D binding protein variants and the risk of COPD. Am J Respir Crit Care Med 157:957–961
98. Ishii T, Keicho N, Teramoto S et al (2001) Association of Gc-globulin variation with susceptibility to COPD and diffuse panbronchiolitis. Eur Respir J 18:753–757
99. Ito I, Nagai S, Hoshino Y et al (2004) Risk and severity of COPD is associated with the group-specific component of serum globulin 1F allele. Chest 125:63–70
100. Sandford AJ, Chagani T, Weir TD, Connett JE, Anthonisen NR, Pare PD (2001) Susceptibility genes for rapid decline of lung function in the lung health study. Am J Respir Crit Care Med 163: 469–473

101. Gombart AF, Borregaard N, Koeffler HP (2005) Human cathelicidin antimicrobial peptide (CAMP) gene is a direct target of the vitamin D receptor and is strongly up-regulated in myeloid cells by 1,25-dihydroxyvitamin D3. FASEB J 19:1067–1077

102. Liu PT, Stenger S, Li H et al (2006) Toll-like receptor triggering of a vitamin D-mediated human antimicrobial response. Science 311:1770–1773

103. Liu PT, Stenger S, Tang DH, Modlin RL (2007) Cutting edge: vitamin D-mediated human antimicrobial activity against Mycobacterium tuberculosis is dependent on the induction of cathelicidin. J Immunol 179:2060–2063

104. Yim S, Dhawan P, Ragunath C, Christakos S, Diamond G (2007) Induction of cathelicidin in normal and CF bronchial epithelial cells by 1,25-dihydroxyvitamin D(3). J Cyst Fibros 6:403–410

105. Hiemstra PS (2007) The role of epithelial beta-defensins and cathelicidins in host defense of the lung. Exp Lung Res 33:537–542

106. Weber G, Heilborn JD, Chamorro Jimenez CI, Hammarsjo A, Torma H, Stahle M (2005) Vitamin D induces the antimicrobial protein hCAP18 in human skin. J Invest Dermatol 124:1080–1082

107. Levy O (2007) Innate immunity of the newborn: basic mechanisms and clinical correlates. Nat Rev Immunol 7:379–390

108. Griffin MD, Xing N, Kumar R (2003) Vitamin D and its analogs as regulators of immune activation and antigen presentation. Annu Rev Nutr 23:117–145

109. Lin R, White JH (2004) The pleiotropic actions of vitamin D. Bioessays 26:21–28

110. Gregori S, Giarratana N, Smiroldo S, Uskokovic M, Adorini LA (2002) 1alpha,25-dihydroxyvitamin D(3) analog enhances regulatory T-cells and arrests autoimmune diabetes in NOD mice. Diabetes 51:1367–1374

111. Meehan MA, Kerman RH, Lemire JM (1992) 1,25-Dihydroxyvitamin D3 enhances the generation of nonspecific suppressor cells while inhibiting the induction of cytotoxic cells in a human MLR. Cell Immunol 140:400–409

112. Schwartz RH (2005) Natural regulatory T cells and self-tolerance. Nat Immunol 6:327–330

113. Matheu V, Back O, Mondoc E, Issazadeh-Navikas S (2003) Dual effects of vitamin D-induced alteration of TH1/TH2 cytokine expression: enhancing IgE production and decreasing airway eosinophilia in murine allergic airway disease. J Allergy Clin Immunol 112:585–592

114. Cantorna MT, Zhu Y, Froicu M, Wittke A (2004) Vitamin D status, 1,25-dihydroxyvitamin D3, and the immune system. Am J Clin Nutr 80:1717S–1720S

115. Pichler J, Gerstmayr M, Szepfalusi Z, Urbanek R, Peterlik M, Willheim M (2002) 1 alpha,25(OH)2D3 inhibits not only Th1 but also Th2 differentiation in human cord blood T cells. Pediatr Res 52:12–18

116. Xystrakis E, Urry Z, Hawrylowicz CM (2007) Regulatory T cell therapy as individualized medicine for asthma and allergy. Curr Opin Allergy Clin Immunol 7:535–541

117. Annesi-Maesano I (2002) Perinatal events, vitamin D, and the development of allergy. Pediatr Res 52:3–5

118. Sampson HA, Munoz-Furlong A, Campbell RL et al (2006) Second symposium on the definition and management of anaphylaxis: summary report – second National Institute of Allergy and Infectious Disease/Food Allergy and Anaphylaxis Network symposium. Ann Emerg Med 47:373–380

119. Hahn EL, Bacharier LB (2005) The atopic march: the pattern of allergic disease development in childhood. Immunol Allergy Clin North Am 25:231–246

120. Hata TR, Kotol P, Jackson M et al (2008) Administration of oral vitamin D induces cathelicidin production in atopic individuals. J Allergy Clin Immunol 122:829–831

121. Weiland SK, Husing A, Strachan DP, Rzehak P, Pearce N (2004) Climate and the prevalence of symptoms of asthma, allergic rhinitis, and atopic eczema in children. Occup Environ Med 61:609–615

122. Staples JA, Ponsonby AL, Lim LL, McMichael AJ (2003) Ecologic analysis of some immune-related disorders, including type 1 diabetes, in Australia: latitude, regional ultraviolet radiation, and disease prevalence. Environ Health Perspect 111:518–523

123. Oren E, Banerji A, Camargo CA Jr (2008) Vitamin D and atopic disorders in an obese population screened for vitamin D deficiency. J Allergy Clin Immunol 121:533–534

124. Sidbury R, Sullivan AF, Thadhani RI, Camargo CA Jr (2008) Randomized controlled trial of vitamin D supplementation for winter-related atopic dermatitis in Boston: a pilot study. Br J Dermatol 159:245–247

125. Wjst M, Hypponen E (2007) Vitamin D serum levels and allergic rhinitis. Allergy 62:1085–1086

126. Camargo CA Jr, Clark S, Kaplan MS, Lieberman P, Wood RA (2007) Regional differences in EpiPen prescriptions in the United States: the potential role of vitamin D. J Allergy Clin Immunol 120: 131–136

127. Camargo CA Jr, Clark S, Pearson JF, Kaplan MS, Lieberman P, Wood RA (2009) Latitude, UVB exposure, and EpiPen prescriptions in 38 urban areas [abstract]. J Allergy Clin Immunol 123(suppl):S109

128. Mullins RJ, Camargo CA Jr (2008) Childhood anaphylaxis in Australia: geographic and socio-economic influences [abstract]. Intern Med J 38(suppl 6):A161.

129. Simons FE, Peterson S, Black CD (2002) Epinephrine dispensing patterns for an out-of-hospital population: a novel approach to studying the epidemiology of anaphylaxis. J Allergy Clin Immunol 110:647–651

130. Schauber J, Gallo RL (2008) Antimicrobial peptides and the skin immune defense system. J Allergy Clin Immunol 122:261–266

131. Scragg R, Camargo CA Jr (2008) Frequency of leisure-time physical activity and serum 25-hydroxyvitamin D levels in the US population: results from the Third National Health and Nutrition Examination Survey. Am J Epidemiol 168:577–586; discussion 87–91

132. Bischoff-Ferrari HA, Giovannucci E, Willett WC, Dietrich T, Dawson-Hughes B (2006) Estimation of optimal serum concentrations of 25-hydroxyvitamin D for multiple health outcomes. Am J Clin Nutr 84:18–28

133. Barger-Lux MJ, Heaney RP, Dowell S, Chen TC, Holick MF (1998) Vitamin D and its major metabolites: serum levels after graded oral dosing in healthy men. Osteoporos Int 8:222–230

134. Green D, Carson K, Leonard A et al (2008) Current treatment recommendations for correcting vitamin D deficiency in pediatric patients with cystic fibrosis are inadequate. J Pediatr 153:554–559

135. Stein RT, Holberg CJ, Morgan WJ et al (1997) Peak flow variability, methacholine responsiveness and atopy as markers for detecting different wheezing phenotypes in childhood. Thorax 52:946–952

136. Legg JP, Warner JA, Johnston SL, Warner JO (2005) Frequency of detection of picornaviruses and seven other respiratory pathogens in infants. Pediatr Infect Dis J 24:611–616

137. Kusel MM, de Klerk NH, Kebadze T et al (2007) Early-life respiratory viral infections, atopic sensitization, and risk of subsequent development of persistent asthma. J Allergy Clin Immunol 119: 1105–1110

138. Mansbach JM, McAdam AJ, Clark S et al (2008) Prospective multicenter study of the viral etiology of bronchiolitis in the emergency department. Acad Emerg Med 15:111–118

VII CLINICAL USES OF VITAMIN D ANALOGUES

56 Treatment of Immunomediated Diseases by Vitamin D Analogs

Luciano Adorini

Abstract   Cells involved in innate and adaptive immune responses, including macrophages, dendritic cells, T and B cells, express the vitamin D receptor (VDR), can produce 1,25-dihydroxyvitamin D_3 [1,25(OH)$_2$D], the biologically active form of vitamin D_3, and respond to this hormone. Thus, 1,25(OH)$_2$D, a secosteroid hormone essential for bone and mineral homeostasis that regulates growth and differentiation of multiple cell types, can display exquisite immunoregulatory and anti-inflammatory properties. The net effect of the vitamin D system on the immune response results in enhancement of innate immunity and in a multi-faceted regulation of adaptive immune responses. Epidemiological evidence indicates a significant association of vitamin D deficiency with increased frequency of several immunomediated and autoimmune diseases, and preclinical models show that the anti-proliferative, pro-differentiative, anti-bacterial, immunomodulatory, and anti-inflammatory properties of synthetic VDR agonists could be exploited to treat a variety of immunomediated diseases. Understanding the physiological role of endogenous VDR agonists in the regulation of immune responses will contribute to guide the clinical translation of these agents.

Key Words: Inflammation; immunomodulation; autoimmune diseases; vitamin D receptor; dendritic cells; T cells; prostatitis; vitamin D analogs; 1,25-dihydroxyvitamin D; VDR agonist; lymphocytes

1. INTRODUCTION

1,25-Dihydroxyvitamin D_3 [1,25(OH)$_2$D] is a secosteroid hormone with pronounced immunoregulatory properties *(1–5)*. Vitamin D_3, as well as the closely related vitamin D_2, can be obtained from diet, but the main source of vitamin D_3 is biosynthesis from 7-dehydrocholesterol in UV-exposed skin *(6)*. Vitamin D_3 is hydroxylated in the liver to 25(OH)D, a reliable indicator of vitamin D status *(6)* and further hydroxylated in the kidney to the active hormone, 1,25(OH)$_2$D *(6)*.

The biological effects of 1,25(OH)$_2$D are mediated by the vitamin D receptor (VDR), a member of the superfamily of nuclear hormone receptors, which includes many ligand-regulated, DNA-binding transcription factors able to both activate and repress gene expression *(7)*. The VDR is expressed in most mammalian cell types, accounting for the pleiotropic activities of VDR agonists. Thousands of different genes are modulated by

From: *Nutrition and Health: Vitamin D*
Edited by: M.F. Holick, DOI 10.1007/978-1-60327-303-9_56,
© Springer Science+Business Media, LLC 2010

VDR agonists, explaining the wide variety of biological effects mediated by this class of agents *(8)*. Ligand binding induces conformational changes in the VDR, which promote heterodimerization with the retinoid X receptor (RXR) and recruitment of corepressor and coactivator proteins. Thus, the VDR functions as a ligand-activated transcription factor that binds to vitamin D responsive elements in the promoter region of vitamin D responsive genes and ultimately influences their rate of RNA polymerase II-mediated transcription *(9)*.

The vitamin D endocrine system is involved in several biological processes able to modulate immune responses and plays an important role in the pathogenesis of autoimmune diseases *(3–5, 10)*. In addition to exerting direct modulatory effects on T and B cell function, VDR agonists shape the phenotype and function of dendritic cells (DCs), promoting tolerogenic properties that favor the induction of regulatory rather than effector T cells *(11)*. Beneficial actions of VDR agonists have been demonstrated in several experimental models of autoimmune diseases and other immunomediated pathologies characterized by chronic inflammatory responses *(12)*. Accumulating data also document the capacity of 1,25(OH)$_2$D, which is produced by macrophages *(13, 14)*, DCs *(15, 16)*, T *(17)*, and B *(18)* cells, to physiologically contribute to regulate, via autocrine and paracrine effects, both innate and adaptive immune responses. The strict control of bioactive hormone production by cells of the immune system itself further supports the relevance of the vitamin D endocrine system in the modulation of immune responses in health and disease.

2. ENHANCEMENT OF INNATE IMMUNE RESPONSES BY VDR AGONISTS

The capacity of the vitamin D system to enhance innate immune responses has recently been firmly established *(19)*, emphasizing the capacity of vitamin D metabolites to act as intracellular regulators of the synthesis and action of naturally occurring defensin molecules against bacterial antigens *(20)*. 1,25(OH)$_2$D has been found to represent a key link between toll-like receptor (TLR) activation and anti-bacterial response in innate immunity against tuberculosis, via induction of the anti-microbial peptide cathelicidin *(14)*, which is encoded by a VDR primary response gene that is strongly upregulated by 1,25(OH)$_2$D *(21)*. A clinical correlate of this important link is provided by the observation that sera from African-American individuals, known to have increased susceptibility to tuberculosis, have reduced levels of 25(OH)D and are inefficient in cathelicidin mRNA induction, suggesting that 1,25(OH)$_2$D sufficiency contributes to decreased susceptibility to tuberculosis *(14)*. However, the anti-bacterial properties of the vitamin D system need to be balanced with its overall inhibitory effects on adaptive immune responses, as reviewed below.

In addition, levels of CYP27B1, which converts 25(OH)D into active 1,25(OH)$_2$D, are increased in wounds and are induced in keratinocytes by TGFβ1, triggering their production of 1,25(OH)$_2$D, which in turn increases expression of cathelicidin and induces the pattern-recognition receptors TLR2 and CD14 *(22)*. These data show a novel role of 1,25(OH)$_2$D in wound repair, enabling keratinocytes to protect wounds against infection and rendering this hormone a key component of innate immunity in anti-microbial response during injury.

Another novel aspect of the immunoregulatory properties of 1,25(OH)$_2$D in innate immunity is provided by the observation that vitamin D$_3$ induced by sunlight in the skin is hydroxylated by local DCs into the active hormone, which in turn upregulates on activated T cells the epidermotropic chemokine receptor CCR10, a primary VDR-responsive gene, enabling them to migrate in response to the epidermal chemokine CCL27 *(16)*. Thus, the capacity of DCs to generate 1,25(OH)$_2$D, able to imprint on T cells skin tropism, has an important physiological role in the control of lymphocyte homing *(23)*. In addition, autocrine 1,25(OH)$_2$D-induced homing of skin-associated T cells, which could include regulatory T cells able to counteract the proinflammatory effects induced in the skin by sun exposure, supports the intriguing hypothesis that pleiotropism of 1,25(OH)$_2$D activities represents the adaptation of the skin to ultraviolet light exposure, coupling sun-induced vitamin D$_3$ synthesis with protection of skin integrity.

3. MODULATION OF ADAPTIVE IMMUNE RESPONSES BY VDR AGONISTS

VDR agonists shape adaptive immune responses via different mechanisms. Several cell types in the immune system express the VDR and can produce 1,25(OH)$_2$D *(13–18)*. Thus, macrophages, DCs, as well as T and B lymphocytes can be considered as extrarenal sites of 1,25(OH)$_2$D synthesis, and their capacity to synthesize active hormone could physiologically contribute to regulate immune responses. Leukocyte-produced 1,25(OH)$_2$D, in addition to exerting autocrine effects, can also modulate immune responses in a paracrine fashion. VDR agonists have been shown to target different cell types participating to adaptive immune responses, but two are most relevant: dendritic cells and T lymphocytes.

3.1. Induction of Pro-tolerogenic Properties in Myeloid Dendritic Cells

VDR agonists arrest the differentiation and maturation of DCs, downregulate expression of the costimulatory molecules CD40, CD80, CD86, and markedly decrease IL-12 while enhancing IL-10 production *(24)*, favoring the induction of DCs with tolerogenic properties *(11)*. In addition, DCs treated with VDR agonists upregulate the expression of ILT3, an inhibitory molecule associated with tolerance induction, although ILT3 expression is dispensable for the capacity of 1,25(OH)$_2$D-treated DCs to induce T regulatory (Treg) cells *(25)*. Tolerogenic DCs induced by VDR agonists have been demonstrated in models of allograft rejection by direct administration to the recipient *(26)* or by adoptive transfer of in vitro-treated DCs *(27)* and have been implicated in the induction of CD4$^+$CD25$^+$ Treg cells able to arrest the development of autoimmune diabetes *(11)*.

DC differentiation is associated with active synthesis of 1,25(OH)$_2$D, which appears to exert an autoregulatory function by inhibiting differentiation of monocyte precursors into immature DCs and their subsequent ability to undergo terminal differentiation in response to maturation stimuli *(15)*. Local generation of 1,25(OH)$_2$D in DCs during inflammatory processes negatively regulates, in a paracrine fashion via ligand-induced recruitment of HDAC3, expression of the NF-κB component RelB *(28)*, thus explaining the inhibitory effect of 1,25(OH)$_2$D on DC maturation, and on their production of proinflammatory mediators. Cell type-specific, ligand-enhanced negative transcriptional regulation represents a novel important paradigm for VDR-controlled genes *(29)*.

Analysis of immunomodulatory effects exerted by 1,25(OH)$_2$D on human blood myeloid (M-DCs) and plasmacytoid (P-DCs) DCs demonstrates a differential capacity of this hormone to modulate cytokine and chemokine production in DC subsets, showing marked effects in M-DCs and negligible ones in P-DCs *(30)*. In addition, inhibition of Th1 development and enhancement of Treg cells are selectively induced by 1,25(OH)$_2$D in M-DCs but not P-DCs *(30)*. P-DCs, characterized by an intrinsic ability to prime naïve CD4$^+$ T cells to differentiate into IL-10-producing T cells and CD4$^+$CD25$^+$ Treg cells and to suppress immune responses, may represent naturally occurring regulatory DCs *(31)*, and the lack of P-DC modulation by 1,25(OH)$_2$D would thus leave their tolerogenic potential unmodified.

3.2. Modulation of Effector Lymphocytes

VDR agonists modulate DC function, thus shaping T-cell activation and development, but they can also have direct effects on T and B cells. VDR agonists have been found to be selective inhibitors of Th1 cell development *(32, 33)* and to inhibit directly Th1-type cytokines such as IL-2 and IFN-γ *(34, 35)*. 1,25(OH)$_2$D has also been shown to enhance the development of Th2 cells via a direct effect on naïve CD4$^+$ cells *(36)*. Thus, both Th1 and Th2 cells can be targets of VDR agonists, depending on their activation and differentiation status *(37)*.

1,25(OH)$_2$D has also potent direct effects on B-cell responses, inducing apoptosis and inhibiting proliferation, generation of memory B cells, plasma cell differentiation, and Ig production *(18)*. B cells, like macrophages, DCs, and T cells, express the VDR and can synthesize 1,25(OH)$_2$D able to exert autocrine and paracrine regulatory actions.

Treatment with VDR agonists inhibits also T-cell production of IL-17 *(38, 39)*, a proinflammatory cytokine recently shown to be produced by pathogenic T cells in various models of organ-specific autoimmunity in the brain, heart, synovium, and intestines *(40)*. Interestingly, IL-17 production is sustained by IL-23, an IL-12 family member formed by p19 and p40 chains, the latter being strongly inhibited by VDR agonists *(32)*. In addition, development of Th17 cells requires interferon regulatory factor 4 (IRF-4) *(41)*, a transcription factor that is strongly inhibited by 1,25(OH)$_2$D in DCs *(42)*.

In conclusion, VDR agonists appear primarily to inhibit proinflammatory, pathogenic T cells like Th1 and Th17 cells and, under appropriate conditions, may favor a deviation to the Th2 pathway. These effects are partly due to direct T-cell targeting, but modulation of DC function by VDR agonists plays an important role in directing T-cell responses. Thus, VDR agonists can target T cells both directly and indirectly, selectively inhibiting T-cell subsets able to mediate chronic inflammation and tissue damage.

3.3. Enhancement of Regulatory T Cells

Tolerogenic DCs induced by a short treatment with $1,25(OH)_2D$ or its analogs can induce $CD4^+CD25^+$ Treg cells that are able to mediate transplantation tolerance (26) and to arrest the development of autoimmune diabetes (43). However, tolerogenic DCs may not always be necessarily involved in the generation of Treg cells by VDR agonists, and a combination of $1,25(OH)_2D$ and dexamethasone has been shown to induce naive $CD4^+$ T cells to differentiate in vitro into IL-10-producing Treg cells, even in the absence of APCs (44).

VDR agonists not only favor induction of $CD4^+CD25^+$ Treg cells (43) and enhance their suppressive activity (45) but can also promote their recruitment at inflammatory sites. Blood-borne M-DCs constitutively produce high levels of CCL17 and CCL22, which are further enhanced by CD40 ligation (30). CCL22, a CCR4 agonist, selectively recruits $Foxp3^+CCR4^+CD4^+CD25^+$ Treg cells (46). Thus, the high constitutive and inducible production of CCL22 by immature M-DCs could lead to the preferential attraction of $CCR4^+CD4^+CD25^+$ Treg cells. Intriguingly, the production of CCL22 is markedly enhanced by $1,25(OH)_2D$ in blood M-DCs (30), indicating that VDR agonists can favor the recruitment of Treg cells by this DC subset.

4. VITAMIN D DEFICIENCY IN AUTOIMMUNE AND OTHER IMMUNOMEDIATED DISEASES

Research on the relation between vitamin D exposure and disease in population-based studies is increasing exponentially (47). Although there is no consensus on optimal serum levels of 25(OH)D, the accepted biomarker of vitamin D status, vitamin D deficiency is usually defined by 25(OH)D levels below 20 ng/ml (48, 49). Mounting evidence indicates a high prevalence of vitamin D deficiency in the general population, and this has been linked to increased frequency of autoimmune diseases, in addition to bone diseases and cancer (6, 50). Epidemiologic analysis shows strong ecologic and case–control evidence that vitamin D reduces the risk of several autoimmune diseases, including multiple sclerosis (MS), type 1 diabetes (T1D), inflammatory bowel disease (IBD), rheumatoid arthritis (RA), osteoarthritis, and SLE (51).

A large prospective study has recently confirmed that high circulating levels of 25(OH)D are associated with a lower risk of MS (52), suggesting that dietary vitamin D supplementation may help prevent the development of MS and may represent a useful addition to therapy in this indication (53, 54). Consistent with this possibility, vitamin D supplementation during infancy significantly reduced T1D incidence evaluated 30 years later (55), as recently confirmed by a meta-analysis of data from case–control studies (56). In addition, vitamin D intake has been found to be inversely correlated with the risk of developing RA (57), and case–control studies have shown significantly lower 25(OH)D levels in SLE patients (58). However, in large prospective cohorts of women, increasing levels of vitamin D intake were not associated with the relative risk of developing either SLE or RA (59). Vitamin D deficiency has been reported to contribute to IBD development (60), and lower amounts of $1,25(OH)_2D$ are synthesized from sunlight exposure in areas in which IBD occurs most often, such as North America and Northern Europe, a situation common to other autoimmune diseases *(10)*.

Collectively, these data indicate vitamin D status as a key environmental factor affecting autoimmune disease prevalence. VDR polymorphisms have also been repeatedly correlated with increased frequency of autoimmune diseases, but so far no association has been described with functional phenotypes *(61)*.

5. ANTI-INFLAMMATORY AND IMMUNOMODULATORY EFFECTS OF VDR AGONISTS IN AUTOIMMUNE AND IMMUNOMEDIATED DISEASES

The capacity of VDR agonists to inhibit autoimmune diseases has been studied in different experimental models, including collagen-induced arthritis, Lyme arthritis, SLE in MRL $^{lpr/lpr}$ mice, T1D in non-obese diabetic (NOD) mice, experimental allergic encephalomyelitis (EAE), and colitis *(3, 4, 10)*. VDR agonists are able not only to prevent but also to treat ongoing autoimmune diseases, as demonstrated by their ability to inhibit T1D development in adult NOD mice, and to ameliorate established chronic relapsing EAE *(3)*. In addition, additive and even synergistic effects have been observed between VDR agonists and immunosuppressive agents, such as cyclosporin A and sirolimus, in autoimmune diabetes and EAE models *(4)*.

Distinct regulatory mechanisms induced by VDR agonists may predominate in different autoimmune disease models but a common pattern characterized by induction of tolerogenic DCs, inibition of Th1 and Th17 cell development, and enhancement of $CD4^+CD25^+$ Treg cells has been frequently observed (Fig. 1). In the following sections, only recent developments in selected autoimmune diseases will be reviewed. More analytical reviews on treatment of autoimmune diseases by VDR agonists can be found in references *(3–5, 10, 62)*.

5.1. Rheumatoid Arthritis

Rheumatoid arthritis (RA) is an immune-mediated disease characterized by articular inflammation and subsequent tissue damage leading to severe disability and increased mortality. VDR agonists have been tested in two RA models, namely Lyme arthritis and collagen-induced arthritis *(63)*. Infection of mice with *Borrelia burgdorferi*, the causative agent of human Lyme arthritis, produces acute arthritic lesions with footpad and ankle swelling. Supplementation with 1,25(OH)$_2$D to mice infected with *B. burgdorferi* minimized or prevented these symptoms, and the same treatment could also inhibit collagen-induced arthritis, preventing the progression to severe arthritis when given to mice with early symptoms *(63)*. In a separate study, VDR agonists displayed a similar capacity to prevent and to suppress already established collagen-induced arthritis without inducing hypercalcemia *(64)*.

1,25(OH)$_2$D contributes to the regulation of MMPs and PGE2 production by human articular chondrocytes in osteoarthritic cartilage *(65)*, suggesting immunomodulatory effects also in human RA. VDR agonists show indeed potential in RA treatment *(66)*, as

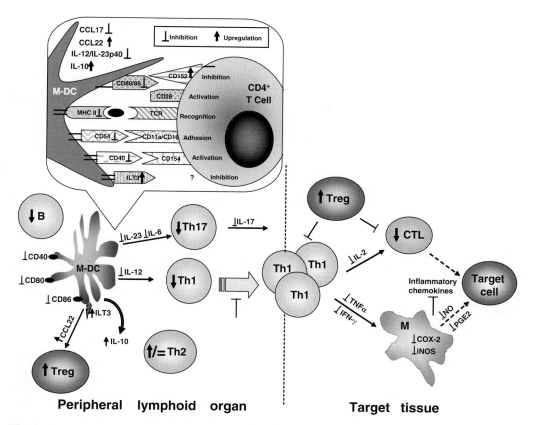

Fig. 1. Mechanisms involved in the regulation of autoimmune responses by VDR agonists. VDR agonists inhibit in myeloid dendritic cells (M-DC), but not in plasmacytoid DCs, expression of surface costimulatory molecules, e.g., CD40, CD80, CD86, as well as MHC class II and CD54 molecules (see *inset*). Production of cytokines affecting T-cell differentiation into Th1 and Th17, IL-12 and IL-23, respectively, is also inhibited. Conversely, expression of surface inhibitory molecules like ILT3 and of secreted inhibitory cytokines like IL-10 is markedly upregulated. Chemokines potentially able to recruit CCR4$^+$ regulatory T cells like CCL22 are also upregulated, whereas the CCR4 ligand CCL17 is downregulated. Upon interaction with M-DCs, CD4$^+$ T cells upregulate expression of the inhibitory molecule CD152 (CTLA-4). DCs expressing low levels of costimulatory molecules, secreting IL-10, and expressing high levels of inhibitory molecules (e.g., ILT3) favor the induction and/or the enhancement of Treg cells. VDR agonists can modulate the inflammatory response via several mechanisms in secondary lymphoid organs and in target tissues. M-DC modulation by VDR agonists inhibits development of Th1 and Th17 cells while inducing CD4$^+$CD25$^+$Foxp3$^+$ Treg cells and, under certain conditions, Th2 cells. VDR agonists can also inhibit the migration of Th1 cells, and they upregulate CCL22 production by M-DC, enhancing the recruitment of CD4$^+$CD25$^+$ Treg cells and of Th2 cells. In addition, VDR agonists exert direct effects on T cells by inhibiting IL-2 and IFN-γ production and on B cells inducing apoptosis and inhibiting proliferation, generation of memory B cells, plasma cell differentiation, and Ig production. In target tissues, pathogenic Th1 cells, that can damage target cells via induction of cytotoxic T cells (CTL) and activated macrophages (MΦ), are reduced in number and their activity is further inhibited by CD4$^+$CD25$^+$ Treg cells and by Th2 cells induced by VDR agonists. IL-17 production by Th17 cells and Th17 cell development are also inhibited. In MΦ, important inflammatory molecules like cyclo-oxygenase 2 (COX-2) and inducible nitric oxide synthase (iNOS) are inhibited by VDR agonists, leading to decreased production of nitric oxide (NO) and prostaglandin E2 (PGE2). MΦs, as well as DCs, T and B cells, can synthesize 1,25(OH)$_2$D which contributes to the regulation of local immune responses. *Blunted arrows* indicate inhibition and *broken arrows* cytotoxicity.

indicated by the beneficial effects of alphacalcidol in a 3-month open-label trial on 19 RA patients *(67)*.

5.2. Systemic Lupus Erythematosus

Systemic lupus erythematosus (SLE) is a T-cell-dependent antibody-mediated autoimmune disease, and the mouse strain MRL$^{lpr/lpr}$ spontaneously develops a SLE-like syndrome sharing many immunological features with human SLE *(68)*. Administration of VDR agonists significantly prolonged the average life span of MRL$^{lpr/lpr}$ mice and induced a significant reduction in proteinuria, renal arteritis, granuloma formation, and knee joint arthritis *(69, 70)*. In addition, dermatological lesions, like alopecia, necrosis of the ear, and scab formation, were also completely inhibited by 1,25(OH)$_2$D therapy *(70)*.

Preclinical models and epidemiological data suggest a beneficial role of VDR agonists in the treatment of human SLE *(71)*. Indeed, VDR agonists can significantly reduce cell proliferation and IgG production, both polyclonal and anti-dsDNA, while enhancing B-cell apoptosis in lymphocytes from SLE patients *(72)*. Vitamin D deficiency is a risk factor for SLE *(58, 73)*, and reduced levels of 1,25(OH)$_2$D in SLE patients may contribute to B-cell hyperactivity in this disease *(18)*.

5.3. Type 1 Diabetes

Vitamin D deficiency predisposes individuals to type 1 and type 2 diabetes *(74)*. VDR agonists have been extensively tested for their capacity to inhibit T1D in the NOD mouse *(75, 76)*, and 1,25(OH)$_2$D itself has been found to reduce the incidence of insulitis and to prevent T1D development but only when administered to NOD mice starting from 3 weeks of age, before the onset of insulitis, whereas a combined treatment of 8-week-old NOD mice with the VDR agonist MC1288 and cyclosporine A reduced T1D incidence *(4)*.

In contrast, the VDR agonist BXL-219 is able, as a monotherapy at non-hypercalcemic doses, to treat T1D in the adult NOD mouse, arresting its immunological progression and preventing clinical onset *(43)*. This VDR agonist significantly down-regulates in vitro and in vivo proinflammatory chemokine production by islet cells, inhibiting T-cell recruitment into the pancreatic islets and T1D development *(77)*. The inhibition of CXCL10, observed at both transcript and secreted protein levels, may be particularly relevant, consistent with the decreased recruitment of Th1 cells into sites of inflammation by treatment with an anti-CXCR3 antibody and with the substantial delay of T1D development observed in CXCR3-deficient mice. The inhibition of islet chemokine production by BXL-219 treatment in vivo persists after restimulation with TLR agonists and is associated with upregulation of IκBα transcription, an inhibitor of NF-κB, and with arrest of NF-κBp65 nuclear translocation in islet cells *(77)*. These findings highlight the capacity of VDR agonists to inhibit secretion of proinflammatory chemokines by target organs of autoimmune attack, a potentially relevant mechanism for the treatment of T1D and other chronic inflammatory conditions.

However, VDR disruption does not alter T1D presentation in NOD mice, in contrast to the more aggressive disease observed in vitamin D-deficient NOD mice *(78)*. Thus, although 1,25(OH)$_2$D is a pharmacological and possibly a physiological immunomodulator in T1D development, its function appears to be redundant to control this disease in the NOD mouse.

5.4. Experimental Allergic Encephalomyelitis

Experimental allergic encephalomyelitis (EAE), an autoimmune disease model resembling MS, is characterized by Th1- and Th17-type cells specific for myelin antigens (40) and can be ameliorated by treatment with VDR agonists (10, 33, 79). The mechanisms involved have been shown to include modulation of JAK–STAT signaling pathways in the IL-12/IFN-γ axis leading to reduced Th1 differentiation (80). IL-10 signaling is also essential for 1,25(OH)$_2$D-mediated inhibition of EAE, as shown by the hormone capacity to significantly inhibit EAE in wild-type mice, but not in mice with disrupted IL-10 or IL-10R genes (81). Thus, a functional IL-10–IL-10R pathway appears to be essential for 1,25(OH)$_2$D-mediated EAE inhibition. Activated inflammatory cells in EAE produce 1,25(OH)$_2$D, which enhances the apoptotic death of inflammatory CD4$^+$ T cells, thus dampening the driving force for continued inflammation (82). Strikingly, 1,25(OH)$_2$D treatment significantly reduced clinical EAE severity within 3 days, preceded by sharp declines in chemokines, inducible iNOS, and CD11b$^+$ monocyte recruitment into the central nervous system (82).

5.5. Inflammatory Bowel Disease

Epidemiological evidence supports a pathogenetic link between vitamin D deficiency and the risk of inflammatory bowel disease (IBD) (60). Vitamin D deficiency may compromise the mucosal barrier, leading to increased susceptibility to mucosal damage and increased risk of IBD, as suggested by studies in VDR-deficient mice indicating a critical role for the VDR in mucosal barrier homeostasis, by preserving the integrity of junction complexes and the healing capacity of the colonic epithelium (83, 84). Vitamin D-deficient in contrast to vitamin D-sufficient IL-10 KO mice, in which IBD symptoms occur spontaneously, rapidly develop diarrhea and a severe wasting disease (10). Administration of 1,25(OH)$_2$D significantly ameliorates IBD symptoms in IL-10 KO mice, and treatment for as little as 2 weeks blocks disease progression and ameliorates symptoms in mice with already established IBD (10). VDR expression is required to control inflammation in the IL-10 KO, because colitis is exacerbated in IL-10/VDR double-deficient mice associated with high local expression of IL-2, IFN-γ, IL-1β, TNFα, and IL-12, and VDR-deficient mice are extremely sensitive to dextran sodium sulfate (DSS)-induced colitis (83). These results point to an important role for the vitamin D system in the control of innate immunity and gastrointestinal homeostasis. Dietary calcium and 1,25(OH)$_2$D treatment directly and indirectly inhibit the TNFα pathway reducing colonic inflammation in IL-10-deficient mice (85), and 1,25(OH)$_2$D delivered rectally can decrease the severity and extent of DSS-induced colitis in wild-type mice (83). In addition, treatment of TNBS-induced colitis with the VDR agonist 22-ene-25-oxa-vitamin D (ZK156979) inhibited disease at normocalcemic doses accompanied by downregulation of myeloperoxidase activity, TNFα, IFN-γ, and T-bet expression, whereas local tissue IL-10 and IL-4 protein levels increased (86).

TNFα represents a validated target in IBD, since this cytokine plays an important role in the initiation and perpetuation of intestinal inflammation in IBD, and anti-TNFα antibodies are approved therapies also for this indication. The VDR agonist TX527 [19-nor-14,20-bisepi-23-yne-1,25(OH)$_2$D] inhibits proliferation and TNFα production by peripheral blood mononuclear cells from CD patients (87). In addition, the activation of NF-κB stimulated by TNFα and its nuclear translocation and the degradation of IKB-α were blocked by TX527, further indicating anti-inflammatory properties of VDR agonists in this indication.

5.6. Immunomediated Prostatic Diseases

The prostate has been recognized as a target organ of VDR agonists and represents an extrarenal synthesis site of 1,25-dihydroxyvitamin D$_3$ (88), but its capacity to respond to VDR agonists has, so far, been probed clinically only for the treatment of prostate cancer (89). Neurological, immunological, and endocrine dysfunctions have been proposed to be involved in the pathogenesis of chronic prostatitis/chronic pelvic pain syndrome (CP/CPPS), with emerging evidence for a primary role of immune-mediated components (90).

Based on the marked inhibitory activity of the VDR agonist elocalcitol on basal and growth factor-induced proliferation of human prostate cells (91), we have tested its anti-inflammatory properties in the treatment of experimental autoimmune prostatitis (EAP) induced by injection of prostate homogenate-CFA in NOD male mice (38). Administration of elocalcitol, at normocalcemic doses, for 2 weeks in already established EAP inhibits significantly the intra-prostatic cell infiltrate, with reduced cell proliferation and increased apoptosis of resident and infiltrating cells (38). Th1-cell responses are decreased as well as production of IL-17 (38), a cytokine involved in prostate inflammation (92), emphasizing the potential of VDR agonists in the treatment of immunomediated diseases of the prostate (93).

We have analyzed the capacity of VDR agonists to treat benign prostatic hyperplasia (BPH), a complex syndrome characterized by a static component related to prostate overgrowth, a dynamic component responsible for urinary irritative symptoms, and an inflammatory component (93). VDR agonists, and notably elocalcitol, reduce the static component of BPH by inhibiting the activity of intra-prostatic growth factors downstream the androgen receptor (91) and the dynamic component by targeting bladder cells (94). In addition, elocalcitol inhibits production of proinflammatory cytokines and chemokines by human BPH cells (93). These data have led to a proof-of-concept clinical study that has successfully shown arrest of prostate growth in BPH patients treated with elocalcitol (95). Ongoing clinical studies will assess the capacity of this VDR agonist to reduce symptoms and ameliorate flow parameters in BPH-affected individuals. The pronounced effects of elocalcitol on bladder smooth muscle cells (Fig. 2) indeed anticipate beneficial effects also on BPH-related lower urinary tract symptoms (96, 97). In particular, its ability to inhibit in the bladder RhoA/Rho-kinase signaling, a calcium-sensitizing pathway, suggests the possible clinical use of elocalcitol in the treatment of altered bladder contractility often associated with BPH-induced lower urinary tract symptoms (96, 97).

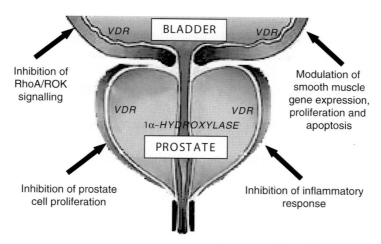

Fig. 2. Prostate and bladder as target organs of VDR agonists. Prostate and bladder cells express the VDR, and prostate cells express the enzymes required for 1,25(OH)₂D production. Key effects of VDR agonists in these target organs are indicated.

6. DEVELOPMENT OF VDR AGONISTS FOR THE TREATMENT OF AUTOIMMUNE AND OTHER IMMUNOMEDIATED DISEASES

VDR agonists have currently widespread clinical applications, notably in the treatment of osteoporosis *(98)*, secondary hyperparathyroidism *(99)*, and psoriasis *(100)*, with considerable efforts ongoing to translate their anti-proliferative properties into treatment of cancer *(89, 101)*. Observational studies in hemodialysis patients report improved cardiovascular and all-cause survival among those receiving VDR agonist therapy, and paricalcitol has been associated with greater survival than calcitriol *(102)*. The survival benefits of paricalcitol appear to be linked, at least in part, to "nonclassical" actions of VDR agonists *(102, 103)*. In cardiovascular tissues, these agents are reported to have several beneficial effects such as anti-inflammatory and antithrombotic effects, inhibition of vascular smooth muscle cell proliferation, inhibition of vascular calcification and stiffening, and regression of left ventricular hypertrophy *(104)*. VDR agonists are also reported to negatively regulate the renin–angiotensin system *(105)*, which plays a key role in hypertension, myocardial infarction, and stroke.

In addition to these indications, VDR agonists, by promoting innate immunity and regulating adaptive immune responses, have the potential to be developed as real immunomodulators in a variety of immunomediated diseases, in particular those mediated by chronic inflammatory responses. However, the exquisite anti-inflammatory, protolerogenic, and immunoregulatory properties exerted by VDR agonists have been clinically exploited so far only partially in the treatment of immunomediated conditions. This is also due to their calcemic liability, the major side effect of this class of agents potentially leading to hypercalciuria and hypercalcemia and in severe cases eventually to tissue calcification *(106)*. The accumulating evidence for multiple immunomodulatory mechanisms regulated by VDR agonists is nevertheless prompting a further exploration of their potential in the development of therapies for immunomediated disorders. Toward this goal, hypocalcemic, tissue-selective VDR agonists, with a wider therapeutic index compared to the natural hormone 1,25(OH)$_2$D, have been identified *(107)*. Elocalcitol, a VDR agonist active in autoimmune disease models*(38)*, has been shown to induce tissue and cell type-selective VDR-mediated activation, likely via differential recruitment of coactivators and corepressors by the VDR–elocalcitol complex in different cell types *(108)*.

Combinations of VDR agonists with other immunomodulators may offer novel treatment perspectives. The 1,25(OH)$_2$D analog TX527, a VDR agonist with an interesting dissociation profile between immunomodulatory and calcemic effects that can attenuate EAE, TX527 combined with IFN-β, an agent currently used in MS treatment, resulted in significant disease protection which was superior to the effect of either treatment separately *(109)*. Thus, adding a VDR agonist like TX527 to IFN-β could represent a potentially interesting combination for the treatment of MS.

The mechanisms of action of VDR agonists extend beyond anti-inflammatory and immunoregulatory properties to non-traditional targets, such as muscle *(110)*. Smooth muscle cell regulation by VDR agonists could be applied to the treatment of overactive bladder *(94, 96)*, while cardiac muscle could be a target in several conditions *(104, 111, 112)*, and skeletal muscle strength has been implicated in the reduction of falls by vitamin D supplementation *(113, 114)*. These beneficial effects exerted by the vitamin D system could also be enhanced by specific VDR agonists, thus expanding considerably their field of potential applications.

REFERENCES

1. Deluca HF, Cantorna MT (2001) Vitamin D: its role and uses in immunology. FASEB J 15: 2579–2585
2. Griffin MD, Xing N, Kumar R (2003) Vitamin D and its analogs as regulators of immune activation and antigen presentation. Annu Rev Nutr 23:117–145
3. Adorini L (2005) Intervention in autoimmunity: the potential of vitamin D receptor agonists. Cell Immunol 233:115–124
4. van Etten E, Mathieu C (2005) Immunoregulation by 1,25-dihydroxyvitamin D3: basic concepts. J Steroid Biochem Mol Biol 97:93–101
5. Arnson Y, Amital H, Shoenfeld Y (2007) Vitamin D and autoimmunity: new aetiological and therapeutic considerations. Ann Rheum Dis 66:1137–1142
6. Holick MF (2007) Vitamin D deficiency. N Engl J Med 357:266–281
7. Evans RM (2005) The nuclear receptor superfamily: a rosetta stone for physiology. Mol Endocrinol 19:1429–1438
8. Wang TT, Tavera-Mendoza LE, Laperriere D et al (2005) Large-scale in silico and microarray-based identification of direct 1,25-dihydroxyvitamin D3 target genes. Mol Endocrinol 19:2685–2695
9. Carlberg C (2003) Current understanding of the function of the nuclear vitamin D receptor in response to its natural and synthetic ligands. Recent Results Cancer Res 164:29–42
10. Cantorna MT (2006) Vitamin D and its role in immunology: multiple sclerosis, and inflammatory bowel disease. Prog Biophys Mol Biol 92:60–64
11. Adorini L, Giarratana N, Penna G (2004) Pharmacological induction of tolerogenic dendritic cells and regulatory T cells. Semin Immunol 16:127–134
12. Mathieu C, Adorini L (2002) The coming of age of 1,25-dihydroxyvitamin D(3) analogs as immunomodulatory agents. Trends Mol Med 8:174–179
13. Overbergh L, Decallonne B, Valckx D et al (2000) Identification and immune regulation of 25-hydroxyvitamin D-1-alpha-hydroxylase in murine macrophages. Clin Exp Immunol 120:139–146

14. Liu PT, Stenger S, Li H et al (2006) Toll-like receptor triggering of a vitamin D-mediated human antimicrobial response. Science 311:1770–1773

15. Hewison M, Freeman L, Hughes SV et al (2003) Differential regulation of vitamin D receptor and its ligand in human monocyte-derived dendritic cells. J Immunol 170:5382–5390

16. Sigmundsdottir H, Pan J, Debes GF et al (2007) DCs metabolize sunlight-induced vitamin D3 to 'program' T cell attraction to the epidermal chemokine CCL27. Nat Immunol 8:285–293

17. Cadranel J, Garabedian M, Milleron B et al (1990) 1,25(OH)2D3 production by T lymphocytes and alveolar macrophages recovered by lavage from normocalcemic patients with tuberculosis. J Clin Invest 85:1588–1593

18. Chen S, Sims GP, Chen XX et al (2007) Modulatory effects of 1,25-dihydroxyvitamin D3 on human B cell differentiation. J Immunol 179:1634–1647

19. Adams JS, Chen H, Chun R et al (2007) Substrate and enzyme trafficking as a means of regulating 1,25-dihydroxyvitamin D synthesis and action: the human innate immune response. J Bone Miner Res 22(Suppl 2):V20–V24

20. Adams JS, Liu PT, Chun R et al (2007) Vitamin D in defense of the human immune response. Ann NY Acad Sci 1117:94–105

21. Wang TT, Nestel FP, Bourdeau V et al (2004) Cutting edge: 1,25-dihydroxyvitamin D3 is a direct inducer of antimicrobial peptide gene expression. J Immunol 173:2909–2912

22. Schauber J, Dorschner RA, Coda AB et al (2007) Injury enhances TLR2 function and antimicrobial peptide expression through a vitamin D-dependent mechanism. J Clin Invest 117:803–811

23. Mebius RE (2007) Vitamins in control of lymphocyte migration. Nat Immunol 8:229–230

24. Penna G, Adorini L (2000) 1alpha,25-dihydroxyvitamin D3 inhibits differentiation, maturation, activation, and survival of dendritic cells leading to impaired alloreactive T cell activation. J Immunol 164:2405–2411

25. Penna G, Roncari A, Amuchastegui S et al (2005) Expression of the inhibitory receptor ILT3 on dendritic cells is dispensable for induction of CD4+Foxp3+ regulatory T cells by 1,25-dihydroxyvitamin D3. Blood 106:3490–3497

26. Gregori S, Casorati M, Amuchastegui S et al (2001) Regulatory T cells induced by 1,25-Dihydroxyvitamin D_3 and mycophenolate mofetil treatment mediate transplantation tolerance. J Immunol 167:1945–1953

27. Griffin MD, Lutz W, Phan VA et al (2001) Dendritic cell modulation by 1alpha,25 dihydroxyvitamin D3 and its analogs: a vitamin D receptor-dependent pathway that promotes a persistent state of immaturity in vitro and in vivo. Proc Natl Acad Sci USA 22:22

28. Dong X, Lutz W, Schroeder TM et al (2005) Regulation of relB in dendritic cells by means of modulated association of vitamin D receptor and histone deacetylase 3 with the promoter. Proc Natl Acad Sci USA 102:16007–16012

29. Griffin MD, Dong X, Kumar R (2007) Vitamin D receptor-mediated suppression of RelB in antigen presenting cells: a paradigm for ligand-augmented negative transcriptional regulation. Arch Biochem Biophys 460:218–226

30. Penna G, Amuchastegui S, Giarratana N et al (2007) 1,25-Dihydroxyvitamin D3 selectively modulates tolerogenic properties in myeloid but not plasmacytoid dendritic cells. J Immunol 178:145–153

31. Liu YJ (2005) IPC: professional type 1 interferon-producing cells and plasmacytoid dendritic cell precursors. Annu Rev Immunol 23:275–306

32. Lemire JM, Archer DC, Beck L et al (1995) Immunosuppressive actions of 1,25-dihydroxyvitamin D3: preferential inhibition of Th1 functions. J Nutr 125:1704S–1708S

33. Mattner F, Smiroldo S, Galbiati F et al (2000) Inhibition of Th1 development and treatment of chronic-relapsing experimental allergic encephalomyelitis by a non-hypercalcemic analogue of 1,25-dihydroxyvitamin D(3). Eur J Immunol 30:498–508

34. Alroy I, Towers T, Freedman L (1995) Transcriptional repression of the interleukin-2 gene by vitamin D_3: direct inhibition NFATp/AP-1 complex formation by a nuclear hormone receptor. Mol Cell Biol 15:5789–5799

35. Cippitelli M, Santoni A (1998) Vitamin D_3: a transcriptional modulator of the IFN-γ gene. Eur J Immunol 28:3017–3030

36. Boonstra A, Barrat FJ, Crain C et al (2001) 1alpha,25-Dihydroxyvitamin D3 has a direct effect on naive CD4(+) T Cells to enhance the development of Th2 cells. J Immunol 167: 4974–4980

37. Mahon BD, Wittke A, Weaver V et al (2003) The targets of vitamin D depend on the differentiation and activation status of CD4 positive T cells. J Cell Biochem 89:922–932

38. Penna G, Amuchastegui S, Cossetti C et al (2006) Treatment of experimental autoimmune prostatitis in nonobese diabetic mice by the vitamin D receptor agonist elocalcitol. J Immunol 177: 8504–8511

39. Daniel C, Sartory NA, Zahn N et al (2008) Immune modulatory treatment of trinitrobenzene sulfonic acid colitis with calcitriol is associated with a change of a T helper (Th) 1/Th17 to a Th2 and regulatory T cell profile. J Pharmacol Exp Ther 324:23–33

40. Steinman L (2007) A brief history of T(H)17, the first major revision in the T(H)1/T(H)2 hypothesis of T cell-mediated tissue damage. Nat Med 13:139–145

41. Brustle A, Heink S, Huber M et al (2007) The development of inflammatory T(H)-17 cells requires interferon-regulatory factor 4. Nat Immunol 8:958–966

42. Gauzzi MC, Purificato C, Conti L et al (2005) IRF-4 expression in the human myeloid lineage: upregulation during dendritic cell differentiation and inhibition by 1{alpha},25-dihydroxyvitamin D3. J Leukoc Biol 77:944–947

43. Gregori G, Giarratana N, Smiroldo S et al (2002) A 1α,25-Dihydroxyvitamin D_3 analog enhances regulatory T cells and arrests autoimmune diabetes in NOD mice. Diabetes 51:1367–1374

44. Barrat FJ, Cua DJ, Boonstra A et al (2002) In vitro generation of interleukin 10-producing regulatory CD4(+) T cells is induced by immunosuppressive drugs and inhibited by T helper type 1 (Th1)- and Th2-inducing cytokines. J Exp Med 195:603–616

45. Gorman S, Kuritzky LA, Judge MA et al (2007) Topically applied 1,25-dihydroxyvitamin D3 enhances the suppressive activity of CD4+CD25+ cells in the draining lymph nodes. J Immunol 179:6273–6283

46. Curiel TJ, Coukos G, Zou L et al (2004) Specific recruitment of regulatory T cells in ovarian carcinoma fosters immune privilege and predicts reduced survival. Nat Med 10:942–949

47. Millen AE, Bodnar LM (2008) Vitamin D assessment in population-based studies: a review of the issues. Am J Clin Nutr 87:1102S–1105S

48. Bischoff-Ferrari HA, Giovannucci E, Willett WC et al (2006) Estimation of optimal serum concentrations of 25-hydroxyvitamin D for multiple health outcomes. Am J Clin Nutr 84:18–28

49. Holick MF (2009) Vitamin D status: measurement, interpretation, and clinical application. Ann Epidemiol 19:73–78

50. Cantorna MT, Mahon BD (2004) Mounting evidence for vitamin D as an environmental factor affecting autoimmune disease prevalence. Exp Biol Med (Maywood) 229:1136–1142

51. Grant WB (2006) Epidemiology of disease risks in relation to vitamin D insufficiency. Prog Biophys Mol Biol 92:65–79

52. Munger KL, Levin LI, Hollis BW et al (2006) Serum 25-hydroxyvitamin D levels and risk of multiple sclerosis. JAMA 296:2832–2838

53. Brown SJ (2006) The role of vitamin D in multiple sclerosis. Ann Pharmacother 40:1158–1161

54. Niino M, Fukazawa T, Kikuchi S et al (2008) Therapeutic potential of vitamin D for multiple sclerosis. Curr Med Chem 15:499–505

55. Hypponen E, Laara E, Reunanen A et al (2001) Intake of vitamin D and risk of type 1 diabetes: a birth-cohort study. Lancet 358:1500–1503

56. Zipitis CS, Akobeng AK (2008) Vitamin D supplementation in early childhood and risk of type 1 diabetes: a systematic review and meta-analysis. Arch Dis Child 93:512–517

57. Merlino LA, Curtis J, Mikuls TR et al (2004) Vitamin D intake is inversely associated with rheumatoid arthritis: results from the Iowa Women's Health Study. Arthritis Rheum 50:72–77

58. Kamen DL, Cooper GS, Bouali H et al (2006) Vitamin D deficiency in systemic lupus erythematosus. Autoimmun Rev 5:114–117

59. Costenbader KH, Feskanich D, Holmes M et al (2008) Vitamin D intake and risks of systemic lupus erythematosus and rheumatoid arthritis in women. Ann Rheum Dis 67:530–535

60. Pappa HM, Grand RJ, Gordon CM (2006) Report on the vitamin D status of adult and pediatric patients with inflammatory bowel disease and its significance for bone health and disease. Inflamm Bowel Dis 12:1162–1174

61. Valdivielso JM, Fernandez E (2006) Vitamin D receptor polymorphisms and diseases. Clin Chim Acta 371:1–12

62. Nagpal S, Na S, Rathnachalam R (2005) Noncalcemic actions of vitamin D receptor ligands. Endocr Rev 26:662–687

63. Cantorna MT, Hayes CE, DeLuca HF (1998) 1,25-Dihydroxycholecalciferol inhibits the progression of arthritis in murine models of human arthritis. J Nutr 128:68–72

64. Larsson P, Mattsson L, Klareskog L et al (1998) A vitamin D analogue (MC 1288) has immunomodulatory properties and suppresses collagen-induced arthritis (CIA) without causing hypercalcaemia. Clin Exp Immunol 114:277–283

65. Tetlow LC, Woolley DE (1999) The effects of 1 alpha,25-dihydroxyvitamin D(3) on matrix metalloproteinase and prostaglandin E(2) production by cells of the rheumatoid lesion. Arthritis Res 1:63–70

66. Cutolo M, Otsa K, Uprus M et al (2007) Vitamin D in rheumatoid arthritis. Autoimmun Rev 7:59–64

67. Andjelkovic Z, Vojinovic J, Pejnovic N et al (1999) Disease modifying and immunomodulatory effects of high dose 1 alpha (OH) D3 in rheumatoid arthritis patients. Clin Exp Rheumatol 17:453–456

68. Santiago-Raber ML, Laporte C, Reininger L et al (2004) Genetic basis of murine lupus. Autoimmun Rev 3:33–39

69. Abe J, Nakamura K, Takita Y et al (1990) Prevention of immunological disorders in MRL/l mice by a new synthetic analogue of vitamin D3: 22-oxa-1 alpha,25-dihydroxyvitamin D3. J Nutr Sci Vitaminol (Tokyo) 36:21–31

70. Lemire JM, Ince A, Takashima M (1992) 1,25-Dihydroxyvitamin D3 attenuates the expression of experimental murine lupus of MRL/l mice. Autoimmunity 12:143–148

71. Cutolo M, Otsa K (2008) Review: vitamin D, immunity and lupus. Lupus 17:6–10

72. Linker-Israeli M, Elstner E, Klinenberg JR et al (2001) Vitamin D(3) and its synthetic analogs inhibit the spontaneous in vitro immunoglobulin production by SLE-derived PBMC. Clin Immunol 99:82–93

73. Ruiz-Irastorza G, Egurbide MV, Olivares N et al (2008) Vitamin D deficiency in systemic lupus erythematosus: prevalence, predictors and clinical consequences. Rheumatology (Oxford) 47:920–923

74. Mathieu C, Gysemans C, Giulietti A et al (2005) Vitamin D and diabetes. Diabetologia 48:1247–1257

75. Mathieu C, Badenhoop K (2005) Vitamin D and type 1 diabetes mellitus: state of the art. Trends Endocrinol Metab 16:261–266

76. Adorini L, Penna G, Giarratana N et al (2005) Inhibition of type 1 diabetes development by vitamin D receptor agonists. Curr Med Chem 4:645–651

77. Giarratana N, Penna G, Amuchastegui S et al (2004) A vitamin D analog down-regulates proinflammatory chemokine production by pancreatic islets inhibiting T cell recruitment and type 1 diabetes development. J Immunol 173:2280–2287

78. Gysemans C, van Etten E, Overbergh L et al (2008) Unaltered diabetes presentation in NOD mice lacking the vitamin D receptor. Diabetes 57:269–275

79. Lemire JM, Archer C (1991) 1,25-Dihydroxyvitamin D3 prevents the in vivo induction of murine experimental autoimmune encephalomyelitis. J Clin Invest 87:1103–1107

80. Muthian G, Raikwar HP, Rajasingh J et al (2006) 1,25 Dihydroxyvitamin-D3 modulates JAK-STAT pathway in IL-12/IFNgamma axis leading to Th1 response in experimental allergic encephalomyelitis. J Neurosci Res 83:1299–1309

81. Spach KM, Nashold FE, Dittel BN et al (2006) IL-10 signaling is essential for 1,25-dihydroxyvitamin D3-mediated inhibition of experimental autoimmune encephalomyelitis. J Immunol 177:6030–6037

82. Pedersen LB, Nashold FE, Spach KM et al (2007) 1,25-dihydroxyvitamin D3 reverses experimental autoimmune encephalomyelitis by inhibiting chemokine synthesis and monocyte trafficking. J Neurosci Res 85:2480–2490

83. Froicu M, Cantorna MT (2007) Vitamin D and the vitamin D receptor are critical for control of the innate immune response to colonic injury. BMC Immunol 8:5

84. Kong J, Zhang Z, Musch MW et al (2008) Novel role of the vitamin D receptor in maintaining the integrity of the intestinal mucosal barrier. Am J Physiol Gastrointest Liver Physiol 294:G208–G216

85. Zhu Y, Mahon BD, Froicu M et al (2005) Calcium and 1 alpha,25-dihydroxyvitamin D3 target the TNF-alpha pathway to suppress experimental inflammatory bowel disease. Eur J Immunol 35: 217–224

86. Daniel C, Radeke HH, Sartory NA et al (2006) The new low calcemic vitamin D analog 22-ene-25-oxa-vitamin D prominently ameliorates T helper cell type 1-mediated colitis in mice. J Pharmacol Exp Ther 319:622–631

87. Stio M, Martinesi M, Bruni S et al (2007) The Vitamin D analogue TX 527 blocks NF-kappaB activation in peripheral blood mononuclear cells of patients with Crohn's disease. J Steroid Biochem Mol Biol 103:51–60

88. Flanagan JN, Young MV, Persons KS et al (2006) Vitamin D metabolism in human prostate cells: implications for prostate cancer chemoprevention by vitamin D. Anticancer Res 26:2567–2572

89. Deeb KK, Trump DL, Johnson CS (2007) Vitamin D signalling pathways in cancer: potential for anticancer therapeutics. Nat Rev Cancer 7:684–700

90. Rivero VE, Motrich RD, Maccioni M et al (2007) Autoimmune etiology in chronic prostatitis syndrome: an advance in the understanding of this pathology. Crit Rev Immunol 27:33–46

91. Crescioli C, Ferruzzi P, Caporali A et al (2004) Inhibition of prostate cell growth by BXL-628, a calcitriol analogue selected for a phase II clinical trial in patients with benign prostate hyperplasia. Eur J Endocrinol 150:591–603

92. Steiner GE, Djavan B, Kramer G et al (2002) The picture of the prostatic lymphokine network is becoming increasingly complex. Rev Urol 4:171–177

93. Adorini L, Penna G, Amuchastegui S et al (2007) Inhibition of prostate growth and inflammation by the vitamin D receptor agonist BXL-628 (elocalcitol). J Steroid Biochem Mol Biol 103:689–693

94. Crescioli C, Morelli A, Adorini L et al (2005) Human bladder as a novel target for vitamin D receptor ligands. J Clin Endocrinol Metab 90:962–972

95. Colli E, Rigatti P, Montorsi F et al (2006) BXL628, a novel vitamin D3 analog arrests prostate growth in patients with benign prostatic hyperplasia: a randomized clinical trial. Eur Urol 49:82–86

96. Morelli A, Vignozzi L, Filippi S et al (2007) BXL-628, a vitamin D receptor agonist effective in benign prostatic hyperplasia treatment, prevents RhoA activation and inhibits RhoA/Rho kinase signaling in rat and human bladder. Prostate 67:234–247

97. Morelli A, Squecco R, Failli P et al (2008) The vitamin D receptor agonist elocalcitol upregulates L-type calcium channel activity in human and rat bladder. Am J Physiol Cell Physiol 294: C1206–C1214

98. Cheskis BJ, Freedman LP, Nagpal S (2006) Vitamin D receptor ligands for osteoporosis. Curr Opin Investig Drugs 7:906–911

99. Brancaccio D, Bommer J, Coyne D (2007) Vitamin D receptor activator selectivity in the treatment of secondary hyperparathyroidism: understanding the differences among therapies. Drugs 67: 1981–1998

100. Segaert S, Duvold LB (2006) Calcipotriol cream: a review of its use in the management of psoriasis. J Dermatolog Treat 17:327–337

101. Campbell MJ, Adorini L (2006) The vitamin D receptor as a therapeutic target. Expert Opin Ther Targets 10:735–748

102. Wolf M, Thadhani R (2007) Vitamin D in patients with renal failure: a summary of observational mortality studies and steps moving forward. J Steroid Biochem Mol Biol 103:487–490

103. Andress D (2007) Nonclassical aspects of differential vitamin D receptor activation: implications for survival in patients with chronic kidney disease. Drugs 67:1999–2012

104. Levin A, Li YC (2005) Vitamin D and its analogues: do they protect against cardiovascular disease in patients with kidney disease? Kidney Int 68:1973–1981

105. Li YC, Kong J, Wei M et al (2002) 1,25-Dihydroxyvitamin D(3) is a negative endocrine regulator of the renin-angiotensin system. J Clin Invest 110:229–238

106. Takahashi T, Morikawa K (2006) Vitamin D receptor agonists: opportunities and challenges in drug discovery. Curr Top Med Chem 6:1303–1316
107. Ma Y, Khalifa B, Yee YK et al (2006) Identification and characterization of noncalcemic, tissue-selective, nonsecosteroidal vitamin D receptor modulators. J Clin Invest 116:892–904
108. Peleg S, Ismail A, Uskokovic MR et al (2003) Evidence for tissue- and cell-type selective activation of the vitamin D receptor by Ro-26-9228, a noncalcemic analog of vitamin D3. J Cell Biochem 88:267–273
109. van Etten E, Gysemans C, Branisteanu DD et al (2007) Novel insights in the immune function of the vitamin D system: synergism with interferon-beta. J Steroid Biochem Mol Biol 103:546–551
110. Demay M (2003) Muscle: a nontraditional 1,25-dihydroxyvitamin D target tissue exhibiting classic hormone-dependent vitamin D receptor actions. Endocrinology 144:5135–5137
111. Wang TJ, Pencina MJ, Booth SL et al (2008) Vitamin D deficiency and risk of cardiovascular disease. Circulation 117:503–511
112. Michos ED, Melamed ML (2008) Vitamin D and cardiovascular disease risk. Curr Opin Clin Nutr Metab Care 11:7–12
113. Broe KE, Chen TC, Weinberg J et al (2007) A higher dose of vitamin d reduces the risk of falls in nursing home residents: a randomized, multiple-dose study. J Am Geriatr Soc 55:234–239
114. Fosnight SM, Zafirau WJ, Hazelett SE (2008) Vitamin D supplementation to prevent falls in the elderly: evidence and practical considerations. Pharmacotherapy 28:225–234

57 Clinical Utility of 1,25-Dihydroxyvitamin D₃ and Its Analogues for the Treatment of Psoriasis

Jörg Reichrath and Michael F. Holick

Abstract   The skin is responsible for the sun-induced production of vitamin D. Keratinocytes not only produce vitamin D but are also a target tissue for its active form 1,25-dihydroxyvitamin D₃ [1,25(OH)₂D]. This is due to the fact that keratinocytes have a vitamin D receptor and when incubated with [1,25(OH)₂D] the cells stopped proliferating and began to mature. The potent antiproliferative activity of [1,25(OH)₂D] has been used to develop it and its analogues as a novel treatment for psoriasis. Other skin disorders including scleroderma, vitiligo, and ichthyosis have been successfully treated with active vitamin D analogues. This chapter will review the rationale for the use of active vitamin D analogues from various skin disorders and clinical trials, demonstrating efficacy especially for psoriasis.

Key Words: Psoriasis; vitamin D analogues; 1,25-dihydroxyvitamin D; skin diseases; 25-hydroxyvitamin D; PASI score; calcitriol; HIV; scleroderma

1. THE VITAMIN D SYSTEM IN THE SKIN

Vitamin D is photochemically synthesized by UVB action in the skin *(1, 2)*. However, the skin not only is the site of vitamin D production but also represents a target tissue for the secosteroid hormone 1α,25-dihydroxyvitamin D₃ (1,25(OH)₂D, calcitriol), the biologically active vitamin D metabolite *(3, 4)*. 1,25(OH)₂D exerts genomic and non-genomic effects. Non-genomic effects of calcitriol and analogues are related to effects on intracellular calcium *(5, 6)*. In keratinocytes and other cell types, calcitriol increases rapidly free cytosolic calcium levels *(5, 6)*. Genomic effects of 1,25(OH)₂D are mediated via binding to a nuclear receptor protein that is present in target tissues and binds calcitriol with high affinity (K_D 10^{-9}–10^{-10} M) and low capacity *(7, 8)*. The human vitamin D receptor (VDR) has been cloned *(9)* and sequence analysis demonstrated that this protein belongs to the superfamily of trans-acting transcriptional regulatory factors, which includes the steroid and thyroid hormone receptors and the retinoic

From: *Nutrition and Health: Vitamin D*
Edited by: M.F. Holick, DOI 10.1007/978-1-60327-303-9_57,
© Springer Science+Business Media, LLC 2010

acid receptors *(9)*. Interaction of calcitriol with VDR results in the phosphorylation of the receptor complex that in turn activates the transcription of calcitriol-sensitive target genes, especially genes involved in cellular differentiation and proliferation. It was previously shown that VDR requires auxillary factors for sufficient DNA binding *(10)*. These auxillary proteins were identified as the retinoid-X receptors (RXR-α,-β,-γ) that were demonstrated to heterodimerize with VDR, thus increasing both the transcriptional function and DNA binding to the respective vitamin D response elements (VDRE) in the promoter region of target genes *(10, 11)*. In the skin, both VDR (Fig. 1) and RXR-α are expressed in keratinocytes, fibroblasts, Langerhans cells, sebaceous gland cells, endothelial cells, and most cell types related to the skin immune system *(12, 13)*.

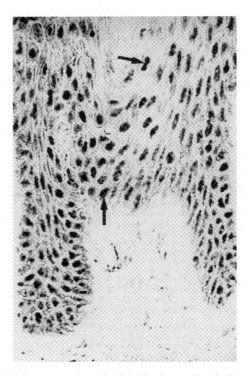

Fig. 1. Immunohistochemical demonstration of 1,25-dihydroxyvitamin D_3 receptors (VDR) in human skin. Notice strong nuclear VDR immunoreactivity in cells of all layers of the viable epidermis (*arrows*). Labeled avidin–biotin technique using mAb 9A7 directed against VDR. Original magnification ×400.

In vitro studies revealed that $1,25(OH)_2D$ is extremely effective in inducing the terminal differentiation and in inhibiting the proliferation of cultured human keratinocytes in a dose-dependent manner *(14–16)*. Additionally, $1,25(OH)_2D$ acts on many cell types involved in immunologic reactions, including lymphocytes, macrophages, and Langerhans cells *(17, 18)*.

Data on the effects of $1,25(OH)_2D$ on the melanin pigmentation system are still conflicting, but most studies do not support the possibility that calcitriol might regulate melanogenesis in human skin *(19)*.

2. PSORIASIS: PATHOGENESIS, IMMUNOLOGY, AND HISTOLOGY OF SKIN LESIONS

Psoriasis is a chronic recurring dermatosis of unknown etiology characterized by hyperproliferation and inflammation of the skin. Psoriasis may affect the skin, nails, and joints in 1–3% of the world's population with significant impacts on quality of life. The disease varies in severity from just a few lesions to involvement of the entire body surface area. Several different types of psoriasis can be discriminated, with psoriasis vulgaris (plaque psoriasis) being the most common form. Psoriasis vulgaris is characterized by sharply demarcated, raised, inflamed, red lesions (plaques) covered by silvery white scales. Although it can occur on any area of the skin, plaque psoriasis typically can be found on the scalp, elbows, knees, and lower back. The peak age of onset for this psychologically debilitating and disfiguring disease is the second decade, but psoriasis may first appear at any age from infancy to the aged (20). It is considered a multifactorial disease and has a prevalence of about 1–2% in the United States. Population, family, and twin studies clearly demonstrate that there is a strong but very complex genetic component leading to the development of psoriatic skin lesions (21). Most likely, multiple genes are involved in the pathogenesis of psoriasis. During the last years, molecular biology techniques that allow studies to analyze psoriasis susceptibility genes have been developed, but until today, no specific genetic marker of the disease is known. Psoriasis has long been known to be associated with certain HLA antigens, particularly HLA-Cw6, although there is no evidence that a psoriasis susceptibility gene exists at this locus (22). Until today, it is still unknown what cell types in human skin are primarily affected by the disease. Recent studies support the hypothesis that epidermal hyperproliferation in psoriasis may be mediated by cells of the immune system, most likely T lymphocytes (23). Activated T cells in psoriatic lesions express HLA-DR, the interleukin-2 receptor (CD 25), and secrete specific immune mediators and cytokines, such as IL-2 and interferon-γ (24–26). Thus, psoriasis represents a so-called Th1 profile disease (characterized by T-lymphocyte secretion of IL-2, IL-12, and interferon-γ) (27). In contrast, atopic dermatitis represents a so-called Th2 profile disease, which is characterized by T-cell secretion of IL-4, IL-5, and IL-10 (28). The activation signal for the development of psoriatic lesions is still unknown, although there is increasing evidence that superantigens such as the N-terminal component of bacterial M-proteins may be of importance for the initiation of T-cell proliferation in psoriasis (29). However, the hypothesis that immune cells may represent the primarily affected cell type in psoriasis has recently been challenged again when it was shown in adult mice that inducible epidermal deletion of Jun proteins resulted in psoriasis-like skin disease and arthritis (30).

The precise appearances of histology will depend upon the age of the psoriatic lesion and the site of the biopsy. In general, epidermal hyperplasia, in which the granular layer may be lost and the stratum corneum shows parakeratosis, can be found (Fig. 2). Typical lesions will histologically show elongation of the dermal papillae, with a relatively thin epidermis at the top of the papillae. Epidermis may show intercellular edema in suprapapillary compartments and infiltration with T lymphocytes and neutrophils, which can extend into spongiform pustules of Kogoj or Munro microabscesses (31).

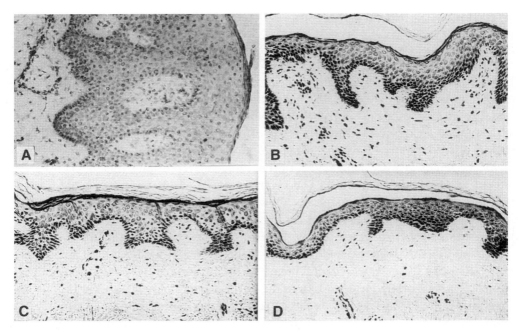

Fig. 2. Histological demonstration of morphological changes in lesional psoriatic skin after 6 weeks of topical treatment with calcitriol (15 µg/g, **b**) and calcipotriol (50 µg/g, **c**). **a**, lesional psoriatic skin before treatment; **d**, nonlesional psoriatic skin. Notice strong reduction of epidermal thickness after topical treatment with vitamin D analogues. Hematoxylin–eosin staining. Original magnification ×200.

3. CLINICAL STUDIES OF VITAMIN D AND ANALOGUES IN PSORIASIS AND OTHER SKIN DISEASES

For there is currently no cure for psoriasis, therapeutic strategies aim at achieving and maintaining remission, i.e. reducing the extent (percentage of the body area involved) and severity (degree of erythema, scaling, and plaque elevation) of the disease, while minimizing adverse events. The use of $1,25(OH)_2D$ and its analogues for the treatment of psoriasis resulted from two independent lines of investigation. Since psoriasis is a hyperproliferative skin disorder, it seemed reasonable that the antiproliferative effects of calcitriol could be used for the treatment of this disease. However, before launching clinical trials in 1985, MacLaughlin and associates reported the observation that psoriatic fibroblasts are partially resistant to the antiproliferative effects of $1,25(OH)_2D$ *(32)*. This observation prompted MacLaughlin and associates to speculate that calcitriol may be effective in the treatment of the hyperproliferative skin disease psoriasis. The other line of investigation resulted from a clinical observation. In 1985, Morimoto and Kumahara reported that a patient who was treated orally with 1α-hydroxyvitamin D_3 for osteoporosis had a dramatic remission of psoriatic skin lesions *(33)*.

Morimoto et al. reported a follow-up study, demonstrating that almost 80% of 17 patients with psoriasis who were treated orally with 1α-hydroxyvitamin D_3 at a dose of 1.0 µg/day for up to 6 month showed clinically significant improvement *(35)*.

Until now, numerous studies reported that various vitamin D analogues, including calcitriol, calcipotriol, tacalcitol, maxacalcitol, and becocalcidiol, are effective and safe in the topical treatment of psoriasis *(35–39)*. It was shown that topical calcitriol is very effective and safe in the long-term treatment of psoriasis vulgaris *(40)*. After treatment for 6–8 weeks in clinical trials, calcipotriol ointment and cream reduced mean PASI (Psoriasis Area and Severity Index) scores by 55–72% and 49–50%, respectively *(41)*. Applied twice daily topically in amounts of up to 100 g ointment (50 μg calcipotriol/g ointment) per week, calcipotriol, the synthetic analogue of calcitriol, was shown to be slightly more effective in the topical treatment of psoriasis than was betamethasone 17-valerate ointment *(42)*. Topical calcipotriol has been compared with other topical treatments for psoriasis. Calcipotriol cream has been reported to be as effective but more cosmetically acceptable than coal tar in a small, observer-blinded trial in patients with psoriasis *(43)*. Recently, twice-daily calcipotriol ointment was compared with once-daily short-contact dithranol cream therapy in a randomized controlled trial of supervised treatment of psoriasis in a day-care setting *(44)*. In that multicentre randomized controlled trial that was performed in six centers in the Netherlands, 106 patients with chronic plaque psoriasis were included, 54 receiving calcipotriol ointment twice daily and 52 receiving dithranol cream once daily in a 12-week intensive treatment program. Patients were treated at the day-care centre, using the care instruction principle of daily visits during the first week and twice-weekly visits subsequently for up to 12 weeks. Quality of life was assessed with the Skindex-29 and the Medical Outcomes Study 36-Item Short-Form General Health Survey (SF-36). At the end of treatment, no statistically significant differences were found between the calcipotriol and the dithranol group in any of the quality-of-life domains or scales of the Skindex-29 and the SF-36. Calcipotriol is available in three different formulations: cream, ointment, and solution. In a large, randomized, double-blind, controlled trial, twice-daily calcipotriol cream was significantly more effective than once-daily calcipotriol cream in terms of the mean percentage reduction in PASI from baseline (48.3% vs. 40.6%, $P = 0.006$) *(45)*. In that study, the reduction in PASI with twice-daily calcipotriol cream did not differ from that with calcipotriol cream in the morning plus clobetasone butyrate cream in the evening (53.7%) and was significantly lower than that with calcipotriol cream in the morning plus betamethasone valerate cream in the evening (57.5%).

Recently, an investigator-masked, randomized, multicenter comparison of the efficacy and safety of twice-daily applications of calcitriol 3 μg/g ointment vs. calcipotriol 50 μg/g ointment in subjects with mild to moderate chronic plaque-type psoriasis has been reported *(46)*. In that study, a total of 250 patients of both gender with mild to moderate chronic plaque-type psoriasis received either calcitriol or calcipotriol ointment twice daily for 12 weeks. Efficacy evaluations comprised global improvement [on a four-point scale from 0 (no change or worse) to 3 (clear or almost clear)] assessed by the investigator and by the subject. Efficacy further included the "dermatological sum score" at each study visit. Safety evaluations including adverse event reporting, cutaneous safety assessment by the investigator, and cutaneous discomfort assessment by the subject [both on a five-point scale from 0 (none) to 4 (very severe)] were recruited. At week 12, the LS_{mean} score of global improvement rated by the investigator was 2.27 for calcitriol and 2.22 for calcipotriol. This difference was not statistically significant, with calcitriol demonstrating to be non-inferior to calcipotriol for global improvement. This same parameter was scored by the subject, with a mean of 2.12 for calcitriol and 2.09 for calcipotriol. The percentage of patients with at least marked improvement tended to be in favor of calcitriol (95.7% vs. 85% for calcipotriol). However, differences were not statistically significant. The mean worst score for the cutaneous safety assessment was higher in the calcipotriol group (0.3 vs. 0.1 and 0.4 vs. 0.2, by the investigator and the patient, respectively). These differences were statistically significant in favor of a better safety profile for calcitriol ($P = 0.0035$). Fourteen dermatological and treatment-related adverse events were reported with calcipotriol vs. only five with calcitriol for a total of 22 adverse events reported throughout the study. The authors concluded that calcitriol administered twice daily over a 12-week treatment period demonstrated similar efficacy as compared to calcipotriol, while showing a significantly better safety profile. It has been reported that a mild dermatitis can be seen in about 10% of patients treated with calcipotriol (50 μg/g), particularly on the face (47). This side effect (mild dermatitis on the face) is not reported after topical treatment with calcitriol.

Recently, a randomized, placebo-controlled, double-blind, multicentre study, analyzing the efficacy and safety of topical becocalcidiol for the treatment of psoriasis vulgaris, has been reported (39). Becocalcidiol is a vitamin D analogue which has not caused hypercalcemia or significant irritation in preclinical trials. In that study, the efficacy and the safety of two dosing regimens of becocalcidiol ointment (low dose, 75 μg/g once daily for 8 weeks; high dose, 75 μg/g twice daily for 8 weeks) in the treatment of plaque-type psoriasis have been evaluated. One hundred and eighty-five subjects with chronic plaque-type psoriasis affecting 2–10% of their body surface area took part in a multicentre, double-blind, parallel-group, vehicle-controlled, randomized controlled trial comparing topical application of placebo, becocalcidiol 75 μg/g once daily (low dose), and becocalcidiol twice daily (high dose) for 8 weeks. Main outcomes included Physician's Static Global Assessment of Overall Lesion Severity (PGA) score; Psoriasis Symptom Severity (PSS) score; adverse events; and laboratory assessment. In that study, in the intent-to-treat population at week 8, high-dose becocalcidiol was statistically superior to vehicle [$P = 0.002$; 95% confidence interval (CI) 6.7–32.2], with 16 of 61 (26%) subjects achieving a PGA score of clear or almost clear. Greater improvement in PSS score was seen with high-dose becocalcidiol than with vehicle, but this result did not quite achieve statistical significance ($P = 0.052$; 95% CI –16.2 to 0.1). In all groups, therapy was safe and well tolerated, with fewer subjects experiencing irritation, than is reported in studies using calcipotriol. The authors concluded that treatment with high-dose topical becocalcidiol for 8 weeks led to almost or complete clearing of moderate plaque-type psoriasis in over a quarter of patients and that the therapy was safe and well tolerated (39).

In 1996, a long-term follow-up study demonstrated the efficacy and safety of oral calcitriol in the treatment of psoriasis (40). Of the 85 patients included in that study that received oral calcitriol, 88.0% had some improvement in their disease after 36 months and 26.5, 263, and 25.3% had complete, moderate, and slight improvement in their disease, respectively. Serum calcium concentrations and 24-h urinary calcium excretion increased by 3.9 and 148.2%, respectively, but were not outside the normal range. Bone mineral density of these patients remained unchanged. A very important consideration for the use of orally administered calcitriol is the dosing technique. To avoid its effects on enhancing dietary calcium absorption, it is very important to provide calcitriol at nighttime. Perez et al. *(40)* showed that as a result of this dosing technique, doses of 2–4 μg/night are well tolerated by psoriatic patients. Recently, the combination of acitretin and oral calcitriol for successful treatment of plaque-type psoriasis has been reported *(48)*.

Patients with psoriasis may need intermittent treatment for the whole of their lives. Vitamin D analogues have been shown not to exhibit tachyphylaxis during treatment of psoriatic lesions and can be continued indefinitely. They are effective and safe for the treatment of skin areas that are usually difficult to treat in psoriatic patients and that respond slowly. Additionally, vitamin D analogues are effective in the treatment of psoriatic skin lesions in children and in HIV patients.

3.1. Treatment of Scalp Psoriasis

It was demonstrated in a double-blind, randomized multicentre study that calcipotriol solution is effective in the topical treatment of scalp psoriasis *(49)*. Forty-nine patients were treated twice daily over a 4-week period. Sixty percent of patients on calcipotriol showed clearance or marked improvement vs. 17% in the placebo group. No side effects were reported.

3.2. Treatment of Nail Psoriasis

The occurrence of nail psoriasis has been reported in up to 50% of patients. Nails in general are very difficult to treat and respond slowly. Up to now, there has been no consistently effective treatment for psoriatic nails. Recently, it was shown that calcipotriol ointment is effective in the treatment of nail psoriasis *(50)*.

3.3. Treatment of Face and Flexures

Although the use of calcipotriol ointment is not recommended on face and flexures due to irritancy, most patients tolerate vitamin D analogues on these sites. Recently, the tolerability and efficacy of calcitriol (3 μg/g) and tacrolimus (0.3 mg/g) ointment in chronic plaque psoriasis affecting facial and genitofemoral regions was analyzed *(51)*. In this double-blind, parallel, 6-week study, 50 patients were randomized in a 1:1 ratio to apply calcitriol or tacrolimus twice daily. The primary efficacy variable was the mean reduction of the target area score (TAS), and the secondary efficacy variable was the percentage of patients with the Physician's Global Assessment (PGA) score of 5 (clear) and 4 (almost clear) at the end of the study. Both calcitriol and tacrolimus were well tolerated. Although calcitriol induced perilesional erythema in a statistically significant higher proportion of patients than did tacrolimus (55% vs. 16% at week 6; $P < 0.05$), it did not necessitate treatment discontinuation. At the end of the study, tacrolimus was significantly more effective than calcitriol based on a significant reduction of mean TAS (67% vs. 51%; $P < 0.05$) as well as more patients achieving complete or almost complete clearance by PGA (60% vs. 33%; $P < 0.05$). The authors concluded that both calcitriol (3 μg/g) and tacrolimus (0.3 mg/g) are safe and well-tolerated therapeutic agents in the treatment of psoriasis in sensitive areas *(51)*. However, tacrolimus demonstrated a slightly more effective clinical outcome compared with calcitriol *(51)*.

3.4. Treatment of Skin Lesions in Children

During the last years it has been shown that topical application of calcitriol ointment (3 μg of calcitriol per gram of petrolatum) is an effective, safe, and reliable therapy to cure psoriatic skin lesions in children *(52, 53)*.

3.5. Treatment of Psoriatic Lesions in HIV Patients

We have treated an HIV-positive patient suffering from psoriatic skin lesions with topical and oral calcitriol. The patient responded well and there was no evidence of enhancement in HIV disease activity or alterations in the number of T lymphocytes or $CD4^+$ and $CD8^+$ cells. Other case reports also demonstrate the efficacy and safety of vitamin D analogues in the treatment of psoriasis *(54)*.

3.6. Combination of Vitamin D Analogues with Other Therapies

Recently, it was reported that efficacy of topical treatment with vitamin D analogues in psoriasis can be increased by combination with other therapies, including very-low-dose oral cyclosporine (2 mg/kg per day), oral acitretin, topical dithranol, topical steroids, and UV-B or narrow-band UV-B phototherapy *(55–59)*. Complete clearing or 90% improvement in PASI was observed in 50% of patients treated with calcipotriol/cyclosporine vs. 11.8% in the placebo/cyclosporine group. No difference was found in that study between the groups in side effects.

Addition of calcipotriol ointment to oral application of acitretin (a vitamin A analogue) was shown to produce a significantly better treatment response achieved with a lower cumulative dose of acitretin in patients with severe extensive psoriasis vulgaris, as compared with the group of patients treated with oral acitretin alone. The number of patients reporting adverse events was similar between the two treatment groups *(57)*.

Combined topical treatment with calcipotriol ointment (50 μg/g) and betamethasone ointment was recently shown to be slightly more effective and caused less skin irritation than did calcipotriol used twice daily *(58)*. As a consequence, calcipotriol is now also available in a combined formulation with betamethasone dipropionate. The efficacy and safety of the combined formulation of calcipotriol and betamethasone dipropionate when used over a 4-week period is well documented. Recently, several publications report on and discuss the safety of this product when used for 52 weeks, representing an option for maintenance therapy for psoriatic patients *(60)*. In a recent investigation using an economic model to simulate the costs and benefits *(61)*, the cost effectiveness of the two-compound formulation calcipotriol and betamethasone dipropionate was compared with other topical treatments commonly used in the management of moderately severe plaque psoriasis in Scotland. In that study, the two-compound formulation calcipotriol and betamethasone dipropionate was associated with reduced costs and superior outcomes as compared to other topical treatments.

Kragballe and coworkers reported that efficacy of topical calcipotriol treatment in psoriasis can be improved by simultaneous ultraviolet B phototherapy. Recently, combination therapy of psoriasis with topical calcipotriol and narrow-band UV-B has been shown to be very effective for the treatment of psoriatic plaques *(59)*. Experimental investigations demonstrated that calcipotriol is degraded by UV radiation and suggested that calcipotriol should be applied after phototherapy but not immediately before. However, it was reported recently that the clinical effect of vitamin D analogues is not inactivated by subsequent UV exposure *(62)*. In that investigation, calcipotriol or maxacalcitol ointment was topically applied to psoriatic plaques of six patients immediately before or after phototherapy on the right or the left side of the body, respectively. The topical application of vitamin D₃ analogues either before or after irradiation by psoralen and UVA radiation (PUVA) or narrow-band (NB)-UV-B showed exactly similar effects in all patients. The authors concluded that therapeutic effects of vitamin D analogues are not clinically inactivated by subsequent irradiation with PUVA or NB-UV-B phototherapy.

4. TREATMENT OF OTHER SKIN DISORDERS WITH VITAMIN D ANALOGUES

Earlier in this century, vitamin D₃ was used in dermatology in huge pharmacological doses for the treatment of scleroderma, psoriasis, lupus vulgaris, and atopic dermatitis. But these first attempts of vitamin D treatment in dermatology were abandoned because of severe vitamin D intoxications that caused hypercalcemia, hypercalciuria, and kidney stones and because other new treatments were introduced for the treatment of these diseases.

4.1. Vitamin D and Ichthyosis

A double-blind, bilaterally paired, comparative study showed the effectiveness of topical treatment with calcipotriol ointment on congenital ichthyoses *(63)*. Reduction in scaling and roughness on the calcipotriol-treated side was seen in all patients with lamellar ichthyosis and bullous ichthyotic erythroderma of Brocq. The only patient treated with Comel–Netherton syndrome showed mild improvement, while the only patient suffering from ichthyosis bullosa of Siemens that was treated with calcipotriol did not show any change in severity on the calcipotriol-treated as compared to the vehicle-treated side.

4.2. Vitamin D and Scleroderma

Some case reports point at the efficacy of vitamin D analogues for the treatment of scleroderma. Humbert et al. *(64)* reported that oral administration of 1.0–2.5 μg/day calcitriol improves skin involvement, probably via inhibition of fibroblast proliferation and dermal collagen deposition.

4.3. Vitamin D and Vitiligo

A large variety of therapeutic agents are being used for the treatment of vitiligo, but treatment remains a challenge. Recent investigations indicate that vitamin D analogues may be efficient and safe in the treatment of vitiligo. While some studies report about a successful therapy with topically applied vitamin D analogues such as calcipotriol *(65)* or tacalcitol alone, others report about a successful therapy with topically applied vitamin D analogues in combination with UV *(66)* or laser therapy *(67)*.

4.4. Vitamin D and Skin Cancer

In vitro studies have demonstrated strong antiproliferative and prodifferentiating effects of vitamin D analogues in many VDR-expressing tumor cell lines, including malignant melanoma, squamous cell carcinoma, and leukemic cells *(18, 68)*. In vivo studies supported these results and showed that active vitamin D analogues block proliferation and tumor progression of epithelial tumors in rats *(69)*. Additionally, it was shown that administration of calcitriol reduced the number of lung metastases after implantation of lung carcinoma cells in mice *(70)*. Inhibition of tumor growth of human malignant melanoma and colonic cancer xenografts was also demonstrated in immune-suppressed mice, but only a high doses of calcitriol *(71)*. Little is known regarding the effects of calcitriol on the formation of metastases in patients with malignant melanoma or squamous cell carcinoma of the skin.

4.5. Vitamin D and Other Skin Diseases

A number of case reports demonstrate positive effects of topical treatment with vitamin D analogues in a variety of skin diseases such as transient acantholytic dermatosis (Grover's disease), inflammatory linear verrucous epidermal nevus (ILVEN), disseminated superficial actinic porokeratosis, pityriasis rubra pilaris, epidermolytic palmoplantar keratoderma of Vorner, and Sjögren–Larsson Syndrome. These promising observations will have to be further evaluated in clinical trials.

5. BIOLOGICAL EFFECTS OF VITAMIN D AND ANALOGUES IN PSORIASIS

Until today, the mechanisms underlying the therapeutic effectiveness of vitamin D analogues in psoriasis are still not completely understood. Results from immunohistochemical and molecular biology studies indicate that the antiproliferative effects of topical calcitriol on epidermal keratinocytes are more pronounced as compared to effects on dermal inflammation. Modulation of various markers of epidermal proliferation (proliferating cell nuclear antigen, Ki-67 antigen) and differentiation (involucrin, transglutaminase K, filaggrin, cytokeratins 10,16) in lesional psoriatic skin after topical application of vitamin D analogues was shown in situ *(72)* (Fig. 3). Interestingly, effects of topical treatment with vitamin D analogues on dermal inflammation are less pronounced (CD-antigens, cytokines, HLA-DR, etc.). One reason for this observation may be that the bioavailability of this potent hormone in the dermal compartment may be markedly reduced as compared to the epidermal compartment *(72)*.

Molecular biology studies demonstrated that clinical improvement in psoriatic lesions treated with calcitriol correlated with an elevation of VDR mRNA *(73)*. It is known that some patients suffering from psoriasis are resistant to calcitriol treatment. It was

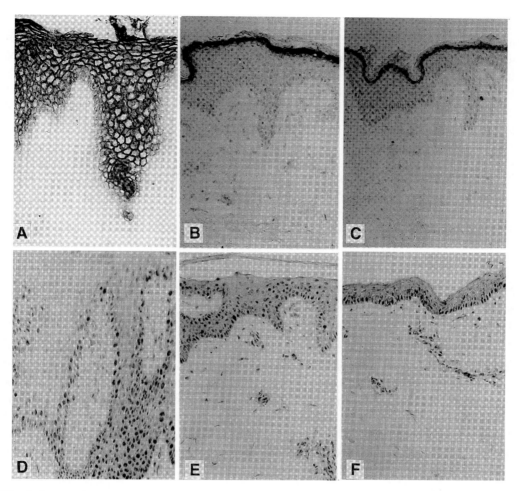

Fig. 3. Immunohistological detection of transglutaminase K (**a, b, c**) and proliferating cell nuclear antigen (**d, e, f**) in lesional psoriatic skin before treatment (**a, d**), lesional psoriatic skin after 6 weeks of topical treatment with calcipotriol (**b, e**), and in nonlesional psoriatic skin (**c, f**). Labeled avidin–biotin technique. Original magnification ×160 (**a, b, c**) and ×400 (**d, e, f**).

recently demonstrated that responders can be distinguished from the nonresponders on the molecular level since nonresponders show no elevation of VDR mRNA in skin lesions along with the treatment. These data suggest that the ability of calcitriol to regulate keratinocyte growth is closely linked to the expression of VDR. The target genes of topical calcitriol that are responsible for its therapeutic efficacy in psoriasis are still unknown. Mayor candidates for calcitriol target genes that are responsible for the calcitriol-induced terminal differentiation in keratinocytes are distinct cell cycle-associated proteins (i.e., INK4 family), including p21/WAF-1 (74). Other possible target pathways include the ERK- and TNF-signaling pathways. Since activation of extracellular signal-regulated kinase (ERK) promotes keratinocyte proliferation and mediates epidermal inflammation, the effect of calcitriol on ERK activation in HaCaT keratinocytes exposed to the ubiquitous inflammatory cytokine TNF has been studied recently *(75)*. By using the EGF receptor (EGFR) tyrosine kinase inhibitor, AG1487 and the Src family inhibitor, PP-1, the authors demonstrated that TNF activated ERK in an EGFR- and Src-dependent and an EGFR- and Src-independent modes. In that study, EGFR-dependent activation resulted in the upregulation of the transcription factor c-Fos, while the EGFR-independent activation mode was of a shorter duration and did not affect c-Fos expression but induced IL-8 mRNA expression. Pretreatment with calcitriol enhanced TNF-induced EGFR–Src-dependent ERK activation and tyrosine phosphorylation of the EGFR but abolished the EGFR–Src-independent ERK activation. These effects were mirrored by enhancement of c-Fos and inhibition of IL-8 induction by TNF. Treatment with calcitriol increased the rate of the dephosphorylation of activated ERK, accounting for the inhibition of EGFR–Src-independent ERK activation by TNF. These findings indicate that effects on the ERK cascade contribute to the effects of calcitriol and its synthetic analogues on cutaneous inflammation and keratinocyte proliferation *(75)*.

6. PERSPECTIVES FOR THE EVALUATION OF NEW VITAMIN D ANALOGUES WITH LESS CALCEMIC ACTIVITY THAT CAN BE USED FOR THE TREATMENT OF HYPERPROLIFERATIVE SKIN DISORDERS

The use of vitamin D analogues in dermatology and other medical fields was shown to be limited, since serious side effects, mainly on calcium metabolism, may occur at supraphysiological doses needed to reach clinical improvement. The evaluation of new vitamin D compounds with strong immunosuppressive, antiproliferative, and differentiating effects but only marginal effects on calcium metabolism introduces new important therapies for the treatment of various skin diseases. The goal to create new vitamin D analogues with selective biological activity and no undesirable side effects is still not reached, but recent findings introduce new and promising finding concepts.

Calcipotriol (MC 903), a synthetic vitamin D analogue with similar VDR-binding properties as compared to calcitriol but low affinity for the vitamin D-binding protein (DBP), was shown to be effective and safe in the topical treatment of psoriasis *(45)*. In vivo studies in rats showed that effects of calcipotriol on calcium metabolism are 100–200 times lower as compared to calcitriol, while in vitro effects on proliferation and differentiation on human keratinocytes are comparable *(76)*. These differential effects are probably caused by the different pharmacokinetic profiles of calcipotriol and calcitriol (different affinity for DBP). Serum half-life of these vitamin D analogues in rats was shown to be 4 min after treatment with calcipotriol in contrast to 15 min after treatment with calcitriol *(76)*. However, one has to mention that most of the calcium studies comparing calcitriol and calcipotriol were done in vivo, while most studies analyzing proliferation or differentiation were done in vitro.

A different approach to create new vitamin D analogues that are effective in the topical treatment of hyperproliferative or inflammatory skin diseases is the goal to create new synthetic compounds with a high degree of dissociation that are metabolized in the skin and therefore exert only little systemic side effects. New analogues of vitamin D obtained by a combination of the 20-methyl modification with biologically interesting artificial side chain subunits *(77)*, 2β-substituted calcitriols *(78)*, or C-2-substituted

19-nor-1α,25-dihydroxyvitamin D3 *(79)* analogues are promising candidates for this strategy.

Another interesting approach to locally enhance the concentration of calcitriol in the skin without obtaining systemic side effects are attempts to inhibit specifically the activity of vitamin D-metabolizing enzymes, i.e., various hydroxylases (catabolic D$_3$-(OH)ases, i.e., 24-hydroxylase for calcitriol) that are present in the skin and that are responsible for the catabolism of calcitriol *(80)*. It is known that various pharmacologic active compounds, including other steroidal hormones but also cytochrome P450 inhibitors such as ketoconazole, inhibit specifically the activity of D$_3$-(OH)ases in the skin and other cell types *(81, 82)*. It may be possible to enhance locally the concentration of endogenous calcitriol in the skin by the topical application of these compounds without obtaining systemic side effects. It can be speculated that the therapeutic effects of various antimycotic compounds including ketoconazole in the treatment of seborrheic dermatitis may at least in part be due to this mechanism.

It is now known that VDR requires nuclear accessory proteins for efficient binding to vitamin D response elements in promoter regions of target genes, thereby inducing VDR-mediated transactivation *(83)*. As a consequence, different vitamin D analogues may have (depending on their chemical structure) different affinities for the various homo- or heterodimers of VDR and nuclear cofactors including RXR-α *(84)*. The synthesis of new vitamin D analogues that activate different vitamin D signaling pathways may lead to the introduction of new therapeutics for the topical or the oral treatment of various skin diseases. These new drugs may induce strong effects on target cell proliferation and differentiation in the skin or the immune system, but only marginal effects on calcium metabolism.

Another approach to enhance the therapeutic effects of orally or topically administered calcitriol may be the combination with synergistic acting drugs. The recent discovery of different vitamin D signaling pathways that are determined and regulated by cofactors of VDR including RXR-α and their corresponding ligands suggests that 9-*cis* RA or all-*trans* RA may act synergistically with vitamin D analogues in inducing VDR-mediated transactivation and regulating the transcriptional activity of distinct gene networks. Only little is known about the effects of the combined application of vitamin D and vitamin A analogues under physiological or pathophysiological conditions in vivo. This combination may selectively enhance or block different biological effects of vitamin D analogues that are mediated by different vitamin D signaling pathways. In conclusion, it can be speculated that new vitamin D analogues will introduce new alternatives for the treatment of various skin disorders. If the final goal to create strong antiproliferative and anti-inflammatory vitamin D analogues with only little calcemic activity is reached, these new agents may herald a new era in dermatologic therapy, which can possibly be compared with the introduction of synthetic corticosteroids or retinoids. These new drugs that may activate selective vitamin D signaling pathways but may exert only little calcemic activity may also be effective in the systemic treatment of various cutaneous malignancies, including lymphomas, malignant melanoma, squamous cell carcinoma, or basal cell carcinoma. In conclusion, it can be summarized that vitamin D analogues have a great potential for the future systemic use as antiproliferative or anticancer agents *(84)*.

REFERENCES

1. Holick MF, MacLaughlin JA, Clark MB, Holick SA, Potts JT, Anderson RR, Blank IH, Parrish JA, Elias P (1980) Photosynthesis of previtamin D_3 in human skin and the physiological consequences. Science 210:203–205

2. Holick MF, MacLaughlin JA, Anderson RR, Parrish J (1982) Photochemistry and photobiology of vitamin D. In: Regan JD, Parrish JA (eds) Photomedicine. Plenum Press, New York, 195–218

3. Holick MF, Smith E, Pincus S (1987) Skin as the site of vitamin D synthesis and target tissue for 1,25-dihydroxyvitamin D_3. Arch Dermatol 123:1677–1682

4. Holick MF (1991) Photobiology, physiology and clinical applications for Vitamin D. In: Goldsmith LA (ed) Physiology, biochemistry and molecular biology of the skin, 2nd edn. Oxford University Press, New York, 928–956

5. Bittiner B, Bleehen SS, Mac Neil S (1991) 1α-25-$(OH)_2$ Vitamin D_3 increases intracellular calcium in human keratinocytes. Br J Dermatol 124:12230–12235

6. MacLaughlin JA, Cantley LC, Holick MF (1990) 1,25$(OH)_2D_3$ increases calcium and phosphatidylinositol metabolism in differentiating cultured human keratinocytes. J Nutr Biochem 1:81–87

7. Haussler MR (1986) Vitamin D receptors: nature and function. Annu Rev Nutr 6:527–562

8. Haussler MR, Mangelsdorf DJ, Komm BS, Terpening CM, Yamaoka K, Allegretto EA, Baker AR, Shine J, McDonnell DP, Hughes M, Weigel NL, O'Malley BW (1988) Molecular biology of the vitamin D hormone. Recent Prog Horm Res 44:263–305

9. Baker AR, Mc Donnell DP, Hughes M, Crisp TM, Mangelsdorf DJ, Haussler MR, Pike JW, Shine J, O'Malley BW (1988) Cloning and expression of full-length cDNA encoding human vitamin D receptor. Proc Natl Acad Sci USA 85:3294–3298

10. Yu VC, Deisert C, Andersen B, Holloway JM, Devary OV, Näär AM, Kim SY Boutin JM, Glass CK, Rosenfeld MG (1991) RXRβ: a coregulator that enhances binding of retinoic acid, thyroid hormone and vitamin D receptors to their cognate response elements. Cell 67:1251–1266

11. Leid M, Kastner P, Lyons R, Nakshatri H, Saunders M, Zacharewski T, Chen J, Staub A, Garnier J, Mader S, Chambon P (1992) Purification, cloning, and RXR identity of the HeLa cell factor with which RAR or TR heterodimerizes to bind target sequences efficiently. Cell 68:377–395

12. Milde P, Hauser U, Simon R, Mall G, Ernst V, Haussler MR, Frosch P, Rauterberg EW (1991) Expression of 1,25-dihydroxyvitamin D_3 receptors in normal and psoriatic skin. J Invest Dermatol 97: 230–239

13. Reichrath J, Münssinger T, Kerber A, Rochette-Egly C, Chambon P, Bahmer FA, Baum HP (1995). In situ detection of retinoid-X receptor expression in normal and psoriatic human skin. Br J Dermatol 133:168–175

14. Smith EL, Walworth NC, Holick MF (1986) Effect of 1α-25-dihydroxyvitamin D_3 on the morphologic and biochemical differentiation of cultured human epidermal keratinocytes grown under serum-free conditions. J Invest Dermatol 86:709–714

15. Hosomi J, Hosoi J, Abe E, Suda T, Kuroki T (1983) Regulation of terminal differentiation of cultured mouse epidermal cells by 1-alpha 25-dihydroxy-vitamin D_3. Endocrinol 113:1950–1957

16. Gniadecki R, Serup J (1995) Stimulation of epidermal proliferation in mice with 1 alpha, 25-dihydroxyvitamin D_3 and receptor-active 20-EPI analogues of 1 alpha, 25-dihydroxyvitamin D_3. Biochem Pharmacol 49:621–624

17. Rigby WFC (1988) The immunobiology of vitamin D. Immunol Today 9:54–58

18. Texereau M, Viac J, Vitamin D (1992) immune system and skin. Europ J Dermatol 2:258–264

19. Ranson M, Posen S, Mason RS (1988) Human melanocytes as a target tissue for hormones: in vitro studies with 1α,25-dihydroxyvitamin D_3, alpha-melanocyte stimulating hormone, and beta-estradiol. J Invest Dermatol 91:593–598

20. Christophers E, Henseler T (1985) Psoriasis of early and late onset: characterization of two types of psoriasis vulgaris. J Am Acad Dermatol 13:450–456

21. Christophers E, Henseler T (1989) Patient subgroups and the inflammatory pattern in psoriasis. Acta Derm Venereol (Stockh) 69:88–92

22. Elder JT, Henseler T, Christophers E, Voorhees JJ, Nair RP (1994) Of genes and antigens: the inheritance of psoriasis. J Invest Dermatol 103:150S–153S

23. Valdimarsson H, Baker BS, Jonsdittir I, Fry L (1986) Psoriasis: a disease of abnormal keratinocyte proliferation induced by T lymphocytes. Immunol Today 7:256–259

24. Lee RE, Gaspari AA, Lotze MT, Chang AE, Rosenberg SA (1988) Interleukin 2 and psoriasis. Arch Dermatol 124:1811–1815

25. Barker JN, Jones ML, Mitra RS, Crockett Torab E, Fantone JC, Kunkel SL, Warren JS, Dixit VM, Nickoloff BJ (1991) Modulation of keratinocyte derived interleukin-8 which is chemotactic for neutrophils and T lymphocytes. Am J Pathol 139:869–876

26. Gottlieb AB (1990) Immunologic mechanisms in psoriasis. J Invest Dermatol 95:18S–19S

27. Schlaak JF, Buslau M, Jochum W, Hermann E, Girndt M, Gallati H, Meyer zum Büschenfelde KH, Fleischer B (1994). T cells involved in psoriasis vulgaris belong to the Th1 subset. J Invest Dermatol 102:145–149

28. van Reijsen FC, Druijnzeel-Koomen CAFM, Kalthoff FS, Maggi E, Romagnani S, Westland JKT, Mudde GC (1992) Skin-derived aeroallergen-specific T-cell clones of Th2 phenotype in patients with atopic dermatitis. J Allergy Clin Immunol 90:184–192

29. Leung DY, Walsh P, Giorno R, Norris DA (1993) A potential role for superantigens in the pathogenesis of psoriasis. J Invest Dermatol 100:225–228

30. Zenz R, Eferl R, Kenner L, Florin L, Hummerich L, Mehic D, Scheuch H, Angel P, Tschachler E, Wagner EF (2005) Psoriasis-like skin disease and arthritis caused by inducible epidermal deletion of Jun proteins. Nature 437(7057):369–375

31. Chowaniec O, Jablonska S, Beutner EH, Proniewska M, Jarzabek Chorzelska M, Rzesa G (1981) Earliest clinical and histological changes in psoriasis. Dermatologica 163:42–51

32. MacLaughlin JA, Gange W, Taylor D, Smith E, Holick MF (1985) Cultured psoriatic fibroblasts from involved and uninvolved sites have partial but not absolute resistance to the proliferation-inhibition activity of 1,25-dihydroxyvitamin D₃. Proc Natl Acad Sci USA 82:5409–5412

33. Morimoto S, Kumahara Y (1985) A patient with psoriasis cured by 1α-hydroxyvitamin D₃. Med J Osaka Univ 35(3–4):51–54

34. Morimoto S, Yochikawa K, Kozuka T, Kitano Y, Imawaka S, Fukuo K, Koh E, Kumahara Y (1986) An open study of vitamin D₃ treatment in psoriasis vulgaris. Br J Dermatol 115:421–429

35. Holick MF, Chen ML, Kong XF, Sanan DK (1996) Clinical uses for calciotropic hormones 1,25-dihydroxyvitamin D₃ and parathyroid hormone related peptide in dermatology: a new perspective. J Invest Dermatol (Symp Proc) 1:1–9

36. Perez A, Chen TC, Turner A, Raab R, Bhawan J, Poche P, Holick MF (1996) Efficacy and safety of topical calcitriol (1,25-dihydroxyvitamin D₃) for the treatment of psoriasis. Br J Dermatol 134:238–246

37. Kragballe K, Beck HI, Sogaard H (1988) Improvement of psoriasis by topical vitamin D₃ analogue (MC 903) in a double-blind study. Br J Dermatol 119:223–230

38. van de Kerkhof PCM, van Bokhoven M, Zultak M, Czarnetzki BM (1989) A double-blind study of topical 1α-25-dihydroxyvitamin D₃ in psoriasis. Br J Dermatol 120:661–664

39. Helfrich YR, Kang S, Hamilton TA, Voorhees JJ (2007) Topical becocalcidiol for the treatment of psoriasis vulgaris: a randomized, placebo-controlled, double-blind, multicentre study. Br J Dermatol 157(2):369–374

40. Perez A, Raab R, Chen TC, Turner A, Holick MF (1996) Safety and efficacy of oral calcitriol (1,25-dihydroxyvitamin D₃) for the treatment of psoriasis. Br J Dermatol 134:1070–1078

41. Van de Kerkhof PC, Vissers WH (2003) The topical treatment of psoriasis. Skin Pharmacol Appl Skin Physiol 16:69–83

42. Kragballe K, Gjertsen BT, de Hoop D, Karlsmark T, van de Kerhof PCM, Larko O, Nieboer C, Roed-Petersen J, Strand A, Tikjob B (1991) Double-blind right/left comparison of calcipotriol and betamethasone valerate in treatment of psoriasis vulgaris. Lancet 337:193–196

43. Tzaneva S, Hönigsmann H, Tanew A (2003) Observer-blind, randomized, intrapatient comparison of a novel 1% coal tar preparation (Exorex R) and calcipotriol cream in the treatment of plaque type psoriasis. Br J Dermatol 149:350–353

44. De Korte J, van der Valk PG, Sprangers MA, Damstra RJ, Kunkeler AC, Lijnen RL, Oranje AP, de Rie MA, de Waard-van der Spek FB, Hol CW, van de Kerkhof PC (2008) A comparison of twice-daily

calcipotriol ointment with once-daily short-contact dithranol cream therapy: quality-of-life outcomes of a randomized controlled trial of supervised treatment of psoriasis in a day-care setting. Br J Dermatol 158(2):375–381

45. Kragballe K, Barnes L, Hamberg KJ, Hutchinson P, Murphy F, Moller S et al (1998) Calcipotriol cream with or without concurrent topical corticosteroid in psoriasis: tolerability and efficacy. Br J Dermatol 139:649–654

46. Zhu X, Wang B, Zhao G, Gu J, Chen Z, Briantais P, Andres P (2007) An investigator-masked comparison of the efficacy and safety of twice daily applications of calcitriol 3 μg/g ointment vs. calcipotriol 50 μg/g ointment in subjects with mild to moderate chronic plaque-type psoriasis. J Eur Acad Dermatol Venereol 21(4):466–472

47. Serup J (1994) Calcipotriol irritation: mechanism, diagnosis and clinical implication. Acta Derm Venereol (Stockh) Abstr 186:42S

48. Ezquerra GM, Regana MS, Millet PU (2007) Combination of acitretin and oral calcitriol for treatment of plaque-type psoriasis. Acta Derm Venereol 87(5):449–450

49. Green C, Ganpule M, Harris D, Kavanagh G, Kennedy C, Mallett R, Rustin M, Downes N (1994) Comparative effects of calcipotriol (MC 903) solution and placebo (vehicle of MC 903) in the treatment of psoriasis of the scalp. Br J Dermatol 130:483–487

50. Petrow W (1995) Treatment of a nail psoriasis with calcipotriol. Akt Dermatol 21:396–400

51. Liao YH, Chiu HC, Tseng YS, Tsai TF (2007) Comparison of cutaneous tolerance and efficacy of calcitriol 3 μg/g ointment and tacrolimus 0.3 mg/g ointment in chronic plaque psoriasis involving facial or genitofemoral areas: a double-blind, randomized controlled trial. Br J Dermatol 157(5): 1005–1012

52. Saggese G, Federico G, Battini R (1993) Topical application of 1,25 dihydroxyvitamin D_3 (calcitriol) is an effective and reliable therapy to cure skin lesions in psoriatic children. Eur J Pediatr 152: 389–392

53. Perez A, Chen TC, Turner A, Holick MF (1995) Pilot study of topical calcitriol (1,25-dihydroxyvitamin D_3) for treating psoriasis in children. Arch Dermatol 131:961–962

54. Gray JD, Bottomley W, Layton AM, Cotterill JA, Monteiro E (1992) The use of calcipotriol in HIV-related psoriasis. Clin Exp Dermatol 17(5):342–343

55. Grossman RM, Thivolet J, Claudy A, Souteyrand P, Guilhou JJ, Thomas P, Amblard P, Belaich S, de Belilovsky C, de la Brassinne M et al (1994) A novel therapeutic approach to psoriasis with combination calcipotriol ointment and very low-dose cyclosporine: a result of a multicenter placebo-controlled study. J Am Acad Dermatol 31:68–74

56. Kerscher M, Volkenandt M, Plewig G, Lehmann P (1993) Combination phototherapy of psoriasis with calcipotriol and narrow band UVB. Lancet 342:923

57. Cambazard F, van de Kerkhof PCM, Hutchinson PE, and the Calcipotriol Study Group (1996) Proceedings of the 3rd International Calcipotriol Symposium, Munich Germany, 23 March 1996.

58. Ortonne JP (1994) Calcipotriol in combination with betamethasone dipropionate. Nouv Dermatol 13:736–751

59. Kragballe K (1990) Combination of topical calcipotriol (MC 903) and UVB radiation for psoriasis vulgaris. Dermatologica 181:211–214

60. Toole JW (2007) Calcipotriol and betamethasone dipropionate for the treatment of psoriasis: a 52-week study. Skin Therapy Lett 12(4):1–3

61. Bottomley JM, Auland ME, Morais J, Boyd G, Douglas WS (2007) Cost-effectiveness of the two-compound formulation calcipotriol and betamethasone dipropionate compared with commonly used topical treatments in the management of moderately severe plaque psoriasis in Scotland. Curr Med Res Opin 23(8):1887–1901

62. Adachi Y, Uchida N, Matsuo T, Horio T (2008) Clinical effect of vitamin D3 analogues is not inactivated by subsequent UV exposure. Photodermatol Photoimmunol Photomed 24(1):16–18

63. Lucker GP, van de Kerkhof PC, van Dijk MR, Steijlen PM (1994) Effect of topical calcipotriol on congenital ichthyosis. Br J Dermatol 131:546–550

64. Humbert P, Dupond JL, Agache P, Laurent R, Rochefort A, Drobacheff C, de Wazieres B, Aubin F (1993) Treatment of scleroderma with oral 1,25-dihydroxyvitamin D_3: evaluation of skin involvement

using non-invasive techniques. Results of an open prospective trial. Acta Derm Venereol (Stockh) 73:449–451

65. Kumaran MS, Kaur I, Kumar B (2006) Effect of topical calcipotriol, betamethasone dipropionate and their combination in the treatment of localized vitiligo. J Eur Acad Dermatol Venereol 20(3): 269–273

66. Amano H, Abe M, Ishikawa O (2008) First case report of topical tacalcitol for vitiligo repigmentation. Pediatr Dermatol 25(2):262–264

67. Goldinger SM, Dummer R, Schmid P, Burg G, Seifert B, Läuchli S (2007) Combination of 308-nm xenon chloride excimer laser and topical calcipotriol in vitiligo. J Eur Acad Dermatol Venereol 21(4):504–508

68. Koeffler HP, Hirji K, Itri L (1985) 1,25-dihydroxyvitamin D$_3$: in vivo and in vitro effects on human preleukemic and leukemic cells. Cancer Treat Rep 69:1399–1407

69. Colston KW, Chander SK, Mackay AG, Coombes RC (1992) Effects of synthetic vitamin D analogues on breast cancer cell proliferation in vivo and in vitro. Biochem Pharmacol 44: 693–702

70. Franceschi RT, Linson CJ, Peter CT, Romano PR (1987) Regulation of cellular adhesion and fibronectin synthesis by 1α,25-dihydroxyvitamin D$_3$. J Biol Chem 262:4165–4171

71. Eisman JA, Barkla DH, Tutton PJM (1987) Suppression of in vivo growth of human cancer solid tumor xenografts by 1α,25-dihydroxyvitamin D$_3$. Cancer Res 47:21–25

72. Reichrath J, Müller SM, Kerber A, Baum HP, Bahmer FA (1997) Biological effects of topical calcipotriol (MC 903) treatment in psoriatic skin. J Am Acad Dermatol 36(1):19–28

73. Chen ML, Perez A, Sanan DK, Heinrich G, Chen TC, Holick MF (1996) Induction of vitamin D receptor mRNA expression in psoriatic plaques correlates with clinical response to 1,25-dihydroxyvitamin D$_3$. J Invest Dermatol 106:637–641

74. Missero C, Calautti E, Eckner R, Chin J, Tsai LH, Livingston DM, Dotto GP (1995) Involvement of the cell-cycle inhibitor Cip1/WAF1 and the E1A-associated p300 protein in terminal differentiation. Proc Natl Acad Sci USA 92:5451–5455

75. Ziv E, Rotem C, Miodovnik M, Ravid A, Koren R (2008) Two modes of ERK activation by TNF in keratinocytes: different cellular outcomes and bi-directional modulation by vitamin D. J Cell Biochem 104(2):606–619

76. Binderup L, Latini S, Binderup E, Bretting C, Calverley M, Hansen K (1991) 20-epi-vitamin D$_3$ analogues: a novel class of potent regulators of cell growth and immune response. Biochem Pharmacol 42:1569–1575

77. Neef G, Kirsch G, Schwarz K, Wiesinger H, Menrad A, Fähnrich M, Thieroff-Eckerdt R, Steinmeyer A (1994) 20-methyl vitamin D analogues. In: Norman AW, Bouillon R, Thomasset M (eds.) Vitamin D. A pluripotent steroid hormone: structural studies, molecular endocrinology and clinical applications. Walter de Gruyter, Berlin, pp. 97–98

78. Schönecker B, Reichenbächer M, Gliesing S, Prousa R, Wittmann S, Breiter S, Thieroff-Eckerdt R, Wiesinger H, Haberey M, Scheddin D, Mayer H (1994) 2β-substituted calcitriols and other A-ring substituted analogues – synthesis and biological results. In: Norman AW, Bouillon R, Thomasset M (eds) Vitamin D. A pluripotent steroid hormone: structural studies, molecular endocrinology and clinical applications. Walter de Gruyter, Berlin, pp. 99–100

79. Chen TC, Persons KS, Zheng S, Mathieu J, Holick MF, Lee YF, Bao B, Arai MA, Kittaka A (2007) Evaluation of C-2-substituted 19-nor-1alpha,25-dihydroxyvitamin D3 analogs as therapeutic agents for prostate cancer. J Steroid Biochem Mol Biol 103(3–5):717–720

80. Schuster I, Herzig G, Vorisek G(1994) Steroidal hormones as modulators of vitamin D metabolism in human keratinocytes. In: Norman AW, Bouillon R, Thomasset M (eds) Vitamin D. A pluripotent steroid hormone: structural studies, molecular endocrinology and clinical applications. Walter de Gruyter, Berlin, pp. 184–185

81. Zhao J, Marcelis S, Tan BK, Verstuaf A, Boillon R (1994) Potentialisation of vitamin D (analogues) by cytochrome P-450 enzyme inhibitors is analog- and cell-type specific. In: Norman AW, Bouillon R, Thomasset M (eds) Vitamin D. A pluripotent steroid hormone: structural studies, molecular endocrinology and clinical applications. Walter de Gruyter, Berlin, pp. 97–98

82. Segersten U, Björklund P, Hellman P, Akerström G, Westin G (2007) Potentiating effects of non-active/active vitamin D analogues and ketoconazole in parathyroid cells. Clin Endocrinol (Oxf) 66(3):399–404

83. Carlberg C, Bendik I, Wyss A, Meier E, Sturzenbecker LJ, Grippo JF, Hunziker W (1993) Two nuclear signalling pathways for vitamin D. Nature 361:657–660

84. Schräder M, Müller KM, Becker-Andre M, Carlberg C (1994) Response element selectivity for heterodimerization of vitamin D receptors with retinoic acid and retinoid X receptors. J Mol Endocrinol 12:327–339

85. Deeb KK, Trump DL, Johnson CS (2007) Vitamin D signalling pathways in cancer: potential for anticancer therapeutics. Nat Rev Cancer 7(9):684–700

58 Affinity Alkylating Vitamin D Analogs as Molecular Probes and Therapeutic Agents

Rahul Ray

Abstract   Affinity alkylating derivatives of naturally occurring molecules were originally developed to target and probe the substrate/ligand-binding sites of enzymes and receptors. These reagents, by virtue of their ability to covalently attach to the substrate/ligand-binding pockets of target enzymes/receptors, provide valuable structural information of the host and the guest, focusing on the dynamic interaction between the two. Our laboratory developed affinity and photoaffinity labeling derivatives of 1,25-dihydroxyvitamin D_3 [1,25(OH)$_2$D] to probe the ligand-binding domain of nuclear vitamin D receptor (VDR). With these studies we obtained crucial information about the ligand contact points and three-dimensional geometry of the ligand-binding domain of VDR, in relation to the biological properties of 1,25(OH)$_2$D. 1,25(OH)$_2$D is an anticancer agent, albeit with severe toxicity at pharmacological doses. We observed that an affinity alkylating derivative of 1,25(OH)$_2$D strongly inhibits the growth of various cancer cells and causes these cells to undergo apoptosis, demonstrating strong potential of this affinity analog of 1,25(OH)$_2$D as an anti-cancer agent. This article summarizes the results of various studies involving affinity alkylating derivatives of vitamin D metabolites to investigate structural–functional aspects of VDR, as well as development of these reagents as potential therapeutic agents in cancer.

Key Words: Vitamin D-binding protein; vitamin D analogs; 25-hydroxyvitamin D; VDR-alkylation; vitamin D receptor; 1,25-dihydroxyvitamin D; 1,25-dihydroxyvitamin D_3-bromoacetate

1. AFFINITY AND PHOTOAFFINITY LABELING REAGENTS AS MOLECULAR PROBES OF ENZYMES AND RECEPTORS

Affinity alkylating derivatives of naturally occurring molecules are classical molecular probes to target the substrate/ligand-binding sites of enzymes and receptors. By virtue of their ability to covalently label the substrate/ligand-binding sites, these high-affinity substrate–analogs/ligand–mimics provide valuable information about the three-dimensional geometries of the substrate/ligand-binding pockets of enzymes/receptors including identity of key amino acids (contact points), orientation of the substrates/ligands inside the pocket, as well as general electronic and steric environments

From: *Nutrition and Health: Vitamin D*
Edited by: M.F. Holick, DOI 10.1007/978-1-60327-303-9_58,
© Springer Science+Business Media, LLC 2010

inside the binding pocket *(1)*. Affinity labeling process involves interaction of a nucleophilic amino acid in close proximity to the electrophilic group in the affinity reagent so that a covalent bond can be formed. On the other hand, in the photoaffinity labeling method, covalent labeling process is initiated by flashing light which degenerates the photoreactive groups in the photoaffinity derivative of the ligand/substrate to produce a highly reactive nitrene or carbene intermediate that indiscriminately inserts into any neighboring amino acid (as shown in Fig. 1a, b).

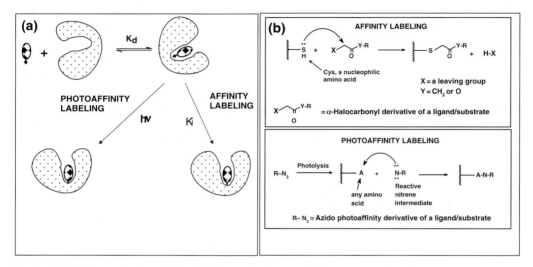

Fig. 1. (**a**) Affinity and photoaffinity labeling of proteins. (**b**) Mechanism of affinity and photoaffinity labeling emphasizing juxtaposition of interacting groups.

Utility of affinity/photoaffinity labeling methods, coupled with mutational analysis and functional assays, has been amply demonstrated to provide a dynamic picture of the binding pocket of macromolecules. These results often complement structural information obtained by X-ray crystallographic and NMR studies. The advantage of affinity/photoaffinity methods is that they are not restricted by the unavailability of the macromolecule in question in large quantities in pure form, which often poses problems for crystallographic and NMR methods of structure determination. However, affinity/photoaffinity labeling processes are restricted by the limited quantity of information that can be obtained by these methods.

2. AFFINITY ALKYLATING COMPOUNDS AS THERAPEUTIC AGENTS

Alkylating compounds that covalently attach to proteins and other macromolecules are important components in the standard cancer chemotherapeutic regimen. Estramustine, lomustine, procarazine, busulfan, cyclophosphamide, chlorambucil, temozolomide, platinum coordination complexes, etc. are members of this class of alkylating chemotherapeutic agents. Each of these compounds contains a chemically reactive group (an electrophile) which reacts with a nucleophilic amino acid component of a protein to establish a covalent bond between the two. This process is similar to the affinity labeling process depicted in Fig. 1b, except that proteins are alkylated randomly, which triggers a signaling cascade that ultimately leads to the death of the alkylated/modified cells. Since the alkylation process is indiscriminate in nature, cancer cells, as well as normal healthy cells, die in the process. This is the root cause of systemic toxicity associated with most alkylating chemotherapeutic agents.

Affinity alkylating agents, on the other hand, alkylate specifically targeted proteins and other macromolecules. In essence, affinity alkylating agents are target-specific "smart molecules." It was recognized early on that some of these agents may act as suicide inhibitors of target enzymes involved in the carcinogenesis process. But, translation of this property to therapeutic agents has remained largely elusive. Recently, aromasin, an affinity alkylating agent/suicide inactivator of aromatase responsible for the conversion of DHEA to estrogen, has been developed to prevent endogenous production of estrogen in breast cancer patients (Fig. 2). Aromasin is currently a first-/second-line therapeutic agent in breast cancer.

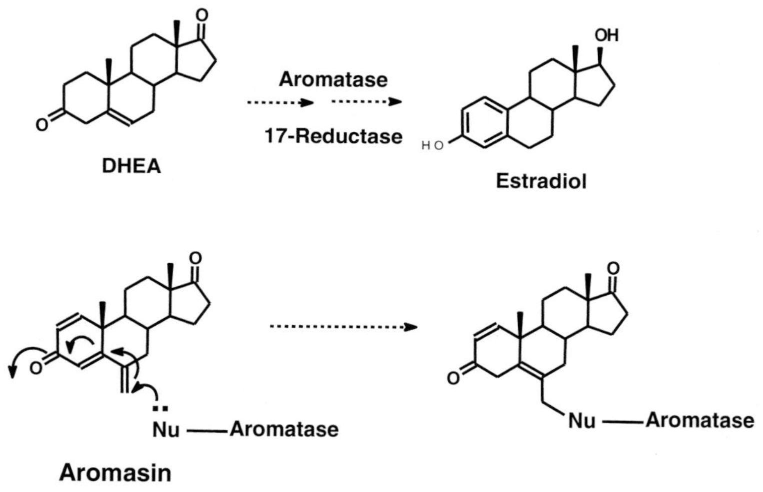

Fig. 2. Suicide inactivation of aromatase by aromasin, blocking endogenous production of estradiol.

3. VITAMIN D AND ITS METABOLITES: BIOSYNTHESIS, PROPERTIES, AND MECHANISM OF ACTION

Vitamin D is a naturally occurring molecule that is biosynthesized in the skin by the interaction of sunlight with 7-dehydrocholesterol in the epidermis. Vitamin D_3, after entering circulation, is scavenged by serum vitamin D-binding protein (DBP) which transports vitamin D to the liver where a hydroxyl group is introduced at the 25-position to produce 25-hydroxyvitamin D_3 [25(OH)D]. 25(OH)D, the most abundant metabolite of vitamin D in circulation, binds to DBP with high affinity and specificity. In turn, 25(OH)D is transported to kidney, where renal 25-hydroxyvitamin D_3-1α-hydroxylase introduces a hydroxyl group at the 1-position with high degree of regio- and stereoselectivity to produce 1α,25-dihydroxyvitamin D_3 [1,25(OH)$_2$D]. 25(OH)D, with serum concentration of 40–100 ng/ml, is considered to be largely biologically inert, while 1,25(OH)$_2$D (serum concentration 8–10 pg/ml) is biologically the most active metabolite of vitamin D. Multiple biological properties of 1,25(OH)$_2$D (*vide infra*) are manifested via its interaction with vitamin D receptor (VDR), a protein present in the nucleus of target cells. Biosynthesis of vitamin D$_3$ and its metabolites, and the role of DBP and VDR are shown in the cartoon in Fig. 3.

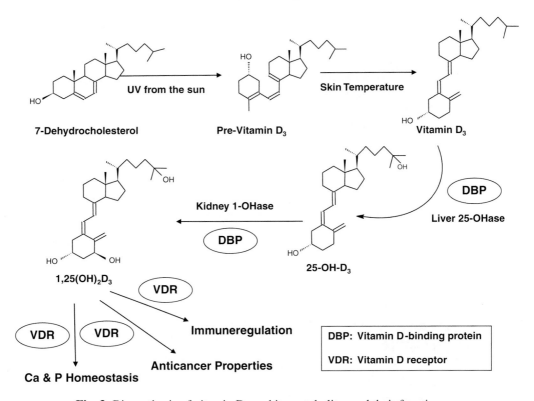

Fig. 3. Biosynthesis of vitamin D$_3$ and its metabolites and their functions.

4. MINERAL HOMEOSTATIC AND CELL-REGULATORY PROPERTIES OF 1,25(OH)$_2$D

1,25(OH)$_2$D is an essential nutrient for our skeletal health due to its crucial role in calcium and phosphorus homeostasis. In addition, several epidemiological studies have strongly suggested that dietary vitamin D has chemopreventive effect in various cancers *(2–9)*. Furthermore, 1,25(OH)$_2$D has profound effect on the growth and maturation of malignant cells. Numerous studies have demonstrated that 1,25(OH)$_2$D and many of its synthetic analogs inhibit the proliferation of neoplastic cells and cause them to differentiate, strongly demonstrating the potential of vitamin D-based compounds for cancer therapy *(10)*. However, clinical use of 1,25(OH)$_2$D has been limited by severe hypercalcemia, hypercalciuria, and considerable loss of body weight. Thus development of 1,25(OH)$_2$D analogs with potent antiproliferative activities and less calcemic toxicity has become an active area of research.

5. TRANSCRIPTIONAL MECHANISM OF 1,25(OH)$_2$D

Modulation of growth, progression, and maturity of cells by 1,25(OH)$_2$D is manifested by a stepwise process that is initiated by strong and highly specific binding of 1,25(OH)$_2$D to VDR, its receptor in the nucleus of the target cells ($K_d = 10^{-9-10}$ mol/l). The VDR–hormone complex heterodimerizes with retinoid X receptor (RXR) and induces cooperative binding of the trimeric complex to the vitamin D response element of chromatin. Such a binding facilitates recruitment of co-activators for transcription to activate transcription and ultimately translational processes. This stepwise mechanism is depicted in the cartoon in Fig. 4. This is a rather simplistic picture, and the actual processes related to the cell-regulatory mechanisms are immensely complex and may involve multiple pathways and numerous players (*vide infra*).

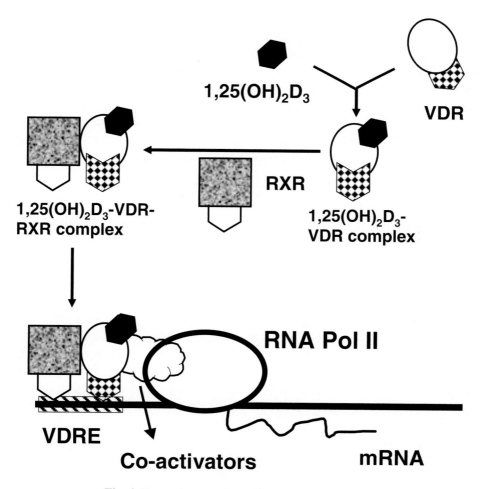

Fig. 4. Transcriptional mechanism of 1,25(OH)$_2$D.

6. VDR AND ITS LIGAND-BINDING DOMAIN (VDR-LBD)

Detailed discussion on the multi-faceted nature of vitamin D-related mechanism is beyond the scope of this review article. But, it is noteworthy that similar mechanism (depicted in Fig. 4) holds good for all nuclear receptors/transcriptional factors (e.g., estrogen receptor, progesterone receptor, retinoic acid receptor); the most unique process in this multi-step mechanism is highly specific interaction between the nuclear receptor and its cognate ligand. In other words, properties of $1,25(OH)_2D$ that set it apart from other steroid hormones depend critically on the highly specific binding between $1,25(OH)_2D$ and VDR. It is well recognized that a specific area of VDR, called its ligand-binding domain (VDR-LBD), is responsible for the high-fidelity molecular recognition between $1,25(OH)_2D$ and VDR and their highly specific binding. This molecular recognition process is critically guided by the three-dimensional geometries of VDR-LBD and $1,25(OH)_2D$. Therefore, it was recognized early on that knowledge of the structural elements in VDR-LBD is crucial for a better understanding of the biological functions of $1,25(OH)_2D$ as well as designing $1,25(OH)_2D$ analogs with better therapeutic index rather than the parent hormone (*vide infra*).

Initially two different approaches were adopted to study the structural–functional aspects of VDR-LBD in the absence of any structural information of VDR. One approach included modeling of VDR-LBD based on the crystal structure coordinates of homologous nuclear receptors. Based on this homology modeling approach, two different models of VDR-LBD were proposed *(11, 12)*. Our laboratory adopted a combinatorial approach of affinity labeling, mutational analysis, and homology modeling to construct a three-dimensional model of VDR-LBD (*vide infra*).

7. AFFINITY LABELING OF VDR FROM ENDOGENOUS SOURCES

We synthesized a simple derivative of $1,25(OH)_2D$, namely $1\alpha,25$-dihydroxyvitamin D_3-3β-(2)-bromoacetate ($1,25(OH)_2D$-3-BE) that became pivotal for our studies. In this compound the 3-hydroxyl group of $1,25(OH)_2D$ is modified with bromoacetic acid. The bromoacetate group acts as a "hook" which can react with a nucleophilic amino acid (such as cysteine) and this compound can get covalently attached to a protein. Therefore, when $1,25(OH)_2D$-3-BE occupies the binding cavity in VDR-LBD, the bromoacetate group (in $1,25(OH)_2D$-3-BE) can react with a neighboring nucleophilic amino acid (in the binding cavity) to form a covalent bond (as shown in Fig. 5). Furthermore, co-treatment with an excess of $1,25(OH)_2D$, the natural ligand, is expected to compete out majority of the labeling reagent from the binding pocket and reduce labeling intensity *(13)*.

We demonstrated that when nuclear extracts of calf thymus cytosol or ROS 17/2.4 cells were treated with ^{14}C-$1,25(OH)_2D$-3-BE, a radiolabeled variety of $1,25(OH)_2D$-3-BE, VDR was dominantly labeled; and significant amount of this labeling was reduced by co-treatment with an excess of $1,25(OH)_2D$ (Fig. 6). Identity of VDR was confirmed by Western blot analysis. These results unequivocally confirmed a direct and a dominant interaction between $1,25(OH)_2D$-3-BE and VDR from natural sources *(14)*.

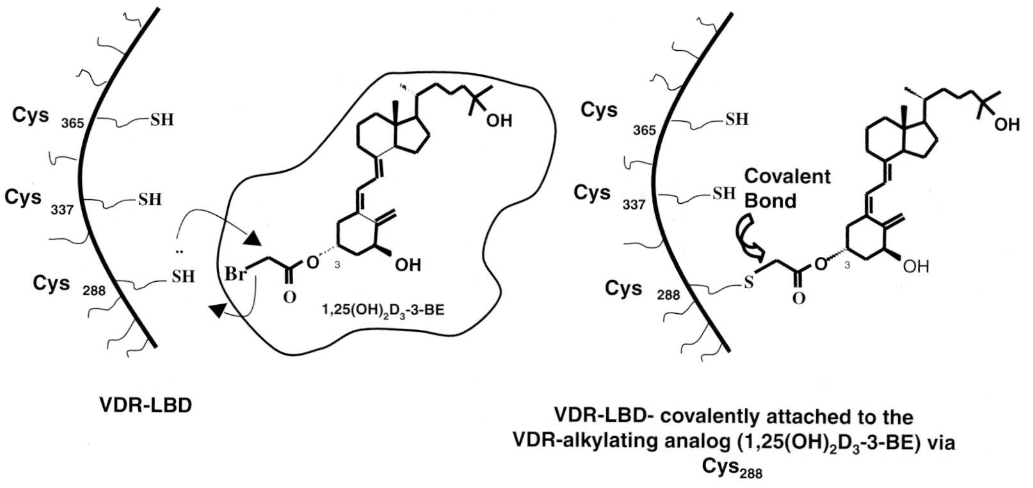

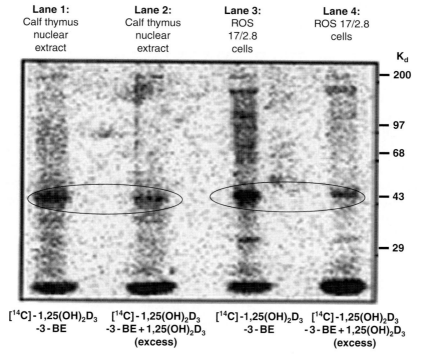

Fig. 5. Affinity labeling of VDR-LBD With 1,25(OH)$_2$D-3-BE.

Fig. 6. Affinity labeling of VDR from natural sources with ^{14}C-1,25(OH)$_2$D-3-BE.

8. AFFINITY LABELING OF RECOMBINANT VDR AND IDENTIFICATION OF A CONTACT POINT

In the next study we affinity labeled a sample of recombinant VDR-LBD with ^{14}C-1,25(OH)$_2$D-3-BE, followed by proteolysis and amino acid sequencing to identify a single cysteine residue (Cys$_{288}$) that is covalently labeled by 1,25(OH)$_2$D-3-BE *(15)*. Mutation of this residue to Gly completely obliterated 1,25(OH)$_2$D binding, demonstrating the functional significance of affinity labeling by ^{14}C-1,25(OH)$_2$D-3-BE. These results confirmed an earlier observation by Haussler's group that Cys$_{288}$ is crucial for 1,25(OH)$_2$D binding *(16)*. It is noteworthy that there are three (3) cysteine residues in VDR-LBD, and 1,25(OH)$_2$D-3-BE specifically labeled a single Cys residue (Cys$_{288}$) (Fig. 5). These results indicated that VDR-LBD has a high steric demand, and once inside the binding pocket, 1,25(OH)$_2$D cannot move or wiggle, otherwise specificity of labeling would have been lost.

Table 1
Role of Met$_{284}$, Cys$_{288}$, and Trp$_{286}$ and in ligand binding by VDR

Mutation	Percentage of loss of ligand binding
Met$_{284}$ – Ala/Ser	70
Trp$_{286}$ – Ala/Phe	>99
Cys$_{288}$ – Gly	>99

Anchoring the 3-OH group of 1,25(OH)$_2$D (containing the bromoacetate moiety that forms covalent bond with Cys$_{288}$) allowed us to identify other contact points. We mutated two amino acids, i.e., Met$_{284}$ and Trp$_{286}$ adjacent to Cys$_{288}$, and observed strong reduction of hormone binding by mutated VDRs as shown in Table 1. These results strongly suggested that a contiguous region containing Met$_{284}$-Ser-Trp-Thr-Cys$_{288}$ plays an indispensable role in the ligand binding by VDR. Importance of Trp$_{286}$ toward 1,25(OH)$_2$D binding is particularly noteworthy, because we believe that delocalized π electrons of Trp interact strongly with the π electrons of the conjugated triene of 1,25(OH)$_2$D to "clamp" the hormone molecule strongly inside the binding pocket (as shown in Fig. 7). It is interesting to note that our earlier work showed that serum DBP,

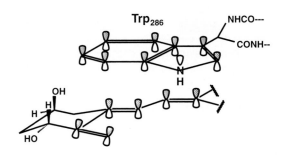

Fig. 7. Probable interaction between the conjugated triene of 1,25(OH)$_2$D and delocalized π electrons of Trp$_{286}$.

the other vitamin D-binding protein, also has a Trp residue in its vitamin D-binding pocket that is essential for binding vitamin D metabolites *(17)*.

Based on this study, combining affinity labeling and mutational analysis, as well as other mutational data published earlier by other groups, we developed a homology model for VDR-LBD *(15)* that corroborated well with the X-ray crystal structure of VDR-LBD *(18)*.

Identification of Cys_{288} as the "anchoring point" of the 3-hydroxyl group of $1,25(OH)_2D$ by affinity labeling with $1,25(OH)_2D$-3-BE is an important observation, because Cys_{288} is present in an unstructured region (β-hairpin loop) of VDR-LBD. Several studies have emphasized the essential role of helix-12, the C-terminal helical part of VDR toward transcription and transactivation *(19)*. In addition, important role of helix-3 to this end is also recognized *(20)*. But the role of other structural elements of VDR-LBD has largely remained unexplored. Our studies clearly demonstrated the essential role of β-hairpin loop area of VDR-LBD toward ligand binding and related properties *(15, 21)*.

In summary, the most important discovery of our VDR-affinity labeling studies are (a) importance of the unstructured β-hairpin region of VDR-LBD, where Cys_{288} is located, and (b) identification of $1,25(OH)_2D$-3-BE as a unique derivative of $1,25(OH)_2D$ with potential therapeutic properties (*vide infra*).

9. VDR-AFFINITY ALKYLATING DERIVATIVES OF $1,25(OH)_2D$ AND $25(OH)D$ AS POTENTIAL THERAPEUTIC AGENTS

It was mentioned earlier that numerous studies have registered strong promise of $1,25(OH)_2D$ as a therapeutic agent in various cancers and other diseases. But its potential as a therapeutic agent is thwarted by its inherent systemic toxicity related to hypercalcemia and hypercalciuria, requiring that sub-optimal doses be used to avoid toxicity, but that deprives the patient of desired effect. Consequently, a strong effort has been underway to develop analogs of $1,25(OH)_2D$ with potent antiproliferative activities and less calcemic toxicity. In a recent phase II clinical trial, effect of seocalcitol (EB-1089), an analog of $1,25(OH)_2D$, was evaluated in 33 patients with inoperable hepatocellular cancer with encouraging results *(22, 23)*. Other analogs with strong therapeutic potential in several cancers are currently under various stages of development. Despite this effort, it is now well recognized that $1,25(OH)_2D$, the natural hormone, probably has the best therapeutic potential, if its toxicity can be mitigated. As a result, toxicity issue is currently being addressed by combining $1,25(OH)_2D$ with standard chemotherapeutic agents such as paclitaxel or dexamethasone with encouraging results *(24–28)*.

10. PHARMACOKINETIC PROPERTY OF $1,25(OH)_2D$ AS IT RELATES TO ITS THERAPEUTIC ACTIVITY

Biological effects of $1,25(OH)_2D$ is governed by its stability in circulation and its bioavailability. $1,25(OH)_2D$ is known to be catabolized very rapidly by CYP450 enzymes, thereby reducing its half-life and potency (as shown in the top half of the cartoon in Fig. 8). However, if catabolic degradation is prevented or slowed down, half-life

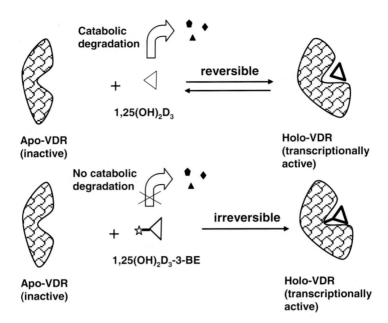

Fig. 8. *Top*: Cartoon depicting reversible nature of 1,25(OH)$_2$D binding by VDR and ensuing catabolism reducing its effective dose. *Bottom*: Irreversible covalent labeling of VDR-LBD by 1,25(OH)$_2$D-3-BE and probability of avoiding/eliminating catabolic degradation.

of 1,25(OH)$_2$D can be enhanced significantly by lowering the dose of 1,25(OH)$_2$D for obtaining optimal effect, but with less toxicity.

11. 1,25(OH)$_2$D-3-BE, A DERIVATIVE OF 1,25(OH)$_2$D, WITH A POTENTIAL OF LESS CATABOLIC DEGRADATION: OUR HYPOTHESIS

1,25(OH)$_2$D-3-BE is a simple derivative of 1,25(OH)$_2$D that allows covalent attachment of 1,25(OH)$_2$D (in its 3-acetate form: minimum structural change) into VDR-LBD (Fig. 5). We hypothesized that 1,25(OH)$_2$D-3-BE, once attached to VDR-LBD by an irreversible process, may be protected from interacting with catabolic enzymes (sheltered by parts of VDR molecule). Thus, its half-life can be boosted significantly (as shown in the bottom half of the cartoon in Fig. 8). According to this hypothesis, less amount of 1,25(OH)$_2$D-BE (1,25(OH)$_2$D in disguise) will be required to attain optimal effect with less toxicity.

12. KINETIC AND STOICHIOMETRIC STUDIES WITH 1,25(OH)$_2$D-BE AND VDR

In order for our above-stated hypothesis to work, 1,25(OH)$_2$D-3-BE must interact and covalently label VDR in target cells rapidly. Furthermore, such an interaction should have a 1:1 stoichiometry, because it is well established that a VDR molecule contains a single 1,25(OH)$_2$D-binding site. In order to demonstrate this, we incubated samples of recombinant VDR-LBD with ^{14}C-1,25(OH)$_2$D-3-BE and removed aliquots from the incubate at specified times. These samples were analyzed by SDS-PAGE–autoradiography. In an accompanying study we incubated samples of VDR-LBD with various amounts of ^{14}C-1,25(OH)$_2$D-3-BE for a fixed time period, followed by analysis of the samples by SDS-PAGE–autoradiography. As shown by the results of this assay (Fig. 9a), covalent labeling of VDR-LBD was extremely rapid (<0.5 min) and stoichiometry of labeling was 1:1 (Fig. 9b) (29). These results showed that 1,25(OH)$_2$D-3-BE may have the properties emulated in our hypothesis (Fig. 8).

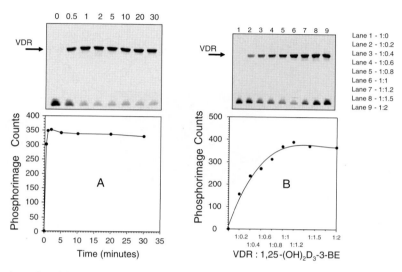

Fig. 9. Kinetic and stoichiometric evaluation of VDR labeling by 1,25(OH)$_2$D-3-BE. (**a**) VDR-LBD treated with ^{14}C-1,25(OH)$_2$D-3-BE, aliquots removed at different times, and analyzed by SDS-PAGE–autoradiography. (**b**) VDR-LBD treated with various amounts of ^{14}C-1,25(OH)$_2$D-3-BE for a fixed time and analyzed by SDS-PAGE–autoradiography.

13. ANTIPROLIFERATIVE EFFECTS OF 1,25(OH)$_2$D-BE IN VARIOUS CANCER CELLS

In an earlier study we observed that 1,25(OH)$_2$D-3-BE possesses a significantly stronger antiproliferative activity compared with equimolar doses of 1,25(OH)$_2$D in human keratinocytes (29). This observation set our goal to investigate the effect of 1,25(OH)$_2$D-3-BE in cancer cells.

We observed that 1,25(OH)$_2$D-3-BE strongly inhibited the growth of many cancer cell lines including LNCaP, PC-3, DU-145 (prostate cancer) (30), Caki-1, A 498 (kidney cancer), T24 (bladder cancer), BxPC3, hs700, hs766 (pancreatic cancer), MDA-MB-231 (estrogen-insensitive breast cancer) cells in a dose-dependent manner. Representative examples are given in Figs. 10 and 11.

Several points are noteworthy from Figs. 10 and 11. In each case antiproliferative effect of 1,25(OH)$_2$D-3-BE is significantly stronger than an equimolar amount of 1,25(OH)$_2$D. Second, growth-inhibitory effect of 1,25(OH)$_2$D-3-BE is strongest with 10^{-6} M of the compound. This is in contrast with published reports where effect of 1,25(OH)$_2$D is reported even at the nanomolar level. Initially it was puzzling for us,

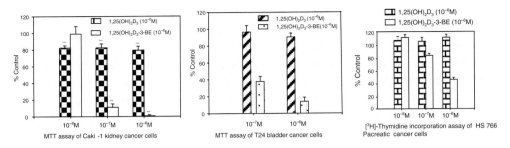

Fig. 10. MTT and [^{3}H]-thymidine incorporation assays of Caki-1 kidney cancer, T24 bladder cancer, and HS766 pancreatic cancer cell lines with 1,25(OH)$_2$D or 1,25(OH)$_2$D-3-BE.

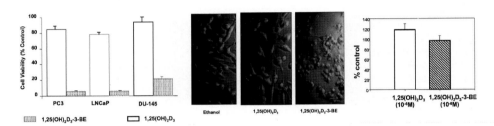

Fig. 11. (**a**) MTT assay of prostate cancer cells treated with 10^{-6} M of 1,25(OH)$_2$D or 1,25(OH)$_2$D-3-BE. (**b**) LNCaP cells treated for 16 h with EtOH (vehicle) or 10^{-6} M of 1,25(OH)$_2$D-3-BE or 1,25(OH)$_2$D (Swamy et al. *(30)*). (**c**) [^{3}H]-Thymidine incorporation assay of HeLa cervical cancer cells.

because we failed to observe any effect of 1,25(OH)$_2$D even at 10^{-6} M level. We believe that fetal bovine serum (FBS), commonly added to the media, may be the reason for such apparent discrepancy. In all the reports, almost without fail, cells are treated in media without FBS, because FBS is known to scavenge small hydrophobic molecules (such as 1,25(OH)$_2$D) and mute their activity. Instead, in all our assays we treated our cells with media containing 5–10% FBS. We believe that treatment of cells with reagents in media containing FBS is much closer to a real physiological situation instead of withholding it.

The third point is that we observed that the effect of 1,25(OH)$_2$D-3-BE in cancer cells is not global, i.e., in certain cell lines (e.g., MCF-7 breast cancer, Caco-2 colon cancer, and HeLa cervical cancer cell lines), antiproliferative effect is considerably muted, as shown in a representative example in Fig. 11c. We also observed considerable difference in the activity (of 1,25(OH)$_2$D-3-BE) among various cell lines of the same cancer type. The cell-specific/cancer-specific property of 1,25(OH)$_2$D-3-BE raises the possibility that its cell-regulatory effect may include VDR-dependent as well as VDR-independent pathways.

We observed that in some cells, 1,25(OH)$_2$D-3-BE induced growth inhibition by triggering cell death. For example, when LNCaP cells were treated with 10^{-6} M of 1,25(OH)$_2$D-3-BE for 16 h, they appeared to be rounding up and dying, typical of apoptosis as shown in Fig. 11b. This effect was even stronger in Caki-1 and A 498 kidney cancer cells (results not shown). In support of this observation we found that 1,25(OH)$_2$D-3-BE strongly induced caspase 3/7 activity in several prostate and kidney cancer cells. A representative example, related to upregulation of caspase 3/7 activity in A 498 kidney cancer cells, is shown in Fig. 12. Caspase activation is a hallmark of apoptosis. Therefore, these results strongly indicated that growth inhibition by 1,25(OH)$_2$D-3-BE in certain cancer cells includes an apoptotic mechanism (*vide infra*).

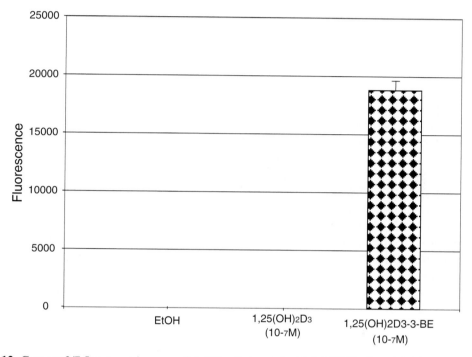

Fig. 12. Caspase 3/7 fluorometric assay of A 498 cells treated with 1,25(OH)$_2$D or 1,25(OH)$_2$D-3-BE.

14. MOLECULAR MECHANISMS RELATED TO THE GROWTH-INHIBITORY EFFECT OF 1,25(OH)$_2$D-3-BE

We studied the enhanced growth-inhibitory property of 1,25(OH)$_2$D-3-BE from two mechanistic standpoints: (1) VDR-related mechanisms and (2) VDR-unrelated mechanisms.

14.1. VDR-Related Mechanisms

VDR-mediated transcriptional process has been studied extensively with the identification of the key events and players, and it is duly recognized that VDR plays an essential role in the manifestation of cellular and biological properties of 1,25(OH)$_2$D and its analogs. However, 1,25(OH)$_2$D-3-BE is an alkylating agent, and in a cellular system it can potentially alkylate irrelevant targets; it can be argued that such a process might be responsible for its enhanced growth-inhibitory and cell-killing properties.

However, as discussed earlier, 1,25(OH)₂D-3-BE is capable of directly interacting with incipient VDR in cellular systems (Fig. 6), which indicates that VDR may be involved in manifesting its growth-inhibitory properties. We further elaborated this by the following assays.

A. *VDR-antisense method to determine the role of VDR in prostate cancer cells:* We probed the connection between VDR and cellular properties of 1,25(OH)₂D-3-BE further by antisense technique. Recently we and others have demonstrated that 25-hydroxyvitamin D_3-3-bromoacetate (25(OH)D-3-BE), a prototype of 1,25(OH)₂D-3-BE without the 1-hydroxyl group, has strong growth-inhibitory and apoptotic activities in prostate cancer cells *(31, 32)*.

In this study we stably transfected ALVA-31 human prostate cancer cells with an antisense VDR expression vector and an empty vector, and assayed for their response to 1,25(OH)₂D or 25(OH)D-3-BE in a cell-proliferation assay. As shown in Fig. 13, growth of VDR-sense cells (empty vector) is strongly inhibited by 10^{-7} to 10^{-6} M of 1,25(OH)₂D. Conversely, the growth of antisense cells treated with 10^{-7} to 10^{-6} M of 1,25(OH)₂D is similar to that of ethanol control, confirming the requirement of VDR in the antiproliferative activity of 1,25(OH)₂D in ALVA-31 cells. In the case of 25(OH)D-3-BE, 10^{-6} M of this compound strongly inhibited the growth of empty vector (sense cells), while growth of antisense cells is significantly weaker. However, with 10^{-7} M of 25(OH)D-3-BE, the growth of both sense and antisense cells are similar to that of ethanol control. This result is in accordance with our previous studies where we observed that the antiproliferative effect of 25(OH)D-3-BE is strongest at 10^{-6} M dose and decreased significantly at lower doses *(31)*. Overall, the result of this assay strongly emphasizes the requirement of VDR in mediating the antiproliferative effect of 25(OH)D-3-BE in prostate cancer cells *(33)*.

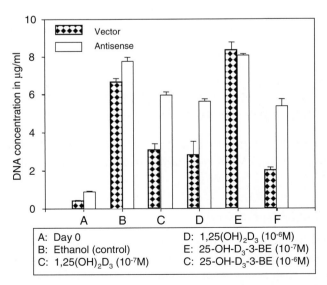

A: Day 0	D: 1,25(OH)₂D₃ (10⁻⁶M)
B: Ethanol (control)	E: 25-OH-D₃-3-BE (10⁻⁷M)
C: 1,25(OH)₂D₃ (10⁻⁷M)	C: 25-OH-D₃-3-BE (10⁻⁶M)

Fig. 13. VDR-antisense assay in ALVA-31 cells. ALVA-31 control (vector) and ALVA-31 VDR "null" cells (antisense) were treated with the indicated doses of 1,25(OH)₂D, 25(OH)D-3-BE and ethanol control and harvested for DNA quantification.

B. *Induction of 1α,25-dihydroxyvitamin D₃-24-hydroxylase (24(OH)ase, CYP24) gene by 25(OH)D-3-BE and 1,25(OH)₂D-3-BE*: 1α,25-Dihydroxyvitamin D$_3$-24-hydroxylase (24(OH)ase, *CYP24*) gene is a 1,25(OH)$_2$D-inducible and VDR target gene. It is known to be strongly and predictably modulated by 1,25(OH)$_2$D and its analogs. We carried out a study to evaluate the effect of 1,25(OH)$_2$D and 25(OH)D-3-BE at various doses on the *CYP24* promoter activity in COS-7 cells that were transfected with a VDR construct, tagged with a chloramphenicol acetyltransferase (CAT) reporter gene. Results of this assay, shown in Fig. 14, demonstrated that CYP24 promoter activity was strongly upregulated by 10^{-8}, 10^{-7}, and 10^{-6} M of 1,25(OH)$_2$D. In contrast, strong promoter activity was displayed only with 10^{-6} M of 25(OH)D-3-BE, and such activity declined almost to the basal level with 10^{-7} M of 25(OH)D-3-BE *(31)*. Very similar results were obtained in A 498 kidney cancer cells, where we observed that approximately one log scale higher concentration of 1,25(OH)$_2$D-3-BE (compared with 1,25(OH)$_2$D) is required to induce equivalent amount of *CYP24* gene expression (unpublished data). In addition, we observed that *CYP24* gene expression is significantly slower with 1,25(OH)$_2$D-3-BE than 1,25(OH)$_2$D in A 498 cells (unpublished data).

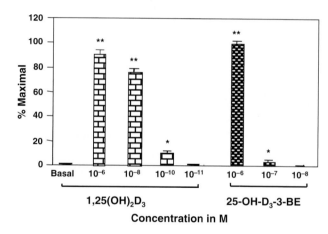

Fig. 14. Analysis of 1,25-dihydroxyvitamin D$_3$-24-hydroxylase (24(OH)ase) promoter activity in COS-7 cells, transiently transfected with a 24(OH)ase-construct, tagged with a chloramphenicol (CAT) reporter gene, and hVDR expression vector. Cells were treated with various doses of 25(OH)D-3-BE or 1,25(OH)$_2$D and CAT activity was determined.

The *CYP24* gene product 1α,25-dihydroxyvitamin D$_3$-24-hydroxylase catalyzes the introduction of a hydroxyl group at the 24-position in 1,25(OH)$_2$D, followed by multiple oxidations of the side chain leading to calcitroic acid, the final catabolite that is excreted *(34, 35)*. Therefore, *CYP24* is the initiator of the catabolic degradation of 1,25(OH)$_2$D. We hypothesized that covalent attachment of 1,25(OH)$_2$D-3-BE to the ligand-binding domain of VDR will decrease its catabolism. In other words, it would require more 1,25(OH)$_2$D-3-BE to induce the same level of CYP24 message as 1,25(OH)$_2$D. Additionally, it is expected that induction of CYP24 mRNA will also be delayed when cells are treated with 1,25(OH)$_2$D-3-BE compared to treatment with 1,25(OH)$_2$D. Results of our CYP24 assay with 25(OH)D-3-BE (Fig. 14) and 1,25(OH)$_2$D-3-BE (in kidney cancer cells, unpublished data) are in concurrence with our hypothesis. Furthermore, these results strongly suggest that cellular actions of 1,25(OH)$_2$D-3-BE (and 25(OH)D-3-BE) follow a VDR-mediated pathway and confirm our earlier finding where we demonstrated by affinity labeling and Western blot analysis that 1,25(OH)$_2$D-3-BE directly interacts with VDR in ROS 2.7 cells and a calf thymus nuclear extract, as shown in Fig. 6 *(14)*.

14.2. Additional VDR-Related and VDR-Unrelated Mechanisms

There are growing evidences to suggest that regulation of cell growth by 1,25(OH)$_2$D may be initiated directly in the cytosol or indirectly in the cytoplasmic membrane by modulation of growth factors, phospholipid metabolism, stimulation of mitogen-activated protein kinases, and regulation of cell-adhesion molecules by VDR-dependent as well as VDR-independent pathways *(36, 37)* as shown in the cartoon in Fig. 15.

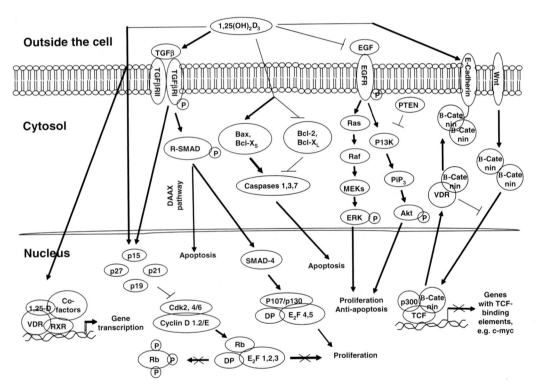

Fig. 15. 1,25(OH)$_2$D signaling pathways.

Signaling via cell cycle regulatory proteins: In a direct effect, 1,25(OH)$_2$D induces expression of cyclin-dependent kinase inhibitors (p15, p19, p21, p27) and inhibits Cdk/cyclin interaction, leading to cell cycle inhibition and initiation of apoptosis *(38, 39)*. Recently we observed that 1,25(OH)$_2$D-3-BE strongly suppressed cyclin D1 activity and retinoblastoma (Rb) phosphorylation in two kidney cancer cell lines (unpublished data). Rb protein is a tumor suppressor that regulates cell cycle progression at the G1–S transition, and its phosphorylation atserine 780 by active Cdk/cyclin complexes relieves its inhibition of E2F-driven gene expression and cell cycle progression *(40)*. The cyclins are positive regulators of cell cycle progression and are necessary for activation of CDK complexes (and inactivation of Rb) to allow cell cycle transition. Therefore, our preliminary results suggest that inhibition of cell growth by 1,25(OH)$_2$D-3-BE may involve this pathway in kidney cancer cells.

Cell signaling via TGFβ–TGFβR axis: In certain cells, inhibition of growth by 1,25(OH)$_2$D involves transforming growth factor TGFβ–TGFβR pathway *(41–44)*. It is an anti-growth pathway and is initiated by interaction of TGFβ with 1,25(OH)$_2$D to stimulate phosphorylation of TGFβR1 in the cytoplasmic membrane. This process inhibits growth either via cyclin-dependent kinase inhibitor pathway or via SMAD proteins feeding into E2F-driven gene expression.

Induction of PDF gene expression by 25(OH)D-3-BE: Prostate-derived factor (PDF) is a member of TGF superfamily; and its expression has been correlated with various cell-regulatory processes. For example, PDF is highly expressed in the prostate, and it is regulated by androgens *(45)*. Li et al. showed that PDF overexpression leads to an 80% reduction in cell viability of MDA-MB-468 breast cancer and a 50–60% reduction in other human breast cancer cells *(46)*. In correlation with the above studies, Tan et al. showed that PDF expression is induced by p53 activation and that conditioned medium from PDF-overexpressing cells, but not from control cells, suppresses tumor cell growth in DU-145 prostate cancer cell line *(47)*. Recently Nazarova et al. reported that 1,25(OH)$_2$D induced PDF expression in LNCaP cells in a dose- and time-dependent manner, and human recombinant PDF inhibited the growth of LNCaP cells *(48)*.

Therefore, we investigated the effect of 25(OH)D-3-BE, in comparison with 1,25(OH)$_2$D, in modulating the expression of PDF in androgen-sensitive LNCaP prostate cancer cells. PDF is synthesized as a pre-protein, cleaved and secreted as a 14-kDa mature form, as designated in Fig. 16. As shown in this figure, strong induction of PDF is observed with 10^{-7} to 10^{-6} M of 1,25(OH)$_2$D (lanes 2 and 3) and 10^{-6} M of

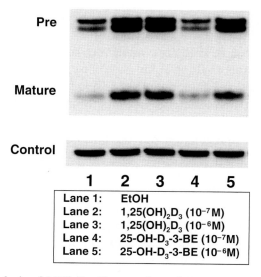

Fig. 16. Western blot analysis of LNCaP cells to evaluate the expression of prostate-derived factor (PDF) by 1,25(OH)$_2$D and 25(OH)D-3-BE.

25(OH)D-3-BE (lane 5). But, with 10^{-7} M of 25(OH)D-3-BE, intensity of the PDF band (lane 4) is similar to ethanol control (lane 1). Therefore, these results demonstrate that 25(OH)D-3-BE is approximately one log scale less efficient than 1,25(OH)$_2$D toward inducing PDF in LNCaP cells, mirroring its effect in modulating message for 24(OH)ase (Fig. 14) and inducing interaction of VDR with RXR GRIP-1 transcription factors, as reported by us earlier *(31)*.

Apoptotic mechanism of 1,25(OH)$_2$D: Apoptotic mechanism of 1,25(OH)$_2$D in certain cells involves an intrinsic pathway, including Bcl family of proteins, with activation of caspases 1, 3, and 7, leading to apoptosis *(49–52)*. Bcl-2/Bcl-X$_L$ are anti-apoptotic, while Bax/Bcl-X$_S$ are pro-apoptotic proteins. We observed that 1,25(OH)$_2$D-3-BE failed to modulate Bcl-2 and Bax proteins in pancreatic, kidney, and prostate cancer cells, yet 1,25(OH)$_2$D-3-BE activated caspases 3 and 7 in prostate cancer (Fig. 12) and kidney cancer cells (unpublished result), as well as in prostate cancer cells with 25(OH)D-3-BE *(30)*, suggesting that a Bcl-independent apoptotic pathway may be operative with 1,25(OH)$_2$D-3-BE, mirroring a recent report by Mullauer et al., who demonstrated that apoptosis by betulinic acid in cancer cells involves Bcl-independent pathways *(53)*.

Cell signaling via MAP-kinase pathway: In certain cells, signal transduction by 1,25(OH)$_2$D involves a mitogen-activated protein kinase pathway that is initiated by the interaction between 1,25(OH)$_2$D and EGF. Such an interaction inhibits phosphorylation of EGFR in the cytoplasmic membrane which triggers a mitogen-activated protein kinase (MAPK) cascade including Ras, Raf, MEKs, and ultimately ERK *(54–56)*. Overall, EGF/EGFR-mediated pathway leads to pro-growth and anti-apoptosis. Recently, we observed that 1,25(OH)$_2$D-3-BE strongly inhibited ERK phosphorylation in DU-145 prostate cancer cells (unpublished data), strongly suggesting that growth inhibition by 1,25(OH)$_2$D-3-BE may involve MAPK pathway.

Cell signaling via PI3K/Akt pathway: EGF/EGFR-mediated downstream signaling is also linked to P13K/Akt pathway, leading to enhanced proliferation and inhibition of apoptosis *(57–59)*. Akt phosphorylation is a key step in stimulating multiple downstream effectors in modulating cell growth and apoptosis.

Akt (*aka* protein kinase B, PKB) is a serine/threonine kinase that is involved in signal transduction by phosphoinositol-3'-kinase/Akt pathway. Akt is involved in a variety of normal cellular functions. In addition, Akt has profound effects in tumorigenesis, cell proliferation, growth, and survival. Recently it has been shown that Akt regulates G(1) cell cycle progression and cyclin expression in prostate cancer cells *(60, 61)*. Another study showed upregulation of Akt and other growth-promoting signaling molecules in malignant prostate epithelial cells *(62)*. In general, Akt phosphorylation is a key event for the survival of cells, and Akt expression is upregulated in malignant cells, while expression of PTEN (phosphatase and tensin homolog deleted on chromosome 10), the upstream modulator of Akt, is downregulated. In an earlier study we reported that 25(OH)D-3-BE induced nuclear DNA fragmentation and activated caspases 3, 8, and 9, hallmarks of apoptosis, in PC-3 cells, while an equimolar concentration of 1,25(OH)$_2$D and 25(OH)D failed to do so *(31)*. Induction of caspases and fragmentation of nuclear DNA represent downstream signaling markers of apoptosis; these markers are regulated by their upstream modulators such as Akt kinase. We postulated that induction of apoptosis by 25(OH)D-3-BE might be mediated by the downregulation of Akt activity, resulting in the observed upregulation of pro-apoptotic proteins.

Our results of the Akt phosphorylation assay are shown in Fig. 17. Akt phosphorylation was barely changed from the ethanol control by 10^{-6} M of 1,25(OH)$_2$D, but an equimolar amount of 25(OH)D-3-BE strongly reduced the level of Akt phosphorylation. In an earlier report we demonstrated that 25(OH)D-3-BE induced apoptosis in PC-3 cells by fragmenting nuclear DNA and activating caspases *(31)*. Results, shown in Fig. 17, suggest that the apoptotic property of 25(OH)D-3-BE may be related to the suppression of a pro-survival event involving Akt phosphorylation. Recently we have observed that 1,25(OH)$_2$D-3-BE strongly inhibits Akt phosphorylation in kidney, prostate, and pancreatic cancer cells (unpublished data), suggesting that growth inhibition by 1,25(OH)$_2$D-3-BE and similar compounds, in general, may follow cell signaling via PI3K/Akt pathway.

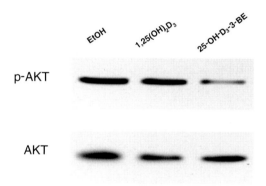

Fig. 17. 25(OH)D-3-BE inhibits Akt phosphorylation in PC-3 prostate cancer cells. PC-3 cells were treated with equimolar (10^{-6} M) amounts of 1,25(OH)$_2$D or 25(OH)D-3-BE.

Cell signaling via E-cadherin–β-catenin pathway: 1,25(OH)$_2$D signaling is also known to involve E-cadherin–β-catenin pathway (*Wnt* signaling). Understimulation from β-catenin, a cell-adhesion molecule and a regulatory molecule for TCF gene expression, accumulates in the cytoplasm and translocates into the nucleus, leading to the formation of β-catenin–TCF complex and transcription of target genes (*Wnt* signaling) *(63)*. A recent publication by Palmer et al. demonstrates that 1,25(OH)$_2$D inhibits transcriptional activity of β-catenin by increasing the amount of VDR bound to β-catenin and subsequent expression of E-cadherin and re-distribution of β-catenin into the plasma membrane *(64)*. Recently it was observed that chemopreventive effect of 1α-hydroxyvitamin D$_5$, a 1,25(OH)$_2$D-like compound with modification in the steroid side chain and without the 25-hydroxyl group, is manifested in chemically induced mammary and colonic carcinogenesis via *Wnt*-signaling pathway *(65)*. Currently we are in the process of evaluating this signaling pathway as a determinant in the effect of 1,25(OH)$_2$D-3-BE in cancer cells.

15. THERAPEUTIC POTENTIAL OF 1,25(OH)$_2$D-3-BE AND RELATED COMPOUNDS IN CANCER

Results discussed above strongly suggest potential therapeutic application of 1,25(OH)$_2$D-3-BE and similar compounds in several cancers including prostate, kidney, and pancreatic cancers. However, druggability of these compounds demands certain pharmacokinetic/pharmacodynamic evaluation.

Determination that intact form of 1,25(OH)$_2$D-3-BE is responsible for its cellular properties: The first concern is the hydrolytic potential of 1,25(OH)$_2$D-3-BE in a live system because 1,25(OH)$_2$D-3-BE contains an ester moiety. To address this issue we carried out a [3H]-thymidine assay in MCF-7 and PC-3 cells, treated with 1,25(OH)$_2$D, 1,25(OH)$_2$D-3-BE, bromoacetic acid (hydrolytic product of 1,25(OH)$_2$D-3-BE) individually or an equimolar mixture of 1,25(OH)$_2$D and bromoacetic acid.

We observed that 10^{-6} M of bromoacetic acid had no effect on the proliferation of MCF-7 and PC-3 cells (Fig. 18). Furthermore, antiproliferative effect of a combination of 1,25(OH)$_2$D and bromoacetic acid (10^{-6} M each) was very similar to that of 1,25(OH)$_2$D (10^{-6} M) alone in both the cells (30). These results strongly suggest that observed effects of 1,25(OH)$_2$D-3-BE are due, at least in large part, to the unhydrolyzed molecule. It is noteworthy that therapeutic agents containing hydrolysable bonds, such as aspirin, are fairly common.

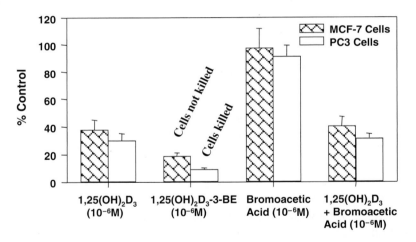

Fig. 18. [3H]-thymidine incorporation assay of MCF-7 and PC-3 cells, treated with 10^{-6} M of 1,25(OH)$_2$D, 1,25(OH)$_2$D, or bromoacetic acid individually or a combination of 1,25(OH)$_2$D and bromoacetic acid (10^{-6} M each).

Serum stability of 25(OH)D-3-BE: Serum stability is an important aspect of a potential therapeutic agent, because it determines the availability of the molecule in its intact and bioactive form. In order to determine potential hydrolysis of 25(OH)D-3-BE in serum, we synthesized [3H]-25(OH)D-3-BE in which the radiolabel (3H) is in the 25(OH)D part of the molecule. Hence its hydrolysis should produce [3H]-25(OH)D and unlabeled bromoacetic acid (as noted in Fig. 19, top panel); organic extract should contain [3H]-25(OH)D and [3H]-25(OH)D-3-BE if any hydrolysis takes place.

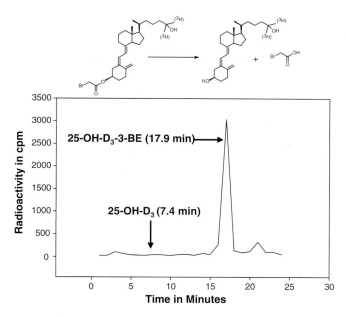

Fig. 19. Stability study of [³H]-25(OH)D-3-BE in human serum. [³H]-25(OH)D-3-BE was incubated with human serum, followed by extraction with an organic solvent and HPLC analysis of the organic extract. Fractions from HPLC were mixed with scintillation cocktail and counted for radioactivity. *Upper panel*: Representation of the hydrolysis of [³H]-25(OH)D-3-BE leading to the production of [³H]-25(OH)D and unlabeled bromoacetic acid. *Lower panel*: Peak indicating intact [³H]-25(OH)D-3-BE. Position of the 25(OH)D peak (retention time 7.4 min) is shown with an *arrow*.

We incubated [³H]-25(OH)D-3-BE in human serum followed by organic extraction, HPLC analysis, and radioactivity counting in HPLC fractions. Results of this assay (Fig. 19) show that majority of radioactivity is concentrated in a single peak corresponding to [³H]-25(OH)D-3-BE. Absence of a radioactive peak corresponding to [³H]-25(OH)D (hydrolysis product) strongly suggests that [³H]-25(OH)D-3-BE is fully stable under the conditions of our experiment (1 h at 37°C).

Cellular uptake of ¹⁴C-25(OH)D-3-BE in DU-145 cells: Cellular uptake analysis is an important aspect of drug development. Therefore, in this study, growing DU-145 cells were incubated with ¹⁴C-25(OH)D-3-BE at 37°C for 60 min, followed by removal of the incubation media, thorough washing of the cells with phosphated saline, and lysis of the cells with methanol. Media and cell extracts were lyophilized and re-dissolved/suspended in 3 ml of water. The aqueous mixtures from cells and media were extracted with ethyl acetate. The organic extract from each fraction was analyzed by HPLC and online radioactivity analysis. As shown in Fig. 20, both media and cellular extracts consist of the radioactive peak for ¹⁴C-25(OH)D-3-BE, strongly suggesting that 25(OH)D-3-BE is taken up by these cells in its intact form.

In vivo activity of 1,25(OH)₂D-3-BE: Recently we carried out an in vivo study of 1,25(OH)₂D-3-BE in an athymic mouse model of DU-145 androgen-insensitive human tumor. We observedthat 1,25(OH)₂D-3-BE is approximately seven times more

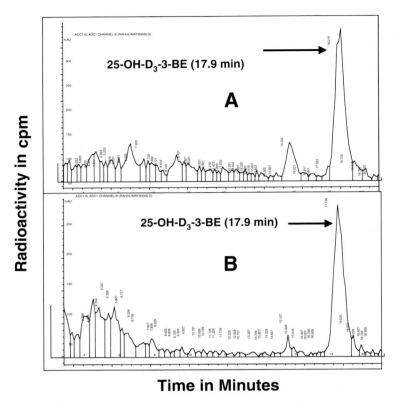

Fig. 20. 25(OH)D-3-BE is taken up by DU-145 prostate cancer cells in its intact form. DU-145 prostate cancer cells were treated with [14]C-25(OH)D-3-BE, and cells and media were extracted with an organic solvent. Extracts were analyzed in HPLC and fractions were counted for radioactive content. (**a**) Media extract; (**b**) cellular extract.

effective in reducing tumor size than is an equimolar dose of $1,25(OH)_2D$ with significantly less toxicity (manuscript in preparation). These results strongly emphasize the potential of $1,25(OH)_2D$-3-BE as a therapeutic agent in prostate cancer, and possibly other cancers.

16. SUMMARY

We have developed hypothesis-driven derivatives of $1,25(OH)_2D$ and $25(OH)D$ that affinity alkylate VDR-LBD. We probed the ligand-binding pocket of VDR with this compound and identified key ligand contact points within VDR-LBD. In addition, we have demonstrated that this compound has a strong translation potential as a therapeutic agent in several cancers.

ACKNOWLEDGMENT

Many people have contributed significantly to the pool of work that culminated into this review. However, the author would like to especially acknowledge contributions from late Dr. Narasimha Swamy (Boston University School of Medicine) and Dr. James Lambert (University of Colorado Health Science Center). Furthermore, this work is supported by grants from National Institute of Diabetes, Digestive and Kidney Diseases (RO1DK044337, RO1DK47418), National Cancer Institute (R21CA127629), Department of Defense (PC 051136), and Community Technology Fund, Boston University.

REFERENCES

1. Sweet FW, Murdock GL (1987) Affinity labeling of hormone-specific proteins. Endo Rev 8:154–184
2. Mizoue T, Kimura Y, Toyomura K, Nagano J, Kono S, Mibu R, Tanaka M, Kakeji Y, Maehara Y, Okamura T, Ikejiri K, Futami K, Yasunami Y, Maekawa T, Takenaka K, Ichimiya H, Imaizumi N (2008) Calcium, dairy foods, vitamin D, and colorectal cancer risk: the Fukuoka Colorectal Cancer Study. Cancer Epidemiol Biomarkers Prev 17:2800–2807
3. Davis CD (2008) Vitamin D and cancer: current dilemmas and future research needs. Am J Clin Nutr 88:565S–569S
4. Ng K, Meyerhardt JA, Wu K, Feskanich D, Hollis BW, Giovannucci EL, Fuchs CS (2008) Circulating 25-hydroxyvitamin d levels and survival in patients with colorectal cancer. J Clin Oncol 26: 2984–2991
5. Giovannucci E, Liu Y, Rimm EB, Hollis BW, Fuchs CS, Stampfer MJ, Willett WC (2006) Prospective study of predictors of vitamin D status and cancer incidence and mortality in men. J Natl Cancer Inst 98:451–459
6. Giovannucci E, Liu Y, Stampfer MJ, Willett WC (2006) A prospective study of calcium intake and incident and fatal prostate cancer. Cancer Epidemiol Biomarkers Prev 15:203–210
7. Garland C, Shekelle RB, Barrett-Connor E, Criqui MH, Rossof AH, Oglesby P (1985) Dietary vitamin D and calcium and risk of colorectal cancer: a 19-year prospective study in men. Lancet 9: 307–309
8. Garland CF, Garland FC, Gorham ED (1991) Can colon cancer incidence and death rates be reduced with calcium and vitamin D?. Am J Clin Nutr 54:193S–201S
9. Garland CF, Garland FC, Shaw EK, Comstock GW, Helsing KJ, Gorham ED (1989) Serum 25-hydroxyvitamin D and colon cancer: eight-year prospective study. Lancet 18:1176–1178
10. Matsuda S, Jones G (2006) Promise of vitamin D analogues in the treatment of hyperproliferative conditions. Mol Cancer Therap 5:797–808
11. Yamada S, Yamamoto K, Masuno H, Choi M (2001) Three-dimensional structure–function relationship of vitamin D and vitamin D receptor model. Steroids 66:177–187
12. Norman AW, Adams D, Collins ED, Okamura WH, Fletterick RJ (1999) Three-dimensional model of the ligand binding domain of the nuclear receptor for 1alpha,25-dihydroxy-vitamin D(3). J Cell Biochem 74:323–333
13. Ray R, Ray S, Holick MF (1994) 1a,25-dihydroxyvitamin D_3-3-deoxy-3 -bromoacetate, an affinity labeling analog of 1a,25-dihydroxyvitamin D_3. Bioorg Chem 22:276–283
14. Ray R, Swamy N, MacDonald PN, Ray S, Haussler MR, Holick MF (1996) Affinity labeling of 1,25-dihydroxyvitamin D_3 receptor. J Biol Chem 271:2012–2017
15. Swamy N, Xu W, Paz N, Hsieh J-C, Haussler MR, Maalouf GJ, Mohr SC, Ray R (2000) Molecular modeling, affinity labeling and site-directed mutagenesis define the key points of interaction between the ligand-binding domain of the vitamin D nuclear receptor and 1,25-dihydroxyvitamin D_3. Biochemistry 39:12162–12171
16. Nakajima S, Hsieh JC, Jurutka P, Galligan MA, Haussler CA, Whitfield GK, Haussler MR (1996) Examination of the potential functional role of conserved cysteine residues in the hormone binding domain of the human 1,25-dihydroxyvitamin D3 receptor. J Biol Chem 271:5143–5149
17. Swamy N, Brisson M, Ray R (1995) Trp 145 is essential for the binding of 25-hydroxyvitamin D_3 to human serum vitamin D-binding protein. J Biol Chem 270:2636–2639
18. Rochel N, Wurtz JM, Mitschler A, Klaholz B, Moras D (2000) The crystal structure of the nuclear receptor for vitamin D bound to its natural ligand. Mol Cell 5:173–179

19. Väisänen S, Peräkylä M, Kärkkäinen JI, Steinmeyer A, Carlberg C (2002) Critical role of helix 12 of the vitamin D(3) receptor for the partial agonism of carboxylic ester antagonists. J Mol Biol 315: 229–238

20. Jiménez-Lara AM, Aranda A (1999) Lysine 246 of the vitamin D receptor is crucial for ligand-dependent interaction with coactivators and transcriptional activity. J Biol Chem 274:13503–13510

21. Kaya T, Swamy N, Persons KS, Ray S, Mohr SC, Ray R (2009) Covalent labeling of nuclear vitamin D receptor with affinity labeling reagents containing a cross-linking probe at three different positions of the parent ligand: structural and biochemical implications. Bioorg Chem; Epub ahead of print 37: 57–63

22. Dalhoff K, Dancey J, Astrup L, Skovsgaard T, Hamberg KJ, Lofts FJ, Rosmorduc O, Erlinger S, BachHansen J, Steward WP, Skov T, Burcharth F, Evans TR (2003) A phase II study of the vitamin D analogue Seocalcitol in patients with inoperable hepatocellular carcinoma. Br J cancer 89:252–257

23. Evans TR, Colston KW, Lofts FJ, Cunningham D, Anthoney DA, Gogas H, de Bono JS, Hamberg KJ, Skov T, Mansi JL (2002) A phase II trial of the vitamin D analogue Seocalcitol (EB1089) in patients with inoperable pancreatic cancer. Br J Cancer 86:680–685

24. Light BW, Yu WD, McElwain MC, Russell DM, Trump DL, Johnson CS (1997) Potentiation of cis-platin antitumor activity using a vitamin D analogue in a murine squamous cell carcinoma model system. Cancer Res 57:3759–3764

25. Yu WD, McElwain MC, Modzelewski RA, Russell DM, Smith DC, Trump DL, Johnson CS (1998) Enhancement of 1,25-dihydroxyvitamin D3-mediated antitumor activity with dexamethasone. J Natl Cancer Inst 90:134–141

26. Smith DC, Johnson CS, Freeman CC, Muindi J, Wilson JW, Trump DL (1999) A Phase I trial of calcitriol (1,25-dihydroxycholecalciferol) in patients with advanced malignancy. Clin Cancer Res 5:1339–1345

27. Hershberger PA, Yu W-D, Modzelewski RA, Rueger RM, Johnson CS, Trump DL (2002) Calcitriol (1,25-dihydroxycholecalciferol) enhances paclitaxel antitumor activity in vitro and in vivo and accelerates paclitaxel-induced apoptosis. Mol Cancer Therap 1:821–829

28. Muindi JR, Peng Y, Potter DM, Hershberger PA, Tauch JS, Capozzoli MJ, Egorin MJ, Johnson CS, Trump DL (2003) Pharmacokinetics of high-dose oral calcitriol phase I trial of calcitriol and paclitaxel. Clin Pharmacol Therap 72:648–659

29. Chen ML, Ray S, Swamy N, Holick MF, Ray R (1999) Anti-proliferation of human keratinocytes with 1, 25-dihydroxyvitamin D_3-3-bromoacetate, an affinity labeling analog of 1,25-dihydroxyvitamin D_3: mechanistic studies. Arch Biochem Biophys 370:34–44

30. Swamy N, Persons KS, Chen TC, Ray R (2003) 1α,25-Dihydroxyvitamin D_3-3β-(2)-bromoacetate, an affinity labeling derivative of 1,25-dihydroxyvitamin D_3 displays strong antiproliferative and cytotoxic behavior in prostate cancer cells. J Cell Biochem 89:909–916

31. Swamy N, Chen TC, Peleg S, Dhawan P, Christakos S, Stewart LV, Weigel NL, Mehta RG, Holick MF, Ray R (2004) Inhibition of proliferation and induction of apoptosis by 25-hydroxyvitamin D_3-3-bromoacetate in prostate cancer cells. Clin Cancer Res 10:8018–8027

32. Lange TS, Singh RK, Kim KK, Zou Y, Kalkunte SS, Sholler GL, Swamy N, Brard L (2007) Anti-proliferative and pro-apoptotic properties of 3-bromoacetoxy calcidiol in high-risk neuroblastoma. Chem Biol Drug Des 70:302–310

33. Lambert JL, Young CD, Persons KS, Ray R (2007) Mechanistic and pharmacodynamic studies of a 25-hydroxyvitamin D3 derivative in prostate cancer cells. Biochem Biophys Res Comm 361:189–195

34. Makin G, Lohnes D, Byford V, Ray R, Jones G (1989) Target cell metabolism of 1,25-dihydroxyvitamin D3 to calcitroic acid. Evidence for a pathway in kidney and bone involving 24-oxidation. Biochem J 262:173–180

35. Rao LS, Ray R, Holick MF, Horst RL, Uskokovic MR, Reddy GS (2000) Metabolism of [3-alpha-^{3}H] 25-hydroxyvitamin D_2 in kidneys isolated from normal and vitamin D2-intoxicated rats. J Nutr Sci Vaminol (Tokyo) 46:222–229

36. Johnson CS, Hershberger PA, Bernardi RJ, Mcguire TF, Trump DJ (2002) Vitamin D receptor: a potential target for intervention. Urology 60(suppl 3A):123–130

37. Bouillon R, Carmeliet G, Verlinden L, van Etten E, Verstuyf A, Luderer HF, Lieben L, Mathieu C, Demay M (2008) Vitamin D and human health: lessons from vitamin D receptor null mice. Endocr Rev 6:726–776

38. Wu W, Zhang X, Zanello LP (2007) 1alpha,25-Dihydroxyvitamin D(3) antiproliferative actions involve vitamin D receptor-mediated activation of MAPK pathways and AP-1/p21(waf1) upregulation in human osteosarcoma. Cancer Lett 254:75–86

39. Rao A, Coan A, Welsh JE, Barclay WW, Koumenis C, Cramer SD (2004) Vitamin D receptor and p21/WAF1 are targets of genistein and 1,25-dihydroxyvitamin D$_3$ in human prostate cancer cells. Cancer Res 64:2143–2147

40. Knudsen ES, Wang JY (1997) Dual mechanisms for the inhibition of E2F binding to RB by cyclin-dependent kinase-mediated RB phosphorylation. Mol Cell Biol 17:5771–5783

41. Defacque H, Piquemal D, Basset A, Marti J, Commes T (1999) Transforming growth factor-beta1 is an autocrine mediator of U937 cell growth arrest and differentiation induced by vitamin D3 and retinoids. J Cell Physiol 178:109–119

42. Yanagisawa J, Yanagi Y, Masuhiro Y, Suzawa M, Watanabe M, Kashiwagi K, Toriyabe T, Kawabata M, Miyazono K, Kato S (1999) Convergence of transforming growth factor-beta and vitamin D signaling pathways on SMAD transcriptional coactivators. Science 283:1317–1321

43. Yanagi Y, Suzawa M, Kawabata M, Miyazono K, Yanagisawa J, Kato S (1999) Positive and negative modulation of vitamin D receptor function by transforming growth factor-beta signaling through smad proteins. J Biol Chem 274:12971–12974

44. Lyakh LA, Sanford M, Chekol S, Young HA, Roberts AB (2005) TGF-beta and vitamin D3 utilize distinct pathways to suppress IL-12 production and modulate rapid differentiation of human monocytes into CD83+ dendritic cells. J Immunol 174:2061–2070

45. Paralkar VM, Vail AL, Grasser WA, Brown TA, Xu H, Vukicevic S, Ke HZ, Qi H, Owen TA, Thompson DD (1998) Cloning and characterization of a novel member of the transforming growth factor-beta/bone morphogenetic protein family. J Biol Chem 273:13760–13767

46. Li PX, Wong J, Ayed A, Ngo D, Brade AM, Arrowsmith C, Austin RC, Klamut HJ (2000) Placental transforming growth factor-beta is a downstream mediator of the growth arrest and apoptotic response of tumor cells to DNA damage and p53 overexpression. J Biol Chem 275:20127–20135

47. Tan M, Wang Y, Guan K, Sun Y (2000) PTGF-beta, a type beta transforming growth factor (TGF-beta) superfamily member, is a p53 target gene that inhibits tumor cell growth via TGF-beta signaling pathway. Proc Natl Acad Sci USA 97:109–114

48. Nazarova N, Qiao S, Golovko O, Lou YR, Tuohimaa P (2004) Calcitriol-induced prostate-derived factor: autocrine control of prostate cancer cell growth. Int J Cancer 112:951–958

49. Guzey M, Luo J, Getzenberg RH (2004) Vitamin D3 modulated gene expression patterns in human primary normal and cancer prostate cells. J Cell Biochem 93:271–285

50. Narvaez CJ, Byrne BM, Romu S, Valrance M, Welsh J (2003) Induction of apoptosis by 1,25-dihydroxyvitamin D3 in MCF-7 Vitamin D3-resistant variant can be sensitized by TPA. J Steroid Biochem Mol Biol 84:199–209

51. Blutt SE, McDonnell TJ, Polek TC, Weigel NL (2000) Calcitriol-induced apoptosis in LNCaP cells is blocked by overexpression of Bcl-2. Endocrinology 141:10–17

52. Mathiasen IS, Lademann U, Jäättelä M (1999) Apoptosis induced by vitamin D compounds in breast cancer cells is inhibited by Bcl-2 but does not involve known caspases or p53. Cancer Res 59: 4848–4856

53. Mullauer FB, Kessler JH, Medema JP (2009) Betulinic acid induces cytochrome c release and apoptosis in a Bax/Bak-independent, permeability transition pore dependent fashion. Apoptosis 14: 191–202

54. Arcidiacon MV, Sato T, Alvarez-Hernandez D, Yang J, Tokumoto M, Gonzalez-Suarez I, Lu Y, Tominaga Y, Cannata-Andia J, Slatopolsky E, Dusso AS (2008) EGFR activation increases parathyroid hyperplasia and calcitriol resistance in kidney disease. J Am Soc Nephrol 19:310–320

55. González EA, Disthabanchong S, Kowalewski R, Martin KJ (2002) Mechanisms of the regulation of EGF receptor gene expression by calcitriol and parathyroid hormone in UMR 106-01 cells. Kidney Int 61:1627–1634

56. Pepper C, Thomas A, Hoy T, Milligan D, Bentley P, Fegan C (2003) The vitamin D3 analog EB1089 induces apoptosis via a p53-independent mechanism involving p38 MAP kinase activation and suppression of ERK activity in B-cell chronic lymphocytic leukemia cells in vitro. Blood 101: 2454–2460

57. Zhang X, Zanello LP (2008) Vitamin D receptor-dependent 1 alpha,25(OH)2 vitamin D3-induced anti-apoptotic PI3K/AKT signaling in osteoblasts. J Bone Miner Res 23:1238–1248

58. Hughes PJ, Lee JS, Reiner NE, Brown G (2008) The vitamin D receptor-mediated activation of phosphatidylinositol 3-kinase (PI3Kalpha) plays a role in the 1alpha,25-dihydroxyvitamin D3-stimulated increase in steroid sulphatase activity in myeloid leukaemic cell lines. J Cell Biochem 103:1551–1572

59. Marcoux N, Vuori K (2003) EGF receptor mediates adhesion-dependent activation of the Rac GTPase: a role for phosphatidylinositol 3-kinase and Vav2. Oncogene 22:6100–6106

60. Gao N, Zhang Z, Jiang BH, Shi X (2003) Role of PI3K/AKT/mTOR signaling in the cell cycle progression of human prostate cancer. Biochem Biophys Res Commun 310:1124–1132

61. Gao H, Ouyang X, Banach-Petrosky WA, Gerald WL, Shen MM, Abate-Shen C (2007) Combinatorial activities of Akt and B-Raf/Erk signaling in a mouse model of androgen-independent prostate cancer. Proc Natl Acad Sci USA 103:14477–14482; Erratum in: Proc Natl Acad Sci USA (2007) 104:17554

62. Uzgare AR, Isaacs JT (2004) Enhanced redundancy in Akt and mitogen-activated protein kinase-induced survival of malignant versus normal prostate epithelial cells. Cancer Res 64:6190–6199

63. Pendás-Franco N, Aguilera O, Pereira F, González-Sancho JM, Muñoz A (2008) Vitamin D and Wnt/beta-catenin pathway in colon cancer: role and regulation of DICKKOPF genes. Anticancer Res 28:2613–2623

64. Pálmer HG, González-Sancho JM, Espada J, Berciano MT, Puig I, Baulida J, Quintanilla M, Cano A, de Herreros AG, Lafarga M, Muñoz A (2001) Vitamin D(3) promotes the differentiation of colon carcinoma cells by the induction of E-cadherin and the inhibition of beta-catenin signaling. J Cell Biol 154:369–387

65. Murillo G, Mehta RG (2005) Chemoprevention of chemically-induced mammary and colon carcinogenesis by 1alpha-hydroxyvitamin D5. J Steroid Biochem Mol Biol 97:129–136

59 Anti-inflammatory Activity of Calcitriol That Contributes to Its Therapeutic and Chemopreventive Effects in Prostate Cancer

Aruna V. Krishnan and David Feldman

Abstract   Calcitriol exerts anti-proliferative and pro-differentiating effects in a number of tumors and malignant cells and its use as an anti-cancer therapy is currently being evaluated. Many molecular pathways are involved in the growth inhibitory effects of calcitriol resulting in cell cycle arrest, induction of apoptosis, and the inhibition of invasion, metastasis, and angiogenesis. Our recent research reveals that calcitriol exhibits several anti-inflammatory actions that we believe contribute to its anti-cancer effects. In normal and malignant prostate epithelial cells calcitriol inhibits the synthesis and biological actions of pro-inflammatory prostaglandins (PGs) by three actions: (i) the inhibition of the expression of cyclooxygenase-2 (COX-2), the enzyme that synthesizes PGs, (ii) the up-regulation of the expression of 15-prostaglandin dehydrogenase (15-PGDH), the enzyme that inactivates PGs, and (iii) the decreasing of the expression of EP and FP PG receptors that are essential for PG signaling. The combination of calcitriol and non-steroidal anti-inflammatory drugs (NSAIDs) results in a synergistic inhibition of the growth of prostate cancer (PCa) cells and offers a potential therapeutic strategy for PCa. The results of our clinical trial in men with early recurrent PCa indicate that the combination of high-dose weekly calcitriol with the non-selective NSAID naproxen slows the rate of rise (doubling time) of serum prostate specific antigen (PSA) levels in most patients indicating a slowing of disease progression. Calcitriol also increases the expression of mitogen-activated protein kinase phosphatase 5 (MKP5) in prostate cells resulting in the subsequent inhibition of p38 stress kinase signaling and the attenuation of the production of pro-inflammatory cytokines. There is also considerable evidence for an anti-inflammatory role for calcitriol in PCa through the inhibition of NFκB signaling in PCa cells. The discovery of these novel calcitriol-regulated molecular pathways reveal that calcitriol has anti-inflammatory actions, which in addition to its other anti-cancer effects, may play an important role in the prevention and/or treatment of cancer in general and PCa specifically.

Key Words: Inflammation; prostate cancer; calcitriol; 1,25-dihydroxyvitamin D; anti-cancer; vitamin D receptor; animal models; angiogenesis; COX-2; prostaglandins

1. INTRODUCTION

Calcitriol (1,25-dihydroxyvitamin D$_3$), the biologically most active form of vitamin D, is an important regulator of calcium homeostasis and bone metabolism

From: *Nutrition and Health: Vitamin D*
Edited by: M.F. Holick, DOI 10.1007/978-1-60327-303-9_59,
© Springer Science+Business Media, LLC 2010

through its actions in intestine, bone, kidney, and the parathyroid glands. However, extensive research demonstrates that calcitriol also exerts anti-proliferative and pro-differentiating effects in a number of tumors and malignant cells raising the possibility of its use as an anti-cancer agent. Many studies have investigated the actions of calcitriol to inhibit the proliferation of malignant cells in culture, revealing several molecular pathways that are involved in these effects. In vivo studies have also demonstrated an anti-cancer effect of calcitriol to retard the development and growth of tumors in animal models. Recent research, including observations from our laboratory, suggests that calcitriol exhibits anti-inflammatory actions that may contribute to its inhibitory effects in several cancers. In this chapter we explore the key molecular pathways underlying the potential chemopreventive and therapeutic utility of calcitriol in prostate cancer (PCa) with emphasis on the contribution of the recently discovered anti-inflammatory actions of calcitriol to its anti-cancer effects.

2. CALCITRIOL AND PROSTATE CANCER

PCa is the most commonly diagnosed malignancy and the third leading cause of cancer death among men in the United States (1). Androgens promote PCa growth and androgen deprivation is the most useful therapy for men who fail primary therapy with surgery or radiation (2, 3). However, many patients eventually fail androgen deprivation therapy and develop androgen-independent PCa (AIPC) and metastatic disease for which there is no effective treatment currently available. One of the goals of current PCa research on early (androgen-dependent) and advanced (androgen-independent) disease is the identification of new, less toxic agents that could be used alone or in combinations to prevent the development and/or slow the progression of the disease. Among these calcitriol has emerged as a promising therapeutic agent. Epidemiological studies suggest that vitamin D deficiency increases the risk of PCa and that mortality rates due to PCa in the United States are inversely related to sunlight exposure (discussed in detail in Chapters 42 and 43).

2.1. Anti-proliferative Effects of Calcitriol in PCa Cells

A number of studies have demonstrated the anti-proliferative and pro-differentiating effects of calcitriol in primary prostatic epithelial cells and human PCa cell lines (4–9). In the normal prostate, $25(OH)D_3$-1α hydroxylase converts $25(OH)D$ to calcitriol suggesting that local production of calcitriol may play an important role in normal growth and differentiation of the prostate (10–12). In prostate cells, the degree of growth inhibition due to calcitriol appears to be inversely proportional to the level of expression and activity of 24-hydroxylase, the enzyme that initiates calcitriol catabolism (13, 14). The co-addition of 24-hydroxylase inhibitors enhances the growth inhibitory activity of calcitriol in these cells (15, 16). These anti-proliferative effects in cultured cells have been observed at high concentrations of calcitriol that in vivo may cause hypercalcemia and hypercalciuria, which may be associated with renal stone formation (17). Many academic investigators and pharmaceutical companies have undertaken intense research to develop calcitriol analogs/derivatives that exhibit increased anti-proliferative activity and reduced tendency to cause hypercalcemia (18–20).

2.2. Tumor Inhibitory Effects of Calcitriol in Animal Models of PCa

Although several rodent models of PCa have been developed *(21)*, there is still a lack of an ideal model for human PCa. Many studies have investigated the establishment and growth of human PCa xenografts in immuno-compromised mice and showed significant reductions in tumor size and volume following treatment with calcitriol or its analogs *(4–9)*. Transgenic models of PCa have also been developed in animals to study the effects of calcitriol and its analogs to prevent or delay the development and progression of PCa *(21)*. These in vivo models provide a valuable tool to study the tumor inhibitory effects of calcitriol and analogs while monitoring their tendency to cause hypercalcemia and to validate their potential use in clinical trials.

3. MECHANISMS OF THE ANTI-PROLIFERATIVE EFFECTS OF CALCITRIOL

Several important mechanisms have been implicated in the anti-proliferative effects of calcitriol in PCa cells. As described below, calcitriol exerts multiple and diverse actions, often cell-specific, including effects on cell cycle arrest, apoptosis, and inhibition of metastasis and angiogenesis *(22, 23)*.

3.1. Growth Arrest

In PCa cells such as LNCaP, calcitriol treatment results in the accumulation of cells in the G_0/G_1 phase of the cell cycle *(24, 25)*. There appear to be multiple mechanisms by which calcitriol causes cell cycle arrest. Calcitriol exerts its effects on some of the key steps in G_1 to S phase transition and causes an increase in the expression of the cyclin-dependent kinase (CDK) inhibitor p21, a decrease in CDK2 activity leading to a decrease in the phosphorylation of the retinoblastoma (Rb) protein, and the repression of E2F transcriptional activity resulting in G_1 arrest of the cells *(26, 27)*. Calcitriol has been shown to directly up-regulate *p21* gene expression acting through a vitamin D response element (VDRE) in the promoter of the p21 gene *(28)* as well as indirectly through the induction of the *insulin-like growth factor binding protein-3 (IGFBP-3)* gene which in turn increases p21 expression *(29)*. The presence of a functional p53 appears to be essential for calcitriol-mediated G_0 arrest of PCa cells *(30)*.

3.2. Apoptosis

There is evidence that calcitriol induces apoptosis in LNCaP cells through the down-regulation of the anti-apoptotic protein Bcl-2 *(31)*, an effect that is caspase- but not p53-dependent *(30)*. Calcitriol or its analogs also decrease Bcl-2 expression in other PCa cells such as ALVA-31 *(32)* and DU 145 cells *(33)*. IGFBP-3 whose expression is increased by calcitriol in some PCa cells *(29, 34, 35)* is regarded as a promoter of apoptosis in breast cancer cells via the activation of caspases 8 and 9 *(36)*. Calcitriol induction of apoptosis, however, appears to be cell specific as it is not uniformly evident in all PCa cells.

3.3. Differentiation

Calcitriol has been shown to induce the differentiation of a number of normal and malignant cells. Histological examination of the prostate tissue revealed a greater degree of epithelial cellular differentiation in rats treated with testosterone and calcitriol compared to rats treated with testosterone alone *(37)*. In some androgen receptor positive PCa cells calcitriol increases the expression of prostate-specific antigen (PSA) *(38, 39)*, which is regarded as a differentiation marker for prostatic epithelial cells. However, the effect appears to be much smaller compared to the induction of PSA expression by androgens. As yet there is no compelling additional evidence supporting a role for calcitriol as a differentiation-promoting agent in PCa.

3.4. Modulation of Growth Factor Actions

Growth factors play an important role in the regulation of prostate epithelial cell growth in the normal prostate by autocrine and paracrine mechanisms. Prostatic stromal cells are capable of modifying the epithelial cell environment through the production of peptide growth factors that can act in a paracrine manner on the basal epithelium *(40, 41)*. In PCa, however, growth factors expressed by the epithelial cells can drive their growth in an autocrine manner leading to the development of independence from epithelial–stromal interactions and contributing to the progression of PCa. Some of these growth factors also appear to play an important role in the establishment and growth of metastatic PCa cells in bone *(42, 43)*. Calcitriol regulates the expression and signaling of some of these key growth factors such as insulin-like growth factor-I (IGF-I) and transforming growth factor β (TGFβ), which play important roles in the regulation of the growth and differentiation of prostate cells. In several PCa cells, calcitriol decreases the availability of IGF by increasing the expression of its binding proteins such as IGFBP-3 *(29, 44, 45)*. IGFBP-3 also exerts important IGF-independent actions that mediate growth inhibitory effects through the up-regulation of p21 expression *(29)* and anti-inflammatory effects through the interference of nuclear factor κB (NFκB) signaling cascade (see Section 4.3 below). Increases in the expression of several IGFBPs including IGFBP-3 expression are associated with the regression of the prostate gland seen in rats following the administration of the vitamin D analog EB1089 *(46)*. In PC-3 cells TGFβ increases the expression of IGFBP-3 leading to growth arrest and apoptosis *(47, 48)*. Calcitriol increases IGFBP-3 expression by a direct transcriptional up-regulation of the *IGFBP-3* gene acting through a well-characterized VDRE sequence in the IGFBP-3 promoter *(35)*. The presence of VDRE sequences has also been demonstrated in the promoter of the human *TGFβ 2* gene *(49)*.

3.5. Inhibition of Invasion, Metastasis, and Angiogenesis

In addition to the inhibition of proliferation of malignant cells, calcitriol also plays an inhibitory role in tumor invasion, metastasis, and angiogenesis. In vitro studies demonstrate that calcitriol and its analogs markedly inhibit the invasiveness and metastatic potential of PCa cells by altering the secretion of matrix metalloproteinases *(50)*, down-regulating the expression of α6 and β4 integrins *(51)*, and increasing the expression of

E-cadherin *(25)*. Several studies have demonstrated that calcitriol inhibits tumor cell-induced angiogenesis in mice *(52, 53)* and modulates angiogenic signaling in tumor-derived endothelial cells *(54)* supporting the hypothesis that the inhibition of angiogenesis plays a role in the antitumor effects of calcitriol *(55)*.

3.6. Novel Molecular Pathways of Calcitriol Actions in PCa Cells

We used cDNA microarrays as a means to achieve the major research goal of gaining a more complete understanding of the molecular pathways by which calcitriol mediates its anti-proliferative and pro-differentiation effects in prostate cells. Our results have revealed that calcitriol regulates the expression of genes involved in the metabolism and signaling of prostaglandins (PGs). PGs are pro-inflammatory molecules that promote prostate tumorigenesis and growth *(56, 57)*. The regulatory actions of calcitriol include the inhibition of the expression of cyclooxygenase-2 (COX-2), stimulation of 15-hydroxyprostaglandin dehydrogenase (15-PGDH) expression, and down-regulation of the levels of the PG receptors EP2 and EP4 *(34, 58)*. We also have shown that calcitriol up-regulates the expression of mitogen-activated protein kinase phosphatase-5 [MKP5; also known as dual specificity phosphatase-10 (DUSP10)] and thereby promotes down-stream anti-inflammatory effects *(59)*. Recent research also indicates that calcitriol interferes with the activation and signaling of NFκB, a transcription factor that regulates the expression of numerous genes involved in inflammatory and immune responses and cellular proliferation *(60)*. In the following sections we will discuss these novel anti-inflammatory effects of calcitriol and their role in the inhibition of the development and progression of PCa.

4. ANTI-INFLAMMATORY EFFECTS OF CALCITRIOL IN PROSTATE CANCER

4.1. Regulation of Prostaglandin Metabolism and Signaling

PGs have been shown to play a role in the development and progression of many cancers including PCa *(56, 57)* by stimulating cellular proliferation, inhibiting apoptosis, promoting angiogenesis, and by activating carcinogens *(61, 62)*. We have recently discovered that calcitriol regulates the expression of several key genes involved in the PG pathway causing a decrease in PG synthesis, an increase in PG catabolism, and the inhibition of PG signaling through their receptors in PCa cells *(63)*.

4.1.1. COX-2 AND PROSTATE CANCER

Cyclooxygenase (COX)/prostaglandin endoperoxidase synthase is the rate-limiting enzyme that catalyzes the conversion of arachidonic acid to PGs and related eicosanoids. COX exists as two isoforms, COX-1, which is constitutively expressed in many tissues and cell types and COX-2, which is inducible by a variety of stimuli. *COX-2* is regarded as an immediate-early response gene whose expression is rapidly induced by mitogens, cytokines, tumor promoters, and growth factors *(57)*. Genetic and clinical studies indicate that increased COX-2 expression is one of the key steps in carcinogenesis *(64)*. Several studies suggest a positive role for COX-2 in prostate tumorigenesis and demonstrate its over-expression in prostate adenocarcinoma *(65, 66)*. However, not all PCa are associated with elevated COX-2 expression *(67, 68)*. Zha et al. *(68)* did not find consistent over-expression of COX-2 in established PCa. However they detected appreciable COX-2 expression in areas of proliferative inflammatory atrophy (PIA), lesions that have been implicated in prostate carcinogenesis. All investigators agree that local production of PGs by the infiltrating inflammatory cells increases the risk of carcinogenesis and/or progression *(68–71)*. Silencing of *COX-2* in metastatic PCa cells induces cell growth arrest and causes morphological changes associated with enhanced differentiation highlighting the role of COX-2 in prostate carcinogenesis *(72)*. At the cellular level, both arachidonic acid, the substrate for COX, and the product prostaglandin E_2 (PGE_2) stimulate proliferation by regulating the expression of genes that are involved in growth regulation including *c-fos (73)*. COX-2 protein expression in prostate biopsy cores and PCa surgical specimens is inversely correlated with disease-free survival *(74)*. A recent analysis of archival radical prostatectomy specimens also found COX-2 expression in PCa cells, adjacent normal glands, and in specimens from patients who exhibited disease progression and concluded that COX-2 expression was an independent predictor of recurrence *(75)*.

4.1.2. 15-PGDH

15-PGDH is the enzyme that catalyzes the conversion of PGs to their corresponding 15-keto derivatives, which exhibit greatly reduced biological activity. Therefore 15-PGDH can be regarded as a physiological antagonist of COX-2. *15-PGDH* has been described as an oncogene antagonist in colon cancer by Yan et al. *(76)*. Their results show that 15-PGDH is universally expressed in normal colon but is routinely absent or severely reduced in cancer specimens. Most importantly, stable transfection of a 15-PGDH expression vector into cancer cells greatly reduces the ability of the cells to form tumors and/or slows tumor growth in nude mice demonstrating that 15-PGDH functions as a tumor suppressor *(76)*. Another study in mice also demonstrates that 15-PGDH acts in vivo as a highly potent suppressor of colon neoplasia development *(77)*. Studies in breast cancer cells demonstrating a suppression of cell proliferation in vitro and decreased tumorigenicity in vivo following the over-expression of 15-PGDH also support a tumor suppressor role for 15-PGDH in breast cancer *(78)*.

4.1.3. PG RECEPTORS

PGE and PGF are the major PGs stimulating the proliferation of PCa cells and they act by binding to G-protein coupled membrane receptors (prostanoid receptors). There are eight members in the prostanoid receptor subfamily and they are distinguished by their ligand-binding profile and the signal transduction pathways that they activate upon ligand binding and account for some of the diverse and often opposing effects of PGs *(79)*. PGE acts through four different receptor sub-types (EP1–EP4), while PGF acts through the FP receptor. PCa cells express both EP and FP receptors *(63, 73)*.

4.1.4. CALCITRIOL EFFECTS ON THE PG PATHWAY IN PROSTATE CELLS

Our studies demonstrate that calcitriol regulates the expression of PG pathway genes in multiple PCa cell lines as well as primary prostatic epithelial cells established from surgically removed prostate tissue from PCa patients *(63)*. We found measurable amounts of COX-2 mRNA and protein in various PCa cell lines as well as primary prostatic epithelial cells derived from normal and cancerous prostate tissue, which were significantly decreased by calcitriol treatment. We also found that calcitriol significantly increased the expression of 15-PGDH mRNA and protein in various PCa cells. We further showed that by inhibiting COX-2 and stimulating 15-PGDH expression, calcitriol decreased the levels of biologically active PGs in PCa cells, thereby reducing the growth stimulation due to PGs. Our data also revealed that calcitriol decreased the expression of EP and FP PG receptors. The calcitriol-induced decrease in PG receptor levels resulted in the attenuation of PG-mediated functional responses even when exogenous PGs were added to the cell cultures. Calcitriol suppressed the induction of the immediate-early gene *c-fos* and the growth stimulation seen following the addition of exogenous PGs or the PG precursor arachidonic acid to PCa cell cultures *(63)*. Thus, as illustrated in Fig. 1, calcitriol inhibits the PG pathway in PCa cells by three separate mechanisms: decreasing COX-2 expression, increasing 15-PGDH expression, and reducing PG receptor levels. We believe that these actions contribute to the suppression of the proliferative stimulus provided by PGs in PCa cells. The regulation of PG metabolism and biological actions

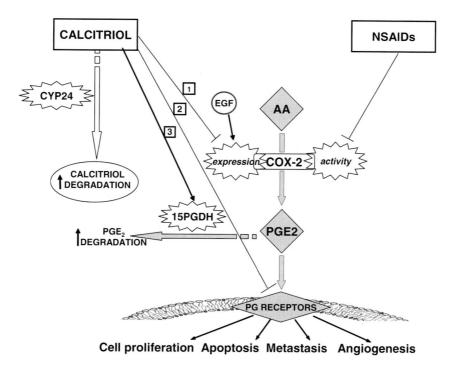

Fig. 1. Calcitriol Inhibits the PG Pathway in Prostate Cells.

constitute an additional novel pathway of calcitriol action mediating its anti-proliferative effects in normal and malignant primary prostate cells and in multiple PCa cell lines.

4.1.5. Combination of Calcitriol and Non-steroidal Anti-inflammatory Drugs (NSAIDs) as a Therapeutic Approach in Prostate Cancer

NSAIDs are a class of drugs that decrease PG synthesis by inhibiting COX-1 and COX-2 enzymatic activities. Several NSAIDs inhibit both the constitutively expressed COX-1 and the inducible COX-2, while others have been shown to be more selective in inhibiting the COX-2 enzyme activity. Our hypothesis is that the action of calcitriol at the genomic level to reduce COX-2 expression leading to decreased COX-2 protein levels will allow the use of lower concentrations of NSAIDs to inhibit COX-2 enzyme activity (63). In addition, an increase in the expression of 15-PGDH due to calcitriol action will lower the levels of biologically active PGs and enhance the NSAID effect. Therefore, we hypothesized that the combination of calcitriol and NSAIDs would exhibit additive/synergistic effects to inhibit PCa cell growth. In cell culture studies we examined the growth inhibitory effects of the combinations of calcitriol with the COX-2-selective NSAIDs NS398 and SC-58125 or the non-selective NSAIDs, naproxen and ibuprofen. The combinations caused a synergistic enhancement of the inhibition of PCa cell proliferation, compared to the individual agents (63). These results led us to further hypothesize that the combination of calcitriol and NSAIDs may have clinical utility in PCa therapy (63).

Preclinical (80) and clinical studies (81) on colon and other cancers have successfully used the strategy of combining low doses of two active drugs to achieve a more effective chemoprevention and therapeutic outcome than those using the individual agents (82). The combination approach would also minimize the toxicities of the individual drugs by allowing them to be used at lower doses while achieving a significant therapeutic effect. Based on our preclinical observations, we proposed that a combination of calcitriol with a NSAID would be a beneficial approach in PCa therapy. The combination strategy allows the use of lower concentrations of NSAIDs thereby minimizing their undesirable side effects. It has very recently become clear that long-term use of COX-2-selective inhibitors such as rofecoxib (Vioxx) causes an increase in cardiovascular complications in patients (83–86). In comparison, non-selective NSAIDs such as naproxen may be associated with fewer cardiovascular adverse effects (87). Our preclinical data show that the combinations of calcitriol with non-selective or selective NSAIDs are equally effective in inducing synergistic growth inhibition (63). We therefore proposed that the combination of calcitriol with a non-selective NSAID would be a useful therapeutic approach in PCa that would allow both drugs to be used at reduced dosages leading to increased safety (88).

Calcitriol, in fact is already being used in combination therapy with other agents that may enhance its anti-proliferative activity while reducing its tendency to cause hypercalcemia (89). The results of the ASCENT I clinical trial in advanced PCa patients who failed other therapies demonstrated that the administration of a very high dose (45 µg) of calcitriol (DN101, Novacea, South San Francisco, CA) once weekly along with the regimen of the chemotherapy drug docetaxel (Taxotere) in use at the time of that trial (once weekly) caused a very significant improvement in overall survival and time to progression, providing evidence indicating that calcitriol could enhance the efficacy of active drugs in cancer treatment (90). The ASCENT I trial did not meet its primary endpoint, i.e., a lowering of serum PSA. However, based on the promising survival results, (16.4 months in the docetaxel arm versus 24.5 months in the docetaxel plus calcitriol arm) a larger, phase III trial (ASCENT II) with survival as an endpoint was initiated. A new, improved docetaxel regimen (every 3 week dosing) was used in the control arm of the ASCENT II trial, which was compared to DN101 plus the older docetaxel dosing regimen (once a week), resulting in an asymmetric study design. Unfortunately the improved survival due to the combination demonstrated in the ASCENT I trial could not be confirmed in the ASCENT II trial (91). In fact, the trial was prematurely stopped by the data safety monitoring committee after 900 patients were enrolled, when an excess number of deaths were noted in the study arm (DN101 plus old docetaxel regimen) versus the control arm (new docetaxel regimen). Since the trial was stopped, further analysis (92) suggests that the increased deaths in the treatment arm compared to the control arm were not due to calcitriol toxicity but due to better survival in the control arm that received the new and improved docetaxel regimen.

Based on our preclinical observations, we recently carried out a single arm, open label phase II study evaluating the combination of the non-selective NSAID naproxen and calcitriol in patients with early recurrent PCa (93). Patients in our study had no evidence of metastases. All the patients received 45 μg of calcitriol (DN101) orally once a week and 375 mg naproxen twice a day for 1 year. The trial was prematurely stopped when the FDA put a temporary hold on DN101 based on the data from the ASCENT II trial described above. We monitored serum PSA levels every 8 weeks. Bone scans were done every 3 months along with ultrasound of the kidney to assess asymptomatic renal stones. Serum testosterone levels were not affected by the therapy and there were no sexual side effects. There was mild gastro-intestinal toxicity in three patients presumably from the naproxen and one patient had to be removed from the study. One patient developed a small asymptomatic renal stone and was removed from the study. He required no intervention for his renal stone. Changes in PSA doubling time (PSA-DT) post intervention were compared to baseline PSA-DT values. The therapy was well tolerated by most patients and a prolongation of the PSA-DT was achieved in 75% of the patients suggesting a beneficial effect of the combination therapy (88, 93).

4.2. Induction of MKP5 and Inhibition of Stress-Activated Kinase Signaling

Our cDNA microarray analysis in prostate cells (58) revealed another novel calcitriol responsive gene, *MKP5*, also known as *DUSP10*. Calcitriol up-regulates MKP5 expression leading to down-stream anti-inflammatory responses in cells derived from normal prostatic epithelium and primary, localized adenocarcinoma, supporting a role for calcitriol in the prevention and early treatment of PCa (59). In primary cultures of normal prostatic epithelial cells from the peripheral zone, calcitriol increased *MKP5* transcription (59). We identified a putative positive VDRE in the *MKP5* promoter mediating this calcitriol effect *(59)*. Interestingly, calcitriol up-regulation of MKP5 was seen only in primary cells derived from normal prostatic epithelium and primary, localized adenocarcinoma but not in the established PCa cell lines derived from PCa metastasis such as LNCaP, PC-3, or DU145. MKP5 is a member of the dual specificity MKP family of enzymes that dephosphorylate, and thereby inactivate, mitogen-activated protein kinases (MAPKs). MKP5 specifically dephosphorylates p38 MAPK and the stress-activated protein kinase Jun-N-terminal Kinase (JNK), leading to their inactivation. Calcitriol inhibited the phosphorylation and activation of p38 in normal primary prostate cells in a MKP-5-dependent manner as MKP5 siRNA completely abolished p38 inactivation by calcitriol *(59)*. A consequence of p38 stress-induced kinase activation is an increase in the production of pro-inflammatory cytokines that sustain and amplify the inflammatory response *(94)*. As interleukin-6 (IL-6), a p38-regulated pleiotropic cytokine, is known to be associated with PCa progression *(95)*, we investigated the effect of calcitriol on IL-6 production. Stimulation of primary prostate cells with the pro-inflammatory factor, tumor necrosis factor α (TNFα), increased IL-6 mRNA stability and concentrations of IL-6 in the conditioned media. Pre-treatment of the cells with calcitriol significantly attenuated the increase in IL-6 production following TNFα treatment *(59)*.

Our data suggest that the ability of calcitriol to inhibit p38 signaling and reduce the subsequent production of pro-inflammatory cytokines, via MKP5 up-regulation contributes to the cancer preventive effects of calcitriol. Since established metastasis-derived PCa cell lines exhibited low levels of MKP5 and were unable to induce MKP5 in response to calcitriol, we speculate that a loss of MKP5 might occur during PCa progression as a result of selective pressure to eliminate the tumor suppressor activity of MKP5 and/or calcitriol.

4.3. Inhibition of NFκB Activation and Signaling

NFκB is a family of inducible transcription factors ubiquitously present in all cells. In contrast to normal cells many cancer cells have elevated levels of active NFκB *(96, 97)*. Constitutive activation of NFκB has been observed in androgen-independent PCa *(98–100)*. The NFκB protein RelB is uniquely expressed at high levels in PCa with high Gleason scores *(101)*. NFκB plays a major role in the control of immune responses and inflammation and promotes malignant behavior by increasing the transcription of the anti-apoptotic gene *Bcl2 (102)*, cell cycle progression factors such as c-myc and cyclin D1, proteolytic enzymes such as matrix metalloproteinase 9 (MMP-9) and urokinase-type plasminogen activator (uPA), and angiogenic factors such as vascular endothelial growth factor (VEGF) and interleukin-8 (IL-8) *(100, 103)*. IL-8, an angiogenic factor and a down-stream target of NFκB, is a potent chemotactic factor for neutrophils that is associated with the initiation of the inflammatory response *(104)*. Calcitriol is known to directly modulate basal and cytokine-induced NFκB activity in many cells including human lymphocytes *(105)*, fibroblasts *(106)*, and peripheral blood monocytes *(107)*. A reduction in the levels of the NFκB inhibitory protein IκBα has been reported in mice lacking the VDR *(108)*. Addition of a VDR antagonist to colon cancer cells up-regulates NFκB activity by decreasing the levels of IκBα *(109)*.

There is considerable evidence for an anti-inflammatory role for calcitriol in PCa through the inhibition of NFκB signaling in PCa cells. Calcitriol decreases the levels of the angiogenic and pro-inflammatory cytokine IL-8 in immortalized normal human prostate epithelial cell lines (HPr-1 and RWPE-1) and established PCa cell lines (LNCaP, PC-3, and DU145) *(110)*. The suppression of IL-8 by calcitriol appears to be due to the inhibition of NFκB signaling. Calcitriol reduces the nuclear translocation of the NFκB subunit p65, thereby inhibiting the NFκB complex from binding to its DNA response element and consequently suppressing the transcription of down-stream targets such as IL-8 *(110)*. Thus calcitriol could delay the progression of PCa by suppressing the expression of angiogeneic and pro-inflammatory factors such as VEGF and IL-8. In addition calcitriol also indirectly inhibits NFκB signaling by up-regulating the expression of IGFBP-3 (see Section 3.4), which has been shown to interfere with NFκB signaling in PCa cells by suppressing p65 NFκB protein levels and the phosphorylation of IκBα *(111)*.

NFκB also provides an adaptive response to PCa cells against cytotoxicity induced by redox active therapeutic agents and is implicated in radiation resistance of cancers *(112, 113)*. A recent study shows that calcitriol significantly enhances the sensitivity of PCa cells to ionizing radiation by selectively suppressing radiation-mediated RelB activation *(114)*. Thus calcitriol may serve as an effective agent for sensitizing PCa cells to radiation therapy via suppression of the NFκB pathway.

5. THE ROLE OF ANTI-INFLAMMATORY EFFECTS OF CALCITRIOL IN PROSTATE CANCER PREVENTION AND TREATMENT

5.1. Inflammation and Prostate Cancer

Current perspectives in cancer biology suggest that inflammation plays a role in the development of cancer *(3, 70, 115)*. Inflammation may contribute to carcinogenesis by several mechanisms including the elaboration of cytokines and growth factors that favor tumor cell growth, induction of COX-2, which leads to the synthesis of PGs that promote tumor proliferation and generation of mutagenic reactive oxygen and nitrogen species *(116)*. Epidemiological studies show that there is decreased risk of PCa associated with the intake of antioxidants and NSAIDs *(117, 118)*. De Marzo et al. *(119)* have proposed that the PIA lesions in the prostate, which are associated with acute or chronic inflammation, are precursors of prostatic intraepithelial neoplasia (PIN) and PCa. The epithelial cells in PIA lesions have been shown to exhibit many molecular signs of stress including elevated expression of COX-2 *(68, 115)*. Based on our recent research demonstrating (i) the repression of COX-2 expression in normal and malignant prostatic epithelial cells by calcitriol *(63)* and (ii) the inhibition of pro-inflammatory cytokine production by calcitriol due to MKP5-mediated p38 inactivation *(59)* as well as (iii) the ability of calcitriol to interfere with NFκB signaling and the resultant attenuation of pro-inflammatory responses (Section 4.3), we postulate that calcitriol may exert anti-inflammatory effects and thereby play a role in delaying or preventing the development and/or progression of PCa.

5.2. Calcitriol and Prostate Cancer Chemoprevention

PCa generally progresses very slowly, likely for decades, before symptoms become obvious and diagnosis is made *(120)*. Recently, inflammation in the prostate has been proposed to be an etiological factor in the development of PCa *(70)*. The observed latency in PCa provides a long window of opportunity for intervention by chemopreventive agents. Dietary supplementation of COX-2 selective NSAIDs such as celecoxib has been shown to suppress prostate carcinogenesis in the Transgenic Adenocarcinoma of the Mouse Prostate (TRAMP) model *(121)*. Our studies on the inhibitory effects of calcitriol on COX-2 expression and the PG pathway and MKP5 induction with the resultant stress kinase inactivation and inhibition of pro-inflammatory cytokine production as well as published observations of calcitriol actions to inhibit NFκB signaling suggest that calcitriol exhibits anti-inflammatory effects in vitro. Therefore we hypothesize that calcitriol has the potential to be useful as a chemopreventive agent in PCa. Recently, Foster and coworkers have demonstrated that administration of high dose of calcitriol (20 μg/kg), intermittently 3 days per week for up to 14–30 weeks, suppresses prostate tumor development in TRAMP mice *(122, 123)*. The efficacy of calcitriol as a chemopreventive agent has also been examined in Nkx3.1; Pten mutant mice, which recapitulate stages of prostate carcinogenesis from PIN lesions to adenocarcinoma *(124)*. The data reveal that calcitriol significantly reduces the progression of PIN from a lower to a higher grade. Calcitriol is more effective when administered before, rather than subsequent to, the initial occurrence of PIN. These animal studies as well as our in vitro observations suggest that clinical trials in PCa patients with PIN or early disease evaluating calcitriol and its analogs as agents that prevent and/or delay progression are warranted.

6. SUMMARY AND CONCLUSIONS

Our recent research has identified several new calcitriol target genes revealing novel molecular pathways of calcitriol action in prostate cells. The data suggest that calcitriol has anti-inflammatory actions that contribute to its anti-proliferative and cancer preventive effects in PCa. Calcitriol reduces both PG production (by suppressing COX-2 and increasing 15-PGDH expression) and PG biological actions (by PG receptor downregulation). We propose that calcitriol inhibition of the PG pathway contributes significantly to its anti-inflammatory actions. Combinations of calcitriol with NSAIDs exhibit synergistic enhancement of growth inhibition in PCa cell cultures suggesting that they may have therapeutic utility in PCa. The results of our recent clinical trial in patients with early recurrent PCa indicate that the combination of a weekly high-dose calcitriol with the non-selective NSAID naproxen has activity to slow the rate of rise of PSA in most patients. Another novel molecular pathway of calcitriol action involves the induction of MKP5 expression and the subsequent inhibition of p38 stress kinase signaling, resulting in the attenuation of the production of pro-inflammatory cytokines in prostate cells. There is also considerable evidence for an anti-inflammatory role for calcitriol in PCa through the inhibition of NFκB signaling in PCa cells. The discovery of these novel calcitriol-regulated pathways suggests that calcitriol has anti-inflammatory actions, in addition to its other anti-cancer effects, that may play an important role in the prevention and/or treatment of PCa. Recent studies in animal models of PCa reveal that calcitriol inhibits the progression of PCa in TRAMP and Nkx3.1; Pten mice. We conclude that calcitriol and its analogs may have utility as chemopreventive agents and should be evaluated in clinical trials in PCa patients with early or precancerous disease.

REFERENCES

1. Ward E, Halpern M, Schrag N et al (2008) Association of insurance with cancer care utilization and outcomes. CA Cancer J Clin 58:9–31
2. Hellerstedt BA, Pienta KJ (2002) The current state of hormonal therapy for prostate cancer. CA Cancer J Clin 52:154–179
3. Nelson WG, De Marzo AM, Isaacs WB (2003) Prostate cancer. N Engl J Med 349:366–381
4. Chen TC, Holick MF (2003) Vitamin D and prostate cancer prevention and treatment. Trends Endocrinol Metab 14:423–430
5. Deeb KK, Trump DL, Johnson CS (2007) Vitamin D signalling pathways in cancer: potential for anticancer therapeutics. Nat Rev Cancer 7:684–700
6. Fleet JC (2008) Molecular actions of vitamin D contributing to cancer prevention. Mol Aspects Med 29:388–396
7. Krishnan AV, Peehl DM, Feldman D (2003) Inhibition of prostate cancer growth by vitamin D: regulation of target gene expression. J Cell Biochem 88:363–371
8. Krishnan AV, Peehl DM, Feldman D (2005) Vitamin D and prostate cancer. In: Feldman D, Pike JW, Glorieux FH (ed) Vitamin D, 2nd edn. Elsevier Academic Press, San Diego, pp. 1679–1707
9. Stewart LV, Weigel NL (2004) Vitamin D and prostate cancer. Exp Biol Med (Maywood) 229: 277–284
10. Chen TC, Wang L, Whitlatch LW et al (2003) Prostatic 25-hydroxyvitamin D-1alpha-hydroxylase and its implication in prostate cancer. J Cell Biochem 88:315–322
11. Hsu JY, Feldman D, McNeal JE et al (2001) Reduced 1alpha-hydroxylase activity in human prostate cancer cells correlates with decreased susceptibility to 25-hydroxyvitamin D3-induced growth inhibition. Cancer Res 61:2852–2856
12. Schwartz GG, Whitlatch LW, Chen TC et al (1998) Human prostate cells synthesize 1, 25-dihydroxyvitamin D3 from 25-hydroxyvitamin D3. Cancer Epidemiol Biomarkers Prev 7: 391–395
13. Miller GJ, Stapleton GE, Hedlund TE et al (1995) Vitamin D receptor expression, 24-hydroxylase activity, and inhibition of growth by 1alpha,25-dihydroxyvitamin D3 in seven human prostatic carcinoma cell lines. Clin Cancer Res 1:997–1003
14. Skowronski RJ, Peehl DM, Feldman D (1993) Vitamin D and prostate cancer: 1,25 dihydroxyvitamin D3 receptors and actions in human prostate cancer cell lines. Endocrinology 132:1952–1960
15. Ly LH, Zhao XY, Holloway L et al (1999) Liarozole acts synergistically with 1alpha, 25-dihydroxyvitamin D3 to inhibit growth of DU 145 human prostate cancer cells by blocking 24-hydroxylase activity. Endocrinology 140:2071–2076
16. Peehl DM, Seto E, Hsu JY et al (2002) Preclinical activity of ketoconazole in combination with calcitriol or the vitamin D analogue EB 1089 in prostate cancer cells. J Urol 168: 1583–1588
17. Gross C, Stamey T, Hancock S et al (1998) Treatment of early recurrent prostate cancer with 1, 25-dihydroxyvitamin D3 (calcitriol). J Urol 159:2035–2039; discussion 2039–2040
18. Bouillon R, Okamura WH, Norman AW (1995) Structure-function relationships in the vitamin D endocrine system. Endocr Rev 16:200–257
19. Ma Y, Khalifa B, Yee YK et al (2006) Identification and characterization of noncalcemic, tissue-selective, nonsecosteroidal vitamin D receptor modulators. J Clin Invest 116:892–904
20. Posner GH, Kahraman M (2005) Overview: rational design of 1a,25-dihydroxyvitamin D3 analogs (deltanoids). In: Feldman D, Pike JW, Glorieux FH (ed) Vitamin D, 2nd edn. Elsevier Academic Press, San Diego, pp. 1405–1422

21. Navone NM, Logothetis CJ, von Eschenbach AC et al (1998) Model systems of prostate cancer: uses and limitations. Cancer Metastasis Rev 17:361–371

22. Krishnan AV, Moreno J, Nonn L et al (2007) Novel pathways that contribute to the anti-proliferative and chemopreventive activities of calcitriol in prostate cancer. J Steroid Biochem Mol Biol 103:1 694–702

23. Krishnan AV, Moreno J, Nonn L et al (2007) Calcitriol as a chemopreventive and therapeutic agent in prostate cancer: role of anti-inflammatory activity. J Bone Miner Res 22(Suppl 2):V74–V80

24. Blutt SE, Allegretto EA, Pike JW et al (1997) 1,25-dihydroxyvitamin D3 and 9-cis-retinoic acid act synergistically to inhibit the growth of LNCaP prostate cells and cause accumulation of cells in G1. Endocrinology 138:1491–1497

25. Campbell MJ, Elstner E, Holden S et al (1997) Inhibition of proliferation of prostate cancer cells by a 19-nor-hexafluoride vitamin D3 analogue involves the induction of p21waf1, p27kip1 and E-cadherin. J Mol Endocrinol 19:15–27

26. Moffatt KA, Johannes WU, Hedlund TE et al (2001) Growth inhibitory effects of 1alpha, 25-dihydroxyvitamin D(3) are mediated by increased levels of p21 in the prostatic carcinoma cell line ALVA-31. Cancer Res 61:7122–7129

27. Zhuang SH, Burnstein KL (1998) Antiproliferative effect of 1alpha,25-dihydroxyvitamin D3 in human prostate cancer cell line LNCaP involves reduction of cyclin-dependent kinase 2 activity and persistent G1 accumulation. Endocrinology 139:1197–1207

28. Liu M, Lee MH, Cohen M et al (1996) Transcriptional activation of the Cdk inhibitor p21 by vitamin D3 leads to the induced differentiation of the myelomonocytic cell line U937. Genes Dev 10:142–153

29. Boyle BJ, Zhao XY, Cohen P et al (2001) Insulin-like growth factor binding protein-3 mediates 1 alpha,25-dihydroxyvitamin d(3) growth inhibition in the LNCaP prostate cancer cell line through p21/WAF1. J Urol 165:1319–1324

30. Polek TC, Stewart LV, Ryu EJ et al (2003) p53 Is required for 1,25-dihydroxyvitamin D3-induced G0 arrest but is not required for G1 accumulation or apoptosis of LNCaP prostate cancer cells. Endocrinology 144:50–60

31. Blutt SE, McDonnell TJ, Polek TC et al (2000) Calcitriol-induced apoptosis in LNCaP cells is blocked by overexpression of Bcl-2. Endocrinology 141:10–17

32. Guzey M, Kitada S, Reed JC (2002) Apoptosis induction by 1alpha,25-dihydroxyvitamin D3 in prostate cancer. Mol Cancer Ther 1:667–677

33. Crescioli C, Maggi M, Luconi M et al (2002) Vitamin D3 analogue inhibits keratinocyte growth factor signaling and induces apoptosis in human prostate cancer cells. Prostate 50:15–26

34. Krishnan AV, Shinghal R, Raghavachari N et al (2004) Analysis of vitamin D-regulated gene expression in LNCaP human prostate cancer cells using cDNA microarrays. Prostate 59:243–251

35. Peng L, Malloy PJ, Feldman D (2004) Identification of a functional VITAMIN D response element in the human insulin-like growth factor binding protein-3 promoter. Mol Endocrinol 18: 1109–1119

36. Kim HS, Ingermann AR, Tsubaki J et al (2004) Insulin-like growth factor-binding protein 3 induces caspase-dependent apoptosis through a death receptor-mediated pathway in MCF-7 human breast cancer cells. Cancer Res 64:2229–2237

37. Konety BR, Schwartz GG, Acierno JS Jr (1996) The role of vitamin D in normal prostate growth and differentiation. Cell Growth Differ 7:1563–1570

38. Zhao XY, Ly LH, Peehl DM et al (1999) Induction of androgen receptor by 1alpha, 25-dihydroxyvitamin D3 and 9-cis retinoic acid in LNCaP human prostate cancer cells. Endocrinology 140:1205–1212

39. Zhao XY, Peehl DM, Navone NM et al (2000) 1alpha,25-dihydroxyvitamin D3 inhibits prostate cancer cell growth by androgen-dependent and androgen-independent mechanisms. Endocrinology 141:2548–2556

40. Craft N, Sawyers CL (1998) Mechanistic concepts in androgen-dependence of prostate cancer. Cancer Metastasis Rev 17:421–427

41. Isaacs JT (1999) The biology of hormone refractory prostate cancer. Why does it develop? Urol Clin North Am 26:263–273

42. Bogdanos J, Karamanolakis D, Tenta R et al (2003) Endocrine/paracrine/autocrine survival factor activity of bone microenvironment participates in the development of androgen ablation and chemotherapy refractoriness of prostate cancer metastasis in skeleton. Endocr Relat Cancer 10: 279–289

43. Gleave ME, Hsieh JT, von Eschenbach AC et al (1992) Prostate and bone fibroblasts induce human prostate cancer growth in vivo: implications for bidirectional tumor-stromal cell interaction in prostate carcinoma growth and metastasis. J Urol 147:1151–1159

44. Drivdahl RH, Loop SM, Andress DL et al (1995) IGF-binding proteins in human prostate tumor cells: expression and regulation by 1,25-dihydroxyvitamin D3. Prostate 26:72–79

45. Huynh H, Pollak M, Zhang JC (1998) Regulation of insulin-like growth factor (IGF) II and IGF binding protein 3 autocrine loop in human PC-3 prostate cancer cells by vitamin D metabolite 1,25(OH)2D3 and its analog EB1089. Int J Oncol 13:137–143

46. Nickerson T, Huynh H (1999) Vitamin D analogue EB1089-induced prostate regression is associated with increased gene expression of insulin-like growth factor binding proteins. J Endocrinol 160: 223–229

47. Hwa V, Oh Y, Rosenfeld RG (1997) Insulin-like growth factor binding protein-3 and -5 are regulated by transforming growth factor-beta and retinoic acid in the human prostate adenocarcinoma cell line PC-3. Endocrine 6:235–242

48. Rajah R, Valentinis B, Cohen P (1997) Insulin-like growth factor (IGF)-binding protein-3 induces apoptosis and mediates the effects of transforming growth factor-beta1 on programmed cell death through a p53- and IGF-independent mechanism. J Biol Chem 272:12181–12188

49. Wu Y, Craig TA, Lutz WH et al (1999) Identification of 1 alpha,25-dihydroxyvitamin D3 response elements in the human transforming growth factor beta 2 gene. Biochemistry 38:2654–2660

50. Schwartz GG, Wang MH, Zang M et al (1997) 1 alpha,25-Dihydroxyvitamin D (calcitriol) inhibits the invasiveness of human prostate cancer cells. Cancer Epidemiol Biomarkers Prev 6:727–732

51. Sung V, Feldman D (2000) 1,25-Dihydroxyvitamin D3 decreases human prostate cancer cell adhesion and migration. Mol Cell Endocrinol 164:133–143

52. Majewski S, Skopinska M, Marczak M et al (1996) Vitamin D3 is a potent inhibitor of tumor cell-induced angiogenesis. J Invest Dermatol Symp Proc 1:97–101

53. Mantell DJ, Owens PE, Bundred NJ et al (2000) 1 alpha,25-dihydroxyvitamin D(3) inhibits angiogenesis in vitro and in vivo. Circ Res 87:214–220

54. Bernardi RJ, Johnson CS, Modzelewski RA et al (2002) Antiproliferative effects of 1alpha, 25-dihydroxyvitamin D(3) and vitamin D analogs on tumor-derived endothelial cells. Endocrinology 143:2508–2514

55. Tosetti F, Ferrari N, De Flora S et al (2002) 'Angioprevention': angiogenesis is a common and key target for cancer chemopreventive agents. Faseb J 16:2–14

56. Badawi AF (2000) The role of prostaglandin synthesis in prostate cancer. BJU Int 85:451–462

57. Hussain T, Gupta S, Mukhtar H (2003) Cyclooxygenase-2 and prostate carcinogenesis. Cancer Lett 191:125–135

58. Peehl DM, Shinghal R, Nonn L et al (2004) Molecular activity of 1,25-dihydroxyvitamin D3 in primary cultures of human prostatic epithelial cells revealed by cDNA microarray analysis. J Steroid Biochem Mol Biol 92:131–141

59. Nonn L, Peng L, Feldman D et al (2006) Inhibition of p38 by vitamin D reduces interleukin-6 production in normal prostate cells via mitogen-activated protein kinase phosphatase 5: implications for prostate cancer prevention by vitamin D. Cancer Res 66:4516–4524

60. McCarty MF (2004) Targeting multiple signaling pathways as a strategy for managing prostate cancer: multifocal signal modulation therapy. Integr Cancer Ther 3:349–380

61. Dubois RN, Abramson SB, Crofford L et al (1998) Cyclooxygenase in biology and disease. Faseb J 12:1063–1073

62. Hawk ET, Viner JL, Dannenberg A et al (2002) COX-2 in cancer – a player that's defining the rules. J Natl Cancer Inst 94:545–546

63. Moreno J, Krishnan AV, Swami S et al (2005) Regulation of prostaglandin metabolism by calcitriol attenuates growth stimulation in prostate cancer cells. Cancer Res 65:7917–7925

64. Markowitz SD (2007) Aspirin and colon cancer – targeting prevention? N Engl J Med 356: 2195–2198

65. Gupta S, Srivastava M, Ahmad N et al (2000) Over-expression of cyclooxygenase-2 in human prostate adenocarcinoma. Prostate 42:73–78

66. Yoshimura R, Sano H, Masuda C et al (2000) Expression of cyclooxygenase-2 in prostate carcinoma. Cancer 89:589–596

67. Wagner M, Loos J, Weksler N et al (2005) Resistance of prostate cancer cell lines to COX-2 inhibitor treatment. Biochem Biophys Res Commun 332:800–807

68. Zha S, Gage WR, Sauvageot J et al (2001) Cyclooxygenase-2 is up-regulated in proliferative inflammatory atrophy of the prostate, but not in prostate carcinoma. Cancer Res 61:8617–8623

69. De Marzo AM, Platz EA, Sutcliffe S et al (2007) Inflammation in prostate carcinogenesis. Nat Rev Cancer 7:256–269

70. Nelson WG, De Marzo AM, DeWeese TL et al (2004) The role of inflammation in the pathogenesis of prostate cancer. J Urol 172:S6–S11; discussion S11–S12

71. Wang W, Bergh A, Damber JE (2005) Cyclooxygenase-2 expression correlates with local chronic inflammation and tumor neovascularization in human prostate cancer. Clin Cancer Res 11: 3250–3256

72. Narayanan BA, Narayanan NK, Davis L et al (2006) RNA interference-mediated cyclooxygenase-2 inhibition prevents prostate cancer cell growth and induces differentiation: modulation of neuronal protein synaptophysin, cyclin D1, and androgen receptor. Mol Cancer Ther 5:1117–1125

73. Chen Y, Hughes-Fulford M (2000) Prostaglandin E2 and the protein kinase A pathway mediate arachidonic acid induction of c-fos in human prostate cancer cells. Br J Cancer 82:2000–2006

74. Rubio J, Ramos D, Lopez-Guerrero JA et al (2005) Immunohistochemical expression of Ki-67 antigen, Cox-2 and Bax/Bcl-2 in prostate cancer; prognostic value in biopsies and radical prostatectomy specimens. Eur Urol 31:31

75. Cohen BL, Gomez P, Omori Y et al (2006) Cyclooxygenase-2 (cox-2) expression is an independent predictor of prostate cancer recurrence. Int J Cancer 119:1082–1087

76. Yan M, Rerko RM, Platzer P et al (2004) 15-Hydroxyprostaglandin dehydrogenase, a COX-2 oncogene antagonist, is a TGF-beta-induced suppressor of human gastrointestinal cancers. Proc Natl Acad Sci USA 101:17468–17473

77. Myung SJ, Rerko RM, Yan M et al (2006) 15-Hydroxyprostaglandin dehydrogenase is an in vivo suppressor of colon tumorigenesis. Proc Natl Acad Sci USA 103:12098–12102

78. Wolf I, O'Kelly J, Rubinek T et al (2006) 15-hydroxyprostaglandin dehydrogenase is a tumor suppressor of human breast cancer. Cancer Res 66:7818–7823

79. Breyer RM, Bagdassarian CK, Myers SA et al (2001) Prostanoid receptors: subtypes and signaling. Annu Rev Pharmacol Toxicol 41:661–690

80. Torrance CJ, Jackson PE, Montgomery E et al (2000) Combinatorial chemoprevention of intestinal neoplasia. Nat Med 6:1024–1028

81. Meyskens FL, McLaren CE, Pelot D et al (2008) Difluoromethylornithine plus sulindac for the prevention of sporadic colorectal adenomas: a randomized placebo-controlled, double-blind trial. Cancer Prev Res 1:32–38

82. Sporn MB, Hong WK (2008) Clinical prevention of recurrence of colorectal adenomas by the combination of difluoromethylornithine and sulindac: an important milestone. Cancer Prev Res 1:9–11

83. Antman EM, DeMets D, Loscalzo J (2005) Cyclooxygenase inhibition and cardiovascular risk. Circulation 112:759–770

84. Graham DJ, Campen D, Hui R et al (2005) Risk of acute myocardial infarction and sudden cardiac death in patients treated with cyclo-oxygenase 2 selective and non-selective non-steroidal anti-inflammatory drugs: nested case-control study. Lancet 365:475–481

85. Ray WA, Stein CM, Daugherty JR et al (2002) COX-2 selective non-steroidal anti-inflammatory drugs and risk of serious coronary heart disease. Lancet 360:1071–1073

86. Solomon SD, Pfeffer MA, McMurray JJ et al (2006) Effect of celecoxib on cardiovascular events and blood pressure in two trials for the prevention of colorectal adenomas. Circulation 114: 1028–1035

87. Bombardier C, Laine L, Reicin A et al (2000) Comparison of upper gastrointestinal toxicity of rofecoxib and naproxen in patients with rheumatoid arthritis. VIGOR Study Group. N Engl J Med 343:1520–1528; 1522 p following 1528

88. Krishnan AV, Srinivas S, Feldman D (2009) Inhibition of prostaglandin synthesis and actions contributes to the beneficial effects of calcitriol in prostate cancer. Dermato Endocrinol 1:1–5

89. Johnson CS, Hershberger PA, Trump DL (2002) Vitamin D-related therapies in prostate cancer. Cancer Metastasis Rev 21:147–158

90. Beer TM, Ryan CW, Venner PM et al (2007) Double-blinded randomized study of high-dose calcitriol plus docetaxel compared with placebo plus docetaxel in androgen-independent prostate cancer: a report from the ASCENT Investigators. J Clin Oncol 25:669–674

91. Release NP. Novacea announces preliminary findings from data analysis of Ascent-2 Phase 3 trial. June 04, 2008 [cited; Available from: http://www.novacea.com]

92. Release NP. Novacea update on Asentar (TM). September 11, 2008 [cited; Available from: http://www.novacea.com/]

93. Srinivas S, Feldman D (2008) A phase II trial of calcitriol and naproxen in recurrent prostate cancer. Anticancer Res 28:1611–1626

94. Park JI, Lee MG, Cho K et al (2003) Transforming growth factor-beta1 activates interleukin-6 expression in prostate cancer cells through the synergistic collaboration of the Smad2, p38-NF-kappaB, JNK, and Ras signaling pathways. Oncogene 22:4314–4332

95. Culig Z, Steiner H, Bartsch G et al (2005) Interleukin-6 regulation of prostate cancer cell growth. J Cell Biochem 95:497–505

96. Palayoor ST, Youmell MY, Calderwood SK et al (1999) Constitutive activation of IkappaB kinase alpha and NF-kappaB in prostate cancer cells is inhibited by ibuprofen. Oncogene 18:7389–7394

97. Sovak MA, Bellas RE, Kim DW et al (1997) Aberrant nuclear factor-kappaB/Rel expression and the pathogenesis of breast cancer. J Clin Invest 100:2952–2960

98. Gasparian AV, Yao YJ, Kowalczyk D et al (2002) The role of IKK in constitutive activation of NF-kappaB transcription factor in prostate carcinoma cells. J Cell Sci 115:141–151

99. Ismail HA, Lessard L, Mes-Masson AM et al (2004) Expression of NF-kappaB in prostate cancer lymph node metastases. Prostate 58:308–313

100. Suh J, Rabson AB (2004) NF-kappaB activation in human prostate cancer: important mediator or epiphenomenon? J Cell Biochem 91:100–117

101. Lessard L, Begin LR, Gleave ME et al (2005) Nuclear localisation of nuclear factor-kappaB transcription factors in prostate cancer: an immunohistochemical study. Br J Cancer 93:1019–1023

102. Catz SD, Johnson JL (2001) Transcriptional regulation of bcl-2 by nuclear factor kappa B and its significance in prostate cancer. Oncogene 20:7342–7351

103. Huang S, Pettaway CA, Uehara H et al (2001) Blockade of NF-kappaB activity in human prostate cancer cells is associated with suppression of angiogenesis, invasion, and metastasis. Oncogene 20:4188–4197

104. Ferrer FA, Miller LJ, Andrawis RI et al (1998) Angiogenesis and prostate cancer: in vivo and in vitro expression of angiogenesis factors by prostate cancer cells. Urology 51:161–167

105. Yu XP, Bellido T, Manolagas SC (1995) Down-regulation of NF-kappa B protein levels in activated human lymphocytes by 1,25-dihydroxyvitamin D3. Proc Natl Acad Sci USA 92:10990–10994

106. Harant H, Wolff B, Lindley IJ (1998) 1Alpha,25-dihydroxyvitamin D3 decreases DNA binding of nuclear factor-kappaB in human fibroblasts. FEBS Lett 436:329–334

107. Stio M, Martinesi M, Bruni S et al (2007) The Vitamin D analogue TX 527 blocks NF-kappaB activation in peripheral blood mononuclear cells of patients with Crohn's disease. J Steroid Biochem Mol Biol 103:51–60

108. Sun J, Kong J, Duan Y et al (2006) Increased NF-kappaB activity in fibroblasts lacking the vitamin D receptor. Am J Physiol Endocrinol Metab 291:E315–E322

109. Schwab M, Reynders V, Loitsch S et al (2007) Involvement of different nuclear hormone receptors in butyrate-mediated inhibition of inducible NF kappa B signalling. Mol Immunol 44:3625–3632

110. Bao BY, Yao J, Lee YF (2006) 1alpha, 25-dihydroxyvitamin D3 suppresses interleukin-8-mediated prostate cancer cell angiogenesis. Carcinogenesis 27:1883–1893

111. Han J, Jogie-Brahim S, Oh Y (2007) New paradigm foe antitumor action of IGF binding protein-3 (IGFBP-3): novel NF-kB inhibitory effect of IGFBP-3 in prostate cancer. In: Proceedings from the Endocrine Society Meeting, Toronto, Canada, pp. P3–P370
112. Criswell T, Leskov K, Miyamoto S et al (2003) Transcription factors activated in mammalian cells after clinically relevant doses of ionizing radiation. Oncogene 22:5813–5827
113. Kimura K, Bowen C, Spiegel S et al (1999) Tumor necrosis factor-alpha sensitizes prostate cancer cells to gamma-irradiation-induced apoptosis. Cancer Res 59:1606–1614
114. Xu Y, Fang F, St Clair DK et al (2007) Suppression of RelB-mediated manganese superoxide dismutase expression reveals a primary mechanism for radiosensitization effect of 1alpha,25-dihydroxyvitamin D(3) in prostate cancer cells. Mol Cancer Ther 6:2048–2056
115. De Marzo AM, Marchi VL, Epstein JI et al (1999) Proliferative inflammatory atrophy of the prostate: implications for prostatic carcinogenesis. Am J Pathol 155:1985–1992
116. Lucia MS, Torkko KC (2004) Inflammation as a target for prostate cancer chemoprevention: pathological and laboratory rationale. J Urol 171:S30–S34; discussion S35
117. Clark LC, Dalkin B, Krongrad A et al (1998) Decreased incidence of prostate cancer with selenium supplementation: results of a double-blind cancer prevention trial. Br J Urol 81:730–734
118. Nelson JE, Harris RE (2000) Inverse association of prostate cancer and non-steroidal anti-inflammatory drugs (NSAIDs): results of a case-control study. Oncol Rep 7:169–170
119. De Marzo AM, DeWeese TL, Platz EA et al (2004) Pathological and molecular mechanisms of prostate carcinogenesis: implications for diagnosis, detection, prevention, and treatment. J Cell Biochem 91:459–477
120. Tsai CJ, Cohn BA, Cirillo PM et al (2006) Sex steroid hormones in young manhood and the risk of subsequent prostate cancer: a longitudinal study in African-Americans and Caucasians (United States). Cancer Causes Control 17:1237–1244
121. Gupta S, Adhami VM, Subbarayan M et al (2004) Suppression of prostate carcinogenesis by dietary supplementation of celecoxib in transgenic adenocarcinoma of the mouse prostate model. Cancer Res 64:3334–3343
122. Alagbala A, Moser MT, Johnson CS et al (2005) Prevention of prostate cancer progression with vitamin D compounds in the transgenic adenocarcinoma of the mouse prostate (TRAMP) model. In: Fourth annual AACR International Conference on Frontiers in Cancer Prevention Research, Baltimore, MD, p. 107
123. Alagbala AA, Moser MT, Johnson CS et al (2006) 1a, 25-dihydroxyvitamin D3 and its analog (QW-1624F2-2) prevent prostate cancer progression. In: 13th Workshop on Vitamin D, Victoria, BC, Canada, p. 95
124. Banach-Petrosky W, Ouyang X, Gao H et al (2006) Vitamin D inhibits the formation of prostatic intraepithelial neoplasia in Nkx3.1;Pten mutant mice. Clin Cancer Res 12:5895–5901

Subject Index

Note: The letters 'f' and 't' following locators refer to figures and tables respectively.

A

AAP, *see* American Academy of Pediatrics (AAP)
ACE, *see* Angiotensin converting enzyme (ACE)
ACEI, *see* ACE inhibitors (ACEI)
ACE inhibitors (ACEI), 943, 944f, 945
Active vitamin D, 226
  high doses, causes, 227
  therapy, 223–225, 228
  treatments, effects, 228
Acute exacerbations of COPD (AECOPD), 1009
Acute laryngitis, 999
Acute lower respiratory infections, 999
  bronchiolitis, 999
  bronchitis, 999
Acute nasopharyngitis, 999
Acute pharyngitis, 999
Acute promyelocytic leukemia, 158
Acute sinusitis, 999
Acute tracheitis, 999
Acute upper respiratory infections
  laryngitis, 999
  nasopharyngitis, 999
  pharyngitis, 999
  sinusitis, 999
  tracheitis, 999
ADAM17, *see* Tumor necrosis factor converting enzyme (TACE)
Adaptive immunity and vitamin D, 1010–1011
  adaptive immunity, CYP27b1
    autoimmune diseases, VDD, 288
    1,25(OH)$_2$D$_3$ administration in experimental models, benefits, 285
  macrophages, 1010
  pulmonary eosinophilic inflammation, 1010
  Th1–Th2 regulation, 1010

Adipocytes
  apoptosis, 348–349, 349f
  glucocorticoid production, 349–350
  inflammatory cytokine production, 352–353
  oxidative stress, 352
Adipogenesis, 193, 347
Adipokine, 350–351, 353, 754
Adynamic bone disease, 957
  alfacalcidol, 957
  biopsy-proven osteitis fibrosa, 957
  disease associated, 957
  level of PTH, 957
  VDRA, bone formation, 958
AECOPD, *see* Acute exacerbations of COPD (AECOPD)
AEV, *see* Avian erythroblastosis virus (AEV)
Affinity/photoaffinity labeling methods, 1062
Africa, VDD and health consequences in, 505–523
  continent of Africa, 506–511
    demography, 507
    diet and foodstuffs, 508
    geography, topography, climate, 506–507
    UV radiation and skin pigmentation, 507–508
    WHO sub-regions, 511t–512t
  health consequences of poor vitamin D, 513–515
    HIV/AIDS, 515
    rickets and osteomalacia, 514
    tuberculosis, 514–515
  impact on immune system, 514
  indicators of health and burden of disease, 511–513
    fluorosis, 512
    HIV/AIDS, 513–514
    low calcium intake, 511–512

From: *Nutrition and Health: Vitamin D*
Edited by: M.F. Holick, DOI 10.1007/978-1-60327-303-9,
© Springer Science+Business Media, LLC 2010

About the Editor

Michael F. Holick, Ph.D, M.D., is professor of medicine, physiology, and biophysics; director of the General Clinical Research Unit; and director of the Bone Health Care Clinic and the Heliotherapy, Light, and Skin Research Center at Boston University Medical Center. After earning a Ph.D. in biochemistry, a medical degree, and completing a research postdoctoral fellowship at the University of Wisconsin, Madison, Dr. Holick completed a residency in medicine at the Massachusetts General Hospital in Boston.

Dr. Holick has made numerous contributions to the field of biochemistry, physiology, metabolism, and photobiology of vitamin D for human nutrition. He determined the mechanism for how vitamin D is synthesized in the skin and demonstrated the effects of aging, obesity, latitude, seasonal change, sunscreen use, skin pigmentation, and clothing on this vital cutaneous process. Dr. Holick has established global recommendations advising sunlight exposure as an integral source of vitamin D. He has also helped increase awareness in the pediatric and medical communities regarding the vitamin D deficiency pandemic and its role in causing not only metabolic bone disease and osteoporosis in adults but increasing risk of children and adults developing common deadly cancers, autoimmune diseases, including type 1 diabetes, and multiple sclerosis as well as heart disease.

Dr. Holick is a diplomate of the American Board of Internal Medicine, a fellow of the American College of Nutrition, and a member of numerous organizations, including the American Academy of Dermatology, American Society for Bone and Mineral Research, and the American Association of Physicians. He is the recipient of numerous awards and honors, including the American Skin Association's Psoriasis Research Achievement Award in 2002, the American College of Nutrition Award in 2002, the Robert H. Herman Memorial Award in Clinical Nutrition from the American Society for Clinical Nutrition in 2003, the Annual General Clinical Research Centers' Program Award for Excellence in Clinical Research in 2006, the Linus Pauling Functional Medicine Award from the Institute for Functional Medicine in 2007, and the Eli Lilly Award from the

Canadian Endocrine Society in 2007. Linus Pauling Prize in human nutrition and the DSM Innovation Nutrition Award in 2009. Dr. Holick serves on a number of national committees, including NIH and NASA and several editorial boards. He has organized and/or co-chaired several international symposia and is editor-in-chief for the *Journal for Clinical Laboratories and Laboratories Related to Blood Transfusion* and Associate editor for Dermato-Endocrinology. He has authored more than 300 peer-reviewed publications and written more than 200 review articles as well as numerous book chapters. He has served as editor and/or co-editor on 11 books including the first edition of Vitamin D and Nutrition and Bone Health-both volumes were developed is part of the Nutrition and Health Series.

About the Series Editor

Dr. Adrianne Bendich is clinical director, Medical Affairs at GlaxoSmith-Kline (GSK) Consumer Healthcare where she is responsible for leading the innovation and medical programs in support of many well-known brands including TUMS and Os-Cal. Dr. Bendich had primary responsibility for GSK's support for the Women's Health Initiative (WHI) intervention study. Prior to joining GSK, Dr. Bendich was at Roche Vitamins Inc. and was involved with the groundbreaking clinical studies showing that folic acid-containing multivitamins significantly reduced major classes of birth defects. Dr. Bendich has co-authored over 100 major clinical research studies in the area of preventive nutrition. Dr Bendich is recognized as a leading authority on antioxidants, nutrition and immunity and pregnancy outcomes, vitamin safety and the cost-effectiveness of vitamin/mineral supplementation.

Dr. Bendich is the editor of nine books including "*Preventive Nutrition: The Comprehensive Guide For Health Professionals*" co-edited with Dr. Richard Deckelbaum, and is series editor of "*Nutrition and Health*" for Humana Press with 35 published volumes including "*Probiotics in Pediatric Medicine*" edited by Dr. Sonia Michail and Dr. Philip Sherman; "*Handbook of Nutrition and Pregnancy*" edited by Dr. Carol Lammi-Keefe, Dr. Sarah Couch, and Dr. Elliot Philipson; "*Nutrition and Rheumatic Disease*" edited by Dr. Laura Coleman; "*Nutrition and Kidney Disease*" edited by Dr. Laura Byham-Grey, Dr. Jerrilynn Burrowes, and Dr. Glenn Chertow; "*Nutrition and Health in Developing Countries*" edited by Dr. Richard Semba and Dr. Martin Bloem; "*Calcium in Human Health*" edited by Dr. Robert Heaney and Dr. Connie Weaver and "*Nutrition and Bone Health*" edited by Dr. Michael Holick and Dr. Bess Dawson-Hughes.

Dr. Bendich served as an associate editor for "*Nutrition*" the International Journal, served on the editorial board of the *Journal of Women's Health and Gender-Based Medicine*, and was a member of the Board of Directors of the American College of Nutrition.

Dr. Bendich was the recipient of the Roche Research Award, is a *Tribute to Women and Industry* Awardee, and was a recipient of the Burroughs Wellcome Visiting Professorship in Basic Medical Sciences, 2000–2001. In 2008, Dr. Bendich was given the Council for Responsible Nutrition (CRN) Apple Award in recognition of her many contributions to the scientific understanding of dietary supplements. Dr Bendich holds academic appointments as adjunct professor in the Department of Preventive Medicine and Community Health at UMDNJ and has an adjunct appointment at the Institute of Nutrition, Columbia University P&S and is an adjunct research professor, Rutgers University, Newark Campus. She is listed in Who's Who in American Women.

Printed in the United States of America